3rd International Conference on Mathematical Modeling in Physical Sciences

AA003429

(IC-MSQUARE 2014)

Journal of Physics: Conference Series Volume 574

Madrid, Spain
28 – 31 August 2014

Editors:

E. C. Vagenas
D. S. Vlachos

ISBN: 978-1-63439-907-4
ISSN: 1742-6588

Printed from e-media with permission by:

Curran Associates, Inc.
57 Morehouse Lane
Red Hook, NY 12571

proceedings.com

Some format issues inherent in the e-media version may also appear in this print version.

Printed by Curran Associates, Inc. (2015)

For permission requests, please contact the Institute of Physics
at the address below.

Institute of Physics
Dirac House, Temple Back
Bristol BS1 6BE UK

Phone: 44 1 17 929 7481
Fax: 44 1 17 920 0979

techtracking@iop.org

Additional copies of this publication are available from:

Curran Associates, Inc.
57 Morehouse Lane
Red Hook, NY 12571 USA
Phone: 845-758-0400
Fax: 845-758-2634
Email: curran@proceedings.com
Web: www.proceedings.com

TABLE OF CONTENTS

WORKSHOP: SILICENE AND GERMANENE: NOVEL SYNTHETIC 2D MATERIALS. CHAIRS: PAOLA DE PADOVA, GUY LE LAY

WORKSHOP: DYNAMICAL PROBABILITY: SIMULATIONS AND COMPUTING. CHAIR: CLEMENT AMPADU

WORKSHOP: NONLINEAR WAVES AND SOLITONS: THEORY AND APPLICATIONS. CHAIR: YANNIS KOMINIS

WORKSHOP: BLACK HOLE PHYSICS. CHAIR, ELIAS VAGENAS

WORKSHOP: CORE WORKSHOP. CHAIRS: ELIAS VAGENAS, DIMITRIOS VLACHOS

Editors:

Dr Elias C. Vagenas, Kuwait University, Kuwait
Prof Dimitrios S. Vlachos, University of Peloponnese, Greece

Co-Editors:

Catarina Bastos, Instituto Plasmas e Fus„o Nuclear, Portugal.
Thomas Hofer, University of Innsbruck, Austria.
Yannis Kominis, National Technical University of Athens, Greece.
Odysseas Kosmas, University of Erlangen-Nuremberg, Germany.
Guy Le Lay, CNRS-PIIM UMR, France.
Paola De Padova, Consiglio Nazionale delle Ricerche-Istituto di Struttura della Materia, Italy.
Bernd Rode, University of Innsbruck, Austria.
Eric Suraud, Universite Paul Sabatier, France.
Kalman Varga, Vanderbilt, United States.

Journal of Physics: Conference Series **574** (2015) 011001 doi:10.1088/1742-6596/574/1/011001

Preface

The third International Conference on Mathematical Modeling in Physical Sciences (IC-MSQUARE) took place at Madrid, Spain, from Thursday 28 to Sunday 31 August 2014.

The Conference was attended by more than 200 participants and hosted about 350 oral, poster, and virtual presentations. More than 600 pre-registered authors were also counted. The third IC-MSQUARE consisted of different and diverging workshops and thus covered various research fields where Mathematical Modeling is used, such as Theoretical/Mathematical Physics, Neutrino Physics, Non-Integrable Systems, Dynamical Systems, Computational Nanoscience, Biological Physics, Computational Biomechanics, Complex Networks, Stochastic Modeling, Fractional Statistics, DNA Dynamics, Macroeconomics etc.

The scientific program was rather heavy since after the Keynote and Invited Talks in the morning, three parallel oral sessions and one poster session were running every day. However, according to all attendees, the program was excellent with high level of talks and the scientific environment was fruitful, thus all attendees had a creative time.

We would like to thank the Keynote Speaker and the Invited Speakers for their significant contribution to IC-MSQUARE. We also would like to thank the Members of the International Advisory and Scientific Committees as well as the Members of the Organizing Committee.

The Conference Chairmen

Elias Vagenas
Kuwait University

Dimitrios Vlachos
University of Peloponnese

International Advisory Committee

- **Professor Constantino Tsallis,** Centro Brasileiro de Perquisas Fisicas, Brazil
- **Professor Athanasios S. Fokas,** University of Cambridge, UK
- **Professor Hentrik J. Jensen** , Imperial College London, UK
- **Professor George E. Karniadakis,** Brown University, USA
- **Professor John E. Straub** , Boston University, USA
- **Professor Jorge Zanelli** , Centro de Estudios Científicos (CECs), Chile

International Scientific Committee

- **Bari Ferenc,** University of Szeged, Hungary
- **Mark Cross,** Swansea University, UK
- **Morten Hjorth-Jensen,** University of Oslo, Norway
- **Nathan A Baker,** Pacific Northwest National Laboratory, Richland, WA, USA
- **Nikolas Markatos,** National Technical University of Athens, Greece
- **Shiyi Chen,** Peking University, People's Republic of China
- **Kostas Kokkotas,** Eberhard Karls University of Tübingen, Germany
- **Sergey Nazarenko,** University of Warwick, UK
- **Rho Shin Myong,** Gyeongsang National University, Republic of Korea
- **Michel A. Van Hove,** City University of Hong Kong, Hong Kong
- **Jacob Barhen,** Oak Ridge Natl. Lab, Oak Ridge, USA
- **Javier Hernández-Andrés,** University of Granada, Spain
- **Maia Martcheva,** University of Florida, USA
- **Emilio Elizalde,** CSIC, Barcelona, Spain
- **Angelo Tartaglia,** Politecnico di Torino, Italy
- **Yoshiharu Yamamoto,** The University of Tokyo, Japan
- **Amaury Mouchet,** Université de Tours, France
- **Mairi Sakellariadou,** King's College of London, UK
- **Kostas Skenderis,** University of Amsterdam, The Netherlands
- **Orfeu Bertolami,** University of Porto, Portugal
- **Saurya Das,** University of Lethbridge, Canada
- **Theocharis Kosmas,** University of Ioannina

Keynote Speaker

- **George Karniadakis**, Charles Pitts Robinson and John Palmer Barstow Professor of Applied Mathematics Brown University

Invited Speakers

- **Yamir Moreno**, Department of Theoretical Physics, Faculty of Sciences, University of Zaragoza, Spain
- **Ofer Lahav**, Perren Chair of Astronomy, Astrophysics Group, University College London
- **Thomas S. Hofer**, Institute of General, Inorganic and Theoretical Chemistry University of Innsbruck, Austria

Peer review statement

All papers published in this volume of *Journal of Physics: Conference Series* have been peer reviewed through processes administered by the proceedings Editors. Reviews were conducted by expert referees to the professional and scientific standards expected of a proceedings journal published by IOP Publishing.

Journal of Physics: Conference Series **574** (2015) 012001 doi:10.1088/1742-6596/574/1/012001

Any ℓ-state solutions of the Feynman propagator for the q-deformed Woods-Saxon potential in arbitrary dimensions

A. Diaf[1], **Dj. Roubehie**[1]

[1]Laboratoire de l'Énergie et des Systèmes Intelligents, Université de Khemis Miliana, Route de Thénia, Khemis Miliana, Algérie.

E-mail: s_ahmed_diaf@yahoo.fr

Abstract. Using the space-time transformations, approximate analytical solutions of the D-dimensional propagator in the presence of q-deformed Woods-Saxon potential are obtained. The analytical expression of the energy eigenvalues is given for various quantum numbers and the corresponding normalized eigenfunctions are obtained in terms of hypergeometric function. Our results are compared with those given by The Nikivorov-Uvarov method.

1. Introduction

It is well known that the analytical solutions of the Schrödinger equations with some exponential-type potentials are impossible for any arbitrary l-wave states, approximation schemes such as the Pekeris-type (or a new improved approximation scheme) need to be employed to deal with the centrifugal term [1, 2]. These exponential-type potentials include the Pöschl?Teller potential [3, 4], Manning-Rosen potential [5, 6], Rosen-Morse-type potentials [7].

The deformed Woods-Saxon (dWS) potential is a short range potential and is given by

$$V^{dWS}(r) = -\frac{V_0}{q + e^{\left(\frac{r-R}{a}\right)}}. \tag{1}$$

where $R = R_0 A_0^{\frac{1}{3}}$, $V_0 = (40.5 + 0.13A_0)MeV$, $r >> a$, $q > 0$. where V_0 is the depth of potential, q is a real parameter which determines the shape (deformation) of the potential, a is the diffuseness of the nuclear surface, R is the width of the potential, A_0 is the atomic mass number of target nucleus and r_0 is radius parameter.

This potential has been used in nuclear, particle, atomic, condensed matter and chemical physics [8-14]. It is known that for the Woods-Saxon potential, the Schrödinger equation can be solved exactly when $\ell = 0$ [15]. Unfortunately, for arbitrary ℓ-state, the radial Schrödinger equation does not admit exact solutions. In this case, some authors have used an approximation scheme to study analytically arbitrary ℓ-wave bound states of the Schrödinger equation for the Woods-Saxon potential [16]and the deformed Woods-Saxon [17].

This paper is organized as follows. In Sec. 2 we derive arbitrary ℓ-states solutions of the deformed Woods-Saxon potential within the Feynman path integrals formalism by approximate method. In Sec.3 the analytical expressions are given to compare with those obtained by other method. The concluding remarks are given in Sec. 4.

2. Feynman Propagator

Since the potential (1) is the deformed Woods-Saxon potential, we have a spherically symmetric system which may be conveniently described in polar coordinates. The propagator is written as [18]:

$$K(\vec{r}'',t'';\vec{r}',t') = \frac{1}{4\pi r''r'}\sum_{l=0}^{\infty}(2l+1)K_l(r'',t'';r',t')P_l(\cos\theta). \tag{2}$$

Where $P_l(\cos\theta)$ is the Legendre polynomial with $\theta \equiv (\vec{r}'',\vec{r}')$ and

$$K_l(r'',t'';r',t') = \lim_{N\to\infty}\int\prod_{j=1}^{N}\exp\left[\frac{i}{\hbar}S_j\right]\prod_{j=1}^{N}\left[\frac{m}{2\pi i\hbar\varepsilon}\right]^{\frac{1}{2}}\prod_{j=1}^{N-1}dr_j. \tag{3}$$

With the sliced action $S_j = \frac{m}{2\varepsilon}(\Delta r_j)^2 - \varepsilon V_{eff}(r_j)$, where

$$V_{eff}(r_j) = \frac{\hbar^2}{2m}\frac{(l+D/2-1)^2-1/4}{r_j r_{j-1}} - \frac{V_0}{q+e^{(\frac{r_j-R}{a})}} \tag{4}$$

with the notations $\Delta r_j = r_j - r_{j-1}$, $\varepsilon = t_j - t_{j-1}$, $t' = t_0$, $t'' = t_N$,$T = t'' - t'$ and for any variable r: $r_j = r(t_j)$

If we take the special case D=3, we get

$$V_{eff}(r_j) = \frac{\hbar^2}{2m}\frac{l(l+1)}{r_j r_{j-1}} - \frac{V_0}{q+e^{(\frac{r_j-R}{a})}} \tag{5}$$

Since the Feynman propagator, with dWS effective potential (3), has no analytical solution for $l \neq 0$ states, an approximation to the centrifugal term has to be made. The good approximation for the centrifugal barrier is taken as [17] :

$$\frac{l(l+1)}{r_j{}^2} \approx \frac{l(l+1)}{R^2}\left[D_0 - \frac{D_1}{q+e^{\nu x}} + \frac{D_2}{(q+e^{\nu x})^2}\right], \tag{6}$$

where

$$D_0 = 1 - \left(\frac{1+e^{-aR}}{aR}\right)^2\left(\frac{4aR}{1+e^{-aR}} - 3 - aR\right),$$

$$D_1 = 2\left(1+e^{aR)}\right)\times\left[3\left(\frac{1+e^{-aR}}{aR}\right) - (3+aR)\left(\frac{1+e^{-aR}}{aR}\right)\right],$$

$$D_2 = \left(1+e^{aR}\right)^2\left(\frac{1+e^{-aR}}{aR}\right)^2\times\left(3+aR-\frac{2aR}{1+e^{-aR}}\right).$$

By substituting the centrifugal term in (5) by its approximation (6), we get

$$V_{eff}(r_j) = A\tanh(\alpha r_j) - \frac{B}{\cosh^2(\alpha r_j)} + C, \tag{7}$$

In the following, let us apply the procedure of path reparametrization [19]. The path integral for the potential (7) is

$$K_l(r'',r';T) = \int\mathcal{D}r(t)\exp\left[\int_{t'}^{t''}\left(\frac{m}{2}\dot{r}^2 - V_{eff}(r)\right)dt\right]. \tag{8}$$

Journal of Physics: Conference Series **574** (2015) 012001 doi:10.1088/1742-6596/574/1/012001

A convenient way to derive the corresponding transformation formula is to use the energy-dependent Green function G of the propagator K_l. By achieving the point canonical transformation and the time transformation,

$$\begin{cases} r &= f(q) \\ dt &= [f'(q)]^2 d\mathfrak{s}, \end{cases} \tag{9}$$

we obtain the following Fourier transform formula

$$K_l(r'', r'; T) = \frac{1}{2\pi i} \int G(r'', r', E) e^{-iET/\hbar} dE, \tag{10}$$

where

$$G(r'', r', E) = \frac{i}{\hbar} \left[f'(q')f'(q'')\right]^{1/2} \int_0^{\infty} \widehat{K}_l(q'', q'; s'') ds'', \tag{11}$$

Here

$$\widehat{K}_l(q'', q'; s'') = \int \mathcal{D}q(\mathfrak{s})$$
$$\exp\left[\frac{i}{\hbar}\int_0^{s''} \left(\frac{m}{2}\dot{q}^2 - f'^2(q)\left[V(f(q)) - E\right] - \Delta V(q)\right) d\mathfrak{s}\right], \tag{12}$$

and

$$\Delta V(q) = \frac{\hbar^2}{8m}\left[3\frac{[f''(q)]^2}{[f'(q)]^2} - 2\frac{f'''(q)}{f'(q)}\right]. \tag{13}$$

The functions f', f'' and f''' are the successive derivatives of the function f.

$$\begin{aligned} \widehat{K}_l(q'', q'; s'') = \exp\left[\frac{is''(E + C - A)}{\hbar\alpha^2}\right] \times \\ \int \mathcal{D}q(\mathfrak{s}) \exp\left[\frac{i}{\hbar}\int_0^{s''} \frac{m}{2}\dot{q}^2 - \frac{\hbar^2}{2m}\left[\frac{\eta(\eta-1)}{\sinh^2 q} - \frac{\upsilon(\upsilon-1)}{\cosh^2 q}\right] d\mathfrak{s}\right] \\ = \exp\left[\frac{is''(E + C - A)}{\hbar\alpha^2}\right] K_l^{MPT}(q'', q'; s'') \end{aligned} \tag{14}$$

with

$$\eta = \frac{1}{2} \pm \sqrt{\frac{-2m(E + C + A)}{\alpha^2\hbar^2}},$$
$$\upsilon = \frac{1}{2} \pm \sqrt{1 + \frac{8mB}{\alpha^2\hbar^2}}$$

and K_l^{MPT} is the path integral of the modified Pöschl-Teller potential which is a known solved problem [?] for witch the bound states wave functions are explicitly given by :

$$\chi_{l,n}^{(k_1,k_2)}(q) = N_n^{(k_1,k_2)}(\sinh q)^{2k_2-1/2}(\cosh q)^{-2k_1+3/2} \times {}_2F_1(-k_1+k_2+k, -k_1+k_2-k+1; 2k_2; -\sinh^2 q), \tag{15}$$

where $k = k_1 - k_2 - n$, and

$$N_n^{(k_1,k_2)} = \frac{1}{\Gamma(2k_2)} \left(\frac{(2k-1)\Gamma(k_1+k_2-k)\Gamma(k_1+k_2+k-1)}{\Gamma(k_1-k_2+k)\Gamma(k_1-k_2-k+1)} \right)^{1/2}, \tag{16}$$

The eigenvalues are expressed by

$$E_n^{MPT} = -\frac{\hbar^2}{2m} \left[2(k_1 - k_2 - n) - 1\right]^2. \tag{17}$$

Calculating $K_l(q'', q'; s'')$ allows us to obtain the Green function. Since this is known, the whole energy spectrum is obtained from its poles, and from the residues at the poles we obtain the corresponding wave functions.

Substituting (15) in (14), we get

$$\widehat{K}_l(q'', q'; s'') = \sum_{n=0}^{N_m} \exp\left\{ \frac{i}{\hbar} s'' \left[\frac{E+C-A}{\alpha^2} - E_n^{MPT} \right] \right\} \chi_{l,n}^{(k_1,k_2)}(q'') \chi_{l,n}^{*(k_1,k_2)}(q') + \int_0^\infty dK \exp\left\{ \frac{i}{\hbar} s'' [\frac{E+C-A}{\alpha^2} - \frac{\hbar^2 K^2}{2m})] \right\} \chi_K^{(k_1,k_2)}(q'') \chi_K^{*(k_1,k_2)}(q'). \tag{18}$$

Then, by integrating this latter over the pseudo-time parameters s'', the Green function G is written as:

$$G_l(r'', r'; E) = \sum_{n=0}^{N_m} \frac{\chi_{l,n}^{dWS(k_1,k_2)}(r'') \chi_{l,n}^{dWS*(k_1,k_2)}(r')}{E_{n,l}^{dWS} - E} + \int_0^\infty dK \frac{\chi_{l,K}^{dWS}(r'') \chi_{l,K}^{dWS*}(r')}{(\hbar^2 K^2 \alpha^2/2m) - A - C - E}. \tag{19}$$

Following [20], the values of k_1, k_2, satisfying the appropriate boundary conditions for $r \to 0$ and $r \to \infty$, are respectively: $k_1 = \frac{1}{2}\left(1 + \sqrt{1 + \frac{8mB}{\alpha^2\hbar^2}}\right) \equiv \frac{1}{2}(1+s)$,
$k_2 = \frac{1}{2}\left(1 + \frac{1}{2}(s - 2n - 1) - \frac{2mA}{\alpha^2(s-2n-1)}\right)$.

3. Energy spectrum and wave functions

Substituting the values of k_1, k_2 in (15) and (17), and by considering $u = \frac{1}{2}[1 + \tanh \alpha r]$, we get

$$\begin{aligned} \chi_{l,n}^{dWS(k_1,k_2)}(r) &= \sqrt{\alpha} N_n^{(k_1,k_2)} u^{k_2-(1/2)} (1-u)_2^{(s/2)-k_2-n} \times F_1(-n, s-n; 2k_2; u) \\ &= \left[\left(1 - \frac{4mA}{\hbar\alpha^2(s-2n-1)^2}\right) \frac{(s-2k_2-2n)\, n!\,\Gamma(s-n)}{\Gamma(s+1-n-2k_2)\Gamma(2k_2+n)} \right]^{\frac{1}{2}} \\ &\quad \times 2^{n+(1-s)/2} (1 - \tanh \alpha r)^{s/2-k_2-n} \\ &\quad \times (1 + \tanh \alpha r)^{k_2-1/2} P_n^{(s-2k_2-2n, 2k_2-1)}(\tanh \alpha r), \end{aligned} \tag{20}$$

where

$$\begin{aligned} E_{n,l}^{dWS} &= \alpha^2 E_n^{MPT} + A - C, \\ &= -\frac{\hbar^2(s-2n-1)^2\alpha^2}{8m} - \frac{2mA^2}{\hbar^2\alpha^2(s-2n-1)^2} + C. \end{aligned} \quad (21)$$

This expression is perfectly identical to that given in refs.[17], obtained with the framework of the Schrödinger formalism.

4. Concluding remarks

We have obtained the bound state solutions of the Feynman Propagator for the deformed Woods-Saxon potential using space-time transformations. The eigenvalue have been obtained from the poles of the Green function for various n and ℓ quantum numbers. We find that our results are in excellent agreement with the findings of other methods.

To conclude, our method is efficient in solving this type of potentials. We intend in a near future, to expand the path integral method to a more general form of potentials and relativistic cases.

References

[1] Ikhdair S M 2011 Phys. Scr. 83 015010.
[2] Zhang L H, Li X P and Jia C S 2011 Int. J. Quantum Chem.111 1870
[3] Eshghi M and Mehraban H 2011 Few-Body Syst. 52 41.
[4] Agboola D 2011 Pramana J. Phys. 76 875.
[5] Ikhdair S M 2011 J. Math. Phys. 52 052303.
[6] Diaf A and Chouchaoui A 2011 Phys. Scr. 84 015004.
[7] Diaf A, Hachama M, Can. Jour. Phys., 91(12) 1081 (2013)
[8] Woods R D and Saxon D B 1954 Phys. Rev. 95 577
[9] Williams W S C 1996 Nuclear and Particle Physics, Clarendon, Oxford
[10] Garcia F et al. 1999 Eur. Phys. J. A 6 49
[11] Goldberg V et al. 2004 Phys. Rev. C 69 031302
[12] Syntfelt A et al. 2004 Eur. Phys. J. A 20 359
[13] Diaz-Torres A and Scheid W 2005 Nucl. Phys. A 757 373
[14] Guo J Y and Sheng Q 2005 Phys. Lett. A 338 90
[15] S. Flügge, Practical Quantum Mechanics, Vol. 1 (Springer, Berlin, 1994)
[16] V.H.Badalov, H.I.Ahmadov, A.I.Ahmadov, Int.J.Mod.Phys.E18, No 3, 631-641 (2009)
[17] S. M. Ikhdair, B. J. Falaye, M. Hamzavi, CHIN. PHYS. LETT. Vol. 30, No. 2 (2013) 020305.
[18] D.C. Khandekar, S.V. Lawande, K.V. Bhagwat, Path Integral Methods and their Applications, World Scientific, Singapore, 1986.
[19] H. Kleinert, Path integrals in Quantum Mechanics, Statistics and Polymer Physics, World scientific, Singapore, 2009.
[20] C. Groshe J. Phys. A. Math. Gen. 22, (1989) 5073.

Family of higher order exponential variational integrators for split potential systems

Odysseas Kosmas and Sigrid Leyendecker
Chair of Applied Dynamics, University of Erlangen-Nuremberg, Germany
E-mail: odysseas.kosmas@ltd.uni-erlangen.de
E-mail: sigrid.leyendecker@ltd.uni-erlangen.de

Abstract. In the present work, we derive a family of higher order exponential variational integrators for the numerical integration of systems containing slow and fast potential forces. To increase the order of variational integrators, first the discrete Lagrangian in a time interval is defined as a weighted sum of the evaluation of the continuous Lagrangian at intermediate time nodes while expressions for configurations and velocities are obtained using interpolating functions that can depend on free parameters. Secondly, in order to choose those parameters appropriately, exponential integration techniques are embedded. When the potential can be split into a fast and a slow component, we use different quadrature rules for the approximation of the different parts in the discrete action. Finally, we study the behavior of this family of integrators in numerical tests.

1. Introduction

In many Hamiltonian systems, the potential part of the corresponding energy function is composed of different parts with strongly varying dynamics. As an example, in classical molecular dynamics, the potential energy includes contributions of several types of atomic interaction. These interactions lead to extremely stiff potentials which force the solution of the equations of motion to oscillate on a very small time scale [1].

Following [1, 2, 3], in order to derive numerical integration schemes for these systems, we split their potential energy into a fast and a slow component. For the resulting Hamilton function, we derive a family of higher order exponential variational integrators that use different quadrature rules in the discrete action corresponding to the different potentials.

2. Review of variational integrators using interpolation techniques

For the derivation of high order variational integrators, we need to recall discrete variational calculus [4]. A discrete Lagrangian is a map $L_d : Q \times Q \to \mathbb{R}$ which may be considered as an approximation of a continuous action with Lagrangian $L : TQ \to \mathbb{R}$, i.e. $L_d(q_k, q_{k+1}) \approx \int_{t_k}^{t_{k+1}} L(q, \dot{q})dt$. The action sum $S_d : Q^{N+1} \to \mathbb{R}, N \in \mathbb{N}$ corresponding to the Lagrangian L_d is defined as $S_d(\gamma_d) = \sum_{k=0}^{N-1} L_d(q_k, q_{k+1})$, with $\gamma_d = (q_0, \ldots, q_N)$ representing the discrete trajectory. The discrete Hamilton principle states that a motion γ_d of the discrete mechanical system extremizes the action sum, i.e. $\delta S_d = 0$. By differentiation and rearrangement of the terms and having in mind that both q_0 and q_N are fixed, the discrete Euler-Lagrange equations (DEL) are obtained [4]

$$D_2 L_d(q_{k-1}, q_k) + D_1 L_d(q_k, q_{k+1}) = 0, \qquad k = 1, \ldots, N-1 \tag{1}$$

where the notation $D_i L_d$ indicates the slot derivative with respect to the i-th argument of L_d.

The definition of the discrete conjugate momentum at time steps k and $k+1$ reads $p_k = -D_1 L_d(q_k, q_{k+1})$, $p_{k+1} = D_2 L_d(q_k, q_{k+1}), k = 0, \ldots, N-1$. The latter equations, also known as position-momentum form of a variational integrator, can be used when an initial condition (q_0, p_0) is known, to obtain (q_1, p_1).

To construct high order methods, we approximate the action integral along the curve segment between q_k and q_{k+1} using a discrete Lagrangian that depends only on the end points. We obtain expressions for configurations q_k^j and velocities $\dot{q}_k^j$ for $j = 0, \ldots, S-1$, $S \in \mathbb{N}$ at time $t_k^j \in [t_k, t_{k+1}]$ by expressing $t_k^j = t_k + C_k^j h$ for $C_k^j \in [0,1]$ such that $C_k^0 = 0, C_k^{S-1} = 1$ using

$$q_k^j = g_1(t_k^j)q_k + g_2(t_k^j)q_{k+1}, \qquad \dot{q}_k^j = \dot{g}_1(t_k^j)q_k + \dot{g}_2(t_k^j)q_{k+1}, \tag{2}$$

where $h \in \mathbb{R}$ is the time step. We choose functions

$$g_1(t_k^j) = \sin\left(u - \frac{t_k^j - t_k}{h}u\right)(\sin u)^{-1}, \quad g_2(t_k^j) = \sin\left(\frac{t_k^j - t_k}{h}u\right)(\sin u)^{-1}, \tag{3}$$

to represent the oscillatory behavior of the solution, see [5, 6]. For continuity, $g_1(t_{k+1}) = g_2(t_k) = 0$ and $g_1(t_k) = g_2(t_{k+1}) = 1$ is required.

For any choice of interpolation used, we define the discrete Lagrangian by the weighted sum $L_d(q_k, q_{k+1}) = h\sum_{j=0}^{S-1} w^j L(q_k^j, \dot{q}_k^j)$, where it can be easily proved that for maximal algebraic order $\sum_{j=0}^{S-1} w^j (C_k^j)^m = \frac{1}{m+1}$, where $m = 0, 1, \ldots, S-1$ and $k = 0, 1, \ldots, N-1$ see [5, 6].

Applying the above interpolation technique with the trigonometric expressions (3), following the phase lag analysis of [5, 6], the parameter u can be chosen as $u = \omega h$. For problems that include a constant and known domain frequency ω (such as the harmonic oscillator) the parameter u can be easily computed. For the solution of orbital problems of the general N-body problem, where no unique frequency is given, a new parameter u must be defined by estimating the frequency of the motion of any moving point mass [6].

3. Exponential integrators

We now consider the Hamiltonian systems

$$\ddot{q} + \Omega q = g(q), \qquad g(q) = -\nabla U(q), \tag{4}$$

where Ω is a diagonal matrix and $U(q)$ is a smooth potential function. We are interested in the long time behavior of numerical solutions when ωh is not small.

Since $q_{k+1} - 2\cos(h\omega)q_k + q_{k-1} = 0$ is an exact discretisation of (4), see [10, 11, 12], we can consider the numerical scheme

$$q_{k+1} - 2\cos(h\omega)q_k + q_{k-1} = h^2\psi(\omega h)g(\phi(\omega h)q_k), \tag{5}$$

where the functions $\psi(\omega h)$ and $\phi(\omega h)$ are even, real-valued functions satisfying $\psi(0) = \phi(0) = 1$, see [1]. The resulting methods using the latter numerical scheme are known as exponential integrators (for some examples of those integrators see [1, 7, 8, 9]).

3.1. Exponential high order variational integrators

If we now use the phase fitted variational integrator in (4) , the discrete Euler-Lagrange equations (1) read

$$q_{k+1} + \Lambda(u, \omega, h, S)q_k + q_{k-1} = h^2\Psi(\omega h)g(\Phi(\omega h)q_k), \tag{6}$$

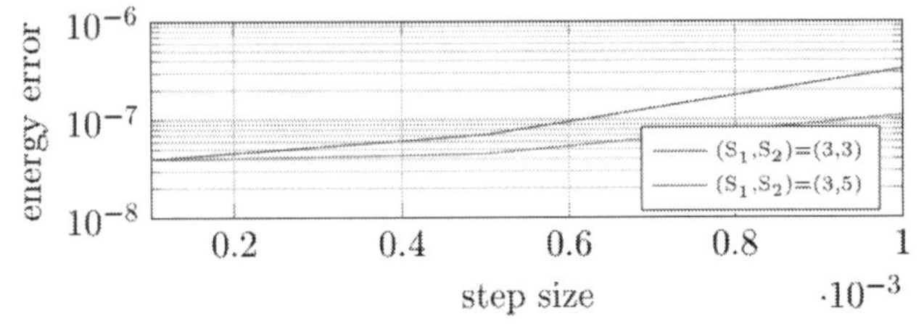

Figure 1: Modified satellite solar system. Global error for the energy error using the proposed family of exponential methods (red line) for $S_1 = 3$, $S_2 = 5$ and the ones that use $S = 3$ (blue line).

where

$$\Lambda(u, \omega, h, S) = \frac{\sum_{j=0}^{S-1} w^j \left[\dot{g}_1(t_k^j)^2 + \dot{g}_2(t_k^j)^2 - \omega^2 (g_1(t_k^j)^2 + g_2(t_k^j)^2) \right]}{\sum_{j=0}^{S-1} w^j \left[\dot{g}_1(t_k^j) \dot{g}_2(t_k^j) - \omega^2 g_1(t_k^j) g_2(t_k^j) \right]}. \tag{7}$$

Exponentially fitted methods using phase fitted variational integrators can be derived when

$$\Lambda(u, \omega, h, S) = -2\cos(\omega h). \tag{8}$$

holds. Thus, phase fitted variational integrators using trigonometric interpolation can be considered as exponential integrators.

4. Family of higher order exponential variational integrators for split potential systems

We focus in the derivation of a family of exponential high order variational integrators for systems containing slow and fast potentials as described in Section 1. Following [1, 2], we split the potential energy of the Lagrangian to fast and slow terms, i.e.

$$L(q, \dot{q}) = T(\dot{q}) - V^f(q) - V^s(q), \tag{9}$$

where $T(\dot{q}) = \frac{1}{2}\dot{q}^T M \dot{q}$ is the kinetic energy of the system (M is a constant mass matrix). In order to use different quadrature rules to approximate the contribution of each potential to the action, we use different numbers of intermediate points for each potential. Denoting the numbers of intermediate points for the slow and the fast potential by S_1 and S_2 respectively, we restrict ourselves to choices that $S_1 < S_2$ (the choice $S_1 = S_2$ creates the exponential integrators of Section 2).

Following Section 2 (see also [5, 6]) the discrete Lagrangian corresponding to (9) is

$$L_d^{S_1,S_2}(q_k, q_{k+1}) = h \left[\sum_{j=0}^{S_1-1} w_1^j \left[T(\dot{q}(t_k^j)) - V^s(q(t_k^j)) \right] - \sum_{j=0}^{S_2-1} w_2^j V^f(q(t_k^j)) \right], \tag{10}$$

where intermediate positions and velocities are given in (2) and (3), for $S = S_1$ and $S = S_2$ respectively.

5. Satellite solar system

As a numerical example, we choose the modified solar system with two planets (with masses $m_1 = 1$ and $m_2 = m_3 = 10^{-2}$) and a satellite ($m_4 = 10^{-4}$) which moves rapidly around the mass m_2, see [1]. Considering initial configurations $q_1 = (0,0)$, $q_2 = (1,0)$, $q_3 = (4,0)$, $q_4 = (1.01,0)$ and initial velocities $\dot{q}_1 = (0,0)$, $\dot{q}_2 = (0,1)$, $\dot{q}_3 = (0,0.5)$, $\dot{q}_4 = (0,0)$, the motion of the two planets is nearly circular with periods close to 2π and 14π, respectively [1]. The problem is then described by the Lagrange function

$$L(q,\dot{q}) = \frac{1}{2}\sum_{i=1}^{4} m_i \dot{q}_i^2 - V^f(q) - V^s(q), \tag{11}$$

where

$$V^s(q) = \sum_{i<j,(i,j)\neq(2,4)}^{4} \frac{m_i m_j}{||q_i - q_j||}, \qquad V^f(q) = \frac{m_2 m_4}{||q_2 - q_4||}. \tag{12}$$

Figure 1 shows the comparison of errors in total energy of the system at $t = 1$ for time steps $h \in \{10^{-4}, 5 \cdot 10^{-4}, 10^{-3}\}$ for the proposed family of exponential methods (red line) for $S_1 = 3$, $S_2 = 5$ and the ones that use $S = 3$ (blue line), see [6]. In that, it is clear that for all the step sizes that are tested, the smallest energy errors are obtained when using the splitting technique of Section 4.

6. Conclusions

A family of higher order exponential variational integrators for the numerical integration of systems containing slow and fast potential forces is presented. For these methods, the discrete Lagrangian in any time interval is defined as a weighted sum of the evaluation of the continuous Lagrangian at intermediate time nodes. In order to use different quadrature rules for the different potential terms, a splitting of their potential energy into a fast and a slow component is addressed, and different numbers of intermediate points for each potential is used. Preliminary numerical results for the case of the modified solar system, show better behavior of the proposed simulation technique, when compared to the one derived without potential energy splitting.

[1] E. Hairer, C. Lubich, G. Wanner, Geometric numerical integration illustrated by the Störmer-Verlet method, Acta Numerica 12 (2003) 399.
[2] A. Stern and E. Grinspun. Implicit-explicit integration of highly oscillatory problems. SIAM Multiscale Modeling and Simulation 7 (2009) 1779.
[3] S. Leyendecker, S. Ober-Blöbaum, A variational approach to multirate integration for constrained systems Multibody Dynamics, Computational Methods in Applied Sciences 28 (2013) 97.
[4] J. Marsden, M. West, Discrete mechanics and variational integrators, Acta Numerica 10 (2001) 357.
[5] O.T. Kosmas, D.S. Vlachos, Phase-fitted discrete Lagrangian integrators, Computer Physics Comm. 181 (2010) 562.
[6] O.T. Kosmas, S. Leyendecker, Phase lag analysis of variational integrators using interpolation techniques, PAMM Proc. Appl. Math. Mech. 12 (2012) 677.
[7] W. Gautschi, Numerical integration of ordinary differential equations based on trigonometric polynomials, Numer. Math. 3 (1961) 381.
[8] P. Deuflhard, A study of extrapolation methods based on multistep schemes without parasitic solutions, Z. Angew. Math. Phys. 30 (1979) 177.
[9] B. García-Archilla, M.J. Sanz-Serna, R.D. Skeel, Long-time-step methods for oscillatory differential equations, SIAM J. Sci. Comput. 20 (1999) 930. SIAM J. Sci. Comput.
[10] R.P. Agarwal, Difference equations and inequalities, Marcel Dekker, New York 2000.
[11] J.L.Cieśliński, B.Ratkiewicz, On simulations of the classical harmonic oscillator equation by difference equations, Adv. Difference Eqs. (2006) 40171.
[12] R.B.Potts, Differential and difference equations, Am. Math. Monthly 89 (1982) 402.

Percolation of interdependent networks with degree-correlated inter-connections

Akito Igarashi and Tomohiro Kuse

Department of Applied Mathematics and Physics, Kyoto university, Kyoto, 606-8501 Japan

E-mail: igarashi@i.kyoto-u.ac.jp

Abstract. In interdependent networks, failures of nodes in one constituent network lead nodes in another network to fail. This happens recursively and leads to a cascade of failures. It is known that the interdependent networks with random inter-connections have weaker robustness than the individual networks. However, if the interdependent networks have degree correlations between the networks constructing them as in the actual cases, the robustness of the interdependent networks may be changed. In this paper, we perform numerical simulations on interdependent networks and obtain the giant cluster sizes after the cascade of failures in order to evaluate the robustness. We show that when a interdependent network has a positive degree inter-correlation, it has the stronger robustness than that for the networks with no degree correlation. We show not only the numerical simulation results but theoretical ones for the robustness of the interdependent networks.

1. Introduction

Most modern systems are not simple. They are complex and construct large networks. Then, it is important that the networks are robust, in other words whether the networks are functional after random removal of a fraction [1]. Furthermore, several modern systems are more complex. They are not isolated networks but connected to one another. They are called interdependent networks. For example, the infrastructures such as water supply, transportation, and fuel and power stations are coupled together mutually. Even if systems are well robust individually, the whole coupled system may have weaker robustness than the individual systems [2]. Random removal of a small fraction of nodes from a constituent network can produce an iterative cascade of failures in total interdependent networks. Electrical blackouts frequently result from a cascade of failures in interdependent networks with the power grids and their communication support systems. For example, the blackout that affected much of Italy on 28 September 2003 occurred by the cascade in the electric network system constituted by the power and communication networks.

In this paper, we consider correlations between the networks that construct interdependent networks. The purpose of this paper is to find how the correlation or the degree distribution affects the robustness of the whole network. We perform numerical simulations for Erdős-Rényi (ER) networks and Barabási-Albert (BA) networks. ER networks have Poissonian degree distributions, and BA networks have power-law degree distributions. Therefore, BA networks have nodes with very large degree that are called hubs. Both of the ER networks and the BA networks are robuster when they have degree inter-correlations, and the difference between the robustness with no correlation and strong correlations in the network constructed by the BA networks is larger than that in the network constructed by the ER networks. We show not only numerical simulation results but theoretical ones. Although several workers [3][4][5] theoretically investigate the robustness of interdependent networks with degree inter-correlations, there have not been any theoretical research for the robustness of interdependent networks consisting of random and scale-free networks with arbitrary inter-correlation. The theory can be used for networks with any degree distributions and any correlations between two constituent networks. The theoretical results fit well with those of numerical simulation results.

2. Model

2.1. Interdependent networks

As an interdependent network, we consider a network constructed from connected two networks, A and B with the same number of nodes. Each node A_i in network A depends on a node B_i in network B and vice versa. That is, A_i and B_i are mutually connected. If node A_i stops functioning owing to an attack or a failure, node B_i also stops functioning. Similarly, if node B_i stops functioning then node A_i stops functioning, since each node in one network can depend only on one node in the other network. In the interdependent networks, we begin by randomly removing a fraction $(1-p)$ of nodes in network A, and failures cascade as in Figure 1. A group that nodes are mutually connected with links is called a cluster. Therefore, if the number of the nodes that the largest or giant cluster contains is large, the network is robust.

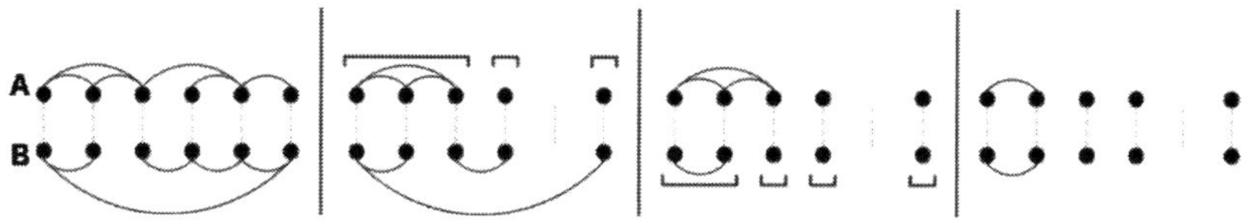

Figure 1. A model of a cascade of failures

1. Remove failures. We remove a fraction $(1-p)$ of A-nodes, where p is the percentage of remaining nodes. where A-nodes mean the nodes in the network A. Then we remove the nodes in the network B, B-nodes, that are connected to the removed A-nodes, as shown in the second figure of Figure 1.

2. Remove the links in the network B across the different clusters in the network A. For example, the network A is divided into three clusters in the second figure of Figure 1, and we remove two links in the network B that connect nodes which are, respectively, in the different clusters in the network A. As the result, we obtain the third figure of Figure 1.

3. Remove the links in the network A across the different clusters in the network B. It is the similar procedure to step 2, as shown in the fourth figure of Figure 1.

4. Repeat steps 2 and 3 until the convergence.

After the convergence, if the percentage μ_∞ of nodes in the giant cluster is high, the network is robust because many nodes are connected and the network is functional. This model shows a percolation phase transition where p equals a threshold p_c. That is, p is smaller than p_c, then $\mu_\infty = 0$. Therefore the network has no cluster. Otherwise, p is larger than p_c, then $\mu_\infty > 0$. That is, the network has a cluster. The smaller p_c is, the robuster the network is.

2.2. Correlation between networks

There is a inter-correlation between two networks in the real world. Therefore it is important that we consider the networks with not only no degree inter-correlation but a degree inter-correlation. We make a interdependent network with a degree correlation between constituent networks by connecting a node in a network to one in another network sampled according to its degree. Actually a node is sampled with a probability proportional to k^{α} from a network and pairs with a similarly sampled node from another network, where k is the degree of the node and α is a parameter for the correlation. When α is 0, the networks have no inter-correlation because the connected nodes are sampled at random. On the other hand, when α is ∞, the networks have a very strong correlation. This procedure makes a positive correlation or no correlation. It is, however, difficult to make a negative correlation.

3. Simulation

We perform numerical simulations for interdependent networks constructed by two Erdős-Rényi (ER) networks and ones constructed by two Barabási-Albert (BA) networks with the total nodes of each constituent networks $N = 25600$ and the average degree $\langle k \rangle = 4$ for various inter-correlation coefficient of degrees. For both ER and BA networks, we find that the stronger the correlation is, the robuster the networks are. Threshold p_c is approximately a linear function of the correlation coefficient r. In individual networks, large degree nodes are hard to be removed from a cluster because they have many links. However, when the interdependent networks have no correlation, large degree nodes easily remove from clusters because the nodes in the other network inter-connected to the large degree nodes may have small degrees and be easily removed. When the networks have strong positive correlations, large degree nodes usually connect to large degree nodes. Therefore both nodes are hard to be removed from the clusters. Then, the networks that have strong correlations are robuster than those with no correlation. Comparing BA networks with ER networks, we find that the difference between robustness with no correlation and the strong correlations for BA networks is larger than one for ER networks, because BA networks have more hubs than ER networks. If the hubs are removed, they greatly affect the cascade. Moreover, for no inter-correlation networks constructed by BA networks, such an iterative process of cascading failures leads to a first-order phase transition. When the correlation is strong, a second-order phase transition is, however, shown, although a first-order phase transition is observed for non- and inter-correlated networks with ER networks.

4. Theory

As in [2], we use the generating function method [8] for evaluating robustness of interdependent networks. Although the degree distribution of each network is unchanged in the cascading processes for randomly inter-connected interdependent networks, one for interdependent networks with the degree inter-correlation is varying in the cascading process. We take into account the variation of the distribution theoretically. In the theory, we use the results of [6] and [7] for the degree distribution of nodes which is not in the giant cluster and one for the constituent network in which some nodes which are linked with failed nodes in another network are removed. The robustness of the networks are theoretically calculated and compared with simulation results in the next section.

5. Comparison between theoretical and simulation results

Figure 2 shows the comparison between the theoretical and numerical simulation results for the networks with BA networks where degree distributions are $P(k) \propto k^{-3}$. The difference between the theoretical and simulation results is small for $\mu_{\infty} > 0.2$, but it is large for $\mu_{\infty} < 0.2$. However, the value for $\mu_{\infty} > 0.2$ is more important than the value for $\mu_{\infty} < 0.2$ because when $\mu_{\infty} < 0.2$, the network is almost broken and the functional network needs a certain value of μ_{∞}. The theoretical results for ER networks and BA networks where the degree distribution are $P(k) \propto k^{-2.7}$ agree well with the numerical simulation results similarly. Therefore the theoretical results reproduce simulation results very well for both ER networks and BA networks, with any correlations.

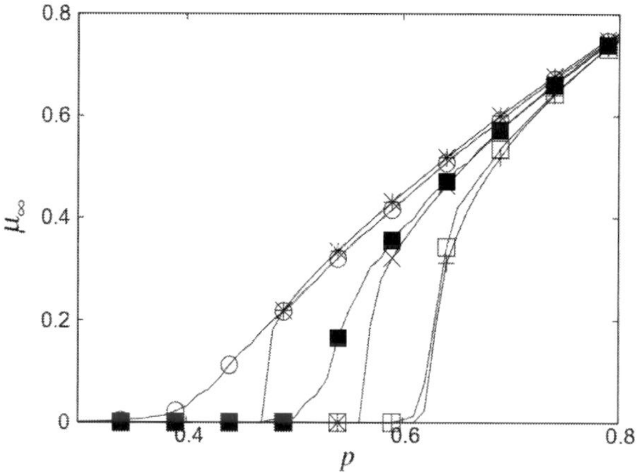

Figure 2. Theoretical and simulation results for μ_∞ as a function of p for BA networks. Circles, filled squares and open squares are simulation results for r=0.986, 0.365 and 0 and asterisks, crosses and pluses are theoretical ones for r=0.986, 0.365 and 0, respectively.

6. Conclusion

In this paper, we study interdependent networks with degree-correlated inter-connections. Although interdependent networks generally have weak robustness, they are robuster if the networks have the strong degree correlations between constituted networks. The robustness with the strong correlation is higher when the network has large hubs. Specifically BA networks is robuster than ER networks, and BA networks with degree distribution $P(k) \propto k^{-2.7}$ is robuster than BA networks with degree distribution $P(k) \propto k^{-3.0}$. Furthermore we propose a simple theoretical method for this model. Then, the theoretical results almost fit to the numerical simulation ones and we can use the theory for networks with any degree distribution and any correlation. It is very useful to treat modern complex systems with the theory.

Acknowledgment

This work has been partially supported by a Grant-in-Aid for Scientific Research from the Japan Society for the Promotion of Science (No. 25390151).

[1] D. S. Callaway, M. E. J. Newman, S. H. Strogatz, and D. J. Watts, Network Robustness and Fragility: Percolation on Random Graphs, Phys. Rev. Lett. **85**, 25 (2000).
[2] S. V. Buldyrev, R. Parshani, G. Paul, H. Eugene Stanley, S. Havlin, Catastrophic cascade of failures in interdependent networks, Nature **464**, 1025 (2010).
[3] S. V. Buldrev, N. W. Shere and G. A. Cwilich, Interdependent networks with identical degrees of mutually dependent nodes, Phys. Rev. E **83**, 016112 (2011).
[4] S. Watanabe and Y. Kabashima, Cavity-based robustness analysis of interdependent networks: Influences of intranetwork and internetwork degree-degree correlations, Phys. Rev. E **89**, 012808 (2014).
[5] B. Min, Su Do Yi, Kyu-Min Lee and K. -I. Goh, Network robustness of multiplex networks with interlayer degree correlations, Phys. Rev. E **89**, 042811 (2014).
[6] J. Shao, S. V. Buldyrev, L. A. Braunstein, S. Havlin, and H. E. Stanley, Structure of shells in complex networks, Phys. Rev. E **80**, 036105 (2009)
[7] Xuqing Huang, Jianxi Gao, S. V. Buldyrev, S. Havlin, and H. E. Stanley, Robustness of interdependent networks under targeted attack, Phys. Rev. E **83**, 065101 (2011)
[8] M. E. J. Newman, S. H. Strogatz, and D. J. Watts, Random graphs with arbitrary degree distributions and their applications, Phys. Rev. E **64**, 026118 (2001)

Critical phenomena in communication/computation networks with various topologies and suboptimal to optimal resource allocation

Marco Cogoni, Giovanni Busonera, Paolo Anedda, Gianluigi Zanetti
CRS4
E-mail: {marco.cogoni, giovanni.busonera, paolo.anedda, gianluigi.zanetti}@crs4.it

Abstract. We generalize previous studies on critical phenomena in communication networks [1,2] by adding computational capabilities to the nodes. In our model, a set of tasks with random origin, destination and computational structure is distributed on a computational network, modeled as a graph. By varying the temperature of a Metropolis Montecarlo, we explore the global latency for an optimal to suboptimal resource assignment at a given time instant. By computing the two-point correlation function for the local overload, we study the behavior of the correlation distance (both for links and nodes) while approaching the congested phase: a transition from peaked to spread $g(r)$ is seen above a critical (Montecarlo) temperature T_c. The average latency trend of the system is predicted by averaging over several network traffic realizations while maintaining a spatially detailed information for each node: a sharp decrease of performance is found over T_c independently of the workload. The globally optimized computational resource allocation and network routing defines a baseline for a future comparison of the transition behavior with respect to existing routing strategies [3,4] for different network topologies.

1. Introduction

Critical phenomena in random networks are an active research field not confined to theoretical physics or mathematics but their application to computer science, biology, traffic engineering and social sciences is increasingly common [5–7]. In this work the problem of congestion in computer networks is addressed, but it is easy to recognize the analogies with traffic flow, container logistics and energy distribution, just to name a few [5].

Ohira and Sawatari [1] first studied the onset of a phase transition within a simple model of communication network under different rates of traffic generation. Their model is defined by a set of nodes on a two dimensional lattice and controlled by a set of parameters such as packet generation rate and routing rules. The expected phase transition is observed by means of the average latency by which packets arrive at their destination during a dynamical simulation over time. In 2001, Solè and Valverde [2] slightly extended the model and studied the time-series dynamics of the number of packets at individual routers, and found a set of power laws at the critical point.

The present model tries to build on the previous attempts, in which only the network communication process has been considered, by (i) seamlessly including the computation time (information processing in general) into the total latency and (ii) avoiding the explicit temporal evolution by considering an average traffic over a given instant. In this way the typical problem of the queue size of dynamic simulations is entirely avoided [8] and the focus is set on the "instantaneous" behavior of the system. In a sense, the model computes the finite-difference load derivative during the time window for each computational node and network link. The model allows an extremely fast evaluation of the latency and this unfolds the possibility of finding the network's optimal resource allocation. The optimization is performed after setting the network topology and the random traffic distribution [9,10]. Here we try to set a baseline for comparing any routing algorithm with the best solution found by a Simulated Annealing (SA) process. Moreover, by varying the temperature in the Metropolis Montecarlo (MC), we estimate the sensitivity of the system performance with respect to the "amount of non-optimality" in the allocation of resources (both routing and processing).

We present the behavior of the average latency with respect to load and resource optimization for two different topologies: a two-dimensional lattice and a Barabási-Albert (BA) random network. We find a transition to the overloaded phase both for a critical load and for a critical temperature T_c. Finally, the two-point correlation function of the local overload for nodes and links is used to show how the system behaves while going from a free to a congested regime: the $g(r)$ for the nodes is found to sharply change its shape from localized to spread when the (MC) temperature is raised.

2. The Model

The model considers three basic components: (i) the physical support for the communication/computation process, (ii) the discrete tasks that are transported/executed, and (iii) the limited capability of the links/nodes to handle each subtask. The communication network is mapped onto a graph where nodes mimic the communicating agents (i.e. routers and servers in a computer network), and the links between them represent communication lines. Each node is characterized by its computational power and each link has its own nominal latency and bandwidth. The computational capability is measured by the number of executable task requests allowable in unit time. This should be intended in two ways: each node is characterized by its own "speed" and by the maximum number of concurrent unitary tasks which can be executed at nominal speed. With respect to the network, each link transports the information at nominal speed when its bandwidth is larger than or equal to the total requests of the tasks running on it (this is equivalent to the betweenness centrality measure of the link). In both cases of networking and computation, whenever the load surpasses the available capabilities, a linear slowdown/overload factor is introduced to take into account the increased time of execution of each subtask. Each link is modeled to be able to support several concurrent communications without overloading, while each computational node can execute more than one subtask at the same time.

We consider two topologies: a simple 2D lattice and a Barabási-Albert random network, with the same number of nodes, but different connectivity.

From the initial node through each stage and until the final node, the information is passed to the next step via a sequence of network links (see Fig.1): the length of each sequence is, in principle, bounded only by the total number of links in the network (no loops are permitted).

The time (latency) needed to complete the k-th task T^k is the sum of two components:

$$L_k^{total} = L_k^{CPU} + L_k^{net},$$

- the network latency L_k^{net} is the sum of the delays $L_{k,(n,m)}^{net}$ introduced by the direct hops $n \Leftrightarrow m$ (with nominal delay D_{nm}^{Net}) within the paths connecting each pair of subtask;
- the computation latency L_k^{CPU} is the sum of the delays $L_{k,n=1,2}^{CPU}$ due to all subtasks.

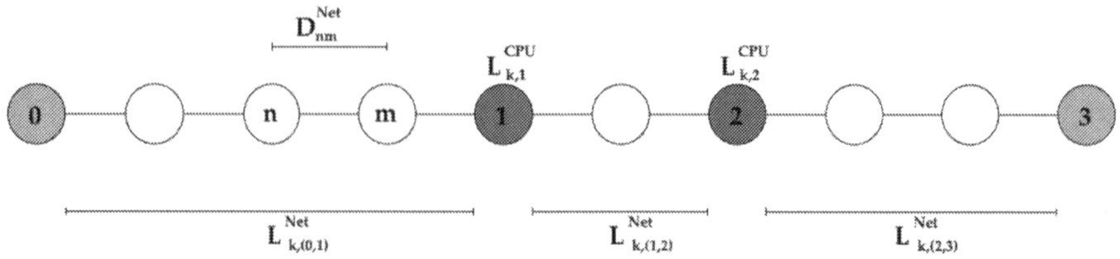

Figure 1. Task structure. Green circles are assigned as the input and output of the task. Red circles represent the processing stages. White circles are nodes used as simple routers between elementary network hops. The basic symbols used to compute the latency are depicted.

The optimizer tries to find the best possible combined performance for the tasks. After initializing the topology of the network and its capabilities, a random, uniform, distribution of traffic is generated. The SA procedure starts from a random allocation of the tasks over the network and typically has a poor performance since it is very unlikely that the computing nodes and the links will have a "good" load distribution (this could be easily estimated since it is equivalent to a balls-in-bins statistical problem).

Being able to simulate the system at constant temperature (T) is equivalent, in a sense, to having a parameter with which to set the amount of "errors" in the total resource allocation. At $T \approx 0$, the simulation ideally leads to an optimal routing and node allocation. Slowly increasing T drives the system through more realistically accessible states and finally to totally random resource allocations at high T.

Results

Simulations were run using several instances of BA networks and 2D lattices with 100 nodes and random traffic. Each node is assigned the same unit power. Each link capability is proportional to the sum of the node degrees. Tasks are modeled with 2 processing stages, each with a computational request of one fourth of unit power. This ratio sets the theoretical maximum computational requests before CPU overload to 200 tasks. Each point on the graphs is obtained by averaging over 10 runs. As shown in Fig.2 (left panel, CPU), the curve labeled with $T = 10^{-5}$ represents the optimal latency at all loads and the transition sharply occurs with a load of $nt = 150$ tasks (note that the maximum theoretical CPU capacity would allow $nt = 200$ with no overload), then a linear latency growth follows with no saturation at high loads. In the central panel (NET), the curve labeled with $T = 10^{-5}$ (optimal), starts a smooth transition at $nt = 175$. then the slope increases up to $nt = 225$ and finally saturates for $nt > 250$. Up to $T = 2.0 \cdot 10^{-3}$ nothing changes, then (solid violet curve) the latency worsens for low loads and grows from about $nt = 75$. The very high load region $nt > 250$ does not suffer at all from resource allocation errors up to $T = 2.0 \cdot 10^{-2}$. Another transition is observed for $T \geq 2.0 \cdot 10^{-2}$ (red curve): the whole curve grows (low and high loads) and has a steep slope at $nt = 25$. For $T > 2.0 \cdot 10^{-2}$ the latency curve shows an increasing vertical offset, even for $nt = 25$. Summarizing, the critical load for both topologies (see Fig.3) starts well before the theoretical maximum CPU load of 200 tasks, but the average latency for the lattice is much larger than for the BA even though both share the same connectivity: The BA is able to resist higher loads than the lattice since the average path is much shorter, thus the network has a lesser impact on the global latency. The resource allocation quality (T) on network performance has a very strong impact near the critical load ($nt \approx 150$). Finally, the $g(r)$ of the CPU overload at $nt \approx 200$ (Fig.4), is a power law for $T < T_c$ but becomes uniformly spread at high values, while for the network starts at zero everywhere and at high T rises especially for distant links. In general for high T we observe an exponential growth of the average number of internal task hops that is only limited by the network size.

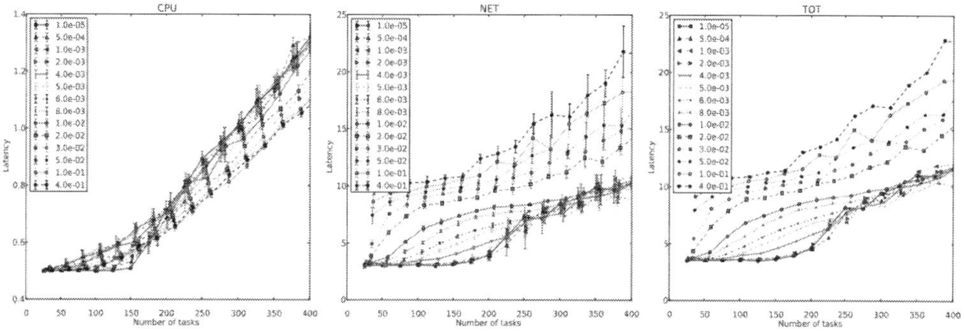

Figure 2. Barabasi random network: Average latency (right) and its components (computation on the left and networking in the center box) dependence on the load computed for different Montecarlo temperatures $(10^{-5}, 4 \cdot 10^{-3}, 10^{-2}, 10^{-1})$.

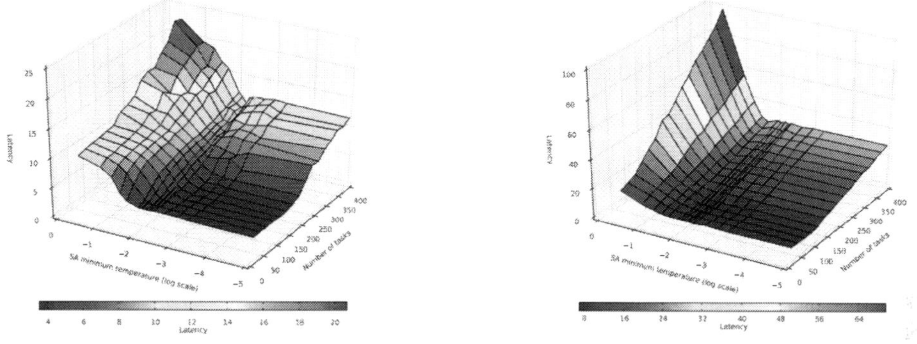

random network 2D lattice

Figure 3. 3D representation of how the global average latency depends on the traffic and on the Montecarlo temperature: The deep blue basin with low latency can be reached with low traffic and moderately bad routing or with high traffic but optimal routing/resource allocation

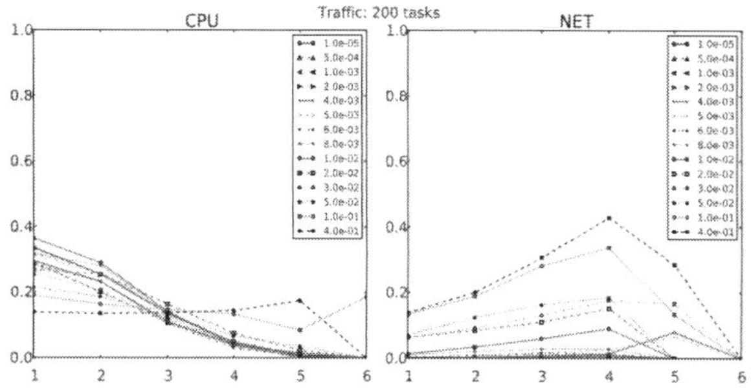

Figure 4. BA: Two-point correlation of the overload (nodes and links) with a traffic of 200 tasks for several temperatures. Once $T > 5 \cdot 10^{-2}$, the CPU overload spreads over all the network.

References

[1] Ohira T and Sawatari R 1998 *Physical Review E* **58** 193–195 URL http://journals.aps.org/pre/abstract/10.1103/PhysRevE.58.193
[2] Solé R and Valverde S 2001 *Physica A: Statistical Mechanics and its Applications* **289** 595–605 URL http://www.sciencedirect.com/science/article/pii/S0378437100005367
[3] Wang W, Wang B, Yin C, Xie Y and Zhou T 2006 *Physical Review E* 1–7 URL http://journals.aps.org/pre/abstract/10.1103/PhysRevE.73.026111
[4] Yan G, Zhou T, Hu B, Fu Z and Wang B 2006 *Physical Review E* 1–5 (*Preprint* 0505366v2) URL http://journals.aps.org/pre/abstract/10.1103/PhysRevE.73.046108
[5] Dorogovtsev S N, Goltsev A V and Mendes J F F 2008 *Rev. Mod. Phys.* **80**(4) 1275–1335 URL http://link.aps.org/doi/10.1103/RevModPhys.80.1275
[6] Barabási A L, Albert R and Jeong H 2000 *Physica A: Statistical Mechanics and its Applications* **281** 69–77 ISSN 03784371 URL http://linkinghub.elsevier.com/retrieve/pii/S0378437100000182
[7] Barabási A L 2009 *Science (New York, N.Y.)* **325** 412–3 ISSN 1095-9203 URL http://www.ncbi.nlm.nih.gov/pubmed/19628854
[8] Arrowsmith D K, Woolf M and Monfragon R J 2004 *Complex Dynamics in Communication Networks: Data Traffic, Topology and Congestion*
[9] Danila B, Sun Y and Bassler K 2009 *Physical Review E* 1–7 (*Preprint* arXiv:0902.2415v2) URL http://journals.aps.org/pre/abstract/10.1103/PhysRevE.80.066116
[10] Guimerà R, Díaz-Guilera A, Vega-Redondo F, Cabrales A and Arenas A 2002 *Phys. Rev. Lett.* **89**(24) 248701 URL http://link.aps.org/doi/10.1103/PhysRevLett.89.248701

A mathematical characterization of the gel point in sol-gel transition

A H Bilge[1], **O Pekcan** [2]

[1] Kadir Has University, Faculty of Engineering and Natural Sciences, Cibali, Istanbul, Turkey
[2] Kadir Has University, Faculty of Engineering and Natural Sciences, Cibali, Istanbul, Turkey

E-mail: ayse.bilge@khas.edu.tr

Abstract. We model the sol-gel transition in terms of Susceptible-Infected-Removed (SIR) and Susceptible-Exposed-Infected-Removed (SEIR) models and compare with experimental results. We show, numerically, that the "gel point" described as the onset of the gelation phenomena and measured experimentally, corresponds to an accumulation point of the extreme values of the derivatives of the gelation curve. We define the "critical point of a sigmoidal curve" as the limit of the points where the derivatives reach their extreme values, provided that this limit exists.

1. Introduction

Physical phenomena involving a phase transition are likely to be described by different mathematical models before and after the transition point. The polymerization of a monomer solution is such an example that displays the passage from the *sol* state to the *gel* state. In the sol state, monomers start to agglomerate, forming small clusters. There is a specific instant, called the *gel point* at which these clusters consolidate to form a giant network. In the steady-state, the gel fraction stabilizes to a value that may be less than 1, i.e, the sol and gel states may coexist. These transitions have been monitored and recorded by Pekcan and co-workers by a fluorescence technique for a variety of chemical and physical gels [1]. In most experiments, the sol-gel transition follows a sigmoidal curve with a sharp rise between the two stable states. At the exact instant of phase transition, physical properties of the monomer solution change; the gel point is either monitored by an independent experiment [2], or by finding the base point that gives the best fit to well-known power low growths [3]. The location of the gel point with respect to the inflection point of the sigmoidal curve is not fixed; in some types of experiments it is closer to the inflection point, in some other it is located earlier in time. Furthermore its location varies also with activation levels in the same experiment.

The Susceptible-Infected-Removed (SIR) and Susceptible-Exposed-Infected-Removed (SEIR) [8] epidemic models describe the spread of an infectious disease in a closed society. We have used these systems to model the gelation of sodium alginate-carragenan solution by fitting the curve for removed individuals to the gelation curve [4] for various levels of activation. We have noticed that the gelation curves obeying the "percolation model" and the "classical model" were described by the SIR and SEIR epidemic models respectively. This approach was motivated by the modeling of the sol-gel transition as a percolation model [5], [6], [7], providing a common basis with the spread of an epidemic in a closed society and the polimerization and gelation phenomena.

In the search for a description of the gel point in terms of the SIR and SEIR systems, we noticed that the points where the derivatives of the sigmoidal curve reach their global extremum formed a convergent sequence. We have used numerical techniques to compute the location of the limit point and we have observed that the location of the critical point agreed qualitatively with the gel point observed by dilatometric techniques [9]. Although we observed the existence of a limit point for the global extremum of the derivatives of a sigmoidal function for a variety of functions, the proof of this fact was extremely challenging. In [10] we have used Fourier and Hilbert transform techniques to prove the existence of a limit point of the global extreme values of a sigmoidal function under fairly general assumptions.

2. Epidemic models and the gelation mechanism

The Susceptible-Infected-Removed (SIR) and Susceptible-Exposed-Infected-Removed (SEIR) models describe the spread of an epidemic in a closed society. These models are based on a subdivision of the population into compartments, as indicated by the names of the models. It is assumed that the passage among the compartments are one-directional, i.e. in epidemiological framework, immunity, once acquired, cannot be lost. The strength of the epidemic depends both on the activity of the virus and on the level of interpersonal relations in the society. If the combination of these effects is strong enough, the number of infected individuals make a peak and then decrease steadily, as there are less and less susceptible individuals around [8].

In the Susceptible-Infected-Removed (SIR) model the time evolution of the number of "Susceptible (S)", "Infected (I)" and "Removed (R)" individuals is given by

$$\frac{dS}{dt} = -\beta IS, \quad \frac{dI}{dt} = \beta IS - \eta I, \quad \frac{dR}{dt} = \eta I, \tag{1}$$

where β and η are constants. In the SEIR model, the "Exposed (E)" individuals act as an intermediate step as seen from the equations below

$$\frac{dS}{dt} = -\beta IS, \quad \frac{dE}{dt} = \beta SI - \epsilon E, \quad \frac{dI}{dt} = \epsilon E - \eta I, \quad \frac{dR}{dt} = \eta I, \tag{2}$$

where β, ϵ and η are constants.

The link between epidemic spread and the gelation is the interpretation of both phenomena by "percolation" that typically represents the passage of gas or fluids trough porous media [11]. For the spread of a disease, susceptible individuals are analogous to empty sites while removed individuals are interpreted as filled sites of a percolating network. The passage from one state to the other is governed by the transition rules. Percolation models are built via simulations and take spatial evolutions into consideration. The SIR and SEIR models being systems of ODE's ignore spatial evolution of the disease and give only cumulative numbers.

3. The gelation of the polyacrilamide-sodium alginate composite

The sol-gel phase transition of polyacrilamide-sodium alginate composite with low and high Sodium Alginate concentrations is studied in [2]. The gelation curves for various levels of activation are presented in Figure 1. The steep, low amplitude curves correspond to low SA concentrations obeying percolation model. The slower rising high amplitude curves obey classical model.

In these experiments, the gel points are determined by a dilatometric technique. The critical point is located at the left of the inflection point of the gelation curve; it is farther away from the inflection point for high activation (low SA concentrations) levels).

Journal of Physics: Conference Series **574** (2015) 012005 doi:10.1088/1742-6596/574/1/012005

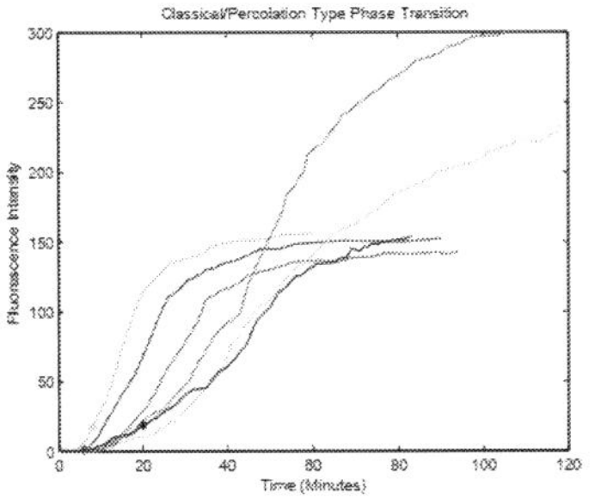

Figure 1. The gelation of the polyacrilamide-sodium alginate composite.

4. Derivatives of the gelation curve

In the search for a mathematical characterization of the gel point, we observed that successive derivatives of the sigmoidal curve representing the number of removed individuals seem to "diverge" near a point that agrees qualitatively with the "gel point" that has been experimentally determined. The mathematical set-up for this phenomenon is described below.

Let $y(x)$ be a smooth sigmoidal curve, i.e, monotone increasing with horizontal asymptotes as $t \to \pm\infty$ and let $\{x_{m,i}\}$ and $\{x_{a,i}\}$, $i = 1, 2, \ldots$, be the set of points where the derivatives of odd and even order reach their extreme values. If the sequences $\{x_{m,i}\}$ and $\{x_{a,i}\}$ are both convergent and they have a common limit x_c, this limit is called "the critical point of the phase transition".

The ith zero of $y^{(n)}$ is x_n^i. The inflection point of $y(x)$ is the unique zero of its second derivative $y^{(2)}$, denoted by x_2^1. x_2^1 lies in between the two zeros of the third derivative x_3^1 and x_3^2, and we have the order relations $x_3^1 < x_2^1 < x_3^2$. The fourth derivative has three zeros, satisfying the order relations $x_4^1 < x_3^1 < x_4^2$, $x_4^2 < x_3^2 < x_4^3$. Although the zeros of the third and fourth derivatives alternate, we can't say anything about the relative positions of x_2^1 and x_4^2. Thus, the observed regular behavior of the zeros near the "critical point" is not a straightforward consequence of the alternation of zeros of successive derivatives. The extrema of odd derivative correspond to the zeros of even derivatives and vice versa. Thus, if the critical point is to represent some type of break-point, the best that we can expect is that the sequences, $\{x_{m,i}\}$ and $\{x_{a,i}\}$, converge to a common limit point x_c.

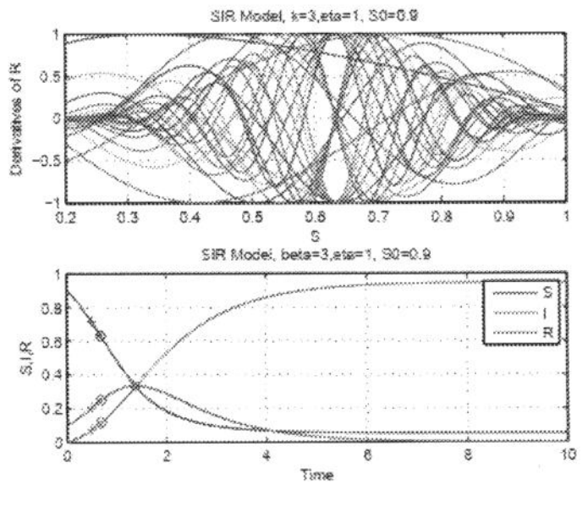

Figure 2. Derivatives of $R(t)$ for $k = \beta/\eta = 3$.

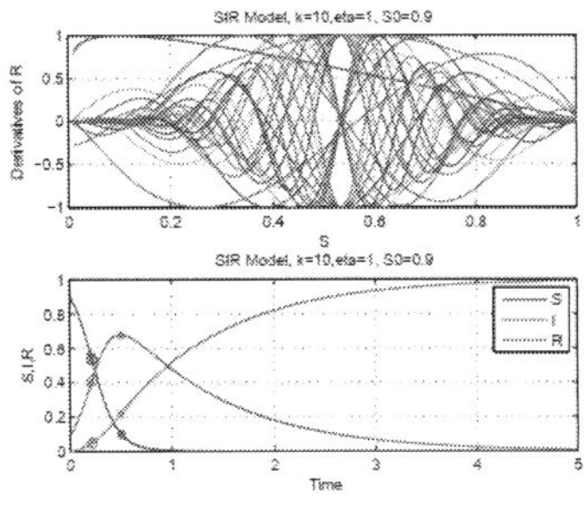

Figure 3. Derivatives of $R(t)$ for $k = \beta/\eta = 10$.

The first 24 derivatives of $R(t)$ normalized to 1 for $\eta = 1$, $S_0 = 0.9$ and $k = \beta/\eta = 3$ and $k = 10$ are plotted against $S(t)$, respectively in Figures 2 and 3. The phase transition points are indicated by $(*)$ in the upper graphs. The lower graphs display the time domain plots of the solution curves S, I, R. $S(t)$ is monotone decreasing while $R(t)$ is monotone increasing. The sequence of points converging to the phase transition point t_c are shown on each curve by (.). The phase transition point t_c indicated by (o), is located between the maximum of zero I, t_m denoted by $(*)$ and the inflection point of I, t_a denoted by $(+)$. For $k = 3$, t_c is located at 78% left of t_m in the interval (t_a, t_m) The first 24 derivatives of $R(t)$ normalized to 1 for $k = 10$, $\eta = 1$ and $S_0 = 0.9$.

5. Dependency of the critical point on the system parameters

The relative position of the critical in the interval bounded by the zeros of the third derivative and the inflection point is shown in Figure 4, for $k = 3, 4, 5, 10$. In these figures, the inflection point, the zero of the third derivative and the critical point are shown respectively by $(*)$, $(+)$ and $(\circ)$. It can be seen that, as k increases, the relative position of the critical point in between the zero of the third derivative and the inflection point is moving towards the zero of the third derivative, as the parameter k increases. For example for $k = 10$, the critical point t_c is located at 94% left of t_m in the interval (t_a, t_m).

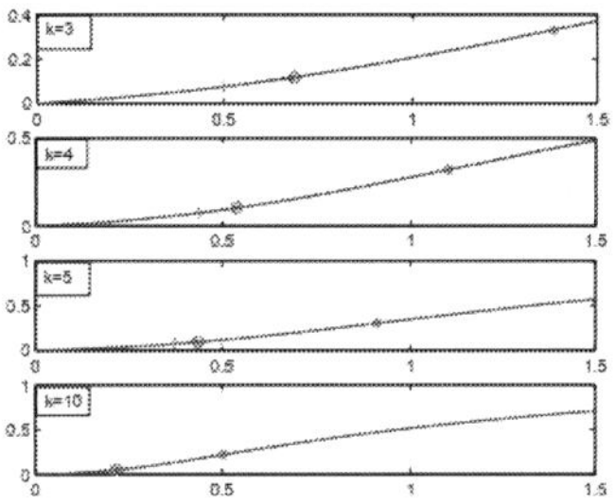

Figure 4. The location of the critical point.

References

[1] O.Pekcan and S. Kara, "Gelation Mechanisms", *Modern Physics Letters B*, 26 (27), 1230019 (2012).
[2] G.A. Evingur, F. Tezcan, F.B. Erim and O. Pekcan, *Monitoring the gelation of polyacrylamide-sodium alginate composite by fluorescence technique*, Phase Transitions, vol. 85, Issue 6, pp. 530-541, (2012)
[3] Y. Yilmaz, A. Erzan and O. Pekcan, Critical exponents and fractal dimension at the sol-gel phase transition via in situ flourescence eperiment, Phys. Rev. E, Third series, vol. 58, Issue 6 Part B, pp. 7487-7491, (1998).
[4] A.H. Bilge, O. Pekcan and V. Gurol, *Application of epidemic models to phase transitions*, Phase transitions, Volume 85, Issue 11, 2012 DOI:10.1080/01411594.2012.672648, pp 1009-1017.
[5] D. Stauffer, Introduction to Percolation Theory (Taylor and Francis, London, 1985).
[6] P. J. Flory, J. Am. Chem. Soc. vol. 63, p. 3083, (1941).
[7] W. H. Stockmayer, J. Chem. Phys. vol. 11, p. 45, (1943).
[8] H.W. Hethcote, *The mathematics of infectious diseases*, SIAM Rev. vol.42, pp.599-653, (2000).
[9] A.H. Bilge and O. Pekcan, *A mathematical Description of the Critical Point in Phase Transitions*, Int. J. Mod. Phys. C, vol. 24, 2013. International Journal of Modern Physics C, Volume 24, Issue 10, id. 1350065
[10] A. H. Bilge and Y. Ozdemir, Preprint: arXiv:1407.4839v1 [math.CA].
[11] D. Stauffer and A. Aharony, Introduction to Percolation Theory (Taylor and Francis, 43 London, 1992).

A mathematical recursive model for accurate description of the phase behavior in the near-critical region by Generalized van der Waals Equation

Jibeom Kim[1,a] and Joonhyeon Jeon[1,b,‡]
[1] Department of Fusion Energy and Advanced Material Engineering, Dongguk University-Seoul, 26, 3-ga, Pil-dong, Jung-gu, Seoul 100-715, Korea

E-mail: [a] jjanggu@dgu.edu, [b] memory@dgu.edu

Abstract. Recently, related studies on Equation Of State (EOS) have reported that generalized van der Waals (GvdW) shows poor representations in the near critical region for non-polar and non-sphere molecules. Hence, there are still remains a problem of GvdW parameters to minimize loss in describing saturated vapor densities and vice versa. This paper describes a recursive model GvdW (rGvdW) for an accurate representation of pure fluid materials in the near critical region. For the performance evaluation of rGvdW in the near critical region, other EOS models are also applied together with two pure molecule group: alkane and amine. The comparison results show rGvdW provides much more accurate and reliable predictions of pressure than the others. The calculating model of EOS through this approach gives an additional insight into the physical significance of accurate prediction of pressure in the near-critical region.

1. Introduction

Even before the GvdW EOS [1], there were a large number of studies [2-6] have been proposed for accuracy of fluids. In those, "Modified Cubic Equation of States (MC EOS)" is the first advanced theory, which based on the van der Waals (vdW) EOS [7]. However, these approaches have limitations on accuracy for near critical region.

As recently published, GvdW EOS is based on the interaction of particles and has potential by setting the thermodynamic theory based on the physical unit. GvdW EOS induced highly simplified EOS which builds on the inter-molecular interactions in Dieterici's form. Nevertheless the results of, extrapolated to other fluids, the GvdW EOS's accuracy declines [8].

For the reason that declines accuracies with GvdW EOS, this paper shows a mathematical recursive model for accurate description in the critical region which named rGvdW EOS. In the supposed process, EOSs tested with various substances for comparing results with well-known EOSs called GvdW, PR (Peng Robinson) [9], MPR (Modified Peng Robinson) [9] EOS. Verifying the accuracy of them, saturated alkane(C_nH_{2n+2}) and saturated amine(RNH_2) are selected for experiment, respectively.

We first describe the theoretical background about rGvdW EOS for the analysing. In the experimental section, we show the volume change with supposed rGvdW EOS for each experiment to confirm the accuracy with various molecules that has linearity and polarity, and finally concluded.

‡ To whom any correspondence should be addressed.

2. Theoretical background

2.1. Mathematical recursive model for GvdW EOS

In this section, introducing a method that has a recursive process to get more accurate performance and optimized results which named rGvdW EOS. As mentioned, the fore study noted that iteration processes can be omitted, but that makes large errors in estimation for critical point. In rGvdW EOS, the recursive processes are performed automatically and exactly by PC with each equation (eq. 1~4).

By the process, calculating cycle begins with Z_{cc} data of each substance. After that, whole calculating processes are started and being terminated with c_c parameter which can be seen milestone for three parameters. In the iteration process, from initialize to complete values which more than 50 times of each parameter are saved and compared values to find a repeated period. And finally, determines to optimize three values with only averaged repeated parameters more than 10 times.

Finally, confirmed parameter values by mean calculating procedure, that can present most precise critical values and those results can be found in section 3. The whole process for accurate calculation starts with Z_{cc} value to find initialized c_c value (k=0) with the equation in eq. (1) in only the first time.

$$c_{c,0} = 2Z_{cc} + \sqrt{4Z_{cc}^{2} + 1} \tag{1}$$

After the initializing c_c, it can be possible to find a_c and b_c which expressible by k in eq. (2) and (3).

$$a_c = \left(\frac{c_{c,k}+1}{c_{c,k}-1}\right)^{c_{c,k}+1} \times P_c, \quad k = 0,1,2... \tag{2}$$

$$b_c = \left(\frac{c_{c,k}-1}{c_{c,k}+1}\right)^{c_{c,k}+1} \times V_c, \quad k = 0,1,2... \tag{3}$$

At that time, the definition process for initial values is done to start recursive calculation for accurate a_c, b_c and c_c with substitution a_c and b_c values in eq. (4).

$$c_c = \frac{\ln\left(\frac{RT_c}{V_c - b_{c,k-1}} - P_c\right) - \ln a_{c,k-1}}{\ln b_{c,k-1} - \ln V_c}, \quad k = 1,2,3... \tag{4}$$

And the whole process for calculating accurate parameter values to estimate critical point is performed with a PC which instructed compare values for every procedure to clustering statistical results.

3. Simulation results and discussion

3.1. Characteristis for alkane group which have linearity the optimized parameters

Normal saturated alkanes can be a milestone for other hydrocarbon-based materials.

From the results in Table 2, alkanes, there are obvious differences between GvdW and rGvdW EOS nonetheless the same formula. In the result of GvdW EOS, there was no iteration process and can make a large error contrary to rGvdW EOS for near the critical point.

Nevertheless the differences in results, GvdW and rGvdW EOS have a similar characteristic which the linearity increases in molecules, Z_{cc} and c_c parameters are decreased and increasing number of carbon particles in each molecule are make grows of b_c parameter by 12~13 cm^3 which be setted void volume.

3.2. Characteristics of alkyl amine group that have a complex intermolecule action

Ammonia, the basic material of amine, has non-shared electron pair which does not participate in the combination with hydrogen, and that makes fast (23.79 GHz) nitrogen inversion due to the instability.

Because of the thermodynamic data obtained by these conditions, the impact of dipole moment may be ignored and it is possible to apply to general liquid with defined as the materials have nonideality only. And the frequency declines after ethylamine in the series of unsaturated amines and be known that the dipole moment are reflected in thermodynamic characteristic, directly [11]. Therefore, the access of these materials is confined only between the liquid region and the critical point [12, 13].
Representatively for many other molecules, Figure 1 shows the comparative results with real gas phase of methylamine and ethylamine on the near critical point for many EOS's estimation results.

Table 1. Parameters used in GvdW and rGvdW EOS for some molecules in alkane group.

Molecule	CP	GvdW EOS	Parameters for GvdW EOS			rGvdW EOS	Parameters for rGvdW EOS		
			c_c	a_c(bar)	b_c(cm^3)		c_c	a_c(bar)	b_c(cm^3)
Methane	46.04	45.97709273	1.7366	1668.9000	26.9159	46.04005852	1.731579	1682.733521	26.594791
Ethane	48.80	48.80075209	1.7127	1829.2000	38.6428	48.80008059	1.718714	1816.825317	39.098545
Propane	42.49	42.02750340	1.6914	1628.6000	51.7242	42.48999198	1.707249	1608.767822	53.006143
Butane	37.97	46.36485079	1.6911	1472.8000	62.7483	37.97007322	1.687660	1480.938477	65.218254
Methylamine	74.58	74.58248992	1.8310	2397.0310	45.2045	74.58002757	1.830993	2397.050537	45.204254
Ethylamine	56.24	56.22631679	1.6761	2233.5314	45.9812	56.23985481	1.676138	2233.395264	45.983124
Propylamine	47.42	47.46031094	1.7614	1663.2420	71.6897	47.42005870	1.761324	1663.621765	71.684521
Butylamine	42.00	42.02503523	1.7580	1480.0643	86.0239	42.00001466	1.757885	1480.293825	86.014461

Table 2. Calculated values for alkane and amine group with each EOSs on critical point

Molecule	Formula	CP	GvdW EOS	rGvdW EOS	PR EOS	MPR EOS
Methane	CH_4	46.04	45.97709273	46.04005852	46.05212995	45.05205110
Ethane	C_2H_6	48.80	48.80075209	48.80008059	48.82483887	48.82475463
Propane	C_3H_8	42.49	42.02750340	42.48999198	42.52569942	42.52562560
Butane	C_4H_{10}	37.97	46.36485079	37.97007322	38.03609805	38.03603144
Methylamine	CH_5N	74.58	74.58248992	74.58002757	74.57463383	74.57451539
Ethylamine	C_2H_7N	56.24	56.22631679	56.23985481	56.38225744	56.38215833
Propylamine	C_3H_9N	47.42	47.46031094	47.42005870	47.42129416	47.42121458
Butylamine	$C_4H_{11}N$	42.00	42.02503523	42.00001466	42.00162568	42.00155502

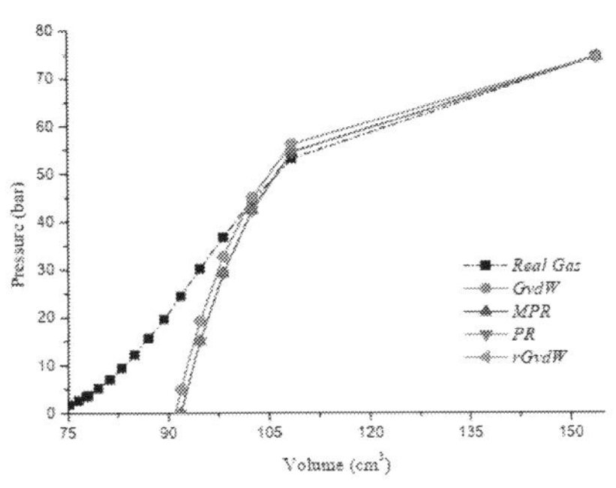

(a)

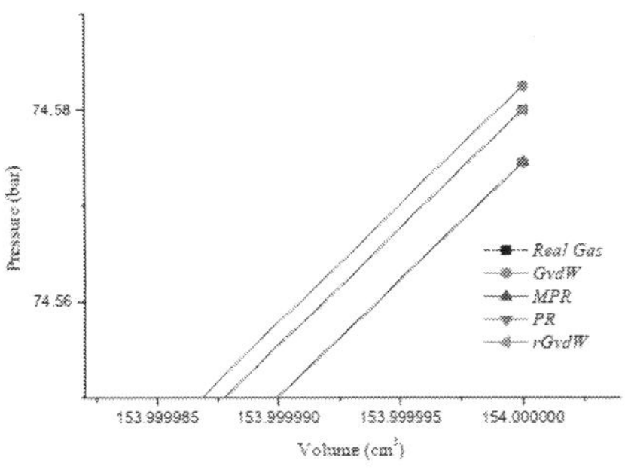

(b)

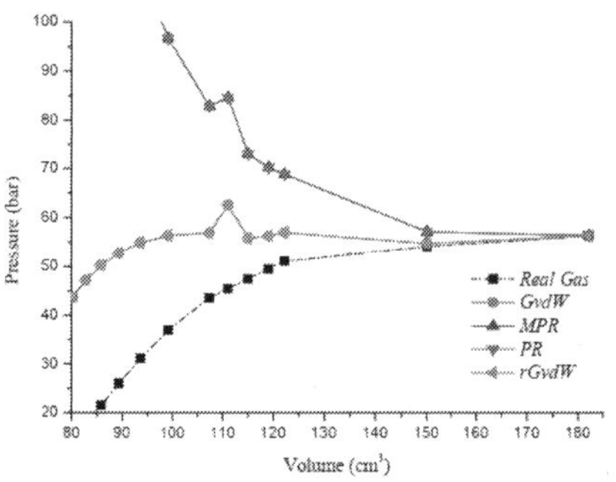

(c)

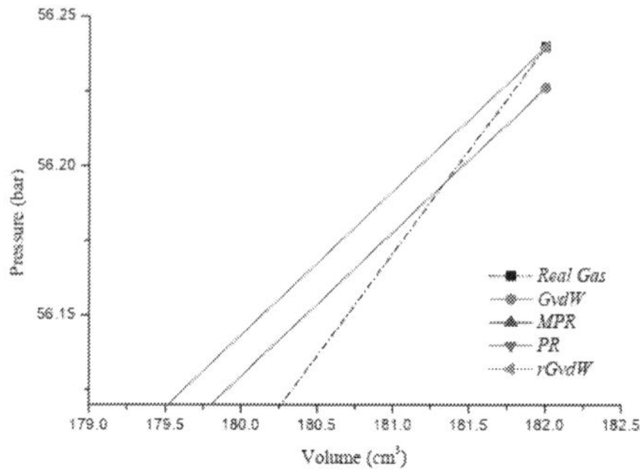

(d)

Figure 1. Result graphs for GvdW, rGvdW, PR and MPR with real gas data in near critical point. (a) Methylamine; (b) Magnifying methylamine; (c) Ethylamine; (d) Magnifying ethylamine

4. Conclusion

In this paper, to improve performance of the GvdW EOS on near the critical region that be known, it has an accuracy to represent without correction of three parameters a_c, b_c and c_c, the recursive mathematical calculation method(rGvdW EOS) is suggested which can be based on computer that automatically compensates parameters on it is dealt with. To compare the accuracy of the suggested rGvdW EOS with others which are well known we use some molecules that have single, double and triple carbon bond. Especially complex interactions also applies to a linear saturated derivative at the critical region than in the other state equations could be seen that virtually no errors. Thus, using rGvdW EOS, it can provide accurate characteristics in the near critical point by the auto-calibration of a_c, b_c and c_c parameters.

Acknowledgments

This work was supported by the Technology Innovation Program (10043787, Development of Ion Selective Membranes for redox Flow Battery) funded By the Ministry of Trade, industry & Energy (MI, Korea).

References

[1] Lee S, Jeon J, KimW and Chair T 2008 J. Phys. Chem. B 112(49) 15725
[2] Fatiha S, Salim S, Samia A, Fairuz K and Chems E 1998 Fluid Phase Equilibria 153(1) 73
[3] Otilio H, Fernando G, David A and Richart V 2002 Fluid Phase Equilibria 198 195
[4] Martin C and Jorgen M 2005 Fluid Phase Equilibria 232 74
[5] Belyakov M and Gorodetskii E 2006 Int'l. J. Thermophysics 27(5) 1387
[6] Lee Y, Shin M, Yeo J and Kim H 2007 J. Chem. Thermodynamics. 39(9) 1257
[7] Johannes D and Waals V 1873 On the Continuity of the Gaseous and Liquid States (doctoral dissertation) *Universiteit Leiden*
[8] Yong W, Guo-bo H and Chang-hua G 2011 Bull. Chem. Soc. Ethiop 25(2) 287
[9] Chorng H, John E and John R 1995 Fluid Phase Equilibria 105(1) 49
[10] Chorng H, John E and John R 1995 Fluid Phase Equilibria 105(1) 61
[11] Lambert J, Roberts G, Rowlinson J and Wilkinson V 1949 Proc. R. Soc. 196 113
[12] Yaws C 1999 *Chemical Properties Handbook* (McGRAW-HILL)
[13] Lide D 1995 *Handbook of Chemistry and Physics* 76th (CRC Press)

Quantum mechanics study on the nonlinear optical responses of a series of 4-hydroxy, 4'-cyanoazobenzene dimers

S N Derrar[1,3], M Sekkal-Rahal[1] and M Springborg[2]
[1]Laboratoire de Microscopie, Microanalyse de la matière et Spectroscopie Moléculaire, Université Djilali Liabès, B.P. 89, Sidi Bel Abbès, 22000, Algérie
[2]Physikalische und Theoretische Chemie, Universitaet des Saarlandes, Postsfach 151150, Saarbruecken 66041, Germany

E-mail: derrarsiham@yahoo.fr

Abstract. Nonlinear optical (NLO) properties are mainly measured by the first hyperpolarizability $\beta(0)$ magnitudes. Besides analyzing the NLO behaviours of monomers, dimerization of chromophores has been found of a huge interest for the application in several optical areas. In the present work, a theoretical study has been carried out on a series of dimers constituted by the azo-chromophore 4-hydroxy, 4'-cyanoazobenzene (HCAB). Combinations of vinylic and styrenic oligomers as well as of 1,3-diphenyl propnane (1,3-DPP) to HCAB have been also investigated. No influence of geometrical parameters has been detected. Indeed, after establishing a comparison with $\beta(0)$ of monomeric entities, results show that the NLO response of these dimers is not guided by any additivity rule.

1. Introduction

Nonlinear optical (NLO) devices are of a huge interest in the technology of photonics and optoelectronics as well as in several telecommunication applications [1-3]. For any system, the capacity of responding to the nonlinear optical standard is, mainly, measured by the first hyperpolarizability magnitude [4-5]. For this aim, the system should be noncentrosymmetric, and a charge transfer should take place from an electron donor to an electron acceptor through a π-spacer.
Organic chromophore systems represent potent candidates in nonlinear optics. Besides their monomers, it would be interesting to analyze dimers and trimers of these chromophores, mainly, to elucidate their NLO behavior.

In the present work, several ways of dimerization of the azo-chromophore 4-hydroxy, 4'-cyanoazobenzene (HCAB) are reported. Static first hyperpolarizability $\beta(0)$ calculations have been carried out, by means of quantum mechanics (QM) and a comparison has been established between the monomer and the different modeled dimers.

Generally, to prepare photoconductor films, chromophores are dissolved in thermoplastic amorphous matrix. This latter contains NLO-phores either as dispersed guest molecules or covalently attached to the polymeric systems. We reported, previously, a case of chromophores covalently grafted on methacrylate copolymers [6]. However, we present here, dimerization of the chromophore model with vinylic and styrenic oligomers, and with 1,3-diphenyl propane (1,3-DPP) which represents a

[3] To whom any correspondence should be addressed.

covalent combination of two styrene molecules, in order to simulate the dispersion of the chromophore in the polymeric matrix.

2. Static first hyperpolarizability *β*

When a strong electric field *E* is applied on a molecule, there is no linearity between the dipoles and the incident field. The polarization *P* is expressed as a Taylor series expansion as following

$$P = P_0 + \chi_1 E + \chi_2 E^2 + \chi_3 E^3 + ... \tag{1}$$

where χ_n is the *n* order susceptibility

At the microscopic level, the polarization can be expressed as

$$P_i = \sum_j \alpha_{ij} E_j + \frac{1}{2}\sum_{j,k} \beta_{ijk} E_j E_k + \frac{1}{6}\sum_{j,k,l} \gamma_{ijkl} E_j E_k E_l + ... \tag{2}$$

where α_{ij} is the linear polarization, β_{ijk} and γ_{ijkl} are first and second hyperpolarizabilities, respectively. With regard to dipole moment, equation (2) can be expressed as

$$\mu(E_i) = \mu_i^0 + \sum_j \alpha_{ij} E_j + \frac{1}{2}\sum_{j,k} \beta_{ijk} E_j E_k + \frac{1}{6}\sum_{j,k,l} \gamma_{ijkl} E_j E_k E_l + ... \tag{3}$$

where μ_i^0 is the dipole moment in the absence of the electric field. Consequently:

$$\alpha_{ij} = \left.\frac{\partial \mu_i}{\partial E_j}\right|_{E=0} \tag{4}$$

$$\beta_{ijk} = \left.\frac{\partial^2 \mu_i}{\partial E_j \partial E_k}\right|_{E=0} \tag{5}$$

$$\gamma_{ijkl} = \left.\frac{\partial^3 \mu_i}{\partial E_j \partial E_k \partial E_l}\right|_{E=0} \tag{6}$$

In centrosymmetric systems, β_{ijk} term vanishes. This tensor is the third derivative of energy:

$$\beta_{ijk} = -\left.\frac{\partial^3 U}{\partial E_i E_j E_k}\right|_{E=0} \tag{7}$$

3. Results

Calculations have been carried out with Gaussian03 program package [7]. A Full optimization in gas phase has been performed at B3LYP/6-31+G(d,p) level of theory, on several dimers of 4-hydroxy, 4'-cyanoazobenzene (Figure 1). Subsequently, their static first hyperpolarizabilities have been calculated at HF/6-311+G(d) level.

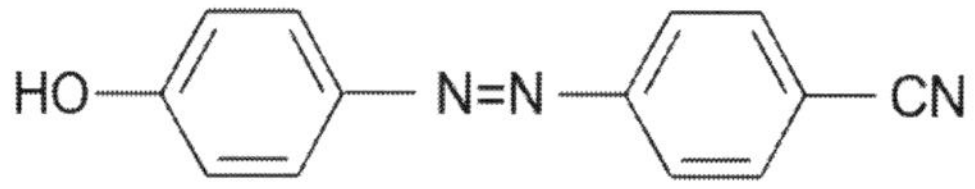

Figure 1. Structure of 4-hydroxy, 4'-cyanoazobenzene (HCAB).

With the Kleinman symmetry [8] contribution, total static first hyperpolarizability $\beta(0)$ is calculated by the combination of nine tensor elements provided by the program as

$$\beta_{tot} = [(\beta_{xxx} + \beta_{xyy} + \beta_{xzz})^2 + (\beta_{yyy} + \beta_{yzz} + \beta_{yxx})^2 + (\beta_{zzz} + \beta_{zxx} + \beta_{zyy})^2]^{1/2} \quad (8)$$

Dipole moments μ and first static hyperpolarizabilities $\beta(0)$ magnitudes of the different HCAB dimers and of dimers constituted by HCAB combined to styrene, to 1,3-DPP and to a vinylic octamer, respectively, before and after optimizations are regrouped in Table 1. Dimers geometries have been built by positioning the two molecules one parallel to the other (Par), in the same plane (Plan), and one perpendicular to the other (Per), respectively.

Table 1. Dipole moment (μ, D), initial and final static first hyperpolarizabilities ($\beta(0)$, X 10^{-30} esu) and HOMO-LUMO energy gap (ΔE_{H-L}, a.u.) of the different dimers studied.

	μ	$\beta_{initial}(0)$	$\beta_{final}(0)$	ΔE_{H-L}
HCAB-HCAB-Par	13.27	25.59	34.48	0.126
HCAB-HCAB-plan	12.84	33.42	34.16	0.126
HCAB-HCAB-Per	11.18	37.79	41.65	0.103
HCAB-Styrene	6.95	21.41	21.37	0.121
HCAB-1,3-DPP	6.83	11.41	18.59	0.130
HCAB-Oct-Par	6.63	17.18	23.33	0.096
HCAB-Oct-Per	8.08	16.13	25.11	0.071
HCAB	6.87		41.60	
Styrene	0.21		0.54	
1,3-DPP	0.44		0.61	
Octamer	0.0014		0.0078	

The final geometries of the studied dimers reveal a perpendicular shape. An enhancement of the $\beta(0)$ magnitudes is noticed from the initial to the final geometries, especially, for the HCAB dimer positioned in the plane. The highest $\beta(0)$ value is attributed to the T-shape of HCAB dimer (perpendicular), probably due to the electronic delocalization enhanced by the azo group and the lone pair of nitrogen atom which accentuates the OH donor effect.

Despite the fact of the existence of two styrene molecules in the HCAB-1,3-DPP dimer, its $\beta(0)$ magnitude is smaller than that one containing only one styrene. This is probably due to the existence of a simple bond in 1,3-DPP which reduces the electronic delocalization effect.

The comparison between $\beta(0)$ of the HCAB monomer and those of its three different dimers indicates that dimerization does not bring any amelioration with regard to the NLO response. Indeed, this fact occurs regardless to the position of the dimers entities.

When HCAB is dimerized by an octamer, $\beta(0)$ magnitudes decrease by about 41% for both parallel and perpendicular geometries. It decreases by 48 % for styrene and up to 55% for 1,3-DPP. This shows that the addition of molecules, even with a significant electronic delocalization, to HCAB chromophore does not contribute to the enhancement of $\beta(0)$.

On the other hand, it is shown from Table 1 that although $\beta(0)$ of octamer is smaller than those of styrene and 1,3-DPP, its dimer with HCAB possesses the largest $\beta(0)$ value.

Seemingly, $\beta(0)$ values are decreased by a factor of about 1.2 when HCAB is dimerized with another HCAB and with a factor of 2 once it is dimerized with octamer or styrene.

The energy gap between the frontier orbitals, high occupied molecular orbital (HOMO) and the lowest unoccupied molecular orbital (LUMO) have been analyzed in Table 1. In the series of HCAB-

HCAB dimers, it is shown that the dimer in the perpendicular geometry seems to be the most excitable molecule due to its lowest ΔE_{H-L}. Concomitantly, the highest energy gap is recorded for HCAB-1,3-DPP dimer. So far, these conclusions are in good agreement with the behavior of $\beta(0)$, where the lower is ΔE_{H-L}, the larger should be $\beta(0)$. However, some exceptions in this classification are noticed concerning dimers containing octamer. In fact, their energy gaps are smaller than that one of HCAB-HCAB-per but their $\beta(0)$ are still smaller than the latter system. It should be mentioned that in the case of HCAB-Oct dimers and unlike all the rest of the other dimers, LUMO are concerned by the octamer rather by HCAB molecule. This can be the reason for the weak character of ΔE_{H-L} for dimers holding the octamer.

4. Conclusion

In the present work, we reported the NLO behavior of a series of dimers composed of HCAB chromophore. It has been noticed that $\beta(0)$ of these dimers are not affected by the geometrical position of the constituents within the dimer. Compared to the NLO response of HCAB monomer, no valuable amelioration has been detected after dimerization.

References

[1] Chemla D S and Zyss J, *Nonlinear Optical Properties of Organic Molecules and Crystals*, Academic Press, Orlando, 1987
[2] Williams D J, *Nonlinear Optical Properties of Organic and Polymer Materials*, ACS Symposium Series 233, American Chemical Society, Washington, DC, 1985
[3] Kobayashi T, *Nonlinear Optics of Organic and Semiconductors*, Springer, Berlin, 1989
[4] Zyss J, *Molecular Nonlinear Optics*, Academic Press, Boston, 1994
[5] Prasad P N and Williams D J, *Introduction to Nonlinear Optical Effects in Molecules and Polymers*, Wiley, New York, 1990
[6] Derrar S N, Sekkal-Rahal M, Guemra K and Derreumaux P 2012 *Int. J. Quant. Chem.* **112** 2735
[7] Frisch M J, Trucks G W, Schlegel H B, Scuseria G E, Robb M A, Cheeseman J R, Montgomery J A Jr, Vreven T, Kudin K N, Burant J C, Millam J M, Iyengar S S, Tomasi J, Barone V, Mennucci B, Cossi M, Scalmani G, Rega N, Petersson G A, Nakatsuji H, Hada M, Ehara M, Toyota K, Fukuda R, Hasegawa J, Ishida M, Nakajima T, Honda Y, Kitao O, Nakai H, Klene M, Li X, Knox J E, Hratchian H P, Cross J B, Adamo C, Jaramillo J, Gomperts R, Stratmann R E, Yazyev O, Austin A J, Cammi R, Pomelli C, Ochterski J W, Ayala PY, Morokuma K, Voth G A, Salvador P, Dannenberg J J, Zakrzewski V G, Dapprich S, Daniels A D, Strain M C, Farkas O, Malick D K, Rabuck A D, Raghavachari K, Foresman J B, Ortiz J V, Cui Q, Baboul A G, Clifford S, Cioslowski J, Stefanov B B, Liu G, Liashenko A, Piskorz P, Komaromi I, Martin R L, Fox D J, Keith T, Al-Laham M A, Peng C Y, Nanayakkara A, Challacombe M, Gill P M W, Johnson B, Chen W, Wong M W, Gonzalez C, Pople J A (2003) Gaussian 03 Revision D.02. Gaussian, Inc, Pittsburgh, PA
[8] Kleinman D A 1962 *Phys. Rev.* **26** 1977

Hydrodynamics studies of cyclic voltammetry for electrochemical micro biosensors

B J Adesokan[1], X Quan[2], A Evgrafov[1], M P Sørensen[1], A Heiskanen[2] and A Boisen[2]

[1]Department of Applied Mathematics and Computer Science, Technical University of Denmark, Richard Petersens Plads, Building 324, 2800 Kgs. Lyngby, Denmark
[2]Department of Micro and Nanotechnology, Technical University of Denmark, Building 345E, Ørsteds Plads, 2800 Kgs. Lyngby, Denmark

E-mail: bojad@dtu.dk

Abstract

We investigate the effect of flow rate on the electrical current response to the applied voltage in a micro electrochemical system. To accomplish this, we considered an ion-transport model that is governed by the Nernst-Planck equation coupled to the Navier-Stokes equations for hydrodynamics. The Butler-Volmer relation provides the boundary conditions, which represent reaction kinetics at the electrode-electrolyte interface. The result shows that convection drastically affects the rate of surface kinetics. At a physically sufficient high flow rates and lower scan rates, the current response is limited by the convection due to fresh ions being brought to the electrode surface and immediately taken away before any surface reaction. However, at high flow and scan rates, the Faradaic current overrides current due to convection. The model also allows predicting the effect of varying electrolyte concentration and scan rates respectively.

I Introduction

Micro biosensors have been wide spread used in biomedical applications and as an analytical device for detection of biochemical substances [1, 2, 3]. It is essential that these devices recognise and translate biochemical properties of a substance into electrical signals. Therefore, the devices must be coupled with an electroanalytical technique to characterise such signals. One of the most powerful analytical techniques used for such a task is cyclic voltammetry (CV) [4, 5, 6]. CV is a widely used technique to probe the physics and chemistry at the electrode-electrolyte interface thereby used as the main mode of operation for sensing devices [7, 8]. This study focuses on simulating CV using the finite element method in a microfluidic cell under flow conditions. We investigate the effect of flow rates, the scan rates and the electrolyte concentration and compare to experimental measurements.

II Experiment: The electrode chip and the microfluidic cell

The electrode chip has dimension $12mm \times 11.3mm$ and was fabricated on a $375\mu m$ thick Si substrate. Four cantilevers were placed in the middle of the chip, each having the dimensions $100\mu m$(width) $\times$ 550nm(thickness) $\times 400\mu m$(length). Each cantilever was capable of functioning as an independent working electrode (WE). The reference electrode with dimensions $100\mu m \times 200\mu m$ was placed on the right side of cantilevers and the counter electrode with dimensions $2700\mu m \times 400\mu m$ on the left side. All the electrodes were coated using e-beam evaporation with a $2nm$ chrome adhesion layer followed by $30nm$ of gold. The microfluidic chamber encapsulates the ECC biosensor with a volume of approximately $105\mu L$ and creates a flow path leading into the chip and through the channel past the cantilevers. The system was connected to a syringe pump (PHD 2000 Infuse/Withdraw syringe pump, Harvard Apparatus, USA) and set the flow rates of up to $600\mu L/min$. The electrolyte with active redox probes used in the CV measurements consists of 200mM KNO3 solutions with 2mM of potassium ferricyanide(III) in Milli-Q water. CV was performed using a computer controlled CHI1030 Potentiostat.

III Mathematical Modeling

The conservation of mass for the ionic species j in the electrolyte bulk solution for a pure electrochemical reaction is governed by

$$\frac{\partial C_j}{\partial t} = \nabla \cdot (z_j u_j F C_j \nabla \phi + D_j \nabla C_j - C_j v), \tag{1}$$

where C_j, D_j, z_j and u_j are the ionic concentration, the diffusion constant, charge and the mobility of ionic species j,. F, ϕ, v represent the Faraday's constant, the electrolyte potential and the convective velocity of the solution, respectively [5, 6].
The electrolyte solution is electrically neutral except at a very thin electric double layer close to the electrode so that ($\sum_j z_j C_j = 0$). The electric potential equation

$$\nabla \cdot (\kappa \nabla \phi) + F \sum_j z_j \nabla \cdot D_j \nabla C_j = 0, \tag{2}$$

can be easily derived from equation (1) by using the electro neutrality condition where $\kappa = F^2 \sum_j z_j^2 u_j C_j$ is the electrolyte conductivity. Finally, the velocity field v of the flowing electrolyte is solved by the Navier-Stokes equations for incompressible fluid flow

$$\nabla \cdot v = 0, \tag{3}$$

$$\rho \left(\frac{\partial v}{\partial t} + v \cdot \nabla v \right) = -\nabla p + \mu \nabla^2 v + s_f . \tag{4}$$

The fluid density is denoted ρ, p is the pressure, μ is the dynamic viscosity and finally s_f denotes an external force per unit volume [9]. We solved equations (1), (2), (3) and (4) with appropriate boundary conditions. Table 1 shows the prescribed boundary conditions. The applied potential as a function of scanning time can be modeled as follows,

$$\phi_{app}(t) = \begin{cases} \phi_{min} + vt & \text{if } 0 \le t \le \mathcal{T}, \\ \phi_{max} - vt & \text{if } \mathcal{T} \le t \le 2\mathcal{T}. \end{cases}$$

where v is the scan rate and $\mathcal{T} = \dfrac{\phi_{max} - \phi_{min}}{v}$ is the half cycle period.

Variables	Insulating boundary $\partial\Omega_{ins}$	Electrode surface $\partial\Omega_e$	Inlet	Outlet
C_j	$\partial C/\partial n = 0$	$\kappa\frac{\partial\phi}{\partial n} + z_j D_j \frac{\partial C_j}{\partial n} = i_n(C_j, \phi, v; \phi_{app})$	C_{in}	$\partial C/\partial n = 0$
ϕ	$\kappa\frac{\partial\phi}{\partial n} = -z_j D_j \frac{\partial C_j}{\partial n}$	$\kappa\frac{\partial\phi}{\partial n} + z_j D_j \frac{\partial C_j}{\partial n} = i_n(C_j, \phi, v; \phi_{app})$		
v	0	0	v_{in}	
p				p_{out}

Table 1: *Specified boundary conditions.*

From Table 1, i_n is the local current density that implicitly models the electrode kinetics via the Butler-Volmer relation, see for example [4, 5, 6]. In a cyclic voltammetry curve, we compute the total current response to the applied potential on the electrode surface by integrating the current density across the surface of the electrode

$$i_{cv} = \int_{\partial\Omega_e} i_n dS. \tag{5}$$

IV Numerical method

We numerically solved our test problem using the finite element commercial software package, COMSOL multiphysics [10] on a Windows work station (Core i7 CPU, 4 cores, 2.40GHz, 16.0GB RAM). We employed a two step approach for solving the coupled fluid and ion-transport problems. First, we solve for the steady state solution to the incompressible Navier-Stokes equations and then substituted the solution of the velocity field into the ion transport problem. The fluid flow problem was solved by discretizing the computational domain using the linear Lagrange finite elements for the pressure field and quadratic elements for the velocity field. For the ion transport problem we employed the quadratic elements while linear Lagrange element was chosen for the potential field. For the time integration, we have used the implicit Euler method that is stable for the numerical computations.

V Results and Discussion

Figure 1 shows series of voltammograms with varying concentrations. Here, there is no flow in the system. The results show that the peak current varies linearly with the concentrations of bulk electrolyte as expected. The simulation was carried out at the scan rate of 50mV/s and electrolyte concentrations of $50\mu M - 2mM$. There is a very good agreement between the simulated and the experimental results. In what follows, we introduce electrolyte flow into the system. Figure 2 shows the effect of flow rates on CV curves. At lower flow rates of about $50\mu L/min$ - $150\mu L/min$ peak currents are observed due to diffusion dominated mass transport. Hence, surface reaction is fast enough for electron transfer at the electrode surface. However, as the flow rate increases to about $250\mu L/min$, voltammograms becomes thinner and the peak current diminishes. This is due to the rapid inflow of the reactants and fast removal of the products induced by the convection. The diffusion layer of ionic species disappears and mass transport is only by the convection. Finally, Figure 3 shows CV curves at different scan rates of 50mV/s - 250mV/s and at volumetric flow rate of $250\mu L/min$. At high scan rate of 250mV/s, peak currents are observed despite the high flow rate. This is due to the fact that the electrode surface kinetics is fast enough for the electron exchange before the reactants are washed away from the electrode surface.

Journal of Physics: Conference Series **574** (2015) 012008 doi:10.1088/1742-6596/574/1/012008

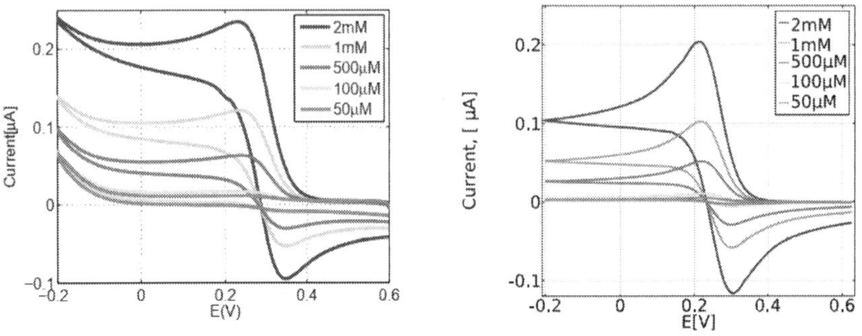

(a) *Experimental data for concentration. The scan rate used is 50mV/s.*

(b) *Simulation result for concentration. The scan rate used is 50mV/s.*

Figure 1: *Cyclic voltammetry results as a function of concentration.*

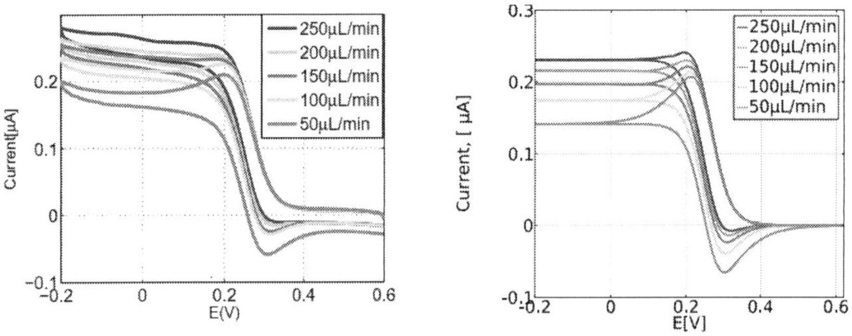

(a) *Experimental data for flow rates. The scan rate used is* $50mV/s$.

(b) *Simulation result for flow rates. The scan rate used is* $50mV/s$.

Figure 2: *Effect of flow rates on cyclic voltammetry.*

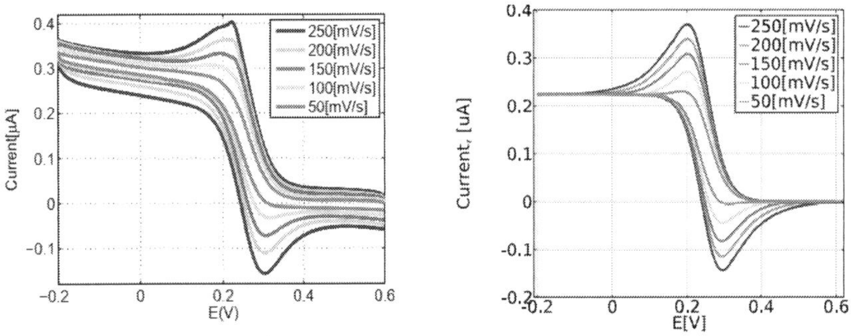

(a) *Experimental data for scan rates.*

(b) *Simulation results for scan rates.*

Figure 3: *Effect of scan rates on cyclic voltammetry. Flow rate of* $250\mu L/min$.

However, at low scan rate of 50mV/s and at high flow rate of $250\mu L/min$ the CV curves reaches saturation and the width of the voltammogram becomes thinner. The reason is due to the fact that fresh reactants are brought very fast to the electrode surface while products are quickly flushed away from the electrode surface. In addition, because of the slow scan rate electrode kinetics is not fast enough for the electron exchange to record peak currents and hence the narrow and flattened voltammograms. Comparison between the experimental data and the simulation results show a very good agreement. See Figure 3a.

VI Conclusions

A two step numerical approach is used to simulate cyclic voltammetry measurement under hydrodynamic flow. We found that the flow rates has a great influence on the current response recorded during cyclic voltammetry. At fast flow rate, the peak current vanishes while at slow rate steady current response with peak current is recorded. We were able to identify characteristics of hydrodynamic effect on the dynamics of current response to applied voltage in cyclic voltammetry measurement. In addition, we predicted the effect of varying electrolyte concentration and scan rates, respectively.

VII Acknowledgement

We aknowledge financial support from the Villum Kann Rasmussen centre of Excellence: Nano Mechanical Sensors and Actuators, fundamentals and new directions (NAMEC).

References

[1] Jianhong P, Fang T, and Thomas T. Glucose biosensor based on the microcantilever. *Ana. Chem.*, 76(2):292–297, 2004.

[2] Cruys-Bagger N, Badino S F, Tokin R, Gontsarik M, Fathalinejad S, Jensen K, Toscano M D, Sørensen T H, Borch K, Tatsumi H, Väljamäe P, and Westh P. A pyranose dehydrogenase-based biosensor for kinetic analysis of enzymatic hydrolysis of cellulose by cellulases. *Enzy. Micro. Tech.*, 58:68 – 74, 2014.

[3] Fischer L M, Pedersen C, Elkjær K, Noeth N N, Dohn S, Boisen A, and Tenje M. Development of a microfabricated electrochemical-cantilever hybrid platform. *Sens. Act. B: Chem.*, 157(1):321–327, 2011.

[4] Compton R G and Banks C E. *Understanding Voltammetry*. World Scientifice, 2 edition, 2011.

[5] Bard A J and Faulkner L R. *Electrochemical Methods: Fundamentals and Applications*. John Wiley & Sons, 2 edition, 2001.

[6] Newmann J and Thomas-Alyea K E. *Electrochemical Systems*. Wiley-Interscience, 3 edition, 2004.

[7] Chevion S, Roberts M A, and Chevion M. The use of cyclic voltammetry for the evaluation of antioxidant capacity. *F. Rad.l Bio. & Med.*, 28(6):860–870, 2000.

[8] Masaya T, Shin-ichi T, Koji T, and Yoji Si. Reaction behavior of LiFePO4 as a cathode material for rechargeable lithium batteries. *Sol. Sta. Ion.*, 148:283–289, 2002.

[9] Batchelor G K. *An Introduction to fluids dynamics*. Cambridge University Press, 1967.

[10] Comsol multiphysics®finite element analysis software https://www.comsol.com/.

[11] Bird R B and Lightfoot W E, Stewart andE N. *Transport Phenomena, Revised 2nd Edition*. John Wiley & Sons, Inc., 2nd edition, 2006.

Calculating Conductance of Ion Channels - Linking Molecular Dynamics and Electrophysiology

Michael A. Wilson[1,2] **and Andrew Pohorille**[1,2]

[1] Dept. Pharmaceutical Chemistry, University of California, San Francisco, CA 94132 USA
[2] Exobiology Branch, MS 239-4, NASA Ames Research Center, Moffett Field, CA 94035 USA

E-mail: `Michael.A.Wilson@nasa.gov`

Abstract. Molecular dynamics computer simulations were combined with an electrodiffusion model to compute conduction of simple ion channels. The main assumptions of the model, and the consistency, efficiency and accuracy of the ion current calculations were tested and found satisfactory. The calculated current-voltage dependence for a synthetic peptide channel is in agreement with experiments and correctly captures the asymmetry of current with respect to applied field.

1. Introduction

Ion channels are pore-forming assemblies of transmembrane proteins that mediate and regulate ion transport through cell membranes [1]. They are ubiquitous to all life. In humans and other higher organisms, they are essential for conducting nerve impulses, cardiac processes, muscle contraction and epithelial transport. In lower organisms, they can act as toxins or antimicrobial agents, and are involved in infectious diseases. Because of their important and diverse biological functions they are frequent targets of drug action. Ion channels also have numerous applications in biotechnology. For these reasons, studies of ion channels are at the forefront of biophysics, structural biology and cellular biology.

The availability of X-ray structures has greatly advanced our understanding of ion channels [2]. However, their mechanism of action remains elusive: ion channels are dynamic by nature, but X-ray crystallography captures the channel in a single, sometimes non-native state. To explain how ion channels work, X-ray structures have to be supplemented with dynamic information that, in principle, can be obtained from molecular dynamics (MD) simulations[3]. However, MD simulations suffer from their own problems, such as inability to access sufficiently long time scales or limited accuracy of force fields. A direct way to assess their reliability is to compare calculated ionic conductances with electrophysiological measurements obtained under similar conditions.

In MD simulations, channel conductance, defined as the ratio of ionic current through the channel to applied voltage, can be calculated once the current, the number of ions that traverse the channel per unit time when an external electric field is applied to the system, has been determined [3]. If the conductance is small, a voltage significantly higher than the experimental one needs to be applied to collect sufficient statistics of ion crossing events. Then, the calculated conductance is extrapolated to the experimental voltage using procedures of unknown accuracy. We propose an alternative approach in which MD simulations at a single voltage are combined with an electrodiffusion (ED) model to recover the full current/voltage (I-V) dependence. Here, we first test the assumptions of the ED model and determine the reliability of the calculated conductance in a model channel built of trichotoxin (TTX), a peptide containing 18 amino acids [4]. The backbone of the TTX was restrained to limit channel fluctuations.

Next, we investigate without any restrains a hexametric channel, LS3, in which each monomer is built of two amino acids, leucine (L) and serine (S) in the sequence $(LSSLLSL)_3$. For this channel we compare calculated and experimentally measured [5] conductance at several applied fields.

2. Theory and Methods

We assume that ion transport through a transmembrane channel can be described as a diffusive process in the presence of an external electric field and the intrinsic potential of mean force (PMF) due to the channel, the membrane, water and other ions. This implies that the channel structure does not undergo substantial changes when an electric field is applied. The transport of ions across channels in the presence of an electric field can be obtained from the time evolution of the probability density of ions in the system, which is governed by the Smoluchowski equation, obtained from the Fokker-Planck equation with the condition that the equilibrium probability distribution is the Boltzmann distribution:

$$\frac{\partial P(z,t)}{\partial t} = -\nabla J(z), \quad J(z) = -\nabla[D(z)P(z,t)] + \frac{D(z)}{\mathrm{k_B T}}F(z)P(z,t) \tag{1}$$

where $J(z)$ is the current, $D(z)$ is the diffusivity, $P(z,t)$ is the probability density of ions, $F(z)$ is the mean force acting on the ion at position z in the channel along the normal to the membrane, $\mathrm{k_B}$ is the Boltzmann constant and T is temperature. This 1-dimensional representation is justified if equilibration of ion positions along directions parallel to the membrane is markedly faster than transport across the channel. This means, for example, that there are no strong ion binding sites along the channel.

If the bulk concentrations of ions on both sides of the membrane and the applied field remain constant over time, then the transport of ions through a channel is a stationary process with absorbing boundary conditions. In other words, the system reaches a steady state and the Fokker-Planck equation reduces to the ED equation, which is a generalized form of the Nernst-Planck (NP) equation [1]. If we assume that that the diffusivity in the channel is constant and known (see Results) integration of this equation yields:

$$\frac{J}{D} = \frac{\rho(z_1)\exp(E(z_1)/\mathrm{k_B T}) - \rho(z_2)\exp(E(z_2)/\mathrm{k_B T})}{\int_{z_1}^{z_2}\exp(E(z)/\mathrm{k_B T})dz}, \tag{2}$$

Here, the number density per unit length, $\rho(z)$, is proportional to the steady-state probability distribution. The total potential of mean force, $E(z)$, is the sum of the intrinsic PMF of an ion crossing the channel in the absence of an electric field, $A(z)$, and the electrostatic potential energy of the applied field, $qV(z)$. z_1 and z_2 define two planes inside the channel. In principle, their choice is arbitrary, but generally, the accuracy improves if they are well separated. Eq. 2 is not the only way to solve the ED equation for J, but it is the only expression that does not require knowledge of the full density profile, $\rho(z)$, inside the channel. For a given applied voltage, only the boundary densities $\rho(z_1)$ and $\rho(z_2)$ need to be calculated, which should be markedly more efficient than determining $\rho(z)$. This makes calculations of the full I-V dependence facile.

The equilibrium PMF, $A(z)$, can be obtained from separate simulations of the channel in the absence of an electric field [6]. Alternatively, it can be determined through integration of Eq. 1, which yields:

$$A(z_2) - A(z_1) = \mathrm{k_B}\left[\ln\rho(z_1) - \ln\rho(z_2) - \frac{J}{D}\int_{z_1}^{z_2}\frac{1}{\rho(z)}dz\right] - q\mathcal{E}_{el}(z_2 - z_1), \tag{3}$$

where $\mathcal{E}_{el}(z_2 - z_1)$ is the change in electric field between z_1 and z_2. This means that $A(z)$ can be determined from a non-equilibrium simulation that is sufficiently long to calculate J and $\rho(z)$ with satisfactory accuracy. Then, only the boundary terms are needed to calculate currents at other applied fields, as described above.

To initiate our MD simulations, both the TTX and LS3 channels were placed in a phospholipid bilayer that was in contact with water lamellae containing 1 M KCl on both sides. A schematic of the system is shown in Fig. 1. Simulations were carried out at several different applied fields for times of 0.5–2 μsec. The simulation protocol resembled that described in our earlier study on ion channels[7]. From these simulations the ion currents, density profiles and/or the density boundary terms were obtained.

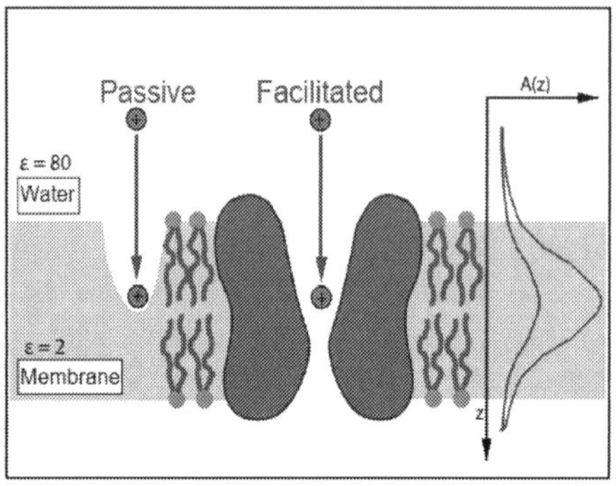

Figure 1. Schematic of an ion channel in a membrane. The water-filled pore significantly lowers the free energy barrier to ion transport, as shown on the right side of the figure.

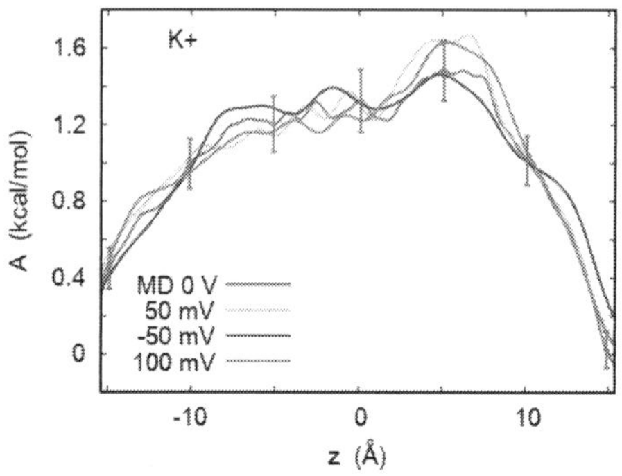

Figure 2. Free energy profile of K^+ in TTX at 0 V (red) and reconstructed from MD at 50 mV (green), -50 mV (blue), and 100 mV (magenta). $z = 0$ is the center of the membrane.

3. Results

In the ED equation, it is assumed that the probability of ion crossing is stationary and the numbers of crossing events in disjoined intervals are independent of each other, which implies that ion crossing is a Poisson process. However, if motions of ions in a channel are correlated or a channel undergoes slow conformational changes these assumptions would not be satisfied. Thus, we constructed semi-log plots of the cumulative probability distribution of waiting times between consecutive ion crossing events vs. time. For a Poisson process, they should yield straight lines with the negative slope equal to the inverse of the average waiting time, which is precisely what we found from our simulations.

The diffusivity was calculated by way of the Einstein relation from short simulations in which an ion was placed at different positions along z, as described previously [7]. Its values were found to be constant to within the statistical errors, so the dependence of $D(z)$ on z was suppressed in the ED equation.

The free energy profiles for K^+ in TTX reconstructed from Eq. 3 are shown in Fig. 2. We see that $A(z)$ obtained from simulations at ± 50 mV and 100 mV are in good agreement with $A(z)$ calculated from the equilibrium simulation. Similarly, the reconstructions for the LS3 channel (not shown) are very good for voltages of ± 100 mV, but not for -200 mV. This indicates that the ED model begins to break down at larger voltages.

For a stationary process, J is independent of the limits of integration in Eq. 2. Thus, calculating J for different ranges of z provides a valuable consistency test. If J is not approximately constant something is amiss; either $A(z)$ or the densities at the endpoints, $\rho(z_1)$ and $\rho(z_2)$, are inaccurate, or the assumption that the drift term in the Fokker-Planck equation is only due to the equilibrium free energy profile and applied field is incorrect. As shown in the example of K^+ in the TTX channel (see Fig. 3), the ED currents integrated in the range $[z_1(= -z_2), z_2]$ are quite stable for z_2 between 10 and 15 Å. For smaller integration ranges, the errors increase, most likely due to the larger relative contributions from the density profiles deep inside the channel, which are known less accurately.

For the LS3 channel, in addition to long MD trajectories at -200, -100 and 100 mV, we also generated shorter trajectories at -50, 50 and 200 mV, from which we extracted $\rho(z_1)$ and $\rho(z_2)$ and computed the current by way of the procedure described in the previous section. In Fig. 4, the results are compared with the experimentally determined I-V curve. The K^+ current calculated directly from MD simulations is in excellent agreement with the total experimental current. The corresponding ED currents are very similar with the exception of that at -200 mV, indicating that the ED equation might become less accurate at large applied fields. Our calculations correctly capture the non-Ohmic character of the currents, which are smaller for the positive voltages than for negative voltages. There is, however, one discrepancy between

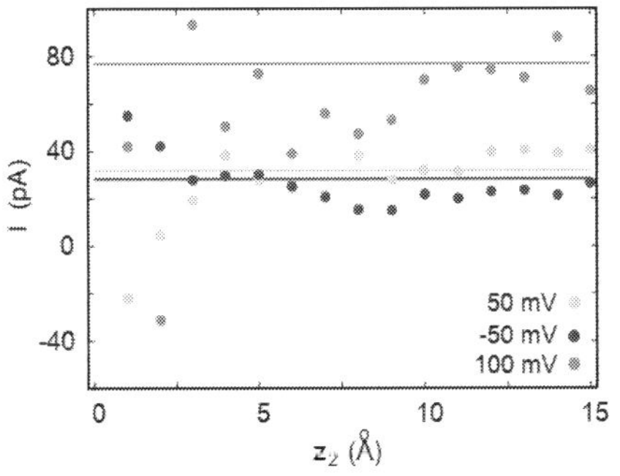

Figure 3. K^+ ED model currents for TTX at 50 mV (green), -50 mV (blue) and 100 mV (magenta). Solid lines are the currents observed in the simulations.

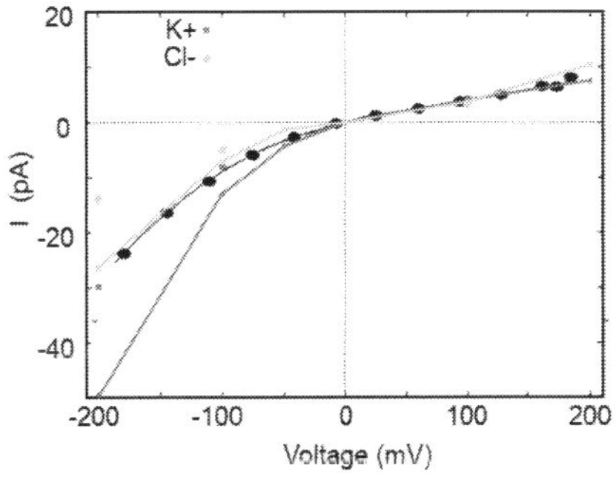

Figure 4. The I-V curves for K^+ (red) and Cl^- (green) in the LS3 channel. Experimental data extracted from [5] are shown as black circles.

the calculated and experimental currents. The channel is selective for K^+ over Cl^- by a factor of 10, which means that Cl^- currents should be quite small. This is not the case in the simulations, as currents of K^+ and Cl^- are comparable. The most likely reason for this outcome is the lack of balance between force fields describing interactions of Cl^- with bulk water and the environment inside the channel.

4. Conclusions

Ion currents were calculated directly from MD simulations and computed from a 1-dimensional ED model, and were found to be in good agreement for a model system based on a heptameric TTX channel structure. The main assumptions underlying the ED model have been shown to be valid: the distribution of ion-crossing events follows Poisson statistics and the thermodynamic force on an ion can be obtained from the intrinsic (zero-voltage) PMF and the applied electric field. The intrinsic PMF can be obtained equally accurately from equilibrium simulations at 0 V and from non-equilibrium density profiles. Once this PMF is known, the current at a desired voltage can be calculated efficiently only from the knowledge of ion densities near the ends of the channel. The extension of this approach to the LS3 channel yields an I-V curve that reproduces the experimentally observed non-Ohmic behavior of the channel. Capturing this feature, called rectification, is a valuable and sensitive test of reliability of computer simulations. The simulations also revealed that standard force fields for Cl^- are, most likely, improperly balanced.

Acknowledgments

This work was supported by the NASA Exobiology Program. Simulations were performed at the NAS Division at NASA Ames and on the Anton computer at the Pittsburgh Supercomputer Center.

References

[1] Hille B 2001 *Ion Channels of Excitable Membranes (3rd edition)* (Sunderland, MA: Sinauer Associates Inc.)
[2] Kew J and Davies C (eds) 2010 *Ion Channels: From Structure to Function* (Oxford: Oxford University Press)
[3] Maffeo C, Bhattacharya S, Yoo J, Wells D and Aksimentiev A 2012 *Chemical Reviews* **112** 6250–6284
[4] Chugh J K, Brückner H and Wallace B A 2002 *Biochemistry* **41** 12934–12941
[5] Lear J D, Schneider J P, Kienker P K and DeGrado W F 1997 *J. Am. Chem. Soc.* **119** 3212–3217
[6] Chipot C and Pohorille A 2007 *Free Energy Calculations. Theory and Applications to Chemistry and Biology* (Berlin: Springer-Verlag)
[7] Wilson M A, Wei C, Bjelkmar P, Wallace B A and Pohorille A 2011 *Biophys. J.* **100** 2394–402

Mathematical Modelling on Chemically Reacting Boundary Type Flow under Reduced Gravity Conditions

Hui Ying WANG[1] **and Thomas HERVIOU**
Institut Pprime (C.N.R.S.-U.P.R. 3346)
Département Fluide-Thermique-Combustion
ENSMA - BP 40109
Téléport 2, 1 av Clément ADER
F-86961 FUTUROSCOPE CHASSENEUIL Cedex
FRANCE

E-mail: wang@ensma.fr

Abstract. A numerical study was performed to give a quantitative description of a heavily sooting, non-premixed laminar flame established in a shear boundary layer in microgravity. The competition between fuel pyrolysis rate, radiation loss due to soot formation and oxygen-side diffusion creates a number of unusual phenomena. A three-dimensional, laminar diffusion flame over plate in low-speed concurrent flow was formulated and solved. The model consists of full Navier-Stokes equations for mass, momentum, energy and species. Gas radiation associated with soot formation is included. The PAH inception model, which is based on the formation of two and three-ringed aromatic species, reproduces correctly the measured soot from a laminar ethylene diffusion flame. The flame standoff distance is beyond the boundary layer thickness, while soot resides always within the boundary layer. The fraction of the total energy released by the flame to the surface increases with a rise of pyrolysis rate due to increase in the flame volume. The zone of the high heat flux increases noticeably when the flame becomes most obviously dual on the cross-stream section due to development of the counter-rotating vortex.

1. Introduction

A fire on board spacecraft is, though unlikely owe to strict safety measures, one of the scenarios with the highest damage potential for hardware and crew. In spacecraft, where buoyancy is negligible, the flow is limited to that induced by the ventilation system with a characteristic velocity of 0.2 m/s. The most likely scenario of a flame incidentally initiated and spread on board of a spacecraft is a wall of combustible material adjacent to a low Reynolds number gas flow and exposed to a potential ignition source. When analysing a combustion process, transport equations can be scaled and characteristic parameters can be extracted. In normal-gravity, a buoyantly induced velocity generates an upper limit for all time scales and dominates all forms of transport. All forms of transport slower than natural convection can thus be neglected. In micro-gravity, this upper limit disappears and different phenomena, negligible in normal gravity, become relevant and sometimes of great importance. All characteristic time scales increase, the natures of heat/mass transfers, species (i.e. fuel and soot) generation, degradation of material change. Ignition includes solid and gas phase processes. Proper selection of materials to minimize the potential of a fire is labelled material flammability. The processes related to the gas phase are solely affected by buoyancy and thus require particular attention. The gas phase induction period can be reduced to negligible time by introducing a pilot.

Under microgravity conditions, as a major portion of heat released from the igniting stage of combustion is transported by forced convection in the flow direction, the superior mode of spread is of co-current (forward) type. Such flame propagation in a microgravity environment is of scientific interest and also essential for the improvement of fire safety in spacecraft and space station. The work of Olson et al. [1] showed that for a forward diffusion flame, radiation becomes the predominant mode of heat transfer. Pyrolysis kinetics have an important role on the onset temperature for pyrolysis of condensed fuel but rapidly fuel generation is limited by transport of oxygen to the surface and of fuel from the interior of the material [2]. A proper evaluation of material flammability requires a detailed understanding of the effect of pyrolysis rate on soot formation, to which the structure of the flame connects [2]. Konsur et al. [3] have attempted to describe the sooting behaviour of non-buoyant jet diffusion flame by using the Smoke-Point concept proposed originally by Markstein et al. [4]. Co-current flame spread over a solid plate could then be linked to a critical soot concentration [2] controlling the trailing edge quenching. Several experimental investigations [5, 6] have demonstrated how changes in blowing affect soot formation and thus, flame structure in the absence of natural convection. Numerical simulations have been performed [7] to determine two regimes (attached and detached flows) of a low Reynolds number flow in the absence of natural convection, function of blowing and thermal expansion, but without radiation.

This is because the ratio of flame length to flame width is usually greater than unity, which implies that the flame is three-dimensional in nature. Many flame spread studies in microgravity were intended to observe the flame behaviour in a two-dimensional (2D) configuration [5, 6]. Besides, by necessity, the experiments were generally conducted in spatially limited experimental volumes which resulted in relatively narrow samples (3-5 cm) despite the original intention of a 2D experimental configuration [5, 6]. It is, therefore, critically important to determine the three-dimensional (3D) effects on flame behaviour and radiation flux as a function of pyrolysis rate, as in this paper. Currently, for predicting soot concentration, extensive efforts are devoted in 3D CFD codes to the semi-empirical soot models [8-12], involving the inception, coagulation, surface growth and oxidation processes. Non-buoyant ethylene laminar jet diffusion flames have been modelled by assuming the nucleation and growth of soot are first-order functions of acetylene concentrations [8, 9]. Recent research suggests that the growth of aromatics is the rate limiting step rather than the formation of the first ring [10]. Vovelle et al. [11] showed that in decane flames, aromatic hydrocarbons are formed from acetylene. Anderson et al. [12] have shown that even a small quantity of aromatic hydrocarbons in the fuel can significantly increase soot emissions. In the present work, the soot model, developed by Hall et al. [13], is applied to a boundary type laminar flame under microgravity. In such soot model, the soot inception rate is based on the formation of two and three-ringed aromatic species from single ring aromatic species from a detailed kinetic reaction mechanism [14].

2. Theoretical analysis

The strong coupling between soot and radiation are studied in three-dimensional flames by solving the Navier-Stokes fluid dynamic equations of elliptic, reacting flow. Binary diffusion coefficients, thermal conductivity and viscosity coefficients are all temperature and composition dependent. This set consists of the conservation equations of mass, momentum, energy and species. The precise formulation of the differential equations describing the dynamic model and the numerical technique can be found elsewhere [15], and will not be repeated here.

2.1. Combustion model

For providing the local concentration of soot precursory species, such as acetylene (C_2H_2), benzene (C_6H_6) and phenyl (C_6H_5), a detailed kinetic reaction [14], including 736 reversible reactions and 99 species has to be used to model the combustion for ethylene/air. This reaction mechanism consists of a set of elementary reactions with corresponding rate parameters. The rate coefficient for reaction is calculated from an Arrhenius expression. These rate parameters, together with the elementary reactions and their stoichiometric coefficients, constitute the reaction mechanism. The rate laws for the species can be calculated by using a time consuming stiff ODE solvers. However, any attempt to use Arrhenius expression for a detailed kinetic [14] coupled with a three dimensional elliptic, reacting flow is discouraged due to a problem of computational cost.

In combustion, the mixture fraction is a conserved quantity traditionally defined as the (mass) fraction of the gas mixture that originates in the fuel stream. The mixture fraction is a function of space and time, commonly denoted $Z(x_i, t)$. Therefore, many combustion models employ a detailed, cost effective kinetic reaction by tabulating the kinetics in terms of mixture fraction and only the mixture fraction is tracked in the calculation [16]. This implies that all species of interest of combustion products can be described in terms of the mixture fraction alone. The correspondence between the mass fraction of an individual species and the mixture fraction is called its "state relation". This state relation can be obtained by using the OPPDIF (Opposed Diffusion Flames) computer code [17] which is performed by using subroutines from the chemical program library Chemkin.

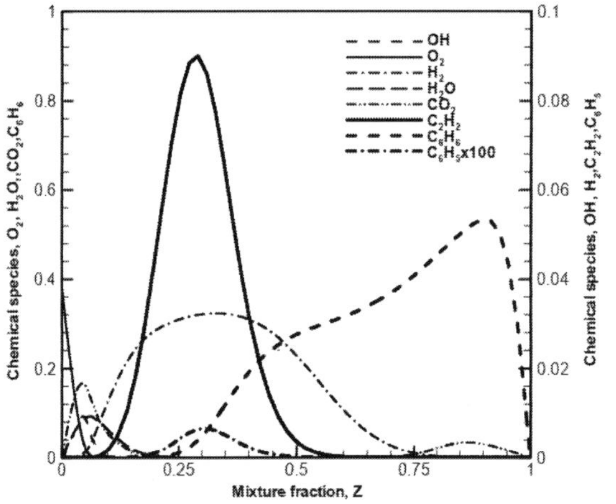

Figure 1. State relation between chemical species and mixture fraction

For obtaining such state relation for combustion of fuel/air, OPPDIF [17] requires the chemical and physical input parameters, such as the fuel type, the initial temperature values of fuel and oxygen, the initial values of fuel and oxygen mass fraction, pressure and strain rate. While this state relation is independent of reactant and oxidizer flow rates. A small value of the strain rate (<0.0003) which is much lower than the critical extinction value, is assigned, corresponding to the chemical equilibrium conditions. At the start of a simulation, the detailed chemical species involved are tabulated as a function of mixture fraction, as shown in figure 1. During the simulation, the local value of all the species is found by table-lookup according to the local mixture fraction value.

Based on the assumption that combustion is mixing-controlled, a linear combination of the conservation equation for fuel and oxidizer leads to the following transport equation for the mixture fraction, Z, a conserved quantity from a normalized mass ratio between fuel and oxidizer.

$$\frac{\partial \rho Z}{\partial t}+\frac{\partial(\rho u_i Z)}{\partial x_j}-\frac{\partial}{\partial x_j}\left(\frac{\mu}{Sc}\frac{\partial Z}{\partial x_j}\right)=0 \quad (1)$$

The oxygen mass conservation equation is given as,

$$\frac{\partial \rho Y_o}{\partial t}+\frac{\partial(\rho u_i Y_o)}{\partial x_j}-\frac{\partial}{\partial x_j}\left(\frac{\mu}{Sc}\frac{\partial Y_O}{\partial x_j}\right)=-\frac{dY_O}{dZ}\bigg|_{Z<Z_{st}}\frac{\mu}{Sc}\nabla Z\cdot\vec{n} \tag{2}$$

where x_j represents the co-ordinate system, and μ denotes the dynamic viscosity, Sc Schmidt number equal to 0.7. For simulation of a laminar diffusion flame, the grid resolution should be sufficient fine (about 1 mm near the reaction zone) to resolve the flame sheet. The heat release rate is directly proportional to the oxygen consumption rate [15] derived from the source term of Eq.(2) at grid cell as a function of the mixture fraction, Z.

2.2. Radiative heat transfer

For a heavily sooting flame such as fire, radiation is a crucial aspect of combustion, and can dominate other modes of heat transfer. A radiative transfer equation (RTE) is solved by using a discrete expression adapted to a finite volume method [15].

$$\vec{\nabla}.\vec{\Omega}I+\kappa I=\kappa\frac{\sigma T^4}{\pi} \tag{3}$$

For simplicity, six bands are selected to give an accurate representation of the most important radiation bands of CO_2 and H_2O (RadCal) [18]. The effect of soot concentration on radiation is included by adding the radiation coefficient of soot κ_s, into that of gas κ_g. As the radiation spectrum of soot is continuous, it is assumed that the mixture of soot and gas behaves as a gray medium with a mean absorption coefficient, $\kappa=\kappa_s+\kappa_g$. The soot absorption coefficient is calculated as a function of the temperature, T, and soot mass density, M :

$$\kappa_s=1225TM/\rho_{soot} \tag{4}$$

2.3. Soot formation model

The two time-dependent conservation equations for soot particles number, N (particles/m^3) and soot mass density, M (kg/m^3) are incorporated in FDS [15].

$$\frac{\partial \rho N}{\partial t}+\frac{\partial[\rho N(u_i+u_i^{th})]}{\partial x_j}-\frac{\partial}{\partial x_j}\left(\frac{\mu}{Sc}\frac{\partial N}{\partial x_j}\right)=\dot{\omega}_{N,inc}+\dot{\omega}_{N,coa} \tag{5}$$

$$\frac{\partial \rho M}{\partial t}+\frac{\partial[\rho M(u_i+u_i^{th})]}{\partial x_j}-\frac{\partial}{\partial x_j}\left(\frac{\mu}{Sc}\frac{\partial M}{\partial x_j}\right)=\dot{\omega}_{M,inc}+\dot{\omega}_{M,gro}+\dot{\omega}_{M,oxi} \tag{6}$$

where u_i^{th} is the thermophoretic velocity components :

$$u_i^{th}=-0.54\frac{\mu}{\rho}\frac{\partial \ln T}{\partial x_i} \tag{7}$$

Several theoretical and experimental works [10, 11, 13], highlight the importance of poly-aromatics species (PAH) as result of intermediary reactions in soot inception processes. Hall et al. [13] developed a soot inception model, which is based on the formation rates of two and three ringed aromatics ($C_{10}H_7$ and $C_{14}H_{10}$) from acetylene (C_2H_2), benzene (C_6H_6) and the phenyl radical (C_6H_5). The growth mechanism of PAH is composed of consecutive repeated reactions that can be classified in two steps mechanism, called HACA (H-Abstraction-C_2H_2-Addition). The first one is an abstraction mechanism of hydrogen.

$$2C_2H_2 + C_6H_5 \rightarrow C_{10}H_7 + H_2 \tag{8}$$

The second one is the addition reaction of acetylene.

$$C_2H_2 + C_6H_6 + C_6H_5 \rightarrow C_{14}H_{10} + H_2 + H \tag{9}$$

The first term in Eq.(5) represents the rate of soot particle inception, and the resulting expression is as,

$$\dot{\omega}_{N,inc} = \frac{8N_A}{M_p}\left\{ c_2\left[\left(\frac{\rho Y_{C_2H_2}}{W_{C_2H_2}}\right)^2 \frac{Y_{C_6H_6} W_{H_2}}{W_{C_6H_6} Y_{H_2}}\right] e^{-\frac{4378}{T}} + c_3\left[\frac{\rho^2 Y_{C_2H_2} Y_{C_6H_6} Y_{C_6H_5} W_{H_2}}{W_{C_2H_2} W_{C_6H_6} W_{C_6H_5} Y_{H_2}}\right] e^{-\frac{6390}{T}}\right\} \tag{10}$$

$$\dot{\omega}_{M,inc} = \frac{M_P}{N_A}\dot{\omega}_{N,inc} \tag{11}$$

where c_2 and c_3 are the scaling factors, given as $c_2 = 127\times10^{8.88}, c_3 = 178\times10^{9.50}$, M_p $(= 12011\,kg/kmol)$ is molar mass of a soot nucleus and N_A represents the Avogadro's number.

The second term in Eq.(5) represents the soot coagulation rate which is determined by Lindstedt [19] from the collision frequency (Puri et al.[20]).

$$\dot{\omega}_{N,coa} = -\left(\frac{24R}{\rho_{soot} N_A}\right)^{1/2}\left(\frac{6}{\pi\rho_{soot}}\right)^{1/6} T^{1/2} M^{1/6} N^{11/6} \tag{12}$$

The second term in Eq.(6) represents the surface growth of soot which is proposed by Frenklach et al. [21].

$$\dot{\omega}_{M,gro} = c_4\left(\rho\frac{Y_{C_2H_2}}{W_{C_2H_2}}\right) e^{-\frac{12100}{T}}\left[(\pi N)^{1/3}\left(\frac{6M}{\rho_{soot}}\right)^{2/3}\right] \tag{13}$$

where c_4 is a scaling factor, given as c_4=9000.6 kg.m.kmol^{-1}.s^{-1}, ρ_{soot} density of a soot particle, R the universal gas constant and W_i molecular weight of species, i.

The last term in Eq.(6) is the soot oxidation rate which is assumed to proceed through two reaction steps,

$$\frac{1}{2}O_2 + C_s \rightarrow CO \tag{14}$$

$$C_s + OH \rightarrow CO + H \tag{15}$$

Lee et al. [22] measured and modelled the O_2 and OH dependence of the soot oxidation in a laminar diffusion flame. By assuming that the kinetics of surface reactions are the limiting mechanism, the specific rate of soot oxidation is expressed by,

$$\dot{\omega}_{M,oxi} = -c_5\tau\rho\frac{Y_{OH}}{W_{OH}} T^{1/2}(\pi N)^{1/3}\left(\frac{6M}{\rho_{soot}}\right)^{2/3} - c_6\rho\frac{Y_{O_2}}{W_{O_2}} T^{1/2}(\pi N)^{1/3}\left(\frac{6M}{\rho_{soot}}\right)^{2/3} e^{-\frac{19778}{T}} \tag{16}$$

where $c_5 = 105.81\,kg.m.kmol^{-1}.K^{-1/2}.s^{-1}$, $c_6 = 18903.51\,kg.m.kmol^{-1}.K^{-1/2}.s^{-1}$ and a value of 0.13 is assigned to the collision efficiency, τ.

2.4. Numerical technique

Low speed solvers used in FDS [15] explicitly eliminate compressibility effects that give rise to acoustic (sound) waves for low-Mach number flows. The Boussinesq approximation invoked – the fluid is still considered thermally-expandable through the divergence of the flow. The temperature is extracted from the density via the equation of state. The fidelity of the numerical solution of the entire system of equations is tied to the pressure/velocity coupling thanks to a direct Poisson solver [15]. All spatial derivatives are approximated by second-order finite differences and the flow variables are updated in time using an explicit second-order predictor-corrector scheme. The convective terms are written as upwind-biased differences in the predictor step and downwind-biased differences in the corrector step. The thermal and material diffusion terms are pure central differences in both the predictor and corrector steps [15]. In this study, the fuel temperature of 300 K and the atmospheric pressure (P=1 atm) are imposed at the porous burner boundaries, conforming to the experimental conditions [5, 6]. At the free boundaries, zero gradient conditions are used for the farfield boundary values of the mass fractions.

3. Results and discussions

The sketch of the experimental setup [5, 6] with the relevant burner dimensions is displayed in figure 2(a, b). In the experiment, microgravity conditions are attained by means of parabolic flights that provide a gravity level of 10^{-3}g during 22 seconds. The experimental set-up corresponds to a square porous burner (L_p=W_p=5 cm, cf. figure 2b) which is embedded in the plane at z=0, centred in the y direction and at 5 cm from the inlet of a large enclosed combustion chamber. Porous burners have been used regularly in an attempt to simplify experiments by avoiding the coupling between heat feedback from the flame and fuel supply so that longer experimentation time can be achieved. The pyrolysis products of a condensed fuel are simulated by injecting ethylene with a constant velocity, V_F, through a porous burner. Due to the limited space and amount of feed gases available in parabolic flight facilities, only small fuel rate can be accommodated. The dimensions of the computational domain, as shown in figure 2(a,b), are 8Lp in the windward direction, and 4Lp in both the lateral and normal directions so that boundary conditions can be set to conform as much as possible to the experimental configuration. The grid system contains 240 x 80 x 60 cells, and cells of 1 mm are concentrated around the reacting zone, giving the best trade-off between accuracy and cost for a three-dimensional reacting flow simulation. The resolution tests indicated that the computations are grid independent when the cell dimensions are lower than 2 mm. A further reduction in the grid size results in a significant reduction in the time step ($\Delta t < 0.0005$ s) for satisfying the CFL stability condition [15]. As the experiment, an uniform oxidizer flow at the inlet with a composition of 35% oxygen and 65% nitrogen flows parallel to the surface with a constant velocity, V_{ox}=0.2 m/s. For describing the effects of pyrolysis rate in boundary layer with blowing, a dimensionless volume coefficient, $C_q=V_F/V_{ox}$, defined by Schlichting [23], is currently employed. It appears that three-dimensional phenomena in flame structure in low-speed concurrent flow are very rich in content as a function of the C_q.

This type flame can be described by the ratio between buoyant force and inertia one. This ratio is defined as follows, $\xi = \frac{Gr_x}{Re_x^{5/2}}$, where Re_x is the Reynolds number and Gr_x the Grashof number [24]. The experimental work [24] showed that in normal gravity, vigorous ejection of pyrolyzed fuel in the pyrolysis zone forms a vertical plume (cf. figure 3a) in the quiescent environment, indicating a predominant natural convection ($\xi > 1$). While microgravity reduces buoyancy enough to enable a much lower ratio ($\xi \leq 1$) even for a low Reynolds number flow (Re<100). This induces a boundary layer diffusion flame (cf. figure 3b, c) with a significant change in the flame orientation and streamlines. The appearance of the predicted flame (cf. figure 4) is qualitatively similar to that described by experiment for laminar diffusion flame (cf. figure 3c). The base of the flame is not attached to the gas fuel supply surface but is stabilized at significant stand-off height downstream. The total flame includes a blue flame plus orange zone. The leading edge portion of the flame consists of a blue region which appears more diffuse than the trailing one. Downstream of this initial zone, the remainder of the flame is orange-red streaks appearance, indicating a presence of great amounts of soot, and par consequent, a significant contribution to the heat flux on the surface.

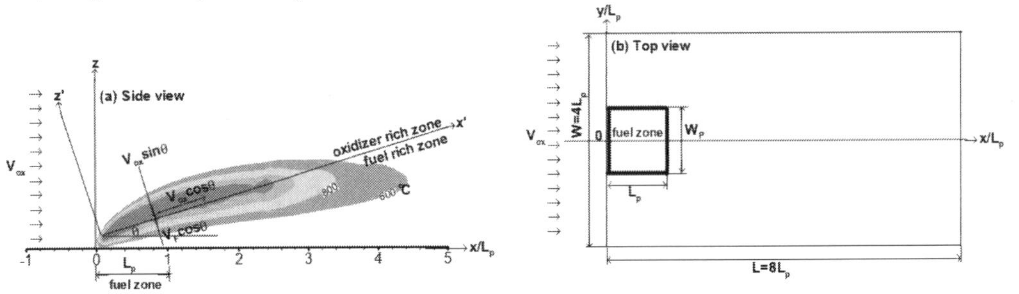

(a) Iso-contours of the temperature (side view) (b) Disposition of the fuel zone (top view)

Figure 2. Computational domain and the coordinate system

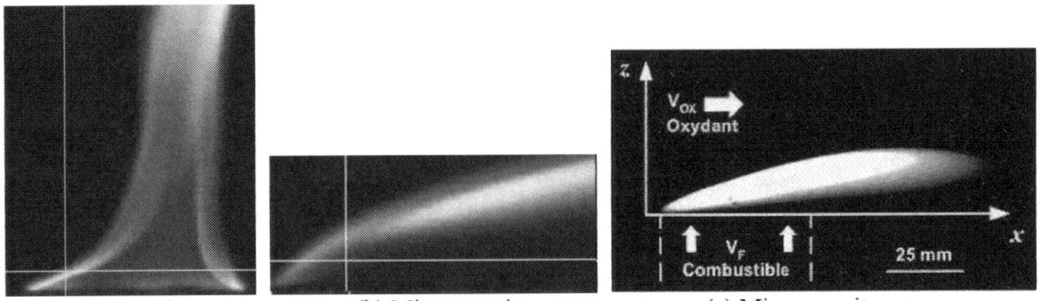

(a) Normal gravity (b) Micro-gravity (c) Micro-gravity

Figure 3. Structure of non premixed reacting flow from experiment

The predicted velocity field in vector form, is shown in figure 5. Energy release from an exothermic chemical reaction enhances the flow perturbations at the leading edge enough to enable a velocity overshoot along the line where fuel and oxidizer are delivered in stoichiometric proportions. The flow is controlled mainly by the cross-flow despite the surface blowing and the energy release due to combustion. The velocity profiles display a pronounced overshoot close to the flame zone at a low free stream velocity of 0.2 m/s. In normal gravity, the experimental work of Andreussi [25] indicated that a similar boundary layer velocity field is formed over a blowing surface only when the oxidizer stream velocity is larger than 1.2 m/s.

Reduced-gravity measurements are generally difficult to obtain, and so usually only the most easily measured quantities for a particular experimental configuration can be performed. This case is selected in this demonstration to clearly illustrate the peak temperature dependence on C_q at the trailing edge for the oxygen concentrations of 0.35. Both these measured and predicted values are presented in figure 6. The authors [5, 6] argued that thermocouples significantly perturb the flame in a low Reynolds number flow (Re<100), and thus, the measured flame temperature of 650°C clearly deviate from real flame temperature ranging from 1200-1600°C. In fact, an accurate determination of the flame temperature and heat flux could not be made due to the limited amount of microgravity time available. The prediction suggests that this characteristic temperature at the trailing edge is strongly dependent of C_q, and a temperature drop of approximately 450°C occurs with an increase of C_q from

0.02 to 0.08. The general characteristics of the flame temperature at the trailing edge include an initially gradual reduction in temperature level with C_q, followed by a rapid decrease in temperature until quenching. That means that rapid pyrolysis of the condensed fuel contributes to reducing the flame temperature below the adiabatic stoichiometric value.

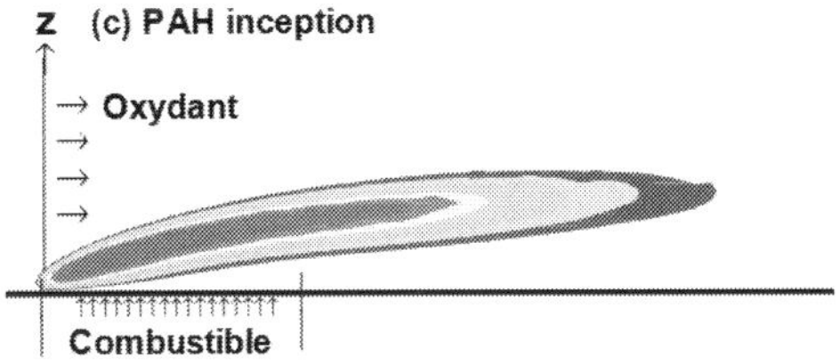

Figure 4. Predicted iso-contours of temperature

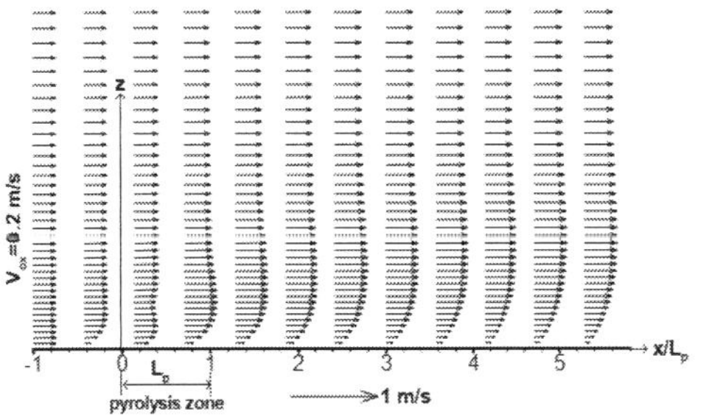

Figure 5. Predicted velocity vectors

The sooting behaviour of such laminar diffusion flame is characterized by using the laser induced incandescence technique (LII). The basic principle of LII relies on the detection of the thermal radiation from the soot particles that have been heated up to vaporization temperature by means of high energy laser pulse. Given in figures 7 and 8 is the comparison between the predicted and measured soot volume fractions across the height, z, at various axial locations x/L_p=0.5-1.75 for V_{ox}=0.25 m/s and V_F=0.005 m/s. By looking at their locations, the measured peak is placed just above the plate surface at z=3 mm, while the calculated one occurs at z=5 mm, with an over-prediction of the vertical stratification of the soot layer. A global description of the integrated soot production, F_{soot}, along x-axis as a function of V_F at an oxidizer velocity of 0.25 m/s is presented in figure 9. Soot volume fraction is integrated only over the soot layer in the z direction, as follows : $F_{soot}(x) = \int_{z=0}^{z_{max}(x)} f_{soot}(x,z)dz$, where Z_{max} (x) is the height so that f_{soot} (x,z)=0 for z>Z_{max}(x) at a given x. It seems that the PAH inception model reproduces the aforementioned sensitivity of the experimental soot formation to V_F. An enhancement of F_{soot} (cf. figure 9) from 20 to 38 (ppm)mm occurs with an increase of the fuel injection velocity, V_F, from 0.003 to 0.005 m/s. Thus, the shortest blue flame length and the longest yellow flame one (cf. figure 3b) are reached when the flame quenching occurs due to radiative loss with the highest fuel injection rate. The peak values of the integrated soot volume fraction are located far away from the trailing edge (x/L_p>1). It is found that the extent of the measured soot is always within the zone of x/L_p<3. It should be noted that the measured soot volume fraction distribution is converted from LII intensity image, and its accuracy depends on the accuracy of the proportionality constant required for correction due to attenuations of both the laser beam and the collected signal. Besides, for the low fuel injection velocity of V_F=0.003 m/s, the flames considered have the particularity of being mostly two-dimensional. While for high V_F, the flames considered develop strongly three-dimensional features far away from the trailing edge, affecting the collected signal for LII. This analyse is supported by the fact that the predicted soot profile follows closely the experimental trend for the low fuel injection velocity of 0.003 m/s, because the soot formation occurs within the extent of the collected signal for LII. While for the fuel injection velocity beyond 0.004 m/s, general shape of the predicted soot profiles has qualitative agreement with the experimental data. It is felt that the difference is due to a combination of experimental uncertainties and the error in the numerical simulation.

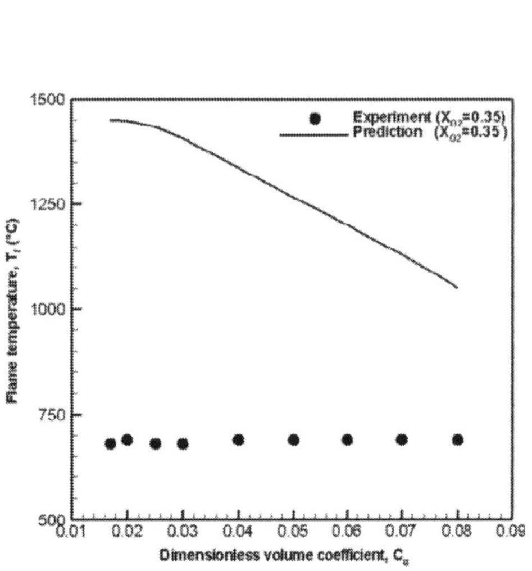

Figure 6. Effect of soot formation on the flame temperature as a function of C_q

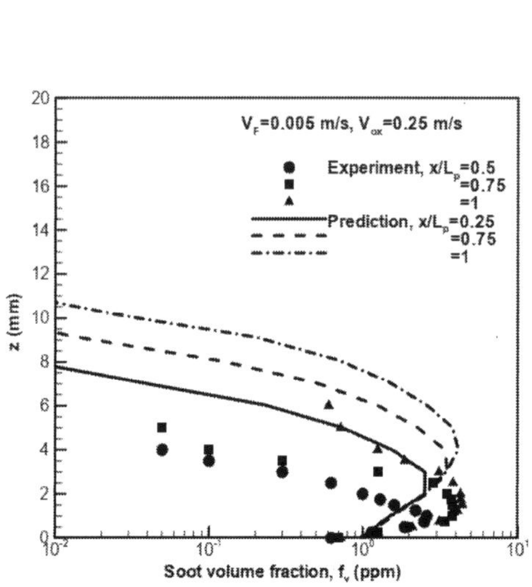

Figure 7. Comparison between computation and experiment for the soot volume fraction at the axial locations x/L_p=0.5-1

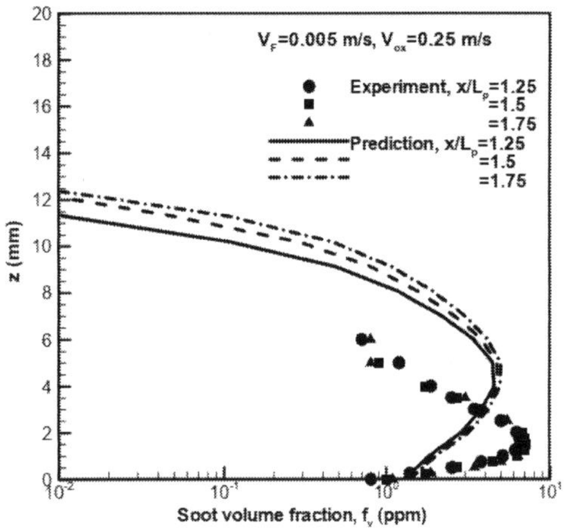

Figure 8. Comparison between computation and experiment for the soot volume fraction at the axial locations x/L_p=1.25-1.75

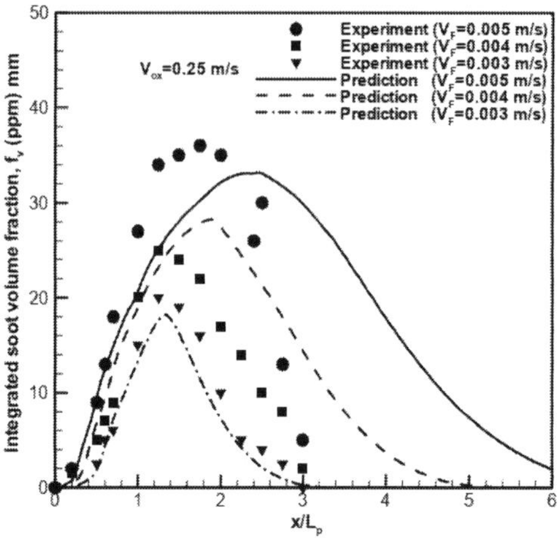

Figure 9. Comparison between computation and experiment for the integrated soot volume fraction at different fuel injection velocity

The trajectories of the jet diffusion flame and the velocity overshoots, characterized respectively by the flame stand-off distance d_T and the boundary layer thickness, d_V, in a cross-flow as a function of C_q are illustrated in figures 10 and 11. Because of radiative heat loss from this heavily sooting flame, the flame temperature does not surpass 1300°C and decreases along the x-axis. The flame stand-off distance, d_T, is the distance of the 600°C contour from the plate (cf. figure 4), and this definition appears to be very close to the soot-radiating zone [5, 6]. The boundary layer thickness, d_V, is defined as the distance above the surface where the parallel component of velocity reaches 99% of the main flow velocity. In considering a wind-blown jet diffusion flame, it is convenient to identify various regions of flow within the jet flame. If the extent of each of these regions is considered to depend on C_q, then three flow regimes can be identified. In the first region, which extends from the fuel exit to the point where the wind forces first start to bend the jet diffusion flame, the momentum flux of the jet flame itself is dominant and the physical characteristics of the jet flame are primarily a function of fuel injection velocity. The second transitional region extends from the end of the fuel injection velocity dominated regime to a point where wind effects begin to prevail. In the final region, the jet flame characteristics are dominated by the wind, with jet flame motion normal to the cross-wind being small. The diffusion flame stand-off distance, d_T, is an order of magnitude larger than that predicted by a Shvab-Zeldovich analysis, as described by Emmons [26] for a reacting boundary layer flow with blowing. The diffusion flame is shorter for the weak pyrolysis rate, and the isotherms are closer to the pyrolysis surface than those for the high one. Since the oxygen-side diffusion is the limiting factor in low-speed diffusive flames, an increase of C_q extends the reaction contour in the downstream direction and pushes the reaction zone far away from the plate surface due to the stoichiometric requirement. The flame stand-off and the boundary layer approach an asymptotical line once the C_q surpasses a critical value of 0.16, and reach a plateau as the flame quenching starts.

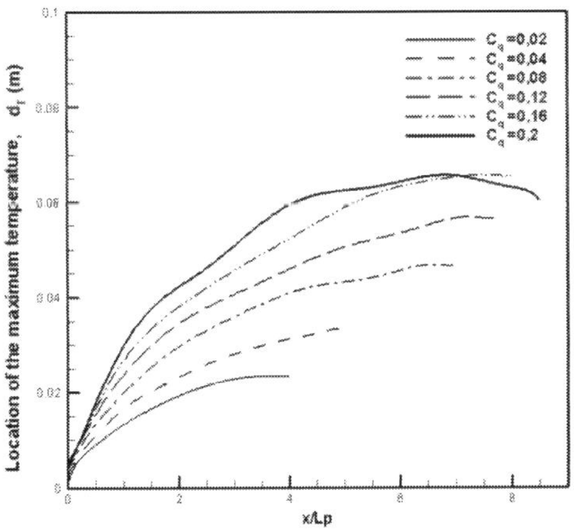

Figure 10. Evolution of the flame standoff distance for different C_q

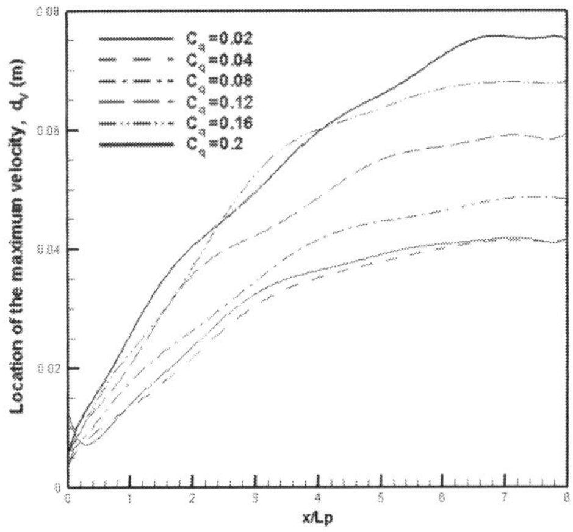

Figure 11. Evolution of the boundary layer thickness for different C_q

Given in figure 12 is the ratio, d_T/d_V, between the flame standoff distance, d_T, and the boundary layer thickness, d_V. Near the leading edge, the flame is positioned very close to the porous plate edge, and the flame establishes inside the viscous boundary layer ($d_T/d_V<1$), denoting the importance of oxygen-side diffusion which strengthens the flame. While soot formation shortens and weakens the blue flame near the leading edge through gas radiation loss with C_q. As x increases, both d_T and d_V move away from the plate into an area of higher velocity, and the ratio, d_T/d_V, converges to one single curve and toward a constant value of approximately 1.2, implying that the flame lies entirely outside the velocity boundary layer. The length of the soot-radiating zone attains a peak until blowoff. Once the flame quenching occurs, since most of the fuel remains in the core and is not participating in the combustion process, the trajectory of the maximum temperature is lower than the velocity one. An asymptotical solution established in natural convection from the reactive boundary layer model [26] suggests that the flame resides always within the boundary layer.

Figure 13 illustrates the ratio, d_H/d_T, between the trajectory of heat generation due to combustion and the flame standoff distance. The trajectory of heat release corresponds to an infinitely thin zone where fuel and oxidizer are delivered in stoichiometric proportions. In agreement with the findings of experiment [5, 6], the maximum temperature achieves on the side of the highly reacting fuel rich region ($d_H/d_T>1$) due to radiation loss. With an increase of C_q, the highest temperatures within the flame are found to approach the reacting zone ($d_H/d_T \approx 1$) due to blowing and thermal expansion. Thus under a zero gravity, the fluid and combustion are strongly coupled through the contributions of the heat release and the thermal radiation associated with soot formation, inducing a major influence on the trajectory of the diffusion flame.

The trajectory of soot particles is obtained from the loci of the profile maxima of soot concentration. Contrary to the flame standoff distance, the maximum soot formation occurs always inside the boundary layer ($d_S/d_V<1$) far away from the stoichiometric line, as shown in figure 14. Soot particles trajectories inside a laminar boundary layer diffusion flame in microgravity do not follow exactly the gas movement due to thermophoretical effects, consistent with the observation of Fujita et al. [27]. This is evidenced by a significant deviation of soot particles from the stoichiometric line due to the dominant thermal expansion which prevents soot particles approaching the flame. The peak soot production is placed just over the plate surface in the fuel rich zone (cf. figures 7 and 8) for the small value of C_q=0.02. The trajectory where soot is delivered in abundance follows the response of flame standoff to changes in the main parameter of C_q.

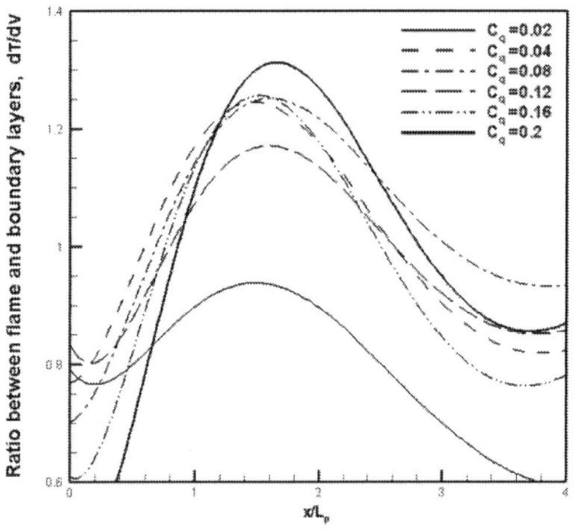

Figure 12. Ratio between flame standoff distance and boundary layer thickness for different C_q

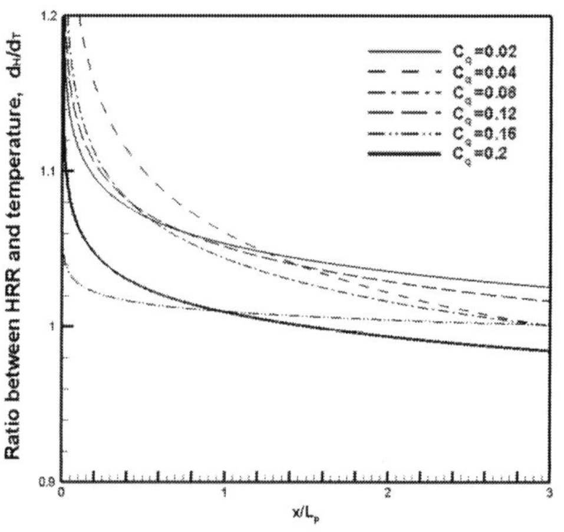

Figure 13. Ratio between the heat release trajectory and flame standoff distance for different C_q

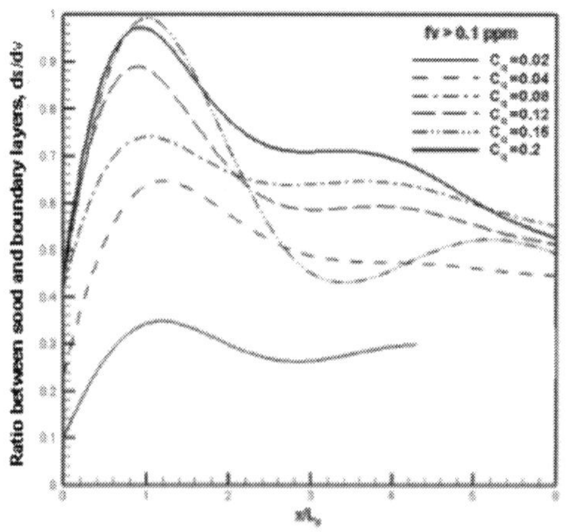

Figure 14. Ratio between soot and boundary layer thicknesses for different C_q

Contours of mean gas temperature on the symmetry plane are shown in figure15. A stable symmetrical flame and counter-rotating vortex in the cross-section are generated under microgravity conditions even with small value of C_q=0.02. The extent of the flame in cross-stream plane surpasses significantly the pyrolysis zone. Development of the counter-rotating vortex leads to the formation of the kidney-shape structure on the cross-stream plane for the temperature. This allows soot particle for long residence times before crossing the flame sheet at fuel-rich conditions, and enhances soot formation. This kidney shaped cross-section can be also characterized by the distance between the peak temperatures S_T, as shown in figure 16. Near the leading edge, the action of the counter-rotating vortices has not led to significant distortion of the temperature field with the contours appearing as its original single-flame. As downstream distance increases (x/L_p>2), the thermal expansion due to an exothermic chemical reaction enhances the air entrainment enough to enable the formation of the double-vortex motions. The action of these vortices is sufficiently strong to lead to two isolated regions with a peak temperature that are remote from the symmetry plane of the flame. The cross-stream contours of figure 15 shows that the main feature of the temperature field of the flame is again its kidney shaped cross-section, caused by the double-vortex motions. The distance S_T extends from the symmetry plane across the flame cross-section, but gradually migrated to the centre-line of the flame as downstream distance increases. Moreover, transition from its original single-flame to the kidney shape dual flame occurs earlier at low C_q due to the higher temperature level. Thus this is attributed to the competing effects between the flame temperature and the soot formation with C_q on the oxygen-side diffusion into the 3D flame.

In concurrent fire growth for a thick solid in microgravity condition, the fire growth rate correlates well with the flame heat feedback in the preheat zone. In microgravity, heat transferred by radiation ahead of a spreading flame is responsible for causing the flame to propagate. After a threshold pyrolysis temperature is attained, all heat flux to the material surface is used entirely for pyrolysis. The radiation heat flux in the preheat zone depends mainly on three factors: 1) the flame volume and its temperature level; 2) the concentrations of gaseous and particulate soot emitting species; 3) the view factor of the flame to the exposed material surface.

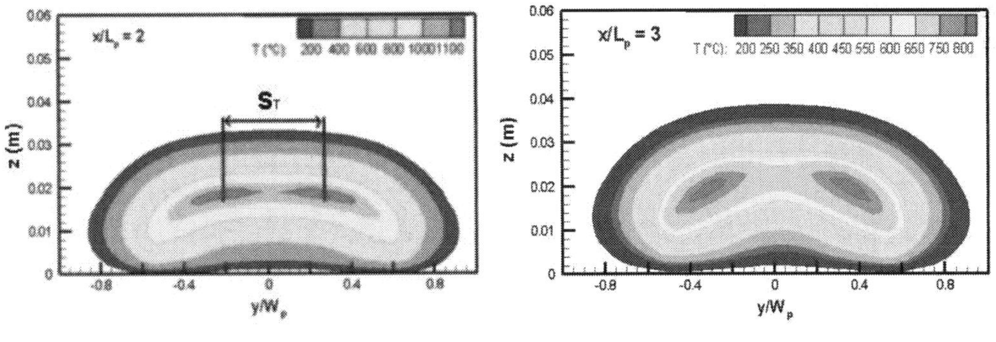

Figure 15. Cross-stream contours of the temperature at the axial locations x/L_p=2 and x/L_p=3

Given in figure 17 is the normalized total heat flux over a plate surface, where q_{pyro} is the heat release rate per unit area (kW/m^2) from a pyrolysis zone. Flame structure is of great importance on the heat flux over a material surface. The peak of the heat flux profile defines the location where the flame approaches to surface most closely. Microgravity flames are typically smooth (cf. figure 3b) with a large heat feedback just downstream the leading edge. The peak in the fraction of the total energy released by the flame to the surface increases from 5% to 15% with a rise of C_q due to increase in the flame volume. As downstream distance increases, the flame standoff distance becomes significant (cf. figure 10). Consequently, the maximum temperature moves away from the surface, implying a reduction in the view factor from flame to the material surface. This leads to a decrease of the fraction of the total energy to the surface, approaching a value of about 3% in spite of C_q.

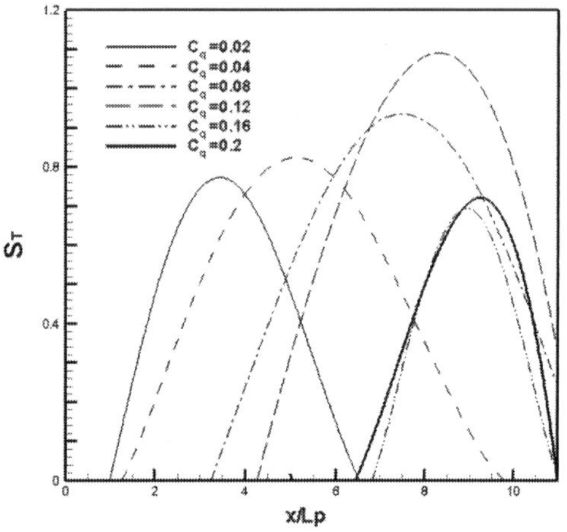

Figure 16. Evolution of the distance between the highest temperature position at a cross-stream section for different C_q

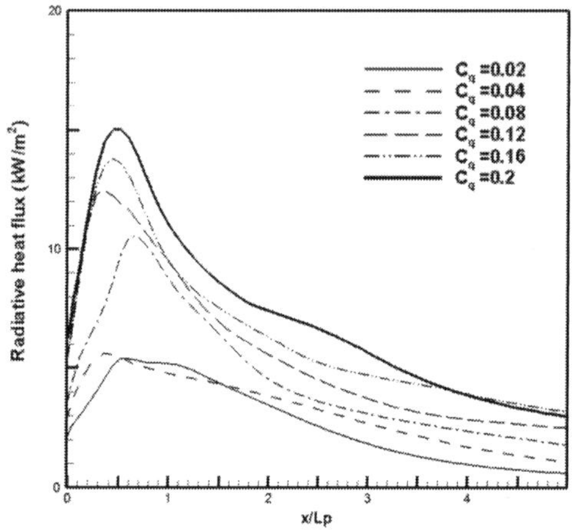

Figure 17. Normalized total heat flux over the horizontal surface for different C_q

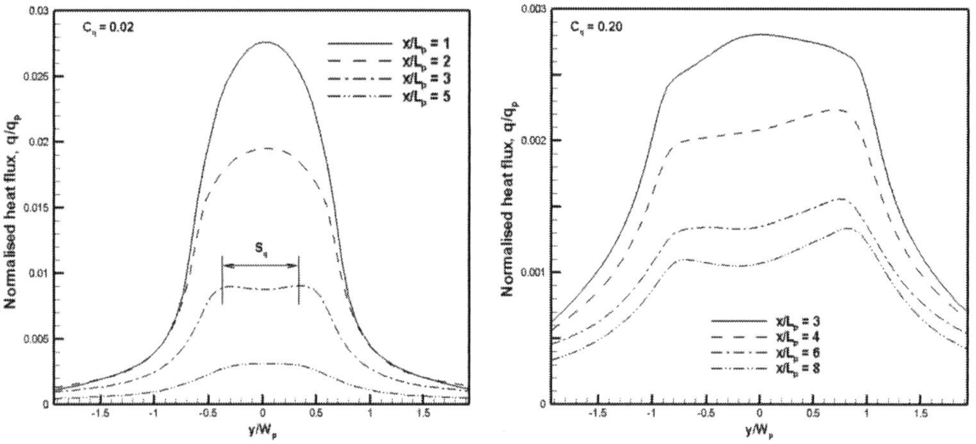

(a) C_q=0.02 (b) C_q=0.2

Figure 18. Evolution of the width of the high heat flux at different cross-stream sections

Turning finally to the thermal radiation profiles in a cross-stream section at different downwind stations, as shown in figure 18(a, b). The peak in the heat flux profile is consistent with the proximity of the flame to the surface. Gas temperatures are higher near the leading edge (x/L_p<1) of the flame, but the higher heat flux distribution is narrower for the small value of C_q. Downwind of the peak, the received heat flux falls off rapidly beyond the flame tip. The width scale S_q (cf. figure 18a) of the higher radiation flux as a function of C_q is illustrated in figure 19. The width S_q of the high heat flux increases noticeably when the flame becomes most obviously dual (cf. figure 15). Increase in C_q from 0.02 to 0.2 widens the soot-radiating orange zone, and consequently, the width of the higher radiation flux increases almost four fold. However, the peak in the heat flux remains almost the same, independent of C_q. Obviously, the lower flame temperature and the increased soot formation with a rise of C_q are the competing effects to determine the magnitude of the flame heat feedback to the exposed material surface.

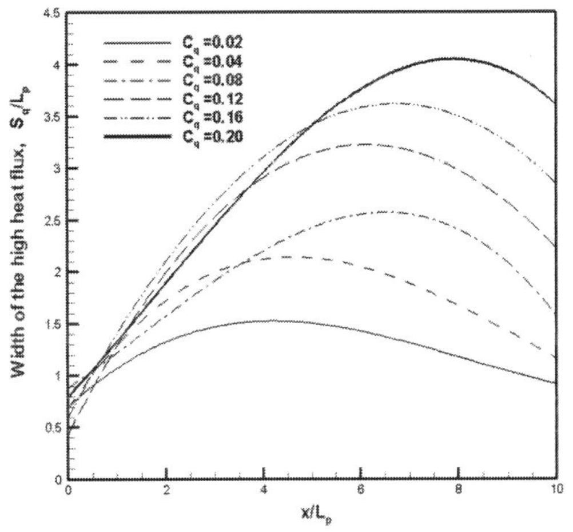

Figure 19. Ratio between the width of the high heat flux and the pyrolysis length for different C_q

4. Conclusions

This numerical study highlights the importance of soot formation on the structure of a microgravity diffusion flame established in a shear boundary layer. The behaviour of the soot production is reasonably well predicted by using PAH inception model. Both the prediction and the experiment suggest that soot production is enhanced with increasing the pyrolysis rate, further emphasizing the role of radiation on the geometric characteristics of a co-current, boundary layer laminar flame. It can be concluded that a large dimensionless volume coefficient favours soot formation, which shortens the blue flame near the leading edge due to radiation loss and strengthens the soot-radiating, orange-red flame. For a low Reynolds number reacting flow in microgravity, the flame establishes outside the viscous boundary layer far away from the leading edge, while the soot resides within the boundary layer. This implies the less importance of oxygen-side diffusion as compared to expansion and convection. A maximum in the flame length for oxidizer velocities below those encountered in natural convection, represents a greater hazard for fire safety in spacecraft.

These results are valid only for gaseous fuel and oxidizer, and should not be directly extrapolated to condensed fuel burning. Ongoing work should be focused on a proper definition of the transport processes for an accurate modelling of the solid phase using a pyrolysis reaction.

5. References

[1] Olson S L and T'ien J S 2000 *Combustion and Flame* **121(3)** 439-452
[2] T'ien J S 1990 *Combustion and Flame* **80** 335-357
[3] Konsur B, Megaridis C M and Griffin D W 1999 *Combustion and Flame* **118 (4)** 509-520
[4] Markstein G H and De Ris J 1985 *20th Symposium (international) on Combustion* (Pittsburgh: *The Combustion Institute*) 1637-46
[5] Legros G, Joulain P, Vantelon J P, Fuentes A and Torero J L 2006 *Combustion Science and Technology* 178-189

[6] Fuentes A, Legros G, Claverie A, Joulain P, Vantelon J P and Torero J L 2007 *31th Symposium (international) on Combustion (*Pittsburgh: *The Combustion Institute*) 2685-92
[7] Rouvreau S, Torero J L and Joulain P 2005 *Combustion Theory and Modelling* **9:2** 137-158
[8] Lui F, Guo H S, Gregory J, Smallwood J and Ömer L 2002 *J. of Quantitative Spectroscopy & Radiative Transfer* **73** 409-421
[9] Leung K M, Lindstedt R P and Jones W P 1991 *Combustion and Flame* 289-305
[10] Wen Z, Yun S, Thomson M J and Lightstone M F 2003 *Combustion and Flame* **135** 323-340
[11] Vovelle C, Delfan J L and Reuillon M 1994 *Soot Formation in Combustion* Springer-Verlag, Berlin 50-71
[12] Anderson H, McEnally C S and Pfefferle L D 2000 *28th Symposium (international) on Combustion (*Pittsburgh: *The Combustion Institute*) 2577-85
[13] Hall R J, Smooke M D and Colket M D 1997 *in Physical and Chemical Aspects of Combustion: A Tribute to Irvine Glassman, F.L Dryer and R.F Sawyer* (Ed.) Gordon & Breach 189-201
[14] Dagaut P and Nicolle A 2005 *Combustion and Flame* **140** 161-171
[15] Mcgrattan K B, Glenn P F and Jason E F 2010 *Fire Dynamics Simulator*
[16] Poinsot T and Veynante D 2005 *Theoretical and Numerical Combustion* second ed. Edwards
[17] Lutz A E, Kee R J, Grcar J F and Rupley F M 1997 *OPPDIF: A fortran computer program for computing opposed-flow diffusion flames* Sandia National Laboratories-Technical Rep. SAND96-8243
[18] Grosshandler W 1993 *National Institute of Standards and Thchnology* Maryland USA
[19] Lindstedt P R 1994 *Soot formation in Combustion, Springer-Verlag* 417-420
[20] Puri R, Richardson T F and Santoro R J 1993 *Combustion and Flame* **92 (3)** 320-333
[21] Frenklach M, Clary D W, Gardier J and Stein S E 1984 *20th Symposium (international) on Combustion (*Pittsburgh: *The Combustion Institute*) 887-901
[22] Lee K B, Thring M W and Beer J M 1962 *Combustion and Flame* **6** 137-145
[23] Schlichting H 1979 *Boundary layer theory* Seventh Edition McGraw-Hill
[24] Torero J L, Bonneau L, Most J M and Joulain P 1994 *25th Symposium (international) on Combustion (*Pittsburgh: *The Combustion Institute*) 1701-09
[25] Andreussi P 1982 *Combustion and Flame* **45** 1-6
[26] Emmons H 1956 *The film combustion of liquid fuel* Angews Z. Math. Mech. **36** 60-71
[27] Fujita O, Ito K, Ito H and Takeshita Y 1997 *4th NASA International Microgravity Combustion Workshop* 217-222

Born Oppenheimer Molecular Dynamics calculation of the ν_{O-H} IR spectra for acetic acid cyclic dimers

Mohamed El Amine Benmalti

University Abdelhamid Ibn Badis-Mostaganem, Laboratoire de Structure Elaboration et Application aux Materiaux Moleculaires, 27000, Mostaganem, Algeria.

E-mail: `malti23@yahoo.fr`

Abdelghani Krallafa

University of Oran, Laboratoire de Physico-Chimie Macromolculaire, LPCM, Oran 31000, Algeria

Marie-Pierre Gaigeot

University of Evry val d Essonne, LAMBE Laboratoire Analyse et Modelisation pour la Biologie et l'Environnement Blvd Franois Mitterrand Bat Maupertuis - 91025 EVRY Cedex France.

Abstract. Both *ab initio* molecular dynamics simulations based on the Born-Oppenheimer approach calculations and a quantum theoretical model are used in order to study the IR spectrum of the acetic acid dimer in the gas phase. The theoretical model is taking into account the strong anharmonic coupling, Davydov coupling, multiple Fermi resonances between the first harmonics of some bending modes and the first excited state of the symmetric combination of the two ν_{O-H} modes and the quantum direct and indirect relaxation. The IR spectra obtained from DFT-based molecular dynamics is compared with our theoretical lineshape and with experiment. Note that in a previous work we have shown that our approach reproduces satisfactorily the main futures of the IR experimental lineshapes of the acetic acid dimer [Mohamed el Amine Benmalti, Paul Blaise, H. T. Flakus, Olivier Henri-Rousseau, Chem Phys, 320(2006) 267-274.].

1. Introduction

The linear infrared (IR) ν_{X-H} spectroscopy of absorption spectra is an important method to understand the physical informations on the dynamics of weak hydrogen bonds (H-bonds). This spectroscopy has been the object of considerable experimental and theoretical works in the literature [1, 2] Numerous studies on hydrogen bonded systems such as water molecule, biological molecules and carboxylic acid dimers is now studied using Density Functional Theory (DFT) techniques [3, 4]. There is also new methods such as Car Parrinello Molecular Dynamics (CPMD) program which is used to understand the dynamics of the hydrogen bond in cyclic dimers and to obtain vibrational spectra of such molecules. The calculation of infrared spectra through Molecular Dynamics (MD) relies on dipole time correlation functions recorded along the trajectory [5]. This is well established in the context of classical MD [6], but is much more recent in the ab initio MD community.

The vibrational properties of acetic acid dimer (A.A.D.) were studied many times, either experimentally [7, 8] or theoretically [9]. In a recent paper Piotr Durlak and Zdzisław Latajka [10] have studied Acetic acid monomer, dimer and its deuterated using ab initio Car-Parrinello molecular dynamics. They have demonstrated that the IR spectra calculated by CPMD method show a good agreement with the experimental Infrared spectrum and reproduces well isotopic effects. We have shown in a recent work [11, 12] how quantum theories are used to explain the changes in the infrared spectrum induced by the formation of the H-bond bridge, i.e. the: (i) large increase in the band width, (ii) band asymmetry, (iii) fine structure, (iv) temperature and isotopic effects. These theories were able to reproduce the IR spectra from some physical basic accessible parameters.

Recently [9, 11], we have proposed a quantum approach of the $\nu_s(X-H)$ IR spectra of the acetic acid dimer. This approach involves (i) the adiabatic approximation (allowing to separate the high frequency mXH motion from that of the H-bond bridge), for each separate part of the dimer, (ii) a strong non-adiabatic correction via the resonant exchange between the excited states of the two fast mode moieties (Davydov coupling), (iii) quantum direct damping taking into account the relaxation of the high frequency modes, (iv) quantum indirect damping taking into account the relaxation of the Hbond bridges.

The present paper aims at demonstrating the applicability of *ab initio* molecular dynamics simulations based on the Born Oppenheimer approach for the calculation of infrared spectra of the A. A. D in the gaz phase. This method and its application to vibrational infrared spectroscopy has been described in details in previous papers [13, 14]. The main advantages of the molecular dynamics approach for the calculation of infrared spectra in comparison with a static *ab initio* calculations is: i) Dynamics simulations are performed at finite temperature. (ii) When calculating IR spectrum from the dipole time correlation function, all anharmonic effects are naturally described. The calculated spectrum from the dynamics is compared with experiment typically done at room temperature.

2. Computational methods

The initial structure of the A.A.D. (figure.1) for wich the trajectories are performed in the present work has been taken from optimized geometry obtained at the B3LYP with sddall (5d,7f) level by using Gaussian 03 program package [15]. DFT-based Born Oppenheimer Molecular Dynamics (BOMD) simulation were performed on this initial structure as a starting for the dynamics with Cp2k package. The calculated parameters concern an isolated dimer of acetic acid.

The dimer was located in a simple cubic box of dimension 13.0 Å. Gaussian basis sets and pseudo-potentials are used. We have used the Becke, Lee, Yang and Parr (BLYP) gradient-corrected functional [16, 17] for the exchange and correlation terms. The calculations are performed with Goedecker-Teter-Hutter (GTH) pseudopotentials [18] and large Gaussian basis sets (in this work (aug-TZV2P) from the CP2K library). We have performed five trajectories (250K, 270K, 300K, 302K and 305K). The average temperature of the dynamics is 285,4K. Our dynamics are strictly microcanonical (NVE ensemble) with a time step of 0,4 fs. In order to eliminate the effect of the periodic images of the charge density we have used the decoupling technique of Martyna and Tuckerman [19]. The data was visualized using the VMD [20] program. The vibrational spectrum has been calculated using the Fourier transformation of the dipole autocorrelation function obtained from dipole trajectories generated by the Cp2k simulation through the following expression.

$$I(\omega) = \frac{2\pi\beta\omega^2}{3cV}\int_{-\infty}^{\infty} dt \; < M(t).M(0) > exp(i\omega\, t) \tag{1}$$

Where $\beta = 1/kT$, c is the speed of light in vaccum, V is the volume. M is the total dipole moment of the system.

Journal of Physics: Conference Series **574** (2015) 012011 doi:10.1088/1742-6596/574/1/012011

3. Results and discussion

Figure 2 presents the comparison between the experimental and theoretical IR spectrum extracted from Ref [12] and the present final IR spectrum extracted from an average of five trajectories of A. A. D. Typical broad band shape for the $\nu(O-H)$ mode is presented. One can see that the Born-Oppenheimer Molecular Dynamics (BOMD) simulated IR spectrum in the range of O–H stretching mode reproduce correctly complicated bands shape observed in both experimental [21] and theoretical spectrum for the nondeuterated acetic acid dimer [9]. The main reason of it is that during the *ab initio* molecular dynamics simulations all possible coupling of modes are taken into accounts due to the oscillations of atoms in molecules and in this approach there is no limitation due to inclusion of only specific coupling between modes. It should be noted that no scaling factors to the vibrational band positions are applied in our BOMD calculations. The sampling of anharmonicities (potential energy surface, dipole anharmonicity, mode couplings, anharmonic modes) being included in our simulations. It is clear that the lineshape obtained by our BOMD calculated spectrum (black line in figure 2) reproduce satisfactorily almost all the experimental features. The shoulder seen at 2900 cm^{-1} is present in the calculation but smaller. The shoulder located at 3150 cm^{-1} is also present in both experimental and theoretical spectrum. As expected the vibrational spectrum obtained by our simulation is very sensitive to the temperture effect and the anharmonicities, all these parameters are responsible for band broadenings. This result is in agreement with the band shapes obtained in the experimental spectra.

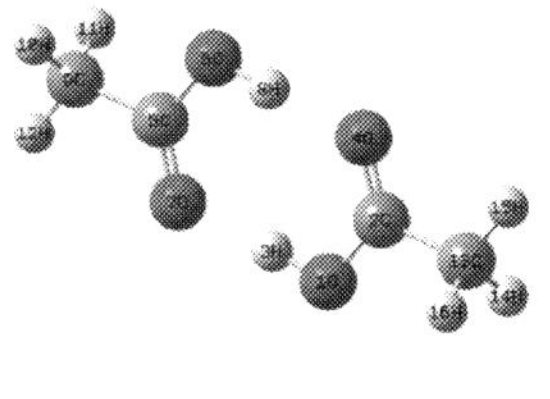

Figure 1. Structure of the acetic acid dimer.

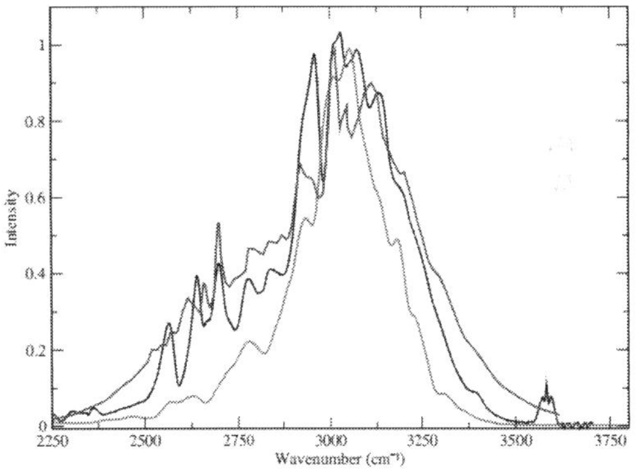

Figure 2. Comparison between the infrared spectra calculated through our calculations (red line) with both the theoretical (blue line) and experimental (black line) infrared spectra.

In table 1 are given the calculated bonds lenghts as a function of the temperature of the simulation. These values are taken for only one moiety of the dimer, the values of the other moiety of the dimer are almost the same. The average distance between the two oxygens O....O of the A. A. D in the gaz phase is 2,66Å. The calculated values are in a good agreement with experimental data 2,684 Å[22]. It should be noted that the hydrogen bond distance is modulate by the coupling mechanism between the bending modes of the hydrogen bond and the stretching

Table 1. Calculated average bond lengths at each trajectories for acetic acid dimer at different temperature.

T(K)	250	270	300	302	305
d_{O-H}(Å)	1,00	1,05	0,99	0,99	0,99
$d_{O....O}$(Å)	2,64	2,72	2,66	2,67	2,61
$d_{H....O}$(Å)	1,67	1,67	1,66	1,69	1,63

modes

In figure 3 are represented the time evolution of interatomic distances of atoms involved in intramolecular H-bonds of A. A. D. at 300K from the BOMD calculation. Analysis of the trajectories of the atoms during the BOMD simulation shows that there is a proton transfer, which is occuring at 16000 step (6,4 Ps) of the simulation. This proton transfer is also observed by Durlak [23] in the case of propinoic acid dimer in a PIMD (Path Integral Molecular Dynamics).

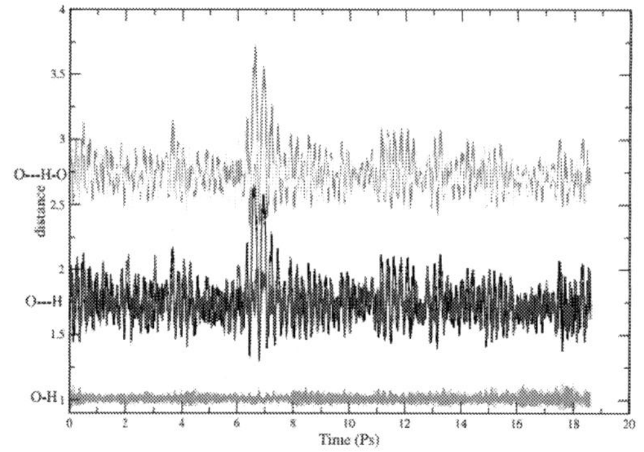

Figure 3. Time evolution of interatomic distances of atoms involved in intramolecular H-bonds of A. A. D. at 300K

4. Conclusion

In this paper, we report *ab initio* molecular dynamics simulations of acetic acid cyclic dimer in order to reproduce experimental infrared spectra. We have shown that vibrationnal properties can be obtained by using Molecular Dynamics within a DFT framework. The results we have obtained are in a good agreement with experiments. Such agreement can not be obtained in the case of static harmonic calculations. We think that our MD IR spectrum will match the experimental one if we increase the temperature of the dynamics and better resolution can be reached if we perform more tarjectories. The MD calculated spectra is better in comparison to that calculated with CPMD method by Durlak et al [23]. If we compare our MD spectrum with that obtained by the quantum approach the calculations reproduces satisfactorily the main features of the experimental spectrum. However, the theoretical model reproduce the main features of the lineshape of the $\nu(OH)$, but the use of approximations in the quantum approach such as adiabatic separation between the vibrational modes (high frequency and the H-bond bridge), Davydov coupling and Fermi resonnances make reaching the experimental IR spectra a very hard task.

Acknowledgments

Acknowledgments: This study was granted access to the HPC ressources of IDRIS made by GENCI (Grand Equipement National de Calcul Intensif) in France.

References

[1] C. Sandorfy, Vibrational spectra of hydrogen bonded systems in the gas phase, Topics in Current Chemistry 120, Springer, Berlin (1984) p. 41.
[2] Y. Maréchal, edited by H. Ratajczak, W. Orville-Thomas, Wiley, New York, 1980, p. 230.
[3] Jens Dreyer, J. Chem. Phys. 122, 184306 (2005)
[4] R.E. Asfin, G.S. Denisov, K.G. Tokhadze, Journal of Molecular Structure 790 (2006) 11–17
[5] D. A. McQuarrie, Statistical Mechanics, Harper-Collins Publishers, New York, 1976.
[6] R. Vuilleumier and D. Borgis, J. Chem. Phys., 1999, 111, 4251.
[7] E.T.J. Nibbering, T. Elsaesser, Chem. Rev. 104 (2004)1887.
[8] K. Heyne, N. Huse, J. Dreyer, E.T.J. Nibbering, T. Elsaesser, S. Mukamel, J. Chem. Phys. 121 (2004) 902.
[9] P. Blaise, M.J. Wójcik and O. Henri-Rousseau, J. Chem. Phys. 122 (2005), p. 64306.
[10] Piotr Durlak, Zdzisław Latajka Chemical Physics Letters 477 (2009) 249-254.
[11] P. Blaise, M.E.-A. Benmalti, O. Henri-Rousseau, J. Chem. Phys 124 (2006), p. 24514.
[12] M.E.-A. Benmalti, P. Blaise, H.T. Flakus,O. Henri-Rousseau, Chem. Phys. 320 (2006) 267-274.
[13] M. P. Gaigeot, R. Vuilleumier, M. Sprik and D. Borgis, J. Chem. Theory Comput., 2005, 1, 772.
[14] C. Marinica, G. Grégoire, C. Desfrancois, J. P. Schermann, D. Borgis and M. P. Gaigeot, J. Phys. Chem. A, 2006, 110, 8802.
[15] M.J. Frisch, G.W. Trucks, H.B. Schlegel, G.E. Scuseria, M.A. Robb, J.R. Cheeseman,J.A. Montgomery, Jr., T. Vreven, K.N. Kudin, J.C. Burant, J.M. Millam, S.S. Iyengar, J.Tomasi, V. Barone, B. Mennucci, M. Cossi, G. Scalmani, N. Rega, G.A. Petersson, H.Nakatsuji, M. Hada, M. Ehara, K. Toyota, R. Fukuda, J. Hasegawa, M. Ishida, T.Nakajima, Y. Honda, O. Kitao, H. Nakai, M. Klene, X. Li, J.E. Knox, H.P. Hratchian, J.B.Cross, C. Adamo, J. Jaramillo, R. Gomperts, R.E. Stratmann, O. Yazyev, A.J. Austin, R.Cammi, C. Pomelli, J.W. Ochterski, P.Y. Ayala, K. Morokuma, G.A. Voth, P. Salvador,J.J. Dannenberg, V.G. Zakrzewski, S. Dapprich, A.D. Daniels, M.C. Strain, O. Farkas,D.K. Malick, A.D. Rabuck, K. Raghavachari, J.B. Foresman, J.V. Ortiz, Q. Cui, A.G.Baboul, S. Clifford, J. Cioslowski, B.B. Stefanov, G. Liu, A. Liashenko, P. Piskorz, I.Komaromi, R.L. Martin, D.J. Fox, T. Keith, M.A. Al-Laham, C.Y. Peng, A. Nanayakkara,M. Challacombe, P.M.W. Gill, B. Johnson, W. Chen, M.W. Wong, C. Gonzalez,J.A. Pople, Gaussian 03, Revision D.01, Gaussian, Inc., Wallingford, CT 06492, 2004,USA.
[16] A. Becke, Phys. Rev. A: At., Mol., Opt. Phys., 1988, 38, 3098.
[17] C. Lee, W. Yang and R. G. Parr, Phys. Rev. B: Condens. Matter, 1988, 37, 785.
[18] S. Goedecker, M. Teter and J. Hutter, Phys. Rev. B: Condens.Matter, 1996, 54, 1703.
[19] G. J. Martyna and M. E. Tuckerman, J. Chem. Phys., 1999, 110, 2810.
[20] W. Humphrey, A. Dalke, K. Schulten, J. Molec. Graphics 14 (1996) 33.
[21] M. Haurie, A. Novak, J. Chim. Phys. 62 (1965) 146.
[22] J.L. Derissen, J. Mol. Struct. 7 (1971) 67.
[23] Piotr Durlak and Zdzislaw Latajka, J Mol Model (2011) 17, 2159-2168.

Journal of Physics: Conference Series **574** (2015) 012012 doi:10.1088/1742-6596/574/1/012012

Computational analysis of difenoconazole interaction with soil chitinases

D. L. VLĂDOIU, M.N. FILIMON, V. OSTAFE, A. ISVORAN

Department of Biology-Chemistry, West University of Timisoara, 16 Pestalozzi, 300115, Timisoara, Romania

E-mail: adriana.isvoran@e-uvt.ro

Abstract. This study focusses on the investigation of the potential binding of the fungicide difenoconazole to soil chitinases using a computational approach. Computational characterization of the substrate binding sites of *Serratia marcescens* and *Bacillus cereus* chitinases using Fpocket tool reflects the role of hydrophobic residues for the substrate binding and the high local hydrophobic density of both sites. Molecular docking study reveals that difenoconazole is able to bind to *Serratia marcescens* and *Bacillus cereus* chitinases active sites, the binding energies being comparable.

1. Introduction

Soil fertility is usually described in terms of the chemical, physical and biological properties such as: pH, cation exchange capacity, texture, structure, water holding capacity, diversity of soil microorganism, etc. All these properties are correlated and impact the crop yields.

The researches of the last years proved the involvement of the soil microorganisms in soil fertility and there are divergent opinions in the specific literature concerning the beneficial and/or detrimental impact of fertilizers and pesticides to soil biology. There are numerous available products and their effects on the soil biology are dependent on the types of products used and also on the application rates adopted [1-4]. There also are human health and environmental impact concerns caused by the fertilizers and pesticides use in agriculture.

Fungicides enter is soil, they are the subject of degradation processes and they also are indicated to have a greater impact on soil microorganisms than other pesticides [5]. Within the present study we focus on the assessment of the effect of the fungicide difenoconazole on soil chitinolytic activity using a molecular docking approach.

Difenoconazole (DFC) is a fungicide belonging to the triazole family and managing to control a comprehensive range of fungi causing diseases of field crops [6]. It usually comes into contact with soil, where it may undergo a variety of transformations and/or causes the loss of soil fertility or environmental damages [7].

Chitinases belong to the family 18 and 19 of the enzymes glycosyl hydrolases being present in a wide range of organisms (fungi, bacteria, plants, humans) that require the reshaping of their own chitin or dissolve and digest the chitin of fungi or animals. [8]. Soil bacteria are usually known to possess chitinolytic activity. Soil chitinases are mainly used for degradation of chitinous fungal cell walls (the main source of the soil chitin) and its utilization as an energy source [8]. They are also used for antagonistic interactions with living fungi, this function being particularly important in the rhizosphere where bacteria and fungi compete for root exudates [9]. Among species of soil bacteria that produce chitinolytic enzymes are *Serratia sp* [10] and *Bacillus sp* [11]. Consequently, we use the molecular docking approach to investigate the possible interaction of DFC with chitinase A from *Seratia marcescens* and from *Bacillus cereus* respectively.

2. Methodology

There are 13 entries in the Protein Data Bank (PDB) [12] concerning crystallographic structures of the native, mutants and complexes with substrates of the *Seratia marcescens* [13] and *Bacillus cereus* chitinases A [14], the corresponding codes entry being: 1CTN, 1EDQ, 1EIB, 1EHN, 1FFR and 1FFQ for the *Seratia marcescens* chitinase A [13] and 3N11, 3N12, 3N13, 3N15, 3N17, 3N18 and 3N1A for the *Bacillus cereus* chitinase A [14], respectively. The structural files of the *Bacillus cereus* chitinase A do not contain the chitin binding and insertion domains [14] but the structural files of *Seratia marcescens* chitinase A contain these domains [13].

The comparison of the considered structures is performed using structures superposition tool under Chimera package [15] and expressed as the root mean square deviation (RMSD) between the equivalent alpha carbon (CA) atoms of the superposed structures [16].

Identification and characterization of the active site cavities of the two enzymes have been performed using Fpocket tool [17].

The computation of the interactions of DFC with *Seratia marcescens* and *Bacillus cereus* chitinases A is performed using SwissDock tool [18], a web-based interface that outcome the most favourable binding modes of the ligand on the protein surface ranked upon the interaction energy expressed as FullFitness score. Preparation of both the protein and ligand for docking in addition to visualization of the binding modes acquired using the SwissDock web-server have been also performed using Chimera package. Selected docking type was „accurate" and we have performed flexible docking.

3. Results and discussions

We have considered all the structural files mentioned above, except the 3N1A structural file because it corresponds to the same mutation of the enzyme as the 3N18 file. For every considered structural file the ligand (if present) has been removed for molecular docking studies.

There are very small differences between the considered structures for every enzyme: the RMSD values are in the interval (0.365÷0.460)Å for the superposition of 538 equivalent CA atoms of *Seratia marcescens* chitinases A and in the interval (0.216÷0.278)Å for the superposition of 320 equivalent CA atoms of *Bacillus cereus* chitinases A, respectively. The superposition of *Seratia marcescens* and *Bacillus cereus* chitinases A gives RMSD values in the interval (1.118÷1.163)Å for 73 equivalent CA atoms.

The active sites cavities of the two enzymes are identified and characterized using the Fpocket tool. These cavities are highly hydrophobic, the local hydrophobic density being 59.76 for the *Seratia marcescens* chitinase A and 59.67 for the *Bacillus cereus* chitinase A respectively. The high local hydrophobic character of these cavities correlates to the hydrophobic character of DFC with the partition coefficient $logP$=4.92 (Pesticide Properties DataBase, http://sitem.herts.ac.uk/aeru/ppdb/en/).

SwissDock outcomes for the interactions of the DFC with *Seratia marcescens* and *Bacillus cereus* chitinases A as well as with their mutants, expressed as the active site binding modes over the total binding modes of the DFC to the protein, the FullFitness scores and the solvation energies are presented in the table 1. All these data reflect that DFC is able to bind to the active site of *Seratia marcescens* and *Bacillus cereus* chitinases A, the interaction with chitinase A from *Seratia marcescens* being stronger. Also, zinc ions presence in the structural file favorites the interaction of DFC with *Bacillus cereus* chitinase A.

Table 1. Outcomes of SwissDock server for the interactions of DFC with *Seratia marcescens* and *Bacillus cereus* chitinases A

Enzyme	PDB code entry	active site binding poses / total binding poses	FullFitness score [kcal/mol]	ΔG [kcal/mol]
Seratia	1CTN	16/37	-1965.38	-8.09
marcescens	1EDQ	3/42	-1961.94	-7.72
chitinase A	1FFQ	4/42	-1928.21	-7.16
	1FFR	3/35	-1950.75	-7.15
	1EIB	3/35	-1924.21	-7.91
	1EHN	17/40	-1925.46	-8.10
Bacillus	3N11	21/34	-1263.84	-7.13
cereus	3N12	6/34	-1863.48	-6.86
chitinase A	3N13	43/44	-1222.91	-7.63
	3N15	26/34	-1235.51	-7.88
	3N17	18/36	-1228.74	-7.69
	3N18	46/49	-1180,76	-7.98

In addition, the punctual mutations of a few amino acids belonging to the active site of considered chitinases (ASP143, GLU145 and TYR227 for the *Bacillus cereus* chitinase A and ASP311, GLU315 and TYR390 for the *Seratia marcescens* chitinase A) negatively affect their interactions with DFC. Experimental data prove that these mutations also have a negative effect for the substrates binding to these enzymes [13, 14].

Previous computational studies reflected that DFC is also able to bind to the active site of the *Bacillus pasteurii* urease, another important enzyme in maintaining the biological properties of soil [19]. Dose dependent response to the DFC application has been also observed for soil dehydrogenase activity, high doses of DFC having inhibitory effect [20].

4. Conclusions

The fungicide difenoconazole provides the ability to bind to the active sites of the *Seratia marcescens* and *Bacillus cereus* chitinases A suggesting its possible inhibitor effect on the chitinolytic activity of soil bacteria. Zinc ions are known to have an inhibitor effect on the GH18 family of glycosyl hydrolases [21] and their presence in the structure of the *Bacillus cereus* chitinase A results in a stronger interaction of the DFC with the enzyme and consequently the inhibitory effect of the chitinolytic activity is stronger.

The results of this study provide insight in the possible inhibitory effect of the fungicide difenoconazole on the chitinolytic activity of soil bacteria affecting the different ecological functions of bacterial chitinase in soil.

Acknowledgements
Diana-Larisa VLĂDOIU would like to thank the project POSDRU/159/1.5/S/137750 - "Doctoral and Postdoctoral programs support for increased competitiveness in Exact Sciences research" cofinanced

by the European Social Found within the Sectorial Operational Program Human Resources Development 2007 – 2013 because this work was supported by this strategic grant.
Nicoleta FILIMON and Adriana ISVORAN would like to thank the grant of the Romanian Ministry of Education, CNCS-UEFISCDI, project number PN-II-RU-PD-2012-3-0220, "Metabolization of difenoconazole by crop plants and fungi communities from soil" for supporting this work.

References

[1] Biederbeck VO, Champbell CA, Ukrainetz H, Curtin D and Bouman OT 1996, *Canadian J Soil Sci* **76**:7-14
[2] Bunemann EK and McNeil A 2004 in *Soil Biology in Agriculture, Proc Workshop Curr Res Soil Biol Agricult*, R. Lines-Kelly, Tamworth, NSW Department of Primary Industries 64-71
[3] Kiney CA, Mandernack KW and Mosier AR *Soil Biol Biochem* **37**:837-50
[4] Smith MD, Hartnett DC and Rice CW 2000 *Soil Biol Biochem* **32**:935-46
[5] Bunemann EK, Schwenke GD and Van Zwieten L 2006 *Aust J Soil Res* 44:379–406
[6] Dahmen H and Staub T 1992 *Plant Disease* **76**(8):774-7
[7] Guo C, Li JZ, Guo BY and Wang HL 2010 *Bull Env Contamin Toxic* **85**(4):427-31
[8] Bhattachrya D, Nagpure A and Gupta RK 2007 *Crit Rev Biotech* **27**:21–8
[9] Hamid R, Khan MA, Ahmad M, Ahmad MM, Abdin MZ, Musarrat J and Javed S 2013 *J Pharm BioAllied Sci*, **5**(1): 21–9
[10] Watanabe T, Kimura K, Sumiya T, Nikaidou N, Suzuki K and Suzuki M 1997 *J Bacteriol* 179:7111–7
[11] Pleban S, Chernin L and Chet I 1997 *Let Appl Microbiol* **25:**284-8
[12] Berman HM, Westbrook J, Feng Z, Gilliland G, Bhat TN, Weissig H, Shindyalov IN and Bourne PE 2000 *Nucl Acids Res* **28**:235-40
[13] Papanikolau Y, Prag G, Tavlas G, Vorgias CE, Oppenheim AB and Petratos K 2001 *Biochem* **40**:11338-43
[14] Hsieh YC, Wu YJ, Chiang TY, Kuo CY, Shrestha KL, Chao CF, Huang YC, Chuankhayan P, Wu WG, Li YK and Chen CJ 2010 *J Biol Chem* **285**:31603-15
[15] Pettersen EF, Goddard TD, Huang CC, Couch GS, Greeblatt DM, Meng EC and Ferrin TE 2004 *J Comp Chem*, **25**:1605-12
[16] Marti-Renon MA, Capriotti E, Shindyalov IN and Bourne PE 2009 *Structural Bioinformatics* Gu J and Bourne PE (Eds),. John Wiley & Sons, Inc.
[17] Le Guilloux V, Schmidtke P and Tuffery P 2009 *BMC Bioinformatics*, **10**:168
[18] Grosdidier A, Zoete V and Michielin O 2011 *Nucl Acids Res*, **39**:W270-7
[19] Vladoiu DL, Filimon MN, Ostafe V and Isvoran A 2014 *The 14th International Balcan Workshop on Applied Physics*, 2-4 July, Constanta, Romania
[20] Madakka M, Sake PK and Rangaswamy V 2010 *Eco Env Cons* **16**(3):369-75
[21] Dahiya N, Tewari R and Hoondal GS 2006 *App Micro Biotech*, **71**(6):773-2.

Journal of Physics: Conference Series **574** (2015) 012013 doi:10.1088/1742-6596/574/1/012013

The reaction of aluminium clusters with water: a quantum chemical investigation

Jerzy Moc[1]
Faculty of Chemistry, Wroclaw University, F. Joliot-Curie 14,
50-383 Wroclaw, Poland

E-mail: jerzy.moc@chem.uni.wroc.pl

Abstract. The reaction of charged aluminium clusters with water in the gas phase is investigated theoretically. To this end, the doublet potential energy surface for the $Al_n^+ + H_2O \rightarrow Al_nO^+ + H_2$ reaction with n=6 is explored using density functional theory. The calculated reaction pathways include the initial adduct ion formation, the O-H bond activation and H_2 elimination steps, and are consistent with the recent experimental report (Ref.2). The results of the current quantum-chemical study are relevant to the issues of catalytic role of the main group metal clusters and H_2 generation.

1. Introduction

Recently, a series of small aluminium cluster cations Al_n^+ were detected in the mass spectrometric study of Al in helium droplets by Spence et al [1]. In another recent mass spectrometric study involving small aluminium cluster cations, Arakawa et al [2] investigated a size-dependency of the reactivity of the Al_n^+ (n=3-19) species towards water in the gas phase. Under single collision conditions, the latter authors observed the Al_nO^+ and/or $Al_n(H_2O)^+$ clusters as major reaction products. More specifically, for $6 \leq n \leq 13$, either the former or the latter species were selectively observed, while the two products were detected for $14 \leq n \leq 19$. This means that the Al_n^+ cluster cations with n=3-5 were found to be unreactive towards H_2O and that aluminium hexamer (n=6) was revealed to be the smallest reactive cluster, producing Al_nO^+ and hydrogen [2]. The earlier experimental and theoretical studies of size-selective reactivity of aluminium cluster anions Al_n^- with water in the gas phase by Castleman and co-workers [3-4] demonstrated that only the relatively large cluster anions (n=16-18) reacted strongly with $2H_2O$ to yield $Al_n(OH)_2^- + H_2$. The proposed mechanism [3-4] for the $Al_n^- + 2H_2O$ elementary reactions was the dissociative chemisorption of water molecules on the metal cluster's neighbouring sites acting as a Lewis acid and a Lewis base, which facilitated the O-H bonds' breakage. As noticed by Arakawa et al [2], compared to the Al_n^- cluster anions, the size range of the Al_n^+ cluster cations upon which the dissociation of water takes place (followed by the H_2 release) is significantly larger. It was also suggested [2] that in the case of the Al_n^+ cations, a different mechanism for the H_2O dissociation/H_2 elimination operates, which involves only one H_2O molecule for the elementary process: $Al_n^+ + H_2O \rightarrow Al_nO^+ + H_2$. However, the actual mechanism of the $Al_n^+ + H_2O$ reactions in the gas phase have not been elucidated yet.

[1] To whom any correspondence should be addressed.

Here, the gas-phase computational scrutiny of the mechanistic pathways of dissociation and dehydrogenation of water on the aluminium hexamer cation Al_6^+ is presented. As mentioned above, Al_6^+ is the smallest Al cluster cation proved experimentally to react with H_2O to form H_2 under thermal conditions [2]. It is known that in order to model adequately the gas-phase reactions of metal clusters with small molecules, an extensive probing of the underlying potential energy surfaces is indispensable [5]. The reported herein computational results are relevant to the issues of catalytic role of the main group metal clusters and H_2 generation.

2. Computational methods

Density functional theory using B3LYP functional [6-7] and the aug-cc-pVTZ basis set [8] was employed to extensively probe the potential energy surface involved in the $Al_6^+ + H_2O$ reaction. All stationary points were subjected to frequency analysis to ensure that structures correspond to local minima (all positive eigenvalues) or transition states (TSs, one negative eigenvalue). Connectivity of the transition states with the respective local minima was assured by calculating the intrinsic reaction coordinates (IRCs) [9]. All the energetics discussed below has been determined at the zero-point vibrational energy (ZPE) corrected B3LYP/aug-cc-pVTZ//B3LYP/aug-cc-pVTZ level. The Gaussian 09 package was employed in the calculations [10].

3. Results and discussion

3.1. Reactants, initial adduct ion and products of the gas-phase reaction $Al_6^+ + H_2O \rightarrow Al_6O^+ + H_2$

First, we calculated the lowest energy geometry of the bare Al_6^+ cluster cation reactant, derived from the prism-like structure and termed **R** (C_{2v},$^2B_1$) (Figure 1). The alternative Al_6^+ geometry derived from the distorted octahedron (C_{2h},$^2A_{1g}$) (not shown) appeared to be 3.4 kcal/mol less stable at the B3LYP/aug-cc-pVTZ//B3LYP/aug-cc-pVTZ + ZPE level. Next, we predicted the structure of the species formed in the initial step of the $Al_6^+ + H_2O$ reaction, which is the $Al_6^+...OH_2$ adduct ion **COM** (C_s, $^2A'$), wherein the water molecule is bound to the aluminium cluster cation through the oxygen atom (see Figure 1). The relatively short Al...O distance of 1.97 Å in the adduct ion **COM** is coupled with the large binding energy of the former species of 19.4 kcal/mol relative to $Al_6^+ + H_2O$. In other words, the formation of the initial adduct ion is a significantly exothermic step. As confirmed by the IRC following, for the two predicted paths of the $Al_6^+ + H_2O \rightarrow Al_6O^+ + H_2$ reaction described below, the formation of **COM** is a shared first step. Here, we also mention the Al_6O^+ product denoted **P** (C_{2v},$^2A_1$), formed along with H_2. As can be seen from Figure 1, Al_6O^+ is the oxygen-bridged cluster.

3.2. PATH A

Starting from the adduct ion **COM**, the O-H bond activation occurs via the four-centered transition state **TS1O-H** ($^2A'$) (see Figure 2). At the B3LYP/aug-cc-pVTZ//B3LYP/aug-cc-pVTZ + ZPE level, the latter transition state is calculated to lie 1.6 kcal/mol below the Al_6^+(**R**) + H_2O asymptote, implying a facile water molecule dissociation on the charged Al cluster (Figure 4). The resulting $Al_6(OH)H^+$ dissociation intermediate (**I1**) features the terminal Al-H bond and the O-bridged hydroxyl group, and is stabilized by 53.6 kcal/mol with respect to the Al_6^+(**R**) + H_2O. The next step along *PATH A* is H_2 elimination involving transition state **TS1H$_2$**. Although **TS1H$_2$** is predicted to lie 25 kcal/mol above **I1**, the former species is placed 28.6 kcal/mol below $Al_6^+ + H_2O$ (Figure 4). The H_2 elimination is followed by the formation of the weak exit channel dihydrogen complex $Al_6O^+...H_2$ (**EXT**) with the concominant rearrangement of the Al_6O^+ core. The *PATH A* is ended with the release of H_2. The overall H_2 elimination reaction Al_6^+(**R**) + $H_2O \rightarrow Al_6O^+$ (**P**) + H_2 is calculated to be exothermic by 52.4 kcal/mol (Figure 4).

3.3. PATH B

For the $Al_6^+ + H_2O \rightarrow Al_6O^+ + H_2$ reaction, we have identified another low energy path involving the prism-like aluminium cluster, termed *PATH B*. With *PATH B* (Figure 3), the H_2O dissociation takes place through transition state **TS2O-H** which connects to the adduct Al_6^+...OH_2 (**COM**) and $Al_6(OH)H^+$ dissociation intermediate (**I2**). *PATH B* is similar to *PATH A* in that it leads to the $Al_6(OH)H^+$ species in a barrier-less O-H bond dissociation step (at the B3LYP/aug-cc-pVTZ//B3LYP/aug-cc-pVTZ + ZPE level) (Figure 4). However, unlike **I1** intermediate of *PATH A*, **I2** intermediate of *PATH B* is an 'open-type' species, with the terminally bound dissociated H and OH (Figure 3). We have found that the next step along *PATH B* involves the isomerization transition state **TSISO**. According to our calculations, this is an easy interconversion with the energy barrier of 6.5 kccal/mol relative to **I2**, and with **TSISO** lying 35.4 kcal/mol below the Al_6^+ + H_2O asymptote. Via transition state **TSISO**, **I2** rearranges to the octahedral-like $Al_6(OH)H^+$ intermediate **I2-1** which features the O-bridged hydroxyl group and terminal Al-H bond.

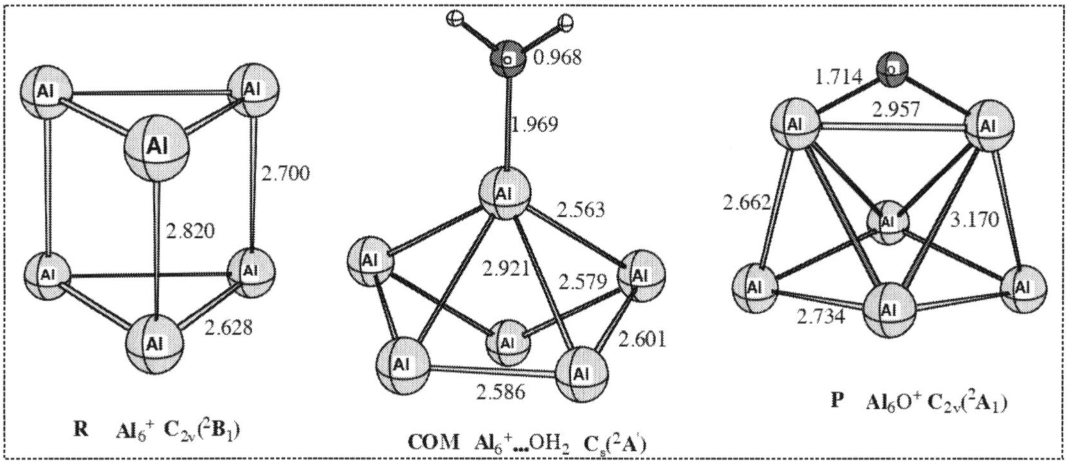

Figure 1. Reactant Al_6^+ (**R**), initial adduct ion Al_6^+...OH_2 of the gas-phase reaction of Al_6^+ with H_2O (**COM**) and product of this reaction, Al_6O^+ (**P**) as optimized at the B3LYP/aug-cc-pVTZ level (bond distances are in Å). The **R**, **COM** and **Al_6O^+** species are common for **PATH A** and **PATH B**.

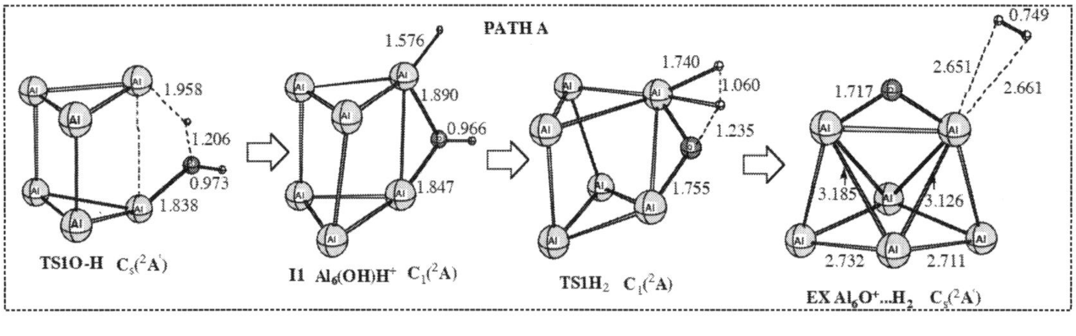

Figure 2. **PATH A:** Transition state for dissociation of the O-H bond of H_2O on Al_6^+ (**TS1O-H**), $Al_6(OH)H^+$ reaction intermediate (**I1**), transition state for H_2 elimination (**TS1H$_2$**) and Al_6O^+...H_2 exit channel dihydrogen complex (**EX**) optimized at the B3LYP/aug-cc-pVTZ level (bond distances are in Å).

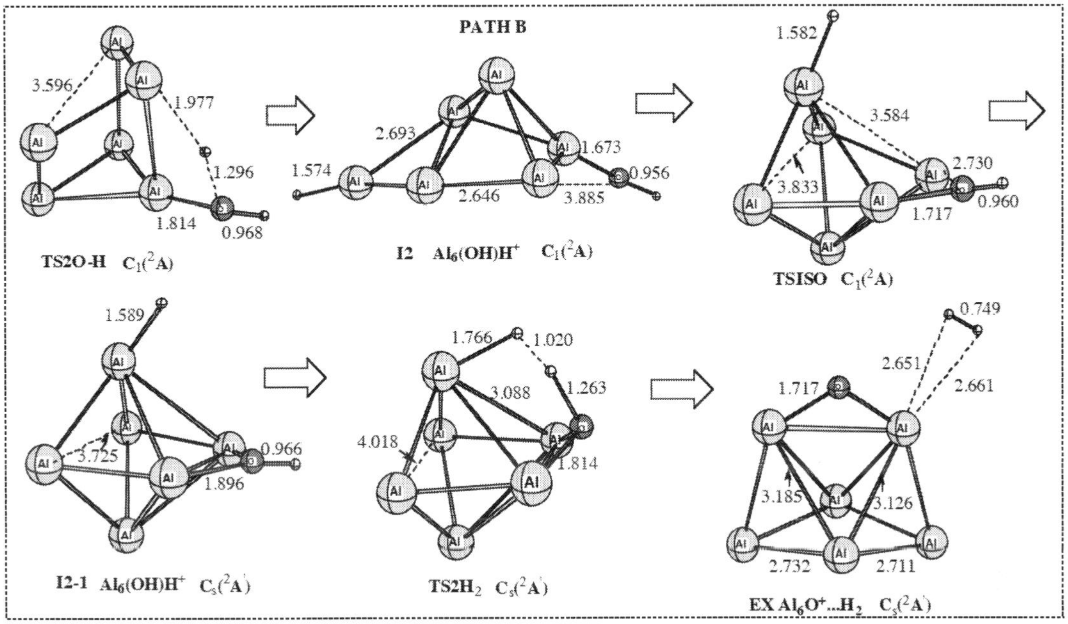

Figure 3. PATH B: Transition state for dissociation of the O-H bond of H_2O on Al_6^+ (**TS2O-H**), $Al_6(OH)H^+$ reaction intermediate (**I2**), isomerization transition state (**TSISO**), $Al_6(OH)H^+$ reaction intermediate (**I2-1**), transition state for H_2 elimination (**TS2H$_2$**) and Al_6O^+...H_2 exit channel dihydrogen complex (**EX**) optimized at the B3LYP/aug-cc-pVTZ level (bond distances are in Å).

According to Figure 3, the latter intermediate is a starting point for the H_2 elimination step taking place through the transition state **TS2H$_2$**. Though this TS for H_2 elimination requires overcoming an energy barrier of 19.7 kcal/mol with respect to **I2-1**, it is located well below the initial reactants, i.e., by 22.4 kcal/mol (Figure 4). As was seen for *PATH A*, the *PATH B* is next continued with the formation of the exit channel dihydrogen complex Al_6O^+...H_2 (**EXT**), and is ended with the H_2 expelling.

In summary, using density functional theory we have identified two low-energy paths for the gas-phase reaction of the Al_6^+ cluster cation with H_2O molecule to yield Al_6O^+ and H_2. Detailed reaction paths comprising the initial adduct ion formation, O-H bond dissociation and H_2 elimination major steps are described. The Al_6^+...OH_2 intial adduct ion formation is predicted to be a significantly exothermic step (by ca. 20 kcal/mol at the B3LYP/aug-cc-pVTZ//B3LYP/aug-cc-pVTZ + ZPE computational level). The relatively deep potential well of the adduct ion along the reaction coordinate contributes to the favourable energetic location of the O-H bond dissociation transition states (related to *PATH A* and *PATH B*) with respect to the entrance channel. The overall H_2 elimination reaction is calculated to be exothermic by 52.4 kcal/mol. The calculated energy profile for the $Al_n^+ + H_2O \rightarrow Al_nO^+ + H_2$ reaction with n=6 is consistent with the experimental observation [2].

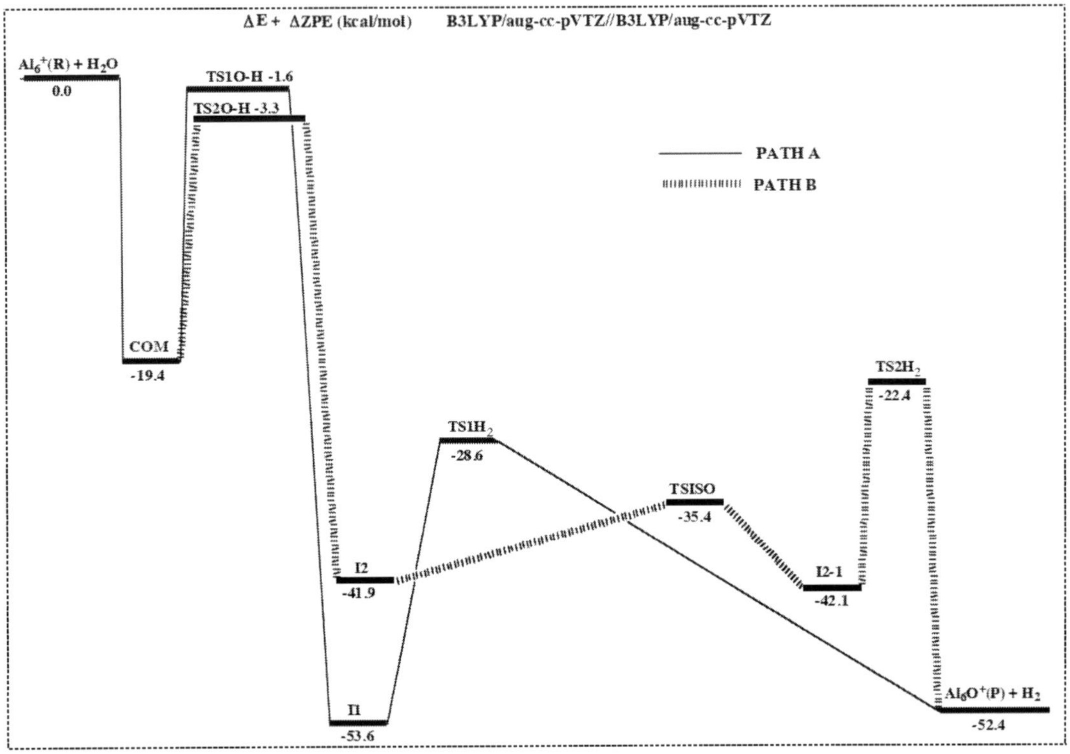

Figure 4. The schematic energy profile for the $Al_6^+ + H_2O \rightarrow Al_6O^+ + H_2$ reaction calculated at the ZPE corrected B3LYP/aug-cc-pVTZ//B3LYP/aug-cc-pVTZ level (note that the exit channel complex is not included).

Acknowledgement
The author acknowledges a generous support of computer time at the Wroclaw Center for Networking and Supercomputing, WCSS.

References

[1] Spence D, Latimer E, York W, Boatwright A, Feng C, Yang S and Ellis A M 2014 *Int. J. Mass Spectrom.* **365-366** 86
[2] Arakawa M, Kohara K, Ito T and Terasaki A 2013 *Eur. Phys. J D* **67** 80
[3] Roach P J, Woodward W H, Castleman Jr A W, Reber A C and Khanna S N 2009, *Science*, **323** 492
[4] Reber A C, Khanna S N, Roach P J, Woodward W H and Castleman Jr A W 2010, *J. Phys. Chem. A* **114** 6071
[5] Moc J, Musaev D G and Morokuma K 2000 *J. Phys. Chem. A* **104** 11606
[6] Becke A D 1993 *J. Chem. Phys.* **98** 5648
[7] Lee C, Yang W and Parr R G 1988 *Phys. Rev. B* **37** 785
[8] Dunning T H 1989 *J. Chem. Phys.* **90** 1007
[9] Gonzalez C and Schlegel H B 1989 *J. Chem. Phys.* **90** 2154
[10] Frisch M J *et al* 2009 *Gaussian 09 revision B.01* (Gaussian Inc, Wallingford, CT.)

Studying the lateral chain packing in a ceramide bilayer with molecular dynamics simulations

N I Papadimitriou[1], S N Karozis, M E Kainourgiakis, G Ch Charalambopoulou

Environmental Research Laboratory, National Center for Scientific Research "Demokritos", Agia Paraskevi 15310, Greece

E-mail: nikpap@ipta.demokritos.gr

Abstract. In this work, we present a novel technique, based on molecular dynamics simulations, that allows the study of the lateral chain packing in a lipid bilayer. It utilizes the radial distribution function of the alkyl chains to determine the arrangement of the chains along the bilayer plane. The positions of the mass centres of the chains are projected onto the bilayer plane and a 2D radial distribution function is calculated for these projections. The proposed technique can be particularly useful for lipid bilayers in the gel (solid) phase where the chains present a limited degree of mobility. As a case study, we have examined a bilayer that consists of ceramide NS 24:0. Ceramide bilayers can be found in the lipid domain of the skin where they have a significant role in its barrier function. The specific bilayer was found (at 300 K) to adopt a strictly hexagonal chain packing with a separation distance between the chains of 0.466 nm, in good agreement with the available experimental data.

1. Introduction

Stratum corneum is the outermost layer of the skin and it is responsible for the mechanical properties of the skin as well as for its barrier function (i.e. control of the flow of substances into or out of the body). It mainly consists of dead cells (corneocytes) that have an elongated shape and are filled with keratin. Corneocytes are surrounded by a lipid domain forming a "brick-and-mortar" structure. This lipid domain comprises of lipid bilayers consecutively placed the one next to the other. In this way, the lipid phase is contiguous and offers the most probable route for a substance to penetrate the stratum corneum.

Concerning its composition, the skin lipid domain is an approximately equimolar mixture of ceramides, cholesterol and free fatty acids. Ceramides are a special class of lipids that contain sphingosine, an unsaturated aminoalcohol. In ceramides, sphingosine is bonded to a fatty acid via a peptide bond. As a result, each ceramide molecule contains a polar head with a protein-like chemical structure and two alkyl chains with unequal lengths (Figure 1). Various ceramide types exist depending on the exact geometry of the polar head, the length of the alkyl chains and the presence of polar groups or unsaturated bonds in them.

At room temperature, ceramide bilayers form a gel phase where each chain may oscillate around a fixed position but without diffusing among the other chains. This is the main difference between ceramide bilayers and phospholipid bilayers that are found in the cell membranes. Since the alkyl

[1] To whom any correspondence should be addressed.

chains are in a solid phase, their positions follow a specific crystal lattice (usually hexagonal or orthorhombic). Temperature changes or other reasons may induce phase transitions from one crystal structure to the other. Such phase transitions dramatically affect the transport properties and the overall functionality of the skin.

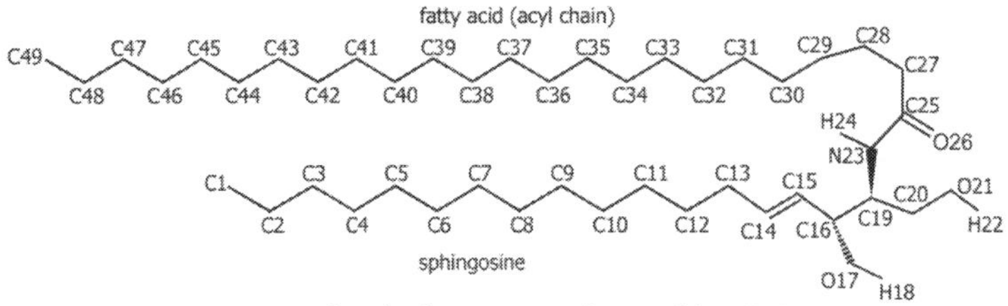

Figure 1. Chemical structure of ceramide NS 24:0.

2. Methodology

For the study of ceramide bilayers we have performed a series of molecular dynamics simulations. The examined system is a lipid bilayer that contains 128 ceramide molecules. Only one ceramide type is used, namely NS 24:0 (Figure 1). The bilayer is confined between two water slabs each containing 2493 water molecules. The initial structure is geometrically optimized and then it is allowed to equilibrate for 140 ns. In this equilibration stage an annealing process is followed: the system is heated from 300 to 360 and then cooled back to 300 K in steps of 20 K with a relaxation time of 20 ns per step. The production run lasts for 20 ns. Atomistic interactions are described with the approach proposed by Berger et al. [1]. More detailed description of the simulations can be found in Ref. [2].

Among several properties examined in the aforementioned system, the lateral packing of the alkyl chains is one of the most interesting and main the subject of this work. For the rigorous study of this property, we have developed a novel technique that uses the radial distribution functions (RDFs) of the chains to quantitatively describe their relative positioning. The RDF shows the probability of finding two objects at a specific distance. In our case, these objects are the projections of the mass centers of the alkyl chains on the bilayer plane. The overall shape of the RDF reveals the type of arrangement (e.g. liquid, hexagonal, or orthorhombic) while the positions of the peaks can be used for the estimation of the characteristic distances of the crystal lattice.

For the RDF calculation, first the position of the mass center of each chain is calculated. Since lipids have limited mobility in the lamellar direction (z-axis) and we are interested in the lateral structure, the positions of the chains are calculated on a 2D basis by considering only the x and y coordinates. Formally, this RDF is calculated as:

$$g(r) = \frac{1}{2\pi r N}\left\langle S(t)\cdot\frac{\delta N(r,t)}{\delta r}\right\rangle \tag{1}$$

where N is the total number of chains, $S(t)$ the xy cross-section area of the simulation box as a function of time t, and $\delta N(r,t)$ is the number of chains located in a circular shell of radius from r to $r+\delta r$. The angle brackets $\langle\rangle$ denote averaging over the total number of chains and total simulation time.

3. Results and discussion

3.1. Lateral chain packing from the RDF

The calculated RDF for the examined system is shown in Figure 2. The first observation on this RDF is that the first peak is sharp and symmetric without any indication of overlapping peaks (e.g. shoulder). This indicates that the chains are placed on a hexagonal lattice. In an ideal hexagonal lattice, each chain is surrounded by 6 other chains (first neighbor shell of chains) and all of them are located at the same distance *d*. This arrangement gives rise to a unique peak in the RDF. In any other lattice type, some of the surrounding chains would be at different distance from the central chain than some others. In this case, we would expect more than one peaks in the first neighbor shell which is clearly contradicted by the calculated RDF. Moreover, the rest of the peaks that correspond to more distant neighbor shells also follow the same hexagonal pattern.

In the next step, the exact position of each peak is calculated with fitting of a Gaussian function. For peaks that are closely placed, a sum of two Gaussian functions is used so as to more accurately determine the positions of the overlapping peaks. The calculated positions for the 5 first peaks (3 first neighbor shells of chain) are shown in Table 1. With simple geometry calculations (and assuming ideal hexagonal geometry), we can calculate the separation distance *d* between two chains that correspond to each peak. These results are also shown in Table 1. It must be noted that all of the peaks give essentially the same distance with a maximum deviation less than 0.5%. This result not only confirms that lateral chain packing is strictly hexagonal but also allows the separation distance between the chains to be calculated with high accuracy. For the specific system, this value is found to be 0.466 nm.

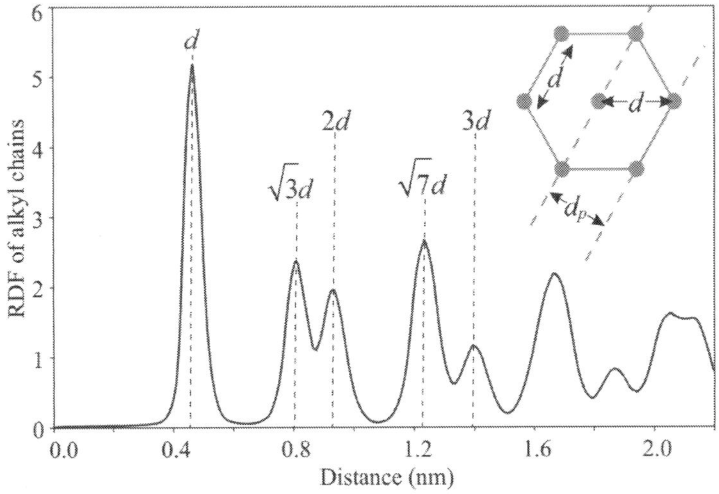

Figure 2. RDF of alkyl chains in a ceramide bilayer. Each peak is associated with the corresponding distance in an ideally hexagonal lattice.

Table 1. Positions of the peaks in the RDF and calculated distances between two chains. The average distance is 0.4661 ± 0.0009 nm.

Peak No.	Shell	Peak position (nm)	Distance (nm)
1	1	0.465	0.4650
2	2	0.808	0.4665
3	2	0.931	0.4656
4	3	1.236	0.4672
5	3	1.399	0.4662

3.2. Comparison with experiments

The RDF function of the alkyl chains provides structural information similar to the X-ray or electron diffraction measurements. In these experimental techniques the distance d_p between the parallel planes of chains is used instead of the distance *d* between the chains. In the examined system, this distance is $d_p = \sqrt{3}/2\, d = \sqrt{3}/2 \cdot 0.4661\,\text{nm} = 0.404\,\text{nm}$. Moreover, the experimentally determined distance is the perpendicular distance d_t between the planes while the RDF refers to the projection of this distance on the bilayer plane. These two distances are equal only in the case when the chains are perpendicular to the bilayer plane. However, in ceramide bilayers the angle between the chains and the bilayer normal (tilt angle) is usually not zero. In this case, the two distances are related by the Equation:

$$d_t = d_p \cos(\theta_t) \qquad (2)$$

where θ_t is the tilt angle. For the calculation of the tilt angle, orthogonal regression is used to find the best-fit line of every chain and then these direction vectors are averaged over the total number of chains and the entire simulation time. In this way, the tilt angle for the examined lipid system is found 9.8° and the corrected distance d_t is 0.398 nm.

Several experimental studies on ceramide bilayers have been reported that examine the chain arrangement but there is no full consensus about the type of lateral packing. This seems to depend on the sample preparation, the exact lipid composition (real specimens usually contain a mixture of ceramides) and the source of the lipids (e.g. synthetic, isolated from human or bovine skin etc.). However, most of the results from different experimental works based on various techniques [3-6] converge to the conclusion that the hexagonal arrangement is the most stable phase of a bilayer consisting of pure CERs. In the work of Pilgram et al [5], the characteristic distance was accurately measured with electron diffraction and it was found 0.399 ± 0.007 nm. This result is in very good agreement with the value (0.398 nm) calculated from our calculations. The quantitative (in terms of the geometry of lateral chain packing) and the qualitative (in terms of distance between the planes of chains) agreement between shows that the method presented above for the study of the lateral structure of a lipid bilayer is accurate and it can be reliably used for the study of similar systems.

4. Conclusions

We have proposed a computational technique that uses the results of molecular dynamics simulations to provide important information about the lateral chain arrangement in lipid bilayers that are in the gel phase. It is based on the RDF function of the alkyl chains. The calculated RDF allows the determination of the lattice type of the lateral chain packing as well as the calculation of the characteristic distance between the parallel planes of chains. The technique was applied on a lipid bilayer comprised of ceramide NS 24:0. The specific system was found to adopt a strictly hexagonal geometry with a characteristic distance of 0.398 nm. The good agreement between the calculated results and the experimental data verifies the reliability of the technique that can be used for other similar systems with limited lipid mobility.

References

[1] Berger O, Edholm O, Jähnig F 1997 *Biophys J* **72** 2002.
[2] Papadimitriou N I, Karozis S N. Kainourgiakis M E, Charalambopoulou G Ch 2014 Studying the structure of single-component ceramide bilayers with molecular dynamics simulations using different force fields *DOI: 10.1080/08927022.2014.947483*.
[3] Shah J, Atienza J M, Rawlings A V, Shipley G G 1995 *J Lipid Res* **36** 1945.
[4] Bouwstra J A, Gooris G S, Cheng K, Weerheim A, Bras W, Ponec M 1996 *J Lipid Res* **37** 999.
[5] Pilgram G S K, Engelsma-van Pelt A M, Bouwstra J A, Koerten H K 1999 *J Invest Dermatol* **113** 403.

Interaction of silicene and germanene with non-metallic substrates

M Houssa[1], E Scalise[2], B van den Broek[1], A Lu[1,3], G Pourtois[3], V V Afanas'ev[1], and A Stesmans[1]
[1] Department of Physics and Astronomy, University of Leuven, B-3001 Leuven, Belgium
[2] Max Planck Institut für Eisenforschung, 40237 Düsseldorf, Germany
[3] imec, 75 Kapeldreef, B-3001 Leuven, Belgium

Abstract. By using first-principles simulations, we investigate the interaction of silicene and germanene with various non-metallic substrates. We first consider weak van der Waals interactions between the 2D layers and dichalcogenide substrates, like MoX_2 (X=S, Se, Te). The buckling of the silicene or germanene layer is correlated to the lattice mismatch between the 2D material and the MoX_2 template. The electronic properties of silicene or germanene on these different templates then largely depend on the buckling of the 2D material layer: highly buckled silicene or germanene on MoS_2 are predicted to be metallic, while low buckled silicene on $MoTe_2$ is predicted to be semi-metallic, with preserved Dirac cones at the K points. We next study the covalent bonding of silicene and germanene on (0001) ZnS and ZnSe surfaces. On these substrates, silicene or germanene are found to be semiconducting. Remarkably, the nature and magnitude of their energy band gap can be controlled by an out-of-plane electric field.

1. Introduction

Very recently, the formation of silicene, the silicon counterpart of graphene, was reported on (111)Ag surfaces [1-3], as well as on (0001)ZrB_2 [4] and (111)Ir [5] surfaces. The electronic properties of the silicene layer on (111)Ag was investigated using angle-resolved photoemission spectroscopy. These measurements revealed the presence of a linear dispersion in the band structure of silicene (so called Dirac cones) with a Fermi velocity of about 1.3×10^6 m/s [3], as theoretically predicted for free-standing silicene [6].

The possible existence of silicene was so far reported on these metallic substrates. However, the characterization of the electronic and electrical properties of silicene and germanene on metallic substrates is very challenging, since these properties are then largely dominated by the metal. The growth of silicene and germanene on semiconducting or insulating substrates is required for their firm identification and complete characterization. In addition, potential applications of these 2D materials in nanoelectronic devices will also require their growth on non-metallic substrates.

We first theoretically study the weak (van der Waals) interaction of silicene and germanene with layered dichalcogenide substrates [7]. On these templates, silicene and germanene are predicted to be either metallic or semi-metallic (with preserved Dirac cones at the k-points), depending on the buckling of the 2D layer.

We next investigate the covalent bonding of silicene and germanene on (0001)ZnS and ZnSe surfaces [8,9]. The charge transfer occurring at the silicene/(0001) ZnS and germanene/(0001) ZnSe interface leads to the opening of an indirect energy band gap in silicene or germanene. Very interestingly, it is found that the nature (indirect or direct) and magnitude of their energy band gap can be controlled by an external electric field.

2. Theoretical details

First-principles simulations were performed using density functional theory (DFT). The generalized gradient approximation (GGA) was used for the exchange-correlation functional [10], as implemented in the Siesta package [11] or in the plane-wave Quantum Espresso package [12]. The core electrons were implicitly treated by using norm-conserving pseudopotentials [13] and the valence electrons were described by using double zeta singly polarized basis sets (Siesta) or plane-wave basis sets (Quantum Espresso). An energy cutoff of 300 Ry and a (10x10x1) k-point mesh were used for the computations, allowing convergence of the total energy of the systems below typically 10 meV. The atomic structure optimization was carried out by relaxing the forces on all atoms until a 0.05 eV/Å force tolerance was reached, using a conjugate gradient method. The computations were performed on periodic slab models, with about 15 Å of vacuum between the periodic cells, in order to minimize the interaction between neighboring cells. To study the effect of an external electric field on the electronic properties of the silicene/(0001) ZnS and silicene/(0001) ZnSe interfaces, a periodic zigzag electric potential was applied in the direction perpendicular to the interface.

3. Silicene and germanene on MoX_2 substrates

We first considered the interaction of silicene and germanene with different semiconducting dichalcogenide substrates, namely MoX_2 with X=S, Se, Te [7]. The computed in-plane lattice parameters and energy band gap of these substrates are given in Table 1.

Table 1. Computed in-plane lattice parameters and energy band gaps of MoX_2.

	In-plane lattice parameter (Å)	Energy band gap (eV)
MoS_2	3.16	1.2
$MoSe_2$	3.30	1.1
$MoTe_2$	3.52	1.0

The initial atomic configuration consists in a supercell with a flat silicene or germanene layer on top of "bulk" MoX_2 (which includes 4 atomic layers). The two bottom MoX_2 layers were kept fixed during relaxation and the initial distance between the 2D material and the top MoX_2 layer equals 4.5 Å. The cell parameter was kept to the one of the template, the silicene or germanene layer being compressively strained, as compared to the free-standing case. We studied three different possible arrangements of the silicon or germanium atoms with respect to the underlying Mo and X atoms, as shown Fig.1: (a) Si (Ge) hexagons placed on top of the MoX_2 hexagons – so called AAA stacking like in h-BN, (b) Si (Ge) hexagons shifted with respect to the MoX_2 hexagons by half a unit cell, and (c) Bernal-like arrangement (ABA stacking), like in graphite.

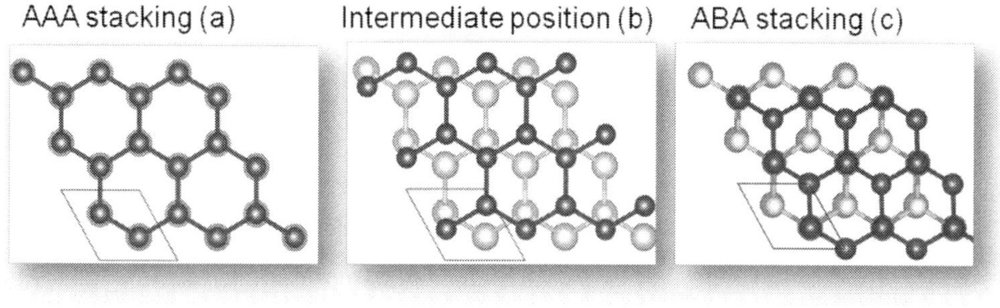

Figure 1. Three different atomic arrangements of the silicene layer on top of MoS_2. Blue, yellow and green spheres are Si, S and Mo atoms, respectively.

After energy relaxation, (a) and (c) structures kept their initial configuration, while structure (b) relaxed to configuration (c). The energy difference between the various configurations is typically less than 5 meV/atom, these atomic configurations being thus equally stable (degenerate). This indicates that the interaction between silicene or germanene layer with the dichalcogenide substrate is very weak. On the other hand, the difference in energy between the initial (before relaxation) and the final (after relaxation) structures is more than 0.1 eV/atom. This energy difference is mainly due to the buckling of the silicene or germanene layer. In all the three structures, the 2D material indeed buckled after relaxation; the buckling distance is given in Table 1 for the silicene/MoX_2 system, together with the in-plane lattice parameter mismatch. From these results, the buckling distance is clearly correlated to the in-plane lattice mismatch. Note that the typical silicene-MoX_2 interlayer distance lies between 3 and 3.5 Å.

Table 2. Silicene buckling distance and in-plane lattice mismatch between silicene and the dichalcogenide substrate.

	Buckling distance (Å)	In-plane lattice mismatch (%)
MoS_2	1.9	18
$MoSe_2$	1.0	14
$MoTe_2$	0.7	9

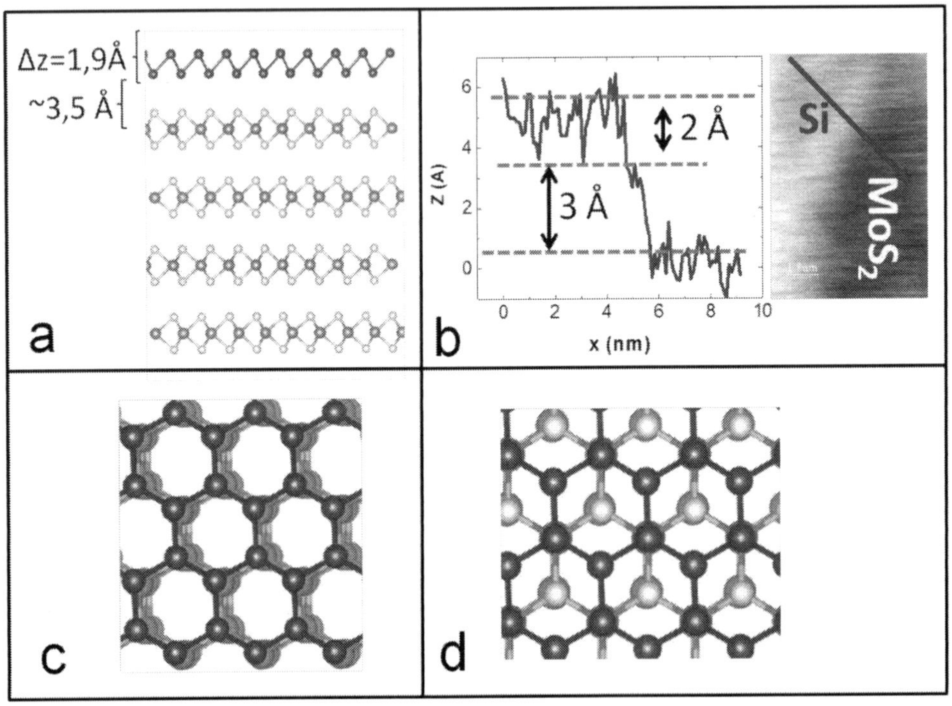

Figure 2. (a) Relaxed atomic configuration of silicene on bulk MoS_2 (side view). In (b) the STM image (VB=0.2V, IS=2nA) of a partially covered MoS_2 surface by Si atoms is shown (right side). A line profile taken across the two terraces allows measuring the amplitude of the step which amounts to about 5 Å (left side of b). In (c) and (d), the top view of *a* and *c* configurations from Fig. 1 (obtained after atomic relaxation) are shown. Blue, yellow and green spheres are Si, S and Mo atoms, respectively. The experimental results are from ref. 14.

Both the Si buckling distance and the Si-MoS_2 interlayer distance are in very good agreement with recently reported experimental STM results [14], see Fig. 2.b, where the step profile between a Si domain and the MoS_2 substrate amounts to 3 Å and exhibits a feature at about 2Å, consistently with the highly buckled silicene arrangement (see the picture in Fig. 2.a). The relatively good agreement between the experimental and computed structural properties of silicene on MoS_2 is supporting the validity of our theoretical approach.

The predicted electronic structure of the silicene and germanene layer on the MoX_2 template largely depends on the buckling parameter in the 2D material. Highly buckled silicene or germanene are predicted to be metallic, as illustrated in Fig. 3 for the case of silicene on MoS_2. On the other hand, the low buckled silicene layer on $MoTe_2$ is predicted to preserve its Dirac cones at the K points, as illustrated in Fig. 4. Silicene on $MoTe_2$ is thus predicted to be a gapless semiconductor, similar to free standing silicene. By increasing the in-plane lattice parameter of the dichalcogenide substrates, we thus found out that the buckling distance in the silicene or germanene layer can be reduced and the 2D material can eventually preserves its gapless semiconducting behavior, i.e. the partial sp^2 hybridization of the Si and Ge atoms is preserved.

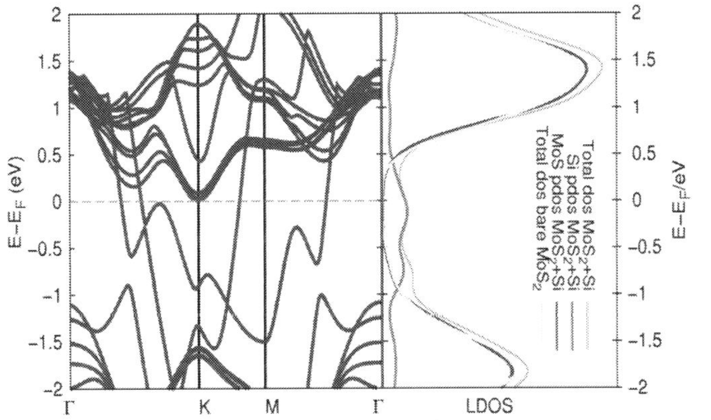

Figure 3. Energy band structure and local density of states (LDOS) of the silicene/MoS_2 structure. The LDOS shows that the density of states of the MoS_2 substrate still preserve a gap very close to that of the bare MoS_2, while all the electronic states close to the Fermi level are due to the contribution of Si atoms, confirming that almost no ''interaction'' (e.g. hybridization) between Si and Mo/S atomic orbitals is induced.

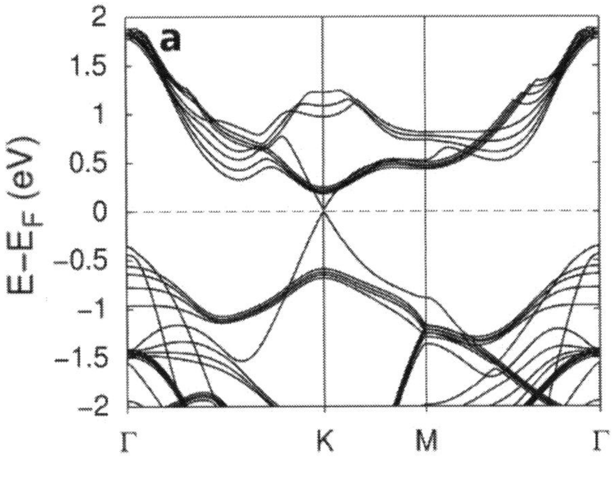

Figure 4. Energy band structure of the silicene/$MoTe_2$ system.

4. Silicene/ZnS and germanene/ZnSe interfaces

ZnS and ZnSe are semiconductors with a large direct energy band-gap of about 3.8 eV and 2.9 eV, respectively [15,16]. Very interestingly, the in-plane lattice constants of ZnS (3.81 Å) and ZnSe (3.98 Å) are very close to the computed ones of free-standing silicene and germanene, i.e. about 3.9 Å and 4.0 Å, respectively; ZnS and ZnSe thus appear as semiconducting templates with adequate symmetry and in-plane lattice parameters for the growth of silicene and germanene, respectively. Since the structural and electronic properties of the silicene/(0001) ZnS and germanene/(0001)ZnSe structures are qualitatively very similar, we only discuss here in details the results pertaining to the germanene/ZnSe interface; results about the silicene/(0001) ZnS interface can be found elsewhere [8].

We first computed the lattice parameters of bulk (wurtzite) ZnSe, which were found to be a=b=3.94 Å and c=6.54 Å, typically 1% smaller than the experimental values. The calculated energy gap is 2.4 eV, i.e. about 17% smaller than its experimental value. We next constructed (0001) polar ZnSe surfaces. A slab model with 8 atomic layers was considered (with a total of 64 atoms), with a 15 Å vacuum layers. After energy relaxation, the atoms of the top and bottom ZnSe layers are displaced, leading to a surface reconstruction very similar to the one of the non-polar (1010) ZnS surface [17,18]; this surface reconstruction is discussed in detail elsewhere [8]. A very similar surface reconstruction was reported recently on (0001) ZnS surfaces, and predicted to be more stable that the non-reconstructed polar ZnS surface for layers up to about 6.6 nm [19]. The reconstructed (0001) ZnSe surface is semiconducting, with a computed energy gap of about 2.1 eV. Note that polar (non-reconstructed) ZnS and ZnSe surfaces are metallic, due to the pining of the Fermi level by the anion surface states, like in ZnO [20,21]. On such polar surfaces, both silicene and germanene were also found to present a metallic character [22].

We next investigated the interaction of germanene with the reconstructed (0001) ZnSe surface. A flat germanene layer was placed on top of the reconstructed (0001) ZnSe surface, followed by geometry relaxation. As discussed in more details in ref. 8 for the case of silicene on ZnS, we studied different possible arrangements of the Ge atoms on the (0001) ZnSe surface. The most energetically stable structure corresponds to an hexagonal arrangements of the Ge atoms placed at intermediate positions between top and hollow sites of the ZnSe hexagons, as shown in Fig. 5. In this geometry, 2 Ge-Se (2.58 Å bond length) and 2 Ge-Zn bonds (2.49 Å bond length) are formed; the electronic charge transfer essentially occurs between the $4p_z$ orbitals of the Ge atoms with the 4s states of Zn and 4p states of Se, the bonded Ge atoms thus adopting an sp^3-like character. Four other Ge atoms from the silicene layer are not bonded to the substrate: two of these atoms are lying at about 2.9 Å from the ZnSe surface, and are marked "intermediate" on Fig. 5 and two other atoms are lying at about 3.71 Å from the ZnSe surface, and are marked "top" on Fig. 5. From the analysis of the partial Mulliken charges on the Ge atoms, the charge transfer occurring at the germanene/(0001) ZnSe interface leads to an excess of negative charge of about 0.15 e on the top Ge atoms, with respect to the intermediate Ge atoms, thus leading to the formation of a dipole at this interface. The average Ge-Ge distance is calculated to be 2.49 Å, very similar to the one of free-standing germanene.

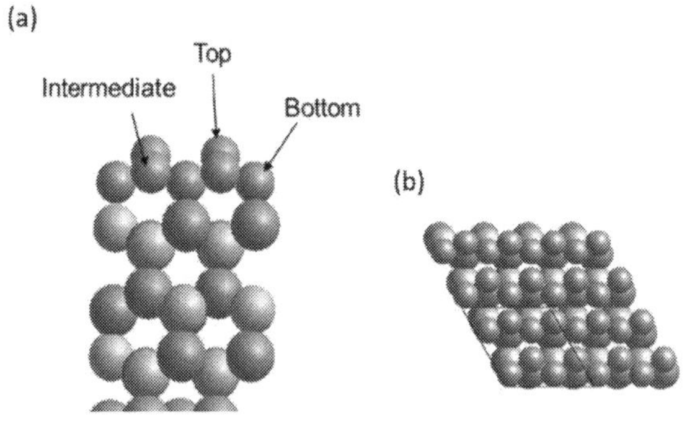

Figure 5. Side view (a) and top view (b) of the relaxed germanene/(0001) ZnSe slab model. Yellow, gray and blue spheres are Se, Zn and Ge atoms, respectively.

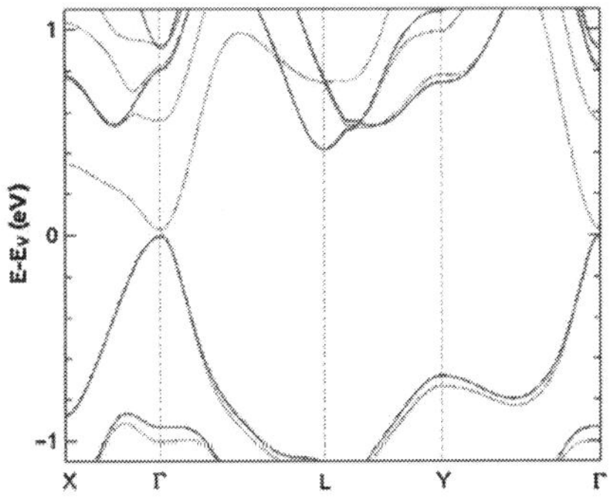

Figure 6. Computed energy band structure of the germanene/(0001) ZnSe slab model, without (solid blue lines) and with (dashed red lines) an external electric field of 0.6 V/ Å in the direction perpendicular to the interface. The reference (zero) energy level corresponds to the top of the valence band E_V of germanene.

The computed energy band structure of the germanene/(0001)ZnSe slab model is shown in Fig. 6. The system is predicted to be semiconducting, with an indirect energy band gap of about 0.4 eV. The opening of an energy gap in the electronic structure of germanene is related to the charge transfer and partial sp^3 hybridization of the Ge atoms bonded to the Zn or Se atoms on the surface. As a matter of fact, the computation of the electronic DOS of a germanene layer with the same atomic arrangement than observed on the ZnSe surface, but with the ZnSe substrate removed, presents a gapless-semiconducting behaviour (not shown).

We next studied the effect of an external electric field on the electronic properties of the germanene/(0001) ZnSe interface. The energy band structure of the system is shown in Fig. 6, without (solid lines) and with (dashed lines) an external electric-field of 0.6 V/Å. Very interestingly, the applied electric field has a substantial effect on the conduction band near the Γ point, leading to a transition from an indirect (Γ to L point) to direct (at Γ point) energy band gap in germanene. The computed direct and indirect energy band gaps are shown in Fig. 7 as a function of the external electric field. For electric fields lower than about 0.4 V/Å, the germanene/(0001) ZnSe interface has an indirect energy band gap, which is almost unaffected by the external electric field. For larger electric fields, the energy band gap becomes direct, and its value decreases approximately linearly with the external electric field, a semiconductor to metal transition being predicted for electric fields higher than 0.6 V/Å. The electric field dependence of the energy gap of germanene/(0001)ZnSe is very similar to the one of the silicene/(0001) ZnS system, which has been discussed in more detail elsewhere [8]; it is related to the reduced dipole at the silicene/ZnS interface in presence of an external electric field.

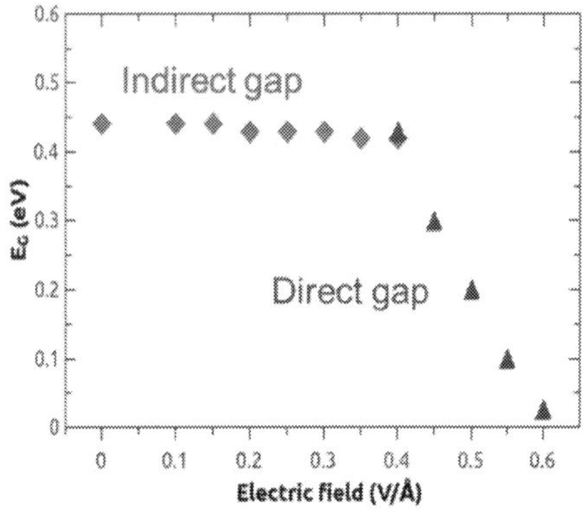

Figure 7. Computed direct (filled triangles) and indirect (filled diamonds) energy band gaps of the germanene/(0001) ZnSe slab model, as a function of the external electric field applied to the system.

5. Conclusions

The interaction of silicene and germanene with several non-metallic substrates has been theoretically investigated. On dichalcogenides, the 2D material is weakly bonded to the substrate, via van der Waals forces. The buckling distance in silicene or germanene is correlated to the lattice parameter of the dichalcogenide template. On MoS_2, silicene and germanene are highly buckled, and are metallic. On $MoTe_2$, silicene is weakly buckled and its (partial) sp^2 hybridization is preserved, the material being a gapless semiconductor. Our theoretical findings thus suggest that the electronic properties of silicene and germanene can be engineered by properly tuning the in-plane crystal parameters "matching" between the 2D material and the dichalcogenide substrate. These non-metallic compounds thus appear as potential templates for the growth of silicene and germanene, potentially enabling their use in future nanoelectronic devices.

On reconstructed (0001) ZnS and ZnSe surfaces, the silicene and germanene layers adopt the surface reconstruction of the underlying ZnS or ZnSe layer, and part of the Si (Ge) atoms are forming bonds with Zn or S (Se) atoms. The bonded Si (Ge) atoms present a sp^3-like hybridization, leading to the formation of silicene (germanene) with an indirect energy band-gap. Remarkably, the value and nature (direct or indirect) of the energy gap of the silicene/(0001) ZnS and germanene/(0001) ZnSe interfaces can be controlled by an external electric field. This predicted electric-field modulation of the electronic properties of silicene and germanene is potentially very interesting for logic applications.

Acknowledgements

Part of this work has been financially supported by the European Project 2D-NANOLATTICES, within the Future and Emerging Technologies (FET) program of the European Commission, under the FET-grant number 270749 as well as the Research Fund of KU Leuven, project GOA/13/011.

References

[1] Vogt P, De Padova P, Quaresima C, Avila J, Frantzeskakis E, Asensio M C, Resta A, Ealet B and Le Lay G 2012 *Phys. Rev. Lett.* **108** 155501
[2] Feng B, Ding Z, Meng S, Yao Y, He X, Cheng P, Chen L and Wu K 2012 *Nano Lett.* **12** 3507
[3] Chiappe D, Grazianetti C, Tallarida G, Fanciulli M and Molle A 2012 *Adv. Mat.* **24** 5088
[4] Fleurence A, Friedlein R, Ozaki T, Kawai H, Wang Y and Yamada-Takamura Y 2012 *Phys. Rev. Lett.* **108** 245501
[5] Meng L, Wang Y, Zhang L, Du S, Wu R, Li L, Zhang Y, Li G, Zhou H, Hofer W A and Gao M J 2013 *Nano Lett.* **13** 685
[6] Houssa M, Pourtois G, Heyns M M, Afanas'ev V V and Stesmans A 2011 *J. Electrochem. Soc.* **158** H107
[7] Scalise E, Houssa M, Cinquanta E, Grazianetti C, van den Broek B, Pourtois G, Stesmans A, Fanciulli M and Molle A 2014 *2D Mater.* **1** 011010
[8] Houssa M, van den Broek B, Scalise E, Pourtois G, Afanas'ev V V and Stesmans A 2013 *Phys. Chem. Chem. Phys.* **15** 3702
[9] Houssa M, van den Broek B, Scalise E, Ealet B, Pourtois G, Chiappe D, Cinquanta E, Grazianetti C, Fanciulli M, Molle A, Afanas'ev VV and Stesmans A 2014 *Appl. Surf. Sci.* **291** 98
[10] Perdew J, Burke K and Ernzerhof M 1996 *Phys. Rev. Lett.* **77** 3865
[11] Soler J M, Artacho E, Gale J D, García A, Junquera J, Ordejón P and Sánchez-Portal D 2002 *J. Phys. C: Cond. Matter* **14** 2745
[12] Giannozzi P *et al.* 2009 *J. Phys.: Cond. Matter* **21** 395502
[13] Troullier N and Martins J L 1991 *Phys. Rev. B* **43** 1993
[14] Chiappe D, Scalise E, Cinquanta E, Grazianetti C, van den Broek B, Fanciulli M, Houssa M and Molle A 2014 *Adv. Mat.* **26** 2096
[15] Weber M J 1986 *Handbook of Laser Science and Technology* (Cleveland: CRC Press)
[16] Xu Y N and Ching W Y 1993 *Phys. Rev. B* **48** 4335
[17] Northrup J E and Neugebauer J 1996 *Phys. Rev. B* **53** R10477
[18] Filippetti A, Fiorentini V, Cappellini G and Bosin A 1999 *Phys. Rev. B* **59** 8026
[19] Zhang X, Zhang H, He T and Zhao M 2010 *J. Appl. Phys.* **108** 064317
[20] Wander A, Schedin F, Steadman P, Norris A, McGrath R, Turner T S, Thornton G and Harrison N M 2001 *Phys. Rev. Lett.* **86** 3811
[21] Meyer B and Marx D 2003 *Phys. Rev. B* **67** 035403
[22] Houssa M, van den Broek B, Scalise E, Pourtois G, Afanas'ev V V and Stesmans A 2013 *ECS Trans.* **53** 51

Journal of Physics: Conference Series **574** (2015) 012016 doi:10.1088/1742-6596/574/1/012016

Kinematics and Simulation of a Hybrid Mechanism: MATLAB/SimMechanics

M E Kütük, R Halicioglu, L C Dulger
Gaziantep University, Faculty of Engineering, Mechanical Engineering Department, Gaziantep 27310, TURKEY

E-mail: mekutuk@gantep.edu.tr, rhalicioglu@gantep.edu.tr, dulger@gantep.edu.tr

Abstract. Kinematic analysis and simulation of hybrid drive system are addressed in this study. A seven link mechanism with two degrees of freedom is selected as the configuration of the system. Kinematic analysis is performed by loop closure equations and required inputs of servo motor are given to get desired ram motion scenario. MATLAB/SimMechanics platform is used to model the hybrid driven mechanical system mechanism characteristics. The simulation results are presented herein.

1. Introduction

The interest in hybrid driving systems is started many years ago [1]. The main principle in hybrid driven systems is to use the conventional motors with a flywheel and the servo motors via a mechanical linkage mechanism. A 2DOF (degree of freedom) link mechanism is driven by a combined input; an ordinary motor with a mechanism coupled to a servo motor (SM) adjusting ram motion. While ordinary motor provides high torque, SM performs the desired motion scenarios. This is presented as an example in flexible manufacturing systems under the assumption of power optimization which is important for selection of useful mechanism. Many different studies are then seen in hybrid configurations in industry. Some related with press operations are taken here [2-5]. The hybrid systems include the advantages of both driving systems. While the existence of conventional motor is to obtain power and torque requirement, SM is used to make system as a flexible motion provider. The basic principle of hybrid systems is to bring together the motion of a large constant velocity motor (CM) with a small SM via a mechanical linkage mechanism. In these systems the CM provides the main torque and motion requirements while the SM assists the modulations on the present motion. The CM undertakes a big amount of workload and the SM is like a real time regulator to change the task. Hybrid systems can be classified as hybrid actuator and hybrid mechanism. An intelligent box having two inputs and one output is considered as the hybrid actuator. The box is driven by a SM and a CM and it provides a programmable output which may be coupled to a mechanism or a load. However, a hybrid mechanism is one that has at least two degrees of freedom. The mechanism has two input links which are coupled to a SM and a CM [6].

The main purpose of this study is to present hybrid mechanism idea with related kinematics and simulation. Reviews on the related works are given as background. A hybrid seven bar (2 DOF) mechanism is taken here. Its kinematics is studied with inverse issues. One of the required ram (slider) motion is explained. A hybrid press alternative is presented. A SimMechanics model is built with the chosen configuration. Simulation results are then given for further studies.

2. Background of the Hybrid Machine

Hybrid machine systems are currently presented many ongoing studies [7-8]. Hybrid machine (HM) systems are employed in common industrial applications including cutting, printing, and stamping. Since the idea is providing a hybrid actuation, this is called hybrid machines. Different studies are seen in hybrid configurations with a five bar, slider crank, and seven bar for example. Motion ranges can be changed and more complicated paths may be produced. A recent study on hybrid systems are summarized in a thesis [8]. Studies on the subject more than two decades are searched. They are given with remarks and things to be added for coming studies. It is composed of three main parts; hybrid machine configurations, hybrid machine synthesis, modelling and control and hybrid machine applications. Optimal synthesis of possible configurations was carried out with the system control.

In hybrid machine applications, control of the machine is carried out in two stages. The first one is the control of SM, the second one is the control of CM to provide its actuation synchronously with the SM. In the studies completed, the proposed configurations are modelled mathematically and different PID control techniques; adaptive control, genetic algorithms (GA), particle swarm optimization (PSO) and sliding mode control (SMC) techniques are applied on them. Main concentration is done its modelling and simulation issues in this study.

3. Hybrid Configuration

A hybrid driven mechanical system is presented in figure 1. The system is composed of a planar five bar mechanism ABDEF and a dyad of CD. All joints except one are revolute joints. The exception is a prismatic joint used in translational motion between the slider and ground. The lengths of the links are represented by r_1, r_2, r_3, r_4, r_5 and r_6 as shown in figure 1. The crank r_5 is driven by a CM. The crank r_2 is driven by a SM. The link r_1 is the ground link and Θ is the orientation angle of it. The coordinate axes are fixed at point F. The BDC is the Bottom Dead Center point of the slider in which r_4, r_5 and r_6 are lined up in a straight line. The starting point of the slider is assumed as BDC. In addition to all, Top Dead Center of the slider is TDC point. In this mechanism, while EF link is driven by a CM, AB link is driven by a SM.

4. Related Kinematics

Kinematics analysis and main configuration of a 2-DOF controllable hybrid driven mechanism are presented in this section. A seven bar configuration is taken here. The inverse kinematics analysis is included. Linkage synthesis criteria's are given. The equations of inverse kinematics are derived. From the sketch of the mechanism given in figure 1, two loop closure equations [7-8] are written. Loops 1 and 2 are given as;

$$\vec{r_5} + \vec{r_6} = \overrightarrow{FG} + \overrightarrow{GC} + \vec{r_4} \tag{1}$$

$$\vec{r_5} + \vec{r_6} = \vec{r_1} + \vec{r_2} + \vec{r_3} \tag{2}$$

These two equations can be written in Euler form [8-9]. When Equations of (1) and (2) are written in Euler form and separated into two parts, Loop 1 and Loop 2 can be written as below;

$$r_5 \sin\theta_5 + r_6 \sin\theta_6 = S_0 + s + r_4 \sin\theta_4 \tag{3}$$

$$r_5 \cos\theta_5 + r_6 \cos\theta_6 = e + r_4 \cos\theta_4 \tag{4}$$

$$r_5 \cos\theta_5 + r_6 \cos\theta_6 = X_A + r_2 \cos\theta_2 + r_3 \cos\theta_3 \tag{5}$$

$$r_5 \sin\theta_5 + r_6 \sin\theta_6 = Y_A + r_2 \sin\theta_2 + r_3 \sin\theta_3 \tag{6}$$

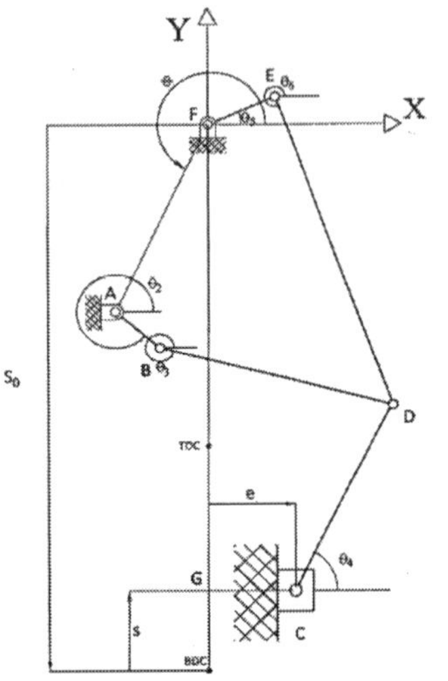

Figure 1 Schematic representation of hybrid driven mechanical system.

$|\overrightarrow{AF}| = r_1,\ |\overrightarrow{AB}| = r_2,\ |\overrightarrow{BD}| = r_3,\ |\overrightarrow{CD}| = r_4,\ |\overrightarrow{FE}| = r_5,\ |\overrightarrow{ED}| = r_6$

By taking 1st and 2nd derivatives of equations (3-6), the angular velocities ad accelerations of each linkage are found [8-10]. Angular displacement, velocity and acceleration expressions are included in a thesis [8]. In order to satisfy different motion characteristics of the slider, the input link r_2 and r_5 must make complete rotations. In a previous study, Li [3] has given the classifications of hybrid five-bar linkages by using the unrestraint double-crank type. The input link r_2 and r_5 are unrestraint cranks. Therefore, the inequality constraints are defined as follows:

$$r_1 + r_2 + r_5 < r_3 + r_6 \tag{7}$$

$$r_2 + r_5 + r_6 < r_1 + r_3 \tag{8}$$

$$r_2 + r_3 + r_5 < r_1 + r_6 \tag{9}$$

The following inequality constraints should be satisfied to provide slider mobility condition in figure 1. It is required that $r_4 > X_D - e$, where e is the coordinate of point D in x direction. Link length condition; in order to have appropriate link lengths, the link lengths inequality constraints are given as follows:

$$r_{i\,\min} \le r_i \le r_{i\,\max} \qquad r_i = 1,\ 2,\ \dots\ 6 \tag{10}$$

When these constraints of mechanism are taken into account, the link dimensions of the hybrid driven mechanism are found and tabulated in table 1. All dimensions are in mm. The orientation angle, Θ is given in degrees.

Table 1. Link dimensions of the hybrid driven mechanism.

r_1	r_2	r_3	r_4	r_5	r_6	e	Θ
530	200	650	900	170	800	6.73	245

5. Motion Applications in Industry

Hybrid machine systems are being used in cutting, stamping and pressing applications in industry. By changing the motion profiles, it is possible to obtain different trajectories and use in various applications in industry. Most of studies are taken on press application according to years of studies. The fifth order polynomials are used to design the motion. Curve Fitting Toolbox in Matlab© is used as a tool in this study to design the motion characteristics of the slider. The Curve Fitting Toolbox is not directly used in motion design operation [8, 11].

5.1. Ram Motion Scenarios

Figure 2 are the selected CM's angular displacement profile. Angular velocity of CM is a constant value (positive direction 1.0472 radian/sec).

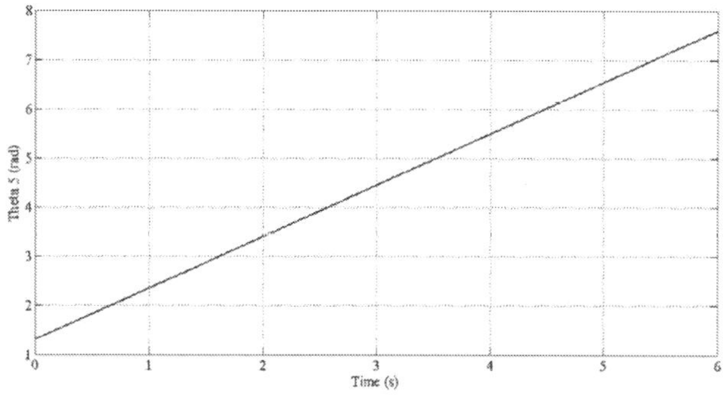

Figure 2. Angular displacement of CV motor.

The designed ram motion is a slow return-dwell and quick rise motion. Stroke of the slider is 687 mm and stroke/min is taken as 10. Figure 3 shows the designed ram motion which is composed of three segments. The kinematics specifications of each segment are illustrated in table 2. It is formed of segment number, time interval and initial and final values of position, velocity and acceleration of each segment.

When the kinematics data of the slider link (ram motion) and the fifth link driven by the CM are known, it is possible to reach the kinematics data of SM (Figure 1) by using the inverse kinematics analysis. The kinematic characteristics of the SM obtained are given in figure 4 (signal building view).

Table 2. Kinematics specifications of the segments

	Motion Time(s)	**Position(mm)**	**Velocity(mm/s)**	**Acceleration (mm/s2)**
I	$0 \leq t \leq 3.0$	687 , 0	0 , 0	0 , 0
II	$3 \leq t \leq 3.5$	0 , 0	0 , 0	0 , 0
III	$3.5 \leq t \leq 6.0$	0 , 687	0 , 0	0 , 0

6. Modelling MATLAB/SimMechanics

A multibody simulation environment is provided by using SimMechanics. It is an important simulation environment for solving many practical problems in industry. With this environment, forward dynamics, inverse dynamics, and kinematics can be done. Mechanisms can be modeled by solving motion equations in MATLAB/Simulink toolbox [12]. In SimMechanics, mechanism's bodies are defined as their coordinates for a one position. Its simulation can be shown directly as 2D or 3D. SimMechanics produces system model without deriving the equations of motion. The input parameters related systems are necessary [13]. SimMechanics model of the hybrid system is given in Figure 5 [11]. SimMechanics structure is designed for the position when Θ_5 is 30 deg. Input parameters are adapted according to that position. Body 7 indicates the slider of the mechanism and the model results of slider are obtained in two types that are "to file: as numerical" and "to scope: as visual". Figure 2 and figure 4 give the signal buildings of CM and SM respectively. Figure 6 presents the 3D simulation view of hybrid mechanism and figure 7 is kinematic outputs of slider link obtained by SimMechanics.

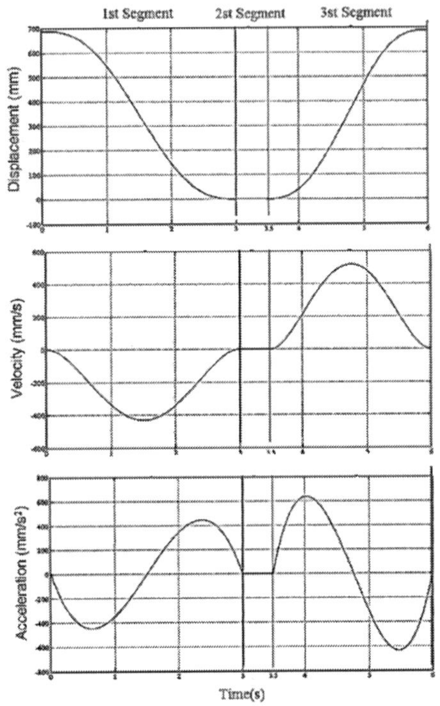

Figure 3. Desired and designed displacement, velocity and acceleration ram motion.

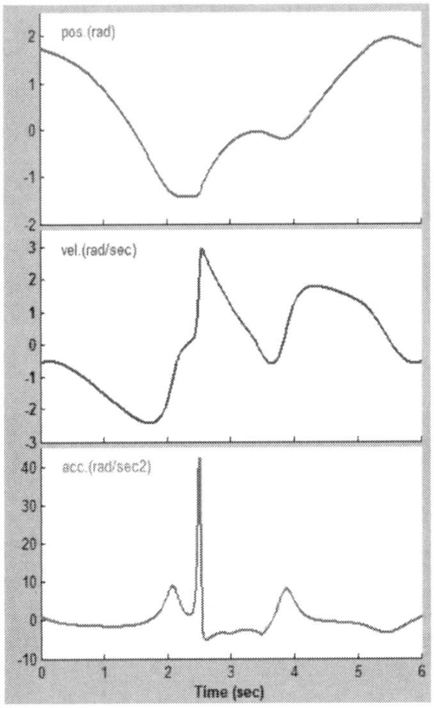

Figure 4. Angular displacement, velocity and acceleration of SM

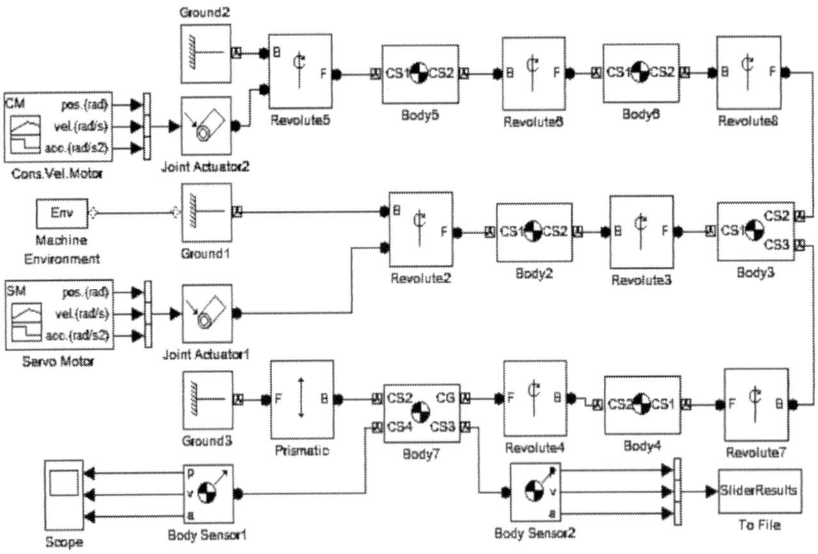

Figure 5. Hybrid mechanism kinematics model by Matlab/SimMechanics

7. Conclusions

A seven bar mechanism is given for hybrid system explanation. It is studied with its inverse kinematics issues. This configuration has advantages of traditional systems with flexible motion output by using a servo motor. The kinematic modelling is facilitated by using SimMechanics and the results are interpreted easily. The results obtained from inverse kinematics are used as input parameters in SimMechanics. The scope of the model gives the expected ram motion output. Different ram motion scenarios can be implemented.

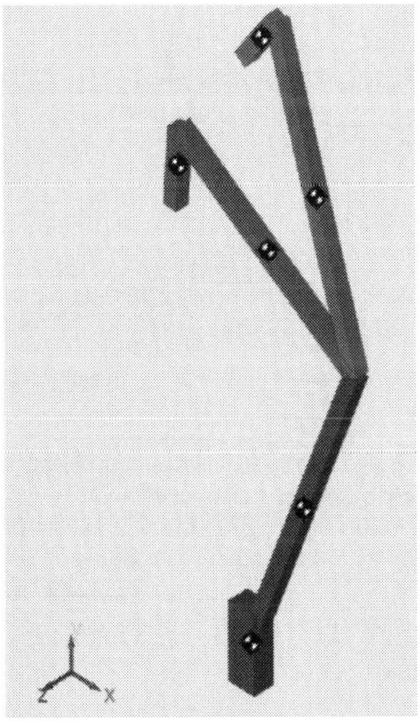

Figure 6. 3D SimMechanics simulation of the hybrid mechanism

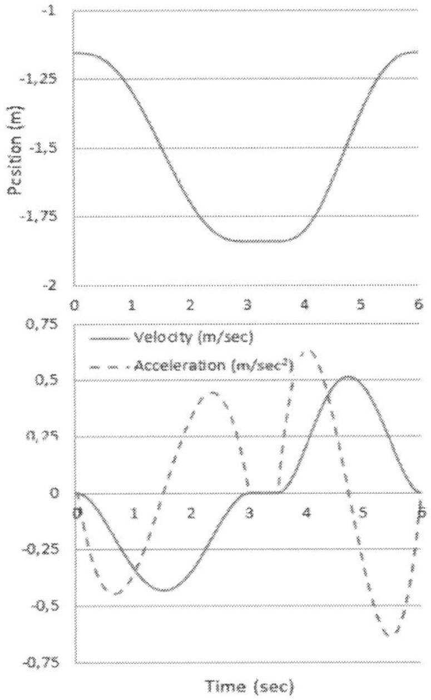

Figure 7. Displacement, velocity and acceleration of the slider (press ram)

8. Acknowledgment
The authors would like to thank the Ministry of Science, Industry and Technology for the financial support under SANTEZ project (Number: 01422.STZ.2012-I)

9. References
[1] Tokuz L C and Jones J R 1992 *Hybrid Machine Modelling and Control Ph.D. Dissert.* (Liverpool: Polytechnic)
[2] Seth B 2004 Programmable Hybrid Mechanisms *11th World Cong. in Mech. and Machine Science* (Tianjin, China).
[3] Ouyang P R, Li Q and Zhang W J 2004 *Mechatronics* **14** 1197-1217
[4] He K, Luo Y, Kong C T and Du R 2009 *WSEAS Transactions on Systems* **8** 614-627
[5] Tso P L 2010 *Journal of Mechanical Design* **132(3)**: 034503
[6] Dulger L C, Kirecci A 1999 *Mechanism and Machine Theory* **35** 1141-1149
[7] Kai H 2008 *Design and Control of a Controllable Hybrid Mechanical Metal Forming Press. PhD Dissert.* (Hong Kong: The Chinese University)
[8] Kütük M E and Dulger L C 2013 *Hybrid Machine Systems: Analysis and Control Ms. Dissert.* (Turkey: Gaziantep University)
[9] Soylemez E 1999 *Mechanisms* (Ankara,Turkey; M.E.T.U)
[10] Norton N R 2004 *Design of Machinery: An Introduction to the Synthesis and Analysis of Mechanisms and Machines* (3th Ed; McGraw Hill)
[11] MathWorks Inc. 2011 *Matlab* (MathWorks)
[12] Halicioglu R, Dulger L C and Bozdana A T 2014 *J. Phys.: Conf. Ser.* **490** 012053
[13] Halicioglu R, Dulger L C 2013*UMTS* (September, Erzurum) *p 451-458*

Journal of Physics: Conference Series **574** (2015) 012017 doi:10.1088/1742-6596/574/1/012017

Approximate method of free energy calculation for spin system with arbitrary connection matrix

Boris Kryzhanovsky and Leonid Litinskii

Center for Optical Neural Technologies, Scientific Research Institute for System Analysis of RAS, Vavilov str., 44/2, off. 216, Moscow, Russia, 119333

kryzhanov@mail.ru, litin@mail.ru

Abstract. The proposed method of the free energy calculation is based on the approximation of the energy distribution in the microcanonical ensemble by the Gaussian distribution. We hope that our approach will be effective for the systems with long-range interaction, where large coordination number q ensures the correctness of the central limit theorem application. However, the method provides good results also for systems with short-range interaction when the number q is not so large.

1. Introduction

We examine a system of N spins $s_i = \{\pm 1\}$, whose state is determined by the configuration vector $\mathbf{s} = (s_1, ..., s_N)$ and the energy $E(s) = -\frac{1}{2}\sum_{i=1}^{N} s_i h_i$. Here $h_i = \sum_{i=1}^{N} T_{ij} s_j$ is the local field acting on the i-s spin. We suppose that $\mathbf{T} = (T_{ij})$ is a symmetric matrix; the diagonal elements of this matrix are equal to zero ($T_{ii} = 0$) and in average its rows contain a large number of nonzero elements $q >> 1$. The inequality $q >> 1$ means that correlation between $s_i h_i$ and $s_j h_j$ is negligible small (it behaves as q^{-1}).

Let us describe our method of calculation of the partition function $Z_N = \sum_{\mathbf{s}} \exp[-\beta E(\mathbf{s})]$, where β is the inverse temperature. Let $\mathbf{s}_0$ be the ground state. The class of all states $\mathbf{s}$ differing from $\mathbf{s}_0$ by n opposite spin signs we define as Ω_n. In the general case we do not know the form of the energy distribution for all $\binom{N}{n}$ states of the class Ω_n. However, the exact expressions of the mean value E_n and dispersion σ_n^2 can be calculated (see [1], [2]). Further on, it can be shown that for large values of n and N the distribution of the energies for the class Ω_n can be approximated reasonable well by the Gaussian distribution with the parameters E_n and σ_n^2 [1].

We verified this assertion for some types of the matrices $\mathbf{T}$, the dimensionalities $N \sim 10^3 - 10^4$ and different values of $n >> 1$. At least if $E \in [E_n \pm 3\sigma_n]$, the experimental density of the distribution for large number of random configurations $K \geq 10^6$ coincides perfectly with the density of the Gaussian distribution (see figure 1). Outside of this interval the densities differ, however for our purposes these distinctions on the tails are not pivotal. Then the summation over the class Ω_n can be replaced by integration over the Gaussian measure, and the expression for the partition function takes the form

$$Z_N=\sum_{n=0}^{N}\sum_{\Omega_n}\exp[-\beta E(\mathbf{s})]\approx\sum_{n=0}^{N}\binom{N}{n}\int_{E_{\min}}^{E_{\max}}\exp(-\beta E)\exp\left(-\frac{(E-E_n)^2}{2\sigma_n^2}\right)\frac{dE}{\sqrt{2\pi}\sigma_n},\qquad(1)$$

where the minimal and maximal values of the energy $E(\mathbf{s})$ are taken as the inferior and superior limits of integration. This technique is essential for our approach. Now the calculation of the partition function reduces to the calculation of the integral with the aid of the saddle-point method.

2. The Main Equation

In the asymptotic limit $N\to\infty$ the expressions for E_n and σ_n^2 take the form:

$$E_n=NE_x,\ \sigma_n^2=N\sigma_x^2,\ E_x=E_0(1-2x)^2,\ E_0=E(\mathbf{s}_0)/N,\ x=n/N,$$
$$\sigma_x^2=\left(N^{-1}\sum\nolimits_{i\neq j}^{N}T_{ij}^2\right)\cdot 2x(1-x)\left[4x(1-x)(1-\varepsilon_0^2)+2N(d_0-\varepsilon_0^2)(1-2x)^2\right]\qquad(2)$$
$$\varepsilon_0^2=4E_0^2\left(\sum\nolimits_{i\neq j}^{N}T_{ij}^2\right)^{-1},\quad d_0=\|\mathbf{Ts}_0\|^2\left(N\sum\nolimits_{i\neq j}^{N}T_{ij}^2\right)^{-1}.$$

Using the Stirling asymptotic and replacing the summation over n with the integration over the variable $x=n/N$, we obtain:

$$Z_N\sim\sqrt{\frac{N}{2\pi}}\int_0^1 e^{-Nf(x)}dx,\qquad f(x)=L(x)+\beta E_x-\tfrac{1}{2}\beta^2\sigma_x^2-N^{-1}\ln\Phi(x)\qquad(3)$$

where $L(x)=x\ln x+(1-x)\ln(1-x)$. Finally, we set $E'_{\max}=E_{\max}/N$, and then

$$\Phi(x)=\frac{1}{\sqrt{2\pi}}\int_{A_x}^{B_x}e^{-\frac{1}{2}t^2}dt,\ B_x=\sqrt{N}\left(\beta\sigma_x+\frac{E'_{\max}-E_x}{\sigma_x}\right),\ A_x=\sqrt{N}\left(\beta\sigma_x+\frac{E_0-E_x}{\sigma_x}\right).$$

Let us calculate the limits of integration B_x and A_x as well as the function $\Phi(x)$. Since $E'_{\max}-E_x>0$, the numerator in the expression for B_x is always positive, and $B_x\to+\infty$ when $N\to\infty$. The numerator in the expression for A_x can be both negative and positive. In the first case $A_x\to-\infty$ when $N\to\infty$, and then $\Phi(x)\to1$. But if the numerator of A_x is positive, the value of $A_x\to+\infty$ and $\Phi(x)\sim\exp\left(-\tfrac{1}{2}A_x^2\right)$. Substituting these expressions for $\Phi(x)$ in (3), we finally obtain:

$$f(x)=\begin{cases}f^{(1)}(x)=L(x)+\beta E_x-\tfrac{1}{2}\beta^2\sigma_x^2, & \text{when}\ \ \beta\sigma_x^2+E_0-E_x<0\\ f^{(2)}(x)=L(x)+\beta E_0-\dfrac{1}{2}\left(\dfrac{E_0-E_x}{\sigma_x}\right)^2, & \text{when}\ \ \beta\sigma_x^2+E_0-E_x>0\end{cases}\qquad(4)$$

The free energy per site is equal to $f(\beta)=\beta^{-1}\min_x f(x)$.

3. The Ising Model on a Hypercube

Our approach is theoretically based for $q>>1$. Nevertheless, rather good results can also be obtained for comparatively small values of q.

For D-dimensional Ising model in each row of the matrix $\mathbf{T}$ there are only $q=2D$ nonzero elements, and they all are equal to J. Here the ground state is the configuration $\mathbf{s}_0=(1,1,...,1)$. It is easy to obtain that $E_0=-\tfrac{1}{2}qJ$, $\varepsilon_0^2=q/N$, $d_0-\varepsilon_0^2=0$, $\sigma_x^2=8qJ^2x^2(1-x)^2$. The function $f(x)$ (4) is a symmetrical one with respect to the middle of the interval $x\in[0,1]$, and this is the reason why we present the expression for $f(x)$ only inside the interval $0\le x\le\tfrac{1}{2}$. Let us introduce dimensionless parameter $\bar{\beta}=\beta J$. Then we obtain for (4):

$$f(x)=\begin{cases} f^{(1)}(x)=L(x)-\frac{1}{2}q\bar{\beta}\left[(1-2x)^2+8\bar{\beta}x^2(1-x)^2\right], & \text{for } 0\le x<x_{\bar{\beta}} \\ f^{(2)}(x)=L(x)-\frac{1}{2}q(\bar{\beta}-\frac{1}{2}), & \text{for } x_{\bar{\beta}}\le x\le\frac{1}{2} \end{cases} \quad (5)$$

Here $x_{\bar{\beta}}=1/2$, when $\bar{\beta}<1$, and $x_{\bar{\beta}}=\left(1-\sqrt{1-1/\bar{\beta}}\right)/2$, when $\bar{\beta}>1$.

It is easy to solve the equation $df/dx=0$ and to examine the properties of the solution (see [7]). At first let us present the final result for $q>16/3$. When $\bar{\beta}$ is not very large, the function $f(x)$ has only one minimum at the point $x_0=1/2$ (figure 2a). When $\bar{\beta}$ becomes larger than the critical value

$$\bar{\beta}_c=\frac{1-\sqrt{1-4/q}}{2}, \quad (6)$$

the minimum point $x_1(\bar{\beta})$ differs from x_0 (figure 2b). The phase transition of the second kind takes place when $\bar{\beta}=\bar{\beta}_c$. When $\bar{\beta}\to\infty$, the minimum point $x_1(\bar{\beta})\to 0$, and the state of the system tends to the ground state $\mathbf{s}_0=(1,1,...,1)$ (figure 2c). This corresponds to the physical description of the spin system: it passes into the ground state when temperature tends to zero.

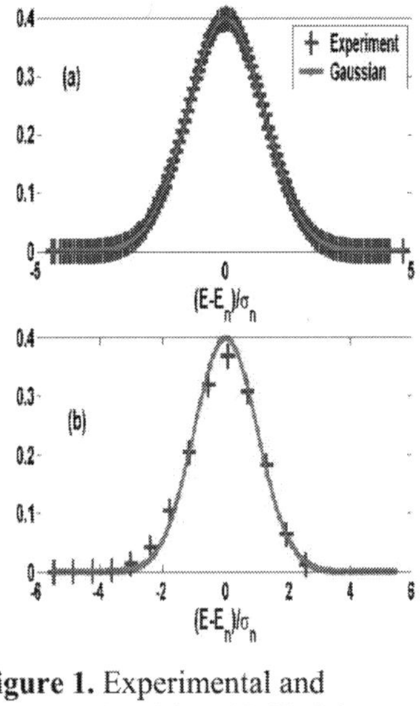

Figure 1. Experimental and Gaussian densities: (a) 3D-Ising, $N=10^3, n=500, K=10^6$; (b) 2D-Ising, $N=10^4, n=15$, $K=32\cdot10^6$.

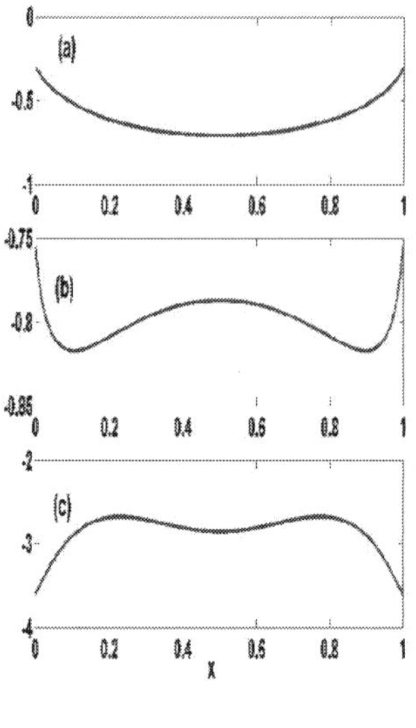

Figure 2. The function $f(x)$ (5) for 3D-Ising: (a) $\bar{\beta}<\bar{\beta}_c$, (b) $\bar{\beta}>\bar{\beta}_c$, and (c) $\bar{\beta}>>\bar{\beta}_c$.

For 3D-Ising model ($q=6$) from Eq.(6) we obtain $\bar{\beta}_c\approx 0.2113$, which is close to the value 0.2224 obtained in [4] with the aid of computer simulations. For 4D-Ising model ($q=8$) we have $\bar{\beta}_c\approx 0.146$, and the difference between this value and the estimate 0.149 obtained in [5] is even less. In general, when q is large enough, from Eq.(6) we obtain $\bar{\beta}_c\approx 1/q$. This is in agreement with the supposition that for large dimensionalities D the Ising model shows indices typical for the mean field model [5], [6].

Now let us show the results for $q<16/3$. For 2D-Ising model ($q=4$) there are some notable distinctions in the scenario described above. (Because of the length of this paper it is impossible to present them in this paper.) Here no analytic expression for the critical value $\bar{\beta}_c$ can be obtained; $\bar{\beta}_c\approx 0.3912$ is a solution of a transcendental equation. This value differs from the known exact solution 0.4407 [3]. For 1D-Ising model ($q=2$) our approach does not work at all: the state of the spin system does not tend to the ground state when $\bar{\beta} \to \infty$. It is not surprisingly, since our approach can be justified only when $q >> 1$.

A magnetic field H can be added to our model for E_n and σ_n^2 as functions of H (see Appendix). In this case an additional term $-\beta H(1-2x)$ appears in Eq.(3) for $f(x)$. Near the critical point $\bar{\beta}_c$ expressions for spontaneous magnetization m_0, magnetic susceptibility χ and specific heat can be obtained [7]. The corresponding critical indices are the classical ones: if $t = 1 - \bar{\beta}_c / \bar{\beta} << 1$, then $m_0 \sim t^{1/2}$, $\chi(t) \sim t^{-1}$, and a specific heat discontinuity corresponds to $\alpha = 0$.

The analysis of this section can be applied to any other model when in each row of the matrix **T** there are only q nonzero, equal to each other elements. Then the configuration $\mathbf{s}_0 = (1,1,...,1)$ is the ground state again, and Eq.(5) is still true. In particular, it works for Ising model on the Bete lattice [3], when each spin interacts with q nearest neighbors. Equation (6) is in qualitative accord with the exact expressions obtained for the Bete lattice. Both results coincide, when $q \to \infty$.

When for any n we have $\sigma_n \to 0$, the expression (1) takes the form well-known in the mean field theory [3]: $Z_N = \sum_{n=0}^{N} \binom{N}{n} e^{-\beta E_n}$. From Eq.(2) we can see that the dispersion σ_x^2 vanishes only if all non-diagonal matrix elements are equal to each other: $q = N-1$, $T_{ij} \equiv J$ $(i \neq j)$. In this case $\mathbf{s}_0 = (1,1,...,1)$, $d_0 - \varepsilon_0^2 = 0$, and $\varepsilon_0^2 = 1$. This simplifies all the calculations, and we easily obtain the classical Bragg-Williams equation [3]. For all other matrices the dispersion $\sigma_x^2 \neq 0$, and our approach allows one to obtain more reasonable estimates than the mean field model.

The work is supported by the Russian Basic Research Foundation (grants 12-07-00295 and 13-01-00504).

Appendix

For the sake of completeness we present the expressions for the mean value $E_n(\mathbf{H})$ and the dispersion $\sigma_n^2(\mathbf{H})$ from the class Ω_n in the presence of a nonuniform magnetic field $\mathbf{H} = (H_1, H_2, ..., H_N)$:

$$E_n(\mathbf{H}) = E_n - \left(1 - \frac{2n}{N}\right) \cdot (\mathbf{s}_0, \mathbf{H}),$$
$$\sigma_n^2(\mathbf{H}) = \sigma_n^2 + \frac{4n(N-n)}{N(N-1)} \left[\|\mathbf{H}\|^2 - \frac{(\mathbf{s}_0, \mathbf{H})^2}{N} + 4\frac{N-2n}{N-2} \left(\frac{(\mathbf{T}\mathbf{s}_0, \mathbf{H})}{2} + \frac{E_0 \cdot (\mathbf{s}_0, \mathbf{H})}{N} \right) \right]. \quad (A.1)$$

The exact expressions for E_n, σ_n^2 in the absence of the magnetic field can be find in [2]. Equations (A.1) are significantly simpler when $\mathbf{H}$ is collinear to $\mathbf{s}_0$: $\mathbf{H} = H \cdot \mathbf{s}_0$. Then it is easy to show that $E_n(\mathbf{H}) = E_n - (1 - 2n/N) NH$, $\sigma_n^2(\mathbf{H}) = \sigma_n^2$.

References

[1] Kryzhanovsky B V, Kryzhanovsky V M 2009 *Lecture Notes in Electrical Engineering* **24** 51-61
[2] Karandashev Ya M, Kryzhanovsky B V 2010 *Optical Memory & Neural Networks* **19** 110
[3] Baxter R J 1982 *Exactly Solved Models in Statistics Mechanics* (London: Academic Press)
[4] Blote H W J, Shchur L N, Talapov A L 1999 *Int. J. Mod. Phys.* C **10** 1137
[5] P.H.Lundow, K.Markstrom 2004 The critical behaviour of the Ising model on the 4-dimensional lattice *Preprint* arXiv:1202.3031v1
[6] Aizenman M 1982 Commun. Math. Phys. 86 1-48
[7] Kryzhanovsky B V, Litinskii L B (in preparation)

An application of Prekopa's theorem on the logconcavity of warranty reserves in reliability engineering

Eva Maria Ortega[1] **and Jose Alonso**[2]

1. Main Researcher MTM2009-13433, Ministerio Economia Competitividad. Res. Universit Oblats. 30001 Murcia, SPAIN. 2. Researcher Hospital Univ. Virgen Arrixaca, 30120 Murcia, SPAIN

E-mail: emariaop@gmail.com

Abstract. The log-concavity of the expected refund rate for an item with a nonrenewing pro-rata rebate warranty policy, and of the expected total reserve to service the warranty over the item life cycle under the same policy are proved. Upper bounds of warranty reserves are derived. The results are extended to other measures for spares and repairs under other warranty policies. Applications are given to ranking items' prices and production and inventory control.

1. Introduction

Log-concave densities have played a prominent role in applied probability for both discrete and absolutely continuous univariate distributions, since log-concave functions have a large number of good mathematical properties. A starting point in this article is to emphasize on some properties of log-concave functions in connection with Polya frequency functions (see Karlin (1968)). The aim of this note is to prove the log-concavity of the expected refund rate for an item and of the expected total reserve to service the warranty over the item's life-cycle with a nonrenewing pro-rata rebate warranty policy. The log-concavity of these functionals is applied for obtaining their upper bounds being useful for optimal pricing under the same warranty policy. Similar results are given for the expected demand for spares per unit time over the warranty period and expected total number of spares over the item's life-cycle, and measures of repairs under a nonrenewing free replacement warranty policy. Throughout the article, given a measure space (Ω, A, λ) we denote by f a probability density function on A with respect to λ, and assume f belongs to the space of functions defined on $A \subset \mathbb{R}$, whose derivatives of all orders $n \leq 2$ exist and are continuous in A, whenever required.

Acknowledgement: We thank Prof. Franco Pellerey from Politecnico di Torino for overall comments on a draft of this paper, and also Prof. Laureano Escudero and Prof. German Badia from University of Zaragoza for their first approval of the main result in this paper.

2. Preliminaries

A non-negative function $f : \boldsymbol{A} \longrightarrow [0, +\infty)$, for $\boldsymbol{A} \subset \mathbb{R}^n$ is said to be log-concave (log-convex) in $x \in \boldsymbol{A}$, if for all x_1, $x_2 \in \boldsymbol{A}$ and $\alpha \in [0, 1]$ such that $\alpha x_1 + (1-\alpha)x_2 \in \boldsymbol{A}$, we have that

$$f(\alpha x_1 + (1-\alpha)x_2) \geq (\leq) f(x_1)^\alpha f(x_2)^{1-\alpha}. \tag{2.1}$$

Taking $log(0) = -\infty$, then an equivalent condition is that $log(f)$ is concave (convex). For the basic properties of log-concave functions, we may refer to Eaton (1987).

Log-concavity can be also characterized in terms of Polya frequency of order 2 functions (see Karlin (1968) for a comprehensive study). A function $g : A \longmapsto [0, +\infty)$, for $\boldsymbol{A} \subseteq \mathbb{R}^2$ is said to be total positive of order 2, for short $g(x, y)$ is TP_2 in (x, y), if for all if for all $x_1 < x_2 \in \boldsymbol{A}$ and $y_1 < y_2 \in \boldsymbol{A}$, then

$$g(x_1, y_2)g(x_2, y_1) \leq g(x_1, y_1)g(x_2, y_2). \tag{2.2}$$

Also, a function $f : A \longmapsto [0, +\infty)$, for $A \subseteq \mathbb{R}(\mathbb{Z})$ is said to be Polya frequency of order 2, for short $f \in PF_2$, if $g(x, y) = f(x - y)$ is TP_2 in $(x, y) \in A^2$. From nomenclature by Karlin (1968), a PF_2 density is distinghuished from a PF_2 frequency function by the additional condition that f is integrable, in this event f is usually normalized by a multiplicative constant, so that $\int_{-\infty}^{+\infty} f(x)dx = 1$. The log-concavity of a measurable function f is equivalent to f being a PF_2 density function vanishing on the negative axis. Several techniques have been developed to construct log-concave functions, as Prekopa theorem in Prekopa (1973), that we recall next.

Proposition 1. *Let $\phi(t, \boldsymbol{x})$ be a convex function in $\mathbb{R}^{n+1}$ and let $\psi(\boldsymbol{x})$ be a convex function in $\mathbb{R}^n$. Assume that*

$$\int_{\mathbb{R}^n} e^{\psi(\boldsymbol{x})-\phi(0,\boldsymbol{x})} d\boldsymbol{x} = 1.$$

Then the function $f(t) = \int_{\mathbb{R}^n} e^{-\phi(t,\boldsymbol{x})} d\boldsymbol{x}$ is log-concave.

The structure of Polya densities of order 2 that vanish on the negative axis provides their moment inequalities and bounds, and they can be characterized by Laplace transforms that exist in an open strip of the complex plane containing the imaginary axis, by different inversion and representation theorems in the literature (see Karlin (1968) and references therein). Finally, we recall how computing Polya densities of order 2 from their Laplace transforms. Let $\Im$ denote the class of entire functions

$$\Im = \{\varphi(s) = Cs^k e^{\delta s} \prod_{i=1}^{\infty} (1 + \lambda_i s) | \varphi(0) = 1, \delta \geq 0, k \in \mathbb{Z}_+, \lambda_i \geq 0, \sum_{i=1}^{\infty} \lambda_i < \infty\}. \tag{2.3}$$

A neccessary and sufficient condition that a density function f for which $f(u) = 0$, for $u < 0$ be PF_2 is that the reciprocal of its Laplace transform $\phi(s)$ be an entire function in the class $\Im$ with $\sum_{i=1}^{\infty} \lambda_i > 0$. Then consider

$$\frac{1}{\phi(s)} = \varphi(s) = Ce^{\delta s - \gamma s^2} s^r \prod_{i=1}^{\infty} (1 + a_i s) e^{-a_i s} \tag{2.4}$$

where γ, δ, a_i and C are real, $\gamma > 0$, and r is a nonnegative integer. Assume that $\phi(s)$ is positive for $\alpha < s < \beta$ where α and β denote consecutive zeros of φ. By the Laplace inversion formula, for $\alpha < \sigma < \beta$

$$h(x) = \frac{1}{2\pi i} \int_{\sigma - i\infty}^{\sigma + i\infty} e^{xs} \phi(s) ds \tag{2.5}$$

and $h(x) = Ce^{\sigma x} f(x)$, where f is a PF_2 density and C is a positive normalizing constant.

3. Log-concavity of warranty reserves

In this section, we obtain the main results on the log-concavity of the behaviour of some warranty expenses' measures. Warranty assures the buyer that the manufacturer will either repair or replace items that do not perform satisfactorily or refund a fraction or the whole of the sale price. This is an important issue in terms of pricing the items. Different policies have been defined in the literature, with criteria as the coverage (or remaining warranty time from the time of the item is replaced, leading to renewing or nonrenewing warranties); the dimensions of the warranty (age and usage of the item); the form of refund or reimbursement to the customer on the failure of the item or dissatisfaction with service; the requirements for the manufacturer, and other diverse factors. For a comprehensive study, we may refer to Blischke and Murthy (2000), Chapter 17. The warranties determined from the reimbursement's point of view, can be classified in: 1. A lump-sum rebate, e.g., 'money-back guarantee', which is usually assigned for a relatively small time interval immediately after the purchase of the item or service. 2. A free repair of the failed item. It is called a free repair warranty (FRW). 3. A repair provided at reduced cost to the buyer. The cost reduction is usually a decreasing function of the time to failure. It is called pro-rata warranty (PRW). 4. A combination of the preceding terms that starts with an FRW up to a specified time and switches to repair at pro-rated cost for the remainder of the warranty period. It is called FRW/PRW. On the other hand, there are two types of warranty from the coverage's point of view: 1. Non-renewing warranty (NR): A newly sold item is covered by a warranty for some calendar time of duration w, called the warranty period, which usually starts ate the time of the purchase of the item. For this policy, during the warranty period, the warrantor assumes all, which is a non-renewing free repair warranty (NRFRW), or a portion of the expenses which is a non-renewing pro-rata repair warranty (NRPRW), associated with the failure of the item. 2. Renewing warranty (R): The warrantor repairs any faulty item from the time of the purchase up to time w, at the time of each repair within an existing warranty, the item is warranted anew for a period of lenght w, and the warranty coverage expires when the lifetime of the item (the original one or its repaired version) exceeds w. It can be RFRW or RPRW.

Recall again that under a NRPRW policy, the manufacturer is required to refund a fraction of the sale price on failure of an item in the warranty period, from the time of the initial purchase. To carry this out, the manufacturer must set aside a fraction of the sale price, that constitutes a warranty reserve. Formally, let w denote the length of the warranty period, c denote the unit sale price, l denote the item's life-cycle, and $\chi(t)$ denote the sales rate (sales per unit time), for $0 \le t \le l$. Let T be the failure time of the item, with probability density function h and distribution function H. Furthermore, we assume that $\chi(t)$ is a measurable log-concave function for $0 \le t \le l$. Consider that the item is sold with a NRPRW policy with linear refund function, such that the expected refund rate (the amount refunded per unit time) at t is given by

$$v(t) = c\int_b^t \chi(x)\frac{t-x}{w}h(t-x)dx \tag{3.1}$$

where $b = \max\{0, t-w\}$, for $0 \le t \le w+l$. We assume that $\int_0^{w+l}\chi(x) = 1$, then χ can be considered the probability density function of a random variable related to the sales intensity.

The log-concavity of the expected refund rate v in (3.1) with respect to $(0, w+l)$ for an item can be stated when χ is a measurable log-concave function. Formally, consider the failure time T of an item with log-concave probability density function h. Assume that a NRPRW policy is applied for the item with warranty time w, life-cycle l and sales rate $\chi(t)$, for any $0 \le t \le l$. If

χ is a measurable log-concave function, then the expected refund rate v in (3.1) is log-concave in $(0, w+l)$. Therefore, the previous result means that the log-concavity of this density function implies the log-concavity of the expected refund rate.

Given an item with lifetime T with a NRPRW policy as above with expected refund rate v given in (3.1) for any $t \in (0, w+l)$, then the expected total reserve needed to service the warranty over the item's life-cycle is given by

$$ETR = ETR(w+l) = \int_0^{w+l} v(t)dt. \quad (3.2)$$

From the earlier result, and from Prekopa's theorem, and under the same assumptions, we obtain that the expected total reserve to service the warranty over the item's life-cycle is log-concave in $w+l$. Furthermore, some upper bounds for the expected refund rate and the expected total reserve to service the warranty over the item's life-cycle can be stated.

Given a nonrepairable item that is sold under a non-renewing free replacement warranty (NRFRW) policy, the manufacturer is required to replace all items that fail within the warranty period w. The demand for spares in the interval $[t, t+\delta t)$ is due to failure of items sold in the period $[b, t)$ where $b = \max\{0, t-w\}$, and is useful for production and inventory control. With the same notation for the parameters of the earlier warranty expenses's measures, again let $\chi(t)$ denote the sales rate (sales per unit time), for $0 \leq t \leq l$; let T denote the failure time of the item, with probability density function h and distribution function H, and assume that $\chi(t)$ is a measurable log-concave function for $0 \leq t \leq l$. The expected demand rate for spares at time t is given by

$$\rho(t) = \int_b^t \chi(x)m(t-x)dx \quad (3.3)$$

where m is the renewal density function associated with a renewal process with inter-arrival times that have distribution function H that is defined by

$$m(t) = h(t) + \int_0^t m(t-x)h(x)dx. \quad (3.4)$$

The expected total number of spares to service the warranty over the item's life-cycle is given by

$$ETS = l + \int_0^{w+l} \rho(t)dt. \quad (3.5)$$

We provide similar conclusions, i.e, log-concavity, as in the main result in this Section, under NRFRW policy, for the demand for spares, the expected total number of spares required to service the warranty over the item's life-cycle, the demand for repairs, and the total expected demand for repair over the warranty period.

The performance of warranty policies is an important issue in terms of pricing the items for quality improvement. From our results on log-concavity that are described above, we derive some consequences on ranking items' prices, and production and inventory control.

4. References

[1] Blischke, W.R. and Murthy, D.N.P. 2000. *Reliability. Modeling, Prediction and Optimization*. Wiley, New York.

[2] Eaton, M.L. 1987. *Lectures on Topics in Probability Inequalities*. CWI Tract 35. Centrum voor Wiskunde en Informatica, Stichting Mathematics Centrum. Amsterdam.

[3] Karlin, S. 1968. *Total Positivity*. Stanford University Press. Stanford.

[4] Marshall, A.W. and Olkin, I. 1979. *Inequalities: Theory of Majorization and Its Applications*. Academic Press, Inc. New York.

[5] Prekopa, A. 1973. On logarithmic concave measures and functions. *Acta Scient Mat*, 34, 335-343.

Journal of Physics: Conference Series **574** (2015) 012019 doi:10.1088/1742-6596/574/1/012019

Transient and Post-irradiation Response of Optoelectronic Devices to Ionizing Radiation

F A A Haider[1] and **F P Chee**[2]
[1,2] Department of Physics with Electronics, Faculty of Science and Natural Resources, University Malaysia Sabah, Kota Kinabalu, Malaysia

E-mail: haider@ums.edu.my[1], fpchee06@ums.edu.my[2]

Abstract. Space and ground level electronic equipment with semiconductor devices are subjected to the deleterious effects by radiation. This paper is attempted to present the transient and post-irradiation response of optoelectronic devices to gamma (γ) rays utilizing cobalt-60. In situ measurements were made on the devices under test (DUTs) up to a total dose of 60 krad followed by a post-irradiation not in-flux test for eight hours. Current transfer ratio (CTR) with is the vital merit of the optoelectronic system is found to decrease remarkably with the absorbed dose. This degradation is induced by the interaction of the energetic photons from gamma rays via two main mechanisms. The dominant effect is the mechanism by ionization while the secondary is by displacement. This radiation effect is found to arouse either a permanent or temporarily damage in the DUTs depending on their current drives and also the Total Ionizing Dose (TID) absorbed. The TID effects by gamma rays are cumulative and gradually take place throughout the lifecycle of the devices exposed to radiation. The full damage cascade phenomenon in the DUTs is calculated via the simulation.

1. Introduction

Studies for introduction of atoms into a solid substrate by bombardment of the solid with ions in the electron-volt (eV) to mega-electron-volt (MeV) energy range have always received great interest [1]. These interests have been stimulated by the feasibilities of synthesizing novel materials with potential applications in the semiconductor, mineral processing and metallurgy, corrosion and optical fields. Moreover, operating environment of the equipment and devices for the technologies on weather forecasting, remote sensing, global navigating position, satellite television and data broadcasting, telecommunication system as well as our surveillance and army based national security technologies are far from stable and not without interference [2, 3]. The physical properties for a particular solid substrate are always sensitive to the presence of a trace amount of foreign atoms. Mechanical, electrical, optical, magnetic and superconducting properties are all affected and sometimes may even be dominated by the existence of such foreign atoms [4, 5].

Gamma rays are photons which are electromagnetic radiation with zero mass, zero charge, and travel with the speed of light, *c*. Since these photons are electrically neutral, they do not steadily lose energy via columbic interactions with atomic electrons as what happened in charged particles. Instead, they travel some considerable distance before undergoing an interaction. Penetration of

[2] To whom any correspondence should be addressed.

gamma rays can be governed statistically via interaction probabilities, per unit distance travelled, which depend on energy and material. This probability is known as the linear attenuation coefficient and has the dimensions of inverse length (eg. *cm-1*). Gamma rays have higher penetration intensity compared to the charged particles of similar energy. Gamma rays can interact with matters primarily in three processes. These are the photoelectric effect, Compton scattering, and pair production [6]. Therefore, the semiconductor device that is exposed to an ionizing radiation environment typically endures degradation in one or more on its performance parameters.

The parameters that give effect to the formation of radiation damage are the ion mass, the target temperature during irradiation, ion flux (number of implanted ions per unit area and time), the ion's energy and the ion's fluence (number of ions per unit area) [7]. The energy transferred to the crystal lattice is strongly determined by the ion mass. The ions can lose their energy by two ways which are the electronic energy loss through the excitation of electronic systems and nuclear energy loss through nuclear collisions with the target atoms. The incident ions will finally come to rest at a certain depth, z.

This paper presents the analytical and numerical changes in optoelectronic devices as a result of gamma radiation. Since the previous study which was modelled on the optoelectronic material had been presented [8, 9] and therefore this paper will further the simulation study to damage extents at different energy level in the structure of the device.

2. Research Methodology

Basically, the research work was divided into two main parts; the first part was the test of gamma radiation effects on the optoelectronic devices while the second part was the Monte Carlo simulation of the resulting mechanism in the model of the DUTs.

2.1. Specialized Devices

Optoelectronic devices consisting of a Plastic IRLED (QEE113) coupled to a Plastic Silicon Infrared Phototransistor (QSE113) is being studied in this research. The QEE113 is a 940 NM GaAs LED, encapsulated in a medium wide angle and clear epoxy plastic side looker package. The QSE113, however, is an NPN silicon phototransistor encapsulated in a wide angle, infrared transparent and black plastic side looker package.

Optoelectronic devices are being studied as they play an important role in optical satellite telecommunications and sensing technology. They are widely used especially in providing electrical isolation between circuits such as a sub-system to sub-system interfaces for space light designing's. These photonics systems are ideal for applications in space system due to their high bandwidth and speed of operation, the immunity to electromagnetic interference, low power consumption and cost, yet, for high reliability. Thus, the understanding of the processes which leads to device failure is crucial in order to meet the radiation hardness requirements.

For the IRLED and phototransistor, two of them are electrically isolated. The phototransistor is a photodetector which is capable of converting the received infrared light into current or voltage. Therefore, the current change in the phototransistor is depending on the forward current, I_F, of the IRLED. In this study, I_F is set at 25, 50, 75 and 100 mA . The I_F of the IRLED can be raised up to 1000 mA, yet, for ideal performance of this device, the I_F range is 10 to 100 mA.

2.2. Radiation Exposure

The radiation testing on the electronic devices consist of multi-parameter test with different exposure levels. Ambient temperature throughout the test was 25±3°C. Before the irradiation process, a control test of 72 hours is performed on the DUTs. This is known as the pre-irradiation testing. Only devices which are within the specific working range and the testing results which reveal that they are within the standard deviation of 2 sigma are submitted for radiation testing. The radiation procedure was referred from the Military Standard (MIL-STD) 883G Method 1019.7, Ionizing Radiation (Total Dose) Test Procedure [10].

The output collecting current, I_C of the phototransistor is measured for bias forward current (I_F) of 25 mA, 50 mA, 75 mA and 100 mA. The schematic of experimental setup for the optoelectronic device is as shown in Figure 1.

The input voltage of the particular device could be changed from a distance of approximately 15 m in a control room during irradiation and the effect could be observed directly using in-situ method. The information and status of the device under test (DUT) will be transmitted through the driver circuit based on an ADC circuit into the PC. The temperature dependence of the charge deposition in the sampling device would be monitored with an electronic thermometer using LM35 precision centigrade temperature sensor. All the data presented to the computer will then be processed and analyzed by the MATLAB program.

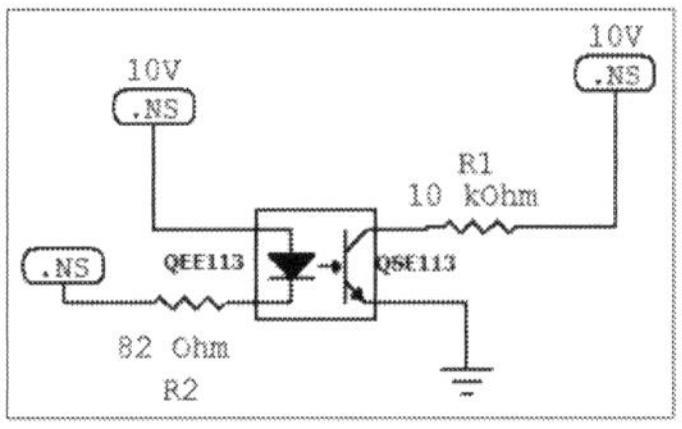

Figure 1. Circuit of experimental setup for the optoelectronic device.

Exposure on the DUTs is utilizing the cobalt-60 (^{60}Co) gamma rays from RAYMINTEX Plant, Malaysian Nuclear Agency. Gamma rays are substantial portion of the photon output of a nuclear detonation [11]. Fissions occurring in the nuclear detonation will provide prompt gamma rays and X-rays that are expelled from the burst as part of the menagerie of particles and radiation energy. Gamma rays are also emitted from radioactive fission fragment components of the bomb fuel debris, as well as from the neutron activated atomic nuclei of weapon [12]. The nature and behavior of gamma rays are in marked contrast to those of neutrons. They have zero rest mass which means that their existence ceases by annihilation when they are brought to rest, as when absorbed by matter [13].

The DUTs are exposure up to a total dose 60 krad. This value is chosen with reference to the NASA report from Poivey [14]. In this research paper, a more accurate estimation of the radiation level using 3D Monte Carlo code is utilized to calculate the total dose level received by the part at geostationary orbit. According to the analysis, the highest 15 years dose level simulated from Monte Carlo code is 24.0 krad whereas those simulated from sectorial analysis- shell sphere/norm model and solid sphere/slant model is 28.5 krad and 18.6 krad respectively. From this data, the total dose of 60 krad is sufficient to cater to a most stringent condition in the proposed space environment.
The dose rate at the selected point could be determined based on Equation 1.

$$DoseRate(rad/s) = \frac{DoseAbsorbed(rad)}{ExposureTime(s)} \tag{1}$$

Since the dose rate at that particular point is known, the exposure time to radiate the DUT up to a total absorbed dose of 60 krad can be calculated. This special configuration and positioning of the coordinated table mode irradiation controlled the distance between the DUT and ^{60}Co sources. This allows equal exposure of radiation at a pre-determined constant dose rate. The dose rate of the activity shows a variation at different time due to the half-life of ^{60}Co. However, this variation is not significant for tests which were conducted at a same period.

2.3. Monte Carlo Simulation
Monte Carlo method is a class of computational algorithms that rely on repeated random sampling and probability statistics to compute their results [15]. This method is ideal for calculation by a computer and tends to be used when it is impracticable to compute an exact result with a deterministic

algorithm. It is applicable in ion-solid interactions and has a number of distinct advantages over analytical calculations based on transport theory [16].

The SRIM-TRIM simulation results are important as they explain the transport of the ion in the target at different energy levels, whereas for the real irradiation exposure, the energy of the incidence ion is constant. Besides, this simulation is used to calculate the range of ions in matter using collisions of ions-atoms. It also can calculate the 3D spreads of ions as well as all the kinetic phenomena that are related to the loss of energy: damage to the target, sputtering, ionization and phonon production. The enumeration of the range of ions in matter and the damage event in the target during the slowing-down process can be done using TRIM [17].

This paper will discuss and analyze the interaction between ^{60}Co ions with the optoelectronic device at different energy level using the SRIM-TRIM. The energies of the incidence ^{60}Co ions are varying from 100 keV up to 10 MeV. The number of incidence ions for each simulation is 2,000 ions.

By the simulation of the detailed calculation with full damage cascades, the plots of ion trajectories, depth vs. Y-Axis, depth vs. Z-Axis, transverse view, ionization, phonons, collision events, atom distributions and energy to recoil can be obtained.

3. Results and Discussions

Current Transfer Ratio, *CTR* is defined as the ratio of I_C to I_F. The rate of change in the current transfer ratio, $\frac{\text{CTR}}{\text{CTR}_0}$ on the optoelectronic device tested is being monitored and recorded. Figure 2 shows the relationship between $\frac{\text{CTR}}{\text{CTR}_0}$ of optoelectronic devices with the total dose absorbed. CTR_o is the reading at pre-irradiation whereas CTR_1 is the reading after being exposed to irradiation.

Drop off for the I_C parameter results in a linearly decreasing *CTR* across the gamma radiation period for different forward biased current applied at a step rate of 25 mA. From this graph, it is illustrated that a higher operating I_F has led to lesser depletion in the *CTR* of the optoelectronic devices.

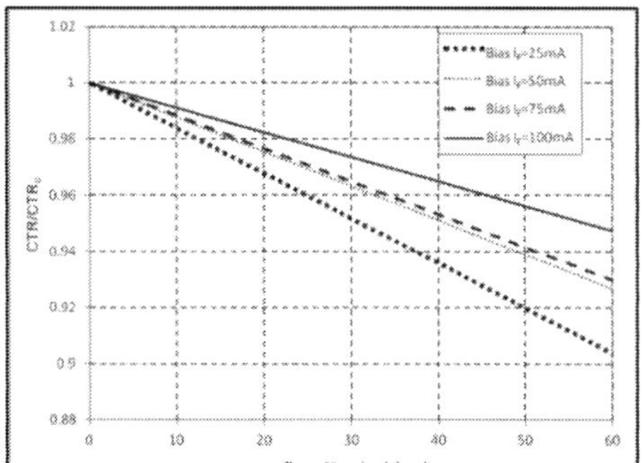

Figure 2. $\frac{\text{CTR}}{\text{CTR}_0}$ during irradiation for different bias I_F

The line graph in Figure 3 indicates the $\frac{\text{CTR}}{\text{CTR}_0}$ of optoelectronic devices during the post-irradiation. After the removal of the radiation, the *CTR* recovers gradually for different forward biased current applied at a step rate of 25 mA. In addition, it is observed that the greater percentage of increase in CTR occurs at a lower bias I_F. The percentage of recovery at post-irradiation is lesser at I_F = 100 mA because the high I_F yields a saturation condition to the operating system.

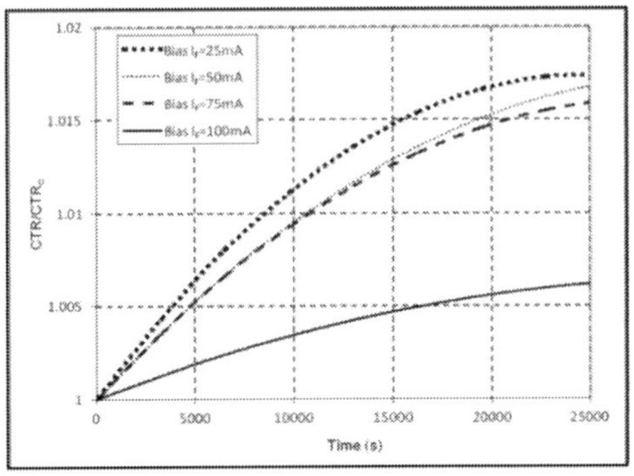

Figure 3. $\dfrac{CTR}{CTR_0}$ at post-irradiation for different bias I_F.

When a flux of energetic ^{60}Co ion traverses the target of thickness dx and containing a total of N target atoms per unit volume, transferring of energy occur during this scattering event. The product $\sigma N dx$ represent the total fraction of the target surface area which acts as an effective scattering center to the ^{60}Co ion. The probability of projectiles with energy E undergoing a scattering event or a collision with a target nucleus while traversing a thickness dx is as shown in following:

$$P(E) = N\sigma(E)dx \qquad (2)$$

The total cross-section gives a measure of the probability for any type of collision to occur where energy E transfers are possible. The software simulation SRIM-TRIM has simulated the trajectories and collision evernts of ^{60}Co ion in LED and phototransistor based on this theory.

Figure 4 to Figure 7 show the trajectories at incidence energy level, E_{Co} = 500 keV and 10 MeV respectively. The red dots indicate the distribution of ^{60}Co ion while travelling through the target atom which leads to lattice displacement. The other colours of dot clusters, however, indicate the displacement of target atom which forms a vacancy due to the recoiling of the other atoms. At E_{Co} = 500 keV, the lattice damage is barely on the surface of the target. As the energy is increased, the projected range started to increase and cause a deeper penetration into the device simulated.

A high energy particle loses its energy in small steps through interactions with electrons in the materials. However, once these particles have lost enough energy, the nuclear collision tends to dominate. As the particle slows down, it captures electrons to form a neutral atom. The high energy particle slows down almost entirely due to Coulomb interactions with the atomic electrons. Due to the large numbers of interaction, the slowing down procedure was nearly continuous and along a straight - line path. The trapped by defects in the optical medium, giving rise to more stable secondary defects and leads to the formation of colour centers. The creation of colour centers absorbs signal photons and therefore degrading the light transmission efficiency. This will result in a decrease in the phototransistor output current for current transfer applications.

The operation of these optoelectronic devices is based on the creation or annihilation of electron-hole pairs. When the energetic particles that interact with the optoelectronic device impart energy which is equal or more than the bandgap energy to a valence band electron, pair production can occur. The bandgap energy of the GaAs semiconductor material is 1.424 eV. This pair production phenomenon can raise an electron in the valence band to the conduction band, leaving a hole behind the valence band. There is a probability that a reverse process might occur, that of electron and hole recombination, is associated with the pair releases all its excess energy after recombination. The process that occurs might be radiative recombination or non-radiative recombination. Radiative recombination is important for the luminescent process. The non-radiative recombination, however, can reduce the radiative efficiency of the optoelectronic material.

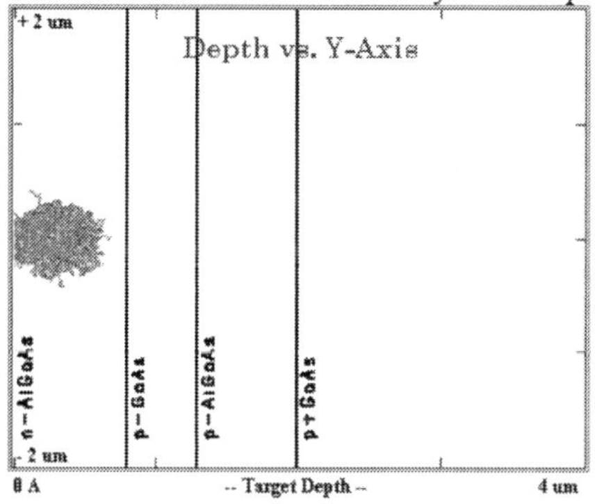

Figure 4. Plot of depth versus Y-axis at E_{Co} = 500 keV in LED

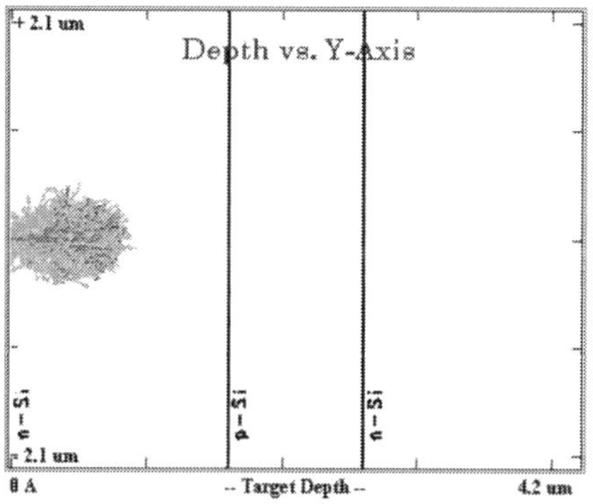

Figure 5. Plot of depth versus Y-axis at E_{Co} = 10MeV in LED.

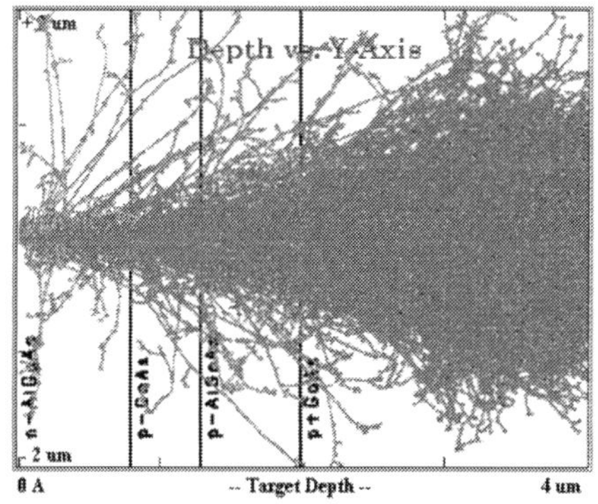

Figure 6. Plot of depth versus Y-axis at E_{Co} = 500 keV in phototransistor.

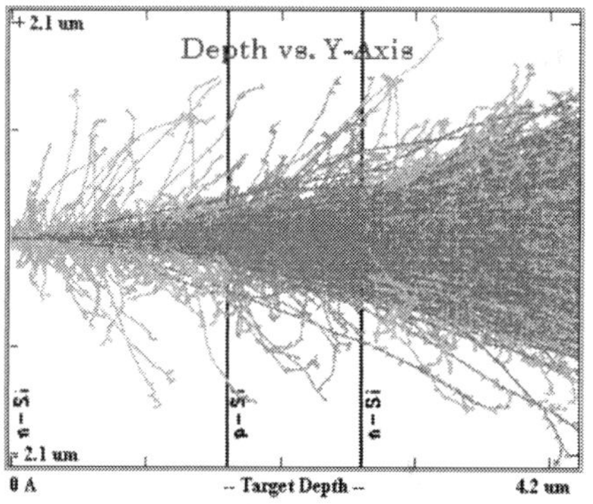

Figure 7. Plot of depth versus Y-axis at E_{Co} = 10 MeV in Phototransistor.

IRLEDs emit light by means of radiative recombination of injected minority carriers with majority carriers in the depletion region. The light output is proportional to the radiative recombination current and consequently an increase of the non-radiataive recombination current causes the degradation in the electrical performance of IRLED. This in turn causes degradation in the output current of the phototransistor. However, as IRLED exhibits short minority carrier lifetime as well as high current densities, thus, it exhibits higher radiation hardness.

Operating parameters can affect the level of degradation of optoelectronic devices in a given application. For the IRLED operating at a high current (I_F =100 mA), the degradation formed at both the I_C and I_F are found to be minimized significantly. This is as when the IRLED is operated at high I_F, a certain amount of power that dissipated in the resistive part of the device leads to the self-heating effect.

4. Conclusion

When significant *CTR* degradation is anticipated for a specific mission application, the effect of degradation can be eventually alleviated in certain cases. This can be done by decreasing the rate of *CTR* or by regulating the application bias condition in order to reduce the harshness of the degradation. In other words, this anticipation is useful in the modelling of the device and also circuit degradation in a radiation environment, without the knowledge of the exact details of the microscopic defects formed during the exposure.

From the results obtained, it is found that the degree of defect and damage produced in the optoelectronic devices highly depend on their operating current and also its mode of operation. The ionization and displacement damage induced by the incident photons have a bigger impact at these devices of low operating currents which is related to the self-annealing effect of the devices.

5. Reference

[1] Avasthi D and Mehta G 2011 Engineering of Materials by Swift Heavy Ion Beam Mixing *Springer* 86-108
[2] Haider F A A and Chee F P 2013 Evaluation on Diffusion of Bipolar Junction Transistor (BJT) Charge-Carrier and its Dependency on Total Dose Irradiation *Advanced Material Research* **701** 71-76
[3] Chee F P, Haider F A A and Saafie S 2014 Defect Generation in Bipolar Devices by Ionizing Radiation *IOSR J. App. Phys.* **6(3)** 92-101
[4] Nastasi M. and Mayer W J 2006 Ion Implantation and Synthesis of Materials *Springer*
[5] Ng K K 2002 Complete Guide to Semiconductor Devices *John Wiley & Sons*
[6] Murray R L 2008 Nuclear Energy: An Introduction to the Concepts, Systems, and Applications of Nuclear Processes *Elsevier*
[7] Wendler E 2009 Mechanisms of Damage Formation in Semiconductors *Journal of Nuclear Instrument and Methods in Physics Research B* **267** 2680-2689
[8] Chee F P Haider F A and Saafie S 2011 Simulation of ionizing radiation induced effects in nanoscale semiconductor material *Micro and Nanoelectronics (RSM), 2011 IEEE Regional Symposium* (IEEE) pp 245-249
[9] Haider F A A and Chee F P 2012 Monte Carlo Simulation of Alpha (α) Particles Penetration in Nanoscale Silicon Semiconductor Materials *Advanced Material Research* **528** 237-240
[10] Department of Defense 2006 MIL-STD-883 Test Method 1019.7 Ionizing radiation (Total Dose) Test Procedure. *Test Method Standard Microcircuits*
[11] Lewis J S 2004 Astronomical perspective *International Geophysics Academic Press* **87** 7-49
[12] Medalia J 2010 Detection of Nuclear Weapons and Materials: Science, Technologies, Observations. *Congressional Research Service*
[13] Buenker R J 2005 Evidence from the Special Relativity and Blackbody Radiation Theories for the Existence of Photons Possessing Zero Kinetic-Energy *Molecular Physics* **76 (2)** 277-291
[14] Poivey C 2009 Radiation Hardness Assurance for Space System *NASA GSFC*
[15] Corke P I 2011 Robotics, Vision and Control: Fundamental Algorithms in MATLAB *New York: Springer Tracts in Advanced Robotics*
[16] Nastasi M, Mayer W J and Hirnoven K J 2004 Ion-solid Interaction Fundamentals and Applications *UK: Cambridge University Press*
[17] Ziegler J F, Ziegler M D and Biersack J P 2010 SRIM-The Stopping and Range of Ions in Matter *Nuclear Instruments and Methods in Physics Research B* **268** 1818-1823

Acknowledgments
The authors are thankful to Fundamental Research Grant Scheme (FRGS) 2013, Project No.: FRG0318-SG-1/2013, with title "Evaluation on Diffusion of Charge-Carrier in Semiconductor and nanostructure Devices and its dependency on Nuclear Radiation".

Computational Simulation of Semiconductor Laser with Optical Injection

G Baimensina[1]
Kazakh National Technical University, Department of Applied Physics, Kazakhstan

E-mail:gulnaz.baimenshina@sdu.edu.kz

Abstract. In this paper we study the dynamics of a semiconductor laser with optical injection. The time behaviour of solutions of a system of three coupled nonlinear rate equations, describing the electric field amplitude and the carrier concentration and the phase difference within the resonator, is discussed both qualitatively and numerically. We then concentrate on the periodic orbits that emanate from Hopf bifurcations. Depending on the injection strength and the phase difference two types of oscillations can be found, such as relaxation and periodic oscillations

1. Introduction

The nonlinear processes plays important role in the development of new communication systems, computing elements. Understanding of laser instabilities is necessary for developing control techniques. From a dynamical systems point of view semiconductor laser systems are very attractive because they show an intriguing variety of complicated dynamics. Lasers including semiconductor, solid-state with the additional external perturbation produce temporal and spatial instability. An optical injection is an example of such perturbation. External perturbations may destabilize or induce the sustain intensity oscillation.

The first unstable fluctuations instead of continuous wave emission were observed in a continuously pumped maser in 1958[1]. The connection between laser and instabilities was published in 1975 by Haken [2]. It was shown that a set of nonlinear differential equations from Maxwell- Bloch equations for a laser model resembles the Lorentz equations that are the basic model for deterministic chaos. Instabilities in lasers have become broad subject in quantum optics. In the present paper we consider the laser rate equations with an optical injection. One of the first studies on the effects of the laser with optical injection was carried out in 1992 [3]. Since then, several papers have discussed the instabilities associated with optical injection [4-8]. The recent books chapters [9-11] give an excellent overview of the dynamical instabilities in lasers.

The objective of this work is to investigate the effects of main parameters, such as detuning and the injection level, on dynamical behavior of laser system using numerical and nonlinear dynamics methods.

This paper is organized as follows. In Section 2 and 3 we consider the model of semiconductor laser with optical injection based on the rate equations and its linear stability analysis. Section 4 discusses

[1] gulnaz.baimenshina@sdu.edu.kz

the numerical results and compares them with obtained analytical results. The time series and phase portraits are plotted for the solutions of the given system. Section 5 concludes the paper.

2. Model

The emission in semiconductor lasers results from electron – hole recombination between energy bands. The dynamics of the semiconductor laser model with optical injection can be studied by three-dimensional set of ordinary differential equations. These equations describe the electric field and the number of electron-hole pairs in the laser active medium. The semiconductor laser is injected with light from an external laser source (master laser). The laser that is influenced by the master laser is called the 'slave' laser.

The rate equations that describe the dynamics of the slave semiconductor laser in response to the injected signal from a master laser can be written as [9]:

$$\frac{dR}{dt} = ZR + \eta\cos\psi \tag{1}$$

$$\frac{d\psi}{dt} = \Omega - \alpha Z - \eta\sin\psi / R \tag{2}$$

$$T\frac{dZ}{dt} = P - Z - (1+2Z)R^2 \tag{3}$$

, where R- the electric field amplitude, Z- carrier concentration over threshold, , α - the linewidth enhancement factor, Ω - detuning frequency (the frequency difference between the master and the slave lasers), T- is the ratio of the photon lifetime to the carrier lifetime, P - pumping current, η - the injection strength, ψ -the phase difference.

Thus, equations (1), (2) and (3) simulate the evolution of the amplitude of the electric field, phase difference, carrier concentration respectively.

Solutions of nonlinear differential equations depend on their parameters. There may be critical parameter values, at which the character of the solution changes completely. For this reason, it is extremely important to know the qualitative behavior of the solutions before starting a numerical calculation.

Let us determine the stability of the steady states solution of the system.

The steady state solutions for R_0, Z_0 and ψ_0 have the following implicit form:

$$R_0^2 = (P - Z_0)/(1+2Z_0) \tag{4}$$

$$Z_0^3(1+\alpha^2) + Z_0^2(-P-\alpha^2 P - 2\alpha\Omega) + Z_0(2\eta^2 + 2\alpha\Omega P + \Omega^2) + \eta^2 - P\Omega^2 = 0 \tag{5}$$

$$\eta\sqrt{1+\alpha^2}\sin(\psi_0 - \arctan\alpha) = \Omega R_0 \tag{6}$$

Steady state solution of the cubic equation (5) Z_0 can be calculated, and then we can obtain the corresponding solutions for R_0 and ψ_0.

To analyze equilibrium points, let's linearize the system (1), (2), (3) near equilibrium points.

The Jacobian matrix of the system (1), (2), (3) near equilibrium point is given by

$$|J| = \begin{vmatrix} Z_0 & R_0(\alpha Z_0 - \Omega) & R_0 \\ \dfrac{\Omega - \alpha Z_0}{R_0} & Z_0 & -\alpha \\ -2\gamma(1+2Z_0)R_0 & 0 & -\gamma(1+2R_0^2) \end{vmatrix}, \tag{7}$$

Here γ=1/T. The characteristic polynomial of matrix (7) has the following form:

$$p(\lambda)=\left|J-\lambda I\right|=\begin{vmatrix} Z_0-\lambda & R_0(\alpha Z_0-\Omega) & R_0 \\ \dfrac{\Omega-\alpha Z_0}{R_0} & Z_0-\lambda & -\alpha \\ -2\gamma(1+2Z_0)R_0 & 0 & -\gamma(1+2R_0^2)-\lambda \end{vmatrix} \quad (8)$$

From (8):

$$-p(\lambda)=\lambda^3+C_1\lambda^2+C_2\lambda+C_3, \quad (9)$$

where the coefficients C_1, C_2, C_3 are equal to the following expressions:

$$C_1=\gamma\frac{1+2P}{1+2Z_0}-2Z_0$$

$$C_2=Z_0^2+(\Omega-\alpha Z_0)^2+2\gamma(P-Z_0)-2\gamma Z_0\frac{1+2P}{1+2Z_0}$$

$$C_3=\gamma Z_0^2\frac{1+2P}{1+2Z_0}+2\gamma\alpha(P-Z_0)(\Omega-\alpha Z_0)-2\gamma Z_0(P-Z_0)+\gamma\frac{1+2P}{1+2Z_0}(\Omega-\alpha Z_0)^2.$$

The eigenvalues of the polynomial (9) can be calculated from the following characteristic equation) to identify the type and stability of the equilibrium points:

$$-p(\lambda)=\lambda^3+C_1\lambda^2+C_2\lambda+C_3=0 \quad (10)$$

According to the theorem of Hurwitz Eq.(5) has only roots with negative real part, which implies asymptotic stability of Z_0, if and only if $C_1>0$, $C_3>0$, $C_1C_2-C_3=0$. The condition $C_1C_2-C_3=0$ implies that a complex pair of eigenvalues is crossing over the imaginary axis, so that the Hopf bifurcation is occurred [9].

3. Discussion and results

Let's now examine how the change in the main control parameter, such as injection strength and detuning frequency affects the system dynamics behaviour. Using the approximate numerical methods, in particular the Runge - Kutta fourth order method, we present the results of numerical solutions of the rate equations (1), (2), (3) and the corresponding phase portraits.

Consider the values of the system parameters Ω = -0.1, η = 0.01. In this case, the cubic equation (5) has one real root or one equilibrium point of the original differential equations. The solution is unstable, since the real parts of the eigenvalues of a complex - conjugate pair of the characteristic polynomial (10) are positive. The phase space trajectory and time responses confirm these analytical results (see Fig. 1) as well.

Raise the value of the injection η = 0.03. In this case the cubic equation has three real roots. According to the eigenvalues of a polynomial, two equilibrium points are stable whereas the third point is unstable. The saddle-node bifurcation is occurred. The system performs the relaxation oscillations.

Consider the value of the injection force η = 0, 04. For a given value of the parameter the pair of complex eigenvalues crosses the imaginary axis. There is a qualitative change in the phase portrait. The system goes to the periodical mode. The curve is a limit cycle in the phase space of R-ψ-Z. The Hopf bifurcation is occurred.

Let us raise the value of the injection to the value η = 0.9. The detuning parameter is left unchanged. The saddle node bifurcation is occurred for given values of the parameters. The stability of the equilibrium point turns back.

Consider the positive values of the detuning Ω, in particular Ω = 0.1 and η = 0.01. In this case, the system has one unstable solution. The real parts of the eigenvalues of a complex - conjugate pair of the characteristic polynomial are positive. The time series and phase portraits for positive detuning value are plotted in Fig. 2.

Let's see what happens if we raise the value of η = 0.04. According to their eigenvalues all three solutions are unstable again. The reason is that in this case, the leading coefficient of characteristic equation for positive values of detuning is always negative. (C_1 =-1.9984 < 0). According to the criterion of Routh – Hurwitz, this condition is not satisfied with the condition of stability of equilibrium point. Thus, the transitions between relaxation and periodic oscillations (see Fig. 1) can occur only for negative detuning values.

4. Conclusions

We have studied the model of laser system with optical injection using numerical analysis. The results from numerical calculations show different types of laser dynamical behaviors. The transitions from relaxation oscillations to a state of periodic oscillations are observed. The behavior of system is sensitive to variations in the main parameters. A slight raise in the parameter η leads to a qualitative change in the behavior system dynamics. A fourth order Runge-Kutta method is used to plot the time integration and phase portraits of the system. It was observed that for certain parameters, raising the value of η can bring stability back. As a result of the simulation was also point out that the saddle node and Hopf bifurcations are occurred only for negative detuning values.

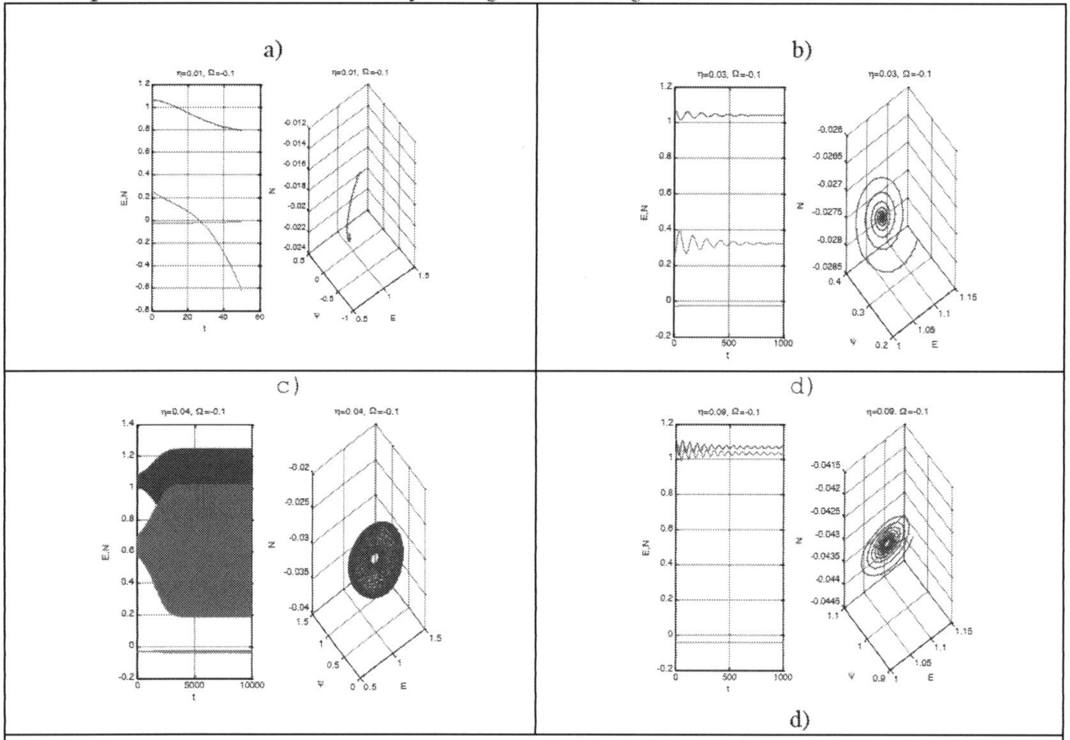

Figure 1. The time series and phase spaces of laser rate equations ($\alpha = 4$, P=1) for negative detuning.

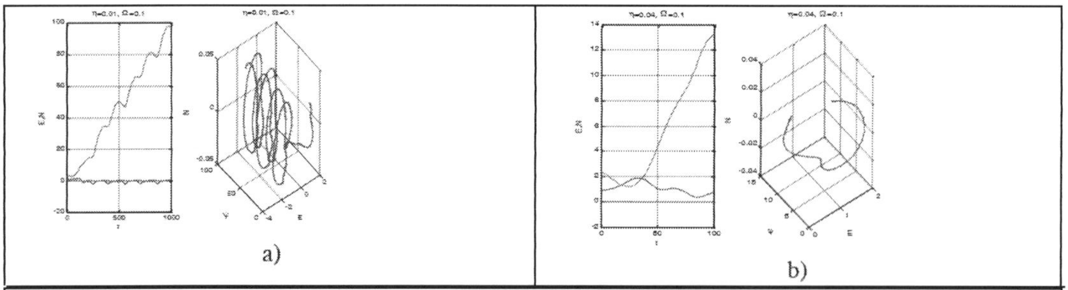

Figure 2. The time series and phase spaces of laser rate equations ($\alpha = 4$, P=1) for positive detuning

References

[1] Makhov G, Kikuchi C, Lambe J, Terhune R W 1958 *Phys. Rev.* 109 1399-1400
[2] Haken H 1975 *Phys Lett.* 53A 77–78
[3] Sacher J, Baums D, Panknin P, Elsasser W, Gobel E O 1992 *Phys. Rev.* A 45 1893-1905
[4] Lee E K, Pang H S, Park J D, Lee H 1993 *Phys.Rev.* A 47 736-739
[5] Annovazzi –Lodi V, Donati S, Manna M 1994 J.Quantum Electron. 30 1537-154
[6] Simpson T.B, Liu J M, Gavrielides A, Kovanis V , Alsing P M 1995 *Phys. Rev.* A 51 4181
[7] Kovanis V, Gavrelides A, Simpson T B, Liu J M 1995 *Appl.Phys.Lett.* 67 2780-2782
[8] Gavrielides A, Kovanis V, Erneux T 1997 Opt. Commun. 136 253-256
[9] Erneux T Glorieux P 2010 *Laser dynamics* (New York: Cambridge University press) p 217-234
[10] Uchida A 2012 *Optical communication with chaotic Lasers, Applications of Nonlinear Dynamics and Synchronization* (Weinheim: Wiley) p 59-143
[11] Ohtsubo J 2008 *Semiconductor lasers* (Berlin: Springer) p 147-176

Dependence of maxima in space

H. Ferreira and L. Pereira

Department of Mathematics, University of Beira Interior, Covilhã, Portugal

E-mail: helenaf@ubi.pt, lpereira@ubi.pt

Abstract. We evaluate the dependence among large values of a spatial process of maxima trough a coefficient that can be applied in natural, technical and societal extreme phenomena. Its main properties are: a) k locations can be taken into account; b) it takes values in $[0,1]$ and higher values indicate stronger dependence; c) it is independent of the univariate marginal distributions of the random field; d) it can be related with other coefficients in the literature such as the tail dependence and the extremal coefficients; e) it agrees with the concordance property for multivariate distributions; f) it has as a particular case the variogram from geostatistics; g) it can be easily estimated. We illustrate the application of the introduced coefficient with the estimation of the degree of dependence among large values of the amount of tritium in drinking water for tree cities.

1. Introduction

Unpleasant extremes of variables like wind, temperature and precipitation, can affect anybody at any place. This thought is applicable also to stock markets fluctuations, given the markets' s globalization.

Despite the negative connotation of extremes, who can imagine the world without extreme events like the Big bang, the extinction of some species, earthquakes or simply the hormonal chaos of passions? In short, a world where there was never large deviations from the average, the usual events or no novel opportunities?

Large fluctuations connected with phase transitions, critical phenomena or catastrophic events can be viewed as extreme or rare events. The mathematical tools for the modeling of the extremal behavior can be provided by the Theory of Extreme Values, a branch of the classical Statistics, where we find the properties of the max-stable processes and max-stable random fields.

Here we will focus on the dependence of the variables of max-stable random fields, which are the natural models for spatial extremes. They can be, for instance, good approximations for annual maxima of daily spatial rainfall and have been widely applied to real data in environmental, atmospheric and geological sciences. The Pearson correlation coefficient proves to be an inadequate measure in contexts of assessment of dependence in which the extreme values are important because it assigns equal weight to all observations when the structure of dependence between extreme observations may differ from other (Resnick (2007)).

We introduce a dependence coefficient, called multivariate variogram, to measure the dependence among extreme values of a spatial process of maxima. We justify its advantages by presenting the main properties and we illustrate its applications to real questions. In particular, we evaluate the dependence among large values of the amount of tritium (pCi/L) in drinking water, for three cities in Alabama State (USA). The first two are relatively close to a nuclear power plant in northern Alabama and its tritium data in drinking water show greater levels in the period 1997-2013. We estimate the dependence of the annual maxima at these locations throughout the estimation of the multivariate variogram.

2. Definition of the multivariate variogram

Here $\{Z(\mathbf{x}), \mathbf{x} \in \mathbb{R}^d\}$ denotes a random field of maxima. That means that, for each choice of $\mathbf{x}_1, ..., \mathbf{x}_k$, the distribution of $(Z(\mathbf{x}_1), ..., Z(\mathbf{x}_k))$ is a Multivariate Extreme Value (MEV) distribution $G_{\mathbf{x}_1,...,\mathbf{x}_k}$. Its tail dependence function

$$l_{\mathbf{x}_1,...,\mathbf{x}_k}(t_1, ..., t_k) = \lim_{u \downarrow 0} \frac{1 - G_{\mathbf{x}_1,...,\mathbf{x}_k}\left(G_{\mathbf{x}_1}^{-1}(1 - ut_1), ..., G_{\mathbf{x}_k}^{-1}(1 - ut_k)\right)}{u},$$

$(t_1, ..., t_k) \in \mathbb{R}_+^k$, characterizes fully the dependence among its marginals distributions $G_{\mathbf{x}_j}$, $j = 1, ..., k$ (Resnick (1987), Beirlant et al. (2004)). The superscripts -1 denote the inverses of the marginal distributions.

Since this function cannot be easily inferred from data, the dependence coefficients are useful, despite the fact that one coefficient cannot preserve all the information about this function.

Several dependence coefficients have been considered in order to resume the dependence among the marginals of $G_{\mathbf{x}_1,...,\mathbf{x}_k}$: extremal coefficients, tail dependence coefficients and madogram (Li (2009), Smith (1990), Schlather, M. and Tawn, J. A. (2003), Cooley et al. (2006), Naveau et al. (2009)).

Next definition introduces our proposal and then we explore its advantages and relations with the previous coefficients. In the following, $\bigvee_{j=1}^{k} a_j$ and $\bigwedge_{j=1}^{k} a_j$ denote the maximum and the minimum of $\{a_1, ..., a_k\}$, respectively.

Definition 2.1. *The variogram of* $(Z(\mathbf{x}_1), ..., Z(\mathbf{x}_k))$ *is the coefficient*

$$v(\mathbf{x}_1, ..., \mathbf{x}_k) = 1 - \frac{k+1}{k-1} E\left(\bigvee_{j=1}^{k} G_{\mathbf{x}_j}(Z(\mathbf{x}_j)) - \bigwedge_{j=1}^{k} G_{\mathbf{x}_j}(Z(\mathbf{x}_j))\right).$$

We remark that

$$v(\mathbf{x}_1, ..., \mathbf{x}_k) = 1 - \frac{k+1}{k-1} E\left(\bigvee_{\{i,j\} \subset \{1,...,k\}} | G_{\mathbf{x}_i}(Z(\mathbf{x}_i)) - G_{\mathbf{x}_j}(Z(\mathbf{x}_j)) |\right).$$

In particular, for two locations $\mathbf{x}_i$ and $\mathbf{x}_j$, it holds $v(\mathbf{x}_i, \mathbf{x}_j) = 1 - 6\nu(| \mathbf{x}_i - \mathbf{x}_j |)$, where

$$\nu(| \mathbf{x}_i - \mathbf{x}_j |) = \frac{1}{2} E\left(| G_{\mathbf{x}_i}(Z(\mathbf{x}_i)) - G_{\mathbf{x}_j}(Z(\mathbf{x}_j)) |\right)$$

is the first-order variogram or madogram (Matheron (1987), Cooley et al. (2006)). Therefore, the previous definition extends the rescaled bivariate variogram, in order to consider several locations.

3. Main properties

We state in this section some properties of the multivariate variogram and its proofs are presented in the appendix. The first proposition is a technical result that relates the introduced coefficient with the previous coefficients in the literature.

Proposition 3.1. *The coefficient $v(\mathbf{x}_1, ..., \mathbf{x}_k)$ is independent of the univariate distributions $G_{\mathbf{x}_j}$, $j = 1, ..., k$ and it holds*

$$v(\mathbf{x}_1, ..., \mathbf{x}_k) = 1 - \frac{k+1}{k-1}\left(\frac{l_{\mathbf{x}}(1, ..., 1)}{1 + l_{\mathbf{x}}(1, ..., 1)} - \sum_{\emptyset \neq I \subseteq \{1,...,k\}} (-1)^{|I|+1} \frac{l_{\mathbf{x}_I}(1, ..., 1)_I}{1 + l_{\mathbf{x}_I}(1, ..., 1)_I}\right),$$

where $\mathbf{x} = (\mathbf{x}_1, ..., \mathbf{x}_k)$ and $\mathbf{x}_I$ denotes the sub-vector of $\mathbf{x}$ with indices in I.

The above proposition enhances that

a) the tail dependence function $l_{\mathbf{x}_1,...,\mathbf{x}_k}$ of $G_{\mathbf{x}_1,...,\mathbf{x}_k}$ determines $v(\mathbf{x}_1, ..., \mathbf{x}_k)$;
b) the extremal coefficients $\epsilon_{\mathbf{x}_I} = l_{\mathbf{x}_I}(1, ..., 1)_I$, for $I \subseteq \{1, ..., k\}$ determine $v(\mathbf{x}_1, ..., \mathbf{x}_k)$;
c) when applied for the particular case of $k = 2$, we obtain $\nu(| \mathbf{x}_i - \mathbf{x}_j |) = \frac{l_{\mathbf{x}_i,\mathbf{x}_j}(1,1)}{1+l_{\mathbf{x}_i,\mathbf{x}_j}(1,1)} - \frac{1}{2}$, which is the equation (14) in Cooley et al. (2006) for the relation between the bidimensional variogram and the extremal coefficient.

We will now see that our dependence coefficient increases with respect to the multivariate concordance ordering. More concordance among the margins of the vector means that extreme values are more likely. In this case, we expect a smaller discrepancy between the maximum and the minimum of the components of the vector and thus a higher value for the coefficient.
The random vector $(Z(\mathbf{x}_1), ..., Z(\mathbf{x}_k))$ is more concordant than another k-dimensional random vector $(Z(\mathbf{y}_1), ..., Z(\mathbf{y}_k))$ (Joe (1997)) if, for each $(u_1, ..., u_k) \in \mathbb{R}^k$, it holds that

$$P\left(\bigcap_{i=1}^{k}\{Z(\mathbf{x}_i) > u_i\}\right) \geq P\left(\bigcap_{i=1}^{k}\{Z(\mathbf{y}_i) > u_i\}\right)$$

and

$$P\left(\bigcap_{i=1}^{k}\{Z(\mathbf{x}_i) \leq u_i\}\right) \geq P\left(\bigcap_{i=1}^{k}\{Z(\mathbf{y}_i) \leq u_i\}\right).$$

Next proposition states that for sets of locations $(\mathbf{x}_1, ..., \mathbf{x}_k)$ with more concordant values for $Z(\mathbf{x}_i)$, $i = 1, ..., k$, we find higher values for the coefficient $v(\mathbf{x}_1, ..., \mathbf{x}_k)$.

Proposition 3.2. *If $(Z(\mathbf{x}_1), ..., Z(\mathbf{x}_k))$ is more concordant than $(Z(\mathbf{y}_1), ..., Z(\mathbf{y}_k))$ then $v(\mathbf{x}_1, ..., \mathbf{x}_k) \geq v(\mathbf{y}_1, ..., \mathbf{y}_k)$.*

If $(Z(\mathbf{x}_1), ..., Z(\mathbf{x}_k))$ has MEV distribution then it is associated and then (Esary et al. (1967))

$$P\left(\bigcap_{i=1}^{k}\{Z(\mathbf{x}_i) > u_i\}\right) \geq \prod_{i=1}^{k} P\left(Z(\mathbf{x}_i) > u_i\right)$$

and

$$P\left(\bigcap_{i=1}^{k}\{Z(\mathbf{x}_i) \leq u_i\}\right) \geq \prod_{i=1}^{k} P\left(Z(\mathbf{x}_i) \leq u_i\right).$$

In other words, a vector $(Z(\mathbf{x}_1), ..., Z(\mathbf{x}_k))$ with MEV distribution is more concordant than the corresponding vector $(\hat{Z}(\mathbf{x}_1), ..., \hat{Z}(\mathbf{x}_k))$ with independent margins.
Otherwise, for the above multivariate concordance ordering the maximal distribution is the Frechét upper bound, corresponding to margins totally dependent. We have then the following consequence of the above proposition.

Proposition 3.3. *a) If $(Z(\mathbf{x}_1), ..., Z(\mathbf{x}_k))$ has independent margins then $v(\mathbf{x}_1, ..., \mathbf{x}_k) = 0$.*
b) If $(Z(\mathbf{x}_1), ..., Z(\mathbf{x}_k))$ has totally dependent margins then $v(\mathbf{x}_1, ..., \mathbf{x}_k) = 1$.
c) $0 \le v(\mathbf{x}_1, ..., \mathbf{x}_k) \le 1$.

4. Estimation

Let $(Z^{(i)}(\mathbf{x}_1), ..., Z^{(i)}(\mathbf{x}_k))$, $i = 1, ..., n$, independent copies of $(Z(\mathbf{x}_1), ..., Z(\mathbf{x}_k))$ and $\hat{G}_{\mathbf{x}_j}$ the empirical distribution function provided by $Z^{(i)}(\mathbf{x}_j)$, $i = 1, ..., n$, $j = 1, ...k$. The natural estimator for the variogram $v(\mathbf{x}_1, ..., \mathbf{x}_k)$ is its empirical counterpart

$$\hat{v}(\mathbf{x}_1, ..., \mathbf{x}_k) = 1 - \frac{k+1}{k-1} \times \frac{1}{n} \sum_{i=1}^{n} \left(\bigvee_{j=1}^{k} \hat{G}_{\mathbf{x}_j}(Z^{(i)}(\mathbf{x}_j)) - \bigwedge_{j=1}^{k} \hat{G}_{\mathbf{x}_j}(Z^{(i)}(\mathbf{x}_j)) \right).$$

The estimator is strongly consistent and its asymptotic normality can be deduced from the Theorem 6 in Fermanian et al. (2004).

5. Application

We compute the estimates for the variogram of the amount $Z(\mathbf{x})$ of tritium (pCi/L) in drinking water, for Muscle Shoals, Scottsboro and Montgomery, three locations $\mathbf{x}$ in Alabama State (USA).
The first two cities are relatively close to a nuclear power plant in northern Alabama (Browns Ferry reactors at Decatur): Muscle Shoals is about 20 miles west of the plant, Scottsboro is about 60 miles east of the plant and both are situated on the banks of the Tennessee River. Montgomery is far from any nuclear plant.

Measurements (quarterly) can be accessed at http://oaspub.epa.gov/enviro/erams. The annual maximum values are summarized in the following graph where one can observe greater concordance between the values of Muscle Shoals and Scottsboro.

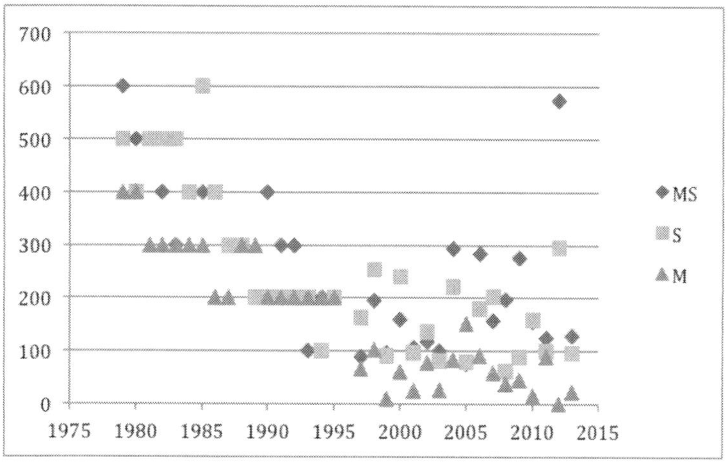

The tritium data in drinking water show both Muscle Shoals and Scottsboro have greater levels than Montgomery in the period 1997-2013, and we evaluate the dependence of the annual maxima at these locations throughout the estimated variogram $\hat{v}$.

location **x**	MS	S	M
MS	$\hat{v}$(Z(MS),Z(MS))=0	$\hat{v}$(Z(MS),Z(S))=0,4	$\hat{v}$(Z(MS),Z(M))=0,13
S	$\hat{v}$(Z(S),Z(MS))=0,4	$\hat{v}$(Z(S),Z(S))=0	$\hat{v}$(Z(S),Z(M))=0,22
M	$\hat{v}$(Z(M),Z(MS))=0,13	$\hat{v}$(Z(M),Z(S))=0,22	$\hat{v}$(Z(M),Z(M))=0

The estimate for the dependence among the three locations presents a low value which can be explained by the week dependence for the pairs $(Z(MS), Z(M))$ and $(Z(S), Z(M))$. The estimated dependence for $(Z(MS), Z(S))$ presents greater value as expected from the locations of these two cities.

6. Appendix: Proofs

Proof of Proposition 3.1 Let $\mathbf{Z}$ be a random vector with MEV distribution G, unit Fréchet F margins and tail dependence function l. It is sufficient to prove that if $\mathbf{Y}$ has MEV distribution with marginal distributions F_j, $j = 1, ..., d$, and the same copula as $\mathbf{Z}$, then it holds that

$$E\left(\bigvee_{j=1}^{k} F_j(Y_j)\right) = E\left(\bigvee_{j=1}^{k} F(Z_j)\right) = \frac{l(1,...,1)}{1+l(1,...,1)}.$$

We first deduce the distribution of $\bigvee_{j=1}^{k} F_j(Y_j)$. Denoting the copula of $\mathbf{Z}$ by $C_{\mathbf{Z}}$, we have, for each $u \in [0,1]$,

$$P\left(\bigvee_{j=1}^{k} F_j(Y_j) \leq u\right) = C_{\mathbf{Z}}(u,...,u) = G\left(-\frac{1}{\log u},...,\frac{1}{\log u}\right) =$$

$$\exp\left(-l\left((-\log u),...,(-\log u)\right)\right) = u^{l(1,...,1)}.$$

Then

$$E\left(\bigvee_{j=1}^{k} F_j(Y_j)\right) = \int_0^1 u^{l(1,...,1)} l(1,...,1)\,du = \frac{l(1,...,1)}{1+l(1,...,1)}.$$

The result then follows by relating the minimum and maximum through

$$\bigwedge_{j=1}^{k} F(X_j) = \sum_{\emptyset \neq I \subseteq \{1,...,k\}} (-1)^{|I|+1} \bigvee_{j \in I} F(X_j).$$

Proof of Proposition 3.2 The concordant order implies (Shaked and Shanthikumar (2007)) that

$$E\left(f\left(\bigwedge_{i=1}^{k} Z(\mathbf{x}_i)\right)\right) \geq E\left(f\left(\bigwedge_{i=1}^{k} Z(\mathbf{y}_i)\right)\right)$$

and

$$E\left(f\left(\bigvee_{i=1}^{k} Z(\mathbf{x}_i)\right)\right) \leq E\left(f\left(\bigvee_{i=1}^{k} Z(\mathbf{y}_i)\right)\right),$$

for all non-decreasing functions f. The result then follows by taking, without loss of generality, $G_{\mathbf{x}_j} = G = f$.

Proof of Proposition 3.3 For $\mathbf{Z} = (Z(\mathbf{x}_1), ..., Z(\mathbf{x}_k))$ let

$$R(\mathbf{Z}, \mathbf{x}) = E\left(\bigvee_{j=1}^{k} G_{\mathbf{x}_j}(Z(\mathbf{x}_j)) - \bigwedge_{j=1}^{k} G_{\mathbf{x}_j}(Z(\mathbf{x}_j))\right).$$

We will prove that if $\bar{\mathbf{Z}} = (\bar{Z}(\mathbf{x}_1), ..., \bar{Z}(\mathbf{x}_k))$ has totally dependent margins then it holds that $R(\bar{\mathbf{Z}}, \mathbf{x}) = 0$ and if $\hat{\mathbf{Z}} = (\hat{Z}(\mathbf{x}_1), ..., \hat{Z}(\mathbf{x}_k))$ has independent margins then it holds that $R(\hat{\mathbf{Z}}, \mathbf{x}) = \frac{k-1}{k+1}$. We need to apply the following equalities to obtain these results:

$$\sum_{\emptyset \neq I \subseteq \{1,...,k\}} (-1)^{|I|+1} = 1, \quad \sum_{i=1}^{k+1} (-1)^{i+1} \binom{k+1}{i} i = 0, \quad \sum_{i=0}^{k+1} (-1)^{i} \binom{k+1}{i} = 0.$$

From the proposition 3.1 we have $R(\bar{\mathbf{Z}}, \mathbf{x}) = \frac{1}{2} - \sum_{\emptyset \neq I \subseteq \{1,...,k\}} (-1)^{|I|+1} \frac{1}{2} = 0$ and

$$R(\hat{\mathbf{Z}}, \mathbf{x}) = \frac{k}{k+1} - \sum_{\emptyset \neq I \subseteq \{1,...,k\}} (-1)^{|I|+1} \frac{|I|}{1+|I|} =$$

$$\frac{k}{k+1} - \sum_{j=1}^{k} (-1)^{j+1} \binom{k}{j} \frac{j}{j+1} = \frac{k}{k+1} + \frac{1}{k+1} \sum_{j=1}^{k} (-1)^{j+2} \binom{k+1}{j+1} j =$$

$$\frac{k}{k+1} + \frac{1}{k+1} \sum_{j=1}^{k} (-1)^{j+2} \binom{k+1}{j+1} (j+1) - \frac{1}{k+1} \sum_{j=1}^{k} (-1)^{j+2} \binom{k+1}{j+1} =$$

$$\frac{k}{k+1} + \frac{1}{k+1}\left(0 - \binom{k+1}{1}\right) + \frac{1}{k+1}\left(0 - \binom{k+1}{0} + \binom{k+1}{1}\right) = \frac{k-1}{k+1}.$$

The statement in c) follows from a), b) and the previous proposition.

Acknowledgments: We acknowledge the valuable suggestions and corrections from the referee. This work was partially supported by the research project PEst-OE/MAT/UI0212/2014.

References

[1] Beirlant J, Goegebeur Y, Segers J and Teugels J 2004 *Statistics of Extremes: Theory and Application.* John Wiley.

[2] Cooley K, Naveau P and Poncet P 2006 Variograms for max-stable random fields. *Lecture Notes in Statistics* **187** 373-390, New York, NY: Springer.

[3] Esary J D, Proschan F and Walkup D W 1967 Association of Random Variables, with Applications. *The Annals of Mathematical Statistics* **38** 1466-74.

[4] Fermanian J D, Radulović D and Wegkamp M 2004 Weak convergence of empirical copula processes Bernoulli **5**, 847-60.

[5] Joe H. 1997 Multivariate Models and Dependence Concepts. Chapman & Hall, London.
[6] Li H 2009 Orthant tail dependence of multivariate extreme value distributions *J. Multivariate Anal.*, **100** 243-56.
[7] Matheron G 1987 Suffit-il, pour une covariance, d'être de type positif? *Sci. Terre. Info. Geol.* **26** 51-66.
[8] Naveau P, Guillou A, Cooley D and Diebolt J 2009 Modelling pairwise dependence of maxima in space, *Biometrika* **96** 1-17.
[9] Resnick S 1987 *Extreme Values, Regular Variation and Point Processes* Springer-Verlag, New York.
[10] Resnick S 2007 *Heavy-Tail Phenomena* Springer-Verlag, New York.
[11] Shaked M and Shanthikumar J.G. 2007 *Stochastic orders.* Springer-Verlag, New York.
[12] Smith R.L. 1990 Max-stable processes and spatial extremes. Preprint, Univ. North Carolina, USA.

Arbitrary ℓ-state solutions of the Feynman propagator with the Deng-Fan molecular potential

Ahmed Diaf

Laboratoire de l'Énergie et des Systèmes Intelligents, Université de Khemis Miliana, Route de Thénia, Khemis Miliana,44225, Algérie

E-mail: s_ahmed_diaf@yahoo.fr

Abstract. The bound state solutions of the Feynman propagator with the rotating Deng-Fan molecular potential are presented approximately. An approximation of the centrifugal potential is used and nonlinear space-time transformations are applied. A relation between the original path integral and the Green function of a new quantum soluble system is derived. The energy spectrum and the normalized eigenfunctions are both obtained for the application of this technique to the Deng-Fan molecular potential. Our results are in very good agreement with those found by using numerical and other methods.

1. Introduction

The exact analytical solution of the Schrödinger equations are only possible with the angular momentum $\ell = 0$ for some potential models. However, when $\ell \neq 0$ [1, 2], the Schrödinger equation can only be solve approximately for different suitable approximation scheme [3, 4]. This is due to the presence of the centrifugal term $(1/r^2)$. Motivated by the success in obtaining approximately the bound state solutions of the Feynman propagator with the Manning-Rosen and the Rosen-Morse potentials [5, 6], we attempt here to solve the arbitrary ℓ-wave propagator with the Deng-Fan potential [7]. An improved approximation scheme for the centrifugal term is used as in [4] and non linear space-time transformations are applied.

The Deng-Fan $(D-F)$ potential is given by

$$V^{D-F}(r) = D(1 - \frac{be^{-\alpha r}}{1-e^{-\alpha r}})^2. \tag{1}$$

where $0 \leq r < \infty$, $b = e^{\alpha r_e} - 1$, the three parameters D, b, and α denote the dissociation energy, the position of the minimum r_e, and the range of the potential, respectively. This potential has been one of the most useful and convenient models to study the energy eigenvalues and electromagnetic transitions of diatomic molecules. It is known that for this potential, the Schrödinger equation can be solved exactly when $\ell = 0$ either directly [8] or by using SO(2, 1) algebra method[9]. Unfortunately, for arbitrary ℓ-state, the radial Schrödinger equation does not admit exact solutions. In this case, some authors have used the approximation scheme proposed by Lu [3] to study analytically arbitrary ℓ-wave bound states of the Schrödinger equation for the Deng-Fan potential [11].

This paper is organized as follows. In Sec. 2 we derive arbitrary ℓ-states solutions of the Deng-Fan potential within the Feynman path integrals formalism by approximate method. In Sec.3

the numerical calculations are given to compare with those obtained by other method. The concluding remarks are given in Sec. 4.

2. Green's function

The propagator for a particle of mass m in the spherically symmetric potential, can be developed into a sum of partial waves of the form [10]:

$$K(\vec{r}'',t'';\vec{r}',t') = \frac{1}{4\pi r''r'}\sum_{l=0}^{\infty}(2l+1)K_l(r'',t'';r',t')P_l(\cos\theta). \tag{2}$$

$P_l(\cos\theta)$ is the Legendre polynomial with $\theta \equiv (\vec{r}'',\vec{r}')$ and

$$K_l(r'',t'';r',t') = \lim_{N\to\infty}\int\prod_{j=1}^{N}\exp\left[\frac{i}{\hbar}S_j\right]\prod_{j=1}^{N}\left[\frac{m}{2\pi i\hbar\varepsilon}\right]^{\frac{1}{2}}\prod_{j=1}^{N-1}dr_j. \tag{3}$$

Here $S_j = \frac{m}{2\varepsilon}(\Delta r_j)^2 - \varepsilon V_{eff}(r_j)$, where

$$V_{eff}(r_j) = \frac{\hbar^2}{2m}\frac{l(l+1)}{r_jr_{j-1}} - \frac{V_1}{\cosh^2(r_j/a)} + V_2\tanh(r_j/a) \tag{4}$$

and $\Delta r_j = r_j - r_{j-1}$, $\varepsilon = t_j - t_{j-1}$, $t' = t_0$ and $t'' = t_N$. Moreover, setting $T = t'' - t'$, we can write:

$$K_l(r'',r',T) := K_l(r'',t'';r',t'). \tag{5}$$

To calculate the propagator K_l given by equation (3) for $\ell \neq 0$ states, we apply the following approximate scheme to the centrifugal term [4] :

$$\frac{1}{r^2} \approx 4\alpha^2\left[C_0 + \frac{e^{-2\alpha r}}{1-e^{-2\alpha r}} + \left(\frac{e^{-2\alpha r}}{1-e^{-2\alpha r}}\right)^2\right], \tag{6}$$

where the parameter $C_0 = 0.0823058167837972$ is a shift determined by the expansion procedures. By substituting the centrifugal term in (4) by its approximation (6), we get

$$V_{eff}(r_j) = A\tanh(2\alpha r_j) - \frac{B}{\cosh^2(2\alpha r_j)} + C, \tag{7}$$

with $A = bD + \frac{Db^2}{2}$, $B = \frac{l(l+1)\alpha^2}{8m} + \frac{Db^2}{4}$ and $C = \frac{C_0 l(l+1)\alpha^2}{2m} + (b+1)D + \frac{Db^2}{2}$.

In the following, we study the potential (7) with the Duru-Kleinert method [12]. The path integral for the potential (7) is

$$K_l(r'',r';T) = \int \mathcal{D}r(t)\exp\left[\int_{t'}^{t''}\left(\frac{m}{2}\dot{r}^2 - V_{eff}(r)\right)dt\right]. \tag{8}$$

By achieving the point canonical transformation and the time transformation [8],

$$\begin{cases} r &= f(q) \\ dt &= [f'(q)]^2 d\mathfrak{s}, \end{cases} \tag{9}$$

In our case, we choose

$$r = \left(\frac{1}{\alpha}\right) arccoth(2\coth^2 q - 1), \tag{10}$$

The propagator expression becomes [11]:

$$\begin{aligned}\widehat{K}_\ell(q'',q';s'') &= \exp\left[\frac{i}{\hbar}s''\frac{E-C)+A}{\alpha^2}\right]\\ &\times\int\mathcal{D}q(\mathfrak{s})\exp\left[\frac{i}{\hbar}\int_0^{s''}\left(\frac{m}{2}\dot{q}^2-\frac{\hbar^2}{2m}\left[\frac{\eta(\eta-1)}{\sinh^2 q}-\frac{v(v-1)}{\cosh^2 q}\right]\right)d\mathfrak{s}\right]\\ &= \exp\left[\frac{i}{\hbar}s''\frac{E-C+A}{\alpha^2}\right]K_\ell^{MPT}(q'',q';s'') \qquad (11)\end{aligned}$$

with $\eta=\frac{1}{2}\pm\sqrt{1+\frac{8mB}{\alpha^2\hbar^2}}$, $v=\frac{1}{2}\pm\sqrt{-\frac{2m(E-C-A)}{\alpha^2\hbar^2}}$ and K_ℓ^{MPT} is the path integral of the modified Pöschl-Teller potential which is a known solved problem [13] for witch the bound states wave functions are explicitly given by :

$$\chi_{l,n}^{(k_1,k_2)}(q)=N_n^{(k_1,k_2)}(\sinh q)^{2k_2-1/2}(\cosh q)^{-2k_1+3/2}\times{}_2F_1(-k_1+k_2+k,-k_1+k_2-k+1;2k_2;-\sinh^2 q), \qquad (12)$$

where $k=k_1-k_2-n$, and

$$N_n^{(k_1,k_2)}=\frac{1}{\Gamma(2k_2)}\left(\frac{(2k-1)\Gamma(k_1+k_2-k)\Gamma(k_1+k_2+k-1)}{\Gamma(k_1-k_2+k)\Gamma(k_1-k_2-k+1)}\right)^{1/2}, \qquad (13)$$

The eigenvalues are expressed by

$$E_n^{MPT}=-\frac{\hbar^2}{2m}\left[2(k_1-k_2-n)-1\right]^2. \qquad (14)$$

Calculating $K_l(q'',q';s'')$ allows us to obtain the Green function. Since this is known, the whole energy spectrum is obtained from its poles, and from the residues at the poles we obtain the corresponding wave functions.

Substituting (12) in (11), we get

$$\begin{aligned}\widehat{K}_l(q'',q';s'') &= \sum_{n=0}^{N_m}\exp\left\{\frac{i}{\hbar}s''\left[(E-C+A)(2a)^2-E_n^{MPT}\right]\right\}\chi_{l,n}^{(k_1,k_2)}(q'')\chi_{l,n}^{*(k_1,k_2)}(q') \qquad (15)\\ &+\int_0^\infty dK\exp\left\{\frac{i}{\hbar}s''[(E-C+A)(2a)^2-\frac{\hbar^2K^2}{2m})]\right\}\chi_K^{(k_1,k_2)}(q'')\chi_K^{*(k_1,k_2)}(q')\end{aligned}$$

Then, by integrating this latter over the pseudo-time parameters s'', the Green function G is written as:

$$\begin{aligned}G_l(r'',r';E) &= \sum_{n=0}^{N_m}\frac{\chi_{l,n}^{DF(k_1,k_2)}(r'')\chi_{l,n}^{DF*(k_1,k_2)}(r')}{E_{n,l}^{DF}-E}\\ &+\int_0^\infty dK\frac{\chi_{l,K}^{DF}(r'')\chi_{l,K}^{DF*}(r')}{(\hbar^2K^2/2m(2a)^2)-A+C-E} \qquad (16)\end{aligned}$$

Following [13], the values of k_1,k_2, satisfying the appropriate boundary conditions for $r\to 0$ and $r\to\infty$, are respectively:

$$\begin{aligned}k_1 &= \frac{1}{2}\left[\left(1+\frac{1}{2}(s+2n+1)\right)+\frac{2mA}{\alpha^2\hbar^2(s+2n+1)}\right]\\ k_2 &= \frac{1}{2}\left(1+\sqrt{1+\frac{8mB}{\alpha^2\hbar^2}}\right)\equiv\frac{1}{2}(1+s) \qquad (17)\end{aligned}$$

Substituting these values of k_1, k_2 in (12) and (14), and by considering $u = \frac{1}{2}\left[1 - \tanh(2\alpha r)\right]$, we get

$$\begin{aligned} \chi_{n,\ell}^{DF(k_1,k_2)}(r) &= \sqrt{\alpha} N_n^{(k_1,k_2)} (1-u)^{\frac{1}{2}-k_1+n} u^{k_1-1-\frac{s}{2}-n} \\ &\times \ {}_2F_1\left(-n, 2k_1 - n - 1; s+1; \frac{1}{1-u}\right) \\ &= \left[\frac{\alpha\,(2k_1-1)\,n!\Gamma\,(2k_1-n-1)}{\Gamma\,(n+s+1)\,\Gamma\,(2k_1-s-n-1)}\right]^{1/2} \left(1-e^{-2r}\right)^{k_2} \\ &\times \exp\left\{-2r\left[k_1 - \frac{s}{2} - n - 1\right]\right\} P_n^{(2k_1-2n-s-2,s)}(1-2e^{-2r}) \,. \end{aligned} \quad (18)$$

The energy spectrum is obtained from the poles,

$$E_{n,l}^{DF} = \alpha^2 E_n^{MPT} - A + C \quad (19)$$

$$= -\left[\frac{\hbar^2\alpha^2(s+2n+1)^2}{8m} + \frac{2mA^2}{\hbar^2\alpha^2(s+2n+1)^2}\right] + C \quad (20)$$

3. Results and discussions

Energies calculated by using the path integral method described above are reported in the Table. the results, Eq. (20) are displayed in Table hereafter, where a comparison is made with the calculations of Dong et al [14], who solved analytically the Schrödinger equation, but using another approach for the centrifugal barrier. All results are compared to the numerical data [1]. Our results are almost indistinguishable from the numerical ones [1]. Nevertheless, except in a few cases, we get results closer to the numerical estimates compared to the method of Dong [14]. This is due to our choice of the centrifugal barrier.

4. Concluding remarks

In this paper, we have presented a path integral treatment for the Deng-Fan potential. By Using an approximation scheme for the centrifugal term and a nonlinear space-time transformations in the radial path integral, We were able to derive explicit expressions of the wave functions in terms of Jacobi polynomials and the energy eigenvalues in function of the potential parameters. The barrier approximation being fixed, we note that the Feynman integral method gives results closer to numerical one than solving the Schrödinger equation.

To conclude, our method is efficient in solving this type of potentials. We intend in a near future, to expand the path integral method to a more general form of potentials and relativistic cases.

References

[1] L. D. Landau and E. M. Lifshitz (1977) Quantum Mechanics, Non-Relativistic Theory, Pergamon, Oxford.
[2] L. I. Schiff (1955) Quantum Mechanics 3rd edition, McGraw-Hill, New York.
[3] R. L. Greene and C. Aldrich, Phys. Rev. A 14 (1976)2363.
[4] S. M. Ikhdair and R. Sever, Ann. Phys. (Berlin) 18, No. 4 (2009)189 197.
[5] A. Diaf, A. Chouchaoui and R.L. Lombard, Ann. Phys. 317, (2005) 354.
[6] A. Diaf and A Chouchaoui, Phys. Scr. 84, (2011) 015004.
[7] Deng, Z. H., Fan, Y. P. Shandong Univ J 7(1957) 162.
[8] Mesa, A. D. S.; Quesne, C.; Smirnov, Y. F. J Phys A: Math Gen, 31(1998), 321.
[9] Codriansky, S.; Cordero, P.; Salamo , S. J Phys A: Math Gen, 32(1999)6287.
[10] D. C. Khandekar, S.V. Lawande, K.V. Bhagwat, Path Integral Methods and their Applications (World Scientific, Singapore,1986)
[11] Dong, S. H.; Gu, X. Y. J Phys: Conf Ser 2008, 96, 012109.

Table 1. Eigenvalues $-E_{n,l}$ of the Deng-Fan potential in atomic units ($\hbar = m = 1$) with $D = 15$ and $r_e = 0.4$.

States	α	Present	Dong et al[14]	Lucha et al[15]
2p	0.05	7.86080	7.86060	7.8628
	0.1	7.95329	7.95247	7.95537
	0.15	8.04508	8.04322	8.04724
	0.2	8.13613	8.13287	8.13842
	0.25	8.22655	8.22142	8.22892
	0.3	8.31629	8.30889	8.31874
3p	0.05	10.99776	10.9976	10.9998
	0.1	11.16255	11.1617	11.1647
	0.15	11.32422	11.3224	11.32647
	0.2	11.48280	11.4795	11.48513
	0.25	11.63827	11.6331	11.64068
	0.3	11.79066	11.7833	9.67565
3d	0.05	10.21597	10.2154	10.21651
	0.1	10.35350	10.3510	10.35409
	0.15	10.48928	10.4837	10.48992
	0.2	10.62334	10.6135	10.62403
	0.25	10.75571	10.7403	10.75645
	0.3	10.88644	10.8642	10.88719
4p	0.05	12.49760	12.4974	12.4992
	0.1	12.69678	12.6960	12.69851
	0.15	12.88832	12.8865	12.8901
	0.2	13.07220	13.0689	13.07400
	0.25	13.24840	13.2433	13.2501
4d	0.05	12.09828	12.0977	12.0989
	0.1	12.28498	12.2825	12.2857
	0.15	12.46635	12.4608	12.46715
	0.2	12.64244	12.6326	12.64324
4f	0.05	11.82077	11.8195	11.8209
	0.1	11.99790	11.9930	11.9981
	0.15	12.17156	12.1604	12.1718
	0.2	12.34182	12.3221	12.3421
5p	0.1	13.54213	13.5413	13.5434
	0.2	13.92894	13.9257	13.9301

[12] H. Kleinert, Path integrals in Quantum Mechanics, Statistics and Polymer Physics, World scientific, Singapore, 2009.
[13] C. Groshe J. Phys. A. Math. Gen. 22, (1989) 5073.
[14] Dong, S. H.; Gu, X. Y. J Phys: Conf Ser , 96(2008)012109.
[15] W. Lucha and F.F. t. J. Mod. Phys. C 10, (1999)607.

Interaction of spatial solitons with a gapless stripe embedded into a Bragg-grating area

Thawatchai Mayteevarunyoo[1] and Boris A. Malomed[2]

[1]Department of Telecommunication Engineering, Mahanakorn University of Technology, Bangkok 10530, Thailand

[2]Department of Physical Electronics, School of Electrical Engineering, Faculty of Engineering, Tel Aviv University, Tel Aviv 69978, Israel

E-mail: thawatch@mut.ac.th

Abstract. We introduce a model in which the grating is absent in a finite-width stripe in the waveguide, thus creating a gapless channel in the gapped medium. Two semi-infinite grating separated by the plain stripe may have a relative phase shift. This system modifies the Bragg bandgap, creating intra-gap defect modes (DFs) which are pinned to the gapless channel. A DF solution in the linear system is found analytically. Further, numerical analysis of the full nonlinear system demonstrates that the shape and stability of Bragg solitons are also strongly affected by the presence of the gapless channel, and by the possible phase shift between the two semi-infinite gratings. In particular, asymmetric and flat-top solitons appear.

1. Introduction and the model

Bragg gratings (BGs) are photonic structures produced by a periodic modulation of the refractive index along an optical fiber or waveguide. Gap solitons (GSs) in gratings [1] are supported through the balance between the BG-induced dispersion or diffraction, which incorporates the bandgap in the system's linear spectrum, and the Kerr nonlinearity of the fiber or waveguide. Analytical solutions for BG solitons in the standard BG model are well known [2, 3]. Slightly more than half of the soliton family is stable [4], [5]. GSs were created in the experiment, using a relatively short BG written in the cladding of an optical fiber [6]. GSs are one of fundamental species of solitons in optics, as well as in other nonlinear media.

An issue of great significance is the development of schemes for the creation of new varieties of BG solitons. One of them is *apodization* [7], i.e., making the Bragg reflectivity gradually varying along the fiber (or waveguide). Also available are BGs with periodic apodization, i.e., an effective *superlattice* built on top of the BG [8]. Analysis of the light propagation in the so developed *supergratings* was performed too [9, 10].

In this work, we address quite a simple scheme which, nevertheless, was not explored before: a structure formed by two semi-infinite BG areas, with possible phase shift α between them, which are separated by a flat (gapless) stripe of width L. The objective is to find the linear spectrum of this system and GSs in its nonlinear version.

The model is based in the following coupled-mode equations:

$$
\begin{aligned}
i\frac{\partial u}{\partial t}+i\frac{\partial u}{\partial x}+\kappa(x)v+\left(|v|^2+\frac{1}{2}|u|^2\right)u &= 0,\\
i\frac{\partial v}{\partial t}-i\frac{\partial v}{\partial x}+\kappa^*(x)u+\left(|u|^2+\frac{1}{2}|v|^2\right)v &= 0,
\end{aligned}
\tag{1}
$$

which may also be realized in the spatial domain, replacing time t by propagation distance z. The local Bragg reflectivity is chosen as

$$
\kappa(x)=\begin{cases} 1 \text{ at } x>0,\\ 0 \text{ at } -L<x<0,\\ \exp(i\alpha) \text{ at } x<-L, \end{cases}
\tag{2}
$$

where α takes values $0\leq\alpha\leq\pi$, and κ^* is the complex conjugate of κ.

Stationary solutions with frequency ω are looked for as $\{u(x,t),v(x,t)\} = e^{-i\omega t}\{U(x),V(x)\}$, where complex functions U and V satisfy equations

$$
\begin{aligned}
+i\frac{dU}{dx}+\omega U+\kappa(x)V+\left[\left(|V|^2+\frac{1}{2}|U|^2\right)\right]U=0,\\
-i\frac{dV}{dx}+\omega V+\kappa^*(x)U+\left[\left(|U|^2+\frac{1}{2}|V|^2\right)\right]V=0,
\end{aligned}
\tag{3}
$$

which are compatible with the relation that reduces the two stationary fields to one, $U^*(x) = -V(x)$. The stationary solutions are characterized by the total energy (or total power, in terms of the spatial-domain model), $E=\int_{-\infty}^{+\infty}\left(|U(x)|^2+|V(x)|^2\right)dx$.

2. The linear spectrum and defect mode in the linear system

First, we explore the spectrum of the linearized version of system (1), the objective being to find the bandgap in which solitons may exist. The computation produces the bandgap which is shown, as a function of α, and for different values of L, in Fig. 1. It is seen that the bandgap shrinks with the increase of α, vanishing at $\alpha=\pi$. At $\alpha=0$, the width of the bandgap is smaller than in the usual uniform Bragg grating (the one corresponding to $L=0$), where the bandgap is $-1<\omega<+1$.

The linear version of Eq. (3), subject to reduction $U^*(x)=-V(x)$, reduces to the single linear equation,

$$
i\frac{dU}{dx}+\omega U-\kappa(x)U^*=0.
\tag{4}
$$

Equation (4) gives rise an exact solution for a *defect mode* (DM) pinned to the central stripe where the grating is absent:

$$
U(x)=\begin{cases}
\exp\left(\frac{i}{2}\arccos\omega-\sqrt{1-\omega^2 x}\right), \text{ at } x\geq 0,\\
A\exp\left(i\omega\left(x+\frac{L}{2}\right)\right), \text{ at } -L\leq x\leq 0,\\
\exp\left(\frac{i}{2}\alpha-\frac{i}{2}\arccos\omega+\sqrt{1-\omega^2}\,(x+L)\right), \text{ at } x\leq -L.
\end{cases}
\tag{5}
$$

The eigenfrequency of this localized mode is determined by the condition of the continuity of $U(x)$ at $x=0$ and $x=-L$, which yield $A=\exp(i\alpha/4+i\pi n)$,

$$
\omega_{\text{eigen}}L+\alpha/2=\arccos(\omega_{\text{eigen}})+2\pi n,
\tag{6}
$$

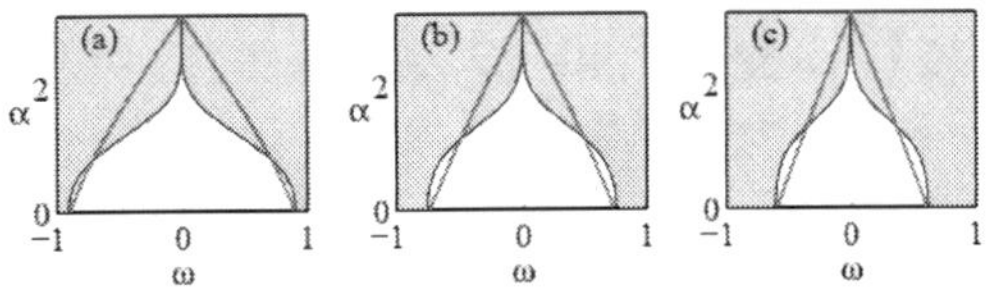

Figure 1. The bandgap structure found from the linearization of Eq. (1) for (a) $L = 0.5$, $L = 1.0$ and $L = 1.5$. Shaded areas are occupied by Bloch bands. The red solid lines correspond to the analytical solutions for the defect modes, see Eq. (6)

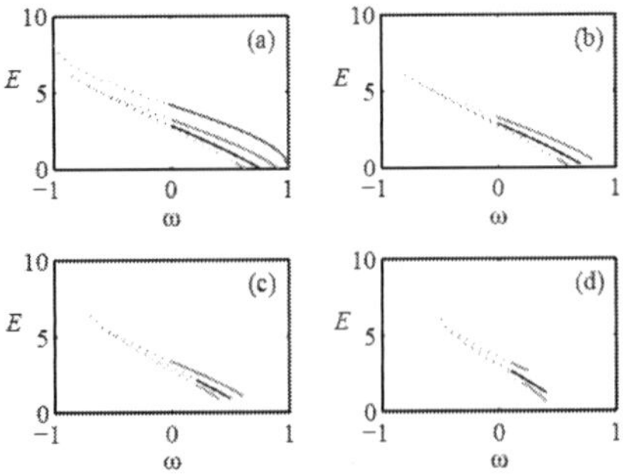

Figure 2. Total energy E of gap solitons versus ω for different L. (a) $\alpha = 0$ (b) $\alpha = \pi/8$ (c) $\alpha = \pi/4$ (d) $\alpha = \pi/3$. Continuous and dashed segments depict stable and unstable solitons, respectively. Different colors correspond to $L = 0$ (blue); $L = 0.5$ (red); $L = 1.0$ (black); $L = 1.5$ (magenta).

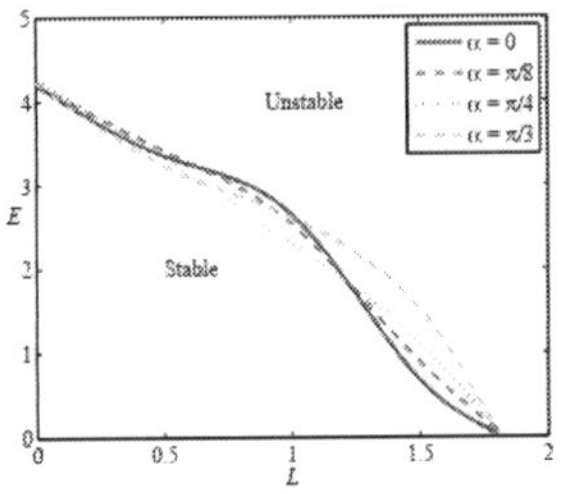

Figure 3. Stability regions for the solitons in the plane of (L, E) at different fixed values of α. In the case of $\alpha = 0$, the result for $L = 0$ pertains to the uniform nonlinear Bragg grating.

where n is an arbitrary integer. In particular, numerical solution of Eq. (6) for $\alpha = 0$ and $n = 0$ yields

$$\omega_{\text{eigen}}(L = 0.5) \approx 0.90, \omega_{\text{eigen}}(L = 1.0) \approx 0.74, \omega_{\text{eigen}}(L = 1.5) \approx 0.61, \tag{7}$$

which is close to the values at which the bandgap closes at $\alpha = 0$ in Fig. 1. The respective eigenvalues at negative frequencies correspond to the defect modes subject to reduction $U(x) = +V^*(x)$, instead of that given by $U^*(x) = -V(x)$.

It follows from Eq. (6) that, for any L, the eigenfrequency shifts to $\omega_{\text{eigen}} = 0$ at $\alpha = \pi$, which is consistent with the picture in Fig. 1. At large values of L, new DMs appear, corresponding to $n > 0$. In particular, the ones corresponding to $\omega = 1$ appear, for each $n \geq 1$, at $L_{\text{cr}}^{(n)} = 2\pi n - \alpha/2$. Therefore, for given large L, the total number of DMs is $n_{\text{DM}} = 1 + \left[\frac{L+(\alpha/2)}{2\pi}\right]$, where [...] stands for the integer part.

3. Numerical Results

The existence and stability of numerically found GS solutions of the full system (3) is summarized in Fig. 2, which displays typical dependences $E(\omega)$ for different L. The stability was identified by means of systematic simulations of Eqs. (1). The stability region is additionally shown in Fig 3 in the plane of (L, E) for different values of α. It is clearly seen that the region shrinks

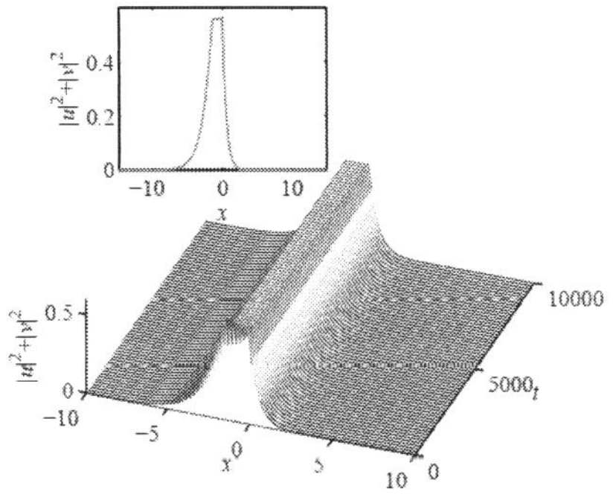

Figure 4. (a) An example of a stable "flat-top" soliton found close to the existence boundary (which is $\alpha = \pi/2.8$) at $\alpha = \pi/3, L = 1.5, \omega = 0.2$. The inset illustrates a the cross section of the final profile.

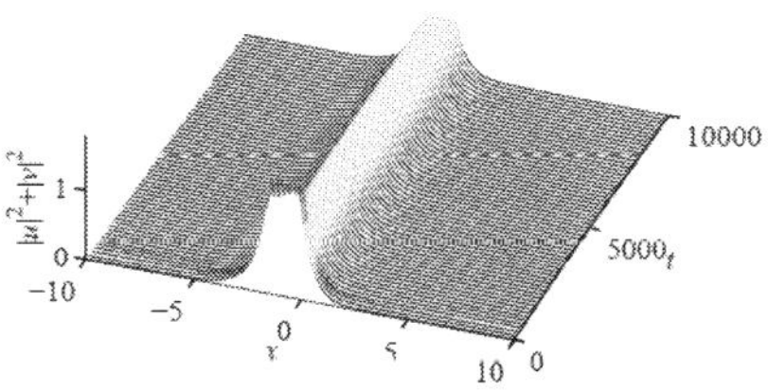

Figure 5. A typical example of the evolution of unstable solitons, for $\omega = -0.3$, $L = 1.5$, $\alpha = \pi/4$.

and vanishes at $L \approx 1.8$. Close to their existence boundary, the solitons, while remaining stable, acquire a specific flat-top shape, see an example in Fig. 4 for $\alpha = \pi/3$. The power profile shown in Fig. 4 is not only flat-top, but also features conspicuous spatial asymmetry, while the profiles of the two components, $|u|^2$ and $|v|^2$, remain mutually identical. A typical example of the evolution of unstable solitons is shown in Fig. 5. It is seen that unstable solitons are not destroyed. Instead, they shed off a part of their energy and thus transform into stable solitons with smaller energy.

4. Summary

We have introduced the model of a complex built of two nonlinear Bragg gratings separated by a flat segment of the waveguide. The linear defect mode was found in the exact form, while gap solitons were found numerically. The solitons feature asymmetry and the flat-top shape. It may also be possible to simulate collisions of moving solitons with the flat segment.

5. Acknowledgment

The work of T.M. was supported by the Thailand Research Fund through Grant No. RSA5780061.

6. References

[1] C. M. de Sterke and J. E. Sipe: Progr. Opt. Vol. 33 (1994), p. 203–260.
[2] A. B. Aceves and S. Wabnitz: Phys. Lett. A Vol. 141 (1989), p. 37–42.
[3] D. N. Christodoulides and R. I. Joseph: Phys. Rev. Lett. Vol. 62 (1989), p. 1746–1749.
[4] B. A. Malomed and R. S. Tasgal: Phys. Rev. E Vol. 49 (1994), p. 5787–5796.
[5] I. V. Barashenkov, D. E. Pelinovsky, and E. V. Zemlyanaya: Phys. Rev. Lett. Vol. 80 (1998), p. 5117–5120.
[6] B. J. Eggleton, R. E. Slusher, C. M. de Sterke, P. A. Krug, and J. E. Sipe: Phys. Rev. Lett. Vol. 76 (1996), p. 1627–1630.
[7] R. E. Slusher, B. J. Eggleton, T. A. Strasser and M. de Sterke: Opt. Exp. Vol. 3 (1998), p. 465–475.
[8] P. St. J. Russell: J. Appl. Phys. Vol. 59 (1986), p. 3344–3355.
[9] N. G. R. Broderick and M. de Sterke: Phys. Rev. E Vol. 55 (1997), p. 3634–3646.
[10] T. Mayteevarunyoo and B. A. Malomed: Opt. Exp. Vol. 16 (2008), p. 7767-7777.

Journal of Physics: Conference Series **574** (2015) 012024 doi:10.1088/1742-6596/574/1/012024

Oscillations at the Si/electrolyte contact: Relation to Quantum Mechanics

J Grzanna[1] and H J Lewerenz[1,2]
[1] Helmholtz Zentrum Berlin for Materials and Energy
Hahn-Meitner-Platz 1, D-14109 Berlin, Germany
[2] California Institute of Technology
1200 E. California Blvd, Pasadena, CA 91125, USA

E-mail: grzanna@helmholtz-berlin.de

Abstract. The basic process at the surface of the Si electrode is characterized by a cyclic oxidation of a thin silicon layer and the subsequent removal of the oxide by etching. Here, the oxide thickness evolves not uniformly due to cracks and nanopores. The mathematical model used to describe the phenomenon is based on a sequence of time dependent (oxide thickness) oscillator density functions that describes the passing of the (infinitesimal) oscillators through their minimum at each cycle. Two consecutive oscillator density functions are connected by a second order linear integral equation representing a Markov process. The kernel of the integral equation is a normalized Greens Function and represents the probability distribution for the periods of the oscillators during a cycle. Both, the oscillator density function and the two-dimensional probability density for the periods of the oscillators, define a random walk. A relation between the oscillator density functions and solutions of the Fokker-Planck equation can be constructed. This allows a connection of the oscillations, originally considered only for the description of a photo-electrochemical observation, to the Schrödinger equation. In addition, if the trajectory of a virtual particle, located at the silicon oxide electrode surface, is considered during one oscillatory cycle, then it can be shown that the displacement of the particle measured at the electrode surface performs a Brownian motion.

1. Introduction

Oscillation phenomena of silicon electrodes in electrochemical systems have been known for a long time [1]. In photo-electrochemistry and electrochemistry, oscillatory behaviour has been extensively studied on Si photo-electrodes [2]. Various models for the oscillation phenomena were discussed based on self-oscillating domains [3], the so-called current bursts [4], and oxide-induced interfacial stress [5] where the latter model explains sustained current oscillations with the existence of two types of oxides with different nanopore densities. The observation of nano-dimensioned pores, fluctuating with the phase of the oscillating (photo) current of Si electrodes immersed in dilute ammonium fluoride solutions motivated the further development [6-8] of the original stress model, also to spatial resolution, by applying cellular automates [7]. If the model is presented by a special phase space analysis based on the holographic principle [9-10], then an analogy between the spectral energy densities of the black body radiation and that obtained from the model for current oscillations can be shown [10]. We first describe (section 2) the mathematics of the model and the relation to the Fokker-Planck and Schrödinger equation. The Brownian motion of a virtual particle, which position is defined by an infinitesimal point located at the silicon surface, is discussed in section 3.

2. Mathematical model and the Fokker-Planck equation

The basic process at the electrode surface domain $\Omega \subseteq R^2$ is characterised by a cyclic oxidation of a small silicon layer and the following removal of the oxide by etching (figure 1). Lattice mismatch between silicon and its oxide leads to stress, cracks, and nanopores resulting in a not uniformly evolution of the oxide thickness. The oxide thickness $d(t,x,y)$ is considered at a fixed point *(x,y)* at the electrode surface in dependence on the time *t*. Here the phase $\varphi = \varphi(t,x,y)$ propagates nonlinearly

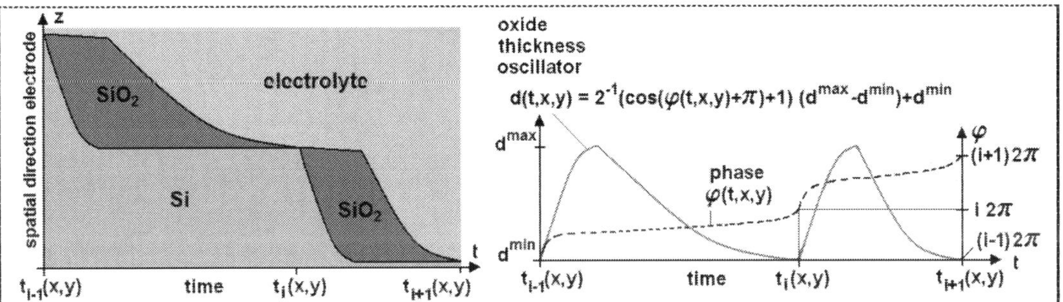

Figure 1. Schematic representation of the oxide thickness evolution at a fixed point *(x,y)* at the electrode surface (left) and the corresponding oxide thickness oscillator and phase (right).

and monotonically increasing for $i2\pi \le \varphi(t,x,y) \le (i+1)2\pi$, $i = 0,1,2,\ldots$. The so called phase oscillator φ is discretized by using the holographic principle [9-10]. Therefore the oxide thickness is registered periodically only once during one cycle. In the phase space $\Phi = \{(x,y,\varphi) \mid (x,y) \in \Omega, \varphi \in R_+\}$ that corresponds to a registration of the phase φ at the so-called snap-shot (holographic) screens $S_i^\phi = \{(x,y,\phi + i\,2\pi) \mid (x,y) \in \Omega\}$ for $i = 0,1,2,\ldots$. The result is the time dependent oscillator density function $p_i^\phi(t)$ [9] defining the differential number of phase oscillators φ passing S_i^ϕ at the time *t* at the *i*-th time. Here the fixed parameter ϕ $(0 \le \phi < 2\pi)$ determines the position of the S_i^ϕ in the phase space. For instance, $\phi = 0$ $(\phi = \pi))$ corresponds to the registration of the oxide thickness at its minimum $d^{\min}$ $(d^{\max})$ at the times $t_i(x,y)$, $i = 0,1,2,\ldots$(figure 1 right) . Assuming that all oscillators φ pass S_i^ϕ during the time interval $[t_{i,\phi}^{\min}, t_{i,\phi}^{\max}]$ at the *i*-th cycle, then two consecutive oscillator density functions are connected by a Markov process represented by a linear integral equation (figure 2 left)

$$p_i^\phi(t) = \int_{t_{i-1,\phi}^{\min}}^{t_{i-1,\phi}^{\max}} p_{i-1}^\phi(s)\, q_{i,s}^\phi(t-s)\, ds \qquad \text{for } 0 \le \phi < 2\pi,\ t_{i,\phi}^{\min} \le t \le t_{i,\phi}^{\max} \text{ and } i = 1,2,3,\ldots \qquad (1)$$

The kernel $q_{i,s}^\phi(t-s)$ of the integral equation is a normalized Greens Function [8] and represents the probability density for the periods $t-s$ of the oscillators in dependence on the starting times s of the oscillators during the *i*-th cycle. Both together, the oscillator density function $p_i^\phi(s)$ and the two-dimensional probability density $q_{i,s}^\phi(t-s)$ for the periods of the oscillators, define a random walk [10]. By equation (1), a temporal evolution of the probability density function $p_i^\phi(t)$ is given (figure 2 right). Moreover, this probability density function is also a solution of the Fokker-Planck equation [11]

$$\frac{\partial}{\partial t} p_i^\phi(t) = -\frac{\partial}{\partial \phi}\left[A(\phi)\, p_i^\phi(t)\right] + \frac{1}{2}\frac{\partial^2}{\partial \phi^2}\left[B(\phi)\, p_i^\phi(t)\right] \quad \text{for } 0 \le \phi < 2\pi,\ t_{i,\phi}^{\min} \le t \le t_{i,\phi}^{\max},\ i = 0,1,2,\ldots \quad (2)$$

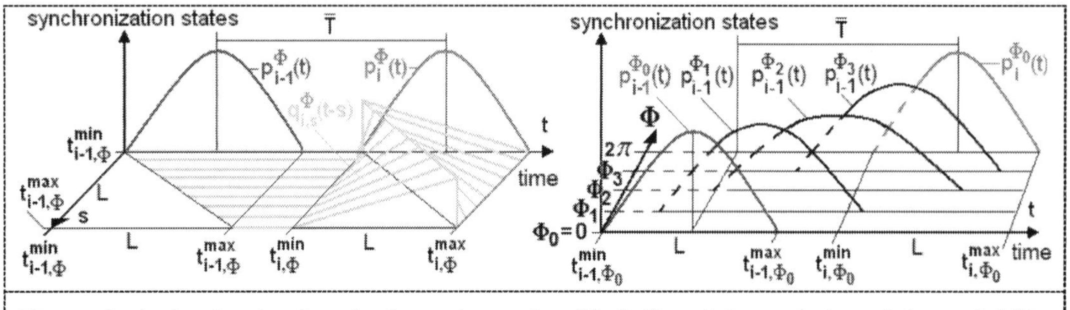

Figure 2. A sketch related to the integral equation (1) (left) and the evolution of the probability density function $p_{i-1}^{\phi}(t)$ shown for $0=\phi_0<\phi_1<\phi_2<\phi_3<2\pi$ (right).

with time-independent drift coefficient $A(\phi)$ and diffusion coefficient $B(\phi)$. In this case the Fokker-Plank equation can be transformed to the Schrödinger equation [11].

3. Brownian motion of a virtual particle

In the following the trajectory of a virtual particle located at the silicon oxide electrode surface is considered during one oscillatory cycle. The trajectory is determined (figure 3 left) using the microscopic nanopore model [5, 9] at which in dependence on the starting time s the evolution of a nanopore as the mean of the evolution of real nanopores during one cycle is regarded. For simplicity, we set $d_{\min}=0$, $\phi=0$ and can omit ϕ in the corresponding notations. In the case of sustained oscillations we choose for q a normalized Greens Function [8, 10] (figure 2 left) i.e.

$$q_{i,s}(t-s)=\begin{cases}2(t-\bar{T})(Ls)^{-1}, & for\ \ \bar{T}\le t\le \bar{T}+s\\ 2(\bar{T}+L-t)L^{-1}(L-s)^{-1}, & for\ \ \bar{T}+s\le t\le \bar{T}+L\end{cases} \tag{3}$$

for $s\in[t_{i-1}^{\min},t_{i-1}^{\max}]$, $L=t_{i-1}^{\max}-t_{i-1}^{\min}$ and $\bar{T}$ denotes the macroscopic oscillation period. Based on the oxide etching process [7], for a fixed $s\in[t_{i-1}^{\min},t_{i-1}^{\max}]$ and $t\in[t_i^{\max},t_i^{\min}]$ the hyperbolic oxide surface $d_{t,s}(r)$ and the always perpendicularly crossing hyperbolic particle trajectory $d_{t,s}^{\perp}(r)$ are defined by

$$d_{t,s}(r)=(1-r_s(t)\,r^{-1})\,d^{\max} \qquad \text{for all } r \text{ with } r_s(t)\le r\le r_{i,s} \tag{4}$$

and

$$d_{t,s}^{\perp}(r)=d^{\max}-((d^{\max})^2+r^2-r_s^2(t))^{1/2} \qquad \text{for all } r \text{ with } \bar{r}_s(t)\le r\le r_s(t). \tag{5}$$

Here $\bar{r}_s(t)=(r_s^2(t)-(d^{\max})^2)^{1/2}$ for $d^{\max}\le r_s(t)\le r_{i,s}$ and $\bar{r}_s(t)=0$ for $0\le r_s(t)<d^{\max}$ determine the radius of the starting point of the particle at the oxide surface. $r_s(t)=r_{i,s}\,(\int_0^t q_{i,s}(\tau-s)\,d\tau)^{1/2}$ defines the radius of the end point of the particle at the bottom of the nanopore where the oxide is removed at the time t and $r_{i,s}=r_{\min}+(r_{\max}-r_{\min})(1-((s-t_{i-1}^{\min})(t_{i-1}^{\max}-t_{i-1}^{\min})^{-1})^{1/2})$ reflects the decreasing of the (half) nanopore distance towards the end of the cycle [9]. Hence, the displacement Δr of a particle is defined by

$$\Delta r_s(t)=r_s(t)-\bar{r}_s(t) \qquad \text{for } s\in[t_{i-1}^{\min},t_{i-1}^{\max}] \text{ and } t\in[t_i^{\max},t_i^{\min}]. \tag{6}$$

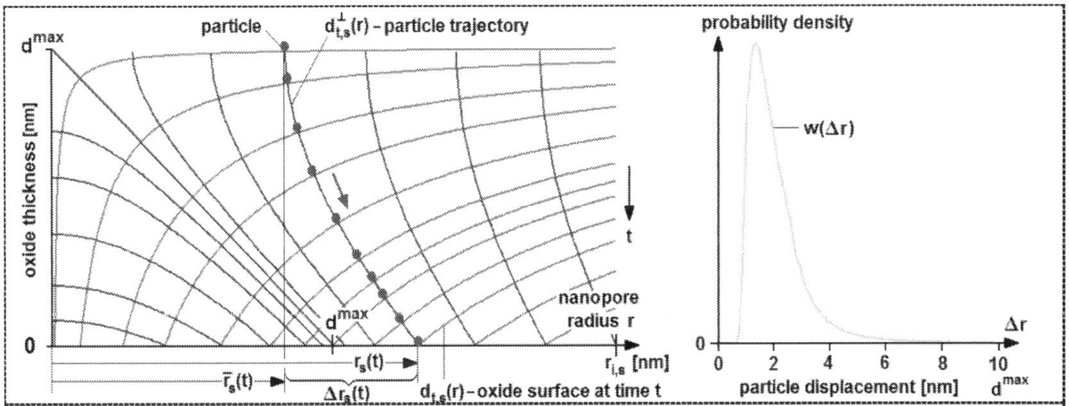

Figure 3. For a fixed oxide growth starting time s of a nanopore, the trajectory (blue line) of a virtual particle crosses the t-dependent oxide surfaces (red lines) always perpendicularly. Here, the oxide is etched away and hence the oxide surface decreases during the time t (left). The probability density for the displacement Δr of a virtual particle shows a Lévy like distribution (right).

Finally, using the nanopore model the probability density w for a displacement Δr of a particle located anywhere at the oxide surface during a cycle is given by

$$w(\Delta r) = c \iint\limits_{\{(s,t)\,|\,\Delta r_s(t)=\Delta r,\, s\in[t_{i-1}^{\min},t_{i-1}^{\max}],\, t\in[t_i^{\min},t_i^{\max}]\}} \bar{r}_s(t)\, q_{i,s}(t-s)\, dt\, ds \qquad \text{for } 0 \le \Delta r \le d^{\max} \tag{7}$$

Here, for a fixed Δr, the path of integration is given by the contour lines of $\Delta r_s(t) = \Delta r$ for $t_{i-1}^{\min} \le s \le t_{i-1}^{\max}$, $t_i^{\min} \le t \le t_i^{\max}$ and c is a normalization factor. Figure 3 right shows the probability density for a particle displacement for $L = \bar{T} = 30s$, $d^{\max} = 10nm$, $r_{\min} = 20nm$, and $r_{\max} = 60nm$.

4. Conclusions and Outlook

The mathematical concept of the discretization of phase oscillators is based on the holographic principle and leads to a linear integral equation representing a Markov process and a random walk. The solution of the integral equation is an oscillator density function whose relation to the Fokker-Planck equation and hence to the Schrödinger equation is shown. Further, a virtual particle located at the silicon oxide surface at the electrode performs a Brownian motion during one cycle. In summary, the model shows analogies to quantum mechanics. Because of the link to the Schrödinger equation and to particle trajectories, the relation to the Bohmian Mechanics [12] should be investigated in future.

5. References

[1] Gerischer H, Lübke M 1988 *Ber. Bunsenges. Phys. Chem.* **92** 573
[2] Aggour M, Giersig M and Lewerenz H J 1995 *J. Electroanal. Chem.* **383** 67
[3] Ozanam F, Blanchard N and Chazalviel J N 1993 *Electrochim. Acta* **38** 1627
[4] Foca E, Carstensen J and Föll H 2007 *J. Electroanal. Chem.* **603** 175
[5] Grzanna J, Jungblut H and Lewerenz H J 2000 *J. Electroanal. Chem.* **486** 181;190
[6] Grzanna J, Jungblut H and Lewerenz H J 2007 *Phys. Stat. Sol. (a)* **204** 1245
[7] Grzanna J, Notz T and Lewerenz H J 2009 *Phys. Stat. Sol. (c)* **6** 1639
[8] Grzanna J, Notz T, Stempel T and Lewerenz H J 2011 *Phys. Stat. Sol. (c)* **8** 1734
[9] Grzanna J and Lewerenz H J 2013 *Journal of Physics: Conference Series* **410** 012160
[10] Grzanna J and Lewerenz H J 2014 *Journal of Physics: Conference Series* **490** 012119
[11] Risken H 1989 *The Fokker Planck Equation*, 2. Edition Springer Verlag
[12] Dürr D, Goldstein S, Tumulka R and Zhanghi N 2004 *Phys. Rev. Letters* **93** 090402

Journal of Physics: Conference Series **574** (2015) 012025 doi:10.1088/1742-6596/574/1/012025

Investigation of nonlinear transformations of the signal spectrum

A Kosarev, E Apushkinskiy, K Maximov, E Ripchinskaya, P Shamray

St. Petersburg Polytechnical University, 195251 St. Petersburg, Russia

sash778@gmail.com, apushkinsky@hotmail.com

Abstract. We consider various echo-processes having different nature and discuss their mathematical models based on the special nonlinear transformations in the spectral domain. It is shown that the presence of a power polynomial in the spectral domain results in a delay of the external perturbation (signal). The correspondence between the degree of nonlinearity and arising delayed signals is established. The agreement between the numerical results and experimental data is demonstrated for several external signals.

1. Introduction

Nonlinear transformation of signal spectrum is a transformation, wherein the output signal depends nonlinearly on the input one. It can be expected that the supplementary signals may occur at time axis in addition to the base (base baud) signals. The echo-signals can be considered as examples of such supplementary pulses [1]. The corresponding interpretation of echo-signals as a result of the nonlinear transformation of input signals spectrum can be found in [2,3].

2. Theory

Let us describe what happens if we apply a nonlinear transformation to a physical signal limited in time. Notice, that physically realizable input signal $s(t)$ is determined only at positive semi-axis of time, while the Fourier transformation is defined for $-\infty < t < \infty$. Because of this, let us redefine the function $s(t)$ along negative time axis by the even reflection. Hence, the spectrum of signal $S(t)$ defined along whole time axis is denoted by expression $\hat{S}(\omega) = \hat{s}(\omega) + \hat{s}^*(\omega)$, where $\hat{s}(\omega) = \int\limits_0^{+\infty} s(t)\mathrm{e}^{-\mathrm{i}\omega t}\mathrm{d}t$. Let us also designate the spectrum of output signal $U(t)$ by $\hat{U}(\omega)$. Let us now suppose that the frequency conversion of input signal by some device (quadrupole) can be expressed by

$$\hat{U}(\omega) = \sum_{n=0}^{N}\left\{a_n[\hat{S}(\omega)]^n\right\}, \tag{1}$$

where a_n are factors determining the level of the corresponding components.

The term with $n = 0$ in formula (1) can be interpreted as the own thermal noise of our quadrupole. The term with $n = 1$ corresponds to the standard linear transformation of input signal (e.g. filtration), and all the subsequent terms with $n \geq 2$ are nonlinear frequency transformations.

Let us consider the sequence of M non-overlapping pulses (see figure 1) as input signal, i.e.,

$$s(t)=\sum_{m=1}^{M}s_m(t-\tau_m). \tag{2}$$

Here, τ_m stands for the beginning of the m-th pulse.

If all the functions $S_m(t-\tau_m)$ are real ones, we have $\hat{S}(\omega)=\hat{S}^*(-\omega)$. It means that the spectrum of the signal in the negative frequency range can be represented by the complex conjugation of spectrum of the same signal in the positive frequency range. So, we obtain the following expansion:

$$\hat{S}(\omega)=\sum_{m=1}^{M}\hat{S}_m(\omega)=\sum_{m=1}^{M}\left[\hat{s}_m(\omega)\mathrm{e}^{-\mathrm{i}\omega\tau_m}+\hat{s}_m^*(\omega)\mathrm{e}^{\mathrm{i}\omega\tau_m}\right], \tag{3}$$

where i is an imaginary unit.

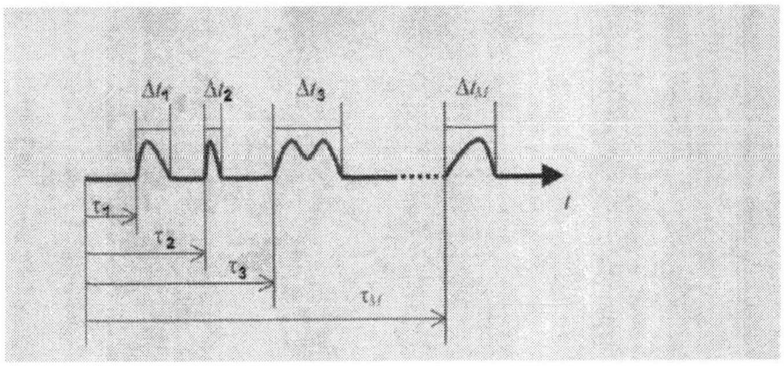

Figure 1. Input signal (see formula (2)).

This signal acts on the quadrupole, which transforms the frequencies in a nonlinear manner. Substituting (3) into (1) we obtain that the spectrum of output signal after M pulses can be expresses by

$$\hat{U}(\omega)=\sum_{n=0}^{N}[a_n\sum_{m=1}^{\frac{(M+n-1)!}{(M-1)!n!}}C_m\prod_{k=1}^{M}\hat{S}_k^{q_k}]. \tag{4}$$

Here, $N\geq 2$, C_m are numerical coefficients, and exponents q_k take the all possible values from 0 to n such that $q_1+q_2+\ldots+q_k=n$.

The signal emitted by the quadrupole can be obtained from the formula (4) with a use of the inverse Fourier transform. If there are several quadrupoles, the overall signal is a sum of the responses from all of them taking into account the distribution function of quadrupoles over resonance frequencies. Assuming that this distribution is continuous and described by distribution function $F(\omega_l)$, we can replace summation by integration and obtain the following formula for the signal emitted by material or system of quadrupoles after irradiation by the sequence of M pulses:

$$U(t)=\frac{1}{2\pi}\int_0^{\infty}F(\omega_l)\int_{-\infty}^{\infty}\left[\sum_{n=0}^{N}a_n\sum_{m=1}^{\frac{(M+n-1)!}{(M-1)!n!}}C_m\prod_{k=1}^{M}\hat{S}_k^{q_k}\right]\mathrm{e}^{\mathrm{i}\omega t}\mathrm{d}\omega\mathrm{d}\omega_l. \tag{5}$$

The output signal formed by contributions from individual quadrupoles can change its intensity with time if the radiation from individuals becomes in-phase one at certain moments.

Now, using (5) we can show that echo-signals can be generated in the systems of quadrupoles with frequency nonlinearity. Let us consider the simplest case, where the quadrupole is undergone by action of two pulses, $s_1(t)$ starting at the moment τ_1 and $s_2(t)$ starting at the moment τ_2. It is obvious that there is no echo if we set $N = 2$ in formula (4). Otherwise, if we take $N = 3$, the output signal emitted by quadrupoles can be expressed by

$$U(t) = \frac{1}{2\pi}\int_0^\infty F(\omega_l)\int_{-\infty}^{\infty}[a_0 + a_1(\hat{S}_1 + \hat{S}_2) + a_2(\hat{S}_1^2 + 2\hat{S}_1\hat{S}_2 + \hat{S}_2^2) + a_3(\hat{S}_1^3 + 3\hat{S}_1^2\hat{S}_2 + 3\hat{S}_1\hat{S}_2^2 + \hat{S}_2^3)]e^{i\omega t}d\omega d\omega_l. \quad (6)$$

Substituting representation $\hat{S}_m = \hat{s}_m e^{-i\omega\tau_m} + \hat{s}_m^* e^{i\omega\tau_m}$ in (6), we obtain that the echo-signals can be provided only by the following terms:

$$\frac{1}{2\pi}\int_0^\infty F(\omega_l)\int_{-\infty}^{\infty} 3a_3[\hat{s}_1\hat{s}_2^2 e^{i\omega(t-2\tau_2-\tau_1)} + \hat{s}_1^*\hat{s}_2^2 e^{i\omega(t-2\tau_2+\tau_1)} + \hat{s}_1^2\hat{s}_2 e^{i\omega(t-\tau_2-2\tau_1)}]\,d\omega d\omega_l. \quad (7)$$

The signal arising during the doubled time interval between the first and second exciting pulses gives the second term in (7). Thus, the primary echo can be expressed by

$$\frac{1}{2\pi}\int_0^\infty F(\omega_l)\int_{-\infty}^{\infty} 3a_3\hat{s}_1^*\hat{s}_2^2 e^{i\omega(t-2\tau_2+\tau_1)}d\omega d\omega_l. \quad (8)$$

3. Calculation / Experiment

With a use of the above described algorithm, we performed the calculations for echo-signals occurring from NMR after excitation of substance (system of quadrupoles) by two or four radio-frequency pulses. We also compared the results of calculations with the experimental data obtained during the excitation of echoes in thin ferromagnetic cobalt films [4,5] and lithium ferrite [6]. The corresponding results are shown in figures 2 and 3. The simulations and experiments are in good agreement.

4. Discussion

It is of interest to compare the results obtained by our method with the spectra of two-pulse echo-signals of a different nature. These spectra can be determined as approximate solutions of the nonlinear partial differential equations describing the relevant phenomena. It is shown in [5] that the approximate solution of the Bloch equations gives at $\tau_1 = 0$ the following formula for the spectrum of the primary nuclear spin echo arising at the moment $t = 2\tau_2$: $\hat{S}_{2\tau_2} = \hat{s}_1^*\hat{s}_2^2$. Similar expressions are also found for the spectra of the primary echo-signals generated in high-temperature superconductors by solving the corresponding nonlinear equations (see [7]). So, the suggested algorithm is in good agreement with the theoretical and experimental results obtained earlier.

Based on the above arguments, one can conclude that in the case of excitation by pulses with equal carrier frequencies, the minimum value of parameter N in formula (4) guaranteeing the existence of the echo-signal is $N = 3$. For higher degree of nonlinearity (i.e., $N \geq 4$), the additional or secondary echo-signals can be registered at the moments $t = k\tau_2$. Notice, for the special case, where the frequency of the second pulse is two times more than the frequency of the first one, the so-called parametric echo [8] appears for $N \geq 2$.

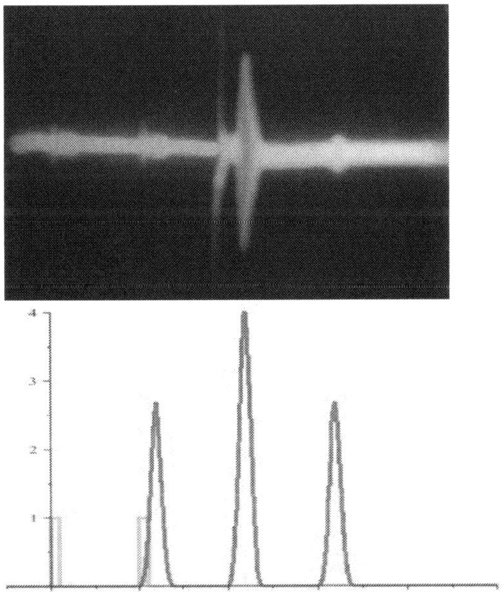

Figure 2. Experimentally observed signals of single-pulse echo, primary and secondary two-pulse echo and their calculations on Maple 14.

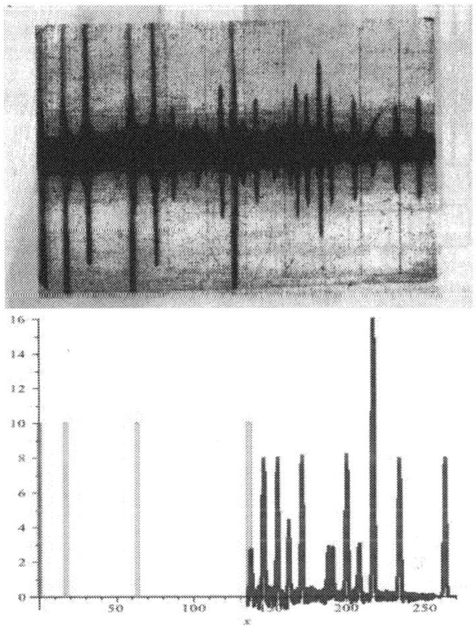

Figure 3. Experimentally observed echo signals from four excitation pulses [6] and their calculations on Maple 14.

5. Conclusion

The suggested algorithm can especially be attractive for echo-signal analysis upon application of the complicated and noise external actions since it is extremely difficult to obtain the approximate solutions of nonlinear equations.

References

[1] Petrov M P, Paugurt A P, Pleshakov I V and Ivanov A V 1985 *Technical Physics Letters* **11**, 1204-07

[2] Korpel A and Chatterjee M 1981 *Proc. IEEE* **69** 1539-56

[3] Ryzak I S 2000 *Radiotechnika i electronika* **45** 5-38

[4] Apushkinsky E and Astrov M 2002 *Physica B: Condensed Matter* **321(1-4)** 408-13

[5] Apushkinsky E G, Kazak A B and Nesterov O A 1982 *Voprosy radioelektroniki, series: Technologia proizvodstva i oborudovanija* (Leningrad) issue 1 13-22

[6] Tarhanov V I 2002 *Geometric algebra, and NMR data processing* (Publishers of Saint-Petersburg Polytechnical University)

[7] Apushkinskii E G, Astrov M S and Sobolevskii V K 2011 *Tech. Phys.* **56** 788-96

[8] Nesterov M M, Pleshakov I V and Fofanov Y A 2006 *Nauchnoe priborostroenie* **16** 64-71

Nonlinear waves on the surface of non-isothermal fluid film

L A Prokudina

South-Ural State University, Chelyabinsk, Russia

E-mail: prokudina-la@mail.ru

Abstract. A mathematical model of non-isothermal flow of a fluid film is presented. In conditions of Marangoni instability wave characteristics are calculated, identify areas of instability of a fluid film. It shows the influence of thermocapillary forces and forces of surface viscosity on the form of waves. Nonlinear development of perturbations belonging to a continuous band of wave numbers on the surface of a thin fluid layer is investigated within the framework of a nonlinear parabolic equation.

1.Introduction

The investigation of a thin layer of a viscous fluid (a fluid film) that combines thin and large contacted surfaces connected with the implementation of the trends in heat and mass transfer apparatus, for example, thermal power, chemical, food, pharmaceutical industry. The study of the wave flow of a fluid film started by P. Kapitsa [1, 2] and further developed in numerous works of scientists, for example, L. Kholpanov [3], V. Shkadov [4], V. Levich [5], V. Nakoryakov, S. Alekseenko [6], L. Jones, S. Whitaker [7], F. Stainthorp [8]C. Massot, F. Irani, E. Lightfoot [9], V. Penev [10], B. Gjevik [11], C. Yih [12] from different countries. On regimes of the flow of a fluid film are influenced of various physical and chemical factors. The use of additives of surface-active substances qualitatively changes the character of the flow of a fluid film. The gradients of temperature and concentration cause the heterogeneity of the surface tension and the appearance of the surface tension gradients, resulting in the interphase surface arise force (thermocapillary), the nature of which is determined by the real physical process.

Various heat-mass exchange processes occur in conditions of interfacial instability of the interface, caused by a variety of effects, factors. Special attention attracts researchers interfacial instability associated with the Marangoni effect. The Marangoni effect related to the change at the interface of surface tension from point to point. Emerging Marangoni instability manifests itself in changes of the wave characteristics [13].

2. Mathematical model of non-isothermal liquid film

Consider for a thin layer of viscous incompressible fluid (fluid film) thick under the action of gravity on solid inclined plane in a rectangular coordinate system OXYZ.

The mathematical model of the flow of a three-dimensional non-isothermal fluid film of thickness δ is a system of Navier–Stokes equations and continuity equations with boundary conditions, taking into account the effect of heat – mass – transfer processes [13]:

$$\begin{cases} \frac{\partial u_+}{\partial t_+} + u_+\frac{\partial u_+}{\partial x_+} + v_+\frac{\partial u_+}{\partial y_+} + w_+\frac{\partial u_+}{\partial z_+} = -\frac{\partial P_+}{\partial x_+} + F_x + \frac{1}{\mathrm{Re}}\left(\frac{\partial^2 u_+}{\partial x_+^2} + \frac{\partial^2 u_+}{\partial y_+^2} + \frac{\partial^2 u_+}{\partial z_+^2}\right) \\ \frac{\partial v_+}{\partial t_+} + u_+\frac{\partial v_+}{\partial x_+} + v_+\frac{\partial v_+}{\partial y_+} + w_+\frac{\partial v_+}{\partial z_+} = -\frac{\partial P_+}{\partial y_+} + F_y + \frac{1}{\mathrm{Re}}\left(\frac{\partial^2 v_+}{\partial x_+^2} + \frac{\partial^2 v_+}{\partial y_+^2} + \frac{\partial^2 v_+}{\partial z_+^2}\right) \\ \frac{\partial w_+}{\partial t_+} + u_+\frac{\partial w_+}{\partial x_+} + v_+\frac{\partial w_+}{\partial y_+} + w_+\frac{\partial w_+}{\partial z_+} = -\frac{\partial P_+}{\partial z_+} + F_z + \frac{1}{\mathrm{Re}}\left(\frac{\partial^2 w_+}{\partial x_+^2} + \frac{\partial^2 w_+}{\partial y_+^2} + \frac{\partial^2 w_+}{\partial z_+^2}\right) \\ \frac{\partial u_+}{\partial x_+} + \frac{\partial v_+}{\partial y_+} + \frac{\partial w_+}{\partial z_+} = 0, \end{cases} \quad (1)$$

$$y = 0: \quad u_+ = w_+ = 0, \qquad v_+ = V_0 \quad (2)$$

$$y = \delta: \quad \frac{1}{\mathrm{Re}}\left[2\frac{\partial u_+}{\partial x_+}\frac{\partial \delta_+}{\partial x_+} - 2\frac{\partial v_+}{\partial y_+}\frac{\partial \delta_+}{\partial x_+} - \left(\frac{\partial v_+}{\partial x_+} + \frac{\partial u_+}{\partial y_+}\right) + \frac{\partial \delta_+}{\partial z_+}\left(\frac{\partial u_+}{\partial z_+} + \frac{\partial w_+}{\partial x_+}\right)\right]$$

$$+M\frac{\partial \delta_+}{\partial x_+} + N\left(\frac{\partial^2 u_+}{\partial x_+^2} + \frac{\partial^2 w_+}{\partial x_+ \partial z_+}\right) + \tau_x = 0 \quad (3)$$

$$\frac{1}{\mathrm{Re}}\left[2\frac{\partial w_+}{\partial z_+}\frac{\partial \delta_+}{\partial z_+} - 2\frac{\partial v_+}{\partial y_+}\frac{\partial \delta_+}{\partial z_+} - \left(\frac{\partial w_+}{\partial y_+} + \frac{\partial v_+}{\partial z_+}\right) + \frac{\partial \delta_+}{\partial x_+}\left(\frac{\partial u_+}{\partial z_+} + \frac{\partial w_+}{\partial x_+}\right)\right]$$

$$+M\frac{\partial \delta_+}{\partial z_+} + N\left(\frac{\partial^2 w_+}{\partial z_+^2} + \frac{\partial^2 u_+}{\partial x_+ \partial z_+}\right) + \tau_z = 0 \quad (4)$$

$$P_+ = \frac{2}{\mathrm{Re}}\left[\frac{\partial v_+}{\partial y_+} - \frac{\partial \delta_+}{\partial x_+}\left(\frac{\partial u_+}{\partial y_+} + \frac{\partial v_+}{\partial x_+}\right) - \frac{\partial \delta_+}{\partial z_+}\left(\frac{\partial w_+}{\partial y_+} + \frac{\partial v_+}{\partial z_+}\right)\right] - \sigma_+\left(\frac{\partial^2 \delta_+}{\partial x_+^2} + \frac{\partial^2 \delta_+}{\partial z_+^2}\right)$$

$$+P_0 - \mathrm{sgn}\,\Delta T \frac{\frac{\rho_1}{\rho_2} - 1}{(\mathrm{Re}\cdot \mathrm{Pr}\cdot \mathrm{Ku})^2}\frac{1}{\delta^2} \quad (5)$$

$$\frac{\partial \delta}{\partial t_+} = v_+ - u_+\frac{\partial \delta}{\partial x_+} - w_+\frac{\partial \delta}{\partial z_+} + \frac{1}{\mathrm{Re}\cdot \mathrm{Pr}\cdot \mathrm{Ku}}\frac{1}{\delta}. \quad (6)$$

In formulas (1–6), dimensionless quantities: $u_+ = \frac{u}{u_0}$, $v_+ = \frac{v}{u_0}$, $w_+ = \frac{w}{u_0}$ are projections of velocity onto the corresponding coordinate axes; $t_+ = \frac{tu_0}{\delta_0}$ is time; $x_+ = \frac{x}{\delta_0}$, $y_+ = \frac{y}{\delta_0}$, $z_+ = \frac{z}{\delta_0}$ are variables;

$\delta_+ = \dfrac{\delta}{\delta_0}$ is the thickness of the fluid film; $\mathrm{Re} = \dfrac{u_0 \delta_0 \rho}{\mu}$ is the Reynolds number; $F_x = \dfrac{g_x \delta_0}{u_0^2}$, $F_y = -\dfrac{g_y \delta_0}{u_0^2}$, $F_z = \dfrac{g_z \delta_0}{u_0^2}$ are the projections of the Froude number onto the corresponding coordinate axes; $P_+ = \dfrac{P}{\rho u_0^2}$ is pressure; $N = \dfrac{k+e}{\rho \delta_0^2 u_0}$ is the surface viscosity parameter; $\bar{\tau}_x = \dfrac{\tau_x}{\rho u_0^2}$, $\bar{\tau}_z = \dfrac{\tau_z}{\rho u_0^2}$ are the projections of tangential stress; $M = M_T + M_K$ is the Marangoni parameter, where $M_T = \dfrac{\partial \sigma}{\partial T^0}\left(\dfrac{\partial T^0}{\partial y}\right)_{y=\delta} \dfrac{1}{\rho u_0^2}$, $M_K = \dfrac{\partial \sigma}{\partial K}\dfrac{\partial K}{\partial \delta}\dfrac{1}{\rho u_0^2}$; $\mathrm{Pr} = \dfrac{\rho c_p \upsilon}{\lambda}$ is the Prandtl number; $\mathrm{Ku} = \dfrac{r'}{c_p \Delta T}$ is the number of a phase transition; $\sigma_+ = \dfrac{\sigma}{\rho u_0^2 \delta_0}$ is the surface tension parameter, where u_0 is the average velocity of the main flow of the fluid film, δ_0 is the thickness of the film in the unperturbed state. During condensation, $\mathrm{sgn}\,\Delta T = 1$, whereas, during evaporation, $\mathrm{sgn}\,\Delta T = -1$.

The projections of the velocity of (1)–(5) we substitute in equation (6) and get a nonlinear differential equation for the deviation $\psi(x,t)$ of the two-dimensional free surface of the fluid film in the form:

$$\left(\frac{1}{\mathrm{Re}\,\mathrm{Pr}\,Ku} - \frac{1}{\mathrm{Re}\,\mathrm{Pr}\,Ku}\psi\right) + \left(a_7 \frac{\partial}{\partial x} + a_{13}\right)\frac{\partial \psi}{\partial t} + a_1 \frac{\partial^4 \psi}{\partial x^4} + a_4 \frac{\partial^3 \psi}{\partial x^3} + a_6 \frac{\partial^2 \psi}{\partial x^2}$$
$$+ a_{11}\frac{\partial \psi}{\partial x} + a_{14}\psi \frac{\partial \psi}{\partial x} + a_{16}\psi \frac{\partial^2 \psi}{\partial x^2} + a_{17}\psi \frac{\partial^2 \psi}{\partial x \partial t} + a_{21}\frac{\partial \psi}{\partial x}\frac{\partial \psi}{\partial t} + a_{22}\left(\frac{\partial \psi}{\partial x}\right)^2$$

$$+ a_{26}\frac{\partial \psi}{\partial x}\frac{\partial^2 \psi}{\partial x^2} + a_{28}\psi \frac{\partial^3 \psi}{\partial x^3} + a_{30}\frac{\partial \psi}{\partial x}\frac{\partial^3 \psi}{\partial x^3} + a_{34}\psi \frac{\partial^4 \psi}{\partial x^4} + a_{37}\psi^2 \frac{\partial \psi}{\partial x} + a_{39}\psi^2 \frac{\partial^2 \psi}{\partial x^2} + a_{40}\psi^2 \frac{\partial^2 \psi}{\partial x \partial t}$$
$$+ a_{44}\psi \frac{\partial \psi}{\partial x}\frac{\partial \psi}{\partial t} + a_{45}\psi \left(\frac{\partial \psi}{\partial x}\right)^2 + a_{49}\psi \frac{\partial \psi}{\partial x}\frac{\partial^2 \psi}{\partial x^2} + a_{51}\psi \frac{\partial \psi}{\partial x}\frac{\partial^3 \psi}{\partial x^3} + a_{55}\psi^2 \frac{\partial^4 \psi}{\partial x^4} + a_{58}\psi^2 \frac{\partial^3 \psi}{\partial x^3} = 0. \quad (7)$$

The numbering of the coefficients of the equation (7) is more complete his mind [14].
The coefficients of Eq. (7) are given by

$$a_1 = -\frac{\mathrm{Re}\,\sigma}{3},\; a_4 = -\frac{\mathrm{Re}^2 F_x N}{2},\; a_6 = -\frac{\mathrm{Re}\,F_y}{3} - \frac{\mathrm{Re}\,M}{2} + \frac{3}{40}\mathrm{Re}^3 F_x\left(\tau_x + F_x\right) + \frac{2}{3} sgn\Delta T \frac{\mathrm{Re}}{\left(\mathrm{Re}\,\mathrm{Pr}\,Ku\right)^2},$$

$$a_7 = \frac{5}{24}\mathrm{Re}^2 F_x,\; a_{11} = -\mathrm{Re}\,F_x - \mathrm{Re}\,\tau_x,\; a_{13} = -1,\; a_{14} = -2\mathrm{Re}\,F_x - \mathrm{Re}\,\tau_x,\, a_{21} = a_{17},\; a_{22} = a_{16},$$

$$a_{16} = -\mathrm{Re}\,F_y - \mathrm{Re}\,M + \frac{3}{8}\mathrm{Re}^3 F_x \tau_x + \frac{9}{20}\mathrm{Re}^3 F_x^2,\; a_{17} = 4a_7,\; a_{18} = \frac{3}{8}\mathrm{Re}^3 (F_z \tau_x + F_x \tau_z) + \frac{9}{10}\mathrm{Re}^3 F_x^2,$$

3. The surface condition of non-isothermal liquid film

Negative temperature gradients occur in the flow of a fluid film on a heated surface. In this case significance $M > 0$ is for the Marangoni parameter.

Under large temperature gradients thermocapillary forces can destroy a fluid film that can cause local overheating cooled film wall and create an emergency situation when working heat and mass– transfer apparatus. The process of development of disturbances in time simulating the destruction of a film (with significant temperature gradients) presented in Figure 1.

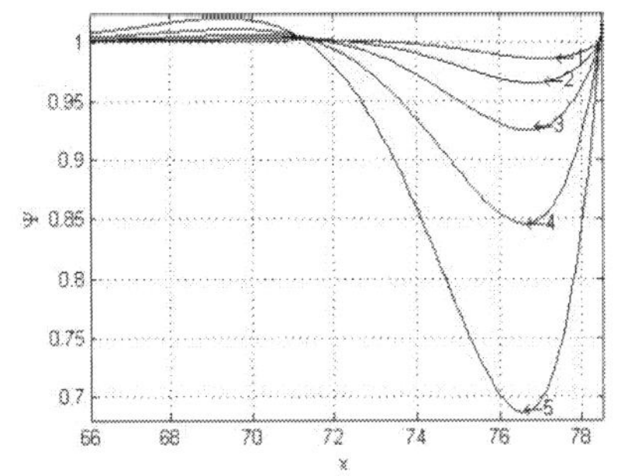

Figure 1. Surface conditions of a vertical film for Re=10, M=1: 1 - t=0; 2 - t=0.0125; 3 - t=0.025; 4 – t=0.0375; t=0.05

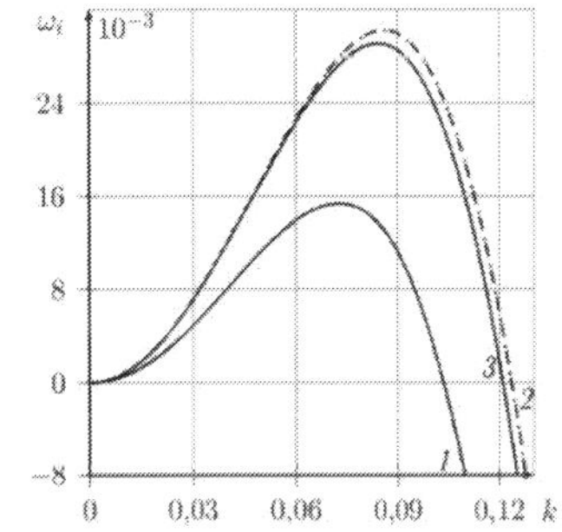

Figure 2. Increment as a function of wave number:

1–M=0, 2–M=1, 3–M=1, N=1

4. Nonlinear parabolic equation

On the surface of a fluid film developing disturbances belonging continuous band of wave numbers (wave packet), under the following assumptions: the wave number band is $\Delta k_{\Sigma} = o(\varepsilon)$, and the increment is $\omega_i = \varepsilon^2 \bar{\omega}_i = o(\varepsilon^2)$,where ε is a small parameter. The equation for the amplitude A envelope of a wave packet is a nonlinear parabolic equation of the form [15, 16]:

$$\frac{\partial A}{\partial t_2} + \frac{i}{\varepsilon}\frac{\partial \omega_i}{\partial k_x}\frac{\partial A}{\partial x_1} + \frac{i}{\varepsilon}\frac{\partial \omega_i}{\partial k_z}\frac{\partial A}{\partial z_1} - \frac{i}{2}\left(\frac{\partial^2 \omega_r}{\partial k_x^2} + i\frac{\partial^2 \omega_i}{\partial k_x^2}\right)\frac{\partial^2 A}{\partial x_1^2}$$

$$-\frac{i}{2}\left(\frac{\partial^2 \omega_r}{\partial k_z^2} + i\frac{\partial^2 \omega_i}{\partial k_z^2}\right)\frac{\partial^2 A}{\partial z_1^2} - i\left(\frac{\partial^2 \omega_r}{\partial k_x \partial k_z} + i\frac{\partial^2 \omega_i}{\partial k_x \partial k_z}\right)\frac{\partial^2 A}{\partial x_1 \partial z_1} = \bar{\omega}_i A - (\beta_1 + i\beta_2)|A|^2 A. \qquad (8)$$

The coefficients of the nonlinear term in equation (8) characterize the nonlinear damping of perturbations (β_1) and the dependence of phase (frequency) on the amplitude (β_2). The first- and second-order derivatives in Eq. (8) are calculated using the dispersion equation

$$\omega(a_7 k_x + a_9 k_z + i) + a_1 k_x^4 + a_2 k_x^2 k_z^2 + a_3 k_z^4 - a_4 i k_x^3 - a_5 i k_z^3 - -a_6 k_x^2 - a_8 k_x k_z$$

$$-a_{10} k_z^2 + a_{11} i k_x + a_{12} i k_z - \frac{1}{\mathrm{Re} \cdot \mathrm{Pr} \cdot \mathrm{Ku}} = 0, \qquad (9)$$

where $\omega = \omega_r + i\omega_i$, ω_r – frequency, ω_i – increment.

The work presents the results of computational experiments for vertical film of water for Reynolds number $\mathrm{Re} = 5$ as in free dripping wet film, and considering thermocapillary forces and forces of surface viscosity: increment (Figure 2).

The waves are wave numbers $k_{\omega_{i\max}}$ with the highest probability are realized in experiments. Wave characteristics are shown in Table 1, where:

$c_r = \dfrac{\omega_r}{k}$ is the phase velocity, $V_{gr} = \dfrac{d\omega_r}{dk}$ is the group velocity.

Table 1. Wave characteristics

Re=5	$k_{\omega_{i\max}}$	$\omega_{i\max}$	ω_r	c_r	V_{gr}	β1	β2
M=0, N=0	0,0725	0,0154	0,2139	2,9520	2,9520	1,6896	-0,0333
M=1, N=0	0,0858	0,0302	0,2493	2,9055	2,9055	1,3955	-0,2014
M=1, N=1	0,0842	0,0291	0,2405	2,8560	2,7503	1,2897	-0,1159

5. Conclusions

1. We have presented a nonlinear mathematical model of a nonisothermal three-dimensional fluid film for Reynolds number of $\mathrm{Re} \leq 20$.

2. We have presented a nonlinear mathematical model for the state of the free surface of a nonisothermal fluid film.

3. Using a dispersion equation, we have calculated the wave characteristics: the frequency, the increment and the phase velocity.

4. We have calculated the forms of waves on the surface of a fluid film under inhomogeneous surface tension.

5. We have derived a nonlinear parabolic equation for the amplitude of the envelope of a wave packet, which is one of the basic models for nonlinear media (a Ginzburg–Landau –type model).

References

[1] Kapitsa P L 1948 Zh. Eksp. Teor. Fiz. **18** 3
[2] Kapitsa P L and Kapitsa S P 1949 Zh. Eksp. Teor. Fiz. **19** 105
[3] Kholpanov L P 1965 Inzh._Fiz. Zh. **8** 488
[4] Shkadov V YA 1967 Izv. Akad. Nauk SSSR, Mekh. Zhidk.Gaza No. 1 43
[5] Levich V G, Krylov V S and Vorotilin V P 1968 Dokl. Akad. Nauk SSSR **183** 6
[6] Alekseenko S V, Nakoryakov V E, Pokusaev B G 1979 Prikl. Mekh. Tekh. Fiz **20** (6) 77
[7] Jones L O and Whitaker S 1966 AIChE J **12** 525
[8] Stainthorp F P and AllenJ M 1965 Trans. Inst. Chem.Eng. **43** 85
[9] MassotC, Irani F and Lightfoot E N 1966 AIChE J **12** 445
[10] Penev V, Krylov V S, Boyadjiev Ch, and Vorotilin V P 1972 Int. J. Heat Mass Transfer **15** 7
[11] Gjevik B 1970 Phys. Fluids **13** 1918
[12] Yih C S 1963 Phys. Fluids **6** 321
[13] Prokudina L A and Vyatkin G P 1998 Dokl. Phys **43**(10) 652
[14] Prokudina L A 2014 Influence of Surface Tension inhomogeneity on the Wave Flow of a liquid Film J. Eng. Phys. and Thermophys. **87** 165
[15] Prokudina L A and Vyatkin G P 2011 Self-organization of perturbations in fluid films Dokl. Phys. **56**(8) 444
[16] Prokudina L A 2014 Nonlinear Evolution of Perturbations in a Thin Fluid Layer during Wave Formation J. Exper. and Theor. Phys. **118**(3) 480

Homoclinic chaos in a pair of parametrically-driven coupled SQUIDs

M. Agaoglou[1], V. M. Rothos[1], H. Susanto[2]
[1] Department of Mechanical Engineering, Faculty of Engineering, Aristotle University of Thessaloniki, Thessaloniki 54124, Greece
[2] Department of Mathematical Sciences, University of Essex, Wivenhoe Park, Colchester CO4 3SQ, United Kingdom
E-mail: rothos@auth.gr

Abstract. An rf superconducting quantum interference device (SQUID) consists of a superconducting ring interrupted by a Josephson junction (JJ). When driven by an alternating magnetic field, the induced supercurrents around the ring are determined by the JJ through the celebrated Josephson relations. This system exhibits rich nonlinear behavior, including chaotic effects. We study the dynamics of a pair of parametrically-driven coupled SQUIDs arranged in series. We take advantage of the weak damping that characterizes these systems to perform a multiple-scales analysis and obtain amplitude equations, describing the slow dynamics of the system. This picture allows us to expose the existence of homoclinic orbits in the dynamics of the integrable part of the slow equations of motion. Using high-dimensional Melnikov theory, we are able to obtain explicit parameter values for which these orbits persist in the full system, consisting of both Hamiltonian and non-Hamiltonian perturbations, to form so-called Silnikov orbits, indicating a loss of integrability and the existence of chaos.

1. Introduction

We consider a series of electrical circuits in a line and we could derive the following nonlinear lattice equation [1],

$$\ddot{f}_n + \gamma\dot{f}_n + f_n + \hat{\beta}\sin(2\pi f_n) - \lambda(f_{n+1} - 2f_n + f_{n-1}) = 0 \qquad (1.1)$$

The coefficient of the nonlinearity, which corresponds to the so-called Josephson critical current, is modulated in time, which can be realized experimentally by modulating the surrounding temperature. As the temperature also influences the damping due to the unpaired electron, additionally we also consider time-periodically modulated dissipation. Compared to [5], we consider softening nonlinearity as opposed to stiffening one, i.e. the nonlinearity has different sign.

Here, we wish to study the possibility of homoclinic chaos near resonances in typical SQUID lattice, which is described by eq. (2.1). In particular, we consider a specific dimer for the lattice equation (2.1) and show that the unperturbed system has a homoclinic orbit in their collective dynamics. Applying the geomertical method of singular perturbation theory near the resonaces and Melnikov theory of near integrable Hamiltonian system to predict the chaotic behavior near resonances [3]. Moreover, we state the theorem for the exisence of multi-homoclinic orbits near resonance following the references [3, 2]. In Section 2, we perform a multi-scale analysis for the dimer case and obtain amplitude equations to describe the slow dynamics of the system of dimer SQUID. It is in the slow dynamics that the homoclinic orbits are found. In Section 3, we perform the analytical method for the existence of homoclinic orbits in the perturbed system. In particular we calculate the Energy-difference function using the Melnikov integral evaluated on the homoclinic solutions and applying the singular perturbation theory we study the dynamics near resonances. In Section 4, we illustrate numerically our analytical results for the SQUID model.

2. Normal mode amplitude equations

We consider the 1D case of eq (6) in [1]

$$\ddot{f}_n + \gamma \dot{f}_n + f_n + \hat{\beta}\sin(2\pi f_n) - \lambda(f_{n+1} - 2f_n + f_{n-1}) = 0 \tag{2.1}$$

for n=1,2 and $f_0 = f_3 = 0$ we obtain

$$\begin{aligned} \ddot{f}_1 + \gamma \dot{f}_1 + f_1 + \hat{\beta}\sin(2\pi f_1) - \lambda f_2 &= 0 \\ \ddot{f}_2 + \gamma \dot{f}_2 + f_2 + \hat{\beta}\sin(2\pi f_2) - \lambda f_1 &= 0 \end{aligned} \tag{2.2}$$

We could consider a small parameter $\gamma_0 \to \epsilon\hat{\gamma}_0$, $\hat{\beta} = \beta + f_p\cos(\omega_p t)$, $\gamma = \gamma_0$ and drive amplitude $f_p = \epsilon h$. The normal mode frequencies for the linear system are:

$$\omega_1^2 = 1 + 2\pi\beta - \lambda, \omega_2^2 = 1 + 2\pi\beta + \lambda \tag{2.3}$$

We use multiple time scales to express the displacements:

$$f_{1,2}(t) = \frac{\sqrt{3\epsilon}}{2}(A_1(T)\mathrm{e}^{i\omega_1 t} \pm A_2(T)\mathrm{e}^{i\omega_2 t} + c.c) + \epsilon^{3/2} f_{1,2}^{(1)}(t) + \dots \tag{2.4}$$

where f_1 is taken with the positive sign and f_2 with the negative sign, slow time $T = \epsilon t$. The drive frequency ω_p is related with ω_2: $\omega_p = 2\omega_2 + \epsilon\Omega$, $\omega_1 = \omega_2 + 2\epsilon\Omega_1$. Substituting multiple time scale expression to the system of f_1, f_2 generates secular terms that yield two coupled equations for the complex amplitudes $A_{1,2}$. Express the complex amplitudes using real amplitudes and phases as

$$A_1(T) = a_1(T)\mathrm{e}^{i(x_1(T)+(\Omega/2-2\Omega_1)T)}, \quad A_2(T) = a_2(T)\mathrm{e}^{i(x_2(T)+\Omega T/2)} \tag{2.5}$$

the real and imaginary parts of the two secular amplitude equations become:

$$\begin{aligned} \frac{\mathrm{d}a_1}{\mathrm{d}T} &= \frac{3\pi^3\beta}{2\omega_1}a_1a_2^2\sin(2(x_2 - x_1) + \frac{\pi\hat{h}}{2\omega_1}\sin(2x_1)a_1 - \frac{1}{2}a_1\hat{\gamma}_0 \\ \frac{\mathrm{d}x_1}{\mathrm{d}T} &= -\frac{1}{2}\Omega + 2\Omega_1 - \frac{3\pi^3\beta}{2\omega_1}(a_1^2 + 2a_2^2 + a_2^2\cos2(x_2 - x_1)) + \frac{\pi\hat{h}}{2\omega_1}\cos(2x_1) \\ \frac{\mathrm{d}a_2}{\mathrm{d}T} &= \frac{3\pi^3\beta}{2\omega_1}a_2a_1^2\sin(2(x_1 - x_2) + \frac{\pi\hat{h}}{2\omega_2}\sin2x_2a_2 - \frac{1}{2}a_2\hat{\gamma}_0 \\ \frac{\mathrm{d}x_2}{\mathrm{d}T} &= -\frac{\Omega}{2} - \frac{3\pi^3\beta}{2\omega_2}(a_2^2 + 2a_1^2 + a_1^2\cos2(x_1 - x_2)) + \frac{\pi\hat{h}}{2\omega_2}\cos(2x_2) \end{aligned} \tag{2.6}$$

2.1. Analytical expressions for homoclinic orbits

We performed a multi-scale analysis [4] for the dimer case and obtained amplitude equations to describe the slow dynamics of the system of dimer SQUID. It is in the slow dynamics that the homoclinic orbits are found. The homoclinic orbits given by

$$B^h(T,I) = \frac{2\delta a^2}{q\cosh(2aT)+p} \tag{2.7}$$

$$\theta^h(T,I) = \tan^{-1}\Big(\sqrt{\frac{I(\delta-3)+\Omega_1}{I(1-\delta)-\Omega_1}}\tanh(aT)\Big) \tag{2.8}$$

$$\chi_1^h(T,I) = \frac{-a(\delta^2-1)}{\sqrt{p^2-q^2}}\tan^{-1}[-\sqrt{\frac{p-q}{p+q}}\tanh(aT)] + (\delta I - \frac{\Omega}{4})T + x_1(0) \tag{2.9}$$

$$\phi^h(T,I) = \chi_1^h(T,I) - \theta^h(T,I) \tag{2.10}$$

where $q = I(\delta^2-1)+2\delta\Omega_1$, $p = -\Omega_1(\delta^2-4\delta+1) - I(\delta^3-6\delta^2+7\delta-2)$ and $a^2 = \Omega_1^2 - 2I\Omega_1(\delta-2) - I^2(\delta-3)(\delta-1)$.

The orbits given by Eqs.2.7-2.10 are homoclinic to Π_0. In Fig. 2 we can see some of them. At resonance, for $I = I_0 = -\Omega/4\delta$, the orbits are heteroclinic, connecting fixed points that are $\Delta\phi$ apart, where $\Delta\phi = \Delta x_1 - \Delta\theta$. Therefore

$$\Delta\phi = \frac{2a(\delta^2-1)}{\sqrt{p^2-q^2}}\tanh^{-1}[\sqrt{\frac{p-q}{p+q}}] - 2\tan^{-1}\sqrt{\frac{I_0(\delta-3)+\Omega_1}{I_0(1-\delta)-\Omega_1}}$$

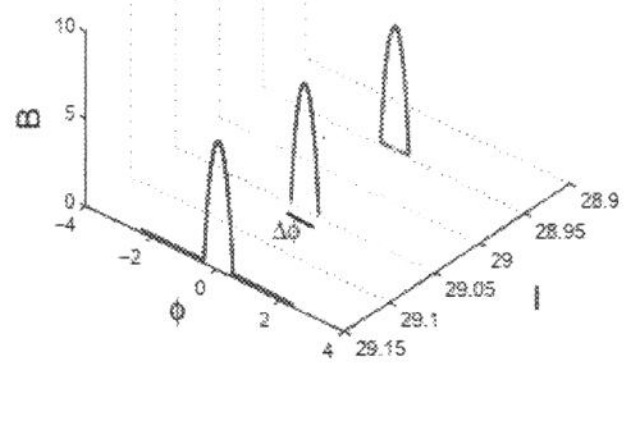

Figure 1. Orbits homoclinic to $\Pi_0 = \{(x,y,I,\phi) : x = y = 0, 1 < \delta < 3, I > \frac{\Omega_1}{3-\delta}\}$. For $I = I_0$ (the orbit in the middle), $d\phi/dT = 0$ on Π_0, and the orbit is heteroclinic, connecting fixed points on Π_0 that are $\Delta\phi$ apart. For $I \lesseqgtr I_0, d\phi/dT \lesseqgtr 0$ on Π_0. The parameters are $\delta = 1.49704$, $\Omega = -110$, $\Omega_1 = 27.6085$.

3. Homoclinic intersections in the perturbed system

In this section we calculate the Energy-difference function using the Melnikov integral evaluated on the homoclinic solutions. We compute that the energy-difference function is

$$\begin{aligned}\Delta^N\mathcal{H}(\phi_0) &= \cos(2(\phi_0+N\Delta\phi)) - \cos(2\phi_0) + \\ &\frac{\sin\frac{(N\Delta\phi)}{2}}{\sin\frac{\Delta\phi}{2}}2(\sin 2\phi_0)\sin(N\Delta\phi) - \\ &\frac{2\xi N}{h\delta}(-2\Delta\phi - \Delta\theta - \frac{\tilde{L}}{I_0})\end{aligned} \tag{3.1}$$

When the dissipation parameter $\xi < \frac{\sin N\Delta\phi(1+2\frac{\sin N\Delta\phi/2}{\sin\Delta\phi/2})}{\frac{2N}{h\delta}(-2\Delta\phi-\Delta\theta-\frac{\tilde{L}}{I_0})}$ then the energy function admits transverse zeros.

3.1. Dynamics near resonance

Now applying the singular perturbation theory we study the dynamics near resonances. The equations that describe the dynamics on $M_\varepsilon = \{(x,y,I,\phi) : x = x^\varepsilon(I,\phi), y = y^\varepsilon(I,\phi)\}$ near the resonace $I = I_0$ are

$$\frac{dI}{dT} = \varepsilon hI\delta\sin(2\phi) - \varepsilon\gamma_0 I, \frac{d\phi}{dT} = -\delta I - \frac{\Omega}{4} + \frac{\varepsilon h\delta}{2}\cos(2\phi) \tag{3.2}$$

According to [4] and [3] in order to study the slow dynamics, which is induced by the perturbation on M_ε near resonance, we set a slow variable $I = I_0 + \sqrt{\varepsilon}J$ into Eqs. (3.2) along with a slow time scale $\tau = \sqrt{\varepsilon}T$, and we obtain the Hamiltonian system in (J, ϕ)

$$H(J, \phi) = -\frac{\delta J^2}{2} + I_0\gamma_0\phi + \frac{1}{2}I_0 h\delta\cos(2\phi) \tag{3.3}$$

3.2. A homoclinic connection to the sink p_ε

We are in a position to show the existence of an orbit homoclinic to the sink p_ε. Here we illustrate numerically our analytical results for the SQUID model. In Fig.2a we show that the unperturbed heteroclinic orbit, which asympotes to p_c as $T \to \infty$, returns back to a point on the circle of fixed points that is inside the homoclinic separatrix loop. In order to have Silnikov Chaos [4] the condition

$$\phi_s < \phi_c + \Delta\phi < \phi_m \tag{3.4}$$

has to be satisfied. The ϕ values of the fixed points are shown in Fig.2b along with $\phi_c + \Delta\phi$ and ϕ_m for a particular value of Ω. The parameter values for which these ϕ values satisfy the condition 3.4 are displayed in Fig.2c.

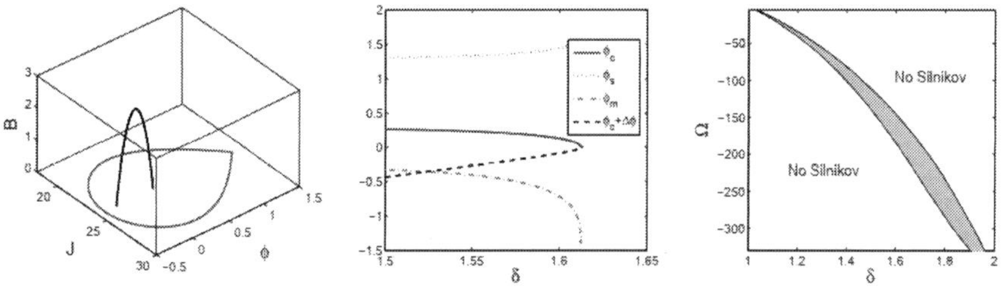

Figure 2. a) The heteroclinic orbit given by Eqs. 2.7 and 2.10 with $I = I_0$, superimposed with the phase portrait of the unperturbed scaled system on Π_ε near resonance. The parameters are $\delta = 1.55, \Omega = -150, \Omega_1 = 35, \zeta = 0.45, h = 1$ ($\xi = \gamma/h$ and $\zeta = \xi/\delta$). This value of ζ sets $\phi_c = 0.233383$ and we can see from the figure that $\phi_s < \phi_c + \Delta\phi < \phi_m$. b) The values of $\phi_s, \phi_c, \phi_c + \Delta\phi$ and ϕ_m as functions of δ , for $\Omega = -150$. For $\delta > 1.524$ the condition is satisfied. c) In gray are indicated the parameter values for which the condition 3.4 is satisfied. In figures b and c: $\varepsilon = 0.01$

Summary: In this work we studied chaotic dynamics of a pair of parametrically-driven coupled SQUIDs arranged in series and we found under which conditions it can exist by using high dimentional Melnikov theory [4].

Acknowledgments: The work of M.A has been co-financed from resources of the operational program "Education and Lifelong Learning" of the European Social Fund and the National Strategic Reference Framework (NSRF) 2007-2013. The work of V.R. has been co-financed by the European Union (European Social Fund ESF) and Greek national funds through the Operational Program Education and Lifelong Learning of the National Strategic Reference Framework (NSRF) Research Funding Program: THALES Investing in knowledge society through the European Social Fund.

References

[1] N.Lazarides, G.P. Tsironis and N. Eleftheriou, *Nonlinear Phenomena in Complex Systems*,11,250, 2008.
[2] G. Mennon, G. Haller and V. Rothos, *Phys Lett A*,263,175, 1999.
[3] G. Haller, *Chaos near Resonances*, Springer, New York, 1999.
[4] G. Kovacic and S. Wiggins, *Physica D*, 57, 185, 1992; G. Kovacic, *SIAM J Math Anal*, 26, 1611, 1995.
[5] Eyal Kenig, Yuriy A. Tsarin, Ron Lifshitz, *Phys. Rev. E* 84 (2011) 016212.

Soliton dynamics in complex potentials

Yannis Kominis

School of Applied Mathematical and Physical Science, National Technical University of Athens, Zographou GR-15773, Greece

E-mail: gkomin@central.ntua.gr

Abstract. Soliton propagation dynamics under the presence of a complex potential are investigated. Cases of both symmetric and non-symmetric potentials are studied in terms of their effect on soliton dynamics. The existence of an invariant of soliton propagation under specific symmetry conditions for the real and the imaginary part of the potential is shown. The rich set of dynamical features of soliton propagation include dynamical trapping, periodic and non-periodic soliton mass variation and non-reciprocal dynamics. These features are systematically investigated with the utilization of an effective particle phase space approach which is shown in remarkable agreement with direct numerical simulations. The generality of the results enables the consideration of potential applications where the inhomogeneity of the gain and loss is appropriately engineered in order to provide desirable soliton dynamics.

1. Introduction

Soliton formation and dynamics in spatially inhomogeneous structures is a subject of intense research interest with applications to many branches of physics, including optical waves in nonlinear photonic structures [1] and matter waves in Bose-Einstein Condensates (BEC), [2]. Spatial modulations of the linear or the nonlinear refractive index of an optical medium have been shown to result in the formation of self-localized waves that have no counterpart in homogeneous systems. [3, 4, 5] Lattice solitons have been shown to exist in a large variety of configurations. [6, 7] In the case of strong spatial modulations, soliton profiles can be interestingly complex but wave dynamics are rather restricted due to the deep soliton trapping and the resulting transverse immobility. Contrarily, in the case of rather weak modulations, soliton profiles remain simple but soliton dynamics can be quite rich and have interesting features with great potential for applications. In such cases solitons move actually as effective particles in a potential, with the form of the latter depending strongly on the characteristics of the soliton. Therefore, different solitons may undergo qualitatively different dynamics in the same inhomogeneous structure. [6, 7]

The consideration of spatial modulation of the material gain and losses appears naturally as a next step for engineering the soliton formation and dynamics and opens new possibilities for applications. The formation of gap solitons has been investigated in periodic lattices with inhomogeneous [8, 9, 10, 11] gain and loss properties. Even for cases of homogeneous gain/loss it has been shown that the interplay between the dynamical soliton power variation and the refractive index modulation results in a rich set of soliton dynamical features. [12, 13, 14]

In this work we study soliton dynamics under the presence of relatively weak symmetric and non-symmetric complex potentials. The presence of gain and loss not only affects the soliton mass (power) but also the effective potential under which the soliton is moving due to the spatial modulation of the refractive index. Soliton dynamics are studied in the three-dimensional phase space of an effective particle of varying mass and the role of spatial symmetries as well as deviations from symmetry is investigated.

2. Model

Soliton propagation in the presence of a complex potential is described by the inhomogeneous NLS equation:

$$iu_z + u_{xx} + [V(x) + iW(x)]\, u + 2|u|^2 u = 0 \tag{1}$$

where u is the wave field envelope, z the normalized propagation distance, and x the scaled transverse coordinate. $V(x)$ and $W(x)$ are the real and imaginary parts of the complex potential. The soliton can be treated as an effective particle [15] of variable mass $m = \int |u|^2 dx$ and momentum $p = i\int (uu_x^* - u_x u^*)dx = mv$ at a position x_0, corresponding to soliton's center, moving with velocity v in an effective potential U_{eff} due to the actual complex potential, according to the equations

$$\frac{dm}{dz} = 2\int_{-\infty}^{+\infty} |u|^2 W(x)dx \tag{2}$$

$$m\frac{dv}{dz} = -\frac{\partial}{\partial x_0}\left[2\int_{-\infty}^{+\infty} |u|^2 V(x)dx\right] \equiv -\frac{\partial U_{eff}}{\partial x_0} \tag{3}$$

$$\frac{dx_0}{dz} = v \tag{4}$$

The dynamical system defined by Eqs. (2)-(4) determines soliton dynamics. In the case of real potential ($W = 0$), soliton moves with a constant mass, the system has fixed points at the extrema of the effective potential (U_{eff}) and the total energy of the effective particle $H = mv^2/2 + U_{eff}(x_0)$ is conserved. The presence of a nonzero part of the potential ($W \neq 0$) introduces an additional degree of freedom related to the particle mass variation and causes the destruction of the conserved quantity of total energy. These features result in drastic qualitative changes of soliton dynamics in comparison to cases of real potentials.

From Eqs. (2),(3) we can obtain

$$m\frac{dv}{dm} = \frac{-\frac{\partial}{\partial x_0}\left[\int_{-\infty}^{+\infty} |u|^2 V(x)dx\right]}{\int_{-\infty}^{+\infty} |u|^2 W(x)dx} \equiv I(m, v, x_0) \tag{5}$$

with the quantity $I(m, v, x_0)$ depending, in general, on all soliton parameters. However, it is readily seen that under the condition

$$\frac{\partial V(x)}{\partial x} = CW(x) \tag{6}$$

with C being a constant, we have $I(m, v, x_0) = -C$ resulting in a conserved quantity of the effective particle motion given by

$$K(m, v) = C \ln m + v = \text{const.} \tag{7}$$

that restricts soliton dynamics in two-dimensional submanifolds of the phase space (x_0, v, m). Moreover, this condition implies the existence of a stable/unstable fixed point at the minima/maxima of the real part of the potential.

The above conditions and discussion are generic with respect to the amplitude and the profile of the complex potential, since the equations (2) and (3) are exact equations for the soliton mass and velocity variation under propagation when u is an exact solution of eq. (1). However, in this work we are mostly interested in soliton dynamics that occur in relatively weak potentials where the solitons are quite mobile. In this case the equations (2) and (3) can be treated perturbatively and provide analytical results by utilizing in the respective integrals the well known soliton solution of the homogeneous NLS equation ($V = W \equiv 0$) that is given by $u = \eta \mathrm{sech}[\eta(x - x_0)] \exp[i(vx/2 + 2\phi)]$ with x_0 and $v = dx_0/dz$ being the position and the velocity of the soliton center and $d\phi/dz = \eta^2/2 - v^2/8$. The soliton mass is $m = 2\eta$.

In the following, we focus on periodic potentials and we investigate soliton dynamics for cases where symmetry conditions or the the condition (6) are either fulfilled or violated. In all cases, the amplitude of the various potentials are of the order of 10^{-2} so that the perturbative approach is valid.

3. Results

A characteristic periodic profile of the complex potential is the sinusoidal profile

$$V(x) = V_0 \cos(K_0 x + \Delta x), \qquad W(x) = W_0 \sin(L_0 x) \tag{8}$$

with V_0, W_0 being the amplitudes and K_0, L_0 the wavenumbers of the real and imaginary parts of the potential. The real part of the potential is an even function for $\Delta x = 0$. The complex potential is known to have a purely real spectrum under the additional condition $W_0 < V_0$ for $K_0 = L_0$. [16] The condition (6) for the existence of the invariant quantity (7) requires both $\Delta x = 0$ and $K_0 = L_0$ but does not restrict the relative amplitude of the real and imaginary parts. For the potential (8), Eqs. (2), (3) provide

$$\frac{dm}{dz} = -\frac{2\pi W_0 L_0}{\sinh(L_0\pi/m)} \sin(L_0 x), \qquad m\frac{dv}{dz} = -\frac{\partial U_{eff}}{\partial x_0} \tag{9}$$

with

$$U_{eff} = -\frac{2\pi V_0 K_0}{\sinh(K_0\pi/m)} \cos(K_0 x + \Delta x) \tag{10}$$

Soliton moves as a particle of varying mass in a potential having a constant spatial period but dynamically varying amplitude due to its strong dependence on the particle mass. The topology of the orbits in the three-dimensional phase space (x_0, v, m) depend strongly on the parameters of the potential as shown in Fig. 1. The case of an even real part and an odd imaginary part with equal periods ($L_0 = K_0$) is shown in Fig. 1(a) for soliton initial conditions corresponding to $m = 1$, positive and negative velocities (v) and various positions (x_0). It is obvious that, in contrast to the conservative case $W_0 = 0$, initial conditions with x_0 and v of opposite sign do not follow the same orbit. Moreover, all orbits with the same initial mass and velocity are restricted on the two-dimensional invariant manifold (7), due to the fulfillment of the condition (6) as shown in Fig. 1(b). Characteristic cases of trapped and traveling soliton propagation are shown in Fig. 2(a) and (b). It is worth emphasizing that in the case of a conservative potential the soliton amplitude and width oscillate in such a way that the soliton mass remain constant, whereas in the dissipative case the soliton mass undergoes oscillations.

Phase space orbits for the case of a potential with an even real and an odd imaginary part, but with different periods of a rational ratio, are depicted in Figs. 1(c) and (d) for positive and negative initial velocities, respectively. In this case the condition (6) is not fulfilled and orbits are not restricted in a two-dimensional manifold. Moreover, as shown in Fig. 1(d), the soliton mass variation can be nonperiodic. Soliton propagation for such a characteristic case of continuous mass increasing is depicted in Fig. 2(c). The case of real and imaginary parts with

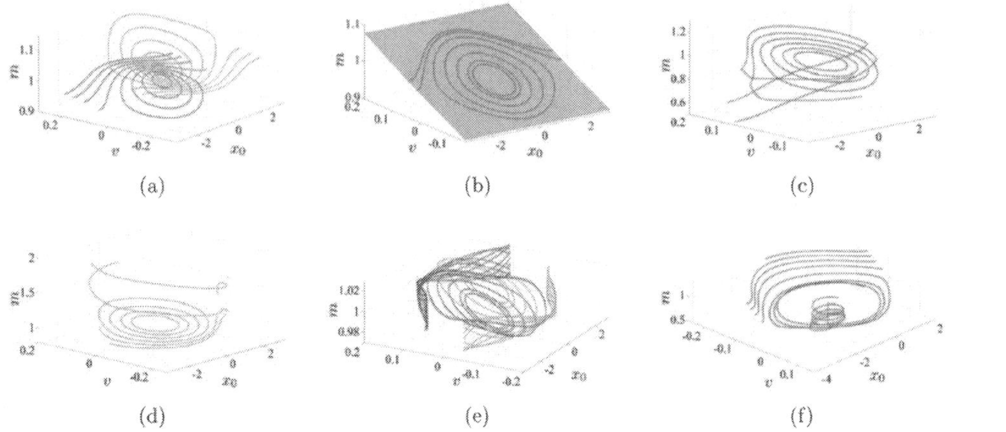

Figure 1. Phase space orbits of the effective particle model for a soliton with initial mass $m(0) = 1$ in the periodic potential (8) with $V_0 = 0.01$, $W_0 = V_0/2$ and $K_0 = 1$. (a) Potential: $L_0 = 1$, $\Delta x = 0$, Initial conditions: $x_0(0) = 0$, $v(0) > 0$ (red / dark gray), $v(0) < 0$, (cyan / light gray); (b) Potential: $L_0 = 1$, $\Delta x = 0$, Initial conditions: $v(0) = 0.05$ (the two-dimensional surface (7) is also shown); (c) Potential: $L_0 = 1/3$, $\Delta x = 0$, Initial conditions: $v(0) = 0.05$; (d) Potential: $L_0 = 1/3$, $\Delta x = 0$, Initial conditions: $v(0) = -0.05$; (e) Potential: $L_0 = \sqrt{2}$, $\Delta x = 0$, Initial conditions: $v(0) > 0$ (red / dark gray), $v(0) < 0$, (cyan / light gray); (f) Potential: $L_0 = 1$, $\Delta x = -\pi/3$, Initial conditions: $v(0) = 0$, $x_0(0) = \pi/3, 1.5\pi/3, 2.8\pi/3$.

spatial periods of an irrational ratio is depicted in Fig. 1(e). It is shown that in addition to trapped orbits, we also have orbits corresponding to traveling solitons with quasiperiodic mass oscillations, each one densely filling a two-dimensional surface. Soliton propagation for such a characteristic case is shown in Fig. 2(d), where the inset shows the details of the quasiperiodic mass and amplitude oscillations.

Finally, a case where neither a spatial symmetry exist nor the condition (6) is fulfilled is shown in Fig. 1(e), where the real part is not an even function whereas the imaginary part is an odd function. In this case, there exist an initial condition for which $x_0 = -\Delta x$ and $v = 0$ remain constant but the local loss is nonzero, resulting to a soliton evolution where the soliton mass continuously decreases and no transverse soliton motion takes place. Such a characteristic case is depicted in Fig. 2(e). Other initial conditions can result to traveling solitons with increasing mass or trapped solitons with decreasing mass, as also shown in Fig. 2(f).

Acknowledgments

The author acknowledges support by the research project NWDCCPS implemented within the framework of the Action "Supporting Postdoctoral Researchers" of the Operational Program "Education and Lifelong Learning" (Actions Beneficiary: General Secretariat for Research and Technology), and is co-financed by the European Social Fund (ESF) and the Greek State.

References

[1] Lederer F, Stegeman G I, Christodoulides D N, Assanto G, Segev M and Silberberg Y 2008 *Phys. Rep.* **463** 1-126
[2] Khaykovich L, Schreck F, Ferrari G, Bourdel T, Cubizolles J, Carr L D, Castin Y and Salomon C 2002 *Science* **296** 1290-1293
[3] Efremidis N K and Christodoulides D N 2003 *Phys. Rev. A* **67** 063608

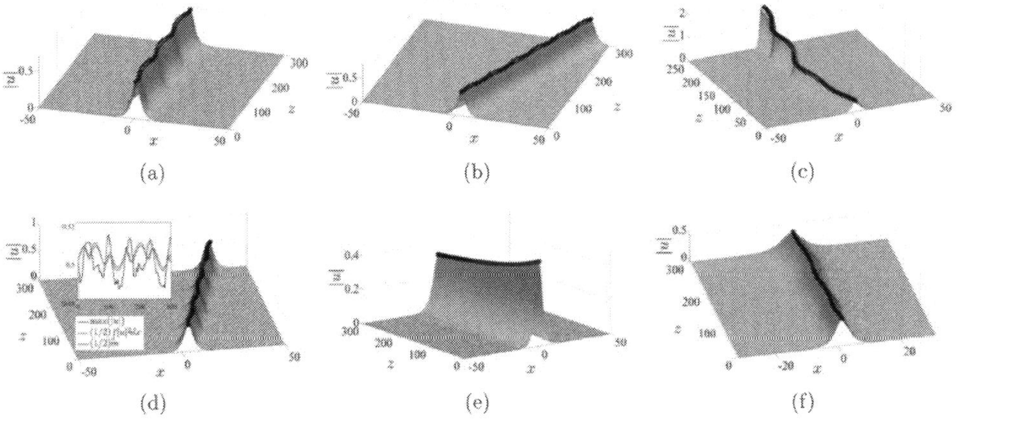

Figure 2. Soliton evolution under propagation as obtained from numerical simulations of the NLS equation (1) and the effective particle model. The thick black line depicts the effective particle orbit $(x_0, v, m/2)$. (a) Potential: as in Fig. 1(a),(b), Initial conditions: $m(0) = 1$, $x_0(0) = 1$, $v(0) = 0.02$; (b) Potential: as in Fig. 1(a),(b), Initial conditions: $m(0) = 1$, $x_0(0) = \pi$, $v(0) = 0.08$; (c) Potential: as in Fig. 1(c),(d), Initial conditions: $m(0) = 1$, $x_0(0) = \pi$, $v(0) = -0.05$; (d) Potential: as in Fig. 1(e), Initial conditions: $m(0) = 1$, $x_0(0) = \pi$, $v(0) = 0.05$, The inset depicts $\max_x(|u|)$ (black), $(1/2)\int |u|^2 dx$ (red) and $m/2$ (blue); (e) Potential: as in Fig. 1(f), Initial conditions: $m(0) = 1$, $x_0(0) = \pi/3$, $v(0) = 0$; (e) Potential: as in Fig. 1(f), Initial conditions: $m(0) = 1$, $x_0(0) = 1.3\pi/3$, $v(0) = 0$.

[4] Louis P J Y, Ostrovskaya E A, Savage C M and Kivshar Y S 2003 *Phys. Rev. A* **67** 013602
[5] Kartashov Y V, Malomed B A, Torner L 2011 *Rev. Mod. Phys.* **83** 247-305
[6] Kominis Y and Hizanidis K 2008 *Opt. Express* **16** 12124-12138
[7] Kominis Y and Hizanidis K 2009 *Phys. Rev. Lett.* **102** 133903
[8] Musslimani Z H, Makris K G, El-Ganainy R and Christodoulides D N 2008 *Phys. Rev. Lett.* **100** 030402
[9] Abdullaev F K, Konotop V V, Salerno M and Yulin A V 2010 *Phys. Rev. E* **82** 056606
[10] He Y, Zhu X, Mihalache D, Liu J and Chen Z 2012 *Phys. Rev. A* **85** 013831
[11] He Y and Mihalache D 2013 *Phys. Rev. A* **87** 013812
[12] Sakaguchi H and Malomed B A 2008 *Phys. Rev. E* **77** 056606
[13] Kominis Y, Droulias S, Papagiannis P and Hizanidis K 2012 *Phys. Rev. A* **85** 063801
[14] Kominis Y, Papagiannis P and Droulias S 2012 *Opt. Express* **20** 18165-18172
[15] Kaup D J and Newell A C 1978 *Proc. R. Soc. London, Ser. A* **361** 413-446
[16] Makris K G, El-Ganainy R, Christodoulides D N and Musslimani Z H 2008 *Phys. Rev. Lett.* **100** 103904

Modeling Mild Collisions

I-Sheng Yang
IOP and GRAPPA, Universiteit van Amsterdam,
Science Park 904, 1090 GL Amsterdam, Netherlands
E-mail: I.S.Yang@uva.nl

Abstract. It is a surprisingly common phenomenon that two objects collide with each other and emerge only mildly altered. We motivate a dynamic-independent, analytical framework to study these mild collisions through two specific examples: (1) Head-on collision between two non-integrable solitons, and (2) Gravitational self-interaction for a collapsing shell of radiation.

1. Introduction

Scattering theory connects detail dynamics to observables with clear physical meanings. Such connection can be particularly sharp for a special class of scattering events: when two localized objects collide and end up almost unaffected. These mild collisions support an *ex post facto* perturbative expansion. As soon as one observes a mild collision, one can immediately write down an abstract expansion $\phi_0 + \phi_1$. Here ϕ_0 describes two colliding objects being completely unaltered, and ϕ_1 describes small corrections. The dynamical reason why ϕ_1 is small will later emerge, and sometimes one can answer important physical questions independent of that.

Although we call them special events, mildness of collisions is a signature behavior of solitons which arises in many branches of physics [1, 2]. Thus we will first review the recent progress made by Amin, Lim and myself regarding collisions of solitons [3, 4]. We demonstrated that the most natural reason why soliton collisions are mild is very different from the common impression of (near) integrability.

Mild collisions are also common in gravitational physics. In our second example, we will demonstrate how to map a problem of gravitational self-interaction of a radiation shell into a problem of mild collisions. We exploit the fact that during mild collisions, the small correction ϕ_1 contains all the information describing how are the objects slightly modified. It can be expanded into various components with clear physical meanings, and employing these projections appropriately is the key feature of our approach.

2. Fast colliding solitons

The complete description of a scattering theory for fast (therefore mild) collisions between solitons can be found in [3]. Here we review the key properties by a simple example given in [4]: a $(1+1)$ dimensional theory of a scalar field with a periodic potential. $\partial_t^2\phi - \partial_x^2\phi + V'(\phi) = 0$, $V(\phi) = V(\phi+\Delta\phi)$. This theory supports a simple soliton which interpolates between two nearby "vacuum", $V(N\Delta\phi) = \mathrm{Min}(V) = 0$, which is usually called a "kink", $\phi_K''(x) = V'[\phi_K(x)]$. This field profile exponentially approaches vacuum values beyond some distance $\sim (L/2)$ away from its center $x = 0$, so we can treat it as a localized object of size L. A reflection takes this to an "antikink" $\phi_K(-x)$, and Lorentz symmetry allows us to describe moving (anti)kinks, $\phi_{K,v}(x,t) = \phi_K\left[\gamma(x - vt)\right]$, $\gamma = (1 - v^2)^{-1/2}$.

A mild collision between two kinks can be easily setup in the center of mass frame, $\phi_0(x,t) = \phi_{K,v}(x,t) + \phi_{K,-v}(x,t)$. This describes two kinks coming toward each other at some $t_i < 0$, colliding at $t = 0$ but nothing happens. They simply pass through each other and continue their merry ways for $t_f > 0$. Assuming that the collision is indeed mild, we can expand true solution as $\phi = \phi_0 + \phi_1$ in which ϕ_1 is small. Applying this assumption to the equation of motion and keep only the leading order in ϕ_1, we get

$$\partial_t^2\phi_1 - \partial_x^2\phi_1 + V''(\phi_0)\phi_1 = V'(\phi_{K,v} + \phi_{K,-v}) - V'(\phi_{K,v}) - V'(\phi_{K,-v}) \equiv S(x,t)\ . \tag{1}$$

The initial condition of two unaltered kinks is $\phi_1 = \partial_t\phi_1 = 0$ at some $t_i < 0$. Solving for ϕ_1 then involves an $x - t$ integral of some Green's function with the source term $S(x,t)$ in Eq. (1), and the smallness of such space-time integral guarantees a small ϕ_1. This is easily arranged. A moving kink only have a nontrivial field profile within a region $\sim \gamma^{-1}L$, and outside such region it approaches the vacuum value $n\Delta\phi$. That means at any time when the two kinks do not overlap, $V'(\phi_{K,v} + \phi_{K,-v}) = V'(\phi_{K,v}) + V'(\phi_{K,-v})$ so the source term is zero. It is only nonzero in the diamond shape region where the two kinks overlaps, as shown in Fig.1. The space-time area of such region is $L^2/(4\gamma^2 v)$, thus a large γ always justifies a small ϕ_1 expansion[1], and at leading order we expect $\phi_1 \propto \gamma^{-2}$.

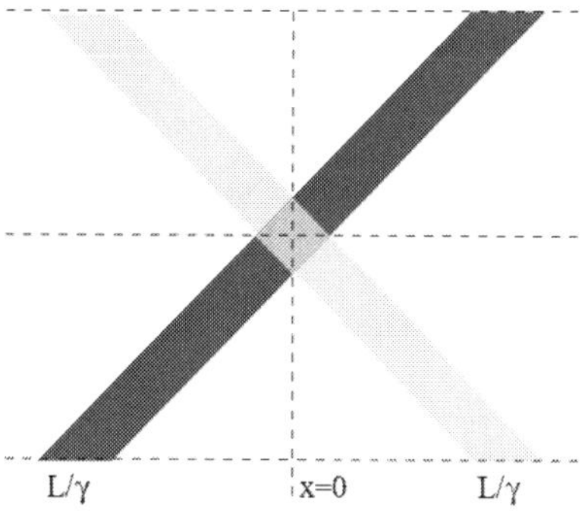

Figure 1. The collision between two fast-moving solitons. They only interact in the center (green) diamond region.

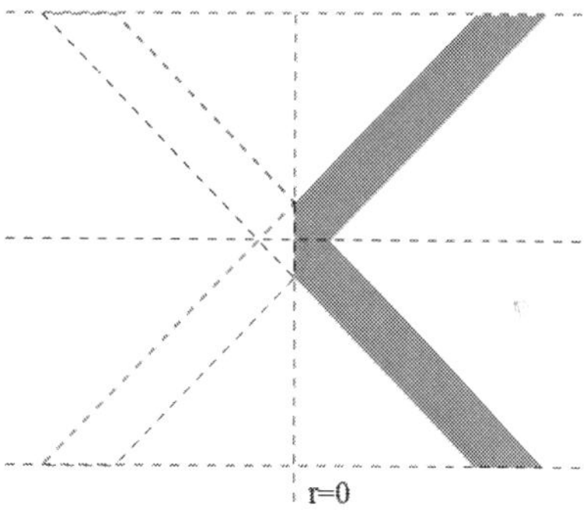

Figure 2. A shell that shrinks and bounces back (shaded region) can be matched to its inverse image and becomes a collision problem.

Before solving Eq. (1) explicitly, it is very helpful to first anticipate what it means by breaking it up into modes. The zero mode is special because it corresponds to the shift symmetry of the kink, $f_0[\gamma(x - vt)] \propto \partial_x\phi_K[\gamma(x - vt)]$. Its coefficient takes the general form of $A_0 + B_0\gamma(t - vx)$. A nonzero A_0 means that the kink acquires a constant position shift during this collision, and a nonzero B_0 means that the kink is moving at a different velocity after the collision. They can be evaluated into simple close-form expressions as shown in [4], but here we will focus on another striking property. We know $B_0 = 0$ even without evaluating it. In other words, we know that solitons do not slow down at the leading order of $1/\gamma^2$.

In order to see that, take energy conservation and expand it in the power of ϕ_1. $E[\phi_0 + \phi_1]|_{t=t_f>0} - E[\phi_0]|_{t=t_i<0} = \frac{\delta E}{\delta\phi_0}\phi_1 + \frac{1}{2}\frac{\delta E}{\delta\phi_0}\phi_1^2 + ...$ A few integration tricks show that a nonzero B_0

[1] The full justification to all order and the expected convergence is described in [3].

is the only possible contribution at the first order of ϕ_1, therefore the only term at the leading order of the γ^{-2} expansion. Since there is nothing else to cancel with, it must be zero on its own [3]. When two colliding kinks slow down, the missing energy must go to other "stuffs". That energy is further suppressed as $\phi_1^2 \propto \gamma^{-4}$, which bounds how much energy the kinks can lose. The center of mass energy of this collision is $E_{\rm CoM} \propto \gamma$, and almost all of those energy will stay with the kinks, only a small amount of energy will be lost, $E_{\rm loss} \propto (\gamma^{-4}E_{\rm CoM}) \propto \gamma^{-3}$.

3. Self-gravitation of a radiation shell

Consider a thin-shell profile of massless scalar field in a Minkowski space background. When the field amplitude ϵ is small, we can adopt the post-Minkowski expansion [5].

$$\sim \mathcal{O}(\epsilon^0) \quad , \quad ds^2 = -dt^2 + dr^2 + r^2 d\Omega^2 \ ; \quad \sim \mathcal{O}(\epsilon^1) \ , \quad \partial_t^2\phi_0 - \partial_r^2\phi_0 - \frac{2}{r}\partial_r\phi_0 = 0 \ ; \tag{2}$$

$$\sim \mathcal{O}(\epsilon^2) \quad , \quad \frac{2\partial_r M}{r^2} = \frac{2}{r}\left(-\frac{M}{r^2} + \partial_r V\right) = 8\pi\frac{(\partial_t\phi_0)^2 + (\partial_r\phi_0)^2}{2} \ , \tag{3}$$

$$ds^2 = -[1 + 2V(r,t)]dt^2 + \left[1 + \frac{2M(r,t)}{r}\right] dr^2 + r^2 d\Omega^2 \ ; \tag{4}$$

$$\sim \mathcal{O}(\epsilon^3) \quad , \quad \partial_t^2\phi_1 - \partial_r^2\phi_1 - \frac{2}{r}\partial_r\phi_1 = C\left(\partial_t^2\phi_0 + \partial_r^2\phi_0 + \frac{2}{r}\partial_r\phi_0\right) + \partial_t C\partial_t\phi_0 + \partial_r C\partial_r\phi_0 \ , \tag{5}$$

where $C(r,t) = [V(r,t) - M(r,t)/r]$. A free field modifies the metric, which eventually back-reacts on the field. One can treat the gravity calculation at the second order as an intermediate step. From that point of view, this entire process is analogous to a mild collision expansion of $\phi = \phi_0 + \phi_1$. The smallness of ϕ_1 is guaranteed by staying in the regime of weak gravity.

In fact, we can further improve the analogy by a change of variable $u(r,t) = r\phi(r,t)$. First of all, the unaltered field profile of a shrinking shell looks like two colliding objects if we take the unphysical $r < 0$ region into account. $u_0(r,t) = r\phi_0(r,t) = \epsilon\,[g(r-t) - g(-r-t)]$, where $g(x)$ is a localized profile centered around $x = 0$. The equation of motion for u_1 becomes

$$\partial_t^2 u_1 - \partial_r^2 u_1 = C(\partial_t^2 u_0 + \partial_r^2 u_0) + \partial_t C\partial_t u_0 + \partial_r C\left(\partial_r u_0 - \frac{u_0}{r}\right) \equiv S(r,t) \ . \tag{6}$$

This is basically identical to Eq. (1). As shown in Fig.2, we have mapped the problem of the shell profile $\phi(r,t)$ in the physical $r > 0$ region into a collision problem. The definition of M and V can be naturally extended to the $r < 0$ region [7]. Everything boils down to solving Eq. (6), and the validity of this expansion seems to follow directly from the weak gravity assumption.

Similar to the previous case, we set $u_1 = \partial_t u_1 = 0$ at some $t_i = -r_0 < 0$ and solve Eq. (6). It is natural to ask what happens to the shell at $t_f = r_0 > 0$ when it is expected to come back to the same radius, and the answer is given by a simple integral.

$$u_1(r,t_f) = \frac{1}{2}\int_{-r_0}^{r_0} dt \int_{|r-r_0+t|}^{r+r_0-t} d\tilde{r}\ S(\tilde{r},t) \ . \tag{7}$$

This, despite being small $\sim \epsilon^3$, diverges as $\ln r_0$. Unlike solitons which only interact when they overlap, gravity is a long range force. The north and south poles still attract each other when shell is large, and such attraction happens to have a diverging effect on u_1.

Nevertheless, this $\ln r_0$ divergence has a clear physical meaning. After simplifying several boundary terms in the integral, Eq. (7) comes to a more tale-telling expression.

$$u_1(r,t_f) = -\epsilon g'(r-r_0)\int_{-r_0}^{r_0} dt\ C\Big(|r-r_0+t|,t\Big) - \frac{1}{2}\int_{-r_0}^{r_0} dt \int_{r-r_0+t}^{r+r_0-t} d\tilde{r}\ (\partial_r C)\frac{u_0}{r} + \mathcal{O}(r_0^{-1}) \ . \tag{8}$$

The divergence only comes from the integral in the first term, and g' shows that it must be a position shift. A massless field profile is supposed to follow a null geodesic. When the metric is modified like in Eq. (4), $|dr| = dt$ is no longer a null geodesic, and a shell starting at $r = r_0$, $t_i = -r_0$ is not coming back to $r = r_0$ on time. Tracking the appropriate null geodesic can give us a position shift, $\Delta r = \int_{-r_0}^{r_0} dt\, C\left(|r - r_0 + t|, t\right) \sim \ln r_0$. The integral here is identical to that in Eq. (8), and it is the only diverging term there. After projecting out an overall position shift, everything else is finite. Thus indeed, gravity is a $1/r^2$ long-range force, but its long range effect here is a boring position shift. After removing that, the next order $1/r^3$ tidal force creates finite "tidal deformations" of this shell.

If we put this shell inside a reflective box, such as a large AdS space, it will be reflected back from the boundary and repeat this self-interaction. One naïve intuition is that since gravity is an attractive force, the shell should becomes narrower from its own gravity, and it will eventually become narrow enough to form a black hole [6]. Using our method, we can show that a shell is equally likely to become wider, since tidal effect tends to pull things apart. Therefore a black hole formation in AdS space is not a completely generic outcome [7].

4. Discussion

We presented two separate cases with very different dynamics, and one method that addresses the outcome of mild collisions in both of them. In the case of soliton collisions, our method provides new insight into the reason why they tend to be mild, and it is very different from the common impression of something related to integrability [3, 4]. The more center of mass kinetic energy enters the collision, the less amount will be lost. In a collision with center of mass energy $E_{\rm CoM} \propto \gamma$ where γ is the usual boost factor, the kinetic energy loss is negligible $E_{\rm loss} \propto (\gamma^{-4} E_{\rm CoM}) \propto \gamma^{-3}$. The solitons pass through each other with all of their original energy.

In the case of gravitational self-interaction, we demonstrated a clean way to separate a naïve divergence in the post-Minkowski expansion: a large position shift due to the change of geodesics. Recognizing this divergence allows us to focus on other finite outcomes and answer important questions such as the AdS stability problem [6, 7].

We are quite used to scattering theories formulated in the language of asymptotic plane-waves, which are convenient since they form a complete basis to describe all possible outcomes. In the case of mild collisions, small changes to the colliding objects should effectively form a complete basis to describe all possible outcomes. It is natural to expect that a fully developed scattering theory in this context can be useful in answering many physical questions.

Acknowledgments

We thank M.A. Amin, F. Dimitrakopoulos, B. Freivogel, E.A. Lim and M. Lippert for discussions. This work is supported in part by the Foundation for Fundamental Research on Matter (FOM), which is part of the Netherlands Organization for Scientific Research (NWO).

References

[1] Scott A C, Chu F and McLaughlin D W, 1973, *IEEE Proc.* **61**, 28.
[2] Malomed B A, 1985, *Physica D: Nonlinear Phenomena* **15**, 385, ISSN 0167-2789.
[3] Amin M A, Lim E A and Yang I S, 2013, *Phys. Rev. D* **88**, 105024 [arXiv:1308.0606 [hep-th]].
[4] Amin M A, Lim E A and Yang I S, 2013, *Phys. Rev. Lett.* **111**, 224101, [arXiv:1308.0605 [hep-th]].
[5] Bizon P, Chmaj T and Rostworowski A, 2009, *Class. Quant. Grav.* **26**, 175006, [arXiv:0812.4333 [gr-qc]].
[6] Bizon P, 2014, *Gen. Rel. Grav.* **46**, 1724, [arXiv:1312.5544 [gr-qc]].
[7] Dimitrakopoulos F, Freivogel B, Lippert M and Yang I S, *to appear*.

Bundle structures of stretching-energy and nonlinear interactions among modes in elastic wave turbulence

Masanori Takaoka[1] **and Naoto Yokoyama**[2]

[1] Department of Mechanical Engineering, Doshisha University, Kyotanabe 610-0394, Japan
[2] Department of Aeronautics and Astronautics, Kyoto University, Kyoto 615-8540, Japan

E-mail: [1]mtakaoka@mail.doshisha.ac.jp
[2]yokoyama@kuaero.kyoto-u.ac.jp

Abstract.
The real-space dynamics and the nonlinear interactions among Fourier modes in elastic wave turbulence are investigated by simulating the Föppl-von Kármán equation. We find that the bundle structures of ridges appear intermittently in the time evolution of the stretching energy field. The time-evolution of the nonlinearity indicates the existence of active and moderate phases in turbulent state. Conditional sampling analysis reveals that the bundle structure, which is the embodiment of the strong nonlinear interactions among modes, induces the energy supply from an external force to the system.

1. Introduction

Elastic wave turbulence, which is governed by the Föppl-von Kármán (FvK) equation [1], has been studied theoretically, numerically and experimentally, and exhibited rich phenomena. Various energy spectra in statistically steady states have been reported. While the energy spectrum $\mathcal{E}(k) \propto k[\log(k_*/k)]^{1/3}$ predicted by the weak turbulence theory (WTT) is successfully reproduced by simulating the FvK equation in the pioneering work [2], the experiments in [3, 4] show different powers of energy spectra $\mathcal{E}(k) \propto k^0 \sim k^{-0.2}$, where $\mathcal{E}(k)$ is an azimuthally-integrated spectrum and $k = (k_x^2 + k_y^2)^{1/2}$. The dimensional analysis suggests other powers of energy spectrum: $\mathcal{E}(k) \propto k^{-1}$ for energy cascade and $\mathcal{E}(k) \propto k^{-1/3}$ for wave action cascade.

In our previous works [5, 7], we have systematically reproduced the variability of the energy spectra by changing the magnitudes of external forces. We have also found the coexistence of the weak and strong turbulence spectra which exists universally for sufficiently wide inertial range. The analytical expression of the separation wavenumber between them has been proposed and numerically validated for various values of parameters. Recently in Ref. [8], we analyzed energy budget by using energy decomposition that is enabled by a single-wavenumber representation of nonlinear energy spectrum. All the above analyses were done in the Fourier space.

We will report here the properties of elastic wave turbulence in the real space. Examined are the real-space structures and the probability density functions of the decomposed energies in addition to the lateral displacement of a plate. We also performed conditional sampling analysis to distinguish active and moderate phases in turbulent state, which reveals the relation between the bundle structure in the real space and the nonlinear interactions among Fourier modes.

2. Formulation

2.1. Basic equations

To investigate the dynamics of elastic waves, we numerically solved the FvK equation

$$\frac{\partial p}{\partial t} = -\frac{Eh^2}{12(1-\sigma^2)}\Delta^2\zeta + \{\zeta,\chi\}, \quad \frac{\partial \zeta}{\partial t} = \frac{p}{\rho}, \quad \Delta^2\chi = -\frac{E}{2}\{\zeta,\zeta\}, \tag{1}$$

where η, p and χ are respectively the lateral displacement, the momentum and the Airy stress potential. The Young's modulus E, the Poisson ratio σ and the density ρ are the physical properties of the elastic plate. The thickness of the plate is expressed by h, and $\Delta = \partial_{xx} + \partial_{yy}$ and $\{f,g\} = \partial_{xx}f\partial_{yy}g + \partial_{yy}f\partial_{xx}g - 2\partial_{xy}f\partial_{xy}g$ are the Laplace operator and the Monge-Ampère operator, respectively. We adopted the parameter values of the steel plate used in the experiment which are shown in Ref. [3].

The pseudo-spectral method is employed, since the plate is supposed to have the periodic boundary of 1m×1m. The number of the aliasing-free modes is 256×256, though 512×512 mode is used in the calculation of the convolution. The fourth-order Runge-Kutta method is used for the time integration. The external force and the dissipation, which are respectively localized in low-, $k \leq 8\pi$, and high-, $k \gtrsim 256\pi$, wavenumber ranges, are added to achieve statistically-steady non-equilibrium states. The details of the numerical method are explained in Ref. [5].

2.2. Energy decomposition

The energy decomposition is convenient to investigate the energy budget in detail as pointed out in our recent manuscript [8]. The decomposed energies are the kinetic energy K, the bending energy V_{b} and the stretching energy V_{s}:

$$K = \frac{p^2}{2\rho}, \quad V_{\mathrm{b}} = \frac{Eh^2\left((\Delta\zeta)^2 - (1-\nu)\{\zeta,\zeta\}\right)}{24(1-\nu^2)}, \quad V_{\mathrm{s}} = \frac{(\Delta\chi)^2 - (1+\nu)\{\chi,\chi\}}{2E}. \tag{2}$$

Fourier spectral representation is suitable for the analysis of homogeneous turbulence. We here introduce the Fourier coefficients of ζ, p and χ as $\zeta_{\boldsymbol{k}}$, $p_{\boldsymbol{k}}$ and $\chi_{\boldsymbol{k}}$ respectively. Adoption of $\zeta_{\boldsymbol{k}}$, $p_{\boldsymbol{k}}$ and $\chi_{\boldsymbol{k}}$ as elementary waves enables the single-wavenumber representations of these decomposed energies under the periodic boundary condition:

$$K_{\boldsymbol{k}} = \frac{|p_{\boldsymbol{k}}|^2}{2\rho}, \qquad V_{\mathrm{b}\boldsymbol{k}} = \frac{\rho\omega_{\boldsymbol{k}}^2|\zeta_{\boldsymbol{k}}|^2}{2}, \qquad V_{\mathrm{s}\boldsymbol{k}} = \frac{k^4|\chi_{\boldsymbol{k}}|^2}{2E}, \tag{3}$$

where the frequency $\omega_{\boldsymbol{k}}$ is determined by the linear dispersion relation. We here categorize the former two energies (latter one energy) as linear (nonlinear) energy, since their (its) order of complex amplitudes is quadratic (quartic). Note that adoption of complex amplitudes as elementary waves makes nonlinear energies the convolution of four waves as known in WTT.

3. Numerical Results

3.1. Energy spectra and real-space structures

In Fig. 1 shown are the ensemble averaged azimuthally-integrated spectra of the total energy $\mathcal{E}$, the kinetic energy $\mathcal{K}$, the potential energy $\mathcal{V}$, the bending energy $\mathcal{V}_{\mathrm{b}}$, the stretching energy $\mathcal{V}_{\mathrm{s}}$, and the linear energy $\mathcal{E}_{\mathrm{L}} = \mathcal{K} + \mathcal{V}_{\mathrm{b}}$. Here, the total energy is decomposed as $\mathcal{E} = \mathcal{K} + \mathcal{V}_{\mathrm{b}} + \mathcal{V}_{\mathrm{s}}$. (See Ref. [8] for details.) Note that the potential energy at each wavenumber $\mathcal{V}$ is decomposed into linear part, $\mathcal{V}_{\mathrm{b}}$, and nonlinear part, $\mathcal{V}_{\mathrm{s}}$, as $\mathcal{V} = \mathcal{V}_{\mathrm{b}} + \mathcal{V}_{\mathrm{s}}$, and the former (latter) takes large values in low- (high-)wavenumber range. The kinetic energy spectrum has flexion downward around $k \approx 200$. As a result, $\mathcal{E}$ as well as $\mathcal{E}_{\mathrm{L}}$ consists of two types of power law spectra with positive and negative exponents.

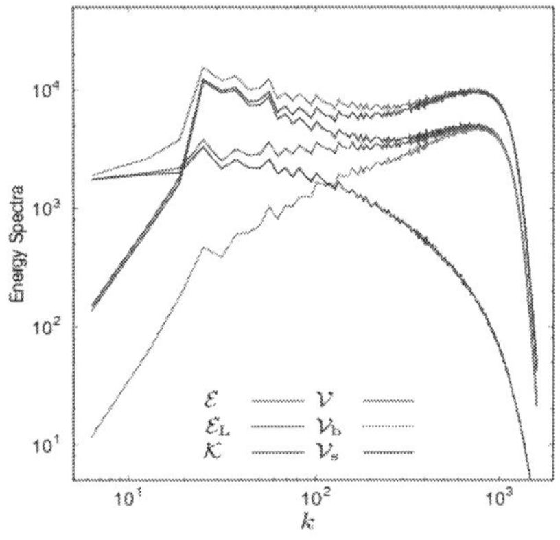

Figure 1. Azimuthally-integrated spectra of total and decomposed energies.

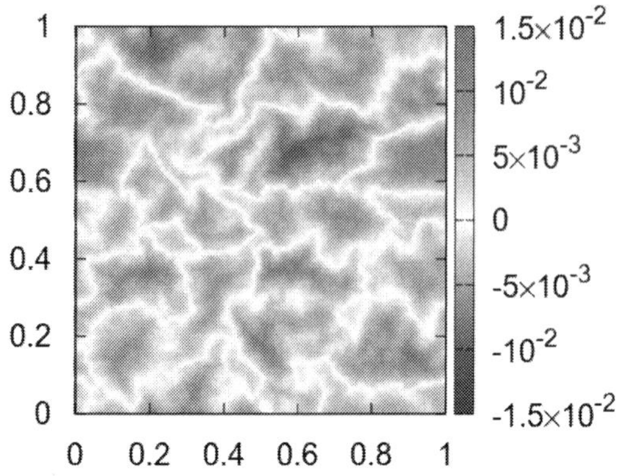

Figure 2. Field of lateral displacement.

Fig. 2 shows the real-space field of the lateral displacement ζ, whose structure has relatively large-scale. One might think that the field structure of ζ contradicts to $\mathcal{V}_b$ in Fig. 1, since the Fourier coefficient of the displacement $\zeta_{\boldsymbol{k}}$, whose dominant scale is large, determines the bending energy spectrum $V_{b\boldsymbol{k}} = \rho\omega_{\boldsymbol{k}}^2|\zeta_{\boldsymbol{k}}|^2/2$ as well as its azimuthally-integrated one $\mathcal{V}_b$, whose dominant scale is small. This is due to the fact that $\omega_{\boldsymbol{k}}$ is proportional to k^2,i.e., $V_{b\boldsymbol{k}} \propto k^4|\zeta_{\boldsymbol{k}}|^2$. The field looks like an intermediate state between the two fields reported in Fig. 2 of Ref. [9], which supports the adequacy of the energy level of the present simulation.

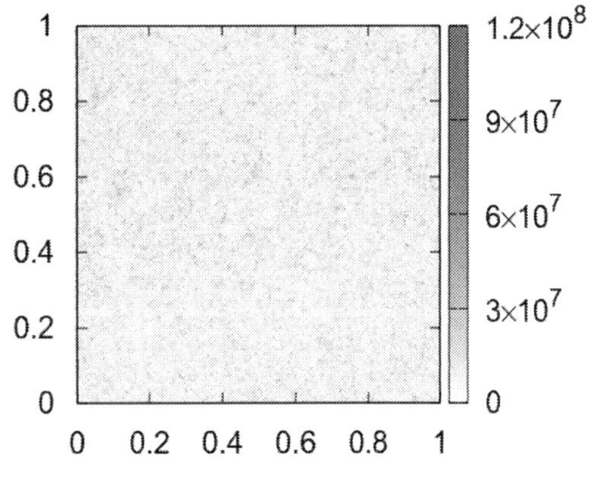

Figure 3. Field of total energy density.

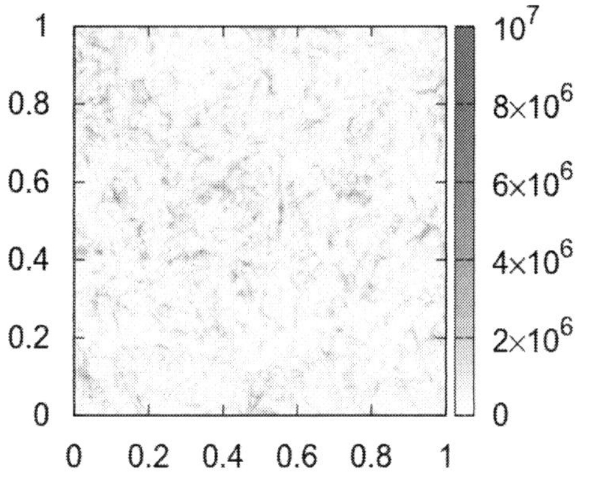

Figure 4. Field of stretching energy density.

On the contrary, the total energy density field shown in Fig. 3 has non-uniform distribution that consists of fine point-like structures. The total energy spectrum shown in Fig. 1 has two local maxima: one is located in low wavenumber $k \approx 8\pi$ and the other in high wavenumber $k \approx 256\pi$. Most of the energy is possessed by the latter high-wavenumber modes, since the number of the high-wavenumber modes is much larger than that of the low-wavenumber ones as known from the logarithmic scale of the horizontal axis. This fact as well as the random phase of weak turbulence in high wavenumber leads the fine-scale structures in the real space.

We also examined the fields of the kinetic energy K and the bending energy V_b, though the graphs are omitted. Both fields have similar fine-scale structures, while the dominant scales of their spectra are different. It may be interesting to point out that the dominant scales of the spectra do not necessarily correspond to the characteristic scale in the real space. We must stress the importance of taking the phase correlations as well as the number of modes into account.

In Fig. 4, we find the ridge and the d-cone structures in the field of the stretching energy. It should be noted that the stretching energy stems from the nonlinearity of the FvK equation, which may induce the phase correlation of modes. It is also consistent with the stretching energy spectrum $\mathcal{V}_s$ in Fig. 4, whose dominant scale is in low-wavenumber range.

3.2. Bundle structure and nonlinearity

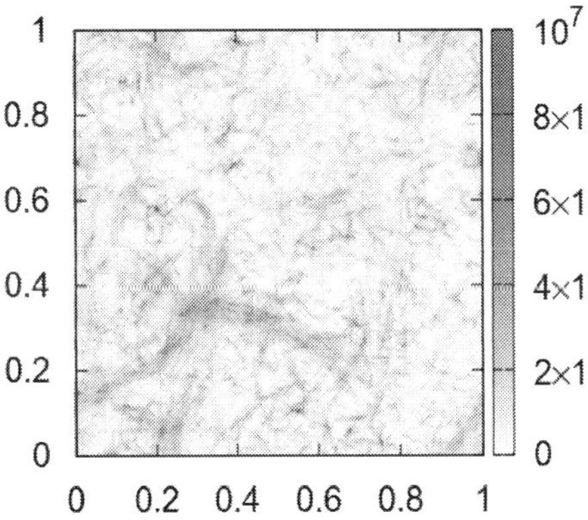

Figure 5. Bundle structure in the field of stretching energy density.

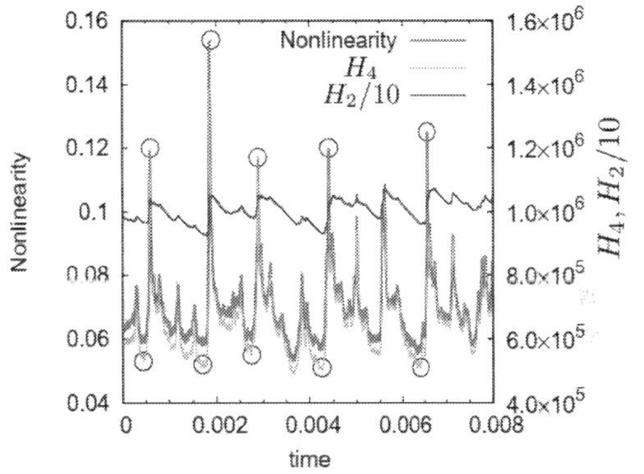

Figure 6. Time evolutions of nonlinearity (red), nonlinear (green) and linear (blue) energies. Circles show representative times.

We furthermore find that the *bundle* structures of the ridges, which are shown in Fig. 5, appear intermittently in the time evolution of the stretching energy field. The time evolution suggests the existence of active and moderate phases in the turbulent state. In Fig. 6 shown is the time evolution of nonlinearity, which is estimated by the ratio of the nonlinear energy to the linear energy, $\int \mathcal{V}_s dk / \int \mathcal{E}_L dk$, corresponding to the ratio of the quartic to quadratic order of the complex amplitude in Hamiltonian, H_4/H_2. The time evolutions of the nonlinear (green) and linear (blue) energies are also drawn in Fig. 6, where the linear energy is shifted downward by one decade. Quasi-periodic sawtooth behavior with mutually-synchronized phase is observed. Strong nonlinearity appears intermittently, while the nonlinearity always fluctuates randomly with relatively small amplitude. The bundle structures are observed at such active phases with the strong nonlinearity.

3.3. Probability distribution functions

Probability distribution functions (PDFs) are examined to characterize the statistical properties. Fig. 7, where the Gaussian PDF is also drawn for reference, shows the normalized PDFs of the lateral displacement ζ at five representative times respectively in active and moderate phases. As the representative times of active phase chosen are the times when the nonlinearity takes top five local maxima in Fig. 6, and the times 1.5×10^{-3} prior to these times are chosen as that of moderate phase. Although the fluctuation of the PDFs of ζ are large, one can safely say that the PDFs are almost Gaussian independent of the phases. The PDFs of p have little fluctuation and are well approximated by the Gaussian distribution, though the graphs are omitted here. The standard deviation of the lateral displacement ζ (the momentum p) in active phase is almost

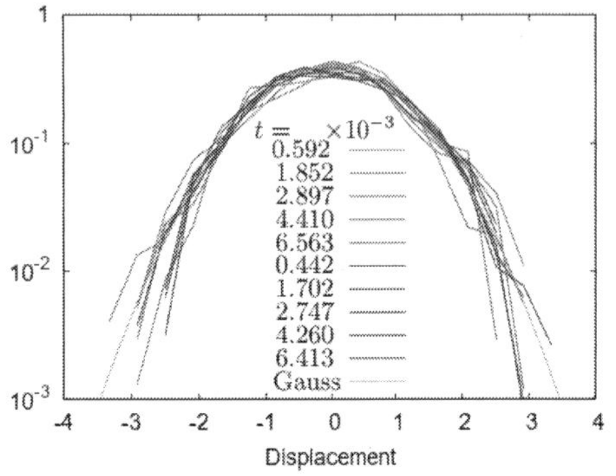

Figure 7. PDF of $\zeta(\boldsymbol{x})$ at active (red) and moderate (blue) phases and Gaussian (green).

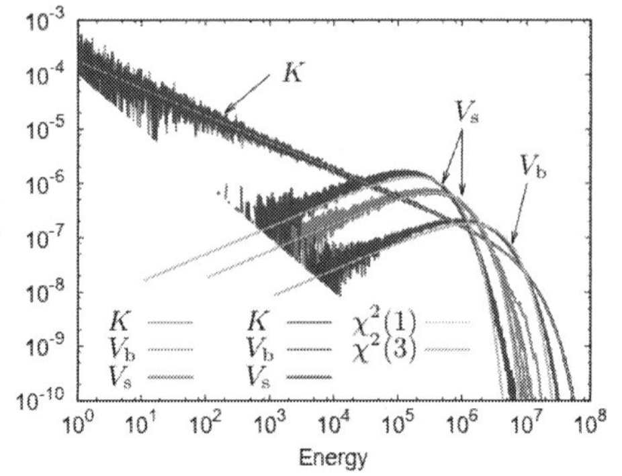

Figure 8. PDFs of $K(\boldsymbol{x})$, $V_{\rm b}(\boldsymbol{x})$ and $V_{\rm s}(\boldsymbol{x})$ at active (red) and moderate (blue) phases.

twice (the same as) that in moderate phase, which is consistent with the fact that most energy of the lateral displacement ζ (the momentum p) is in low- (high-)wavenumber range.

Fig. 8 shows the PDFs of the decomposed energies, K, $V_{\rm b}$ and $V_{\rm s}$ at the representative times. Since the momentum p is regarded as a random variable that obeys the Gaussian distribution with zero mean, the PDF of K is expected to obey the χ^2-distribution with 1-degree of freedom, $\chi^2(1)$, following from $K \propto p^2$. Since the expanded expressions of $V_{\rm b}$ and $V_{\rm s}$ in Eq. (2) respectively contain three quadratic terms and one cross term as $V_{\rm b} \propto \zeta_{xx}^2 + \zeta_{yy}^2 + 2(1-\nu)\zeta_{xy}^2 + 2\nu\zeta_{xx}\zeta_{yy}$ and $V_{\rm s} \propto \chi_{xx}^2 + \chi_{yy}^2 + 2(1+\nu)\chi_{xy}^2 - 2\nu\chi_{xx}\chi_{yy}$, the PDFs of $V_{\rm b}$ and $V_{\rm s}$ may be approximated by χ^2-distribution with 3-degrees of freedom, $\chi^2(3)$, under the assumption of the independency of the derivatives whose supports exist in high-wavenumber range. Although the distributions of linear energies, K and $V_{\rm b}$, are well approximated, that of $V_{\rm s}$ slightly deviate from $\chi^2(3)$ in the vicinity of $V_{\rm s} \sim 10^7$ where the nonlinearity is large. While the distributions of K and $V_{\rm b}$ are independent of the phases, that of $V_{\rm s}$ shifts according to the phases as seen in Fig. 8.

3.4. Nonlinear interactions among modes

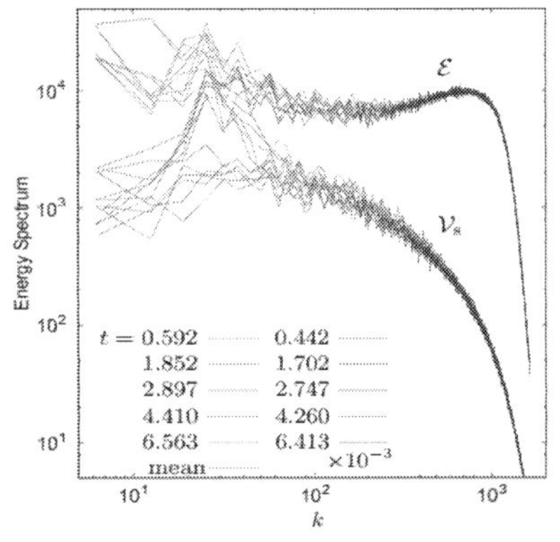

Figure 9. Representative energy spectra in active (red) and moderate (blue) phases.

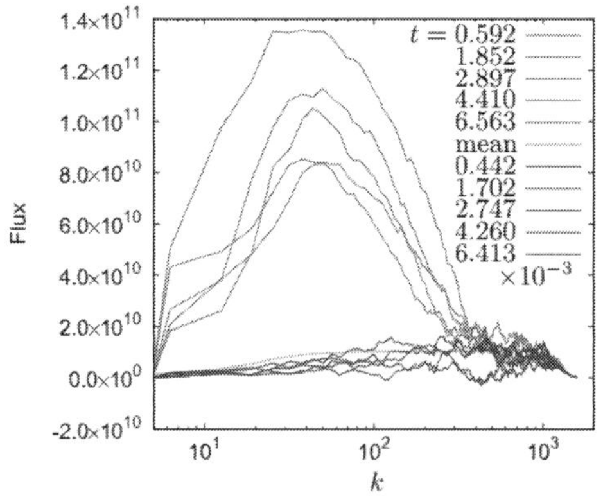

Figure 10. Representative energy fluxes in active (red) and moderate (blue) phases.

Fig. 9 shows the representatives of the total energy spectra $\mathcal{E}$ and the stretching energy $\mathcal{V}_s$ in both active and moderate phases. In low-wavenumber range, $k \lesssim 300$, the magnitudes of $\mathcal{E}$ in the active phases are relatively larger than those in the moderate phases, though they coincide in high-wavenumber range. The result is consistent with the PDF in the real space, shown in Fig. 8, since the linear energy dominates most of the total energy and remains unchanged in both phases. The difference appears more clearly in the curves of $\mathcal{V}_s$: the magnitudes of $\mathcal{V}_s$ in active phases in the forcing scale are almost one digit larger than those in moderate phases.

The fluxes of the total energy are also examined and shown at the representative times, since the nonlinear interaction among Fourier modes governs them. It should be noted here that these fluxes are different from the fluxes of linear energy which are conventionally used in WTT. The difference between those in the active and moderate phases is clearly seen in Fig. 10. The fluxes in the moderate phases can happen to take negative values in some wavenumber range. The fluxes in the active phases increase rapidly in low-wavenumber range, $k \lesssim 8\pi$, and decrease to around common value in relatively high-wavenumber range, $k \gtrsim 100\pi$, corresponding to the coincidence of the energy spectra in both phases as seen in Fig. 9. It means that the bundle structure observed in the real space is the embodiment of the strong nonlinear interactions among Fourier modes, which cause the energy input from the external force into the system.

4. Concluding Remark

We have investigated the real-space dynamics and the nonlinear interactions among Fourier modes in elastic wave turbulence by numerically simulating the Föppl-von Kármán equation. We find that the bundle structures of ridges appear intermittently in the time evolution of the stretching energy field, which is the nonlinear part of the total energy. The time evolution of the nonlinearity indicates the existence of active and moderate phases in turbulent state. Conditional sampling analysis to distinguish these phases reveals that the nonlinear interactions among modes behave differently depending on whether they are in active or moderate phases. The bundle structure, which stems from the strong nonlinear interactions, plays the essential role in supplying energy into the system. Some of the results, such as the bundle structures and the sawtooth behavior in Fig. 6, may depend on the forcing term adopted in the present work. It is our future work to examine their universality.

Acknowledgments

Numerical computation in this work was carried out at the Yukawa Institute Computer Facility. This work was partially supported by KAKENHI Grant No. 25400412.

References

[1] Landau L D and Lifshitz E M 1986 *Theory of Elasticity* (Butterworth-Heinemann, Oxford)
[2] Düring G, Josserand C and Rica S 2006 *Phys. Rev. Lett.* **97** 025003
[3] Boudaoud A, Cadot O, Odile B and Touzé C 2008 *Phys. Rev. Lett.* **100** 234504
[4] Mordant N 2008 *Phys. Rev. Lett.* **100** 234505
[5] Yokoyama N and Takaoka M 2013 *Phys. Rev. Lett.* **110** 105501
[6] Nazarenko S 2011 *Wave Turbulence* (Springer, Heidelberg)
[7] Yokoyama N and Takaoka M 2014 *Phys. Rev.* E **89** 012909
[8] Yokoyama N and Takaoka M (submitted to) *Phys. Rev.* E (arXiv:1407.1582 [nlin.CD])
[9] Miquel B, Alexakis A, Josserand C and Mordant N 2013 *Phys. Rev. Lett.* **111** 054302

IC-MSQUARE 2014 IOP Publishing
Journal of Physics: Conference Series **574** (2015) 012031 doi:10.1088/1742-6596/574/1/012031

Hierarchy of solutions to the NLS equation and multi-rogue waves

P. Gaillard

Université de Bourgogne, 9 Avenue Alain Savary, Campus de Mirande, 21000 Dijon, France

E-mail: pierre.gaillard@u-bourgogne.fr

Abstract. The solutions to the one dimensional focusing nonlinear Schrödinger equation (NLS) are given in terms of determinants. The orders of these determinants are arbitrarily equal to $2N$ for any nonnegative integer N and generate a hierarchy of solutions which can be written as a product of an exponential depending on t by a quotient of two polynomials of degree $N(N+1)$ in x and t. These solutions depend on $2N-2$ parameters and can be seen as deformations with $2N-2$ parameters of the Peregrine breather P_N : when all these parameters are equal to 0, we recover the P_N breather whose the maximum of the module is equal to $2N+1$. Several conjectures about the structure of the solutions are given.

PACS numbers :
33Q55, 37K10, 47.10A-, 47.35.Fg, 47.54.Bd

1. Introduction

Here, we consider the one dimensional focusing nonlinear Schrödinger equation (NLS) to describe the phenomena of rogue waves. Recently, many works about NLS equation have been published using different methods. Rational solutions to the NLS equation were written in 2010 as a quotient of two wronskians [1]; the present author constructed in [2] another representation of the solutions to the NLS equation in terms of a ratio of two wronskians of even order $2N$ composed of elementary functions using truncated Riemann theta functions in 2011; Guo, Ling and Liu found in 2012 another representation of the solutions as a ratio of two determinants [3] using generalized Darboux transformation; a new approach was proposed by Ohta and Yang in [4] using Hirota bilinear method; finally, the present author has obtained in 2013 rational solutions in terms of determinants which do not involve limits in [5].

The present paper presents multi-parametric families of quasi rational solutions to NLS of order N in terms of determinants (determinants of order $2N$) depending on $2N-2$ real parameters. With this method, a hierarchy of solutions to the NLS equation is obtained. With this representation, at the same time, the well-known ring structure, but also the triangular shapes also given by Ohta and Yang [4], Akhmediev et al. [6], Matveev and Dubard [7] are found.

The aim of this paper is to summarize results on solutions to the NLS equation depending on $2N-2$ parameters and try to classify them. Solutions depending on $2N-2$ parameters give the (analogue) Peregrine breather P_N of order N as a particular case when all the parameters are equal to 0 : for this reason, these solutions will be called $2N-2$ parameters deformations of the Peregrine P_N.

The paper is organized as follows. First, we recall the representation of the solutions to the NLS equation in terms of wronskians. This representation allows to obtain quasi rational solutions to NLS equation, when some parameter tends towards 0. Quasi rational solutions depending a priori on $2N-2$ parameters at the order N are constructed.
Moreover we give a theorem which states the structure of the quasi-rational solutions to the NLS equation. Families depending on $2N-2$ parameters for the N-th order as a ratio of two polynomials of degree $N(N+1)$ of x and t multiplied by an exponential depending on t are obtained. We state that the highest amplitude of modulus of the Peregrine breather of order N is equal to $2N+1$. Then we try to classify the hierarchy of solutions to the NLS equation in function of the order N and the parameters $\tilde{a}_j$ and $\tilde{b}_j$.

2. Families of multi-parametric solutions to NLS equation in terms of a ratio of two determinants depending on $2N-2$ parameters

We consider the focusing NLS equation

$$iv_t + v_{xx} + 2|v|^2 v = 0. \tag{1}$$

We recall the main result obtained in [2]. In the following, we need to define some notations. The parameters $-1 < \lambda_\nu < 1$, $\nu = 1, \dots, 2N$, are real numbers such that

$$\begin{array}{l} -1 < \lambda_{N+1} < \lambda_{N+2} < \dots < \lambda_{2N} < 0 < \lambda_N < \lambda_{N-1} < \dots < \lambda_1 < 1 \\ \lambda_{N+j} = -\lambda_j, \quad j = 1, \dots, N. \end{array} \tag{2}$$

The terms κ_ν, δ_ν and γ_ν are defined by

$$\begin{array}{l} \kappa_j = 2\sqrt{1-\lambda_j^2}, \quad \delta_j = \kappa_j \lambda_j, \quad \gamma_j = \sqrt{\frac{1-\lambda_j}{1+\lambda_j}}, \\ \kappa_{N+j} = \kappa_j, \quad \delta_{N+j} = -\delta_j, \quad \gamma_{N+j} = 1/\gamma_j, \quad j = 1 \dots N. \end{array} \tag{3}$$

We choose the parameters a_j and b_j in the form

$$a_j = \sum_{k=1}^{N-1} \tilde{a}_k j^{2k+1} \epsilon^{2k+1}, \quad b_j = \sum_{k=1}^{N-1} \tilde{b}_k j^{2k+1} \epsilon^{2k+1}, \quad 1 \le j \le N. \tag{4}$$

Complex numbers e_ν $1 \le \nu \le 2N$ are defined in the following way :

$$e_j = ia_j - b_j, \quad e_{N+j} = ia_j + b_j, \quad 1 \le j \le N, \quad a, b \in \mathbf{R}. \tag{5}$$

The terms $x_{r,\nu}$ $(r = 3, 1)$ are defined by

$$x_{r,\nu} = (r-1) \ln \frac{\gamma_\nu - i}{\gamma_\nu + i}, \quad 1 \le j \le 2N. \tag{6}$$

We consider X_ν and Y_ν defined by

$$\begin{array}{l} X_\nu = \kappa_\nu x/2 + i\delta_\nu t - ix_{3,\nu}/2 - ie_\nu/2, \\ Y_\nu = \kappa_\nu x/2 + i\delta_\nu t - ix_{1,\nu}/2 - ie_\nu/2, \end{array}$$

for $1 \le \nu \le 2N$, with κ_ν, δ_ν, $x_{r,\nu}$ defined in (3), (6); parameters e_ν are defined by (5).
We define the functions $\varphi_{j,k}$ for $1 \le j \le 2N$, $1 \le k \le 2N$ by

$$\begin{array}{ll} \varphi_{4j+1,k} = \gamma_k^{4j-1} \sin X_k, & \varphi_{4j+2,k} = \gamma_k^{4j} \cos X_k, \\ \varphi_{4j+3,k} = -\gamma_k^{4j+1} \sin X_k, & \varphi_{4j+4,k} = -\gamma_k^{4j+2} \cos X_k, \end{array} \tag{7}$$

for $1 \leq k \leq N$, and

$$\begin{array}{ll} \varphi_{4j+1,N+k} = \gamma_k^{2N-4j-2} \cos X_{N+k}, & \varphi_{4j+2,N+k} = -\gamma_k^{2N-4j-3} \sin X_{N+k}, \\ \varphi_{4j+3,N+k} = -\gamma_k^{2N-4j-4} \cos X_{N+k}, & \varphi_{4j+4,N+k} = \gamma_k^{2N-4j-5} \sin X_{N+k}, \end{array} \quad (8)$$

for $1 \leq k \leq N$.
We define the functions $\psi_{j,k}$ for $1 \leq j \leq 2N$, $1 \leq k \leq 2N$ in the same way, the term X_k is only replaced by Y_k.

$$\begin{array}{ll} \psi_{4j+1,k} = \gamma_k^{4j-1} \sin Y_k, & \psi_{4j+2,k} = \gamma_k^{4j} \cos Y_k, \\ \psi_{4j+3,k} = -\gamma_k^{4j+1} \sin Y_k, & \psi_{4j+4,k} = -\gamma_k^{4j+2} \cos Y_k, \end{array} \quad (9)$$

for $1 \leq k \leq N$, and

$$\begin{array}{ll} \psi_{4j+1,N+k} = \gamma_k^{2N-4j-2} \cos Y_{N+k}, & \psi_{4j+2,N+k} = -\gamma_k^{2N-4j-3} \sin Y_{N+k}, \\ \psi_{4j+3,N+k} = -\gamma_k^{2N-4j-4} \cos Y_{N+k}, & \psi_{4j+4,N+k} = \gamma_k^{2N-4j-5} \sin Y_{N+k}, \end{array} \quad (10)$$

for $1 \leq k \leq N$.
All the functions $\varphi_{j,k}$ and $\psi_{j,k}$ and their derivatives depend on ϵ and can all be prolonged by continuity when $\epsilon = 0$.
Then we get the following result [5] :

Theorem 2.1 *The function v defined by*

$$v(x,t) = \exp(2it - i\varphi) \times \frac{\det((n_{jk})_{j,k\in[1,2N]})}{\det((d_{jk})_{j,k\in[1,2N]})} \quad (11)$$

is a quasi-rational solution to the NLS equation (1)

$$iv_t + v_{xx} + 2|v|^2 v = 0,$$

quotient of two polynomials $N(x,t)$ and $D(x,t)$ depending on $2N-2$ real parameters $\tilde{a_j}$ and $\tilde{b_j}$, $1 \leq j \leq N-1$.
N and D are polynomials of degrees $N(N+1)$ in x and t, where

$$\begin{array}{ll} n_{j1} = \varphi_{j,1}(x,t,0),\ 1 \leq j \leq 2N & n_{jk} = \frac{\partial^{2k-2}\varphi_{j,1}}{\partial\epsilon^{2k-2}}(x,t,0), \\ n_{jN+1} = \varphi_{j,N+1}(x,t,0),\ 1 \leq j \leq 2N & n_{jN+k} = \frac{\partial^{2k-2}\varphi_{j,N+1}}{\partial\epsilon^{2k-2}}(x,t,0), \\ d_{j1} = \psi_{j,1}(x,t,0),\ 1 \leq j \leq 2N & d_{jk} = \frac{\partial^{2k-2}\psi_{j,1}}{\partial\epsilon^{2k-2}}(x,t,0), \\ d_{jN+1} = \psi_{j,N+1}(x,t,0),\ 1 \leq j \leq 2N & d_{jN+k} = \frac{\partial^{2k-2}\psi_{j,N+1}}{\partial\epsilon^{2k-2}}(x,t,0), \\ 2 \leq k \leq N,\ 1 \leq j \leq 2N & \end{array}$$

The functions φ and ψ are defined in (7),(8), (9), (10).

Moreover, we have the following result which gives the highest amplitude of the module of the Peregrine breather of order N :

Theorem 2.2 *The function v_0 defined by*

$$v_0(x,t) = \exp(2it - i\varphi) \times \left(\frac{\det((n_{jk})_{j,k\in[1,2N]})}{\det((d_{jk})_{j,k\in[1,2N]})} \right)_{(\tilde{a_j}=\tilde{b_j}=0,\ 1\leq j\leq N-1)} \quad (12)$$

is the Peregrine breather of order N solution to the NLS equation (1) whose highest amplitude in module is equal to $2N+1$.

The two previous results has been proven by the author. We postpone to give their demonstrations to another publication in order not to weigh down the text of this article.

3. Hierarchy of solutions to NLS equation depending on $2N-2$ parameters

For each nonnegative integer N, we have constructed solutions to NLS equation depending on $2N-2$ parameters.
The case $N=1$ correspond to the classical breather, first constructed by Peregrine [8] in 1983; the analogue breather in the case $N=2$ was built for the first time by Akhmediev, Eleonski and Kulagin [9, 10] in 1986. Other analogues of the Peregrine breathers of order 3 and 4 were constructed in a series of articles by Akhmediev et al. [11, 12] using Darboux transformations. It is only in 2012 that solutions of order 3 and 4 with respectively 4 and 6 parameters were first explicitly found by Matveev using another method based on results of [13], but only published in 2013 in [7].
The solutions for orders 3 and 4 have also been explicitly found by the present author [14, 15]. The equivalence between with two types of solutions was made in [7] for the order 3; the equivalence between these solutions for the order 4 was made by the present author in [15].
In a series of articles, we have studied other higher orders. We have also explicitly found the solutions at order 5 with 8 parameters [16] : these expressions are too extensive to be presented : it takes 14049 pages! For orders 6, 7 the solutions are also explicitly found but are more complicated and cannot be published in any review; the analysis has been done in all these cases in respectively [17, 18]. The solutions for order 8, with 14 parameters are also found and submitted to a review.
The cases of orders 9 and 10 have just been finished and the solutions with respectively 16 and 18 parameters are explicitly found.
From these various studies, it arises that the solutions have quite particular structures depending on the parameters $\tilde{a}_j$ and $\tilde{b}_j$. The parameters $\tilde{a}_j$ and $\tilde{b}_j$ play a similar role in obtaining the structures of the solutions. One can thus establish a certain number of conjectures about these solutions at the order N.

3.1. Case $a_1 \neq 0$ (or $b_1 \neq 0$)

For $\tilde{a}_1 \neq 0$ or $\tilde{b}_1 \neq 0$ and other parameters equal to 0, one obtains a triangle with $\frac{N(N+1)}{2}$ peaks. It is important to note that we obtain triangle only in this case; in all the other cases for only one parameter non equal to 0, we obtain rings.
We do not have the space to give the figures in the present text (see the figures on line).

3.2. Case $a_{N-1} \neq 0$ (or $b_{N-1} \neq 0$), $N \geq 3$

For $\tilde{a}_{N-1} \neq 0$ or $\tilde{b}_{N-1} \neq 0$ and other parameters equal to 0, one obtains only one ring of $2N-1$ peaks with in the center Peregrine P_{N-2} of order $N-2$ [1]; here, $N \geq 3$.
Again, we cannot give the figures in the present text; it can be seen on line.

3.3. Case $a_{N-2} \neq 0$ (or $b_{N-2} \neq 0$), $N \geq 5$

For $\tilde{a}_{N-2} \neq 0$ or $\tilde{b}_{N-2} \neq 0$ and other parameters equal to 0, one obtains two concentric rings of $2N-3$ peaks with in the center Peregrine P_{N-4} of order $N-4$; here $N \geq 5$.
We also have not the space to give the figures here (see the corresponding figures on line).

3.4. Case $a_{N-3} \neq 0$ (or $b_{N-3} \neq 0$), $N \geq 7$

We have the same type of figures in this case, but we don't have the space to give in this publication : namely, at order N, we get 3 rings of $2N-5$ peaks with the P_{N-6} breather in the center (the figures can be seen on line).

[1] This conjecture was already formulated by different authors, in particular by Akhmediev et al. [19]

3.5. General case

In general, we can conjecture that :
For $\tilde{a}_{N-k} \neq 0$ or $\tilde{b}_{N-k} \neq 0$ and other parameters equal to 0, for $N > 0$ and $2k \leq N - 1$, one obtains k concentric rings of $2N - 2k + 1$ peaks with in the center Peregrine P_{N-2k} of order $N - 2k$.
It would be relevant to study the cases for the integers k such that $k > \frac{N}{2}$ and the parameters $a_{N-k} \neq 0$ or $b_{N-k} \neq 0$; the structure seems to be more complicated and would be clarified.

4. Conclusion

Here we have given the structure of quasi-rational solutions to the one dimensional focusing NLS equation at the order N. They can be expressed as a product of an exponential depending on t by a ratio of two polynomials of degree $N(N+1)$ in x and t.
If we choose $\tilde{a}_i = \tilde{b}_i = 0$ for $1 \leq i \leq N$, we obtain the classical (analogue) Peregrine breather. Thus these solutions appear as $2N-2$-parameters deformations of the Peregrine breather P_N of order N; this P_N breather has an highest amplitude in module equal to $2N+1$.
These results have been recently proven by the author and their proofs will be given in another publication.
The method described in the present paper provides a powerful tool to get explicit solutions to the NLS equation and to understand the behavior of rogue waves. There are currently many applications in these fields as recent works by Akhmediev et al. [20] or Kibler et al. [21] attest it in particular.
This study leads to a better understanding of the phenomenon of rogue waves, and it would be relevant to go on with the higher orders.
A beginning of classification of the solutions to NLS equation was started with Akhmediev et al. [22]. It would be important in the future to prove the conjectures given in this paper and to give a complete classification for the order N of the quasi rational solutions to the NLS equation.

[1] Dubard P Gaillard P Klein C Matveev V B 2010 Eur. Phys. J. Special Topics **185** 247
[2] Gaillard P 2011 J. Phys. A : Meth. Theor. **44** 1
[3] Guo B Ling L Liu Q P 2012 Phys. Rev. E **85** 026607-1
[4] Ohta Y Yang J 2012 Pro. R. Soc. A **468** 1716
[5] Gaillard P 2013 Jour. Of Math. Phys. **54** 013504-1
[6] Kedziora D J Ankiewicz A Akhmediev N 2012 Phys. Rev. E **86** 056602-1
[7] Dubard P Matveev V B 2013 Nonlinearity **26**
[8] Peregrine H. 1983 J. Austral. Math. Soc. Ser. B **25** 16
[9] Akhmediev N Eleonski V Kulagin N 1985 Sov. Phys. J.E.T.P.
[10] Akhmediev N Eleonski V Kulagin N 1987 Th. Math. Phys. **72**183
[11] Akhmediev N Ankiewicz A Soto-Crespo J M 2009 Physical Review E **80** 026601-1
[12] Akhmediev N Ankiewicz A Clarkson P A 2010 J. Phys. A : Math. Theor.
[13] Eleonski V Krichever I Kulagin N 1986 Sov. Dokl. Sect. Math. Phys.
[14] Gaillard P 2013 Phys. Rev. E **88** 042903-1
[15] Gaillard P 2014 J. Math. Phys. **54** 073519-1
[16] Gaillard P 2014 Commun. Theor. Phys. **61** 365
[17] Gaillard P 2014 Physica Scripta V. **89** 015004-1
[18] Gaillard P 2014 Jour. Of Phys. : conferences Series **482** 012016-1
[19] Kedziora D J Ankiewicz A Akhmediev N 2011 Phys. Review E **84** 056611-1
[20] Chabchoub A Hoffmann N Akhmediev N 2011 Phys. Rev. Lett. **106** 204502-1
[21] Kibler B Fatome J Finot C Millot G Dias F Genty G Akhmediev N Dudley J M 2010 Nature Physics **6** 790
[22] Kedziora D J Ankiewicz A Akhmediev N 2013 Phys. Review E **88**, 013207-1

Soliton in an extended nonlinear Schrödinger equation with stimulated scattering on damping low-frequency waves and nonlinear dispersion

E M Gromov

National Research University Higher School of Economics,
25/12 Bolshaja Pecherskaja Ulitsa, Nizhny Novgorod, 603155, Russia

E-mail: egromov@hse.ru

Abstract. Dynamics of solitons is considered in an extended nonlinear Schrödinger equation, including a pseudo-stimulated-Raman-scattering (pseudo-SRS) term (scattering on damping low-frequency waves, nonlinear dispersion and inhomogeneity of the spatial second-order dispersion (SOD). It is shown that wave-number downshift by the pseudo-SRS may be compensated by upshift provided by spatially increasing SOD with taking into account nonlinear dispersion. The equilibrium state is stable for negative parameter of nonlinear dispersion and unstable for positive one. The analytical solutions are verified by comparison with numerical results

1. Introduction

The great interest to the dynamics of solitons is motivated by their ability to travel long distances keeping the shape and transferring the energy and information. Soliton solutions are relevant to nonlinear models in various areas of physics which deal with the propagation of intensive wave fields in dispersive media: optical pulses and beams in fibers and spatial waveguides, electromagnetic waves in plasma, surface waves on deep water, etc. [1-7].

Dynamics of long high-frequency (HF) wave packets is described by the second-order nonlinear dispersive wave theory. The fundamental equation of the theory is the nonlinear Schrödinger equation (NLSE) [8,9], which includes the second-order dispersion (SOD) and cubic nonlinearity. Soliton solutions in this case arise as a result of the balance between the dispersive stretch and nonlinear compression of wave packets.

To solve many applied problems there is the necessary decreasing of solitons`s space size. Such decreasing is accompanied, as usually, by stimulated scattering on low-frequency (LF) media perturbations. To this time stimulated scattering on spatially homogeneous LF time modes (stimulated Raman scattering (SRS)) was considered in details [1]. SRS is described in extended NLSE by term with time delay of nonlinear kerr response. For localized nonlinear wave packets (solitons), the SRS gives rise to the downshift of the soliton frequency [10]. The compensation of the SRS was studied to this time distally [10-21].

For a series media the propagation of short solitons is accompanied by arising of damping LF waves. These LF modes are internal waves in the stratified fluid and ion-sound waves in the plasma.

Model for describing of stimulated scattering of HF waves on damping LF waves, named as pseudo-stimulated-Raman-scattering (pseudo-SRS), was proposed in [22-24]. Taking into account the wave factor of stimulated LF perturbations significantly varieties the dynamics of short HF solitons. The pseudo-SRS leads to the self-wavenumber downshift, similar to what is well known in the temporal domain [1,10-21]. The model equation elaborated in [22-24] also included smooth spatial variation of the SOD, accounted for by a spatially decreasing SOD coefficient, which leads to an increase of the soliton's wave-number, making it possible to compensate the effect of the pseudo-SRS on the soliton by the spatially inhomogeneous SOD. The equilibrium between the pseudo-SRS and decreasing SOD gives rise to stabilization of the soliton's wave-number spectrum. However, the consideration was carried out in disregard of the nonlinear dispersion.

In this work the soliton dynamics is considered in the frame of extended NLSE with a pseudo-SRS, decreasing dispersion and with taking into account nonlinear dispersion. Shows that equilibrium state between the pseudo-SRS and decreasing SOD is stable focus for negative nonlinear dispersion, and unstable focus for positive nonlinear dispersion.

2. The basic equation and integral relations

Let's consider the dynamics of the HF wave field $U(\xi,t)\exp(i\omega t - i\kappa\xi)$ in the frame of extension NSE with pseudo-SRS, nonlinear dispersion and inhomogeneous SOD:

$$2i\frac{\partial U}{\partial t}+\frac{\partial}{\partial \xi}\left[q(\xi)\frac{\partial U}{\partial \xi}\right]+2U|U|^2+2i\beta\frac{\partial\left(U|U|^2\right)}{\partial\xi}+\mu U\frac{\partial\left(|U|^2\right)}{\partial\xi}=0, \qquad (1)$$

where $q(\xi)$ is the SOD, μ is the pseudo-SRS, β is the nonlinear dispersion. Equation (1) with zero conditions on infinity $U|_{\xi\to\pm\infty}\to 0$ has the following integrals:

$$\frac{\mathrm{d}N}{\mathrm{d}t}\equiv\frac{\mathrm{d}}{\mathrm{d}t}\int_{-\infty}^{+\infty}|U|^2 d\xi=0, \qquad (2)$$

$$2\frac{\mathrm{d}}{\mathrm{d}t}\int_{-\infty}^{+\infty}K|U|^2 d\xi=-\mu\int_{-\infty}^{\infty}\left[\frac{\partial\left(|U|^2\right)}{\partial\xi}\right]^2 d\xi-\int_{-\infty}^{\infty}\frac{\mathrm{d}q}{\mathrm{d}\xi}\left|\frac{\partial U}{\partial\xi}\right|^2 d\xi, \qquad (3)$$

$$N\frac{\mathrm{d}\bar{\xi}}{\mathrm{d}t}\equiv\frac{\mathrm{d}}{\mathrm{d}t}\int_{-\infty}^{\infty}\xi|U|^2 d\xi=\int_{-\infty}^{+\infty}qK|U|^2 d\xi+\frac{3}{2}\beta\int_{-\infty}^{+\infty}|U|^4 d\xi, \qquad (4)$$

where $U\equiv|U|\exp(i\phi)$, $K\equiv\partial\phi/\partial\xi$ is the local wave-number of wave packet.

3. Analytical results

For analysis of the system (2)-(4) let use the adiabatic approximation, presenting solution in sech-like form

$$U(\xi,t)=A(t)\,\mathrm{sech}\left(\frac{\xi-\bar{\xi}(t)}{\Delta(t)}\right)\exp\left(-ik(t)\xi-i\frac{3}{2}\frac{\beta}{\Delta}\tanh\left(\frac{\xi}{\Delta}\right)-i\int\Omega(t)dt\right), \qquad (5)$$

where $\Delta(t)=\sqrt{q(\bar{\xi}(t))}/A(t)$, $\Omega(t)=A^2(t)/2$, $A^2(t)\Delta(t)=\mathrm{const}$. With taking into account (2), (5) and system (2)-(4) is reduced to:

$$2q_0\frac{\mathrm{d}k}{\mathrm{d}t}=-\frac{8}{15}\frac{\mu A_0^4}{n^3}-q'\left(\frac{A_0^2}{3n^2}+q_0k^2+\frac{2\beta A_0^2k}{n^2}\right),\ \frac{\mathrm{d}n}{\mathrm{d}t}=q'nk, \qquad (6)$$

where $q'(\bar{\xi}) = (dq/d\xi)_{\bar{\xi}}$, $n = q(\bar{\xi})/q_0$, $q_0 = q(0)$, $A_0 = A(0)$. Equilibrium state of Eqs.(6) is $k_* = 0, n_* = -8\mu A_0^2/(5q')$. In particular for $\mu = \mu_* \equiv -5q'/(8A_0^2)$ equilibrium state achieved for $n_* = n_0 \equiv 1$, coinciding with initial wave-packet parameter. For linearly decreasing SOD, $q = q_0(1 - \xi/D)$, after replacements, $\theta \equiv t\sqrt{q_0}A_0/(D\sqrt{3})$, $y \equiv k\sqrt{3q_0}/A_0$, system (6) is reduced to

$$2\frac{\mathrm{d}y}{\mathrm{d}\theta} = -\frac{\lambda}{n^3} + \frac{1}{n^2} + y^2 + \delta\frac{y}{n^2}, \quad \frac{\mathrm{d}n}{\mathrm{d}\theta} = -ny, \tag{7}$$

where $\lambda \equiv 8\mu A_0^2 D/(5q_0)$, $\delta \equiv \beta A_0\sqrt{3/q_0}$. For $\delta < 0$ the equilibrium state of Eqs.(7) is stable focus, $\delta = 0$: center, $\delta > 0$: unstable focus. The trajectories following from Eqs. (7) on plane (n, y) are shown in Figure 1 for initial conditions $y_0 = 0$, $n_0 = 1$ for $\lambda = 3/4$ with different δ.

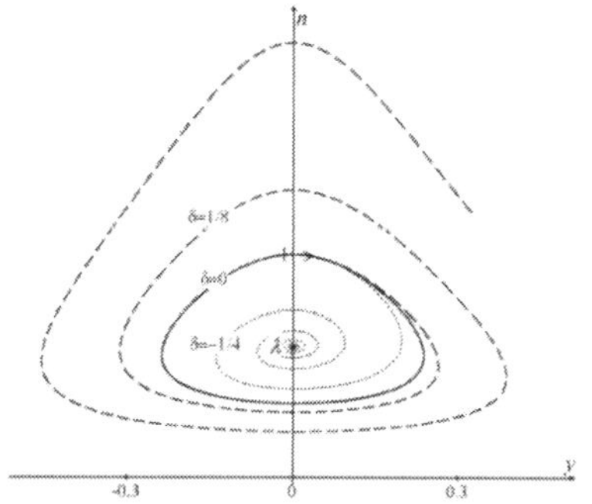

Figure 1. The trajectories (7) on plane (y, n) for $\lambda = 3/4$, initial conditions $y_0 = 0$, $n_0 = 1$ and different δ.

4. Numerical results

We simulated solutions of the initial-value problem for the wave packet, $U(\xi, t = 0) = \exp[-i(3/2)\beta \tanh \xi] \mathrm{sech}\xi$ in the framework of Eq. (1) for $q(\xi) = 1 - \xi/10$, and different values of μ, β. The analytically predicted equilibrium value of the pseudo-SRS coefficient for this initial pulse is $\mu_* = 1/16$. In direct simulations, the initial pulse for $\mu = 1/16$ and different β is a stationary localized distribution (figure 2).

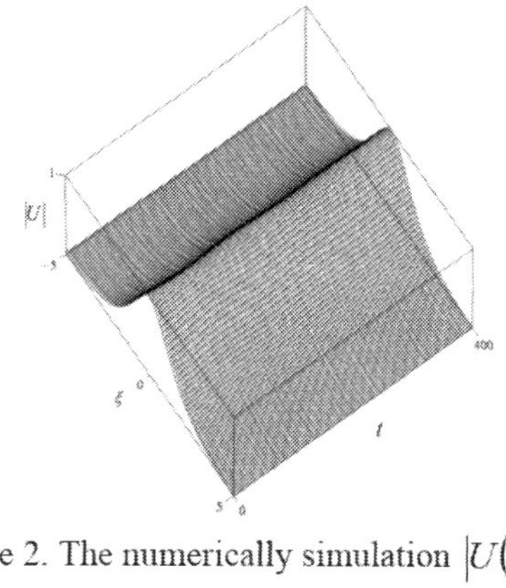

Figure 2. The numerically simulation $|U(\xi, t)|$ for $\mu = 1/16 \equiv \mu_*$ with different values of β.

At values of the pseudo-SRS coefficient different from μ_*, the simulations produce nonstationary solitons, see an example for $\mu=(3/4)\mu_*\equiv 3/64$ in figures 3 and 4.

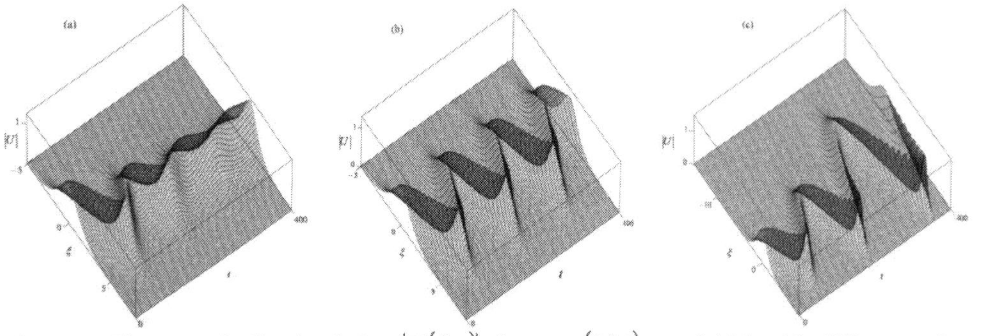

Figure 3. The numerically simulation $|U(\xi,t)|$ for $\mu=(3/4)\mu_*\equiv 3/64$ with different values of β [(a): $\beta=-1/16$, (b): $\beta=0$, (c): $\beta=1/32$].

In figure 4, numerical results produced, as functions of time, by the simulations for the value of point coordinate of the maximum modulus of the wave-packet's shape ξ_{m} ($\max|U(\xi,t)|=|U(\xi_{\text{m}},t)|$), are compared with the analytical counterparts of the mass-center wave-packet envelope $\bar{\xi}\equiv q_0(n-1)/q'$ obtained from Eqs. (6) for $\mu=(3/4)\mu_*\equiv 3/64$ and different values of β.

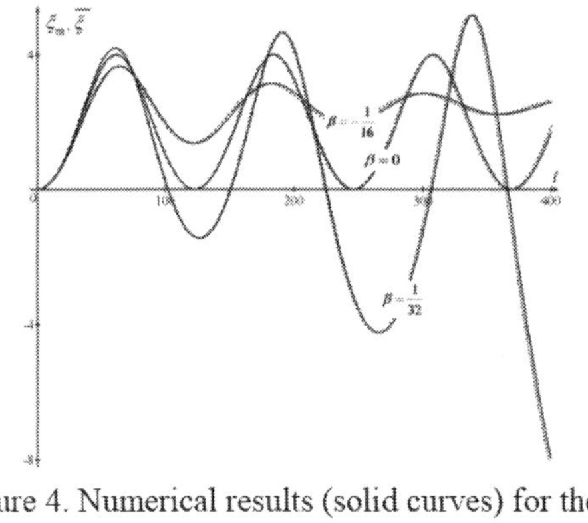

Figure 4. Numerical results (solid curves) for the value of point coordinate of the maximum modulus of the wave-packet's shape ξ_{m} and analytical results (dashed curves) for the mass-center wave-packet envelope $\bar{\xi}$ for $\mu=(3/4)\mu_*\equiv 3/64$, and different values of β.

5. Conclusion

In this work, we studied the soliton dynamics in the framework of the extended inhomogeneous NLSE, includes the pseudo-SRS effect, the lineally decreasing SOD and nonlinear dispersion. The results were obtained by means of analytical method, based on evolution equations for the field moments, and verified by direct simulations. The stationary solitons exist due to the balance between the self-wavenumber downshift, caused by the pseudo-SRS, and the upshift induced by the decreasing SOD. The analytical solutions are close to their numerically found counterparts.

The present model does not take into regard the third-order linear dispersion. The compensation of the pseudo-SRS in the model of inhomogeneous media which includes this higher-order term will be considered elsewhere.

Acknowledgements
This study (research grant No 14-01-0023) was supported by The National Research University Higher School of Economics' Academic Fund Program in 2014/2015.

References

[1] Infeld E and Rowlands G 2000 *Nonlinear Waves, Solitons and Chaos* (Cambridge: Cambridge University Press)
[2] Agrawal G P 2001 *Nonlinear Fiber Optic* (San Diego: Academic Press)
[3] Yang J 2001 *Solitons in Field Theory and Nonlinear Analysis* (New York: Springer)
[4] Kivshar Y S and Agraval G P 2003 *Optical Solitons: From Fibers to Photonic Crystals* (San Diego: Academic)
[5] Dickey L A 2005 *Soliton Equation and Hamiltonian Systems* (New York: World Scientific)
[6] Malomed B A 2006 *Soliton Management in Periodic Systems* (New York: Springer)
[7] Dauxois T and Peyrard M 2006 *Physics of Solitons* (Cambridge: Cambridge University Press)
[8] Zakharov V E and Shabat A B 1972 *Sov. Phys. JETP.* **34** 62-9
[9] Hasegawa A and Tappert F 1973 *Appl. Phys. Lett.* **23** 142-4
[10] Gordon J P 1986 *Opt. Lett.* **11** 662-4
[11] Kodama Y 1985 *J. Stat. Phys.* **39** 597-614
[12] Kodama Y and Hasegawa A 1987 *IEEE J. Quant. Electron.* **23** 510-24
[13] Hong B and Lu D 2009 *Inter. J. Nonlin. Science* **7** 360-7
[14] Obregon M A and Stepanyants Yu A 1998 *Phys. Lett. A* **249** 315-23
[15] Kivshar Y S 1990 *Phys. Rev. A* **42** 1757-61
[16] Malomed B A and Tasgal R S 1998 *J. Opt. Soc. Am. B* **15** 162-70
[17] Biancalama F, Skrybin D V and Yulin A V 2004 *Phys. Rev. E* **70** 011615
[18] Essiambre R-J and Agrawal G P 1997 *J. Opt. Soc. Am. B* **14** 314-22
[19] Essiambre R-J and Agrawal G P 1997 *J. Opt. Soc. Am. B* **14** 323-30
[20] Andrianov A, Muraviev S, Kim A and Sysoliatin A 2007 *Laser Phys.* **17** 1296-302
[21] Chernikov S, Dianov E, Richardson D and Payne D 1993 *Opt. Lett.* **18** 476-8.
[22] Gromov E M and Malomed B A 2013 *J. Plasma Phys.* **79** 1057-62
[23] Gromov E M and Malomed B A 2014 *Opt. Comm.* **320** 88-93
[24] Aseeva N V, Gromov E M and Tyutin V V 2013 *Radiophys. Quant. Electr.* **56** 157-66

Supersymmetric extension of a coupled Korteweg-de Vries system

Adrián Sotomayor[1], **Alvaro Restuccia**[2]

[1] Departament of Mathematics, Universidad de Antofagasta, Antofagasta, Chile
[2] Physics Department, Universidad de Antofagasta, Antofagasta, Chile
[2] Physics Department, Universidad Simón Bolívar, Caracas, Venezuela

E-mail: `adrian.sotomayor@uantof.cl,arestu@usb.ve`

Abstract. We introduce a supersymmetric extension for a parametric coupled Korteweg-de Vries system. The supersymmetric system has two real hamiltonian functionals and two associated basic Poisson structures. The basic Poisson structures allows the construction of a pencil of Poisson brackets and associated to them a hamiltonian functional depending on the parameter of the pencil. The two basic Poisson brackets are compatible.

1. Introduction

It was shown in [1] that among the one parameter supersymmetric extensions of the Korteweg-de Vries (KdV) equation there is a special one, for a particular value of one of the coefficients on the equations, that has an infinite number of conservation laws. It was called the $N = 1$ supersymmetric extension of KdV equation. It is equivalent to the super KdV equation obtained in [2] by reduction from the super Kadomtsev-Petviashvili hierarchy. Other supesymmetric extensions of the KdV equations and interesting analysis of them can be found in [3, 4, 5, 6, 7].

The $N = 1$ supersymmetric extension of KdV equation is a system of coupled equations formulated in terms of fields u, ξ valued on the even and odd part of a given Grassmann algebra respectively.

The system is

$$\begin{aligned} u_t &= -u''' + 6uu' - 3\xi\xi'', \\ \xi_t &= -\xi''' + 3(\xi u)'. \end{aligned} \tag{1}$$

The Grassmann algebra may have any number of generators. In particular, if we consider only one generator e, then

$$u = u_0\mathbf{1}$$
$$\xi = \xi_0 e$$

since $\mathbf{1} \circ \mathbf{1} = \mathbf{1}, \mathbf{1} \circ e = e \circ \mathbf{1} = e, e \circ e = 0$.

u_0, ξ_0 are real fields $u_0 = u_0(x,t), \xi_0 = \xi_0(x,t)$.

In this particular case (1) reduces to ninth Hirota-Satsuma coupled KdV system [8, 9]. Other extensions are obtained by considering more generators of the Grassmann algebra. In all these cases the equation for u_0 is the KdV equation.

The super KdV system (1) has one local hamiltonian structure obtained via a super Miura transformation. The associated Poisson algebra is equivalent to the super Virasoro algebra with central terms. It is consequently related to conformal field theory and string theory.

There exists also a super Gardner transformation [1] which allows to obtain in a direct way an infinite sequence of local conserved quantities. Moreover, in [10] it was found an infinite sequence of non-local conserved quantities of odd weight. In [11] this infinite sequence was derived from a non-local conserved quantity of the super Gardner system and later on, using a supersymmetric cohomology, a new infinite sequence of non-local conserved quantities of even weight was derived [12].

In the present work we consider a supersymmetric extension of a parametric coupled KdV system which we have analyzed recently [13]. For $\lambda = -1$ the system is equivalent to the complexification of the KdV equation. For $\lambda = +1$ the system is equivalent to two decoupled KdV equations. For $\lambda = 0$ the system corresponds to the ninth Hirota-Satsuma coupled system [9].

2. The supersymmetric extension of a coupled KdV system

The parametric coupled KdV system was formulated in terms of two real fields

$$\begin{aligned} u_t &= -u''' + 6uu' + 6\lambda vv' \\ v_t &= -v''' + 6(uv)'. \end{aligned} \tag{2}$$

We found [13] a Bäcklund transformation, proved the permutability theorem and obtained the Gardner transformation, which allows to obtain directly infinite local conserved quantities for the system. We found new solutions for the system, a class of them describes multi-solitonic solutions while another class describes periodic solutions. We also found the complete hamiltonian and Poisson structure for the system [14].

We now consider a supersymmetric extension of the system (2). We start from the supersymmetric extension [1] of KdV equation and consider the corresponding fields to be valued on a $\mathcal{Z}^\lambda$ algebra [15]. If we denote the generators of the associative and commutative algebra over $\mathbb{R}$ by e_1 and e_2 we then have

$$\begin{aligned} U &= ue_1 + ve_2 \\ \xi &= \mu e_1 + \nu e_2 \end{aligned} \tag{3}$$

where u, v are even and μ, ν are odd Grassmann valued fields respectively. The generators of $\mathcal{Z}^\lambda$ satisfy $e_1 \circ e_1 = e_1, e_1 \circ e_2 = e_2, e_2 \circ e_2 = \lambda e_1$.

The resulting supersymmetric extension of the coupled KdV system (2) becomes

$$\begin{aligned} u_t &= -u''' + 6uu' + 6\lambda v'v - 3\mu\mu'' - 3\lambda\nu\nu'' \\ v_t &= -v''' + 6(vu)' - 3\mu\nu'' - 3\nu\mu'' \\ \mu_t &= -\mu''' + 3(u\mu + \lambda v\nu)' \\ \nu_t &= -\nu''' + 3(u\nu + v\mu)' \end{aligned} \tag{4}$$

which are invariant under the supersymmetric transformation

$$\begin{aligned} \delta u &= \epsilon_1\mu' + \lambda\epsilon_2\nu', \; \delta v = \epsilon_1\nu' + \epsilon_2\mu' \\ \delta\mu &= \epsilon_1 u + \lambda\epsilon_2 v, \; \delta\nu = \epsilon_1 v + \epsilon_2 u, \end{aligned} \tag{5}$$

where ϵ_1 and ϵ_1 are the odd Grassmann supersymmetric parameters.

We notice that for $\mu = \nu = 0$ the system (4) reduces to (2). If $\lambda = 0$ the system becomes

$$\begin{aligned} u_t &= -u''' + 6uu' - 3\mu\mu'' \\ \mu_t &= -\mu''' + 3(u\mu)' \\ v_t &= -v''' + 6(vu)' - 3\mu\nu'' - 3\nu\mu'' \\ \nu_t &= -\nu''' + 3(u\nu + v\mu)', \end{aligned} \tag{6}$$

where the first two equations define the $N = 1$ supersymmetric extension of KdV equation [1], the third and fourth equations couple the u and μ fields to v and ν. This system is the $N = 1$ supersymmetric extension of the ninth Hirota-Satsuma coupled KdV system [8, 9]. This bosonic system has all the integrability properties of KdV systems, in particular integrable hierarchies related to polynomial Lie algebras [16].

3. The Miura transformation, Poisson structure and Gardner transformation

We consider the supersymmetric Miura transformation

$$\begin{aligned} U &= V' + V^2 - \Upsilon\Upsilon' \\ \xi &= \Upsilon' + V\Upsilon, \end{aligned} \tag{7}$$

where V and Υ are valued on the $\mathcal{Z}^\lambda$ algebra with coefficients on the even and odd parts of a Grassmann algebra respectively.

The hamiltonians of the system are

$$\begin{aligned} H_1 &= \int_{-\infty}^{+\infty} \left[u^2 + \lambda v^2 + \mu'\mu + \lambda\nu'\nu\right] dx \\ H_2 &= \int_{-\infty}^{+\infty} \left[2uv + \mu'\nu + \nu'\mu\right] dx. \end{aligned} \tag{8}$$

Associated to each hamiltonian there is a Poisson structure which can be deduced in terms of the Miura fields V, Υ by following a Dirac approach on the constrained phase space obtained from two real singular lagrangians. The two Poisson structures are compatible. Moreover one may construct a pencil of Poisson brackets from these basic Poisson structures. The resulting Poisson algebra is a deformation of the super Virasoro algebra which is the Poisson algebra of the supersymmetric extension of the KdV equation.

The supersymmetric system (4) has, for any λ, an infinite sequence of conserved quantities. They can be obtained in a direct way, from the associated super Gardner system which follows from (4) via a Gardner transformation. The few basic conserved quantities of the super Gardner system provides the infinite sequence of conserved quantities of (4).

The super Gardner system is the supersymmetric extension of the Gardner system in terms of the two real fields presented in [13].

4. Conclusions

We have introduced a supersymmetric extension of a coupled KdV system. For particular values of the parameter it corresponds to supersymmetric of interesting coupled KdV systems. The supersymmetric extension has a sequence of infinite conserved quantities and a natural extension of the hamiltonian structure of the supersymmetric KdV equation.

Acknowledgments

A. R. and A. S. are partially supported by Projects Fondecyt 1121103 and Mecesup ANT398 (Chile).

References

[1] Mathieu P 1988 *J. Math. Phys.* **29** 2499
[2] Manin Yu I and Radul A O 1985 *Commun. Math. Phys* **98**, 65
[3] Labelle P and Mathieu P 1991 *J. Math. Phys.* **32** 923
[4] Belucci S, Ivanov E and Krivonos S 1993 *J. Math. Phys.* **34** 3087
[5] Delduc F and Ivanov E 1993 *Phys. Lett.* B **309** 312-319
[6] Delduc F, Ivanov E and Krivonos S 1996 *J. Math. Phys.* **37** 1356

[7] Popowicz Z 1999 *Phys. Lett.* B **459** 150-158
[8] Hirota R and Satsuma J 1981 *Phys. Lett.* A **85** 407
[9] Sakovich S Yu 1999 *J. Nonlin. Math. Phys.* **6** N3 255-262
[10] Dargis P and Mathieu P 1993 *Phys. Lett.* A **176** 67-74
[11] Andrea S, Restuccia A and Sotomayor A 2005 *J. Math. Phys.* **46** 103517
[12] Andrea S, Restuccia A and Sotomayor A 2012 *Phys. Lett.* A **376** 245-251
[13] Cortés Vega L , Restuccia A and Sotomayor A 2014 *Preprint* arXiv: math/ph 1407.7743 v1
[14] Restuccia A and Sotomayor A 2014 *Preprint* arXiv: math/ph 1409.2418 v1
[15] Zuo D 2014 *Preprint* arXiv: math/ph 1403.0027 v1
[16] Casati P and Ortenzi G 2006 *J. Geom. and Phys.* **56** 418-449

Journal of Physics: Conference Series **574** (2015) 012034 doi:10.1088/1742-6596/574/1/012034

Poisson structure on the phase space associated to the hamiltonian dynamics of coupled Korteweg-de Vries type equations

Adrián Sotomayor[1], **Alvaro Restuccia**[2]

[1] Departament of Mathematics, Universidad de Antofagasta, Antofagasta, Chile
[2] Physics Department, Universidad de Antofagasta, Antofagasta, Chile
[2] Physics Department, Universidad Simón Bolívar, Caracas, Venezuela

E-mail: `adrian.sotomayor@uantof.cl,arestu@usb.ve`

Abstract. We present the hamiltonian structures for a wide class of coupled Korteweg-de Vries systems, including the Gear and Grimshaw system that models the strong interaction of internal waves in a stratified liquid and the system of Lou, Tong, Hu and Tang that describes a two layer fluid model. Among the hamiltonian structures of these systems we found new Poisson brackets which define consistent algebras of observables.

1. Introduction

Since Hirota-Satsuma [1], coupled Korteweg-de Vries (KdV) systems have been object of a extensive study in view of their intrinsically relevant mathematical properties and of its relationship with several physics models. In particular it was derived in [2] a system which describes linearly stable internal waves in a density stratified fluid and in [3] it was analyzed a system which under some physical reductions can be used in the description of the atmospheric and oceanic phenomena. An interesting study of coupled KdV systems related with its stability properties was presented in [4].

Among the properties of such systems one has the existence of associated Lax pairs and Bäcklund transformations as well as the existence of multi-solitonic solutions and the Painlevé property [5].

In this work we present a lagrangian approach to the Gear and Grimshaw coupled KdV system [2, 6], which includes two different lagrangians with two associated hamiltonians describing the coupled system. Through that description is possible to obtain the complete hamiltonian structure of the system with his inherited Poisson mathematical description. We notice that this procedure give rise in a natural way to a pencil of Poisson structures for the system.

2. Coupled KdV systems

The system introduced by Gear and Grimshaw [2] and analyzed in [6] is

$$\begin{aligned} &u_t + uu_x + u_{xxx} + a_3 v_{xxx} + a_1 vv_x + a_2(uv)_x = 0 \\ &b_1 v_t + rv_x + vv_x + v_{xxx} + b_2 a_3 u_{xxx} + b_2 a_2 uu_x + b_2 a_1 (uv)_x = 0, \end{aligned} \tag{1}$$

where a_1, a_2, a_3, b_1, b_2 and r are real constants with b_1 and b_2 positive. u and v are real valued functions, $u = u(x,t), v = v(x,t)$.

In [6] the Cauchy problem was shown to be well posed. A hamiltonian functional for this system was also obtained. The hamiltonian density is

$$\mathcal{H} = b_2(u_x)^2 + (v_x)^2 + 2b_2a_3u_xv_x - b_2\frac{u^3}{3} - b_2a_2u^2v - b_2a_1uv^2 - \frac{v^3}{3} - rv^2. \tag{2}$$

If we consider x to be of dimension 1, then u and v must have dimension -2 in order that the different terms on the above expression have the same dimension. From the field equation (1) we conclude that t must have dimension -3. We then notice that the term rv^2 has different dimension unless the constant r has dimension -2. This feature distinguish this term from the others where the constant coefficients a_a, a_2, a_3, b_1, b_2 are dimensionless.

The hamiltonian function introduced in [4] has the same terms as $\mathcal{H}$ in (2) together with $-r_1v^2 - r_2u^2 - kvu$ which have coefficients with dimension -2.

Similar terms appear in the field equations in [3] where in order to consider a hamiltonian system we must take $\alpha_1 = \delta_1 = 0$ (in their notation).

The hamiltonian functional in all these cases contains terms of weight -6, essentially they contain all possible terms of that weight.

3. Hamiltonian structure of the coupled KdV system

The field equations (1) and the hamiltonian density $\mathcal{H}$ in (2) arise from a lagrangian density $\mathcal{L}$ formulated in terms of the fields w, y defined as

$$\begin{aligned} u &= w_x \\ v &= y_x. \end{aligned} \tag{3}$$

The lagrangian density is given by

$$\mathcal{L} = -\frac{1}{2}w_xw_t - \frac{1}{2}y_xy_t - \mathcal{H}. \tag{4}$$

By taking independent variations of $L = \int_{t_1}^{t_2} dt \int_{-\infty}^{+\infty} \mathcal{L}dx$ with respect to w and y one obtains the field equations (1).

The interesting point of the coupled system is that there exists a second lagrangian from which we can deduce, by taking independent variations with respect to w and y, the same field equations. The new lagrangian density gives rise, via a Legendre transformation, to a new hamiltonian structure of the system.

Let us consider the first basic lagrangian density

$$\hat{\mathcal{L}} = \frac{1}{2}w_xw_t - \frac{1}{2}y_xy_t - \hat{\mathcal{H}}, \tag{5}$$

where

$$\begin{aligned} \hat{\mathcal{H}} = {} & \alpha_1(w_{xx})^2 + \alpha_2(y_{xx})^2 + \alpha_3w_{xx}y_{xx} + \alpha_4(w_x)^3 + \\ & +\alpha_5(w_x)^2y_x + \alpha_6(y_x)^2w_x + \alpha_7(y_x)^3 + \alpha_8(w_x)^2 + \alpha_9w_xy_x + \alpha_{10}(y_x)^2. \end{aligned} \tag{6}$$

The corresponding fields equations are

$$\begin{aligned} w_{xt} &= 2\alpha_1w_{xxxx} + \alpha_3y_{xxxx} - 3\alpha_4\left((w_x)^2\right)_x - 2\alpha_5(w_xy_x)_x - \alpha_6\left((y_x)^2\right)_x - 2\alpha_8w_{xx} - \alpha_9y_{xx} \equiv \frac{\delta H}{\delta w} \\ y_{xt} &= 2\alpha_2y_{xxxx} + \alpha_3w_{xxxx} - \alpha_5\left((w_x)^2\right)_x - 2\alpha_6(y_xw_x)_x - 3\alpha_7\left((y_x)^2\right)_x - \alpha_9w_{xx} - 2\alpha_{10}y_{xx} \equiv \frac{\delta H}{\delta y}. \end{aligned} \tag{7}$$

Now we introduce the second basic lagrangian density,

$$\tilde{\mathcal{L}} = -\frac{1}{2}w_x y_t - \frac{1}{2}y_x w_t - \tilde{\mathcal{H}}. \tag{8}$$

The corresponding field equations are

$$\begin{aligned} w_{xt} &= \frac{\delta \tilde{H}}{\delta y} \\ y_{xt} &= \frac{\delta \tilde{H}}{\delta w} \end{aligned} \tag{9}$$

where $\widetilde{H} = \widehat{\mathcal{H}}(\tilde{\alpha_1}, \tilde{\alpha_2}, \ldots, \tilde{\alpha_{10}}) = \int_{-\infty}^{+\infty} \widetilde{\mathcal{H}}dx$, $\widetilde{\mathcal{H}}$ has the same general form (6) but now in terms of the coefficients $\tilde{\alpha_1}, \tilde{\alpha_2}, \ldots, \tilde{\alpha_{10}}$. In order to obtain the same field equations we get a relation between the coefficients. For example: $\tilde{\alpha_3} = 2\alpha_1, \tilde{\alpha_2} = \frac{1}{2}\alpha_3$ and so on.

We thus have two different lagrangians and two different hamiltonian densities which give rise to the same field equations. The two lagrangians provides two different basic Poisson bracket structures. From the two basic Poisson structures one may construct a pencil of Poisson brackets with an associated hamiltonian for each value of the parameter of the pencil.

The field equations for suitable values of the constants $\alpha_i, i = 1, \ldots, 10$ describe the Grimshaw equations in [4].

4. Conclusions

We presented the most general hamiltonian approach for the Gear and Grimshaw coupled KdV system which contains several physical models, in particular the one considered in [3]. We introduce two lagrangians and associated hamiltonians for the systems and discuss the Poisson pencil structure generated by them.

Acknowledgments

A. R. and A. S. are partially supported by Projects Fondecyt 1121103 and Mecesup ANT398 (Chile).

References

[1] Hirota R and Satsuma J 1981 *Phys. Lett.* A **85** 407
[2] Gear J A and Grimshaw R 1984 *Stud. Appl. Math.* **70**, 235 ; Gear J A 1985 *Stud. Appl. Math.* **72**, 95
[3] Lou S Y, Tong B, Hu H C and Tang X Y 2006 *J. Phys. A: Math. Gen.* **39** 513-527
[4] Grimshaw R 2013 *Without Bounds: A Scientific Canvas of Nonlinearity and Complex Dynamics, Understanding Complex Systems* Rubio R G et al (eds.) (Springer-Verlag Berlin Heidelberg)
[5] Wang D S 2010 *Appl. Math. Comp.* **216** 1349-1354
[6] Bona J L, Ponce G, Saut J C and Tom M N 1992 *Commun. Math. Phys.* **143** 287-313

Dynamic of HNLS Solitons using Compact Split Step Padé Scheme

Moussa Smadi[1] and Derradji Bahloul[2]

[1,2]Département de Physique, Faculté des Sciences, Université de Batna
Laboratoire de Physique des Rayonnements et de leurs Interactions avec la Matière
LPRIM, 5000 Avenue Chahid Boukhlouf, 05000 Batna, Algeria

[1]msmadi@univ-batna.dz
[2]bahloul@univ-batna.dz

Abstract: In this communication, we use a compact split step Padé scheme (CSSPS) to solve the scalar higher-order nonlinear Schrödinger equation (HNLS) with higher-order linear and nonlinear effects. The second part, consisting of two sections, the first section is dedicated to the study numerically the stabilization of high-order solitons dynamic in optical fibers by compensation or by the interplay of higher order nonlinearity - especially quintic nonlinearity- and the self-steepening. In the second section we study also numerically the propagation of conventional chirped or unchirped solitons in optical fibers with to the management of the non-linearity, dispersion and loss (gain).

Keywords: higher order optical solitons, compact split step Padé scheme, higher-order nonlinear Schrödinger equation (HNLS), quintic nonlinearity, dispersion managed.

1. Introduction:

Propagation of solitons in optical fibers has attracted a great attention of many authors [1-3]. The balance between nonlinear Kerr effect and chromatic dispersion is the clue of the stability of optical NLS solitons [2]. The propagation of ultra-short and ultra-intense optical solitons in optical fibers is not so obvious and requires efficient and fast numerical methods in order to be investigated. At high power higher order nonlinearities may alter the propagation of the solution especially with the interplay of higher order dispersion effects such as third or fourth ordered dispersion. Quintic nonlinearity [4] is the most important nonlinear effect due to the saturation of optical field [5] and must be taken into account in the study of ultra-intense optical solitons. In this communication we use a compact split step Padé scheme (CSSPS) [6] to solve the higher-order nonlinear Schrödinger (HNLS) equation with power law nonlinearity [7,8] and higher order dispersion effects, in order to study numerically the impact of the combined nonlinear effects, such self-steepening and quintic nonlinearity on the propagation of the higher-order soliton in optical fibers, and the evolution of conventional unchirped or chirped solitons in optical fibers with to the management of the non-linearity, dispersion and loss (gain).

2. Theory:

The higher-order nonlinear Schrödinger HNLS equation with power law nonlinearity and higher order dispersion:

$$\frac{\partial A(\mathrm{Z,T})}{\partial \mathrm{Z}} = -\frac{\alpha}{2}A + i\sum_{n=2}^{N=4}\frac{i^n}{n!}\beta_n\frac{\partial^n A}{\partial T^n} + i\gamma\left[|A|^2A + \gamma'|A|^{2m}A + \frac{i}{\omega_0}\frac{\partial}{\partial T}(|A|^2A) - T_R A\frac{\partial |A|^2}{\partial T}\right] \qquad (1)$$

Where $i = \sqrt{-1}$ and $\mathrm{A(Z,T)}$ is the slowly varying envelope of the optical pulse and $T = t_{lab} - \mathrm{Z}/v_g$ the temporal coordinate in retarded frame that moves at the group velocity v_g of the pulse and Z the spatial coordinate representing the distance of transmission β_2, β_3, β_4 and Represents dispersion parameters of the second, third and fourth orders. α is the attenuation factor. For silica fibers $\beta_2 \sim 20ps^2/km$ at $\sim 1.55\,\mu m$ [1] $\gamma = n_2\omega_0/cA_{eff}$ and γ' are respectively the nonlinear Kerr effect coefficient and the power law nonlinearity coefficient. n_2 is the nonlinear index coefficient, for silica fibers $n_2 = 2.2 \times 10^{-20} m^2/W$ at the same wave length[1]. The term proportional to $1/\omega_0$ governs the Kerr dispersion that is responsible of self-steepening and shock formation. The last term is responsible of self-frequency shift induced by intrapulse Raman scattering and T_R is related to the Raman response function at wave length $\lambda_p \sim 1.55\,\mu m$ $T_R \approx 3fs$. [1]

3. Numerical scheme

We first normalize equation (1) in the following way:

$$t = \frac{T}{T_0}, z = \frac{Z}{L_D}, L_D = \frac{T_0^2}{|\beta_2|}, L_D' = \frac{T_0^3}{|\beta_3|}, L_D'' = \frac{T_0^4}{|\beta_4|}, L_{NL} = \frac{1}{\gamma P_0}, \kappa = \frac{1}{\gamma' P_0^{m-1}}, s = \frac{1}{\omega_0 T_0}, \tau_R = \frac{T_R}{T_0}$$

$$and\ u(\mathrm{z,t}) = \frac{\mathrm{A(z,t)}}{\sqrt{\mathrm{P_0}}}$$

Where T_0 is the initial pulse width, L_D, L'_D, L''_D and L_{NL} are respectively the dispersion lengths for the different orders and the nonlinear length. The parameters s and τ_R govern the effects of self-steeping and intrapulse Raman scattering respectively. Both of these effects are quite small for picoseconds pulses but must be considered for ultra-short pulses with $T_0 < 0.1ps$ [1]. In our simulation, we suppose the parameter $(\alpha = 0)$ for lossless fibers. We choose to rewrite the equation (1) in the form (2) because it is more convenient for the numerical calculations.

$$\frac{\partial u}{\partial z} = i\alpha_2\frac{\partial^2 u}{\partial t^2} + \alpha_3\frac{\partial^3 u}{\partial t^3} + i\alpha_4\frac{\partial^4 u}{\partial t^4} + i\left[\alpha_5|u|^2u + \alpha_6|u|^{2m}u - i\alpha_7\frac{\partial}{\partial t}(|u|^2u) + \alpha_8 u\frac{\partial |u|^2}{\partial t}\right] \qquad (2)$$

Where:

$$\alpha_2 = -\frac{\mathrm{sgn}(\beta_2)}{2}, \quad \alpha_3 = \frac{\mathrm{sgn}(\beta_3)}{6}\frac{L_D}{L_D'}, \alpha_4 = \frac{\mathrm{sgn}(\beta_4)}{24}\frac{L_D}{L_D''}, \quad \alpha_5 = \frac{L_D}{L_{NL}}, \ \alpha_6 = \frac{L_D}{L_{NL}}\kappa, \quad \alpha_7 = -s\frac{L_D}{L_{NL}},$$

$$and\ \alpha_8 = -\tau_R\frac{L_D}{L_{NL}}$$

This equation (2) can be written formally in the form:

$$\frac{\partial u}{\partial z} = (\hat{L} + \hat{N})u \qquad (3)$$

Where $\hat{L}$ is a linear operator that accounts for all the linear effects, and $\hat{N}$ is a nonlinear operator that governs the effect of all the fiber nonlinearities.

$$\hat{L}u = \left(i\alpha_2\frac{\partial^2}{\partial t^2} + \alpha_3\frac{\partial^3}{\partial t^3} + i\alpha_4\frac{\partial^4}{\partial t^4}\right)u \qquad (3-a)$$

$$\hat{N}u = i\alpha_5|u|^2u + i\alpha_6|u|^{2m}u + \alpha_7\frac{\partial}{\partial t}(|u|^2u) + i\alpha_8 u\frac{\partial |u|^2}{\partial t} \qquad (3-b)$$

The exact solution is approximated by solving for the half step size $\Delta z/2$ separately and alternatively the purely linear equation

$$\frac{\partial u}{\partial z} = \hat{L}u \qquad (3-c)$$

And purely nonlinear equation

$$\frac{\partial u}{\partial z} = \hat{N}u \qquad (3-d)$$

The solution of one sub problem is employed as an initial condition for the next sub problem. The validity of this approximation is discussed in reference [1] using Baker–Hausdorff formula . The linear equation is solved using the compact Padé scheme algorithm an implicit scheme that is unconditionally stable and well adapted for the different derivative orders for more detail see for example [6]. Whereas the nonlinear equation is solved using a fourth order Runge-Kutta scheme (RK4) that satisfies the CFL condition.

4. Simulation and discussion

4.1. Study of the combined effects self-steepening and quintic nonlinearity on the propagation of high order soliton in optical fiber.

The remarkable effect of the self-steepening on the propagation of higher order-solitons in optical fibers is the breakup of such solitons into their constituents, a phenomenon referred to as soliton fission [9]. In the first case we study numerically by means of (CSSPS) method the impact of the self-steepening governed by the parameter α_7 in the equation (2). In the second case we study the impact of combined effects the self-steepening and the quintic effect governed by the parameter α_6 in the equation (2) on the propagation of higher order-solitons. Pulse evolution inside fibers is then governed by equation (4):

$$\frac{\partial u}{\partial z} = i\alpha_2\frac{\partial^2 u}{\partial t^2} + i\left[\alpha_5|u|^2u + \alpha_6|u|^{2m}u - i\alpha_7\frac{\partial}{\partial t}(|u|^2u)\right] \qquad (4)$$

Figure 1, shows this comportment for a second order soliton with $\alpha_7 = 0.2$. We note that both solitons propagate at different speeds, as a result, they separate from each other, and the separation increases in a linear manner as a function of distance this result is in good agreement by comparison with literature [10]. Figure 2, shows the temporal and amplitude evolution of a second order soliton ($N = 2$) over ten dispersion lengths. One can conclude that the simultaneous presence of the self-steepening effect and quintic effect allows the high-order soliton to keep its shape longer without significant breaking or deformation on a longer propagation distance, when the quintic parameter increases. It can be concluded as that the existence of the quintic effect plays an important role in the removal of the breakup caused by the self-steepening effect. This is almost similar to the phenomenon of cancellation effects caused by the dispersion of the group velocity (GVD) by the Kerr effect (SPM) for a fundamental soliton.

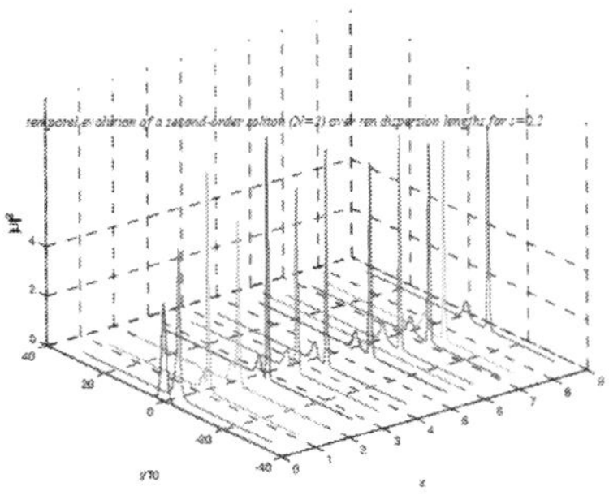	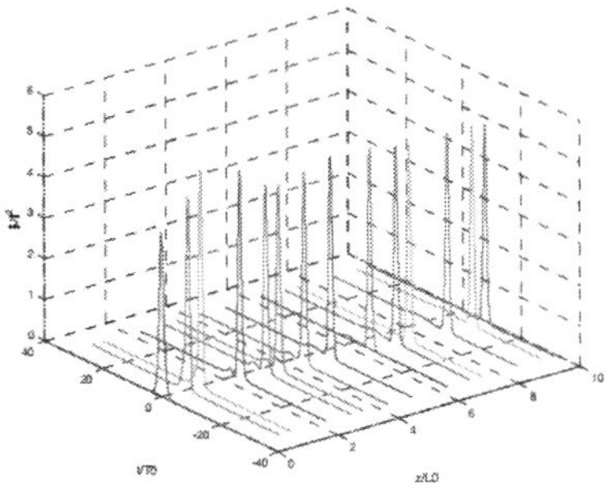
Figure 01: temporal evolution of a second order soliton (N=2) over ten dispersion lengths for, $\alpha_3 = 0, \alpha_4 = 0, \alpha_6 = 0, \alpha_7 = 0.2 and\ \alpha_8 = 0$	**Figure 02**: temporal evolution of a second order soliton (N=2) over ten dispersion lengths for $\alpha_2 = 0.5, \alpha_3 = 0, \alpha_4 = 0, \alpha_6 = -0.10, \alpha_7 = 0.2,\ \alpha_8 = 0\ and\ m = 2$

4.2 Study of conventional unchirped or chirped soliton propagation in optical fiber in the presence of management of dispersion, nonlinearity and gain (loss).

The concept of dispersion management technique is to compensate these effects by using a periodic dispersion map of combining fibers with different characteristics [1]. However in real soliton application systems, the dispersion, nonlinearity, gain and loss are generally varied with the propagation distance. In the following we present the results of our study obtained by application of the (CSSPS) in the case of the propagation of conventional unchirped or chirped soliton in an optical fiber when the dispersion, non-linearity, gain and loss are generally varied with the propagation distance. In many references [11-12-13-14] intensive studies have been devoted to this situation. The Schrödinger equation model which governs the dynamics of solitons in this type of fiber is given by the equation (5):

$$i\frac{\partial A(Z,T)}{\partial Z}+\frac{1}{2}D(z)\frac{\partial^2 A}{\partial T^2}+\gamma(z)|A|^2A=i\alpha(z)A \qquad (5)$$

Where $D(z)$ represents the variation of the dispersion of the group velocity (GVD) as a function of the distance, $\gamma(z)$ represents the variation of the Kerr non-linearity parameter and $\alpha(z)$ represent the variation of the gain (loss) as a function of propagation distance also. Note that the nonlinear Schrödinger equation with variable coefficients (6) is not integrable except for a few very special cases of combinations of variables carefully chosen $D(z)$, $\gamma(z)$ and $\alpha(z)$ [15]. In our numerical simulations we considered that the variation of the dispersion of the group velocity (GVD) is given by [16].

$$D(z)=\frac{1}{D_0}exp(\sigma z)\gamma(z), \quad \gamma(z)=\gamma_0+\gamma_1\sin(g.z) \quad , \quad \alpha(z)=\frac{\sigma}{2} \qquad (5-1)$$

The peak power of the initial pulse takes into account by the D_0 factor. γ_0, $\gamma_1$and g parameters represents the Kerr nonlinearity. We take $\sigma < 0$ for a system having a dispersion decreasing and loss (absorption), $\sigma > 0$ for the reverse case. The initial pulse injected in the form of a conventional chirped soliton given by equation [1]:$u(0,T)=sech(T)\,exp\left(-\frac{i.C.T^2}{2}\right)$ $(5-2)$, Where C is the chirp parameter. And for the others parameters we took the following values [14]:$\gamma_0=0, D_0=1; \gamma_1=1; g=1$ $(5-3)$, The results obtained are shown by curves (4) (5) and (6) below:

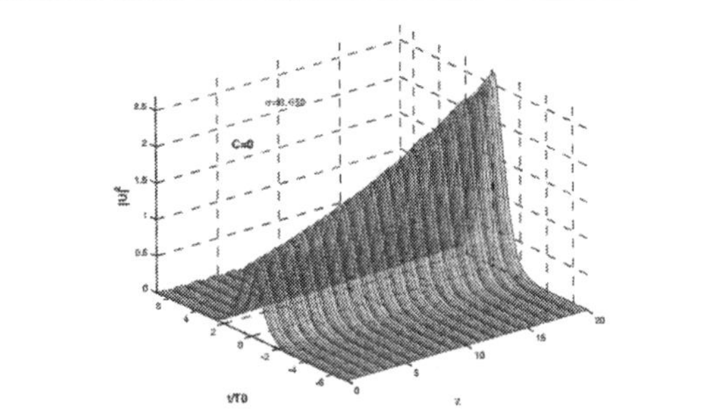	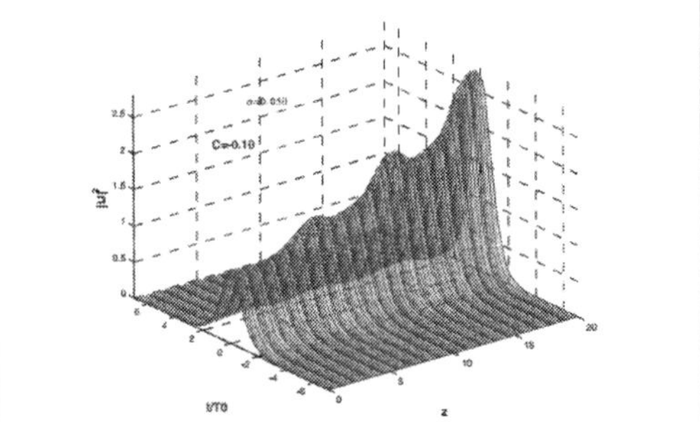
Figure 04: Evolution of the intensity profile of a conventional unchirped soliton with propagation distance	**Figure 05**: Evolution of the intensity profile of a conventional chirped soliton with propagation distance for $c=-0.10, \sigma=0.050$.

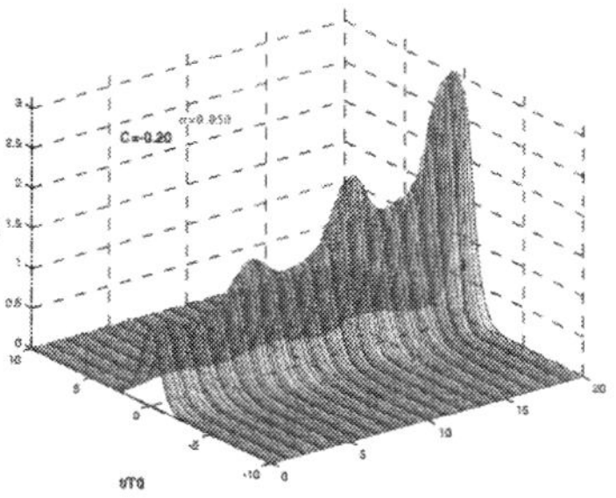

Figure 06: Evolution of the intensity profile of a conventional chirped soliton "with propagation distance $c = -0.20, \sigma = 0.050$.

We noticed that the amplitude of a conventional unchirped solitons have an increase in a linear manner as a function of distance when the parameter σ is positive, in the opposite case σ negative the amplitude of a conventional unchirped solitons have an decrease in a linear manner as a function of distance. While the amplitude of the conventional chirped soliton oscillates (increase / decrease) depending on the module of the "chirped" parameter and it can be seen that the chirped soliton is obviously and periodically compressed and broadening with the increase of propagation distance. These results are in good agreement with the model presented in reference [17]

5-Conclusion

The compact split step Padé Scheme has been implemented to investigate the propagation of ultra-intense and ultra-short optical solitons under the effect of power law nonlinearity and high order dispersion effect. This scheme is more efficient, rapid and takes less memory space. It is also well adapted to the higher order derivatives. We applied this method for the case of combined effects such self-steepening quintic nonlinearity and concluded that the quintic effect plays an important role in the removal of the breakup caused by the self-steepening effect on the propagation of optical second-order solitons. In the presence of management of dispersion, nonlinearity and gain (loss) the amplitude of a conventional unchirped solitons have a decrease in a linear manner as a function of distance of propagation. While the amplitude of the conventional chirped soliton oscillates (increase / decrease) depending on the module of the "chirped" parameter.

Acknowledgments

This work is supported by the post-graduation of The *Département de Physique*, *Faculté des Sciences* of the *Université de Batna*, Algeria.

References:

[1] G. P. Agrawal, *"Nonlinear Fiber Optics"*, fourth edition, Academic Press, Boston, 2007.

[2] A. Hasegawa and Y. Kodama *Solitons in Optical communications*,"Clarendon Press; Oxford (1995)

[3] A. Hasegawa Theory of information transfer in optical fibers*: A tutorial review*," *Opt. Fiber Technol*, (2004), 10, pp-50-170.

[4] J. Sonesone and A. Peleg, The effect of quintic nonlinearity on the propagation characteristics of dispersion managed optical solitons", *Chaos Solitons and Fractals*. (2006), 29, pp823-828

[5] S Konar, M. Mishra, S. Jana, Effect of quintic nonlinearity on soliton collisions in fibers,"*Physica D*. (2004), 195, pp123-140.

[6] Moussa. Smadi and Derradji. Bahloul "A compact split step Padé scheme for higher-order nonlinear Schrödinger equation (HNLS) with power law nonlinearity and fourth order dispersion", *Computer Physics Communications*, Volume 182, Issue 2, February 2011, Pages 366-371

[7] A.M. Wazwas, Exact solutions for the fourth order nonlinear Schrödinger equations with cubic and power law nonlinearities", *Math. Comput. Modell.* (2006), 43 (7-8):802-8.

[8] A. Biswas and D. Milovic Optical solitons in power law media with fourth order dispersion," *Commun. Nonlinear Sci. Numer. Simulat.* 14 (2009), 1834-1837.

[9] A. Hasegawa, M. Matsumoto, *Optical Solitons in Fibers*, Springer, New York, 2002.

[10] K Ohkuma, Y H Ichikawa, Y Abe, Opt. Lett. 12 (1987) 516.

[11]. James Wilson, G. I. Stegeman, E. M. Wright, "Soliton Switching in an erbium doped nonlinear fiber coupler", *Opt, Lett,* 16 (1991) 1653-1655.
[12]. Serkin V. N, A. Hasegawa, "Novel soliton solution of the nonlinear Schrodinger equation model" *Phys. Rev. Lett.* 85 (2000) 4502-4505.
[13]. Serkin V. N, A. Hasegawa, "*Soliton management in the nonlinear Schrodinger equation model with varying dispersion, nonlinearity, and gain" JETP. Lett.* 72(2) (2000)89-92.
[14].Partha P. Banerjee "*Nonlinear optics –Theory, Numerical Modeling, and applications*-2004 by Marcel Dekker, New York, U S A, pp164-171
[15] J. Wang, L. Li, S. Jia, "Exact chirped gray soliton solutions of the nonlinear Schrödinger equation with variables coefficients" , *Optics Communications* 274(1) (2007)223-230.
[16] R. Hao, L. Li, X. Shi, W. Xue, G. Zhou, "A new approach to exact soliton solutions and soliton interaction for the nonlinear Schrödinger equation with variables coefficients" *Optics Communications* 236(1-3) (2004)79-86
[17] H Zheng, C Wu, Z Wang, H Yu, S Liu and X Li "Propagation Characteristics of chirped soliton distributed amplification systems with variable coefficients", *Optik*, 123, 2012, pp 818-822.

The Solutions of the equations of pulse propagation in optical fibers

Y Z Pchelkina[1,2,3], **I V Alimenkov**[1]

[1] Samara State Aerospace University (SSAU), 34, Moskovskoye shosse, Samara, 443086, Russia
[2] 148-32, Unyh Pionerov prospect, Samara, 443105, Russia
[3] musina@yandex.ru

We have found a solution of the optical pulses propagation equation in optical fibers in quadratures. We have derived an expanded equation of optical pulses propagation in silica optical fibers and found its localized solution. The derived solution of the extended equation of optical pulses propagation in optical fibers for arbitrary function of the nonlinear medium response to an external harmonic disturbance has been found in quadratures.

1. Introduction.

The optical pulse field which is distributed in a single-mode optical fiber supporting the linear polarization condition has the form [1]:

$$\boldsymbol{E}(\boldsymbol{r},t) = \boldsymbol{e}_x F(x,y) A(z,t) \exp\{i(\beta_o z - \omega_o t)\}. \tag{1}$$

For this function we have derived a propagation equation [1], [2]:

$$\frac{\partial A}{\partial z} + \beta_1 \frac{\partial A}{\partial t} + i\frac{\beta_2}{2}\frac{\partial^2 A}{\partial t^2} = i(\Delta\beta)A, \tag{2}$$

where $\beta_1 = 1/v_g$ - is the value of the reverse group velocity. n the field of waveguide transparency $\Delta\beta$ is the real function of $|A|^2$.

The equation (2) holds true for pulses with duration $> 0{,}1nc$ which corresponds to quasi-monochromatic approximation approximation. This equation belongs to Ginsburg – Landau class of equations. In this abridged form (without the second spatial derivative) it can be reduced to a nonlinear Schrodinger equation, in which the coordinate and time switch places. Most of the known publications about optics use a standard approach to the solution of the nonlinear Schrodinger equation. The aim of this work is to apply a non-standard approach to the equation solution (2).

2. Method for solving equation of optical fibers pulses propagation in quadratures.

The solution will be conducted in the laboratory frame of reference without specifying the nonlinear response functions $\Delta\beta$. Substitution into (2)

$$A(z,t) = \sqrt{I}\,\exp\{i\,qz\}, \tag{3}$$

were q - is a small correction to the central wave number β_o in (2) after separating the real and imaginary parts will result in the following two equations for envelope optical intensity $I(z,t) = |A(z,t)|^2$:

$$\frac{\partial I}{\partial z} + \beta_1 \frac{\partial I}{\partial t} = 0 \tag{4}$$

$$\beta_2 \left[2I \frac{\partial^2 I}{\partial t^2} - \left(\frac{\partial I}{\partial t} \right)^2 \right] = -8qI^2 + 8I^2 \Delta\beta(I). \tag{5}$$

The general solution of the linear homogeneous equation (4) is an arbitrary differentiable function $I = I(s)$, где $s(z,t) = z - z_o - v_g t$. Going over in (5) to a form of the differential equation, we get:

$$2I \frac{d^2 I}{ds^2} - \left(\frac{dI}{ds} \right)^2 = \frac{8I^2}{\beta_2 v_g^2} \left[\Delta\beta(I) - q \right]. \tag{6}$$

A quadratic equation (6) is an equation of Euler - Lagrange for mechanical analogue-particle. We have found the solution in quadratures:

$$\int \frac{dI}{I\sqrt{(G(I)/I - q)/\beta_2}} = \frac{2\sqrt{2}}{v_g} s. \tag{7}$$

Thus, the solution of the problem of optical pulses propagation in single-mode optical fibers described by the differential equation in partial derivatives (2) is reduced to the calculation of the primitive in the left-hand side of equation (7) and to the treatment of the obtained expression that is certainly an easier task.

3. The solution of the extended equation of optical fibers pulses propagation

Let γ – be the nonlinearity coefficient, then, from (2):

$$\frac{\partial A}{\partial z} + \frac{1}{v_g} \frac{\partial A}{\partial t} + \frac{i\beta_2}{2} \frac{\partial^2 A}{\partial t^2} = i\gamma |A|^2 A, \tag{8}$$

All the above parameters are known for single mode quartz fibers and their numerical values are given in [1], [2], [6]. But the solution (7) is flawed because the distribution of the pulse velocity has the form of a Delta-function, and because this rate fixed for all impulses does not depend on the pulse peak intensity.

We have derived an extended equation of pulses propagation and we propose its solution to solve this problem. We have considered an equation for the field strength in a spectral representation [1].

$$\nabla^2 \tilde{E}(\vec{r}, \omega - \omega_o) + \varepsilon(\omega) k_o^2 \tilde{E}(\vec{r}, \omega - \omega_o) = 0. \tag{9}$$

where $\tilde{E}(\vec{r}, \omega - \omega_o) = \int_{-\infty}^{\infty} E(\vec{r}, t) \exp\{i[\omega - \omega_o]t\} dt$, $\varepsilon(w) = (n + \Delta n)^2 \cong n^2 + 2n\Delta n$ - is dielectric permeability.

The equation (4) is solved by the standard method of separation of variables, by substituting equation (9) into the solution. Let $\bar{\beta}^2$ - be a separation constant, then we obtain two equations:

$$\frac{\partial^2 F}{\partial x^2} + \frac{\partial^2 F}{\partial y^2} + \left[\varepsilon(\omega) k_o^2 - \bar{\beta}^2 \right] F = 0, \tag{10}$$

$$\frac{\partial^2 \tilde{A}}{\partial z^2} + 2i\beta_o \frac{\partial \tilde{A}}{\partial z} + \left(\bar{\beta}^2 - \beta_o^2 \right) \tilde{A} = 0. \tag{11}$$

(10) determines the distribution of the mode field $F(x,y)$ and the correction $\Delta\beta$ to the propagation constant in linear approximation $\overline{\beta}(\omega)=\beta(\omega)+\Delta\beta$.

While considering (11) we will not follow instructions [1], [2], [6] to neglect the second derivative in the equation (11) because of assumptions of the slow function changeability $\widetilde{A}(z,\omega-\omega_o)$, so we get the complete solution from the equation (11)

$$i\left(\frac{\partial A}{\partial z}+\beta_1\frac{\partial A}{\partial t}\right)+\frac{1}{2\beta_o}\frac{\partial^2 A}{\partial z^2}-\frac{\beta_2}{2}\frac{\partial^2 A}{\partial t^2}+\gamma|A|^2A=0. \qquad (12)$$

It is interesting to find solutions to this equation and study various limiting cases. We have derived the solution of this equation:

$$A(z,t)=\frac{E_o\exp\{iz\gamma E_o^2/2\}}{ch\left[(z-z_o-vt)\sqrt{\frac{\gamma E_o^2\beta_o}{1+|\beta_2|\beta_o v^2}}\right]}. \qquad (13)$$

This solution represents a wave packet moving with a constant speed which depends on the tension peak values and it has a central wave number. This radically distinguishes it from formula (3). Within the limit formula (13) goes over into (3), i.e. it is more general.

4. The solution of the extended equation of optical fibers pulses propagation at arbitrary nonlinearity in quadratures

Now let us consider the solution of equation (12) in quadratures for an arbitrary function of the nonlinear response. The function $\Delta\beta(|A|^2)$ describing the nonlinear response of the medium to external harmonic perturbation, belongs to one of three classes: competing, saturating and transitional nonlinearity [7]. Let in (12):

$$A(z,t)=R(z,t)\exp\{\pm iqz\} \qquad (14)$$

were R - the real function, q - arbitrary nonnegative parameter, which is associated with field peak strength.

Thus, the envelope is a running wave of unchanged profile that moves with constant speed $v=v_g(1\pm q/\beta_o)$. Profile of this wave is determined by the equation

$$\left(\frac{1}{2\beta_o}-\frac{\beta_2 v_g^2}{2}\right)\frac{d^2R}{ds^2}=\left[\pm q+\frac{q^2}{2\beta_o}-\Delta\beta(R^2)\right]R, \qquad (15)$$

which is soluble in quadratures:

$$\sqrt{\frac{1-\beta_o\beta_2 v_g^2}{2\beta_o}}\int\frac{dR}{\sqrt{(\pm q+q^2/2\beta_o)R^2-B(R^2)+C_1}}=z-z_o-vt, \qquad (16)$$

where $(-z_o)$ is an arbitrary constant of integration. Constant C_1 is determined by the behavior of the envelope at infinity. We should notice that parameter q is a correction to the wave number β_o, which should not exceed half of the value, so the sign of the expression inside the parentheses under the radical is determined by the sign before q.

5. Conclusions

The solution of the problem of optical pulses propagation in single-mode optical fibers that support the status of linear polarization is reduced to the calculation of the primitive in the left-hand side of (16). It is certainly technically easier than the solution of equations in partial derivatives. As graphical analysis does not require treatment of expression which is obtained from (16), the envelope graph can be constructed in the case when the left part of the formula (16) is not integrable in elementary functions.

Acknowledgements
This work was supported by the Ministry of education and science of the Russian Federation in the framework of the implementation of the Program of increasing the competitiveness of SSAU among the world's leading scientific and educational centers for 2013-2020 years.

References
[1] Agrawal, G.P. Nonlinear Fiber Optics. – Moscow: " Mir", 1996. – 324p. – (in Russian).
[2] Agrawal, G.P. Nonlinear Fiber Optics. – Academic Press, 2007. – 478p.
[3] Rajaraman, R. Solitons and instantons in quantum field theory. – Moscow:"Mir" Publisher, 1985. - 416p. – (in Russian).
[4] Stepanov, V.V. Course of differential equations. – Moscow: " GITTL" Publisher, 1953. – 468p. – (in Ruaaian)
[5] Dwight, H.B. Tables of integrals and other mathematical data. – Moscow: " Nauka" Publisher, 1977. – 224p. – (in Russian).
[6] Agrawal, G.P. Nonlinear Fiber Optics. – Academic Press. – 2013. – 629p.
[7] Kivshar, Y.S. Optical solitons. From Fibers to Photonic Grystals / Y.S. Kivshar, G.P. Agrawal – Moscow: "Fizmatlit" Publisher, – 2005. – 648p. – (In Russian).

Frequency comb generation for wave transmission through the nonlinear dimer

K.N. Pichugin and A.F. Sadreev
L.V. Kirensky Institute of Physics, 660036, Krasnoyarsk, Russia
E-mail: almas@tnp.krasn.ru

Abstract. We study dynamical response of the nonlinear dimer relative to monochromatic wave injected via the waveguide. We show existence of a domain in space of frequency and injected amplitude where the stationary solutions of the time evolution equations do not exist. We present time dependent solutions which show that scattering waves carry multiple harmonics with frequencies spaced equidistantly.

In a series of publications [1, 2, 3, 4, 5, 6] time evolution of quantum nonlinear dimer was explicitly considered. The dimer is the simplest realization of discrete nonlinear Schrödinger equation and consists of two nonlinear sites. That case is exactly solvable with the solutions in the form of elliptic functions [2]. Observation of the closed nonlinear dimer implies application

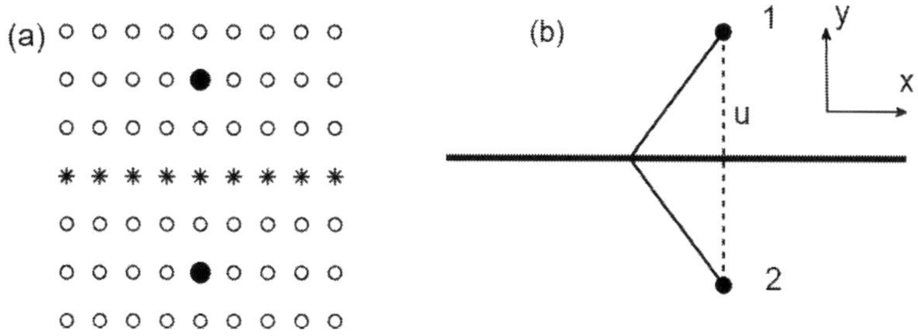

Figure 1. (a) Two identical microcavities made from a Kerr media marked by filled circles are inserted into the square lattice photonic crystal of dielectric rods with dielectric constant ϵ_0. The 1D waveguide is formed by substitution of linear chain of rods by the rods with dielectric constant $\epsilon_W + \epsilon_0$ marked by stars. (b) Schematic system consisting of a waveguide coupled with two off-channel microcavities.

of probing signal onto the system that opens the dimer. Respectively, dynamical equations describing the opened nonlinear dimer become non integrable that constitutes a main difference between the closed and opened nonlinear dimer. The opened quantum nonlinear dimer can be realized in the form of directional linear photonic crystal waveguide with two in-channel or

Published under licence by IOP Publishing Ltd

off-channel Kerr microcavities as shown in Fig. 1 [7]. As dependent on engineering of openness the stationary states of the opened nonlinear dimer were studied in Refs. [8, 9, 10, 11, 12] to show a symmetry breaking phenomenon. There is a domain in space of frequency and injected power where the stationary solutions of the time evolution equations do not exist [7, 10]. Then one can expect that dynamical response of the nonlinear dimer will display features which can not be described by the stationary scattering theory. In particular injection of a monochromatic symmetric wave onto the nonlinear plaquette gives rise to emission of anti-symmetric satellite waves with frequencies different from the frequency of the incident wave [10]. This phenomenon is determined as frequency comb (FC) generation and studied in many linear and nonlinear systems [13, 14].

We write the equations for the dynamical evolution of the microcavity modes in the framework of the coupled mode theory (CMT) [7, 15] for the dimer shown in Fig. 1 (b):

$$
\begin{aligned}
i\dot{A}_1 &= (\omega_0 + \lambda|A_1|^2)A_1 - i\gamma A_1 + (u - i\gamma)A_2 + i\sqrt{\gamma}E_{in}e^{i\omega t},\\
i\dot{A}_2 &= (\omega_0 + \lambda|A_2|^2)A_2 - i\gamma A_2 + (u - i\gamma)A_1 + i\sqrt{\gamma}E_{in}e^{i\omega t}.
\end{aligned} \tag{1}
$$

where $I_1 = |A_1|^2, I_2 = |A_2|^2$, $\sqrt{\gamma}$ is the coupling constant of the microcavities with probing wave with the amplitude E_{in} and frequency ω injected through the directional waveguide, u is the coupling constant between the microcavities due to overlapping. In the present paper we consider the case of uncoupled microcavities $u = 0$. Even in that simplest case dynamics of the nonlinear dimer effected by injected monochromatic wave remains cardinally different from the case of closed dimer because of interaction of the microcavities via the continuum. An amplitude of outgoing (transmitted) wave is given by equation [7, 15]

$$
E_{out} = E_{in} - \sqrt{\gamma}(A_1 + A_2). \tag{2}
$$

If to assume that the solution is stationary $A_j(t) = A_{j0}e^{i\omega t}$ the CMT equations (1) are simplified to be equal

$$
\begin{aligned}
(\nu - \lambda|A_{10}|^2)A_{10} + i\gamma(A_{10} + A_{20}) &= i\sqrt{\gamma}E_{in}\\
(\nu - \lambda|A_{20}|^2)A_{20} + i\gamma(A_{10} + A_{20}) &= i\sqrt{\gamma}E_{in}
\end{aligned} \tag{3}
$$

where $\nu = \omega - \omega_0$. It was shown the system bifurcates from the symmetry preserving state $A_{10} = A_{20}$ to the symmetry breaking state $|A_{10}| \neq |A_{20}|, \theta_1 - \theta_2 = 0, \pi$ where $A_{j0} = |A_{j0}|\exp(i\theta_j), j = 1, 2$ with variation of E_{in} [7], similar to the closed nonlinear dimer [1, 3, 4]. The symmetry preserving solution has the form

$$
A_{10} = A_{20} = A_0 = \frac{i\sqrt{\gamma}E_{in}}{\nu - \lambda|A_0|^2 + 2i\gamma} \tag{4}
$$

that defines the equation for the stationary intensity $I_0 = |A_0|^2$

$$
I_0[(\nu - \lambda I_0)^2 + 4\gamma^2] = \gamma E_{in}^2. \tag{5}
$$

However as different from the closed nonlinear dimer the opened dimer can transit into the phase symmetry breaking state with $|A_{10}| = |A_{20}|$ but $\theta_1 - \theta_2 \neq 0, \pi$ [7, 10].

Analysis of stability of the stationary solutions relative to small perturbations was performed in Ref. [7] and revealed a domain in space of parameters ω and E_{in} where *all* stationary solutions are unstable. Similar result was found in the opened nonlinear off-channel microcavity [16] and of four nonlinear microcavities [10]. We apply a standard procedure of small perturbations to find domains in space of physical parameters where the stationary solutions defined by Eq. (3) are stable [17]:

$$A_j(t) = [A_{j0} + a_j(t)]e^{i\omega t}, \quad j = 1, 2. \tag{6}$$

Presenting small perturbations $a_j(t) = (x_j + iy_j)e^{-\mu t}$ we obtain the following equations for eigenfrequencies μ

$$(\mu - 2\gamma)^2 = -(\nu - \lambda I_0)(\nu - 3\lambda I_0), \ \mu^2 = -(\nu - \lambda I_0)(\nu - 3\lambda I_0). \tag{7}$$

Therefore the boundary where the symmetry preserving family of solutions becomes unstable is defined by equations $\nu - \lambda I_c = 0$ and $\nu - 3\lambda I_c = 0$. Substituting these boundary values I_c into Eq. (5) we find a domain where stationary solutions do not exist

$$E_{in}^2 = \frac{4\gamma\nu}{\lambda}, \ E_{in}^2 = \frac{4\nu}{3\gamma\lambda}[\gamma^2 + \frac{\nu^2}{9}]. \tag{8}$$

shown in Fig. 2.

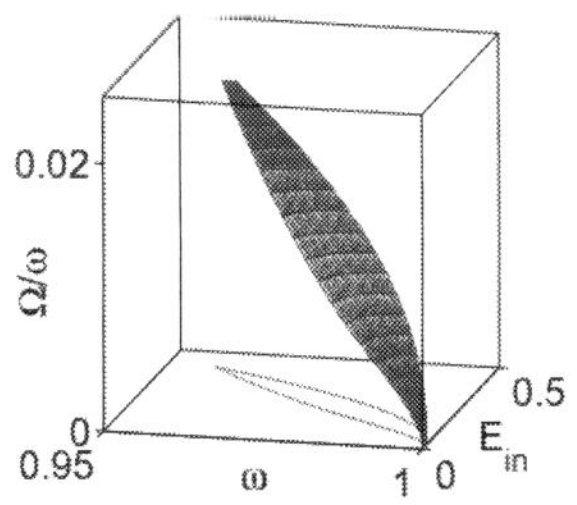

Figure 2. (Color online) The frequency comb generated frequency Ω vs amplitude E_{in} and frequency ω of injected monochromatic wave. The parameters of the dimer: $\omega_0 = 1, \gamma = 0.01, \lambda = -0.01$.

Numerics shows that in this domain the solution of the CMT equations (1) evolves into the periodical one after long time. One can see from Eq. (1) that there is a symmetry relative to a half of period and permutation of the microcavities:

$$A_1(t + T/2) = A_2(t), A_2(t + T/2) = A_1(t). \tag{9}$$

In Fig. 3 (a) and Fig. 3 (b) we present results of numerical simulations of Eq. (1) in the domain of unstable stationary solutions which indeed demonstrate the symmetry (9). Fig. 3 (c) shows that the oscillations of the cavities are nonlinear and include many Fourier harmonics $A_j(t) \approx \exp(i\omega t)\sum_n a_n \exp(in\Omega_c t)$ that demonstrates the FC effect in Fig. 3 (d). Moreover one can see that the time evolution of amplitudes $a_j(t)$ are not symmetrical in time, i.e., there is no any time t_0 that $A_j(t_0 + t) = A_j(t_0 - t)$. Absence of the time reversal symmetry is also demonstrated in Fig. 3 (c) where the oscillations of the real and imaginary parts of the amplitudes $A_j(t)$ show nonsymmetric behavior.

We have chosen the parameters listed in caption of Fig. 3 in correspondence to the photonic crystal system considered in Ref. [7]. Fig. 3 (e) and Fig. 3 (f) demonstrate similar FC effect in the transmittance amplitude $E_{out}(t)/E_{in} \approx \exp(i\omega t)\sum_n t_n \exp(2in\Omega_c t)$ calculated by Eq. (2). The Fourier components for the site amplitudes and transmittance in Figs. 3(d) and 3 (f) show that $a_n \neq a_{-n}$ and $t_n \neq t_{-n}$ that is the result of breaking of the time reversal symmetry considered above. Respectively Fig. 3 (c) shows breaking of symmetry relative to the complex conjugation of the site amplitudes.

Thus, probing of the nonlinear quantum dimer by monochromatic wave with the frequency ω and the amplitude E_{in} revealed a domain where the stationary solutions does not exist. In this domain numerical solution of the CMT equations show the following results: (i) the time reversal

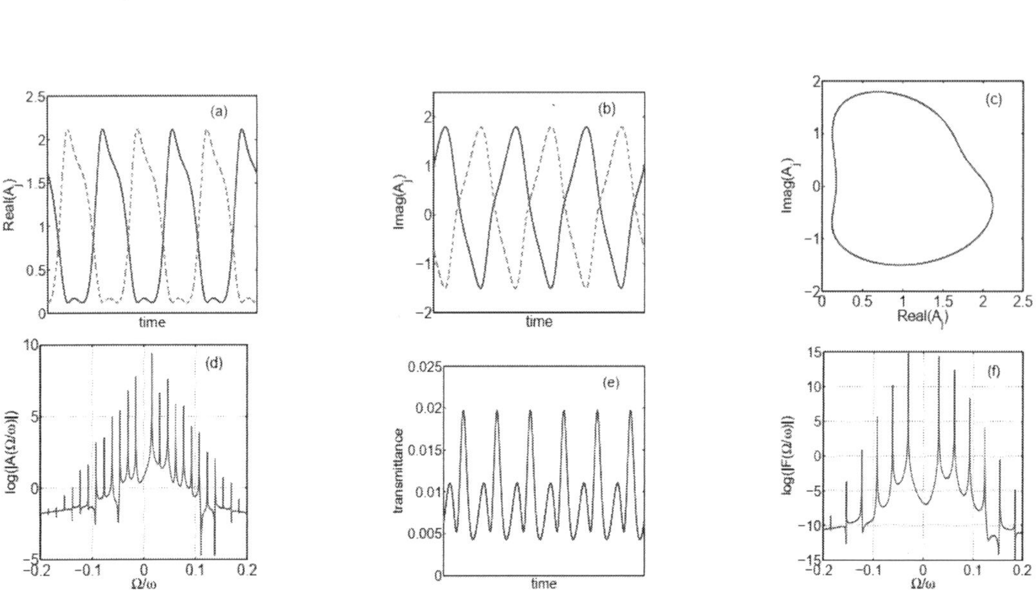

Figure 3. (Color online) Time evolution of site amplitudes $A_j(t), j = 1, 2$ (a), (b) and (c). The Fourier transform $|A(\Omega/\omega)|$ in logarithmic scale (d) with the characteristic frequency $\Omega_c \approx 0.0154\omega$. Time evolution of the transmittance $|E_{out}(t)/E_{in}|^2$ (e) and the Fourier transform $F(\Omega/\omega)$ of the transmittance amplitude $[E_{out}(t)/E_{in}]\exp(-i\omega t)$ in logarithmic scale (f). The parameters: $\omega_0 = 1, \omega = 0.98, \gamma = 0.04, E_{in} = 0.4, \lambda = -0.01$

symmetry is broken to give rise to the ratchet effect, (ii) the time evolution of the sites displays the frequency comb (FC) effect, i.e., probing wave transmitted through the nonlinear dimer acquires many harmonics spaced equidistantly by the frequency $\Omega_c \ll \omega$, (iii) the frequency of harmonics Ω_c is goes down with decreasing of the probing signal amplitude E_{in} as shown in Fig. 2. Therefore the FC frequency generated by the dimer can be governed by injected wave.

The work was supported by RFBR grant 03-02-00497. A.S acknowledges fruitful discussions the problem of frequency comb with Lushuai S. Cao.

References

[1] Eilbeck J C, Lomdahl P S and Scott A C 1985 *Physica* D **16** 318
[2] Kenkre V M and Campbell D K 1986 *Phys. Rev.* B **34** 4959
[3] Tsironis G P and Kenkre V M 1988 *Phys. Lett.* A **127** 209
[4] Kenkre V M and Kuś M 1994 *Phys. Rev.* B **49** 5956
[5] Molina M I 1999 *Mod. Phys. Lett.* **13** 225
[6] Xu H, Kevrekidis P G and Saxena A 2014 arXiv:1404.4382
[7] Bulgakov E N, Pichugin K N and Sadreev A F 2011 *Phys. Rev.* B **83** 045109; *J.Phys.: Cond. Mat.* 2011 **23** 065304
[8] Maes B, Soljačić M, Joannopoulos J D, Bienstman P, Baets R, Gorza S -P and Haelterman M 2006 *Opt. Express* **14** 10678
[9] Brazhnyi A and Malomed B A 2011 *Phys. Rev.* A **83** 053844
[10] Maksimov D N and Sadreev A F 2013 *Phys. Rev.* E **88** 032901
[11] Miroshnichenko A E, Malomed B A and Kivshar Yu S 2011 *Phys. Rev.* A **84** 012123
[12] Xu Yi and Miroshnichenko A E 2014 *Phys. Rev.* B **89** 134306
[13] Kippenberg T J, Holzwarth R and Diddams S A 2011 *Science* **332** 555
[14] Cao L S, Qi D X, Peng R W, Wang Mu and Schmelcher P 2014 *Phys. Rev. Lett* **112** 075505
[15] Suh W, Wang Z and Fan S 2004 *IEEE J. of Quantum Electronics* **40** 1511
[16] Miroshnichenko A E, Kivshar Yu, Etrich C, Pertsch T, Iliew R and Lederer F 2009 *Phys. Rev.* A **79** 013809
[17] Cowan A R and Young J F 2003 *Phys.Rev.* E **68** 046606

Journal of Physics: Conference Series **574** (2015) 012038 doi:10.1088/1742-6596/574/1/012038

Some More Solutions of Burgers' Equation

Mukesh Kumar & Raj Kumar
Department of Mathematics,
Motilal Nehru National Institute of Technology Allahabad–211004, India.
E-mail: mukeshkumar12@rediffmail.com

Abstract. In this work, similarity solutions of viscous one–dimensional Burgers' equation are attained by using Lie group theory. The symmetry generators are used for constructing Lie symmetries with commuting infinitesimal operators which lead the governing partial differential equation (PDE) to ordinary differential equation (ODE). Most of the constructed solutions are found in terms of Bessel functions which are new as far as authors are aware. Effect of various parameters in the evolutional profile of the solutions are shown graphically and discussed them physically.

Introduction

Burgers' equation is widely used in the physical problems to model various phenomenon such as modelling of gas dynamics [1], traffic flow, investigating the shallow water waves [2, 3], examining the chemical reaction diffusion model [4], turbulent flow patterns of soil water [5], shock waves in hydrodynamics turbulence [6]. First time, it was also used in the context of a statistical theory of turbulent motion of fluids [7, 8].

One–dimensional viscous Burgers' equation has a non–linear convection term and a second order viscous diffusion term so it becomes a simplified form to one–dimensional analogue of the Navier–Stokes equations. Initially, Burgers' equation was presented by Bateman [9] and later describing a mathematical model of turbulence [1, 8, 10]. Due to extensive work [1, 8] of Burgers it is known as Burgers' equation. Besides its fundamental non–linearity, exact solutions of the Burgers' equation have been obtained for a wide range of initial and boundary conditions [6]. Numerical solutions of Burgers' equation were found impractical for small viscosity due to slow convergence of solutions [11].

Although the literature devoted to Burgers' equation is indeed enormous [12, 13]. For the equation, A. Ouhadan et al. [13] obtained some exact solutions namely exponential, rational and periodic solutions forming a new class of Lie point symmetries through its maximal sub algebras. Furthermore, exact solutions of the forced Burgers' equation are constructed [14] making use of the modified Hopf–Cole transformation for stationary and transient external forces. Motivation of these researches leads us to find the exact solutions of Burgers' equation by using similarity transformations method (STM).

Nevertheless, due to a lot of physical significance of Burgers' equation in real life problems, we have an unquestionable interest to get exact solutions by using STM. The STM of differential equations depends on the Lie groups theory of invariance [15–16]. Indeed, the invariance enables us to reduce the number of variables by one. The applications of the method for various problems of fluid dynamics can be studied [17–20].

The goal of this research is to get the exact solutions of one dimensional Burgers' equation by using STM. In this work, authors have been focused on the exact solutions in terms of Bessel functions. Due to arbitrary choice of parameters, exact solutions are shown graphically and discussed them physically as far as possible. Some results [12, 13, 21] are deduced as special cases of our results.

Similarity Solutions and Discussions

Authors considered the Burgers' equation in the form

$$u_t + u\,u_x = \nu\, u_{xx}, \qquad \nu > 0, \tag{1}$$

where $u(x,t)$ is velocity and ν is kinematic viscosity of fluid. Eq. (1) is a quasilinear parabolic PDE. To apply STM on (1) one can consider one–parameter (ϵ) Lie group transformations

$$x^* = x + \epsilon\,\xi(x,t,u) + O(\epsilon^2), \quad t^* = t + \epsilon\,\tau(x,t,u) + O(\epsilon^2), \quad u^* = u + \epsilon\,\eta(x,t,u) + O(\epsilon^2), \tag{2}$$

where ξ, τ and η are the infinitesimals corresponding to x, t and u respectively which are to be determined in such a way that the Eq. (1) is invariant under the Lie group of transformations (2). Existence of such group allows the number of independent variables to be reduced by one.

Applying Lie group theory [15–16] on (1) yields the following infinitesimals

$$\xi = a_1\,xt - a_2\,x + a_3\,t + a_4, \quad \tau = a_1\,t^2 - 2a_2\,t + a_5, \quad \eta = a_1\,(x - ut) + a_2\,u + a_3, \tag{3}$$

where a_1, a_2, a_3, a_4 and a_5 are five arbitrary parameters. Further, Eq. (1) admits five dimensional Lie algebra generated by

$$X_1 = (x-ut)\,\frac{\partial}{\partial u} + xt\,\frac{\partial}{\partial x} + t^2\,\frac{\partial}{\partial t}, \quad X_2 = u\,\frac{\partial}{\partial u} - x\,\frac{\partial}{\partial x} - 2t\,\frac{\partial}{\partial t}, \quad X_3 = \frac{\partial}{\partial u} + t\,\frac{\partial}{\partial x},$$
$$X_4 = \frac{\partial}{\partial x}, \quad X_5 = \frac{\partial}{\partial t}. \tag{4}$$

The generators X_1, X_2, X_3, X_4 and X_5 are linearly independent. In order to obtain exact solutions, one has to solve the following Lagrange system

$$\frac{dx}{a_1\,xt - a_2\,x + a_3\,t + a_4} = \frac{dt}{a_1\,t^2 - 2a_2\,t + a_5} = \frac{du}{a_1\,(x-ut) + a_2\,u + a_3}. \tag{5}$$

To get similarity solutions of Eq. (1) following cases can be raised:

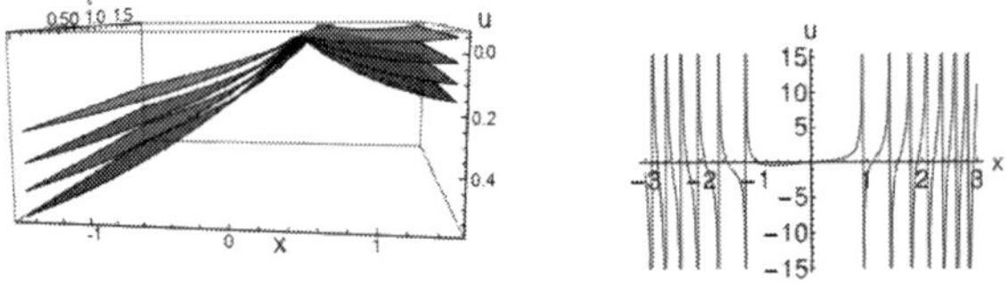

(a) (b)

Figure 1. Variation in spatial structure of velocity profile for (10). (a) At $0 \le t \le 2$ for $\nu = 0.10$ (magenta), 0.20 (red), 0.30 (blue), 0.40 (black) ; (b) At $t = 0.5$ for $\nu = 0.10$.

Case (I): If $a_1 \neq 0$ and treating $A_1 = -\frac{a_2}{a_1}$, $B_1 = \frac{a_3}{a_1}$, $C_1 = \frac{a_4}{a_1}$ and $D_1 = \frac{a_5}{a_1}$ leads to the following sub cases:

(Ia): If $D_1 \neq A_1^2$ then similarity form of the solution can be furnished as

$$u = \alpha + \frac{Xt}{\sqrt{t^2 + 2A_1 t + D_1}} + \frac{F(X)}{\sqrt{t^2 + 2A_1 t + D_1}}, \tag{6}$$

where $X = \frac{x+B_1-\alpha(A_1+t)}{\sqrt{t^2+2A_1t+D_1}}$ is a similarity variable.
Substitution of (6) into (1) results the following equation after integration

$$2\nu F'(X) = F^2 - 2A_1 XF + D_1 X^2 + b_1, \tag{7}$$

where b_1 is an arbitrary constant of integration.

Eq. (7) can be transformed into an ODE of second order by taking $F = -\frac{2\nu}{v}\frac{dv}{dX}$

$$v''(X) + \frac{A_1 X}{\nu} v' + (\frac{D_1 X^2 + b_1}{4\nu^2})v = 0. \tag{8}$$

Eq. (8) becomes Bessel's Equation [22] under the parametric restrictions $b = -\frac{A_1}{4\nu}$, $b_1 = 2A_1\nu$ and $d^2 = \frac{D_1 - A_1^2}{16\nu^2}$.

If $D_1 < A_1^2$, then real solutions of Eq. (8) are not possible. On the other hand, if $D_1 > A_1^2$ then real solutions of (8) can be written as

$$v = X^{\frac{1}{2}} e^{bX^2} \Big[k_1 J_{\frac{1}{4}}(X^2 d) + k_2 J_{-\frac{1}{4}}(X^2 d) \Big], \tag{9}$$

where k_1 and k_2 are arbitrary constants of integration and $J_n(.)$ is the first kind Bessel function of order n.

Assuming $k = \frac{k_2}{k_1}$ $(k_1 \neq 0)$, solution of (1) is furnished by

$$u(x,t) \quad = \alpha + \frac{X}{\sqrt{t^2+2A_1t+D_1}} \Big[(t + A_1) - 4\nu d \frac{J_{-\frac{3}{4}}(X^2 d) - kJ_{\frac{3}{4}}(X^2 d)}{J_{\frac{1}{4}}(X^2 d) + kJ_{-\frac{1}{4}}(X^2 d)} \Big] \tag{10}$$

Solution curves for expression (10) are shown in Fig. 1. In Fig. 1(a), the viscous damping effect is obvious because u decreases as ν increases from 0.10 to 0.40 at time t showing that (1) experiences a loss of momentum hence, loss of kinetic energy with respect to time. Also for all t solutions (10) show the degenerate nature near to $x = 0.06143$, $u = 0.4424$.

Apart from it, solution profiles in Fig. 1(a) at small viscosity $\nu = 0.10$ are smooth in the range $-1.50 \leq x \leq 1.50$ so these solutions lead to some physical applications as discussed in Miller [11]. This happens due to the presence of the term νu_{xx} in right hand side of Eq. (1). Furthermore, if such initial data for Eq. (1) is used which has finite discontinuity for small values of ν then u_{xx} becomes very large, and the viscous term becomes important whenever the discontinuity begins to form. This term keeps the solution smooth for all time, and determines the correct physical nature of the system.

Solution curves represented in Fig. 1(b) at $t = 0.5$ for $\nu = 0.1$ support to the fact that the solutions of Burgers' Eq. (1) are impractical for the small values of viscosity due to slow convergence of solutions [11]. It can also be observed that the number of shocks increases as ν decreases. So it is physically meaningful leading to a new research scope to capture the shocks and resolving them analytically or numerically.

(Ib): If $D_1 = A_1^2$, solution is given by

$$u \quad = \frac{(C_1 - A_1 B_1)(2t + A_1)}{2(t+A_1)^3} + \frac{(x+B_1)}{(t+A_1)^2} t + \frac{F_1(X_1)}{(t+A_1)}, \tag{11}$$

with $X_1 = \frac{C_1-A_1B_1}{2(t+A_1)^2} + \frac{(x+B_1)}{t+A_1}$ as the similarity variable, while k_1 and k_2 are constants of integration, $b = -\frac{A_1}{4\nu}$ and $\alpha_1^2 = \frac{2}{9}\frac{(A_1B_1-C_1)}{\nu^2}$ are parameters.
So we obtain

$$F_1(X_1) \;= A_1\,X_1 - 3\nu\alpha_1\sqrt{X_1}\Big[\frac{J_{-\frac{2}{3}}(\alpha_1 X_1^{3/2}) - kJ_{\frac{2}{3}}(\alpha_1 X_1^{3/2})}{J_{\frac{1}{3}}(\alpha_1 X_1^{3/2}) + kJ_{-\frac{1}{3}}(\alpha_1 X_1^{3/2})}\Big]. \tag{12}$$

We have taken $k = \frac{k_2}{k_1}$ (follows $k_1 \neq 0$). Thus by the help of Eqs. (11) and (12), the solution of (1) can be expressed as

$$\begin{aligned} u(x,t) \;&= \tfrac{(x+B_1)}{(t+A_1)} + \tfrac{1}{(t+A_1)^2}\Big[C_1 - A_1B_1 - \tfrac{3\nu\alpha_1}{\sqrt{2}}\sqrt{2(x+B_1)(t+A_1)+(C_1-A_1B_1)} \\ &\quad\times\Big(\frac{J_{-\frac{2}{3}}(\alpha_1 X_1^{3/2}) - kJ_{\frac{2}{3}}(\alpha_1 X_1^{3/2})}{J_{\frac{1}{3}}(\alpha_1 X_1^{3/2}) + kJ_{-\frac{1}{3}}(\alpha_1 X_1^{3/2})}\Big)\Big]. \end{aligned} \tag{13}$$

Case (II): If $a_1 = 0$ and $a_2 \neq 0$ then one can obtain the similarity variable $X_2 = \frac{(x-A_2t+l)}{\sqrt{C_2+2t}}$ (where $l = B_2 - A_2C_2$, $A_2 = -\frac{a_3}{a_2}$, $B_2 = -\frac{a_4}{a_2}$ and $C_2 = -\frac{a_5}{a_2}$) with similarity form

$$u = A_2 + \frac{F_2(X_2)}{\sqrt{C_2+2t}}. \tag{14}$$

and corresponding ODE after integration can be written as

$$2\,\nu\,F_2^{'}(X_2) = F_2^2(X_2) - 2X_2\,F_2(X_2) + b_3, \tag{15}$$

Eq. (15) can be transformed into the following with help of $F_2(X_2) = -\frac{2\nu}{v}\frac{dv}{dX_2}$

$$v^{''}(X_2) + \frac{X_2}{\nu}\,v^{'}(X_2) + \frac{b_3}{4\nu^2}\,v(X_2) = 0. \tag{16}$$

Setting $b_3 = -4\,\nu$ and applying the transformation $v(X_2) = X_2\,P(X_2)$ on (16) yields the solution of (1) as follows

$$u(x,t) \;= A_2 - \frac{2\nu}{X_2^2\sqrt{C_2+2t}}\Big[X_2 + \frac{\exp\big(-\frac{X_2^2}{2\nu}\big)}{k + \int \frac{1}{X_2^2}\exp\big(-\frac{X_2^2}{2\nu}\big)\,dX_2}\Big]. \tag{17}$$

On the other hand if $\mathbf{b_3 = 0}$ then particular solution is given by

$$u(x,t) = A_2 + \frac{1}{\sqrt{C_2+2t}}\Big[\frac{\exp\big(-\frac{X_2^2}{2\nu}\big)}{c_2 - \frac{1}{2\nu}\int \exp\big(-\frac{X_2^2}{2\nu}\big)\,dX_2}\Big] \tag{18}$$

It can also be noticed that solutions for $\nu = 0.10, 0.50, 0.70$ degenerate while coincide for $\nu = 0.10$ and 0.70. Solution (13) is an explicit expression involving Bessel functions. Multisolitons occur at $x = -0.25$ for (13) and represented by Fig. 2 taking $a_1 = 1.3517$, $a_2 = 0.8308$, $a_3 = 0.5853$, $a_4 = 0.5497$, $k = 0.9172$, $b = 1.5366$ and $b_2 = -0.1229$. Nature of solution curves have been represented for viscosity $\nu = 0.10$.

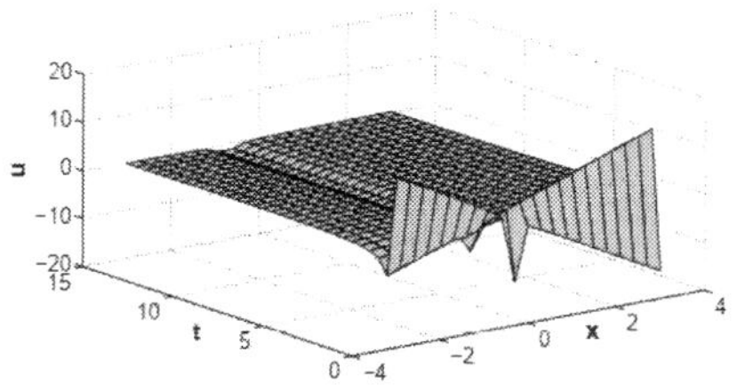

Figure 2. Multisoliton profile of (13) for $\nu = 0.10$

Conclusions

In summary, Lie groups theory reduces the one-dimensional Burgers' equation into an ODE which is solved analytically. Since we have adequate options to select different arbitrary constants applying the STM thus three exact solutions have been obtained explicitly. Solutions in the case (Ia) show viscous damping effect with time evolution which are of most physical significance and support to Miller [11] for a very small value of viscosity. Solutions in the case (Ib) are useful for shocks capturing of Burgers' Eq. (1). The quadrature form solutions are obtained in the case (II). Solutions in the cases (Ia), (Ib) and (II) are constructed in terms of Bessel functions. Three new solutions in the cases (Ia), (Ib) and (II) have been obtained. The exact solutions can serve as benchmark in analysis, accuracy testing of numerical results and their comparison.

1. References

[1] Burgers J M 1974 *The Non-Linear Diffusion Equation: Asymptotic Solutions and Statistical Problems* (New York: Springer)
[2] Logan J D 1994 *An Introduction to Non-Linear Partial Differential Equations* (New York: Wiley)
[3] Debtanth L 1997 *Non-Linear Partial Differential Equations for Scientist and Engineers* (Boston: Birkhauser)
[4] Adomian G 1995 *Comput. Math. Appl.* **29** 1
[5] Su N, Jim P C W, Keith W V, Murray E C and Mao R, 2004 *Aust. J Soil Res.* **42** 9
[6] Cole J D 1951 *Quart. Appl. Math.* **9** 225
[7] Burgers J M 1940 *Proc. K. Ned. Akad. Wet.* **43** 2
[8] Burgers J M 1948 *Adv. Appl. Mech.* **1** 171
[9] Bateman H 1915 *Mon. Wea. Rev.* **43** 163
[10] Blackstock D T 1966 *J Acoust. Soc. Am.* **39** 411
[11] Miller E L 1966 MS Thesis University of Delaware Newark DE
[12] Polyanin A D and Zaitsev V F 2004 *Handbook of Non-Linear Partial Differential Equations* (Boca Raton FL: Chapman and Hall/CRC) p 10)
[13] Ouhadan A, Mekkaoui T and Kinani E H El 2011 *Int. J Nonlinear Sci.* **11** 51
[14] Petrovskiĭ S V 1999 *Tech. Phys.* **44** 878
[15] Bluman G W and Cole J D 1974 *Similarity Methods for Differential Equations* (New York: Springer)
[16] Olver P J 1993 *Application of Lie Groups to Differential Equation: Graduate Texts in Mathematics* (New York: Springer) p 30–130
[17] Kumar M and Kumar R 2014 *Meccanica* **49** 335
[18] Lie S 1881 *Arch. Math.* **6** 328
[19] Zhi H Y 2013 *Commun. Theor. Phys.* **59** 263
[20] Kumar M, Kumar R and Kumar A 2014 *Comput. Math. Appl.* **68** 454
[21] Fletcher C A J 1984 *Computational Galerkin Methods* (New York: Springer)
[22] Kamke E 1959 *Differentialgleichungen Löungsmethoden Und Lösungen* (New York: Chelsea Publishing Co) p 441

Solitonics effects in supercontinuum generation in highly nonlinear fibers

M Ferreira[1]
I3N- Institute of Nanostrcutures, Nanomodelling and Nanofabrication
Department of Physics, University of Aveiro, 3810-193 Aveiro, Portugal

E-mail: mfernando@ua.pt

Abstract. When ultrashort pulses with large enough power are launched into highly nonlinear fibers, soliton fission gives origin to multiple fundamental solitons of different widths and peak powers. Two soliton related effects become then particularly important in the context of supercontinuum generation: the soliton self-frequency shift and the emission of dispersive radiation in the normal dispersion region. The dispersive characteristics of the optical fiber assume play a key role in these circumstances.

1.Introduction

Supercontinuum generation (SCG) occurs when narrow-band incident pulses undergo extreme nonlinear spectral broadening to yield a broadband spectrally continuous output. It results generally from the synergy between several fundamental nonlinear processes, such as self-phase modulation, cross-phase modulation, stimulated Raman scattering, and four-wave mixing [1]. Soliton related effects also play a significant role whenever light with sufficient power propagates in the anomalous dispersion regime [2].

The first observation of supercontinuum was realized in 1970 by Alfano and Shapiro in bulk borosilicate glass [3]. SCG in fibers occurred for the first time in a 1976 experiment, when 10 ns pulses with more than 1-kW peak power were launched in a 20-m long fiber, producing a 180-nm-wide spectrum [4]. Several types of highly nonlinear fibers, namely microstructured optical fibers (MOFs) with optimized designs, have been developed during the recent years in order to enhance the supercontinuum generation [2,5].

Pumping with ultrashort pulses in the anomalous dispersion region of the fiber, the soliton fission process gives origin to multiple fundamental solitons of different widths and peak powers [6]. Among the host of soliton related effects, there are two which become then particularly important: the soliton self-frequency shift [7] induced by intrapulse Raman scattering (IRS) and the emission of dispersive (or "Cherenkov") radiation, which comes from the overlap of a soliton spectrum with the normal dispersion range [8]. The peculiar dispersive characteristics of highly nonlinear fibers play a key role in these circumstances.

[1] Corresponding author

2. Modelling of the supercontinuum generation

Modelling the supercontinuum generation can be realized considering a generalized nonlinear Schrödinger equation that includes higher-order dispersion effects, as well as IRS. Such equation can be written as [1,2,9]:

$$\frac{\partial U}{\partial z} - i\sum_{k\geq 2}\frac{i^k \beta_k}{k!}\frac{\partial^k U}{\partial \tau^k} + \frac{\alpha(\omega)}{2}U = i\gamma\left(1+\frac{i}{\omega_0}\frac{\partial}{\partial t}\right)\left(U(z,t)\int_{-\infty}^{+t} R(t')\left|U(z,t-t')\right|^2 dt'\right) \quad (1)$$

where $U(z,t)$ is the electric field envelope, ω_0 is the center frequency, $\alpha(\omega)$ is the frequency-dependent fiber loss, γ is the nonlinear parameter, and $\beta_k = \left(d^k \beta / d\omega^k\right)$, ($k$ = 0, 1, 2...) are the dispersion coefficients at the center frequency, $\beta(\omega)$ being the mode-propagation constant. The group-velocity dispersion (GVD) is characterized by the parameter β_2. The nonlinear response function $R(t)$ in Eq. (1) can be written as $R(t) = (1 - f_R)\delta(t) + f_R h(t)$, where the δ-function represents the instantaneous electron response (responsible for the Kerr effect), $h(t)$ represents the delayed ionic response (responsible for the Raman scattering) and f_R is the fractional contribution of the delayed Raman response to the nonlinear polarization, in which a value $f_R = 0.18$ is often assumed. It is common to approximate $h(t)$ in the form [9]:

$$h(t) = \frac{\tau_1^2 + \tau_2^2}{\tau_1 \tau_2^2}\exp(-t/\tau_2)\sin(t/\tau_1) \quad (2)$$

where $\tau_1 = 12.2$ fs and $\tau_2 = 32$ fs.

3. Solitonic effects

Eq. (1) can be used to describe the propagation of femtosecond pulses in optical fibers, in both the normal and anomalous dispersion regimes. The soliton order is given by $N = \sqrt{L_D / L_{NL}}$, where $L_D = T_0^2 / |\beta_2|$ is the dispersion distance and $L_{NL} = 1/\gamma P_0$ is the nonlinear length. In the presence of higher-order dispersion, an Nth-order soliton gives origin to N fundamental solitons whose widths and peak powers are given by $T_k = T_0/(2N + 1 - 2k)$ and $P_k = P_0 (2n + 1 - 2k)^2 / N^2$, respectively, for k = 1,..., $\tilde{N}$, where $\tilde{N}$ is the integer closest to N when N is not an integer. Soliton fission occurs generally after a propagation distance $L_{fiss} \sim L_D / N$, at which the injected higher-order soliton attains its maximum bandwidth.

The fission phenomenon can be observed in Fig. 1, which shows the temporal and spectral evolution of an optical pulse along a highly nonlinear MOF with hole diameter d=1.4 μm and pitch $\Lambda = 1.6$ μm. In this case, the dispersion curve has only one zero dispersion wavelength, located at 780 nm. We consider a pump in the anomalous dispersion region at 835 nm, where the nonlinear parameter is $\gamma = 11 \mathrm{W}^{-1}\mathrm{km}^{-1}$. An input pulse $U(0,\tau) = \sqrt{P_0}\,\mathrm{sech}(\tau / T_0)$ is assumed, where $P_0 = 10$kW and $T_0 = 28.4$ fs, which corresponds to an intensity full width at half maximum (FWHM) of 50 fs. For the assumed parameters, we have $N = 8.5$ and $L_{fiss} \sim 2.3$ cm.

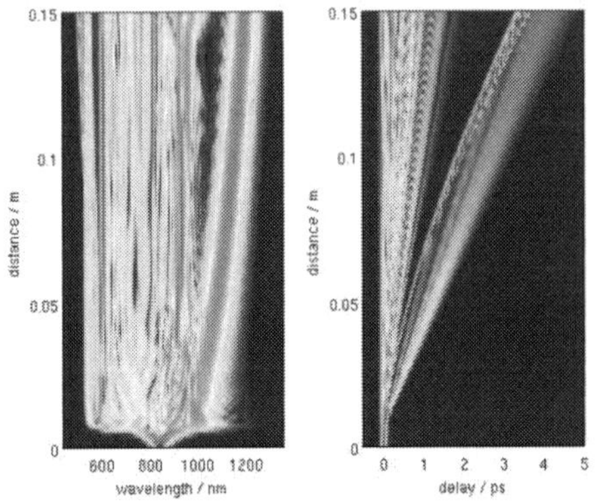

Figure 1 – (a) Spectral and (b) temporal evolution of an optical pulse along a MOF obtained from Eq. (1). The input sech pulse at 835 nm has a peak power of 10 kW and 50 fs FWHM.

Besides higher-order dispersion, another main effect affecting the dynamics of a higher-order soliton fission is the intrapulse Raman scattering. This phenomenon leads to a continuous red-shift of the soliton carrier frequency, an effect known as the soliton self-frequency shift (SSFS) [7]. The rate of frequency shift of a sech pulse propagating as a fundamental soliton is given by [1]

$$\frac{df}{dz} = -\frac{4t_R|\beta_2|}{15\pi T_s^4} = -\frac{4t_R(\gamma P_s)^2}{15\pi|\beta_2|} \tag{3}$$

where $t_R \equiv f_R \int_{-\infty}^{+\infty} t h(t) dt \approx 5\,\text{fs}$ is the Raman parameter, whereas T_s and P_s are the soliton width and peak power, respectively. Since the SSFS effect is proportional to $(\gamma P_s)^2$, it will be enhanced if short pulses with high peak pulses are propagated in highly nonlinear fibers.

The red-shift of the solitons created by the fission process is clearly seen in Fig. 1. As a consequence of the SSFS, such solitons separate from each other. Since the SSFS is the largest for the shortest soliton, its spectrum shifts the most toward the red side. The change of the soliton's frequency determines a reduction in the soliton's speed because of dispersion. This deceleration appears as a bending of the soliton trajectory in the time domain.

Besides the fission phenomenon and the resulting Raman-induced spectral shifts of constituent fundamental solitons, Fig. 1 shows also the appearance of spectral components on the short wavelength side of the pulse spectrum. Such spectral components correspond to *nonsolitonic radiation* (NSR), also called *Cherenkov radiation* [8], which is emitted by the solitons resulting from the fission process in the presence of higher-order dispersion. NSR is emitted at a frequency ω such that its phase velocity matches that of the soliton. This situation occurs when the phase-matching condition $\beta(\omega) = \beta(\omega_s) + \beta_1(\omega - \omega_s) + \gamma P_s / 2$ is satisfied, where P_s and ω_s are the peak power and the frequency, respectively, of the fundamental soliton formed after the fission process. Expanding $\beta(\omega)$ around ω_s, this phase-matching condition reduces to

$$\sum_{n\geq 2} \frac{\Omega^n}{n!} \beta_n(\omega_s) = \frac{\gamma P_s}{2} \tag{4}$$

where $\Omega = \omega - \omega_s$ is the frequency shift between the dispersive wave and the soliton. Neglecting the fourth- and higher-order dispersions, this frequency shift is given approximately by [8]:

$$\Omega \approx -\frac{3\beta_2}{\beta_3} + \frac{\gamma P_0 \beta_3}{3\beta_2^2} \tag{5}$$

For solitons propagating in the anomalous-GVD regime we have $\beta_2 < 0$. In these circumstances, Eq. (5) shows that $\Omega > 0$ when $\beta_3 > 0$, in which case the NSR is emitted at wavelengths shorter than that of the soliton, as observed in Fig. 1. On the contrary, if $\beta_3 < 0$ the NSR is emitted at a longer wavelength than that of the soliton and it can result in the suppression of the SSFS effect [1].

In the special case that the third-order dispersion vanishes and the soliton is perturbed by the fourth-order alone, Eq. (4) gives the result:

$$\Omega_{\pm} \approx \pm\sqrt{-12\beta_2 / \beta_4} \tag{6}$$

where the last term on the right-hand side of Eq. (4) has been neglected. In these circumstances, since solitons propagate in the anomalous-GVD regime ($\beta_2 < 0$) both solutions in Eq. (6) become imaginary and no dispersive waves are produced if $\beta_4 < 0$. However, if $\beta_4 > 0$, both solutions are real and have opposite signs, indicating that NSR occurs on both sides of the soliton frequency.

4. Conclusions

In this paper we discussed the main solitonic effects contributing to the supercontinuum generation process in highly nonlinear fibers. Such effects become important whenever light propagates in the anomalous dispersion regime. Soliton fission occurs if ultrashort pulses with large enough power are launched into highly nonlinear fibers. Multiple fundamental solitons of different widths and peak powers are generated in this case. Among the host of soliton related effects, there are two which become particularly important: the soliton self-frequency shift and the emission of dispersive radiation in the normal dispersion region. For typical dispersion conditions, solitons dominate the long-wavelength edge of the supercontinuum, while the dispersive radiation determines the short wavelength edge.

5. References

[1] Ferreira M F 2011 *Nonlinear Effects in Optical Fibers* (John Wiley & Sons)
[2] Dudley J and Taylor J (Eds) 2010 *Supercontinuum Generation in Optical Fibers* (Cambridge University Press, Cambridge, UK)
[3] Alfano R R and Shapiro S L 1970 Phys. Rev. Lett. **24** 584
[4] Lin C and Stolen R H 1976 Appl. Phys. Lett. **28** 216
[5] Russel P 2006 J. Lightwave Technol. **24** 4729
[6] Kodama Y and Hasegawa A 1987 *IEEE J. Quant. Elect.* **QE23** 510
[7] Mitschke F and Mollenauer L 1986 Opt. Lett. **11** 659
[8] Akhmediev N and Karlsson M 1995 *Phys. Rev. A* **51** 2602
[9] Agrawal G P 2007 *Nonlinear Fiber Optics*, 4th ed. (San Diego: Academic Press)

IC-MSQUARE 2014 IOP Publishing
Journal of Physics: Conference Series **574** (2015) 012040 doi:10.1088/1742-6596/574/1/012040

One dark matter mystery: halos in the cosmic web

Jose Gaite
Instituto de Microgravedad IDR, ETSIAE,
Universidad Politécnica de Madrid, E-28040 Madrid, Spain
E-mail: jose.gaite@upm.es

Abstract. The current cold dark matter cosmological model explains the large scale cosmic web structure but is challenged by the observation of a relatively smooth distribution of matter in galactic clusters. We consider various aspects of modeling the dark matter around galaxies as distributed in smooth halos and, especially, the smoothness of the dark matter halos seen in N-body cosmological simulations. We conclude that the problems of the cold dark matter cosmology on small scales are more serious than normally admitted.

1. Introduction

Cold dark matter, which consists of weakly interacting particles whose velocity dispersion in the early universe was not sufficient to prevent the formation of structure on galactic and sub-galactic scales, is fundamental in modern cosmology, since the early 1980s. Naturally, the full standard cosmological model, namely, the ΛCDM model, presently includes a cosmological constant and inflationary initial conditions. However, just the cold dark matter (CDM) hypothesis, with the consequent richness of structure on very small scales, does not fit well the indirect observations of dark matter. To be precise, the results of N-body simulations of the evolution of collisionless CDM show that too much structure is produced on galactic scales. This excess of small-scale structure manifests itself in two problems: the core-cusp problem [1] and the missing satellites problem (or dwarf galaxy problem) [2, 3].

Observations of galactic rotation velocities seem to indicate an approximately constant dark matter density in the inner parts of galaxies, whereas cosmological N-body simulations indicate that the dark matter radial density profile has a singularity of power-law type. This discrepancy has been named "the core-cusp problem". It is illustrated in Fig. 1, which shows the results of Kuzio de Naray et al [4] for a particular galaxy (F568-3): the points with error bars are rotation velocity measurements and are well fitted by the solid curve, which corresponds to an isothermal dark matter halo with a constant density core. In contrast, the dashed curve, corresponding to a Navarro-Frenk-White (NFW) singular density profile, clearly overpredicts the rotation velocity near the center. Any other power-law like profile would also induce too high rotation velocities near the center.

The "missing satellites" problem refers to the appearance of a large number of subhalos (satellites) close to every single halo in cosmological N-body simulations, without observational counterparts. Put in other words, the number of dwarf satellite galaxies around a normal-sized galaxy is much smaller than the number predicted by collisionless CDM N-body simulations. For example, only about 11 dwarf galaxies orbiting the Milky Way have been observed, whereas the number of dark matter subhalos produced in, for example, the "Via Lactea" cosmological

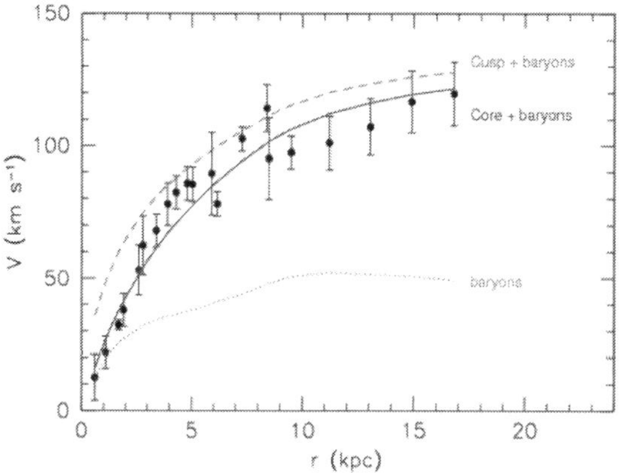

Figure 1. Measured rotation curve of a galaxy (points with error bars) compared to model fits assuming a cored dark matter halo (solid curve) or a cuspy dark matter halo with an NFW profile (dashed curve). The dotted curve shows the contribution of baryons (stars+gas). From Kuzio de Naray et al.

Figure 2. The dark matter distribution from the "Via Lactea II" cosmological simulation of a Milky Way-mass cold dark matter halo, from Diemand et al. (http://www.ics.uzh.ch/~diemand/vl). The cube has a size of 800 kpc.

simulation of a Milky Way-mass cold dark matter halo is much larger (the simulation has been made by Diemand et al [3]). This is easily perceived in Fig. 2. Naturally, this problem is intrinsic to the CDM model, since cold dark matter clusters hierarchically from the smallest scales (from the bottom up). If dark matter is composed of neutralinos, as is likely, the smallest clumps of dark matter can range from $10^{-11}\, M_\odot$ to $10^{-3}\, M_\odot$ [5].

Both problems suggest that the real dark matter distribution, on small scales, is smoother than the one produced by collisionless CDM N-body simulations: the real distribution is less grainy and has no density singularities. On the other hand, on medium or large scales, the results of N-body simulations agree with observations: the large scale structure of the Universe is arranged as a "cosmic web" constituted by matter sheets, filaments, and nodes, plus the cosmic voids that these objects leave in between. This type of structure was initially proposed in connection with a simplified model of the cosmic dynamics, namely, the adhesion model [6], and both galaxy surveys and cosmological N-body simulations have confirmed it.

On scales smaller than those corresponding to the cosmic-web structure in N-body simulations, the dark matter distribution transforms into a distribution of relatively smooth dark matter clusters or *halos* that have a limited range of sizes. These dark matter halos are to be identified with the ones surrounding galaxies, detected by rotation velocity measurements. Halo models of the large scale structure of matter [7] have become very popular. The distribution of dark matter inside an N-body halo is smooth, save for a concentration at its center, presumably constituting a density singularity, and for the presence of subhalos close to it. The existence of a density singularity and of an excess of subhalos are the already mentioned problems of the CDM model on small scales.

Although the existence of smooth halos is rarely questioned, they really are not part of the CDM model. In fact, their presence in N-body simulations can be a consequence of numerical artifacts [8]. From a theoretical viewpoint, the continuum CDM dynamics is ruled by the Vlasov-Poisson equations, which give rise to the same type of gravitational clustering on ever decreasing scales. Therefore, assuming that the cosmic web geometry is the attractor solution of those equations, there is no reason why this solution should become relatively smooth from some galactic or super-galactic scale downwards. On the contrary, it would be natural that the cosmic web structure in collisionless N-body simulations were a self-similar multifractal, displaying the same features, namely, sheets, filaments, nodes, and cosmic voids, on ever decreasing scales, down to the scale of the smallest dark matter clumps (with mass $\ll M_\odot$). For example, the cosmic web produced by the adhesion model [6] is of that type [9].

With the above in mind, we study the smoothness of halos in CDM N-body simulations, to determine the range of scales where the distribution of matter is smooth and the connection of halos with the non-smooth cosmic web structure [10]. We also consider how the abundance of halos depends on the mass resolution and on the discreteness parameters pertaining to N bodies. Our conclusions question the consistency of the CDM-halo model with observations much more seriously than the well-known problems, namely, the core-cusp problem and the missing satellites problem.

2. Smoothness of halos in N-body simulations

A dark matter halo consists of a spherical like distribution of dark matter particles with a density that is generally smooth but singular at the center [7]. Most models of formation of halos are motivated by the spherical collapse model and its variants. Naturally, in a halo, smoothness and isotropy tend to diminish away from the center. If we imagine a spherical surface with origin on the halo center and with increasing radius, initially the dark matter distribution over the surface is mildly anisotropic (triaxial) but, as the radius increases, the anisotropy increases and, eventually, the spherical surface intersects a distribution that is essentially indistinguishable from a non-smooth cosmic web structure.

To measure anisotropy and non-smoothness, we need to define them in mathematical terms [10]. In mathematics, a function is smooth if it has an infinite number of derivatives. However, we cannot just use derivatives of the density field, on account of the density singularities at halo centers. Therefore, a more useful distinguishing feature is the change from mild to strong anisotropy on increasing scales: on small scales, namely, close to halo centers, the anisotropy is caused by triaxiality and, possibly, subhalos, whereas, on larger scales, the anisotropy patterns are much more complex. If we consider a spherical shell centered on a halo, the density in it must be fairly smooth for a small radius (except for subhalo singularities), but the density must become increasingly non-smooth as the radius grows.

To find how uniform a distribution of k particles in a spherical shell is, we can compare the angular spherical coordinates of the particles, $\{\phi_i, \theta_i\}_{i=1}^k$, with the coordinates corresponding to a uniform distribution of the particles. To do this, we must consider that the variables that must be uniformly distributed are not ϕ and θ but ϕ and $\cos\theta$. For convenience, we define coordinates $\phi/(2\pi)$ and $(1+\cos\theta)/2$ over the unit square $[0,1]\times[0,1]$, corresponding to a cylindrical equal-area projection on it. To analyze if the particles are uniformly distributed, we can use the Rényi entropies

$$S_q(\{p_i\}) = \frac{\log_2(\sum_{i=1}^M {p_i}^q)}{1-q}, \quad q \neq 1, \tag{1}$$

where $\{p_i = n_i/N\}_{i=1}^M$, with $N = \sum_{i=1}^M n_i$, and $\{n_i\}_{i=1}^M$ are the counts-in-cells corresponding to the distribution of N particles in M spherical cells (set in the given coordinates). The limit of S_q as $q \to 1$ yields the standard Boltzmann-Gibbs-Shannon entropy. The Rényi entropies with $q \geq 0$ fulfill $0 \leq S_q(\{p_i\}) \leq \log_2 M$, so we divide them by $\log_2 M$ to obtain numbers between 0 and 1: zero corresponds to maximum order (or dispersion of counts-in-cells) and one to minimum order. The entropic coefficients $S_q(\{p_i\})/\log_2 M$ are just coarse Rényi dimensions divided by the dimension of the ambient space, which is, in the present case, a two-dimensional spherical surface.

Journal of Physics: Conference Series **574** (2015) 012040 doi:10.1088/1742-6596/574/1/012040

When we observe how the $q \geq 1$ entropic coefficients of a shell in a given halo change with the radius of the shell, we expect that they start at values close to one and have a generally decreasing trend, in accord with the transition from small-scale smoothness to large-scale graininess.

2.1. Halos in the Bolshoi and Via Lactea II simulations

The Bolshoi simulation [11] assumes cosmological parameters $\Omega_\Lambda = 0.73$, $\Omega_M = 0.27$, Hubble parameter $h = 0.70$, and initial spectral index $n = 0.95$, while the "Via Lactea II" (VL2) simulation [3] takes $\Omega_\Lambda = 0.76$, $\Omega_M = 0.24$, Hubble parameter $h = 0.73$, and initial spectral index $n = 0.95$. The comoving size of the simulation boxes are 250 h^{-1} Mpc and 40 h^{-1} Mpc, respectively. The number of Bolshoi particles is $N = 2048^3$, yielding a mass resolution of $1.35 \cdot 10^8\, h^{-1}\, M_\odot$. The VL2 simulation focuses on the formation of a single, Milky-Way size CDM halo, using the method of refinement. The VL2 dark matter halo is refined with more than 10^9 high-resolution particles, yielding a mass resolution of $4098\, M_\odot$.

Our procedure consists in adjusting spherical shells to have 1024 particles and then setting a 8×8 mesh in the unit square corresponding to every shell (in our coordinates), obtaining $M = 64$ cells per shell. Since subhalo singularities strongly affect smoothness measures, we avoid them by removing the four most populated cells of every shell, reducing M to 60. This operation hardly alters the overall smoothness properties. In consequence, the entropic coefficients are given by $S_q/\log_2 60 = S_q/5.9$. The variation of the $q = 1$ and $q = 2$ coefficients with radius for two Bolshoi halos, on the one hand, and the VL2 halo, on the other, are plotted in Fig. 3 and Fig. 4, respectively.

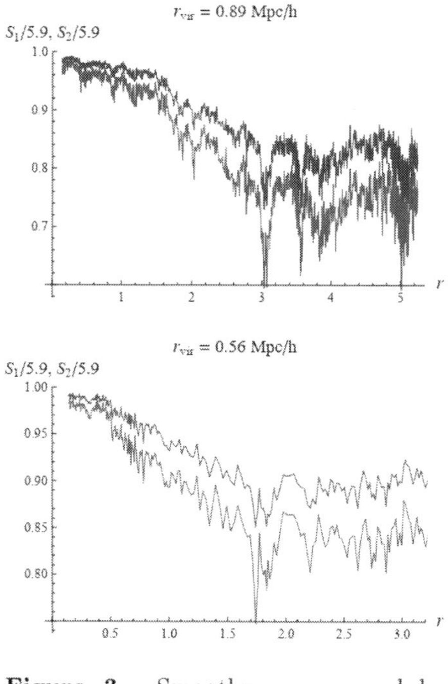

Figure 3. Smoothness measured by two Rényi entropies in two Bolshoi halos, showing the decrease with radius (in h^{-1} Mpc).

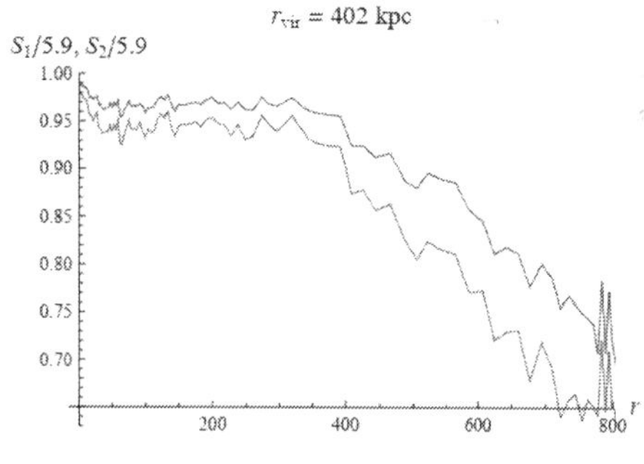

Figure 4. The variation with radius of the same two Rényi entropies in the "Via Lactea II" halo. The radius is measured in kpc.

We see in all the plots the progressive decrease of smoothness with increasing radius, from values close to one to values that approximately tend to the asymptotic values corresponding to the multifractal cosmic-web structure, namely, 0.8 and 0.6, for $q = 1$ and $q = 2$, respectively [12, 13]. The considerable fluctuations superposed on the decreasing trend are due to subhalos, whose effect is not totally eliminated in our procedure. Remarkably, nothing special happens at the virial radius.

2.2. *Subhalo abundance*

The excellent mass resolution of the VL2 simulation reflects on the large number of subhalos, easily perceived in Fig. 2. Actually, there are more than 9000 subhalos within the virial radius of the Milky Way VL2 halo; or a few thousands, at any rate, if we suitably restrict the minimal halo size. In contrast, the halos in the Bolshoi simulation contain much fewer subhalos. For example, the Bolshoi simulation halo with $r_{\text{vir}} = 0.56$ Mpc/h and whose entropies are plotted at the bottom of Fig. 3 contains only 22 subhalos within the virial radius (with more than 50 particles each).

This remarkable difference is a consequence of the different mass resolution of both simulations: the particle mass ratio between the two simulations is 33000, approximately. An improved resolution must bring out unresolved substructure. Presumably, as suggested by Diemand et al [3], "at infinite resolution one would find a long nested series of halos within halos within halos etc."

However, a ratio of 33000 is considerably larger than the ratio between the numbers of subhalos of the respective halos. In fact, the definitions of subhalo in the two simulations do not coincide, but this is only a part of the cause for the discrepancy: the influence of the gravity softening on smoothness properties must be taken into account as well. We discuss both the discreteness parameters next.

3. Influence of discreteness on smoothness

The nature of halos and, in particular, their smoothness are influenced by the discreteness parameters, namely, total number of particles N and softening length ϵ. The optimal choice of softening is still a debated issue [14, 15]. In a CDM simulation with N particles that produces objects of size R, namely, halos of size R, the optimal softening length ϵ has to be determined in terms of N and R. Here, R is the average radius of halos, for a given simulation, where the radius of an individual halo is defined as its smoothness radius, in accord to the preceding section. Since the size of the simulation box L and, hence, the total number of particles N, must not have any effect on the nature and size of individual halos, we can express the softening length in terms of local quantities, namely, the discretization length, $\ell = LN^{-1/3}$, and R. In principle, the function $\epsilon(\ell, R)$ can be arbitrary. Besides, considering that ℓ and ϵ are discreteness parameters, fixed at the start of the cosmological simulation, whereas the sizes of halos belong to the result of the simulation, it is preferable to write the relation among the three quantities as $R(\ell, \epsilon)$.

The evidence suggests that the function $R(\ell, \epsilon)$ greatly depends on the variables ℓ and ϵ [8]. In this regard, we would like to know the separate dependence of R on ℓ and ϵ. While a finite ℓ (a finite N) introduces corrections to the mean-field Vlasov-Poisson dynamics, partially remedied by the gravity softening, this softening also perturbs the Vlasov-Poisson dynamics. Joyce et al [15] point out that the rigorous Vlasov-Poisson limit, $\ell \to 0$, is to be taken at *fixed* finite ϵ, obtaining a smoothed version of the Vlasov-Poisson equations. One may ask what $\lim_{\ell \to 0} R(\ell, \epsilon)$ is: dimensional analysis demands that $\lim_{\ell \to 0} R(\ell, \epsilon) \sim \epsilon$. In words, the softening on a scale ϵ must produce a smoothing on the same scale. Therefore, if halo sizes are determined by smoothness, these sizes must be of the order of ϵ. The subsequent $\epsilon \to 0$ limit that leads to the actual Vlasov-Poisson equations makes the halo sizes vanish and the halo number density diverge. This implies that the relevant solutions of these equations correspond to a self-similar cosmic web structure that is present on ever decreasing scales. Naturally, the limit $\ell \to 0$ at *fixed* finite ϵ or, in other words, the domain of discreteness parameters such that $\epsilon \gg \ell$, is not studied by cosmological N-body simulations. Nevertheless, its effect on halos can be understood by analogy with the adhesion model [9].

4. Conclusions

The often mentioned problems of the CDM model on small scales, namely, the core-cusp problem and the missing satellites problem, reflect just the bottom-up formation of structure in the CDM model, which necessarily gives rise to a considerable amount of structure on galactic scales. The amount of structure produced on small scales is usually inferred from the results of collisionless CDM N-body simulations, which actually show that the distribution of matter on those scales can be described as a distribution of *smooth* halos with a limited range of sizes. The core-cusp problem arises because the halos in simulations are singular at the center, whereas the observations of galactic rotation velocities are better fitted by totally smooth halos, namely, with soft cores. The missing satellites problem arises because there are too many small halos and subhalos in simulations. In summary, the dark matter distribution on small scales in N-body simulations is a distribution of halos, which are relatively smooth but not as smooth as required by indirect observations of the dark matter distribution.

However, as we have seen in this paper, the existence of halos in the cosmic web is surely an aspect of the small-scale smoothening produced by N-body discreteness effects. Our analysis of the smoothness and abundance of halos corroborates this conclusion. The analysis of smoothness is based on the application of robust entropic measures to halos in two recent CDM N-body simulations with different resolutions, finding a progressive and essentially monotonic growth of halo graininess or anisotropy with radius, which stops at some radius, for every halo. The radius at which the limit graininess or anisotropy are attained can be defined as the final end of the smooth halo and the beginning of the cosmic web structure, which merge seamlessly. The removal of discreteness parameters should make all the smoothness radii vanish and therefore remove the halos and their relative smoothness altogether. This makes the problems of the CDM hypothesis on small scales somewhat different and much more serious than they are currently known to be.

References

[1] de Blok W J G 2010 The Core-Cusp Problem, *Advances in Astronomy* 2010, article id. 789293
[2] Moore B, Ghigna S, Governato F, Lake G, Quinn T, Stadel J and Tozzi P 1999 *Astrophys. J. Lett.* **524** L19-L22
[3] Diemand J, Kuhlen M, Madau P, Zemp M, Moore B, Potter D and Stadel J 2008 *Nature* **454** 735–38
[4] Kuzio de Naray R, McGaugh S S, de Blok W J G 2008 *Astrophys. J.* **676** 920–43
[5] Berezinsky V S, Dokuchaev V I and Eroshenko Yu N 2014 *Phys.-Usp.* **57** 1–36
[6] Gurbatov S.N., Saichev A.I. and Shandarin S.F. 2012 *Phys.-Usp.* **55** 223–49.
[7] Cooray A and Sheth R 2002 *Phys. Rep.* **372** 1–129.
[8] Gaite J. 2013 Halo Models of Large Scale Structure and Reliability of Cosmological N-Body Simulations, *Galaxies* **1** 31–43.
[9] Vergassola M., Dubrulle B., Frisch U. and Noullez A. 1994 *Astron. & Astrophys.* **289** 325–56
[10] Gaite J 2014 Smooth halos in the cosmic web *Preprint* arXiv:1407.6197
[11] Klypin A A, Trujillo-Gomez S and Primack J, 2011 *Astrophys. J.* **740** 102
[12] Gaite J. 2007 *Astrophys. J.* **658** 11–24
[13] Gaite J 2010 Fractal analysis of the dark matter and gas distributions in the Mare-Nostrum universe *J. Cosmol. Astropart. Phys.* **03** 006
[14] Splinter R J, Melott A L, Shandarin S F and Suto Y 1998 *Astrophys. J.* **497** 38–61.
[15] Joyce M, Marcos B and Baertschiger T 2009 *MNRAS* **394** 751–773

Combining a reactive potential with a harmonic approximation for molecular dynamics simulation of failure: construction of a reduced potential

I G Tejada[1], L Brochard[1], G Stoltz[2,3], F Legoll[1,3], T Lelièvre[2,3] and E Cancès[2,3]

[1] Université Paris-Est. Laboratoire Navier (UMR 8205), CNRS, ENPC, IFSTTAR. 6 & 8 avenue Blaise Pascal 77455 Marne-la-Vallée, France.

[2] Université Paris-Est. CERMICS, École des Ponts ParisTech (ENPC). 6 & 8 avenue Blaise Pascal 77455 Marne-la-Vallée, France.

[3] INRIA Rocquencourt, MATHERIALS research-team, 78153 Le Chesnay Cedex, France.

E-mail: ignacio.tejada@enpc.fr

Abstract. Molecular dynamics is a simulation technique that can be used to study failure in solids, provided the inter-atomic potential energy is able to account for the complex mechanisms at failure. Reactive potentials fitted on ab initio results or on experimental values have the ability to adapt to any complex atomic arrangement and, therefore, are suited to simulate failure. But the complexity of these potentials, together with the size of the systems considered, make simulations computationally expensive. In order to improve the efficiency of numerical simulations, simpler harmonic potentials can be used instead of complex reactive potentials in the regions where the system is close to its ground state and a harmonic approximation reasonably fits the actual reactive potential. However the validity and precision of such an approach has not been investigated in detail yet.

We present here a methodology for constructing a reduced potential and combining it with the reactive one. We also report some important features of crack propagation that may be affected by the coupling of reactive and reduced potentials. As an illustrative case, we model a crystalline two-dimensional material (graphene) with a reactive empirical bond-order potential (REBO) or with harmonic potentials made of bond and angle springs that are designed to reproduce the second order approximation of REBO in the ground state. We analyze the consistency of this approximation by comparing the mechanical behavior and the phonon spectra of systems modeled with these potentials. These tests reveal when the anharmonicity effects appear. As anharmonic effects originate from strain, stress or temperature, the latter quantities are the basis for establishing coupling criteria for on the fly substitution in large simulations.

1. Introduction

As many macroscopic failure mechanisms in materials, such as fracture or plastic deformation, originate at the atomic scale (bond breaking, vacancies, dislocations) it is natural to study failure at such scale [1]. However, atomistic modeling is computationally expensive because models usually contain an extremely large number of particles and because events at atomic scale require complex modeling. Ab initio approaches (i.e. based on first principles) have a good predictive ability, but a high computational cost which limits their use to small systems and time scales. Molecular dynamics (MD) is an alternative technique that makes it possible to simulate larger systems. It is used to calculate the motion of atoms (e.g. following the Hamiltonian dynamics) once a model for their mutual interactions is postulated. The interatomic potential used is what mostly determines the physical validity and the computational cost of the simulation.

Reactive potentials [2, 3, 4, 5, 6] are classical potentials that allow bonds to break and form. This is done by empirically modeling the effect of the current atomic environment on interaction forces, which is essential at failure when the atomic structure is strongly rearranged. However the reactive ability is useless in regions where the atomic structures are close to the ground state and the material behaves elastically. Substituting there the reactive potential by a non-reactive approximation makes it possible to get the same macroscopic response of the model while considerably reducing the computation time. This technique has been used, for instance, to study dynamic crack propagation in a silicon single crystal [7], but the existing approaches lack a detailed methodological investigation and a proper validation. To make further progress in this direction we discuss here a coupling method between a reactive bond-order potential and a harmonic potential.

2. Theoretical background

Graphene is a material made of carbon atoms arranged in a two-dimensional honeycomb lattice with unique physical properties. In this work, we use it as a simple case study to illustrate our methodology. In this respect, we do not pay attention to out-of-plane bending or vibration modes, and we use the 2nd-generation reactive empirical bond-order potential [8], REBO, as implemented in [5]. It accounts for the angular forces and gives the correct cohesive energy and equilibrium lattice constants of graphene, but is known to only imperfectly reproduce its elastic constants and phonon spectra [9]. REBO is a reformulation of Brenner-Tersoff potentials [3, 4], that are in turn based on Abell's formalism [2] in which the binding energy is produced by pairwise nearest-neighbor interactions that depend on the local atomic environment.

In the ground state (i.e., the system unstressed at 0 K), the potential energy is minimum and it can be expanded in terms of the powers of the atomic displacements around their equilibrium positions. The harmonic approximation consists in truncating the Taylor expansion after quadratic terms. The Hessian of the potential energy is the so-called force constants matrix [10], Φ.

We seek a substitute for REBO interactions such that, close to the ground state, both the original potential and its harmonic approximation provide the same (or very close) forces. Accordingly, their Hessians must be the same, or very close. We consider harmonic potentials made of harmonic springs (bond interactions between pairs of atoms) and angles (angle interactions between triplets of atoms), widely implemented in MD codes. The analytical expressions for the potential energies respectively are: $E_s = (1/2)\,K_s\,(r_s - r_{\mathrm{eq},s})^2$ and $E_a = (1/2)\,G_a\,(\theta_a - \theta_{\mathrm{eq},a})^2$. The topology of the network of springs and angles as well as the values of the parameters K_s, $r_{\mathrm{eq},s}$, G_a and $\theta_{\mathrm{eq},a}$ are fitted so that the Hessian of that harmonic potential is exactly equal to that of REBO in the ground state. Note that the harmonic potential is not a quadratic function in terms of the current positions of the atoms but in terms of bond lengths r_s and angles θ_a.

The use of a substitute potential affects the bulk properties of graphene. As we focus on fracture tests, we analyze the mechanical behavior, i.e. the static strain-stress relation, and the phonon dispersion curves, i.e. the dynamics of the atomic vibrations, which are both essential to the failure mechanisms of materials. If the substitute potential reproduces the Hessian of REBO in the ground state, then their mechanical behaviors and phonon spectra deviate only when the second order expansion fails approximating the REBO potential, e.g. because of excessive anharmonicity due to large strains and/or large temperatures.

The desired hybrid model consists of a reactive zone close to the crack tip (REBO interactions), a non-reactive zone (harmonic interactions) far from the crack tip and a transition zone which serves for the computation of the REBO potential (because of the range of the interactions) and for a smooth transition between the potentials.

3. Methodology

In order to set the parameters of springs and angles, we follow the algebraic method proposed by Mounet [11] to enforce the acoustic sum rules and index symmetries on the Hessian of carbon allotropes obtained with density-functional perturbation theory. First a dot product between two matrices is defined as $\Phi \cdot \Psi = \sum_{i\alpha j\beta} \Phi_{i\alpha j\beta} \Psi_{i\alpha j\beta}$, where i, j label atoms $(1, \cdots, N)$ and α, β represent the in-plane coordinates (x, y). The associated distance is $\mathrm{d}(\Phi, \Psi) = \sqrt{(\Phi - \Psi) \cdot (\Phi - \Psi)}$, so that minimizing such distance is equivalent to applying a least squares method. The methodology for finding the optimal potential parameters for a simplified harmonic potential (valid for non-crystalline structures too) is as follows:

- Compute the force constants matrix with REBO potentials in the ground state: Φ^{REBO}.
- Use the same geometry of the system in its ground state for the harmonic approximation (e.g. a honeycomb lattice of constant $a \simeq 2.4$Å for graphene).
- Prescribe a topology of harmonic interactions between atoms produced by springs and angles.
- Construct a basis for the approximated solutions. The contribution of each prescribed interaction to the whole approximated Hessian Φ^{HARMO} is known and depends on the corresponding constant, in such a way that we can write: $\Phi^{\mathrm{HARMO}} = \sum_s K_s \Phi_s^{\mathrm{spring}} + \sum_a G_a \Phi_a^{\mathrm{angle}}$. Thus $\left\{\Phi_s^{\mathrm{spring}}, \Phi_a^{\mathrm{angle}}\right\}$ is the basis of the subspace of approximated solutions.
- Orthonormalize the basis $\left\{\Phi_s^{\mathrm{spring}}, \Phi_a^{\mathrm{angle}}\right\} \to \{U_I\}$.
- Take Φ^{HARMO} as the projection of Φ^{REBO} onto the orthonormalized basis. $\Phi^{\mathrm{HARMO}} = \sum_I (\Phi^{\mathrm{REBO}} \cdot U_I)\, U_I$ is the element closest to Φ^{REBO} in the subspace of approximated models. If $\mathrm{d}\left(\Phi^{\mathrm{REBO}}, \Phi^{\mathrm{HARMO}}\right) = 0$, then the prescribed interactions are able to exactly reproduce the original force constants matrix.
- Express Φ^{HARMO} in the original basis, and get the constants of the springs to use.
- Optional: if the system is a crystal, the approximated model can be symmetrized to use fewer constants.

Once the approximated model is known, we use MD simulation to compare the mechanical behavior of both models. The graphene sheet is deformed in one direction (zig-zag or armchair) while the perpendicular one is left undeformed. At each deformed configuration, the energy is minimized and the stress is computed as $\sigma_{\alpha\beta} = \sum_i^N r_{i\alpha} f_{i\beta}/A$, where A is the area of the sheet, $f_{i\beta}$ the β-component of the total force per unit of length acting on particle i, and $r_{i\alpha}$ the α-component of its position vector. The engineering strain is defined as $\varepsilon_\alpha = (L_\alpha - L_{\alpha,0})/L_{\alpha,0}$, where $L_{0,\alpha}, L_\alpha$ are the initial and final dimensions of the sheet in α direction.

By applying lattice dynamics we get their phonon spectra. We use a Langevin thermostat to get these curves at different temperatures following the so-called quasi-harmonic approximation (i.e. assuming that the anharmonicity is restricted to thermal expansion, so that the temperature dependence of the phonon frequencies arises only from the dependence on crystal structure and volume [10]). We have used LAMMPS [12] and our own MD code.

4. Results and discussion

The Hessian of two-dimensional graphene at ground state modeled with REBO can be exactly reproduced with a system of reduced potentials based on four classes of harmonic interactions, as

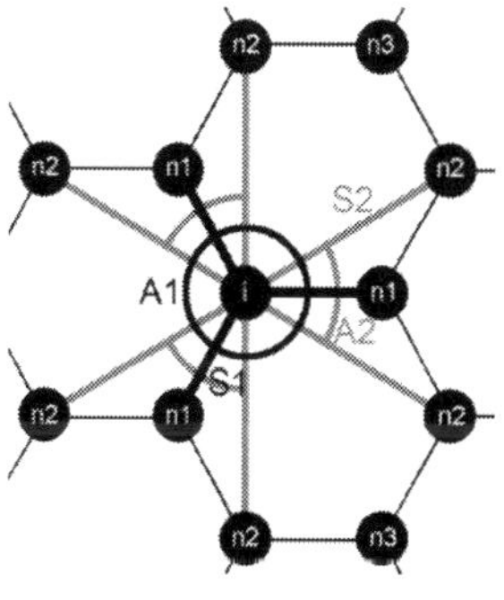

Figure 1. Schematic of the topology of 3+6 harmonic springs, of types S1 (——) and S2 (——), and 3+3 angles, of types A1 (⊲) and A2 (⊲), that connect atom i to its 1st- and 2nd order neighbors (n1,n2) in the approximated model.

Table 1. Harmonic springs and angles used in the approximated model

Interaction	Atom/s in interaction (figure 1)	Constant	Equilibrium
Spring 1 (S1)	(i, n1)	$K_1 = 39.76$ eV/ Å^2	$r_{\mathrm{eq},1} = 1.3977$ Å
Spring 2 (S2)	(i, n2)	$K_2 = 1.21$ eV/ Å^2	$r_{\mathrm{eq},2} = 2.4208$ Å
Angle 1 (A1)	(i, n1, n1')	$G_1 = 1.31$ eV	$\theta_{\mathrm{eq},1} = 2\pi/3$
Angle 2 (A2)	(i, n2, n2')	$G_2 = 7.11$ eV	$\theta_{\mathrm{eq},2} = \pi/3$

depicted in figure 1 and detailed in table 1. Accordingly, the phonon dispersion curves obtained with REBO and with the harmonic approximation are indistinguishable at 0 K (figure 2). At nonzero temperatures, these curves change with the former potential, while remaining unchanged with the latter (there is no thermal expansion with harmonic models). Moreover, the mechanical behaviors in tensile tests are the same at small deformations (figure 3).

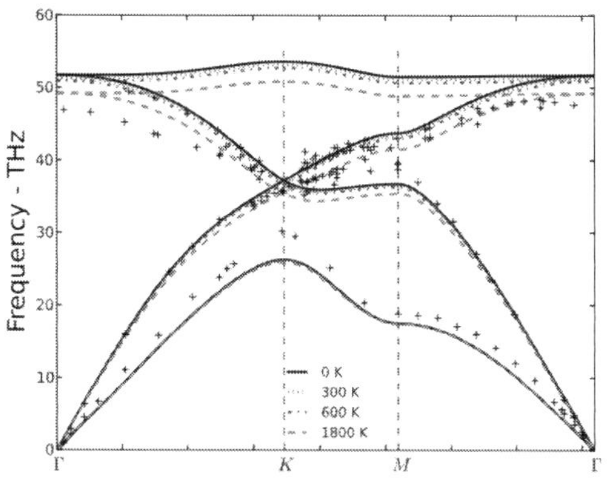

Figure 2. Phonon spectra of graphene obtained with either REBO or a harmonic approximation at $T = 0$ K (——) and with REBO at $T = 300$ (······), 600 (— · —) and 1800 (– – –) K. Experimental data (+) [13] (included to contextualize the curves).

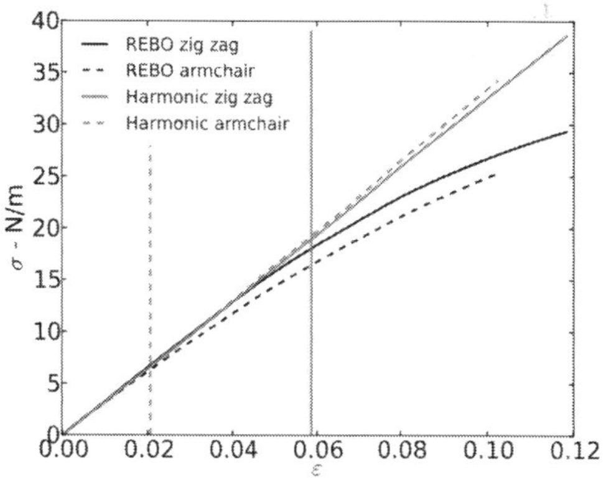

Figure 3. Tensile response of graphene (stress vs. strain) in ziz-zag (——) and arm-chair (– – –) directions obtained from MD simulations with REBO (black) and with a harmonic approximation (red). The vertical lines indicate the strain at which the difference in the stress is larger than 5%.

When atoms move away from their positions in the ground state, the harmonic approximation becomes less and less valid, because of the anharmonic contributions to the reactive potential. It can be due to an increase of the temperature or of the strain, quantities that can be locally estimated around each atom during a simulation, and are therefore suited for constructing indicators for on the fly substitution. The analysis of the bulk behavior far from the ground state enables us to quantify the effects of anharmonicity. For instance, these effects start to be significant for strains between 2% and 6% (see figure 3).

On the other hand, when using a harmonic approximation one has to consider the main macroscopic physical phenomena that originate from anharmonicity and therefore cannot be captured with harmonic approximations. These can be grouped into equilibrium properties (thermal expansion and dependency of elastic constants on temperature and volume) and transport properties (thermal conductivity and almost any of the processes by which the lattice vibrations transmit energy). This is a drawback of the harmonic approximation, which limits the use of the coupling to regions where anharmonicity is small.

Finally, there are various ways of achieving a perfect coupling in the hybrid model to leave the phonon relations unaffected. For instance, the same algebraic methodology explained in section 3 can be used to obtain the spring and angle constants in a representative volume element (RVE) that includes reactive and non-reactive zones. The whole Hessian of this hybrid RVE is the result of the contribution of a reactive and a harmonic part: $\Phi^{\mathrm{HYB}} = \Phi^{\mathrm{HYB(R)}} + \Phi^{\mathrm{HYB(H)}}$. By minimizing the distance $\mathrm{d}\left(\Phi^{\mathrm{REBO}} - \Phi^{\mathrm{HYB(R)}}, \Phi^{\mathrm{HYB(H)}}\right)$, the spring and angles constants of the harmonic part are obtained. The constants near the interface have to be different to those in the pure harmonic zone to guarantee that there are no undesired effects due to the transition between the potentials. However the new values to use there depend on the geometry of the interface in the hybrid RVE. Alternatively, the methodology can be also applied to obtain the set of springs and angles that can substitute a single REBO bond in a full REBO environment. Then, every time a REBO bond is replaced, the corresponding springs and angles are added. Following this approach, a perfect coupling is achieved without prescribing any specific interface, and the substitution can be performed at any time during the simulation, so that the interface can be displaced on the fly. More details on the coupling methodology will be provided in [14].

Acknowledgments

The authors gratefully acknowledge support from Labex MMCD - Modélisation & Experimentation pour la Construction Durable under contract ANR-11-LABX-0022.

References

[1] Holland D and Marder M 1999 *Advanced Materials* **11**(10) 793–806
[2] Abell G C 1985 *Phys. Rev. B* **31**(10) 6184–6196
[3] Brenner D W 1990 *Phys. Rev. B* **42**(15) 9458–9471
[4] Tersoff J 1986 *Phys. Rev. Lett.* **56**(6) 632–635
[5] Stuart S J, Tutein A B and Harrison J A 2000 *The Journal of Chemical Physics* **112** 6472–6486
[6] van Duin A C T, Dasgupta S, Lorant F and Goddard W A 2001 *The Journal of Physical Chemistry A* **105** 9396–9409
[7] Buehler M J, van Duin A C T and Goddard W A 2006 *Phys. Rev. Lett.* **96**(9) 095505
[8] Brenner D W, Shenderova O A, Harrison J A, Stuart S J, Ni B and Sinnott S B 2002 *Journal of Physics: Condensed Matter* **14**(4) 783
[9] Tewary V K and Yang B 2009 *Phys. Rev. B* **79**(7) 075442
[10] Dove M T 1993 *Introduction to Lattice Dynamics* (Cambridge University Press) ISBN 9780511619885
[11] Mounet N 2005 *Structural, vibrational and thermodynamic properties of carbon allotropes from first-principles: diamond, graphite, and nanotubes* Master's thesis Massachusetts Institute of Technology
[12] Plimpton S J 1995 *J. Comput. Phys.* **117** 1–19
[13] Mohr M, Maultzsch J, Dobardžić E, Reich S, Milošević I, Damnjanović M, Bosak A, Krisch M and Thomsen C 2007 *Phys. Rev. B* **76**(3) 035439
[14] Tejada I G, Brochard L, Stoltz G, Legoll F, Lelièvre T and Cancès E *In preparation*

Journal of Physics: Conference Series **574** (2015) 012042 doi:10.1088/1742-6596/574/1/012042

Coherent interactions in nonlinear multilevel media.

E Gazazyan, V Chaltykyan

Institute for Physical Research, Armenian National Academy of Sciences, Ashtarak-2, 0203, Armenia

E-mail: emilgazazyan@gmail.com

Abstract. We propose in this work an alternative method of easier calculation of necessary conditions for lossless propagation of short laser pulses in multilevel atomic media. Method is based on the quasienergy approach and illustrated with example of a five-level system. We also demostrated effective population transfer in this system.

1. Introduction

Coherent interaction of light signals with quantum systems attract considerable interests for their importance in both fundamental science and practical applications. The most observable experimentally and applicable are adiabatic coherent interactions which rely upon sufficiently slow evolution of a quantum system [1-3]. Constructing desired coherent superposition states of atoms interacting with laser pulses in macroscopic media is one of key problems in quantum informatics and related topics.

Microscopic theory of adiabatic interaction between laser pulses and individual multilevel atoms is sufficiently complete [4, 5, 6] but the realization of adiabatic interaction in macroscopic volumes is not completely clear and sufficiently studied as of today. In adiabatic approximation the populations of atomic levels follow the instantaneous values of field envelopes. So, the adiabatic approximation corresponds to a dispersion-free medium. Even in linear optics the approximation of dispersion-free medium is restricted by propagation lengths where the group delay (caused by the group velocity which is different from c) of the pulse in the medium may be neglected. In nonlinear media the situation is more complicated because of the dependence of the group velocity on the intensities of available laser fields. In the simplest, two-level, model more intense portions of the pulse are moving faster, resulting in steepening of the leading edge and formation of a shock wave (see, e.g., [7, 8, 9]). In addition to this mechanism, nonlinearity of medium leads to parametric broadening of spectrum (self-phase modulation [10]) which again shortens the pulse duration during its propagation and also furthers the violation of adiabaticity. So adiabatic processes in medium are quite robust if and only if they are implemented at sufficiently short optical lengths, where adiabaticity of interaction is preserved. However, it was shown as early as in seventies that at interaction of a laser pulse with a V-system the length of adiabaticity of interaction increases essentially, as compared with the two-level system, due to compensation for the medium dispersion [11]. After that the possibility of soliton regime of pulse propagation in such media was studied in detail [12]. This mechanism of compensation for the nonlinear dispersion of medium is based on destructive interference of dipole moments induced by radiation at the adjacent transitions of medium atoms.

Another striking example of bleaching of resonant medium via adiabatic interaction is electromagnetically induced transparency (EIT) [13, 14] which is based on the socalled dark state formed in a three-level system [15]. The dark state does not involve the intermediate state because of quantum interference; due to this fact the field-induced dipole moments vanish leading to medium bleaching.

The problem of medium transparency at certain frequencies which enables distortion-less propagation of laser pulses in media, is of considerable importance because it is concerned directly with such applications as optical information storage in media with its subsequent retrieval and optical communication lines [14], adiabatic population transfer [15], creation of superposition states, control of chemical processes [16], implementation of logic gates and many others.

In this work we propose a relatively simple calculation technique to find the necessery (but not always sufficient) condition for similar bleaching of a medium consisting of atomic systems with more complicated level diagrams. The method relies on the system quasienergies, which allows obtainig analytical expressions describing controllable population transfer and pulse propagation without distostion in media. We also obtain the necessery conditions for the propagation length where the adiabaticity of interaction does not break in the medium. For visualization, we demonstrate these effects in media of M-type and W-type atomic system.

2. Formalism

We consider the multilevel schemes (see, Fig.1) interacting with incident multipulse field $E(z,t) = \sum_i E_i exp(i\omega_i z/c - i\omega_i t) + c.c.$. Frequencies ω_i of pulses may be in resonance with one or several atomic transitions, i.e. $|\omega_i - \omega_{ij}| << \omega_{ij}$, where ω_{ij} are the frequencies of respective atomic transitions. Conditions imposed on intensities of pulses, their on/off switching sequences, ranges of overlapping, choice of polarizations (if magnetic sublevels are employed) etc. are determined while solving the problem (see Section 3) depending on specific configurations. We assume that considered pulses have durations T_i much shorter than all times of relaxation and at the same time much longer than the inverse frequency distance between the closest quasienergies i.e. $|\lambda_i - \lambda_j|T_i >> 1$ ($\lambda_{i,j}$ are eigenvalues of the interaction Hamiltonian), in order to ensure the adiabaticity of interaction. In approximation of non-overlapping atomic wave functions, the polarization of medium may be represented in the form $P = N\langle d\rangle$, with N being the number density of atoms and $\langle d\rangle$ being the atomic dipole moment induced by an external field, $\langle d\rangle = \langle\psi|d|\psi\rangle$, where $|\psi\rangle$ is the state vector of the atom in the external field. Within approximations listed above we have

$$\langle\psi|d|\psi\rangle = \sum b_i^* b_{i+1}\langle i|d|i+1\rangle e^{-i\omega_i t} + c.c. \tag{1}$$

Here $|i\rangle$ is the bare state of atom corresponding to the energy E_i, $\langle i|d|j\rangle = d_{ij}$ is the matrix element of the dipole moment of atomic transition, b_i is the coefficient of bare state $|i\rangle$ in superposition state $|\psi\rangle$ and $b_i^* b_{i+1}\langle i|d|i+1\rangle e^{-i\omega_i t} = \langle d_i\rangle e^{-i\omega_i t}$ is the dipole moment at frequency ω_i.

It is known that in the approximation of slowly varying amplitudes, propagation of pulses in a dielectric homogeneous isotropic medium can be described by the system of truncated equations in running coordinate system $x = z, \tau = t - z/c$ (see, e.g., [1]):

$$\frac{\partial E_i}{\partial x} = -i\frac{2\pi\omega_i}{c}P_i \tag{2}$$

where P_i is the complex amplitude of medium polarization at frequency ω_i, $P_i = N\langle d_i\rangle$. As shown in works [14,15], in the resonant approximation and adiabatic interaction with medium, the system of truncated equations [22,23] may be represented in the form

$$\frac{\partial E_j}{\partial x} = -i\frac{2\pi N\omega_j\hbar}{c}\frac{\partial\lambda_i}{\partial E_j^*} \tag{3}$$

Here λ_i is the eigenvalue of the interaction Hamiltonian corresponding to eigenstate realized in adiabatic interaction. The last equation is correct only at such lengths of propagation where the approximation of dispersion-free medium is valid. Limitations of applicability of this equation are considered in Section 4. If the system has a quasienergy which remains constant during the overall time of interaction, i.e., $\partial\lambda/\partial E_j^* = 0$, then according to[1], the induced dipole moment in such a system is zero and the pulse propagates in such medium without change in shape at the group velocity equal to c. It is known[1] that the quasienergies (eigenvalues) of the system are the roots of characteristic equation

$$\det(H - \lambda I) = 0 \tag{4}$$

where I is the unit matrix and H the interaction Hamiltonian (in frequency units) which in the dipole approximation and the resonance approximation may be represented in the form

$$H = \sum \sigma_{i,i}\delta_{i-1} - (\sum \sigma_{i,i+1}\Omega_i + h.c.) \tag{5}$$

with $\sigma_{i,j}$ being the projection matrices, Ω_i amplitude of Rabi frequencies of the pulse interacting with the transition $i \longrightarrow i+1$: $\tilde{\Omega_i} = \Omega_i e^{i\varphi_i} = -E_i d_{i,i+1}/\hbar$, and δ_{i-1} the (i-1)-photon detuning (δ_0=0). The phases of complex Rabi frequencies $\tilde{\Omega}_i$, which can vary when traveling in the medium, are included into one-photon detunings Δ_i. So the quantities Ω_i are real and positive.(The one photon detunings are defined as $\Delta_i = \omega_{i+1,i} - \omega_i + \dot{\varphi}_i$, if $\omega_{i+1,i} > 0$ and $\Delta_i = \omega_{i,i+1} - \omega_i + \dot{\varphi}_i$, if $\omega_{i+1,i} < 0$, where φ_i are the phases of corresponding Rabi frequencies). Definition of multiphoton detunings depends on the specific scheme of interaction. By differentiating the characteristic equation with respect to E_j^* and setting to zero the derivatives of quasienergy, we obtain a system of algebraic equations, from which the needed values of interaction parameters may be determined. The problem can, however, be simplified. Really, at turning off the fields the interaction Hamiltonian is diagonalized and the roots of [2] go to the following constant values:

$$\lambda_i \longrightarrow \delta_{i-1} \tag{6}$$

The problem of determination of necessary (but not always sufficient) conditions for medium transparency is thus reduced to the search of such parameters of adiabatic interaction for which the system quasienergy remains equal to one of multi-photon detunings always during interaction. By substituting these values successively into equation (2) we obtain the needed conditions for the interaction parameters.

3. Criterion of adiabaticity of interaction

Procedure considered above allows determining only one eigenvalue of the interaction Hamiltonian, but determination of the adiabaticity criterion for a single atom requires knowledge of all other eigenvalues, i.e., finding the algebraic roots of fourth-order equation. This can be done in general only numerically. Since we are interested in the propagation problem, we must determine the adiabaticity criterion for the medium, rather than for a single atom; it may be found directly from the propagation equations. As follows the dipole moment $< d >$ of an atom interacting with electromagnetic field and residing the state $|\psi\rangle$, can be expressed in terms of the coefficients b_i of bare state $|i\rangle$ in superposition state $|\psi\rangle$ and matrix elements of the dipole moment of atomic transition d_{ij}. The coefficients b_i are determined by non-stationary Schroedinger equation with Hamiltonian (3). Separating real and imaginary parts in

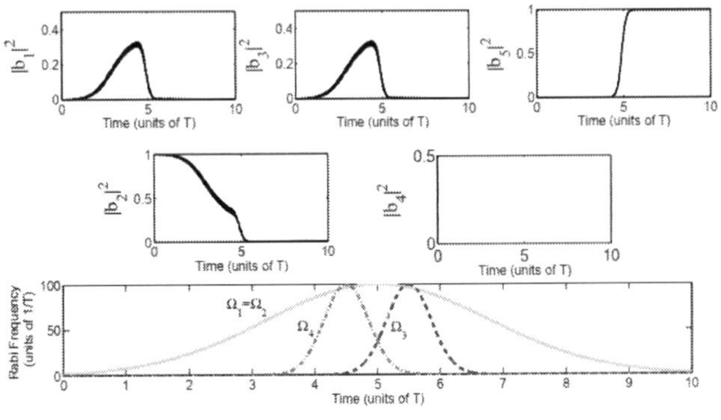

Figure 1. Dynamics of atomic level populations $|b_i|^2$ in W-system and the sequence of pulses.

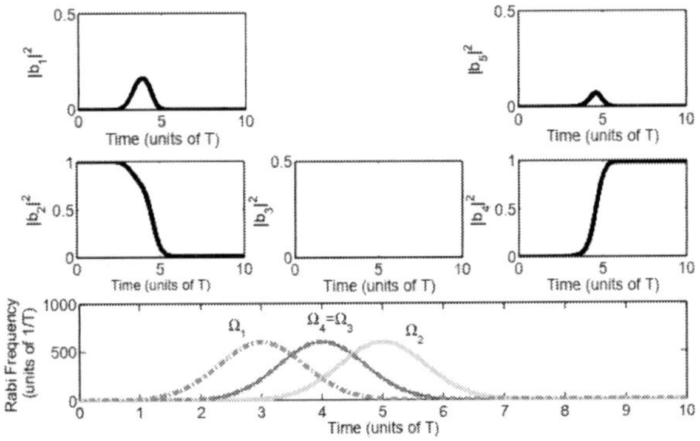

Figure 2. Dynamics of atomic level populations $|b_i|^2$ in M-system and the sequence of pulses.

the truncated equation of propagation, differentiating the equation for the phase with respect to time, and combining the obtained equations with the Schroedinger equation, in general case we obtain a self-consistent system of equations describing variation of frequencies (one-photon detunings) and intensities (Rabi frequencies) of pulses during propagation in medium. For example, in the case of medium consisting of W-type atoms (Fig.1c) we obtain:

$$\begin{aligned}
\frac{\partial \Omega_1^2}{\partial x} &= -q_1 \frac{\partial |b_1|^2}{\partial \tau} \\
\frac{\partial \Omega_2^2}{\partial x} &= q_2 \frac{\partial (|b_1|^2 + |b_2|^2)}{\partial \tau} \\
\frac{\partial \Omega_3^2}{\partial x} &= q_3 \frac{\partial (|b_1|^2 + |b_2|^2 + |b_3|^2)}{\partial \tau} \\
\frac{\partial \Omega_4^2}{\partial x} &= q_5 \frac{\partial |b_5|^2}{\partial \tau} \\
\frac{\partial \Delta_i}{\partial x} &= q_i \frac{\partial}{\partial \tau} \frac{Re(b_i^* b_{i+1})}{\Omega_i}
\end{aligned} \tag{7}$$

where $q_i = 2\pi N \omega_i |d_{i,i+1}|^2/\hbar c$ For other atomic level schemes equations for intensities differ in only the signs of derivatives with respect to populations. As it follows from equations, during propagation in the medium not only the shapes of pulses may vary essentially, but also the conditions for detuning of resonances may be violated.

Since it is the time derivatives that stand in right hand side of these equations, by use for atomic populations the expressions following from the wave functions obtained above, we take into account the first nonadiabatic correction which lead to the change in spectral and temporal characteristics of pulses. Correspondingly, the conditions of smallness of these changes are criteria of adiabaticity in the medium.

4. Conclusion

We have developed and analyzed, for the example of a five-level system, a relatively simple method for finding necessary conditions under which adiabatic short light pulses can travel in the medium without distortion of the shape and phase. It was shown that the necessary conditions for such bleaching of medium (i.e., vanishing of dipole moments induced in the medium at the frequencies of all interacting fields) is the equality of system quasienergies to one of the values of multi-photon detunings of resonances. In this case the medium bleaching may be caused by both interference of quantum states and interference of dipole moments of different transitions if pumping is degenerate. The existence of such regimes means that at the propagation lengths where the interaction adiabaticity does not break, stable superposition states are produced all over the medium and these states may easily be controlled by adjusting the parameters of pulses. Finding of the sufficient conditions requires determination of roots of algebraic equations which may be done only numerically for the order of equation higher than three. However, the sufficiency of found necessary conditions may easily be checked directly by calculation of the induced dipole moments. In addition, we demonstrated for the examples of W-systems the possibility to find a generalized criterion of adiabaticity of interaction in the medium without determination of all eigenvalues of interaction Hamiltonian; we analyzed population transfer in this system and its peculiarities during propagation in the medium.

Acknowledgment

Authors are grateful to Professor G. Grigoryan for multiple fruitful discussions and assistance.

Work was in part supported by IRMAS International Associated Laboratory, and grant of NFSAT YSSP-13-46.

References

[1] Shore B W 1990 *The Theory of Coherent Atomic Exitation* (Wiley-Interscience, New York)
[2] Ter-Mikaelyan M 1997 *Physics-Uspekhi* **40** 1195.
[3] Zheltikov A M and Koroteev N I 1997 *Physics-Uspekhi* **42** 321
[4] Kral P, Thanopulos I and Shapiro M 2007 *Rev. Mod. Phys.* **79** 53
[5] Oreg J, Hioe F and Eberly J H 1984 *Phys.Rev.* A **29** 690
[6] Elk M ,1995 *Phys. Rev.* A **52** 4017
[7] Grischkowsky D, Courtens E and Armstrong J A 1973 *Phys. Rev. Letters* **31** 422
[8] Mazets I E and Matisov B G 1996 *Quantum and Semiclass.Opt.* **8** 909
[9] Chaltykyan V, Grigoryan G and Nikogosyan G 2003 *Phys.Rev.*A **68** 013819
[10] Shen Y R 1984 *The Principles of Nonlinear Optics* (Wiley-Interscience, New York)
[11] Achmanov S, Kovrigin M, Maksimov S and Ogluzdin V 1972 *Pis'ma JETF* **15** 186
[12] Paspalakis E, Kylstra N J and Knight P L 2000 *Phys.Rev.*A **61** 045802
[13] Harris S 1997 *Physics Today* **50(7)** 36
[14] Fleischhauer M, Imamoglu A and Marangos J 2005 *Rev. Mod. Phys.* **77** 633.
[15] Bergmann K, Theuer H and Shore B 1998 *Rev. Mod. Phys.* **70** 1003
[16] Vitanov N, Fleischhauer M, Shore B and Bergmann K 2001 *Adv. Atom. Mol. Opt. Physics* **46** 55

A Stochastic Model of Particle Deposition and Evaporation for Ionic Self-Assembly of Thin Films

E. M. Schwen, I. Mazilu, D. A. Mazilu

Washington and Lee University, 204 W. Washington Street, Lexington, VA 24450, USA

E-mail: schwene15@mail.wlu.edu, mazilui@wlu.edu, mazilud@wlu.edu

Abstract.
We investigate nanoparticle self-assembly using a stochastic model based on *cooperative sequential adsorption with evaporation* mechanisms and aimed specifically at the creation of optical thin films. Applying the mean field approximation, we derive a rate equation for particle density. We solve directly for the particle density in both the steady state and time-dependent cases. The analytical results are compared to Monte Carlo simulations of the self-assembly process and to experimental data for self-assembled thin films. We relate our theoretical model to the final particle density for thin films created under varied nanoparticle suspension concentrations.

1. Introduction

Nanoparticle self-assembly is an interdisciplinary area of research with a wide range of applications throughout the sciences. The different applications have given rise to an extensive array of experimental techniques that make use of the tendency of particles towards organized self-assembly under controlled conditions. Our goal in this paper is to present a general stochastic model which can address the main features of nanoparticle self-assembly including both the time development and the steady state of the system.

Our theoretical model addresses self-assembly as stochastic process using *cooperative sequential adsorption with evaporation* (CSAE) mechanisms. This method accounts for interaction between nearest-neighbors in the deposition process and allows evaporation of particles from the surface. The mean field approximation yields a rate equation for particle density which predicts both the evolution of the system and the equilibrium particle density. We utilized Monte Carlo simulations for a further investigation of the nanoparticle self-assembly process and found excellent agreement between the theoretical and computational results.

We applied our model experimentally to layer-by-layer self-assembly of thin films, which can be used in the creation of antireflective coatings [1]. Using this assembly process, we studied the effects of initial nanoparticle colloidal suspension concentration on the equilibrium particle density of the resulting thin film. We then related our theoretical model parameters to the nanoparticle concentration and found a functional dependence between nanoparticle concentration and the level of interaction between nearest-neighbors in the deposition process.

Our paper is organized as follows. In section 2, we introduce our stochastic particle deposition and evaporation model. In section 3, we solve our model for the steady state and time-dependent particle densities using the mean field approximation. In section 4, we present our Monte Carlo simulation methods and results. In section 5, we outline our experimental technique for thin film production and relate our theoretical model to the experimental data. We conclude in section 6 with an analysis of our results and possible extensions for our work.

2. Model Definition

We define our model on a uniform deposition surface where each site is connected to z other sites, referred to as its nearest-neighbors. We consider singular nanoparticles which can both evaporate from the surface and attach to the surface with rates dependent on the number of occupied nearest-neighbors. The occupation state of site i is determined by the occupation number n_i: $n_i = 0$ represents an empty site and $n_i = 1$ represents an occupied site.

We define the following transition rate for site occupation:

$$c(n_i \to (1-n_i)) = \gamma n_i + (1-n_i)\alpha \left(1 - \beta \frac{1}{z} \sum_{j \in NN} n_j\right) \tag{1}$$

The first term in the transition rate accounts for evaporation: if a particle is present, it will evaporate with the probability γ. The second term addresses deposition: if a site is empty, a particle will attach with the rate $\alpha(1 - \beta\eta/z)$, where $\eta = \sum_{j \in NN} n_j$ is the number of occupied nearest-neighbors of site i. The parameters α and γ set the relative rates for deposition and evaporation and β, restricted to the range $0 \leq \beta \leq 1$, controls the interaction with nearest-neighbors. For self-assembly with charged particles, electrostatic repulsion suggests that the presence of particles in neighboring sites will reduce the deposition rate. Higher values of β increase repulsion effects, while the extreme case of $\beta = 0$ models a situation with no interaction between neighboring particles.

With this transition rate, the number of particles on the lattice changes according to:

$$\frac{\partial n_i}{\partial t} = -\gamma n_i + (1-n_i)\alpha \left(1 - \beta \frac{1}{z} \sum_{j \in NN} n_j\right) \tag{2}$$

3. Mean Field Solution

Our overall goal with this model is to derive an equation for the total particle density on the deposition surface. We take the ensemble average of $\langle n_i \rangle$ and employ the mean field technique [2] to approximate the higher order correlations as $\langle n_i n_j \rangle = \langle n_i \rangle \langle n_j \rangle$. We obtain:

$$\frac{\partial \langle n_i \rangle}{\partial t} = -\gamma \langle n_i \rangle + (1 - \langle n_i \rangle)\alpha \left(1 - \beta \frac{1}{z} \sum_{j \in NN} \langle n_j \rangle\right) \tag{3}$$

Assuming translational invariance across the surface, the average site density $\rho_i = \langle n_i \rangle$ is the same as the total particle density $\rho = \sum_i p_i/N$, where N is the total number of sites. We can therefore write the rate equation in terms of the particle density ρ:

$$\frac{\partial \rho}{\partial t} = -\gamma\rho + (1-\rho)\alpha(1-\beta\rho) \tag{4}$$

In the steady state $\partial\rho/\partial t = 0$, this equation is easily solved for the final particle density:

$$\rho = \frac{(\alpha + \alpha\beta + \gamma) - \sqrt{(\alpha + \alpha\beta + \gamma)^2 - 4\alpha^2\beta}}{2\alpha\beta} \tag{5}$$

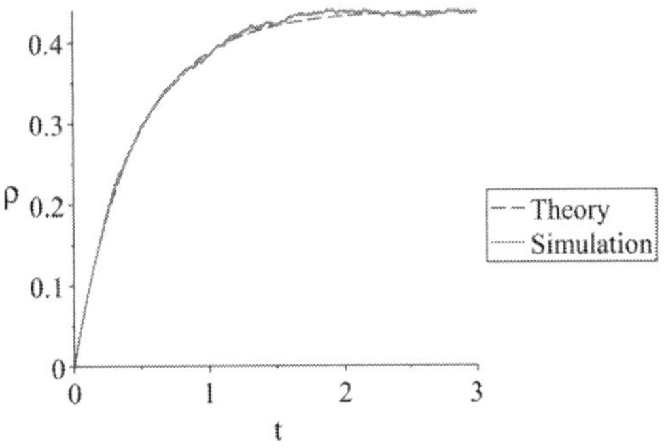

Figure 1. Comparison of simulation (blue solid line) and theoretical (red dashed line) results for particle density over time in arbitrary units. Parameters used: $\alpha = 1$, $\beta = 0.5$, $\gamma = 1$.

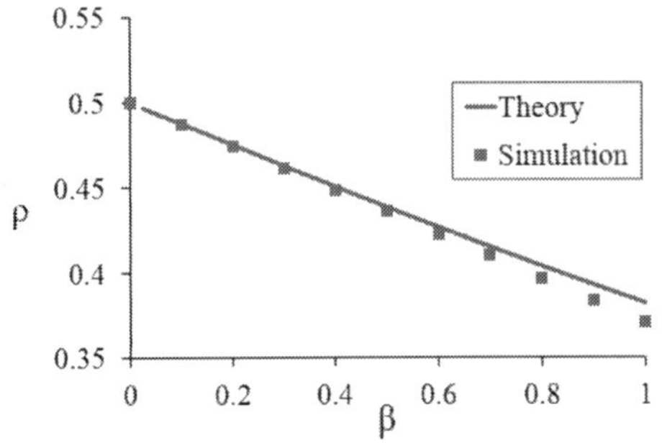

Figure 2. Comparison of simulation (blue squares) and theoretical (red line) results for steady-state particle density as a function of β. Parameters used: $\alpha = 1$, $\gamma = 1$.

The rate equation can also be solved exactly in the general case for the time-dependent particle density. This solution is displayed graphically in comparison with simulation results in Fig. 1.

4. Monte Carlo Simulations

We utilize Monte Carlo simulations in order to further investigate the dynamics of the CSAE process. We model the deposition surface as a 120×120 square lattice. In order to minimize edge effects, we record data for only the 100×100 lattice section at the center of the larger lattice. This choice mimics the experimental process, where only small sections in the center of the optical coating are analyzed and edge effects are not observed.

The simulations proceed according to the transition rate given in Eq. (1) and utilize an event-driven algorithm in order to improve efficiency. Instead of selecting a random site, the simulation selects a random event (evaporation, deposition with zero neighbors, deposition with one neighbor, etc.) weighted by the rate of the event multiplied by the number of qualifying sites. It then selects a random qualifying site and changes the occupation state n_i. Beginning from an empty lattice, we allow the simulation to proceed for 1.44×10^6 site updates in order to ensure that a steady state is reached. We average the particle density results over 100 realizations of the system.

We found excellent agreement between simulations and theoretical results for both the time development of the system and steady-state particle densities as shown in Figs. 1 and 2.

5. Comparison with Experiment

We used a layer-by-layer self-assembly process to create optical thin films. Typically, a substrate is dipped in alternating aqueous solutions of cations and anions. The process has been extended to include colloidal suspensions as well as aqueous solutions. This dipping procedure forms cation/anion bilayers which make up the film.

We used negatively charged glass slides as the substrates for our experiment. We exposed them first to an aqueous solution of the polycation poly(diallyldimethylammonium chloride) (PDDA). After rinsing in deionized water, we dipped the slides in a coloidal suspension of negatively charged spherical silica nanoparticles with a nominal 40-50 nm diameter. We varied the concentration of the silica nanoparticle suspension to observe its effects on the

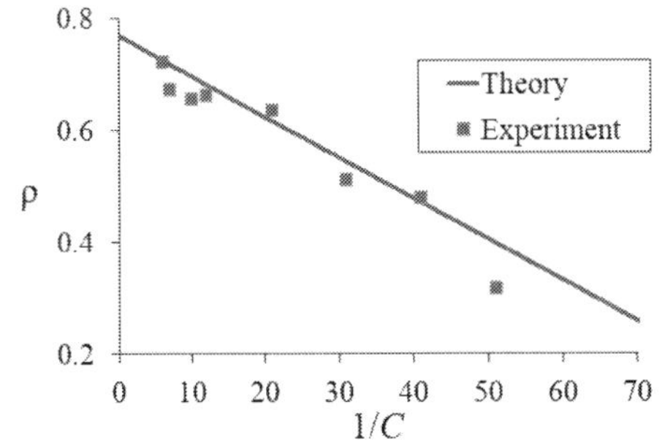

Figure 3. Comparison of experimental data and theoretical results for steady-state particle density as a function of the inverse of nanoparticle concentration in arbitrary units. The linear fit equations are: i) theory (red line), $y = -0.0078x + 0.7582$ with $R^2 = 0.9988$; ii) experiment (blue squares), $y = -0.0078x + 0.7566$ with $R^2 = 0.9468$.

density of particles deposited on the substrate. We analyzed single-bilayer films using SEM micrographs and determined the particle density using an automated pixel-counting process. A comprehensive description of the experimental process can be found in [4].

Comparing our theoretical model with the experimental data, we find the nanoparticle concentration C is a function of the interaction parameter β:

$$C = \frac{1}{23\beta + 5} \tag{6}$$

A graphical comparison of our experimental and theoretical results is presented in Fig. 3.

6. Conclusion

In this paper, we presented a stochastic self-assembly model which we solved for particle density under the mean field approximation. We compared our theoretical results to Monte Carlo simulations and related the deposition rate to nanoparticle concentration for the creation of thin films. While we applied our model specifically to the experimental process of ionic self-assembly of thin films, the mechanisms of our model are quite general and can be effectively extended to other self-assembly processes or even other two-state systems with interactions between neighboring elements.

Our results leave several open questions which could be the subject of further research. Additional investigation into the link between β and nanoparticle concentration could lead to a more thorough understanding of the relationship and a provide general solution that applies over a larger particle density range. Future studies could also investigate the applications of this model to different geometries or physical systems.

Acknowledgments

The authors would like to thank V. O. Kim and B. M. Simpson for film production and SEM imaging. Funding for this research was provided through the Summer Research Scholars Program and the Lenfest Grant at Washington and Lee University.

References

[1] Iler R K, *J. Colloid Interface Sci.* **21**, 569 (1966.)
[2] Krapivsky P L, Redner S and Ben-Naim E V, *A kinetic view of statistical physics*, Cambridge University Press, Cambridge, England (2010.)
[3] Di Ventra M, Evoy S and Heflin J R, *Introduction to Nanoscience and Technology* Springer, New York, (2004.)
[4] Cook L J, Mazilu D A, Mazilu I , Simpson B M, Schwen E M, Kim V O and Seredinski A M, *Phys. Rev. E* (in press), (2014.)

A quantum theory for the excitation spectrum of a rectangular Andreev billiard

J. S. Espinoza Ortiz[1], **R. E. Lagos**[2], **H. Belich**[3], **L. S. Amorim**[3], **F. H. S. Eleutério**[3], **M. T. D. Orlando**[3]

[1]Departamento de Física, Universidade Federal de Goiás, Catalão, Goiás 75704/020, Brasil.
[2]Departamento de Física, IGCE, UNESP, Rio Claro, São Paulo 13500/970, Brasil.
[3]Depto de Física e Química, Universidade Federal do Espírito Santo, ES 29060/900, Brasil.

E-mail: jsespino@physics.org

Abstract. We consider a $\mathcal{N}-\mathcal{S}$ box system consisting of a rectangular conductor coupled to a superconductor. The Green functions are constructed by solving the *Bogoliubov-de Gennes* equations at each side of the interface, with the pairing potential described by a step-like function. Taking into account the mismatch in the Fermi wave number and the effective masses of the normal metal - superconductor and the tunnel barrier at the interface, we use the quantum section method in order to find the exact energy Green function yielding accurate computed eigenvalues and the density of states. Furthermore, this procedure allow us to analyze in detail the *nontrivial* semiclassical limit and examine the range of applicability of the *Bohr-Sommerfeld* quantization method.

1. Introduction

There is an increasing technological interest in the study of normal-conductor($\mathcal{N}$) ballistic quantum dots attached to a superconductor($\mathcal{S}$), giving rise to the coherent scattering of electron into holes and conversely at the superconductor-conductor interface. This peculiar phenomena known as Andreev reflection is an important concept, necessary to understand the properties of nanostructures with $\mathcal{N}-\mathcal{S}$ hybrid structures, commonly called Andreev billiards[1]. In this direction, interesting experiments and theoretical works using graphene are still calling the attention of the scientific community [2, 3] .

The $\mathcal{N}-\mathcal{S}$ box system consists of a rectangular conductor $\mathcal{N}$ of height w and width a attached to a superconductor $\mathcal{S}$. The composed system is integrable and its density of states is gapless. The Bogoliubov-de-Gennes (BdeG) equation describes the physics of the system, solved under suitable boundary conditions enabling us to construct explicit Green functions at each side of the junction, the latter defined as the quantum section, as shown in Figure 1.a. The quantum section method is a tool useful to construct the full energy Green function, thus providing a method to compute the quantum spectrum, and to properly derive the semiclassical limit[4]. In section 2 we apply the quantum section method and obtain the energy Green function eigenvalue condition. Next in section 3, we quantify the role of classical orbits. In section 4 we obtain an expression for the density of states and discuss its leading order approximation. Finally, our conclusions are presented in section 5.

2. The $\mathcal{N}-\mathcal{S}$ Box System

Let us consider the BdeG equation: $\begin{pmatrix} \mathcal{H}_o & \triangle \\ \triangle^\star & -\mathcal{H}_o \end{pmatrix}\Psi = \mathcal{E}\Psi$, where $\mathcal{H}_o = \frac{\mathcal{P}^2}{2M} - \mathcal{E}_{\mathcal{F}}$. $\mathcal{P}$ is the particle momentum, M its mass and the chemical potential at any region is defined as equal to $\mathcal{E}^{\mathcal{N}}_{\mathcal{F}}$ for the $\mathcal{N}$ region and equal to $\mathcal{E}^{\mathcal{S}}_{\mathcal{F}}$ for the $\mathcal{S}$ region. In order to set up the wave functions, we construct the wave functions as a single layer potential density over the quantum section. Thus, considering $\chi_m(x_2)$ an eigenmode function along the section the integral $\int_\Sigma \mu_m \chi_m(x\prime_2)dx_2\prime$ is always a constant. The prescription $|\triangle| \equiv 0$ is imposed inside the normal region $\mathcal{N}$ and the wave function is written as:

$$\Psi^{\mathcal{N}} \;=\; \sum_m \left[a^e_m \begin{pmatrix} 1 \\ 0 \end{pmatrix} \mathcal{G}^{e\,\mathcal{N}}_m(x_1, x\prime_1) + a^h_m \begin{pmatrix} 0 \\ 1 \end{pmatrix} \mathcal{G}^{h\,\mathcal{N}}_m(x_1, x\prime_1)\right] \chi_m(x_2)\,. \tag{1}$$

Similarly, for the superconductor region $\mathcal{S}$, $\mathcal{E} \leq |\triangle| \neq 0$ and

$$\Psi^{\mathcal{S}} \;=\; \sum_{m\prime} \left[c^e_{m\prime} \begin{pmatrix} \gamma_e \\ \frac{1}{\sqrt{2}} \end{pmatrix} \mathcal{G}^{e\,\mathcal{S}}_m(x_1, x\prime\prime_1) + c^h_{m\prime} \begin{pmatrix} \gamma_h \\ \frac{1}{\sqrt{2}} \end{pmatrix} \mathcal{G}^{h\,\mathcal{S}}_m(x_1, x\prime\prime_1)\right] \chi_{m\prime}(x_2)\,. \tag{2}$$

The choice $\chi_m(x_2) = \sqrt{\frac{2}{w}}\sin\left(\frac{m\pi}{w}x_2\right)$ satisfies Dirichlet boundary condition along x_1 for $x_2 = 0$ and w. Next, wave functions above (Eqs. (1)-(2)) are substituted into the BdeG equation to find the appropriated horizontal *Green-mode function.* Inside the $\mathcal{N}$-region, for the m-mode we define $\mathcal{K}^{\mathcal{N}}_{\mathcal{F}} = \sqrt{2\frac{M_{\mathcal{N}}}{\hbar^2}\mathcal{E}^{\mathcal{N}}_{\mathcal{F}}}$ and by exchanging $\mathcal{E}\prime \to \mathcal{E} - \frac{\hbar^2}{2\,M_{\mathcal{N}}}(\frac{m\pi}{w})^2$, we write for electrons (e) $\left[\partial^2_x + \frac{2\,M_{\mathcal{N}}}{\hbar^2}\left(\mathcal{E}^{\mathcal{N}}_{\mathcal{F}} + \mathcal{E}\prime\right)\right]\mathcal{G}^{e\,\mathcal{N}}_m$, similarly for holes ($h$) exchanging $\mathcal{E}\prime \to -\mathcal{E}\prime$. To satisfy Dirichlet's boundary conditions along the vertical axis at $x = a$, we choose $\mathcal{G}^{e(h)\,\mathcal{N}}_m(x, x\prime) = \frac{1}{k^{e(h)}_m} e^{\pm\imath k^{e(h)}_m (a - x\prime)} \sin k^{e(h)}_m\,(a - x)$ with $k^{e(h)}_m = \mathcal{K}^{\mathcal{N}}_{\mathcal{F}}\sqrt{1 \pm \frac{\mathcal{E}}{\mathcal{E}^{\mathcal{N}}_{\mathcal{F}}} - \left(\frac{\pi\, m}{\mathcal{K}^{\mathcal{N}}_{\mathcal{F}}\, w}\right)^2}$. Here, we can observe that these *scattering functions* describe two classical paths: a short path around the neighborhood of the section (from $x\prime$ to x). The other path is longer ($2a$), since it now leaves the section and bounces at the boundary of the conductor ($x = a$) and then goes back to the neighborhood of the section. Observe also that electrons and holes travel opposite classical paths, this classical picture will be of paramount importance later when we derive the semiclassical limit. Then we consider the superconductor region and similarly we test $\mathcal{G}^{e(h)\,\mathcal{S}}_{m\prime}(x, x\prime\prime) = \frac{1}{q^{e(h)}_{m\prime}} e^{\mp\imath\, q^{e(h)}_{m\prime}(x - x\prime\prime)}$, where amplitude damping results in *quasi-short-path* contributions, from the semiclassical point of view. Setting up $\mathsf{K}_{\mp} = \mathcal{K}^{\mathcal{S}}_{\mathcal{F}}\sqrt{\tilde{k}^{e(h)\mathcal{S}}_{m\prime}}$ and $\mathcal{K}^{\mathcal{S}}_{\mathcal{F}} = \sqrt{2\frac{M_{\mathcal{S}}}{\hbar^2}\mathcal{E}^{\mathcal{S}}_{\mathcal{F}}}$, for the $m\prime$−eigenmode, we now have

$$\begin{bmatrix} -\partial^2_x - \mathsf{K}^2_+ & \frac{2\,M_{\mathcal{S}}}{\hbar^2}\triangle \\ \frac{2\,M_{\mathcal{S}}}{\hbar^2}\triangle^\star & -(-\partial^2_x - \mathsf{K}^2_-) \end{bmatrix} \begin{bmatrix} \mathcal{G}^{e\mathcal{S}}_{m\prime} \\ \mathcal{G}^{h\mathcal{S}}_{m\prime} \end{bmatrix} = \begin{bmatrix} q^{e(h)\,2}_{m\prime} - \mathsf{K}^2_+ & \frac{2\,M_{\mathcal{S}}}{\hbar^2}\triangle \\ \frac{2\,M_{\mathcal{S}}}{\hbar^2}\triangle^\star & -\left(q^{e(h)\,2}_{m\prime} - \mathsf{K}^2_-\right) \end{bmatrix} \begin{bmatrix} \gamma_{e(h)} \\ \frac{1}{\sqrt{2}} \end{bmatrix} = 0\,. \tag{3}$$

To find solutions for $\gamma_{e(h)}$, we compute the secular equation yielding (with $|\gamma_{e(h)}| = 1/\sqrt{2}$):

$$\gamma_{e(h)} = \frac{1}{\sqrt{2}}\, e^{\pm\imath \arccos(\mathcal{E}/|\triangle|)}, \qquad q^{e(h)}_m = \mathcal{K}^{\mathcal{S}}_{\mathcal{F}}\sqrt{\tilde{\mathcal{Q}}^{\mathcal{S}}_m}\, e^{\frac{\imath}{2}(\mathsf{q}^{e(h)}_m + \frac{\theta}{2})}, \qquad \tilde{k}^{e(h)\mathcal{S}}_m = \left[1 - (+)\frac{\mathcal{E}}{\mathcal{E}^{\mathcal{S}}_{\mathcal{F}}} - \left(\frac{m\pi}{\mathcal{K}^{\mathcal{S}}_{\mathcal{F}}\, w}\right)^2\right],$$
$$\tilde{\mathcal{Q}}^{\mathcal{S}}_m = \sqrt{\left(\frac{\triangle}{\mathcal{E}^{\mathcal{S}}_{\mathcal{F}}}\right)^2 + \tilde{k}^{e\mathcal{S}}_m\, \tilde{k}^{h\mathcal{S}}_m}, \quad \mathsf{q}^e_m = \arctan\left[\frac{\left(\frac{\triangle}{\mathcal{E}^{\mathcal{S}}_{\mathcal{F}}} - \tilde{k}^{e\mathcal{S}}_m\right)}{\left(\frac{\triangle}{\mathcal{E}^{\mathcal{S}}_{\mathcal{F}}} + \tilde{k}^{e\mathcal{S}}_m\right)}\tan\frac{\theta}{2}\right], \quad \mathsf{q}^h_m = \arctan\left[\frac{\left(\frac{\triangle}{\mathcal{E}^{\mathcal{S}}_{\mathcal{F}}} + \tilde{k}^{h\mathcal{S}}_m\right)}{\left(\frac{\triangle}{\mathcal{E}^{\mathcal{S}}_{\mathcal{F}}} - \tilde{k}^{h\mathcal{S}}_m\right)}\tan\frac{\theta}{2}\right] - \pi\,. \tag{4}$$

Defining $\lambda = \frac{M_{\mathcal{N}}}{M_{\mathcal{S}}}$ and $\mathcal{Z} = 2\frac{M_{\mathcal{N}}}{\hbar^2}U_o$, we now search for eigenvalues imposing the continuity of the wave function and a step-like discontinuity of the wave function derivative, at the section,

$$\left[\Psi^{\mathcal{N}} - \Psi^{\mathcal{S}}\right]_\Sigma = 0 \quad , \quad \left[\partial_x\left(\Psi^{\mathcal{N}} - \lambda\,\Psi^{\mathcal{S}}\right) - \mathcal{Z}\,\Psi^{\mathcal{S}(\mathcal{N})}\right]_\Sigma = 0\,. \tag{5}$$

With $x\prime = x\prime\prime = 0$ on the section, we obtain the energy Green function $\mathcal{G}_m(\mathcal{E})$ imposing that at the eigenenergy:

$$\langle m|\mathcal{G}(\mathcal{E})|m\rangle = \frac{1}{\sqrt{2}}\sum_{mm\prime}\int_{\Sigma}\langle m|y\rangle\left[\gamma_e\,\mathcal{D}^{ee}_{mm}\,\mathcal{D}^{hh}_{m\prime m\prime} - \gamma_h\,\mathcal{D}^{he}_{m\prime m}\,\mathcal{D}^{eh}_{mm\prime}\right]\langle y|m\prime\rangle\,dy \equiv 0\,. \tag{6}$$

Since $\int_{\Sigma}\langle m|x_2\rangle\langle x_2|m\prime\rangle dx_2 = \delta_{mm\prime}$, with $\mathcal{D}^{\alpha\beta}_{mm} = \det\begin{vmatrix} \mathcal{G}^{\alpha\,\mathcal{N}}_m & \mathcal{G}^{\beta\,\mathcal{S}}_m \\ \mathcal{G}^{\alpha\,\mathcal{N}}_{m}{}_\prime & (\mathcal{Z}\mathcal{G}^{\beta\,\mathcal{S}}_n + \lambda\mathcal{G}^{\beta\,\mathcal{S}}_{m}{}_\prime) \end{vmatrix}$; for $\alpha = (e\,,\,h)$ and $\beta = (e\,,\,h)$. The next step consists to rewrite the energy Green function in terms of the scattering matrix. We will see that this strategy is useful when searching for a semiclassical description.

3. The Classical Orbits

We define $\mathcal{P}_m = \imath\frac{\lambda^2\sin\theta}{2\,K_m\Gamma_m^2}\,\mathcal{Z}^e_m\mathcal{Z}^h_m e^{\imath(k^-_m a+[\mathsf{z}^e_m+\mathsf{z}^h_m]-\theta)}\,[\cos(k^-_m a) - \cos(k^+_m a)]$ with $k^{\mp}_m = k^e_m \mp k^h_m$, $\mathsf{z}^{e(h)}_m = \arctan\left\{\left(\frac{\Gamma_m-\mathcal{Z}}{\Gamma_m+\mathcal{Z}}\right)\tan\frac{1}{4}(\mathsf{q}^{e(h)}_m + \frac{\theta}{2} - (+)\pi)\right\}$, $\mathcal{Z}^{e(h)}_m = \sqrt{\mathcal{Z}^2 + \Gamma_m^2 + 2\mathcal{Z}\Gamma_m\cos\frac{1}{2}(\mathsf{q}^{e(h)}_m + \frac{\theta}{2} - (+)\pi)}$, $K_m = k^e_m k^h_m$, $\Gamma_m = \lambda\mathcal{K}^{\mathcal{S}}_{\mathcal{F}}\sqrt{\tilde{\mathcal{Q}}^{\mathcal{S}}_m}$, and rewrite Eq. (6) as a function of the scattering matrix's terms of reflection $(r_m, \tilde{r}_m)$ and transmission$(t_m, \tilde{t}_m)$ [5]

$$\mathcal{G}_m(\mathcal{E}) \;=\; \mathcal{P}_m \det|I - S_m| = [\mathcal{P}_m - \mathcal{P}_m\,(r_m + \tilde{r}_m) + \mathcal{P}_m\,r_m\,\tilde{r}_m - \mathcal{P}_m\,t_m\,\tilde{t}_m]\,, \tag{7}$$

yielding $\mathsf{r}^{-(+)}_m = \arctan\left[\left(\frac{k^e_m+(-)\,k^h_m}{k^e_m-(+)\,k^h_m}\right)\tan\theta\right]$, $\mathcal{R}^{-(+)}_m = \sqrt{[k^e_m]^2 + [k^h_m]^2 - (+)\,2\,K_m\,\cos 2\theta}$,

$\mathsf{d}^{-(+)}_m = \arctan\left[\frac{\left(\Gamma_m\mathcal{R}^{-(+)}_m\cos(\frac{1}{4}[\mathsf{q}^h_m-\mathsf{q}^e_m]+\mathsf{r}^{-(+)}_m)-\mathcal{Z}\,k^{+(-)}_m\sin\theta\right)}{\left(\Gamma_m\mathcal{R}^{-(+)}_m\cos(\frac{1}{4}[\mathsf{q}^h_m-\mathsf{q}^e_m]+\mathsf{r}^{-(+)}_m)+\mathcal{Z}\,k^{+(-)}_m\sin\theta\right)}\tan\frac{\theta}{2}\right]$, $\varpi^{+(-)}_m = \mathcal{Z}\,k^{+(-)}_m\,\sin\theta$,

$\Xi^{-(+)}_m = \Gamma_m\,\mathcal{R}^{-(+)}_m\,\cos(\frac{1}{4}[\mathsf{q}^h_m - \mathsf{q}^e_m] + \mathsf{r}^{-(+)}_m)$, $\mathcal{D}^{-(+)}_m = \sqrt{\left(\varpi^{+(-)}_m\right)^2 + \left(\Xi^{-(+)}_m\right)^2 + 2\,\varpi^{+(-)}_m\,\Xi^{-(+)}_m\,\cos\theta}$,

$$\begin{aligned} -\mathcal{P}_m\,(r_m + \tilde{r}_m) &= \imath\frac{\lambda^2}{2\,K_m\,\Gamma_m^2}\left\{\mathcal{D}^-_m\,e^{\imath(k^-_m a+\mathsf{d}^-_m-\frac{3}{2}\theta)}\sin(k^+_m a) - \mathcal{D}^+_m\,e^{\imath(k^-_m a+\mathsf{d}^+_m-\frac{3}{2}\theta)}\sin(k^-_m a)\right\}, \\ \mathcal{P}_m\,r_m\,\tilde{r}_m - \mathcal{P}_m\,t_m\,\tilde{t}_m &= \imath\frac{\lambda^2\sin\theta}{2\,\Gamma_m^2}\,e^{\imath(k^-_m a-2\theta)}\,[\cos(k^-_m a) + \cos(k^+_m a)]\,. \end{aligned}$$

Here we found that the scattering reflection contributions are diminished by the scattering transmission contributions, taking account of the quantum capability of the wave function to tunnel.

4. Leading order approximations

We consider all the eigenmode contributions to compute the density of states,

$$\begin{aligned} \rho(\mathcal{E}) = -\tfrac{1}{\pi}\textstyle\sum_m\mathcal{G}_m(\mathcal{E}) \;&=\; -\tfrac{1}{\pi}\textstyle\sum_m\frac{\lambda^2\sin\theta}{4\,K_m\,\Gamma_m^2}\,\{\Phi_{1m}\cos(2\theta+\Theta_{1m}) - \Phi_{2m}\cos(2k^-_m a - 2\theta - \Theta_{2m}) \\ &-\; \Phi_{3m}\cos([k^-_m - k^+_m]a - 2\theta - \Theta_{3m}) + \Phi_{4m}\cos([k^-_m + k^+_m]a - 2\theta - \Theta_{4m})\}\,. \end{aligned} \tag{8}$$

We now analyze the phase of the respective contributions in the above expression. Even the short trajectory is affected by quantum tunneling, the phase 2θ appears in all contributing terms in $\rho(\mathcal{E})$. The second term is a long trajectory which is influenced by Andreev retroreflections $2k^-_m a\left(= 2[k^e_m - k^h_m]\,a\right)$, where both electron and hole particles travel the same length $(2a)$ in opposite directions while bouncing inside the conductor, see Figure 1.b. The last two terms are influenced by Andreed retroreflections and specular-reflections represented by $k^+_m a\left(= [k^e_m + k^h_m]\,a\right)$, respectively. In particular, the third term considers contributions of two bouncing electrons traveling the same length a; one in a clockwise and another in a counterclockwise fashion. Meanwhile the hole travels twice the width of the conductor, it is shown in Figure 1.c. Exchanging electron by hole the physical scenario presented before can be applied to the fourth term, as depicted in Figure 1.d. We find that every phase contribution is also affected by a term Θ_{im} taking into account the fact that the waves functions can go further into the superconductor region. In fact this is a correction applied to the usual tunneling term θ, since it depends on it and the superconductor material physical parameters. Finally, the amplitudes Φ_{im} are related to the particle's path length distribution[3, 6, 7],

Aiming at a better understanding of the semiclassical framework, we make the following approximations: $\frac{\tilde{\mathcal{A}}^{\pm}_{2m}\sin\theta}{\tilde{\mathcal{A}}^{\mp}_{2m}\cos(\theta-\tilde{\alpha}^{\mp}_{1m})} < 1$, $\frac{\tilde{\mathcal{B}}^{\pm}_{2m}\sin\theta}{\tilde{\mathcal{B}}^{\mp}_{2m}\cos(\theta-\tilde{\beta}^{\mp}_{1m})} < 1$ and $\mathcal{Z}\left(\frac{\triangle/\mathcal{E}^{\mathcal{S}}_{\mathcal{F}}}{\mathcal{Q}^{\mathcal{S}}_{m}}\right) < 1$, and obtain

$$\tilde{\mathcal{A}}^{\mp}_{1m} \approx \sqrt{\left(\Gamma_m k^{+}_{m}\right)^2 + \left(\mathcal{Z}k^{-}_{m} \mp 2\Gamma^2_m \sin^2\theta\right)^2}, \quad \tilde{\mathcal{A}}^{\pm}_{2m} \approx \Gamma_m k^{+}_{m}\cos\theta \pm \left[\left(\Gamma^2_m + \mathcal{Z}^2\right)\cos 2\theta + K_m\right].$$

$$\tilde{\mathcal{B}}^{\mp}_{1m} \approx \sqrt{\left(\Gamma_m k^{-}_{m}\right)^2 + \left(\mathcal{Z}k^{+}_{m} \mp 2\Gamma^2_m \sin^2\theta\right)^2}, \quad \tilde{\mathcal{B}}^{\pm}_{2m} \approx \Gamma_m k^{-}_{m}\cos\theta \pm \left[\left(\Gamma^2_m + \mathcal{Z}^2\right)\cos 2\theta - K_m\right].$$

$$\tilde{\alpha}^{\mp}_{1m} \approx \arccos\left(\frac{\Gamma_m k^{+}_{m}}{\tilde{\mathcal{A}}^{\mp}_{1m}}\right) \quad , \quad \tilde{\beta}^{\mp}_{1m} \approx \arccos\left(\frac{\Gamma_m k^{-}_{m}}{\tilde{\mathcal{B}}^{\mp}_{1m}}\right).$$

$$\Phi_{1m} \approx \tilde{\mathcal{A}}^{-}_{1m}\cos(\theta-\tilde{\alpha}^{-}_{1m}), \Phi_{2m} \approx \tilde{\mathcal{A}}^{+}_{1m}\cos(\theta-\tilde{\alpha}^{+}_{1m}), \Phi_{3m} \approx \tilde{\mathcal{B}}^{-}_{1m}\cos(\theta-\tilde{\beta}^{-}_{1m}), \Phi_{4m} \approx \tilde{\mathcal{B}}^{+}_{1m}\cos(\theta-\tilde{\beta}^{+}_{1m}).$$

$$\Theta_{1m} \approx \arccos\left(\frac{\tilde{\mathcal{A}}^{+}_{2m}\sin\theta}{\Phi_{1m}}\right), \Theta_{2m} \approx \arccos\left(\frac{\tilde{\mathcal{A}}^{-}_{2m}\sin\theta}{\Phi_{2m}}\right), \Theta_{3m} \approx \arccos\left(\frac{\tilde{\mathcal{B}}^{+}_{2m}\sin\theta}{\Phi_{3m}}\right), \Theta_{4m} \approx \arccos\left(\frac{\tilde{\mathcal{B}}^{-}_{2m}\sin\theta}{\Phi_{4m}}\right).$$

In the strong limit $\mathcal{E} << \mathcal{E}_{\mathcal{F}}$, the phases and the amplitudes are approximately equal and the net contribution of the mixture of Andreev retroreflection and specular reflection are negligible, thus retroreflection constitute the most important contribution, and so the *Bohr-Sommerfeld* (BS) formula quantization rule becomes $2\left(k^{e}_{m} - k^{h}_{m}\right)a - 2\theta - \Theta_{2m} = (2n-1)\pi$, with m, n integers. The Θ_{2m} term is not present in the BS expression obtained in reference[6], where both the mass of the particles and the Fermi energy of the $\mathcal{N}-\mathcal{S}$ system are considered indistinguishable and clean.

5. Conclusions

We presented a full quantum theory for the $\mathcal{N}-\mathcal{S}$ system via the quantum section method. It enable us to properly obtain the semiclassical limit. In general, both Andreev retroreflection and Andreev specular reflections determine the spectrum of the system. Our formulation enable us to analyze the B-S quantization method. We found that in the strong limit $\mathcal{E} << \mathcal{E}_{\mathcal{F}}$ the BS formula applies, including the extra term Θ_{2m}. Numerical work in progress may corroborate Cserti *et al.*[6] conjecture, namely: third and fourth contributions in equation (8) cancel each other in the semiclassical limit.

Acknowledgments

JSEO and REL would like to thank the support of the Goiás Research Foundation - FAPEG.

References

[1] Beenakker C W J (2008) Rev. Mod. Phys. **80** 1337
[2] Castro Neto A H, Guinea F, Peres N M R, Novoselov K S and Geim A K (2009) Rev. Mod. Phys. **81** 109
[3] Cserti J, Hagymási I and Kormányos A (2009) Phys. Rev. **B 80** 073404
[4] Bogomolny E B (1992) Nonlinearity **5** 805
[5] Beenakker C W J (2005) Lect. Notes Phys. **667** 131
[6] Cserti J, Bodor A, Koltai J, and Vattay G (2002) Phys. Rev. **B 66** 064528
[7] Libisch F, Rotter S, Burgdörfer J, Kormányos A and Cserti J (2005) Phys. Rev. **B 72** 075304

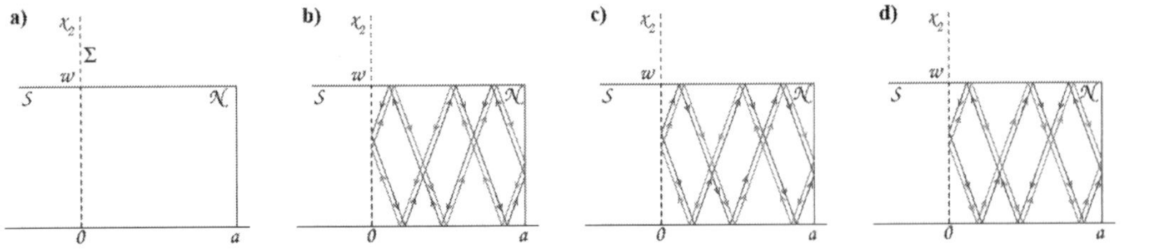

Figure 1. **a)** The Andreev billiard composed of two rectangular regions, one conducting and the another superconducting. The quantum section Σ is defined over the junction at $x_1 = 0$. Different phase contributions of an electron (blue-line) and hole (red-line) into the Andreev billiard. Andreev retroreflections is depicted in **b)**. In **c)**, two electrons traveling in clockwise and counterclockwise directions and one hole. In **d)**, two holes traveling in clockwise and counterclockwise directions and one electron. These two contributions are interpreted as a superposition of Andreev retroreflection $(k_m^e - k_m^h)$ and specular reflections $(k_m^e + k_m^h)$, respectively.

Modeling contact formation between atomic-sized gold tips via molecular dynamics

W Dednam[1,2], C Sabater[1,4], M A Fernandez[1], C Untiedt[1], J J Palacios[3], M J Caturla[1]

1 Departamento de Fisica Aplicada, Universidad de Alicante, San Vicente del Raspeig, E-03690 Alicante, Spain
2 Department of Physics, Science Campus, University of South Africa, Private Bag X6, Florida Park 1710, South Africa
3 Departamento de Fisica de la Materia Condensada, Universidad Autonoma de Madrid, Cantoblanco, Madrid 28049, Spain
[4] Huygens-Kamerlingh Onnes Laboratorium, Leiden University, Niels Bohrweg 2, 2333 CA Leiden, Netherlands

E-mail: wdednam@gmail.com

Abstract. The formation and rupture of atomic-sized contacts is modelled by means of molecular dynamics simulations. Such nano-contacts are realized in scanning tunnelling microscope and mechanically controlled break junction experiments. These instruments routinely measure the conductance across the nano-sized electrodes as they are brought into contact and separated, permitting conductance traces to be recorded that are plots of conductance versus the distance between the electrodes. One interesting feature of the conductance traces is that for some metals and geometric configurations a jump in the value of the conductance is observed right before contact between the electrodes, a phenomenon known as jump-to-contact. This paper considers, from a computational point of view, the dynamics of contact between two gold nano-electrodes. Repeated indentation of the two surfaces on each other is performed in two crystallographic orientations of face-centred cubic gold, namely (001) and (111). Ultimately, the intention is to identify the structures at the atomic level at the moment of first contact between the surfaces, since the value of the conductance is related to the minimum cross-section in the contact region. Conductance values obtained in this way are determined using first principles electronic transport calculations, with atomic configurations taken from the molecular dynamics simulations serving as input structures.

1. Introduction

The microelectronics industry faces a great challenge if it is to continue the trend that has characterized its success over the last few decades [1]. The size of the gate in a transistor is now approaching a few nm in dimension and new strategies must be devised to achieve smaller sizes. Nanotechnology is a promising field and potentially offers a solution to the above-mentioned problem. In this connection, one of the most important tools both in the development and in the characterization of systems at the nano-scale is the scanning tunnelling microscope (STM) [2]. With the aid of an STM or mechanically controllable break-junction (MCBJ), another related technique, we can study the electronic transport in few-atom systems [3]. Such studies typically require two electrodes of the same metal and a piezo system to join and separate the electrodes with subatomic precision. By also connecting a battery across the electrodes, the current through the nano-contact can be measured, usually in units of the quantum of conductance ($G_0 = 2e^2/h$) [4].

The electronic transport through atomic-sized contacts can be interpreted from a plot of the measured conductance versus the displacement between the electrodes. Fig. 1 shows experimental data for gold electrodes at 4.2K; the blue curve corresponds to separation of the electrodes, and the red curve to the process of bringing them into contact. These curves are usually referred to as conductance traces. The insets in the figure help to visualise the contact formation and rupture processes. Moreover, the traces in the figure exhibit a jump in conductance [5] that is close to 1 G_0 in value. In this case, the rupture trace corresponds to a transition from a monatomic contact to a complete loss of contact. In the case of the formation trace we observe a jump in the conductance that corresponds to a transition from tunnelling current to first atomic contact [6].

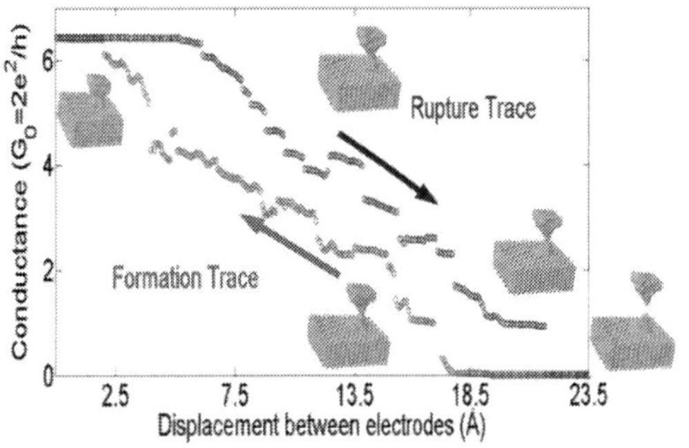

Figure 1. (Colour online) Experimental conductance measurements recorded at 4.2 K using gold electrodes. The insets illustrate the geometry of the electrodes as contact is made or broken.

A disadvantage of the above experimental techniques is the difficulty of imaging the electrodes at atomic resolution. Therefore, theoretical modelling can prove useful in obtaining more information about the geometry of the electrodes and the mechanism of contact formation.

2. Theory

Molecular dynamics (MD) simulations [7] based on empirical potentials can serve to model nano-contact formation and rupture. However, MD relies on interatomic potentials that are fitted to a limited number of parameters of the material. So, doubts remain as to the reliability of the potentials when systems are taken to situations far away from the range where the parameters have been fitted. Also, there is no description of the electrons in MD and it is thus not possible to obtain information such as the conductance of the nano-contacts.

On the other hand, DFT calculations [8] have the advantage of being able to describe both the nuclei and the electrons in the system, with the use of certain approximations, but without the need of fitting parameters. Electronic transport calculations based on DFT permit the conductance to be obtained to a high level of accuracy in materials such as gold [9]. However, these calculations are computationally costly and systems of up to only a few tens of atoms can be conveniently handled.

We use a combination of these two methods to study the dynamical formation and breaking of gold nano-contacts and to calculate the conductance through the most probable structures at the moment of first contact.

The MD code LAMMPS [10, 11] is used to perform the dynamical calculations. The embedding atom potential [12] for gold developed by Zhou *et al.* [13] serves to model the interactions between the atoms. This potential has been fitted to parameters such as lattice constants, elastic constants, bulk moduli, vacancy formation energies, sublimation energies, and heats of solution [13].

Note that it has been necessary to modify the LAMMPS source code to enable continuous breaking and making of nano-contacts. The simulation structure is divided along its length (z-axis) into evenly-spaced bins via a spatial averaging *fix* that comes standard with LAMMPS. Following Ref. [14], the choice of bin size (height) is the interplanar distance between close-packed layers (2.04 Å along (001) and 2.3556 Å along (111)). Special functions have been added to the *variable* subroutine in LAMMPS to find the bin containing the fewest atoms and the number of atoms in it. Each rupture-formation simulation is repeated over twenty cycles and paused every 10000 fs to determine the least-atom bin number and the number of atoms therein.

System sizes ranging from 500 up to almost 3000 atoms and oriented along two different crystallographic directions, (001) and (111), have been considered. The initial configuration consists of a neck type structure (see Fig. 2). In the case of a narrow neck (e.g., Fig. 2(a)), the direction of motion during each rupture-formation cycle is reversed when there are a minimum of 5 atoms in the least-atom bin. In the case of wide neck (e.g., Fig 2(b)), on the other hand, the cycles are reversed when the least-atom bin contains 7 atoms. Each rupture or formation cycle is continued for a further 1000 fs after these conditions have been met to ensure that the contact has actually ruptured or that the least-atom bin indeed contains the predefined minimum number of atoms.

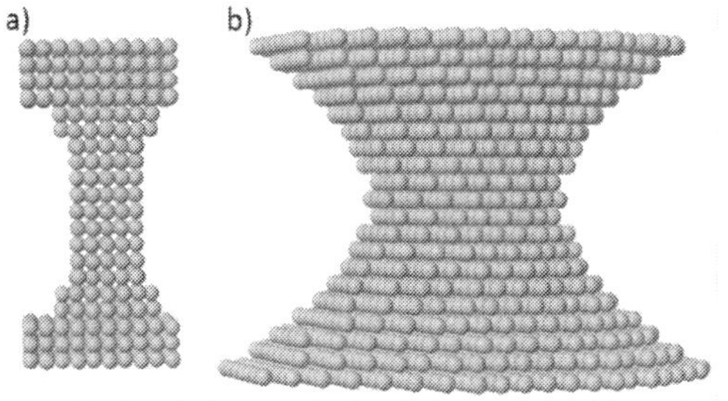

Figure 2. (Colour online) Two of the initial configurations used in the simulations, left panel a) 525 atoms of gold, right panel b) 2804 atoms. Both are oriented along the (001) crystallographic direction in this case.

The first two top and bottom layers of all the simulated structures are frozen and displaced in opposite directions by 0.0041 Å and 0.0047 Å every 1000 fs for structures in the (001) and (111) crystallographic directions, respectively. The velocity of deformation in simulations (0.41 and 0.47 m/s along (001) and (111), respectively) are within the limits that can be computed by MD - about four orders of magnitude below the speed of sound in gold (3000 m/s) (See Ref. [15] and the references therein). In order to keep the temperature constant throughout the calculation, a Nose-Hoover thermostat is applied [16, 17].

For the formation cycles, a Fortran subroutine is used to calculate the minimum distance between the two opposing electrodes. Since the electrodes are allowed to separate to about 5 Å, the atoms belonging to the top fragment are labelled "1" and those belonging to the bottom fragment "2". The first two distinctly labelled atoms ("1" and "2", respectively) to come within a distance halfway between first and second nearest neighbours during contact formation cycles, are then judged to have made contact first.

Subsequently, the conductance of selected structures at the moment of first contact - taken from the MD simulation trajectories - are computed by means of the electronic transport code ALACANT [18, 19, 20], using a minimal basis set that assigns only 1 orbital to each atom. It is well known that conductance values obtained from electronic transport calculations on few-atom nano-contacts, are determined by the few atoms in the minimum cross-section of the contact. Therefore, in order to save time on DFT calculations, we remove a few of the outer layers from each electrode.

Journal of Physics: Conference Series **574** (2015) 012045 doi:10.1088/1742-6596/574/1/012045

3. Results and Discussion

3.1. Molecular Dynamics Simulations

Figures 3(a), (b), (c) and (d) show the MD formation traces of the four structures considered here. In all cases we observe that the first few cycles of formation (labelled "run1", "run2", etc.) require more steps before first contact, with the exception of "run13" in Fig. 3(c). All the traces clearly exhibit a jump in distance at the moment of first contact between the electrodes. This jump is reminiscent of those observed in experimental conductance traces [5]. We also see that after a number of cycles the traces all overlap, indicating that the geometric configuration of the contact repeats in a regular fashion. When the traces repeat in this way we can talk about a process of mechanical annealing or, in other words, that we now have reproducible contacts with sharpened tips.

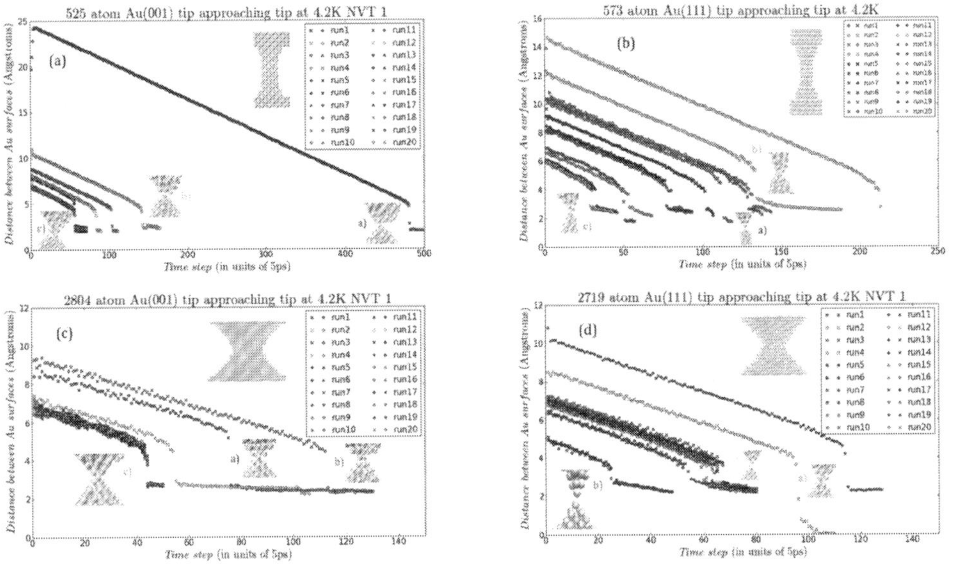

Figure 3. (Color online) Traces of distance between electrodes versus time for a system of: (a) 525 atoms oriented along the (001) direction. Insets a, b and c correspond to the geometries of the electrodes at the moment of first contact for formation cycles labelled run1, run2 and run18, respectively; (b) 573 atoms along (111). Insets a, b and c correspond to the first-contact geometries of formation cycles run1, run4 and run20, respectively; (c) 2804 atoms along (001). Insets a, b and c correspond to the first-contact geometries of formation cycles run1, run3 and run20, respectively; (d) 2719 atoms along (111). Insets a, b and c correspond to the first-contact geometries of formation cycles run4, run8 and run12, respectively.

Traces 1 to 6 in Fig. 3(a) (525 atoms, (001) orientation) differ significantly from each other: the jump occurs at different time steps, and its height also differs from trace to trace. However, after run 7 all traces up to run 20 are nearly the same: the jumps occur at the same time step and are nearly equal in height. The insets in Fig. 3(a) show the atomic configurations right before contact for several traces. In many cases contact is made through a single atom, or *monomer*; in other cases, as in inset c, contact is made through two adjacent atoms, or *double contact*. There is greater diversity among the traces in Fig. 3(b) (573 atoms, (111) orientation), both in terms of the time taken for the jump to occur and its height. Here, it takes at least 10 indentation cycles before the traces start to overlap. When the traces are stable, the jump to contact occurs through a monomer (inset c). However, there are also cases where the jump

Journal of Physics: Conference Series **574** (2015) 012045 doi:10.1088/1742-6596/574/1/012045

occurs through a *dimer* (inset a) and more than one atom (inset b). In Fig. 3(c) (2804 atoms, (001)), the traces overlap after only three cycles. Note that the aspect ratio of this neck (the length of the neck divided by the diameter of its minimum cross-section) is lower ($\sim 2:1$) than that of the structure in Fig. 3(a) ($\sim 5:1$). As in Fig. 3(b), the traces in Fig. 3(d) (2719 atoms, (111)) exhibit greater diversity. The most energetically stable surface layers of a face-centred cubic (FCC) crystal are oriented along the (111) crystallographic direction, since atoms in the exposed layers have more nearest neighbours than in any other direction. Because (111) layers form on the oblique faces of the pyramid shaped electrodes that are oriented along (001), e.g. in Figs. 3(a) and 3(c), it is easier to "cold" anneal surfaces in this direction, requiring less than half the number of cycles than is needed to achieve reproducible structures along (111). The insets in Fig. 3(d) show that contact occurs more often through a vertical dimer than through a monomer, and double atomic contacts also occur.

3.2. Results of the DFT calculations on the MD-generated contacts

Following Refs. [5, 21], we have selected three representative (111) contact geometries from the MD atomic configurations at the moment of first contact: the monomer in inset b and the double contact in inset c in Fig. 3(b) and the dimer in inset c in Fig. 3(d). We have calculated the conductance of these structures since the conductance of their counterparts along (001) have already been calculated elsewhere [21]. The results are shown in Fig. 4 and the conductance values are expressed in units of G_0. The crystallographic orientation and cycle of formation of the three configurations are also indicated, as well as the time step of and distance to first contact.

Type of contact, conductance and distance		
Monomer	Dimer	Double Contact
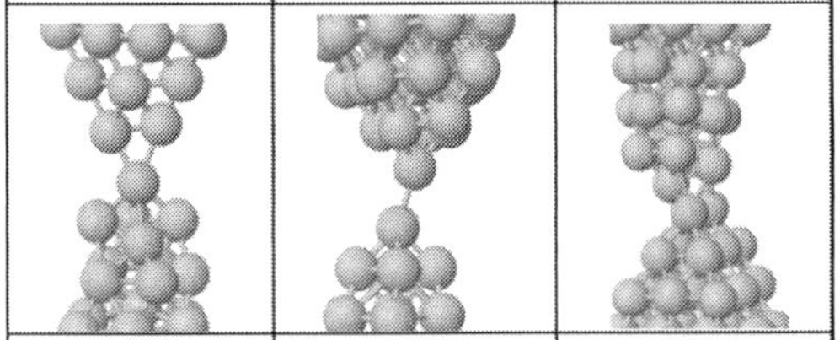		
1.60 G_0	0.87 G_0	2.04 G_0
Au(111) Run20 step 150000	Au(111) Run12 step 320000	Au(111) Run04 step 665000
First contact distance 3.00 Å	First contact distance 2.92 Å	First contact distance 3.34 Å

Figure 4. (Colour online) Different types of contact: monomer, dimer and double contact and their corresponding conductances. The distance to first contact in each case is also shown.

We should expect the conductance of the monomer to be between that of the dimer and the double contact, since the tunnelling current across atoms just above and below the monomer ought to be higher than the dimer's – because of the closer proximity of the atoms around the monomer. It appears that neighbouring atoms in the minimum cross-section make a substantial contribution to conductance in the form of tunnelling current, which could explain the high value of 1.60 G_0 for the monomer. The size of the distance to first contact is also consistent with the type of contact under consideration. Thus, one would expect the jump to be smallest in the case of the dimer – the approach of two single atoms towards each other – since the oncoming atoms "see" fewer atoms, and neighbours on the same electrode effectively keep them from jumping to the opposing electrode for longer than would be expected. This contrasts with the case of a single atom feeling the attractive force of more than one atom on the opposing tip, e.g., the monomer in Fig. 4. The force of attraction between more than one atom on both opposing electrodes, e.g., the double contact in Fig. 4, ought to be the greatest and hence so should the jump to contact, as is indeed the case.

4. Conclusion

We have shown previously [22, 23] that when repeated indentation is performed on structures that are oriented in the (001) direction, traces of *minimum cross-section* versus time overlap perfectly. This has been explained as a process of mechanical annealing of the electrodes that results in very sharp and stable tips [22]. Here we observe the same behaviour, repetition of the traces after several cycles of indentation, but of the *minimum distance* between the electrodes vs time. Moreover, we show that electrodes oriented along (111) can also be mechanically annealed. In contrast to (001) electrodes, the traces of (111) electrodes take longer to repeat in a regular way, apparently due to the stability of the exposed (111) surface layers on the oblique faces of (001) structures. In the case of (111) electrodes we also find that the final jump to contact occurs predominantly through a monomer or a dimer, and to a lesser extent, through a double contact.

The conductance values we obtain are in agreement with previous works [21]. However, in the case of the monomer it is higher than expected. Instead of using a minimal one-electron basis set, it might be necessary to use an eleven-electron basis set to improve this result [22].

Acknowledgments

In compiling this paper, W. Dednam has made extensive use of python scripts written by Prof André E. Botha, who kindly shared them. W. Dednam acknowledges support from the National Research Foundation of South Africa through the Scarce Skills Masters scholarship funding programme. This work is supported by the Generalitat Valenciana through grant reference PROMETEO2012/011 and the Spanish government through grant FIS2010-21883. This work is part of the research programme of the Foundation for Fundamental Research on Matter (FOM), which is financially supported by the Netherlands Organisation for Scientific Research (NWO).

References

[1] Moore G E 1965 *Electronics Magazine* **38** 114
[2] Binnig G, Roher H, Gerber C and Weibel E 1982 *Phys. Rev. Lett.* **49** 57
[3] Agraït N, Levy-Yeyati A and van Ruitenbeek J M 2003 *Phys. Rep.* **377** 81
[4] van Wees B J *et al.* 1988 *Phys. Rev. Lett.* **60** 848
[5] Untiedt C *et al.* 2007 *Phys. Rev. Lett.* **98** 206801
[6] Gimzewski J K, Möller R, Pohl D W and Schlittler R R 1987 *Surface Science* **189–190** 15–23
[7] Allen M P and Tildesley D J 1986 *Computer Simulation of Liquids* (Oxford: Oxford Science Publications)
[8] Hohenberg P and Kohn W 1964 *Phys. Rev.* **136** B864–B871
[9] Palacios J J, Pérez-Jiménez A J, Louis E, SanFabián E and Vergés J A 2002 *Phys. Rev. B* **66** 035322
[10] Plimpton S 1995 *J. Comp. Phys.* **117** 1
[11] Lammps: Large-scale atomic/molecular massively parallel simulator `http://lammps.sandia.gov` accessed: 2014-09-12
[12] Daw M and Baskes M 1983 *Phys. Rev. Lett.* **50** 1285
[13] Zhou X W *et al.* 2001 *Acta Materialia* **49** 4005
[14] Takahashi A, Kurokawa S and Sakai A 2014 *Phys. Status Solidi B* **251** 1363
[15] Dreher M, Pauly F, Heurich J, Cuevas J C, Scheer E and Nielaba P 2005 *Phys. Rev. B* **72** 075435
[16] Nosé S 1984 *Molecular Physics* **52** 255
[17] Hoover W G 1985 *Phys. Rev. A* **31** 1695
[18]
[19] Frisch M J *et al.* Gaussian 09 Revision a.01 gaussian Inc. Wallingford CT 2009
[20] Palacios J J *et al.* Alicante atomistic computation applied to nanotransport `http://alacant.dfa.ua.es` accessed: 2014-09-12
[21] Sabater C, Caturla M J, Palacios J J and Untiedt C 2013 *Nanoscale Research Letters* **8** 257
[22] Sabater C, Untiedt C, Palacios J J and Caturla M J 2012 *Phys. Rev. Lett.* **108** 205502
[23] Sabater C and Scixell Mechanical annealing of metallic electrodes at the atomic scale publicly available at https://www.youtube.com/watch?v=OlZekZNbKec

First-principle studies of phonons and thermal properties of AlN in wurtzite structure

J Q Fu, T L Song[1], X X Liang and G J Zhao

Department of Physics, School of Physical Science and Technology, Inner Mongolia University, Hohhot, 010021, P. R. China

E-mail: songtl@126.com

Abstract. We calculate the band structure, density of states, phonon dispersions and thermodynamic properties of wurtzite Aluminum nitride (AlN) using the local density approximation (LDA) and the generalized gradient approximation (GGA). The results show that wurtzite AlN is a direct gap semiconductor. The phonon, dielectric, and thermodynamic properties are discussed in detail. The calculated values are in agreement with available experimental data.

1. Introduction

Group-III nitrides have attracted considerable attention during the past decade due to technological applications in optoelectronic and electronic devices. Aluminum nitride (AlN) is one of the important materials among them. It has high thermal conductivity, high melting point, large bulk modulus, and large band gap. Attention has also been paid to the lattice vibrations (phonons), which related closely to their electric, optical as well as thermal properties. Some investigations have given the phonon dispersions and phonon frequencies at the Γ point of hexagonal AlN by the first-principles [1, 2, 3].

In this work, we have performed the first-principle calculations for the band structure, density of states, phonon dispersion, dielectric and thermodynamic properties of hexagonal AlN based on two different exchange-correlation potentials (GGA and LDA).

2. Calculation method

Wurtzite AlN belongs to the space group P63mc and its unit cell contains four atoms (see figure 1). In our calculation, the software Quantum-Espresso [4] is used, which is an integrated suite of Open-Source computer codes for electronic-structure calculations and materials modeling at the nanoscale based on density functional theory. Two different exchange-correlation potentials (GGA and LDA) have been adopted and the energy cut-off is chosen as 105Ry for GGA and 75Ry for LDA, respectively, in the band structure calculation. A $10\times10\times10$ Monkhorst-Pack mesh have been used to perform the Brillouin zone (BZ) integration. Phonon calculation is presently a two-step process. First, find the ground-state atomic and electronic configuration; Second, calculate phonons using Density Functional Perturbation Theory. In order to obtain the phonon dispersion curves, we have selected a $4\times4\times4$ q mesh for the force constant calculation.

[1] Corresponding author E-mail address: songtl@126.com

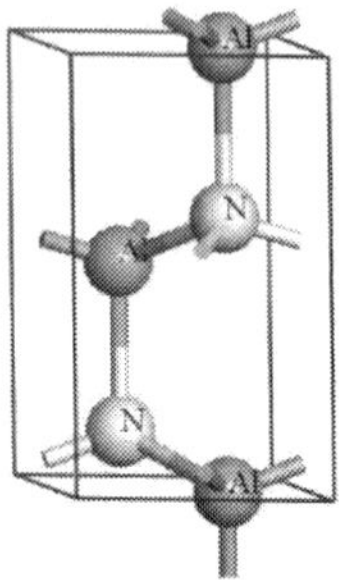

Figure 1. Schematic of the primitive cell of wurtzite AlN after the optimization

3. Result and discussion

3.1. Lattice parameters

It is read from table 1 that our results for the optimized lattice parameters very closer to the previous theoretical [5, 6] and experimental data [7]. However, the corresponding values by LDA is slightly smaller than the experimental values but those by GGA larger.

Table 1. The lattice parameters of wurtzite AlN compare with other theories and the experimental data.

		a(Å)	c(Å)	a/c
AlN	This work^{GGA}	3.120	5.000	1.606
	This work^{LDA}	3.090	4.910	1.589
	Theory[5]	3.100	5.100	1.615
	Theory[6]	3.096	4.959	1.602
	Experiment[7]	3.110	4.980	1.601

3.2. Band structures and density of states

Figures 2 and 3 show wurtzite AlN band structures and density of states (DOS) obtained by GGA and LDA, respectively. The results indicate that the AlN is the direct band gap semiconductor. The band gap for GGA is 4.2eV and for LDA is 4.5eV. Compared with the experimental value 6.2eV, the LDA results are better than the GGA, but they are all lower than experimental value. The band splits up into 3 sub-bands. We can also see that the valence band are mostly dominated by N-2s,2p and A1-3s electronic states, but the conduction band comes from Al-3s, 3p and a little N-2s,2p electron.

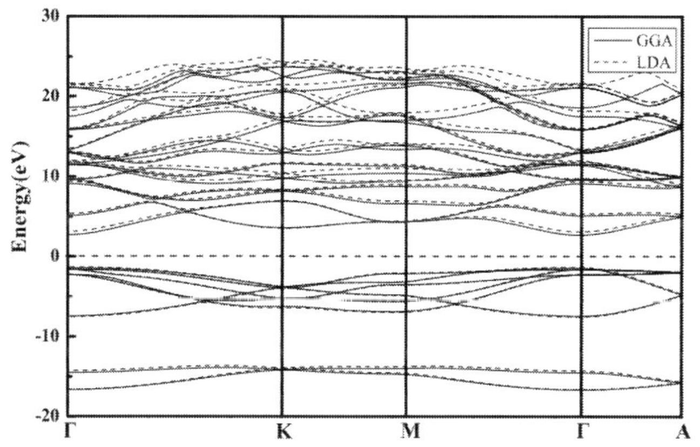

Figure 2. Band structure of wurtzite AlN obtained by GGA (solid line) and LDA (dash line).

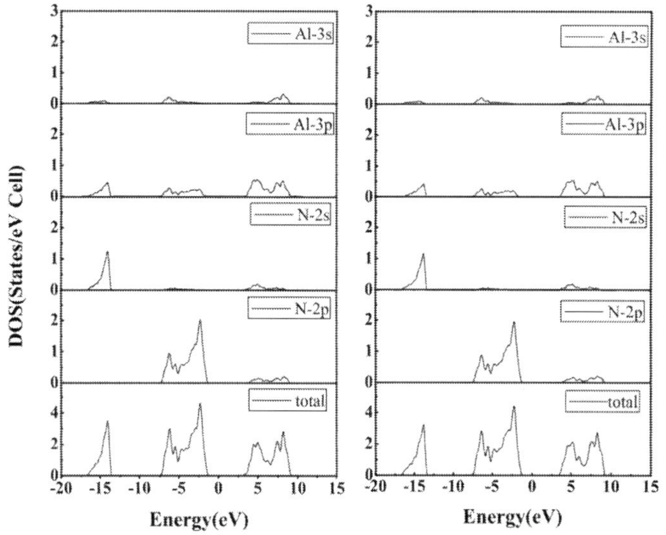

Figure 3. Total and projected DOSs of wurtzite AlN by (a) GGA and (b) LDA

3.3. Phonon dispersion curves

The unit cell of wurtzite AlN contains 4 atoms. We can see from figure 4 that there are totally 12 phonon branches containing 3 acoustic phonon branches and 9 optical phonon branches. The phonon modes degenerate at some high symmetry points of BZ. For example, at A point, only 4 phonon modes can be observed, but 8 modes appear at the Γ point. The calculated phonon frequencies at the Γ point in this work are listed in table 2. The corresponding results given by previous theories are also listed partly for comparison. The calculated results are closer to the available experimental values.

$\{\omega[E_1(\mathrm{TO})]-\omega[A_1(\mathrm{TO})]\}/\omega[E_1(\mathrm{TO})]$ could be used to measure the anisotropy of materials of wurtzite structure. The calculated value is 0.09 for GGA and 0.06 for LDA, respectively. It is seen that the wurtzite AlN displays obviously the anisotropy.

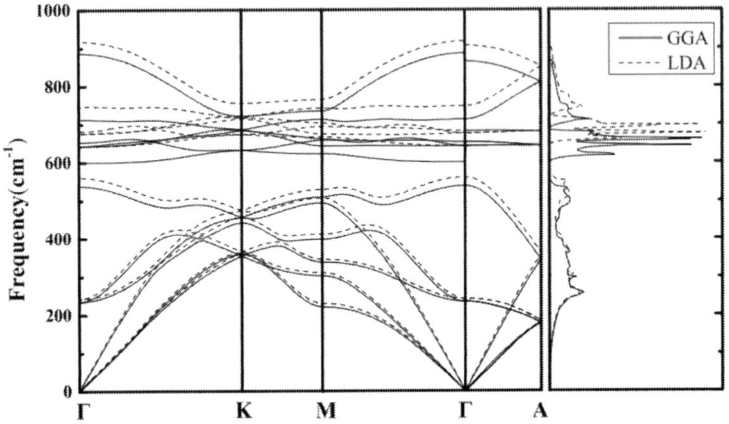

Figure 4. Phonon dispersion curves (left) and DOS (right) of wurtzite AlN by GGA (solid line) and LDA (dash line).

Table 2. The phonon frequencies at the Γ point for wurtzite AlN by GGA and LDA as well as some other theoretical and experimental data.

	E_2^l	B_1^l	A_1(TO)	E_1(TO)	E_2^h	B_1^h	A_1(LO)	E_1(LO)
This work^{GGA}	238	538	599	655	645	711	866	887
This work^{LDA}	241	560	643	681	675	749	905	918
Theory[1]	215	557	640	668	652	764	883	922
Theory[5]	–	–	619	677	–	–	893	918
Experiment[2]	248	–	614	673	–	–	894	917
Experiment[3]	241	–	607	–	660	–	–	924

3.4. Dielectric properties

We have calculated the high-frequency dielectric tensor and the Born effective charge tensors. The calculated results are listed in table 3. For wurtzite AlN, $\varepsilon(\infty)$ is composed of two independent components, corresponding respectively to the directions parallel to the c axis, $\varepsilon_{\parallel}(\infty)=\varepsilon_{zz}(\infty)$, and perpendicular to the c axis, $\varepsilon_{\perp}(\infty)=\varepsilon_{xx}(\infty)=\varepsilon_{yy}(\infty)$. The average dielectric constants $\varepsilon(\infty)=1/3\mathrm{Tr}\varepsilon(\infty)$ is also given. We also consider Born effective charge tensors with two independent components $Z_{\perp}^{*}=Z_{xx}^{*}=Z_{yy}^{*}$ and $Z_{\parallel}^{*}=Z_{zz}^{*}$. It is seen that the wurtzite AlN displays obviously the anisotropy in the dielectric constant and the Born effective charge. The our results for the high-frequency dielectric tensors are in agreement with the experimental data.

Table 3. The born effective charge and high-frequency dielectric tensors

	$Z^*_\perp$	$Z^*_\parallel$	Z^*	$\varepsilon_\perp(\infty)$	$\varepsilon_\parallel(\infty)$	$\varepsilon(\infty)$
This work[GGA]	2.50	2.66	2.56	4.40	4.64	4.47
This work[LDA]	2.53	2.67	2.57	4.30	4.58	4.45
Theory[8]	2.53	2.69	2.58	4.38	4.61	4.46
Theory[9]	–	2.70	–	4.42	4.70	4.51
Theory[10]	–	–	–	4.4	4.8	–
Experiment[11]	–	–	–	–	–	4.68

3.5. Thermodynamic properties

To understand thermodynamic properties we have performed the computations for some main thermodynamic functions, such as the inner energy, free energy, specific heat and entropy for the wurtzite AlN lattice system. The auxiliary code of Quantum Espresso called quasi harmonic approximation (QHA) is used in the calculation. The calculated results for the thermodynamic function are illustrated in figure 5.

The values of ΔE and F_v by GGA are slightly lower than those by LDA, but the values of C_v and S by LDA are slightly lower than those by GGA. From curve of Internal energy, the change of ΔE is more slowly when T below 200K. It is seen that the specific heat presents T^3 at low temperature and tend to a certain value when the temperature is high.

The Debye temperature (Θ_D) is also calculated by QHA at 0 K and room temperature

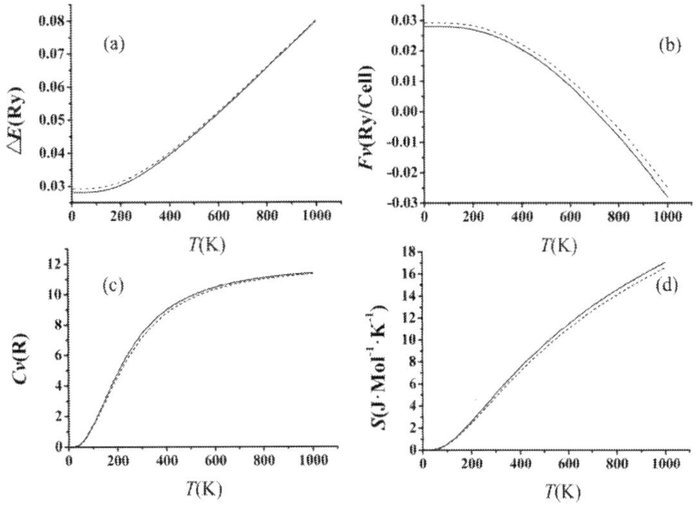

Figure 5. The phonon internal energy ΔE (a), vibrational free energy F_v (b), specific heat C_v (c) and entropy S (d) of AlN as functions of temperature. The specific heat is measured in $R = 8.31\mathrm{JMol}^{-1}\mathrm{K}^{-1}$.

Table 4. Debye temperature (Θ_D) for wurtzite AlN by GGA and LDA

		Θ_D,K	
		$T = 0$K	$T = 298$K
AlN	This work^{GGA}	984	1017
	This work^{LDA}	1026	1063
	Theory[13]	991	1057
	Theory[14]	1028	–
	Experiment[12]	950	1010

(298K). The calculated results of the Debye temperature for wurtzite AlN by GGA and LDA are listed in table 4. Our calculated values by GGA are very close to the experiment data [12] obtained from specific heat measurements, and better than the other theoretical values [13, 14].

4. Conclusion

In this work, we calculate the electronic states and the lattice dynamics of wurtzite AlN with two different pseudo potentials, LDA and GGA. The band structure of electrons and the phonon dispersion characteristics as well as the dielectric and thermal properties are showed. The calculated values are in agreement with available experimental data.

Acknowledgments

This work was supported in part by the PhD Progress Foundation of Higher Education Institutions of China (No.20111501110003), the National Natural Science Foundation of China (No.11347208), the key Educational Science Foundation of Inner Mongolia Autonomous Region of China (Grant NJZZ14004) and the enhancing comprehensive strength project of Inner Mongolia University (No.14020202).

References

[1] Tütüncü H M and Srivastava G P 2000 *Phys. Rev. B* **62** 5028–5035
[2] Davydov V Y, Kitaev Y E, Goncharuk I N, Smirnov A N, Graul J, Semchinova O, Uffmann D, Smirnov M B, Mirgorodsky A P and Evarestov R A 1998 *Phys. Rev. B* **58** 12899–12907
[3] Perlin P, Polian A and Suski T 1993 *Phys. Rev. B* **47** 2874–2877
[4] Giannozzi P, Baroni S, Bonini N, Calandra M, Car R, Cavazzoni C, Ceresoli D, Chiarotti G L, Cococcioni M, Dabo I, Corso A D, Gironcoli S d, Fabris S, Fratesi G, Gebauer R, Gerstmann U, Gougoussis C, Kokalj A, Lazzeri M, Martin-Samos L, Marzari N, Mauri F, Mazzarello R, Paolini S, Pasquarello A, Paulatto L, Sbraccia C, Scandolo S, Sclauzero G, Seitsonen A P, Smogunov A, Umari P and Wentzcovitch R M 2009 *J. Phys.: Condens. Matter* **21** 395502
[5] Bungaro C, Rapcewicz K and Bernholc J 2000 *Phys. Rev. B* **61** 6720–6725
[6] Ferreira da Silva A, Souza Dantas N, de Almeida J S, Ahuja R and Persson C 2005 *Journal of Crystal Growth* **281** 151–160
[7] Wright A F and Nelson J S 1995 *Phys. Rev. B* **51** 7866–7869
[8] Karch K 1997 *Phys. Rev. B* **56** 7404–7415
[9] Christensen N 1994 *Phys. Rev. B* **50** 4397–4415
[10] Adachi S 2005 *Properties of Group-IV, III-V and II-VI Semiconductors* 1st ed (Chichester, England: Wiley)
[11] Akasaki I and Hashimoto M 1967 *Solid State Communications* **5** 851–853
[12] Bruls R J, Hintzen H T, de With G and Metselaar R 2001 *Journal of the European Ceramic Society* **21** 263–268
[13] Marmalyuk A A, Akchurin R K and Gorbylev V A 1998 *High Temperature* **36** 817–819
[14] Peng F, Chen D, Fu H and Cheng X 2008 *Physica B: Condensed Matter* **403** 4259–4263

First-principles study of hydrogen diffusion in transition metal Rhodium

Wulijibilige Bao[1], Xin Cui[1]*, Zhi-Ping Wang[1]

[1]School of Physical Science and Technology, Inner Mongolia University, Hohhot 010021, China

E-mail: pycuixin@imu.edu.cn

Abstract. In this study, the diffuse pattern and path of hydrogen in transition metal rhodium are investigated by the first principles calculations. Density functional theory is used to calculate the system energies of hydrogen atom occupying different positions in rhodium crystal lattice. The results indicate that the most stable position of hydrogen atom in rhodium crystal lattice locates at the octahedral interstice, and the tetrahedral interstice is the second stable site. The activation barrier energy for the diffusion of atomic hydrogen in transition metal rhodium is quantified by determining the most favorable path, i.e., the minimum-energy pathway for diffusion, that is the indirect octahedral-tetrahedral-octahedral (O-T-O) pathway, and the activation energy is 0.8345eV.

1. Introduction

Research into the reaction of hydrogen with metals has attracted much attention because of potential applications as effective hydrogen storage materials [1], Ho K M *et. al* [2] studied the metal-hydrogen interactions in NbH by first-principles total energy calculations. And the transition metal-hydrogen systems have also attracted considerable interest for the unfilled *d* orbitals. Antonov V E [3] discussed the phase transformations, crystal and magnetic structures of high-pressure hydrides of *d*-metals, and obtained the conclusions that the high-pressure technique had been most effective in hydrogenation of the group VI–VIII transition metals, neither of which except Pd forms hydrides at low hydrogen pressures, and the hydrides formed at high pressures were shown to have close-packed metal sublattices with *fcc*, *hcp* or *double hcp* structures, in which hydrogen occupied octahedral interstitial positions. Then Pronsato M E *et. al* [4] investigated the electronic and bonding properties of iron monohydride using tight-binding method. We[5,6] had investigated the structural, electronic and magnetic properties of RhH alloy and PdH_x with both cubic and tetrahedral structures based on *ab-initio* total energy calculations.

Hydrogen diffusion and migration in metals are important aspects to determine the behavior of hydrogen in metals. The diffusion of hydrogen in metals or alloys are also investigated by scientific workers. Diffusion of hydrogen in the α-phase Pd system was investigated by electrochemical methods[7], and the concentration of hydrogen in the metal was gradually increased by control of the electrode potential. The dissolution and diffusion behaviors of hydrogen in copper were investigated based on first-principles calculations in combination[8]. The adsorption and diffusion of hydrogen on ordered $Ni_3Fe(111)$ surface and in the bulk were studied by first-principles calculations based on density functional theory[9]. The hydrides of Rh were synthesized at the Institute of Solid State Physics, Russian Academy of Sciences for the first time [10,11], and RhH had a NaCl-type crystal structure with H atoms occupying every octahedral site in the face - centered cubic of rhodium lattice with the parameter a =4.010 Å. In this work, we investigate the hydrogen diffusion behaviors in face - centered cubic of rhodium structures based on *ab-initio* total energy calculation. And *ab-initio* calculations are very helpful as they are able to provide fundamental understanding of many properties and insight into microscopic details.

2. Computational method

A first principles simulation is carried out using the plane wave ultra-soft pseudo-potentials (USPP) [12] method, which is based on density-functional theory (DFT)[13]. In this method, the ultra-soft pseudo-potentials are used to describe the computationally expensive core-valence interaction, and the Perdew-Wang parameterization (PW91) has been used for the GGA exchange- correlation functional in the calculations. The Monkhort-Pack [14] special *k*-point sampling of the brillouin zone used in the calculations is 8×8×8, the energy cut-off for the plane waves are set to 280eV.

We obtain the possible filling positions of hydrogen in face - centered cubic of rhodium by comparing the binding energy of the hydrogen atom in interstitial sites of Rh. Define a formula to calculate the binding energy of the hydrogen atom:

$$E_{\text{int}} = E_{\text{cell+H}} - E_{\text{cell}} - E_{\text{H}} \quad (1)$$

Where $E_{\text{cell+H}}$ is the total energy of the system when a hydrogen atom embedding to interstitial site, E_{cell} is the energy of the pure face - centered cubic of rhodium cell, and E_{H} is the energy of one hydrogen atom. The energy is obtained in the same calculation method and accuracy for all the systems. The whole system is most stable when the binding energy of the hydrogen atom is the lowest.

The most possible diffusion pathway of hydrogen atom is from an octahedral position to the adjacent octahedral position. We calculate the activation energies for H diffusion in Rh along three possible pathways, as shown in Figure 1. The hydrogen atoms both migrate linearly from an original octahedral site to its neighbor octahedral site (the O-O path) for the first two pathways, and the hydrogen atom moves between two octahedral sites in the same panel via an adjacent tetrahedral site intermediate (the O-T-O path) for the third pathway, as shown in Figure 1(c).

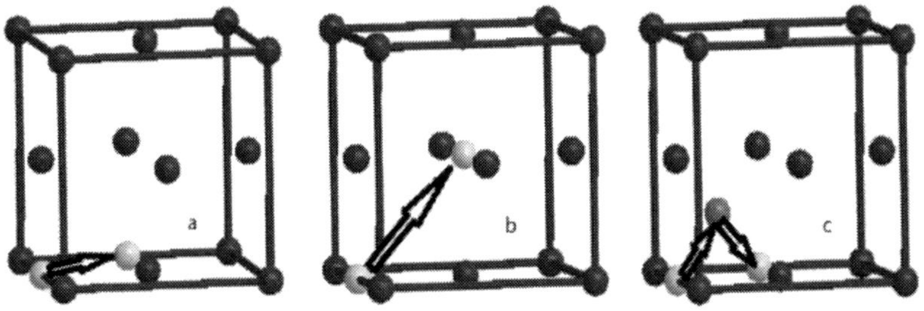

Figure 1. Pathway of H diffusion in fcc Rh from (a) original O site(1/2,0,0) to edge O site(0,1/2,0), (b) original O site(1/2,0,0) to body-centered O site(1/2,1/2,1/2), (c) original O site(1/2,0,0) to edge O site(0,1/2,0) via a T site(1/4,1/4,1/4). The black spheres is Rh atoms, the white and gray spheres are single H atom in octahedral and tetrahedral sites respectively.

H atom must overcome a potential barrier when migrating along the diffusion pathway and the maximum point for the potential barrier is identified as the saddle point, which is generally located at the midpoint of migration path in high-symmetric configurations. The activation energy for H diffusion (Q_{H}) in Rh is calculated as the energy difference[15]:

$$Q_{\text{H}} = E_{\text{sad}} - E_0 \quad (2)$$

where E_{sad} and E_0 are the total energies of the cells for the saddle point and for the equilibrium site, respectively.

3. Results and discussions

In order to find the equilibrium lattice parameters and the most stable adsorption position of this system, we calculate the total energies with both octahedral and tetrahedral structures. By total energy minimization, lattice parameter of the fcc rhodium structure is estimated to be 3.8966 Å, which is in good agreement with the experimental value, 4.010 Å reported by Somenkov *et al.* [16]. The calculated binding energies for hydrogen in different interstitial sites are shown in Table 1. It can be see that the binding energy of hydrogen atom in octahedral site is lower than that in tetrahedral site, that is the octahedral site is the global minimum, the tetrahedral site being higher in energy by 0.5423eV, so the most stable position of hydrogen atom in fcc rhodium locates at the octahedral interstice, and the tetrahedral interstice is the second stable site. Then we study the diffusion path between octahedral sites that is the optimum geometry for the fully lattice for hydrogen.

Table 1. Binding energy for H in different interstitial sites

E/eV		□*E*/eV
octahedral site	tetrahedral site	
-2.0411	-1.4988	0.5423

To estimate the activation energy for hydrogen interstitial diffusion between octahedral sites in fcc rhodium, we calculate the energy variation of the whole system with diffusion distance along the pathway. Nine images are used for both the first direct pathways from original octahedral site(1/2,0,0) to edge octahedral site(0,1/2,0), and the second one from original octahedral site(1/2,0,0) to body-centered octahedral site(1/2,1/2,1/2), as shown in Figure 2 and 3, respectively. We can see that the saddle point is explicitly located at the midpoint of the migration paths for both two paths, which turns out to be a common feature in many high-symmetry configurations[17].

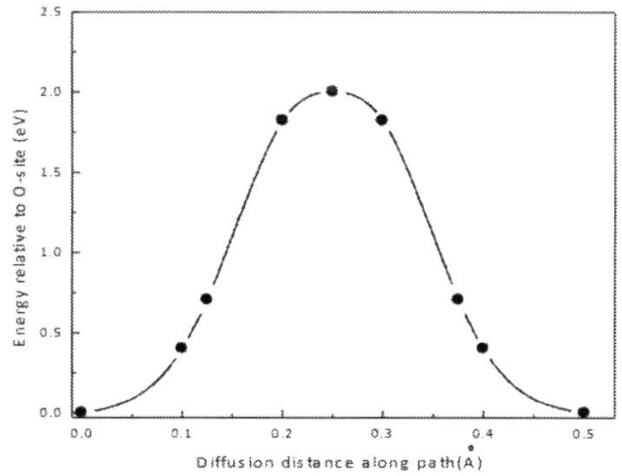

Figure 2. Activation energies for H interstitial diffusion in Rh along the first direct pathway from original O site(1/2,0,0) to edge O site(0,1/2,0).

In the first direct path from original octahedral site(1/2,0,0) to edge octahedral site(0,1/2,0), the calculated activation energy for H diffusion in fcc rhodium is 2.0008 eV, which is lower than the value in the second path from original octahedral site(1/2,0,0) to body-centered octahedral site(1/2,1/2,1/2), indicating that the potential barrier of path 2 is larger than that of path 1, and it is easier for H diffusion in fcc rhodium along the first direct path from original octahedral site(1/2,0,0) to edge octahedral site(0,1/2,0).

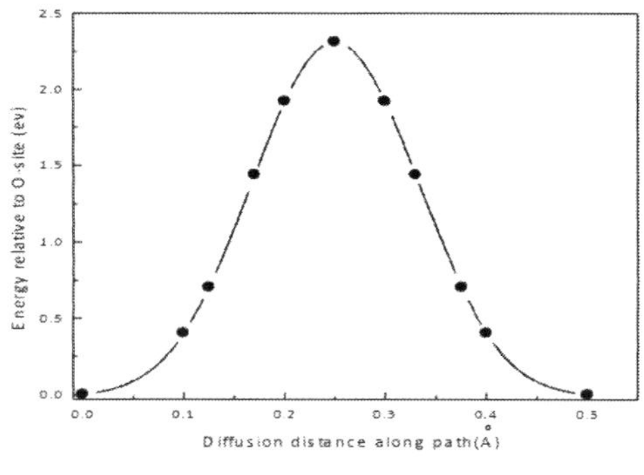

Figure 3. Activation energies for H interstitial diffusion in Rh along the second pathway from original O site(1/2,0,0) to body-centered O

site(1/2,1/2,1/2)

The calculated diffusion activation energies for hydrogen interstitial diffusion in Rh for the first two direct pathways are listed in Table 2.

Table 2. Calculated results of activation energy for H interstitial diffusion in Rh via three different pathways

Diffusion pathway	Activation energy /eV
The first O-O path	2.0008
The second O-O path	2.3105
The third O-T-O path	0.8345

Then we compare the direct and indirect path between the two octahedral sites, For the third path, that is an indirect O-T-O pathway, as many as twenty one images are used to find the saddle point accurately, and the energy variation with diffusion distance along this pathway is calculated, shown in Figure 4.We can see that the midpoint of this diffusion path corresponding to the tetrahedral site is not the saddle point as found in the first two direct O-O paths but the local minimum point, that is, it is locally stable but globally metastable. From the structure of image for the saddle point, we find that it lies symmetrically at the centroid of the triangle of fcc rhodium atoms shared by the octahedron and its adjacent tetrahedron.

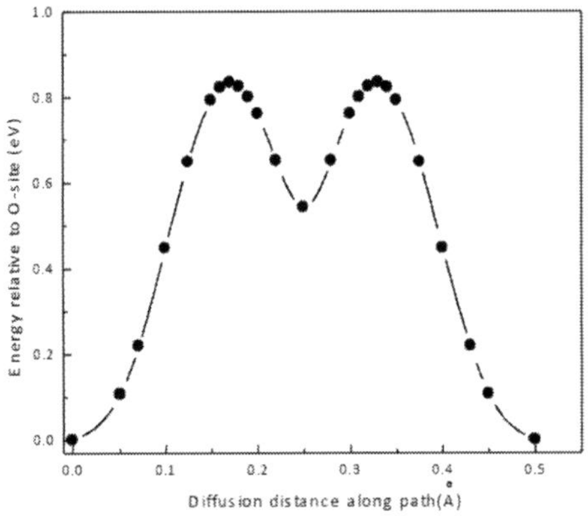

Figure 4. Activation energies for H interstitial diffusion in Rh along the third path from original O site(1/2,0,0) to edge O site(0,1/2,0) via a T site(1/4,1/4,1/4).

The calculated activation energy for H diffusion along the third path is 0.8345 eV, also shown in Table 2, which is much lower than the value along the first path. We compare the activation energies of H diffusion along the first pathway and the third pathway, both of which can be denoted for H diffusion with the same initial octahedral site and the final one. It can be found that the activation energy for H diffusion in the indirect O–T–O pathway is about 1.1663 eV lower than that in the direct O–O path, therefore, the energetically most favorable pathway for H diffusion in fcc rhodium is the indirect O–T–O pathway.

From a chemical point of view, the role of the interstitial atom is to debilitate the Rh-Rh nearest-neighbor (NN) interaction. As noted above, the repulsive forces among Rh atoms induced by the hydrogen atom passing through the saddle point of the indirect O–T–O path are smaller than those in the direct O–O path, so that the indirect O–T–O mechanism for hydrogen diffusion in fcc rhodium is not only energetically but also geometrically more favorable.

4. Conclusion

With the use of the DFT-based first principle method, we have investigated the diffusion behavior of hydrogen atom in face-centered cubic structure of rhodium. The calculated results indicate that the most stable position of hydrogen atom in fcc rhodium locates at the octahedral interstice. By comparing the activation energy of hydrogen diffusion along the different paths, find that the most favorable pathway for H diffusion in fcc rhodium is the indirect O–T–O pathway, with the activation energy of 0.8345 eV.

Acknowledgment
This work has been supported by the National Natural Science Foundation of China (No. 11247209).

References
[1] Boddien A, Mellmann D, Gärtner F, Jackstell R, Junge H, Dyson P J, Laurenczy G, Ludwig R, Beller M 2011 *Science* **333** 1733
[2] Ho K M, Tao H J, and Zhu X Y 1984 *Phys. Rev. Lett.* **53** 1586
[3] Antonov V E 2002 *J. Alloys Comp.* **330–332** 110
[4] Pronsato M E, Brizuela G, Juan A 2003 *J. Phys. Chem. Solids* **64** 593
[5] Cui X, Wang J T, Liang X X, Zhao G Zh 2009 *Solid State Commun* **149** 322
[6] Cui X, Liang X X, Wang J T, Zhao G Zh 2011 *Chin. Phys. B* **20(2)** 026201
[7] Piotr Z and Ewa M 2001 *Phys. Chem. Chem. Phys.* **3** 2935
[8] Zhou H B, ZhangY, Ou X 2013 *Comp. Mater. Sci.* **79** 923
[9] Li J, Xie Y P, Chen Y X, Wang B W, Zhao S J 2014 *Intermetallics* **44** 64
[10] Ponyatovskii E G, Antonov V E, Belash I T 1982 *Sov. Phys. Usp.* **25** 596
[11] Antonov V E, Antonova T E, Chirin N A, Ponyatovsky E G, Baier M, Wagner F E 1996 *Scripta Mater.* **34** 1331
[12] Vanderbilt D 1990 *Phys. Rev. B* **41(11)** 7892
[13] Dreizler R M and Gross E K U 1990 *Density Functional Theory: An Approach to the Quantum Many-Body Problem* (Berlin: Springer-Verlag)
[14] Monkhort H J, Pack J D 1976 *Phys. Rev. B* **13(12)** 5188
[15] Han X L, Wang Q, Sun D L, Sun T, Guo Q 2009 *Int. J. Hydrogen Energy* **34** 3983
[16] Somenkov V A, Glazkov V P, Irodova A V, Shilstein S Sh 1987 *J. Less-Common Met.* **129** 171
[17] Krimmel H, Fähnle M. 2000 *Phys. Rev. B* **62(9)** 5489

Journal of Physics: Conference Series **574** (2015) 012048 doi:10.1088/1742-6596/574/1/012048

First-principle study of the electronic band structure and the effective mass of the ternary alloy $Ga_xIn_{1-x}P$

H Q Yang, T L Song[1]**, X X Liang and G J Zhao**

Department of Physics, School of Physical Science and Technology, Inner Mongolia University, Hohhot, 010021, P. R. China

E-mail: songtl@126.com

Abstract. In this work, the electronic band structure and the effective mass of the ternary alloy $Ga_xIn_{1-x}P$ are studied by the first principle calculations. The software QUANTUM ESPRESSO and the generalized gradient approximation (GGA) for the exchange correlations have been used in the calculations. We calculate the lattice parameter, band gap and effective mass of the ternary alloy $Ga_xIn_{1-x}P$ for the Ga composition x varying from 0.0 to 1.0 by the step of 0.125. The effect of the Ga composition on the lattice parameter and the electronic density of states are discussed. The results show that the lattice parameter varies with the composition almost linearly following the Vegard's law. A direct-to-indirect band-gap crossover is found to occur close to $x = 0.7$. The effective masses are also calculated at $\Gamma(000)$ high symmetry point along the [100] direction. The results show that the band gap and the electron effective mass vary nonlinearly with composition x.

1. Introduction

Extensive experimental investigations of the electronic structure of the group III-V semiconductors and their alloys have been carried out because of their importance in the optoelectronic device technology. Several methods have been developed to calculate the electronic properties of the ternary alloy $Ga_xIn_{1-x}P$. Bouarissa [1] studied the electronic properties of the ternary alloy $Ga_xIn_{1-x}P$ by using the empirical pseudopotential method (EPM) within the virtual crystal approximation (VCA). Onton et al. [2] employed the cathodoluminescence (CL) and photoluminescence (PL) approaches to investigate the conduction-band structure of $Ga_xIn_{1-x}P$. Kumar et al. [3] calculated disorder effects on electronic and optical properties of the ternary alloy $Ga_xIn_{1-x}P$ ($x = 0.25$, $x = 0.5$, and $x = 0.75$) by using the WIEN2k code. The effect of the alloy composition on physical properties is important and should been investigated in detail.

In this paper the electronic band structure and the effective mass of the ternary alloy $Ga_xIn_{1-x}P$ are studied. The calculations are based on first-principle calculations within the framework of the density functional theory.

2. Calculation model and methods

The first principle calculations have been performed by using the software QUANTUM ESPRESSO [4] and the generalized gradient approximation (GGA) for the exchange correlations.

[1] Corresponding author E-mail address: songtl@126.com

Based on the conventional cell illustrated in figure 1(a), we build a supercell ($1\times1\times2$) illustrated in figure 1(b) for substitutional doping. Convergence with respect to k points and energy cutoff is carefully checked. The plane-wave cutoff energy is set to 75 Ry, and the numerical integration of the Brillouin zone (BZ) is performed using a k-mesh of dimensions $6 \times 6 \times 6$.

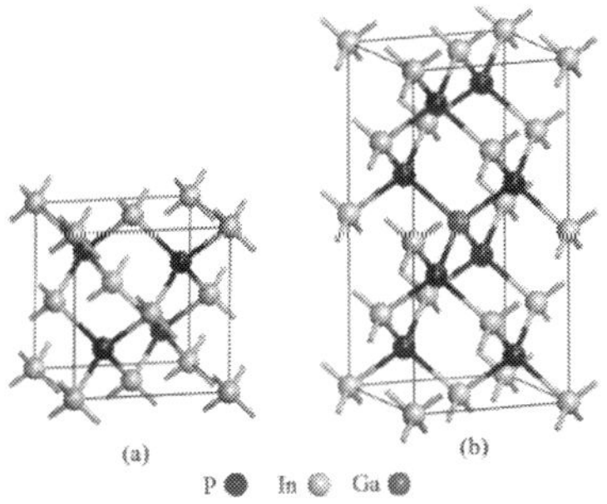

Figure 1. Crystal structures: (a) InP or GaP, (b) $Ga_xIn_{1-x}P$.

3. Result and discussion

3.1. Lattice parameters

In our calculation, the supercell contains sixteen atoms. We perform an iterative process in which the coordinates of the atoms and the cell parameters are adjusted so that the total energy of the structure is minimized. The calculated lattice parameters as functions of the compositions x for the ternary alloy $Ga_xIn_{1-x}P$ are illustrated in the figure 2. From it, we can find that the lattice parameter of the ternary alloy $Ga_xIn_{1-x}P$ decreases with increasing Ga composition x. It is almost linearly following the Vegard's law. The equation of lattice parameter a can be obtained by a linear fitting method as follows:

$$a\left(Ga_xIn_{1-x}P\right) = 5.87063 - 0.45537x. \tag{1}$$

Our conclusions are very close to other theoretical calculations [2].

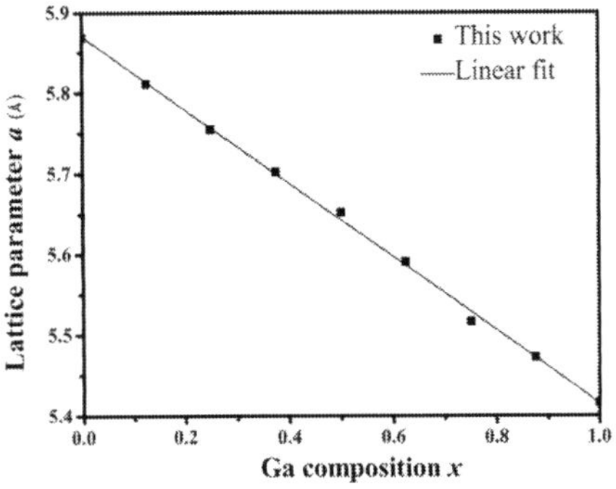

Figure 2. Lattice parameters of the ternary alloy $Ga_xIn_{1-x}P$ for different Ga compositions x.

3.2. Energy band gap

The calculated band gaps are listed in table 1 and compared with other theoretical results [3] as well as the available experimental [5] data.

Table 1. Band gap E_g calculated by GGA for InP, GaP and $Ga_xIn_{1-x}P$ and the corresponding experimental Ref[5] and other theoretical Ref[3] values

x	Γ_v–Γ_c			Γ_v–X_c		
	this work	Ref[5]	Ref[3]	this work	Ref[5]	Ref[3]
InP	0.444	1.34	0.67	1.201	2.19	1.77
$Ga_{0.125}In_{0.875}P$	0.517	–	–	1.83	–	–
$Ga_{0.25}In_{0.75}P$	0.628	–	0.79	1.156	–	1.74
$Ga_{0.375}In_{0.625}P$	0.754	–	–	1.113	–	–
$Ga_{0.5}In_{0.5}P$	0.91	1.846	0.97	1.115	–	1.72
$Ga_{0.625}In_{0.375}P$	1.008	–	–	1.128	–	–
$Ga_{0.75}In_{0.25}P$	1.133	–	1.32	1.115	–	1.68
$Ga_{0.875}In_{0.125}P$	1.288	–	–	1.133	–	–
GaP	1.559	2.78	1.81	1.155	2.26	1.65

We also illustrate the band gap of the ternary alloy $Ga_xIn_{1-x}P$ as functions of Ga composition x in figure 3 (a). It is found that a direct-to-indirect band-gap crossover [3, 6] occur close to $x = 0.7$. The band gaps show nonlinear behavior with the composition x. By fitting the non-linear variation of the calculated band gaps in terms of the composition with polynomial function, the relationship between the band gaps and the composition is obtained as follows:

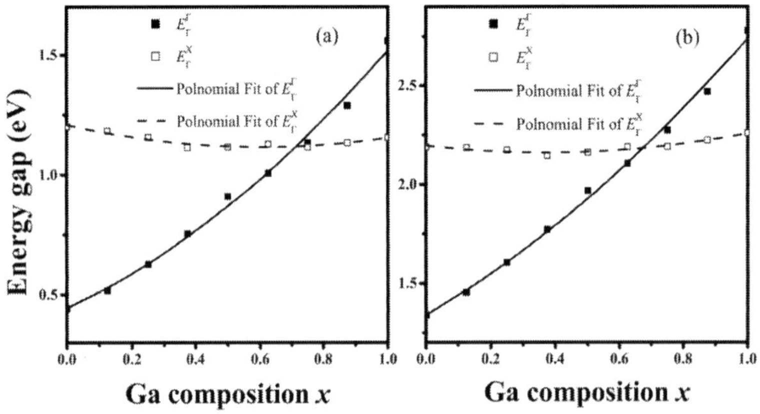

Figure 3. Band gap E_g as a function of the composition x.(a) for calculation.(b)for correction

$$E_\Gamma^\Gamma(x) = 0.44529 + 0.62298x + 0.44862x^2, \quad (2)$$

$$E_\Gamma^X(x) = 1.20758 - 0.30386x + 0.25039x^2. \quad (3)$$

We have computed the Kohn-Sham band structure of the ternary alloy $Ga_xIn_{1-x}P$. As was seen, the calculated results for the band gaps are underestimated in comparison with experimental results. Now let us amend the band gaps by the the following correction formula of ternary alloy:

$$E^{\text{cor}}_{g,\text{Ga}_x\text{In}_{1-x}P} = E^{\text{cal}}_{g,\text{Ga}_x\text{In}_{1-x}\text{P}} + (1-x)\cdot\Delta E_{\text{InP}} + x\cdot\Delta E_{\text{GaP}}, \tag{4}$$

where

$$\Delta E_{\text{InP}} = E^{\text{exp}}_{g,\text{InP}} - E^{\text{cal}}_{g,\text{InP}}, \tag{5}$$

$$\Delta E_{\text{GaP}} = E^{\text{exp}}_{g,\text{GaP}} - E^{\text{cal}}_{g,\text{GaP}}. \tag{6}$$

$E^{\text{cal}}_{g,\text{Ga}_x\text{In}_{1-x}\text{P}}$ are calculation values and $E^{\text{cor}}_{g,\text{Ga}_x\text{In}_{1-x}\text{P}}$ are the correction values. $E^{\text{exp}}_{g,\text{InP}}$ and $E^{\text{exp}}_{g,\text{GaP}}$ are experimental values of InP and GaP. $E^{\text{cal}}_{g,\text{InP}}$ and $E^{\text{cal}}_{g,\text{GaP}}$ are the calculation values of InP and GaP. Finally, the correction band gap of the ternary alloy $Ga_xIn_{1-x}P$ as functions of Ga composition x is showed in figure 3(b). By fitting the non-linear variation of the calculated band gaps in terms of concentration with polynomial function, the relationship between the band gaps and the composition is obtained as follows:

$$E^{\Gamma}_{\Gamma}(x) = 1.34129 + 0.94796x + 0.44864x^2, \tag{7}$$

$$E^{X}_{\Gamma}(x) = 2.19658 - 0.18786x + 0.25039x^2. \tag{8}$$

3.3. Effective mass

The theoretical effective mass is a tensor with nine components defined as:

$$\frac{1}{m^*_{ij}} = \frac{4\pi^2}{h^2}\frac{\partial^2 E(k)}{\partial k_i \partial k_j}, \tag{9}$$

where E is the energy and k_i are the components of the wave vector $\mathbf{k}$ in a periodic potential field of a semiconductor [3]. Here we restrict ourselves to a particular case and regarded the effective mass as a scalar. Then the electrons effective mass can be defined as:

$$\frac{1}{m^*} = \frac{4\pi^2}{h^2}\frac{\partial^2 E(k)}{\partial k^2}. \tag{10}$$

We have calculated the effective mass of the electrons at the conduction-band minima at Γ and along the [100] direction and the results for various composition are listed in table 2. Figure 4 shows the effective mass as a function of the composition x. It is clearly seen form table 2 that the electron effective mass increases monotonously with the Ga composition x. The calculated value of m^* for InP and GaP is in good agreement with the experimental data. The nonlinear behavior of electron effective mass with composition x is also found from figure 4.

By piecewise fitting method, we get the following equation of effective mass m^*:

$$m^* = \begin{cases} 0.0740 + 0.0925x & (0 \le x \le 0.625) \\ 2.0493 - 6.0572x + 4.7890x^2 & (0.625 \le x \le 1). \end{cases} \tag{11}$$

An inflection-point is found to occur around at $x = 0.7$, where a change of the direct-to-indirect band-gap arises for the ternary alloy $Ga_xIn_{1-x}P$.

Journal of Physics: Conference Series **574** (2015) 012048 doi:10.1088/1742-6596/574/1/012048

Table 2. Effective mass m^* calculated according by GGA for InP, GaP and $Ga_xIn_{1-x}P$ and the corresponding experimental Ref[7] and other theoretical Ref[3] values

	x	m^*		
		this work	Ref[7]	Ref[3]
InP	0.000	0.073060	0.077	0.0768
$Ga_{0.125}In_{0.875}P$	0.125	0.089079	–	–
$Ga_{0.25}In_{0.75}P$	0.250	0.096235	–	0.099
$Ga_{0.375}In_{0.625}P$	0.375	0.105914	–	–
$Ga_{0.5}In_{0.5}P$	0.500	0.120593	–	0.121
$Ga_{0.625}In_{0.375}P$	0.625	0.133188	–	–
$Ga_{0.75}In_{0.25}P$	0.750	0.203615	–	0.1938
$Ga_{0.875}In_{0.125}P$	0.875	0.412529	–	–
GaP	1.000	0.782270	0.82	0.6

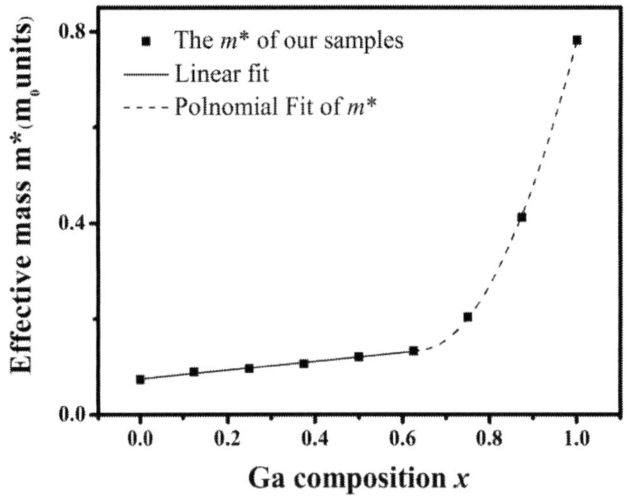

Figure 4. Effective mass as a function of the composition x.

4. Conclusion

In this paper, the band structure and the effective mass of the ternary alloy $Ga_xIn_{1-x}P$ are calculated using the software QUANTUM ESPRESSO and the generalized gradient approximation (GGA). The lattice parameter, band gap, and effective mass of the ternary alloy $Ga_xIn_{1-x}P$ for the Ga composition x varying from 0.0 to 1.0 by the step of 0.125 are presented. The results show that the lattice parameter varies with the composition x almost linearly following the Vegard's law. A direct-to-indirect band-gap crossover is found to occur close to $x = 0.7$. The effective masses are also calculated at $\Gamma(000)$ symmetry point along the [100] direction. The results show that the band gap and the electron effective mass vary nonlinearly with composition x.

Acknowledgments
This work was supported in part by the PhD Progress Foundation of Higher Education Institutions of China (No.20111501110003), the National Natural Science Foundation of China (No.11347208), the key Educational Science Foundation of Inner Mongolia Autonomous Region of China (Grant NJZZ14004) and the enhancing comprehensive strength project of Inner Mongolia University (No.14020202).

References
[1] Bouarissa N 2010 *Materials Chemistry and Physics* **124** 336–341
[2] Onton A, Lorenz M R and Reuter W 1971 *Journal of Applied Physics* **42** 3420–3432
[3] Kumar S, Parashari S S and Auluck S 2009 *phys. stat. sol. (b)* **246** 2294–2300
[4] Giannozzi P, Baroni S, Bonini N, Calandra M, Car R, Cavazzoni C, Ceresoli D, Chiarotti G L, Cococcioni M, Dabo I, Corso A D, Gironcoli S d, Fabris S, Fratesi G, Gebauer R, Gerstmann U, Gougoussis C, Kokalj A, Lazzeri M, Martin-Samos L, Marzari N, Mauri F, Mazzarello R, Paolini S, Pasquarello A, Paulatto L, Sbraccia C, Scandolo S, Sclauzero G, Seitsonen A P, Smogunov A, Umari P and Wentzcovitch R M 2009 *J. Phys.: Condens. Matter* **21** 395502
[5] Levinshtein m, Rumyantsev S and Shur M 1996 *Handbook Series on Semiconductor Parameters* vol 2 (World Scientific)
[6] Adachi S 2009 *Properties of Semiconductor Alloys: Group-IV, III-V and II-VI Semiconductors* 1st ed (Chichester, U.K: Wiley)
[7] Madelung O 2013 *Semiconductors: Group IV Elements and III-V Compounds* softcover reprint of the original 1st ed. 1991 edition ed (Berlin ; New York: Springer)

First-principle Studies of Lattice and Electronic Structure of $Be_xZn_{1-x}O$

Xue Lei, G J Zhao[1], X X Liang and T L Song
Department of Physics, School of Physical Science and Technology, Inner Mongolia University, Hohhot 010021, P. R . China

Email: stzhaogj@imu.edu.cn

Abstract. The lattice and electronic structure of beryllium doped zinc oxide ($Be_xZn_{1-x}O$) ternary mixed crystal are studied by the first-principle calculations within the framework of the density functional theory (DFT). It turned out that the lattice parameter *a* and lattice parameter *c* decrease linearly as Be doping concentration increases, compliance with Vegard's law, the lattice parameters of the $Be_xZn_{1-x}O$ ternary mixed crystal are consistent with experimental results. Band gap increases with increasing Be content, and the band gap values are corrected. As Be doping, Zn 4s states dominate the conduction band and the conduction band bottom position constantly moving to higher energy region. Density of states strength Zn 4s states with decreasing proportion of Zn is constantly reduced, make the band gap width of $Be_xZn_{1-x}O$ increases.

1. Introduction

With the development of optoelectronics industry, II-VI wide-gap semiconductor zinc oxide has become a great importance to the photoelectric material, and it was used widely in ultraviolet light photoelectric devices [1-2]. The causes of study ZnO mixed crystal is due to its easy to doping ternary mixed crystals formation mechanism and it is easy to meet the needs of a variety of optoelectronic devices. Experimental [3-5] studies have found that magnesium doped zinc oxide form $Mg_xZn_{1-x}O$ ternary mixed crystal, as Mg concentrations ($x \leq 0.33$) increases, the band gap of $Mg_xZn_{1-x}O$ ternary mixed crystal increases from 3.3 to 3.99eV, but because of the different crystal structures and large lattice mismatch between ZnO (3.25Å, hexagonal) and MgO (4.22Å, cubic) observed crystal phase segregation between ZnO and MgO as Mg concentrations ($x \geq 0.36$) [6]. Recently used Be instead of Mg in order to expand the band gap, the band gap of BeO is 10.6eV, the lattice structure of BeO as the same as ZnO. Ryu et al. have investigated $Be_xZn_{1-x}O$ ternary mixed crystal, they synthesized BeZnO films on *c*-Al_2O_3 substrates by the method of hybrid beam deposition (HBD) [7-8]. The experimental results indicated that the concentration of Be varied over the entire range from 0 to 100% and no phase segregation between ZnO and BeO was detected, after this, they got ZnO-based ultraviolet light emitting diodes that employ BeZnO/ZnO [9].

[1] Corresponding author E-mail address: stzhaogj@imu.edu.cn

In this paper, in order to obtain a full understanding for the structural and electronic properties of $Be_xZn_{1-x}O$ ternary mixed crystal, we are using first principles method based on the density functional theory (DFT) to calculate $Be_xZn_{1-x}O$, the optimal lattice parameters and the corresponding electronic band structures for the $Be_xZn_{1-x}O$ ($0 \leq x \leq 1$) are obtained and analysed.

2. Calculation Method

The calculations in this paper were performed within the framework of the density-functional theory (DFT) by using the plane-wave pseudopotential approach in the Cambridge Sequential Total Energy Package(CASTEP) codes [10]. We used the generalized gradient approximation (GGA) of PBE scheme [11] to describe the exchange-correlation potential. Energy cutoff of plane-wave was set to be 450eV in the present calculations. The special points sampling integration over the Brillouin zone (BZ) were performed using a k-mesh of dimensions 4×4×4 k-points mesh according to a Monkhorst–Pack scheme [12]. Select Be $2s^2$, Zn $3d^{10}4s^2$, O $2s^22p^4$ configuration as the valence states, the tolerances for geometry optimization were set as the difference in total energy being within 1×10^{-6}eV/atom, the basic model was wurtzite ZnO, figure 1 is model by the 2×2×1 supercell structure.

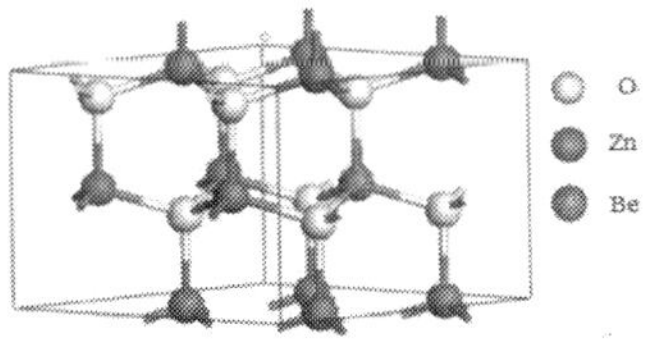

Figure 1. The crystal structure of wurtzite $Be_{0.125}Zn_{0.875}O$ alloys.

3. Results and discussion

3.1. Lattice parameters

Table 1. Calculated lattice parameter *a* and c for $Be_xZn_{1-x}O$ ternary mixed crystal. Compared with other theoretical calculations and experimental results.

x	Lattice parameter *a* (nm)		Lattice parameter *c* (Å)		
	This work	Theoretical	This work	Exp.	Theoretical
0	0.3290	0.3256[a]	5.29	5.20[b]	5.26[a]
0.1				5.11[b]	
0.125	0.3215	0.3190[a]	5.20		5.15[a]
0.25	0.3134	0.3134[a]	5.13		5.08[a]
0.375	0.3083	0.3052[a]	5.06		5.02[a]
0.5	0.3020	0.2972[a]	4.84		4.99[a]
0.6				4.79[b]	
0.625	0.2975	0.2924[a]	4.78		4.83[a]
0.7				4.71[b]	
0.75	0.2891	0.2889[a]	4.68		4.68[a]
0.8				4.61[b]	
0.875	0.2823	0.2808[a]	4.59		4.56[a]
1	0.2753	0.2764[a]	4.47	4.38[b]	4.49[a]

[a] Ref.[14]. [b] Ref.[7].

After the structure optimization, the lattice parameters of $Be_xZn_{1-x}O$ ternary mixed crystal have been calculated, table 1 show that our calculated compared with other theoretical and experimental, our results are in agreement with them.

The lattice parameter *a* and lattice parameter *c* of the $Be_xZn_{1-x}O$ ternary mixed crystal decrease linearly as Be doping concentration increases, shown in figure 2. The result compliance with Vegard's law [13] and consistent with experimental results[7], fitting the calculated results, we get the linear equation as follow,

$$a = 0.328 - 0.052x$$

$$c = 5.313 - 0.839x$$

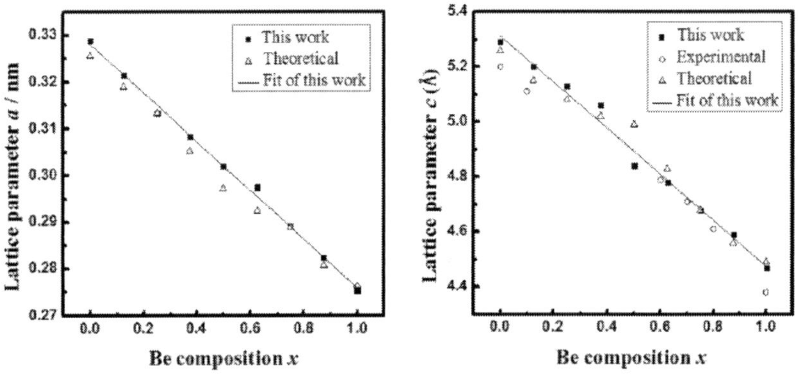

Figure 2. Calculated the lattice parameters of $Be_xZn_{1-x}O$ ternary mixed crystal for different Be concentrations.

3.2. Density of states

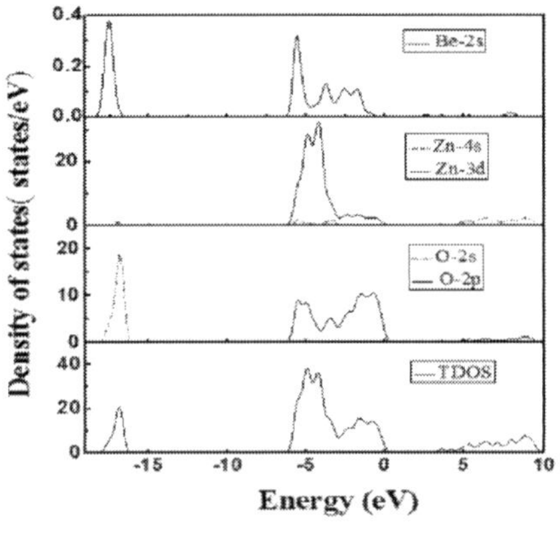

Figure 3. Total and partial density of states for $Be_{0.25}Zn_{0.75}O$.

The partial and total density of states for $Be_xZn_{1-x}O$ are calculated, figure 3 shows the total and partial DOS for the $Be_{0.25}Zn_{0.75}O$, as an example. The results shown that the valence band maximum (VBM) occupied by O 2p states, and the conduction band minimum (CBM) occupied by Zn 4s states. From figure 4, as Be doping, the contribution of zinc and beryllium in the conduction band position constantly moving to higher energy region, density of states strength Zn 4s states with decreasing proportion of Zn is constantly reduced, density of states strength Be 2s states with increasing proportion of Be is constantly increased, make the band gap width of $Be_xZn_{1-x}O$ increases.

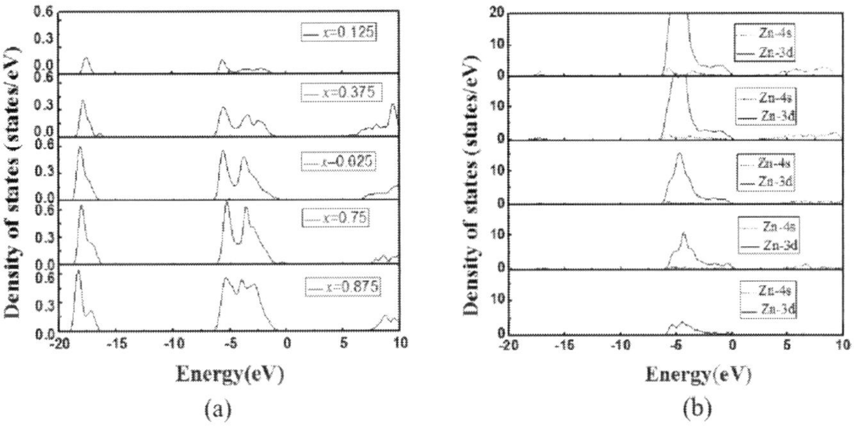

Figure 4. (a) Beryllium PDOS of different components, (b) zinc PDOS of different components.

3.3. Electronic band structure

Table 2. Band gap *Eg* and correction Band gap *Eg* for $Be_xZn_{1-x}O$ ternary mixed crystal. Compared with other theoretical calculations and experimental results.

x	E_g (eV)			
	This work	Theoretical	Correctional	Experimental
0	0.734	0.972[a]	3.3000	3.37[b]
0.1				3.53[b]
0.125	1.083	1.325[a]	3.7777	
0.25	1.568	1.645[a]	4.3913	
0.375	1.816	2.058[a]	4.7679	
0.44				4.63[b]
0.5	2.463	2.381[a]	5.5435	
0.59				4.74[b]
0.6				5.01[b]
0.625	3.041	3.076[a]	6.2502	
0.68				5.45[b]
0.75	4.192	3.799[a]	7.5298	
0.875	4.943	5.236[a]	8.4094	
1	7.004	7.322[a]	10.600	10.60[b]

[a] Ref.[14]. [b] Ref.[7].

The band gap of $Be_xZn_{1-x}O$ ternary mixed crystal is increases with increasing Be concentrations, but our calculated is smaller than experimental, because of the semiconductor based on DFT of GGA exchange-correlation effects between electrons and electron handle deficiency caused errors, but in the same environment, changing the doping concentration for the same structure system material research, the band gap was calculated by DFT is still meaningful. In order to close to the experimental value, Band gap value has been corrected, using the formula[15],

$$E^{cor}_{g,Be_xZn_{1-x}O} = E^{cal}_{g,Be_xZn_{1-x}O} + (1-x)\cdot \Delta E_{ZnO} + x \cdot \Delta E_{BeO}$$

$E^{cor}_{g,Be_xZn_{1-x}O}$ is correction band gap of $Be_xZn_{1-x}O$, $E^{cal}_{g,Be_xZn_{1-x}O}$ is calculation band gap of $Be_xZn_{1-x}O$, x is concentration of Be, ΔE_{ZnO} is experimental band gap minus calculated band gap of ZnO, ΔE_{BeO} is experimental band gap minus calculated band gap of BeO. Correction parameters are experimental of ZnO 3.3eV and experimental of BeO 10.6eV, table 2 and figure 5 shown our calculated compared with experimental and our calculated results are consistent with experimental results, in figure 5 we used the quadratic function shown band gap of $Be_xZn_{1-x}O$ ternary mixed crystal, the solid line is a fit of our calculation data using the quadratic function, as follow,

$$E_g = 3.491 + 1.173x + 5.608x^2$$

and the dotted line is a fit of experimental [16], the formular is

$$E_g = 3.4 - 2.1x + 9.3x^2$$

and our calculated is consistent with experimental.

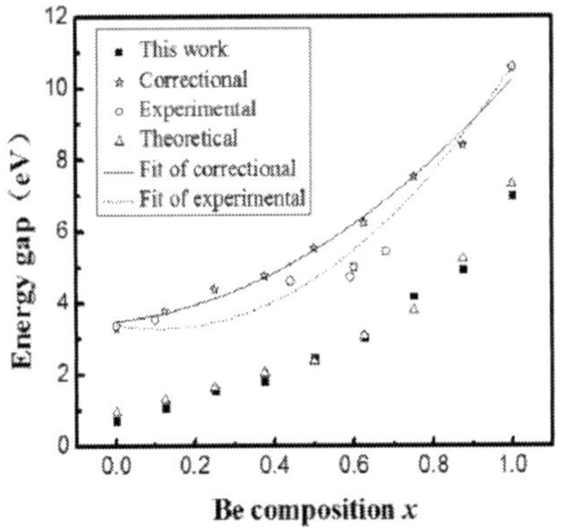

Figure 5. Calculated the Band gap of $Be_xZn_{1-x}O$ ternary mixed crystal for different Be concentrations.

4. Conclusions

In this paper, we have used the first-principle calculations to study the lattice parameters and electronic band structure of $Be_xZn_{1-x}O$ ternary mixed crystals. The structural parameters of $Be_xZn_{1-x}O$ ternary mixed crystals have been optimized. The results show that lattice parameters are almost linearly and obey the Vegard's law. After correction the electronic band gap is consistent with experimental results. The results of the density of states for $Be_xZn_{1-x}O$ show that as Be doping the contribution of zinc and beryllium in the conduction band position constantly moving to higher energy region, make the band gap width of $Be_xZn_{1-x}O$ increases.

Acknowledgments

This work was supported in part by the PhD Progress Foundation of Higher Education Institutions of China (No. 20111501110003), the Key Educational Science Foundation of Inner Mongolia Autonomous Region of China (Grant NJZZ14004), the National Natural Science Foundation of China (No. 11347208) and the Enhancing Comprehensive Strength Project of Inner Mongolia University (No. 14020202).

References

[1] Bagnall D M, Chen Y F, Zhu Z, Yao T, Koyama S, Shen M Y and Goto T 1997 *Appl. Phys. Lett.* **70** 2230
[2] Aoki T, Hatanaka Y and Look D C 2000 *Appl. Phys. Lett.* **76** 3257
[3] Ohtomo A, Kawasaki M, Ohkubo I, Koinuma H, Yasuda T and Segawa Y 1999 *Appl. Phys. Lett.* **75** 980
[4] Makino T, Chia C H, Tuan N T, Segawa Y, Kawasaki M, Ohtomo A, Tamura K and Koinuma H 2000 *Appl. Phys. Lett.* **77** 1632
[5] Ohtomo A, Kawasaki M, Koida T, Masubuchi K and Koinuma H 1998 *Appl. Phys. Lett.* **72** 2466
[6] Madelung O 2003 *Semiconductors: Data Handbook, third ed* (New York: Springer)
[7] Ryu Y R, Lee T S, Lubguban J A, Corman A B, White H W, Leem J H, Han M S, Park Y S, Youn C J and Kim W J 2006 *Appl. Phys. Lett.* **88** 052103
[8] Ryu Y R, Lee T S and White H W 2004 *J. Crystal Growth* **261** 502.
[9] Ryu Y R, Lee T S, Lubguban J A, White H W, Kim B J, Park Y S and Youn C J 2006 *Appl. Phys. Lett.* **88** 241108.
[10] Segall M D, Lindan P J D, Probert M J, Pickard C J, Hasnip P J, Clark S J and Payne M C 2002 *J. Phys.: Condens. Matter* **14** 2717
[11] Perdew J P, Burke K and Ernzerhof M 1996 *Phys. Rev. Lett.* **77** 3865.
[12] Monkhorst H J and Pack J D 1976 *Phys. Rev. B* **13** 5188
[13] Vegard L 1921 *Z. Phys.* **5** 17
[14] Zheng Y P, Chen Z G, Lu Y, Wu Q Y, Weng Z Z and Huang Z G 2008 *Journal of Semiconductors* **29** 2317
[15] Tang X, Lu H F, Zhao J J and Zhang Q Y 2010 *J. Phys. Chem. Solids* **71** 336
[16] Adachi S 2009 *Properties of Semiconductor Alloys: Group-IV, III-V and II-VI Semiconductors* (Japan: John Wiley & Sons, Ltd)

Computer simulation of electron states of the isoelectronic substitutional impurities in quantum wire of KCl with edge dislocation

Yury K Timoshenko[a], **Valentina A Shunina**[b], **and Alexander I Shashkin**[a]

[a] Voronezh State University, Universitetskaya pl., 1, Voronezh, 394006, Russia
[b] Voronezh State Technical University, Moscow Avenue, 14, Voronezh, 394026, Russia

E-mail: `yutim@phys.vsu.ru`

Abstract. The electron states of quantum wires KCl, KCl : Br, and KCl : I with an edge dislocations were investigated. The main problem was in study of dislocation influence on localized electron states connected with substitutional isoelectronic impurities Br^- and I^- in the neighborhood of edge dislocation line. The tight-binding semi-empirical band approximation, semi-empirical and non-empirical cluster approach were used. Semi-empirical calculations were carried out in framework of model [1, 2]. Besides, the algorithms for electronic levels calculations of polar nanosystems with the partial self-consistency [3] were used. The computer simulation results lead to the conclusion that the substitutional isoelectronic impurity anions Br^- and I^- located near the dislocation line capture the holes more efficiently than in bulk of systems without dislocations.

1. Introduction

As it is wellknown the maximal values of the actual length parameters of nanosystems are equal to about 100 nanometers. Such large polyatomic systems are able to have the analogs of extended defects in monocrystals. For example, the edge dislocations. There are some papers devoted to the theoretical investigations of nanosystems with dislocations, e.g. [4, 5]. A more complicated problem is the investigations of nanosystems with dislocations and point defects. In our paper the electron states of potassium chloride nanosystems with substitutional isoelectronic impurities Br^- and I^- have been studied. The theoretical investigations of this sort attract interest in connection with problem of computer simulation of a dislocation influence on the impurity's local levels. The information about positions of the energy levels of complex defects such as "dislocation + impurity" is one of fundamental points of the electron structure theory of defects, which is necessary for the interpretation of experiments and prediction of physical properties of the crystals and nanosystems.

2. Models and methods

The physical and mathematical models based on our works [1–3, 6] have been used. The edge dislocation had Burgers vector $\langle 110 \rangle$ and was located in the glide plane $\{1\bar{1}0\}$ as well as in papers [1, 2]. As a starting point, nanocrystal model without periodic boundary conditions was taken into consideration. The potassium chloride nanocrystal with dislocation have contained $43 \times 8 = 344$ ions. I.e., there was 8 layers along the dislocation line. The $K_{172}Cl_{172}$ nanocrystal relaxation was considered in framework of the model of pointlike polarizable ions. The equilibrium geometry of nanocrystals were calculated by minimization of the total potential energy over the ion coordinates. Naturally, the dipole moments were calculated in a self-consistent way (for details see paper [7]). After the calculations of the equilibrium configurations of the nanocrystal, its electron structure was calculated. We used the Slater-Koster variant of the tight-binding method modified by us earlier. Using the obtained results we have constructed the model of infinite quantum nanowire with one-dimensional periodicity along dislocation line. Namely, the coordinates of ions in a unit cell (86 ions, fig. 1) and correction to diagonal elements of Hamiltonian have been obtained by using the data for central area of nanocrystal. The total density of states (TDOS) and partial local densities of states (PLDOS) for KCl

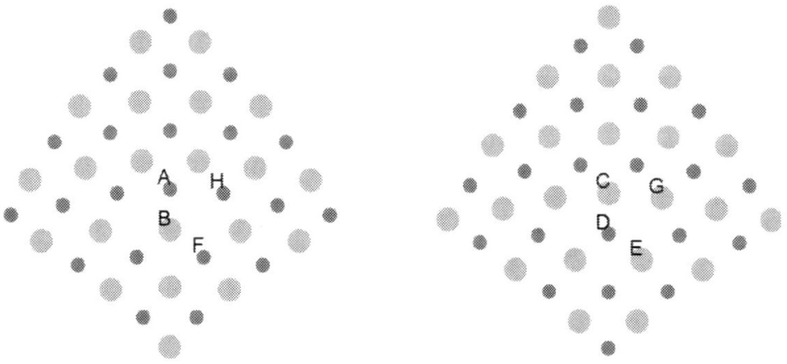

Figure 1. The central layers of nanocrystal $K_{172}Cl_{172}$. Gray disks – cations; yellow disks – anions.

nanowire were calculated. It was found that obtained data for nanowire are in reasonably good agreement (if we restrict our attention to the problems of electron state localization near the dislocation line) with the results of calculations for nanocryatal. By the way, the results of non-empirical calculations (see [6]) are in reasonably *qualitative* good agreement with the results of semiempirical calculations [1, 2]. Non-empirical quantum-chemical calculations were made with the use of the program "Gaussian 03" [8] which was installed on the parallel computer cluster of the Digital Technologies Department of Computer Sciences Faculty, Voronezh State University, and with the use of the program "Firefly" [9, 10] on the supercomputer "Chebyshev" of Moscow State University.

For simplicity we have investigated the electron states of impurities near the dislocation line in cluster approach (in fact, the nanocrystals were considered). According to paper [6] the occupied states are localised to a greater degree near anions C and G (fig. 1). Because of this result we located the substitutional isoelectronic impurities (Br^- or I^-) in C position. Taking into account that a relaxation of ions near substitutional isoelectronic impurity ion is ordinary rather small, we completely neglected this relaxation. Further, for estimations of matrix elements of Hamiltonian we have used the simplification in accordance with paper [11]. For comparison the electronic structure of nanocrystal $K_{196}Cl_{196}$ ($7 \times 7 \times 8 = 392$ ions) without a dislocation was considered also.

The nanocrystals to be studied (without and with substitutional isoelectronic impurity ions) are the systems containing "whole ions". These systems allow to use the partial self-consistency approximation [3]. Three algorithms for calculations with the partial self-consistency have been presented in paper [3]. In particular, the ordinary recursion method may be used in this approximation. In our paper we show the results obtained by recursion method in variant of Nex [12, 13].

3. Results and discussion

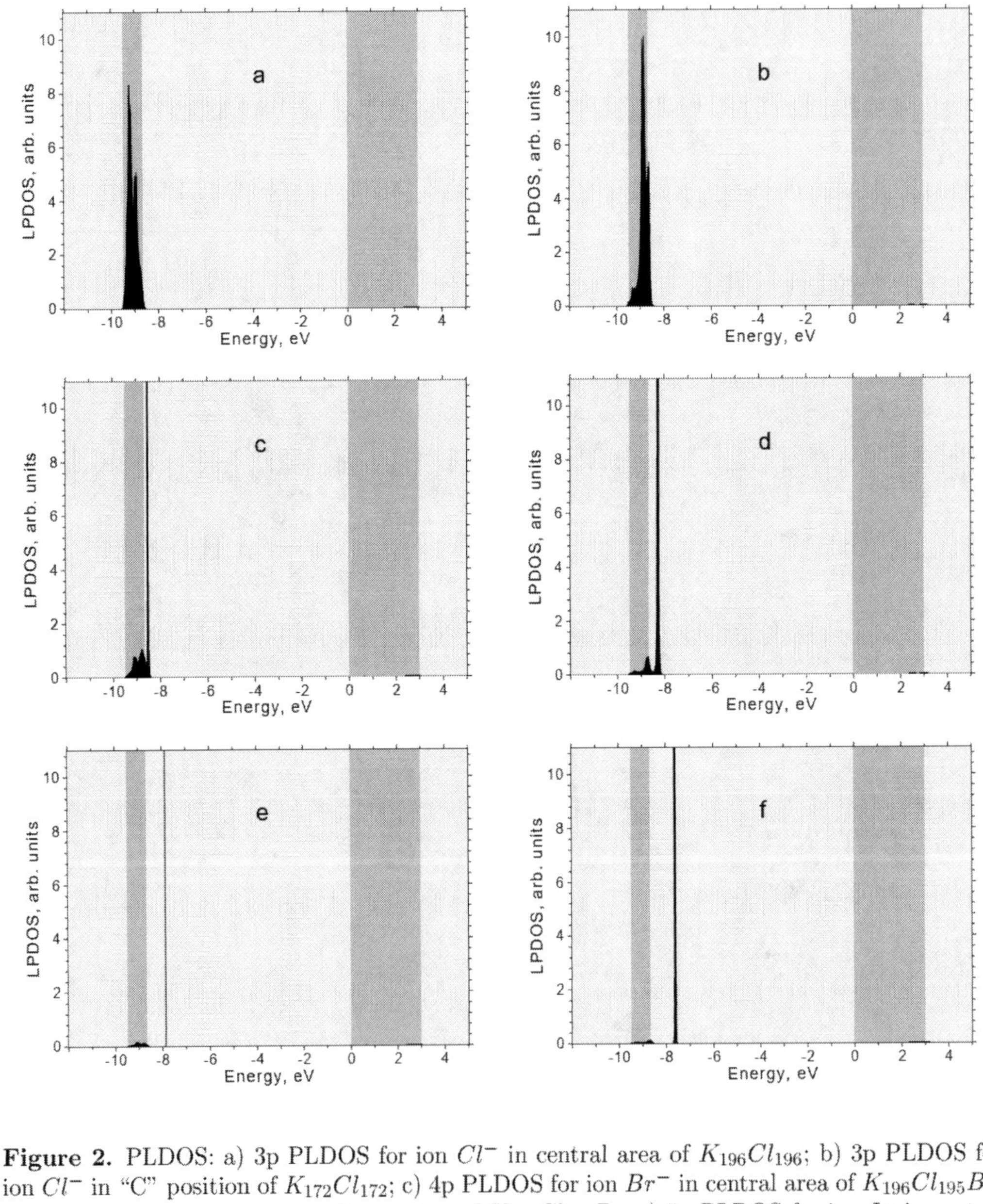

Figure 2. PLDOS: a) 3p PLDOS for ion Cl^- in central area of $K_{196}Cl_{196}$; b) 3p PLDOS for ion Cl^- in "C" position of $K_{172}Cl_{172}$; c) 4p PLDOS for ion Br^- in central area of $K_{196}Cl_{195}Br$; d) 4p PLDOS for ion Br^- in "C" position of $K_{172}Cl_{171}Br$; e) 5p PLDOS for ion I^- in central area of $K_{196}Cl_{195}I$; d) 5p PLDOS for ion I^- in "C" position of $K_{172}Cl_{171}I$.

Some results of calculations are shown in fig. 2. Namely, The PLDOS for central anions of $K_{196}Cl_{196}$, $K_{172}Cl_{172}$, $K_{196}Cl_{195}Br$, $K_{172}Cl_{171}Br$, $K_{196}Cl_{195}I$, and $K_{172}Cl_{171}I$ are presented.

The regions of valence and conduction bands of ideal crystal KCl are marked by light-green color.

Let us extrapolate *in our thoughts* the graphical data of fig. 2a and fig. 2b to monocrystals. Then it becomes evident that largest splitting off for the occupied dislocation energy levels is sufficiently less than 1 eV in spite of results of some publications (see short review of such results in paper [1]). Further, according to fig. 2(c – f) localization of occupied electron states connected with substitutional isoelectronic impurity anion is being increased if impurity location near the dislocation line. The maximal value of PLDOS is being shifted towards the bottom of conduction band on 0.23 eV in case of substitutional isoelectronic impurity anion Br^- location near the dislocation line. In case of substitutional isoelectronic impurity anion I^- energy shift is equals to 0.25 eV.

4. Conclusions

Thus, in present work the electron structure simulations of KCl wires with the edge dislocations and the isoelectronic substitutional impurities were carried out. For simplicity the wires were modelled by the nanocrystals extended along one axis. The plots of PLDOS are presented and discussed. The obtained results show that the substitutional isoelectronic impurity anions Br^- and I^- located near the dislocation line capture the holes more efficiently than in bulk of KCl systems without dislocations.

Acknowledgments

Many thanks to Professor S. D. Kurgalin (Head of the Digital Technologies Department of Voronezh State University) for the possibility of using the computer cluster.

References

[1] Timoshenko Yu K and Shunina V A 2002 *Physics of the Solid State (Russia)* **44** 888
[2] Timoshenko Yu K and Shunina V A 2005 *Phys. Stat. Sol. (c)* **2** 1788
[3] Timoshenko Yu K 2014 *J. Phys.: Conf. Series* **490** 012173
[4] Masuda-Jindo K, Hung V V and Menon M 2005 *Phys. Stat. Sol. (c)* **2** 1781
[5] Omar M S and Taha H T 2009 *Physica B* **404** 5203
[6] Timoshenko Yu K, Shunina V A and Shashkin A I 2014 *J. Phys.: Conf. Series* (in press)
[7] Timoshenko Yu K and Shunina V A 2009 *Surface Sci.* **603** 2564
[8] Frisch M J, Trucks G W, Schlegel H B, Scuseria G E, Robb M A, Cheeseman J R, J A Montgomery J, Vreven T, Kudin K N, Burant J C, Millam J M, Iyengar S S, Tomasi J, Barone V, Mennucci B, Cossi M, Scalmani G, Rega N, Petersson G A, Nakatsuji H, Hada M, Ehara M, Toyota K, Fukuda R, Hasegawa J, Ishida M, Nakajima T, Honda Y, Kitao O, Nakai H, Klene M, Li X, Knox J E, Hratchian H P, Cross J B, Adamo C, Jaramillo J, Gomperts R, Stratmann R E, Yazyev O, Austin A J, Cammi R, Pomelli C, Ochterski J W, Ayala P Y, Morokuma K, Voth G A, Salvador P, Dannenberg J J, Zakrzewski V G, Dapprich S, Daniels A D, Strain M C, Farkas O, Malick D K, Rabuck A D, Raghavachari K, Foresman J B, Ortiz J V, Cui Q, Baboul A G, Clifford S, Cioslowski J, Stefanov B B, Liu G, Liashenko A, Piskorz P, Komaromi I, Martin R L, Fox D J, Keith T, Al-Laham M A, Peng C Y, Nanayakkara A, Challacombe M, Gill P M W, Johnson B, Chen W, Wong M W, Gonzalez C and Pople J A 2004 *Gaussian 03, Revision C.02* (Wallingford CT: Gaussian, Inc.)
[9] Granovsky A A Firefly version 7.1.g http://classic.chem.msu.su/gran/firefly/index.html
[10] Schmidt M W, Baldridge K K, Boatz J A, Elbert S T, Gordon M S, Jensen J H, Koseki S, Matsunaga N, Nguyen K A, Su S, Windus T L, Dupuis M and Montgomery J A 1993 *J. Comput. Chem.* **14** 1347
[11] Timoshenko Yu K and Shununa V A 2008 *Proc. of SPIE (Physical Chemistry of Interfaces and Nanomaterials VII : Conference of SPIE Symposium on NanoScience + Engineering, 10–14 August 2008, San Diego, USA)* **7034** 70340
[12] Nex C M M 1978 *J. Phys. A* **11** 653
[13] Nex C M M 1984 *Comp. Phys. Commun.* **34** 101

On the computer simulations of electron states of ionic nanosystems with extended defects

Yury K Timoshenko
Voronezh State University, Universitetskaya pl., 1, Voronezh, 394006, Russia
E-mail: yutim@phys.vsu.ru

Abstract. The electron states of nanosystems of ionic compounds AgCl and KCl with extended defects (charged defects, edge dislocations) were under consideration. The semi-empirical tight-binding approximation and different calculation methods were used. The obtained results and efficiency of calculation schemes are discussed. This work is an improvement of our early investigations.

1. Introduction

The notion of the "extended defect" means a defect which perturbs extensive region of a crystal. Dislocations and charged defects in ionic systems are extended defects as a rule. The calculation procedures for equilibrium space configurations and electron states of such defects in polar semiconductors and insulators have some particularities caused by long range Coulomb interaction and effects of polarization. Which is why the executions of self-consistent calculations of electron structure of polar systems with extended defects cause serious difficulties. But in special case of polar systems which consist of the "whole ions" it's possible to use the partial self-consistency approximation [1]. Some applications of computational approaches in partial self-consistency approximation are been considered in present paper.

2. Models and methods

Semi-empirical techniques was used. At the first step the equilibrium geometry structure of nanocrystal (NC) was calculated within the model of pointlike polarizable ions by minimization of the total potential energy over the ion coordinates or by molecular dynamics method. Next, after determining of the ions coordinates in the equilibrium NC configuration the electron structure was calculated by semi-empirical self-consistent tight-binding method. If NC had large size and contained "whole ions" only, the partial self-consistency approximation was used. The use of partial self-consistency allow us to apply the effective methods without executing of unitary transformation of Hamiltonian matrixes. For example, numerical investigations of size effect influence on electronic structure of KCl nanocrystal (NC) with charged surface defect were carried out with the use of such algorithms, during which maximal NC under consideration was $(K_{7504}Cl_{7505})^-$. For more details, see the papers [1, 2]. Naturally, for NC with some hundred ions the standard problems of calculations of eigenvalues and eigenvectors are resolved quite efficiently.

In this work we have restricted the study of extended defects by charged defects only. Namely, adsorbed ion Ag^+ on surface of NC $AgCl$ has been considered. This problem was studied by us earlier [2]. In present paper the additional results of computer simulation of the localized electron states and the probabilities of quantum transitions are presented.

Generally speaking, absorbed energy is not equal to a difference of one-particle energies of system in the ground state. However, with a view of the qualitative analysis it is assumed very often [3, 4]. According to these papers, in the present work the simple formulas for calculations of probabilities of transitions and absorption factor in the electric dipole approximation (see, for example, [5]) were used:

$$\mathcal{P}_{n\to n'} = c_1|\,\mathbf{e}\cdot\mathbf{M}_{n'n}|^2\delta\left(E_{n'} - E_n - \hbar\,\omega\right), \tag{1}$$

$$\alpha\left(\hbar\,\omega\right) = \frac{c_2}{\hbar\,\omega}\sum_{n,n'}{}^{'}|\,\mathbf{e}\cdot\mathbf{M}_{n'n}|^2\delta\left(E_{n'} - E_n - \hbar\,\omega\right), \tag{2}$$

where

$$\mathbf{e}\cdot\mathbf{M}_{n'n} = \langle\psi_{n'}| - i\hbar\,\mathbf{e}\cdot\nabla|\psi_n\rangle. \tag{3}$$

Here (3) – matrix element of optical transition; c_1, c_2 –constants; $\mathbf{e}$ – unit vector in direction of an electric field strength; E_n and ψ_n – energy and a wave function n th one-electron state; $\hbar\,\omega$ – energy of quantum.

Following the paper [4], the atomic orbitals (AOs), usually applied in non-empirical quantum chemistry, were used for calculation of matrix elements (3). Such procedure has allowed authors of paper [4] to carry out the successful analysis of dipole matrix elements for the silicon quantum wires in the qualitative agreement with later non-empirical DFT calculations. Further, AOs, generally speaking, do not form orthonormal basis set of functions. Meanwhile, in our variant of tight-binding method it is supposed. Therefore the orthogonalizaiton and normalization of AOs were carry out with use of iterative algorithm [6].

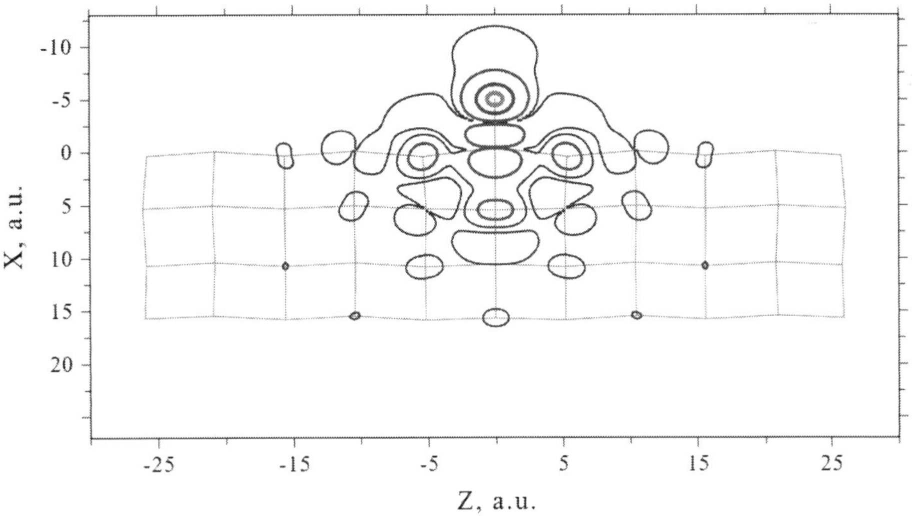

Figure 1. Module of the wave function of AgCl NC with the adsorbed silver ion (485 ions): first unoccupied state.

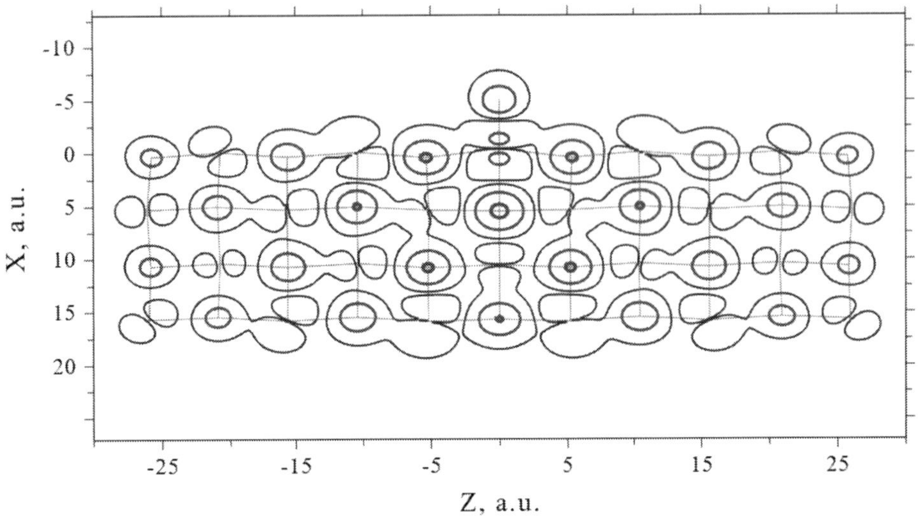

Figure 2. Module of the wave function of AgCl NC with the adsorbed silver ion (485 ions): second unoccupied state.

3. Results

The investigated nanocrystals contained four layers 11 × 11 ions without (484 ions – neutral system) and with adsorbed ion Ag^+ (485 ion – charged system); see fig. 2a, [2]. The linear combinations of Gauss Cartesian functions as the basis orbitals with parameters taken from [7] for 4d and 5s shells of silver and from the tables [8] for 3p shell of chlorine were used.

There is strong localization of first unoccupied state near adsorbed ion, see fig. 1. For comparison the module of a wave function for the second unoccupied state is given, fig. 2. The contours of the wave functions are presented in XZ plane (adsorbed ion's position on vertical axis). The function values that define the contour lines to be calculated are 0.5 (red color), 0.1 (dark-red color), 0.01 (blue color), and 0.001 (dark-blue color).

How much strongly the influences of adsorbed ion on the nanocrystal absorption spectrum? The calculated absorption spectrums and joint density of states are presented on fig. 3. In the formula (2) $|\,\mathbf{e}\cdot\mathbf{M}_{n'n}|^2$ it has been exchanged on $|\,\mathbf{e}_x\cdot\mathbf{M}_{n'n}|^2+|\,\mathbf{e}_y\cdot\mathbf{M}_{n'n}|^2+|\,\mathbf{e}_z\cdot\mathbf{M}_{n'n}|^2$.

The peak value of probability of quantum transition for NC $Ag_{242}Cl_{242}$ at 5.5 eV was accepted equal to unity. Further this normalization was taken into account for plotting of other spectrums. Joint densities of states were calculated with use of formula (2) in which the matrix element of optical transition was equal to unity. According to the practice of calculation of such expressions, the Dirac delta function in the formula (2) was exchanged by a Gaussian. For the best visual perception the joint densities of states were normed so that their maximums coincided with maximums of corresponding absorption spectrums.

The labels E_g^{indir} (the indirect forbidden gap), E_g^{NC} (the difference between lowest unoccupied and highest occupied energy levels of NC), and E_g^{dir} (the direct forbidden gap) are presented in fig 3 also. It is necessary to note, that the direct forbidden gap was found by the calculations of joint density of states for ideal crystal.

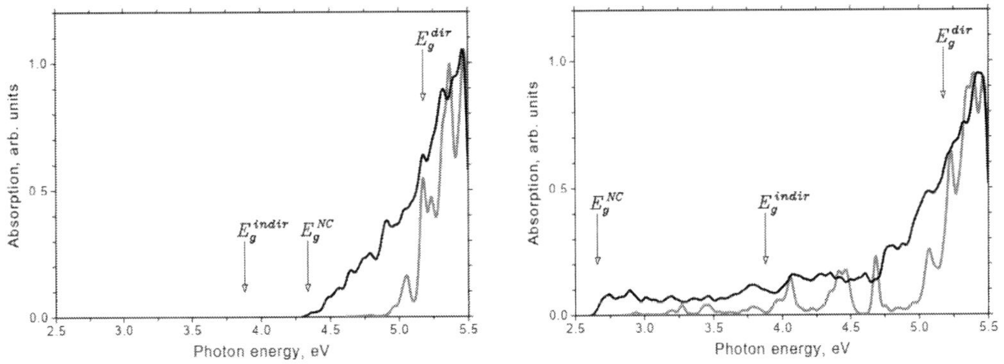

Figure 3. The absorption spectrums (red lines) and joint densities of states for NC $Ag_{242}Cl_{242}$ (left) and $(Ag_{243}Cl_{242})^+$ (right).

4. Conclusions

The absorption spectrum and joint density of states are essentially different near the absorption edge of NC $Ag_{242}Cl_{242}$ (fig. 3, left). The left boundary of joint density of states corresponds to energy gap of NC (energy gap is equal to difference of energy of the lowest unoccupied level and energy of the uppermost occupied level). Meanwhile, appreciable absorption begins with energies which are about the width of direct forbidden gap of ideal crystal $AgCl$!.

Thus, in the present work with use of computer simulation it is shown: 1) presence of the short-range order in silver chloride NC consisting of several hundreds of ions is affected on a absorption spectrum: absorption spectrum of NC is similar to an absorption spectrum of an ideal crystal $AgCl$ near to its edge; 2) extended charge defect leads to appearance of absorption below edge of absorption spectrum of an ideal crystal $AgCl$.

References

[1] Timoshenko Yu K 2014 *J. Phys.: Conf. Series* **490** 012173
[2] Timoshenko Yu K and Shunina V A 2009 *Surface Sci.* **603** 2564
[3] Farberovich O V, Timoshenko Yu K, Bugakov A M and Domashevskaya E P 1981 *Solid State Communications* **40** 559
[4] Kopaev Yu V, Molotkov S N and Nazin S S 1992 *JETP Lett.* **55** 728
[5] Bassani F and Pastori Parravicini G 1975 *Electronic States and Optical Transitions in Solids* (Oxford, New York: Pergamon Press)
[6] Xiang H J, Li Z, Liang W Z, Yang J, Hou J G and Zhu Q 2006 *J. Chem. Phys.* **124** 234108
[7] Friedlander M E, Howell J M and Snyder G 1982 *J. Chem. Phys.* **77** 1921
[8] Huzinaga S 1984 *Gaussian Basis Sets for Molecular Calculations. Physical Sciences Data. Vol. 16* (Amsterdam–Oxford–New York–Tokio: Elsevier)

Accelerating voltage effect on the ablation process during pulsed electron deposition

Muddassir Ali and Redhouane Henda

Bharti School of Engineering, Laurentian University, 935 Ramsey Lake Road, Sudbury, ON P3E2C6, Canada

E-mail: mx1_ali@laurentian.ca

Abstract. In this study, we have analyzed the effect of electron beam accelerating voltage on the maximum temperature of a graphite target during pulsed electron beam ablation (PEBA). To this end, a two stage, one dimensional thermal model is used. The target is subjected to an electron pulse with accelerating voltages of 10, 13, 15, 17 and 18 kV. The surface temperature, vaporization front velocity, ablated depth, and ablation rate are estimated from the solution of the model. Simulation results have shown that target surface temperature is not proportional to the accelerating voltage. It has been found that the surface temperature increases with the accelerating voltage from 10 kV to 15 kV, and reaches a maximum value (7500 K) at 15 kV. After 15 kV, the temperature decreases with increasing accelerating voltage. Similar trends have been observed in the vaporization front velocity, ablated depth and ablation rate, with maximum values (75 m/s, 2.1 μm, and 4 μg/mm^2) at 15 kV. The calculation results are in good agreement with relevant experimental data from the literature.

1. Introduction

Over the past few decades, the interaction of intense pulsed energy beams with solid-state materials has gained growing interest for the fabrication of high quality thin films. Pulsed electron beam ablation has recently emerged as a potential alternative to well-known pulsed laser deposition (PLD) for the fabrication of thin films with superior qualities. PEBA is conceptually similar in overall physics to PLD, with distinctive advantages of efficient energy transfer (electric-electric: ~30 %), lower operating costs, and ease of use [1-3]. Thin film deposition during PEBA is significantly influenced by many factors such as power density, beam energy, background gas pressure, and target to substrate distance. Target ablation is mainly affected by two critical factors, which are beam energy and power density delivered to the target surface. Experimental results have shown that it is not always favorable to operate the electron beam at higher voltage in order to increase the target heating rate. The fundamental principle of PEBA involves the generation of a highly energetic pulsed electron beam with a short pulse width (~100 ns) and high power density (10^8 W/cm^2). The energetic electron beam impinges on the target surface, which leads to rapid sublimation of a fine fraction of the surface and to the ejection and expansion of a plasma plume. The latter will eventually condense as a thin film on a properly placed substrate in a vertical direction to the target. Due to high power density delivered to the target surface (10^8 W/cm^2), fast heating and vaporization take place far from thermodynamic equilibrium. Solid-to-vapor transition is principally independent of the phase diagram and composition of the target material, resulting in congruent sublimation and potentially, target stoichiometry preservation in the deposited thin film [1,4].

Despite the fact that PEBA has been used in the fabrication of thin films for two decades now, very limited attempts have been made to model and optimize the critical parameters involved in the ablation process [5,6]. In an early attempt, Kowalewicz and Redel [5] have proposed a model, which describes irradiation by a polyenergetic electron beam, energy dissipation via conduction and the effect of the vapor pressure of the ablated mass on the temperature of the target. Recently, Tricot et al. [6] have suggested a simple model of heat conduction in a target irradiated by a polyenergetic pulsed electron beam, but does not include phase change during ablation. Both of these models do not account for the receding surface of the target during ablation. In order to thoroughly assess PEBA for the deposition of thin films with superior properties, it is essential to interpret how the critical parameters, particularly the accelerating voltage behaves during the ablation process. As mentioned earlier, beam accelerating voltage is one of the key parameters towards thin film fabrication control. In this study, we present the effect of accelerating voltage on the ablation efficiency of graphite.

2. Mathematical model

The interaction of an electron beam with a target material is quite a complex physical process and takes place in a sequence of phenomena. When the target surface is exposed to electron beam radiation, the incident beam energy is mostly absorbed by the electron subsystem. Later, due to interactions between electrons and phonons, energy is transferred to phonons in the lattice. The energy transferred to the phonons propagates in the target material through the mechanism of heat conduction [7]. The thermal model for electron beam-target interaction in the regime of sublimation can be categorized to occur in two stages: (i) heating of the target surface, and (ii) sublimation of the target surface. Therefore, the temperature distribution inside the target surface along the depth can be described by a two stage one-dimensional heat conduction model. The latter depends on the incident electron beam parameters and the thermophysical properties of the target material. The diameter of the electron beam is normally in the range of several millimetres, which is much greater than the diffusion depth. Therefore, heat conduction can be assumed to take place predominantly in one dimension along the target depth [8]. The heat conduction equation, which allows the calculation of time-dependent temperature distribution along the target depth, $T(x,t)$, before the target surface has started to vaporize is expressed as,

$$\rho_s C_{p_s} \frac{\partial T}{\partial t} = \nabla \cdot (k_s \nabla T) + S(x,t) \qquad (0 < t \le t_s) \tag{1}$$

with the initial condition,

$$T(x,0) = T_0 \tag{2}$$

and boundary conditions,

$$-k_s \nabla T \Big|_{x=0} = Q_s(t) \qquad (0 < t \le t_s) \tag{3}$$

$$-k_s \nabla T \Big|_{x=l} = 0 \qquad (0 < t \le t_s) \tag{4}$$

where ρ_s, C_{p_s}, and k_s are the solid phase density, solid state specific heat capacity, and solid phase thermal conductivity of the target material, respectively. S is the heat source term, T_0 is the initial temperature of the target surface, x is the distance from the target surface along the incident direction, l is the thickness of the target material, Q_s is the surface power density delivered by the heat source, and t_s is the time required by the target surface to attain the sublimation point.

When the target surface temperature increases and exceeds the sublimation temperature, vaporization will start to take place. During this sta. During the vaporization stage, the thermal field in the target can be calculated by introducing the moving vaporization front in the one-dimensional heat conduction equation [9]. The latter takes the form,

$$\rho_v C_{p_v}\left(\frac{\partial T}{\partial t} - u(t)\frac{\partial T}{\partial x}\right) = \nabla \cdot (k_v \nabla T) + S(x,t) \qquad (t_s \leq t \leq \tau) \tag{5}$$

here the initial and boundary conditions are

$$T(0,t) = T_s(t) \tag{6}$$

$$-k_v \nabla T \Big|_{x=0} = Q_s(t) - L_v \rho_v u(t) \qquad (t_s \leq t \leq \tau) \tag{7}$$

$$-k_v \nabla T \Big|_{x=l} = 0 \qquad (t_s \leq t \leq \tau) \tag{8}$$

where ρ_v, C_{p_v}, and k_v are the vapor phase density, vapor phase specific heat capacity, and vapor phase thermal conductivity of the target material, respectively. $u(t)$, L_v, and τ are the vaporization front velocity, latent heat of vaporization, and duration of the electron beam pulse, respectively. The time required by the target surface to attain the sublimation point can be estimated by [10],

$$t_s = \frac{3(T_{sb} - T_0)^2 \rho_s C_{p_s} k_s}{4Q_s(t)^2} \tag{9}$$

During the process of sublimation, the vaporization front velocity can be evaluated assuming that the vaporization flux from the target surface follows Hertz-Knudsen equation [9], and it can be given by,

$$u(t) = 0.82\frac{P_s(T_s)}{\rho}\left(\frac{m}{2\pi k_B T_s}\right)^{1/2} \tag{10}$$

where $P_s(T_s)$ is the saturated vapor pressure at the surface temperature T_s, m is the average mass of target particle, and k_B is the Boltzmann constant. The solid to the vapour phase transition has been assumed to follow the Clausius-Clapeyron equation [9],

$$P_s(T_s) = P_b exp\left[\frac{mL_v}{k_B}\left(\frac{1}{T_{sb}} - \frac{1}{T_s}\right)\right] \tag{11}$$

where P_b, L_v, and T_{sb} are the atmospheric pressure, latent heat of vaporization, and sublimation temperature during the atmospheric pressure P_b.

The source of electrons in PEBA is polyenergetic [5]. In order to describe the heat source appropriately, the distribution of energy must be taken into consideration. Monte Carlo simulations are able to estimate the profile of the energy transferred to the target material by the electron beam. Monte Carlo is based on tracking a primary electron of initial energy E and direction v, and finding where the next collision will occur based on the calculation of the inelastic mean free path. To achieve statistical sampling, this process is repeated for thousands of electrons using CASINO code [11]. The analytical expression of maximum energy deposited by each group of electrons takes a polynomial form as

$$E(x) = f_{A,B,C}(x) = \sum_{i=0}^{7} a_i \times x^i \tag{12}$$

where x is the penetration depth along the target material. In order to account for energy losses, the fraction of beam electrons that are effectively used during ablation has been set to 0.6. This parameter has been assessed in a separate study [12].

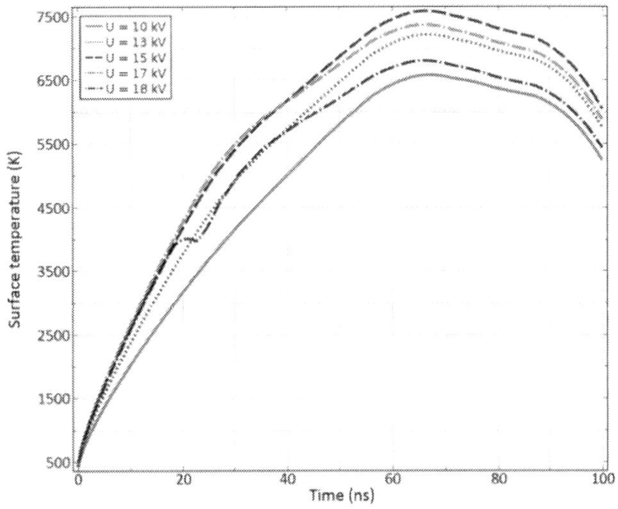

Figure 1. Target surface temperature for various accelerating voltages as a function of pulse duration.

3. Results and discussion

In our calculations, the target is initially at ambient room temperature (300 K). Due to the electron beam energy deposition in the target material, the surface is heated very rapidly. The calculated target surface temperature for various values of the accelerating voltage is depicted for the entire pulse width (~100 ns) in Fig. 1. As anticipated, the maximum temperature is located at the target surface. It can be seen that the surface temperature and heating rate increase rapidly for the accelerating voltage in the range of 10 to 15 kV. This finding is likely due to the polyenergetic nature of the electron beam [5]. The majority of electrons produced by the beam for the voltage in the range of 10 to 15 kV have lower energies. As a result, the maximum energy of the electrons is deposited near the target surface. The highest temperature is observed for an accelerating voltage of 15 kV at about 7500 K between 60 and 70 ns. At higher voltages, the heating rate falls due to an increase in the penetration depth of the electrons, which offsets the intensity of the beam. As the acceleration voltage increases the energy of the electrons in the beam increases while, at the same time, the highly energetic electrons will penetrate deeper in the target. This results in greater losses of the beam energy, which becomes predominate at higher voltages (above 15 kV), and eventually cancels out the effect of the acceleration voltage on the beam energy. For 17 kV, it can be seen that initially the heating rate is about the same as 15 kV, but it slightly decreases after 40 ns yielding a maximum temperature of 7300 K. The evolution of the target surface temperature with time and dependence on voltage seem to have a similar trend for all voltages except for 18 kV. At 18 kV, the temperature profile goes through a short plateau at ~20 ns corresponding to a target surface temperature of 4000 K. This is the result of phase change due to sublimation, which starts to occur around 4000 K. Afterwards, there is a sudden climb in the temperature whereby the surface temperature reaches 6700 K. For lower voltages than 18 kV, sublimation seems to occur much faster so that not plateauing of the temperature can be observed.

The vaporization front velocity from the target surface is illustrated in Fig. 2. As can be seen, at 15 kV the velocity reaches a maximum value of 75 m/s at about 65 ns, i.e., when the surface temperature of the target is at maximum (7500 K). Beyond the latter, the surface front receding velocity decreases as the target surface temperature drops. A similar trend can be observed for the other values of the accelerating voltage.

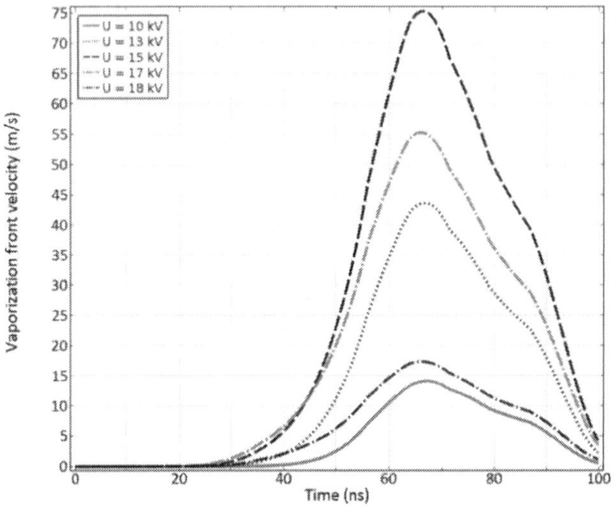

Figure 2. Vaporization front velocity for various accelerating voltages as a function of pulse duration.

The ablated depth and ablation rate can be obtained from the surface front receding velocity. These are plotted as a function of pulse duration in Fig. 3 and 4, respectively. As expected for the case of 15 kV, the calculated ablated depth and ablation rate are larger than at other voltages. The calculated ablated depth is around 2.1 μm and the ablation rate is about 4 μg/mm^2. Similar profiles of both ablated depth and ablation rate for the remaining accelerating voltage values can be observed.

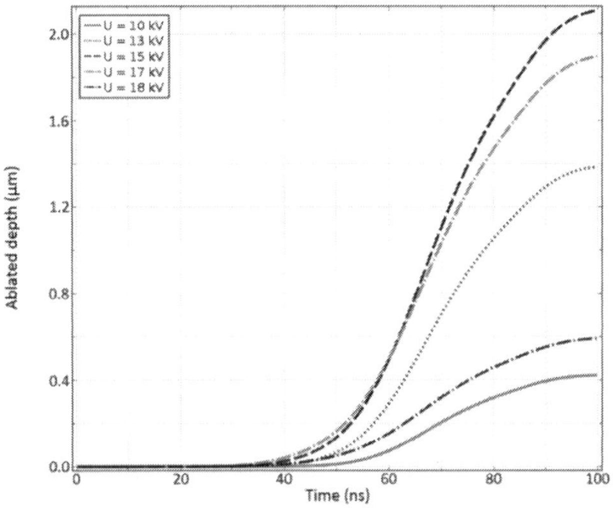

Figure 3. Ablated depth for various accelerating voltages as a function of pulse duration.

4. Model validation

Experimental data on the solid-state target ablation by pulsed electron beams are very scarce and more so on graphite. To validate the model, the estimated parameters for different values of the accelerating voltage are compared with relevant experimental data in the literature [4,5,13-16]. For all voltage values, the ablated rate ranges between 9 and 50 μg/pulse or 50×10^{16} and 250×10^{16} atoms/pulse, which is in good agreement with reported experimental data of a few tens of μg/pulse [15] or some 10^{16} atoms/pulse

[5]. Experimental results have shown that the ablation rate in PEBA is at least 10 times larger than its counterpart in PLD. This is true in our case where the ablation rate in PEBA is 10-20 times the rate in PLD for a graphite target [14,15]. In terms of the heating rate, our results are in qualitative (no absolute values have been provided by the authors) agreement with experimental data [4,13]. The latter indicate that the heating rate is at maximum at a voltage of about 15 kV. Finally, electron beams have been reported to reach a penetration depth in the target of 1-2 μm [15,16]. This is in good accordance with our calculated ablated depth as reported in Fig. 3.

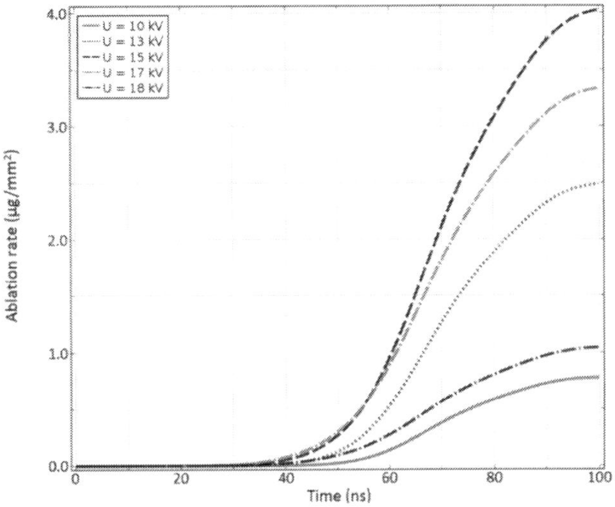

Figure 4. Ablation rate for various accelerating voltages as a function of pulse duration.

5. Conclusion

We have proposed a two stage, one-dimensional heat conduction model to investigate graphite ablation during PEBA at various electron beam accelerating voltages. The temporal dependence of the temperature of the target has been numerically estimated. Vaporization front velocity, ablated depth and ablation rate have also been calculated. The findings have shown that an optimum value of the accelerating voltage is around 15 kV. The latter corresponds to the maximum value of the surface front receding velocity, ablated depth and ablation rate. The calculation results have been compared with available experimental data and have been found to be in overall good agreement.

Acknowledgment

The authors would like to acknowledge the financial support received from the Natural Sciences and Engineering Research Council (NSERC) of Canada.

References

[1] Müller G, Konijnenberg M, Krafft G and Schultheiss C 1995 Thin film deposition by means of pulsed electron beam ablation *Scienc and Technology of Thin Films* ed F C Matacotta and G Ottaviani (Singapore: World Scientific) pp 89-119

[2] Gilgenbach R M, Kovaleski S D, Lash J S, Ang L K and Lau Y Y 1999 *IEEE Trans. Plasma Sci.* **27** 150

[3] Kovaleski S, Gilgenbach R M, Ang L K, Lau Y Y and Lash J S 1998 *Appl. Surf. Sci.* **127-129** 947

[4] Strikovski M, Kim J and Kolagani S H 2010 Plasma energetics in pulsed laser and pulsed electron deposition *Springer Handbook of Crystal Growth* ed G Dhanaraj, K Byrappa, V Prasad and

M. Dudley (Berlin Heidelberg: Springer–Verlag) pp 1193-1211
[5] Kowalewicz R and Redel T 1995 *IEEE Trans. on Plasma Sci.* **25** 270
[6] Tricot S, Semmar N, Lebbah L and Boulmer–Leborgne C 2010 *J. Physs D: Appl. Phys.* **43** 065301
[7] Ptitsin VE 2012 *Phys. Status Solidi C* **9** 15
[8] Steibeck J, Braunstein G, Dresselhaus M S, Venkatesan T and Jacobson D C 1985 *J. Appl. Phys.* **58** 4374
[9] Peterlongo A, Mintello A and Kelly R 1994 *Phys. Rev.* E*: Stat. Nonlinear Soft Matter Phys.* **50** 4716
[10] Zhang D, Liu D, Li Z, Li G, Tan X, Li L, Fang R, Hu D and Liu G 2007 *Appl. Surf. Sci.* **253** 6144
[11] Drouin D, Couture A R, Joly D, Tastet X, Aimez V, and Gauvin R 2007 *Scanning* **29** 92
[12] Ali M, and Henda R 2014 *Science & Applications of Thin Films Conference & Exhibition (Cesme, Turkey, 15-19 September 2014)* vol 1 p 95
[13] Strikovski M and Harshavardhan KS 2003 *Appl. Phys. Lett.* **82** 853
[14] Bulgakova N M and Bulgakov A V 2001 *Appl. Phys. A: Mater. Sci. Process.* **73** 199
[15] Harshavardhan K S and Strikovski M 2005 Pulsed electron–beam deposition of high temperature superconducting films for coated conductor applications *Second Generation HTS Conductors* ed A. Goyal, (Boston: Kluwer Academic Publishers) 109-33
[16] Hobel M, Geerk J, Linker G and Schultheiss C 1990 *Appl. Phys. Lett.* **56** 973

The origin of self-focusing effect in terahertz quantum cascade lasers

Jian Wang[1], **Yanfang Li**[2], **and Ning Yang**[2*]
[1]Department of Physics, Beijing Jiaotong University, 100044, Beijing, China
[2]Institute of Applied Physics and Computational Mathematics, P.O. Box 8009(28), 100088, Beijing, China
*yang_ning@iapcm.ac.cn

Keyword: Quantum cascade lasers, Self-focusing, Kerr lensing

Abstract. The terahertz quantum cascade lasers (THz-QCLs) are the compact and coherent terahertz light sources based on the inter-subband transition and resonant tunneling of carriers in semiconductor superlattice. In recent studies on tapered THz-QCLs, it was found that the self-focusing effect in the active region of the devices may cause the abnormal increase of the far-field divergence of the laser beam. By simulating the propagation of optical mode in QCL waveguide and considering both the nonlinearity effect and thermal accumulation in the active region, we propose that the refractive index change caused by the third-order nonlinearity of multi-quantum-wells in active region may be the key reason for the self-focusing in THz-QCLs. This result indicates that the nonlinear effect may have great impact on the beam quality of QCLs which must be carefully considered in applications of THz-QCLs, such as the TH- imaging systems.

1. Introduction

Quantum cascade lasers (QCLs), whose emission wavelengths can cover the frequency range from mid-infrared to terahertz (THz) regions[1, 2], are light sources based on the inter-subband transitions and resonant tunnelings of carriers in multiple quantum wells. The THz-QCLs have great applications in the fields of security check, free space optical communication, spectroscopy and imaging. However, the small output power and large divergence of the laser beam still limit the applications of THz-QCLs. To improve the output power maintaining the single-mode operation, the tapered gain region had been widely used in near-infrared diode and mid-infrared QCLs[3, 4, 5].

Recently, we developed the THz-QCLs with tapered gain region and demonstrated the power amplification effect[6]. The SEM imaging of the devices is shown in Fig.1a. The devices are composed of a ridged region of 0.5 mm$\times$103 μm (length$\times$width) and a tapered region of 2 mm length and with 0°, 3°, 5°, and 8° tapered angles, respectively. In the experiment, we observed an abnormal increase of far-field divergence of the tapered THz-QCLs. More specifically, the horizontal divergence of a far field laser beam decreases with increasing the tapered angle only when the angle is smaller than 5°, and then it increases with further increasing the angle. Generally speaking, the increase of the aperture of the output facet will decrease the divergence

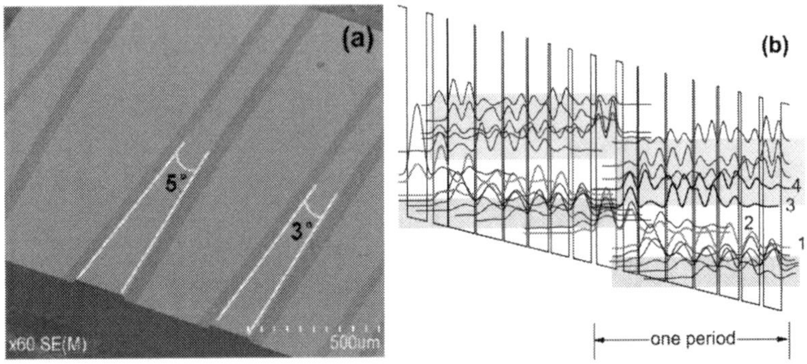

Figure 1. (a) SEM image of the tapered THz QCLs. (b) Conduction band energy diagram for two periods of multi-quantum-wells in the THz-QCL studied in Ref.[6]. The thicknesses of the GaAs quantum wells and $Al_{0.15}Ga_{0.85}As$ barriers of one period are, from left to right, 12.6/**4.4**/9.3/**0.6**/16.9/**0.9**/16.6/**1.0**/14.3/**1.2**/ 12.5/**1.5**/11.4/**2.5**/11.4/**3.3** nm. The barriers are in bold font and the underlined layer are doped with Si to 3×10^{16} cm^{-3}. The designed bias is 2.55 kV/cm. The red subbands are upper and lower laser levels, and the subbands labeled with numbers 1~4 are levels involved in the third order nonlinearity of the medium.

of laser beam, and the abnormal increase of the divergence may be due to the emergence of high-order transverse modes in the waveguide[7]. However, in our experiment, there are absorption stripes on the two sides of the ridges of the device which strongly suppress the high-order transverse modes. Then the observation suggests that there is self-focusing effect in the large tapered angle and high power THz-QCLs. The self-focusing can be caused by either the Kerr nonlinearity of the medium (Kerr lensing effect) or the thermal accumulation (thermal lensing effect). For the former case, if the laser medium possesses large third order susceptibility $\chi^{(3)}$, the refractive index n for a high intensity light beam itself is modified by $\Delta n_{in} = n_2 I$, where I is the optical intensity. For the latter case, the temperature gradient within the medium causes a refractive index change according to the local temperature as $\Delta n_{th} = \alpha_T T$, where α_T is the thermal coefficient and T is the local temperature. Then we want to answer whether the optical nonlinearity or the thermal accumulation accounts for the self-focusing in tapered THz-QCLs.

2. Simulation of nonlinearity and thermal accumulation in THz-QCL

To specify the origin of the self-focusing in THz-QCL, we need to include the index changes caused by the Kerr lensing and the thermal lensing into the simulation of the optical mode propagation. For the Kerr lensing, we first calculate the third order susceptibility of the active region superlattice. The conduction band diagram of the THz-QCLs studied in Ref.[6] is calculated and is shown in Fig.1b. The calculated energy separation between the two laser levels (red lines) is 11.3 meV (~2.75 THz), the observed laser wavelength is 103 μm (~2.9 THz). The subbands labeled with 1~4 are levels give birth to the third order optical nonlinearity. The density matrix equations can be employed to calculate the third order nonlinear susceptibility,

$$\begin{aligned} \frac{d\rho_{mn}}{dt} + (\gamma_{mn} + i\omega_0^{mn})\rho_{mn} &= \frac{ie}{\hbar}\mathbf{E}\sum_q (z_{mq}\rho_{qn} - z_{qn}\rho_{mq}) \\ \frac{d\rho_{mm}}{dt} + \hat{R}_m &= \frac{ie}{\hbar}\mathbf{E}\sum_q z_{mq}\rho_{qm} - z_{qm}\rho_{mq}, \end{aligned} \quad (1)$$

where m, n, q run over 1∼4, γ_{mn} is the relaxation rate of the off-diagonal element of the density matrix, ω_0^{mn} is the transition frequency, z_{mn} is the optical dipole between the subbands m and n, and $\hat{R}_m$ represents all relaxation and pumping terms for the suband m in the absence of the radiation fields. e is the electron charge and $\hbar$ is the reduced Plank constant. The radiation field is considered as a sum of components with slowly varying amplitudes $\mathbf{F}_{mn}$ and corresponding frequencies ω_{mn} near to the transition frequencies ω_0^{mn}. Solving the steady-state solution of Eq.(1), we can calculate the polarization of the system to arbitrary order of the nonlinearity and arbitrary field intensity. Since we are only concerned about the third order term which cause the index change, we expand the polarization in powers of the field amplitudes assuming weak enough field to get a analytical expression for the third order nonlinear susceptibility as

$$\chi^{(3)} \cong \frac{iez_{12}z_{23}z_{34}z_{14}N_e}{\hbar^3\Gamma_{41}}[\frac{1}{\Gamma_{31}}(\frac{n_2-n_3}{\Gamma_{32}}+\frac{n_2-n_1}{\Gamma_{21}}-\frac{1}{\Gamma_{42}}(\frac{n_3-n_4}{\Gamma_{43}}-\frac{n_2-n_3}{\Gamma_{32}}))], \qquad (2)$$

where $\Gamma_{mn} = \gamma_{mn} + i(\omega_0^{mn} - \omega_{mn})$, N_e is the density of electrons in the active region, $n_m = \rho_{mm}$ is the subband population. Then the second order nonlinear refractive index is given by $n_2 = \chi^{(3)}/2n_0$, where n_0 is the linear refractive index.

For the energy diagram shown in Fig.1b, the calculated energy separations ΔE_{12}, ΔE_{23}, and ΔE_{34} are 11.3, 11.1, and 10.5 meV, respectively. The corresponding dipole matrix elements are $z_{12} = 6.46$ nm, $z_{23} = 7.08$ nm, $z_{34} = 15.94$ nm, and $z_{14} = 3.13$ nm. Using $N_e = 5 \times 10^{15}$ cm^{-3}, assuming that all the population is in the upper laser level 2, and all $\gamma_{mn} = 10$ meV except $\gamma_{41} = 20$ meV, and neglecting all detunings, we can get a $\chi^{(3)} \sim 2.2 \times 10^{-5}$ esu and corresponding nonlinear refractive index $n_2 \sim 5.17 \times 10^{-8}$ cm^2/W. Such a large nonlinearity is attributed to the large dipole moments. Of course, this value is an estimated upper limit of the nonlinear refractive index. The real value depends on the subband lifetimes, populations and even the temperature. So we use a smaller value $n_2 \sim 5 \times 10^{-9}$ cm^2/W in our simulation.

To study thermal lensing effect in THz-QCLs, we need simulate the temperature distribution in QCLs. The thermal model of QCLs has been widely studied. Here we just give a brief note on it, the details can be found in Ref.[8, 9] and the references therein. The heat diffusion in QCLs is described by the heat equation $\nabla \cdot [\kappa \nabla T] + Q = 0$, where κ is the thermal conductivity tensor, T is the temperature, and Q is the heat generated per unit volume. The equation can be solved by standard finite-difference approach. Since the active region of the QCLs consist of near one thousand layers of quantum wells, the thermal conductivity is highly anisotropic. In the simulation, the thermal conductivity in the growth direction is significantly lower (typically one order lower) than thermal conductivity parallel to the layers. The detailed parameters of thermal conductivity of QCLs' active region, the waveguide layers, the substrate as well as the heat sink can be also found in Ref.[9].

In Fig.2, we present the simulated temperature distributions of 0°- and 8°-tapered THz-QCLs. It can be seen that the temperature in the center of 8°-device is higher than that of 0°-device due to the larger width. For both devices, the temperature gradient is mainly along the growth direction. In the direction parallel to the layer, the temperature gradient is much smaller. This is because the conduction with the heat sink is the main way of the heat abstraction in QCLs. After the simulation of temperature distribution, the local temperature and the thermal coefficient $\alpha_T = 3.3 \times 10^{-4}$K^{-1} are employed to study the thermal lensing effect in QCLs.

3. The origin of the self-focusing in tapered THz-QCLs

The optical mode within the waveguide of tapered THz-QCLs is described by the standard Helmholtz equation with the index changes due to the Kerr lensing and thermal lensing effects, and is solved with Beam propagation method[10]. Then the far-field distribution of the laser beam can be obtained from the optical mode distribution at the output facet according to the Rayleigh-Sommerfeld diffraction integral theory[11].

IOP Publishing
doi:10.1088/1742-6596/574/1/012053

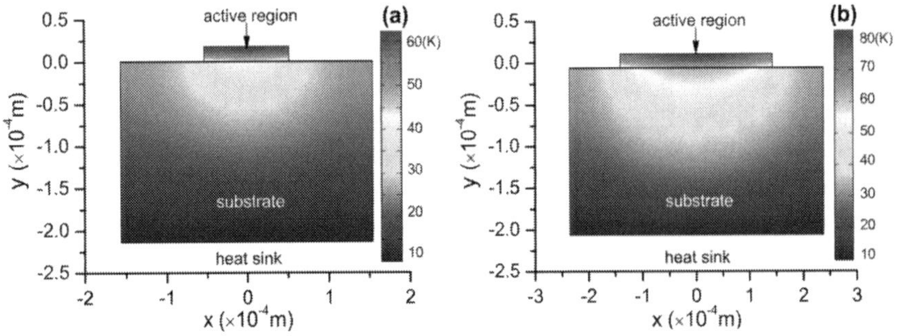

Figure 2. The simulated temperature distribution on the front facet of (a) 0°- and (b)8°-tapered THz-QCLs. The temperatures of the environment and the heat sink are set to 10 K.

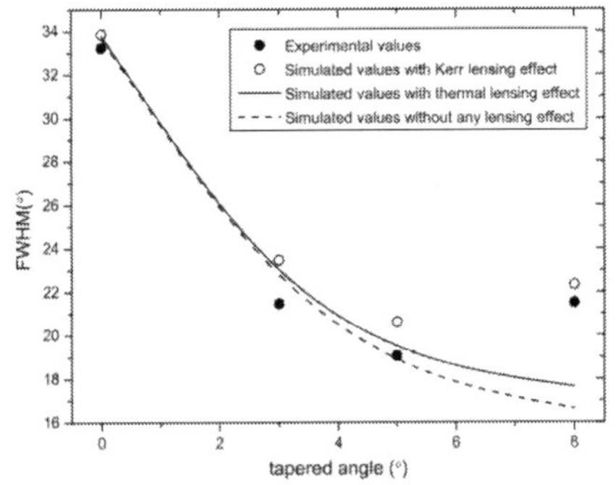

Figure 3. The experimental and simulated angles of full width at half maximum (FWHM) of the horizontal far field laser beam of the tapered THz-QCLs.

In Fig.3, we show the experimental and simulated results of horizontal full width at half maximum (FWHM) of the far field laser beam of the tapered THz-QCLs. As shown by the experimental data, the far-field divergence of tapered THz-QCLs decreases with increasing the tapered angle only when the angle is smaller than 5°. The divergence increase with further increasing the angle, see the experimental FWHM of 8° device. In the simulation side, the calculated divergence always decreases with increasing the angle if non of the lensing effects is considered (see the dashed line in Fig.3). Considering the thermal effect is also helpless. As shown in Fig.2, since the temperature gradient in the direction parallel to the quantum well layers is very small, the thermal lensing effect only slightly increases the far-field divergence of tapered THz-QCLs (the solid line in Fig.3). Even if we use a larger thermal coefficient α_T, the simulated result cannot recover the experimental result. However, if the Kerr lensing effect is considered, the simulated result (data points ○) can accurately recover the experimental result (data points ●). These comparisons suggest that the refractive index change caused by Kerr lensing, i.e. the third order optical nonlinearity of the active region may be the main reason for the abnormal increase of the divergence of tapered THz-QCLs.

4. Conclusions

In summary, we have presented the simulation of the mode propagation and the far-field distribution of THz-QCLs. By including the self-focusing effect caused by the third-order optical nonlinearity of the superlattice in the active region, the simulation can recover the abnormal increase of the far-field divergence of the laser beam. This result suggests that the optical nonlinearity may have a great influence on the far-field beam quality of THz-QCLs, which should be carefully considered in future imaging application based on the THz-QCLs.

Acknowledgments

This work is supported by National Basic Research Program of China (2013CB632801/5) Special-Funded Program on National Key Scientific Instruments and Equipment Development (2011YQ13001802) and CAEP under Grant No. 2012B0401061.

References

[1] J. Faist, F. Capasso,Sivco, D. L. Sivco, C. Sirtori, A. L. Hutchinson, and A. Y. Cho 1994 *Science* **264** 553
[2] B. S. Williams 2007 *Nat. Photonics* **1** 517
[3] D. Vijayakumar, O. B. Jensen, and R.Ostendorf, T. Westphalen, and B. Thestrup 2010 *Opt. Express* **18** 893
[4] S. Menzel, L. Diehl, C. Pflugl, A. Goyal, C. Wang, A. Sanchez, G. Turner, and F. Capasso 2011 *Opt. Express* **19** 16229
[5] A. Lyakh, R. Maulini,A. Tsekoun, R. Go, C. Kumar, and N.Patel 2012 *Opt. Express* **20** 4382
[6] Y. F. Li, J. Wang, Ning Yang, J. Q. Liu, T. Wang, F. Q. Liu, Z. G. Wang, W. D. Chu, and S. Q. Duan 2013 *Opt. Express* **21** 15998
[7] L. Nähle, J. Semmel, W. Kaiser, S. Höfling, and A. Forchel 2007 *Appl. Phys. Lett.* **91** 181122
[8] C. A. Evans, D. Indjin, Z. Ikonic, P. Harrison, M. S. Vitiello, V. Spagnolo, and G. Scamarcio 2008 *IEEE J. Quan. Electron.* **44** 680
[9] K. Pierscinski, D. Pierscinska, M. Iwinska, K. Kosiel, A. Szerling, P. Karbownik, and M. Buga 2012 *J. Appl. Phys.* **112** 043112
[10] C. C. Wang 1966 *Phys. Rev. Lett.* **16** 344C346
[11] J. Wang, W. D. Wu, X. L. Zhang, and S. Q. Duan 2011 *Information and Electronic Engineering* **9** 365

Anatomy of Demagnetizing and Exchange Fields in Magnetic Nano-Dots Influenced by 3D Shape Modifications

T Blachowicz[1,3] **and A Ehrmann**[2, 3]
[1] Institute of Physics – Center for Science and Education, Silesian University of Technology, 44-100 Gliwice, Poland
[2] Faculty of Textile and Clothing Technology, Niederrhein University of Applied Sciences, 41065 Mönchengladbach, Germany
[3] VIARAM – Virtual Institute for Applied Research on Advanced Materials
E-mail: tomasz.blachowicz@polsl.pl

Abstract. Hysteresis loops of 3D ferromagnetic permalloy nano-half-balls (dots) with 100 nm base diameter have been examined by means of LLG micromagnetic simulations and finite element methods. Tests were carried out with two orthogonal directions of the externally applied field at 10 kA/(m.ns) field sweeping speed. The comparison of samples with different 3D modifications at the sub-10nm scale, accessible by nowadays lithographic techniques, enables conclusions about different mechanisms of competition between demagnetizing and exchange fields. Design paradigms provided here can be applied, e.g., in bit-patterned media used as novel magnetic storage systems.

1. Introduction

Analysis of the microscopic structure of magnetic states plays an important role in understanding magnetic states in low dimensional objects, which is necessary for the development of nanostructures with novel features. Round objects, especially nano-dots, are of particular interest due their use in bit-patterned media [1]. Observed in nano-dots and rings, flux-closure states, vortexes, and onion states were extensively investigated in many experimental works [2-6]. First experiments with thin magnetic films on non-magnetic spheres with diameters between 20 and 1000 nm underline the importance of probing 3D nano-magnets [7-8]. The 3D nano-objects can principally be structured with today's techniques inspired by biologically self-assembled processes using proper (smart) components [9].

2. Mathematical Model

The idea of the study presented here is based on the effect that for small magnetic elements the shape modifications – due to dipolar interactions – play a crucial role in magnetization dynamics; however, due to reduced dimensionality, the demagnetizing fields can compete with exchange fields leading to specific magnetic behavior, such as oscillations, rapid transient states, and quasi-static states.

Thus, we theoretically examine 3D ferromagnetic half-balls made of commonly applied permalloy (Py) with different shape modifications in order to support development of new 3D performance paradigms. In our samples the circular base of diameter 100 nm is located in the x-y-plane. In some of the half-balls, a perpendicular hole of diameter 50 nm has been cut in order to extract the core region of a possible magnetic vortex state which is known to have a comparable diameter [10]. In that way the samples represent all basic shape features to be met in 3D-fabricated nanomagnets: planes, curved surfaces, straight and bent edges, as well as intentionally introduced cuts and holes. In order to narrow the analysis to magnetostatic vs. exchange-based competition, magnetocrystalline anisotropy is excluded. All deviations from an ideal half-ball, created in different ways, can be seen in Fig. 1.

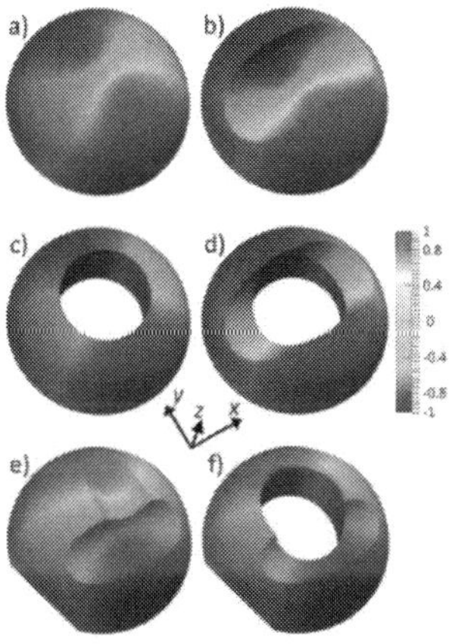

Figure 1. The samples under simulation: (a), (b) and (e) are solid half-balls, while (c) and (d) half-balls possess a 50 nm diameter hole, respectively. The height of the half-ball (a) equals 50 nm. The cylindrical cuts in the x-y plane (b, d) have 50 nm diameter, and the maximum depth of a cut, seen along the half-ball z-axis, equals 25 nm with respect to the top point of the bulk half-ball (a). Samples (e) and (f) possess arbitrary, asymmetrical cuts, and the elliptical hole in (f) case is not parallel to the z-axis. The surface x-components of magnetization, at zero-valued externally applied magnetic field, are color-coded.

Here we report results of hysteretic behavior of the half-balls using micromagnetic simulations [11], allowing for an insight into the magnetic properties, basing on the Landau-Lifshitz-Gilbert (LLG) equation of motion

$$\frac{\partial \vec{J}}{\partial t} = -\frac{\gamma}{1+\alpha^2}\vec{J} \times \vec{H}_{eff} - \frac{\gamma\alpha}{(1+\alpha^2)J_s}\vec{J} \times \left(\vec{J} \times \vec{H}_{eff}\right), \tag{1}$$

where $\vec{J}$ is the magnetic polarization vector, J_s is the magnetization polarization at saturation, γ is the electron gyromagnetic ratio, α is the damping parameter, and $\vec{H}_{eff}$ is the effective magnetic field. The effective field, in general, is a superposition of four contributing fields: the externally applied field, the magnetocrystalline field, the demagnetizing field, and the exchange field, namely

$$\vec{H}_{eff} = \vec{H}_{ext} + \frac{2K_1}{J_s}\vec{n}\left(\vec{n} \cdot \frac{\vec{J}}{J_s}\right) + \vec{H}_{demag} + \frac{2A}{J_s}\Delta\left(\frac{\vec{J}}{J_s}\right), \tag{2}$$

where K_1 is the magnetocrystalline anisotropy constant, $\vec{n}$ is the unit vector pointing in the magnetic easy-axis direction, and A is the exchange constant.

For the simulations, the MagPar [12] LLG micromagnetic solver was used, dynamically integrating the equation of motion. Finite elements were meshed as tetrahedral elements with dimensions of maximal 3.7 nm, which is smaller than the Py exchange length of 5.7 nm [11]. Near the edges, meshing was approximately 10 times denser, to include the influence of demagnetizing fields more exactly. Else, we chose the exchange constant A=1.05 10^{-11} J/m, the magnetic polarization at saturation J_s =1 T, and the Gilbert damping constant α = 0.01 [11].

3. Results and Discussion – Hysteresis Loops

Hysteresis loops have been examined with the external magnetic field H_{ext} applied along the x-axis (in the sample plane) and along the z-axis (perpendicular to the base plane), respectively. In all cases, starting from a random state of magnetization at H_{ext} = 0, the field has been changed with constant speed of 10 kA/(m.ns) up to 450 kA/m (H_{ext} along x-axis) or to 600 kA/m (H_{ext} along z-axis), respectively (this step is not shown in the graphs for clarity). Next, starting from positive saturation, the field has been swept at a constant speed of -10 kA/(m.ns) to -450 kA/m (H_{ext} along x-axis) or to -600 kA/m (H_{ext} along z-axis), respectively. In order to close the hysteresis loops, the field has been swept back from negative to positive saturation with the same constant speed afterwards. This field sweeping speed is correlated to typical values for magneto-electronic applications [11].

3.1 Hysteresis loops for the in-plane applied field (x-direction)

The hysteresis loops obtained for H_{ext} applied along the x-axis are shown in Fig. 2. For the solid half-ball (Fig. 2a), it should be mentioned that the start of the magnetization reversal before $H_{ext} = 0$ is reached. Such a behavior has also been observed, e.g., by He *et al.* and interpreted as the beginning of a magnetic vortex state formation [6]. Our simulation strongly supports this interpretation, as can be seen in Fig. 2a where snapshots of the magnetization in z-direction are presented, showing vortex formations correlated with the start of the oscillations. A detailed analysis of the oscillation frequencies will be given in a future article.

While cutting the top of the solid half-ball suppresses magnetization reversal before $H_{ext} = 0$ (Fig. 2b), a hole in the sphere (Fig. 2c) significantly enhances the coercivity as well as the field region between first and second step of the complete magnetization reversal. This broad step is correlated with a flux-closed vortex state, while the saturation magnetization states and the oscillatory region are associated with two different onion states. The snapshots for the magnetization reversal from positive to negative saturation show the z-components of the magnetization. The oscillation here is not circular but a symmetric oscillation with the two colored domain walls oscillating towards each other and back again. Following the snapshots of the x-component of the magnetization from negative to positive saturation, the magnetization reversal from the original onion state along a second onion state to the vortex state and finally to the reversed saturated onion state is clearly visible.

However, additional top cuts in a sphere with a hole (Fig. 2d) reduce the coercivity again, which is opposite to the finding in solid half-balls. Obviously, both effects have to be taken into account if a certain coercive field is desired.

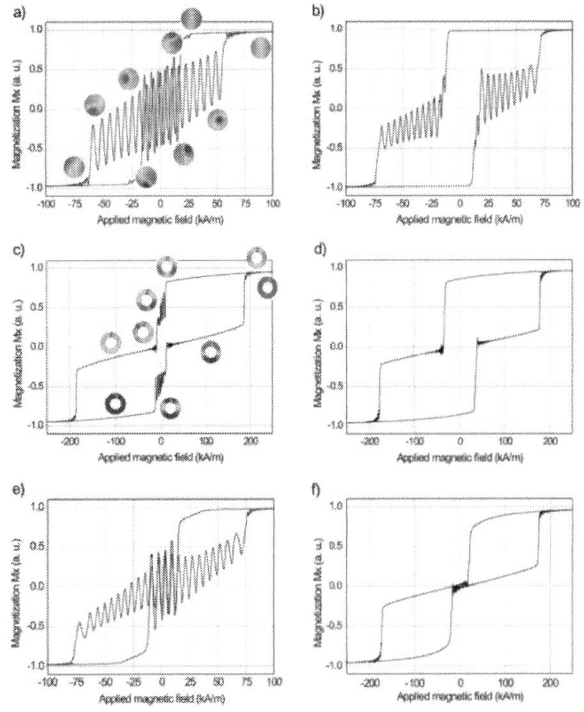

Figure 2. Hysteresis loops, simulated for the samples depicted in figure 1, for the external field applied along the x-axis. The color-coded insets in (a) show the z-component of the magnetization for different external fields, the insets in (c) exhibit the z-components (for field sweep from positive to negative values) and x-components (for field sweep from negative to positive saturation), respectively.

3.2 Hysteresis loops for the field applied perpendicular to a base-plane (z-direction)

Figure 3 shows hysteresis loops for H_{ext} applied along the z-axis. There are two types of reversal processes obtained, those with a step (Figs. 3a, 3b, and 3e) and those with smooth dynamics (Figs. 3c, 3d, and 3f). The insets in Fig. 3a show that in the solid half-ball a vortex core is built [13], which is separated from the outer part of the vortex by a circular domain wall. During the magnetization reversal process, the core region becomes smaller, while the outer ring reverses. The oscillatory step finalizes the reversal by switching the core region. Both reversal processes from positive to negative saturation and vice versa differ by the orientation of the core and the outer region.

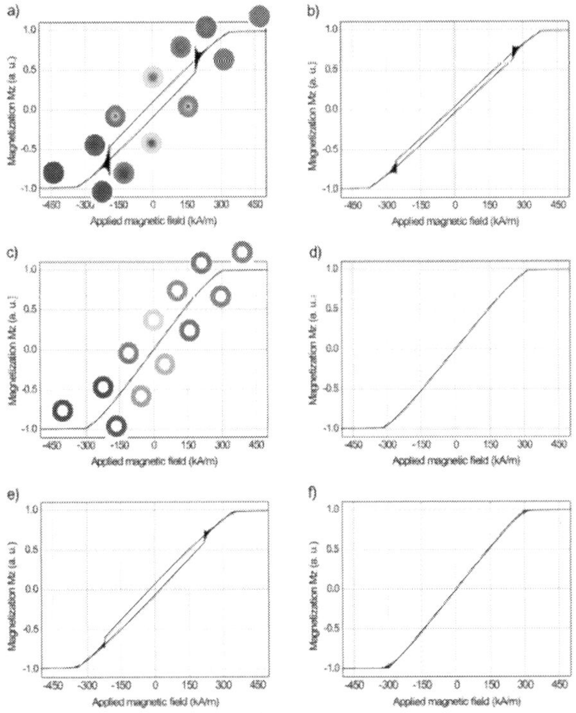

Figure 3. Hysteresis loops, simulated for the samples depicted in figure 1, for the external field applied along the z-axis. The color-coded insets in (a) and (c) show the z-component of the magnetization for different external fields.

Apparently, the cylindrical cuts in the solid half-balls (Figs. 3b and 3e) lead to smaller coercivities and remanences, while the holes (Figs. 3c, 3d, and 3e) produce smooth reversible magnetization curves. Such a reversible magnetization curve of quasi-spherical Fe particles of diameter 200 nm has also been found by Diao *et al.* [14]. However, investigation of the magnetization dynamics [13] and the color-coded insets in Fig. 3c shows that here a flux-closed state is generated, corresponding to two possible states with flux rotational direction clockwise or counter-clockwise, enabling the utilization of these nano-magnets in storage devices as well. During the magnetization reversal process, the magnetization values as well as the color-coded snapshots are identical for all external fields; the process is completely reversible.

Moreover, comparing Figs. 2 and 3, it is obvious that for the Py half-balls, the hard axis is always directed along the z-axis, as in 2D nano-magnets.

All the observed effects can be more deeply understood by considering Fig. 4, where both competing fields are visualized. It can be recognized that an additional horizontal cut increases the in-plane (x-y) components of the demagnetizing field blocking the exchange field precessions. A more detailed examination of demagnetizing and exchange energy is given in [15].

a) b)

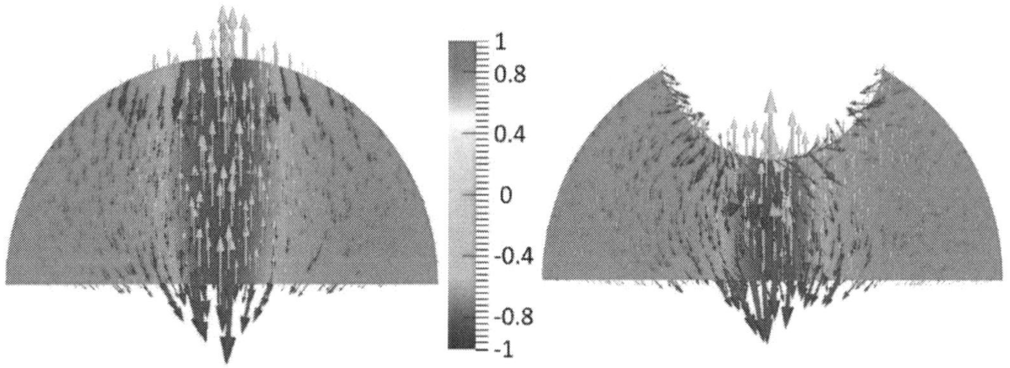

Figure 4. Visualizations of cross-sections of competing demagnetizing (dark lilac arrows) and exchange fields (light grey arrows) for the full solid sample (a) and the sample with a single horizontal cut (b). The colors at cross-sections represent local values of the M_z magnetization component: 1 (red) – vector parallel to z-axis, 0 (green) – vector perpendicular to z-axis, -1 (blue) – vector antiparallel to z-axis. The figure represents zero-valued external-field state (remanence).

4. Conclusion

In summary, we have provided some basic nanomagnetic sample 3D paradigms and given an overview about magnetization characteristics which can be reached by changes in the geometry of magnetic nano-particles, enabling deeper understanding of desired properties in a relatively broad range of modifications to meet the challenges of new 3D technology of magnetic devices. This could give an opportunity for new applications in magneto-electronics.

References

[1] Richter H, Dobin A, Heinonen O, Gao K, Veerdonk R, Lynch R, Xue J, Weller D, Asselin P, Erden M and Brockie R 2006 *IEEE Trans. Magn.* **42** 2255-60
[2] Cowburn R P, Koltsov D K, Adeyeye A O, Welland M E and Tricker D M 1999 *Phys. Rev. Lett.* **83** 1042
[3] Zhu J G, Zheng Y F and Prinz G A 2000 *J. Appl. Phys.* **87** 6668
[4] Zhu F Q, Fan D L, Zhu X C, Zhu J G, Cammarata R C and Chien C L 2004 *Adv. Mater.* **16** 2155
[5] Zhang W and Haas S 2010 *Phys. Rev. B* **81** 064433
[6] He K, Smith D J and McCartney M R 2010 *J. Appl. Phys.* **107** 09D307
[7] Soares M M, de Biasi E, Coelho L N, dos Santos M C, de Menezes F S, Knobel M, Sampaio L C and Garcia F 2008 *Phys. Rev. B* **77** 224405
[8] Amaladass E, Ludescher B, Schütz G, Tyliszczak T, Lee M S and Eimüller T 2010 *J. Appl. Phys.* **107** 053911
[9] Leong T G, Zarafshar A M and Gracias D H 2010 *Small* **6** 792
[10] Chou K W, Puzic A, Stoll H, Dolgos D, Schütz G, van Waeyenberge B, Vansteenkiste A, Tyliszczak T, Woltersdorf G and Back C H 2007 *Appl. Phys. Lett.* **90** 202505
[11] Smith N, Markham D and la Tourette D 1989 *J. Appl. Phys.* **65** 4362
[12] Scholz W, Fidler J, Schrefl T, Suess D, Dittrich R, Forster H and Tsiantos V 2003 *Comp. Mat. Sci.* **28** 366
[13] Blachowicz T, Ehrmann A, Steblinski P and Pawela L 2010 *J. Appl. Phys.* **108** 123906
[14] Diao Z, Abid M, Upadhyaya P, Venkatesan P and Coey J M D 2010 *J. Magn. Magn. Mat.* **322** 1304
[15] Ehrmann A, *Examination and simulation of new magnetic materials for the possible application in memory cells*, Logos Verlag, Berlin / Germany 2014, ISBN 978-3-8325-3772-2

Ab-initio simulation of the ionization and fragmentation of acetylene by strong femtosecond laser pulses

Arthur Russakoff
Department of Physics and Astronomy, Vanderbilt University, Nashville, Tennessee 37235, USA
E-mail: arthur.g.russakoff@vanderbilt.edu

Sergiy Bubin
Department of Chemistry, University of Rochester, Rochester, New York 14627, USA
E-mail: sergiy.bubin@nu.edu.kz

Kálmán Varga
Department of Physics and Astronomy, Vanderbilt University, Nashville, Tennessee 37235, USA
E-mail: kalman.varga@vanderbilt.edu

Abstract. The electron and nuclear dynamics of acetylene when interacting with strong short laser pulses has been simulated in the framework of real–space Time Dependent Density Functional Theory (TDDFT) and molecular dynamics. The stretching and dissociation of individual bonds are reported, and are shown to depend on the laser field intensity and orientation relative to the laser polarization. The ionization dynamics, including ionization from individual Kohn–Sham orbitals, is also reported. The orbital ionization dynamics are shown to vary with an increase in the intensity of the laser field.

1. Introduction

The interaction of strong short laser pulses with small molecules has become an active area of recent research. Many advancements in experimental techniques have allowed the measurement of both electron and nuclear dynamics. These recent advancements include imaging of the momentum distribution of dissociated fragments with the cold target recoil ion spectrometer (COLTRIMS)[1]. The study of aligned molecules has also become possible[2, 3, 4, 5], allowing for comparison between theory and experiment without the need to average over an ensemble of molecular orientations[6]. Also among the recent experimental advancements is the ability to probe the difference in ionization from different molecular orbitals[7], and to study the ionization of individual orbitals and the sequential tunnelling of two electrons from multiple orbitals[8].

Among the ionization and fragmentation processes investigated, the Coulomb explosion has been a topic of particular interest[9, 10]. In the Coulomb explosion, the short intense laser field

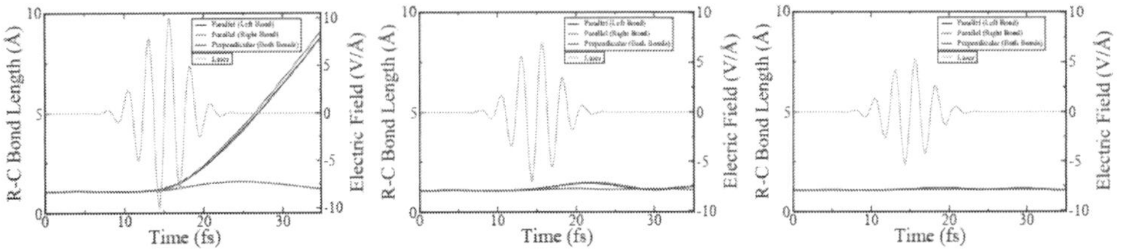

Figure 1. (Color online) Time–dependent C-H bond lengths and the corresponding laser pulses for parallel and perpendicular orienations. The laser pulses are a) $14 \cdot 10^{14} \frac{W}{cm^2}$, b) $7 \cdot 10^{14} \frac{W}{cm^2}$, and c) $4 \cdot 10^{14} \frac{W}{cm^2}$.

ionizes the molecule and the molecular fragments dissociate with high kinetic energy. Earlier studies of the Coulomb explosion have proposed a model of charge localization[9]. More recent studies using momentum coincidence measurements [10] point towards a theory of enhanced ionization[11, 12, 13]. Theoretically, the enhanced ionization mechanism has been investigated within Time Dependent Hartree Fock (TDHF)[14, 15, 16] for the small linear hydrocarbon, acetylene (C_2H_2). In these simulations, the ions were held fixed for the duration of the laser pulse. In this paper we use Time-Dependent Density Functional Theory[17] and Ehrenfest dynamics[18] to investigate the coupling between the electron and nuclear dynamics in acetylene under the influence of an intense short–lived laser pulse.

2. Computational Details

The electron dynamics in the simulations are modeled using Time Dependent Density Functional Theory (TDDFT) on a real–space grid with real–time propagation. Core electrons, which are difficult to handle computationally, are represented using norm-conserving Troullier-Martins pseudopotentials [19].

The initial (ground) state of the system is first prepared by performing the ground-state DFT calculation. Next, the time–dependent Kohn-Sham orbitals, ψ_k, are determined by solving the time–dependent Kohn–Sham equations,

$$i\hbar \frac{\partial \psi_k(\mathbf{r}, t)}{\partial t} = H\psi_k(\mathbf{r}, t), \tag{1}$$

where k is a quantum number labeling the orbital. The Hamiltonian is given by,

$$H = -\frac{\hbar^2}{2m}\nabla^2_{\mathbf{r}} + V_{\rm H}[\rho](\mathbf{r}, t) + V_{\rm XC}[\rho](\mathbf{r}, t) + V_{\rm ext}(\mathbf{r}, t). \tag{2}$$

ρ is the electron (number) density which is defined by a sum over all occupied orbitals,

$$\rho(\mathbf{r}, t) = \sum_{k=1}^{N} 2|\psi_k(\mathbf{r}, t)|^2. \tag{3}$$

V_H is the Hartree potential, defined by

$$V_H(\mathbf{r}, t) = \int d\mathbf{r}' \frac{\rho(\mathbf{r}', t)}{|\mathbf{r} - \mathbf{r}'|}, \tag{4}$$

and accounts for the electrostatic Coulomb interactions between electrons. V_{XC} is the exchange–correlation potential, whose exact form is a complicated functional of the entire history of the

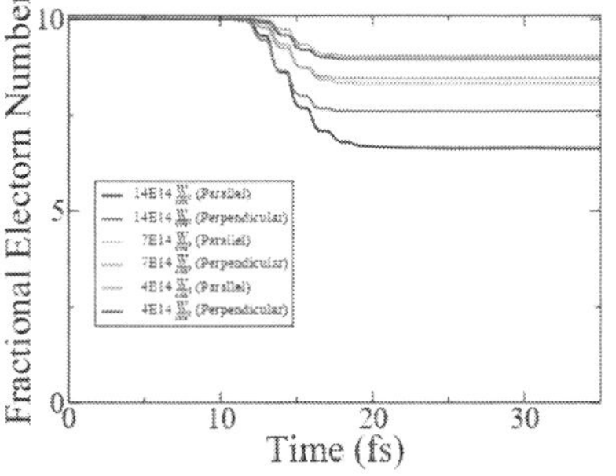

Figure 2. (Color online) Time–dependent fractional electron number all considered laser intensities and geometric orientations.

electron density. It is approximated using the adiabatic local–density approximation (ALDA) with the parameterization of Perdew and Zunger [20]. V_{ext} is the external potential, including the Coulomb potential due to the ions, V_{ions}, and the explicitly time–dependent potential due to the electric field of the laser V_{laser}. V_{ions} is given by,

$$V_{\text{ion}}(\mathbf{r}) = \sum_{i=1}^{N_{\text{ions}}} \frac{-Z_i}{|\mathbf{r} - \mathbf{R}_i|}, \tag{5}$$

where where N_{ions} is the total number of ions, and R_i and Z_i are the position and pseudocharge (valence) of the ith ion, respectively. V_{laser} is described using the dipole approximation, $V_{laser} = \mathbf{r} \cdot \mathbf{E}(t)$, with the time–dependent electric field given by,

$$\mathbf{E}(t) = \hat{\mathbf{k}} E_{max} \exp\left[-\frac{(t-t_0)^2}{2a^2}\right] \sin(\omega t). \tag{6}$$

The parameters a, t_0, and E_{max} define the width, initial position of the center, and the maximum amplitude of the Gaussian envelope, respectively. ω describes the frequency of the laser and $\hat{\mathbf{k}}$ is a unit vector defining the (linear) polarization of the electric field. The Kohn–Sham orbitals are propagated with the time–evolution operator,

$$U(0,t) = \prod_q U(t_q, t_q + \delta t), \quad t_q = q\delta t, \tag{7}$$

where δt is a short time step and t the total simulation time. $U(t_q, t_q + \delta t)$ is approximated with fourth-order Taylor expansion, so that the propagation of the Kohn-Sham orbitals over a short time step, δt, is given by,

$$\psi_k(\mathbf{r}, t_q + \delta t) \approx \sum_{n=0}^{4} \frac{1}{n!} \left(-\frac{i\delta t}{\hbar} H(\mathbf{r}, t_q)\right)^n \psi_k(\mathbf{r}, t_q). \tag{8}$$

The Kohn-Sham orbitals are represented at discrete points in real space which are organized in a uniform rectangular grid. The accuracy of the numerical calculations is controlled by adjusting the grid spacing. At the walls of the simulation cell, a zero–boundary condition is enforced. This can lead to an unphysical reflection of the wavefunction off the walls of the simulation cell. A complex absorbing potential (CAP) at the walls of the simulation cell is therefore implemented, using the form proposed by Manolopoulos[21] with an additional softening parameter which suppresses extremely large CAP magnitudes at the boundaries. As the molecule is ionized by

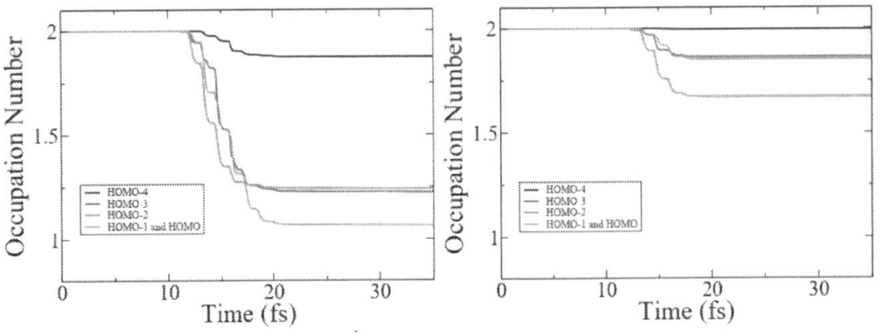

Figure 3. (Color online) Time–dependent occupation number of the Kohn-Sham orbitals for the parallel orientation and laser intensities of a) $14 \cdot 10^{14} \frac{W}{cm^2}$ and b) $4 \cdot 10^{14} \frac{W}{cm^2}$.

the laser field, electron density travels to the edge of the simulation box where it is absorbed by the CAP. The integral of the electron density over the volume of the box,

$$N(t) = \int_V \rho(\mathbf{r}, t) d^3x, \tag{9}$$

therefore diverges from the initial electron number, $N(0)$. Since, in general, $N(t)$ is not an integer, we will refer to it henceforth as the fractional electron number. We interpret $N(0) - N(t)$ as the total ionization of the molecule. We note this method of calculating the total ionization is dependent upon the size of the simulation box. The box size is therefore tuned to be large enough to allow for electron re-collision, but small enough such that the computational complexity remains tractable. Motion of the ions in the simulations were treated semi–classically using Ehrenfest dynamics [18]. The quantum forces on the ions, which are due to the electrons, are given by the derivatives of the expectation value of the total electronic energy with respect to the ionic positions. These forces are then fed into Newton's Second Law, giving

$$M_i \frac{d^2\mathbf{R}_i}{dt^2} = Z_i \mathbf{E}_{\text{laser}}(t) + \sum_{j \neq i}^{N_{\text{ions}}} \frac{Z_i Z_j (\mathbf{R}_i - \mathbf{R}_j)}{|\mathbf{R}_i - \mathbf{R}_j|^3} - \nabla_{\mathbf{R}_i} \int V_{\text{ion}}(\mathbf{r}, \mathbf{R}_i) \rho(\mathbf{r}, t) d\mathbf{r}, \tag{10}$$

where M_i and Z_i are the mass and pseudocharge (valence) of the ith ion, respectively.

3. Results

The interaction of acetylene, C_2H_2, with a linearly polarized laser pulse was investigated with TDDFT. The laser field had a full width half maximum pulse width of 4.5fs and three intensities were used: 4, 7, and $14 \cdot 10^{14} \frac{W}{cm^2}$ (see Figure 1). Parallel and perpendicular orientations of the linear C_2H_2 molecule relative to the laser polarization were considered.

As the acetylene molecule is exposed to a time–dependent laser field, bond stretching or even bond breaking may occur. Bond breaking where the fragments are repelled at high kinetic energy is referred to as Coulomb explosion. Figure 1 shows the distance between the hydrogen atoms and the nearest carbon atom, i.e. the C-H bond lengths. The corresponding laser fields are also shown for reference. At the lowest intensity, $4 \cdot 10^{14} \frac{W}{cm^2}$, the bond lengths are essentially unchanged. As the laser intensity is increased to $7 \cdot 10^{14} \frac{W}{cm^2}$, oscillatory bond stretching is observed when the molecule is parallel to the laser polarization. At the high intensity, $14 \cdot 10^{14} \frac{W}{cm^2}$, some oscillatory bond stretching occurs for the perpendicular orienation and bond breaking via Coulomb explosion is observed in the parallel orientation.

In addition to the bond stretching, ionization dynamics also occur. As shown in Figure 2, as the laser intensity is increased the molecule becomes more ionized. The ionization dynamics of each molecular orbital also changes with laser intensity. As shown in Figure 3, when the molecule is in the parallel orientation and the laser intensity is $4 \cdot 10^{14} \frac{W}{cm^2}$, the most ionized orbital is the highest occupied molecular orbital (HOMO) and the next highest orbital (HOMO-1), which is degenerate with the HOMO. When the laser intensity is increased to $14 \cdot 10^{14} \frac{W}{cm^2}$, the most ionized orbital is the HOMO-2, followed by the HOMO-3.

In conclusion, we have reported on nuclear dynamics and the total and orbital ionization dynamics of Acetylene when subjected to three laser fields of different intensity. The stretching and dissociation of individual C-H bonds were reported, and were shown to depend on the laser field intensity and orientation relative to the laser polarization. The orbital ionization dynamics were shown to vary with an increase in the intensity of the laser field.

4. References

[1] Ullrich J, Moshammer R, Dorn A, Drner R, Schmidt L P H and Schmidt-Bcking H 2003 *Reports on Progress in Physics* **66** 1463 URL `http://stacks.iop.org/0034-4885/66/i=9/a=203`
[2] Kumarappan V, Holmegaard L, Martiny C, Madsen C B, Kjeldsen T K, Viftrup S S, Madsen L B and Stapelfeldt H 2008 *Phys. Rev. Lett.* **100**(9) 093006 URL `http://link.aps.org/doi/10.1103/PhysRevLett.100.093006`
[3] Holmegaard L, Nielsen J H, Nevo I, Stapelfeldt H, Filsinger F, Küpper J and Meijer G 2009 *Phys. Rev. Lett.* **102**(2) 023001 URL `http://link.aps.org/doi/10.1103/PhysRevLett.102.023001`
[4] Bisgaard C Z, Clarkin O J, Wu G, Lee A M D, Gener O, Hayden C C and Stolow A 2009 *Science* **323** 1464–1468 (*Preprint* `http://www.sciencemag.org/content/323/5920/1464.full.pdf`) URL `http://www.sciencemag.org/content/323/5920/1464.abstract`
[5] Meckel M, Comtois D, Zeidler D, Staudte A, Pavii D, Bandulet H C, Ppin H, Kieffer J C, Drner R, Villeneuve D M and Corkum P B 2008 *Science* **320** 1478–1482 (*Preprint* `http://www.sciencemag.org/content/320/5882/1478.full.pdf`) URL `http://www.sciencemag.org/content/320/5882/1478.abstract`
[6] Bubin S, Atkinson M, Varga K, Xie X, Roither S, Kartashov D, Baltuška A and Kitzler M 2012 *Phys. Rev. A* **86**(4) 043407
[7] Akagi H, Otobe T, Staudte A, Shiner A, Turner F, Drner R, Villeneuve D M and Corkum P B 2009 *Science* **325** 1364–1367 (*Preprint* `http://www.sciencemag.org/content/325/5946/1364.full.pdf`) URL `http://www.sciencemag.org/content/325/5946/1364.abstract`
[8] Wu J, Schmidt L P H, Kunitski M, Meckel M, Voss S, Sann H, Kim H, Jahnke T, Czasch A and Dörner R 2012 *Phys. Rev. Lett.* **108**(18) 183001 URL `http://link.aps.org/doi/10.1103/PhysRevLett.108.183001`
[9] Markevitch A N, Romanov D A, Smith S M and Levis R J 2004 *Phys. Rev. Lett.* **92** 063001
[10] Roither S, Xie X, Kartashov D, Zhang L, Schöffler M, Xu H, Iwasaki A, Okino T, Yamanouchi K, Baltuska A and Kitzler M 2011 *Phys. Rev. Lett.* **106**(16) 163001
[11] Zuo T and Bandrauk A D 1995 *Phys. Rev. A* **52**(4) R2511–R2514
[12] Seideman T, Ivanov M Y and Corkum P B 1995 *Phys. Rev. Lett.* **75**(15) 2819–2822
[13] Bocharova I, Karimi R, Penka E F, Brichta J P, Lassonde P, Fu X, Kieffer J C, Bandrauk A D, Litvinyuk I, Sanderson J and Légaré F 2011 *Phys. Rev. Lett.* **107**(6) 063201
[14] Lötstedt E, Kato T and Yamanouchi K 2012 *Phys. Rev. A* **85** 041402 ISSN 1050-2947
[15] Lötstedt E, Kato T and Yamanouchi K 2012 *Phys. Rev. A* **86**(2) 023401
[16] Lötstedt E, Kato T and Yamanouchi K 2013 *J. Chem. Phys.* **138** 104304
[17] Runge E and Gross E K U 1984 *Phys. Rev. Lett.* **52** 997–1000
[18] Ehrenfest P 1927 *Z. Phys.* **45**(7) 455–457
[19] Troullier N and Martins J L 1991 *Phys. Rev. B* **43** 1993–2006
[20] Perdew J P and Zunger A 1981 *Phys. Rev. B* **23** 5048–5079
[21] Manolopoulos D E 2002 *J. Chem. Phys.* **117** 9552

IC-MSQUARE 2014 IOP Publishing
Journal of Physics: Conference Series **574** (2015) 012056 doi:10.1088/1742-6596/574/1/012056

Wormhole Dynamics

Hisa-aki Shinkai[1,2], **Takashi Torii** [3]

[1] Faculty of Information Science and Technology, Osaka Institute of Technology, Hirakata, Osaka 573-0196, Japan
[2] Computational Astrophysics Laboratory, Institute of Physical & Chemical Research (RIKEN), Hirosawa, Wako, Saitama, 351-0198 Japan
[3] Faculty of Engineering, Osaka Institute of Technology, Osaka, Osaka 535-8585, Japan

E-mail: `hisaaki.shinkai@oit.ac.jp`

E-mail: `takashi.torii@oit.ac.jp`

Abstract. Wormholes are theoretical products in general relativity, and are popular tools in science fictions. We know numerically the four-dimensional Ellis wormhole solution (the so-called Morris-Thorne's traversable wormhole) is unstable against an input of scalar-pulse from one side. We investigate this feature for higher-dimensional versions, both in n-dimensional general relativity and in Gauss-Bonnet gravity. We derived Ellis-type wormhole solution in n-dimensional general relativity, and found existence of unstable modes in its linear perturbation analysis. We also evolved it numerically in dual-null coordinate system, and confirmed its instability. The wormhole throat will change into black-hole horizon for the input of (relatively) positive energy, while it will change into inflationary expansion for (relatively) negative energy input. If we add Gauss-Bonnet terms (higher curvature correction terms in gravity), then wormhole tends to expand (or change to black-hole) if the coupling constant α is positive (negative).

1. Introduction

Wormhole is hypothetical object such as a short-cut tunnel connecting two points in space-time. The idea is essential in science fictions as a way for rapid interstellar travel, warp drives, and time machines. However, wormhole is also a theoretical research topic with long history.

The first appearance of a "tunnel structure" was in 1916 by Flamm [1] just after the discovery of Schwarzschild's black-hole solution. Einstein and Rosen [2] proposed a "bridge structure" between black-holes in order to obtain a regular solution without a singularity. The name "wormhole" was coined by John A. Wheeler in 1957, and its fantastic applications are popularized after the influential study of traversable wormholes by Morris and Thorne [3].

They considered "traversable conditions" for human travel through wormholes responding to Carl Sagan's idea for his novel *Contact*, and concluded that such a wormhole solution is available if we allow "exotic matter" (negative-energy matter).

The introduction of exotic matter sounds to be unusual for the first time, but such matter appears in quantum field theory and in alternative gravitational theories such as scalar-tensor theories. The Morris-Thorne solution is constructed with a massless Klein-Gordon field whose gravitational coupling takes the opposite sign to normal, which is found in Ellis's earlier work [4], so that we call it Ellis wormhole, hereafter. (See a review e.g. by Visser [5] for earlier works; See also e.g. Lobo [6] for recent works).

Ellis wormhole solution was studied in many contexts. Among them, we focus on its dynamical features. The first numerical simulation on its stability behavior was reported by one of the authors [7]. They use a dual-null formulation for spherically symmetric space-time integration, and observed that the wormhole is unstable against Gaussian pulses in either exotic or normal massless Klein-Gordon fields. The wormhole throat suffers a bifurcation of horizons and either explodes to form an inflationary universe or collapses to a black hole, if the total input energy is negative or positive, respectively. These basic behaviors were repeatedly confirmed by other groups [8, 9].

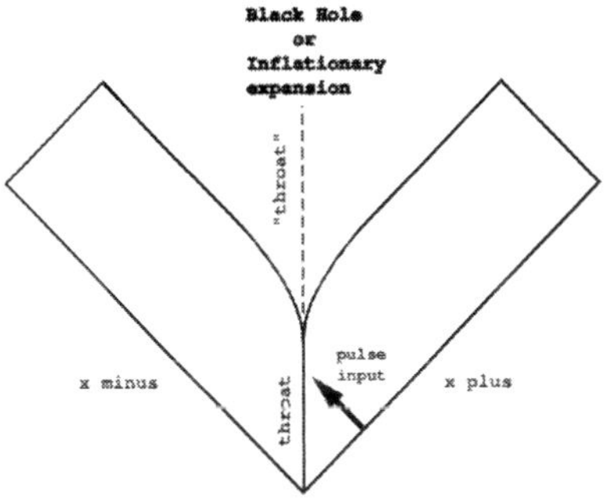

Figure 1. Partial Penrose diagram of the evolved space-time. Suppose we live in the right-side region and input a pulse to an Ellis-wormhole in the middle of the diagram. The wormhole throat suffers a bifurcation of horizons and either explodes to form an inflationary universe or collapses to a black hole, if the total input energy is negative or positive, respectively [7].

The changes of wormhole either to a black hole or an expanding throat supports an unified understanding of black holes and traversable wormholes proposed by Hayward [10]. His proposal is that the two are dynamically interconvertible, and that traversable wormholes are understandable as black holes under the existence of negative energy density.

In this article, we introduce our extensional works of [7], mainly its stability behavior in higher-dimensional space-time. The higher-dimensional theories such as string/M theories are applied for various unsolved problems in gravitational phenomena and cosmology, and gain new insights into them. We believe that wormholes will also give us new fundamental physical landscapes. We therefore demonstrate wormhole dynamics also in Gauss-Bonnet gravity, which is one of the modified gravity theory including higher-order corrections of curvatures, one of the string-motivated gravity theories.

Wormhole study in higher-dimensional space-time is not a new topic. We can find the articles from 80s [12, 13], and the recent studies are including higher-curvature terms (see e.g. [14] and [15] and references therein). Most of the researches concern the solutions and their energy conditions mainly, but to our knowledge there is no general discussion on the stability analysis of the solutions.

The main four contents in this article are: (a) constructing Ellis solutions in higher-dimensional general relativity, (b) stability analysis using linear perturbation method [16], (c) stability analysis using numerical evolution method, and (d) dynamical effects of Gauss-Bonnet coupling in 5-dimensional wormhole solution.

In §2, we derive the simplest wormhole solution with ghost scalar field in spherically symmetric, higher-dimensional space-time. We then study its stability using linear perturbation analysis, and find that there is at least one unstable mode in any dimensional space-time.

In §3, we try to confirm the prediction of instability using numerical evolutions. We implemented numerical code in [7] as it can treat higher-dimensional versions. We also demonstrate the wormhole structure in Gauss-Bonnet gravity theory.

2. Wormhole solutions in higher-dimensional general relativity

2.1. Field equations

We start from the n-dimensional Einstein-Klein-Gordon system

$$S = \int d^n x \sqrt{-g}\left[\frac{1}{2\kappa_n^2}R - \frac{1}{2}\epsilon(\nabla\phi)^2 - V(\phi)\right], \tag{1}$$

where κ_n^2 is a n-dimensional gravitational constant. The scalar field ϕ is the *normal* (or *ghost*) field if $\epsilon = 1$ (-1).

The metric of the space-time is assumed to be

$$ds^2 = -f(t,r)e^{-2\delta(t,r)}dt^2 + f(t,r)^{-1}dr^2 + R(t,r)^2 h_{ij}dx^i dx^j, \tag{2}$$

where $h_{ij}dx^i dx^j$ represents the line element of a unit $(n-2)$-dimensional constant curvature space with curvature $k = \pm 1, 0$ and volume Σ_k. In order to construct a static wormhole solution, we restrict the metric function as $f = f(r)$, $R = R(r)$, $\phi = \phi(r)$, and $\delta = 0$ in this subsection.

The Klein-Gordon equation becomes

$$\frac{1}{R^{n-2}}(R^{n-2}f\phi')' = -\epsilon\frac{dV}{d\phi}. \tag{3}$$

Hereafter, we construct the solution with the massless ghost scalar field ($V(\phi) = 0$ and $\epsilon = -1$). The Klein-Gordon equation (3) is integrated as

$$\phi' = \frac{C}{fR^{n-2}}, \tag{4}$$

where C is an integration constant. The Einstein equations are reduced to

$$\frac{(n-2)R'}{R}\left[\frac{f'}{f} + \frac{(n-3)R'}{R}\right] - \frac{(n-2)(n-3)k}{fR^2} = -\frac{\kappa_n^2 C^2}{f^2 R^{2(n-2)}}, \tag{5}$$

$$\frac{(n-2)R''}{R} = \frac{\kappa_n^2 C^2}{f^2 R^{2(n-2)}}. \tag{6}$$

The throat of the wormhole is at $r = 0$, and a is the radius of the throat, i.e. $R(0) = a$. The regularity conditions at the throat is written as $R(0) = a > 0$, and $f(0) = f_0 > 0$, where f_0 is a constant. Here we can assume $a = 1$ and $f_0 = 1$ without loss of generality, but we keep a in the equations in this section for later convenience. We assume the reflection symmetry with respect to the throat: $R'(0) = 0$, and $f'(0) = 0$. We also impose $\phi(0) = 0$ by a shift symmetry of the scalar field. With these conditions, the integration constant C is determined as

$$\kappa_n^2 C^2 = (n-2)(n-3)ka^{2(n-3)}. \tag{7}$$

This relation implies that there is no wormhole solution for the cases $k = 0$ and $k = -1$. Hence we investigate the spherically symmetric case $k = 1$.

The solution of Eqs. (4) (6) is obtained as

$$f \equiv 1, \tag{8}$$

$$R' = \sqrt{1 - \left(\frac{a}{R}\right)^{2(n-3)}}, \tag{9}$$

$$\phi = \frac{\sqrt{(n-2)(n-3)}}{\kappa_n}a^{n-3}\int\frac{1}{R(r)^{n-2}}dr. \tag{10}$$

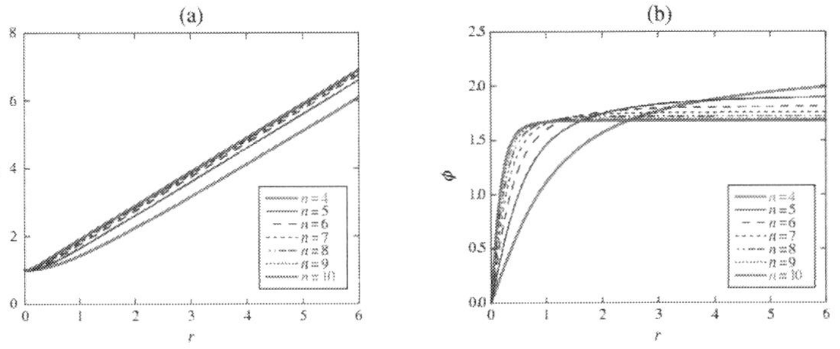

Figure 2. The n-dimensional wormhole solutions; (a) The circumference radius R and (b) the scalar field ϕ are plotted as a function of the radial coordinate r.

The eq. (9) is integrated to give

$$r(R) = -mB_z\Big(-m, \frac{1}{2}\Big) - \frac{\sqrt{\pi}\Gamma[1-m]}{\Gamma[m(n-4)]}, \tag{11}$$

where $m = 1/2(n-3)$ and $z = R^m$. $B_z(p,q)$ is the incomplete beta function defined by $B_z(p,q) := \int_0^z t^{p-1}(1-t)^{q-1}dt$. For $n = 4$, this solution reduces to Ellis's wormhole solution $f \equiv 1$, $R = \sqrt{r^2+a^2}$, $\phi = \sqrt{2}\tan^{-1}(r/a)$.

At the throat

$$R''(a) = \frac{n-3}{a}, \quad \text{and} \quad \phi'(a) = \frac{\sqrt{(n-2)(n-3)}}{\kappa_n a}. \tag{12}$$

These indicate that the throat of the wormhole has larger curvature and the scalar field ϕ becomes steeper as n goes higher. We plotted these behaviors in Figure 2. For $n \to \infty$, the functions have the limiting solution, $R = r + a$ and $\phi = \pi/2$ $(r > 0)$.

2.2. Stability analysis

In this subsection, we investigate the linear stability of the higher-dimensional wormhole. We follow the analysis in [9], where the throat radius is not fixed since it is shown that the instability occurs by resolution of the degeneracy of a double trapping horizon [7].

We focus on the "spherical" modes, where the $(n-2)$-dimensional constant curvature space is not perturbed. In the time-dependent metric ansatz eq. (2) we write the perturbed functions as

$$f(t,r) = f_0(r) + \varepsilon f_1(r)e^{i\omega t}, \tag{13}$$
$$\delta(t,r) = \delta_0(r) + \varepsilon \delta_1(r)e^{i\omega t}, \tag{14}$$
$$R(t,r) = R_0(r) + \varepsilon R_1(r)e^{i\omega t}, \tag{15}$$
$$\phi(t,r) = \phi_0(r) + \varepsilon \phi_1(r)e^{i\omega t}, \tag{16}$$

where ε is an infinitesimal parameter, and the variables with subscript 0 denote the static solution.

By introducing the new variable,

$$\psi_1 = R_0^{\frac{n-2}{2}}\Big(\phi_1 - \frac{\phi_0'}{R_0'}R_1\Big), \tag{17}$$

the first-order equations give the single master equation,

$$-\psi_1'' + V(r)\psi_1 = \omega^2\psi_1, \tag{18}$$

with the potential,

$$V(r) = \frac{n-2}{2}\left[\frac{n-3}{R_0^{2(n-2)}} + \frac{(n-4)R_0'^2}{2R_0^2}\right] + \frac{2(n-3)^2}{R_0^{2(n-2)}R_0'^2}. \tag{19}$$

Here we have assumed $a = 1$. The variable ψ_1 is gauge invariant under the spherically symmetric ansatz. Since R_0' is zero at the throat and the potential V diverges there, we regularize the master equation (18).

Using the 0-mode solution of the master equtaion (18), $\bar{\psi}_1 = \left(R_0^{\frac{n-4}{2}} R_0'\right)^{-1}$, we define a pair of differential operators, $\mathcal{D}_\pm = \pm\frac{d}{dr} - \frac{\bar{\psi}_1'}{\bar{\psi}_1}$. Then the master equation, (18), can be written as

$$\mathcal{D}_-\mathcal{D}_+\psi_1 = \omega^2\phi_1. \tag{20}$$

Operating $\mathcal{D}_+$ from the left and defining the new variable $\Psi_1 = \mathcal{D}_+\psi_1$, we find the regularized master equation

$$-\Psi_1'' + W(r)\Psi_1 = \omega^2\Psi_1, \tag{21}$$

where

$$W(r) = -\frac{1}{4R_0^2}\left[\frac{3(n-2)^2}{R_0^{2(n-3)}} - (n-4)(n-6)\right]. \tag{22}$$

Now the potential function is regular everywhere. For $n = 4$, $W(r)$ has the minimum at the throat and is negative definite. For $n \geq 5$, $W(r)$ has the minimum at the throat, while it increases apart from the throat and becomes positive for large r.

We search eigenfunctions $\Psi_1(r)$ of eq. (21), and find that in any dimension n. There exists one negative eigenvalue for ω^2, which are listed in Table 1. The existence of the eigenfunction with negative ω^2 implies that the solution is unstable. We find large negative ω^2 for higher n, which indicates the timescale of instability becomes shorter.

Table 1. The negative eigenvalues ω^2.

n	ω^2	n	ω^2
4	− 1.39705243371511	10	−12.0442650147438
5	− 2.98495893027790	11	−13.9552091676647
6	− 4.68662054299460	20	−31.5751101285105
7	− 6.46258414126318	50	−91.3457759137153
8	− 8.28975936306259	100	− 191.283017729717
9	− 10.1535530451867		

Journal of Physics: Conference Series **574** (2015) 012056 doi:10.1088/1742-6596/574/1/012056

3. Numerical evolutions of wormhole solutions

The purpose of this section is to confirm the prediction of instability of wormholes using numerical evolutions. We developed our numerical code as it can treat higher-dimensional versions. We also implemented it for studying the wormhole structure in Gauss-Bonnet gravity theory, which is one of the modified gravity theory including higher-order corrections of curvatures.

We begin describing Gauss-Bonnet theory briefly in §3.1, then explain our strategy in §3.2. The results will be shown in §3.3 and §3.4.

3.1. Einstein-Gauss-Bonnet gravity

Einstein-Gauss-Bonnet (EGB) gravity is derived from the superstring theory, with additional higher-order curvature correction terms to general relativity. Such higher-order corrections can be treated as an expansion of $\mathcal{R}$ in the action, but the Gauss-Bonnet term,

$$\mathcal{L}_{\rm GB} = \mathcal{R}^2 - 4\mathcal{R}_{\mu\nu}\mathcal{R}^{\mu\nu} + \mathcal{R}_{\mu\nu\rho\sigma}\mathcal{R}^{\mu\nu\rho\sigma}, \tag{23}$$

has good properties such that it is ghost-free combinations[11] and does not give higher derivative equations but an ordinary set of equations with up to the second derivative in spite of the higher curvature combinations.

The EGB action in n-dimensional space-time $(\mathcal{M}, g_{\mu\nu})$ is described as

$$S = \int_{\mathcal{M}} d^n X \sqrt{-g}\left[\frac{1}{2\kappa^2}\left\{\alpha_{\rm GR}\left(\mathcal{R} - 2\Lambda\right) + \alpha_{\rm GB}\mathcal{L}_{\rm GB}\right\} + \mathcal{L}_{\rm matter}\right], \tag{24}$$

where κ^2 is the n-dimensional gravitational constant, $\mathcal{R}$, $\mathcal{R}_{\mu\nu}$, $\mathcal{R}_{\mu\nu\rho\sigma}$ and $\mathcal{L}_{\rm matter}$ are the n-dimensional scalar curvature, Ricci tensor, Riemann curvature and the matter Lagrangian, respectively. This action reproduces the standard n-dimensional Einstein gravity, if we set the coupling constant $\alpha_{\rm GB}$ equals to zero.

The action (24) gives the gravitational equation as

$$\alpha_{\rm GR}G_{\mu\nu} + \alpha_{\rm GB}H_{\mu\nu} + g_{\mu\nu}\Lambda = \kappa^2 T_{\mu\nu}\,, \tag{25}$$

where

$$G_{\mu\nu} = \mathcal{R}_{\mu\nu} - \frac{1}{2}g_{\mu\nu}\mathcal{R}, \tag{26}$$

$$H_{\mu\nu} = 2\Big[\mathcal{R}\mathcal{R}_{\mu\nu} - 2\mathcal{R}_{\mu\alpha}\mathcal{R}^{\alpha}{}_{\nu} - 2\mathcal{R}^{\alpha\beta}\mathcal{R}_{\mu\alpha\nu\beta} + \mathcal{R}_{\mu}{}^{\alpha\beta\gamma}\mathcal{R}_{\nu\alpha\beta\gamma}\Big] - \frac{1}{2}g_{\mu\nu}\mathcal{L}_{\rm GB}, \tag{27}$$

$$T_{\mu\nu} = -2\frac{\delta\mathcal{L}_{\rm matter}}{\delta g^{\mu\nu}} + g_{\mu\nu}\mathcal{L}_{\rm matter}. \tag{28}$$

The higher-order curvature terms are considered as correction terms from string theory. These terms are known to produce two solution branches normally, only one of which has general-relativity limit. The theory is expected to have singularity-avoidance features in the context of gravitational collapses and/or cosmology, but as far as we know there is no studies so far using fully numerical evolutions. (Numerical studies on critical phenomena are recently reported for small $\alpha_{\rm GB}$ [17, 18, 19]).

Studies on wormholes in Gauss-Bonnet gravity have long histories. Several solutions and their classifications are reported in [20, 21], while their energy conditions are considered in [14]. Similar researches are extended to the Lovelock gravity[22], and also to the dilatonic Gauss-Bonnet system [23].

3.2. Dual-null evolution system

We implemented our evolution code [7] for higher-dimensional space-time, and with Gauss-Bonnet gravity terms. The system we consider is spherical symmetry, and expressed using dual-null coordinate. The use of dual-null coordinate simplifies the treatment of horizon dynamics and radiation propagation clearly.

We adopt the line element

$$ds^2 = -2e^{f(x^+,x^-)}dx^+\,dx^- + r^2(x^+,x^-)\gamma_{ij}dz^idz^j, \tag{29}$$

where the coordinate (x^+, x^-) are along to null propagation directions, and $\gamma_{ij}dx^idx^j$ is the metric of the $(n-2)$-dimensional unit constant curvature space with $k = \pm 1,\ 0$.

For writing down the Einstein equations, we introduced the variables

$$\begin{aligned}
\Omega &= \frac{1}{r}, &(30)\\
\vartheta_\pm &= (n-2)\partial_\pm r &(31)\\
\nu_\pm &= \partial_\pm f &(32)
\end{aligned}$$

which are conformal factor, expansions, and in-affinities, respectively. We also define

$$\eta = \Omega^2\frac{(n-2)(n-3)}{2}\left(ke^{-f} + \frac{2}{(n-2)^2}\vartheta_+\vartheta_-\right). \tag{33}$$

The non-zero Einstein tensor components, then, are

$$\begin{aligned}
G_{++} &= -\Omega(\partial_+\vartheta_+ + \vartheta_+\nu_+), &(34)\\
G_{--} &= -\Omega(\partial_-\vartheta_- + \vartheta_-\nu_-), &(35)\\
G_{+-} &= \Omega\partial_-\vartheta_+ + \eta, &(36)\\
G_{ij} &= \gamma_{ij}\left\{e^f\left[\frac{\partial_+\nu_-}{\Omega^2} - \frac{2(n-3)}{(n-2)\Omega}\partial_-\vartheta_+ - \frac{(n-3)(n-4)}{(n-2)^2}\vartheta_+\vartheta_-\right] - k\frac{(n-3)(n-4)}{2}\right\} &(37)
\end{aligned}$$

The EGB equation, (25), becomes

$$\begin{aligned}
\partial_+\vartheta_+ &= -\vartheta_+\nu_+ + \frac{1}{\alpha_{\rm GR}\Omega}(\alpha_{\rm GB}H_{++} - \kappa^2 T_{++}), &(38)\\
\partial_-\vartheta_- &= -\vartheta_-\nu_- + \frac{1}{\alpha_{\rm GR}\Omega}(\alpha_{\rm GB}H_{--} - \kappa^2 T_{--}), &(39)\\
\partial_-\vartheta_+ = \partial_+\vartheta_- &= -\frac{\eta}{\Omega} + \frac{1}{\alpha_{\rm GR}\Omega}(\kappa^2 T_{+-} - \alpha_{\rm GB}H_{+-} + e^{-f}\Lambda), &(40)\\
\partial_+\nu_- = \partial_-\nu_+ &= -\frac{2(n-3)}{(n-2)}\eta + \frac{(n-3)(n-4)}{(n-2)^2}\Omega^2\vartheta_+\vartheta_- + k\frac{(n-3)(n-4)}{2e^f}\Omega^2\\
&\quad + \frac{2(n-3)}{(n-2)}\frac{1}{\alpha_{\rm GR}}(\kappa^2 T_{+-} - \alpha_{\rm GB}H_{+-} + e^{-f}\Lambda)\\
&\quad + \frac{\Omega^2}{e^f\alpha_{\rm GR}}\left(\kappa^2 T_{zz} - \alpha_{\rm GB}H_{zz} - r^2\Lambda\right), &(41)
\end{aligned}$$

where $H_{\mu\nu}$ components are lengthy and we omit them in this article.

We assume ghost scalar field $\phi(x^+, x^-)$,

$$T_{\mu\nu} = -\partial_\mu\phi\partial_\nu\phi - g_{\mu\nu}\left(-\frac{1}{2}(\nabla\phi)^2 + V(\phi)\right), \tag{42}$$

which obeys the Klein-Gordon equation,

$$\Box\phi = \frac{dV}{d\phi}. \tag{43}$$

If we define scalar momentum as

$$p_\pm = r\partial_\pm\phi = \frac{1}{\Omega}\partial_\pm\phi, \tag{44}$$

then non-zero $T_{\mu\nu}$ components are $T_{++} = -\Omega^2 p_+^2$, $T_{--} = -\Omega^2 p_-^2$, $T_{+-} = e^{-f}V$, and $T_{zz} = -e^f p_+ p_- - \Omega^{-2}V$. The eq. (43) becomes

$$\partial_+ p_- = \left(\frac{1}{n-2} - \frac{1}{2}\right)\Omega\vartheta_+ p_- - \frac{1}{2}\Omega\vartheta_- p_+ - \frac{1}{2e^f\Omega}\frac{dV}{d\phi}, \tag{45}$$

$$\partial_- p_+ = -\frac{1}{2}\Omega\vartheta_+ p_- + \left(\frac{1}{n-2} - \frac{1}{2}\right)\Omega\vartheta_- p_+ - \frac{1}{2e^f\Omega}\frac{dV}{d\phi}. \tag{46}$$

These equations complete the system.

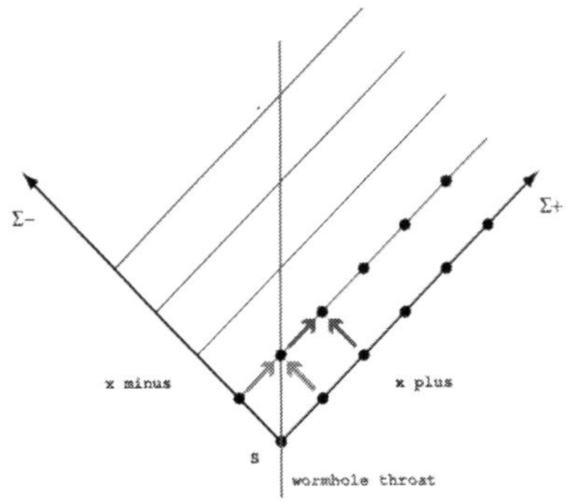

Figure 3. Numerical grid structure. Initial data are given on null hypersurfaces $\Sigma_\pm$ ($x^\mp = 0$, $x^\pm > 0$) and their intersection S. [7].

The basic idea of numerical integration is as follows. We prepare our numerical integration range as drawn in Figure 3. The grid will cover both universes connected by the wormhole throat $x^+ = x^-$. We give initial data on a surface S and the two null hypersurfaces $\Sigma_\pm$ generated from it. Generally the initial data have to be given as

$$(\Omega, f, \vartheta_\pm, \phi) \quad \text{on } S\text{: } x^+ = x^- = 0 \tag{47}$$

$$(\nu_\pm, p_\pm) \quad \text{on } \Sigma_\pm\text{: } x^\mp = 0,\ x^\pm > 0. \tag{48}$$

We then evolve the data $u = (\Omega, \vartheta_\pm, f, \nu_\pm, \phi, p_\pm)$ on a constant-x^- slice to the next.

Due to the dual-null decomposition, the causal region of a grid is clear, and there are in-built accuracy checks: the integrability conditions or consistency conditions $\partial_-\partial_+ u = \partial_+\partial_- u$. In order to update a point N (north), we have two routes from the points E (east) and W (west). The set of equations, (38)-(41) and (45)-(46), gives us x^+-direction (W to N) and x^--direction (E to N) integrations together with the consistency conditions.

As a virtue of the dual-null scheme, we can follow the wormhole throat or black-hole horizons easily. They are both trapping horizons, hypersurfaces where $\vartheta_+ = 0$ or $\vartheta_- = 0$ [24, 25]. Another benefit is the singular point excision technique, since the causal region of each grid point is apparent. When a grid point is inside a black-hole horizon and near to the singularity, we can exclude that point and grid points in its future null cone from further numerical computation.

In order to evaluate the energy, we apply the Misner-Sharp mass in n-dimensional EGB gravity[14],

$$E_n = \frac{(n-2)A_{n-2}}{2\kappa_n^2}\Omega\left[-\frac{1}{\Omega^2}2\frac{\Lambda}{(n-1)(n-2)}+\left(k+\frac{2}{(n-2)^2}e^f\vartheta_+\vartheta_-\right)\right.$$
$$\left.+(n-3)(n-4)\alpha_{\rm GB}\Omega^2\left(k+\frac{2}{(n-2)^2}e^f\vartheta_+\vartheta_-\right)^2\right]. \quad (49)$$

3.3. *Evolutions of 4, 5, and 6-dimensional wormhole solutions in GR*

We checked our numerical code whether it reproduces the static wormhole solution, obtained in §2. We express our n-dimensional Ellis wormhole solutions in dual-null coordinate and evolved. We find that numerical truncation error can quite easily destroy the stability, but this stability can be controlled with fine resolution. All the results below are obtained after we confirmed the resolution of the code which does not destroy the stability of wormhole throat structure by numerical errors within the evolution (in x^--direction) shown in the figure.

We put perturbations of the static wormhole in the form of Gaussian pulses, input from the right-hand universe. The perturbation is put in the scalar field momentum on its initial data on Σ_+

$$\delta p_+ = c_1 \exp(-c_2(l-c_3)^2), \quad (50)$$

with all the other initial data as the the static wormhole solution. Here c_1, c_2, c_3 are parameters, and we show the cases with small amplitude and width $c_1 = \pm 0.01$ and $c_2 = 3$, and the initial location $c_3 = 1$. That is, the pulse will hit the wormhole throat at $x^+ = x^- = 1$. Positive (or negative) c_1 corresponds enhancing (or reducing) the supporting ghost field.

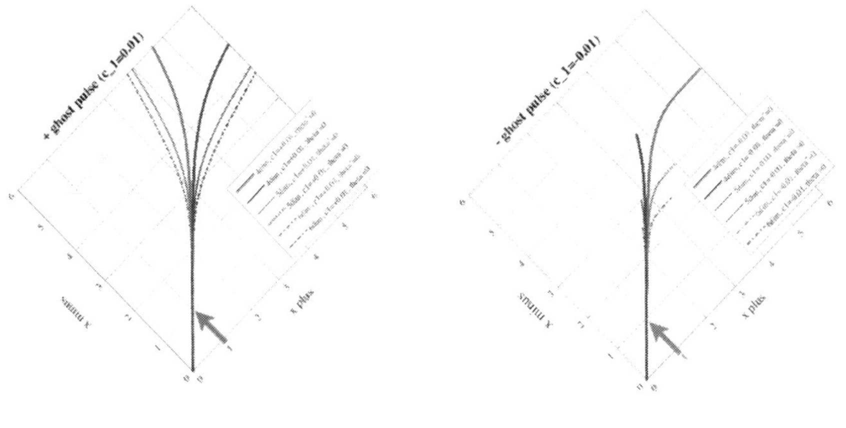

Figure 4. Location of the expansion ϑ_+ (red lines) and ϑ_- (blue lines) for evolutions of a solution in 4, 5, and 6-dimensional General Relativity ($\alpha_{\rm GB} = 0$) as a function of (x_+, x_-). The throat begins expanding if we input negative energy scalar flux (left panel), while the throat turns to be a black hole if we input positive energy scalar flux (right panel).

Figure 4 shows the results of $n = 4, 5$, and 6 dimensional wormhole solution with above perturbation. The plot shows where the vanishing locations of expansions, $\vartheta_\pm = 0$, in (x^+, x^-) plane. We see first the wormhole throat is the location both $\vartheta_\pm = 0$, but after a small pulse hit it, then the throat (or horizon) split into two ($\vartheta_+ = 0$ and $\vartheta_- = 0$), depending on the signature of the energy of pulse.

If the location of ϑ_+ is outer (in x^+-direction) than that of ϑ_-, then the region $\vartheta_- < x < \vartheta_+$ is judged as a black-hole. Otherwise the region $\vartheta_+ < x < \vartheta_-$ can be judged as an expanding throat. The throat begins expanding if we input negative energy scalar flux (left panel in Figure 4), while the throat turns to be a black hole if we input positive energy scalar flux (right panel).

3.4. Evolutions of 5-dimensional wormhole solution in EGB gravity

We also evolve the same initial data with Gauss-Bonnet terms $\alpha_{\rm GB} \neq 0$ and study their effects to the evolutions.

Figure 5 shows the case of 5-dimensional EGB gravity. The initial data of wormhole on $\Sigma_\pm = 0$ are obtained numerically, solving the set of equations. The evolutions with $\alpha_{\rm GB} \neq 0$ are quite unstable, and we are hard to keep its static configurations long enough. We see if $\alpha_{\rm GB} > 0$ the throat begins expansion (left panel in Figure 5). On the contrary, if $\alpha_{\rm GB} < 0$, then the throat turns to be black-hole (right panel).

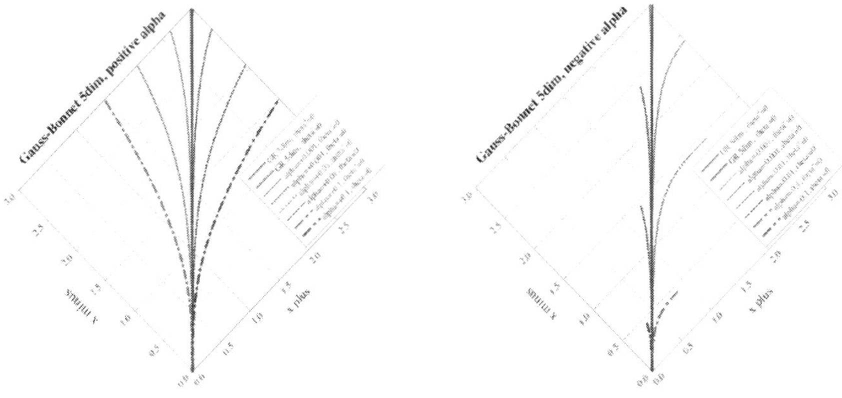

Figure 5. The same with Figure 4, but for 5-dimensional Einstein-Gauss-Bonnet gravity. When $\alpha_{\rm GB} > 0$ (left panel), the throat begins expanding (left panel), while the throat turns to be a black hole for $\alpha_{\rm GB} < 0$ (right panel).

4. Conclusions and Discussions

We studied the simplest wormhole solutions and their stability.

The space-time is assumed to be static and spherically symmetric, has ghost scalar field, and has reflection symmetry at the throat. The four-dimensional version is known to the Ellis (Morris-Thorne) solution, and we derived its extension in the n-dimensional space-time.

Using the linear perturbation technique, we showed that the solutions have at least one negative mode, which concludes that all wormholes are linearly unstable. The time scale of instability becomes shorter as n becomes large.

We, next, confirm the instability with numerical evolutions. Our code uses dual-null coordinate system, which is well-suited for studying horizon structure. At the throat, both the ingoing and outgoing expansions vanish, which means that the throat consists of a degenerate horizon. If we put a small amplitude scalar pulse to the throat, then the wormhole throat bifurcates, its horizon structure changes into a black-hole or an expanding throat depending the pulse energy is positive or negative, respectively. We also investigated the effect of the higher-order curvature corrections using Gauss-Bonnet terms, and found that such corrections do not work for stabilization of wormholes.

All the behavior of wormholes may be explained simply with energy balance. In [7], for small perturbations, an existence of critical solution is suggested. The similar behaviors are also observed in our investigations, which will be reported elsewhere.

We guess a wormhole with exotic matter is a disguise to avoid public notice, and does prefer to appear as a black-hole or an expanding universe.

Acknowledgments
This work was supported in part by the Grant-in-Aid for Scientific Research Fund of the JSPS (C) No. 25400277. Numerical computations were carried out on SR16000 at YITP in Kyoto University, and on the RIKEN Integrated Cluster of Clusters (RICC).

References

[1] L. Flamm, Physik Z. **17**, 448 (1916).
[2] A. Einstein & N. Rosen, Phys. Rev. **48**, 73 (1935).
[3] M. S. Morris & K. S. Thorne, Am. J. Phys. **56**, 395 (1988).
[4] H. G. Ellis, J. Math. Phys. **14**, 395 (1973).
[5] M. Visser, *Lorentzian Wormholes* (AIP Press, 1995).
[6] F. S. N. Lobo, in *Classical and Quantum Gravity Research* (Nova Sci. Pub., 2008). [arXiv:0710.4474]
[7] H. Shinkai & S.A. Hayward, Phys. Rev. D **66**, 044005 (2002).
[8] A. Doroshkevich, J. Hansen, I. Novikov, & A. Shatskiy, Int. J. Mod. Phys. D **18**, 1665 (2009)
[9] J. A. Gonzalez, F. S. Guzman & O. Sarbach, Class. Quant. Grav. **26**, 015010, 015011 (2009); Phys. Rev. D **80**, 024023 (2009). O. Sarbach & T. Zannias, Phys. Rev. D **81**, 047502 (2010).
[10] S A Hayward, Int. J. Mod. Phys. D **8**, 373 (1999).
[11] B. Zwiebach, Phys. Lett. B **156**, 315 (1985).
[12] A. Chodos & S. Detweiler, Gen. Rel. Grav. **14**, 879 (1982).
[13] G. Clément, Gen. Rel. Grav. **16**, 131 (1984).
[14] H. Maeda & M. Nozawa, Phys. Rev. D **78**, 024005 (2008).
[15] P. Kanti, B. Kleihaus & J. Kunz, Phys. Rev. Lett. **107**, 271101 (2011); Phys. Rev. D **85**, 044007 (2012).
[16] T. Torii & H. Shinkai, Phys. Rev. D **88**, 064027 (2013).
[17] S. Golod & T. Piran, Phys. Rev. D **85**, 104015 (2012).
[18] F. Izaurieta & E. Rodriguez, Class. Quant. Grav. **30** 155009 (2013).
[19] N. Deppe, C. D. Leonard, T. Taves, G. Kunstatter, & R. B. Mann, Phys. Rev. D **86**, 104011 (2012).
[20] B. Bhawal & S. Kar, Phys. Rev. D **46**, 2464 (1992).
[21] G. Dotti, J. Oliva & R. Troncoso, Phys. Rev. D **76**, 064038 (2007).
[22] M H Dehghani & Z Dayyani, Phys. Rev. D**79**, 064010 (2009).
[23] P Kanti, B Kleihaus & J Kunz, Phys. Rev. Lett **107**, 271101 (2011); Phys. Rev. D **85**, 044007 (2012).
[24] S A Hayward, Class. Quant. Grav. **15**, 3147 (1998).
[25] S A Hayward, Phys. Rev. D **49**, 6467 (1994).

Geometry Design of an Electron Dump with Simple Geo

M Şekerci[1], F Ç Öztürk[2], N Hafızoğlu[2] and S Özkorucuklu[2]

[1] Süleyman Demirel University, Department of Physics, 32260, Isparta, Turkey
[2] Istanbul University, Department of Physics, 34134, Istanbul, Turkey

mertsekerci@sdu.edu.tr

Abstract. The goal of this study is to design a geometry of an electron dump with Simple Geo code which is a freeware product and provides the ability of designing complex geometric systems easily. Also, Simple Geo can output the designed geometry in many different formats. Desired design of the electron dump is to stop the 40 – 42 MeV electron beams. To reach this aim, requested geometric design with the possible material was done with Simple Geo and a FLUKA output format file created to run the simulations in FLUKA code.

In particle accelerators and research laboratories, the transportation of the high energetic charged particles done by the structures called beam lines. With the help of this structures high energetic charged particle beams could be transported in desired ways to perform the requested experiments and to observe the interactions. That kind of structures are also used in the areas of high energy physics experiments and related facilities such as particle colliders, synchrotron radiation devices for example to produce a synchrotron light source, particle therapy in where that energetic particle beams consisting of usually protons but also neutrons to be used for cancer treatment, astrophysics, military purposes and for increasing the industrial improvements. Used for all kind of desires, high energetic charged particle beams used in the beam lines must be neutralized and ended finally. This is a must property of a suitable beam line since without a neutralization that high energetic charged particle beams should harm even destroy the rest of the system, located after the beam line. Structures called "beam dumps" are used for this purpose [1].

Basically, dump systems are used to stop and neutralize the high energetic charged particle beams without any harm to their organic and inorganic surroundings including the beam line itself. The design studies completed in this study is done for an electron dump system which can be used in a beam line that houses 40 – 42 MeV electron beam produced from a thermionic electron gun and used to create a free electron laser light.

Electron dumps have been designed to have different geometric details and are made of different materials according to the characteristic specifications of the charged particle beams of the systems that they are going to be used [2]. For this study as is mentioned before, the studied design will be able to stop 40 – 42 MeV energetic electron beams. The energy of the charged particle beam is the most important value of that characteristic specifications.

Electron dump systems also requested to minimize the effects of the secondary particles which could be occurred due to the uncontrolled interactions of high energetic charged particle beams and matter inside the electron dump systems. Therefore, neutralization and extermination of high energetic charged particle beams and possible formed secondary particles are done by the absorption of the energies of the beam and secondary particles after hitting the beam to a structure that holds a conical space inside a cylindrical material. Most of the electron dumps which are used for high energetic electron beams like in this particular study, have the same shape which is a conical space inside a cylindrical structure. The reason of that conical space inside the cylindrical structure is to eliminate the possibility of the damage to the beam line which may occur due to the backscattering of the secondary particles and beam itself [3]. This shape will not let the secondary particles creation allowing to scatter directly on the opposite way of the incoming beam.

In Figure 1. mentioned conical space itself is shown inside a cylindrical structure. This figure was obtained during the designing with the Simple Geo [4].

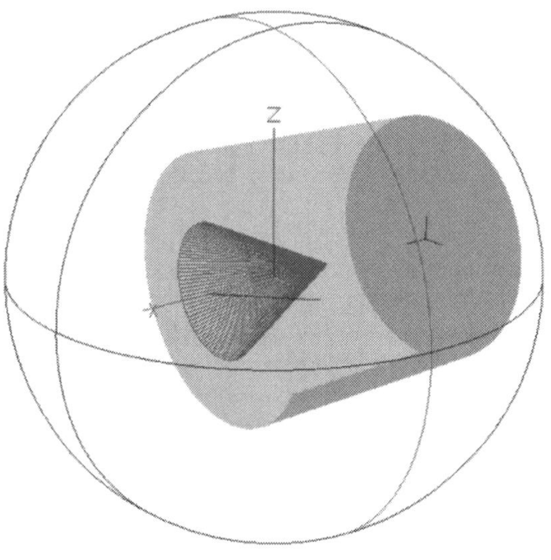

Figure 1. Conical space itself inside the cylindrical structure

Simple Geo is the code used in the overall design and used to create the geometric design. It is possible to create and design different complex geometries and structures with this freeware code. Besides, to create even more complex geometries, the code provides pre-defined geometries and mathematical logic based operators such as addition and subtraction to make the connections between different geometric objects to create the final object. Primary reason of using Simple Geo is its simplicity of geometric design and ability to provide different format output files that could be directly imported by other simulation or design codes. This property of Simple Geo, provides us the simplicity of testing the designed structure with some useful other codes like FLUKA.

In Figure 2. the conical space and the cylindrical structure is shown together. This figure was obtained with the Simple Geo's mentioned mathematical logic subtraction method. In order to create this shape, a cylinder defined first with specific properties. Continuously, the required conical space is created by defining the properties of that conical structure.

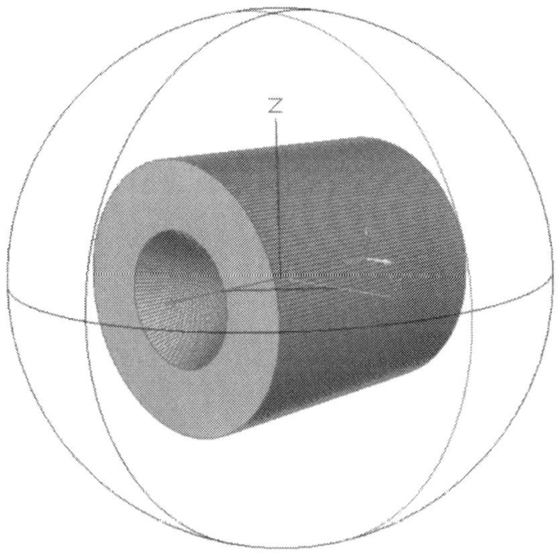

Figure 2. Conical space inside the cylindrical structure together

Cylindrical core of the electron dump system is designed as 24 cm long on the direction of the beam propagation way with a 6 cm radius to provide the requests and be used on a beam line that houses 40 – 42 MeV electron beam as mentioned before. The conical space on the cylindrical core has a 3 cm maximum radius and 0 cm minimum radius and the length of the core is determined as 4 cm. Those dimensions are selected since they are observed to be the most applicable and suitable ones. Some simulations of different dimensional geometries designed with Simple Geo were studied with FLUKA code before giving the final decision. Simulation studies covering this context still continues with the simulation code FLUKA [5].

One other important topic is about the material selection of the production of these kind of devices. The materials used in electron dump systems are usually carbon, aluminum, copper, beryllium, and lead. That diversity to produce an electron dump is due to the usage purposes of the dump system and to be able to produce it with low cost. Some of these materials are able to work with respectively lower energetic high energy charged particle beams.

Carbon – graphite was selected for the electron dump system designed in this study. Primary reason for this is the low neutron production rate of the carbon – graphite. Furthermore, carbon – graphite provides the best results for the mentioned energy range and price-performance rate.

An electron dump structure with these dimensions and with a lead cover has been observed to operate with high efficiency and also incarcerate the secondary particles inside itself.

References

[1] McIntosh D H, 1994 *UCLA Technical Design Report* **3**-12
[2] Colin Smith, Yun He, Charles Sinclair, 2005 *IPAC 5* **1**-3
[3] Scientific Assessment of High-Power Free-Electron Laser Technology, 2009, Committee on a Scientific Assessment of Free-Electron Laser Technology for Naval Applications, Board on Physics and Astronomy, Division on Engineering and Physical Sciences, National Research Council
[4] Theis C, Buchegger K.H, Brugger M, Forkel-Wirth D, Roesler S, Vincke H, Nuclear Instruments and Methods in Physics Research A **562**, 827-829
[5] Ferrari A, Sala P R , Fasso A, Ranft J, 2008 **3**-10, 243-252

Wormholes from Chiral Fields

E. Charalampidis[1†], T. Ioannidou[1*], B. Kleihaus[2‡] and J. Kunz[2+]
[1]School of Civil Engineering, Faculty of Engineering, Aristotle University of Thessaloniki, 54124 Thessaloniki, Greece
[2]Institut für Physik, Universität Oldenburg, Postfach 2503 D-26111 Oldenburg, Germany
E-mail: †echarala@auth.gr, *ti3@auth.gr, ‡b.kleihaus@uni-oldenburg.de, +jutta.kunz@uni-oldenburg.de

Abstract. In this paper, Lorentzian wormholes with a phantom field and chiral matter fields have been obtained. In addition, it is shown that for different values of the gravitational coupling of the chiral fields, the wormhole geometry changes. Finally, the stability of the corresponding wormholes is studied and it is shown that are unstable (eg. Ellis's wormhole instability)

1. Introduction

A wormhole[1] is a topological object of space-time which correspond to a *shortcut* through space-time. Wormholes that could be crossed in both directions, are known as *traversable wormholes*. Transversable wormholes introduced in [1], were obtained when a phantom field was coupled to gravity. Their energy-momentum tensor would violate all (null, weak and strong) energy conditions; while the phantom field is a scalar field with a reversed sign in front of its kinetic term.

The action of the Einstein gravity coupled to a phantom field [2] and ordinary matter fields is of the form

$$S = \int \left[\frac{1}{16\pi G}\mathcal{R} + \mathcal{L}_{\rm ph} + \mathcal{L}_{\rm ch} + \mathcal{L}_{\rm sk}\right]\sqrt{-g}\; d^4x \tag{1}$$

where $\mathcal{R}$ is the curvature scalar, G is the Newton's constant and g is the determinant of the metric. Also, the Lagrangian of the phantom field ϕ is

$$\mathcal{L}_{\rm ph} = \frac{1}{2}\partial_\mu\phi\partial^\mu\phi\ , \tag{2}$$

and the non-linear sigma model Lagrangian is

$$\mathcal{L}_{\rm ch} = \frac{\kappa^2}{4}{\rm Tr}\left\{L_\mu L^\mu\right\} \tag{3}$$

where $L_\mu = \partial_\mu U U^\dagger$ and κ is a coupling constant. The U is the chiral field which is a function on the space-time manifold taking values in the Lie group SU(2). The last term in the action is the Skyrme term defined as

$$\mathcal{L}_{\rm sk} = \frac{1}{32e^2}{\rm Tr}\left\{F_{\mu\nu}F^{\mu\nu}\right\} \tag{4}$$

[1] Based on a talk given by T. Ioannidou

where $F_{\mu\nu} = [L_\mu, L_\nu]$ and e a coupling constant.

The variation of the action with respect to the metric leads to the Einstein equations

$$G_{\mu\nu} = \mathcal{R}_{\mu\nu} - \frac{1}{2}g_{\mu\nu}\mathcal{R} = 8\pi G T_{\mu\nu} \tag{5}$$

where the stress-energy tensor is

$$T_{\mu\nu} = g_{\mu\nu}\mathcal{L}_M - 2\frac{\partial \mathcal{L}_M}{\partial g^{\mu\nu}} \ . \tag{6}$$

Note that $\mathcal{L}_{\rm M} = \mathcal{L}_{\rm ph} + \mathcal{L}_{\rm ch} + \mathcal{L}_{\rm sk}$ is the matter Lagrangian.

We consider static spherically symmetric wormhole solutions by assuming that

$$ds^2 = -A^2 dt^2 + d\eta^2 + R^2 d\Omega^2 \ , \tag{7}$$

where $d\Omega^2 = d\theta^2 + \sin^2\theta d\varphi^2$ denotes the metric of the unit sphere, while A and R are functions of η. Note that the coordinate η takes values in $-\infty < \eta < \infty$. The limits $\eta \to \pm\infty$ correspond to the *disjoint asymptotically flat regions*.

In addition, the chiral field is taken to be of the form

$$U = \cos F + i \sin F \ \vec{e} \cdot \vec{\tau} \ , \tag{8}$$

where $\vec{e}$ is the unit vector field

$$\vec{e} = (\sin\theta\cos\varphi, \sin\theta\sin\varphi, \cos\theta) \ , \tag{9}$$

$\vec{\tau}$ is the Pauli matrices vector and $F(\eta)$ is the profile function.

For simplicity, we introduce the dimensionless quantities

$$\eta = \frac{\tilde{\eta}}{\kappa} \ , \quad R = \frac{\tilde{R}}{\kappa} \ , \quad \phi = \kappa\tilde{\phi} \ , \quad 4\pi G = \frac{\alpha}{\kappa^2} \tag{10}$$

which is equivalent to $\kappa = 1$ and $4\pi G = \alpha$ and we rename $\tilde{\eta} = \eta$, $\tilde{R} = R$, $\tilde{\phi} = \phi$.

2. Wormhole Topology

In order, to obtain wormhole solutions, we assume that the function R does not possess any zero. In particular, we assume that R behaves like $|\eta|$ in the asymptotic regions and possesses (at least) one minimum. Suppose $R'(\eta_0) = 0$ for some η_0 and $R(\eta_0) = r_0$. Then from the equations of motion we get that

$$R''(\eta_0) = \frac{1}{r_0}\left(1 - \alpha \sin^2 F_0 \left[2 + \frac{\sin^2 F_0}{e^2 r_0^2}\right]\right) = \frac{1}{r_0}\left(1 - \alpha/\alpha_{\rm cr}\right) \ , \tag{11}$$

where $F_0 = F(\eta_0)$ and

$$\alpha_{\rm cr} = \frac{1}{\sin^2 F_0 \left[2 + \frac{\sin^2 F_0}{e^2 r_0^2}\right]} \ . \tag{12}$$

Therefore, R has a minimum at η_0 if $\alpha < \alpha_{\rm cr}$ and a maximum when $\alpha > \alpha_{\rm cr}$. The first case corresponds to a wormhole since it describes a surface of minimal area which separates two asymptotically flat regions. In the second case the maximum is a local maximum since in the asymptotic regions $R = |\eta|$. This implies that there are two minima of R, one for $\eta < \eta_0$ and another for $\eta > \eta_0$. For a sequence of alternating minima and maxima the corresponding spatial hyper-surfaces would also have two asymptotically flat regions divided by a throat. However, the geometry of the throat would be more complicated.

For simplicity let us choose $\eta_0 = 0$ which defines the position of the throat. In addition, we impose the following conditions at the throat

$$R(0) = r_0 \ , \quad R'(0) = 0 \ , \quad F(0) = F_0 \ . \tag{13}$$

The first choice correspond to the areal radius of the throat, and the second is the extremum condition. The third choice fixes the value of the chiral profile function at the throat, which is a free parameter. Let us consider the case where $F_0 = n\pi/2$, i.e. the wormholes are symmetric under the interchange of the asymptotic regions.

Finally, the following boundary conditions are considered

$$A(\eta \to \infty) \to 1 \ , \quad F(\eta \to \infty) \to 0 \ , \quad F(\eta \to -\infty) \to n\pi \ . \tag{14}$$

which sets the time scale, and results from requiring finite energy and conserved topological charge respectively.

Let us conclude, by identifying the topological charge of the solutions. Consider the chiral field U as a map of spatial slices of the wormhole space-time to the group manifold $SU(2) \sim S^3$. Since U takes constant values in both asymptotic regions the spatial slices become topologically equivalent to a three dimensional sphere, where the north and south pole correspond to the asymptotic regions $\eta \to +\infty$ and $\eta \to -\infty$, respectively. Thus the chiral field is a map between two three-spheres, and the topological charge is defined as the degree of the map. Due to the the chiral field ansäntz given by (8) and the above asymptotic boundary conditions for the profile function F with $n = 1$ the corresponding topological charge of our solution is one.

The stability of wormholes is crucial for their physical relevance. It is known that the Ellis wormhole possesses an unstable mode [3, 4]. Therefore, we investigate whether our solutions possess the Ellis instability. However, in our work [2], we make a careful choice of the gauge condition in order not to miss the unstable mode present in the Ellis wormhole. Unfortunately, all our wormholes are Ellis unstable, as discussed in the next section.

3. Conclusion

Morris-Thorne wormholes threaded by chiral fields that carry a conserved topological charge have been obtained. We consider static spherically symmetric solutions, which are symmetric under an interchange of the two asymptotically flat universes with respect to the throat.

We observe that when the chiral fields are described by the NLS model, the gravitational coupling constant α is the single free parameter. For small α the chiral fields have no influence upon the wormhole geometry, and the throat is characterized by a single minimal surface. As the gravitational coupling increases, the presence of the chiral fields influence the geometry of the wormhole. At a critical coupling $\alpha_{\rm cr}$, the throat becomes degenerate. Beyond $\alpha_{\rm cr}$ the throat exhibits a set of three extrema, a (local) maximal surface and two minimal surfaces located symmetrically, one on each side. Therefore, the wormhole geometry develops a belly in the interior, which increases in size as α increases.

When the Skyrme term is added, another free parameter appears. Therefore, the gravitational coupling and the throat size can vary independently. As before increasing α influence the matter on the geometry of the throat, and the formation of an inner belly beyond a critical value $\alpha_{\rm cr}$. However, the splitting of the space-time into three parts is not as smooth as the NLS case, since at the maximal coupling the asymptotically flat universes touch the inner universe.

Let us conclude by stating that the chiral fields cannot stabilize the wormhole space-time. The instability of the Ellis solution is inherited by the wormholes threaded by chiral fields, although they *do* carry a topological charge. According to our observations this is not surprising though, since, the topological charge may finally simply reside in a single of several disconnected space-time parts.

Acknowledgement
E.C and T.I. acknowledge support from FP7, Marie Curie Actions, People, International Research Staff Exchange Scheme (IRSES-606096). T.I. also acknowledges support from The Hellenic Ministry of Education: Education and Lifelong Learning Affairs, and European Social Fund: NSRF 2007-2013, Aristeia (Excellence) II (TS-3647). B. Kleihaus and J. Kunz gratefully acknowledge support by the German Research Foundation within the framework of the DFG Research Training Group 1620 *Models of gravity*.

References

[1] M. S. Morris, K. S. Thorne, Am. J. Phys. 56, 395, 1988.
[2] E. Charalampidis, T. Ioannidou, B. Kleihaus and J. Kunz, Phys. Rev. D 87, 084069, 2013
[3] J. A. Gonzalez, F. S. Guzman and O. Sarbach, Class. Quant. Grav. 26, 015010, 2009
[4] K. A. Bronnikov, J. C. Fabris and A. Zhidenko, Eur. Phys. J. C 71, 1791, 2011

Microscopic black hole stabilization via the uncertainty principle

Constantinos G. Vayenas[1,2,*] **and Dimitrios Grigoriou**[1]

[1]School of Engineering, Caratheodory 1, St., University of Patras, Patras GR 26504, Greece
[2]Division of Natural Sciences, Academy of Athens, Panepistimiou 28, Ave., GR-10679 Athens, Greece

E-mail: cgvayenas@upatras.gr

Abstract. Due to the Heisenberg uncertainty principle, gravitational confinement of two- or three-rotating particle systems can lead to microscopic Planckian or sub-Planckian black holes with a size of order their Compton wavelength. Some properties of such states are discussed in terms of the Schwarzschild geodesics of general relativity and compared with properties computed via the combination of special relativity, equivalence principle, Newton's gravitational law and Compton wavelength. It is shown that the generalized uncertainty principle (GUP) provides a satisfactory fit of the Schwarzschild radius and Compton wavelength of such microscopic, particle-like, black holes.

1. Introduction

The Black Hole Uncertainty Principle [1, 2, 3] seeks for a connection between the Heisenberg Uncertainty Principle (HUP) on microscopic scales, expressed via the Compton wavelength, and the event horizon of black holes, commonly referring to macroscopic scales [1, 2, 3]. These considerations have led to the concepts of Generalized Uncertainty Principle and Generalized Event Horizon [1, 2, 3, 4]. One implication is that there could exist sub-Planckian black holes with a size of order their Compton wavelength [1, 2, 3, 4]. It has been thus proposed that there may exist some link between black holes and elementatry particles [2].

In this work we examine the gravitational confinement of two or three relativistic light particles rotating around their center of mass and we use the Schwarzschild geodesics of GR in conjunction with the HUP to investigate the properties of such states. It is found that such rotational states exist and their properties are similar with those computed by treating the two- or three- rotating particle motion via the combination of HUP, special relativity, Newton's gravitational law and the equivalence principle [5, 6, 7]. Interestingly when the rest mass of the two or three light relativistic particles is in the range of neutrino masses ($\sim 5 \cdot 10^{-2}\ eV/c^2$), then the mass of the gravitationally confined microscopic Planckian or sub-Planckian rotating particle states are in the hadron ($1\ GeV$) mass range [5, 6, 7].

2. SR treatment

The SR treatment of the three rotating particle model (Fig.1) has been presented already elsewhere [5, 7]. In brief the special relativity equation of motion for a circular orbit [8, 9] is used in conjunction with Newton's gravitational law to obtain

$$\gamma m_o \frac{\mathrm{v}^2}{r} = \frac{G m_g^2}{\sqrt{3} r^2}, \tag{1}$$

where m_g is the gravitational mass of each light particle, which according to the equivalence principle equals its inertial mass, m_i. The latter is given by

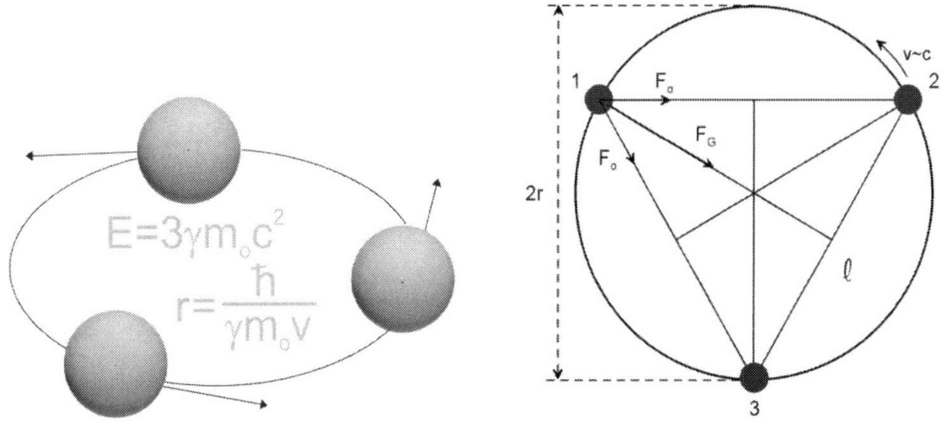

Figure 1. Schematic of the model of three rotating particles, together with the expressions for the rest energy and the radius of the bound state (left). The three particles are moving at a constant tangential velocity **v** in a circle of radius r around their center of mass and are equally spaced. F_o is the attraction force between two particles at distance $\ell = \sqrt{3}r$ and F_G is the resultant radial force (right).

$$m_i = \gamma^3 m_o, \tag{2}$$

where $\gamma(= (1-\mathrm{v}^2/c^2)^{-1/2})$ is the Lorentz-factor, as shown already in 1905 for linear motion [10] and more recently for arbitrary particle motion [5], including circular orbits. It must be noted that the transverse mass γm_o, which is equal to the relativistic mass, is not an inertial mass, since the force computed from the product of mass and acceleration it is not invariant [5, 7, 8]. Combining (1) and (2) one obtains

$$r = \frac{G m_o}{\sqrt{3} c^2} \gamma^5 \frac{\gamma^2}{\gamma^2 - 1}, \tag{3}$$

or, equivalently, for $\gamma >> 1$:

$$r/r_s = (1/2\sqrt{3})\gamma^5, \tag{4}$$

where $r_s(= 2Gm_o/c^2)$ is the Schwarzschild radius.

Upon combining with the de Broglie wavelength equation

$$\lambda\!\!\!^{-} = \frac{\hbar}{\gamma m_o \mathrm{v}}, \tag{5}$$

and assuming $r = \lambda\!\!\!^{-}$ one obtains from (4) and (5) that

$$\gamma^3 m_o = 3^{1/4} m_{Pl}, \tag{6}$$

where $m_{Pl}(= (\hbar c/G)^{1/2})$ is the Planck mass, therefore

$$\gamma^6 = 3^{1/2}\left(\frac{m_{Pl}}{m_o}\right)^2. \tag{7}$$

Very interestingly one observes from (6) that the mass of each rotating particle equals $3^{1/4}m_{Pl}$, regardless of the value of m_o. It also follows from (6)that the gravitational force acting on each of the three rotating particles is given by

$$F = \frac{Gm_o^2\gamma^6}{\sqrt{3}r^2} = \frac{Gm_{Pl}^2}{r^2} = \frac{\hbar c}{r^2}, \tag{8}$$

which is a factor of $\alpha^{-1}(= \epsilon\hbar c/e^2 \approx 137)$ larger than the electrostatic force $e^2/\epsilon r^2$ between two unit electric charges at a distance r. This is the expected value of the strong force at the same distance [11]. The ratio of this force to the nonrelativistic ($\gamma \approx 1$) Newtonian force, F_N, at the same distance is given by

$$\frac{F}{F_N} = \gamma^6 = 3^{1/2}\left(\frac{m_{Pl}}{m_o}\right)^2. \tag{9}$$

Since the mass of composite state formed by the three rotating particles equals $3\gamma m_o$, it follows from (7) that

$$m = 3\gamma m_o = 3^{13/12}m_o^{2/3}m_{Pl}^{1/3}. \tag{10}$$

Solving this equation for m_o one obtains

$$m_o = \frac{(m/3)^{3/2}}{3^{1/8}m_{Pl}^{1/2}}. \tag{11}$$

Considering as the composite particle mass m the mass of a neutron ($939.565\ MeV/c^2$) and as m_o the mass of one of the constituent quarks one finds

$$m_o = 0.0437\ eV/c^2, \tag{12}$$

which, surprisingly, is within the current uncertainty limits ($0.05 \pm 0.01\ eV/c^2$) of the rest mass of the heaviest neutrino [12]. The corresponding relativistic mass γm_o equals 313 MeV/c^2 which falls in the range of effective quark masses [13]. Consequently, as already shown [5, 6, 7], the mass of quarks can be modeled successfully by the mass of gravitationally confined neutrinos.

3. GR treatment

It is interesting to examine if the key results of the SR treatment, e.g. equations (6) and (9) can be obtained using the theory of general relativity (GR) where the masses are considered fixed and not velocity dependent.

In order to apply the Schwarzschild geodesics equation of GR [14] to the rotating neutrino problem one must adjust the physical model of Fig. 1 to the standard geometry of the Schwarzschild metric which involves a light test particle of mass m^* rotating around a central mass M. This can be done via the model shown in Figure 2. First we note that in the three-rotating particle model the force exerted to each particle is given by $F_N = Gm_o^2/(\sqrt{3}r^2)$, therefore the Newtonian potential energy due to the other two particles is given by

$$V_N = -Gm_o^2/(\sqrt{3}r). \tag{13}$$

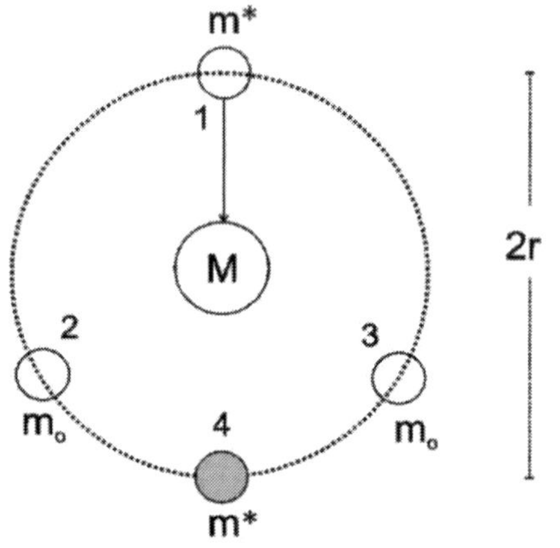

Figure 2. GR model for the three-neutrino state; $m^* = 2^{1/2}3^{-1/4}m_o$.

Since the same cyclic motion of particle 1 due to particles 2 and 3 can be obtained by substituting particle 1 with a new particle of mass m^* and by also substituting particles 2 and 3 by a particle 4 of mass m^* in the antidiametric position of 1, it follows that V_N must also equal the potential energy of 1 due to 4, i.e. $V_N = -Gm^{*2}/2r$. From this and equation (13) it follows $m^* = 2^{1/2}3^{-1/4}m_o$. We then consider the one-dimensional Schwarzschild effective potential, $V_s(r)$ with $M >> m^*$

$$V_s(r) = -\frac{GMm^*}{r} + \frac{2^{-1/2}3^{1/4}L^2}{2m_or^2} - \frac{2^{-1/2}3^{1/4}GML^2}{c^2m_or^3}, \tag{14}$$

where L is the angular momentum. Setting $L = \hbar$ one obtains

$$\frac{V_s(r)}{(m^*c^2/2)} = -\frac{r_s}{r} + \frac{a^2}{r^2} - \frac{a^2r_s}{r^3}, \tag{15}$$

where $r_s(= 2GM/c^2)$ is the Schwarzschild radius of the central mass M and $a(= \hbar/m^*c)$ is the Compton wavelength of the rotating mass m^*.

The effective potential is not the actual potential experienced by the rotating particle. It is the potential of an identical particle in ordinary one-dimensional nonrelativistic mechanics which causes the same one-dimensional motion as the radial motion of the actual rotating particles [14]. The derivative $F = dV_s/dr$, termed effective force, is thus the force acting on this identical particle in ordinary one-dimensional nonrelativistic motion.

Two circular orbits are obtained when the effective force $F = dV_s(r)/dr$ is zero, which upon differentiation of (15) gives

$$r^{\pm} = \frac{a^2}{r_s}\left[1 \pm \sqrt{1 - \frac{3r_s^2}{2a^2}}\right]. \tag{16}$$

The larger one of this roots (a minimum in V_s) is very large ($10^{24}m$) and irrelevant in the present model. The smaller one (a maximum in V_s) is unstable.

When the rotational radius r is smaller than r^-, which is close to the Schwarzschild radius, r_s, then the effective force suggests that the formation of a black hole with an event horizon at r_s is energetically favored (Fig. 3).

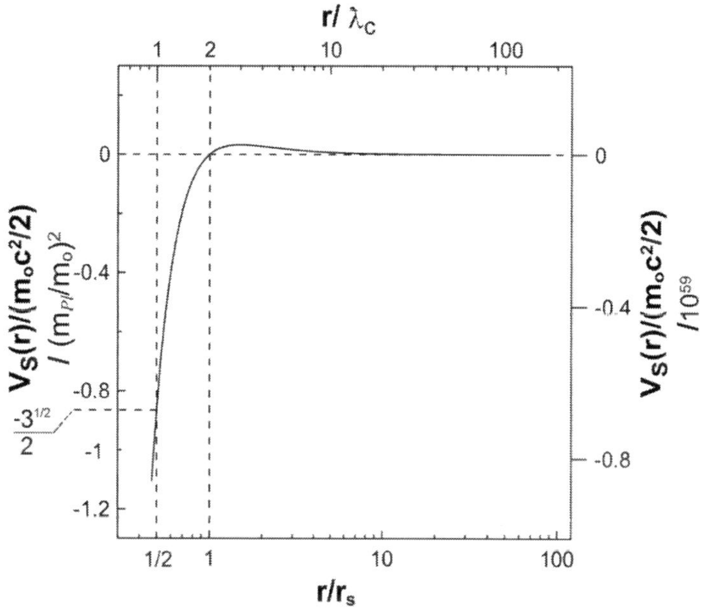

Figure 3. Plot of equation (15).

Differentiation of eq. (15) gives

$$F_s(r) = dV_s(r)/dr = (m^*c^2/2)\left[\frac{r_s}{r^2} - \frac{2a^2}{r^3} + \frac{3a^2 r_s}{r^4}\right]. \tag{17}$$

The first term in (17) corresponds to normal gravitational attraction according to Newton's gravitational law. Indeed, denoting this force by F_N and accounting for $r_s = 2GM/c^2$ one obtains

$$F_N(r) = \frac{m^*c^2}{2} \cdot \frac{2GM}{c^* r^2} = \frac{GMm^*}{r^2}. \tag{18}$$

The ratio $F(r)/F_N(r)$ of the effective force $F(r)$ to the Newtonian force F_N is thus given by:

$$F_s(r)/F_N(r) = 1 + \frac{a^2}{r_s^2}\left[3\left(\frac{r_s}{r}\right)^2 - 2\left(\frac{r_s}{r}\right)\right]. \tag{19}$$

Using the definition of a, r_s and $m_{Pl}(=\hbar c/G)^{1/2}$ and recalling $m^* = 2^{1/2}3^{-1/4}m_o$ we obtain

$$F_s(r)/F_N(r) = 1 + \frac{m_{Pl}^4}{4M^2 m^{*2}}\left[3\left(\frac{r_s}{r}\right)^2 - 2\left(\frac{r_s}{r}\right)\right] = 1 + \frac{\sqrt{3}}{8}\cdot\frac{m_{Pl}^4}{M^2 m_o^2}\left[3\left(\frac{r_s}{r}\right)^2 - 2\left(\frac{r_s}{r}\right)\right]. \tag{20}$$

The minimum r value for which $F_N(r)$ can serve as a Newtonian centripetal force, and thus for which the ratio $F(r)/F_N(r)$ is defined, is the value $r_s/2$. This follows from

$$\frac{GMm^*}{r^2} = \frac{m^* \mathrm{v}^2}{r} \quad thus \quad r = \frac{GM}{\mathrm{v}^2}. \tag{21}$$

Accounting for v $< c$ it follows that indeed $r > r_s/2$. This is the minimum allowed r value which leads (Fig. 3) to the lowest possible value of $V_s(r)$ and which according to eq. (15) is given by

$$V_s(r)_{min} = (m^*c^2/2)\left[-2 - \frac{4a^2}{r_s^2}\right] \approx -(m^*c^2/2)\frac{4a^2}{r_s^2} = -m^*c^2\frac{m_{Pl}^4}{2M^2m^{*2}} = -m^*c^2\frac{\sqrt{3}}{4}\cdot\frac{m_{Pl}^4}{M^2m_o^2} \tag{22}$$

When the limit $r = r_s/2$ coincides with the Compton wavelength limit, $\lambda_c(= \hbar/Mc)$, imposed by the Heisenberg uncertainty principle (HUP), i.e. when

$$r_s/2 = \frac{GM}{c^2} = \frac{\hbar}{Mc} = \lambda_c, \tag{23}$$

it follows

$$M = \left(\frac{\hbar c}{G}\right)^{1/2} = m_{Pl} \tag{24}$$

This result is consistent with that extracted from the SR treatment (eq. 8) which shows that the mass keeping, via its gravitational attraction, each rotating particle to its orbit is the Planck mass m_{Pl}.

Substituting in (19) with $r = r_s/2$ and accounting for $a^2/r_s^2 >> 1$ one obtains

$$F_s(r)/F_N = 3^{1/2}\left(\frac{m_{Pl}}{m_o}\right)^2 \tag{25}$$

which, surprisingly, is the same result as that obtained via the special relativistic treatment (eq. 9).

It thus appears that $r_s/2$ provides a satisfactory generalized event horizon (GEH), at least when used in conjunction with the Heisenberg uncertainty principle (HUP). At $r = r_s/2 = \lambda_c$ both the Schwarzschild geodesics and the Heisenberg uncertainty principle are satisfied for $M = m_{Pl}$.

4. Generalized uncertainty principle

These results are consistent with the generalized uncertainty principle (GUP) discussed by Adler [1], Carr [2, 3], Vagenas [4] and others, i.e.

$$\Delta x \geq \frac{\hbar}{\Delta p} + L_p^2\frac{\Delta p}{\hbar} \tag{26}$$

where Δx is the position uncertainty, p is the momentum and $L_p(= (G\hbar/c^3)^{1/2})$ is the Planck length. Using in (26) $\Delta p = Mc$ one obtains

$$\Delta x \geq \frac{\hbar}{Mc} + \frac{G\hbar}{c^3}\cdot\frac{Mc}{\hbar} = \lambda_c\left(1 + \frac{GM^2}{\hbar c}\right) = \lambda_c\left(1 + \frac{M^2}{m_{Pl}^2}\right) \tag{27}$$

Consequently for $M = m_{Pl}$ one obtains $\Delta x = 2\lambda_c$, thus $\Delta x/2 = \lambda_c$ and, therefore, in view of eq. (23), i.e. $r_s/2 = \lambda_c$, it follows $\Delta x = r_s$. Consequently the uncertainty in position equals r_s, i.e. the event horizon of the black hole and therefore for $M = m_{Pl}$, the use of the GUP in conjunction with $\Delta x = r_s$ leads directly to $\lambda_c = r_s/2$ and to the key result of equation (25) which shows the good agreement between the SR and GR treatments of the rotating particle model. This result, i.e. $\Delta x = r_s = 2\lambda_c$ also shows how the GUP leads to black hole stabilization forming bound states with some interesting properties, similar to those of hadrons.

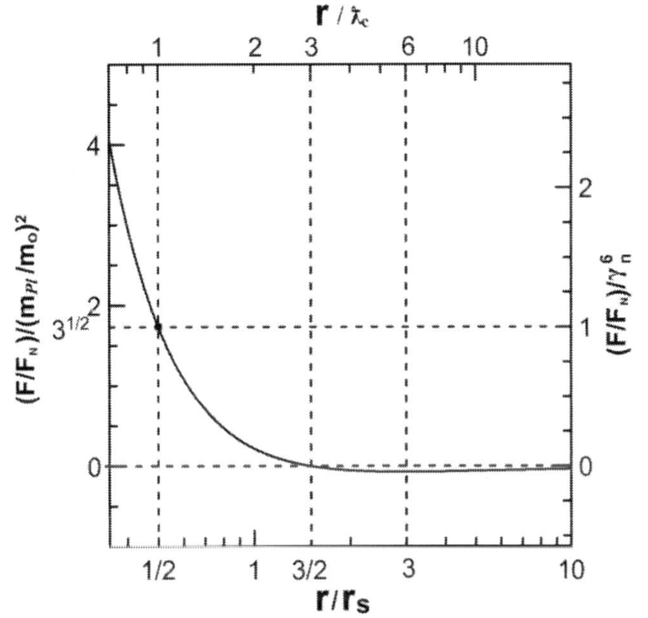

Figure 4. Plot of equation (20) for $M = m_{Pl}$.

For example, the effective force, F, at $r = r_s/2 = \lambda_c$ is equal (eqs 9 and 25) to the actual relativistic gravitational force, $\hbar c/r^2$, which keeps each particle in orbit in the SR treatment (eq. 8) and this force is a factor of $\alpha^{-1}(= \epsilon c\hbar/e^2 \approx 137)$ stronger than the Coulombic force at the same distance, as expected for the strong force [11]. Also if one were to borrow from SR the inertial-gravitational mass expression $\gamma^3 m_o = 3^{1/4} m_{Pl} = 3^{1/4} M$ (eq. 6), then one can compute $\gamma(= 7.163 \cdot 10^9$ for $m_o = 0.04378\ eV/c^2)$ and thus the relativistic mass $3\gamma m_o(\approx 939\ MeV/c^2)$ of the bound state, which is in the baryon mass range. Also upon using the above γ value in the Compton wavelength expression $\lambda_c = \hbar/\gamma m_o c$, one finds $\lambda_c = 0.63 \cdot 10^{-15}\ m$, which is in the baryon radius range. It thus appears that indeed a link exists between microscopic black holes and elementary particles as already suggested [2].

Acknowledgment
We thank Professor A. Fokas from Cambridge University for numerous helpful discussions.

References

[1] Chen P & Adler RJ (2002) Black hole remnants and dark matter. arXiv:gr-qc/0205106v1.
[2] Carr B, Modesto L Premont-Schwarz I (2011) Generalized Uncertainty Principle and Self-dual Black Holes. arXiv:1107.0708v1 [gr-qc].
[3] Carr BJ (2014) The Black Hole Uncertainty Principle Correspondence. arXiv:1402.1427v1 [gr-qc].
[4] Das S & Vagenas EC (2009) Phenomenological Implications of the Generalized Uncertainty Principle. Can. J. Phys. 87:233-240.
[5] Vayenas CG, Souentie S, Fokas A (2014) A Bohr-type model of a composite particle using gravity as the attractive force. Physica A 405:360-379.
[6] Vayenas CG (2014) Mathematical modeling of mass generation via confinement of relativistic particles. Journal of Physics Conference series 490:012084, doi: 10.1088/1742-6596/490/1/012084.
[7] Vayenas CG & Souentie S (2012), Gravity, special relativity and the strong force: A Bohr-Einstein-de-Broglie model for the formation of hadrons. Springer, ISBN 978-1-4614-3935-6

[8] French AP (1968) Special relativity. W.W. Norton and Co., New York
[9] Freund J (2008) Special relativity for beginners. World Scientific Publishing, Singapore
[10] Einstein A (1905) Zür Elektrodynamik bewegter Körper. Ann. der Physik., Bd. XVII, S. 17:891-921; English translation On the Electrodynamics of Moving Bodies (*http : //fourmilab.ch/etexts/einstein/specrel/www/*) by G.B. Jeffery and W. Perrett (1923)
[11] Povh, B., Rith, K., Scholz, Ch. & Zetsche, F.: Particles and Nuclei: An Introduction to the Physical Concepts 5th Ed., Springer-Verlag Berlin Heidelberg, (2006)
[12] Mohapatra RN et al (2007) Rep Prog Phys 70:1757-1867
[13] Griffiths D (2008) Introduction to Elementary Particles. 2nd ed. Wiley-VCH Verlag GmbH & Co. KgaA, Weinheim
[14] Wald RM (1984) General relativity, The University of Chicago Press, Chicago

Perceiving light rays above and below the horizon of a black hole in Schwarzschild spacetime

A Siwek[1], B Kuśmierz[1,2], P Gusin[3], J Masajada[1] and A Radosz[1]

[1] Institute of Physics, Wrocław University of Technology, Wybrzeże Wyspiańskiego 27, 50-370 Wrocław, Poland

[2] Institute of Mathematics, University of Wrocław, Plac Grunwaldzki 2/4, 50-384 Wrocław, Poland

[3] Institute of Theoretical Physics, University of Wrocław, Plac Maxa Borna 9, 50-204 Wrocław, Poland

E-mail: alicja.siwek@pwr.edu.pl

Abstract. We consider a novel kind of plot for presenting the behaviour of null geodesics in Schwarzschild spacetime. Our diagram depicts various features of the light rays as recorded by different classes of observers. The idea stems from the use of phase portraits for illustrating the null geodesics in Schwarzschild spacetime, above and below the horizon of a black hole, but they reveal unphysical characteristics. The plots we propose are free from such anomalies. Moreover they allow us to discover new under-horizon features.

1. Introduction

In general relativity, the notion of an observer becomes crucial when one intends to describe an outcome of an experiment conducted in a strong gravitational field. In our work, we focus on the perception of light rays by different kinds of observers in Schwarzschild spacetime. We present a novel kind of plot for illustrating the behaviour of null geodesics in this geometry. The spatial trajectories of light rays can be presented in a form of a phase portrait [1] but it may reveal unphysical features: the trajectories corresponding to light rays moving in the interior of a black hole away from the singularity, passing through the event horizon and leaving this region of spacetime. We present the plots where only physical null geodesics are illustrated as each point in the diagram represents a light ray that is registered by a local observer where the spatial (the direction in space) as well as temporal aspects (frequency shift) are taken into account.

2. Light rays as measured by different observers in Schwarzschild spacetime

In the static, spherically symmetric spacetime the line element can be expressed using Schwarzschild coordinates $(t,\, r,\, \theta,\, \phi)$ as:

$$ds^2 = \left(1 - \frac{r_S}{r}\right)(cdt)^2 - \left(1 - \frac{r_S}{r}\right)^{-1} dr^2 - r^2 d\theta^2 - r^2 \sin^2\theta d\phi^2 \equiv g_{tt}(cdt)^2 + g_{rr}dr^2 + g_{\theta\theta}d\theta^2 + g_{\phi\phi}d\phi^2\,, \tag{1}$$

where $r_S = 2GM/c^2$ and we shall use the system of units where $G = c = 1$. Due to the spherical symmetry of the spacetime, we restrict our considerations to the $\theta = \pi/2$ plane.We determine the geodesics of massive and massless geodesics, described by the four-velocity and the wave four-vector, respectively, using the conservation laws: $g_{tt}\dot{t}$ = const and $-g_{\phi\phi}\dot{\phi}$ = const, where the dot signifies the derivative with respect to the auxiliary parameter λ. For null geodesics $g_{\mu\nu}\dot{x}^\mu\dot{x}^\nu = 0$ and a four-vector corresponding to a light ray, denoted as $\boldsymbol{k} = k^t\partial_t + k^r\partial_r + k^\phi\partial_\phi$, reads

$$k^t = \frac{\Omega}{g_{tt}}\,,\; k^r = \pm\left(\Omega^2 - g_{tt}\frac{L^2}{r^2}\right)^{1/2}\,,\; k^\phi = \frac{L}{r^2}\,, \tag{2}$$

where Ω and L are constants, related to the energy and angular momentum, respectively, and $\pm$ signs correspond to outgoing/ingoing light rays.

2.1. Light rays as measured by static observers

Let us introduce a static observer SO in Schwarzschild spacetime whose velocity four-vector reads $\boldsymbol{V} = V^t\partial_t = g_{tt}^{-1/2}\partial_t$. The components of a wave vector which are measured by this observer determined within his/her coordinate frame (see e.g. [2]) are:

$$k^0 = g_{tt}^{1/2}\Omega\,,\; k^1 = \pm\left(\frac{\Omega^2}{g_{tt}} - \frac{L^2}{r^2}\right)^{1/2}\,,\; k^3 = \frac{L}{r}\,. \tag{3}$$

The angle of incidence ψ of a light ray for SO can be expressed as

$$\cot\psi = \frac{k^1}{k^3} = \pm r\left(\frac{1}{b^2 g_{tt}} - \frac{1}{r^2}\right)^{1/2}\,, \tag{4}$$

where the parameter describing a specific null geodesics $b = L/\Omega$ (impact parameter) and $+/-$ correspond to in-/outgoing geodesics, respectively. We can illustrate the evolution of light rays, specified by b, as measured by SO at each point along the trajectory, in the diagram depicting the $\psi(r)$ dependence, the ψ-diagram, in Figure 1. The range of the angle ψ is restricted to

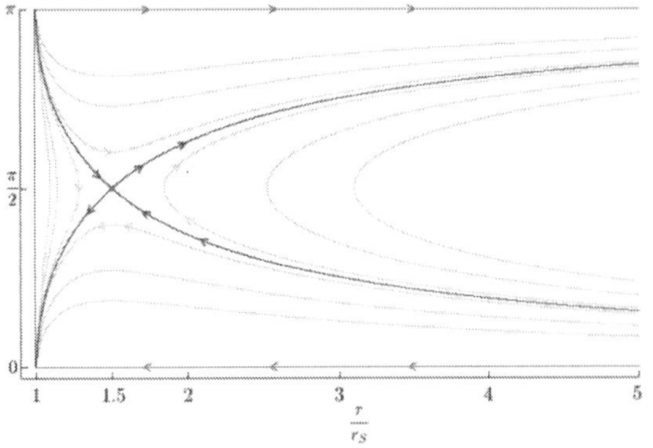

Figure 1. ψ-diagram for static observers. The unstable circular orbit of the photon sphere is represented by an isolated point: $r_{phs} = 3r_S/2$, $\psi_{phs} = \pi/2$. The blue lines, tending asymptotically to this point, are four separate curves corresponding to critical value of b: $b_{cr} = 3\sqrt{3}\,r_S/2$.

$[0, \pi]$ due to the axial symmetry. We notice the symmetry with respect to the line $\psi = \pi/2$: the ingoing trajectories fill the lower part $\psi < \pi/2$ and outgoing trajectories the upper part $\psi > \pi/2$ of the diagram. Approaching the event horizon, all trajectories tend to be oriented radially: $\psi \to 0$ or $\psi \to \pi$.

2.2. Light rays as measured above the horizon by RF – observers in a radial free fall

For an observer falling freely from $r = r_0$, $A = [g_{tt}(r_0)]^{1/2}$ and the four-velocity

$$\boldsymbol{U} = U^t\partial_t + U^r\partial_r = (A/g_{tt})\partial_t - (A^2 - g_{tt})^{1/2}\partial_r\,. \tag{5}$$

In the frame of reference of such an observer the angle of incidence of a light ray is determined as (see also [2, 3])

$$\cot\psi = \frac{k^1}{k^3} = \frac{rA}{g_{tt}b}\left[-\frac{(A^2-g_{tt})^{1/2}}{A} \pm \left(1-g_{tt}\frac{b^2}{r^2}\right)^{1/2}\right] = \frac{rA}{g_{tt}b}\left[-v \pm \left(1-g_{tt}\frac{b^2}{r^2}\right)^{1/2}\right], \quad (6)$$

where $v = [1 - g_{tt}/A^2]^{1/2}$ is velocity of the RF as measured by an SO (see e.g. [4]).

Let us choose a class of observers falling freely from infinity ($A = 1$), registering the light rays of different b, where the wavelength measured at $r = 7\,r_S$ is $\lambda_0 = 590\,\text{nm}$ for all the ingoing light rays. The frequency shift as measured by each RF is described by the ratio of frequencies registered at radial coordinates r_2 and r_1 (see e.g. [5, 6])

$$\frac{\Omega(r_2)}{\Omega(r_1)} = \frac{g_{tt}(r_1)}{g_{tt}(r_2)}\frac{1 \mp v(r_2)\left[1-g_{tt}(r_2)\frac{b^2}{r_2^2}\right]^{1/2}}{1 \mp v(r_1)\left[1-g_{tt}(r_1)\frac{b^2}{r_1^2}\right]^{1/2}}. \quad (7)$$

In Figure 2 a ψ-diagram for RF, falling from infinity, where the differences in wavelength are indicated by the colours of the points in the plot is presented.

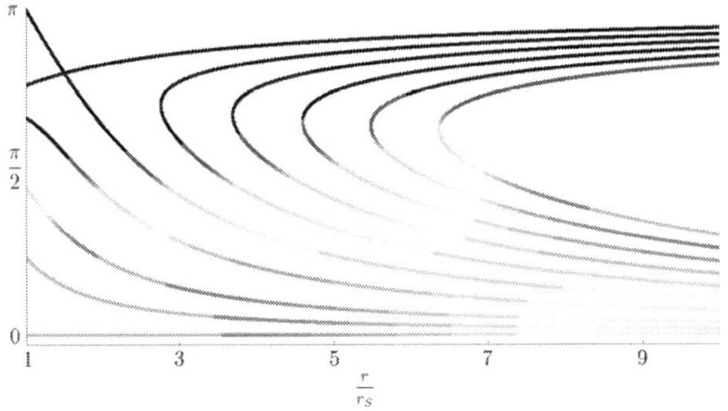

Figure 2. The ψ-diagram for observers falling freely from infinity where for each observer and each ingoing light ray $\lambda(r_1 = 7\,r_S) = 590\,\text{nm}$. Black/grey points indicate ultraviolet/infrared range, respectively.

2.3. Light rays as measured above and below the horizon by RF and NRF (non-radially falling observers)

In the interior of the Schwarzschild black hole, the angle ψ as measured by a radially falling observer satisfies:

$$\cot\psi = \frac{rA}{g_{tt}b}\left[\mp\frac{(A^2-g_{tt})^{1/2}}{A} + \left(1-g_{tt}\frac{b^2}{r^2}\right)^{1/2}\right], \quad (8)$$

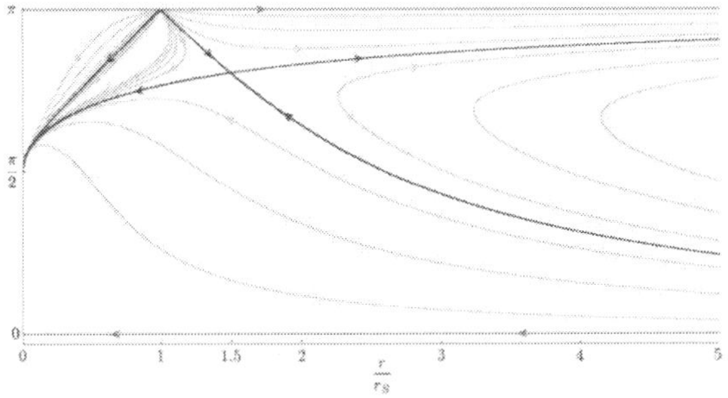

Figure 3. ψ-diagram for observers in the radial free fall from infinity.

where the interchange of roles of time t and radial r coordinates below the horizon (see [7, 8]) has been taken into account. The curves corresponding to different light rays measured by RF, above and below the event horizon are presented in Figure 3.

In this diagram only outgoing geodesics tend to be focused, $\psi \to \pi$, at the horizon. The ingoing rays cover the whole range of $\psi \in [0, \pi)$. The small area in the upper left part of the diagram between the blue lines $b = b_{cr}$ comprises the trajectories corresponding to geodesics characterized by $b > b_{cr}$, i.e. in this part of the diagram big changes in b induce very small changes in ψ.

Another type of observer, whose ψ-diagram can lead to interesting insights, is the observer in a free fall but with a non-zero angular momentum J. The velocity four-vector of such an observer is given by:

$$\boldsymbol{U} = U^t\partial_t + U^r\partial_r + U^\phi\partial_\phi = (A/g_{tt})\partial_t - [A^2 - g_{tt}(1 + J^2/r^2)]^{1/2}\partial_r + J/r^2\partial_\phi \ . \qquad (9)$$

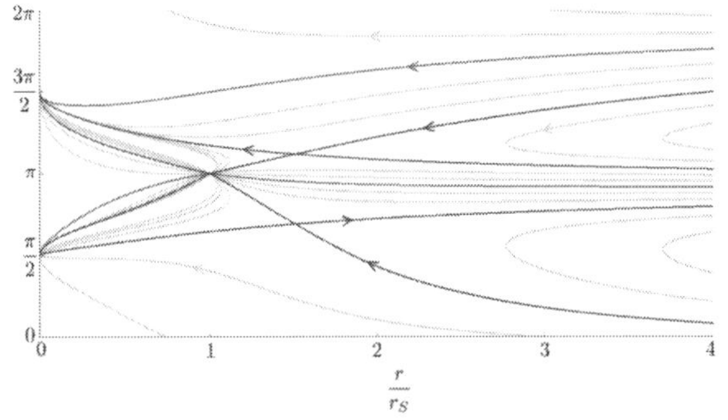

Figure 4. ψ-diagram for a class of observers with $A = 1$ and $J = r_S$.

In this case the range of ψ is extended to $\psi \in [0, 2\pi]$. The diagram is not symmetric with respect to the horizontal line $\psi = \pi$, apart from the curves corresponding to the light rays with $|b| = \infty$ below the horizon. Radial light rays no longer correspond to $\psi = 0/\pi$. On the other hand, all the outgoing trajectories tend to the value $\psi = \pi$ at the horizon, just like in the case of RF.

3. Discussion and conclusions

In this paper we present plots illustrating the trajectories of light rays as registered by different types of observers in Schwarzschild spacetime, both above and below the horizon of the black hole. Each plot, corresponding to a specific class of observers, pictures all the light rays which can be detected by this type of observer, presenting the angle at which a given light ray is registered. The obtained ψ-diagrams reveal the main differences in the perception of light rays by static, radially and non-radially falling observers. The most notable issues are the symmetry of the diagrams and the behaviour of null geodesics in the vicinity and below the horizon of the black hole – for RF and NRF. Originating in the vicinity of the horizon, there exist almost empty, "tear-shaped" areas between the curves $|b| = b_{cr}$. It is characteristic for these parts of the diagrams that in order to fill them up with trajectories one needs curves corresponding to light rays of very large (up to inifinite) angular momentum. This "repulsion" of trajectories around the curves $|b| = \infty$ is reminiscent of the neighbourhood of the photon sphere above the horizon. The possible analogy between these parts of ψ-diagrams will be the subject of further investigations. The important aspect concerning the area beneath the horizon is that the coordinate r describes the evolution in time, $r = 0$ being the ultimate instant. The consequences of this fact will be considered in our future studies. Moreover, the next significant step will be the construction of ψ-diagrams in other geometries – so far the preliminary results have been obtained in the case of a Kerr black hole.

References

[1] Dean B *Am. J. Phys.* 1999 **67** 78
[2] Müller T *Gen. Relat. Gravit.* 2008 **40** 2185-99
[3] Trümper M and Wayland J R *Phys. Rev. D* 1974 **10** 3931
[4] Augousti A T, Radosz A and Ostasiewicz K *Eur. J. Phys.* 2011 **32** 535
[5] Radosz A, Augousti A T and Ostasiewicz K *Phys. Lett. A* 2009 **373** 801
[6] Radosz A, Augousti A T and Siwek A *Gen. Relat. Gravit.* 2013 **45** 705-715
[7] Novikov I D *Astron. Zh.* 1961 **38** 564
[8] Grib A A and Pavlov Y V *Phys. Usp.* 2009 **52** 257

The Geometrical Meaning of Time - Some Cosmological Implications

Asher Yahalom
Ariel University, Ariel 40700, Israel
E-mail: asya@ariel.ac.il

Abstract. It is stated in many text books that the any metric appearing in general relativity should be locally Lorentzian i.e. of the type $g_{\mu\nu} = \text{diag}(1, -1, -1, -1)$ this is usually presented as an independent axiom of the theory, which cannot be deduced from other assumptions. The meaning of this assertion is that a specific coordinate (the temporal coordinate) is given a unique significance with respect to the other spatial coordinates. It was shown that the above assertion is a consequence of requirement that the metric of empty space should be linearly stable and need not be assumed. Some cosmological implications of the above result will be suggested.

1. Introduction

It is well known that our daily space-time is approximately of Lorentz (Minkowski) type that is, it possesses the metric $\eta_{\mu\nu} =$ diag $(1, -1, -1, -1)$. The above statement is taken as one of the central assumptions of the theory of special relativity and has been supported by numerous experiments. But why should it be so?

Furthermore, many textbooks [2] state that in the general theory of relativity any space-time is locally of the type $\eta_{\mu\nu} =$ diag $(1, -1, -1, -1)$, although it can not be presented so globally due to the effect of matter. This is a part of the demands dictated by the well known equivalence principle. The above principle is taken to be one of the assumptions of general relativity other assumption such as diffeomorphism invariance, and the requirement that theory reduce to Newtonian gravity in the proper regime lead to the Einstein equations:

$$G_{\mu\nu} = -\frac{8\pi G}{c^4} T_{\mu\nu} \tag{1}$$

in which $G_{\mu\nu}$ is the Einstein tensor, $T_{\mu\nu}$ is the stress-energy tensor, G is the gravitational constant and c is the velocity of light.

The Principle of Equivalence rests on the equality of gravitational and inertial mass, demonstrated by Galileo, Huygens, Newton, Bessel, and Eötvös. Einstein reflected that, as a consequence, no external static homogeneous gravitational field could be detected in a freely falling elevator, for the observers, their test bodies, and the elevator itself would respond to the field with the same acceleration [2]. This means that the observer will experience himself as free, not feeling the effect of any force at all. Mathematically speaking for the observer space time is locally (but not globally) flat and Minkowskian.

It was shown by several authors that one need not assume that space-time is locally Minkowskian based on an empirical (unexplained) fact, rather this can be justified theoretically.

Greensite [5] and Carlini & Greensite [6, 7] have studied the metric $\eta_{\mu\nu} =$ diag $(e^{i\theta}, -1, -1, -1)$ in which θ the "Wick angle" was treated as a quantum field dynamical variable. They have shown that the real part of the quantum field effective potential is minimized for the Lorentzian metric $\theta = 0$ and for the same case the imaginary part of the quantum field effective potential is stationary. Furthermore they have calculated the fluctuations around this minimal value and have shown them to be of the order $(\frac{l_p}{R})^3$ in which l_p is the Planck length and R is the scale of the universe. Elizalde & collaborators [8] have shown that the same arguments apply to a five dimensional Kaluza-Klein universe of the type $R^4 \times T^1$.

Itin & Hehl [9] have deduced that space time must have a Lorentzian metric in order to support classical electric/magnetic reciprocity.

H. van Dam & Y. Jack Ng [10] have argued that in the absence of a Lorentzian metric one can not obtain an appropriate finite representation of the relevant groups and hence the various quantum wave equations can not be written.

What is common to the above approaches is that additional theoretical structures & assumptions are needed in order to justify what appears to be a fundamental property of space-time. In previous works [1, 11, 12] it was claimed otherwise. It was shown that General relativistic equations and some "old fashioned" linear stability analysis will lead to a unique choice of the Lorentzian metric being the only one which is linearly stable for empty space. Other metrics are also allowed for empty space but are unstable and thus can exist in only a limited region of space-time. It should be mentioned that the choice of coordinates in the Fisher approach to physics can be also be justified using the stability approach [13]. The nonlinear stability question of the Lorentzian metric was settled by D. Christodoulou & S. Klainerman [14].

This paper assumes that space time must have four dimensions, it does not explain why this is so. For a possible explanation derived from string theory one can consult a paper by S. K. Rama [15]. The problem of the "arrow of time" that is time directionality was thought to be explained by Boltzman's H theorem. However, this explanation is thought to be wrong by some researchers who argue that this theorem cannot provide the required result unless one assumes low entropy at the beginning of the universe, for a popular account see a book by S. Carroll [16]. We will not discuss time directionality here.

For non empty space-time the situation can be drastically different. The existence of the intuitive partition of 4 dimensional space into "spatial" space and "temporal" time, is a feature of an almost empty space-time. This does not contradict the fact that such a partition can not be demonstrated in general solutions of equation (1) such as the one discovered by Gödel [26]. But this problem is not a characteristic of exotic space-times rather it is a property of standard cosmological models.

Standard Cosmology has many fundamental problems those include the horizon problem, the flatness problem, the entropy problem and the monopole problem [17]. A possible solution to those problems were suggested by Alan Guth using his famous inflation theory [18]. Entropy problems which plagued the original inflation model has led to a new inflation model suggested by Linde [19] which solve the entropy problem but required fine tuning of parameters. The same criticism holds for chaotic inflation also suggest by Linde [20]. On 17 March 2014, astrophysicists of the BICEP2 collaboration announced the detection of inflationary gravitational waves in the B-mode power spectrum, which if confirmed, would provide clear experimental evidence for the theory of inflation [21]. However, on 19 June 2014, lowered confidence in confirming the findings was reported.

It is the opinion of the author of this paper that a basic flaw in common to all inflation models. All inflation models require to postulate one or more scalar fields which have no function, implication or purpose in nature except for their ad-hoc use in the inflation model. This is in sharp contradiction with the principle of Occam's razor which demand that a minimum number of assumptions will explain a maximum number of phenomena. Postulating a physical field for every phenomena does not serve the purpose of theoretical science. Moreover, it will be shown that a perfectly good explanation within the frame-work of standard Cosmology does exist for the horizon problem if one looks closely at the metric changes of the Friedman-Lemaitre-Robertson-Walker metric.

The plan of this paper is as follows: First we look at non empty space-times and see how generic solutions of the Einstein equations in the presence of matter change signature. Then we will discuss some possible implications of our results, in particular the conditions under which a particle can travel in speeds exceeding the speed of light. Finally some cosmological implications and conclusions would be given.

2. Mechanisms Of Signature Change

It was shown that among the possible flat space metrics only the Lorentzian metric is stable in empty space [1]. Nevertheless one may still inquire if a mechanism exists by which a metric change does occur[1], can we create some how a metric of the type $g_{\mu\nu} =$ diag $(+1, +1, -1, -1)$ in some region of space-time? The answer obviously has to do with the only reason a metric should change according to equation (1) and this is $T_{\mu\nu}$. Looking at available solution of general relativity one finds that metric changes are quite common.

The Schwarzschild square interval (in terms of spherical coordinates t, r, θ, ϕ) is given by:

$$c^2 d\tau^2 = (1 - \frac{r_s}{r})c^2 dt^2 - \frac{dr^2}{1 - \frac{r_s}{r}} - r^2(d\theta^2 + \sin^2\theta d\phi^2) \tag{2}$$

In which τ is the proper time, and r_s is the Schwarzschild radius (in meters) of the massive body, which is related to its mass M by $r_s = \frac{2GM}{c^2}$. It is obvious that while for $r > r_s$ the metric is locally (up to scaling) $g_{\mu\nu} =$ diag $(+1, -1, -1, -1)$. For $r < r_s$ the metric is locally (up to scaling) $g_{\mu\nu} =$ diag $(-1, +1, -1, -1)$. Hence the direction of temporal and (one) spatial axis is exchanged. Notice, however, that although the sign of the eigen-values did change we are still left with a Lorentzian metric.

Another example is the Friedman-Lemaitre-Robertson-Walker square interval which is well known in cosmological models:

$$c^2 d\tau^2 = c^2 dt^2 - a(t)^2 \left(\frac{dr^2}{1 - \kappa r^2} + r^2(d\theta^2 + \sin^2\theta d\phi^2) \right) \tag{3}$$

$a(t)$ is known as the "scale factor" and κ may be taken to have units of length^{-2}, in which case r has units of length and a(t) is unit less. κ is then the Gaussian curvature of the space at the time when $a(t) = 1$. Hence for radial distances such that $r < \frac{1}{\sqrt{\kappa}}$ the metric is locally (up to scaling) $g_{\mu\nu} =$ diag $(+1, -1, -1, -1)$ that is Lorentzian. However, for $r > \frac{1}{\sqrt{\kappa}}$ the metric is locally (up to scaling) $g_{\mu\nu} =$ diag $(+1, +1, -1, -1)$. This means that a particle propagating in a radial direction will experience an Euclidean metric. Of course the critical radius $\frac{1}{\sqrt{\kappa}}$ only exists for closed universes in which $\kappa > 0$. Flat universes with $\kappa = 0$ or open universes with $\kappa < 0$ do not have such a critical radius.

One should notice that in the above cases a signature change is accompanied by a metric singularity while the signature changes considered by Eddington [4] involve zeros. However, metric singularities are not curvature singularities and can be removed by proper choice of coordinates.

[1] In the sense that the eigen-values of the metric change signs.

3. Implications For Particle Trajectories

A particle travelling in a space-time with a constant metric can be described by the Action $\mathcal{A}$ and Lagrangian L:

$$\mathcal{A} = \int L d\tau, \qquad L = \frac{1}{2} m u_\alpha u^\alpha + \frac{q}{c} u_\alpha A^\alpha \tag{4}$$

In the above τ is some parameter along the trajectory, x_α are the particle coordinates, $u_\alpha \equiv \frac{dx_\alpha}{d\tau}$, m is the particle mass, q is the particle charge and A_α are some functions of the particle coordinates (that transform as a four dimensional vector). Basic variational analysis leads to the following equations of motion:

$$m \frac{du^\alpha}{d\tau} = -\frac{q}{c} u^\beta (\partial_\beta A^\alpha - \partial^\alpha A_\beta) \tag{5}$$

It is customary to use as a parameter the length of the trajectory:

$$d\tau^2 = |\eta_{\alpha\beta} dx^\alpha dx^\beta| \tag{6}$$

The analysis of trajectories in different flat metrics is described in [22] and will not be repeated here due to space limitations.

The main results are as follows: a subluminal particle in a Lorentz space-time will remain subluminal for ever (as is well known). A superluminal particle in a Lorentz space-time will remain superluminal for ever. A particle in an Euclidean space-time can attain any velocity.

4. Possible Cosmological Implications

One obvious physical implication of the previous analysis is that a particle can be accelerated to a velocity close to the velocity c in a Lorentz space, enter into an Euclidean space and be accelerated further in this region to velocities above the speed c and emerge in a Lorentz space in which it will remain above the speed c for ever unless it is decelerated in an Euclidean space again.

This certainly may happen to a particle which travels radially in a Friedman-Lemaitre-Robertson-Walker metric passing outwards the critical radius of $r_c = \frac{1}{\sqrt{\kappa}}$ and then coming back at superluminal velocities.

But if such particles do exist how would their existence bear on existing physical and astrophysical problems?

An obvious implication has to do with the homogeneity (horizon) problem. According to an analysis given by Narlikar [17] a proper radius for a particle horizon of a sub luminal particle at the radiation dominated epoch was $R_L = 2ct$, taking into account temperatures of the early universe led him to conclude that this radius was of the order of magnitude of about 1 meter on present day scales whether in reality the cosmological micro wave background is homogeneous on a scale of 10^{26} meters. Notice, however, that superluminal particles are not restricted by the velocity of light and hence can bring a very young universe into thermal equilibrium. Of course a more popular mechanism for achieving this is inflation [24]. However, one should notice that a Higgs type fields do not give the correct density perturbation spectrum [24], hence one is forced to postulate a new field which is not a part of any particle model and thus is a possible but inelegant solution of the homogeneity problem. Alternatively one can speculate that homogeneity is achieved by ordinary matter which can become superluminal as the current analysis shows.

5. Conclusions

Mathematically speaking one of the main differences between time and space is encapsulated in the flat metric of our space-time which is **locally** of the Lorentzian type $\eta_{\mu\nu} =$ diag (1,-$-1, -1, -1$). But this is an empirical fact or a mathematical postulate thus unexplained. One can imagine also other flat metrics such as the Euclidian metric: diag $(1, 1, 1, 1)$. In Euclidian metrics there is no restriction on the speed of any moving body as the speed of light restricts the speed of propagation only in the presence of a Lorentzian metric. Why is our space-time Lorentzian and not Euclidean? The answer is that only the Lorentzian metric is stable [1] for an (almost) empty space. But space-time is not empty and thus the notion of time always progressing forward with the increase of entropy is probably just a consequence of the scales of reality that we are exposed to. In huge cosmological scale the Friedman-Lemaitre-Robertson-Walker universe losses its Lorentzian character. The horizon problem related to the homogeneity of cosmic microwave background can be solved if one takes into account the superluminal motion of particles for $r > r_c$ and the same particles moving into the $r < r_c$ Lorentzian domain. Another scenario that one needs to take into account are particles created such that there initial velocities are super luminal. Statistical physics of such particles is needed an hopefully will be developed in the future.

References

[1] Yahalom, A. (2008). The Geometrical Meaning of Time, Foundations of Physics, Volume 38, Number 6, Pages 489-497.
[2] Weinberg, S. (1972). Gravitation and Cosmology: Principles and Applications of the General Theory of Relativity, John Wiley & Sons, Inc.
[3] Misner, C. W. Thorne, K.S. & Wheeler J.A.(1973). Gravitation W.H. Freeman & Company.
[4] Eddington, A. S. (1923). The mathematical theory of relativity, Cambridge University Press.
[5] Greensite, J. (1992), Los Alamos Archive gr-qc/9210008.
[6] Carlini, A. & Greensite, J. (1993). Los Alamos Archive gr-qc/9308012.
[7] Carlini, A. & Greensite, J. (1994). Phys. Rev. D, Volume 49, Number 2.
[8] Elizalde, E., Odintsov, S. D. & Romeo, A. (1994). Class. Quantum Grav. 11 L61-L67.
[9] Itin, Y. & Hehl, F. W. (2004). Los Alamos Archive gr-qc/0401016.
[10] van Dam, H. & Jack, Y. (2001). Los Alamos Archive hep-th/0108067.
[11] Yahalom, A. (2009). The Gravitational Origin of the Distinction between Space and Time, International Journal of Modern Physics D, Vol. 18, Issue: 14, pp. 2155-2158.
[12] Yahalom, A. (2011). Advances in Classical Field Theory, Chapter 6, Bentham eBooks eISBN: 978-1-60805-195-3.
[13] Yahalom, A. (2010). Gravity and the Complexity of Coordinates in Fisher Information, International Journal of Modern Physics D, Vol. 19, No. 14 1-5, World Scientific Publishing Company.
[14] Christodoulou, D. & Klainerman, S. (1989-1990). The global nonlinear stability of the Minkowski space, Seminaire Equations aux derivees partielles (dit "Goulaouic-Schwartz"), Exp. No. 13, p. 29.
[15] Kalyana Rama, S. (2006). Los Alamos Archive hep-th/0610071 v2.
[16] Carroll, S. (2010). From Eternity to Here: The Quest for the Ultimate Theory of Time, A Plume Book.
[17] Narlikar, J. V. (1993). Introduction to Cosmology, Cambridge University Press.
[18] Guth, A. H. (1981). Inflationary universe: A possible solution to the horizon and flatness problems, Phys. Rev. , D23, 347 (1981).
[19] Linde, A. (1982). A new inflationary universe scenario, Phys. Lett., B108, 389.
[20] Linde, A. (1983). Chaotic inflation, Phys. Lett., B129, 177.
[21] Ade, P. A. R. et al. (BICEP2 Collaboration) (2014). Detection of B-Mode Polarization at Degree Angular Scales by BICEP2, Phys. Rev. Lett. 112, 241101 Published 19 June 2014.
[22] Yahalom, A. (2013). Gravity and Faster than Light Particles, Journal of Modern Physics (JMP), Vol. 4 No. 10 PP. 1412-1416. DOI: 10.4236/jmp.2013.410169.
[23] Weinberg, S. (1995). The Quantum Theory of Fields, Cambridge University Press.
[24] Guth, A. H. (1995). Starting the universe: the Big Bang and cosmic inflation, p. 105 in Bubbles, voids and bumps in time: the new cosmology, Edited by James Cornell, Cambridge University Press.
[25] Rubin, V. C. (1995). "Weighting the universe: dark matter and missing mass" p. 73 in Bubbles, voids and bumps in time: the new cosmology, Edited by James Cornell, Cambridge University Press.
[26] Yourgrau, P. (2006). A World Without Time, Basic Books.

The generalized uncertainty principle as quantum gravitational friction

Pedro Bargueño

Departamento de Física, Universidad de los Andes, Apartado Aéreo 4976, Bogotá, Distrito Capital, Colombia

E-mail: p.bargueno@uniandes.edu.co

Abstract. In this article we present a dissipative Schrödinger–Langevin–like Hamiltonian which incorporates implicitly the deformed commutation relations which are linear in particle momenta due to a generalized uncertainty principle. This result is based on interpreting the deformation parameter as quantum gravitational friction on the configuration space.

1. Introduction

After Mead [1], who was the first who pointed out the role of gravity on the existence of a fundamental measurable length, a considerably ammount of effort has been devoted to study the modification of the Heisenberg uncertainty principle, known as Generalized Uncertainty Principle (GUP), together with the consequences it leads to [2, 3, 4, 5, 6, 7, 8, 9, 10, 11, 12, 13, 14, 15, 16, 17, 18]. The specific form of the GUP and its associated commutation relations, together with the physical consequences they lead to have been recently studied [19, 20]. Moreover, there has been recent interest in a particular version of the GUP [22, 21, 23], which predicts not only a minimum length but also a maximum momentum [14, 15, 16].

As shown in [19], the effects of this GUP can be implemented both in classical and quantum systems by defining deformed commutation relations by means of

$$x_i = x_{0i} \; ; \; p_i = p_{0i} \left[1 - \gamma p_0 + 2\gamma^2 p_0^2\right], \tag{1}$$

where $[x_{0i}, p_{0j}] = i\hbar\delta_{ij}$ and $p_0^2 = \sum_{j=1}^{3} p_{0j}p_{0j}$ and $\gamma = \gamma_0/m_p c$, being γ_0 an adimensional constant.

These GUPs give place to deformed commutation relations which are linear and/or quadratic in particle momenta, depending on the theories which are consistent with the corresponding GUP (theories with a GUP were shown to be equivalent to doubly special relativity theories in certain sense [24]). When acting on massive particles, this minimum–legth and maximum–momentum new form of the GUP has been shown to be equivalent to a gravitationally–induced damping process in an Ohmic environment at zero temperature [25]. Therefore, both minimum length and maximum momentum give place to quantum gravitational friction for massive particles. In this work, a dissipative Schrödinger–Langevin–like Hamiltonian which incorporates implicitly this specific GUP will be introduced.

2. Quantum gravitational friction

Let us consider a non–relativistic Hamiltonian of the form $H = \frac{p^2}{2m} + V(x)$. To simplify notation p_0 will be denoted as p (only 1D systems will be considered). If the deformed commutator

$$[x, p] = i\hbar\left[1 - \frac{\gamma_0}{m_p c}p + \mathcal{O}\left(\gamma_0^2\right)\right] \tag{2}$$

is introduced, the corresponding Heisenberg equations of motion, $\dot{A} = \frac{i}{\hbar}[H, A]$ (where A is any observable) read $\dot{p} = -V'(x)$ and $m\dot{x} = p\left[1 - \gamma_0 \frac{p}{m_p c}\right]$. Therefore, the evolution equation in the configuration space is

$$m\ddot{x} - \frac{2\gamma_0}{m_p c}V'(x)m\dot{x} + V'(x) = 0. \tag{3}$$

Moreover, after defining

$$\bar{\alpha}(x) \equiv -\frac{2\gamma_0}{m_p c}V'(x), \tag{4}$$

Eq. (4) reads

$$m\ddot{x} + m\bar{\alpha}(x)\dot{x} + V'(x) = 0, \tag{5}$$

which resembles a Langevin equation with a position–dependent friction.

This equation can be derived from the so called system + bath approach to open systems [27] in the following way. We split the total Hamiltonian including system, bath and their couplings, in three parts as

$$H = H_s + H_b + H_{sb}, \tag{6}$$

where

$$H_s = \frac{p^2}{2m} + V(x) \tag{7}$$

stands for the Hamiltonian of the isolated system,

$$H_b = \frac{1}{2}\sum_i \left(\frac{p_i^2}{m_i} + m_i\omega_i^2 x_i^2\right) \tag{8}$$

is the Hamiltonian for the bath, which acts as a resevoir, and can be represented as a set of harmonic oscillators, and

$$H_{sb} = \sum_i \left[\frac{f^2(x)d_i^2}{m_i\omega_i^2} - 2d_i f(x)x_i\right] \tag{9}$$

expresses an interaction term between the isolated system and the bath, where d_i are appropriate coupling constants. When the system–bath coupling is linear, that is, for $f(x) = x$, the total Hamiltonian is known as the Caldeira–Leggett model [26, 27], widely used in the field of condensed matter physics. Following a standard approach [27], the bath degrees of freedom can be eliminated. Moreover, if Ohmic dissipation (memory free) and a nonlinear sistem–bath coupling is considered, the corresponding Langevin equation reads (after dropping a term involving noise)

$$m\ddot{x}(t) + m\alpha\left[f'(x)\right]^2 \dot{x}(t) + V'(x) = 0. \tag{10}$$

Therefore, we conclude that

$$-\bar{\alpha}(x) = m\alpha\left[f'(x)\right]^2, \tag{11}$$

or

$$-\frac{2\gamma_0}{m_p c}V'(x) = \alpha\left[f'(x)\right]^2. \tag{12}$$

Thus, one can define a quantum gravitational friction as

$$\alpha_{QG} \equiv \frac{2\gamma_0}{m_p c}. \tag{13}$$

Moreover, after absorbing the negative sign by the d_i constants, we note that the non–linear coupling between the system and the bath is given by

$$f(x) \equiv \int_0^x \sqrt{V'(y)}dy. \tag{14}$$

As the noise function (not considered here) is the responsible of the appearance of thermal effects, we can conclude that the deformed commmutator derived from Eq. (1) can be written as

$$[x, p] = i\hbar \left(1 - \frac{\alpha_{QG}}{2} p\right) \tag{15}$$

and think of the deformation as a consequence of a gravitationally–induced damping process due to an Ohmic environment at zero temperature.

Therefore, from the configuration space perspective, the dynamics driven from H_s using deformed canonical commutation relations (linear in p), is completely equivalent to that obtained from the system + bath approach applied to the total Hamiltonian $H = H_s + H_b + H_{sb}$ when the deformation parameter γ is interpreted as a friction coefficient and a specific non–linear system–bath coupling is introduced.

Moreover, by means of the GUP–induced deformed canonical commutator, a general non–relativistic Hamiltonian of the form $H = \frac{p^2}{2m} + V(x)$ transforms into (see, for example [19])

$$\begin{aligned} H &= \frac{p^2}{2m} + V(x) - \frac{\gamma_0}{m m_p c} p^3 + \mathcal{O}(\gamma_0^2) \\ &= \frac{p^2}{2m} + V(x) - \frac{\alpha_{QG}}{2m} p^3 + \mathcal{O}(\gamma_0^2). \end{aligned} \tag{16}$$

Thus, in light of the previous interpretation, the second term of Eq. (16) can be taken as a dissipative one.

A different (but equivalent in certain sense) formulation of the linear GUP as a quantum gravitational friction effect can be stated as follows. In a recent work [28] we derived a generalization of the so–called Schrödinger–Langevin or Kostin equation for a Brownian particle interacting with a heat bath. This generalization is based on a nonlinear interaction model providing a state-dependent dissipation process exhibiting multiplicative noise. The idea of the GLE is to introduce a dissipative potential, V_d, in the time–dependent Schrödinger equation such that the evolution of the average values for the position, velocity and acceleration observables satisfy the classical equation of motion, which results to be a (generalized) Langevin equation [29]. But, as we shown, the GUP leads to a generalized Langevin equation with a position–dependent coupling that turns out to be $f(x) = \int_0^x \sqrt{V'(y)}dy$. Therefore, the formalism developed in Ref. [28] is suitable for the linear GUP case. In fact, following [28] the following damping potential for the GUP can be derived:

$$V_d = -\frac{\alpha_{QG}}{2}\tilde{S} = -\alpha_{QG}\, p\, V(x), \tag{17}$$

where

$$\tilde{S} \equiv \left(\frac{df}{dx}\right)^2 S - 2\int S \frac{df}{dx}\frac{d^2 f}{dx^2} dx \tag{18}$$

is the coupling–dependent phase of the wavefunction.

Therefore, the GUP–dissipative Hamiltonian equivalent to that of Eq. (16) can be written as

$$H = \frac{p^2}{2m} + V(x)\left[1 - \alpha_{QG}\, p\right] + \mathcal{O}(\gamma_0^2) \tag{19}$$

Notice, that, although Eq. (19) predicts the correct Langevin–like equation derived from the deformed commutation relations, the correct dynamics for the phase space variables can not be obtained from it. This situation is similar to that found within the well known Caldirola–Kanai effective Hamiltonian approach [30].

3. Conclusions

In this work we have argued that the deformation parameter which enters in a generalized uncertainty principle which is linear in particle momenta can be interpreted as a friction coefficient form the configuration space point of view. Moreover, a dissipative Schrödinger–Langevin–like Hamiltonian which incorporates implicitly the deformed commutation relations due to the mentioned generalized uncertainty principle has been introduced.

References

[1] C. A. Mead, Phys. Rev. **135**, B849 (1964).
[2] D. Amati, M. Ciafaloni and G. Veneziano, Phys. Lett. B **216**, 41 (1989).
[3] M. Maggiore, Phys. Lett. B **304**, 65 (1993).
[4] M. Maggiore, Phys. Rev. D **49** 5182 (1994).
[5] M.Maggiore, Phys. Lett. B **319**, 83 (1993).
[6] L.J. Garay, Int. J. Mod. Phys. A **10**, 145 (1995).
[7] F. Scardigli, Phys. Lett. B **452**, 39 (1999).
[8] S. Hossenfelder *et al.*, Phys. Lett. B **575**, 85 (2003).
[9] C. Bambi and F.R. Urban, Class. Quant. Grav. **25**, 095006 (2008).
[10] G.M. Hossain, V. Husain and S.S. Seahra, Class. Quant. Grav. **27**, 165013 (2010).
[11] A. Kempf, G. Mangano and R.B. Mann, Phys. Rev. D **52**, 1108 (1995).
[12] A. Kempf, J. Phys. A **30**, 2093 (1997).
[13] F. Brau, J. Phys. A **32** 7691 (1999).
[14] J. Magueijo and L. Smolin, Phys. Rev. Lett. **88**, 190403 (2002).
[15] J. Magueijo and L. Smolin, Phys. Rev. D **67**, 044017 (2003).
[16] J. Magueijo and L. Smolin, Phys. Rev. D **71**, 026010 (2005).
[17] J . L. Cortes and J. Gamboa, Phys. Rev. D **71**, 065015 (2005).
[18] S. Ghosh and P. Pal, Phys. Rev. D **75**, 105021 (2007).
[19] S. Das and E. C. Vagenas, Phys. Rev. Lett. **101**, 221301 (2008).
[20] S. Das and E. C. Vagenas, Can. J. Phys. **87**, 233 (2009).
[21] S. Das, E. C. Vagenas and A.F. Ali, Phys. Lett. B **690**, 407 (2010) and Erratum ibid B **692**, 342 (2010).
[22] A.F. Ali, S. Das and E. C. Vagenas, Phys. Lett. B **678**, 497 (2009).
[23] A.F. Ali, S. Das and E. C. Vagenas, Phys. Rev. D **84**, 044013 (2011).
[24] S. Hossenfelder, Class. Quantum Grav. **23**, 1815 (2006).
[25] P. Bargueño, Phys. Lett. B **727**, 496 (2013).
[26] A. O.Caldeira and A. J. Leggett, Phys. Rev. Lett., **46**, 211 (1981).
[27] U. Weiss, *Quantum Dissipative Systems*, Series in Modern Condensed Matter Physics **13**, World Scientific (2008).
[28] P. Bargueño and S. Miret–Artés, Ann. Phys. **346**, 59 (2014).
[29] M. Razavi, Classical and Quantum Dissipative Systems, Imperial College Press, London (2005).
[30] A. S. Sanz and S. Miret-Artés, *A Trajectory Description of Quantum Processes. A Bohmian Perspective.*, Lecture Notes in Physics **850** (2012) and **831** (2014), Springer.

IC-MSQUARE 2014 IOP Publishing
Journal of Physics: Conference Series **574** (2015) 012063 doi:10.1088/1742-6596/574/1/012063

Canonical Quantization of the Reissner-Nordström Black Hole via Conditional Symmetries

Evangelos Melas

Technological Educational Institution of Patras, Department of Management, Greece

E-mail: evangelosmelas@yahoo.co.uk

Abstract. We use the conditional symmetry approach to study the $r-$evolution of Reissner–Nordström black hole both at the classical and quantum level. We Dirac–quantize the Reissner–Nordström black hole using the quantum analogues of the classical conditional symmetries, and show that the existence of such symmetries yields solutions to the Wheeler–DeWitt equation which, as a semiclassical analysis shows, exhibit a good correlation with the classical regime. Finally, we use the resulting wave functions to investigate the possibility of removing the classical singularities.

1. Introduction

In this paper we study [1] the behavior of a static, spherically symmetric space–time in the framework of the presence of conditional symmetries in minisuperspace constrained systems. The phase–space variables turn out to correspond to the $r-$dependent metric functions and to an electrostatic field with which the action of the model is augmented.

In section 2 we follow [2] - [4] and construct a minisuperspace Lagrangian using the canonical decomposition along the radial coordinate r which now plays the role of a dynamical variable. We then in section 3 pass to the Hamiltonian formalism and reveal six conditional symmetries and a rheonomic integral of motion. It can be shown [1] that the classical solution of such a system can be identified with the space–time of a Reissner–Nordström (RN) black hole ([5], [6]). In section 4 we Dirac–quantize the model.

2. Classical formulation and conditional symmetries

The general form of a static, spherically symmetric line element is

$$ds^2 = -a^2(r)dt^2 + n^2(r)dr^2 + b^2(r)\left(d\theta^2 + \sin^2\theta\, d\phi^2\right), \tag{1}$$

with $n(r)$ playing the role of the $r-$lapse function, while $a(r)$ and $b(r)$ are the "dynamical" dependent variables in the $r-$foliation. In order to acquire the RN solution we need to consider an electrostatic field minimally coupled to gravity. Thus, the full action is written as

$$S_{g+em} = \int \mathcal{L}_{GR}\, d^4x + \int \mathcal{L}_{EM}\, d^4x = \int \sqrt{-g}\, R\, d^4x - \int \sqrt{-g}\, F_{\mu\nu}F^{\mu\nu}\, d^4x, \tag{2}$$

where $F_{\mu\nu} = A_{\mu,\nu} - A_{\nu,\mu}$ is the antisymmetric electromagnetic tensor and A_μ is the potential with $A_0 = f(r)$ and $A_1 = A_2 = A_3 = 0$. In (2) we have chosen the units $c = 1$, $G = \frac{1}{4\pi}$. The variation of this action with respect to the space−time metric $g_{\mu\nu}$ leads to Einstein's field equations

$$E_{\mu\nu} = 2\,T_{\mu\nu}, \tag{3}$$

where $E_{\mu\nu} = R_{\mu\nu} - \frac{1}{2}R\,g_{\mu\nu}$ is the Einstein tensor and $T_{\mu\nu} = F_{\mu\kappa}F_\nu{}^\kappa - \frac{1}{4}g_{\mu\nu}F_{\kappa\lambda}F^{\kappa\lambda}$ is the stress−energy tensor associated with the electromagnetic field.

Apart from the field theory approach, one can be led to effectively the same equations of motion by integration of the redundant degrees of freedom in action (2), i.e. integrating over t, θ and ϕ and ignoring a multiplicative (infinite) constant. All system information is then contained in a reduced, point−like action $S = \int L(a, b, f, a', b', f', n)\,dr$ with the following Lagrange function:

$$L = \frac{1}{2\,n}\left(8\,b\,a'\,b' + 4\,a\,b'^2 + 4\frac{b^2}{a}f'^2\right) + 2\,n\,a, \tag{4}$$

where $'$ denotes differentiation with respect to the spatial coordinate r. It is easy to verify that the Euler−Lagrange equations ensuing from (4) are equivalent to the reduced Einstein's equations obtained by the substitution of the line element (1) and $A_\mu = (f(r), 0, 0, 0)$ in (3). The Lagrangian (4) belongs to a particular form of singular Lagrangians: $L = \frac{1}{2n}\,G_{\mu\nu}\,q'^\mu\,q'^\nu + n\,V(q)$. If one uses the freedom to reparametrize the lapse, then (4) can be brought to a form in which the potential V is constant. In our case we choose to set $n = \frac{N}{2\,a}$, which leads to

$$L = \frac{1}{2\,N}\left(16\,a\,b\,a'\,b' + 8\,a^2\,b'^2 + 8\,b^2 f'^2\right) + N, \tag{5}$$

or, in a more concise form, $L = \frac{1}{2\,N}\,\overline{G}_{\mu\nu}\,q'^\mu\,q'^\nu + N$ with $q'^\mu = (a', b', f')$ and

$$\overline{G}_{\mu\nu} = \begin{pmatrix} 0 & 8\,a\,b & 0 \\ 8\,a\,b & 8\,a^2 & 0 \\ 0 & 0 & 8b^2 \end{pmatrix}. \tag{6}$$

As shown in [2], it is in this particular lapse parametrization that the conditional symmetries of the phase−space, as defined in [7], become Killing vector fields of the supermetric (6) in the configuration space. As it can be straightforwardly verified, the above given metric $\overline{G}_{\mu\nu}$ is flat and admits the following six Killing vectors:

$$\begin{aligned} &\xi_1 = \partial_f, \quad \xi_2 = \frac{1}{2\,a\,b}\,\partial_a, \quad \xi_3 = \frac{f}{2\,a\,b}\,\partial_a + \frac{1}{2\,b}\,\partial_f, \quad \xi_4 = -a\,\partial_a + b\,\partial_b - f\,\partial_f, \\ &\xi_5 = -\frac{a^2 + f^2}{2\,a\,b}\,\partial_a + \partial_b - \frac{f}{b}\,\partial_f, \quad \xi_6 = -a\,f\,\partial_a + b\,f\,\partial_b - \frac{a^2 + f^2}{2}\,\partial_f. \end{aligned} \tag{7}$$

Additionally, the supermetric $\overline{G}_{\mu\nu}$ exhibits a homothetic symmetry ($\pounds_{\xi_h}\overline{G}_{\mu\nu} = \overline{G}_{\mu\nu}$) generated by

$$\xi_h = \frac{a}{4}\frac{\partial}{\partial a} + \frac{b}{4}\frac{\partial}{\partial b} + \frac{f}{4}\frac{\partial}{\partial f}. \tag{8}$$

3. Hamiltonian formulation

Let us now turn to the Hamiltonian formulation; invoking the usual definition of the momenta

$$p_N := \frac{\partial L}{\partial N'} = 0, \tag{9a}$$

$$p_a := \frac{\partial L}{\partial a'} = \frac{8\,a\,b\,b'}{N}, \tag{9b}$$

$$p_b := \frac{\partial L}{\partial b'} = \frac{8\,a\,(b\,a' + a\,b')}{N}, \tag{9c}$$

$$p_f := \frac{\partial L}{\partial f'} = \frac{8\,b^2\,f'}{N}, \tag{9d}$$

and following Dirac–Bergman's algorithm [8], we acquire one first class primary constraint $p_N \approx 0$, the Hamiltonian

$$H = N\,\mathcal{H} = N\left(-\frac{p_a^2}{16\,b^2} + \frac{p_a\,p_b}{8\,a\,b} + \frac{p_f^2}{16\,b^2} - 1\right), \tag{10}$$

and the first class secondary constraint $\{p_N, H\} \approx 0 \Rightarrow \mathcal{H} \approx 0$. If we associate the phase–space quantities $Q_I := \xi_I^\mu p_\mu$ with the six Killing vector fields (7), we are provided with six linear integrals of motion

$$\begin{aligned} &Q_1 = p_f, \quad Q_2 = \frac{1}{2\,a\,b}\,p_a, \quad Q_3 = \frac{f}{2\,a\,b}\,p_a + \frac{1}{2\,b}\,p_f, \quad Q_4 = -a\,p_a + b\,p_b - f\,p_f, \\ &Q_5 = -\frac{a^2+f^2}{2\,a\,b}\,p_a + p_b - \frac{f}{b}\,p_f, \quad Q_6 = -a\,f\,p_a + b\,f\,p_b - \frac{a^2+f^2}{2}\,p_f, \end{aligned} \tag{11}$$

which form a Poisson bracket algebra with the previously mentioned structure constants. As also stated in [2], under the given lapse parametrization (in which the potential is constant) the Poisson brackets of the Q_I's with the Hamiltonian H are exactly equal to zero and not just weakly vanishing, $\{Q_I, H\} = 0$, for $I = 1, \ldots 6$. Moreover, since $\mathcal{H} \approx 0$, the constancy of the potential part is carried over to the quadratic in the momenta kinetic term, leading inevitably the latter to become a Casimir invariant of the Lie algebra formed by the Q_I's. In the case we are studying this is

$$Q_C = \frac{1}{4}\left(Q_2\,Q_5 + Q_3^2\right) = \mathcal{H} + 1. \tag{12}$$

As it is known, the integrals of motion, Q_I's, become constants, say κ_I's, on the solution space. However, these are not the only existing integrals of motion. As shown in [9], in principle, all conformal Killing vectors of the supermetric define rheonomic integrals of motion. Here we do have such a vector, namely ξ_h. We conclude [1] that the the following seven relations are valid on the solution space:

$$Q_I = \kappa_I, \qquad I = 1, \ldots, 6 \tag{13a}$$

$$Q_h - \int N dr = c_h \Rightarrow \frac{1}{4}(a\,p_a + b\,p_b + f\,p_f) - \int N dr = c_h, \tag{13b}$$

with κ_I's and c_h being constants. It is quite interesting that the above relations completely determine the entire classical solution space along with the two relations quadratic in the κ_I's emanating from the two Casimir invariants of the algebra.

4. Quantization through symmetries

The identification of the linear integrals of motion as physical quantities leads to the need of expressing them as operators. The algebra defined by these operators has to match the classical Lie algebra and, moreover, one has to determine which of them can be applied at the same time on the wave function together with the constraints mentioned in the previous section. These issues have been clearly addressed in [2]. In a quick view, we start with the usual definition of the momenta ($\hbar = 1$) as operators

$$p_\alpha \to \widehat{p}_\alpha := -\mathrm{i}\frac{\partial}{\partial q^\alpha}, \tag{14}$$

where q^α is any one of a, b, f, N. After that, the quantum analogues of the conditional symmetries Q_I, are expressed in the most general form of a linear Hermitian (under an arbitrary measure μ) differential operator of the first order:

$$\widehat{Q}_I := -\frac{\mathrm{i}}{2\mu}\left(\mu\,\xi_I^\alpha\,\partial_\alpha + \partial_\alpha\,\mu\,\xi_I^\alpha\right). \tag{15}$$

It has been proved in [2] that operators $\widehat{Q}_I$ defined as in (15) satisfy the same algebra as do the classical quantities Q_I, i.e. $[\widehat{Q}_I, \widehat{Q}_J]F = C_{IJ}^K\,\widehat{Q}_K F$ for any function F for which the action of the operators is well defined. It is noteworthy that this happens for any arbitrary measure $\mu(a, b, f)$.

Apart from the primary constraint

$$\widehat{p}_N = -\mathrm{i}\frac{\partial}{\partial N}\Psi = 0 \Rightarrow \Psi = \Psi(a, b, f), \tag{16}$$

the main operator one has to apply is the quantum analogue of the Hamiltonian constraint or, equivalently in the particular lapse parametrization, of the Casimir invariant ($\widehat{Q}_C$), since

$$\widehat{\mathcal{H}}\Psi = (\widehat{Q}_C - 1)\Psi = 0. \tag{17}$$

Apart from $\widehat{\mathcal{H}}$, we also have at our disposal the conditional symmetries. The wave function of the system is to be realized as an eigenstate of those physical quantities that can be measured together:

$$\widehat{Q}_I\Psi = \kappa_I\Psi. \tag{18}$$

In [1] it is shown that solutions to the Wheeler–DeWitt equation (17) which at the same solve (18), as a semiclassical analysis shows, exhibit a good correlation with the classical regime. Finally in [1] we use the resulting wave functions to investigate the possibility of removing the classical singularities.

References

[1] T. Christodoulakis, N. Dimakis, Petros A. Terzis, Babak Vakili, E. Melas and Th. Grammenos: *Phys.Rev.* D Vol. 89 (2014) p.044031

[2] T. Christodoulakis, N. Dimakis, P.A. Terzis, G. Doulis, Th. Grammenos, E. Melas and A. Spanou: *J. Geom. Phys.* Vol. 71 (2013) p.127 (arXiv: 1208.0462 [gr-qc])

[3] M. Cavaglià, V. de Alfaro and A.T. Filippov, *Int. J. Mod. Phys.* D Vol. 4 (1995) p.661 (arXiv: gr-qc/9411070)

[4] M. Cavaglià, V. de Alfaro and A.T. Filippov: *Int. J. Mod. Phys.* D Vol. 5 (1996) p.227 (arXiv: gr-qc/9508062)

[5] H. Reissner: *Annalen der Physik* Vol. 50 (1916) p.106

[6] Nordström: *Verhandl. Koninkl. Ned. Akad. Wetenschap., Afdel. Natuurk. Amsterdam* Vol. 26 (1918) p.1201

[7] K.V. Kŭchar: *J.Math.Phys.* Vol. 23 (1982) p.1647

[8] P.A.M. Dirac, *Lectures on Quantum Mechanics*, Yeshiva University, Academic Press, New York (1964)

[9] T. Christodoulakis, N. Dimakis and P.A. Terzis: *Lie-point and variational symmetries in minisuperspace Einstein's gravity* (arXiv: 1304.4359 [gr-qc])

Meteorological time series forecasting based on MLP modelling using heterogeneous transfer functions

C. Voyant[1], ML. Nivet[1], C. Paoli[2], M. Muselli[1] and G. Notton[1]

[1]University of Corsica - SPE CNRS UMR 6134, 20250 Corte - France (Corsica)
[2]University of Galatasaray - Computer Engineering Department, No:36 34357 Ortaköy, Istanbul - Turkey
voyant@univ-corse.fr

Abstract. In this paper, we propose to study four meteorological and seasonal time series coupled with a multi-layer perceptron (MLP) modeling. We chose to combine two transfer functions for the nodes of the hidden layer, and to use a temporal indicator (time index as input) in order to take into account the seasonal aspect of the studied time series. The results of the prediction concern two years of measurements and the learning step, eight independent years. We show that this methodology can improve the accuracy of meteorological data estimation compared to a classical MLP modelling with a homogenous transfer function.

1. Introduction

The time series (TS) formalism is useful in many scientific fields [1,3,4]. In the particular case of the meteorology, the prediction is essential to anticipate weather variation and thus to prevent the population of potential risks, but it can also be used for the energy management (intermittent energy sources switching e.g. global solar radiation or wind speed) [5-8]. The measured series are often seasonal and noised; they have repetitive and more or less predictable fluctuations (24 hours and 1 year). The noise increases when the time step between measurements decreases. It is essential to take into account periodicities and to build a predictor with seasonal adjustments [9]. In this paper, we suggest to study periodic meteorological TS measured during 11 years in Ajaccio (France; at an hourly step) and concerning four kinds of data with different degrees of periodicity: solar irradiation (Wh/m^2, [6]), humidity (%), ground temperature (□C) and wind velocity at 10 meters from the ground (Beaufort scale) [10]. The variation coefficients (VC; standard_deviation/mean) of these series are respectively: 1.5, 0.2, 0.4 and 0.5. This coefficient represents the variability and the seasonality of the series; the high values are often related to bad prediction results. In the next section, we will present the one hour ahead prediction methodology based on the heterogeneous transfer functions multilayer perceptron (MLP) and on a particular time index taken as input. Then the results of predictions and the comparison between different architectures will be exposed; finally in the last part, we will analyze the results and will draw conclusions.

2. Methods

In this section, we first introduce the global methodology based on the heterogeneous transfer functions MLP that is to say the fact to mix transfer functions within on network [2]. Next, we present the different meteorological data estimation methodologies used during the cross-comparison. .

2.1. MLP with heterogeneous transfer functions

A classical MLP with one hidden layer is, in the time series ($x(t)$) prediction context, defined by equation 1 where W^1,$b^1$and W^2,b^2 are the weights and biases matrices of the hidden and output layers and f the transfer function of the hidden layer (note that in our case, the transfer function of the output layer is the identity function). The tapped delay line (p delays) is defined from the lag operator (L) by $TDL = \{L^{i-1}x(t)\} i \in [1, p]$ considering that $L^i x(t) = x(t - i)$. With this notation, the output of the MLP becomes $O = L^{-1}x(t)$ [3].

$$O = W^2 \, f(W^1 TDL + b^1) + b^2 \tag{1}$$

Often a hyperbolic tangent function (or similar function) is used for f. In this study we propose to combine two types of transfer function in the hidden layer: bijective and non-bijective. As the first one allows activating a specific hidden node once the argument is upper than the bias; the second one allows to introduce an excitation/inhibition balance in the system: low and high signals will have an inhibition effect. We choose to test respectively the hyperbolic tangent function (noted tanh) and its derivative. Figure 1 represents these two functions types and the scaled form of the non-bijective transfer function (expanded between -1 and 1).

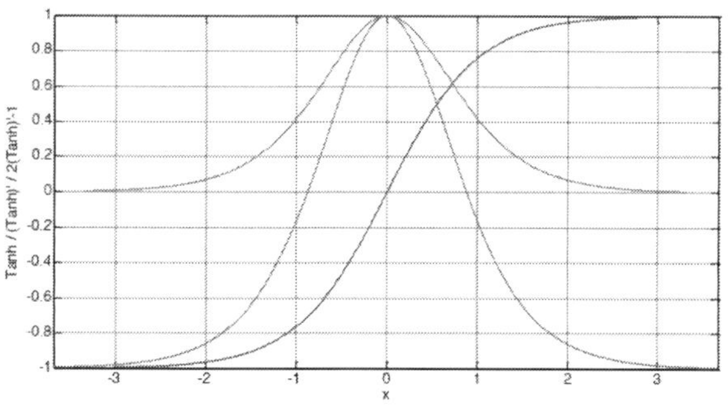

Fig. 1: In blue the tanh function and in red its derivative (= 1-tanh²). The green curve represents the scaled form of (tanh)'

With this modification, the new network is defined by (2), where $W^{1'}$ and $b^{1'}$ define the weights and biases related to the non-bijective nodes.

$$O = W^2 (f(W^1 TDL + b^1) + f'(W^{1'} TDL + b^{1'})) + b^2 \tag{2}$$
$$O = W^2 (\tanh(W^1 TDL + b^1) + 1 - \tanh^2(W^{1'} TDL + b^{1'})) + b^2$$

The difference between the "classic" and the heterogeneous transfer functions MLP is related to the two following expressions:

$$\tanh(W^1 TDL + b^1) \text{ and } \tanh(W^1 TDL + b^1) - \tanh^2(W^{1'} TDL + b^{1'}) + 1 \tag{3}$$

If we pose $W^{1'} TDL + b^{1'} = (W^1 TDL + b^1)\Delta$ where Δ is the asymmetry rate, and $W^1 TDL + b^1 = u$, we obtain that in the classic case the output is related to $\tanh(u)$ while in the heterogeneous transfer functions MLP case it is related to $\tanh(u) - \tanh^2(u\Delta) + 1$. In figure 2 are shown these two types of outputs related to linear combination of inputs (u) and to Δ. If Δ is upper than 0.5 then the global output is non-bijective and the inhibition phenomenon becomes leading. Concerning the low values of Δ the output is not between -1 and 1 but between -0.3 and 1.7 (shift of 0.7).

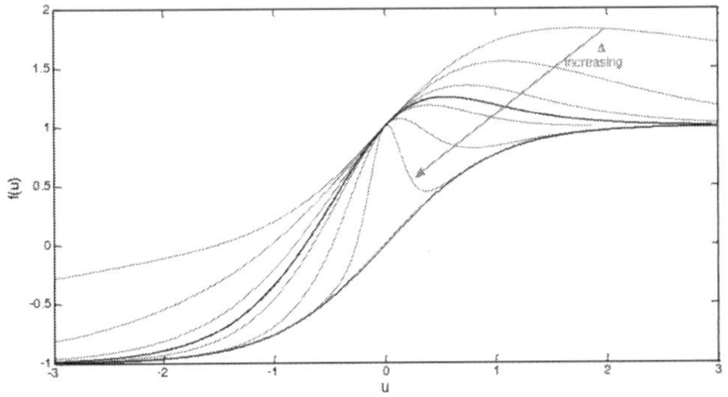

Fig. 2: Output of the classic (blue) and the heterogeneous transfer functions MLP (dotted curve) relating to the *u* parameter (linear combination of inputs) and the asymmetry rate Δ (0.2, 0.5, 0.8, 1.2, 2 and 5). In black, the output concerning Δ=1.

2.2. Meteorological data forecast

The test and validation process of the proposed methodology consists in a cross-comparison with some classical tools usually proposed in literature in order to make Meteorological data forecasting. These different methodologies of prediction are tested through the two last years of measurements, all the others data are related to learning step of the MLP.

2.2.1. Persistence

The first studied prediction methods is the persistence method; the simplest way of producing a forecast. The persistence method assumes that the actual conditions will not change. To improve the results and to take into account the fact that the TS is more or less periodical, we corrected the persistence with a scale term computed from the moving average related to the nine first years [4,9].

2.2.2. *MLP*

In the presented study, the MLPs have been computed with the Matlab© software and its Neural Network toolbox. The characteristics chosen related to previous works are the following: one hidden layer, the activation functions are the continuously and differentiable hyperbolic tangent (hidden) and linear (output), the Levenberg-Marquardt (approximation to the Newton's method) learning algorithm with a maximum fail parameter before stopping training equal to 3. The optimization of the number of input nodes is done with the partial autocorrelation factor, which measures the degree of association between two measures, with the effect of a set of controlling intermediate measures removed. Concerning the four types of TS considered, this number varies between 4 and 9 lags. The number of hidden neurons is taken equal to the input nodes number. The results shown in the following parts are related to the best network among six different trainings coupled with a random weight initialization [1,4,7]. Concerning the MLP with heterogeneous transfer functions, the characteristics are described in the section 2.1.

2.2.3. MLP with a time index input

The last approach consists in adding a time index input to the two MLP previously proposed, that is to say, the classical one and the MLP with the heterogeneous transfer function. An input is added and is related to the time concerning the hour of prediction. This node takes the values i/24 (for i=1 to 24) for each hour of the day [9].

3. Results

The first series of results (Table 1) is related to the persistence (P), normal MLP (noted N-MLP) and the heterogeneous transfer functions MLP (noted HTF-MLP; 50% of tanh' transfer function in the hidden layer). The error metric used is the nRMSE (RMSE/mean). Concerning the use of the MLP, the results presented in the following tables are related to the runs (among six) minimizing the nRMSE.

	P	N-MLP	HTF-MLP
Humidity	**0.070**	0.073	0.073
Solar irradiation	0.319	**0.305**	0.308
Temperature	**0.058**	0.067	0.067
Wind speed	0.394	**0.359**	**0.359**

Table. 1: nRMSE related to the persistence, normal MLP and heterogeneous transfer functions MLP. In bold the lowest nRMSE.

In Table 1, we see that HTF-MLP is not really efficient, it gives the best results only for only one TS (Wind speed) tied with N-MLP. With a MLP, one way to improve the results is often to make the time series stationary and to operate a preprocessing before to choose the inputs nodes. In our case, there are four kinds of data, so we chose to apply a seasonal adjustment by periodic coefficients (computed from moving average during the learning phase). In the following results, when this pre-post processing is used, the suffix –s is added (example N-MLP-s). The Table 2 shows the impact of this processing on the nRMSE value.

	N-MLP-s	HTF-MLP-s
Humidity	**0.070**	**0.070**
Solar irradiation	**0.333**	0.334
Temperature	**0.059**	**0.059**
Wind speed	0.351	**0.350**

Table. 2: nRMSE related to the normal MLP and heterogeneous transfer functions MLP and concerning the time series made stationary. In bold the lowest nRMSE.

The stationary process doesn't improve systematically the nRMSE, in fact, in the solar radiation case the error is increased by more than 2 percentage point. The second tool that should improve results is the time index taken as input of the networks. In the following parts, we will note –t (example N-MLP-s-t) when this index is used. The Table 3 shows the impact of this exogenous data (multivariate analysis).

	N-MLP-t	HTF-MLP-t	N-MLP-s-t	HTF-MLP-s-t
Humidity	**0.069**	0.070	0.071	0.070
Solar radiation	**0.292**	0.298	0.324	0.331
Temperature	0.058	0.060	0.059	**0.057**
Wind speed	0.358	0.357	**0.351**	0.353

Table. 3: nRMSE related to the normal MLP and heterogeneous transfer functions MLP and concerning the time index taken as input. In bold the lowest nRMSE.

Considering the different configurations -s, -t, non-bijective or bijective transfer functions of hidden nodes, a lot of combinations are possible. The next table (4) summarizes for each data the predictor with the lowest nRMSE and the related error metric.

*the hidden node with the (tanh)' transfer function is only linked to the time index input

	VC	Type	nRMSE	Time index	Stationary	Transfer functions
Humidity	0.2	N-MLP-t	0.069	Yes	No	100% tanh
Solar irradiation	1.5	N-MLP-t	0.292	Yes	No	100% (tanh)'
Temperature	0.4	HTF-MLP-s-t	0.057	Yes	Yes	50% tanh 50% (tanh)'
Wind speed	0.5	HTF-MLP-s	0.350	No	Yes	80% tanh 20% (tanh)'*

Table. 3: Best configuration considering nRMSE, for all combinations of tested MLP parameters.

The time index and the (tanh)' transfer function are the tools giving the best results 75% of the cases). The stationary process does not define really the best approaches. For the two time series with the lowest VC the nRMSE is under 10% whereas for the other series (solar irradiation and wind speed) the nRMSE is close to 30%.

4. Conclusion

In this study, a MLP coupling two transfer functions, a time index and a stationary process was studied. If the interest of the time index taken as inputs of the network is proved in this study, future studies expanding the nature or the location related to time series will be necessary to show the interest of the MLP with heterogeneous transfer functions. As conclusion of this paper, for each case concerning the time series forecasting of meteorological data, the possibility to use a time index, a stationary process and an inhibition transfer functions should be considered.

References

[1] M. A. Behrang, E. Assareh, A. Ghanbarzadeh and A. R. Noghrehabadi, The potential of different artificial neural network (ANN) techniques in daily global solar radiation modeling based on meteorological data, Solar Energy, 84:1468–1480, Elsevier, 2010.

[2] W. Duch and N. Jankowski, Transfer functions: hidden possibilities for better neural networks, ESANN'2001 proceedings - European Symposium on Artificial Neural Networks, Bruges (Belgium), 25-27 April 2001, D-Facto public., ISBN 2-930307-01-3, pp. 81-94

[3] S.F. Crone, Stepwise selection of artificial neural networks models for time series prediction, J. Intelligent Syst. 14: 99–122, De Gruyter, 2005.

[4] J. G. De Gooijer and R. J. Hyndman, 25 years of time series forecasting, Int. J. Forecasting, 22: 443–473, Elsevier, 2006.

[5] C. Duchon and R. Hale, Time Series Analysis in Meteorology and Climatology: An Introduction, Wiley-Blackwell, 2012.

[6] F. O. Hocaoglu, Stochastic approach for daily solar radiation modeling, Solar Energy, 85: 278–287, Elsevier, 2011.

[7] S. Kalogirou, Artificial neural networks in renewable energy systems applications: A review, Renew. Sustain. Energy Rev., 5: 373–401, Elsevier, 2001.

[8] A. Mellit and S. Kalogirou, Artificial intelligence techniques for photovoltaic applications: A review, Prog. Energy Combustion Sci., 34: 574–632, Elsevier, 2008.

[9] C. Voyant, C. Paoli, M. Muselli and M.-L. Nivet, Multi-horizon solar radiation forecasting for Mediterranean locations using time series models, Renew. Sustain. Energy Rev., 28: 44–52, Elsevier, 2013.

[10] T.T. Warner, Numerical Weather and Climate Prediction, Cambridge University Press, 2010.

Simulations in Medicine and Biology: Insights and perspectives

George M. Spyrou

Biomedical Research Foundation of the Academy of Athens, Soranou Efessiou 4, 115 27 Athens, Greece

gspyrou@bioacademy.gr

Abstract. Modern medicine and biology have been transformed into quantitative sciences of high complexity, with challenging objectives. The aims of medicine are related to early diagnosis, effective therapy, accurate intervention, real time monitoring, procedures/systems/instruments optimization, error reduction, and knowledge extraction. Concurrently, following the explosive production of biological data concerning DNA, RNA, and protein biomolecules, a plethora of questions has been raised in relation to their structure and function, the interactions between them, their relationships and dependencies, their regulation and expression, their location, and their thermodynamic characteristics. Furthermore, the interplay between medicine and biology gives rise to fields like molecular medicine and systems biology which are further interconnected with physics, mathematics, informatics, and engineering. Modelling and simulation is a powerful tool in the fields of Medicine and Biology. Simulating the phenomena hidden inside a diagnostic or therapeutic medical procedure, we are able to obtain control on the whole system and perform multilevel optimization. Furthermore, modelling and simulation gives insights in the various scales of biological representation, facilitating the understanding of the huge amounts of derived data and the related mechanisms behind them. Several examples, as well as the insights and the perspectives of simulations in biomedicine will be presented.

This paper is dedicated to the memory of Professors George Tzanakos and Evangelia Micheli-Tzanakou.

1. Introduction

There is a wonderful interplay between Physics, Mathematics, Informatics, Medicine and Biology that drives to a continuous fertile explosion producing other fields like Computational Physics and Mathematics, Biophysics, Medical Physics, Medical Informatics, Bioinformatics as well as Nanomedicine and NanoBiotechnology, AstroBiology and NeuroInformatics. All these scientific fields are interconnected belonging to one or more networks. Their interconnections can be of various types: physical, content-based and concept-based, demonstrating various levels of randomness.

Each of the basic sciences has an initial contribution delineating a large space of new questions and specific needs. To achieve information and data understanding in medicine and biology across the axis prediction/prognosis – diagnosis – therapy – integration, state-of-the-art computational methods are applied. Some of the main objectives of computer-aided Medicine are the early diagnosis of diseases, the effective therapy, the minimization of medical error and the connection to biological data towards Molecular Medicine. In parallel, some of the main objectives of computer-aided Biology are the prediction of biomolecules' properties, the prediction of the biological systems' properties as well as the investigation of the molecular basis for various diseases.

The related applications to the aforementioned interdisciplinary fields require high performance computing (perhaps several hundred of Petaflops), especially when we are going to systems levels requiring realistic simulations (e.g. simulations of protein machine interactions and molecule-based cell simulations as well). However, up to now, according to the TOP500 list (June 2013), the best available supercomputers present a maximum performance ranging from 2 to 50 Petaflops.

2. Simulations in general

The simulations' general pipeline starts always by having interesting questions (Figure 1). Afterwards, we have to understand (or believe that we understand) the processes and the phenomena of the systems involved. Then, we make a model that describes our system and we select the proper numerical methods to implement this model. These are translated to code development which is checked for consistency and verification under the prism of a set of predefined expected results. There is a feedback loop here concerning code optimization mainly in technical terms.

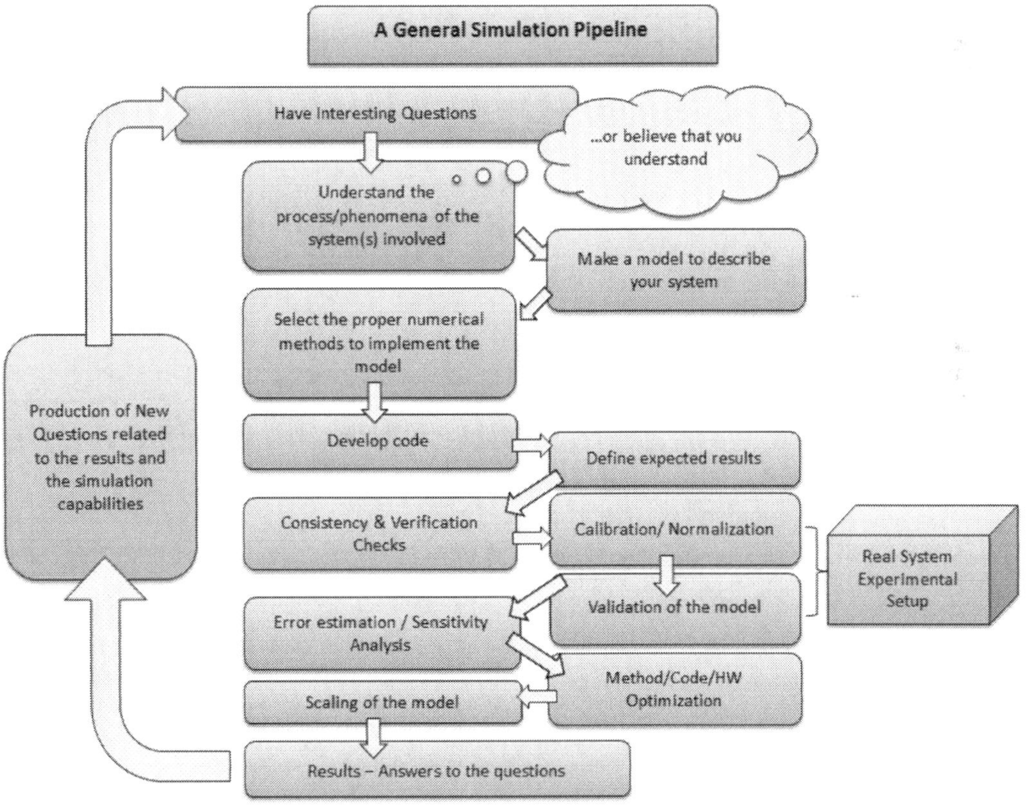

Figure 1. A general pipeline describing simulations' design and validation

In parallel, we try to develop an experimental setup resembling the real system or crucial features of the real system we are simulating. This gives us the opportunity to calibrate our model and normalize properly the results in order to match the scale of the real ones.

Then, the model can be validated via comparisons with experimental results and in the sequel we are able to perform sensitivity analysis and error estimation. Finally, we can proceed to overall optimization by fine tuning the methods, the code and the hardware used for the simulation. By performing scaling of the model we are able to answer whether the more CPUs will proportionally reduce the needed time to complete a long simulation.

Following this pipeline, we reach to results that hopefully help us answer the initial questions. Usually then, we realize the capabilities of the simulation and we haste to put new questions, more exotic than the initial ones.

3. Simulations in Medicine and Biology

Actually, Medicine and Biology are suitable for weird questions. According to Richard Feynman, "everything that living things do can be understood in terms of the jigglings and wigglings of atoms" [1]. As such, simulations in these fields deal with very complex systems, multidimensional feature spaces and multi-scale approaches.

Depending on what is the level of modeling, simulations in medicine and biology can be process driven or data driven. In the first case, the simulations are based on modeling the phenomena/procedures/processes of the real system under investigation. Given a good understanding of the underlying mechanisms, we produce results that should be similar to the experimental ones and vice versa: by tuning our model to fit the experimental results we understand better the underlying mechanisms. In the second case, simulations are based on the experimental data structure and data features, producing similar data to the real ones, preserving the data features and their correlations. In this way, we can populate our samples without necessarily understanding the underlying mechanisms for their production.

Simulations with biomolecules are usually process driven. They generate representative configurations of the systems providing us with reliable values of structural and thermodynamic properties. They also enable the time-dependent behavior of atomic and molecular systems to be determined, providing a detailed picture of the way in which a system changes from one conformation to another. Such simulations help to predict molecular structures, understand interactions and properties, design of bio-nano materials, experiment on what cannot be studied experimentally, obtain movies of the interacting molecules and many other fascinating capabilities. Molecular modeling is a key methodology for research and development in bionanotechnology, since it provides nanoscale images at atomic and electronic resolution, and it predicts the interaction of biological and inorganic materials. Thus, molecular modeling facilitates studies and in silico experiments with exciting objectives, like the use of carbon nanotubes as in situ biosensors and the use of nanopores for DNA sequencing [2]. For this kind of simulations, the two most popular simulation methods are Molecular dynamics, and Monte Carlo.

Molecular Dynamics is a deterministic method where time dependent properties are calculated, configurations in the past and in the future can be predicted, calculations of kinetic energy and potential energy can be performed, and time average can be calculated from simulation trajectories. On the other hand, Monte Carlo is a stochastic method presenting no temporal relationship and no predictive power among the past and the future. Only the potential energy is used and each configuration depends only on its predecessor. A key function in biomolecular simulations is the energy function that is a target function which describes the interaction energies of all atoms and molecules in the system. The energy function comprises energy terms concerning bonded and non-bonded states. The bonded energy terms describe the capabilities of stretching, bending and rotating around a bond, whereas the non-bonded energy terms include the van der Waals and the electrostatic contributions. The landscape of this energy function is downhill explored when we are investigating stable structural instances of molecules.

Of course, biomolecules are not alone. They are inside a solution and their accurate simulation requires realistic models for the interaction between biomolecules and solvent. There are two types of solvent consideration: the explicit solvent models and the implicit solvent models. In the first category, the models take into consideration all the solvent molecules presenting high accuracy but also high complexity, slow convergence and high computing requirements. In the second category, solvent is considered as a continuous medium and average estimations of the interactions are produced. In this direction, very useful is the determination of the molecular surface that can be done as a van der Waals, Connolly or Solvent accessible surface by using a rotating probe over the molecule. The choice of the 'probe radius' does have an effect on the observed surface area, as using a smaller probe radius detects more surface details and therefore reports a larger surface. A typical probe radius value is the one that approximates the radius of a water molecule [3].

However, there are also efforts in presenting data driven simulations in biology concerning especially biological data modelling like the one presented in the BiDaS Web server we have developed. BiDaS is a web-application that can generate massive Monte Carlo simulated sequence or numerical feature data sets (e.g. dinucleotide content, composition, transition, distribution properties) based on small user-provided data sets. BiDaS server enables users to analyze their data and generate large amounts of: (i) Simulated DNA/RNA and aminoacid (AA) sequences following practically identical sequence and/or extracted feature distributions with the original data. (ii) Simulated numerical features, presenting identical distributions, while preserving the exact 2D or 3D between-feature correlations observed in the original data sets [4].

In parallel, at Medicine's side, we have put a major effort to simulate mammographic imaging, focusing on tasks such as: (i) Simulation of x-ray focal spots with finite size and realistic intensity distribution, (ii) Generation of simulated mammographic images, (iii) Dosimetric calculations, (iv) Investigation of various magnification setup geometries, (v) Simulation of digital detectors for mammographic imaging [5-15].

The objective was to set up the design parameters and the functional parameters of a mammographic unit as well as the properties of a 3-D phantom with inhomogeneities of various geometries and compositions, run a simulation in order to generate the corresponding image and make dosimetric calculations as well. This is a process-driven simulation, thus we have simulated the physical phenomena related to the photons travelling inside material, namely the anelastic and elastic scattering as well as photoelectric absorption. Furthermore, the large number of parameters, along with the need for multiparameter studies and unified result management led us to the development of some simulation environment platforms, named MASTOS and DOSIS as the acronyms for "MAmmography Simulation Tool for design Optimization Studies" and "tool for DOsimetry Simulation Studies". Through such powerful tools we managed to perform various investigations concerning the contrast-to-noise ratio in magnification mammography, the influence of various focal spot sizes and magnification geometries, the influence of x-ray spectra on the image quality and the absorbed dose, the behavior of photoconductor materials as detectors for digital mammography. The Monte Carlo generated mammographic images along with the accompanying information concerning the absorbed dose can be used both for the evaluation of new mammographic imaging setups (geometry, materials, and spectra) and for the training and evaluation of Computer Aided Diagnosis (CAD) schemes.

Moving towards molecular imaging, we have started studying another imaging modality, namely Multispectral Optoacoustic Tomography (MSOT), in collaboration with the Institute for Biological and Medical Imaging at the Helmholtz Center in Munich, Germany. This is an improved optoacoustic imaging modality of biological tissues and molecular imaging, used mainly for resolving contrast of highly absorbing tissue components or contrast agents. The development of an MSOT simulator includes the optical and the acoustic forward problem formulation. Simulating the optical and acoustic phenomena in concert facilitates the direct and easy comparison with experimental data, providing better understanding of the optical and acoustic phenomena that are present in experimental tissue imaging. The development of the simulation and the initial validation results has been just presented [16].

Switching to the other kind of simulations, the data driven ones, we are running a project concerning Modeling of Solitary Pulmonary Nodules in PET/CT images using Monte Carlo Methods. The assessment of solitary pulmonary nodules remains a difficult diagnostic task despite the rapid improvement of various imaging modalities. CT and PET are the most common methods for diagnosis of SPN. Physicians and computational systems as well, would gain benefit if trained to a large number of SPN cases with controlled topological and morphological characteristics. The modeling of SPN has been implemented by Monte Carlo methods taking into consideration morphological characteristics, internal features and Standardized Uptake Value activity distribution. Again, the generated images can be used both for the evaluation of the medical experts' diagnostic performance as well as for the training and evaluation of Computer Aided Diagnosis (CAD) schemes [17].

4. Discussion

Concluding, simulations in Medical and Biological research are a very useful and promising tool that may provide us with invaluable information at various levels and scales. The presented research works are only a drop in a wonderful ocean that we are all welcome to sail.

References

[1] Feynman RP, Leighton RB, and Sands ML 1963 The Feynman Lectures on Physics. Reading, MA: Addison-Wesley Publishing Co. vol. 1, p.3-6
[2] Lu D, Aksimentiev A, Shih AY, Cruz-Chu E, Freddolino PL, Arkhipov A, Schulten K 2006 Phys Biol. 2;3(1):S40-53.
[3] Connolly ML 1983 Science. Aug 19;221(4612):709-13.
[4] Paraskevopoulou MD, Vlachos IS, Athanasiadis E, Spyrou G 2013 Nucleic Acids Res.41:W582-6.
[5] Spyrou G, Tzanakos G, Bakas A, Panayiotakis G 1998 Phys Med Biol. 43(11):3341-57.
[6] Spyrou G, Panayiotakis G, Tzanakos G 2000 Med Inform Internet Med. 25(4):275-93.
[7] Spyrou G, Tzanakos G, Nikiforides G, Panayiotakis G 2002 Phys Med Biol. 47(6):917-33.
[8] Delis H, Spyrou G, Panayiotakis G, Tzanakos G 2005 Eur J Radiol 54(3):371-6.
[9] Delis H, Spyrou G, Costaridou L, Tzanakos G, Panayiotakis G 2006 Med Phys. 33(11):4221-35.
[10] Koutalonis M, Delis H, Spyrou G, Costaridou L, Tzanakos G, Panayiotakis G 2006 Phys Med Biol. 51(21):5539-48.
[11] Koutalonis M, Delis H, Spyrou G, Costaridou L, Tzanakos G, Panayiotakis G. 2007 Phys Med Biol. 52(11):3185-99.
[12] Koutalonis M, Delis H, Spyrou G, Costaridou L, Tzanakos G, Panayiotakis G 2008 Phys Med Biol.53(5):1369-84.
[13] Sakellaris T, Spyrou G, Tzanakos G, Panayiotakis G 2005 Phys Med Biol. 50(16):3717-38.
[14] Sakellaris T, Spyrou G, Tzanakos G, Panayiotakis G 2007Phys Med Biol. 52(21):6439-60.
[15] Sakellaris T, Koutalonis M, Spyrou G, Pascoal A 2013 Phys Med. S1120-1797(13)00253-6.
[16] Maneas E, Tzoumas S, Ntziachristos V and Spyrou G 2013Proceedings of the 13th IEEE International Conference on BioInformatics and BioEngineering – BIBE2013.
[17] Tzanoukos G, Gaitanis A, Georgakopoulos A, Chatziioannou A, Chatziioannou S, Spyrou G 2013Proceedings of the 13th IEEE International Conference on BioInformatics and BioEngineering – BIBE2013.

Journal of Physics: Conference Series **574** (2015) 012066 doi:10.1088/1742-6596/574/1/012066

On the generalized Hartman effect for symmetric double-barrier point potentials

Molly A Lee[1], José T Lunardi[2], Luiz A Manzoni[1] and Erik A Nyquist[1]

[1] Department of Physics, Concordia College, 901 8th St. S., Moorhead, MN 56562, USA

[2] Departamento de Matemática e Estatística, Universidade Estadual de Ponta Grossa Av. General Carlos Cavalcanti, 4748, Cep 84030-000, Ponta Grossa, PR, Brazil

E-mail: mlee11@cord.edu, jttlunardi@uepg.br, manzoni@cord.edu, enyquist@cord.edu

Abstract. We consider the scattering of a non-relativistic particle by a symmetrical arrangement of two identical barriers in one-dimension, with the barriers given by the well-known four-parameter family of point interactions. We calculate the phase time and the stationary Salecker-Wigner-Peres clock time for the particular cases of a double δ and a double δ' barrier and investigate the off-resonance behavior of these time scales in the limit of opaque barriers, addressing the question of emergence of the generalized Hartman effect.

Quantum tunneling times have been extensively investigated in the last decades, leading to several alternative definitions of transmission time (see [1] and references therein). A phenomenon that attracted particular interest is the so-called Hartman effect [2], which states that in the opaque limit (low probability of transmission) the tunneling time becomes independent of the barrier width. In a generalized version of the Hartman effect the tunneling time for a particle incident on a double-barrier potential also becomes independent of the inter-barrier distance in the opaque limit [3, 4]. Due to their counterintuitive nature and consequent conceptual difficulties, these effects have been investigated for a variety of tunneling times (phase time, dwell time, Larmor time, etc.). However, most of these investigations are restricted to rectangular barriers or (in the generalized case) double-δ barriers – thus, it is important to verify if such effects hold for other interactions, such as generalized point interactions.

Point interactions in one-dimensional quantum mechanics have been the source of significant theoretical and mathematical interest for providing the opportunity to investigate advanced tools such as regularization and renormalization, generally employed in quantum field theory, in a simpler scenario (see, e.g., [5]). From the point of view of applications, point-like interactions have long provided models of significant experimental relevance for lower dimensional systems in, e.g., atomic physics. The most general point interaction in one-dimensional quantum mechanics is given by a four-parameter family of interactions, as can be shown by either considering self-adjoint extensions (SAE) of the kinetic energy operator (see, e.g., [6, 7, 8]) or by using a distributional approach to the corresponding Schrödinger equation [9]. Recently this general four-parameter family of interactions in one dimension has been used to investigate tunneling times for a non-relativistic particle incident on a single scatterer [10].

In this work we calculate the phase and Salecker-Wigner-Peres (SWP) clock times for a non-relativistic particle tunneling through a *symmetrical* arrangement of two identical potential barriers given by the four-parameter generalized point interaction in one dimension (in this case the SWP clock time is equivalent to the dwell time – see below). Our main interest is in investigating the emergence of the generalized Hartman effect (GHE); thus we restrict ourselves to the δ and δ' interactions and analyze the behavior of those times in the opaque limit. For the double δ-potential the stationary tunneling times have already been addressed in the literature (see [11] and references therein), however, as we will see below it is worth revisiting the discussion of the GHE for this potential.

Let us consider the scattering of a non-relativistic particle of mass μ and energy E by two identical point scatterers localized at $z = 0$ and $z = L$. The point interaction at position z can be characterized by the boundary conditions (b.c.)

$$\Phi(z_+) = \Lambda\, \Phi(z_-), \qquad \text{with} \qquad \Phi(z) \equiv \begin{pmatrix} \psi(z) \\ \psi'(z) \end{pmatrix}, \tag{1}$$

where $\psi(z)$ indicates the wave function, $\psi'(z)$ denotes its space derivative and $\Phi(z_\pm) = \lim_{\epsilon\to 0^+} \Phi(z \pm \epsilon)$; Λ is the matrix

$$\Lambda = e^{i\theta} \begin{pmatrix} a & b \\ c & d \end{pmatrix}, \quad ad - bc = 1, \quad \theta \in [0, \pi), \tag{2}$$

with $a, b, c, d \in \mathbb{R}$.

The four-parameter family of point interactions given by eq. (2) represents the most general point interactions consistent with conservation of the probability current in one dimension, as can be demonstrated by using a rigorous distributional approach to treat the corresponding Schrödinger equation [9], or as obtained by the SAE approach [6, 7, 8]. The interaction (1)-(2) includes the so-called δ and δ' interactions as well as other point interactions which do not have a simple characterization.

We also assume that the particle is coupled to a SWP clock [12, 13] which only runs when the particle is in the region $(0, L)$. As is well known, the effect of the particle-clock coupling is to introduce a small potential barrier of high $\mathcal{V}_m$ in the region $(0, L)$ [13, 14]. Then, for incidence from the left, the general solution of the time-independent Schrödinger equation can be written as (omitting the dependence on $\mathcal{V}_m$, for simplicity)

$$\psi_{\rm I}(z) = e^{ikz} + r(\mathbf{k})e^{-ikz}; \qquad \psi_{\rm III}(z) = t(\mathbf{k})e^{ikz}, \tag{3}$$

where $\psi_{\rm I}$ and $\psi_{\rm III}$ indicate the wave function in the regions $z < 0$ and $z > L$, respectively. In (3) we defined $\mathbf{k} \equiv (k, q)$, with $k \equiv \frac{1}{\hbar}\sqrt{2\mu E}$ and $q \equiv \frac{1}{\hbar}\sqrt{2\mu(E - \mathcal{V}_m)}$ being the respective wave numbers in regions I/III and II $(0 < z < L)$.

It is convenient to express the transmission and reflection coefficients as

$$t(\mathbf{k}) = |t(\mathbf{k})|e^{i[\varphi_t(\mathbf{k}) - kL]} \qquad \text{and} \qquad r(\mathbf{k}) = |r(\mathbf{k})|e^{i\varphi_r(\mathbf{k})}. \tag{4}$$

The phase transmission time for this system can be immediately calculated as $t_p^T(k) = \hbar\frac{\partial}{\partial E}\varphi_t(k)$, where the limit $\mathcal{V}_m \to 0$ is implicit. Similarly, the phase reflection time is given by $t_p^R(k) = \hbar\frac{\partial}{\partial E}\varphi_r(k)$. We are also interested in the dwell time $\tau_D(k)$ (see, e.g., [1] and references therein), which can be obtained from the identity [14]

$$\tau_D(k) = |t(k)|^2 t_c^T(k) + |r(k)|^2 t_c^R(k), \tag{5}$$

where $t_c^{T(R)}(k)$ stands for the transmission (reflection) SWP clock time, which is given by $t_c^T(k) = -\hbar\left[\frac{\partial}{\partial \mathcal{V}_m}\varphi_t(\mathbf{k})\right]_{\mathcal{V}_m=0}$ (with a similar expression for the reflection SWP time).

Journal of Physics: Conference Series **574** (2015) 012066 doi:10.1088/1742-6596/574/1/012066

In this work we are interested in the case in which both the barriers are symmetric (which implies that the whole potential is also symmetric). This corresponds to the restriction $\theta = 0$ and $a = d$ in (2) [9, 15]. In this case, after imposing the b.c. (1)-(2) at $z = 0$ and $z = L$ and taking into account the potential $\mathcal{V}_m$ in the region $(0, L)$ due to the interaction of the particle with the SWP clock, we obtain (from now on we use units such that $\hbar = 2\mu = 1$)

$$\varphi_t(\mathbf{k}) = \varphi_r(\mathbf{k}) - \frac{\pi}{2} = -\tan^{-1}\left\{\frac{2a(c-bk^2)q + \left[c^2 + b^2k^2q^2 - a^2(k^2+q^2)\right]\tan(qL)}{2k\left[(a^2+bc)q + a(c-bq^2)\tan(qL)\right]}\right\} . \quad (6)$$

Thus, it follows from (5), and using conservation of probability, that for these kinds of symmetric potential we have $\tau_D(k) = t_c^T(k) = t_c^R(k)$. It also follows that $t_p^T(k) = t_p^R(k)$. Such result, which holds for any symmetric regular potential, also holds for any symmetric arrangement of singular point interactions (not necessarily identical). Now we restrict ourselves to the double δ and double δ' interactions and focus on the opaque limit, with the purpose of analyzing the GHE.

The double δ-interaction. The (symmetric) double delta potential can be obtained by taking $a = d = 1$, $b = 0$, $c = \gamma$ and $\theta = 0$, with γ characterizing the strength of the interaction [8, 9, 15]. With these substitutions the phase time becomes

$$t_p^T(k)\big|_\delta = \frac{4k^4L + \gamma^3 + 2k^2\gamma(2+L\gamma) - \gamma^3\cos(2kL) + 2k\gamma^2\sin(2kL)}{k\left[8k^4 + 4k^2\gamma^2 + \gamma^4 + \gamma^2(4k^2-\gamma^2)\cos(2kL) + 4k\gamma^3\sin(2kL)\right]} , \quad (7)$$

and the dwell/clock time can be shown to be [11]

$$t_c^T(k)\big|_\delta = \frac{2k\left[2k^2L + \gamma + L\gamma^2 - \gamma\cos(2kL)\right] - \gamma^2\sin(2kL)}{8k^4 + 4k^2\gamma^2 + \gamma^4 + \gamma^2(4k^2-\gamma^2)\cos(2kL) + 4k\gamma^3\sin(2kL)} . \quad (8)$$

In the limit $L \to 0$ the two barriers coincide at the origin and the above times reduce to the ones obtained in reference [10] for a single point scatterer with strength 2γ.

The opaque limit is characterized by very strong interactions, i.e., $\gamma \gg 1$. It is straightforward to show that in the *extreme* opaque limit, $\gamma \to \infty$, both the off-resonance phase and dwell/clock times vanish, regardless of L (assumed to be finite), and this is generally interpreted as indicating the presence of the GHE – see [11] and references therein. However, in the extreme limit $\gamma \to \infty$ there is absolutely no transmitted wave beyond the first barrier and the transmission phase delay φ_t becomes meaningless (in fact, in this situation the delta barrier becomes impenetrable [6, 9]). Accordingly, in this extreme limit the transmission phase and dwell/clock times cannot be understood as traversal times (see [1, 14] for a similar discussion in the context of rectangular barriers). Therefore, we follow the approach of reference [14] and investigate the opaque limit by considering the asymptotic behavior of the tunneling times when γ is very large, but still finite. The leading contributions to the off-resonance phase and dwell/clock times are respectively given by $t_p^T(k)\big|_\delta = \frac{1}{\gamma k}$ and $t_c^T(k)\big|_\delta = \frac{1}{\gamma^2}\left[kL\csc^2(kL) - \cot(kL)\right]$, and we observe that for the phase time it does not depend on the barrier separation L, while for the dwell/clock time there is a dependence on L. Therefore, by taking into account *only* the leading contributions to the transmission times in the opaque regime, when the probability of transmission is small but non-vanishing, it appears that the transmission phase time presents the GHE, whereas the dwell/clock time does not present such effect.

The double δ'-interaction. We consider a (symmetric) double δ' interaction as obtained from (1) and (2) by taking the parameters to be $\theta = 0$, $a = d = 1$, $b = \gamma$, $c = 0$, with γ giving the strength of the interaction [8, 9, 15] (alternative, if less common, definitions for a δ'-interaction exist – see, e.g., [16]).

The phase and the dwell/clock times are respectively given by

$$t_p^T(k)\big|_{\delta'} = \frac{2L(2+k^2\gamma^2)+\gamma(4+k^2\gamma^2)-k\gamma^2\left[k\gamma\cos(2kL)+2\sin(2kL)\right]}{k\left\{8+4k^2\gamma^2+k^4\gamma^4+k^2\gamma^2\left[(4-k^2\gamma^2)\cos(2kL)-4k\gamma\sin(2kL)\right]\right\}}, \tag{9}$$

$$t_c^T(k)\big|_{\delta'} = \frac{2(2L+\gamma+k^2L\gamma^2)-2\gamma\cos(2kL)+k\gamma^2\sin(2kL)}{k\left\{8+4k^2\gamma^2+k^4\gamma^4+k^2\gamma^2\left[(4-k^2\gamma^2)\cos(2kL)-4k\gamma\sin(2kL)\right]\right\}}. \tag{10}$$

Similarly to the case of the double δ barrier, in the *extreme* opaque limit $\gamma \to \infty$ both of the above times converge to zero. But, again, in this extreme limit the interaction corresponds to two impenetrable point barriers, and any quantity calculated in terms of transmitted waves does not have physical meaning. Thus, here we also consider the asymptotic behavior of the off-resonance times in the opaque regime, when γ is very large but finite – still allowing a non-vanishing probability of transmission. The leading contributions to the phase and dwell/clock times are respectively given by $t_p^T(k)\big|_{\delta'} = \frac{1}{\gamma k^3}$ and $t_c^T(k)\big|_{\delta'} = \frac{1}{\gamma^2 k^4}\left[\cot(kL)+kL\csc^2(kL)\right]$. Again, in what concerns *only* the leading contributions to the off-resonance times in the opaque limit, it seems that the GHE occurs for the phase time but not for the dwell/clock time.

It should be noticed, however, that for any finite interaction strength the next-to-leading contribution to the off-resonance phase time (of order $1/\gamma^2$) *does* depend on the barrier separation L for both the δ and δ' interactions[1]. That is, at the same order as the leading contribution for the clock/dwell times, the phase times also present a slow growth with L for any *finite* γ (here, "slow growth" means that it still allows superluminal velocities). This indicates that the GHE is just a mathematical consequence of taking the extreme opaque limit $\gamma \to \infty$, in which case any "transmission" time loses its physical meaning, as mentioned above.

Acknowledgments
This work was partially supported by NASA – Minnesota Space Grant Consortium (M.A.L. and L.A.M.) and NSF-STEP grant (E.A.N. and L.A.M.). L.A.M. also thanks the Office of Undergraduate Research of Concordia College for partial funding.

References

[1] Winful H G 2006 *Phys. Rep.* **436** 1
[2] Hartman T E 1962 *J. Appl. Phys.* **33** 3427
[3] Olkhovsky V S, Recami E and Salesi G 2002 *Europhys. Lett.* **57** 879
[4] Lunardi J T and Manzoni L A 2007 *Phys. Rev.* A **76** 042111
[5] Manuel C and Tarrach R 1994 *Phys. Lett.* A **328** 113
[6] Šeba P 1986 *Czech. J. Phys.* B **36** 667
[7] Carreau M 1993 *J. Phys. A: Math. Gen.* **26** 427
[8] Albeverio S, Gesztesy F, Høegh-Krohn R H and Holden H 2004 *Solvable Models in Quantum Mechanics*, 2nd ed (AMS Chelsea Publishing)
[9] Calçada M, Lunardi J T, Manzoni L A and Monteiro W 2014 *Front. Physics* **2** 23
[10] Kočinc S Lj S and Milanovič V 2012 *Mod. Phys. Lett.* B **26** 1250092
[11] Lunardi J T, Manzoni L A and Nystrom A T 2011 *Phys. Lett.* A **375** 415
[12] Salecker H and Wigner E P 1958 *Phys. Rev.* **109** 571
[13] Peres A 1980 *Am. J. Phys.* **48** 552
[14] Calçada M, Lunardi J T and Manzoni L A 2009 *Phys. Rev.* A **79** 012110
[15] Román J M and Tarrach R 1996 *J. Phys. A: Math. Gen.* **29** 6073
[16] Albeverio S, Dabrowski L and Kurasov P 1998 *Lett. Math. Phys.* **45** 33

[1] A detailed numerical study of the dependence of the tunneling times with L, as well as an investigation of double-barrier point potentials more general than those considered in (6), will appear elsewhere.

Parallelization of shadow projection in synthetic three–dimensional scenes

Javier Hernández–Aceituno
Departamento de Ingeniería Informática, Universidad de La Laguna.
Avda. Astrofísico Fco. Sánchez s/n, 38204 Canary Islands, Spain
E-mail: `javier@isaatc.ull.es`

Abstract. The generation of projected shadows in synthetic three–dimensional scenes is a complex procedure which, because of its computational cost, has yet to be perfected. In this work, we present an algorithm that simplifies this process using the basic three–dimensional geometry of three elements: a point cloud which defines the object that casts the shadow, the position of the light source, and a projection plane. Shadows are generated as irregular polygons with as many vertexes as the size of their corresponding generative point cloud. The parallelization potential of the resulting algorithm is then studied for real–time applications.

1. Introduction

Light effects, such as reflection, refraction or shadow projection, are very difficult to simulate when computing synthetic three–dimensional scenes, because of their computational cost. The accurate reproduction of light behavior for each image pixel – a procedure known as *ray–tracing* [1, 2] – often requires recursion, as each beam may be reflected or refracted on several objects along its path. Therefore, this technique requires a long processing time for each image, and can not be used in real time applications, for which developers favor inaccurate heuristic methods [3] or simplified approximations [4, 5].

In this work, we describe an algorithm that produces the projected shadows of three–dimensional objects, which are defined as point clouds, given the positions of the light sources that may affect them and the geometric planes that are affected by the projection. This procedure is explained in section 2, the computational cost of its application is studied on section 3 and the results of its parallelization are analyzed on section 4.

2. Algorithm definition

According to the basic principles of optical physics, given the three–dimensional positions of a point light source L and a hypothetical point object P, the shadow that the latter casts over any surface must be aligned with the direction of the light beam that produces it, that is, $P - L$. If we name the projection of the object as X, we can deduce that

$$X = P + \alpha \times (P - L), \quad (1)$$

where $\alpha > 0$ depends on the desired projection plane; if $\alpha < 0$, the projection plane is located at a relative position to P where a shadow cannot be casted, such as between P and L, and X would not exist.

A projection plane Z will be defined by its normal vector N and any point Q which belongs to it, such that $N \cdot (Y - Q) = 0$ for every point $Y \in Z$ [6]. Since the projection point X must necessarily belong to the projection plane, using the properties of the dot product [7], the value of α can be obtained as follows:

$$N \cdot (X - Q) = N \cdot (P + \alpha \times (P - L) - Q) = 0 \Rightarrow \alpha = \frac{N \cdot (Q - P)}{N \cdot (P - L)} \quad (2)$$

Notice how, if the light beam $P - L$ is parallel to Z, $N \cdot (P - L) = 0$ and P does not have a valid projection. This calculation must be executed once for each object point, light source, and projection plane. Since any image is bound to contain several of these, we must optimize (2) to reduce its computational cost as much as possible. Calculating α as

$$\alpha = \frac{N \cdot Q - N \cdot P}{N \cdot P - N \cdot L}, \quad (3)$$

allows us to compute $N \cdot Q$ only once per projection plane, $N \cdot L$ once per light source, and $N \cdot P$ once per object point.

Some restrictions of this technique must be considered. Our algorithm considers hypothetically boundless flat surfaces as projection planes, and the projection of points of different polygons of an object may overlap, which would cause undesired effects in case of semitransparent shadows. Users of this method must assess the impact of these limitations on their image generation procedures.

3. Experiments

The point cloud which defines a complex three–dimensional object may contain a very large number of points. For the sake of simplicity, we will study the behavior of our algorithm depending on the amount of points to project, regardless of whether they belong to different objects.

Our method was coded in *C/C++* language [8, 9], using the *OpenGL* three–dimensional graphic library [10] and the *OpenMP* parallelization interface [11], as shown in Fig. 1. We defined a three–dimensional `Point` structure, with all the required mathematical operators (vector addition and subtraction, dot product and scalar multiplication) and a `Plane` structure, which includes two `Point` elements: a normal vector `n` and a plane point `q`. Function `projection` receives three parameters: a vector of light sources `l`, a vector of projection planes `z`, and a vector of object points `p`; it returns a matrix of shadow projections `x`.

The generation time for a single image depends on its target media, but a minimum frame rate of 30 Hz is required, or else the animation would not look right to human viewers [12]. This sets a maximum restriction of 33.3 milliseconds per frame if our algorithm is to be used in real time.

In order to observe the efficiency of our method on slow computers, all experiments were done in a virtual machine. We observed unacceptable generation times for scenes with more than 10000 object points; however, since the shadows of static objects can be calculated offline and complex objects do not usually require more than 10000 points, our algorithm works correctly in real time.

4. Parallelization

Since our function contains three nested loops, it can be easily parallelized. However, although the number of object points in a scene may be very high, there are normally few light sources and projection planes. This may cause the communication between cores and memory accesses to increase the computational cost of the procedure and undo the parallelization enhancement. [13]

```
std::vector<Point> projection(const std::vector<Point> l,
                              const std::vector<Plane> z,
                              const std::vector<Point> p) {
  std::vector<Point> x;
  x.reserve(p.size() * l.size() * z.size());
  int i, j, k;
  float nq, np, nl, alpha;
  for (i = 0; i < z.size(); i++) {       // For each projection plane
    nq = z[i].n * z[i].q;
    for (j = 0; j < l.size(); j++) {     // For each light source
      nl = z[i].n * l[j];
      for (k = 0; k < p.size(); k++) {   // For each object point
        np = z[i].n * p[k];
        if (np == nl) continue;          // p[k] does not cast a shadow
        alpha = (nq - np) / (np - nl);
        if (alpha < 0) continue;         // p[k] does not cast a shadow
        x.push_back(p[k] + alpha * (p[k] - l[j]));
  } } }
  return x;
}
```

Figure 1. Sequential *C++/OpenGL* code for shadow projection.

To prove this point, we tested different configurations: parallelizing all loop levels (projection planes, light sources, and object points), parallelizing only the innermost loop (object points), and, for the sake of comparison, parallelizing only the two outer loops (projection planes and light sources). The time results for a four core virtual machine are shown in Fig. 2.

A full parallelization was found to improve the image generation time for over two projection planes, but to increase it otherwise. As expected, an innermost parallelization solved this problem and produced the best results for most cases, especially for large amounts of object points. An outer parallelization was found less efficient than the rest of options, including sequential executions, for most configurations.

5. Conclusion

The proper parallelization of our algorithm depends on the complexity of the scene to be generated. Parallelization will allow for greater amounts of visible objects, but applying it to all loops will only be useful when more than one light source and one projection plane appear in the image.

A greater number of light sources and projection planes will however reduce the allowable amount of object points, even when parallelizing. We observed that in this situation, full parallelization and innermost parallelization behave similarly. Therefore, an innermost parallelization is advisable for scenes which include unknown amounts of object points.

References

[1] Appel A 1968 *AFIPS Spring Joint Comp. Conf.* (ACM) pp 37–45
[2] Reif J H, Tygar J D and Yoshida A 1994 *Discrete Comput. Geom.* **11** 265–88
[3] Laine S, Aila T, Assarsson U, Lehtinen J and Akenine-Möller T 2005 *ACM Trans. Graph.* **24** 1156–65
[4] Crow F C 1977 *SIGGRAPH Comput. Graph.* **11** 242–48
[5] Porquet D, Dischler J and Ghazanfarpour D 2005 *SIGGRAPH Comput. Graph.* (ACM) pp 213–20
[6] Anton H 2010 *Elementary linear algebra* (John Wiley & Sons)
[7] Istrătescu V I 1987 *Inner product structures: theory and applications* vol 25 (Springer)

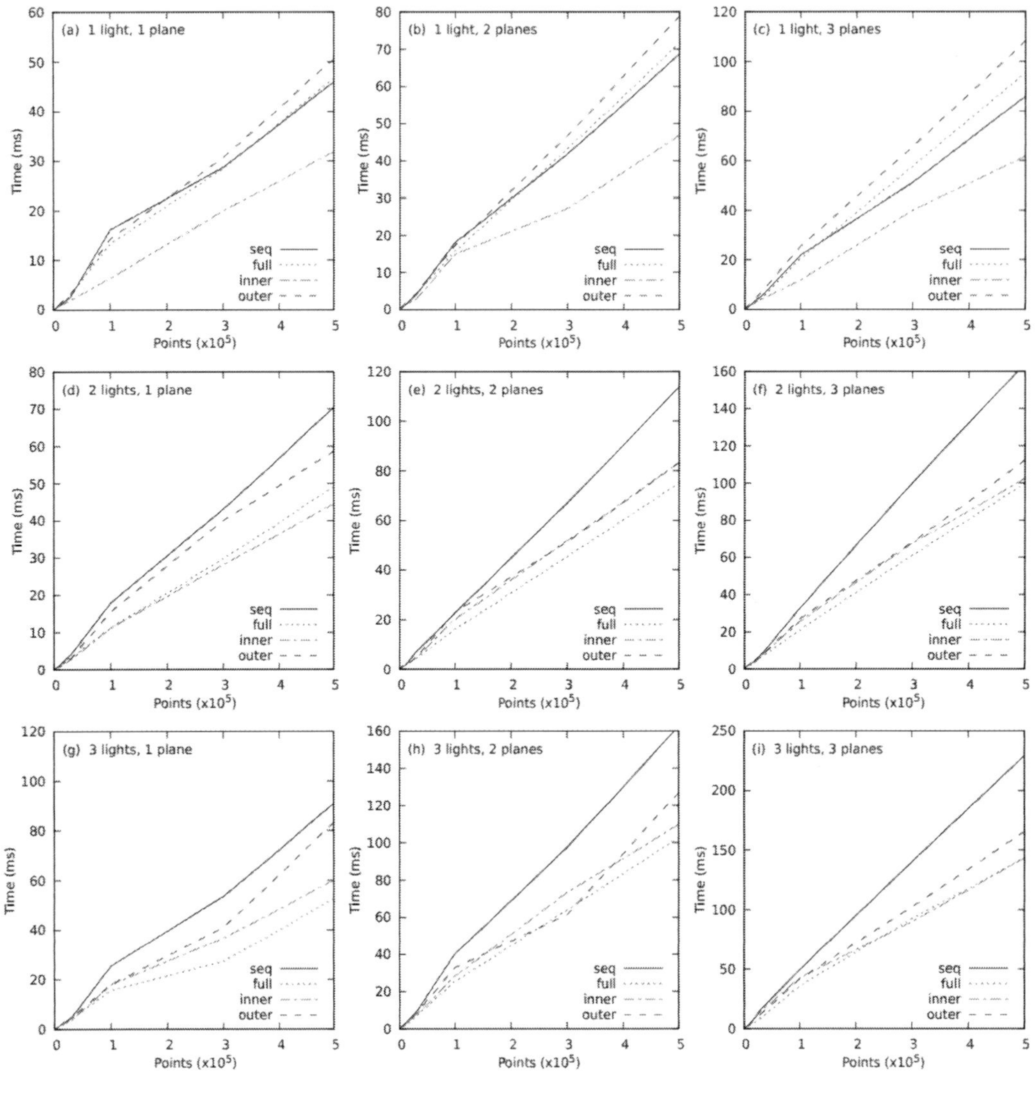

Figure 2. Generation times for different scene configurations, for sequential executions (*seq*), full parallelization (*full*), innermost parallelization (*inner*) and outer parallelization (*outer*).

[8] Ritchie D M 1993 *ACM SIGPLAN Notices* **28** 201–208
[9] Stroustrup B 2000 *The C++ Programming Language* (Addison-Wesley Longman Publishing Co., Inc.)
[10] Woo M, Neider J, Davis T and Shreiner D 1999 *OpenGL Programming Guide: The Official Guide to Learning OpenGL* (Addison-Wesley Longman Publishing Co., Inc.)
[11] Chapman B, Jost G and Pas R 2007 *Using OpenMP* (The MIT Press)
[12] Boff K R, Kaufman L and Thomas J P Cognitive processes and performance Tech. rep.
[13] Rünger G and Rauber T 2013 *Parallel Programming for Multicore and Cluster Systems* 2nd ed (Springer)

Simulation of General Physics laboratory exercise

P Aceituno[1], J Hernández–Aceituno[2] and A Hernández–Cabrera[1]

[1] Departamento de Física, Universidad de La Laguna. La Laguna, 38206 Tenerife, Spain
[2] Departamento de Ingeniería Informática, Universidad de La Laguna.
Avda. Astrofísico Fco. Sánchez s/n, 38204 Canary Islands, Spain

E-mail: paceitun@ull.edu.es

Abstract. Laboratory exercises are an important part of general Physics teaching, both during the last years of high school and the first year of college education. Due to the need to acquire enough laboratory equipment for all the students, and the widespread access to computers rooms in teaching, we propose the development of computer simulated laboratory exercises. A representative exercise in general Physics is the calculation of the gravity acceleration value, through the free fall motion of a metal ball. Using a model of the real exercise, we have developed an interactive system which allows students to alter the starting height of the ball to obtain different fall times. The simulation was programmed in ActionScript 3, so that it can be freely executed in any operative system; to ensure the accuracy of the calculations, all the input parameters of the simulations were modelled using digital measurement units, and to allow a statistical management of the resulting data, measurement errors are simulated through limited randomization.

1. Introduction

A large amount of the teaching hours of Science curricula are dedicated to laboratory experiments. The advantages of including these exercises are clear since the first contact with Physics at high school; they not only aid the understanding of physics concepts, but also help students learn how to properly handle measurement instruments and process experimental data statistically [1, 2].

Motivated by the costs institutes must face to provide all of their students with the required laboratory equipment, we propose a computerized simulation of some characteristic exercises. Since nowadays computer access in teaching is widespread, the execution of these exercises should come at no additional cost [3–6].

Our intention is not to replace all of the current laboratory exercises with simulated alternatives, since the handling of laboratory equipment and measurement instruments should remain a key aspect in the training of future engineers and scientists. We only mean to convert some exercises to a virtual media, in order to make them accessible to all institutes and to give students access to additional exercises they would not be able to do otherwise.

A characteristic General Physics exercise is the calculation of the numerical value of the acceleration of gravity on the surface of the Earth, using the free fall motion of a steel ball. Taking a real exercise (which can be found at the General Physics laboratory of Universidad de La Laguna) as a model, we generated an interactive diagram from which students can obtain experimental data.

The initial height of the ball can be set manually, and the application returns the corresponding fall time; a limited uniform noise is added to simulate measurement errors, so that data can be processed statistically. In order to increase the appeal of our simulation to young students, we have developed a graphical user interface which emulates the appearance of the real laboratory exercise, rather than just return time data for different height input values.

2. Laboratory exercise

The required equipment for the laboratory exercise, shown in Fig. 1 is:

- A metal base which supports a one-meter long vertical metal bar.
- A contact plate, located at the bottom of the metal bar, and an electromagnet, located at adjustable height at the top, fed by a direct current source.
- A digital chronometer, and a Morse switch which activates it and deactivates the DC source.

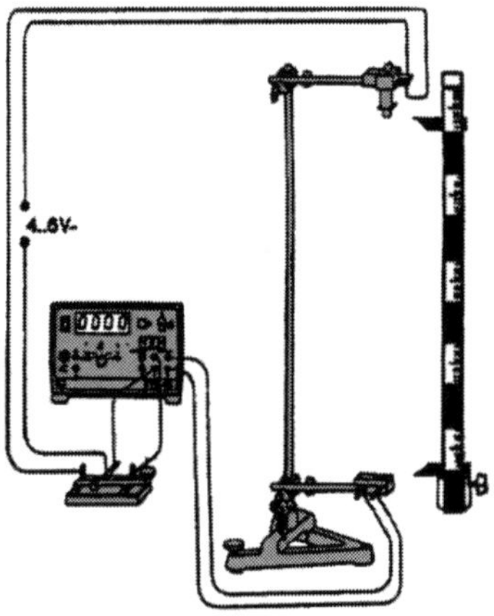

Figure 1. Laboratory equipment required for the exercise.

In this experiment, the steel ball initially hangs motionless from the electromagnet. When the Morse switch is pressed, the current is turned off, the ball falls, and the chronometer starts. When the ball hits the contact plate, the chronometer stops and the fall time in milliseconds is obtained.

The initial height of the ball is named h, and its fall time, t. Considering a free fall motion with negligible air friction, the relation between the travelled space and the fall time is given by $h = \frac{1}{2}gt^2$, where g is the acceleration of gravity. Following this model, the numerical value g can be obtained from the measured values of h and t as

$$g = \frac{2h}{t^2}.$$

At the laboratory, the students set the height of the ball, turns the power source on, and places the ball on the electromagnet. Then, they turn the power off using the Morse switch, which starts the chronometer. When the ball hits the contact plate, the chronometer stops and the students write the fall time down. For each height value h_i, the procedure repeats several times, so as to obtain the mean value of the fall time $\bar{t}_i$, along with its standard deviation. A gravity value can be obtained from these data as

$$g_i = \frac{2h_i}{\bar{t}_i^2}.$$

The student is expected to repeat this process for different height values, in order to obtain a more accurate gravity value: by plotting the values of $2h_i$ versus their corresponding mean time values $\overline{t}_i^2$, the value of g can be obtained as the gradient of the least squares regression line $2h = gt^2$.

3. Simulation

The simulation was programmed in *ActionScript 3.0* [7] and compiled as an *Adobe Flash* [8] file, so that it can be freely executed in most operative systems. The file contains three interactive elements:

- A ball which can be dragged to different heights using the mouse.
- A red button which both starts the execution and resets the system.
- A counter which displays the elapsed time in milliseconds during an execution.

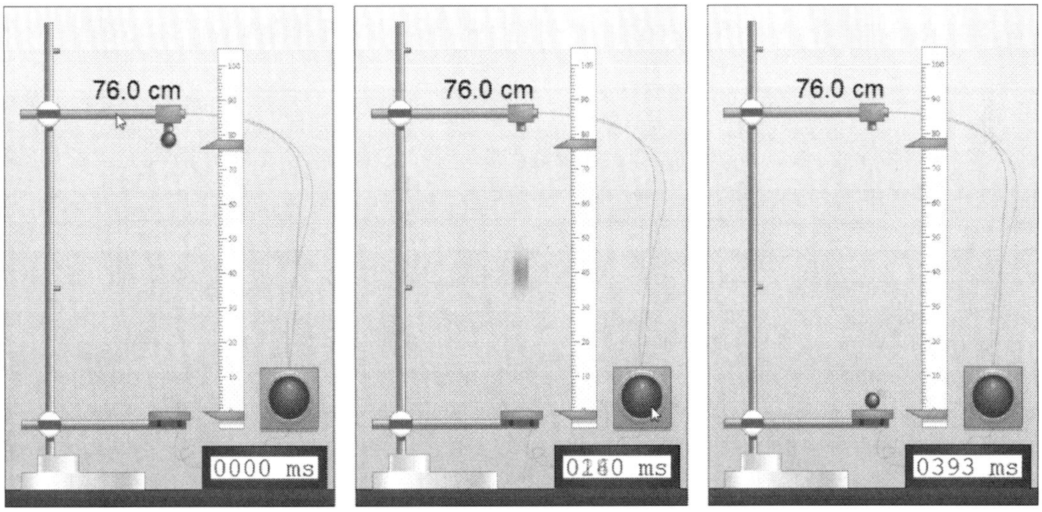

Figure 2. Example of an execution of the simulation.

To create an accurate simulation of the exercise, we required a correspondence between physical and computational scales: the height of the virtual ball can be set between 70 and 310 px (pixels) which correspond to 30 and 90 cm respectively. Therefore, the spatial resolution of our file is

$$\frac{(310-70)\ px}{(90-30)\ cm} = 4\ \frac{px}{cm};$$

the time resolution of our simulation was set manually to 100 fps (frames per second). Using these two data, the acceleration can be recalculated as

$$g = 9.81\ \frac{m}{s^2} \times \frac{4\ px}{10^{-2}\ m} \times \left(\frac{1\ s}{100\ f}\right)^2 = 0.3924\ \frac{px}{f^2}.$$

When the button is pressed, the time counter starts continually displaying the elapsed time, and the starting position of the ball if saved as y_0. For each following frame t, the height of the ball is recalculated as usual,

$$y_t = y_0 - \tfrac{1}{2}gt^2.$$

The counter stops once y_t equals the lowest position. To simulate measurement inaccuracies, a uniform noise of $\pm 10\ ms$ is added to the final time reading.

To improve the appearance of the simulation, procedural motion blurring and physically accurate bouncing have been added to the motion of the ball. These effects however are merely visual and do not affect the calculations at all. An example of an execution of the simulation is shown in Fig. 2.

Students are expected to follow the same steps as during the real laboratory exercise. They must calculate the mean values and standard deviations of the fall time values, and obtain the value of g as the gradient of a least squares regression line. Fig. 3 shows an example of a plot obtained using the simulation file.

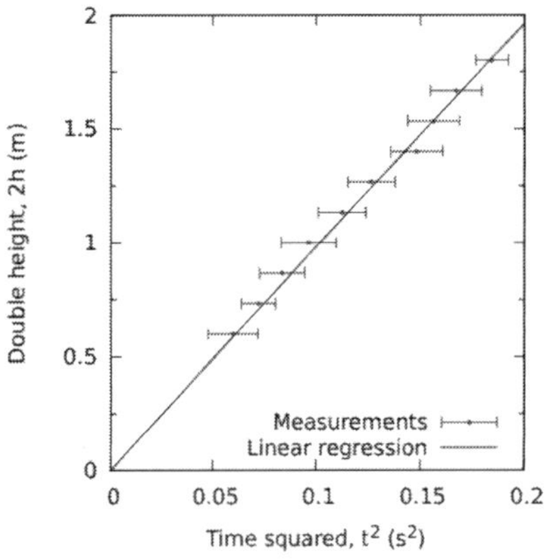

Figure 3. Example of regression line.

4. Conclusion

We have developed a simulation of a General Physics laboratory exercise, at which the value of the acceleration of gravity is calculated from the fall times of a steel ball, initially set at an adjustable height. The exercise has been generated as an *Adobe Flash* file using *ActionScript 3.0* language, so that it can be freely executed under most operative systems. Measurement errors and visual effects have been simulated in order to improve the realism and appearance of the exercise.

This system can be applied to similar laboratory exercises. Future works may include exercises that are too complex to assemble or operate in real life, such as parabolic motions or viscous media which would include a friction factor.

References

[1] Feisel L D and Rosa A J 2005 *Journal of Engineering Education* **94** 121–30
[2] Hofstein A and Lunetta V N 2004 *Science Education* **88** 28–54
[3] Civelek T, Uçar E, Gökçöl O, Üstünel H and Umut İ 2013 *Cypriot Journal of Educational Sciences* **8** 457–71
[4] Ma J and Nickerson J V 2006 *ACM Computing Surveys (CSUR)* **38** 7
[5] Rutten N, van Joolingen W R and van der Veen J T 2012 *Computers and Education* **58** 136–53
[6] Shpresa A, Allegra M and Prenjasi E 2013 *Conf. eLearning and Software for Education* 02 pp 344–49
[7] Adobe Systems Inc 2007 *Programming Actionscript 3.0*
[8] Adobe Systems Inc 2008 *SWF File Format Specification, Version 10*

Journal of Physics: Conference Series **574** (2015) 012069 doi:10.1088/1742-6596/574/1/012069

A modified transfer matrix approach to include Zeeman spin splitting in dispersion relations for coupled quantum wells.

A Hernández-Cabrera and P Aceituno

Departamento de Física, Universidad de La Laguna, La Laguna, 38206-Tenerife, Spain, and Instituto Universitario de Estudios Avanzados (IUdEA) en Física Atómica, Molecular y Fotónica, Universidad de La Laguna, La Laguna, 38206 Tenerife, Spain

E-mail: ajhernan@ull.edu.es

Abstract. In our work, we calculate the dispersion relations for $InGaAs - InAlAs$ based double quantum wells (narrow gap structures). We have developed an improved 4×4 version of the Transfer Matrix Approach, considering contributions from external fields when tunneling through central barrier exists. The transverse electric field is necessary to reach the resonance of electronic levels in asymmetric structures. The in-plane magnetic field induces the Zeeman effect and the spin splitting of the resonant levels. We have also included abrupt barrier effects due to the nature of the interfaces between the above materials.

1. Introduction

The use of spin better than charge in electronic devices is known as spintronics. The low power consumption as well as the slightly dissipative transport are advantages of spin-based electronics that have made it a center of interest in last years. In order to make the most of its applications, the choice of material is a crucial point. Those with a narrow gap but a large Landé factor, which will have a large spin-orbit coupling, seem to be a good option, as $InGaAs - InAlAs$ structures are. To be useful in practice, electronic (or hole) states with different spins must be separated in energy. Moreover, the material must be electrically polarized [1].

In this work we analyze variations of the spin splitting in asymmetric structures, such as asymmetric coupled quantum wells (ACQW), when transverse electric and in plane magnetic fields are applied. We use an improved variant of 8×8 Kane formalism for the dispersion relations in the vicinity of the spin anticrossings together with a 4×4 modified version of the Transfer Matrix Approach . We have included contributions of the Pauli Hamiltonian and its mixing with the effective spìn-orbit interaction, and spin splitting caused by abrupt interfaces.

2. Eigenstate problem

In the parabolic approximation we can write the Schrödinger equation for ACQW in the form [2]:

$$\left(\varepsilon_p^\mu + \frac{\hat{p}_z^2}{2m_\mu} + \varepsilon_c^\mu(z) + \widehat{W}^\mu\right)\Psi^\mu(\mathbf{p},z) = E\Psi^\mu(\mathbf{p},z), \tag{1}$$

for the eigenfunctions $\Psi^{\mu}(\mathbf{p},z)$ and eigenvalues E. Here $\mu = b, w$ denotes de barrier and well regions, respectively. In the above equation ε_p^{μ} is the kinetic energy in the in-plane direction, which includes the effective mass m_{μ}. The $\varepsilon_c^{\mu}(z)$ energy is $U(z)$ in the wells, and $\Delta E_c + U(z)$ in the barriers, where $U(z) \simeq eF_{\perp}z$ for an uniform transverse electric field $F_{\perp}$, and ΔE_c is the band offset for conduction band. Band diagram for ACQW is shown in Fig. 1.

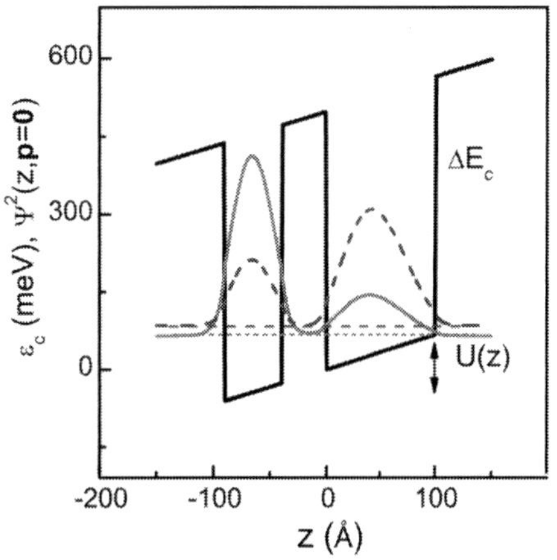

Figure 1. Conduction band diagram for asymmetric coupled quantum well. Horizontal thin lines show electron energy levels and thin curves correspond to squared wave functions close to the resonance.

The magnetic energy is described by $\widehat{W}^{\mu} = \overline{v}^{\mu}\left[\hat{\sigma} \times \mathbf{p}\right]_z + w_H^{\mu}\hat{\sigma}_y$, where $\hat{\sigma}$ is the Pauli matrix. We define the Pauli splitting energy (half of the Zeeman splitting) caused by the magnetic field as $w_H^{\mu} = (\overline{g}^{\mu}/2)\mu_B H$. Here $\overline{g}^{\mu}$ is the effective Landé factor, μ_B is the Bohr magneton, and H is the in plane magnetic field. We can take characteristic spin velocity for each layer as $\overline{v}^{\mu} = eF_{\perp}\hbar/4m_{\mu}\varepsilon_g$, with ε_g the well gap energy.

After some cumbersome algebra [3], we obtain the fundamental solutions of Eq (1):

$$\Psi^{\mu+}(\mathbf{p},z) = \left\{\left[a_{\mu+}A_i(\xi_{\mathbf{p}}^{\mu+}) + b_{\mu+}B_i(\xi_{\mathbf{p}}^{\mu+})\right] + \rho^{\mu-}(\mathbf{p})\left[a_{\mu-}A_i(\xi_{\mathbf{p}}^{\mu-}) + b_{\mu-}B_i(\xi_{\mathbf{p}}^{\mu-})\right]\right\}(1/\sqrt{2})$$
$$\Psi^{\mu-}(\mathbf{p},z) = \left\{\rho^{\mu+}(\mathbf{p})\left[a_{\mu+}A_i(\xi_{\mathbf{p}}^{\mu+}) + b_{\mu+}B_i(\xi_{\mathbf{p}}^{\mu+})\right] + \left[a_{\mu-}A_i(\xi_{\mathbf{p}}^{\mu-}) + b_{\mu-}B_i(\xi_{\mathbf{p}}^{\mu-})\right]\right\}(1/\sqrt{2}), \quad (2)$$

where $\rho^{\mu\pm}(\mathbf{p}) = (\overline{v}^{\mu}p_{\pm} + w_H^{\mu})/iw_{\mathbf{p}}^{\mu}$, with $p_{\pm} = p_x \pm ip_y$, and $w_{\mathbf{p}}^{\mu} = [(\overline{v}^{\mu}p_x + w_H^{\mu})^2 + (\overline{v}^{\mu}p_y)^2]^{1/2}$. In the above equation $A_i(\xi_{\mathbf{p}}^{\mu\pm})$ and $B_i(\xi_{\mathbf{p}}^{\mu\pm})$ are the Airy functions with arguments

$$\xi_{\mathbf{p}}^{\mu\pm} = \frac{z}{l_{\perp}^{\mu}} + \frac{\varepsilon_{\mathbf{p}}^{\mu\pm} - E + \delta_b^{\mu}\Delta E_c}{\varepsilon_{\perp}^{\mu}}, \quad (3)$$

with δ_{μ}^{b} acting as a Kronecker function and $l_{\perp}^{\mu} = \left(\hbar^2/2m_{\mu}eF_{\perp}\right)^{1/3}$, $\varepsilon_{\perp}^{\mu} = \hbar^2/[2m_{\mu}\left(l_{\perp}^{\mu}\right)^2]$, and $\varepsilon_{\mathbf{p}}^{\mu\pm} = \varepsilon_p^{\mu} \pm |w_{\mathbf{p}}^{\mu}|$. Finally, $a_{\mu\pm}$, $b_{\mu\pm}$ are unknown coefficients that we will obtain by means of the boundary conditions, which include abrupt interface parameter $\chi = (2eF_{\perp}\delta + \Delta E_c)/2\varepsilon_g \approx \Delta E_c/2\varepsilon_g$, where δ is the halfwidth of that interface [4].

Now we generate transfer matrices, $M^{\mu}(L_i, E, \mathbf{p})$, which include boundary conditions at interface L_i. To obtain electronic levels for each 2D momentum $\mathbf{p} = (p_x, p_y)$ we introduce a modification of the method used before [3]. The total transfer matrix can be written as:

$$\begin{aligned} S(E,\mathbf{p}) \;=\; & [M^b(L_1,E,\mathbf{p})]^{-1} \cdot M^w(L_1,E,\mathbf{p}) \cdot [M^w(L_2,E,\mathbf{p})]^{-1} \cdot M^b(L_2,E,\mathbf{p}) \cdot \\ & [M^b(L_3,E,\mathbf{p})]^{-1}\; M^w(L_3,E,\mathbf{p})\; [M^w(L_4,E,\mathbf{p})]^{-1}\; M^b(L_4,E,\mathbf{p}). \end{aligned} \quad (4)$$

And the exact solution of the Hamiltonian is obtained from the Wronskyan-like:

$$S_w(E,\mathbf{p}) = S_{11}(E,\mathbf{p}) \cdot S_{33}(E,\mathbf{p}) - S_{31}(E,\mathbf{p}) \cdot S_{13}(E,\mathbf{p}) = 0 \quad (5)$$

Fig. 2 represents $S_w(E,\mathbf{p})$, $S_{11}(E,\mathbf{p})$ and $S_{33}(E,\mathbf{p})$ versus E for a given momentum $\mathbf{p}$. Solutions of Eq. (1) correspond to their roots: $S_w(E,\mathbf{p})$ has four roots for each $\mathbf{p}$ value, corresponding to the two deepest coupled levels of the ACQW, and their respective spin down and spin up sublevels. In order to identify spin orientation, we look for $S_{11}(E,\mathbf{p})$ roots, which correspond to the two coupled levels with spin up. On the other hand, $S_{33}(E,\mathbf{p})$ roots indicate spin down sublevels. Sweeping over a wide range of values of $\mathbf{p}$ we obtain dispersion relations as four paraboloids.

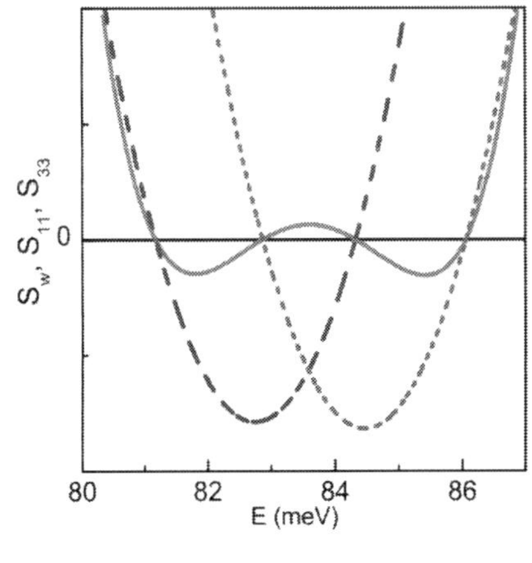

Figure 2. Solutions of the transcendental Eq. (5). $S_w(E,\mathbf{p})$: red solid line. Spin down solutions $S_{33}(E,\mathbf{p})$: blue dashed line. Spin up solutions $S_{11}(E,\mathbf{p})$: green dotted line.

Once energy and coefficients $a_{\mu\sigma}$, $b_{\mu\sigma}$ (Eq. 2) have been obtained, we normalize wave functions for each momentum $\mathbf{p}$. Scheme of Fig. 1 includes the two resonant energy levels and the respective wave functions for $\mathbf{p} = \mathbf{0}$. Spin levels are indistinguishable in this figure.

3. Results

We present here numerical results for $In_{0.53}Ga_{0.47}As - In_{0.52}Al_{0.48}As$-based ACQWs [5] and formed by two $InGaAs$ wells of 50 and 100 Å wide separated by a 40 Å $InAlAs$ barrier. We have applied an electric field of 67 kV/cm to get resonance of the deepest conduction levels. We have used a magnetic field of 0.5 T. We can find for this field an anticrossing between sublevels spin down and up at each electronic level.

The four paraboloids corresponding to the two lower electronic levels after spin splitting are represented in Fig. 3, where spin anticrossing is visible at the right of the two sets of paraboloids. Spin anticrossing occurs around $p_x/p_0 = 30$ for $p_y/p_0 = 0$, where $p_0 = m_w\overline{v}^w$. Note that paraboloids have not rotation symmetry because the magnetic energy $w_{\mathbf{p}}^{\mu}$, which is an essential part of the Airy functions argument, induces a breaking of the p_x momentum symmetry and, as a consequence, the spin anticrossing.

In Fig. 4 we present a $p_y = 0$ section of the former figure where we can get a more accurate 2D description of the anticrossing p_x position and energy splitting.

Interestingly, interface contributions are opposite to the intrinsic spin-orbit effect. The mechanisms contributing to mix Pauli contribution with the other kinds of spin-orbit contributions (low magnetic field or interface effects) are different. As a consequence, numerical

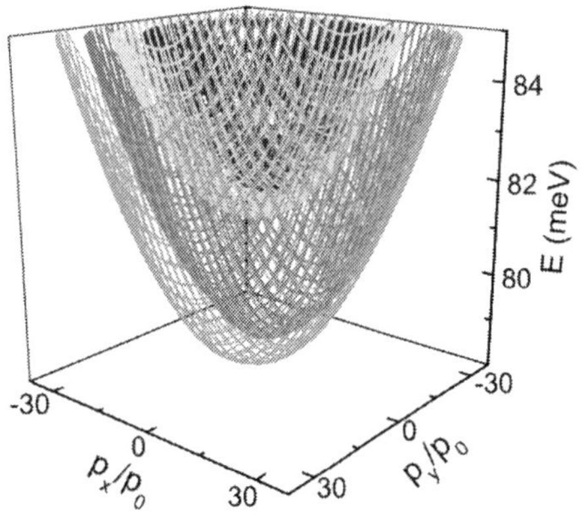

Figure 3. Energy versus p_x and p_y. From bottom to top: first level, spin down and spin up, second level, spin down and spin up.

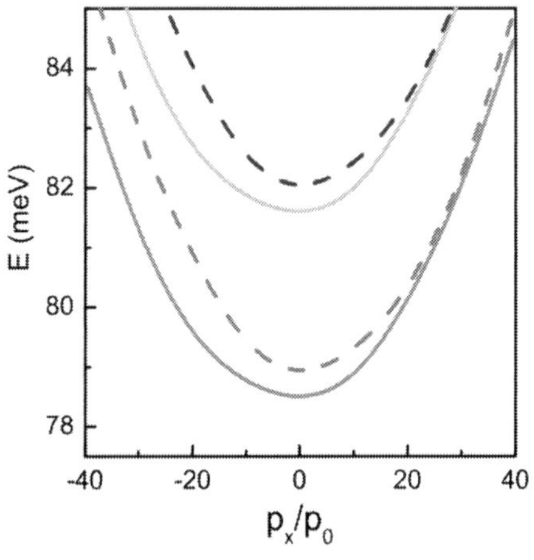

Figure 4. Energy versus p_x for $p_y = 0$. Solid line: spin down, dashed line: spin up.

calculations for $InGaAs - InAlAs$ structures, subjected to very low in-plane magnetic fields, lead to pronounced magnetoinduced changes in the energy dispersion relations, This effect is particularly noticeable for spin anticrossings.

4. Conclusions

In this work we have presented an improved modification of the transfer matrix approximation. So we have raised 4×4 matrices to include spin sublevels from the Pauli (Zeeman) splitting caused by a magnetic field and the effect of abrupt interfaces. Although we have limited ourselves to include these two contributions, the method is capable of handling any other contribution to the spin splitting. We have applied the method to narrow-gap ACQWs under transverse electric and in-plane magnetic fields. We have used non-symmetric contour conditions, caused by the asymmetry of the structure, to solve the eigenvalue problem. The present version of the transfer matrix, based on the Wronskian-like of wave functions space, has proven to be a highly efficient and effective tool to study the physical properties of any layered material when subjected to a wide range of perturbations. In particular, the method is especially useful in analyzing spin peculiarities.

References

[1] Bernevig B A and Zhang S 2006 *IBM J. Res. and Dev.* **50** 1. Kikkawa J M and Awschalom D D 1998 *Phys. Rev. Lett.*. **80** 4313

[2] Hernández-Cabrera A, Aceituno P and Vasko F T 2008 *Journal of Luminescence* **128** 862. Hernández-Cabrera A, Aceituno P and Vasko F T 2006 *Phys. Rev.* B **74** 035330

[3] Hernández-Cabrera A and Aceituno P 2014 *Indian J. Phys* In press.

[4] Vasko F T and Kuznetsov A V 1999 *Electronic States and Optical Transitions in Semiconductor Heterostructures* (New York: Springer)

[5] Vurgaftman I, Meyer J R and Ram-Mohan L R 2001 *J. Appl. Phys* **89** 5815

Modeling and optimizing the design of matrix treatments in carbonate reservoirs with self-diverting acid systems

G T Bulgakova[1], R Ya Kharisov[2], A R Sharifullin[3], A V Pestrikov[4]

[1]Ufa State Aviation Technical University, Ufa, Russian Federation
[2]RN-UfaNIPIneft Limited, Ufa, Russian Federation
[3]BashNIPIneft Limited, Ufa, Russian Federation
[4]Gubkin Russian State University of Oil and Gas, Moscow, Russian Federation

E-mail: math@mail.rb.ru

Abstract. Application of a self-diverting-acid based on viscoelastic surfactant (SDVA) is a promising technology for improving the efficacy of acid treatment in oil and gas–bearing carbonate reservoirs. In this study, we present a mathematical model for assessing SDVA flow and reaction with carbonate rock using the SDVA rheological characteristics. The model calculates the technological parameters for acidizing operations and the prediction of well productivity after acid treatment, in addition to technical and economic optimization of the acidizing process by modeling different acid treatment options with varying volumes, injection rates, process fluids stages and initial economic scenarios.

Introduction

Acidizing carbonate reservoirs is the most common method for bottom-hole zone chemical treatments which are used to enhance oil recovery. Despite the years of experience and substantial volume of studies aimed at improving and enhancing efficiency of this method, a significant number of treatments do not provide positive results. One of the current problems in oil production is that acid treatments are not very efficient, particularly in the mature fields. This could be solved using materials that are capable of transformation and ensuring acid stimulation control [1-3]. Using such methods, more effective materials have been found that are capable of gelation and are resistant to acid.

An example of these improvements in acid viscosity management and diversion is an in-situ gelled acid. The main advantage of this, compared to the conventional acid stimulation, is the minimum number of steps necessary for acid diversion, since it allows simultaneous stimulation and diversion. The gelling material, based on viscoelastic surfactants, is injected and it clogs or reduces the permeability of stimulated reservoir zones. The gelling agent, injected after an acid solution, penetrates into the low permeability oil-saturated zone (as high-permeability zones are generally clogged with gel) reacts with the rocks, and forms new pore channels and expands existing ones. As a result, the permeability of low-permeability zones increases. Thus, self-diverting acid based on viscoelastic surfactant (SDVA) is seen as a promising approach to improving the efficacy of acid treatments. The self-diverting effect is based on the surfactant's ability to form a viscoelastic gel during the acid reaction with carbonate rocks. The resulting gel creates an effective local diversion of new acid portions to untreated low-permeability zones. After the treatment, due to the susceptibility of cylindrical micelles to hydrocarbons, a deflecting gel is destroyed by oil and easily flows out of the wellbore. Thus, the application of a self-diverting-acid based on viscoelastic surfactant provides an even treatment for the entire productive interval in oil reservoirs.

When designing acid treatments with diverters, it is essential to have justified calculations for agent injection rates, injected acid volumes, diverter volumes, the number of injection cycles, the number of diversion stages, and the volume of injected fluid at each stage to enable well performance prediction after treatment. It is also important to calculate the expected economic effect of well acidization. When designing large-scale selective acid treatments, the expected effects should be obtained from models of basic physical and chemical processes using dedicated software products [4]. For wells with a heterogeneous permeability profile, the problem of acid placement in target zones cannot be properly resolved without a numerical simulation. In addition, numerical simulations allow technical-economic optimization of well treatments by modeling scenarios with different volumes, stages of process fluids and initial economic cases.

SDVA rheological model

The rheological tests show that SDVA viscosity is described by a complex function of the spent acid concentration and the shear rate [5]. SDVA viscosity significantly increases with acid spending in a chemical reaction, which leads to flow diversion and, consequently, to more uniform wormhole propagation in carbonates. According to the results of rheological testing, the dependence of dynamic viscosity of the SDVA upon the acid concentration and shear rate [5]:

$$\mu_a = K\dot{\gamma}^{(n-1)}\left[1+\left(\frac{\mu_{\max}}{\mu_0}-1\right)\exp\left(\frac{-\alpha(c-c_{\max})^2}{c(1-c)}\right)\right] \tag{1}$$

where μ_0 is the basic viscosity (viscosity at acid mass concentration c = 0.12 or 0), $\mu_{\max}$ corresponds to the maximum viscosity, defined relative to the reference value because of gelation, $c_{\max}$ is the acid concentration at maximum viscosity, and α measures the acid concentration range for gel formation, which corresponds to the reciprocal variance (width) of the correlation between viscosity and acid concentration. With an α value increase, the range of gelation decreases. In the above equation, K represents fluid consistency (Pa s). A higher viscosity corresponds to a higher consistency, $\dot{\gamma}$ is the shear rate, s^{-1}, n is an indicator of non-Newtonian fluid behaviour [6]. It is necessary to evaluate the following parameters of the model: K, n, $c_{\max}$, α.

The identification of model parameters is related to the solution of inverse problems that are ill-posed [7]. Additional assumptions can turn ill-posed problems into well-posed problems. The use of supplementary (a priori) information of a qualitative nature yields the Ivanov method of quasisolution [8]. In this case, as the inverse problem solution, the parameter set K, n, $c_{\max}$, α. gives a minimum of the residual function (equation 2) that can be considered as a correct solution for the described problem. The model was adjusted against the measured viscosity $\mu^{\exp}$ and calculated viscosity μ^{calc} minimising the corresponding function

$$J(K,n,c_{\max},\alpha) = \sum_i(\mu_i^{\exp} - \mu_i^{calc})^2 \rightarrow \min . \tag{2}$$

The rheological model, describing the apparent SDVA viscosity behavior versus acid concentration and shear rate, is taken as a mathematical model for describing well acidizing with SDVA use in carbonate reservoirs [4]. The mathematical model optimizes well acidizing reactions using SDVA in

Journal of Physics: Conference Series **574** (2015) 012070 doi:10.1088/1742-6596/574/1/012070

carbonate reservoirs and evaluates the effectiveness of acid treatment under specific reservoir conditions.

Flow diversion efficiency

Maintaining acid distribution over the whole producing interval is a major objective of acid treatments in carbonate oil and gas bearing reservoirs. Acid treatment modeling for various types of diverters showed that the diversion efficiency depends not only on properties of diverters, but on the heterogeneity in vertical permeability of stimulated reservoirs. Calculations show that if maximum well production increment is selected as an assessment criterion, an optimal treatment of high-permeability zones is the necessary condition. On the other hand, from the standpoint of effective reservoir depletion, the drainage of low-permeability layers with minimal impact on high-permeability ones is required. If the purpose of acid treatment optimization is efficient reservoir depletion, then, as a criterion of reservoir profile conformance after treatments, the design optimization algorithm takes square dispersion D_q, defined as the specific flow rate standard deviation from the average Q_{av}

$$D_q = \sum_i^n \left(\frac{Q_{av}}{\sum_i^n h_i} - \frac{Q_i}{h_i} \right)^2 \left(\frac{n^2 Q^2}{\sum_i^n h_i^2} \right)^{-1} \tag{3}$$

where Q - total liquid rate, n - number of layers, Q_i, h_i - rate and thickness of i–th layer, respectively. In this case, minimum dispersion indicates a uniform zone treatment by hydrochloric acid. High dispersion points to an uneven reserve depletion or uneven sweep efficiency by waterflooding.

A promising alternative diversion technique is the application of materials capable of altering their properties (in particular, viscosity) directly during the process of acid treatment, thereby providing a certain control over the acidizing processes. An example of such an approach is acidizing with self-diverting acid that is based on viscoelastic surfactant (SDVA). Using the simulation, we compared the following different acid treatment designs for a layered-heterogeneous reservoir: the base case design without diverters (1), the design with viscous liquid application - INERA diverter (2) and a design with SDVA application (3). The total volume of working fluid in all cases corresponded to 36 m^3. When calculating designs (2) and (3) the volume of conventional acid composition was taken as 15 m^3. The volume of inhibited hydrochloric acid in design (1) corresponded to 30 m^3. Table 1 shows the distribution of initial permeability and porosity.

The dispersion, calculated for acidizing designs 1-3, is shown in Figures 1 and 2. Figure 1 shows the dispersion vs. agent injection rate while acidizing without diverters and with diverters INERA and SDVA (designs 1-3 of acidizing, respectively).

The calculations indicate that design (3), i.e. acidizing with SDVA, results in the smallest dispersion. When injecting the INER diverter at an initial rate of 0.1 m^3/min, the dispersion was the same as the design using SDVA. However, after some reduction, the dispersion flattens out later on. This can be explained by the nonlinear relationship between viscosity and injection rate. When the rate increases, the INER viscosity decreases and consequently the diverting properties calm down. When only inhibited acid is injected, the dispersion is higher than the one calculated in either design (3) with SDVA or design (2) with INERA. However, when the injection rate increases, the acid distribution over the interval for a zone increases and well dispersion decreases but remains higher than the dispersion when SDVA is used as a diverter.

Assessment of acidizing efficiency

Well performance after acid treatment is estimated using the Hawkins equation, which defines the equivalent formation damage skin factor s:

$$s = k_0 \int_{r_w}^{r_{wh}} \frac{dr}{r\,k(r)} - \ln \frac{r_{wh}}{r_w} \tag{4}$$

k_0 – untouched zone permeability around the wellbore, $k(r)$ – stimulated zone permeability around the wellbore, r_w – wellbore radius, r_{wh} – stimulated zone radius. For the selected designs we calculated the post-treatment skin values. Figure 2 shows skin values for low-permeability layers versus specific agent injection rate, used in the above mentioned designs. Specific agent consumption was calculated per a length unit of the layer thickness. Initial skin factor was 16. The calculations revealed that after well treatments with SDVA, skin of low-permeability layers decreased to negative values, while in designs (1) and (2), the post-treatment skin in low-permeability layers remained positive. The simulation results proved that SDVA application is highly efficient for acid treatments in carbonate reservoirs.

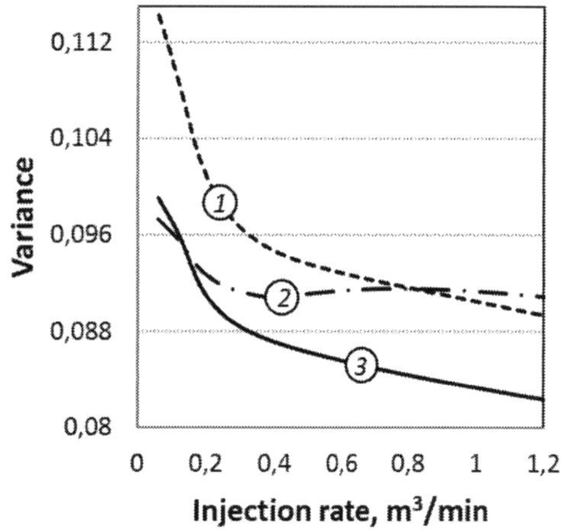

Figure 1. Well dispersion vs. agent injection rate

1 - without diverters
2 - INERA diverter
3 - SDVA

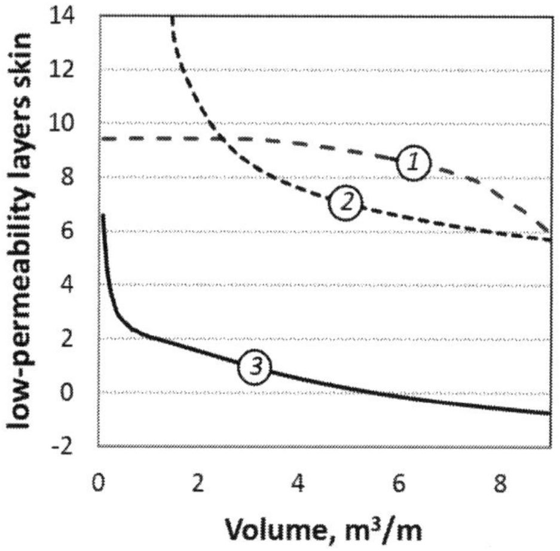

Figure 2. Skin of low-permeability layer vs. injection specific volume

1 - INERA diverter
2 - without diverters
3 - SDVA

Optimization steps

The optimization process is divided into several steps, corresponding to the selected measures. At each stage an optimal ratio between the volumes of acidizing stages and the diverter stages is determined. For each stage, an optimum injection rate is calculated, as well as cost effective volumes. The sequence of stages is fixed, however, one or more stages can be excluded from the overall sequence.

SDVA injection tests on cores have shown that acid treatment can be the most efficient process under certain optimal conditions [9]. Such conditions are described by an optimal Damköhler number. Under the optimal Damköhler number, several long dissolution channels (wormholes) can be created, penetrating into formations and providing the best connectivity between formation and wellbore. In the case of radial SDVA flow from the well into the formation, optimization ensures the optimum Damköhler number with regard to wormholes.

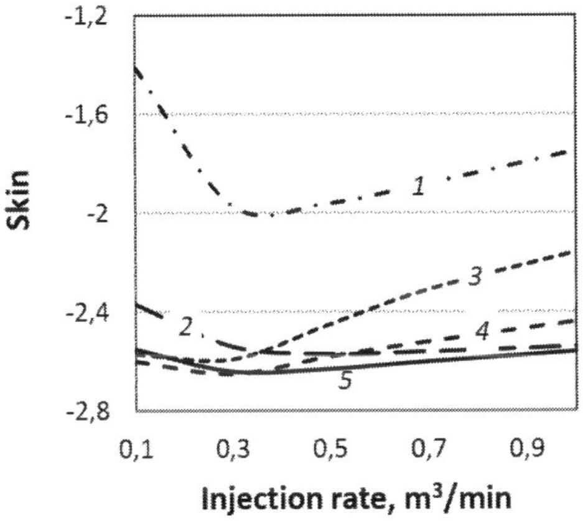

Figure 3. Skin vs. SDVA injection rate at different acid injection rates *u*
1 - u = 0.1 m³/min; 2 - u = 0.3 m³/min
3 - u = 0.5 m³/min; 4 - u = 0.7 m³/min
5 - u = 1 m³/min

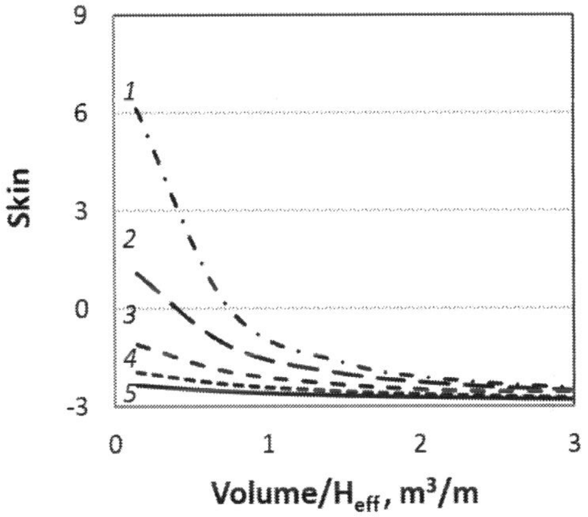

Figure 4. Skin vs. SDVA volume at different acid injection volumes V_a
1 – V_a = 1 m³; 2 – V_a = 5 m³
3 – V_a = 10 m³; 4 – V_a = 15 m³
5 – V_a = 20 m³

To investigate the influence of SDVA injection rate and volume on the inflow profile conformance and acid treatment efficiency, the mathematical model was applied to the design cases for SDVA and inhibited acid injection. As Figure 3 shows, the skin changes versus the SDVA injection rates at different inhibited hydrochloric acid injection rates are non-monotone. The optimal range for the SDVA injection rate is from 0.25 m³/min to 0.4 m³/min. The increase in the acid injection rate increases treatment efficiency.

Figure 4 shows the relationship between the skin and SDVA injection volumes at different acid injection volumes. The dynamic of the skin slows down with the increase in SDVA volume, and respectively decreases in conjunction with an increase in the injected acid volumes. Analysis of the results indicates that the optimum ratio of SDVA volume V_{SDVA} to injected acid volume V_a is related to the interval of $V_{SDVA} = (0.9 - 1.1)V_a$. The same values were obtained earlier [4] when calculating volumes of high viscosity liquid diverters. Analysis of previous well acidizing operations in carbonate reservoirs confirms that the unjustified increase in diverter volumes reduces the whole treatment efficiency.

Conclusions

A mathematical model can be used to effectively plan, design and optimize acidizing operations in carbonate reservoirs. In order to efficiently design acid treatment processes, an accurate simulation of complex acid dissolution and working fluid distribution in carbonate reservoirs is required.

Mathematical modeling also allows calculation of a preliminary injection program to achieve a negative skin, a number of hydrochloric acid injection cycles, a number of diversion stages using self-diverting acid systems, fluid injection volumes and rates at each stage.

This simulation experiment aimed to investigate the effect of SDVA injection rate and volume on injectivity profile conformance and acid treatment performance. The results revealed that combined use of SDVA and conventional acid composition reduces the injectivity profile non-conformance, compared to simple acid and nonlinear viscous liquids such as emulsions. In this case, the production increment has a nonlinear correlation with SDVA and acidic volumes, and their optimal volumes can be determined in the process of economic optimization. A criterion for determining the optimum ratio of SDVA and acid volumes was established, which ensures maximum inflow profile conformance under maximum acid treatment efficiency. The numerical studies revealed a need for simultaneous optimization on injection rates and volumes for both SDVA and consequent conventional HCl composition.

This study shows that mathematical modeling is useful for predicting the economic effects of acid treatments in carbonate reservoirs. Further use of these mathematical techniques in the future will make acidizing operations more efficient and reduce costs.

Acknowledgments
This study was supported by the Ministry of Education and Science of the Russian Federation as the base part of the state orders for higher educational institutions and the Russian Foundation for Basic Research (Project 14-01-97012 r_povolzh'e_a).

References

[1] Nasr-El-Din H A, Al-Mohammad A M, Al-Aamri A D, Al-Fahad M A and Chang F F **2009** *SPE Prod. Oper.* **24** 107–116
[2] Yu M, Mahmoud M A, Nasr-El-Din H A **2011** *SPE J.* **16** 993–1001
[3] Shipilov A I, Krutikhin E V, Kudrevatykh N V and Mikov A I **2012** *Neftyanoe khozyaystvo (Oil Industry)* **2** 80–83 (in Russian)
[4] Bulgakova G T, Kharisov R Ya, Sharifullin A R, Pestrikov A V **2011** *Proc. of the Society of Petroleum Engineers (SPE) European Formation Damage Conference* (Noordwijk, The Netherlands, June 7-10, 2011) (SPE 143959)
[5] Bulgakova G T, Kharisov R Ya, Pestrikov A V, Sharifulin A R **2014** *Energy Fuels* **28** 1674−85
[6] Rao M A 2007 *Rheology of Fluid and Semisolid Foods: Principles and Applications* (New York: Springer)
[7] Tikhonov A N, Arsenin V Y 1977 *Solutions of Ill Posed Problems* (New York: John Wiley and Sons)
[8] Ivanov V K, Vasin V V, Tanana V P 2002 *Theory of Linear Ill- Posed Problems and Its Applications* (Netherlands, Leiden: Brill Academic Publishers)
[9] Fredd C N, Fogler H S 1998 *AIChE J.* **44** 1933–49

IC-MSQUARE 2014 IOP Publishing
Journal of Physics: Conference Series **574** (2015) 012071 doi:10.1088/1742-6596/574/1/012071

Asymptotic saturation of mode frequency for the magnetorotational instability in resistive plasmas

F. E. M. Silveira

Centro de Ciências Naturais e Humanas, Universidade Federal do ABC,
Rua Santa Adélia, 166, Bairro Bangu, CEP 09.210-170, Santo André, SP, Brazil

E-mail: francisco.silveira@ufabc.edu.br

Abstract. Resistive effects are investigated on the magnetorotational instability. The main modifications with respect to the approach to the problem in the ideal limit are the introduction of the diffusive term in the induction equation and of the decay time T of the velocity field perpendicular to the magnetic field in the dispersion relation. As a result, when resistive effects are strong, it is shown that the instability condition in the ideal limit is corrected by the coefficient $\kappa^2 T^2$, where κ denotes the epicyclic frequency. The proposed formulation is then applied to the description of the instability for a Keplerian disk and it is found that the characteristic frequency of the unstable modes saturates asymptotically to a finite value which does not depend on the distance from the central object and is fully determined by the resistivity of the disk.

When a plasma rotates differentially about a fixed center, on a plane perpendicular to a magnetic field, the coupling of the current density flowing in the fluid with the field (the Lorentz force density) may be disruptive. Such a process is commonly referred to as the magnetorotational instability [1, 2]. This phenomenon has been intensively studied in recent years, in connection with the description of the dynamics of accretion disks [3, 4, 5, 6]. These systems are astrophysical structures which are made up of ionized gases and orbit massive objects like black holes and neutron stars. Perhaps, the simplest approach to the problem is to consider an ideal plasma (its resistivity is negligible) subjected to a small perturbation parallel to the equilibrium magnetic field. In this case, it is widely known that the condition for the magnetorotational instability to occur in the neighborhood of a given point over the disk depends on the distance of the point from the central object. However, it can be shown that such a result implies the characteristic frequency of the unstable modes is unbounded. In this work, we derive the condition for the magnetorotational instability to occur in a plasma with a finite resistivity. As a consequence, we find that the characteristic frequency of the unstable modes saturates asymptotically to a finite value which does not depend on the distance from the central object and is fully determined by the resistivity of the disk.

Let us start by considering an incompressible fluid with (constant and uniform) mass density ρ_0. In this case, the principle of matter conservation (the continuity equation) implies the velocity field $\vec{V}$ is divergenceless,

$$\nabla \cdot \vec{V} = 0. \tag{1}$$

Next, by supposing that the fluid is an (electrically neutral) inviscid plasma, subjected to a magnetic field $\vec{B}$, the time evolution of $\vec{V}$ will be governed by Euler's equation,

$$\frac{\partial \vec{V}}{\partial t} + \left(\vec{V} \cdot \nabla\right)\vec{V} = -\frac{\nabla P}{\rho_0} + \frac{\left(\nabla \times \vec{B}\right) \times \vec{B}}{\mu_0 \rho_0}, \tag{2}$$

where P denotes the hydrodynamic pressure and use has been made of Ampère's law (in the hydromagnetic approximation: the displacement current is negligible with respect to the conduction current in the Ampère-Maxwell law), with μ_0 standing for the vacuum magnetic permeability (a non-magnetic medium is assumed). Further, the absence of magnetic monopoles implies

$$\nabla \cdot \vec{B} = 0. \tag{3}$$

Finally, the time evolution of $\vec{B}$ obeys Faraday's law,

$$\frac{\partial \vec{B}}{\partial t} = \nabla \times \left(\vec{V} \times \vec{B}\right) + \frac{\eta}{\mu_0}\nabla^2 \vec{B}, \tag{4}$$

where use has been made of the standard form of Ohm's law for a plasma with finite resistivity η. Eqs. (1) to (4) are the basic equations for the considerations to follow.

We suppose now that the plasma rotates differentially about a fixed center, on a plane perpendicular to a constant and uniform magnetic field. By adopting cylindrical coordinates (r, θ, z), such hypotheses may be expressed through the relations $\vec{V} = \hat{\theta} r \Omega$ and $\vec{B} = \hat{z} B_0$, where $\Omega = \Omega(r)$ and $B_0 =$ constant. In this case, we see at once that Eqs. (1), (3), and (4) are automatically satisfied, and that Eq. (2) implies

$$\frac{1}{\rho_0}\frac{dP}{dr} = r\Omega^2. \tag{5}$$

Eq. (5) describes an equilibrium configuration (actually, a stationary state) of the fluid, for which cylindrical isobaric surfaces are supported by a centrifugal force density.

Let us consider small perturbations $f \sim \exp(\gamma t + \imath kz)$ (expressed in terms of pure Fourier mode decompositions) for the equilibrium quantities F, $F \to F + f$, such that $\mathcal{O}\left(f^2\right), \mathcal{O}\left(f/r^2\right) \sim 0$. This means that we regard linearized disturbances for the equilibrium configuration in the neighborhood of a given radial coordinate, to the first order in $1/r$ (only first order curvature terms are kept). In this case, Eq. (1) leads to

$$\frac{v_r}{r} + \imath k v_z = 0, \tag{6}$$

it follows from Eq. (2) that

$$\begin{aligned} \gamma v_r - 2\Omega v_\theta &= \imath \frac{k B_0}{\mu_0 \rho_0} b_r, \\ \gamma v_\theta + \left(2\Omega + r\Omega'\right) v_r &= \imath \frac{k B_0}{\mu_0 \rho_0} b_\theta, \\ \gamma v_z &= -\imath \frac{k}{\rho_0} p, \end{aligned} \tag{7}$$

Eq. (3) leads to

$$\frac{b_r}{r} + \imath k b_z = 0, \tag{8}$$

and it follows from Eq. (4) that

$$\begin{aligned} \left(\gamma + \frac{\eta k^2}{\mu_0}\right) b_r &= \imath k B_0 v_r, \\ \left(\gamma + \frac{\eta k^2}{\mu_0}\right) b_\theta - r\Omega' b_r &= \imath k B_0 v_\theta, \\ \left(\gamma + \frac{\eta k^2}{\mu_0}\right) b_z &= \imath k B_0 v_z, \end{aligned} \tag{9}$$

where the prime denotes a first order total derivative with respect to r.

Eqs. (6) to (9) seem to show that we have eight equations for seven unknowns. However, with the help of Eqs. (6) and (8), it can be easily checked that the first and third of Eqs. (9) are linearly dependent. This means that, actually, we have seven linearly independent equations for seven unknows. By choosing the third of Eqs. (9) to be a spurious equation, we may determine p in terms of v_z from the third of Eqs. (7), v_z in terms of v_r from Eq. (6), and b_z in terms of b_r from Eq. (8). As a result, we are left with the determination of the subset of four unknowns v_r, v_θ, b_r, and b_θ from the first two of both Eqs. (7) and (9) to fix the whole set of seven perturbative quantities. From the physical perspective, the relevant quantities are the non-trivial solutions for the whole set, which are obtained by requiring that the determinant of the coefficient matrix for the subset vanishes. In this case, the (in principle, complex) time rate γ and (definitely, real) wavenumber k must satisfy the dispersion relation

$$\left(\gamma + \omega_{\mathrm{A}}^2 T\right)^2 \gamma^2 + \left[\kappa^2 \left(\gamma + \omega_{\mathrm{A}}^2 T\right) + 2\omega_{\mathrm{A}}^2 \gamma\right] \left(\gamma + \omega_{\mathrm{A}}^2 T\right) + \left(\omega_{\mathrm{A}}^2 - 2q\Omega^2\right) \omega_{\mathrm{A}}^2 = 0, \tag{10}$$

where we have introduced the usual epicyclic frequency κ, shear parameter q, Alfvén frequency ω_{A}, and decay time T of the velocity field component flowing on the polar plane perpendicular to the externally applied axial magnetic field, which are defined through the formulae

$$\kappa^2 = 2\Omega\left(2\Omega + r\Omega'\right), \qquad q\Omega = -r\Omega', \qquad \omega_{\mathrm{A}}^2 = \frac{k^2 B_0^2}{\mu_0 \rho_0}, \qquad \text{and} \qquad T = \frac{\eta \rho_0}{B_0^2}. \tag{11}$$

The fourth of Eqs. (11) shows that η is contained in T. This means that weak and strong resistive effects on the perturbed equilibrium may be investigated by examining the limiting functional forms acquired by Eq. (10) as the result of assuming small and large values attained by T, respectively.

When resistive effects are weak, Eq. (10) approaches

$$\gamma^4 + \left(\kappa^2 + 2\omega_{\mathrm{A}}^2\right)\gamma^2 + \left(\omega_{\mathrm{A}}^2 - 2q\Omega^2\right)\omega_{\mathrm{A}}^2 = 0. \tag{12}$$

We see that Eq. (12) has a root $\gamma > 0$ if the condition

$$\omega_{\mathrm{A}}^2 < 2q\Omega^2 \tag{13}$$

is satisfied, provided that $q > 0$. Result (13) is nothing but the usual condition for the ideal (the plasma resistivity is negligible) magnetorotational instability to occur and γ denotes its standard growth rate. An important application of result (13) in astrophysics concerns the description of the instability of an accretion disk. By assuming that its differential rotation about a central object of mass M is of gravitational origin, one may easily check that

$$\Omega^2 = \frac{GM}{r^3}, \tag{14}$$

where G stands for Newton's gravitational constant. Eq. (14) is nothing but Kepler's third law and the system is known as a Keplerian disk. Given Eq. (14), condition (13) shows that the wavelength $\lambda = 2\pi/k$ of an unstable mode exhibits a lower limit,

$$\lambda > \frac{2\pi B_0/\Omega}{\sqrt{3\mu_0\rho_0}}, \tag{15}$$

for a Keplerian disk. However, according to Eq. (14), limit (15) is unbounded. Actually, as the central object is approached, more and more monochromatic unstable modes fit within the disk (thickness). Such a result is physically unacceptable because it states that as the radial coordinate decreases, the frequency of the unstable modes increases with no limit. Let us see what is the situation if a finite plasma resistivity is introduced into the problem.

When resistive effects are strong, Eq. (10) approaches

$$\left(\omega_{\rm A}^2 T^2\right)\gamma^2 + \left(2\omega_{\rm A}^2 T\right)\gamma + \left[\left(\kappa^2 T^2\right)\omega_{\rm A}^2 - 2q\Omega^2\right] = 0, \tag{16}$$

such that Eq. (16) has a positive root, provided that the condition

$$\left(\kappa^2 T^2\right)\omega_{\rm A}^2 < 2q\Omega^2 \tag{17}$$

is satisfied for some also positive shear parameter. Quite interestingly, Eq. (14) shows that the lower limit for the wavelength of an unstable mode in a Keplerian disk is bounded,

$$\lambda > \frac{2\pi\eta/B_0}{\sqrt{3\mu_0/\rho_0}}, \tag{18}$$

because it does not depend on the radial coordinate. This means that the frequency of the unstable modes saturates asymptotically to a finite value, which is fully determined by the resistivity of the disk.

Resistive effects have been investigated on the magnetorotational instability. The fundamental modifications with respect to the approach to the problem in the ideal limit have been: *(1)* the introduction of the diffusive term, proportional to η/μ_0, in the induction equation (4) and *(2)* the identification $\eta k^2/\mu_0 = \omega_{\rm A}^2 T$ in the general dispersion relation (10). As a result, when resistive effects are strong, it has been shown that the instability condition in the ideal limit is corrected by the coefficient $\kappa^2 T^2$, see Eq. (17). The proposed formulation has been then applied to the description of the instability for a Keplerian disk and it has been found that the characteristic frequency of the unstable modes saturates asymptotically to a finite value which does not depend on the distance from the central object and is fully determined by the resistivity of the disk, see Eq. (18).

References

[1] Chandrasekhar S 1981 *Hydrodynamic and Hydromagnetic Stability* (New York: Dover)
[2] Velikhov E P 1959 *Soviet Physics JETP* **36** 995
[3] Liu W, Goodman J, Herron I and Ji H 2006 *Phys. Rev. E* **74** 056302
[4] Pessah M E and Chan C-K 2008 *Astrophys. J.* **684** 498
[5] Kunz M W and Lesur G 2013 *Mon. Not. R. Astron. Soc.* **434** 2295
[6] Silveira F E M and Galvão R M O 2013 *Phys. Plasmas* **20** 082126

Simulating CC and MLO compressions with the Surface Evolver

Marcelo Zanchetta do Nascimento[1], **Valério Ramos Batista**[2]
[1]*FACOM-UFU, Av. João Naves de Ávila 2121, Bl.A, 38400-902 Uberlândia-MG, Brazil*
[2]*CMCC-UFABC, R. Sta. Adélia 166, Bl.B, 09210-170 St. André-SP, Brazil*

E-mail: nascimento@facom.ufu.br

Abstract. Mammographies are X-ray images of the breast under external compressions called Craniocaudal (CC) and Mediolateral Oblique (MLO). Together they increase the chances of detecting cancer but the breast is shown in strongly deformed shapes. Cancer location is highly uncertain for the surgery and so the breast is commonly taken out entirely, a serious trauma for the patient. In this paper we present a fully virtual mammography procedure that faithfully reproduces all shapes of the breast and in its inside tracks the cancer at any step. The cancer is then precisely located for the surgery and can be removed through a small incision. So the whole structure is preserved and cured as an integral benefit to the patient.

The 2nd author dedicates this work to his wife Clarice.

1. Introduction

The mammography procedure consists of taking X-ray images of the breast. They can reveal the presence of nodules of tissues. Certain nodules appear as a reaction of the body against instability in health. They are called tumours, which can either stall and be harmless or spread and attack other cells. The first case is called benign and the second malignant, namely *cancer*.

Nodules have a hard consistency but can be pretty small. An X-ray image should capture them with the best distinctness, which also helps classify them as Normal, Benign and Malignant. Because of that the breast is submitted to external compressions of the mammographer called Craniocaudal (CC) and Mediolateral Oblique (MLO).

These two different views increase the chances of detecting and classifying such nodules, but the breast is shown in strongly deformed shapes. Cancer location is highly uncertain for the surgery, and so the breast is commonly taken out entirely, a serious trauma for the patient.

If cancer were located precisely, it could be removed through a small incision. After some slight corrections by plastic surgery the whole breast would remain preserved. As a matter of fact, this is already possible through cutting-edge technologies like tomosynthesis [1, 2] and high-tech goggles for cancer detection [3].

There are may articles about these modern techniques but we cited just three for the sake of brevity. However, such techniques are quite expensive and still under improvement. Hence very few women can afford them in developing countries. Magnetic resonance could be a less dear alternative, but unfortunately not in poor countries yet.

The cheapest techniques are ultrasound and mammography. Both deform the shape of the breast considerably, but the latter has an advantage: not only the breast structure but also the mammography procedure can be modelled and simulated together in a totally virtual environment. Some related to softwares have been developed exactly with the purpose of locating nodules for the surgery. We cite [4–9] and [10].

Unfortunately, they still present relevant limitations already discussed in [11]. To the best of our knowledge, until now, not one of them has been officially approved by a Medical Council as a reliable nodule locator permitting it to become part of surgical preparations.

In [11] we introduced the first part of a long-term research that aims at a full and detailed reproduction of the mammography procedure with the Surface Evolver [12]. Differently from previous works, our approach is to implement nodule trajectories under CC and MLO compressions in our model. These trajectories are studied by means of transparent breast phantoms. See `www.cirsinc.com/products/all/44` for a picture. This present article is a continuation of [11]. Herewith we include the CC and MLO compressions and give some technical details omitted in that first paper.

2. Methodology

As explained in [11], our simulation has 6 main steps: SRG (*surgery*), STU (*stand-up*), LAT (*lay-on-table*), CRC (*cranio-caudal*), LET (*lean-on-table*) and MLO (*medio-lateral-oblique*). Figure 1 shows important measurements of the initial positions. We *always* use the cgs-system. For CRC and MLO, measurements are taken directly from the X-ray images.

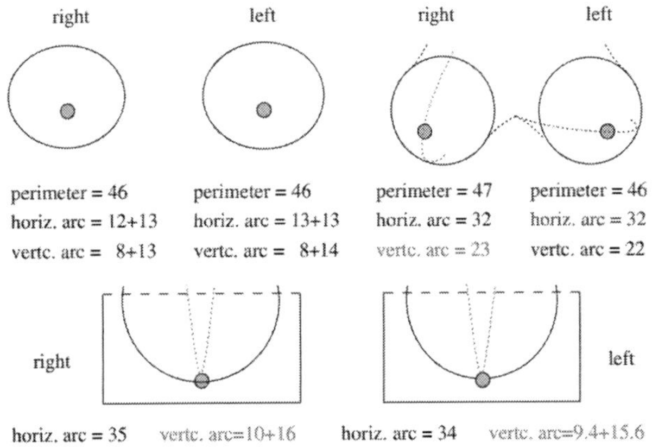

Figure 1: Measurements of SRG, STU and LAT.

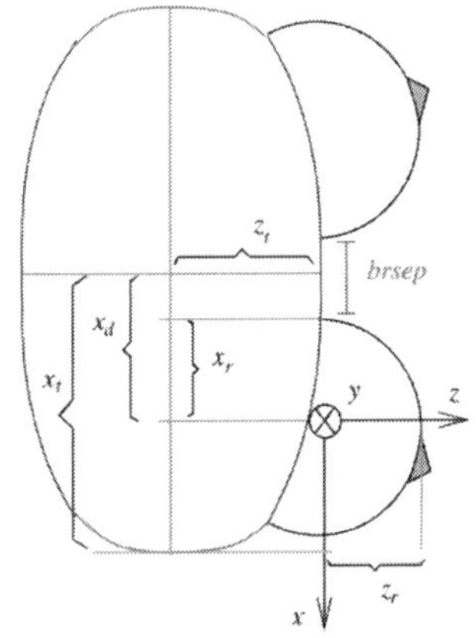

Figure 2: Deduced *unsigned* values.

Now we focus on SRG. Figure 2 is deduced from Figure 1. Empirically, the breasts are within a range of the thorax that can be equated as a cylindrical surface called *bbase*, whose generatrix is a *cubic ellipse* given by $|1 + z/z_t|^3 + |(x + x_d)/x_t|^3 = 1$.

Here the subscript t means "thorax", whereas d stands for "detachment". Namely, the generatrix of *bbase* has a centre with x-coordinate detached from the portion of *bbase* that is actually under the breast. More precisely, $x_d = x_r + brsep/2$.

The breast surface is typically modelled as the upper hemisphere of an ellipsoid with the following radii: x_r, y_r and z_r. It lies on *bbase* and z_r is also the 3rd coordinate of the nipple. In our case measurements indicate that $y_r = z_r = vtarc/\pi$, where $vtarc$ stands for "vertc. arc" in Figure 1. As depicted in Figure 2, the upper blue arc that we call $fthrx$ is almost straight. Therefore we take $x_r = (fthrx - bsep)/4$.

Although we use the half-ellipsoid approach to compute initial values of area and volume, breast and thorax meet at a contour that cannot be an ellipse. Empirically, it is also a cubic ellipse called *bbcrc* in our programmes and equated as $(|x|/x_r)^3 + (|y|/y_r)^3 = 1$. We take $x_t = fthrx/1.7$ and $z_t = lthrx/1.7$, where the factor 1.7 is a rescaling of $\pi/2$. This one would be the factor if the generatrix of *bbase* were a circumference.

Figure 3 shows how we locate the nipple and place the lower plate for CC compressions. The woman places her breast on the lower plate in such a way that most of it will lie thereon. We have observed it as a quite intuitive action. Figure 3 shows the point of contact between the breast and the edge of the plate: it is at the origin of Oxz.

Of course, the edge of the lower plate is tangent to the thorax. In Figure 3 we drew a red vector that is normal to the point of tangency. Let θ be the angle that this vector N makes with Ox. An easy computation gives $\theta = \arccos((x_d/x_t)^{3/2})$, whence

$$N = \left(z_t \cos^{\frac{4}{3}} \theta,\ 0,\ x_t \sin^{\frac{4}{3}} \theta\right). \tag{1}$$

Of course, the corresponding unitary normal is $n = N/||N||$. Now an important remark: *bbcrc* is used as a reference to obtain the values in Figure 1. Namely, the tape-measure has to contour arcs of the breast with endpoints on *bbcrc*. Its considered to be fixed to the woman's thorax. This contour is easily recognised when the woman is lain down. When she stands up, you can use the jugular notch to locate it again.

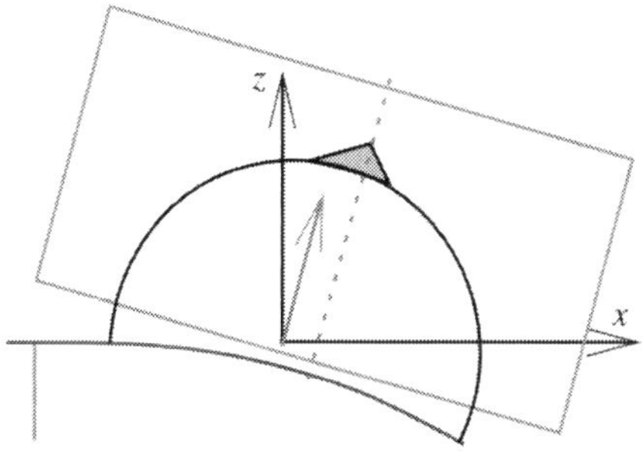

Figure 3: Nipple and plate for CC.

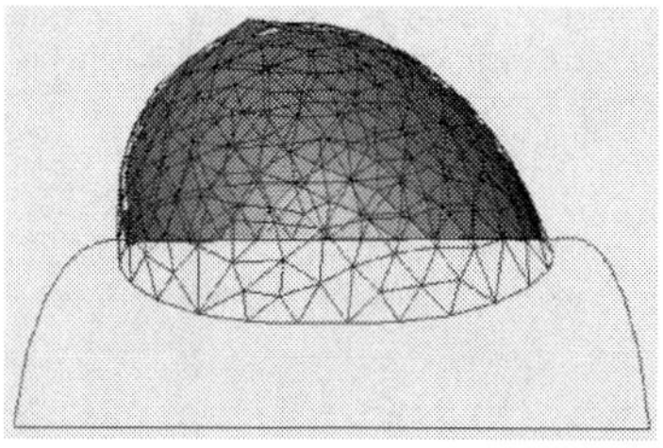

Figure 4: Profile curve.

Figure 3 indicates the nipple at a point of the breast where the unitary normal η to the surface has the following property: its projection onto Oxz is parallel to n. This observation is empirical. Curiously, it holds for *any* of the 6 steps described above (from SRG to MLO).

Now we recall Figure 1. For the left breast at SRG notice that $vtarc = 8 + 14$ from bottom to top. Let $n = (n_1, n_2, n_3)$, where in fact $n_2 = 0$. By taking this example our programme will comb the breast surface until it finds the normal vector

$$\eta = \left(n_1,\ -n_3 \cos(\frac{8+11}{44} \cdot \pi),\ n_3 \sin(\frac{8+11}{44} \cdot \pi) \right). \tag{2}$$

The factor $(8+11)/44$ in (2) is not $8/22$ because the profile of the breast is not quite an upper circumference (see Figure 4), whence the average between $8\pi/22$ and the right angle $11\pi/22$.

We locate our coordinate system by opening the back of the right hand upon the woman's sternum. Technical details can be found in [11, 13].

As Evolver works with surface layers, our first approach is to mark a virtual nodule on a layer inside our model and track its trajectory. In Evolver we represent it by a black triangle. Our programme makes all steps and compressions gradually, and the user can follow the graphical output in real time. Figures 5 and 6 show only the beginning and the end of SRG→STU→LET→MLO.

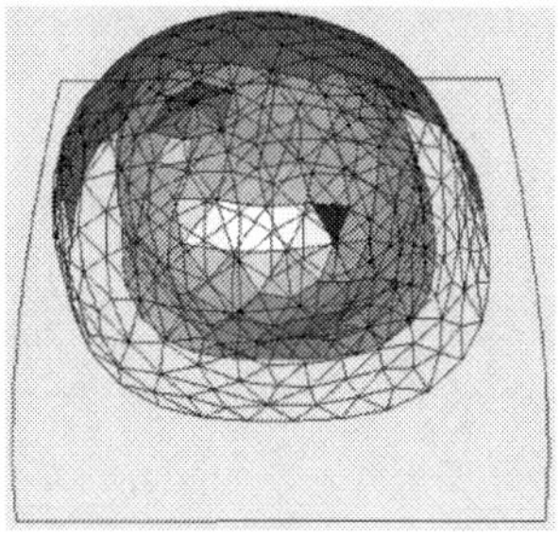

Figure 5: Marking a nodule at SRG.

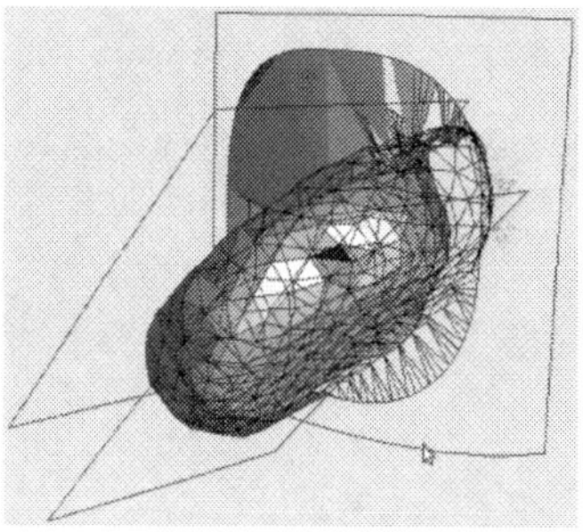

Figure 6: Its position at MLO.

There are many kinds of breast tumours: lipoma (in the fat), carcinoma (in the glands), papilloma (inside the nipple), etc. In all cases they represent an abnormal group of cells of the corresponding component: fat, gland, lactiferous duct, etc. Therefore, a tumour cannot move about as if it were detached from its component. That is why we consider the layers as a reasonable approach.

Although inexact, it is worthwhile to study this approach before adding further complexity to our model. In Section 3 we compare Evolver's virtual displacement with actual displacements inside the phantom when it is compressed by a mammographer. Tests are still at the beginning but we have not found any inconsistency yet.

3. Results

The link *Softwares* of our home page

`http://www.facom.ufu.br/~nascimento`

contains five sample videos of tests with a transparent phantom. A thorough inspection of the videos enabled us to plot the nodule trajectories as shown in Figures *11* and *12* of [13].

Our simulator tracks any chosen virtual nodule as depicted in Figures 5 and 6. Its coordinates are saved in a file named `<patient's number>Xtrj.st`, where `X=L,R` indicates either the left or the right breast, respectively. As an example, we have chosen a virtual nodule at SRG that lead to the following table:

coord.	SRG	STU	LAT	CRC	LET	MLO
x	4.25	3.48	3.37	3.45	5.19	6.15
y	2.74	1.16	0.48	-1.31	1.15	0.65
z	3.12	2.53	2.70	5.00	3.82	4.63

(3)

In Section *5* of [13] we inspect a nodule of the phantom that is the closest to the one of Table 3. In its case, the CC compression gives

$$(4.90, 0.41, 3.67) + \Delta_{\text{phantom}} = (5.71, -0.49, 4.89). \tag{4}$$

Our example in [11] is a breast that weighs 350. Its volume varies from 700 (SRG) to 660 (LAT). But the density of the phantom is nearly the double. Adjustments must be taken into account (see [13] for details). Going back to Table 3, the transition from LAT to CRC gives

$$(3.37, 0.48, 2.70) + \Delta_{\text{breast}} = (3.45, -1.31, 5.00). \tag{5}$$

In practice the x-coordinate of (5) has not changed. Thus we shall not use it for comparisons. But both the y and z coordinates of the displacements Δ in (4) and (5) have grown consistently. Of course, in (5) they are the double of (4) but we recall that the phantom is 1.75 times denser than the breast.

4. Conclusions

This work has many progresses compared with [11]. We have been dealing with the direct virtual mammography. However, the most important is the reverse procedure. In future our software will work as follows: starting from CC and MLO images that show a cancer, it will be located at SRG through the layer-approach. See Section *6* of [13] for details.

5. Acknowledgements

We are grateful to José Artur Quilici Gonzalez, professor at the Federal University of ABC, for his assistance with the state of the art. We thank Dr Ana Cláudia Veronesi for details about gynaecological surgery, and her husband Alejandro Montepeloso for translating some technical terms. We also thank *Ribeirão Pires Hospital (SP - Brazil)* for their assistance and for the concession to use their mammographer in our compression experiments with the phantom. This research is supported by FAPESP proc.No.12/16244-3.

References

[1] Ciatto S, Houssami N, Bernardi D, Caumo F, Pellegrini M, Brunelli S, Tuttobene P, Bricolo P, Fantò C, Valentini M *et al.* 2013 *The lancet oncology* **14** 583–589
[2] Skaane P, Bandos A I, Gullien R, Eben E B, Ekseth U, Haakenaasen U, Izadi M, Jebsen I N, Jahr G, Krager M *et al.* 2013 *Radiology* **267** 47–56
[3] Liu Y, Akers W J, Bauer A Q, Mondal S, Gullicksrud K, Sudlow G P, Culver J P and Achilefu S 2013 *Analyst* **138** 2254–2257
[4] Bliznakova K, Bliznakov Z, Bravou V, Kolitsi Z and Pallikarakis N 2003 *Physics in Medicine and Biology* **48** 3699–3721
[5] Bhatti S N and Sridhar-Keralapura M 2012 *Medical Physics* **39** 1748–1768
[6] Hsu C M, Palmeri M L, Segars W P, Veress A I and Dobbins III J T 2011 *Medical Physics* **38** 5756–5770
[7] Kuhlmann M, Fear E, Ramirez-Serrano A and Federico S 2013 *Medical Engineering & Physics* **35** 470–478
[8] Nishikawa R M 2007 *Computerized Medical Imaging and Graphics* **31** 224–235
[9] Pathmanathan P, Gavaghan D J, Whiteley J P, Chapman S J and Brady J M 2008 *Biomedical Engineering, IEEE Transactions on* **55** 2471–2480
[10] Tang J, Rangayyan R M, Xu J, El Naqa I and Yang Y 2009 *Information Technology in Biomedicine, IEEE Transactions on* **13** 236–251
[11] Nascimento M Z and Ramos Batista V 2014 *Journal of Physics: Conference Series* vol 490 (IOP Publishing) p 012146
[12] Brakke K A 1992 *Experimental Mathematics* **1** 141–165
[13] Nascimento M Z and Ramos Batista V 2014 Simulating CC and MLO compressions with the surface evolver File `preprint.pdf` on URL `https://sites.google.com/site/vramos1970`

Journal of Physics: Conference Series **574** (2015) 012073 doi:10.1088/1742-6596/574/1/012073

An interactive programme for weighted Steiner trees

Marcelo Zanchetta do Nascimento[1], **Valério Ramos Batista**[2], **Wendhel Raffa Coimbra**[3]

[1] *FACOM-UFU, Av. João Naves de Ávila 2121, Bl.A, 38400-902 Uberlândia-MG, Brazil*
[2] *CMCC-UFABC, R. Sta. Adélia 166, Bl.B, 09210-170 St. André-SP, Brazil*
[3] *UFMS, rod. BR 497 km 12, 79500-000 Paranaiba-MS, Brazil*

E-mail: `nascimento@facom.ufu.br`

Abstract. We introduce a fully written programmed code with a supervised method for generating weighted Steiner trees. Our choice of the programming language, and the use of well-known theorems from Geometry and Complex Analysis, allowed this method to be implemented with only 764 lines of effective source code. This eases the understanding and the handling of this beta version for future developments.

1. Introduction

One of the main problems at implementing multicast in Wide Area Networks (WAN) is the high cost of transmissions between terminals. Cost reduction is attained by adding routers to the network, but this increases complexity (see [1, 2]). Steiner trees have long been used in order to optimise routes, aiming at the lowest cost possible (see [3, 4]).

The Steiner Minimal Tree (SMT) problem is NP-hard [5] but can be exactly solved for terminals in thousands. GeoSteiner does it amazingly fast for terminals at random. Its strategy is to look for terminals that (almost) make equilateral triangles, and then to prune sub-optimal trees. See `www.diku.dk/hjemmesider/ansatte/martinz/geosteiner` for details. But when terminals follow a pattern, GeoSteiner can be very slow. For instance, we use 4GB of RAM, microprocessor Intel Core i5 3.2GHz, and operating system Linux Ubuntu 12.04. With this setting GeoSteiner takes 73.02s to generate Figure 1 (compare it with Figure 2). This time drops to only 0.06s when the 31 terminals are at random.

Moreover, the GeoSteiner algorithm cannot be easily adapted to find *weighted* minimal Steiner trees. This fact, together with the slowness in patterned cases, is precisely due to the strategy of looking for equilateral triangles.

Given a graph $G = (V, E, w)$, a subset $S \subset V$ and a weight function w, a *weighted minimal Steiner tree* (WSMT) $T \subset G$ is one that spans all vertices of S and also minimises the total weight. WSMTs have many applications. Among others, they are used in network formation games, computational sustainability and electric power networks [6, 7, 8]. The problem has further variations, like for unity disk graphs and restrictions on w, that have been studied recently [9, 10, 11, 12]. These and other works use heuristics for non-supervised methods. They are fast at generating weighted Steiner Trees with good chances to be minimal. But if one really seeks *the minimum*, non-supervised methods will not be reliable unless applied to few terminals.

In the last paragraph from §6 of [13] the authors commented that skill and good guesses of a trained user can help find the SMT. Their work is previous to GeoSteiner by three decades but

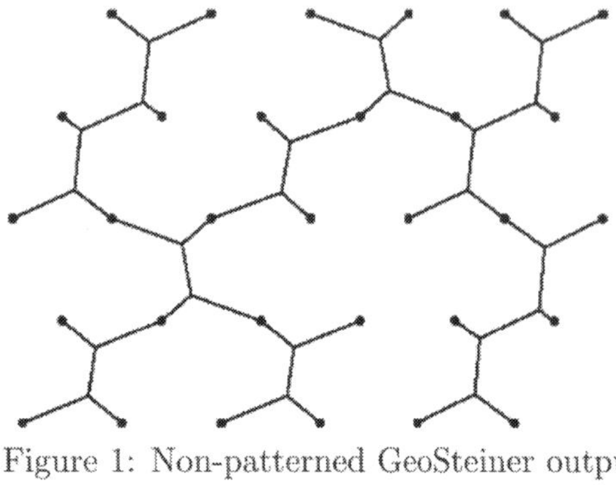

Figure 1: Non-patterned GeoSteiner output.

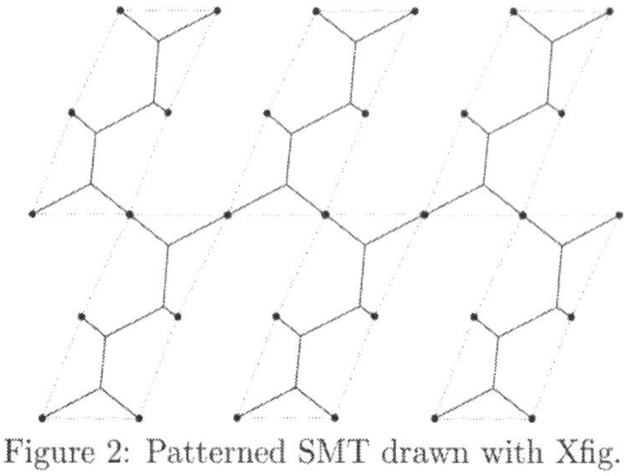

Figure 2: Patterned SMT drawn with Xfig.

their comment is still pertinent. Indeed, one cannot always predict whether a fast algorithm exists to solve a given problem. In that same work, Gilbert and Pollak conjectured that the length L of an SMT always verify $L \geq Lprim \cdot \sqrt{3}/2$, where $Lprim$ is the length of the Minimal Spanning Tree (MST). This one can be obtained by Prim's algorithm [14]. According to [15] their conjecture is true. If so, any Steiner tree obtained from the terminals of an MST can improve it of at most 13.4%.

It seems rather little, but if a connection is used extensively it represents a great saving in the long-term. Moreover, in [16] the authors claim that the Gilbert and Pollak's conjecture is not completely answered yet. Thus it might even exist an SMT of which $L < Lprim \cdot \sqrt{3}/2$.

A supervised programme is not suitable for thousands of terminals, except for a long-term project distributed to a team of users and terminals in hundreds. This is still the case of sound and video cards. Their frequent access makes it desirable to bring delay to a minimum (even a 1% decrease would count). For them we need to focus on rectilinear Steiner trees, a variation we shall work on in future. However, in VLSI-design we have millions of terminals. For such extreme cases supervised methods are unsuitable.

We present `stree`, a fully programmed code to generate WSMTs. It is written in Matlab/Octave to spare the GUI. Moreover, many original ideas are within the code. They are based on theorems from Geometry and Complex Analysis. This produces a very short programme (only 855 lines, or 764 without graphical input/output), easier to understand and to handle for future developments. The present version of our programme has didactic purposes. It is even accessible to undergraduate students. The reader can access the link *Softwares* of our webpage

`http://www.facom.ufu.br/~nascimento`

to download `stree.zip`, which contains the executable and some source codes. The link also contains a *preprint* (this one with all technical detais and a user's manual).

2. Methodology

As we mentioned in the Introduction, the MST is frequently used to construct a Steiner tree. Prim's algorithm is easily adaptable to find the MST for terminals that are weighted as follows:

Definition 2.1. Consider the tree $T = (V, E)$ with weight function $w : V \to \mathbb{R}^+$ and 0-1 adjacency matrix a_{ij}. Then T is an MST if it minimises the total cost C given by

$$C = \sum_{i<j} a_{ij}(w_i + w_j)||V_i - V_j||, \tag{1}$$

where $||V_i - V_j|| = \frac{1}{2}(w_i + w_j)|V_i - V_j|$ is the connection cost between terminals V_i and V_j.

When we have $w : V \to \{1\}$, then it is the Euclidean non-weighted case. For a given set of terminals $V = \{V_1, \ldots, V_n\}$ and a weight function $w : V \to \mathbb{R}^+$, we can find the edges E that minimise the cost C in (1) and result in an MST $T = (V, E)$. By adding extra points $S = \{S_1, \ldots, S_m\}$ to V, namely $\tilde{V} = V \cup S$, we shall find the corresponding $\tilde{T} = (\tilde{V}, \tilde{E})$ such that $\tilde{C} \leq C$. We claim that $\tilde{C} < C$ exactly when $S \neq \emptyset$, providing one chooses a suitable extension $\tilde{w} : \tilde{V} \to \mathbb{R}^+$ of the weight function. The following lemma shows how to make this choice when $n = 3$ and $m = 1$.

Lemma 2.1. *Consider a triangle with vertex weights a, b and c, and suppose it admits a classical (Euclidean) Steiner point. If $s \leq \min\{a, b, c\}$ is the weight of the Steiner point, then its connection with the vertices will cost less than any other connection through the vertices only.*

Its proof and all the others' are in our preprint (see the link *Softwares* of our webpage).

Lemma 2.2. *Consider a full Steiner tree with $n \geq 3$ terminals $A_1, \ldots, A_n$ and Steiner points $S_1, \ldots, S_{n-2}$. Add weights to their respective terminals as a_i, $1 \leq i \leq n$, and to the Steiner points as $s_i = \min\{a_1, \ldots, a_n\} \, \forall i$. The resulting weighted tree is then a local minimum of C.*

The most important consequence of Lemmas 2.1 and 2.2 is that Steiner points can be added exactly as in the Euclidean case for arbitrary $n \geq 3$. The resulting MST $\tilde{T} = (\tilde{V}, \tilde{E})$ coincides with a Euclidean non-weighted Steiner tree, which will not be necessarily the SMT. Anyway, many results proved in [13] still hold for $\tilde{T}$: *Convex hull*, *Maxwell's Theorem*, *Lune* and *Wedge* properties.

3. Results

Our pseudocode in Figure 3 works recursively while the connection matrix is incomplete.

Input: Set of terminals
Output: Steiner tree
1. Pts ← terminals, Connexion_Matriz ← 0
2. Tree_of_Prim ← Compute_Tree_of_Prim(Pts)
3. Steiner_hull ← Compute_Steiner_hull(Pts)
4. While Connexion_Matriz implies Stree non-connected
  i. Subset_Pts ← User's Choice of a Subgroup(Pts)
  ii. Subtree ← Compute_Connection(Subset_Pts)
  iii. If Subtree not OK, redo Step i
  iv. Else Connexion_Matriz ← Connection(Subset_Pts)
5. return Steiner tree (and its length)

Figure 3: The pseudocode of the `stree` algorithm.

The main result is our programme `stree`, of which some outputs are depicted in Figure 4. There we start with a *Weighted* Minimal Spanning Tree (WMST) obtained by the adapted Prim's algorithm (see Definition 2.1). Notice that a WMST can have self-intersections, but this never happens to a WSMT because Lemmas 2.1 and 2.2 imply that any self-intersection can be split into two Steiner points in order to lower the cost.

Basically, one uses the WMST as a reference to draw Steiner subtrees. These will build an entire tree with good chances of being the WSMT. The weighted length is printed on the Matlab terminal window. The user gets some feedback while drawing and can also undo some choices. Sections *3* and *4* of our preprint are a detailed manual with hints, tutorials and illustrations.

4. Conclusions

Differently from trying a fully automated method, we propose to take advantage of the good choices that a user can make. Many attributes like intuition, guess, practice and a bird's-eye view are valuable means that one cannot translate into any programming language. As long as

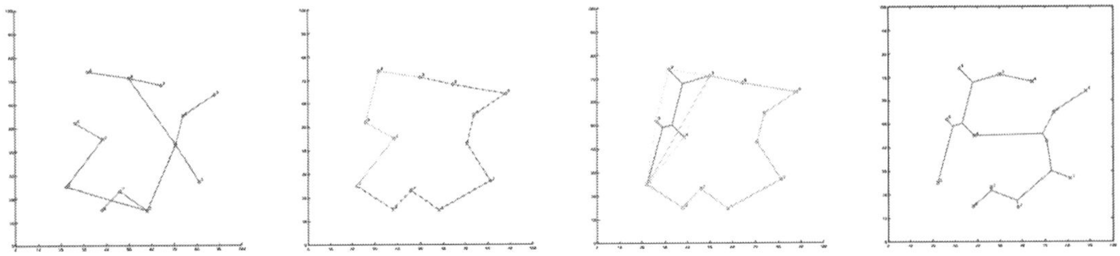

Figure 4: Some steps at running the `stree` programme.

a task is feasible with supervision we suggest taking it into account, besides the fully automated methods. This proposal is not new, but we endeavour to obtain a code that is both easy to run and to understand. The programme `stree` is still in the beta version. Further improvements will include more feedback to the user.

5. Acknowledgements

Many improvements in this paper were due to the careful analyses carried out by referees. We thank them for their valuable help. We are also grateful to Cláudio Nogueira de Meneses, professor at the Federal University of ABC, for his assistance with weighted graphs.

References

[1] Jia X, Hu X-D, Lee M, Du D-Z and Gu J 2001 Optimization of wavelength assignment for QoS multicast in WDM networks *IEEE Trans. on Communications* **49** 341–350
[2] Sahasrabuddhe L H and Mukherjee B 2001 Multicast routing algorithms and protocols: a tutorial *IEEE Network* **14** 90–102
[3] Ausiello G, Crescenzi P, Gambosi G, Kann V, Marchetti-Spaccamela A and Protasi M 1999 *Complexity and approximation: combinatorial optimization problems and their approximability properties* (Berlin: Springer)
[4] Hu X-D, Shuai T-P, Jia X and Zhang M-H 2004 Multicast routing and wavelength assignment in WDM networks with limited drop-offs *INFOCOM 2004: Twenty-third AnnualJoint Conference of the IEEE Computer and Communications Societies* **1** (Hong-Kong: IEEE)
[5] Garey M R, Graham R L and Johnson D S 1977 The complexity of computing Steiner minimal trees *SIAM Journal of Applied Mathematics* **32** 835–859
[6] Dilkina B and Gomes C P 2010 *Integration of AI and OR Techniques in Constraint Programming for Combinatorial Optimization Problems* (Berlin: Springer) p 102–116
[7] Guha S, Moss A, Naor J S and Schieber B 1999 Efficient recovery from power outage *Proceedings of the 31st Annual ACM Symposium on Theory of Computing* (Atlanta: ACM) p 574–582
[8] Anshelevich E, Dasgupta A, Kleinberg J, Tardos E, Wexler T and Roughgarden T 2008 The price of stability for network design with fair cost allocation *SIAM Journal on Computing* **38** 1602–1623
[9] Angelopoulos S 2006 The node-weighted Steiner problem in graphs of restricted node weights *Algorithm Theory-SWAT* (Berlin: Springer) p 208–219
[10] Demaine E D, Hajiaghayi M and Klein P N 2009 Node-weighted Steiner tree and group Steiner tree in planar graphs *Automata, Languages and Programming* (Berlin: Springer) p 328–340
[11] Li X, Xu X-H, Zou F, Du H, Wan P, Wang Y and Wu W 2009 A PTAS for node-weighted Steiner tree in unit disk graphs *Combinatorial Optimization and Applications* (Berlin: Springer) p 36–48
[12] Zou F, Li X, Gao S and Wu W 2009 Node-weighted Steiner tree approximation in unit disk graphs *Journal of Combinatorial Optimization* **18** 342–349
[13] Gilbert E N and Pollak H 0 1968 Steiner minimal trees *SIAM Journal of Applied Mathematics* **16** 1–29
[14] Prim R C 1957 Shortest connection networks and some generalizations *Bell System Technical Journal* **36** 1389–1401
[15] Du D Z and Hwang F K 1992 A proof of the Gilbert-Pollak conjecture on the Steiner ratio *Algorithmica* **7** 121–135
[16] Innami N, Kim B H, Mashiko Y and Shiohama K 2010 The Steiner ratio conjecture of Gilbert-Pollak may still be open *Algorithmica* **57** 869–872

Study on Durability and Stability of an Aqueous Electrolyte Solution for Zinc Bromide Hybrid Flow Batteries

Donghyeon Kim[1, a], **Joonhyeon Jeon**[1, b, ‡]
[1] Department of Energy and Advanced Material Engineering, Dongguk University-Seoul, 30, Pildong-ro 1gil, Jung-gu, Seoul 100-715, Korea

E-mail: [a] allan.k@dgu.edu, [b] memory@dgu.edu

Abstract. Zinc-bromine flow battery using aqueous electrolyte has advantages of cost effective and high energy density, but there still remains a problem improving stability and durability of electrolyte materials during long-time cell operation. This paper focuses on providing a homogeneous aqueous solution for durability and stability of zinc bromide electrolyte. For performance experiments of conventional and proposed electrolyte solutions, detailed cyclic voltammetry (CV) measurements (at a scan rate of 20 mV s^{-1} in the range of -1.5 V~1.5 V) are carried out for 40 cycles and five kinds of electrolytes containing which has one of additives, such as (conventionally) zinc chloride, potassium chloride, (newly) lithium perchlorate, sodium perchlorate and zeolite-Y are compared with the 2.0 M $ZnBr_2$ electrolyte, respectively. Experimental results show that using the proposed three additives provides higher anodic and cathodic peak current density of electrolytes than using other two conventional additives, and can lead to improved chemical reversibility of zinc bromide electrolyte. Especially, the solution of which the zeolite-Y added, shows enhanced electrochemical stability of zinc bromide electrolyte. Consequently, proposed electrolytes have a significant advantage in comparison with conventional electrolytes on higher stability and durability.

1. Introduction

Recently, The flow battery technology has high interest of energy storage system for users who requires cost effective and eliminates the limitation of traditional battery systems that has a decoupling problem with the power and energy components of the system. The zinc bromine hybrid flow battery (ZBB) has characteristic of high power density and can be a useful rechargeable device for energy conversion [1]. The process of charging and discharging energy of ZBB is generally achieved through a reaction of redox couples like Zn^{2+} / $Zn_{(s)}$ and Br^- / Br_2 in zinc bromide electrolyte [2, 3]. The zinc-bromine electrolyte solution has the advantage of providing cost-effective high energy density for battery energy storage, but its distinctive-electrochemical property during cell charging causes a lot of problems, which leads to highly concentrated bromine in catholye and zinc metal deposition on the anode electrode surface. This is due to the fact that the activated bromine can corrode components of the cell and zinc metal can be formed a variety of dendritic morphology. This provides reason enough not only to decreasing the energy efficiency of ZBB but also to reduce the battery life cycle. Thus, zinc-bromine electrolyte solution has a significant influence on the electrochemical performance of ZBB for energy storage [4].

‡ To whom any correspondence should be addressed.

Various additive materials such as quaternary ammonium salts and dendrite inhibitors has been deveopled which combine activated bromine and prevent indiscriminate formation of zinc dendrite, respectively[4]. Improving the electrolyte conductivity has also long been one of the major topics in the development of an electrolytic capacitor. In conventional electrolye solutions, zinc chloride or potassium chloride have generally been used for this purpose [5], but these additives can hinder stability and durability of zinc bromide electrolyte. Hence, there still remains a problem of improving stability and durability of electrolyte materials during long-time cell operation.

This paper focuses on providing additives such as lithium perchlorate, sodium perchlorate and zeolite-Y for durability and stability of the electrolyte. Depending on the kinds of conventional and new additives, electrochemical effects of these electrolytes are investigated by cyclic voltammetry measurement (CV) [6].

2. Experimental Method

2.1. Electrolyte Preparation

Several kinds of additives such as $LiClO_4$, $NaClO_4$, zeolite-Y, $ZnCl_2$ and KCl are used to improve the stability and durability of electrolyte. These compounds have a stronger electronegativity than $ZnBr_2$. Zinc bromide electrolyte solution is prepared by electrolytic dissociation of zinc bromide powder which is supplied by Kanto Chemical Co., Japan. A certain amount of additives, $LiClO_4$, $NaClO_4$, zeolite-Y, $ZnCl_2$ and KCl are prepared into the electrolyte, respectively. The total content of zinc bromide in the electrolyte is determined as 2.0 mol L^{-1}, because minimum resistivity of the electrolyte and maximum solubility of bromine was observed at that concentration [7, 8].

2.2. Cyclic Voltammetry Measurement

The CV is carried out on a potentiostat mode in a three-electrode cell kit which consists of electrolyte bridge and two reservoirs. To investigate the stability and electrochemical characteristic of the zinc bromide electrolyte, the current versus potential curves are recorded in the cell kit with platinum wire electrode as the counter electrode, saturated calomel electrode (SCE) as reference electrode, and a carbon felt electrode as 3D working electrodes, respectively. The potential range of cyclic voltammograms is -1.5 V to 1.5 V and the scan rate is 20 mV s^{-1} [9]. To confirm the durability and electrochemical property of electrolytes, the tests are progressed for 40 cycles.

3. Results and discussion

To investigate influence of additives on the electrochemical properties, anodic peak current (I_{PA}), cathodic peak current (I_{PC}) and peak current density ratio (I_{PA}/I_{PC}) are obtained from the CV experiments. Figure 1 shows the electrochemical property of these electrolytes, and measurement results to compare durability and stability of these electrolytes for 40 cycles. Also, Table 1 indicates the related I_{PA} and I_{PC} data where these values measured on 2nd, 10th and 40th cycle, respectively. It can be seen that the I_{PA} of the electrolytes with each of $LiClO_4$, $NaClO_4$ and zeolite–Y has higher anodic peak current and the I_{PC} of the aqueous solutions with $LiClO_4$ and zeolite-Y slightly decrease. And also the peak current density ratios of these solutions with additives show value close to 1.0, excepting the KCl.

Table 1 also demonstrates that the I_{PA} of the electrolyte can be enhanced by additives as $LiClO_4$, $NaClO_4$ and zeolite–Y, respectively. The I_{PC} of the solution with each of KCl, $LiClO_4$ and zeolite–Y slightly decrease. It is found from these results that I_{PA} of all electrolytes (in comparison to I_{PC}) is continuously decreased in progress up to 40 cycles. In the case of ideal electrolyte can be seen that peak currents have to be undiversified during long time operation. But when additives are added into the electrolytes, anodic and cathodic peak current decrease. Thus, this result indicates that all kinds of additives can hinder durability of electrolyte for a long time. But anodic peak current decreasing rate of solution with additives as zeolite-Y or $LiClO_4$ slightly decline and the cathodic peak current decreasing rate can be improved by additives as KCl or zeolite-Y. The peak current density ratios of zinc bromide electrolyte with $LiClO_4$, $NaClO_4$ or zeolite-Y are closer to 1.0 than that of without additive, and also peak current density decreasing rates of electrolytes with $LiClO_4$ or zeolite-Y can be enhanced.

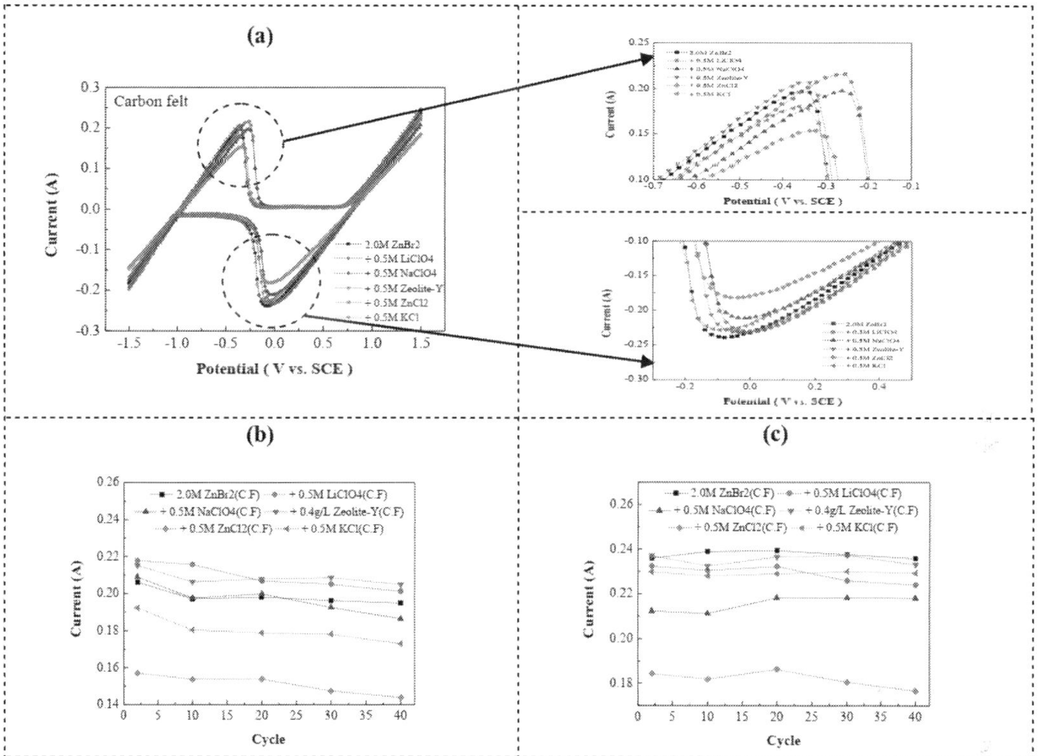

Figure 1. (a) Cyclic voltammograms at 10th cycle for the electrochemical property of electrolytes, (b) the anodic peak currents, and (c) the cathodic peak currents of the zinc bromide electrolyte with various additives such as $LiClO_4$, $NaClO_4$, zeolite-Y, $ZnCl_2$ and KCl.

Table 1. The related data of Figure 1

Sample	2nd cycle			10th cycle			40th cycle		
	I_{PA} (A)	I_{PC} (A)	I_{PA} / I_{PC}	I_{PA} (A)	I_{PC} (A)	I_{PA} / I_{PC}	I_{PA} (A)	I_{PC} (A)	I_{PA} / I_{PC}
2.0 M $ZnBr_2$	0.20613	0.23592	0.874	0.19713	0.23810	0.828	0.19492	0.23585	0.826
2.0 M $ZnBr_2$ + 0.5 M $ZnCl_2$	0.15705	0.18417	0.853	0.15371	0.18153	0.847	0.14391	0.17654	0.815
2.0 M $ZnBr_2$ + 0.5 M KCl	0.19239	0.22977	0.837	0.18041	0.22805	0.791	0.17307	0.22932	0.755
2.0 M $ZnBr_2$ + 0.5 M $LiClO_4$	0.21783	0.23237	0.937	0.21572	0.23044	0.936	0.20114	0.22408	0.898
2.0 M $ZnBr_2$ + 0.5 M $NaClO_4$	0.20897	0.21228	0.984	0.19766	0.21076	0.938	0.18656	0.21804	0.856
2.0 M $ZnBr_2$ + 0.4 g/L Zeolite-Y	0.21519	0.23701	0.908	0.20662	0.23244	0.889	0.20497	0.23307	0.879

4. Conclusions

To improve the stability and durability of zinc bromide electrolyte, the effects of the electrochemical behavior in these additives as $ZnCl_2$, KCl, $LiClO_4$, $NaClO_4$ and zeolite-Y are studied and in order to investigate the electrochemical characteristics of the electrolyte, the carbon felt electrode as 3D working electrode is used. As a result, the carbon felt electrode can be appropriate at the reduction oxidation reaction of bromine and zinc for ZBB. But the creation of zinc dendrite is accelerated by the carbon felt electrode, so the anodic peak current decreasing rate has a higher ratio than the cathodic peak current decreasing rate. The anodic current of electrolyte for ZBB can be improved by additives as $LiClO_4$ and zeolite-Y, and also the cathodic current is slightly diminished by KCl and zeolite-Y. Moreover, the decreasing rates of the electrolyte with experimented additives can show almost considerable changes in comparison with the electrolyte without additive. Also, the experiment results can show that $LiClO_4$ and zeolite-Y are proper additional materials at the anodic peak current decreasing rate, as well as KCl and zeolite-Y appertain to the cathodic peak current decreasing rate. Thus, the electrolyte with zeolite-you can indicate that the peak current decreasing rate of anodic and cathodic slightly decreased, and also anodic and cathodic currents enhanced in comparison with conventional electrolytes. It will be used as a promising additive for the electrolyte of ZBB.

5. Acknowledgments

This work was supported by the Technology Innovation Program (10043787, Development of Ion Selective Membranes for redox Flow Battery) funded By the Ministry of Trade, industry & Energy (MI, Korea)

6. Reference

[1] C. Ponce Leon, A. Frias-Ferrer, J. Gonzalez-Garcia, D. A. Szanto and F. C. Walsh, 2006, *J. Power Sources*, **160**, 716-732.

[2] M. J. Watt-Smith, R. G. A. Wills and F. C. Walsh, 2009, *Encyclopedia of Electrochemical Power Sources*, Jurgen Garche, Amsterdam, The Netherlands, Elsevier, 438-443.

[3] C. Ponce Leon, A. Frias-Ferrer, J. Gonzalez-Garcia, D. A. Szanto and F. C. Walsh, 2006, *Journal of Power Sources*, **160**, 716-732.

[4] M. Skyllas-Kazacos, M.H. Chakrabarti, S. A. Hajimolana, F. S. Mjalli and M. Saleem, 2011, *Journal of The Electrochemical Society*, **158**, 8, R55-R79

[5] H. S. Lim, A. M. Lackner, R. C. Knechtli, 1977, J. Electrochem. Soc., **124**, 8, 1154-1157

[6] M. H. Chakrabarti, R. A. W. Dryfe and E. P. L. Roberts, 2007, *J. Electrochemica Acta*, **52**, 2189-2195.

[7] H.S. Lim, A.M. Lackner, and R.C. Knechtli, 1977, *J. Electrochemical Science and Technology*, **124**, 8, 1154-1157.

[8] Palph Zito, Jr., 1982, Zinc-bromine battery with long term stability, US-4482614-A

[9] H. S. Choi, Y. H. Oh, C. H. Ryu and G. J. Hwang, 2013, Trans. of the Korean Hydrogen ad New Energy Society, **24**, 4, 347-352.

[10] Petr Vanysek, 2009, *Handbook of Chemistry and Physics*, David R. Lide, Boca Raton, Florida, CRC press, 76th, 8-20 – 8-29

[11] Xiaojuan Wu, Suqin Liu, Nanfang Wang, Sui Peng and Zhangxin He, 2012, *J. Electrochemica Acta*, **78**, 475-482.

[12] Sha Li, Kelong Huang, Suqin Liu, Dong Fang, Xiongwei Wu, Dan Lu and Tao Wu, 2011, *J. Electrochemica Acta*, **56**, 5483-5487.

[13] Huang Fei, Zhao Qiang, Luo ChunHui, Wang GuiXin,Yan KangPang and Luo DongMei, 2012, *Chinese Science Bulletin*, **57**, 32, 4237-4243.

[14] Qinzhi Lai, Huamin Zhang, Xianfeng Li, Liqun Zhang and Yuanhui Cheng, 2013, *Journal of Power Sources*, **235**, 1-4.

Modeling indirect detectors for performance optimization of a digital mammographic detector for dual energy applications

N Martini[1], V Koukou[1], N Kalyvas[2], P Sotiropoulou[1], C Michail[2], I Valais[2], A Bakas[3], I Kandarakis[2], G Nikiforidis[1], G Fountos[2]

[1] Department of Medical Physics, Medical School, University of Patras, 265 00 Patras, Greece
[2] Department of Biomedical Engineering, Technological Educational Institute of Athens, Egaleo, 122 10 Athens, Greece
[3] Department of Medical Radiologic Technology, Technological Educational Institute of Athens, Egaleo, 122 10 Athens, Greece

E-mail: valais@teiath.gr

Abstract. Dual Energy imaging is a promising method for visualizing masses and microcalcifications in digital mammography. The advent of two X-ray energies (low and high) requires a suitable detector. The scope of this work is to determine optimum detector parameters for dual energy applications. The detector was modeled through the linear cascaded (LCS) theory. It was assumed that a phosphor material was coupled to a CMOS photodetector (indirect detection). The pixel size was 22.5 μm. The phosphor thickness was allowed to vary between 20mg/cm^2 and 160mg/cm^2. The phosphor materials examined where Gd_2O_2S:Tb and Gd_2O_2S:Eu. Two Tungsten (W) anode X-ray spectra at 35 kV (filtered with 100 μm Palladium (Pd)) and 70 kV (filtered with 800 μm Ytterbium (Yb)), corresponding to low and high energy respectively, were considered to be incident on the detector. For each combination the contrast-to-noise ratio (CNR) and the detector optical gain (DOG), showing the sensitivity of the detector, were calculated. The 40 mg/cm^2 and 70 mg/cm^2 Gd_2O_2S:Tb exhibited the higher DOG values for the low and high energy correspondingly. Higher CNR between microcalcification and mammary gland exhibited the 70mg/cm^2 and the 100mg/cm^2 Gd_2O_2S:Tb for the low and the high energy correspondingly.

1. Introduction

Breast cancer, which is a common cause of death among female population, may manifest as microcalcifications. In X-ray mammography, medical diagnosis (including screening techniques) relies on the detection and visualization of such microcalcifications (μCs) and/or soft tissue masses. The early detection of breast cancer has been shown to decrease breast cancer mortality [1]. Dual-energy subtraction imaging techniques [2-8] offer an alternative approach to the detection and visualization of μCs . With this technique, high- and low-energy images are separately acquired and "subtracted" from each other in a weighted fashion to cancel out the cluttered tissue structure so as to decrease the obscurity from overlapping tissue structures. Current Full-Field Digital Mammography (FFDM) systems, are based on either direct X-ray detectors using amorphous Selenium (a-Se) or indirect X-ray detectors using scintillators coupled to amorphous Silicon (a-Si) sensors [9,10]. Detector modelling has been carried out to determine the optimum detector design for the best detector performance in optimizing X-ray detection and signal generation. One method to determine the optimum detector parameters is linear cascaded systems theory (LCS). This theory calculates the output of a detector as a series of cascaded stages. These stages describe the statistics of signal carrier interactions and are divided into gain stages and blur stages [11-17]. In this study, the aforementioned theory was used in order to calculate the optimum phosphor thickness of an indirect detector for CNR and DOG maximization.

2. Materials and Methods

In this study the LCS theory was used. This theory calculates the output of a detector as a series of cascaded stages. Every stage has a frequency domain input $S_{in}(u)$, where u is the spatial frequency, a mean input value $\bar{x}_{in}$, a frequency domain output $S_{out}(u)$ and a mean output $\bar{x}_{out}$. Every gain stage is characterized by a statistical mean value $\bar{q}$ and variance σ_q^2, while every blur stage is characterized by a Modulation Transfer Function $MTF(u)$. The blur stages are either stochastic or deterministic. The frequency depended output and the mean output signal for each stage can be calculated from the following relationships: (i) $S_{out}(u)=\bar{q}^2 S_{in}(u)+\bar{x}_{in}\sigma_q^2$ and $\bar{x}_{out}=\bar{q}\bar{x}_{in}$, for gain stages, (ii) $S_{out}(u)=(S_{in}(u)-\bar{x}_{in})MTF^2(u)+\bar{x}_{in}$ and $\bar{x}_{out}=\bar{x}_{in}$ for stochastic blur stages and (iii) $S_{out}(u)=S_{in}(u)MTF^2(u)$ and $\bar{x}_{out}=\bar{x}_{in}$ for deterministic blur stages [11-17]. In this work the following stages were considered: the X-ray absorption in the phosphor material, the optical photon production per absorbed X-ray, the optical photon escape and spread to the output, the impingement of the optical photons at the CMOS surface and the production of electrons at the CMOS output. If an X-ray fluence $\bar{\Phi}_o(E)$, of energy E is incident on the detector, then it is absorbed exponentially in different depths of the phosphor. The probability of X-ray interaction is assumed to follow binomial distribution with a mean probability $\bar{M}_x(E,t)$ per incident X-ray photon, where $\bar{M}_x(E,t)=e^{-\mu(E)t}\mu(E)\Delta t$, with $\mu(E)$ being the total attenuation coefficient of the material. The absorbed X-ray energy is transformed into optical photons with a Poisson process with a mean value $\bar{m}_o(E)$, expressing the number of optical photons per absorbed X-ray photon. $\bar{m}_o(E)$ can be calculated as, $\bar{m}_o(E)=n_C E/E_\lambda$, where n_C is the intrinsic conversion efficiency of X-ray power to optical photons power. E_λ is the optical photons energy. A fraction $G(t)$ of these photons, escape to the output. $G(t)$ is assumed to follow binomial distribution with mean value $\bar{G}(t)$ and variance $\bar{G}(t)(1-\bar{G}(t))$. The optical photons are spread to the output. This spread is characterized by $MTF(u,t)$. By combining the above stages through LCS theory the Noise Power Spectrum ($NPS(u)$) and the mean number of optical quanta escaping the detector, M_l, can be calculated [13-16]. The optical photons impinge at the CMOS pixel with size a_{pix}. Only a fraction of these photons are detected. This is a gain stage characterized by a pixel fill factor ff and an effective pixel area, a_{pd}, where $a_{pd}=\sqrt{ffa^2_{pix}}$. This process follows binomial distribution with a mean value $\overline{ff}$ and a variance $\overline{ff}$ (1-$\overline{ff}$). Due to spectral matching a fraction of these photons will be actually detected. This is a gain stage with a probability characterized by a mean value $\bar{a}_s$ and a variance $\bar{a}_s$(1-$\bar{a}_s$). The optical photons are absorbed and electron-hole pairs are created. It was assumed that this process also follows a binomial distribution with a mean probability $\bar{Q}_p$ and a variance $\bar{Q}_p$(1-$\bar{Q}_p$). A

Journal of Physics: Conference Series **574** (2015) 012075 doi:10.1088/1742-6596/574/1/012075

fraction of these pairs will reach the detector output. It was considered that this process follows a Poisson distribution with a mean probability value $\overline{Q}_e$ and a variance $\overline{Q}_e(1-\overline{Q}_e)$. By applying LCS theory in the above stages the total detector NPS ($NPS_{Ia}(u)$) is equal to:

$$NPS_{Ia}(u) = (\overline{ff\, a_s Q_p Q_e})^2 NPS(u) + \overline{M_I\, ff\, a_s Q_p Q_e}(1 - \overline{ff\, a_s Q_p Q_e})$$

The total number of electrons reaching the output ($\overline{X_e}$) equals to $\overline{Xe} = \overline{M_I}\, \overline{ff\, a_s}\, \overline{Q_p}\, \overline{Q_e}$ The electrons are spread to the output. This is a blur stage characterized by CMOS MTF (MTF_{inh}). The final NPS equals to $NPS_I(u) = \left[NPS_{Ia}(u) - \overline{Xe}_I\right] MTF^2_{inh}(u) + \overline{Xe}_I$, where it was assumed that $MTF_{inh}(u) = \mathrm{sinc}(\pi a_{pd} u)$. Finally the signal is integrated due to the finite pixel size. This is a deterministic blur stage characterized by $MTF_{pix}(u)$ where $MTF_{pix}(u) = \mathrm{sinc}(\pi a_{pix} u)$ [13-16]. An additional useful parameter is the number of optical photons per incident X-ray (DOG). DOG expresses the sensitivity of the scintillator and is expressed as the total number of signal carriers produced per incident X-ray photon. With regards to the dual energy optimization, CNR was utilized as an index for image evaluation. CNR was calculated as $CNR = \dfrac{\left|X_{e,bg} - X_{e,micro}\right|}{\sqrt{\sigma^2_{bg} + \sigma^2_{micro}}}$ where $X_{e,bg}$ is the signal passing through 4cm PMMA and $X_{e,micro}$ is the signal passing through microcalcification. In addition, σ^2 is the corresponding signal variance calculated as $\sigma = \sqrt{\sum_{u=0}^{\infty} NPS_T(u)}$. The model was tested with two different CMOS indirect detectors each one comprised of a RadEye CMOS photoreceptor and Gd_2O_2S:Tb or Gd_2O_2S:Eu scintillator. The latter has been evaluated as a probable candidate for imaging detectors [19-23]. Different surfaces densities of the scintillator were considered in the range between 20mg/cm^2 and 160 mg/cm^2. The CNR was tested for microcalcification thicknesses of 100 μm and 200 μm. The low energy spectrum was obtained by filtering a standard 35kV spectrum with Pd of 100 μm thickness and the high energy spectrum was obtained by filtering a standard 70kV spectrum with Yb of 800 μm thickness. The data used for calculating the equations were obtained from literature [13-27].

Results and Discussion

The low and high energy spectra are presented in Figure 1. It may be observed from the calculated data that the addition of Pd and Yb filtration introduces a sharp edge at 25keV and 62keV respectively. due to the k-edge of these two materials at approximately 24.4keV and 61.3keV respectively. These sharp edges can minimize the effect of variance in assessing material differences and composition.

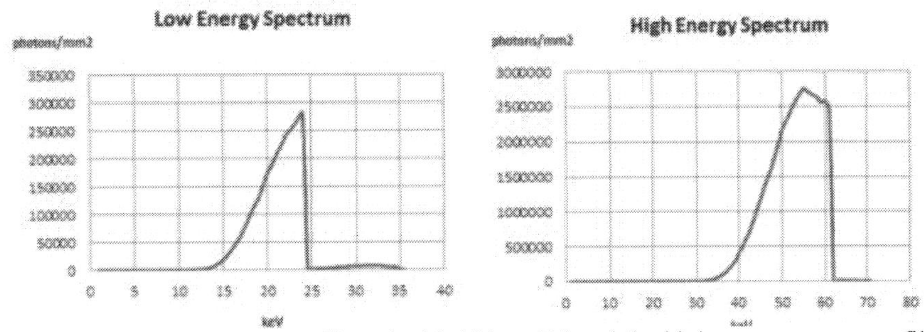

Figure 1. The low energy spectrum filtered with 100μm Pd and the high energy spectrum filtered with 800μm Yb.

Figure 2 present the CNR for the low and the high energy spectra for different Gd2O2S:Tb and Gd2O2S:Eu surface densities respectively. The thickness of the microcalcification was 200 μm . It can be observed from Figure 2 that Gd_2O_2S:Tb phosphor based X-ray detector exhibits better CNR than Gd_2O_2S:Eu phosphor based one. This occurs because CNR is the ratio of signal differences over the total noise variance. The signal in Gd_2O_2S:Tb detector is higher than that of Gd_2O_2S:Eu mainly due to the higher intrinsic conversion efficiency of the former which is 0.15 over 0.06. In addition the total number of electrons produced per keV absorbed (i.e. m_o) equals to 61 for Gd_2O_2S:Tb and 30 for Gd_2O_2S:Eu. Therefore the total optical photon output for Gd_2O_2S:Tb is higher than Gd_2O_2S:Eu. As it may be observed from Figure 2 the optimum phosphor thicknesses for Gd_2O_2S:Tb, are 70 mg/cm^2 and 100 mg/cm^2 for the low and the high X-ray energy respectively, while for Gd_2O_2S:Eu the corresponding thicknesses are 50 mg/cm^2 and 80 mg/cm^2. It is interesting to notice that the surfaces densities corresponding to higher DOG values were calculated similar to Figure 2, but approximately 30 mg/cm^2 lower. This occurs because the screen sensitivity with respect to thickness is affected by the X-ray absorption properties, as well as, the optical photon escape properties from the phosphor. When the thickness is increased the X-ray absorption probability also increases but the optical photon escape probability, as it is calculated by equation 3, decreases. Therefore, for each X-ray photon energy an optimum thickness can be associated. In our case, if detectability (i.e CNR) is considered, the best surface densities are between 70 mg/cm^2 and 100 mg/cm^2 for Gd_2O_2S:Tb and 50 mg/cm^2 and 80 mg/cm^2 for Gd_2O_2S:Eu.

The presented methodology was also tested for microcalcification thickness of 100 μm . The results were qualitatively the same. However the corresponding CNR values were lower due to the smaller difference of X-ray absorption probability between the100μm microcalcification and the mammary gland.

3. Conclusions

In this work, optimum indirect detector parameters, with respect to scintillation thickness for dual energy applications, were studied. The detector was modeled through the LCS theory, assuming Gd_2O_2S:Tb and Gd_2O_2S:Eu scintillators coupled to a CMOS photodetector. The 40mg/cm^2 and the 70mg/cm^2 Gd_2O_2S:Tb exhibited the higher DOG values for the low and the high energy correspondingly. Higher CNR between 200 μm microcalcification and mammary gland exhibited the 70mg/cm^2 and the 100mg/cm^2 Gd_2O_2S:Tb for the low and the high energy respectively. The results were qualitatively the same for 100 μm microcalcification but the corresponding CNR values were lower due to the lower absorption differences between the microcalcification and the mammary gland.

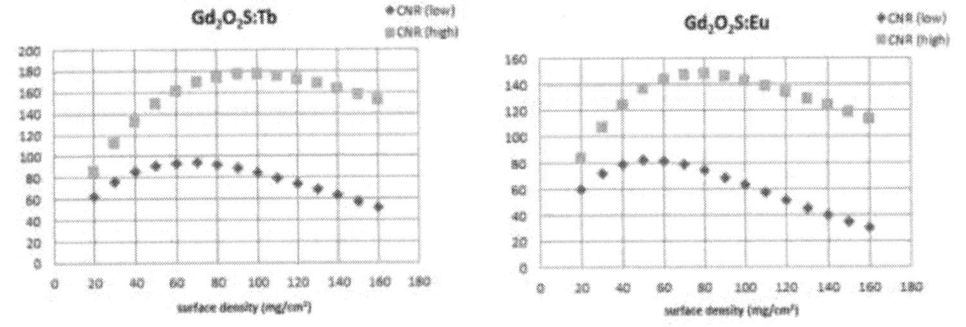

Figure 2. The CNR of Gd2O2S:Tb and Gd2O2S:Eu based detector for various surface densities.

Acknowledgement
This research has been co-funded by the European Union (European Social Fund) and Greek national resources under the framework of the "Archimedes III: Funding of Research Groups in TEI of Athens" project of the "Education & Lifelong Learning" Operational Programme.

4. References

[1] Smigel K 1995 *J. Natl. Cancer Inst.* **87** 1940
[2] Johns P and Yaffe M 1985 *Med. Phys.* **12** 289
[3] Johns P, Drost D, Yaffe M and Fenster A 1985 *Med. Phys.* **12** 297
[4] Boone M, Shaber G and Tecotzky M 1990 *Med. Phys.* **17** 665
[5] Boone M 1991 *Invest. Radiol.* **26** 521
[6] Brettle S and Cowen A 1994 *Phys. Med. Biol.* **39** 1989
[7] Asaga T, Masuzawa C, Yoshida A and Mattsuura H 1995 *J. Digit. Imag.* **8** 70
[8] Shaw C and Gur D 1992 *J. Digit. Imag.* **5** 262
[9] Lemacks M, Kappadath S, Shaw C, Liu X and Whitman G 2002 *Med. Phys.* **29** 1739
[10] Jong R, Yaffe M, Skarpatiotakis M, Shumak R, Danjoux N, Gunesekara A and Plewes D 2003 *Rad.* **228** 842
[11] Meter R and Rabbani M 1990 *Med. Phys.* **17** 65
[12] El-Mohri Y, Antonuk L, Zhao Q, Wang Y, Hong Du Y and Sawant A 2007 *Med. Phys.* **34** 315
[13] Nishikawa R and Yaffe M 1990 *Med. Phys.* **17** 894
[14] Kim H, Jun S, Ko J, Cho G and Graeve T 2008 *IEEE Trans. Nucl. Sci.* **55** 1357
[15] Liaparinos P, Kalyvas N, Kandarakis I and Cavouras D 2013 *Nucl. Instrum. and Meth. A* **697** 87
[16] Michail C, Spyropoulou V, Fountos G, Kalyvas N, Valais I, Kandarakis I and G Panayiotakis 2011 *IEEE Trans. Nucl. Sci.* **58** 314
[17] Lin C, Mathur B and Chang M 2002 *IEEE Trans. Nucl. Sci.* **49** 754
[18] Ji W and Rowlands J 1990 *Med. Phys.* **25** 2148
[19] Michail C, Fountos G, Liaparinos P, Kalyvas N, Valais I, Kandarakis I and Panayiotakis G 2010 *Med. Phys.* **37** 3694
[20] Michail C, Fountos G, Valais I, Kalyvas N, Liaparinos P, Kandarakis I and Panayiotakis G 2011 *IEEE Trans. Nucl. Sci.* **58** 2503
[21] Michail C, Valais I, Toutountzis A, Kalyvas N, Fountos G, David S, Kandarakis I and Panayiotakis G 2008 *IEEE Trans. Nucl. Sci.* **55** 3703
[22] Seferis I, Michail C, Valais I, Fountos G, Kalyvas N, Stromatia F, Oikonomou G, Kandarakis I and Panayiotakis G 2013 *Nucl. Instr. and Meth. Phys. Res.* **729** 307
[23] Seferis I, Michail C, Valais I, Zeler J, Liaparinos P, Fountos G, Kalyvas N, David S, Stromatia F, Zych E, Kandarakis I and Panayiotakis G 2014 *J. of Lum.* **151** 229
[24] https://w9.siemens.com/cms/oemproducts/Home/X-rayToolbox/spektrum/Pages/Default.aspx
[25] Kalyvas N, Liaparinos P, Michail C, David S, Fountos G, Wójtowich M, Zych E and Kandarakis I 2012 *Appl. Phys.* **106** 131
[26] http://www.rad-icon.com/products-software.php
[27] Rad-icon Imaging Corp. Remote Radeye data sheet

Optimization of breast cancer detection in Dual Energy X-ray Mammography using a CMOS imaging detector

V Koukou[1], G Fountos[2], N Martini[1], P Sotiropoulou[1], C Michail[2], N Kalyvas[2], I Valais[2], A Bakas[3], E Kounadi[4], I Kandarakis[2] and G Nikiforidis[1]

[1] Department of Medical Physics, Medical School, University of Patras, 265 00 Patras, Greece
[2] Department of Biomedical Engineering, Technological Educational Institute of Athens, Egaleo, 122 10 Athens, Greece
[3] Department of Medical Radiologic Technology, Technological Educational Institute of Athens, Egaleo, 122 10 Athens, Greece
[4] Ministry of Health, SEYYP, Pireos 205, 11853, Athens, Greece

E-mail: gfoun@teiath.gr

Abstract. Dual energy mammography has the ability to improve the detection of microcalcifications leading to early diagnosis of breast cancer. In this simulation study, a prototype dual energy mammography system, using a CMOS based imaging detector with different X-ray spectra, was modeled. The device consists of a 33.91 mg/cm^2 Gd_2O_2S:Tb scintillator screen, placed in direct contact with the sensor, with a pixel size of 22.5 μm. Various filter materials and tube voltages of a Tungsten (W) anode for both the low and high energy were examined. The selection of the filters applied to W spectra was based on their K-edges (K-edge filtering). Hydroxyapatite (HAp) was used to simulate microcalcifications. Calcification signal-to-noise ratio (SNR_{tc}) was calculated for entrance surface dose within the acceptable levels of conventional mammography. Optimization was based on the maximization of SNR_{tc} while minimizing the entrance dose. The best compromise between SNR_{tc} value and dose was provided by a 35kVp X-ray spectrum with added beam filtration of 100μm Pd and a 70kVp Yb filtered spectrum of 800 μm for the low and high energy, respectively. Computer simulation results show that a SNR_{tc} value of 3.6 can be achieved for a calcification size of 200 μm. Compared with previous studies, this method can improve detectability of microcalcifications.

1. Introduction

Screening and diagnosis using X-ray mammography rely on the early detection of microcalcifications (μCs) and soft-tissue masses [1-4]. The detection and visualization of μCs are often obscured by the overlapping tissue structures. Dual-Energy (DE) imaging technique offers an alternative approach for imaging and visualizing μCs. With this technique, high- and low-energy images are acquired and their differences are used to cancel out the background tissue structures. In the resulting subtracted image, 'signal' is the difference of the means between a target and its background, and 'noise' is the standard deviation of that difference. The SNR is the ratio between signal and noise [1]. For the dual-kVp technique, the SNR depends, among other factors, on the energy spectrum of the original low- and high-energy X-ray beams, the elemental composition of the objects in the image and the efficiency of the image receptor [2].

SNR in DE subtracted images is a powerful parameter to predict the visibility of the detail of interest against the background after the subtraction. Generally, the SNR in the subtracted images will be lower than in the original ones [3]. *Shaw and Gur* analyzed the SNR for various techniques of dual-energy subtraction and concluded that for the dual-kVp method, the SNR in the subtracted image would generally increase as the difference in kVp between the low and the high-energy exposures maximized [3]. *Lemacks et al* presented a numerical framework to calculate the signal-to-noise ratio of the subtracted microcalcification (μC) image for dual-kVp subtraction, and applied it to polyenergetic beams [4]. Their results applicable to Mo/Mo and W/La (anode/filter) beams, showed that for a 5cm thick compressed breast composed of 50% glandular and 50% adipose tissue, a SNR of 3 was achieved for a μC size of 250 μm, for certain combinations of low- and high-energy exposures [4]. However, under the current implementations of Dual Energy Digital Mammography (DEDM) the minimum detectable microcalcification size is in the range of 250 μm and 355 μm [4-8].

In this simulation study, a prototype dual energy mammography system using a CMOS based imaging detector with different X-ray spectra was modeled. The aim of the present study is the detection of the minimum microcalcification size for entrance surface dose within the acceptable levels of conventional mammography. Calcification signal-to-noise ratio (SNR_{tc}) was calculated and optimization was based on the maximization of SNR_{tc} while minimizing the entrance dose. Also, monochromatization of the X-ray spectra with the use of K-edge filtering technique was included.

2. Materials and Methods

An analytical model was developed in order to determine the calcification SNR assuming only Poisson distribution [4]. The breast is composed of adipose tissue of thickness t_a, glandular tissue of thickness t_b, and a cubic microcalcification of thickness t_c.

Considering that polyenergetic X-rays are used, the mean measured signals in the low- and high-energy images, S_l and S_h, can be expressed as:

$$S_{l,h} = \int dE \times R_j \times d^2 \times \Phi_j(E) \times e^{-\frac{\mu}{\rho}\alpha(E)\rho_a t_a - g_r\left(\frac{\mu}{\rho}g\rho_g - \frac{\mu}{\rho}a\rho_a\right)T - \frac{\mu}{\rho}c(E)\rho_c t_c} \times A(E) \times Q(E) \quad (1)$$

The SNR of the subtracted signal, t_c, can be expressed as follows [4]:

$$SNR_{tc} = \frac{tc}{\sqrt{\sigma_{tc}^2}} \quad (2)$$

where $\boldsymbol{\sigma_{tc}}$ represents the noise level in the calcification subtraction image.

The unfiltered spectra were parameterized by Boone *et al* for a Tungsten (W) anode for both low and high energy [9,10]. The maximum tube voltage range was 23-35kVp for the low energy (LE) and 50-70kVp for the high energy (HE). The beams have been attenuated by various filters according to their K-edge (K-edge filtering) from 18keV to 61keV [11]. The filter thicknesses varied from 100-1000 μm for both energies.

The surface dose, after filtration of the X-ray beams, was calculated according to equation (3) [12]. The total exposure is the sum of the low- and high-energy exposures. So the whole process was adapted to all, low- and high-, kVp spectra from the Tungsten (W) anode.

$$Dose(mGy) = 1.83 \cdot 10^{-6} \cdot \Phi_o(E) \cdot E \cdot \left(\frac{\mu_{en}}{\rho}\right)_{air} \cdot 8.77 \cdot 10^{-3} \quad (3)$$

where $\Phi_o(E)$ is the filtered X-ray spectrum value (photons/mm^2) at energy E. μ_{en}/ρ is the X-ray mass energy absorption coefficient of air at energy E obtained from NISTIR (National Institute of Standards and Technology Interagency Report) [13].

The SNR has been evaluated for the subtracted digital images of 5cm thick compressed breast, with composition of 50% glandular and 50% adipose tissue. The microcalcifications tested, were 100, 150, 200, 250 and 300 μm thick [14].

The elemental compositions of adipose and glandular breast tissue, the microcalcifications, and the scintillator (Gd_2O_2S:Tb) were used to calculate the mass-attenuation coefficients using data published by NISTIR with the computer program XMuDat [13,15]. Hydroxyapatite (HAp) was used to simulate microcalcifications. Hydroxyapatite is a calcium-phosphate mineral form ($Ca_{10}(PO_4)_6(OH)_2$) and in crystalline form has a density of 3.18gr/cm^3 [16].Note that, in the calculation of SNR_{tc}, the mass attenuation coefficients were then replaced by *effective mass attenuation coefficients*.

In this study, terbium-doped gadolinium oxysulfide (Gd_2O_2S:Tb) was considered to be the scintillator material coupled to a CMOS photodiode pixel array (RadEye HR). This scintillator was selected due to its efficiency and imaging properties, which are superior in comparison to other detectors [17]. The density of Gd_2O_2S:Tb is 7.34 g/cm^3. The surface density, W, (density multiplied by the thickness of the material) of Gd_2O_2S:Tb used in this study is 0.03394 g/cm^2 (approximately 34 mg/cm^2). The sensor pitch is 22.5 μm. The Quantum Detection Efficiency (QDE), $A(E)$, and the matching factor (α_s), $Q(E)$, were calculated according to Michail *et al* [18].

Computer simulations provided values for the following parameters: (i) mean energy (ME), (ii) Full Width at Half Maximum (FWHM) and (iii) total entrance surface dose. Optimization was based on the minimization of total entrance surface dose, while maintaining SNR_{tc} above the threshold. The SNR_{tc} threshold was set to be 3 ($SNR_{tc} \geq 3$) [4]. Furthermore, minimization of FWHM leading to narrow energy band X-rays was included.

3. Results and Discussion

For low energy filter selection, FWHM was plotted as a function of surface density (density multiplied by the thickness of the material) for all low energy filters tested (Figure 1). Palladium (Pd) filter has the lower values of FWHM in the whole surface density range, leading to narrower spectra. For this reason, Pd was selected as the low energy filter.

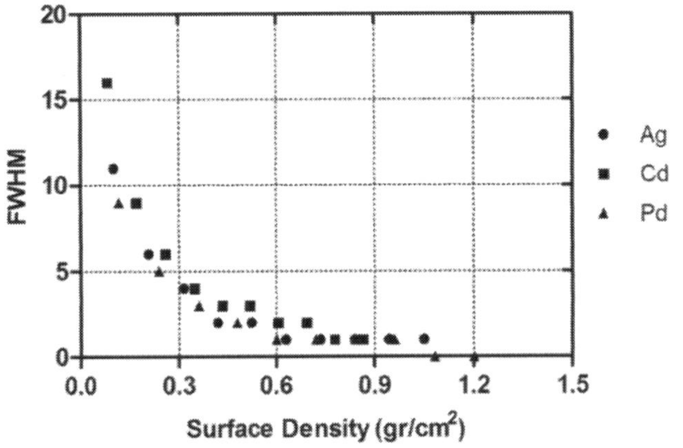

Figure 1. FWHM as a function of surface density for all low energy filters tested at 35kVp.

Figure 2 shows the SNR_{tc} as a function of surface density for low-/ high-energy tungsten spectra at 35/70kVp filtered with Pd-100 μm and Ce, Cu, Eu, Nd, Sm, Ho, Yb respectively, for a μC size of 200 μm. It is obvious that SNR_{tc} values were above the threshold only for Cu, Ho and Yb filters. Maximization of SNR_{tc} was accomplished at the surface density range between 0.4 and 0.55gr/cm^2. An SNR value of 3 is achieved for a μC size of 200 μm. Thus, a 200 μm microcalcification size was considered as the minimum detectable μC size in this study.

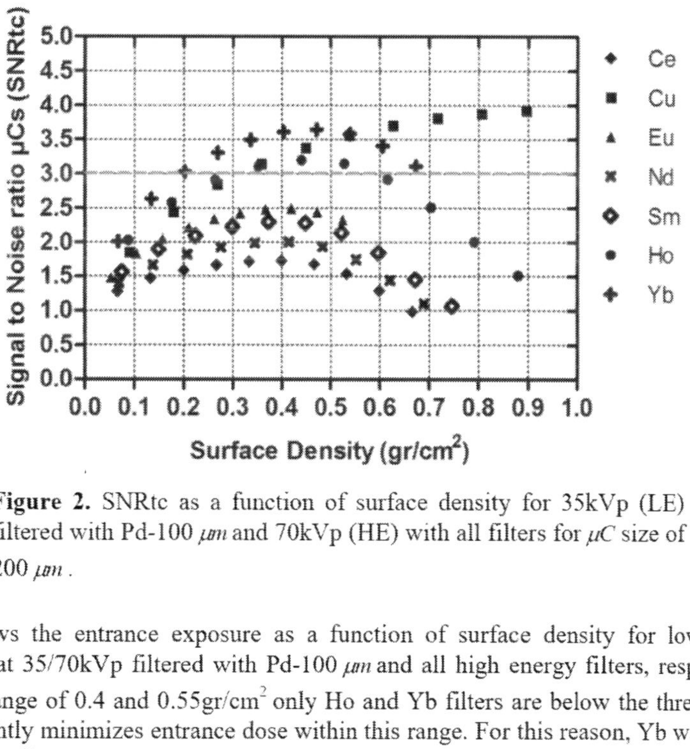

Figure 2. SNRtc as a function of surface density for 35kVp (LE) filtered with Pd-100 μm and 70kVp (HE) with all filters for μC size of 200 μm.

Figure 3 shows the entrance exposure as a function of surface density for low-/ high-energy tungsten spectra at 35/70kVp filtered with Pd-100 μm and all high energy filters, respectively. In the surface density range of 0.4 and 0.55gr/cm^2 only Ho and Yb filters are below the threshold of 6mGy. Note that Yb slightly minimizes entrance dose within this range. For this reason, Yb was considered to be the high energy filter.

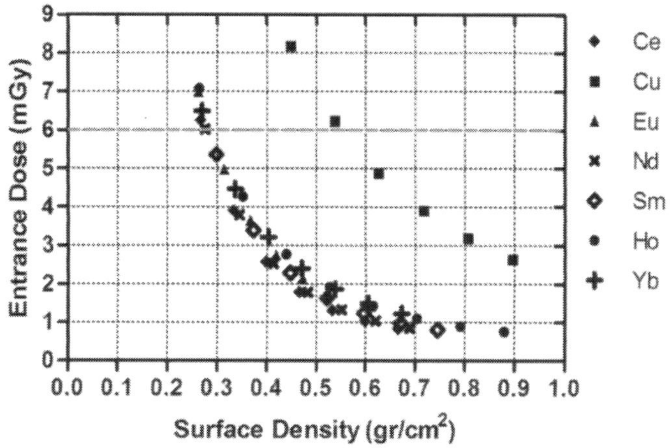

Figure 3. Entrance dose as a function of surface density for 35kVp (LE) filtered with Pd-100 μm and 70kVp (HE) with all filters.

4. Conclusion

In this simulation study, a prototype dual energy mammography system was modelled, using a CMOS based imaging detector with different X-ray tungsten spectra. Optimization was based on the maximization of SNR_{tc} while minimizing the entrance dose. The best compromise between SNR_{tc} value and dose was provided by 35kVp with added beam filtration of 100 μm Pd and 70kVp Yb filtered spectrum of 800 μm for the low and high energy, respectively. Computer simulation results show that a SNR_{tc} value of 3.6 can be achieved for a calcification size of 200 μm and assuming a 50% adipose and 50% glandular tissue composition and a breast thickness of 5 cm. The total entrance dose was 1.85mGy. Compared with previous studies, this method can improve detectability of microcalcifications.

Acknowledgement

This research has been co-funded by the European Union (European Social Fund) and Greek national resources under the framework of the "Archimedes III: Funding of Research Groups in TEI of Athens" project of the "Education & Lifelong Learning" Operational Programme.

5. References

[1] Wagner F and Brown G 1985 *Phys. Med. Biol.* **30** 489
[2] Brandan M and Ramirez V 2006 *Phys. Med. Biol.* **51** 2307
[3] Shaw C and Gur D 1991 *J. Digit. Imaging* **5** 262
[4] Lemacks M, Kappadath S, Shaw C, Liu X, Whitman G 2002 *Med. Phys.* **29** 1739
[5] Kappadath S and Shaw C 2003 *Med. Phys.* **30** 1110
[6] Kappadath S and Shaw C 2004 *Phys. Med. Biol.* **21** 2563
[7] Kappadath S and Shaw C 2008 *Phys. Med. Biol.* **53** 5421
[8] Ducote J and Molloi S 2008 *Med. Phys.* **35** 5411
[9] Boone J, Fewell T and Jennings R 1997 *Med. Phys.* **24** 1863
[10] Boone J and Seibert A 1997 *Med. Phys.* **24** 1661
[11] Koukou V, Martini D, Sotiropoulou P, Fountos G, Michail C, Valais I, Kandarakis I and Nikiforidis G 2012 *e-JST* **7** 79
[12] Michail C, Spyropoulou V, Fountos G, Kalyvas N, Valais I, Kandarakis I and Panayiotakis G 2011 *IEEE Trans. Nucl. Sci.* **58** 314
[13] Hubbell J and Seltzer S 1995 US Department of commerce NISTIR 5632
[14] Koukou V, Martini N, Fountos G, Sotiropoulou P, Michail C, Valais I, Kandarakis I and Nikiforidis G 2014 *IFMBE Proceedings* **41** 459
[15] Hammerstein R, Miller D, White D, Masterson M, Woodward H and Laughlin J 1979 *Radiology* **130** 485
[16] Gong J, Arnold J and Cohn S 1964 *Anat. Rec.* **149** 319
[17] Seferis I, Michail C, Valais I, Fountos G, Kalyvas N, Stromatia F, Oikonomou G, Kandarakis I, Panayiotakis G. 2013 *Nucl. Instrum. Meth. Phys. Res. A.* **729** 307
[18] Michail C, Fountos G, Liaparinos P, Kalyvas N, Valais I, Kandarakis I and Panayiotakis G. 2010 *Med. Phys.* **37** 3694

Crank-Nicholson method for rate equations in powder random lasers

B García-Ramiro[1], M A Illarramendi[2] and J Zubia[3]

[1]Department of Applied Mathematics, E.U. Ingeniería Técnica Industrial de Bilbao, Universidad del País Vasco UPV/EHU, Paseo Rafael Moreno "Pitxitxi" 3, 48013 Bilbao, Spain

[2]Department of Applied Physics I, Escuela Superior de Ingeniería, Universidad del País Vasco UPV/EHU, Alda. Urquijo s/n 48013 Bilbao, Spain

[3]Department of Communications, Escuela Superior de Ingeniería, Universidad del País Vasco UPV/EHU, Alda. Urquijo s/n 48013 Bilbao, Spain

E-mail: mariabegona.garciar@ehu.es

Abstract. In this work, we show the resolution of the rate equations in powder random lasers by using the Crank-Nicholson finite difference method. Light propagation in our powders is described by the model of light diffusion. The generalized time-dependent random laser equations describing our system are formed by three differential coupled equations: two diffusion equations for the pump and emitted light and a rate equation for the density of the dopant molecules in the excited state. The system has been solved for two pumping schemes (one-photon and two-photon excitation) and for a wide range of temporal incident pulses (from femtoseconds to nanoseconds).

1. Introduction

Conventional lasers are usually constructed from two basic components: a gain material that is pumped in order to provide amplification of light and a cavity to provide feedback. However, random lasers replace the traditional laser cavity with a random, multiple-scattering medium. This type of laser becomes a subject of intense theoretical and experimental studies because of its important potential applications [1]. In addition, a major advantage of random lasers over regular lasers is that their production is cheap and the required technology relatively simple, but the laser dynamics are much more complex than that of the conventional laser and there is still much to understand. For example, scientists recently explored the mode-locking of random lasers and investigated how to control their operation.

The random laser equations have been solved by using several numerical methods due to the difficulty of finding the exact solution: method of lines [2], Montecarlo simulation [3] and finite-difference time domain method (FDTD) [4]. In this work, we solve these equations for two pumping schemes, one-photon (OP) and two-photon (TP) excitations, and for a wide range of temporal incident pulses (from femtoseconds to nanoseconds) by using the Crank-Nicholson finite difference method.

2. Theoretical model

Assuming a diffusive propagation of light in powder random lasers, the equations describing our model for one and two-photon excitations are:

$$\frac{\partial W_p(z,t)}{\partial t} = D_p \frac{\partial^2 W_p(z,t)}{\partial z^2} - g(z,t) + p(z,t) \tag{1}$$

$$\frac{\partial W_e(z,t)}{\partial t} = D_e \frac{\partial^2 W_e(z,t)}{\partial z^2} + f\,\mathrm{v}\,\sigma_{em} N(z,t) W_e(z,t) + \gamma \frac{N(z,t)}{\tau_s} \tag{2}$$

$$\frac{\partial N(z,t)}{\partial t} = g(z,t) - f\,\mathrm{v}\,\sigma_{em} N(z,t) W_e(z,t) - \frac{N(z,t)}{\tau_s} \tag{3}$$

where $W_p(z,t)$ and $\partial W_e(z,t)$ are the light densities at the pump and emission wavelengths respectively and $N(z,t)$ is the density of dye molecules in the excited state. The term corresponding to light absorption, $g(z,t)$ is given by $g_{OP}(z,t) = f\,\mathrm{v}\,K_{abs} W_p(z,t)$ and by $g_{TP}(z,t) = f\,\mathrm{v}^2 \beta \hbar \omega_p W_p^2(z,t)$, for one and two- photon excitations respectively. K_{abs} is the one-photon absorption coefficient of the material at the pump wavelength and β is the two-photon one. The volume fraction, *f*, occupied by the scatters has been included in the equations to take into account the effective part of light density which penetrates into the particles. v is the speed of light in the medium. σ_{em} is the stimulated emission cross section, τ_s is the excited state lifetime and , D_p and D_e, are light diffusion coefficient for pump and emitted radiation, respectively. γ is the fraction of spontaneous emission contributing to the laser process. In both pumping schemes, the source of diffuse radiation, $p(z,t)$, is an incoming Gaussian pulse in the *z* direction which is extinguished along its path through the sample. This function $p(z,t)$ is different for each type of excitation. The system of equations is solved with the following boundary and initial conditions [5]:

$$\begin{aligned} &W_p(-l_e,t) = W_p(L+l_e,t) = W_e(-l_e,t) = W_e(L+l_e,t) = 0 && \forall t \\ &W_p(z,0) = W_e(z,0) = N(z,0) = 0 && \forall z \end{aligned} \tag{4}$$

l_e is the extrapolation length and L is the scattering sample thickness.

3. Numerical solution

The set of coupled non linear partial differential equations (1)-(3) are numerically solved by using the Crank-Nicholson method. On the condition that $n = (L+2l_e)/h$ is an integer, the domain (z, t) is discretized by two sizes of step h (spatial) and k (time): $z_i = -l_e + ih \quad 0 \le i \le n$ and $t_j = jk \quad j \ge 0$.

3.1. One- photon excitation

The equation (1) is solved first and the derivatives are approximated at the mesh point by

$$\begin{aligned} &\frac{\partial W_p(z_i,t_j)}{\partial t} \approx \frac{W^p_{i,j+1} - W^p_{i,j}}{k} \quad \text{with} \quad W^p_{i,j} = W_p(z_i,t_j) \\ &\frac{\partial^2 W_p(z_i,t_j)}{\partial z^2} \approx \frac{1}{2}\left(\frac{W^p_{i+1,j} - 2W^p_{i,j} + W^p_{i-1,j}}{h^2} + \frac{W^p_{i+1,j+1} - 2W^p_{i,j+1} + W^p_{i-1,j+1}}{h^2} \right) \end{aligned} \tag{5}$$

Replacing the derivatives (6) in the equation (1), evaluating $g(z,t)$ and $p(z,t)$ in the intermediate step between j and j +1 and using the boundary and initial condition, this equation can be written in matrix form as:

$$A^{OP}W_{j+1}^{p*} = B^{OP}W_j^{p*} + P_j^* \tag{6}$$

with $W_j^{p*} \equiv \left(W_{1,j}^p, W_{2,j}^p, \ldots, W_{n-1,j}^p\right)^T$, $P_j^* = \left(p\left(z_1, t_{j+\frac{k}{2}}\right), p\left(z_2, t_{j+\frac{k}{2}}\right), \ldots, p\left(z_{n-1}, t_{j+\frac{k}{2}}\right)\right)^T$

where $A^{OP} = \left(a_{ij}^{OP}\right)_{1\le i\le n-1}$ and $B^{OP} = \left(b_{ij}^{OP}\right)_{1\le i\le n-1}$ are tridiagonal matrices

$$a_{ij}^{OP} = \frac{1}{k} + \frac{D_p}{h^2} + \frac{f\,\mathrm{v}\,K_{abs}}{2} \quad \text{if } i=j, \qquad a_{ij}^{OP} = -\frac{D_p}{2h^2} \text{ if } |i-j|=1, \qquad a_{ij}^{OP} = 0 \text{ if } |i-j|>1$$

$$b_{ij}^{OP} = \frac{1}{k} - \frac{D_p}{h^2} - \frac{f\,\mathrm{v}\,K_{abs}}{2} \quad \text{if } i=j, \qquad b_{ij}^{OP} = \frac{D_p}{2h^2} \text{ if } |i-j|=1, \qquad b_{ij}^{OP} = 0 \text{ if } |i-j|>1$$

Therefore, after solving the matrix equation (6), the solutions of equations (2) and (3) can be determined by the solution of the following matrix system:

$$\begin{cases} A_1^{OP}\,W_{j+1}^{e*} - B_1^{OP}\,W_j^{e*} + f\,\mathrm{v}\,\sigma_{em}N_j^{e*} * W_j^{e*} + \dfrac{\gamma}{\tau_s}N_j^{e*} \\ N_{j+1}^{e*} = N_j^{e*} + k(f\,\mathrm{v}\,K_{abs}W_j^{p*} - f\,\mathrm{v}\,\sigma_{em}N_j^{e*} * W_j^{e*} - \dfrac{1}{\tau_s}N_j^{e*}) \end{cases} \tag{7}$$

where A_1^{OP} eta B_1^{OP} are the corresponding tridiagonal matrices and $N_j^{e*} * W_j^{e*}$ represents the element by element product of N_j^{e*} and W_j^{e*}. In this case the functions of the coupled equations (2) and (3) are computed at the time gridpoints instead of at midpoints of the temporal subinterval in order to obtain a system of linear equations.

3.2. Two-photon excitation

The same process can be applied for two-photon excitations. Then, the equations (1)-(3) can be written as:

$$A^{TP}W_{j+1}^{p*} = B_1^{TP}W_j^{p*} - f\,\mathrm{v}^2\beta\hbar\omega_p W_j^{p*} * W_j^{p*} + P_j^* \tag{8}$$

$$\begin{cases} A_1^{TP}W_{j+1}^{e*} = B_1 W_j^{e*} + f\,\mathrm{v}\,\sigma_{em}\,N_j^{e*} * W_j^{e*} + \dfrac{\gamma}{\tau_s}N_j^{e*} \\ N_{j+1}^{e*} = N_j^{e*} + k(f\,\mathrm{v}^2\beta\hbar\omega_p W_j^{p*} * W_j^{p*} - f\,\mathrm{v}\,\sigma_{em}N_j^{e*} * W_j^{e*} - \dfrac{1}{\tau_s}N_j^{e*}) \end{cases} \tag{9}$$

4. Numerical results

We have carried out simulations to study the laser-like emission in several powder random lasers for OP [5, 6] and TP [7] excitations. The theoretical results are in a good agreement with the experimental data which validates the Crank-Nicholson method. As an example, we show the theoretical pulse shortening calculated from the reduction of the full width at half maximum (FWHM) of the temporal profiles when increasing the pump energies under TP excitation in a ground powder of a silica gel containing Rhodamine 6G doped silica nanoparticles. As the emitted photons are collected along the backward direction of the incident pump beam, the time evolution of the emitted light has been calculated from $\vec{F}_e(t) = -D_e\dfrac{\partial W_e(z,t)}{\partial z}\hat{z}$ evaluated at the front sample surface($z = 0$). Figure 1 shows the FWHM of the emitted pulse obtained as a function of the pump pulse energy for different temporal incident pulses. Δ represents the temporal half width at half maximum of incident pulse. The data have been fitted to a sigmoid function. The inflection points of these fits represent the threshold energy densities at which the light amplification begins. The dependence of the lasing threshold on the Δ incident pulse has been plotted in Figure 2. As it can be observed, lasing threshold increases very abruptly as the value of Δ is getting closer to the spontaneous lifetime of the material. The input values for these calculations are the material parameters: K_{abs} = 148.5 cm^{-1}, σ_{em} = 2.5x10^{-16} cm^2, τ_s (OP) = 1.65 ns, τ_s (TP) = 2.1 ns, $\beta\hbar\omega_p$ = 3.86x10^{-13} μm·ps, n_{eff} = 1.16, f = 0.43.

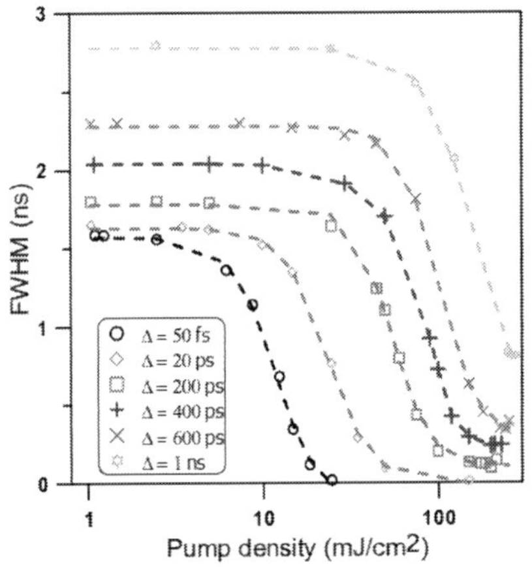

Figure 1. Theoretical FWHM of the pulses as a function of the pump density. (○) Δ = 50 fs, (◊) Δ = 20 ps, (□) Δ = 200 ps, (+) Δ = 400 ps, (×) Δ = 600 ps, (¤) Δ = 1 ns. The dashed lines are the sigmoidal fits to the data.

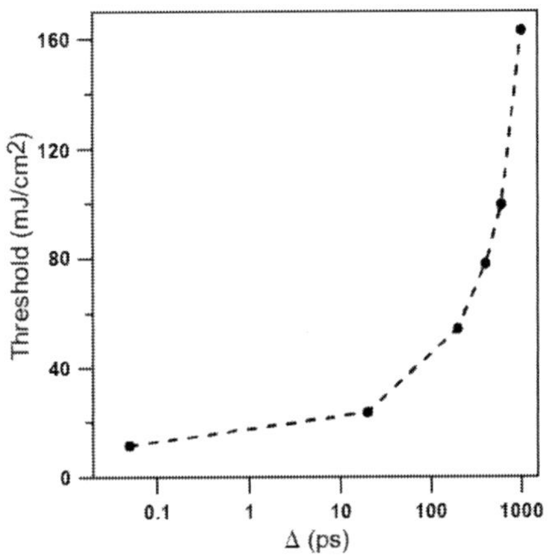

Figure 2. Lasing threshold as a function of the Δ incident pulse. The Δ values are the same as in Figure 1. The dashed line is a guide for the eye.

5. Conclusion

In this work, it is shown that the Crank-Nicholson finite difference method solves satisfactorily the laser rate equations of powder random lasers when a diffusive propagation of light is considered.The method has been applied for one and two-photon excitation. The numerically calculated results in a Rhodamine 6G doped ground powder agree with the corresponding experimental results [5, 7].

References

[1] Wiersma D 2008 *Nat. Phys.* **4** 359
[2] van Soest G, Poelwijk F J, Sprik R and Lagendijk A 2001 *Phys. Rev. Lett.* **86** 1522
[3] Ignesti E, Tommasi F, Fini L, Lepri S, Radhalakshmi V, Wiersma D and Cavalieri S 2013 *Phys. Rev. A* **88** 033820
[4] Andreasen J and Cao H 2010 *Phys. Rev. A* **82** 063835
[5] García-Revilla S , Fernández J , Illarramendi M A, García-Ramiro B, Balda R, Cui H, Zayat M, and Levy D 2008 *Opt. Express* **16**, 12251 (2008)
[6] García-Ramiro B, Illarramendi M A, Aramburu I, Fernández J and Balda R 2010 *Opt. Mater.* **33** 211
[7] García-Revilla S, Solá I, Balda R, Roso L, Levy D, Zayat M and Fernández J 2010 *Proc. SPIE* **7598**, 759804

Stochastic approach to the numerical solution of the non-stationary Parker's transport equation

A. Wawrzynczak[1], **R. Modzelewska**[2], **A. Gil**[2]
[1]Institute of Computer Sciences, Siedlce University, Poland,
[2]Institute of Mathematics and Physics, Siedlce University, Poland.
E-mail: awawrzynczak@uph.edu.pl, renatam@uph.edu.pl, gila@uph.edu.pl,

Abstract. We present the newly developed stochastic model of the galactic cosmic ray (GCR) particles transport in the heliosphere. Mathematically Parker transport equation (PTE) describing non-stationary transport of charged particles in the turbulent medium is the Fokker-Planck type. It is the second order parabolic time-dependent 4-dimensional (3 spatial coordinates and particles energy/rigidity) partial differential equation. It is worth to mention that, if we assume the stationary case it remains as the 3-D parabolic type problem with respect to the particles rigidity R. If we fix the energy/rigidity it still remains as the 3-D parabolic type problem with respect to time. The proposed method of numerical solution is based on the solution of the system of stochastic differential equations (SDEs) being equivalent to the Parker's transport equation. We present the method of deriving from PTE the equivalent SDEs in the heliocentric spherical coordinate system for the backward approach. The advantages and disadvantages of the forward and the backward solution of the PTE are discussed. The obtained stochastic model of the Forbush decrease of the GCR intensity is in an agreement with the experimental data.

1. Introduction

Many problems in physics, finance or biology can be represented as models of the diffusive transport processes described by the Fokker-Planck type equations (FPE). The difficulty of the numerical solution of this type equations increases with the problem dimension. Reason is the instability of the numerical schemes like finite-differences and finite-volume in the higher dimensions. In consequence, to ensure the scheme stability and convergence the density of numerical grid must be improved, increasing the computational complexity. To overcome this problem the stochastic methods can be applied (e.g.[1],[2]). In this approach the individual particle motion is described as a Markov stochastic process, and the system evolves probabilistically. Accordingly, the FPE can be solved by the corresponding stochastic differential equations (SDEs) (e.g. [3]).

We apply stochastic methodology to model the galactic cosmic rays (GCR) transport in the heliosphere. During the propagation through the heliosphere GCR particles are modulated by the solar wind and heliospheric magnetic field (HMF). Modulation of the GCR is a result of action of four main processes: convection by the solar wind, diffusion on irregularities of HMF, particles drifts in the non-uniform magnetic field and adiabatic cooling (e.g. [4]). Transport of the GCR particles in heliosphere can be described by the Parker transport equation (PTE) [5] being the second order parabolic type partial differential equation:

$$\frac{\partial f}{\partial t} = \vec{\nabla} \cdot (K_{ij}^S \cdot \vec{\nabla} f) - (\vec{v}_d + \vec{U}) \cdot \vec{\nabla} f + \frac{R}{3}(\vec{\nabla} \cdot \vec{U})\frac{\partial f}{\partial R}, \tag{1}$$

where $f = f(\vec{r}, R, t)$ is an omnidirectional distribution function of three spatial coordinates $\vec{r} = r(r, \theta, \varphi)$, particles rigidity R and time t; $\vec{U}$ is solar wind velocity, $\vec{v}_d$ the drift velocity, and K_{ij}^S is the symmetric part of the diffusion tensor of the GCR particles.

Based on the stochastic approach we present model of the short-time variation of the GCR intensity, called the Forbush decrease (Fd) [9]. The Fd occurs as an occasional decrease in the GCR intensity recorded on the Earth surface by the neuron monitors. The Fds follow the occurrence of the solar flares and intensive solar coronal mass ejecta (CME) [10]. There can be distinguished two types of the Fds: 1) sporadic- being the result of the shock waves and the magnetic clouds appearing in the interplanetary space, caused by the solar flares on the Sun; 2) recurrent type Fd connected with the corotating interaction regions (CIR) appearing in the interplanetary space in connection with the solar rotation.

2. Stochastic approach

To model the GCR transport by the stochastic methods the corresponding SDEs must be derived. Firstly, the PTE (Eq.1) should be transformed to the standard form of the FPE which, depending on integration's direction, can be generally expressed in two forms [3]:
time-forward:

$$\frac{\partial F}{\partial t} = \sum_i \frac{\partial}{\partial x_i}(A_i \cdot F) + \frac{1}{2}\sum_{i,j} \frac{\partial^2}{\partial x_i \partial x_j}(B_{ij}B_{ij}^T \cdot F), \tag{2}$$

and time-backward:

$$\frac{\partial F}{\partial t} = \sum_i A_i \frac{\partial F}{\partial x_i} + \frac{1}{2}\sum_{i,j} B_{ij}B_{ij}^T \frac{\partial^2 F}{\partial x_i \partial x_j}. \tag{3}$$

Corresponding to Eqs. 2 and 3 SDE can be written as (e.g. [3]):

$$d\vec{r} = \vec{A}_i \cdot dt + B_{ij} \cdot d\vec{W}, \tag{4}$$

where $\vec{r}$ is the trajectory of individual pseudoparticle in the phase space and dW_i is the Wiener process, usually written as $dW_i = \sqrt{dt} \cdot dw_i$, dw_i is the randomly fluctuating term having Gaussian distribution.

In both, forward and backward cases, to solve Eq.4 first we start from some initial position in space and time and integrate along the pseudoparticles trajectories until they reach the boundary. The choice between the forward and backward integration depends on the problem that has to be solved. In the forward approach pseudoparticles should be initialized at various points at the boundary, where the GCR particles enter the heliosphere. Then, its trajectory should be traced until they arrive the point of interest, e.g. Earth orbit. To obtain the reasonable statistic a huge number of pseudoparticles should be initialized because most of them do not reach the Earth orbit. The backward approach is more efficient for the GCR propagation in the heliosphere, because it reduces the number of 'useless' particles. In the backward integration pseudoparticles are initialized at the Earth orbit and traced backward in time until they reach the heliosphere's boundary (in this paper assumed at 100 AU, Fig. 1). The particle distribution function can be obtained by averaging over the entrance points, $f(\vec{r}, R) = \frac{1}{N}\sum_{n=1}^{N} f_{LIS}(R)$, where $f_{LIS}(R)$ is the cosmic ray spectrum at the outer boundary taken as in [14] and R is the rigidity of the nth particle at the entrance point.

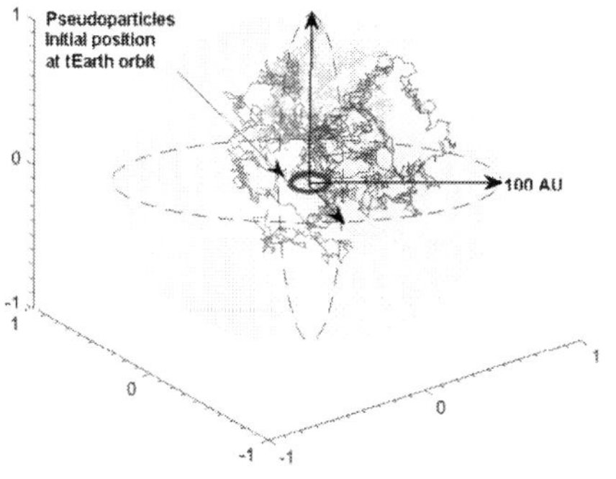

Figure 1. The sample pseudoparticles trajectories within the heliosphere.

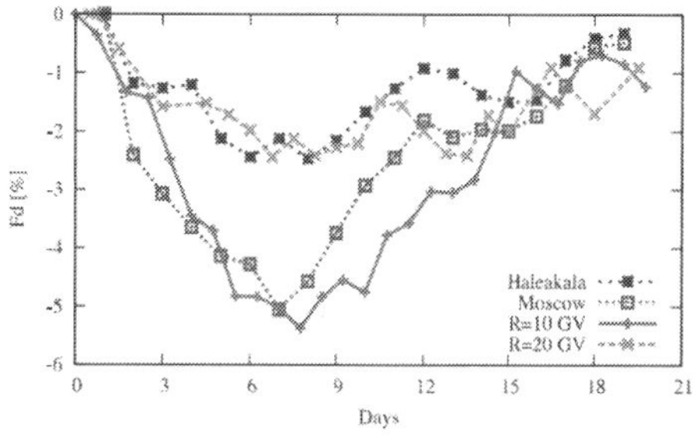

Figure 2. Changes of the expected amplitudes of the Fd of the GCR intensity at the Earth orbit, for the rigidity of 10 and 20 GV based on the solutions of the Eqs.6 in comparison with the GCR intensity registered by Moscow and Haleakala neutron monitors during the Fd in 18 March - 4 April 2002.

The PTE (Eq.1) in the 3-D spherical coordinate system (r, θ, φ) can be written as time-backward FPE diffusion equation:

$$\frac{\partial f}{\partial t} = A_1\frac{\partial^2 f}{\partial r^2} + A_2\frac{\partial^2 f}{\partial \theta^2} + A_3\frac{\partial^2 f}{\partial \varphi^2} + A_4\frac{\partial^2 f}{\partial r\partial \theta} + A_5\frac{\partial^2 f}{\partial r\partial \varphi} + A_6\frac{\partial^2 f}{\partial \theta\partial \varphi} + A_7\frac{\partial f}{\partial r} + A_8\frac{\partial f}{\partial \theta} + A_9\frac{\partial f}{\partial \varphi} + A_{10}\frac{\partial f}{\partial R} \quad (5)$$

with following coefficients:

$A_1 = K^S_{rr}, A_2 = \frac{K^S_{\theta\theta}}{r^2}, A_3 = \frac{K^S_{\varphi\varphi}}{r^2 sin^2\theta}, A_4 = \frac{2K^S_{r\theta}}{r}, A_5 = \frac{2K^S_{r\varphi}}{r sin\theta}, A_6 = \frac{2K^S_{\theta\varphi}}{r^2 sin\theta},$

$A_7 = \frac{2}{r}K^S_{rr} + \frac{\partial K^S_{rr}}{\partial r} + \frac{ctg\theta}{r}K^S_{\theta r} + \frac{1}{r}\frac{\partial K^S_{\theta r}}{\partial \theta} + \frac{1}{r sin\theta}\frac{\partial K^S_{\varphi r}}{\partial \varphi} - U - v_{d,r},$

$A_8 = \frac{K^S_{r\theta}}{r^2} + \frac{1}{r}\frac{\partial K^S_{r\theta}}{\partial r} + \frac{1}{r^2}\frac{\partial K^S_{\theta\theta}}{\partial \theta} + \frac{ctg\theta}{r^2}K^S_{\theta\theta} + \frac{1}{r^2 sin\theta}\frac{\partial K^S_{\varphi\theta}}{\partial \varphi} - \frac{1}{r}v_{d,\theta},$

$A_9 = \frac{K^S_{r\varphi}}{r^2 sin\theta} + \frac{1}{r sin\theta}\frac{\partial K^S_{r\varphi}}{\partial r} + \frac{1}{r^2 sin\theta}\frac{\partial K^S_{\theta\varphi}}{\partial \theta} + \frac{1}{r^2 sin^2\theta}\frac{\partial K^S_{\varphi\varphi}}{\partial \varphi} - \frac{1}{r sin\theta}v_{d,\varphi}, A_{10} = \frac{R}{3}\nabla \cdot U.$

We apply the full 3D anisotropic diffusion tensor of GCR particles $K_{ij} = K^{(S)}_{ij} + K^{(A)}_{ij}$ consisting of the symmetric $K^{(S)}_{ij}$ and antisymmetric $K^{(A)}_{ij}$ parts presented in [7]. The drift velocity of GCR particles is implemented as: $v_{d,i} = \frac{\partial K^{(A)}_{ij}}{\partial x_j}$ [8]. The corresponding to Eq. 5 set of SDEs with matrix B_{ij}, $(i, j = r, \theta, \varphi)$ has a form (the same form can be found in [15]):

$$\begin{aligned} dr &= A_7 \cdot dt + [B \cdot dW]_r \\ d\theta &= A_8 \cdot dt + [B \cdot dW]_\theta \quad (6) \\ d\varphi &= A_9 \cdot dt + [B \cdot dW]_\varphi \\ dR &= A_{10} \cdot dt, \end{aligned} \qquad B_{ij} = \begin{bmatrix} \sqrt{2A_1} & 0 & 0 \\ \frac{A_4}{\sqrt{2A_1}} & \sqrt{2A_2 - \frac{A_4^2}{2A_1}} & 0 \\ \frac{A_5}{\sqrt{2A_1}} & \frac{A_6 - \frac{A_4 A_5}{2A_1}}{B_{\theta\theta}} & \sqrt{2A_3 - B^2_{\varphi r} - B^2_{\varphi\theta}} \end{bmatrix}.$$

We integrate Eqs. 6 backward in time by the Euler–Maruyama scheme. As an initial condition we assume an empty heliosphere, as discussed in [16]. Solving Eqs. 6 in spherical coordinates we apply the following boundary conditions: $\varphi_i < 0 \rightarrow \varphi_i = \varphi_i + 2\pi$, $\varphi_i > 2\pi \rightarrow \varphi_i = \varphi_i - 2\pi$, $\theta_i < 0 \rightarrow \theta_i = \theta_i + 2\pi$ and $\theta_i > \pi \rightarrow \theta_i = \pi - |\theta_i|$. For the inner radial boundary we consider the reflecting boundary, $\frac{\partial f}{\partial r} = 0$ at $r = 0.001$ AU.

3. Model of the short-time variation of the GCR intensity

In this paper we present the model of the recurrent Fd taking place due to established corotating heliolongitudinal disturbances in the interplanetary space. CIR passing the Earth gradually diminishes the diffusion at the Earth orbit, causing larger scattering of the GCR particles, and in effect fewer GCR particles reach the Earth. We simulate this process by the gradual decrease and then the increase of the diffusion coefficient at the Earth orbit with respect the heliolongitudes. The diffusion coefficient K_{II} of cosmic ray particles has a form: $K_{II} = K_0 \cdot K(r) \cdot K(R,\nu)$, where $K_0 = 10^{21} cm^2/s$, $K(r) = 1 + 0.5 \cdot (r/1AU)$ and $K(R,\nu) = R^{2-\nu}$. The exponent ν pronounce the increase of the HMF turbulence in the vicinity of space where the Fd is created (e.g. [11, 12]), and is taken as: $\nu = 0.8 + 0.2sin(\varphi - 90°)$ for $90° \leq \varphi \leq 270°$. In the model we assume the existence of the two dimensional spiral Parker's heliospheric magnetic field B [13] implemented through the angle $\psi = arctan(-B_\varphi/B_r)$ in the 3D anisotropic diffusion tensor K_{ij} of GCR particles [7].

The expected changes of the GCR intensity for the rigidity of 10 and 20 GV during the simulated Fd in comparison with the profiles of the daily GCR intensities recorded by the two neutron monitors with different cut off rigidities in 18 March - 4 April 2002 presents Fig. 2. One can see that the proposed model is in a good coincidence with the experimental data and, as is expected the amplitude of the Fd decreases for higher rigidities. Moreover, the model of the Fd obtained based on the solution of the SDE allows to reflect the stochastic character of the GCR particles distribution in the heliosphere and present the pseudoparticle trajectory thorough the 3D heliosphere (Fig. 1), which is not possible based on the solution of the Parker transport equation by the e.g. finite difference method (e.g. [11]). This feature can allow a thorough analysis of the trajectory of the GCR particles when encountering occasional shock waves accompanying the sporadic Fd. The presented method of solution of the PTE is unconditionally stable in the higher dimensions and doesn't rely on the numerical grid. However, the integration method involves the statistical error, which can be reduced by increasing the number of simulated pseudoparticles and applying the higher order integration method e.g. Milnstein's method. This work is in progress.

Acknowledgments

This work is supported by The Polish National Science Centre grant awarded by decision number DEC-2012/07/D/ST6/02488. We thank the principal investigators of Moscow and Haleakala neutron monitors for the ability to use their data.

References

[1] Zhang M 1999 *Astrophys. J.* **513** 409-420
[2] Gervasi M, Rancoita P G, Usoskin I G and Kovaltsov G A 1999 *Nuclear Physics B (Proc. Suppl.)* **78** 26-31
[3] Gardiner C W 2009 *Handbook of stochastic methods. For physics, chemistry and the natural sciences* Springer Series in Synergetics
[4] Moraal H 2013 *Space Science Reviews* **176** 299-319
[5] Parker E N 1965 *Planetary and Space Science* **13** 9-49
[6] Potgieter M S 2013 *Space Science Reviews* **176** 165-176
[7] Alania M V 2002 *Acta Physica Pol. B* **33** 4 1149-1166
[8] Jokipii J R, Levy E H and Hubbard W B 1977 *Astrophysical Journal* **1 213** 861-868
[9] Forbush S E 1954 *J. Geophys. Res.* **59** 525
[10] Cane H V 2000*Space Sci. Rev.* **93** 49
[11] Wawrzynczak A and Alania M V 2008 *Adv. Space Res.* **41** 325-334
[12] Wawrzynczak A and Alania M V 2010 *Adv. Space Res.* **45** 622-631
[13] Parker E N 1958 *Astrophys. J.* **128** 664
[14] Webber W R and Lockwood J A 2001 *Journal of Geophysical Research* **106** 29323-29332
[15] Kopp A, Bsching I, Strauss R D and Potgieter M S 2012 *Computer Physics Communications* **183** 530542
[16] Pei C, Bieber J W, Burger R A and Clem J 2010 *Journal of Geophysical Research* **115** A12107

Moving particle semi-implicit method for fluid simulation with implicitly defined obstacles

Yasutomo Kanetsuki[1], Yasuaki Sakamoto[1] and Susumu Nakata[2]

[1] Graduate School of Information Science and Engineering, Ritsumeikan University, Japan
1-1-1 Noji-higashi, Kusatsu Shiga 525-8577 JAPAN

[2] College of Information Science and Engineering, Ritsumeikan University, Japan
1-1-1 Noji-higashi, Kusatsu Shiga 525-8577 JAPAN

E-mail: is0061ee@ed.ritsumei.ac.jp, is019084@ed.ritsumei.ac.jp, snakata@is.ritsumei.ac.jp

Keywords: fluid simulation, moving particle semi-implicit method, implicit form

Abstract. We developed a particle-based fluid simulation method with obstacles defined in implicit form. Our fluid simulation is based on the Moving Particle Semi-implicit (MPS) method, a typical particle-based algorithm achieving incompressible flow. In general, the particle-based fluid simulation is performed with obstacles that are defined as a set of particles located along the boundaries. We employ the implicit representation for the geometric information of obstacles which requires a new formulation of particle motion at the vicinity of boundaries. The main di culty of MPS-based simulation with implicit obstacles is the lack of boundary particles that are required for the computation of two quantities: particle number density and particle force determined by pressure field. In our formulation, new definitions of the two quantities giving good estimation of the original ones is developed and incorporated in the MPS algorithm. In addition, we provide a modified algorithm for the construction of discrete linear system specific to the implicit representation for the computation of particle pressure. Our numerical tests show that the proposed approximation techniques provide adequate particle motion at the vicinity of the implicitly defined obstacles.

1. Introduction

In this paper, we present a technique for incompressible fluid simulation with obstacles defined in implicit form. Our method is performed under the following conditions: the incompressible fluid simulation is performed using the Moving Particle Semi-implicit (MPS) method [1] and each obstacle in the domain has implicit representation that defines the geometry, i.e., the boundary of each obstacle is defined as the solution of $f(\boldsymbol{x}) = 0$.

The main problem to be addressed for the purpose above is the formulation of particle motion in the vicinity of obstacles. Since implicitly defined obstacles do not possess boundary particles, a new formulation specific to the implicit representation is required. The quantities such as the particle number density and particle force of each fluid particle in the vicinity of boundaries need to be determined depending on the implicit form. In our formulation, we define the former quantity by assuming hypothetical particles located along the boundaries and the latter one is defined by assuming an artificial repulsive force acting between obstacles and fluid particles.

The particle-based fluid simulation method called the Smoothed Particle Hydrodynamics (SPH) is one of the most popular approaches that has been used for scientific computation [2]

and for computer graphics [3]. The weakly compressible SPH [4] is one advanced method achieving nearly incompressible flow, a key requirement for many practical applications. The MPS is another approach that accomplishes the incompressibility with high accuracy and has been successfully applied to engineering applications.

In many particle-based simulations, the geometry of obstacles is assumed to be represented by a set of boundary particles. Another approach is polygonal representation of boundaries [5]. In our formulation, obstacles are assumed to have geometry defined in implicit form. In computer graphics, a number of techniques for generating implicit surface have been proposed [6, 7]. In addition, the implicit representation is suitable for representation of time-varying shapes [8].

Our method is performed by combining the two techniques: the MPS and the implicit representation. This new formulation helps to simulate incompressible fluid with time-varying smooth obstacles.

2. MPS with boundary particles

The MPS is a method for numerical analysis of incompressible flow with free surfaces. The governing equations consist of the Navier-Stokes equations, mass conservation law and the incompressibility condition:

$$\frac{D\boldsymbol{v}}{Dt} = -\frac{1}{\rho}\nabla P + \nu\nabla^2\boldsymbol{v} + \boldsymbol{g}, \quad \frac{D\rho}{Dt} + \rho\nabla\cdot\boldsymbol{v} = 0, \quad \frac{D\rho}{Dt} = 0,$$

where $\boldsymbol{v}$, ρ, P, ν and $\boldsymbol{g}$ are the velocity, density, pressure, kinematic viscosity coefficient and the gravity, respectively. The pressure term $\boldsymbol{a}_i^{\text{press}}$ is obtained using the following model:

$$\boldsymbol{a}_i^{\text{press}} = -\frac{D}{\rho n^0}\sum_{j\neq i}\left[\frac{P_j^{k+1} - \hat{P}_i^{k+1}}{\left\|\boldsymbol{r}_j^* - \boldsymbol{r}_i^*\right\|^2}\left(\boldsymbol{r}_j^* - \boldsymbol{r}_i^*\right) w\left(\left\|\boldsymbol{r}_j^* - \boldsymbol{r}_i^*\right\|\right)\right],$$

where D, n^0, $\hat{P}_i^{k+1}$ and $\boldsymbol{r}_i^*$ are the number of spatial dimensions, the particle number density n_i at the initial state, the minimum pressure in the effective radius and the particle position, respectively. The particle number density n_i and the weight function $w(r)$ are evaluated as:

$$n_i = \sum_{j\neq i} w\left(\|\boldsymbol{r}_j - \boldsymbol{r}_i\|\right), \quad w(r) = \begin{cases} \frac{r_e}{r} - 1\,(0 \leq r < r_e) \\ 0\,(r_e \leq r) \end{cases},$$

where r_e is the effective radius. The pressure $\{P_i^{k+1}\}$ is obtained as the solution of the linear system:

$$-\frac{\rho_0}{\Delta t^2}\left(\frac{n_i^* - n^0}{n^0} - \alpha P_i^{k+1}\right) = \frac{2D}{\lambda n^0}\sum_{j\neq i}\left[\left(P_j^{k+1} - P_i^{k+1}\right) w\left(\left\|\boldsymbol{r}_j^* - \boldsymbol{r}_i^*\right\|\right)\right],$$

where α is the compressibility of the fluid and λ is the coefficient determined statistically.

3. MPS with implicitly defined obstacles

The standard MPS is formulated assuming a set of particles located inside obstacles and along their boundary. In this study, the particle motion needs to be formulated without using the particles of obstacles and, instead of using them, we divide the computation of the physical quantities at each particle into the quantities determined by surrounding fluid particles and those by obstacles. Assuming a set of boundary particles, the quantities determined by the obstacles defined by the following equations:

$$\begin{cases} \boldsymbol{a}_{i,\text{obstacle}}^{\text{press}} = -\frac{D}{\rho n^0}\sum_{j\in\text{obstacle}}\left[\frac{P_j^{k+1} - \hat{P}_i^{k+1}}{\left\|\boldsymbol{r}_j^* - \boldsymbol{r}_i^*\right\|^2}\left(\boldsymbol{r}_j^* - \boldsymbol{r}_i^*\right) w\left(\left\|\boldsymbol{r}_j^* - \boldsymbol{r}_i^*\right\|\right)\right], \\ \left\langle\nabla^2 P\right\rangle_{i,\text{obstacle}}^{k+1} = \frac{2D}{\lambda n^0}\sum_{j\in\text{obstacle}}\left[\left(P_j^{k+1} - P_i^{k+1}\right) w\left(\left\|\boldsymbol{r}_j^* - \boldsymbol{r}_i^*\right\|\right)\right]. \end{cases} \quad (1)$$

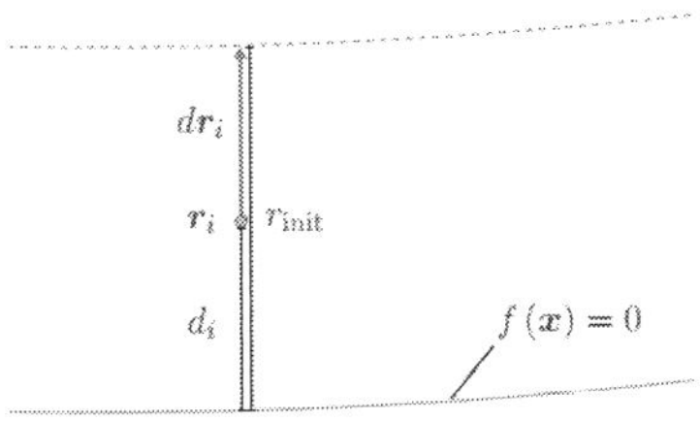

Figure 1. The distance between the fluid particle $\boldsymbol{r}_i$ and the obstacle $f(\boldsymbol{x}) = 0$. The blue circle represents the fluid particle and the brown represents the obstacle.

In order to replace these equations with expressions without boundary particles, we apply the following two models. The first one is to assume hypothetical boundary particles that determine quantities at fluid particles affected by the obstacles. The second one is to define an artificial pressure term affected from obstacles to each fluid particle. In this study, we assume that the hypothetical boundary particles are located along the tangent plane at the closest boundary point on the implicit surface.

By applying the first model to the weight function, we obtain

$$w^{\text{obstacle}}(d_i) \approx \sum_{j \in \text{obstacle}} w\left(\|\boldsymbol{r}_j - \boldsymbol{r}_i\|\right),$$

where d_i is the distance from i-th fluid particle to the closest point on the boundary and $\boldsymbol{r}_j$ is the hypothetical boundary particle. Note that the weight function is free of the direction of the tangent plane and, as a result, the weight function can be determined as a function of $d_i \geq 0$ in advance of the time-step computation in practical computation. For the pressure term in (1), we use the second model, i.e., we define an artificial force that acts on each fluid particle as following:

$$\boldsymbol{a}^{\text{press}}_{i,\text{obstacle}}(d_i) \equiv \beta \frac{n_i w^{\text{obstacle}}(d_i)}{d_i^2} \boldsymbol{dr}_i,$$

where β is the coefficient for force level and $\boldsymbol{dr}_i$ is the direction from the fluid particle to the nearest boundary (see Figure 1). The bottom equation in (1) is obtained by

$$\left\langle \nabla^2 P \right\rangle^{k+1}_{i,\text{obstacle}} \approx \beta \frac{2\rho n_i w^{\text{obstacle}}(d_i) \|\boldsymbol{dr}_i\|}{\lambda d_i}.$$

The quantities undefined in the above discussion are d_i, the distance from a fluid particle to the closest boundary point, and $\boldsymbol{dr}_i$, the normal direction of the corresponding tangent plane at the closest boundary point. We obtain the distance d_i by the Taubin ' s distance approximation. For the normal $\boldsymbol{dr}_i$, we replace the direction with the gradient of the implicit function at the corresponding fluid particle:

$$d_i \approx |f(\boldsymbol{r}_i)| / \|\nabla f(\boldsymbol{r}_i)\|, \quad \boldsymbol{dr}_i \approx (r_{\text{init}} - d_i) \cdot \nabla f(\boldsymbol{r}_i) / \|\nabla f(\boldsymbol{r}_i)\|,$$

where r_{init} is the distance of initial particles.

Figure 2 shows the results of our MPS computation with implicit surfaces. The simulation is performed with 7,900 fluid particles and the obstacles defined by $\left(\sqrt{x^2+y^2}-R\right)^2 - r^2 = 0$. The results show that adequate particle motion can be achieved in the vicinity of the boundary by our MPS-based formulation with implicit expression.

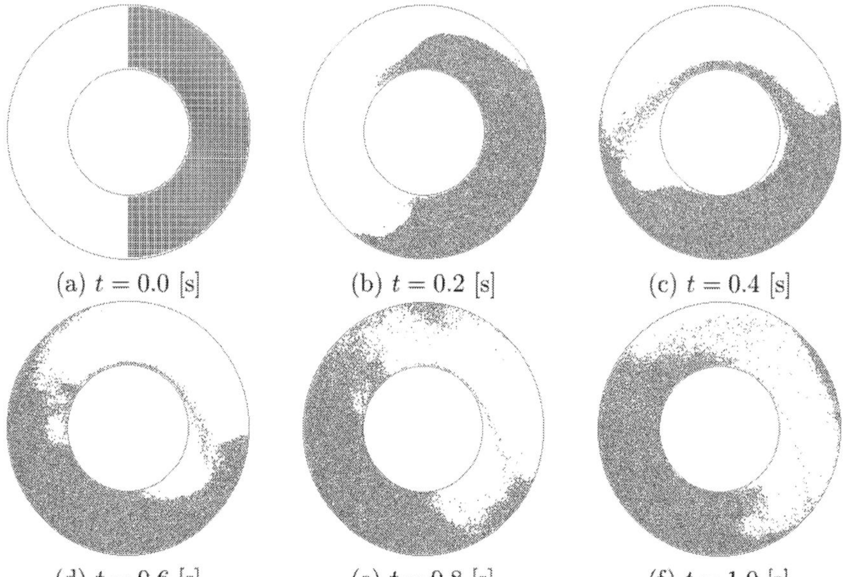

(a) $t = 0.0$ [s] (b) $t = 0.2$ [s] (c) $t = 0.4$ [s]

(d) $t = 0.6$ [s] (e) $t = 0.8$ [s] (f) $t = 1.0$ [s]

Figure 2. Results at every 0.2 [s] starting from initial state.

4. Conclusions

We have presented a fluid simulation method with implicit function form obstacles. The formulation is performed by giving two models specific to implicit representation: one is to assume hypothetical particles along boundaries and the other is to define a new model of the pressure term component affected by obstacles. Some numerical tests have shown that our method gives adequate particle motion with implicitly defined obstacles.

Acknowledgments

This work was supported by JSPS KAKENHI Grant Number 24760066.

References

[1] S. Koshizuka and Y. Oka., "Moving particle semi-implicit method for fragmentation of incompressible fluid," *Nuclear Science and Engineering*, vol.123, no.3, pp.421–434, 1996.

[2] J. J. Monaghan., "Smoothed particle hydrodynamics," *Annual Review of Astronomy and Astrophysics*, vol.30, no.1, pp.543–574, 1992.

[3] M. Müller, D. Charypar and M. Gross., "Particle-based fluid simulation for interactive applications," *In Proceedings of the 2003 ACM SIGGRAPH/Eurographics Symposium on Computer Animation*, pp.154–159, 2003.

[4] M. Becker and M. Teschner., "Weakly compressible SPH for free surface flows," *Proceedings of the 2007 ACM SIGGRAPH/Eurographics Symposium on Computer animation*, pp.209–217, 2007.

[5] T. Harada, S. Koshizuka and Y. Kawaguchi., "Improvement in the boundary conditions of smoothed particle hydrodynamics," *Computer Graphics & Geometry*, vol.9, no.3, pp.2–15, 2007.

[6] M. Berger, J. A. Levine, L. Gustavo Nonato, G.. Taubin and C. T. Silva., "A benchmark for surface reconstrucrion," *ACM Transactions on Graphics*, vol.32, no.2, pp.20:1–17, 2013.

[7] M. Sugihara, B. Wyvill and R. Schmidt., "WarpCurves: A tool for explicit manipulation of implicit surfaces," *Computers & Graphics*, vol.34, no.3, pp.282–291, 2010.

[8] J. Sussmuth, M. Winter and G. Greiner., "Reconstructing animated meshes from time-varying point clouds," *Computer Graphics Forum*, vol.27, no.5, pp.1469–1476, 2008.

Hydrogen-Like Impurities in novel potential barriers: Theoretical development

J D González[1], F J Rondano[1], J Barba-Ortega[2]

[1] Teoría de la materia condensada, Universidad del Magdalena, Santa Marta, Colombia
[2] Departamento de Física, Universidad Nacional de Colombia, Bogotá, Colombia

E-mail: `jedgonzalezac@unal.edu.co`

Abstract. A systematic theoretical study of the binding energy of the ground state of a hydrogenic off-axis donor in a cylindrical quantum wire containing two quantum wells in a section of the tube layer is calculated in the presence of a uniform magnetic field applied parallel. We express the wave function as a product of free electron wave functions and an envelope function that depends only on the electronion separation. By using the variational principle we derive a differential equation for the envelope function, which we solve numerically.

1. Introduction

The study of quasi-one-dimensional system has attracted considerable attention in the multilayer structures [1,2] and nanotubes with those properties of the embedded QWs have been experimentally obtained [3]. In our previous work we show that the low lying 1s and 2px,y subbands of the free electron in nanotube, become almost degenerated as the width of the repulsive core grows and the mixing of these subbands in the presence of the off-axis donor no longer should be depreciated [4].The purpose of this paper is to present a detailed theoretical treatment of the variational method used to determine the ground state energy of the off-axis neutral donor of the impurity states in cylindrical nanotube with the presence of a magnetic field oriented along the growth axis.

2. Theoretical Formalism

Let us consider a nanotube with a off-axes hidrogen-like impurity and its solution is described below. The first theoric study on donor impurity in quantum well was realized by Bastard [5] whose consider a trial function as:

$$\Psi = f_0 e^{-\alpha R} \tag{1}$$

where f_0 is the trial wave function of the ground state for a free electron and $e^{-\alpha R}$ include the electrostatic interaction ion-electron. Later Bajaj [6] analyze the same problem considering additional modification:

$$e^{-\alpha R} \rightarrow \sum_i \beta_i e^{-\alpha i R} \tag{2}$$

However, analyzing the energy binding of a donor in a GaAs/AlGaAs core-shell nanotube mixing subbands take place in this heterostructure due to height of confinement potential V_{Layer} and the layer width, in consequence, The probability distributions of the ground states (1s) and first excited state (2p) are very similar, and the system may become degenerated. Therefore it was necessary to modify the trial function, so that:

$$\Psi = [\alpha f_{1s} + \sqrt{(1-\alpha^2)} f_{2p}] \sum_i \beta_i e^{-\alpha i R} \quad (3)$$

The variational method that we propose change $\sum_i \beta_i e^{-\alpha i R}$ by $\Phi(|\vec{r} - \vec{\xi}|)$ which only depend on the ion-electron separation equation (3). Using the equation (2)

$$H_0 f_m(\rho) = E_m f_m \quad (4)$$

and considering

$$f_{1s}(\rho) = R_{1s}(\rho) \quad \text{and } f_{2p}(\rho) = R_{2p}(\rho) cos\varphi \quad (5)$$

We obtain

$$[-\frac{1}{\rho}\frac{d}{d\rho}\rho\frac{d}{d\rho} + \frac{m^2}{\rho^2} + V(\rho) + \frac{\gamma^2\rho^2}{4} - (E_m - \gamma_m)]R(\rho) = 0 \quad (6)$$

by replacing

$$R(\rho) = \rho^m U(\rho) \quad (7)$$

by substituting in equation (6)

$$-U''(\rho) - (2m+1)\frac{U'(\rho)}{\rho} + [V(\rho) + \frac{\gamma^2\rho^2}{4} - (E_m - \gamma_m)]U(\rho) = 0 \quad (8)$$

to avoid the singularity at ρ use the following approximation

$$U(\rho) = a + b\rho^2; \quad 0 \le \rho \le \xi \quad (9)$$

where

$$U(\rho) = a[1 + \alpha_0\rho^2]; \; \alpha_0 = \frac{V(0) - E_m + \gamma_0}{4(m+1)} \quad (10)$$

The solution of the equation (8) at $\xi \le \rho \le R_{ext}$ by using trigonometric sweep method as folowing:

$$U(\rho) = A(\rho)cos\theta(\rho); U'(\rho) = A(\rho)sin\theta(\rho); U''(\rho) = A'(\rho)sin\theta(\rho) + A(\rho)cos\theta(\rho).\theta'(\rho) \quad (11)$$

substituting (11) in ec. (8) we obtain

$$A(\rho)[[V(\rho) + \frac{\gamma^2\rho^2}{4} - (E_m - \gamma_m)] - cos\theta(\rho) - cos\theta(\rho).\theta'(\rho) - \frac{(2m+1)sin\theta(\rho)}{\rho}] - A'(\rho).sin\theta(\rho) (12)$$

Deriving the first equation in (11) and equating with the second one, we obtain:

$$A(\rho)[sin\theta(\rho) + sin\theta(\rho).\theta'(\rho)] - cos\theta(\rho).A'(\rho) = 0 \quad (13)$$

The system of equations (12) and (13) has the following nontrivial solution

$$\theta'(\rho) = -[sin^2\theta(\rho) + S(\rho)cos^2\theta(\rho) + W(\rho)sin\theta(\rho)cos\theta(\rho)]$$
$$S(\rho) = E_m - \gamma_m - V_{eff}; V_{eff} = V(\rho) + \frac{\gamma^2\rho^2}{4}; W(\rho) = \frac{(2m+1)}{\rho}; \quad (14)$$

Applying the criteria of smoothness $\rho = \xi$ to determine the necessary initial condition to solve equation (14)

$$\frac{R'(\rho)}{R(\rho)}|_{\rho=\xi^-} = \frac{R'(\rho)}{R(\rho)}|_{\rho=\xi^+} \tag{15}$$

where

$$\theta(\xi) = tan^-1(\frac{2\alpha_0\xi}{1+\alpha_0\xi^2}) \tag{16}$$

To obtain the wave function is necessary to determine $A(\rho)$, the from (13)

$$A(\rho) = \exp\int_\xi^\rho (1-s(\rho))sin\theta(\rho)cos\theta(\rho)d\rho - \exp\int_\xi^\rho W(\rho)sin^2\theta(\rho)d\rho \tag{17}$$

Using equations (14) y (17) is posible to obtain the radial part of the wave equation:

$$R(\rho) = \rho^m A(\rho)cos\theta(\rho) \tag{18}$$

In order to find the solution at the region $\rho \geq R_{ext}$, it is necessary to know additional condition to found the solution of equation (14). $\theta(R_{ext})$ which define the energy implicitly. In this region $V(\rho) = V_{ext} = cte$, in addition is possible to consider the following aproximation $\frac{\gamma^2\rho^2}{4} \sim \frac{\gamma^2 R_{ext}^2}{4}$ due to the electron always tend to remain inside the each well and the tunneling will be relatively negligible, that is way the modifications in the potential for $\rho \geq R_{ext}$ not will affect considerably the calculus. The solution of the equation (6) at the region $\rho > \rho_{ext}$ are functions of Bessel modified.

$$R(\rho) = K_m(\sqrt{\lambda}\,\rho); \lambda = V_{ext} + \frac{\gamma^2 R_{ext}^2}{4} + \gamma m - E \tag{19}$$

Using the criteria of smoothness in $\rho = R_{ext}$ we obtain:

$$\theta(R_{ext}) \quad = \quad -atan\left[\frac{\sqrt{\lambda}\,K_{m-1}(\sqrt{\lambda}\,R_{ext})}{K_m(\sqrt{\lambda}\,R_{ext})} + \frac{m}{R_{ext}}\right]; K_{-1}(x) = K_1(x) \tag{20}$$

Starting with the Schrodinger variational principle

$$F[\Psi] = \langle\Psi|\hat{\mathrm{H}} - E|\Psi\rangle \;\rightarrow\; min \tag{21}$$

and choosing the origin of our coordinate system on the impurity donor position, the functional can be written as:

$$F[\phi] = \langle(\alpha f_{1s} + \sqrt{1-\alpha^2}f_{2p})\phi(r'))|\hat{\mathrm{H}}_0 - \frac{2}{r'} - E(D^0)|(\alpha f_{1s} + \sqrt{1-\alpha^2}f_{2p}))\phi(r')\rangle \tag{22}$$

with

$$f_{1s} = f_{1s}(\vec{r}+\vec{\xi}) \quad f_{2p} = f_{2p}(\vec{r}+\vec{\xi}) \tag{23}$$

where $\vec{\xi}$ is impurity possition and $\vec{r}$ is the electron position from the impurity We express

$$F[\psi] = I_1 + I_2 \tag{24}$$

Where

$$I_1 = \langle(\alpha f_{1s} + \sqrt{1-\alpha^2}f_{2p})\phi(r'))|\hat{\mathrm{H}}_0|(\alpha f_{1s} + \sqrt{1-\alpha^2}f_{2p}))\phi(r')\rangle \tag{25}$$

and

$$I_2 = \langle(\alpha f_{1s} + \sqrt{1-\alpha^2} f_{2p})\phi(r'))| - \frac{2}{r'} - E(D^0)|(\alpha f_{1s} + \sqrt{1-\alpha^2} f_{2p}))\phi(r')\rangle \quad (26)$$

working over equation (25) and considering f_{1s}, $f_{2s} = R_{2p}\, cos\varphi$ and $\nabla \Longleftrightarrow \nabla'$ we obtain

$$\begin{aligned} F[\phi] &= \int_0^\infty [[\alpha^2 P_{00} + 2\alpha\sqrt{1-\alpha^2}P_{01} + (1-\alpha^2)P_{11}][\left(\frac{d\phi(r')}{dr'}\right)^2 + \phi(r')^2\left(-\frac{2}{r'} - E(D^0)\right)] \\ &+ [\alpha^2 E_{1s}P_{00} + \sqrt{1-\alpha^2}(E_{12}+E_{2p})P_{01} + (1-\alpha^2)E_{2p}P_{11}]\phi(r')^2]r'^2 dr' \end{aligned} \quad (27)$$

with

$$\begin{aligned} P_{00} &= \int_0^{2\pi}\int_0^{\pi} R_{1s}^2 sin\theta' d\theta' d\varphi'; P_{11} = \int_0^{2\pi}\int_0^{\pi} R_{2p}^2 cos^2\varphi sin\theta' d\theta' d\varphi'; \\ P_{01} &= \int_0^{2\pi}\int_0^{\pi} R_{2p}R_{1s} cos\varphi sin\theta' d\theta' d\varphi' \end{aligned} \quad (28)$$

we obtain

$$F[\Phi] = \int_0^\infty J(r')\left(\frac{d\phi(r')}{dr'}\right)^2 dr' + \int_0^\infty J(r')\left(\frac{\tilde{\mathrm{E}}(r')}{J(r')} - \frac{2}{r'} - E(D^0)\right)\phi(r')^2 dr' \quad (29)$$

with

$$\begin{aligned} J(r') &= \alpha^2 r'^2 P_{00} + 2\alpha\sqrt{1-\alpha^2}r'^2 P_{01} + (1-\alpha^2)r'^2 P_{11}; \quad (30) \\ \tilde{\mathrm{E}}(r') &= \alpha^2 r'^2 E_{1s}P_{00} + \alpha\sqrt{1-\alpha^2}r'^2(E_{1s}+E_{2p})P_{01} + E_{2p}(1-\alpha^2)r'^2 P_{11} \quad (31) \end{aligned}$$

It is necessary to find the function ϕ' that minimize the functional (31) and finally we to obtain the the Euler-Lagrange equation for the correlation function:

$$\frac{-1}{J(r)}\frac{d}{dr}[J(r)\frac{d\Phi(r)}{dr}] + [\tilde{E}(r) - \frac{2}{r}]\Phi = E(D^0)\Phi \quad (32)$$

This equation (34) is solved by using the trigonometric sweep method.

3. Conclusions

We study the mathematical development to determine the binding energy of hydrogen-like impurity of a cylindrical nanotube. Additionally, this paper shows in detail the proposed mathematical solving the variational problem considering confinement potential produced by a square well potential added to potential generated by the applied magnetic field.

Acknowledgments

This work was partially financed by the Universidad del Magdalena (Fonciencias) and the Colombian Agency COLCIENCIAS through doctoral scholarships 567.

References

[1] Branis S V , Li G, and Bajaj K K, 1993 Phys. Rev. B 47, 1316.
[2] Sierra-Ortega J, Mikhailov I D, and Betancur F J, 2004, Physica B 348, 66.
[3] Kubota K, Vaccaroa P O, Ohtania N, Hiroseb Y, Hosodab M, Aida T, 2002 Physica E 13, 313.
[4] Betancur F J, Orozco E A, Gonzalez J D and Mikhailov I D, 2005 Phys. Stat. Sol. (b), 242, 1833.
[5] Bastard G 19981 phys.Rev. B 24 4714.
[6] Greene R L, Bajaj K K 1983 Solid State Comunication 45 825.

Journal of Physics: Conference Series **574** (2015) 012081 doi:10.1088/1742-6596/574/1/012081

Influence of both quantum well thickness and radius on density exciton states on a semiconductor nanoribbon

J D González[1], F J Rondano[1], J Barba-Ortega[2]

[1] Teoría de la materia condensada, Universidad del Magdalena, Santa Marta, Colombia
[2] Departamento de Física, Universidad Nacional de Colombia, Bogotá, Colombia

E-mail: jedgonzalezac@unal.edu.co

Abstract. Simultaneous effects of quantum well thickness and radius on exciton energy and density exciton states on a semiconductor nanoribbon with a finite and rectangular confining potential are studied theoretically. We show that in the structural adiabatic limit, when the width of the pattern of the particles pathways within the ribbon is much smaller than its radius, the wave equations for exciton are separable and their solutions can be found numerically. We present results of calculation of the density exciton states as functions of the wavelength and energy for different values of ribbon radius and quantum well thickness, with potential profiles rectangular and parabolic.

1. Introduction

There is much interest in the physical properties of semiconductor structures, because the geometry and strong dimensional confinement of carriers make them particularly attractive as potential building blocks for nanoscale electronics and optoelectronic devices, including solar cells [1], photodetectors [2], tunnel [3] and light emitting [4] diodes, field effect transistors [5], single electron devices [6] and lasers. In particular, if an exciton is trapped inside a low-dimensional nanostructure, for example inside a quantum well, wire or dot, its binding energy increases considerably, the bound state becomes more stable facilitating in this manner its formation. The stronger the confinement, the larger is the enhancement of the binding energy of the exciton. Therefore, in nanoribbons where the confinement is the strongest, the exciton are more stable than in other types of nanostructures. In this work, we present a theoretical model for exciton confined in a nanoribbon with quantum well thickness and differents radius, which describe the method of solution of the corresponding wave equations. The results of calculation of the exciton energy is used to find the density exciton states with confinement potential rectangular and parabolic.

2. Theoretical Formalism

Let us consider a ribbon with internal radius in order of micrometers (Fig. 1 (left)) and a longitudinal section cuasi-flat considering that the quantum well thickness are in order of nanometers (Fig. 1 (right)). The exciton is moving into the quantum well (QW) with rectangular and parabolic shapes, using different positions of the QW into the ribbon measured

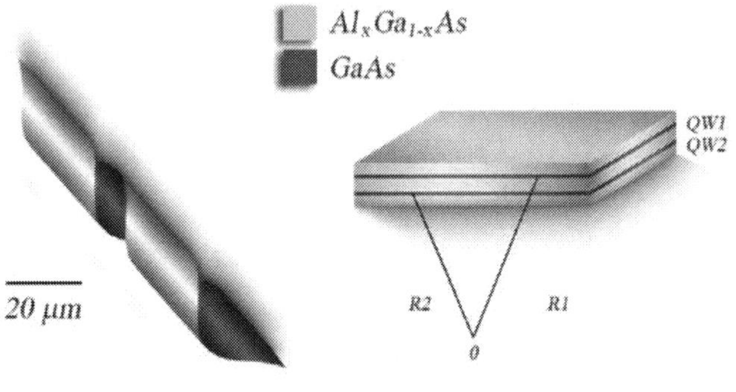

Figure 1. (Left) Schematic figure of a fabricated nanoribbon. (Right) Cross-section of nanoribbon in which QW1 and QW2 are considered quasi-flat.

from its center Fig1.b. The method used is a simple variational procedure and he quantum wells QW1 and QW2 are considered non-interacting due to the separation of 40nm between them. For modeling a nanoribbon containing two quantum wells, we follow the method used in [7] to obtain the desired potential profile. In this work, we consider a single exciton in semiconductor heterostructures within the effective mass approximation by using a spherical hole mass approximation and constant physical parameters throughout the heterostructures. In this case the characteristic energy of an exciton can consequently be defined through the effective Rydberg $Ry_0^* = e^2/2a_0\epsilon$ and its characteristic length through the effective Bohr radius, $a_0^* = \epsilon\hbar/\mu e^2$, with $\mu = m_e^* m_h^*/(m_e^* + m_h^*)$ being the reduced exciton effective mass. Thus the dimensionless Hamiltonian for the electron-hole pair confined in a heterostructure may be written as:

$$H(\mathbf{r_e}, \mathbf{r_h}, \tau) = H_e(\mathbf{r_e}) + H_h(\mathbf{r_h}) - 2\tau/r_{eh} \quad (1)$$

$$H_\alpha = -\eta_\alpha \nabla_\alpha^2 + V_\alpha(\mathbf{r}_\alpha) \quad (2)$$

Where H_e and H_h describe the free motion of electron and hole confined in the heterostructure respectively. $V_\alpha(\mathbf{r}_\alpha)$ with $\alpha = e, h$ is the potential confinement, $2/r_{eh}$ is the energy of the electron-hole interaction, $\mathbf{r_e}$ and $\mathbf{r_h}$ are electron and hole position vectors, and $r_{eh} = |r_e - r_h|$ is the electron-hole separation. The parameter τ in Eq.1 is equal to 1 for exciton and to 0 for uncorrelated electron-hole pair, and the parameters $\eta_\alpha = \mu/m_\alpha^*$ in Eq.2 are characteristics of the corresponding dimensionless effective masses of the hole in units of the reduced exciton effective mass μ respectively. We choose the ground state exciton trial function in the form:

$$\Psi(\mathbf{r_e}, \mathbf{r_h}) = f_0(\mathbf{r_e}, \mathbf{r_h})\Phi(r_{eh}) \quad (3)$$

where $f_0(\mathbf{r_e}, \mathbf{r_h})$ is the product of the one-particle electron and hole ground state wave functions, whereas $\Phi(r_{eh})$ is a variational function that describes the intrinsic properties of the exciton and depends only on the electron-hole separation. The ground state energy of the exciton is found by minimizing a functional which is minimized with respect to Φ and its first derivative yields the Euler-Lagrange equation:

$$\frac{1}{G_0(r)}\frac{d}{dr}\left[G_0(r)\frac{d\Phi}{dr}\right] - \frac{2}{r}\Phi(r) = [E_{ex} - E_0]\Phi(r) \quad (4)$$

The binding energy of the exciton, $E_b = E_0 - E_{ex}$, is obtained by solving numerically Eq. 4.

3. Results and Discussion

In Fig. 2, 3 y 4 we are presenting the results obtained for the energies of the motion along the z-direction corresponding to the lowest electron and ground hole states as functions of wavelength and energies. The rise of the ground hole energy is due to the fact that the increment in the confinement leads to a stronger localization of the hole wave function on parabolic potential barrier opposite on the rectangular case. In order to calculate the density of the energy states (DOS) with respect from the valence band to excitonic state transition we assume that the resulting DOS is an envelope curve of all individual transitions: $\rho(E) = \sum_{i=1,2}\sum_{m=1,10} f(E - E_i(m)$ and $f(x) = 1/\sqrt{2\pi}\sigma \exp(-x^2/2\sigma^2)$ Here σ is a natural width of individual spectral lines and we adopt a gaussian form for them. All Figures shown the variation of the density states considering different position of the quantum well R_i from the nanoribbon axis with profiles of confinement rectangular and parabolic, for quantum well thickness $L = 5a_0^*$ and $L = 8a_0^*$. Increasing the radii as follow $R_1 = 2.0a_0^*, R_2 = 3.0a_0^*, R_3 = 4.0a_0^*$ the peaks reach higher values including when a parabolic potential is used instead of the rectangular. In consequence, this changes gives additional possibilities to obtain a nanoribbon with huge characteristics to manipulate its physical parameters. The Fig.4 (Right) shows the

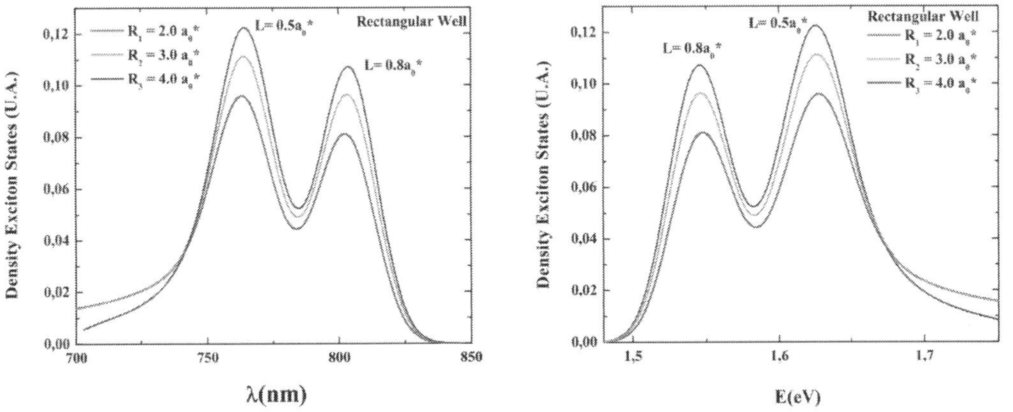

Figure 2. Density exciton states for different values of radii using a rectangular potential profile as a function of wavelength (left) and energy (right).

density of states as a function of the energy of the exciton and with it an investment in the peak position with respect to Fig.5 (Left) for each value of L, due to the inverse relationship between wavelenght and energy, determined by the equation $\lambda = hc/E(m)(nm/eV)$. The curves shown the peak shift when changes related with the potential shape.

4. Conclusions

We study the features of density exciton states in a nanoribbon and novels curves were obtained changing the quantum wells thickness and positions from the axis. Additionally, the study of system with confinement potential rectangular and parabolic is more realistic and the optimal performance of the heterostructure can be obtained. It is hoped that this work will stimulate further experimental activity on nanoribbon.

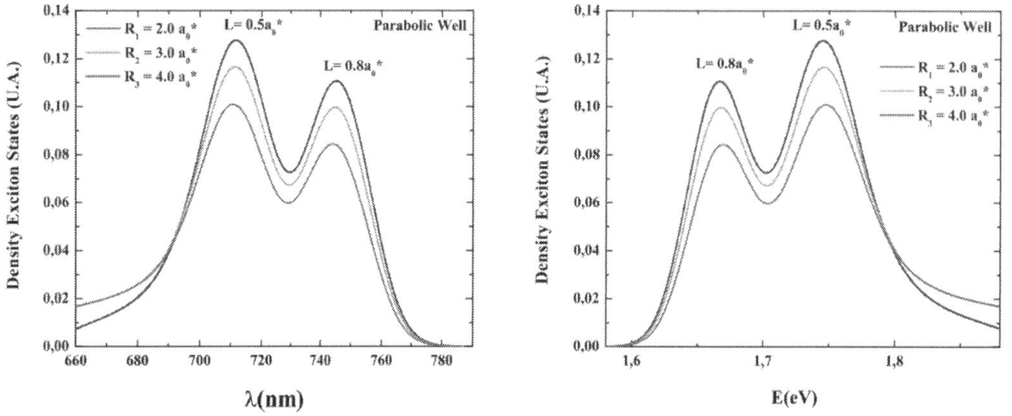

Figure 3. Density exciton states for different values of radii using a parabolic potential profile as a function of wavelength (left) and energy (right).

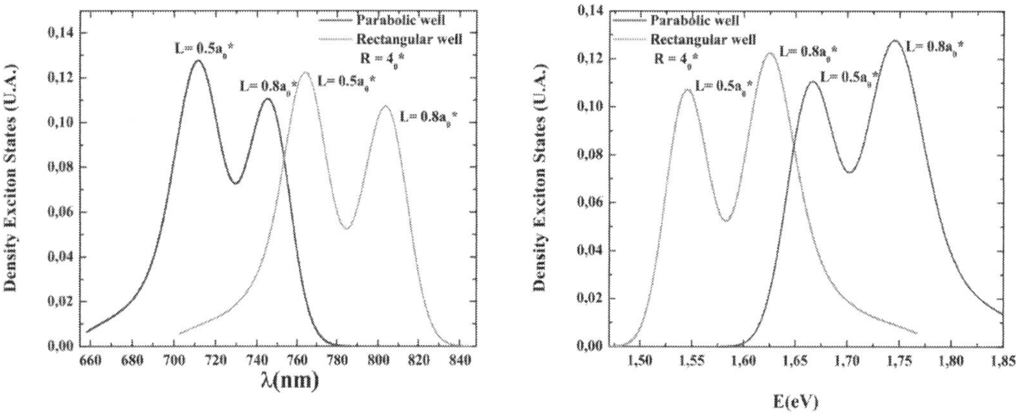

Figure 4. The figure compare the density exciton states using a rectangular or parabolic potential profile as a function of wavelength (left) and energy (right)

Acknowledgments
This work was partially financed by the Universidad del Magdalena (Fonciencias) and the Colombian Agency COLCIENCIAS through doctoral scholarships 567.

References
[1] Czaban J A, Thompson D A, LaPierre R R 2009 *Nano Letters* **9** 148.
[2] Pettersson H, Tragardh J, Persson A I, Landin L, Hessman D, Samuelson L 2006 *Nano Letters* **6** 229.
[3] Wallentin J, Persson J M, Wagner J B, Samuelson L, Deppert K, Borgstrom M T 2010 *Nano Letters* **10** 974.
[4] Qian F, Gradecak S, Li Y, Wen C Y, Lieber C M 2005 *Nano Letters* **5** 2287.
[5] Liang G, Xiang J, Karche N, Klimeck G, Lieber C M, Lundstrom M 2007 *Nano Letters* **7** 642.
[6] Fuhrer A, Fasth C, Samuelson L 2007 *Applied Physics Letters* **91** 52109.
[7] Mikhailov I D, Betancur F J, Escorcia R A, Sierra-Ortega J 2002 *Phys. Stat. Sol. (b)* **234** 590.

Confinement effects on quasi 0D dimensional structure with $Al_xGa_{1-x}As/GaAs/Al_xGa_{1-x}As$ quantum wells

J D González[1,3], **J E González**[2], **J Barba-Ortega**[3]

[1] Teoría de la materia condensada, Universidad del Magdalena, Santa Marta, Colombia
[2] Departamento de Física, Universidad Pontificia Bolivariana, Bucaramanga, Colombia
[3] Departamento de Física, Universidad Nacional de Colombia, Bogotá, Colombia

E-mail: jedgonzalezac@unal.edu.co

Abstract. In this work we study the effects of barrier height and position on symmetric $Al_xGa_{1-x}As/GaAs/Al_xGa_{1-x}As$ quantum wells in absence of external influences. We use a variational method within the effective mass approximation to observe the effects on the structure. The donor trial function is taken as a product of the ground state wave function, with an arbitrary correlation function that depends only on ion-electron separation. It has been observed two peaks in the curves for the dependence of the ground-state binding energies versus the donor distance from the axis and it is shown that the impurity binding energy depends strongly on the impurity position, potential shape on both quantum wells.

1. Introduction

Quantum dots (QDs) have attracted huge attention due to their potential applications in solar cells [1], biomedicine [2], photo-detectors [3], quantum cascade lasers [4], single-photon sources [5, 6], and quantum computing [7]. It has been found out that the electronic and optical properties of QDs are extremely sensitive to size, stoichiometry, and shape which leads to a tunable absorption, emission, and polarization-sensitive-spectra [8]-[10]. This dependence comes from relatively small number of atoms forming such structures. In this work, we propose a simple variational method to analyze the on- and off-center D^0 ground states in which the problem of the donors confined in quasi 0D structure is reduced to the similar one for the hydrogenic analogous in an isotropic space [11]. Taking advantage of the high symmetry in the spherical quantum dots (SQDs) we apply this method to analyze the ground state binding energy as a function of the donor impurity position in the SQD with two quantum well considering different confinement potential height and barrier position.

2. Theoretical Formalism

Within the framework of the effective-mass theory, the dimensionless Hamiltonian in a system of reference $\mathbf{r}' = \mathbf{r} + \boldsymbol{\zeta}$ for a D^0 in a SQD, can be written as [12]:

$$\hat{H}(\mathbf{r} + \boldsymbol{\zeta}, Z) = -\nabla^2 + V(|\mathbf{r} + \boldsymbol{\zeta}|) - \frac{2Z}{r} \tag{1}$$

where the Z parameter is zero for the electron and one for the donor; $\boldsymbol{\zeta}$ and $\mathbf{r}$ designate the 3D-vectors of the ion and electron positions, respectively measured from the center of the structure and $V(r)$ is an arbitrary confinement potential. The effective Bohr radius $a_0^* = \varepsilon\hbar^2/m^*e^2$ and the effective Rydberg, $R_y^* = e^2/2a_0^*$ have been taken as units of length and energy, respectively. A a function for the D^0 ground state, we consider the following trial function

$$\Psi_{D^0}(\mathbf{r}) = f_0(\mathbf{r} + \boldsymbol{\zeta})\Phi_{D^0(r)} \tag{2}$$

Here, $f_0(\mathbf{r} + \boldsymbol{\zeta})$ is the ground state wave function for a free electron in the SQD and the envelope function $\Phi_{D^0(r)}$ depends only the electron-ion separation.

Starting from the variational principle of Schrödinger and using the procedure of the functional derivative developed by us [12] we obtain the following one-dimensional Euler-Lagrange equation for the function, $\Phi_{D^0(r)}$,

$$\frac{1}{J_0(r)}\frac{d}{dr}J_0(r)\frac{d\Phi(r)_{D^0}}{dr} - \frac{2}{r}\Phi(r)_{D^0} = [E_{D^0} - E_0]\Phi(r)_{D^0} \tag{3}$$

where the radial part of the Jacobian volume element $J(r)$ is related to the density of distribution $P(r)$ of the free electron charge over a sphere of the radius r as

$$J(r) = 4\pi r^2 P(r); P(r) = \frac{1}{2}\int_{-1}^{1} f_0^2(\sqrt{r^2 + \zeta^2 - 2r\zeta x})dx \tag{4}$$

In our calculations, to find the D^0 energy, $E(D^0)$, we solve numerically the Eq. (3) by using the trigonometric sweep procedure [13]. Eq. (3) describes the hydrogen-like atom in an effective space with the radial part of the Jacobian volume element $J(r)$ [12].

For modeling a quantum dot containing two quantum wells,we use the empirical relationship $V_i(\rho) = 0.61[1.155x(\rho) + 0.37x^2(\rho)]eV$, to obtain the height of the potential barrier with the aluminum concentration as we can see in Eq. 5. Where $x(\rho)$ represents the aluminum concentration as a function of radial distance r. Thus the confinement potential given by the following expression:

$$V(\rho) = V_1\theta(-\rho, -R_1, W) + V_2\theta(-\rho, R_2, W) - V_3\theta(-\rho, R_3 + W, W) + V_4\theta(\rho, R_4, W) \tag{5}$$

Where

$$\theta(z, z_0, W) = \begin{cases} 0; & z < z_0 - W \\ [(z - z_0)^2/W^2 - 1]^2; & z_0 - W \leq z < z_0 \\ 1; & z \geq z_0 \end{cases} \tag{6}$$

$\theta(-\rho, -R_i, W)$ is soft-edge version of the Heaviside function R_i, V_i represent the radii and heights of the barriers,respectively; W is a parameter related to the width of the transition region. $W = 0.01a_0^*$ for rectangular barrier.

3. Results and Discussion

The Fig. 1 (left) shows the profile of the confinement potential with two wells (QW1 and QW2), finite height and rectangular shape. The internal and external potential remain unchanged $V_{int} = V_{ext} = 40R_y^*$ and only the potential used for the layer vary for values of the $V_{layer} = 10R_y^*, 20R_y^*, 30R_y^*$. The radii are $R_1 = 1a_0^*$,$R_2 = 2a_0^*$,$R_3 = 3a_0^*$ and $R_4 = 4a_0^*$, in consequence the layer width is $1a_0^*$. In Fig. 1 (right) we present the D_0 binding energy dependence on the donor displacement from the center of the $Al_{0.3}Ga_{0.7}As/GaAs/Al_{0.3}Ga_{0.7}As$ SQD with different values of V_{layer}.One can observe a successive displacement of the curve peaks

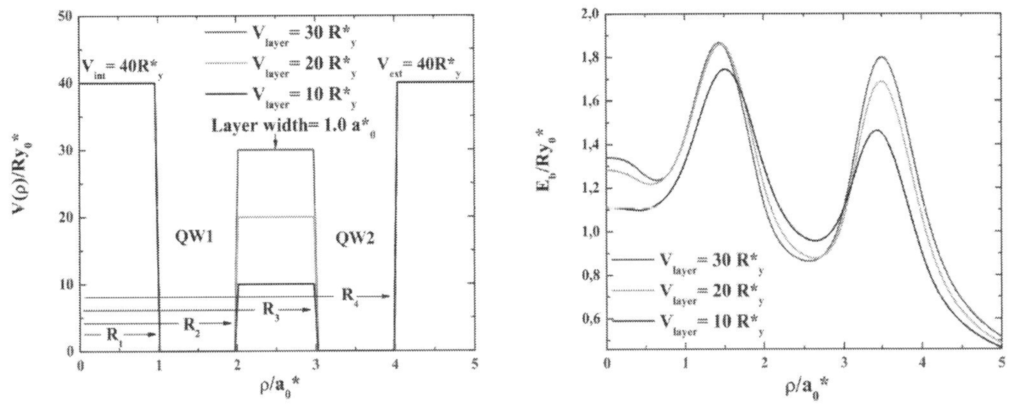

Figure 1. (left) Confinement potential profile containing two $Al_xGa_{1-x}As/GaAs/Al_xGa_{1-x}As$ quantum wells (QWs) and different values of V_{layer}. **Figure 1.** (right) D_0 ground state binding energies as a function of the donor displacement from the center of a spherical quantum dot.

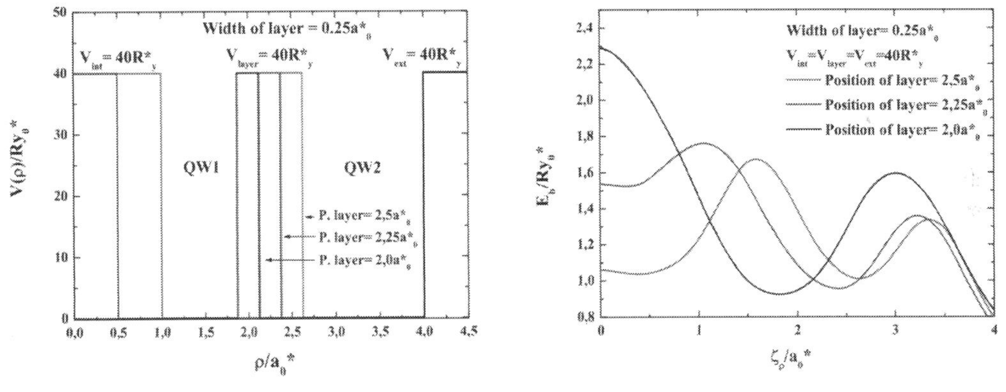

Figure 2. (left) Confinement potential profile containing two $Al_xGa_{1-x}As/GaAs/Al_xGa_{1-x}As$ quantum wells (QWs) and different values of layer position. **Figure 2.** (right) D_0 ground state binding energies as a function of the donor displacement from the center of a spherical quantum dot.

position to the layer with increase of the V_{layer} values, additionally the peaks are coinciding approximately with the midpoint between the interior and exterior radii, $(R_1 + R_2)/2$ for the firts peaks in QW1 whereas for the secund peaks are located approximately $(R_3 + R_4)/2$ in QW2, where the confinement is the largest. Moreover,the curves minimum in Fig. 1 are in the position of the layer, it shown that increasing V_{layer}, the probabilities of electron penetration in the barrier is reduced and he smallest value of the binding energy is achieved using $V_{layer} = 30R_y^*$, in additon, it is interesting that the crossovers of all curves related to the inversion of their order occur approximately at the same point, corresponding to distance at $2a_0^*$ and $3a_0^*$ from the axis of the SQD, in this position of the donor binding energy is almost insensitive to the potential barrier changes. The Fig. 2 (left) also shows the profile of the confinement potential with similar characteristics than Fig. 1 (left), but the potential height are the same $V_{int} = V_{ext} = V_{layer} = 40R_y^*$, whereas the position of potential for the layer vary for values measured from the axis of the SQD $P.layer = 2.0a_0^*, 2.25a_0^*$ and $2.5a_0^*$ and the layer width is used $0.25a_0^*$. It is seen from Fig. 2 (right) that as the position layer is changed from $2.0a_0^*$ to $2.5a_0^*$ the altitude of the peaks in QW1 and QW2 falls as the P.layer increases. It is due to the fact that for those values, the electron mostly is localized in the central region of QW1 and QW2 and therefore, the electron-donor separation increases with the donor displacement from the dot center, whereas, the binding energy decreases when the donor is situated in the same position of the barriers, due to the electron-donor separation becomes very large. Additionally, the Fig.2 (right) shows a successive displacement of the curves peaks position to the right in the QW1 whereas the opposite occurs in QW2 with reducing P.layer in consequence a large curves configurations can be obtained.

4. Conclusions

In this paper, we have presented a simple method of calculation of the binding energy for the lowest states of the off-center D^0 donor in a SQD applicable for arbitrary confinement potential. By using this procedure we calculate the D^0 binding energies in dots with two rectangular wells with different values of potential barrier for the layer, in consequence, a large curves configurations can be obtained which is useful to design electronic and opto-electronic devices.

Acknowledgments

This work was partially financed by the Universidad del Magdalena (Fonciencias) and the Colombian Agency COLCIENCIAS through doctoral scholarships 567.

References

[1] Sargent E H 2012 *Nat. Photonics* **6** 133.
[2] Ma N, Sargent E H, Kelley S O 2008 *Nat. Nanotechnol.* **4** 121.
[3] Konstantatos G, Sargent E H 2010 *Nat. Nanotechnol.* **5** 391.
[4] Williams B S 2007 *Nat. Photonics* **1** 517.
[5] Claudon J,Bleuse J, Malik N S, Bazin M, Jaffrennou P, Gregersen N 2010 *Nat. Photonics* **4** 174.
[6] Kako S, Santori C, Hoshino K, Gtzinger S, Yamamoto Y, Arakawa Y 2006 *Nat.Mater.* **5** 887.
[7] Herz L M, Phillips R T 2002 *Nat. Mater.* **1** 212.
[8] Padilha L A, Nootz G, Olszak P D, Webster S, Hagan D J, Van Stryland E W 2011 *Nano Lett.* **11** 1227.
[9] Tang J, Liu H, Zhitomirsky D, Hoogland S, Wang X, Furukawa M 2012 *Nano Lett.* **12** 4889.
[10] Zibik E, Grange T, Carpenter B, Porter N, Ferreira R, Bastard G 2009 *Nat.Mater.* **8** 803.
[11] Sierra-Ortega J, Mikhailov I D, Betancur F J 2002 *Phys. Stat. Sol. (b)* **230** 469.
[12] Mikhailov I D, Betancur F J, Escorcia R A, Sierra-Ortega J 2002 *Phys. Stat. Sol. (b)* **234** 590.
[13] Betancur F J, Mikhailov I D, Oliveira L E 1998 *J. Appl. Phys. D* 31 3391.

Surface Coverage in Wireless Sensor Networks Based on Delaunay Tetrahedralization

M G Ribeiro[1], L A Neves[1], A R Pinto[2], M Z Nascimento[3], G F D Zafalon[1] and C Valêncio[1]

[1] São Paulo State University (UNESP), Department of Computer Science and Statistics (DCCE), Brazil
[2] Universidade Federal de Santa Catarina, Campus de Blumenau, Brazil
[3] Federal University of Uberlandia, FACOM, Brazil

E-mail: leandro@ibilce.unesp.br

Abstract. In this work is presented a new method for sensor deployment on 3D surfaces. The method was structured on different steps. The first one aimed discretizes the relief of interest with Delaunay algorithm. The tetrahedra and relative values (spatial coordinates of each vertex and faces) were input to construction of 3D Voronoi diagram. Each circumcenter was calculated as a candidate position for a sensor node: the corresponding circular coverage area was calculated based on a radius *r*. The *r* value can be adjusted to simulate different kinds of sensors. The Dijkstra algorithm and a selection method were applied to eliminate candidate positions with overlapped coverage areas or beyond of surface of interest. Performance evaluations measures were defined using coverage area and communication as criteria. The results were relevant, once the mean coverage rate achieved on three different surfaces were among 91% and 100%.

1. Introduction

Wireless Sensor Networks (WSN) are defined as a subclass of Ad hoc networks, aiming to monitor some phenomena. This kind of network is mainly applied on locations of hard access or in dangerous areas, for military purposes or not [1-3]. The main component of this kind of network is the sensor node, which is responsible for detecting a signal of interest. The inappropriate positioning of a node can cause failures on the WSN to monitor the desired event [1-5]. Studies involving WSN are developed to guarantee appropriate distributions of the sensor nodes on different interest areas. Among the motivations, emphasize to obtain the biggest sensing coverage area, with the best possible quality and with the lesser sensor quantity.

Distributions of sensor nodes with the Delaunay triangulation and Voronoi diagram provide relevant results, but there are few approaches exploring more realistic terrains, considering volumes and roughness of natural reliefs [4, 5]. The generally encountered are studies on simplified and bi-dimensional spaces [6]. For 3D proposal, the limitations involve the need of a non-trivial pre-processing step to discretize a single surface, surfaces non-representative of the real reliefs or restriction to increase the number of nodes. Therefore, in this work a new method is presented for sensor deployment on 3D surfaces, considering a combination of Delaunay and 3D Voronoi. The method was structured on representing the surface with the Delaunay algorithm, construction of the 3D Voronoi diagram and definition of the circumcenter from each cell as candidate position for a sensor node. The preliminary results were significant, indicating which the proposed approach is promising and already provides an important contribution for studies involving sensor deployment.

2. Methodology

The first step of the method was to define the boundary of the 3D reliefs for distribution of the sensor nodes. This task was performed applying the Blender graphic package, which is maintained by *Blender Foundation* and available under double license (*BL / GNU General Public License*) [7]. From the boundaries represented, the Delaunay triangulation [6] was applied, aiming to subdivide the geometric domain in simplexes. This was possible considering the limits as a set of vertices V and s as a k-simplex ($0 \leq k \leq n$), composed by the vertices V. The circumcenter of s was comprehended as a circumference with radius r that crosses all vertices from s. If $k=d$, s has a single circumcenter, otherwise, there are infinite circuncemters from s. The simplex s is Delaunay if exists a circumcenter from s which no vertex from V is inside it. Therefore, the lines between the points that satisfy the condition were drawn, constituting the Delaunay triangles.

The next step was to define the Voronoi diagram [5]. Each Voronoi cell V from the diagram was represented by equation (1). The duality was defined as: for each Delaunay's edge, the middle point was determined and a perpendicular line was drawn, obtaining the edges from the Voronoi cell. Then, the circumcenter was calculated as the intersection of the bisector from the sides of a triangle. Therewith, it is possible to draw a circumscribed circumference to the triangle which touches the corresponding vertex, guaranteeing that the cell is Voronoi. The circumcenter of a cell was used as a candidate position of the sensor node on the surface.

$$V(p_i)=\{p \mid d(p,p_i)\leq d(p,p_j),\ j\neq i,\ j=1,\ldots,n\}, \tag{1}$$

where, $p_1,\ldots, p_n$ is a set of nodes (defined with x, y and z coordinates), generated by the Delaunay algorithm; n define the total of nodes which discretize the analyzed surface; $d(p,p_i)$ represents the Euclidian distance between a node p (took as base) and a node p_i, both belonging to a tetrahedron; $V(p_i)$ represents the Voronoi cell associated to the node p_i.

From each candidate position defined, the next step was to calculate performance measures, such as: (a) local and global communications; and, (b) coverage area with quality [4].

Local communication measure was defined by verifying if a candidate position has at least one neighbor for communication. Local communication was determined by the coordinates and the circumcenter radius of interest, checking if the neighbors are within enough distance for communication. This was guaranteed considering that the distance between the analyzed circumcenters is less than the sum of radius, identifying positions which imply in isolated nodes. Global communication was guaranteed with the use of Dijkstra algorithm [8], aiming to guarantee the existence of routes among the nodes of the interest area. The weight between each edge was determined according to the Euclidean distance and using as input method the adjacency matrix.

To define the coverage area with quality was necessary to delete possible circumcenters with identical spatial positions and analyze unwanted intersections. This was performed by verifying the distances among a position taken as reference and the relative neighbor positions: considering a radius range r of the sensor node (which was provided by the user), the positions were deleted when the distance between evaluated positions was greater than the sum of radius. This condition was applied for each candidate position. The coverage area of a node was obtained by equation (2):

$$s=\frac{1}{2}\left(det\begin{vmatrix} y_1 & z_1 & 1\\ y_2 & z_2 & 1\\ y_3 & z_3 & 1\end{vmatrix}^2+det\begin{vmatrix} z_1 & x_1 & 1\\ z_2 & x_2 & 1\\ z_3 & x_3 & 1\end{vmatrix}^2+det\begin{vmatrix} x_1 & y_1 & 1\\ x_2 & y_2 & 1\\ x_3 & y_3 & 1\end{vmatrix}^2\right)^{1/2}, \tag{2}$$

where s is the surface area and x, y and z are vertex coordinates which constitute the tetrahedron of the 3D surface of interest.

Total area was obtained by the sum of individual areas, following the relation $a=\pi r^2$. Therewith, it was necessary to consider areas with circumference intersections of the coverage, analyzing pairs of positions, applying the described method by [4]. The relations were defined by the Euclidean distances c among each position i=1,...,n, equation (3). Values r_0 and r_1 were the corresponding radius range of the centers called A and B. The points C and D are the intersection points between the circumcenters A and B. Therefore it was possible to determine the angles between the points CAD and CBD, which were called respectively by θ and ρ. At first, only half of the angle was calculated, considering the triangle constituted by the point CBA, equation (4). By calculating the angle among the points CBA, it can be concluded that $CAD = 2(CAB)$. For the other half, the process was similar, therefore it can be concluded that $CAD = 2(CAB)$. Finally obtained by equation (5), the angles composed by the points CBD and CAD in radian, it was possible to determine the region of each one of the circles cut by the chord CD. Calculating the sector area of the circle BCD and subtracting the area of the triangle ACD, the desired area was defined without overlap (equation (6)).

$$c = \left(\left(x_{c_i} - x_{c_{i+1}}\right)^2 + \left(y_{c_i} - y_{c_{i+1}}\right)^2 + \left(z_{c_i} - z_{c_{i+1}}\right)^2\right)^{1/2} \quad (3)$$

$${r_0}^2 = {r_1}^2 + c^2 - 2r_1 c \cos(CBA) \Rightarrow \cos(CBA) = \frac{{r_0}^2 + c^2 - {r_1}^2}{2r_0 c} \quad (4)$$

$$\gamma = \mathrm{acos}(CBA), \theta = 2\gamma; \zeta = \mathrm{acos}(CAB), \rho = 2\zeta \quad (5)$$

$$area = \frac{1}{2}\rho {r_1}^2 - \frac{1}{2}{r_1}^2 sen(\rho) + \frac{1}{2}\theta {r_0}^2 - \frac{1}{2}{r_0}^2 sen(\theta) \quad (6)$$

It is important to emphasize that the area (equation 6) is true for intersection in pairs. For others scenarios, the strategy used was broader, aiming to consider two nodes which are neighbors (P and Q) with intersection and involving another intersection with a third node R. In this scenario, the intersection area was PQ and QR. This situation was corrected considering the difference between the intersections PQ and QR, defining the intersection area among the three circumferences PQR and subtracting of the area of the previous calculation from the intersection area two by two. The last step was to sum the intersection value PQR to obtain the final area.

3. Results

The proposed approach was tested on three distinct reliefs: plain, plateau and hilly. These reliefs are commonly found on applications of WSN for forest fire control, for instance. The length of the example surfaces were fixed in 2000 x 2000 m². In figure 1(a) is shown the plain relief, with total area of 4022.23 m². In figure 1(b) is presented the plateau relief, with total area of 4080.24 m² and in figure 1(c) is presented the hilly relief (4405.92 m²). The values were obtained considering peaks and valleys.

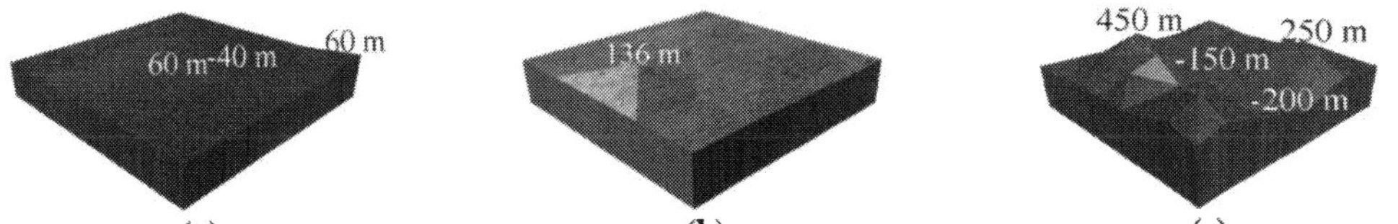

(a) (b) (c)

Figure 1. Representative surfaces of **(a)** plain, **(b)** plateau and **(c)** hilly, which were used as examples for sensor deployment considering the proposed method.

After applying the proposed method, it was obtained: 26 sensor nodes for the plain relief, where 295 ± 56 meters of mean range radius of the sensor, resulting in 100% of coverage area (4022.23 m²); 20 sensor nodes for the plateau relief, with coverage of 98.2% (4006.74 m²). The mean range radius of the sensor was 335 ± 20.7 meters; and, 62 sensor nodes for the hilly relief, representing 91.01% of coverage (4009.65 m²), with mean of 180 ± 4.5 meters from range radius of the sensor.

Another quality measure used was the check on local and global communications, using the Dijkstra shortest path algorithm. As results, the global communication of the distribution was from 18 sensor nodes for the plain terrain, 8 sensor nodes for plateau terrain and 42 sensor nodes for hilly terrain. In all cases, by the minimum limits presented for global communication, local communication was satisfied through Dijkstra algorithm. To ilustrate these distributions, in figures 2(a), 2(b) and 2(c) are presented the positions of the sensor nodes in relation to the relief, by Delaunay triangulation and 3D Voronoi. There is still the representative circumferences from the region covered by the nodes, from the circuncenters (position of the sensor nodes).

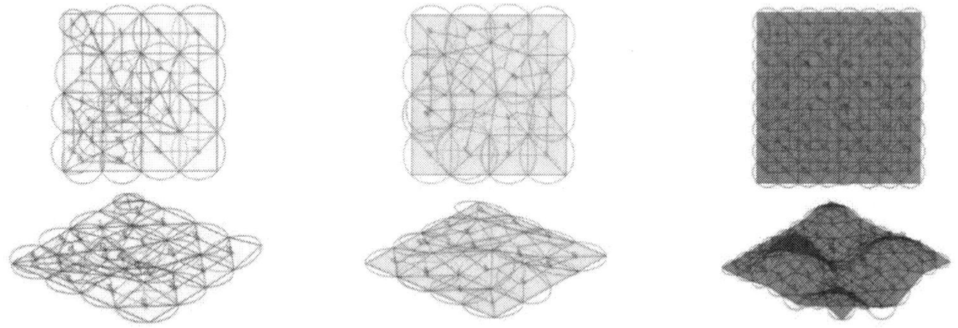

(a) **(b)** **(c)**

Figure 3. Visualization of the candidate positions to all sensor nodes relative to the plain **(a)**, plateau **(b)** and hilly **(c)** surfaces.

4. Conclusion

In this work was presented a new method for sensor deployment on 3D surfaces, which provided relevant candidate positions for the surfaces used as example. The coverage values were above 91% and with guarantees of global and local communications: the 3D approach presented by [5] provided a coverage rate of 76.3%, less than the one achieved here. The developed method still does not handle obstacles or priority regions. However, the preliminary results indicate which the method is promising and can be already used by specialists focused on studies involving sensor deployment and 3D surfaces. In future work, the method will be refined, considering obstacles, deployment on priority regions and inclusion of another quality measures.

Acknowledgements

This work was financially supported by PROPe/UNESP (Pró-Reitoria de Pesquisa/UNESP).

References

[1] Winkler M, Tuchs K D, Hughes K and Barclay G 2008 Theoretical and practical aspects of military wireless sensor networks. *Journal of Telecommunications and Information Technology* **2** 37-45

[2] Lynch J P and Loh K J 2006 A summary review of wireless sensors and sensor networks for structural health monitoring *Shock and Vibration Digest* **38**(2) 91-130

[3] Wang Y, Liu Y and Guo Z 2012 Three-dimensional ocean sensor networks: a survey. *Journal of Ocean University of China* **11**(4) 436-450

[4] Unaldi N, Temel S and Asari V K 2012 Method for optimal sensor deployment on 3D terrains utilizing a steady state genetic algorithm with a guided walk mutation operator based on the wavelet transform *Sensors* **12**(4) 5116-5133

[5] Jin M, Rong G, Wu H, Shuai L and Guo X 2012 Optimal surface deployment problem in wireless sensor networks In Proceedings *IEEE INFOCOM* 2345-2353

[6] Chen L and Holst M 2011 Efficient mesh optimization schemes based on optimal delaunay triangulations *Computer Methods in Applied Mechanics and Engineering* **200**(9) 967-984

[7] Blender, Blender foundation, http://www.blender.org/ (Dec. 2013).

[8] Dijkstra E W 1959 A note on two problems in connection with graphs *Numerische Math* **1** 269-271

The Lorentz Attractor and Other Attractors in the Economic System of a Firm

V I Shapovalov[1,3] and N V Kazakov[2]

[1]The Volgograd Branch of Moscow Humanitarian-Economics Institute, Volgograd, Russia
[2]The Volgograd State Technical University, Volgograd, Russia

[3]E-mail: shavi71@rambler.ru

Abstract. A nonlinear model of the economic system of "a firm" is offered. It is shown that this model has several chaotic attractors, including the Lorentz attractor and a new attractor that, in our opinion, has not yet been described in the scientific literature. The chaotic nature of the attractors that were found was confirmed by computing the Lyapunov indicators. The functioning of our economic model is demonstrated with examples of firm behaviour that change the control parameters; these are well known in practice. In particular, it is shown that changes in the specific control parameters may change the system and avoid bankruptcy for the firm.

Keywords: chaotic attractor, synergetics, nonlinear economic model.

1. Introduction

A change of external influence corresponds to a change of control parameters in nonlinear dynamics. Let us remember that the control parameters are all constants in the evolutionary equation: $dX_i/dt = F_i(X_1, X_2, ..., X_n)$, where X_i are the system variables; F_i are functions of the variables; t is time; and n is the minimum number of variables needed for a description of the system. By changing the control parameter values we can change the current steady state of the system into an unstable state, and vice versa.

In this paper we consider economic systems as self-organizing systems. We can observe the coordinated, collective movement of the elements in such systems. Therefore we understand the X_i variables in (1) to be macroscopic variables corresponding to some generalized characteristics of a collective movement of system elements. We recall that such variables are called order parameters in synergetics [1].

Creating a synergetic model usually starts with the selection of the order parameters, i.e. the macroscopic variables that quantify the main links in the system. The next step consists in drawing up the proportions to form these relationships. (The rule of drawing up proportions is described in detail in [1]. According to this rule, the increase over time in a particular value is eventually proportional to the gain in this value minus its loss). These proportions are then converted into evolutionary equations.

Below we consider a firm for which the number of employees and the working capital are average for a particular region. We also assume that, in order to develop its production base, the firm borrows money. The aim of our work is to create a synergetic model of this system. In particular, we will try to find its attractors.

2. The basic model for the firm system

Selecting the system variables for the firm is easy – they are the number of employees X, the working capital Y and the loan amount Z.

We call our model the *basic* model if it includes the minimum number of relationships between these variables (i.e., only those relationships that are sure to arise if the company operates in the market for a long time). These relationships are the following:

1) The increase in the number of employees over time (dX/dt) is proportional to the number of employees joining, J_1, minus the number leaving, J_2. An increase in the number of employees at the firm depends on the firm having an attractive image. In most cases, to create an attractive image the firm must spend part of its capital Y and a portion of the loan Z. Consequently $J_1 \sim Y,Z$. It is obvious that the number of employees who are lost due to layoffs is proportional to the number of employees. Therefore $J_2 \sim X$.

2) The capital increase over time (dY/dt) is proportional to the capital gains, J_3, minus the capital reduction, J_4. It is obvious that $J_3 \sim (Y+Z)$. The main losses in capital are caused by the payments of the employees' salaries and credit service payments, so $J_4 \sim X,Z$.

3) The increase in the loan over time (dZ/dt) is proportional to the capital of the company ($\sim Y$) (we assume that the larger the firm's capital, the greater the willingness to extend credit to the firm), minus losses that are due to the magnitude of the loans taken ($\sim Z$) (in particular, any bank is less willing to lend to a firm that has many loan obligations).

The above proportions can be represented as a system of three evolution equations [2,3]:

$$\dot{X} = \alpha YZ - \gamma X; \quad \dot{Y} = \mu(Y + Z) - \beta XZ; \quad \dot{Z} = \delta Y - \lambda Z \ , \tag{1}$$

where $\dot{X} = dX/dt$, $\dot{Y} = dY/dt$, $\dot{Z} = dZ/dt$; and α, γ, μ, β, δ, λ are coefficients of proportionality. The values α, γ, μ, β, δ, λ are constants, and hence play a role as control parameters.

The system of equations (1) is similar in its structure to the well-known Lorenz system, which has no analytic solutions. Therefore a numerical investigation of this system is carried out for different values of the control parameters. We choose the values of the control parameters taking into account the restriction imposed by the properties of the real economic system. For example, it is obvious that the terms of each equation in system (1) must be of similar size. Consequently, the pairs of coefficients α and γ, μ and β, and δ and λ cannot be of a different order. If they differ in order (or more) then the terms with very small coefficients can be neglected.

Different types of attractors are found as a result of the calculations. To clarify the type of attractor we apply the Lyapunov indicators (recall that in the three-dimensional coordinate system, the signs of the Lyapunov indicators have the following distribution [4]: "–, –, –" for a point; "0, –, –" for a limit cycle; "0, 0, –" for the torus; and "+, 0, –" for a chaotic attractor). The calculation of these parameters was performed as described in [5], and included the orthogonalization procedure on the Gramm-Schmidt method. Some typical attractors are presented in 'Figure 1'.

The simple attractors shown in 'Figure 1(a)' and 'Figure 1(b)' correspond to the states in which the system is found as λ increases. Recall that this control parameter reflects the factors that have an adverse effect on receiving credit. In other words, rigid actions by the bank force the company to complicate its structure. We suppose the complication of the structure to be shown in the changes to the configuration of the attractors. However, if λ continues to increase the structure shows chaotic attractors ('Figure 1(c)' and 'Figure 1(d)').

In our opinion, the chaotic attractors in 'Figure 1(c)' and 'Figure 1(d)' correspond to the bankruptcy of the firm. However, they appear only for a certain ratio of the control parameters. Therefore, by changing the values of these parameters, we can avoid the unwanted attraction domain of the attractor. For example, we can increase γ (i.e. fire some employees). In this case, the chaotic behaviour of the firm also changes to an ordered behaviour ('Figure 2(e)'). Alternatively, the company can improve its management, i.e. increase μ. As a consequence, it leaves the chaotic attractor system and enters the attractor "limit cycle" ('Figure 1(f)').

The chaotic attractor shown in 'Figure 1(d)' coincides in form with the famous Lorenz attractor. If we calculate the fractal dimension *D*, using a known simplified formula [6,7], we obtain the value coinciding with dimension of the Lorentz attractor: $D = 2 + [L(X) + L(Y)]/|L(Z)| = 2.06$, where *L*(*X*), *L*(*Y*) and *L*(*Z*) are the Lyapunov indicators. However, the chaotic attractor shown in 'Figure 1(c)' is not a Lorenz attractor as it has a tape configuration. The borders of the tape remain unchanged for a long time period. During this period of time the trajectories of the system almost completely fill the space inside the tape. We believe that the attractor 'Figure 1(c)' is a *tape modification* of the Lorenz attractor. This is explained by the fact that this attractor is obtained from the system (1), which is not much different from the Lorenz system.

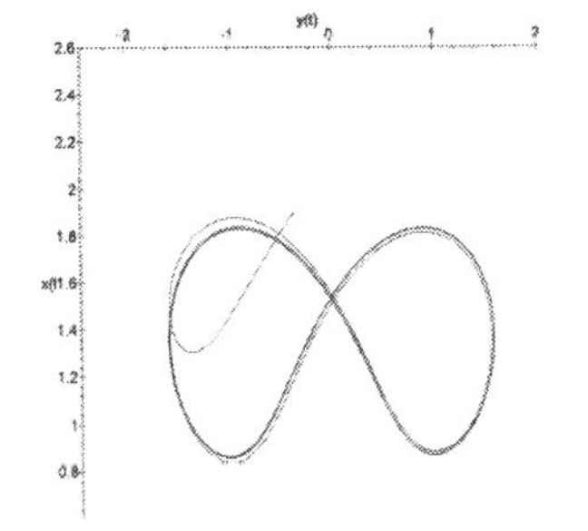 $L(X) = 0.000$; $L(Y) = -0.507$; $L(Z) = -1.393$	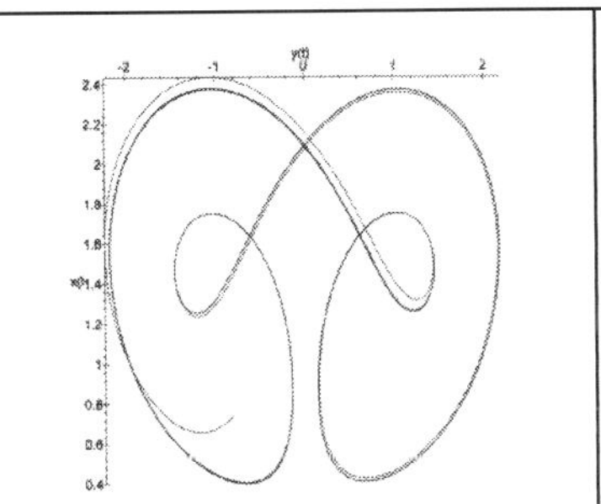 $L(X) = 0.000$; $L(Y) = -0.295$; $L(Z) = -2.205$	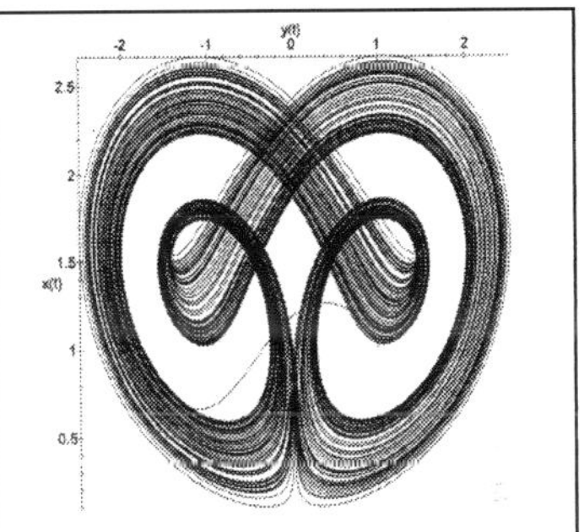 $L(X) = 0.121$; $L(Y) = 0.000$; $L(Z) = -2.721$
(a) simple attractor: λ=3.1	**(b)** simple attractor: λ=3.7	**(c)** chaotic attractor: λ=3.8
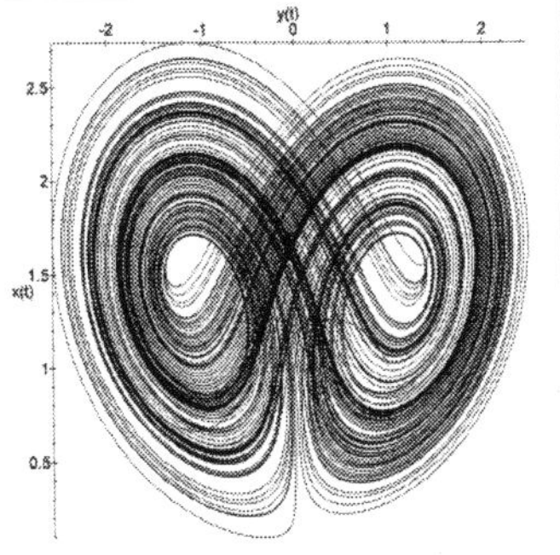 $L(X) = 0.176$; $L(Y) = 0.000$; $L(Z) = -2.976$	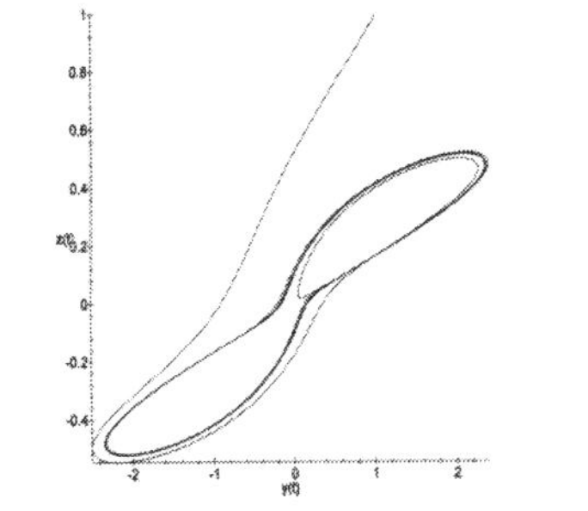 $L(X) = 0.000$; $L(Y) = -0.146$; $L(Z) = -3.654$	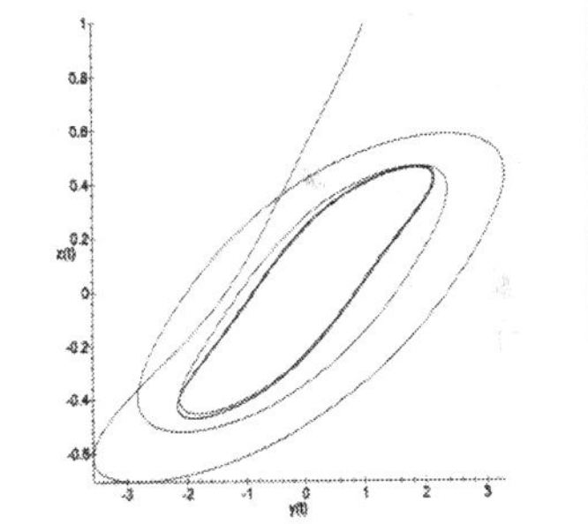 $L(X) = 0.000$; $L(Y) = -0.100$; $L(Z) = -0.900$
(d) chaotic attractor: λ=4	**(e)** simple attractor: λ=4, γ=2	**(f)** simple attractor: λ=4, μ=3.2

Figure 1. Some attractors of the system (1). Changes in the parameter λ are shown: **(a)** – **(d)** (the unchangeable parameters are: α=5, γ=1, μ=2.2, β=8, δ=1). Attractors **(e)** and **(f)** follow on from the chaotic attractor represented in **(d)**. *L*(*X*), *L*(*Y*) and *L*(*Z*) are the Lyapunov indicators.

Attractors shown in 'Figure 1' are selected from a series of attractors resulting from changes of the λ control parameter with step of 0,1. We assume that a smaller step would not be noticeable in a real economic system. However, from a mathematical point of view, such a restriction does not make sense. As an example, we explored the small λ interval from 3.6 to 3.7 in increments of 0.01. As a result, we found that the system (1) has tape-type chaotic attractors in this range. At the same interval simple attractors appear between chaotic attractors. If we reduce the step further, we also see a sequence of simple and chaotic attractors.

The appearance and disappearance of chaotic attractors while the control parameter changes is well known (for example, in the Lorenz system [4]). At the same time, in our opinion, a controlled change in the economic value of as little as 0.01 (and less) is difficult to implement in a real economic system. Therefore, such variations can be neglected for the average firm in practice.

3. Complex model of the firm

The basic model (1) can be complicated, depending on the economic situation. In particular, the firm can create its own attractive image using only capital equity Y or only a loan Z. In order to consider these cases using the evolution equations system we have to add terms with only Y and only Z in the first equation of system (1). In addition, for some periods of time, the company may not have credit obligations ($Z = 0$). Then its basic expenses will only be the cost of the employees X. To account for this, a term containing only X (i.e., excluding loan servicing) should be subtracted in the second equation. As a result the evolutionary equations system for a complex model of the company will take the form:

$$\dot{X} = \varepsilon Y + \nu Z + \alpha YZ - \gamma X; \quad \dot{Y} = \mu(Y + Z) - \beta XZ - \eta X; \quad \dot{Z} = \delta Y - \lambda Z , \qquad (2)$$

where ε, η and ν are coefficients of proportionality. We carried out numerical research for this system by changing the values of the control parameters λ, γ and μ. In general, the economic picture was similar to that of the base model. An increase in λ (i.e., reducing the possibility of receiving credit) led to a change from an ordered movement to a chaotic one ('Figure 2'). An increase in μ (i.e. an increase in the efficiency of the firm in the market) brings the firm out of a chaotic attractor system ('Figure 2'). However, in contrast to the basic model, changing only γ (e.g., by dismissing some of the staff) is not enough to move the company away from a chaotic attractor. At the same time, this is possible by a joint change in γ and μ ('Figure 2(f)').

4. Conclusion

1. In this paper we proposed a nonlinear mathematical model of the economic system of "a firm" (see the systems of equations (1) and (2)). We found simple attractors (limit cycles) that, in our opinion, correspond to stable states of the firm, and chaotic attractors, that correspond to the state of bankruptcy in this model. Also, we have shown that by changing the values of specific control parameters it is possible to avoid a chaotic attractor, i.e. bankruptcy. In particular, by reducing λ (for example, by taking a loan at a lower interest rate), we will transfer the company from a chaotic attractor to a stable state – a limit cycle ('Figure 1' and 'Figure 2'). The configuration of the limit cycles presumably reflects the complexity of the firm's economic activity that is needed to ensure sustainability with varying degrees of harshness in the credit obligations. If a company cannot change its credit obligations, it can change some other control parameters. For example, it can increase the parameter γ (which corresponds to dismissing some employees) and/or increase the parameter μ (i.e. improve the management). In these cases, the chaotic behaviour firm also changes to a stable behaviour ('Figure 1' and 'Figure 2').

We should emphasize that the mathematical models considered in this paper are fairly abstract. They describe a general methodological approach, and reveal new capabilities given by synergetic relationships for the analysis of the activity of a particular enterprise. In our opinion, these models can serve as a good methodological basis for the forecasting of bankruptcy and the search for the optimum way of managing a business.

2. We assume that we have found *a new chaotic attractor*. Its appearance is presented in 'Figure 2(c)'. This attractor is seen if the values of α=6, γ=1, ε=10, ν=6, μ=1, β=8, η=4, δ=1, λ=5 are substituted in the system (2). We also found chaotic attractors, which we called tape modifications of the Lorenz attractor because of their appearance (e.g., 'Figure 1(c)'). A distinctive feature of a tape chaotic attractor is the fact that, over a long time, the trajectory of the system completely fills the space within the boundaries of the tape but does not go beyond them. In our opinion, research into the

attractors mentioned above can be of interest to scientists dealing with problems of nonlinear dynamics.

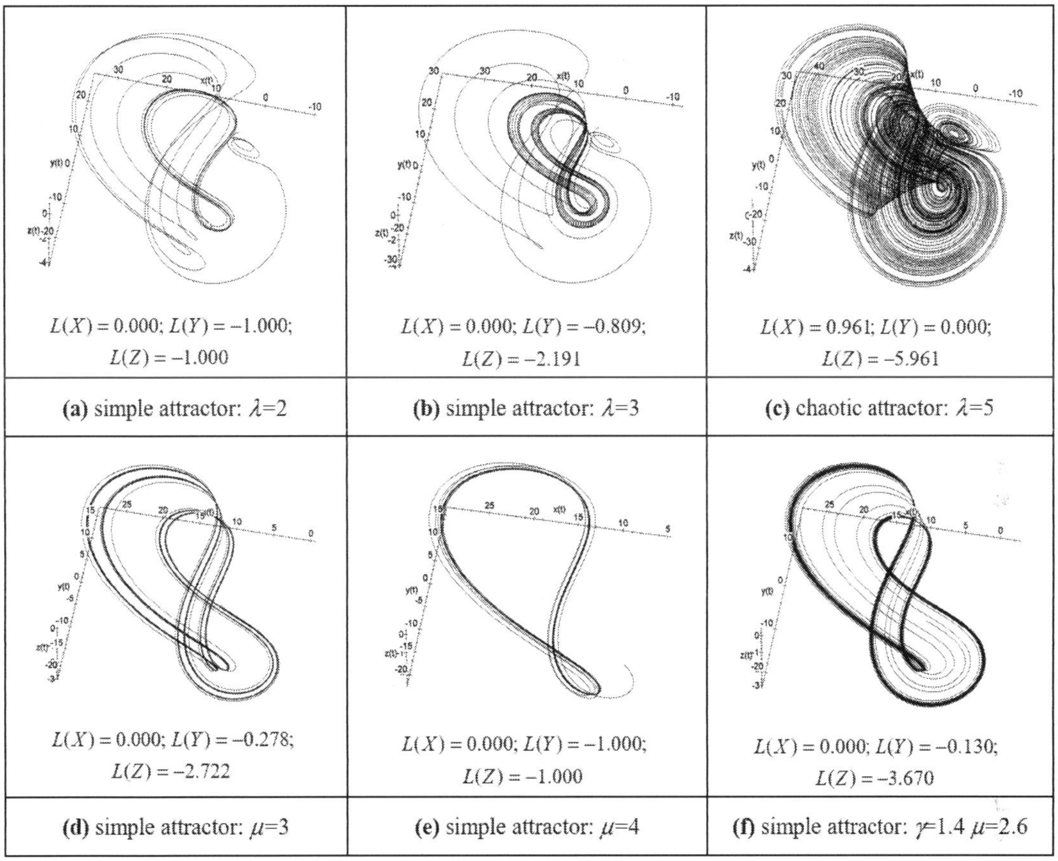

Figure 2. Some attractors of system (2). Changes in the parameter λ are shown: **(a)** – **(c)** (the unchangeable parameters are α=6, γ=1, ε=10, ν=6, μ=1, β=8, η=4, δ=1). Attractors **(d)** – **(f)** follow on from the chaotic attractor represented in **(c)**.

References

[1] Haken H 1978 *Synergetics* (Berlin, Heidelberg, New York: Springer Verlag)
[2] Shapovalov V I, Kablov V F, Bashmakov V A and Avvakumov V E 2004 Synergetic stability model for an average firm *Synergetics and Problems in Control Theory* ed A A Kolesnikov (Moscow: FIZMATLIT Publishers) pp 454–464
[3] Shapovalov V I 2005 *Basis of Ordering and Self-Organization Theory* (Moscow: ISPO-Service)
[4] Berge P, Pomeau Y and Vidal Ch 1988 *L'ordre Dans Le Chaos* (Paris: Hermann)
[5] Naymark Yu I and Landa P S 1987 *Stochastic and chaotic oscillations* (Moscow: Nauka Publishers)
[6] Kaplan J L and Yorke J A 1979 *Lect. Notes in Math.* **730** 228
[7] Ahmed I, Mu Ch and Zhang A 2014 *International Journal of Analysis and Applications* **6** 27

Order and Disorder Evolution Determined by Non-Linear Processes

Yehuda Roth
Oranim Academic College, Q. Tivon 36006, Israel
E-mail: yudroth@gmail.com

Abstract. We show that a low dimensional system can evolve toward an infinite dimensional ergodic system. We also show the existence of the reverse process, that is, a large-dimensional ergodic system that evolves toward a low-dimensional ordered system.

1. Introduction

Order and disorder are related to coherence and de-coherence concepts, respectively (see for example Refs. [1, 2]). Consequently, exploring the de-coherence processes gives rise to the understanding of the entropy increase evolution toward maximum value in thermal equilibrium. In this work, by exploring recursive maps, we discover that in addition to the de-coherence evolution, there is an opposite process referred to as re-coherence. Re-coherence describes a system that originally possessed a disordered component, but evolves toward absolute coherence. A demonstration of a low dimensional system that experiences such a re-coherenne evolution was recently introduced by us [3]. In this paper we show that re-coherence and de-coherence have the same origin, that is, nonlinear maps.
Two contradicting processes determine a system of nonlinear dynamics where if these two processes reach a finite number of equilibrium states the associated map is regarded as a regular map. Otherwise, the map is considered to be chaotic [4]. We show that chaotic maps are associated with ergodicity evolution while regular maps are responsible for the re-coherence evolution.

2. Primary Definitions

2.1. Low Dimensional Complementary Maps

We start with a low dimensional nonlinear map and we show how it expands into an infinite number of degrees of freedom with an ergodic nature.
Let $f_R(x_n)$ be an iterative map with a control parameter R such that

$$x_{n+1} = f_R(x_n), \quad x \in [0,1] \tag{1}$$

where all values of x are within the interval $[0,1]$. We refer it as an **x-map**.
We introduce coherence by introducing a **complementary map**, referred to as the **y-map**, as follows[3]

$$y_{n+1} \equiv 1 - f_R(1 - y_n), \tag{2}$$

where it is easy to see that $\forall n \; y \in [0,1]$.
For example, the y-map for the logistic map

$$x_{n+1} = Rx_n\left(1 - x_n\right) \tag{3}$$

becomes

$$y_{n+1} = 1 - Ry_n\left(1 - y_n\right). \tag{4}$$

In general the complementary maps can evolve independently provided that the initial variables x_0 and y_0 are determined separately. Otherwise, when the initial conditions are coordinated such that $x_0 + y_0 = 1$ the maps are considered to be **initially coherent**.
In a pure mathematical sense it can be shown that coherence, namely the relation $x_n + y_n = 1$, is conserved during the iterations.
Consider initially correlated maps,

$$y_0 = 1 - x_0. \tag{5}$$

Correspondingly, the next iteration yields

$$\begin{array}{l} x_1 = f_R\left(x_0\right) \\ y_1 = 1 - f_R\left(1 - y_0\right) = 1 - f_R\left(x_0\right) = 1 - x_1 \\ \Rightarrow y_1 = 1 - x_1. \end{array} \tag{6}$$

Thus, the initial relation $y_0 = 1 - x_0$ is replicated to the $n = 1$ iteration such that $y_1 = 1 - x_1$. Obviously, in a pure mathematical manner, this correlation is maintained for all n iterations

$$if \;\; y_0 = 1 - x_0 \;\; then \;\; \forall n \;\; y_n = 1 - x_n. \tag{7}$$

2.2. States Representation of the Nonlinear Maps

Suppose that we have some 2-D system with the **generating states** $|1\rangle$ and $|\emptyset\rangle$[5]. These generating states span a primary low dimensional basis that under the influence of chaotic maps, expands toward an infinite dimensional space.
The complementary maps are embedded in the generating states through the states coefficients. Since the coefficients are defined under the complex space we need to define at least two sets of complementary maps: The ordinary maps of eqs. 1 and 2 that defines the coefficients real part and what we refer as **hidden** maps that defines the states phases.
The hidden maps are defined as follows:

$$\begin{array}{l} p_{n+1} = \phi_\rho\left(p_n\right) \\ \tilde{p}_{n+1} = 1 - \phi_\rho\left(p_n\right) \\ \forall\, n \quad 0 \le p_n \le 1. \end{array} \tag{8}$$

The phase θ_n is defined as

$$e^{i\theta_n} = \sqrt{p_n} + i\sqrt{\tilde{p}_n}. \tag{9}$$

Altogether, we define the initial generating states

$$\begin{array}{l} |0\rangle = \sqrt{x_0}e^{i\theta_0}\,|1\rangle + \sqrt{y_0}\,|\emptyset\rangle \\ |\tilde{0}\rangle = -\sqrt{y_0}e^{i\theta_0}\,|1\rangle + \sqrt{x_0}\,|\emptyset\rangle\,, \end{array} \tag{10}$$

where we considered θ_0 as a relative phase. We assume that the initial maps x_0, y_0 and p_0, p_0 are unitary correlated meaning that $x_0 + y_0 = p_0 + \tilde{p}_0 = 1$.
We assume that the states interact with an external environment that modifies the states coefficients in according to the recursive-complementary maps to generate the iterating states

$$\begin{array}{l} |n\rangle = \sqrt{x_n}e^{i\theta_n}\,|1\rangle + \sqrt{y_n}\,|\emptyset\rangle \\ |\tilde{n}\rangle = -\sqrt{y_n}e^{i\theta_n}\,|1\rangle + \sqrt{x_n}\,|\emptyset\rangle \end{array} \tag{11}$$

Journal of Physics: Conference Series **574** (2015) 012085 doi:10.1088/1742-6596/574/1/012085

3. Description of Maps Evolution

Part of the following was published in Ref. [5].
Recursive nonlinear maps have the tendency of reaching constant values, provided that they are not in the chaotic regimes. If the maps reach single values, denoted by x_∞, y_∞ and θ_∞, we obtain that any initial basis of states terminates at the unique basis

$$\begin{aligned} |n \to \infty\rangle &= \sqrt{x_\infty} e^{i\theta_\infty} |1\rangle + \sqrt{y_\infty} |\emptyset\rangle \\ |\tilde{n} \to \infty\rangle &= -\sqrt{y_\infty} e^{i\theta_\infty} |1\rangle + \sqrt{y_\infty} |\emptyset\rangle . \end{aligned} \tag{12}$$

For example, when $R < 3$ the logistic maps reach the single set $x_\infty = \frac{R-1}{R}$ and $y_\infty = \frac{1}{R}$. Consequently, for real coefficients we obtain states that, except for the $x_n = 0, 1$-values, always converge into the unique basis of states

$$\begin{aligned} |\infty\rangle_R &= \sqrt{\tfrac{R-1}{R}} |1\rangle + \sqrt{\tfrac{1}{R}} |\emptyset\rangle \\ |\tilde{\infty}\rangle_R &= -\sqrt{\tfrac{1}{R}} |1\rangle + \sqrt{\tfrac{R-1}{R}} |\emptyset\rangle . \end{aligned} \tag{13}$$

This basis of states spans a 2-D Hilbert space.

3.1. The Maps Dimension Extension Toward Chaos and Ergodicity

The ϕ and the f maps are considered to be independent, meaning, that at the same time one set of maps can evolves in a regular manner, while the other can behave chaotically. For that reason we restrict our discussion only to a single set, that is, the f-maps keeping in mind that the proceeding arguments valid for both maps.
It is known that for a large variety of maps, increasing the strength parameter R such that the map evolves from regular toward chaotic behaviors is followed by universal steps.
The first step (relatively small R) is characterized by maps that are converging into a single pair of stable points "$x^{(1)}$" and "$y^{(1)}$" (or $\theta^{(1)}$) calculated as

$$\begin{aligned} x_{n+1} &= x_n \equiv x^{(1)} \Rightarrow x^{(1)} = f_R(x^{(1)}) \\ y_{n+1} &= y_n \equiv y^{(1)} \Rightarrow y^{(1)} = 1 - f_R(1 - y^{(1)}). \end{aligned} \tag{14}$$

For a higher level of R, the maps no longer converge into single values as they reach a state of two alternating values. Nevertheless, it is possible to redefine a map which retrieves a single stable value as follows

$$\begin{aligned} & x_{n+2} = x_n \equiv x^{(2)} \Rightarrow x^{(2)} = g(R, x^{(2)}) \\ & g \equiv f \circ f \\ & y_{n+2} = y_n \equiv y^{(2)} \Rightarrow y^{(2)} = \tilde{g}(R, 1 - y^{(2)}) \\ & \tilde{g} \equiv (1 - f) \circ (1 - f) . \end{aligned} \tag{15}$$

For higher strength parameter the maps terminate with four alternating values characterized by the maps $G = g \circ g$ and $\tilde{G} = \tilde{g} \circ \tilde{g}$ and the proceeding eight splits are determined by the $G \circ G$ and $\tilde{G} \circ \tilde{G}$ maps. Increasing R, the same splitting continues until it reaches the chaotic regime for which it is impossible to redefine maps that are stabilized at a single value.
Suppose we start with the regular initial conditions for the 2-D Hilbert space. For $l = 1$ the map reaches a final value. For $l = 2$ there are two final values. Thus, in order to retrieve a single value we use the double map $g = f \circ f$. If n represents some time interval that passes between the f-iterations, the composed map g corresponds with a time interval which is twice the original f-time. We now show that this interval doubling corresponds with an extension of the 2-D Hilbert space dimension.

We start with the basic map f (or ϕ) applied in the $l = 2$ - regime where in the steady state it alternates between two values.
Suppose that we start with an initial condition that is one of the two final states. For $n = 0, 1, 2, 3...$ the maps split into two types - even series and odd series iterations. Even series iterations are characterized by

$$\begin{array}{ll} Even\ series & \\ x_{2n+2} = f_R(x_{2n}) & (The\ series\ x_0, x_2, x_4...) \\ y_{2n+2} = 1 - f(y_{2n}) & (The\ series\ y_0, y_2, y_4...) \\ Odd\ series & \\ x_{2n+3} = f_R(x_{2n+1}) & (The\ series\ x_1, x_3, x_5...) \\ y_{2n+1} = 1 - f(y_{2n-1}) & (The\ series\ y_1, y_3, y_5...). \end{array} \tag{16}$$

The $l = 4$ space possesses the following degrees of freedom:

$$\begin{array}{ll} For\ n = 0, 1, 2, 3... & \\ x_{4n+4} = f_R(x_{4n}) & (The\ series\ x_0, x_4, x_8...) \\ x_{4n+5} = f_R(x_{4n+1}) & (The\ series\ x_1, x_5, x_9...) \\ x_{4n+6} = f_R(x_{4n+2}) & (The\ series\ x_2, x_6, x_{10}...) \\ x_{4n+7} = f_R(x_{4n+3}) & (The\ series\ x_3, x_7, x_{11}...). \end{array} \tag{17}$$

Counting the $l = 4$ map degrees of freedom we obtain the four branches of eq. 17. We assume that the three former $l = 2$ branches also participate in the degree of freedom counting, thereby obtaining the $l = 4$ space dimension $d = 2^1 + 2^2 + 2^4 = 26$.
In general we assume that the l-space dimension is $d_l = \sum_{i=0}^{l/2} 2^{2^i}$.
The corresponding states are $|\nu_l\rangle_{i,j}$ and $|\tilde{\nu}_l\rangle_{i,j}$ where

$$\begin{array}{l} |\nu_l, \nu_\lambda\rangle_{i,j} = \sqrt{x_{\nu_l}} \exp\{i\theta_{\nu_\lambda}\} |\emptyset\rangle + \sqrt{y_{\nu_l}} |1\rangle \\ |\tilde{\nu}_l, \nu_\lambda\rangle_{i,j} = -\sqrt{y_{\nu_l}} \exp\{i\theta_{\nu_\lambda}\} |\emptyset\rangle + \sqrt{x_{\nu_l}} |1\rangle \\ \nu_l = ln + i, \quad i = 0, 1...l - 1 \\ \nu_\lambda = \lambda n + j, \quad j = 0, 1, ..\lambda - 1 \end{array} \tag{18}$$

where the symbols i and j identify a states in the extended space. For example $l = 1$ yields the single sequence of iterating states $|n\rangle$ and $|\tilde{n}\rangle$. In that case the two time sequences are the same. For $l = 2$ or $\lambda = 2$, we obtain two sets of iterating states where the even states corresponds with the $i = 0$ states while the $i = 1$ is associated with the odd series.

3.2. Evolution Toward Ergodicity

We show that the transition to a longer iteration time step (such as the time doubling that happens in the transition from the f to the g maps) corresponds with enlarging the system dimension.
In the chaotic regime the time interval is defined to be infinitely long and the space dimension infinitely large. For that everlasting time, the iteration concept becomes meaningless. In other words, at some stage, the x, y values lose the evolution history and thereby become random numbers with the appropriate evolved corresponding distribution. The infinite space dimension defines the associated ensemble. In other words, a single everlasting time step possesses an infinite independent initial possible values and the apparent time evolution calculation is just a new random selection.
We note that although the maps' tendency toward randomness is a primary characteristic, it does not appear at once; it takes some time for the random errors to take control over the

system. Nevertheless, randomness is the final stage in the variables' evolution that can be associated with the coherence loss during the evolution. This chaotic stage can be associated with thermodynamic equilibrium. We note that the final distribution is calculated by considering an ensemble which is composed of all x y values regardless of their iteration order. In that sense, the original 2-D space that evolved into a random infinite random space describes the ergodicity process.

The transition from time evolving system toward random ensemble can be introduced as follows: Suppose that N_{max} is the final iteration being measured. Large values of the scaled time intervals ν_l or ν_{lambda} are crucial for the system to reach a final situation (see eqs. 180. Consequently, small values of l or λ requires large values of n with small iteration steps ln or λn. In addition, the conditions $i = 0, 1...l-1$ or $j = 0, 1, ..\lambda-1$ yields a relatively low dimensional space. However, for large number of l or λ, the values of ν_l or ν_λ can reach high values even for a meaningless time interval ($N_{max} = 1$). These high values are obtained due to the possible high values of i or j which stand for the space high dimensionality.

3.3. The Re-Coherence Process

Suppose that we have an ensemble of variables uniformly distributed in the interval $[0, 1]$. At $t = 0$ ($n = 0$) we randomly pick two variables and expose them to the double nonlinear **regular** maps as defined by eqs. 1,2. At this initial time the chosen random variables serve as initial conditions x_0 and y_0 for the $x - y$ maps. Assuming that the chaotic maps that generated the ensemble are no longer active, these two picked initial variables will iterate towards one pair of the finite number of values as described by eqs. 14-17. We can now repeat this experiment many times but the result will always be the same. For simplicity, we will no focus on maps that converge into the single pair values, x_∞, $y_\infty = 1 - x_\infty$. Regardless of the ensemble chosen member, all the ensemble values will converge into the same pair x_∞, $y_\infty = 1 - x_\infty$. In that sense the infinite dimensional random ensemble is reduced into a 2-D space. In other words, if we wait long enough, every measurement we conduct will end up with the same reading. In the states language the random ensemble will reduce into an infinite number of identical pairs of basis states as described in eqs.12. The fact that every selected pair out of the random ensemble eventually generates a state justifies naming the process **re-coherence**.

4. Summary

System evolution toward Ergodicity is a known fact in statistical mechanics. However, it only describes a partial feature of the word, namely, inanimate systems. In this paper, we showed the existence of an opposite process, namely ergodic systems that evolve toward order. We can say that both the ergodicity and these *reorder* time evolutions are the outcome of the same process namely recursive nonlinear maps. If ergodicity is the outcome of maps defined in the chaotic regime, it is only reasonable to assume that regular maps are also active in nature and, in particular, there are processes that ergodic systems become ordered. Since ergodic systems describe the inanimate word, we cannot avoid the conclusion that the reordered processes describe animate systems creation as was suggested long ago by Schrödinger[6, 7]. In biology such an evolution is referred to a **Biopoiesis** process[8].

[1] R. Alicki, A. ozinski, P. Pakonski and K.l Z. yczkowski, (2004) *J. Phys. A: Math. Gen.* **37** 51575172.
[2] A. N. Soklakov and R. Schack (2002) *Phys. Rev. E*, **66**, 036212.
[3] Y. Roth (2014), *J. Phys.: Conf. Ser.* **490** 012055.
[4] H. Schanz, T. Dittrich, and R. Ketzmerick *Phys. Rev. E* , **71**, 026228 (2005).
[5] Y. G. Roth, **Advances in Pure Mathematics**, **4**, (2014).
[6] E. Schrdinger (1944), *What Is Life?.* (Cambridge University Press 1992).
[7] M. P. Murphy,Luke A. J. O'Ne *What is Life? The Next Fifty Years: Speculations on the Future of Biology*, (Cambridge University Press, 1997).
[8] A. K. Lal, *Astrophysics and Space Science*, **317** 267-278 (2008).

A stochastic model of nanoparticle self-assembly on Cayley trees

I. Mazilu, E. M. Schwen, W. E. Banks, B. K. Pope, D. A. Mazilu
Washington and Lee University, 204 W. Washington Street, Lexington, VA 24450, USA
E-mail: mazilui@wlu.edu, schwene15@mail.wlu.edu, banksw14@mail.wlu.edu, popeb14@mail.wlu.edu, mazilud@wlu.edu

Abstract. Nanomedicine is an emerging area of medical research that uses innovative nanotechnologies to improve the delivery of therapeutic and diagnostic agents with maximum clinical benefit. We present a versatile stochastic model that can be used to capture the basic features of drug encapsulation of nanoparticles on tree-like synthetic polymers called dendrimers. The geometry of a dendrimer is described mathematically as a Cayley tree. We use our stochastic model to study the dynamics of deposition and release of monomers (simulating the drug molecules) on Cayley trees (simulating dendrimers). We present analytical and Monte Carlo simulation results for the particle density on Cayley trees of coordination number three and four.

1. Introduction

Many studies in the field of nanomedicine explore the versatile properties of *dendrimers* and their potential use as a novel drug delivery mechanism via drug attachment and encapsulation [1]. Dendrimers are new synthetic polymers able to carry both targeting molecules and drug molecules to cancerous tumors, minimizing the negative side effects of medications on healthy cells; they are perfect physical examples of Cayley tree structures.

We propose a cooperative sequential adsorption model with evaporation (CSAE) defined on a general Cayley tree. CSAE models are ideal for describing drug encapsulation and release because (i) the deposition process of the drug nanoparticles is stochastic and can be modeled by sequential adsorption models; (ii) the deposited drug nanoparticles are electrically charged, as are the substrate deposition sites, suggesting a cooperative model with deposition rates dependent on nearest-neighbor site occupation; (iii) the drug nanoparticles have a probability of detachment, which is incorporated in the model via an evaporation rate.

We first map our CSAE model onto an Ising model defined on a Cayley tree of coordination number $z = 4$. In section 3, we derive the rate equations for the spin magnetizations and find numerical results for the particle densities for specific attachment and detachment probabilities. In section 4, we compare our analytical results with Monte Carlo simulations modeling particle deposition onto a Cayley tree. Lastly, we present a summary of our work and suggestions for future research in section 5.

2. Model definition and its connection to the Ising model

We model the adsorption surface as a Cayley tree with coordination number z. We consider monomers that attach/detach from the nodes of the tree. We define an occupation number $n_i = 0$ for an empty site and $n_i = 1$ for an occupied site.

We define the following transition rate for the particle occupation:

$$c(n_i \rightarrow (1 - n_i)) = \gamma n_i + (1 - n_i)\alpha\beta^{\sum_{j\in NN} n_j} \tag{1}$$

The first term in the transition rate is the evaporation term: if a particle is present, it will evaporate with probability γ. The second term describes the deposition of monomers. If the lattice cell is empty, a monomer will attach with a rate equal to $\alpha\beta^{\eta}$, where $\eta = \sum_{j\in NN} n_j$ is the number of occupied nearest neighbors.

We map our model onto an Ising model defined on a Cayley tree of coordination number z. The Hamiltonian associated with the d-dimensional Ising model of a system of N spins in an external field is [2]:

$$H = -J \sum_{i,j\in NN} s_i s_j - B_{ext} \sum_{i=1}^{N} s_i \tag{2}$$

The first term describes the interactions between nearest-neighboring spins, and the second term expresses the interaction between each spin and an external magnetic field B_{ext}. The spin numbers are $s_i = 1$ for a spin up, and $s_i = -1$ for for a spin down. In terms of the spin numbers s_i, the particle occupation numbers are: $n_i = \frac{1+s_i}{2}$.

The equilibrium properties of this model can be derived from the partition function (provided that it can be solved), but the non-equilibrium properties depend on the nature of the spin dynamics. The steady-state of our non-equilibrium system is equivalent to the corresponding equilibrium Ising model if the detailed balance condition is satisfied [3]:

$$P_{eq}(s)c_j(s) = P_{eq}(s^j)c_j(s^j) \tag{3}$$

The variable s stands for a configuration of the system, and s^j is the same configuration with the jth spin flipped. P_{eq} is the Boltzmann factor.

The coupling and field constants K and h can be found from the detailed balance condition, and are dimension dependent:

$$K = \frac{J}{kT} = \frac{1}{4}ln(\beta) \tag{4}$$

$$h = \frac{B_{ext}}{kT} = \frac{1}{4}ln(\frac{\alpha^2\beta^z}{\gamma^2}) \tag{5}$$

This mapping onto the Ising model permits the use of known results [5] to analyze our stochastic model.

3. Cayley tree theoretical analysis

Glauber presented the solution for the magnetization of a spin system in one dimension in [4]. We generalize his method for a Cayley tree. We assume translational invariance within each generation of the Cayley tree: all spins within a specific generation are equivalent. We label the central node of the tree as "n", and then each subsequent generation from n to 1, with generation 1 being the outermost generation of the tree. We define the magnetization of generation i as $q_i =< s_i >$. In terms of this magnetization, the particle density of generation i is defined by: $\rho = \frac{1+q_i}{2}$. The time evolution of q_i is derived [6] to be:

Journal of Physics: Conference Series **574** (2015) 012086 doi:10.1088/1742-6596/574/1/012086

$$\frac{dq_i}{dt} = -q_i(t) + B + < \tanh(K \sum_{j \in V(i)} s_j) > + B < s_i \tanh(K \sum_{j \in V(i)} s_j) > \quad (6)$$

where $B = \tanh(h)$ reflects the effect of the external field.

This equation does not have exact solutions in higher dimensions, and in that situation, one has to use different approximations schemes. We present results for a Cayley tree of coordination number $z = 4$ with arbitrary magnetic field h.

In order to be able to get a closed form for the system of equations, we use the series expansion approximation for $\tanh(K \sum_{j \in V(i)} s_j) = C_1(\sum_{j \in V(i)} s_j) + C_2(\sum_{j \in V(i)} s_j)^3$, and find the coefficients $C_1 = \frac{2}{3}\tanh(2K) - \frac{1}{12}\tanh(4K)$, and $C_2 = \frac{1}{48}\tanh(4K) - \frac{1}{24}\tanh(2K)$.

We also use the factorization approximation, $< \prod_{j \in NN} s_j > = \prod_{j \in NN} q_j$, which allows us to remove multi-spin correlations. With these approximations, the system of equations is:

$$\begin{aligned}
\frac{dq_n}{dt} &= -q_n + 4(C_1 + 10C_2)q_{n-1} + 24C_2 q_{n-1}^3 + B(1 + 4(C_1 + 10C_2)q_n q_{n-1} + 24C_2 q_n q_{n-1}^3) \\
\frac{dq_i}{dt} &= -q_i + (C_1 + 10C_2)(3q_{i-1} + q_{i+1}) + 6C_2(3q_{i-1}^2 q_{i+1} + q_{i+1}^3) \\
&\quad + B(1 + (C_1 + 10C_2)(3q_i q_{i-1} + q_i q_{i+1}) + 6C_2(3q_i q_{i-1}^2 q_{i+1} + q_i q_{i-1}^3)) \\
\frac{dq_1}{dt} &= -q_1 + q_2 \tanh(K) + B(1 + \tanh(K) q_1 q_2) \qquad (7)
\end{aligned}$$

This system of equations can be solved numerically, and Figs. 1 and 2 present the associated particle densities per generation and for the entire tree for sample parameters.

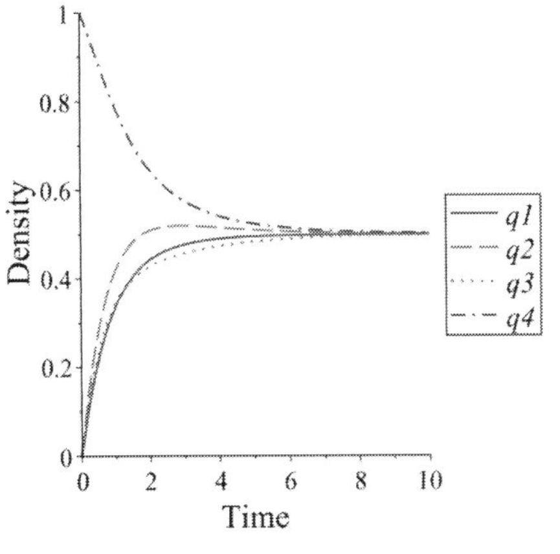

Figure 1. Density vs. time for each generation, with $z = 4$, $\alpha = 1$, $\beta = 0.5$, $\gamma = 0.25$. Time is in arbitrary units.

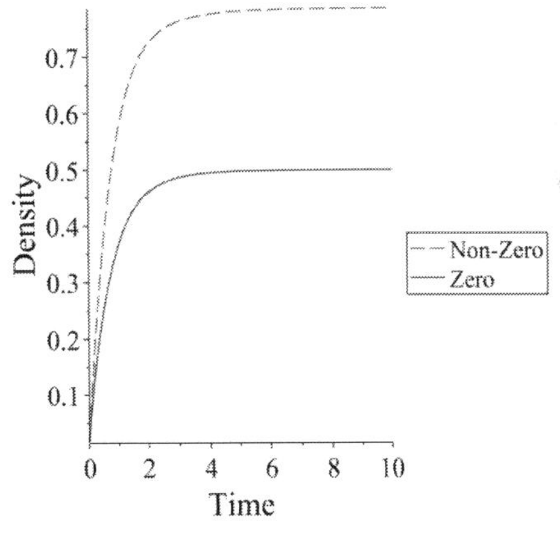

Figure 2. Density vs. time for entire tree for $z = 4$. Comparison for zero ($h = 0$) and nonzero ($h = 1$) external field. Time is in arbitrary units.

We also matched our analytical results with Monte Carlo simulations on Cayley trees with coordination number $z = 4$ (Fig. 3) for a variety of parameter regimes, both with and without the presence of an external field. The good match between simulation results and analytical solutions suggests that simulations can be an effective tool for studying the dynamics of larger tree structures. A larger number of sites also decreases the impact of random variations on total

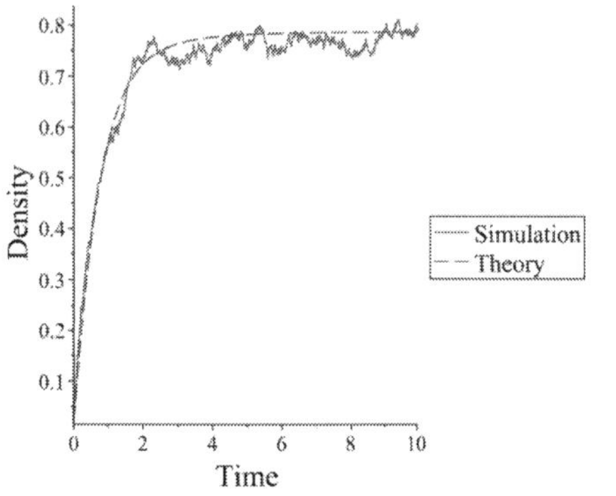

Figure 3. Comparison of theoretical solution (red) and the simulation average over ten trials (green). Parameters used: $z = 4$, $\alpha = 1$, $\beta = e$, $\gamma = 1$. Time is in arbitrary units.

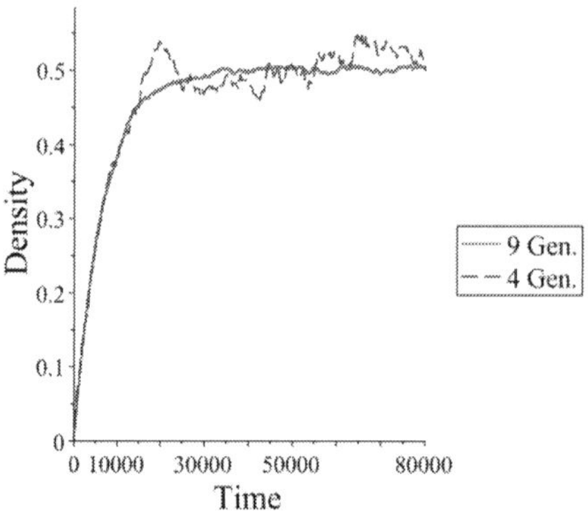

Figure 4. Comparison of simulations for 4-generation tree (red) and 9-generation tree (blue) on an arbitrary time scale, for $z = 4$, $\alpha = 1$, $\beta = 1$, $\gamma = 1$.

particle density, making simulations more effective for larger systems than they are for small systems. We therefore investigate the time evolution of a 9-generation tree and compare it to our results for a 4-generation tree. As seen in Fig. 4, the density plots for the two trees are nearly identical when the arbitrary time for the larger tree is rescaled. This result suggests that our theoretical solutions will apply equally well to trees with more generations.

4. Conclusion

In this article, we presented a model of cooperative sequential adsorption with evaporation for particle deposition on a Cayley tree, which we mapped on an equivalent Ising model. Theoretically, we found systems of differential equations describing the time-development of magnetization for each generation of Cayley trees with coordination number $z = 4$. We related these Ising model magnetization results to the particle density for our model. Computationally, we simulated the CSAE process on a Cayley tree and found excellent agreement between simulation results and theoretical predictions. This agreement validates our analytical solutions and supports the effectiveness of the simulations in mapping the dynamics of the system. Further studies could apply this model more directly to the drug encapsulation process or adapt it to address other systems such as self-assembled thin films, epidemic models, or social networks.

References

[1] Hecht S and Frechet J M J, *Angew. Chem. Int. Ed.* **40**, 74 (2001.)
[2] Ising E, *Z. Phys.* **31** , 25 (1925.)
[3] Krapivsky P L, Redner S, and Ben-Naim E V, *A kinetic view of statistical physics*, Cambridge University Press, Cambridge, England (2010.)
[4] Glauber R J, *J. Math.Phys.* **4**, 294, (1963.)
[5] Baxter R J, *Exactly Solved Models in Statistical Mechanics*, Academic Press, 1982.
[6] Melin R, Angles d'Auriac J C, Chandra P and Doucot B, *J. Phys. A: Math. Gen.* **29**, 5773, (1996.)

An analytical and computational study of a stochastic adsorption model with variable attachment and detachment rates

D. A. Mazilu, E. M. Schwen, I. Mazilu

Washington and Lee University, 204 W. Washington Street, Lexington, VA 24450, USA

E-mail: `mazilud@wlu.edu, schwene15@mail.wlu.edu, mazilui@wlu.edu`

Abstract. We present a stochastic model for adsorption and evaporation of monomers with applications to optical coatings. We consider a general case of attachment and detachment rates dependent on the overall number of particles in the system. The model is applicable to all dimensions and topologies, and can describe a variety of two-state physical systems. We report analytical results for the time-dependent particle density. We compare our analytical results with experimental data and Monte Carlo simulations.

1. Introduction

Analytical and computational models of sequential adsorption [1] have proven successful in describing diverse physical systems ranging from surface deposition and chemisorption on crystal surfaces to epidemics problems and voting behavior. A particular application for such models, the dynamics of nanoparticle deposition, is currently an active area of research in nanotechnology studies [2] that raises interesting questions on the theoretical front.

We present a stochastic model for both adsorption and evaporation of monomers with variable attachment and detachment rates. The attachment and detachment rates depend on the overall number of particles already present in the system. The model is quite general, is applicable to all dimensions and topologies, and can describe a variety of two-state physical systems. We use mean field theory and matrix theory to find solutions for the particle density and probabilities of having a set number of particles present in the system (section 2.)

In section 3, we compare our analytical results to Monte Carlo simulations and to experimental data on the self-assembly of charged nanoparticles on glass substrates [3]. This technique has been used successfully in making antireflective coatings [2, 4] with optical properties dependent on the surface coverage of the substrate. We summarize our results and present possible extensions of the project in section 4.

2. Model description

Our model is defined on a general lattice of arbitrary topology (rectangular grids, Cayley trees, etc.) of N sites. Each site of the grid has two states: empty or filled. Empty sites are filled at a rate α_i; filled sites are emptied at a rate β_i. These rates are functions of the total number i of filled sites in the lattice. In order to mimic the cooperative effects due to electrostatic repulsion during the ionic self-assembly of nanoparticles, we consider the attachment rates to decrease as the number i of filled sites increases, and the detachment rates to increase with i. The functions picked for these rates can be modified easily depending on the physical situation considered. We consider here a linear case:

$$\begin{aligned} \alpha_i &= \alpha(N-i) \\ \beta_i &= \beta i \end{aligned} \tag{1}$$

with α and β positive constants. This model has the virtue of simplicity and can be solved using the mean field approximation, but exhibits sufficient complexity to be useful as a standard of comparison for experimental results as well as analytic and computational models that include more complex rate assumptions.

Using the mean field approximation we can write the equation for the mean density of filled sites $\bar{\rho}$:

$$\frac{d\bar{\rho}}{dt} = \alpha(1-\bar{\rho})^2 - \beta\bar{\rho}^2,$$

with general solution

$$\bar{\rho} = \frac{\alpha - \tanh\left(t\sqrt{\alpha\,\beta} + A\sqrt{\alpha\,\beta}\right)\sqrt{\alpha\,\beta}}{\alpha - \beta} \tag{2}$$

for positive α, β and $\alpha \neq \beta$. When the two rates are equal, $\bar{\rho}\,(t) = 1/2 + A\,\mathrm{e}^{-2\,\alpha\,t}$, and the surface coverage settles at 50%.

If cooperative effects are not being considered (constant attachment and detachment rates), the solution for the particle density is:

$$\bar{\rho} = A\exp\left(-(\alpha+\beta)t\right) + \frac{\alpha}{\alpha+\beta} \tag{3}$$

In each case, the coefficient A is determined by the mean density at $t = 0$.

Moving beyond the mean field result, we let Q_i represent the time-dependent ensemble-average probability that exactly i sites of the lattice are filled. This obeys the master equation:

$$\frac{d}{dt}Q_i = -((N-i)^p\alpha + i^p\beta)Q_i + (N-i+1)^p\alpha Q_{i-1} + (i+1)^p\beta Q_{i+1} \tag{4}$$

where $p = 1$ for constant attachment and detachment rates, and $p = 2$ for variable rates with linear dependence on the number of filled sites at time t. This equation can be solved exactly using matrix theory or the generating function method for $p = 1$.

The general time-dependent solution is given by $Q_i = \sum_{k=0}^{N} c_k E_{ik}\exp(\lambda_k t)$, where λ_k is the k-th eigenvalue of the associated matrix, and E_{ik} is the i-th component of its k-th eigenvector. The derivation of these eigenvalues and eigenvectors for the $p = 1$ case has been presented in a recent article [5]. Unfortunately, the matrix approach is analytically intractable for $p > 1$, and approximation methods become necessary.

We show in Fig. 1 the probability, as a function of time, that exactly i sites are filled for several values of i, in a case with $N = 100$ initially empty sites, using $\eta = \frac{\beta}{\alpha} = 4$, for constant attachment and detachment rates. As time progresses, each Q_i for $0 < i < 20$ at some point becomes the dominant term, spikes, and then diminishes. The expected steady state at $i = 20$ does not spike, but rather levels off as t increases. Q_{20} does not become unity because the deposition and evaporation will continue, meaning that other states will remain possible at all times, with states nearest to $i = 20$ having higher probabilities than others.

Journal of Physics: Conference Series **574** (2015) 012087 doi:10.1088/1742-6596/574/1/012087

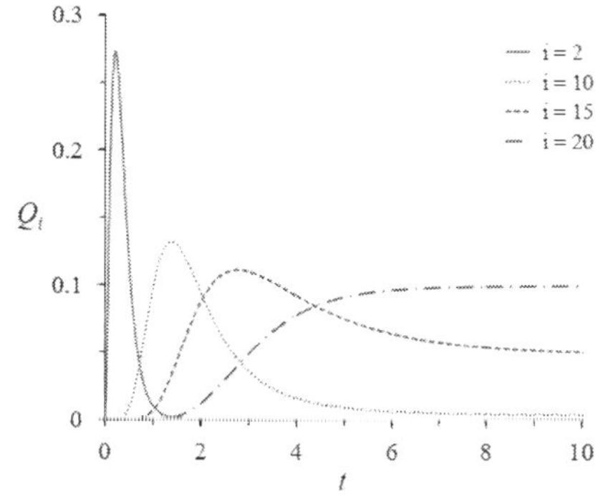

Figure 1. The probability Q_i, as a function of time, that exactly i sites are filled in a case of $N = 100$ initially filled sites and $\eta = 4$.

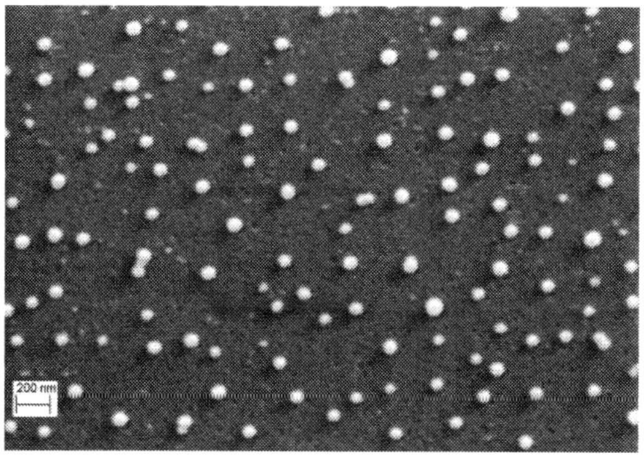

Figure 2. Typical SEM micrograph of a single-bilayer $SiO_2/PDDA$ film deposited on a glass microscope slide.

3. Comparisons: experiment, theory, and computer simulations

3.1. Experiment and theory

Ionic self-assembly allows the creation of highly uniform, conformal coatings of charged nanoparticles on a substrate with relative ease and low cost. In our experiments we deposited negatively charged, spherical silica nanoparticles of nominal 40-50 nm diameter on negatively charged glass slides using poly(diallyldimethylammonium chloride) (PDDA) as polycation. We examined the nanoparticle coverage of the substrate for single-bilayer films using SEM micrographs in which deposited particles appear as light regions on a dark background. We determine the surface coverage by employing a pixel-count comparison of grayscale values. A sample SEM micrograph is presented in Fig. 2.

The theoretical model presented in the previous section, defined for non-interacting particles, approximates well the ionic self-assembly process for the case of low particle density in the steady state. In the analysis of the SEM micrographs, we assume the deposited particles have a circular cross section, and that complete coverage would result in a rectangular grid of particles in contact. In this packing geometry, the maximum fraction of the total area that can be covered is $\pi/4 \approx 0.785$. Correcting for this, the dark pixel coverage of 0.063 translates to a circle coverage of $\bar{\rho} = 0.080 \pm 0.009$ which corresponds to $\eta = 11.5$. This allows an evaluation of adsorption and evaporation rates of $\alpha = 0.0032$ and $\beta = 0.037$. Hand counting of deposited particles in the micrographs produces a mean number of 252, thus a maximum capacity of $N = 3150$.

3.2. Computer simulations and theory

We have created Monte Carlo simulations to further investigate the stochastic particle adsorption process. We present results from simulation runs on a 100×100 two-dimensional grid using two sets of attachment and detachment rates. The first series of simulations uses the variable attachment and detachment rates presented in Eq. (1). The second series, presented for comparison, uses constant attachment and detachment rates.

We utilize an event driven algorithm for both series of simulations. At each timestep, the simulation selects a random event (attachment or detachment) weighted by the chosen attachment and detachment rates. It then randomly selects a qualifying site and performs the selected action. We allow the simulation to proceed for 1.44×10^6 site updates in order to ensure a steady state is reached. We record the particle density over time (Fig. 3) as well as the steady state particle density averaged over 100 realizations of the system (Fig. 4). These plots show the excellent agreement between the analytical solutions and the computer simulations and validate our use of the mean field approximation in deriving the analytical results.

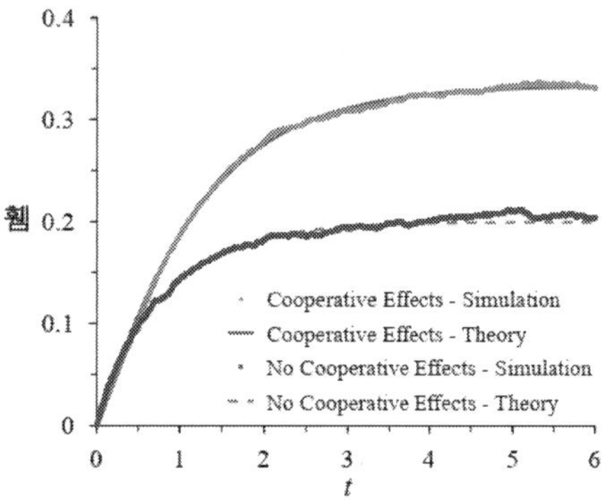

Figure 3. Mean particle density as a function of time for $\eta = 4$, with and without cooperative effects. Time is in arbitrary units.

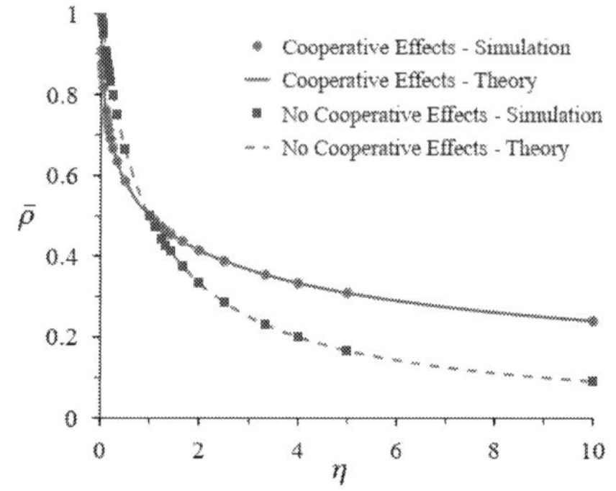

Figure 4. Mean particle density for the steady state as a function of η, with and without cooperative effects.

4. Conclusions

In this paper, we presented exact analytical results for the time-dependent probability that i sites of a d-dimensional lattice are filled, for the case of constant attachment and detachment rates. We also calculated the mean coverage of the lattice for both constant and variable attachment and detachment rates using the mean field approximation. The analytical results matched very well the Monte Carlo simulations. We compared our theory with experimental results obtained for ionic self-assembly of silica nanoparticles. The comparison with the experimental data leads us to believe that cooperative effects (such as electrostatic interactions, for example) are, in fact, present, but are not relevant for low particle density. The model proposed in this paper can be tailored to other physical situations that involve particle attachment and detachment. It can apply to any lattice structure or dimension and can serve as a starting point for other studies.

Acknowledgments

Special thanks to W. E. Banks, B. M. Simpson, V. O. Kim, and A. M. Seredinski for the SEM imaging and analysis. This research was supported by the Washington and Lee University Summer Research Program and the Lenfest grant.

References

[1] Liggett T M *Stochastic Interacting Systems: Contact, Voter, and Exclusion Processes*, Berlin: Springer-Verlag (1999.)
[2] Di Ventra M, Evoy S and Heflin J R *Introduction to nanoscience and technology*, New York: Springer (2004.)
[3] Iler R K *J. Colloid Interface Sci.*, **21**, 569 (1966.)
[4] Yancey S E, Zhong W, Heflin J R and Ritter A L *J. Appl. Phys.* **99**, 034313 (2006.)
[5] Fonseca C M, Mazilu D A, Mazilu I and Williams H T *Applied. Math. Lett.*, **26**, 1206 (2013.)

Any realistic model of a physical system must be computationally realistic

A Bolotin

Ben-Gurion University of the Negev, Beersheba 84105, Israel

E-mail: arkadyv@bgu.ac.il

Abstract. It is argued that any possible definition of a realistic physics theory – i.e., a mathematical model representing the real world – cannot be considered comprehensive unless it is supplemented with requirement of being computationally realistic. That is, the mathematical structure of a realistic model of a physical system must allow the collection of all the system's physical quantities to compute all possible measurement outcomes on some computational device not only in an unambiguous way but also in a reasonable amount of time.

In the paper, it is shown that a deterministic quantum model of a microscopic system evolving in isolation should be regarded as realistic since the NP-hard problem of finding the exact solution to the Schrödinger equation for an arbitrary physical system can be surely solved in a reasonable amount of time in the case, in which the system has just a small number of degrees of freedom. In contrast to this, the deterministic quantum model of a truly macroscopic object ought to be considered as non-realistic since in a world of limited computational resources the intractable problem possessing that enormous amount of degrees of freedom would be the same as mere unsolvable.

1. Introduction

Recently a definition of realistic physics theories – i.e., mathematical models representing the real world – has proposed, according to which a model of a physical system is classified as realistic if and only if, according to the mathematical structure of this model, the collection of all physical quantities written in the system *unambiguously* determines the probabilities of all possible measurement outcomes [1]. Attractive and credible as this definition might appear, it nevertheless fails to mention that before anything else a realistic model has to be *computationally* realistic. This means that the mathematical structure of a realistic model of a physical system must allow the collection of all the system physical quantities to determine all possible measurement outcomes (or their probabilities) not only in an unambiguous way but also in a realistic – that is, *reasonable* – amount of time. Undeniably, the defining characteristic of any mathematical physics model is that it makes falsifiable or testable predictions. So, if the mathematical model were insoluble in a reasonable amount of time (even with access to a supercomputer), then the model would not have a realistically testable predictive content, and therefore the term "realistic" would be hardly applicable to it.

Of course, the requirement of being computationally realistic would be completely redundant for the definition of a realistic physics theory if every physics model were capable of being

Journal of Physics: Conference Series **574** (2015) 012088 doi:10.1088/1742-6596/574/1/012088

solvable in reasonable time as soon as enough computational resources were thrown at the model. The goal of this paper is to show that, in all likelihood, such is not the actual state of things in *quantum theory*. In particular, it will be shown that the deterministic quantum model of a microscopic system evolving in isolation should be regarded as a realistic model, whilst the quantum model of a truly macroscopic object (built around exact solutions to the object's Schrödinger equation) ought to be considered as non-realistic.

2. Schrödinger's equation computational complexity

To demonstrate this, the complexity of Schrödinger's equation, which is one of the fundamental equations of quantum theory that describes how the quantum state of a physical system changes with time, should be analyzed.

Theorem 1 *Let H be the Schrödinger Hamiltonian for a physical system, the vector $|\psi(t)\rangle$ be the state of this system at time t, the wave function $\Psi(\mathbf{r},t)$ be the product $\langle \mathbf{r}\,|\psi(t)\rangle$, where $\mathbf{r}=(\mathbf{r}_1,\mathbf{r}_2,\ldots,\mathbf{r}_N)$ denotes the sets of position vectors such that $\mathbf{r}_n=(x_{n1},x_{n2},x_{n3})$ and N is the system constituent particle number. Then the runtime complexity of the operations needed to verify that any given solution $\Psi(\mathbf{r},t)$ is indeed the solution to the Schrödinger equation $i\hbar\,\partial\Psi(\mathbf{r},t)/\partial t = H\Psi(\mathbf{r},t)$ is polynomially bounded.*

As it can be readily seen, to calculate the complexity of verification (defined as value L) it is sufficient to find the minimal number of elementary operations required to compute the effects of the Schrdinger Hamiltonian H on the solution $\Psi(\mathbf{r},t)$. After been projected into the position basis $\{|\mathbf{r}\rangle\}$ the generic form of the Schrdinger Hamiltonian H can be expressed as the following sum:

$$H = -\frac{\hbar^2}{2}\sum_{n=1}^{N}\frac{1}{m_n}\frac{\partial^2}{\partial \mathbf{r}_n^2} - \frac{\hbar^2}{2M}\sum_{j\neq l}^{N}\frac{\partial}{\partial \mathbf{r}_j}\frac{\partial}{\partial \mathbf{r}_l} + \widehat{\mathrm{H}}_{\mathrm{s}} + V(\mathbf{r}_1,\mathbf{r}_2,\ldots,\mathbf{r}_N,t) \quad , \tag{1}$$

where m_n is the mass of the system constituent particle n, M denotes the mass of the collection of the constituent particles resulting in the extra kinetic energy K (owing to the dependency K on the spatial configuration of the particles), $\widehat{\mathrm{H}}_{\mathrm{s}}$ is the term accounting for the presence of the constituent particles' spins (this terms may include spin-orbit coupling, spin-rotation coupling and spin-spin coupling). Additionally, in the case of the external electromagnetic field existing represented by the vector potential $\mathbf{A}=(A_1,A_2,A_3)$, the interaction of the constituent particle of charge q_n with the electromagnetic field can be taken into account by replacing the partial derivatives $\partial/\partial x_{nj}$ by $\partial/\partial x_{nj} - iq_nA_j/\hbar c$. So, making use of the results of the papers [2, 3], the complexity L can be presented as follows:

$$L(H\Psi) = L\left(\frac{\partial^2\Psi}{\partial \mathbf{r}_1^2},\ldots,\frac{\partial^2\Psi}{\partial \mathbf{r}_N^2},\frac{\partial\Psi}{\partial \mathbf{r}_1},\ldots,\frac{\partial\Psi}{\partial \mathbf{r}_N}\right) \le O\!\left(N^2\right)\cdot cost(\Psi(\mathbf{r},t)) \quad , \tag{2}$$

where only nonscalar arithmetic operations (i.e., binary operations whose both operands involve the solution Ψ) are considered contributed to the complexity of verification L, whereas additions/subtractions and multiplications by arbitrary scalars $k(\mathbf{r},t)$ are allowed for free, $cost(\Psi(\mathbf{r},t))$ denotes the computational cost of the wave function evaluation at particular numerical values $\mathbf{r}=(\mathbf{r}_1,\mathbf{r}_2,\ldots,\mathbf{r}_N)$ and t. In order to the interpretation of solutions to Schrödinger's equation to make sense, it must be *feasible* to evaluate the given solution $\Psi(\mathbf{r},t)$ at arbitrary $\mathbf{r}$ and t because otherwise it would be impossible to use $\Psi(\mathbf{r},t)$ to compute a measurable observable of the quantum system. In conformity with Cobham's thesis [4], the function $\Psi(\mathbf{r}, t)$ can be feasibly evaluated only if it can be evaluated on some computational device in polynomial time. This implies that $cost(\Psi(\mathbf{r}, t)) = poly(N)$, and hence, the verification complexity L is upper-bounded by a polynomial. □

Corollary 1 *The decision problem of Schrödinger's equation "Does the ground state of a physical system* $\Psi_g(\mathbf{r})$ *have energy* $E_g \le 0$*?" is in the NP complexity class of computational problems, whose solutions can be verified in polynomial time.*

Certainly, if the ground state $\Psi_g(\mathbf{r})$ with the energy $E_g \le 0$ is given, then by substituting it back into Schrödinger's equation one can prove that $\Psi_g(\mathbf{r})$ is indeed the solution to this equation in only a polynomial amount of time.

Theorem 2 *Given an exact generic algorithm capable of solving exactly the Schrödinger equation for an arbitrary physical Hamiltonian (that is, for any and all possible physical systems), any problem in the NP complexity class can be solved with only polynomially more work.*

Let $A(\Psi)$ be such an algorithm. As $A(\Psi)$ can solve exactly the Schrödinger equation for all Hamiltonians, it can also solve the Schrdinger equation for Ising Hamiltonians that describe Ising models of a spin glass [5, 6]. On the other hand, the paper [7] explains how "all the famous NP problems" (such as Karp's 21 NP-complete problems [8]) can be written down as Ising models with only a polynomial number of steps (to be exact, with a polynomial number of spins which scales no faster than N^3). Therefore, in just a polynomial number of steps one can get from any NP-complete problem to the Schrödinger equation for the Hamiltonian of an Ising spin glass, whose decision problem "Does the ground state of a glass $\Psi_g(\mathbf{r})$ have energy $E_g \le 0$?" solves the NP-complete problem of interest using the algorithm $A(\Psi)$. From here, the proof of the theorem follows: Since an arbitrary NP problem is polynomial-time reducible to any NP-complete problem and thus to the decision form of the Ising model, any problem in NP can be solved exactly by the algorithm $A(\Psi)$ with only polynomially more work. □

Corollary 2 *The problem of solving exactly Schrödinger's equation for an arbitrary physical system is NP-hard.*

Seeing as the algorithm $A(\Psi)$ that solves this problem can also solve any problem in the NP complexity class, the given problem is NP-hard.

Corollary 3 *If the algorithm* $A(\Psi)$ *were efficient, the class NP would be equal to the class P of computational problems solvable in polynomial time. If P≠NP, the algorithm* $A(\Psi)$ *could not be efficient, i.e., its runtime complexity could not be polynomial in* N.

Theorem 3 *If P≠NP, Schrödinger's equation for a truly macroscopic object would not have an exact solution reachable in polynomial time.*

As an entity of classical realm, a truly macroscopic object contains a large (and essentially unchecked) number of constituent microscopic particles that interact constantly with many different physical systems (of various properties and scales) within the vast causal horizon for the object. As such, its Hamiltonian must be a sum of (practically) all possible physical Hamiltonians, and hence an algorithm capable of solving exactly the Schrödinger equation for this object must be the exact generic algorithm $A(\Psi)$. Therefore, if P≠NP, Schrödinger's equation for a truly macroscopic object could not be solved exactly in polynomial time. □

Assuming that P≠NP, is it possible that the exact generic algorithm $A(\Psi)$ is significantly faster than exhaustive search (an exact generic algorithm as well), which, according to the postulates of quantum mechanics, runs exponentially in N as $O(2^N)$?

Theorem 4 *In all probability, if P≠NP, then the runtime complexity of the exact generic algorithm $A(\Psi)$ could not be sub-exponential in N.*

Suppose that $A(\Psi)$ is a sub-exponential time algorithm. Since any NP-complete problem – including the 3-SAT problem – can be written down as the Schrödinger equation for the Hamiltonian of an Ising spin glass, it follows that the algorithm $A(\Psi)$ can solve any NP-complete problem in sub-exponential time. However, according to the widely believed conjecture (called *the exponential time hypothesis*), the 3-SAT problem does not have a sub-exponential time algorithm [9, 10, 11]. Hence, if the runtime complexity of $A(\Psi)$ were sub-exponential in N, then the exponential time hypothesis could be shown to be false. □

Corollary 4 *In all probability, if P≠NP, then an algorithm solving exactly Schrödinger's equation for a truly macroscopic object could not be significantly faster than exhaustive search.*

3. Discussion

It is clear that (even if P≠NP) the NP-hard problem of solving exactly Schrödinger's equation for an arbitrary physical system could surely be solved in a reasonable amount of time for the instance where the system is composed of just a few constituent microscopic particles. This means that the quantum model of a microscopic system evolving in isolation should be regarded as a realistic model. In contrast to this, the quantum model of a truly macroscopic object (such as a macroscopic detector, Schrödinger's cat, an observer), ought to be considered as a non-realistic model. Indeed, as it follows from Corollary 4, it is quite likely that if P≠NP, then to everyone living in a world of limited in time computational resources, the problem of solving exactly Schrödinger's equation for a truly macroscopic object is the same as a mere unsolvable problem.

On the other hand, if the macroscopic object is considered as a system with only a few controlled or measured degrees of freedom (among many others that are uncontrolled or unmeasured) then the object's quantum model (an inexact, probabilistic one as it can only provide an incomplete description of the system) can surely be solved in reasonable time and thus regarded as a realistic physics model.

References

[1] Gisin N 2014 A possible definition of a Realistic Physics Theory *Preprint* arXiv:1401.0419 [quant-ph]
[2] Paterson M and Stockmeyer L 1973 On the number of nonscalar multiplications necessary to evaluate polynomials *SIAM J. Comput.* **2** 1
[3] Baur W and Strassen V 1983 The complexity of partial derivatives *Theoretical Computer Science* **22** 317-330
[4] Cobham A 1964 The intrinsic computational difficulty of functions *Proc. Int. Cong. for Logic, Methodology and Philosophy of Science II* (North Holland)
[5] Fischer K and Hertz J 1991 *Spin Glasses* (Cambridge University Press)
[6] Guerra F and Toninelli F 2002 The thermodynamic limit in mean field spin glass models *Communications in Math. Phys.* 230 **1** 71-79 (*Preprint* arXiv:0204280 [cond-mat])
[7] Lucas A 2014 Ising formulations of many NP problems *Frontiers in Physics* **2** 5 (*Preprint* arXiv:1302.5843 [cond-mat.stat-mech])
[8] Garey M and Johnson D 1979 *Computers and Intractability: a Guide to the Theory of NP-Completeness* (New York: Freeman & Co)
[9] Impagliazzo R, Paturi R, and Zane F 2001 Which problems have strongly exponential complexity? *J. Comput. System Sci.* **63** 512-530
[10] Woeginger G 2003 *Exact Algorithms for NP-hard Problems: A Survey* Combinatorial Optimization - Eureka, You Shrink! (Springer-Verlag) pp. 185-207
[11] Lokshtanov D, Marx D, Saurabh S 2011 Lower bounds based on the exponential time hypothesis *Bulletin of the EATCS* **84** 41-71

D-dimensional Eckart+deformed Hylleraas potential: Bound state solutions

J J Peña, G Ovando and J Morales
Universidad Autónoma Metropolitana - Azcapotzalco,
CBI - Area de Física Atómica Molecular Aplicada,
Av. San Pablo 180, 02200 México, D.F., MEXICO.
E-mail: jjpg@correo.azc.uam.mx

Abstract. By using an exactly-solvable multiparameter exponential-type potential, the bound-state solutions of the D-dimensional Eckart + Hylleraas potential are directly derived as a particular case. Althought our proposal accepts different approximations to the centrifugal term, its usefulness is exemplified in the frame of the Taskin and Kocak approach. This fact enable us to compare our results with specific potentials found in literature which are obtained here as particular cases from our proposal. That is, instead of solving an specific exponential-type potential, by resorting each time to a specialized method, the energy spectra and wavefunctions are found straightforwardly by means of the simple choice of the involved parameters. Furthermore, our proposal can be used as alternative way in the search of bound-state solutions to new exponential-type potentials besides that one can study different approximations to the term $1/r^2$.

PACS: 0.365Ge, 03.65.Fd, 03.65Ca.
Keywords: Point canonical transformation, Schrödinger equation, Exponential-type potentials. Bound states.

1. Introduction

The search of analytical bound state solutions of the Schrödinger equation for exponential-type potentials, has been of great interest as shown by many cases of specific potentials studied by means of different approaches to the centrifugal term and by using various methods such as the Nikiforov-Uvarov [1], asymptotic iteration [2], supersymmetric quantum mechanics [3], the path integral approach [4], numerical calculations [5] and many others. In particular, the bound state solutions of the $D-$dimensional Eckart-Hylleraas potential has been recently obtained by using the Nikiforov-Uvarov method [6]. Such a procedure, is one of the most popular approaches extensively used with many other specific exponential-type potentials such as the mixed potentials of Deng-Fan + Eckart [7], Hylleraas + Rosen Morse [8] or by other different approaches to the five-parameter exponential-type potential model of Chun-Sheng *et al* [9] as well as for the four parameter molecule potential proposed by Hua [10] and so on. These mixed potential models has been created with the aim to contain some particular cases of single exponential potentials, although they are limited in their extension as will see at the end of section III. For that, reason, the aim of this paper is to show that the bound state solutions of the $D-$dimensional Eckart-Hylleraas potential derives from a particular case of an exactly solvable multiparameter exponential-type potential adapted to the $D-$dimensions case in the frame of the Taskin and Kocak approximation to the centrifugal term [11] such as shown in section III. In section IV, as worked examples, we consider some particular cases directly obtained from the proposal by selecting specific values of the involved parameters. As will be shown, instead of solving a specific exponential-type potential by means of a specialized solution method, a mixed exponential-type potential such as the one considered in this work has the advantage that it can be used to find the energy spectra and eigenfunctions of single potentials in $D-$dimensions.

II. Schrödinger equation in D-dimensions.

The Schrödinger equation in D-dimensions is given, in natural units ($\frac{\hbar^2}{2m}=1$), by [12]

$$\left[r^{1-D}\frac{\partial}{\partial r}(r^{D-1}\frac{\partial}{\partial r})-\frac{\Lambda^2(\Omega)}{r^2}+E-V(r)\right]\psi_{n\ell m}(r,\Omega)=0 \tag{1}$$

where $\psi_{n\ell m}(r,\Omega)$ is the solution in the D-dimensional coordinates $(r,\Omega)=(r,\theta_1,\theta_2,\theta_3,...\theta_{D-2},\phi)$ and $\Lambda^2(\Omega)$ is the D-dimensional angular momentum operator. Assuming the separation of variables $\psi_{n\ell m}(r,\Omega)=R_{n\ell}(r)Y_\ell^m(\Omega)$ Eq.(1) is rewritten as

$$-\frac{d^2R_{n\ell}(r)}{dr^2}-\frac{D-1}{r}\frac{dR_{n\ell}(r)}{dr}+\frac{\ell_D(\ell_D+1)}{r^2}R_{n\ell}(r)+V(r)R_{n\ell}(r)=ER_{n\ell}(r) \tag{2}$$

such that $\Lambda^2(\Omega)Y_l^m(\Omega)=\ell_D(\ell_D+1)Y_l^m(\Omega)$, with $\ell_D=\sqrt{\frac{1}{4}+\ell(\ell+D-2)}-\frac{1}{2}$. Furthermore, if we set $R_{n\ell}(r)=r^{-\frac{D-1}{2}}\psi_{n\ell}(r)$ Eq.(2) becomes

$$-\frac{d^2\psi_{n\ell}(r)}{dr^2}+\left[V(r)+\frac{L_s}{r^2}\right]\psi_{n\ell}(r)=E\psi_{n\ell}(r)(r), \tag{3}$$

where $L_s=L(L+1)=\ell_D(\ell_D+1)-(D-1)(3-D)/4$ and, as expected, the case $D=3$ implies $L=\ell_D=\ell$.

The Schrödinger equation given in Eq.(3) has been solved by different authors by means of the Nikiforov-Uvarov (NU) method [1] applied to specific potentials [13]. With the same purpose, in the next section we will consider a straightforward approach that can be applied to any exponential-type potential. In particular, the usefulness of our proposition is exemplified with the application to the Eckart+deformed Hylleraas potential [6].

III. Bound state solutions of the Eckart+deformed Hylleraas potential.

Recently, by using the canonical transformation method applied to the hypergeometric differential equation (DE) it was obtained a class of multiparameter exponential-type potential given by [14]

$$V(r)=\frac{qA\exp(-r/k)}{1-q\exp(-r/k)}+\frac{qB\exp(-r/k)}{(1-q\exp(-r/k))^2}+\frac{q^2C\exp(-2r/k)}{(1-q\exp(-r/k))^2}, \tag{4}$$

such that for the Schrödinger equation

$$-\frac{d^2\psi_n(r)}{dr^2}+V(r)\psi_n(r)=E_n\psi_n(r), \tag{5}$$

it has the eigenvalues

$$E_n = -\frac{1}{4k^2}\left(\frac{n^2+(h+1)(2n+1)/2+k^2(A+B)}{n+(h+1)/2}\right)^2 \tag{6}$$

and eigenfunctions

$$\psi_n(r) = (q\exp(-r/k))^{(c-1)/2}(1-q\exp(-r/k))^{(h+1)/2}\ {}_2F_1(-n,b;c;q\exp(-r/k)). \tag{7}$$

where a, b and c are given from the hypergeometric DE, with the A, B and C parameters related accordingly to

$$A+B = \frac{2ab-c(a+b+1-c)}{2k^2} \qquad and \qquad A-C = \frac{(c-1)^2-(a-b)^2}{4k^2}, \tag{8}$$

such that

$$b = \frac{(h+1)(h-a)-2k^2(A+B)}{h+1-2a}, \qquad c = \frac{2a(h-a)-2k^2(A+B)}{h+1-2a} \tag{9}$$

and

$$c = a+b-h\ , \qquad with \qquad h = \left(1+4k^2(B+C)\right)^{1/2}. \tag{10}$$

Consequently, in order to take advantage of above result for obtaining bound state solutions of exponential-type potentials in D-dimensions, we propose $q = 1$ and the definitions

$$A = \mathbf{A} + \boldsymbol{\alpha}\frac{L_s}{k^2}, \qquad B = \mathbf{B} + \boldsymbol{\beta}\frac{L_s}{k^2}, \qquad C = \mathbf{C} + \boldsymbol{\gamma}\frac{L_s}{k^2}. \tag{11}$$

in such a way that the potential of Eq.(4) is rewritten as

$$V(r) = \frac{\mathbf{A}\exp(-r/k)}{1-\exp(-r/k)} + \frac{\mathbf{B}\exp(-r/k)}{(1-\exp(-r/k))^2} + \frac{\mathbf{C}\exp(-2r/k)}{(1-\exp(-r/k))^2} + L_s T_c \tag{12}$$

where the term

$$T_c = \frac{(\boldsymbol{\alpha}+\boldsymbol{\beta})\exp(-r/k) + (\boldsymbol{\gamma}-\boldsymbol{\alpha})\exp(-2r/k)}{k^2(1-\exp(-r/k))^2} \tag{13}$$

may be identified with an approximation to $\frac{1}{r^2}$. However, before to do that, we want to notice that, as shown recently [15,16], the choice of the A, B and C parameters leads to many of most popular exponential-type potentials as particular cases of Eq.(4). So, one can make extensive this feature to the $\mathbf{A}, \mathbf{B}$ and $\mathbf{C}$ parameters in the effective potential for which, to fulfill the purpose of this work, let us consider the selection $\mathbf{A} = \frac{V_0(a-1)}{b} - V_1$, $\mathbf{B} = V_2$, $\mathbf{C} = 0$, $k = \frac{1}{2\alpha}$, $\boldsymbol{\alpha} = \omega k^2$, $\boldsymbol{\beta} = \lambda k^2$ *and* $\boldsymbol{\gamma} = 0$ that allows to write Eq.(14) as

$$V_{eff}(r) = \left(\frac{V_0(a-1)}{b} - V_1\right)\frac{\exp(-2\alpha r)}{1-\exp(-2\alpha r)} + V_2\frac{\exp(-2\alpha r)}{(1-\exp(-2\alpha r))^2} + \frac{L_s}{r^2} \tag{14}$$

where the parameters $\boldsymbol{\alpha}, \boldsymbol{\beta}$ and $\boldsymbol{\gamma}$ have permitted to use the approximation $T_c \approx r^{-2}$ of Taskin and Kocak [11].

Straightforwardly, this potential is identified with the *Eckart +deformed Hylleraas potential* given by $V_{EH}(r) = V_{eff}(r) + \frac{aV_0}{b}$, namely

$$V_{EH}(r) = \frac{V_0}{b}\left(\frac{a-\exp(-2\alpha r)}{1-\exp(-2\alpha r)}\right) - V_1\frac{\exp(-2\alpha r)}{1-\exp(-2\alpha r)} + V_2\frac{\exp(-2\alpha r)}{(1-\exp(-2\alpha r))^2} + \frac{L_s}{r^2}. \tag{15}$$

where $\frac{aV_0}{b}$ is a displacement. Consequently, according to our generalized proposal for exponential-type potentials, $V_{EH}(r)$ has, from Eq.(6), the energy spectra

$$E_{EH(n,L)} = -\alpha^2 \left(\frac{n^2 + (2n+1)(h_L+1)/2 + \frac{1}{4\alpha^2}\left(\frac{V_0(a-1)}{b} - V_1 + V_2 + (\omega+\lambda)L_s\right)}{n + (h_L+1)/2} \right)^2 + \frac{aV_0}{b} \tag{16}$$

that is in good agreement with the solution obtained from the Nikiforov-Uvarov method by Ikot *et.al.* [6] matching their parameters

$$\left\{ \begin{array}{l} \gamma = \frac{V_0}{4\alpha^2} \\ \theta = \frac{1}{4\alpha^2}\left[V_1 - \omega\left(\ell(\ell+D-2) + (D-1)(D-3)/4\right)\right] \\ \phi = \frac{1}{4\alpha^2}\left[V_2 + \lambda\left(\ell(\ell+D-2) + (D-1)(D-3)/4\right)\right] \\ \ell(\ell+D-2) + (D-1)(D-3)/4 = L_s \\ \sigma = \frac{1}{2}\left[1 + \sqrt{1+4\phi}\right] = \frac{h_L+1}{2}. \end{array} \right. \tag{17}$$

Also, directly form Eq.(7), the eigenfunctions are given by

$$\psi_{n,L}(r) = (\exp(-2\alpha r))^{(c_L-1)/2}(1 - \exp(-2\alpha r))^{(h_L+1)/2}\ {}_2F_1(-n, b_L; c_L; \exp(-2\alpha r)), \tag{18}$$

where $h_L = \sqrt{1 + \frac{1}{\alpha^2}\left[V_2 + \lambda L_s\right]}$ and

$$b_L = \frac{(h_L+1)(n+h_L) - \frac{1}{2\alpha^2}\left(\frac{V_0(a-1)}{b} - V_1 + V_2 + (\omega+\lambda)L_s\right)}{2n + h_L + 1}, \tag{19}$$

$$c_L = \frac{-2n(h_L+n) - \frac{1}{2\alpha^2}\left(\frac{V_0(a-1)}{b} - V_1 + V_2 + (\omega+\lambda)L_s\right)}{2n + h_L + 1}. \tag{20}$$

IV. Bound state solutions of some specific exponential-type potentials.

As mentioned before, the selection of parameters involved in the proposed multiparameter exponential-type potential of Eq.(12), leads to various specific potentials as particular cases [15]. The same is valid for the Eckart+deformed Hylleraas potential here considered. In fact, as shown in Table I, for specific parameters, it become as particular cases of $V_{EH}(r)$ the standard $V_E(r)$ Eckart, $V_H(r)$ Hylleraas, $V_{Hu}(r)$ Hulthén and $V_{RM}(r)$ Rosen-Morse potentials just for give some examples.

In table I, it should be used $N_X = n^2 + (h_{X(L)} + 1)(2n+1)/2$ with $X = E,\ H,\ Hu$ and RM for Eckart, Hylleraas, Hulthén and Rosen-Morse respectively such that

$$h_{E(L)} = \sqrt{1 + \frac{1}{\alpha^2}(V_2 + \lambda Ls)} \qquad \textit{and} \qquad h_{H(L)} = h_{Hu(L)} = h_{RM(L)} = \sqrt{1 + \frac{\lambda Ls}{\alpha^2}} \tag{21}$$

Also, the corresponding wavefunctions are derived from Eq.(18) as

$$\begin{aligned} \psi_{X(n,L)}(r) &= (\exp(-2\alpha r))^{(c_{X(L)}-1)/2}(1 - \exp(-2\alpha r))^{(h_{X(L)}+1)/2} \\ &\quad {}_2F_1(-n, b_{X(L)}; c_{X(L)}; \exp(-2\alpha r)), \end{aligned} \tag{22}$$

Journal of Physics: Conference Series **574** (2015) 012089 doi:10.1088/1742-6596/574/1/012089

where the corresponding parameters $b_{X(L)}$ and $c_{X(L)}$ are obtained directly from Eqs.(19) and (20) for the respective involved parameters given in the Table I.

Table I. Specific potential models obtained directly as particular cases of $V_{EH}(r)$.

Parameters	Potential	Energy Spectra
$b \to \infty$ or $V_0 = 0$	$V_E(r) = \frac{-V_1 \exp(-2\alpha r)}{1-\exp(-2\alpha r)} + \frac{V_2 \exp(-2\alpha r)}{(1-\exp(-2\alpha r))^2} + \frac{L_s}{r^2}$	$E_E = -\alpha^2 \left(\frac{N_E + \frac{1}{4\alpha^2}(-V_1+V_2+(\omega+\lambda)L_s)}{n+(h_{E(L)}+1)/2} \right)^2$
$V_1 = V_2 = 0$	$V_H(r) = \frac{V_0}{b} \left(\frac{a-\exp(-2\alpha r)}{1-\exp(-2\alpha r)} \right) + \frac{L_s}{r^2}$	$E_H = -\alpha^2 \left(\frac{N_H + \frac{1}{4\alpha^2}\left(\frac{V_0(a-1)}{b}+(\omega+\lambda)L_s\right)}{n+(h_{H(L)}+1)/2} \right)^2 + \frac{aV_0}{b}$
$V_0 = V_2 = a = 0$	$V_{Hu}(r) = \frac{-V_1 \exp(-2\alpha r)}{1-\exp(-2\alpha r)} + \frac{L_s}{r^2}$	$E_{Hu} = -\alpha^2 \left(\frac{N_{Hu} + \frac{1}{4\alpha^2}(-V_1+(\omega+\lambda)L_s)}{n+(h_{Hu(L)}+1)/2} \right)^2$
$V_1 = V_2 = 0$, $b = 1$, $a = -1$	$V_{RM}(r) = \frac{-V_0(1+\exp(-2\alpha r))}{1-\exp(-2\alpha r)} + \frac{L_s}{r^2}$	$E_{RM} = -\alpha^2 \left(\frac{N_{RM} + \frac{1}{4\alpha^2}(-2V_0+(\omega+\lambda)L_s)}{n+(h_{RM(L)}+1)/2} \right)^2 - V_0$

Finally, we want to notice that due to the fact that we assume $\mathbf{C} = 0$ and $\gamma = 0$ other specific D–dimensional potentials such as the Manning-Rosen or Deng-Fan models can not be derived in the present work from $V_{EH}(r)$. However, they are particular cases contained into the multiparameter exponential-type potential (Eq.12) as already published elsewhere [15,16].

V Concluding remarks.

This work has been devoted to find the analytical bound-state solutions of the Schrödinger equation for the Eckart-deformed Hylleraas potential in $D-$dimensions. At difference with the celebrated Nikiforov-Uvarov method, our proposal has been adapted to the $D-$dimensional case of the exactly-solvable solutions of a class of exponential-type potentials. The effective potential that we are proposing accepts different approximations to the centrifugal term as well as different exponential-type potential models depending on the choice of the involved parameters **A**, **B** and **C**. However, we have considered the Taskin - Kocak approximation to $1/r^2$ as well as specific choices of parameters to get the Eckart-deformed Hylleraas potential. From there, the particular cases of bound-state solutions in $D-$dimensions to the Hulthén, Eckart, Rosen-Morse and the single deformed Hylleraas potential are directly derived from the proposal. However, we stand why some other potentials such as the Manning-Rosen or Deng-Fan potential can not be obtained from the Eckart-deformed Hylleraas model. By the contrary, these potentials are particular cases from the proposed multiparameter exponential-type potential in $D-$dimensions. In short, instead of obtaining the analytical bound-state solutions of the Schrödinger equation for an specific exponential-type potential in $D-$dimensions, by means of a specialized method, our proposal gives the general solution from where specific potentials can be obtained as particular cases.

Acknowledgments

This work was partially supported by the projects UAM-A-CBI-2232004 and 009. We are grateful with the SNI - Conacyt - México for the stipend received.

References

[1] Nikiforov A F and Uvarov V B 1988 *Special Functions of Mathematical Physics* (Birkhauser, Bassel)
[2] Cifci H, Hall R L and Saad N 2003 *J. Phys A: Math. Gen.* **36** 1807
[3] Morales D A. 2004 *Chem. Phys. Lett.* **394** 68
[4] Diaf A, Chouchaoui A and Lombard R J 2005 *Annals of Physics* **317** 354
[5] Lucha W and Schöberl F F 1999 *Int. J. of Mod. Phys. C* **10** 607
[6] Ikot A N, Awoga O A and Antia A D 2013 *Chin. Phys. B* **22** 020304
[7] Awoga O A, Ikot A N and Emah J B 2013 *Rev. Mex. Fis.* **59** 229
[8] Ikot A N, Antia A D, Akpan I O and Awoga O A 2013 *Rev. Mex. Fis.* **59** 46
[9] Chun-Sheng Jiaa, Yun Lia, Yu Sunb, Jian-Yi Liuc, Liang-Tian Suna 2013 *Phys. Lett. A* **311** 115
[10] Hua W 1990 *Phys. Rev. A* **42** 2524
[11] Taskin F and Kocak G 2010 *Chin. Phys. B* **19** 090314
[12] Hassanabadi H, Zurrinkamar S and Rajabi A A 2011 *Commun. Theor. Phys.* **55** 541
[13] Awoga O A and Ikot A N 2012 *Pramana J. of Phys.* **79** 345
[14] García-Martínez J, García-Ravelo J, Morales J and Peña J J 2012 *Int. J. Quan. Chem.* **112** 195
[15] Peña J J, García-Martínez J, García-Ravelo J and Morales J. 2014 *IOP J. of Phys.: Conf. Ser.* **490** 012199
[16] Peña J J, García-Martínez J, García-Ravelo J and Morales J. 2014 *Int. J. Quan. Chem.* DOI: 10.1002/qua.24803

Pair correlation functions of FeAs-based superconductors: Quantum Monte Carlo study

V A Kashurnikov, A V Krasavin

National Research Nuclear University MEPhI (Moscow Engineering Physics Institute), 31 Kashirskoye shosse, 115409, Moscow, Russia

avkrasavin@gmail.com

Abstract. The new generalized quantum continuous time world line Monte Carlo algorithm was developed to calculate pair correlation functions for two-dimensional FeAs-clusters modeling of iron-based superconductors within the framework of the two-orbital model. The analysis of pair correlations depending on the cluster size, temperature, interaction, and the type of symmetry of the order parameter is carried out. The data obtained for clusters with sizes up to 10×10 FeAs-cells favor the possibility of an effective charge carrier's attraction that is corresponding the A_{1g}–symmetry, at some parameters of interaction.

1. Introduction

Recently new iron-based high-temperature superconductors (HTSC) [1], both pnictides and chalcogenides, came to be regarded as the subject of intense theoretical and experimental research [2, 3]. The reason is not only the similarity in structural, magnetic, electrophysical, and other properties and features to copper-oxide superconductors but also the presence of a complex band structure, and variety of magnetic and superconducting phases.

Iron pnictides have a layered structure of closely spaced atomic planes consisting of Fe and As atoms. As well as copper-oxide HTSC, they have a pronounced anisotropy which allows description of these systems on the basis of two-dimensional tight-binding models. However, these models have multiband structure and they are much more complex than, for example, the Emery model for copper-oxide HTSC [4]. Most appropriate tight-binding models for these compounds are two- [5, 6], three- [7–9] and the five-orbital ones [9, 10], which are typical generalized Hubbard multiband models. Strong correlations, a multi-gap band structure, features of magnetic and superconducting properties [11–13], the lack of expansion parameters for analytical approximations are the cause of to employ exact quantum methods for the study of iron-based HTSC.

The results known to date on exact calculations of FeAs clusters were obtained mainly using the methods based on the exact diagonalization (ED) of the Hamilton matrix of the model under consideration, and on Monte Carlo (MC) technique.

The properties of the two-orbital model for a cluster of size $\sqrt{8} \times \sqrt{8}$ (the so-called «tilted» cluster) containing eight iron atoms were studied in detail [6, 14] by ED technique. However, the ED method, although giving precise answers for the two-orbital model, is limited by the size of the clusters as the size of the Hamilton matrix grows exponentially with the system size. The largest cluster calculated by ED technique in the case of the two-orbital model took 10 iron atoms that correspond to cluster size 2×5. The power of Earth Simulator supercomputer (Japan) was required for such calculation [15]. The data on pairing correlations and the binding energy obtained in these studies were allowed to identify areas on the phase diagram corresponding to the pairing symmetries A_{1g} and B_{2g} of the superconducting order parameter [6, 16].

As regards cluster calculations of iron-based HTSC by MC methods, it should be noted the following.

In [17] 8×8 and 10×10 clusters were calculated using the variational MC algorithm also in the framework of the two-orbital model. From our point of view the choice of a variational wave function with taking into account antiferromagnetic correlations only leads to results close to the Hartree-Fock (mean-field) approximation. Strictly speaking, this approach may be inadequate to do correct conclusions about such an important issue as the pair correlations in the system.

Calculations of FeAs clusters up to 16×16 size using the determinant MC method were implemented in [18, 19] in the framework of a simplified model that is referred to as the model with S_4-symmetry. In fact, this model is a two-band generalized Hubbard model with diagonal interaction between the bands. The data obtained on pair correlations, nevertheless, show the presence of A_{1g} pairing symmetry.

The three-orbital model was studied in [20, 21] by MC method for clusters of size up to 16×16 in undoped state. On the one hand, this model is more complete in comparison with two-orbital one, and the results allow to describe correctly the experimentally observed photoemission spectra and the spectra of neutron scattering, as well as to obtain data on magnetic ordering and conductivity in the non-superconducting state. On the other hand, the pair correlations were not discussed in these works, and, according to [6], the two-orbital model is sufficient for the analysis of pair correlations in FeAs systems.

There were also approaches describing the features of pnictides by spin degrees of freedom [22] and focusing mainly on the study of antiferromagnetic correlations.

In this paper, the new generalized quantum continuous time world line MC algorithm (the generalized CTWL-algorithm) was developed and adapted to the full two-orbital model. An accurate calculation of non-diagonal terms of the form $a_{i\sigma}^{+}a_{j\sigma'}a_{k\sigma''}^{+}a_{l\sigma'''}$ within the full two-orbital model which may play a leading role for the pairing of charge carriers and, in our view, cannot be neglected, is the key feature of the presented algorithm. The new algorithm is based on the previously developed CTWL-algorithm [23, 24] which allowed a convenient coding of states for MC procedures.

2. Two-orbital model

In the crystal structure of iron pnictides each of Fe atoms is surrounded by a tetrahedron of As atoms, and vice versa. Band structure calculations [25] show that $3d$ –states of iron atoms make the main contribution to the electronic density of states near the Fermi level. Therefore, a basic structural element can be defined as two Fe atoms located in a simple square lattice and two As atoms above and below the Fe-plane. One is responsible for conducting and superconducting properties of the system. Accordingly the tight-binding model of the system is defined [5, 6].

The two-orbital model is quite complex to implement algorithms based on quantum MC or ED technique for modeling of the iron based superconductors. The Hamiltonan of the two-orbital model is given by

$$H = H_{int} + H_{kin},$$

$$H_{int} = U\sum_{i,\alpha} n_{i,\alpha,\uparrow}n_{i,\alpha,\downarrow} + V\sum_{i} n_{i,x}n_{i,y} - \mu\sum_{i} n_i - J\sum_{i}\left(n_{i,x,\uparrow}n_{i,y,\uparrow} + n_{i,x,\downarrow}n_{i,y,\downarrow}\right)$$

$$-J\sum_{i}\left(a_{i,x,\downarrow}^{+}a_{i,x,\uparrow}a_{i,y,\uparrow}^{+}a_{i,y,\downarrow} + a_{i,x,\uparrow}^{+}a_{i,x,\downarrow}a_{i,y,\downarrow}^{+}a_{i,y,\uparrow} + a_{i,x,\uparrow}^{+}a_{i,y,\downarrow}a_{i,x,\downarrow}^{+}a_{i,y,\uparrow} + a_{i,y,\uparrow}^{+}a_{i,x,\downarrow}a_{i,y,\downarrow}^{+}a_{i,x,\uparrow}\right),$$

$$H_{kin} = -t_1\sum_{i,\sigma}\left(a_{i,x,\sigma}^{+}a_{i+x,x,\sigma} + a_{i,y,\sigma}^{+}a_{i+y,y,\sigma} + h.c.\right) - t_2\sum_{i,\sigma}\left(a_{i,y,\sigma}^{+}a_{i+x,y,\sigma} + a_{i,x,\sigma}^{+}a_{i+y,x,\sigma} + h.c.\right)$$

$$-t_3\sum_{i,\sigma}\left(a_{i,x,\sigma}^{+}a_{i+x+y,x,\sigma} + a_{i,x,\sigma}^{+}a_{i+x-y,x,\sigma} + a_{i,y,\sigma}^{+}a_{i+x+y,y,\sigma} + a_{i,y,\sigma}^{+}a_{i+x-y,y,\sigma} + h.c.\right)$$

$$-t_4 \sum_{i,\sigma} \left(a^+_{i,x,\sigma} a_{i+x+y,y,\sigma} + a^+_{i,y,\sigma} a_{i+x+y,x,\sigma} + a^+_{i,x,\sigma} a_{i+x-y,y,\sigma} + a^+_{i,y,\sigma} a_{i+x-y,x,\sigma} + h.c. \right) \quad (1)$$

where operator $a^+_{i,x(y),\sigma}(a_{i,x(y),\sigma})$ creates (annihilates) an electron with spin σ on site $\boldsymbol{i}$ and orbital $x(y)$; t_i, $i = 1, \ldots, 4$ are the hopping amplitudes between d_{xz} and d_{yz} orbitals defined as x and y, respectively; parameters U and V describe the Coulomb interaction within and between orbitals, respectively; J is the exchange integral; μ is the chemical potential.

Experimental studies of iron-based HTSC, as well as band-structure calculations within the two-orbital model (see, e.g., Ref. 6 and references therein) give different possible sets of parameters of the Hamiltonian which allow to describe qualitatively the observed electronic properties of these compounds. The realistic parameters of hopping amplitudes were chosen in the present work (hereinafter all energy quantities are measured in 0.2 eV units):

$$t_1 = 0.058; t_2 = 0.22; t_3 = -0.0208; t_4 = -0.079. \quad (2)$$

The relationship between the parameters of interaction in most calculations was determined as

$$V = 0.5U; J = 0.25U. \quad (3)$$

3. Method

In this paper, the quantum world-line MC algorithm was modified so that the calculation of the Hamiltonian of general form

$$H = \sum_{kln_1n_2n_3n_4} t_{kl}^{n_1n_2n_3n_4} \left(A_k^{n_1n_2} A_l^{n_3n_4} + h.c. \right) + \sum_{kln_1n_2} U_{kl}^{n_1n_2} N_k^{n_1} N_l^{n_2}$$
$$+ \sum_{kn_1n_2} B_k^{n_1n_2} \left(A_k^{n_1n_2} + h.c. \right) + \sum_{kn} C_k^n N_k^n + \sum_{kln_1n_2n_3} D_{kl}^{n_1n_2n_3} N_l^{n_3} \left(A_k^{n_1n_2} + h.c. \right) \quad (4)$$

becomes possible. Here $A_k^{n_1n_2}$ are off-diagonal terms of transitions between states n_1 and n_2 on site k; N_k^n is the particle number operator of a state n on site k; t, B, C, D are the corresponding amplitudes of interactions. The two-orbital model enables the description in the framework of the specified general Hamiltonian [24].

In order for the analysis of the off-diagonal long-range order can be made, the pair correlation functions for different types of symmetry of the pairing of charge carriers should be calculated,

$$Q_\delta = \sum_{ij} \int_0^\beta d\tau \langle \Delta_\delta(i,\tau) \Delta_\delta^+(j,0) \rangle. \quad (5)$$

Here δ denotes the pairing channel corresponding to the type of symmetry of the order parameter. According to the classification [6, 16], the correlator Q_2 is responsible for A_{1g} pairing, and Q_9 for B_{2g} pairing. The data on small clusters obtained by ED technique [6] showed that these types of symmetry appear to be the main ones for FeAs clusters; similar symmetry types were considered in the simplified S_4 −model [18]. The corresponding order parameters are given by

$$\Delta_2^+(i) = \frac{1}{\sqrt{N}} \sum_{\alpha\mu} \left(a^+_{i,\alpha,\uparrow} a^+_{i+\mu,\alpha,\downarrow} - a^+_{i,\alpha,\downarrow} a^+_{i+\mu,\alpha,\uparrow} \right); \quad (6)$$

$$\Delta_9^+(i) = \frac{1}{\sqrt{N}} \sum_{\alpha\mu} \left(a^+_{i,-\alpha,\uparrow} a^+_{i+\mu,\alpha,\downarrow} - a^+_{i,\alpha,\downarrow} a^+_{i+\mu,-\alpha,\uparrow} \right), \quad (7)$$

where N is the number of sites (Fe atoms), α denotes the orbitals x and y, and μ here is a unit vector to one of four neighbor sites of the site i.

For extracting the anomalous terms of the type $\langle aa \rangle$ and $\langle a^+a^+ \rangle$ (5) has to be expanded according to Wick's theorem. Each four-fermion term can be expressed in the following way [26]:

$$\langle T_\tau a_{i\alpha'\sigma'}(\tau) a_{j\alpha\sigma}(\tau) a^+_{k\beta\sigma''}(0) a^+_{l\beta'\sigma'''}(0) \rangle \rightarrow$$
$$\langle T_\tau a_{i\alpha'\sigma'}(\tau) a^+_{l\beta'\sigma'''}(0) \rangle \langle T_\tau a_{j\alpha\sigma}(\tau) a^+_{k\beta\sigma''}(0) \rangle - \langle T_\tau a_{i\alpha'\sigma'}(\tau) a^+_{k\beta\sigma''}(0) \rangle \langle T_\tau a_{j\alpha\sigma}(\tau) a^+_{l\beta'\sigma'''}(0) \rangle. \quad (8)$$

As a result, we have the following sequence of pairwise products of normal terms (Green's functions):

$$G_2 = \frac{1}{N}\sum_{\substack{ij\alpha\beta\\ \mu\nu\sigma}}\int_0^\beta d\tau\,\big(\langle a_{j+\nu,\beta,\sigma}(\tau)a^+_{i+\mu,\alpha,\sigma}(0)\rangle - \langle a_{j,\beta,\sigma}(\tau)a^+_{i,\alpha,\sigma}(0)\rangle - \\ -\langle a_{j+\nu,\beta,\sigma}(\tau)a^+_{i+\mu,\alpha,-\sigma}(0)\rangle + \langle a_{j,\beta,-\sigma}(\tau)a^+_{i,\alpha,\sigma}(0)\rangle\big); \tag{9}$$

$$G_9 = \frac{1}{N}\sum_{\substack{ij\alpha\beta\\ \mu\nu\sigma}}\int_0^\beta d\tau\,\big(\langle a_{j+\nu,s\beta,\sigma}(\tau)a^+_{i+\mu,s\alpha,\sigma}(0)\rangle - \langle a_{j,s\beta,\sigma}(\tau)a^+_{i,s\alpha,\sigma}(0)\rangle - \\ -\langle a_{j+\nu,s\beta,\sigma}(\tau)a^+_{i+\mu,-s\alpha,-\sigma}(0)\rangle + \langle a_{j,-s\beta,-\sigma}(\tau)a^+_{i,s\alpha,\sigma}(0)\rangle\big), \tag{10}$$

where i and j denote the sites of the lattice (Fe atoms), α and β denote the orbitals x and y, μ and ν are unit vectors to one of neighbor sites of the site i and j, respectively, $\sigma = \uparrow, \downarrow$ is spin z-axis projection, $s = 1$ if $\sigma = \uparrow$ and $s = -1$ otherwise. In this case the differences $Q_2 - G_2$ and $Q_9 - G_9$ give information about the anomalous terms $\langle a^+_{i\alpha'\uparrow}a^+_{l\beta'\downarrow}\rangle\langle a_{j\alpha\downarrow}a_{k\beta\uparrow}\rangle$ and about possible pairing of charge carriers in the given symmetry channel of the order parameter.

4. Correlation functions

Pair correlation functions (5) and normal terms (9, 10) were calculated for FeAs clusters of size 4×4, 6×6, 8×8, and 10×10 within the framework of the full two-orbital model (1) by the quantum MC method developed in section 3 in the temperature range $\beta = 5 \div 20$ and the range of the interaction $U = 2 \div 16$ while maintaining the relation (3). Results are presented for the half filling.

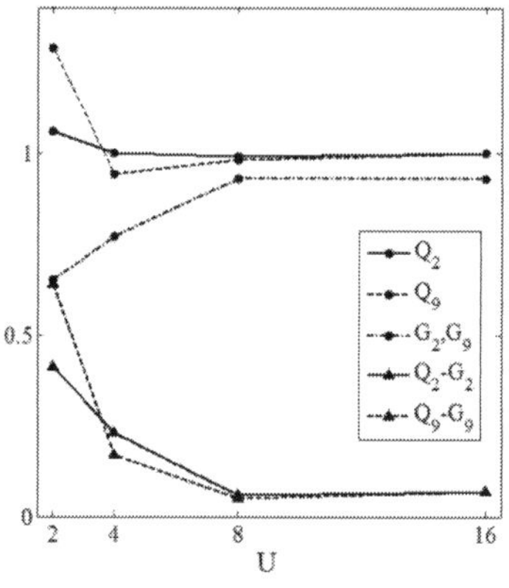

Figure 1. The dependence of the correlation functions Q_2 and Q_9, normal terms G_2 and G_9, and their differences on the interaction parameter for 8×8 cluster. $\beta = 5$.

Figure 1 shows the dependence of the correlation functions Q_2 and Q_9, normal terms G_2 and G_9, and the difference between these values on the interaction parameter for 8×8 cluster. The first thing to note is that in all our calculations G_2 and G_9 coincided within the error of the calculation which was $\sim 10^{-2} - 10^{-3}$ for all data. Hence, the normal terms (9, 10) were insensitive to the type of symmetry. Further, the differences $Q_2 - G_2$, $Q_9 - G_9$ points up the presence of anomalous terms, i.e. the pairing of charge carriers at $U < 8$ for both types of symmetries is possible. For larger values of the interaction the pair correlations were not observed.

The dependence of the same correlation functions on temperature is shown in figure 2 for the largest calculated cluster of size 10×10 and two values of interaction $U = 4;\ 8$. The differences in the behavior of the pair correlations are evident. At $U = 4$ correlations Q_2 take place right up to the smallest temperature $\beta = 20$, and the symmetry channel Q_9 disappears at $\beta > 10$. At larger value $U = 8$ pair correlations continuously decrease at $\beta > 5$.

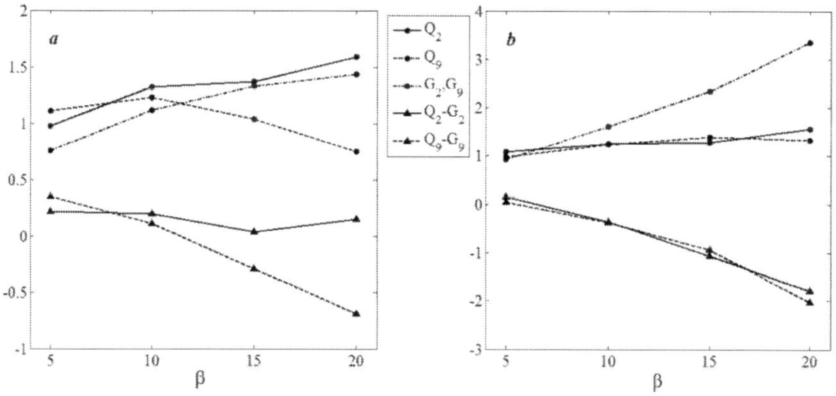

Figure 2. The dependence of the correlation functions Q_2 and Q_9, normal terms G_2 and G_9, and their differences on temperature for 10×10 cluster. (a) $U = 4$; (b) $U = 8$.

Let us consider now the dependence of the pair correlations on the system size. The results for the lowest temperature $\beta = 20$ are presented in figure 3. The scaling of the pair correlation functions shows pronounced A_{1g}-type correlations at $U = 4$ for clusters of size 8×8 and larger; they weakly depend on the cluster size. The data indicate also the disappearance of B_{2g}-type correlations with increasing of system size. The same way as in the case of higher temperature, at greater interaction $U = 8$ the differences $Q_2 - G_2$ and $Q_9 - G_9$ are less than zero for both types of symmetries and clusters of any size.

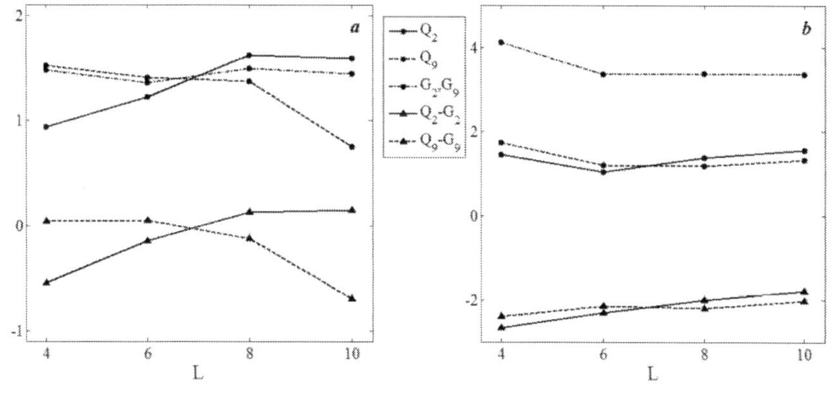

Figure 3. The dependence of the correlation functions Q_2 and Q_9, normal terms G_2 and G_9, and their differences on the system size. (a) $U = 4$; (b) $U = 8$. $\beta = 20$.

Thus, the whole set of obtained data provides the evidence for the possibility of charge carriers pairing at the chosen values of the hopping amplitudes (2) and the interaction parameter $U < 8$ (with relations (3) between the U, V, and J). The type of realized symmetry is A_{1g}. The results indicate also that the normal terms G_2 and G_9 depend on the interaction strength and temperature and almost linearly increase with U, β (for strong interactions), while the dependence of the correlation functions on these parameters is much weaker. The increase of the normal terms leads to effective suppression of pair correlations with increasing of interaction strength and inverse temperature at $U \gtrsim 8$. For 6×6 cluster the values of correlation parameters reach saturation and slightly change in larger systems.

For the analysis of the relation between pair correlations and magnetic ordering, as the authors of [18], we investigated the magnetic correlations using diagonal part (along the z-axis) at zero frequency of the spin susceptibility

$$\chi(q) = \int_0^\beta d\tau \sum_{\alpha\alpha'} \sum_{ij} e^{iq(i-j)} \langle m_{i\alpha}(\tau) m_{j\alpha'}(0) \rangle ; \; m_{i\alpha} = \frac{1}{2}(n_{i\alpha\uparrow} - n_{i\alpha\downarrow}). \quad (11)$$

The data obtained show a clear trend for the displacement of the peak point from (π, π) to $(\pi, 0)$ with increasing of the interaction. Thus, in the region of small U where according to our data pair correlations are observed, a specific antiferromagnetic ordering emerges (spin density wave), which was also noted in [6, 14, 15, 18, 20].

Acknowledgements
The work was supported by Russian Science Foundation (project 14-22-00098).

References

[1] Kamihara Y, Watanabe T, Hirano M, and Hosono H 2008 J. Am. Chem. Soc. **130** 3296.
[2] Isyumov Yu, and Kurmaev E 2010 High-Tc Superconductors Based on FeAs Compounds, Springer Series in Materials Science Vol. 143 (Springer, Berlin).
[3] Dagotto E 2013 *Rev. Mod. Phys.* **85** 849.
[4] Emery V J 1987 *Phys. Rev. Lett.* **58** 2794.
[5] Raghu S, Qi X L, Liu C X, Scalapino D J, and Zhang S C 2008 *Phys. Rev. B* **77** 220503(R).
[6] Moreo A, Daghofer M, Riera J A, and Dagotto E 2009 *Phys. Rev. B* **79** 134502.
[7] Daghofer M, Nicholson A, Moreo A, and Dagotto E 2010 *Phys. Rev. B* **81** 014511.
[8] Yu S L, Knang J, and Li J X 2009 *Phys. Rev. B* **79** 064517.
[9] Luo Q, Martins G, Yao D X, Daghofer M, Yu R, Moreo A, and Dagotto E 2010 *Phys. Rev. B* **82** 104508.
[10] Graser S, Maier T A, Hirschfeld P J, and Scalapino D J 2009 *New J. Phys.* **11** 025016.
[11] Sadovskii M V 2008 *Physics-Uspekhi* **51** 1201.
[12] Ishida K, Nakai Y, and Hosono H 2009 *J. Phys. Soc. Jpn.* **78** 062001.
[13] Sadovskii M V, Kuchinskii E Z, Nekrasov I A 2012 *Journal of Magnetism and Magnetic Materials* **324** 3481.
[14] Nicholson A, Ge W, Zhang X, Riera J, Daghofer M, Oleś A M, Martins G B, Moreo A, and Dagotto E 2011 *Phys. Rev. Lett.* **106** 217002.
[15] Okumura M, Nakai N, Nakamura H, Hayashi N, Yamada S, and Machida M 2009 *Physica C* **469** 932.
[16] Wan Y and Wang Q H 2008 *Preprint* cond-mat/0805.0923.
[17] Kubo K and Thalmeier P 2011 *J. Phys. Soc. Jpn.* **80** SA121.
[18] Ma T, Lin H Q, and Hu J 2013 *Phys. Rev. Lett.* **110** 107002.
[19] Wu Y, Liu G, and Ma T 2013 *EPL* **104** 27013.
[20] Liang S, Alvarez G, Sen C, Moreo A, and Dagotto E 2012 *Phys. Rev. Lett.* **109** 047001.
[21] Liang S, Moreo A, and Dagotto E 2013 *Phys. Rev. Lett.* **111** 047004.
[22] Applegate R, Singh R R P, Chen C C, and Devereaux T P 2012 *Phys. Rev. B* **85** 054411.
[23] Kashurnikov V A, Krasavin A V 2010 *ZhETF* **111** 180.
[24] Kashurnikov V A, Krasavin A V 2013 *JETP Lett.* **97** 333.
[25] Singh D J and Du M H 2008 *Phys. Rev. Lett.* **100** 273003.
[26] Abrikosov A A, Gorkov L P, and Dzyaloshinski I E 1975 Methods of Quantum Field Theory in Statistical Physics, *Dover Books on Physics* (Dover Publications, New York).

The numerical calculation of shear zone refraction's law crossover the different stickiness sand

K. Yamamoto[1], Y. Yamada[2], S. Miyazima[2]

Faculty of Science and Engineering, Setsunan University, Neyagawa, Osaka 572-8508,Japan
Faculty of Engineering, Chubu University, Kasugai, Aichi 487-8501, Japan

E-mail: keizo@ele.setsunan.ac.jp

Abstract. The refraction's law of sheared region near the boundary of two kinds of sand with the different stickiness is investigated by a simulation method. Our experimental apparatus for sand system is one of the simplest sets in order to obtain shear pattern, called linear split-bottom cell. In this paper we give an analysis based on a classical numerical calculation for the above-mentioned system. Although a sand particle has a variety of forms from sphere to needle in general, but here we assume an ensemble of the simplest shape such as cubes in a rectangular parallel pipe which is split into two halves. One of the halves can be moved along the split quasi-statically. We obtain a refraction pattern like an equipotential surface across the interface of different dielectric constants shown in the textbook. Various quantities obtained from our calculation were compared with experiments in the preceding conference. In this conference, we will show the refraction's law at the shear zone boundary crossing over two different sand systems with unequal stickiness.

1. Introduction

Properties in sand system have recently been studied from elementary processes [1]-[3] and also actual applications to technological fields [4]. As is well-known, sand systems consist of sand particles. Generally, forms of sand particles diverge from sphere to needle. Ensemble of such particles forms an example of typical complex system. When stress is given at a site, along a line, along a surface and/or to bulk body, stress and strain propagates and complicate pattern of stress region is formed. It is also important to investigate how the quasi-static propagation of stress and strain in the sand system progresses.

Therefore we built up an experimental apparatus to study how the shear propagates and is formed as simple as possible which is shown in Fig.1. Our results of sheared width which depend on height from the bottom in homogeneous sand system have been shown in Ref. [5]. Reference [6] is improved the insufficient parts of the preceding conference results [5].

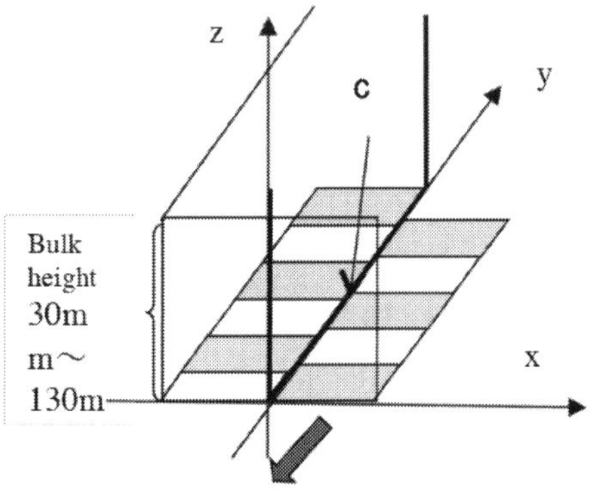

Fig.1 Experiment apparatus which is made of acrylic box with a split along the longest edge of the box. One side of the box is movable with a quasi-static velocity(blue thick arrow).

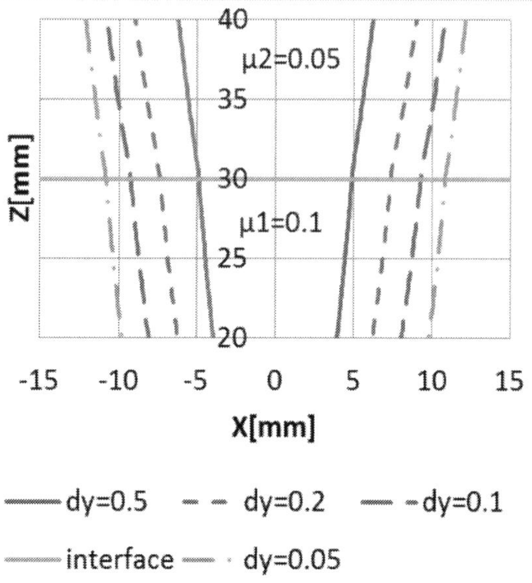

Fig.2 Shear lines across the horizontal interface of two sand layers with two different stickiness.

In Fig.1, the right hand side of box can be moved quasi-statically, where stress propagates from bottom to top. Although a sand particle has a variety of forms from sphere to needle in general, but here we assume an ensemble of the simplest shape such as cubes in a rectangular parallel-pipe split into two halves. One of the halves is moved along the split quasi-statically shown in Fig.1. Figure 2 shows the boundary of shear region which across the horizontal interface of different kinds of sand in our numerical calculation. The longitudinal lines are shear lines which sliding distances from initial positions are 0.5[mm], 0.2[mm], 0.1[mm] and 0.05[mm], respectively. But all incidence angles to the interface are so small that we can't confirm whether the refraction's law is followed Snell's law known for geometric optics [7]. In this paper we give an analysis based on numerical calculation to get the large incident angle for the above-mentioned system. Then the interface of difference sand size is leaned 45 degrees from the level.

2. A simulation corresponding the present experiment

We assume that sand particles are approximated by small cubes and our sand ensemble is regularly piled 340×680 cubes which approximates a random piling. When one side of the box is under the force which is shown by an arrow in Fig.1, the sand particles at the bottom move together with the bottom plate and many particles at the top share the slip between the right and left half boxes. By assuming the friction force between two particles we have obtained the following equation.

$$\left[\mu_x\left(f(x,z)-f(x-1,z)\right)-d_{shr}\right]+\left[\mu_x\left(f(x,z)-f(x+1,z)\right)-d_{shr}\right]$$
$$+\left[\mu_z\left(f(x,z)-f(x,z+1)\right)\left(z_{top}-z\right)-d_{shr}\right]$$
$$+\left[\mu_z\left(f(x,z)-f(x,z-1)\right)\left(z_{top}-z+1\right)-d_{shr}\right]=0 \qquad (1)$$

Here, $f(x,z)$ is displacement of a particle at x and z, $\boldsymbol{\mu_x}\left(\boldsymbol{f(x,z)-f(x-1,z)}\right)$ is friction between the particles at x, z and x-1, z and $\boldsymbol{\mu_z}\left(\boldsymbol{f(x,z)-f(x,z+1)}\right)\left(\boldsymbol{z_{top}-z}\right)$ is friction between the particles at x, z and x, z+1 where $\boldsymbol{\mu_x}$ and $\boldsymbol{\mu_z}$ are the friction coefficients, as shown in Fig.3. A factor $\left(\boldsymbol{z_{top}-z}\right)$ is consideration about the effect of gravity. Then, $[G-d_{shr}]$ means $G-d_{shr}$ if $G-d_{shr}>0$, and 0 if $G-d_{shr}<0$. Once we know the numeric values of four neighbour sites of $f(x, z)$, we can obtain the value of $f(x, z)$ by using this equation.

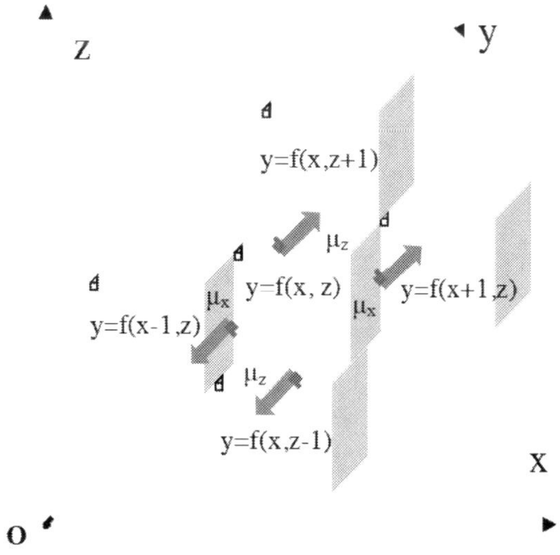

Fig.3. A sand particle is under the force from four nearest neighbour cubes.

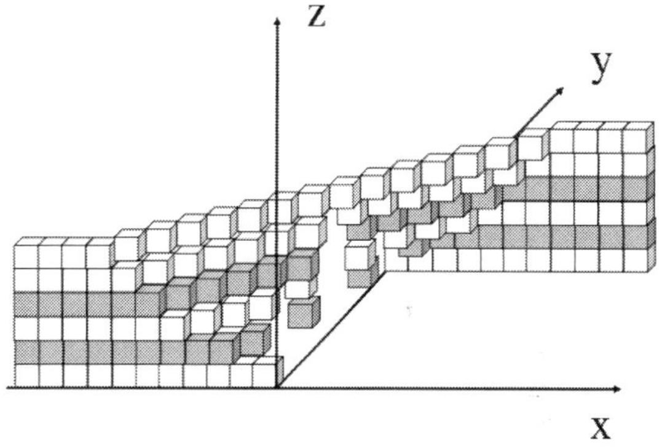

Fig. 4. A sketch of sheared portion.

3. Some results of calculation in heterogeneous sand system

Figure 4 shows a movement of particles at the central area schematically. Here we adopt that a sliding distance is 7mm as experiment where a side box is moved in a quasi-static manner. The particles at the bottom move by a little together with the plate, but next upper particles near the sliding plate move a little less distance. In this heterogeneous sand simulation, the inclination of the interface of different sand size is 45° from the horizon. The graph shown in Fig.5 is an intersection by x-z plane by adopting μ_1=0.1, μ_2=0.05 and d_{shr} =0.01. The straight line of 45° inclination is the interface of different sand size. The curves extended in the lengthwise direction of both sides in Fig.5 are shear lines whose sliding distance is 1.5 mm. The central lines show the position of the maximum slid. In order to confirm the refraction's law, the figure which expanded near a part ① of Fig. 5 is Fig. 6. The incidence angle θ_1 is 32.3° and the refraction angle θ_2 is 50.7°. Therefore μ_1 tanθ_1 =0.0447, μ_2 tanθ_2 =0.0433 and μ_1 sinθ_1 =0.0378, μ_2 sinθ_2 =0.0274. As a result of this, the refraction's law doesn't obey Snell's law known for geometric optics but obeys the refraction law like the equipotential surface across the interface of different dielectric constants.

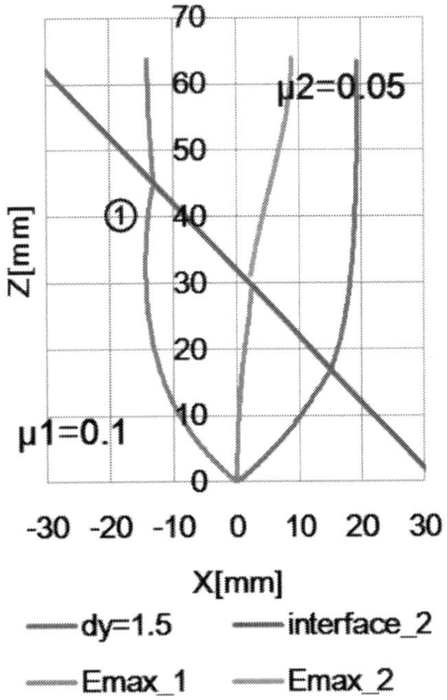

Fig.5.The maximum shear line and the line of the mean value of shear when μ1> μ2.

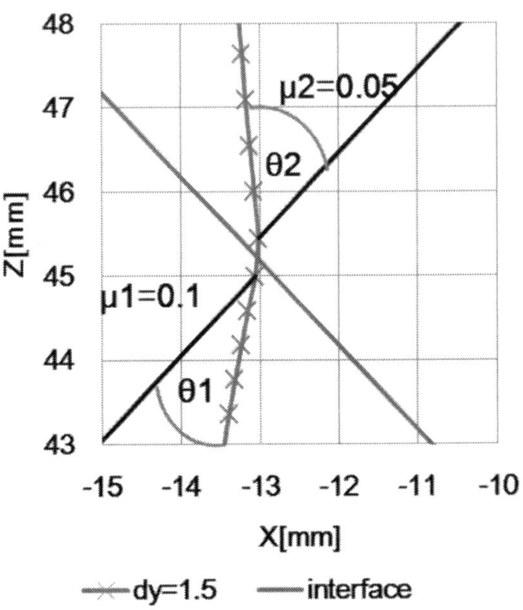

Fig.6. The enlargement shear line near ① shown in Fig.5.

4. Concluding remarks

Shear pattern in the sand system is examined experimentally or numerically by many scientists and here is a trial of simulation for the experiment. Some results of experiment are understood by the present simulation, but some are not. The present calculation shows that the refraction's law of shear line obeys the refraction law like the equipotential surface across the interface of different dielectric constants. The maximum slid line shows a complicated feature. Probably, our model will still be insufficient.

References

[1] Ries A Wolf D E and Unger T 2007 *Physical Review* E **76** 051301.
[2] Fenistein D, van de Meent J W, and van Hecke M 2004 *Physical Review Letters* **92** 094301.
[3] Torok J, Unger T, Kertesz J and Wolf D E 2007 *Physical Review* E **75**, 011305.
[4] For example, Advanced Powder Technology, Elsevier.
[5] S. Miyazima, Y. Yamada and K. Yamamoto 2012 *Journal of Physics:Conference Series* **410** 012069.
[6] K. Yamamoto, Y. Yamada, N. Miyashita and S. Miyazima 2013 IC-MSQUARE 2013 P3P17.
[7] T. Borzonyi, T. Unger and B. Szabo 2009 *Physical Review* E **80**, 060302.

Evaluation of Kirkwood-Buff integrals via finite size scaling: a large scale molecular dynamics study

W Dednam[1,2], A E Botha[2]

1 Departamento de Fisica Aplicada, Universidad de Alicante, San Vicente del Raspeig, E-03690 Alicante, Spain

2 Department of Physics, Science Campus, University of South Africa, Private Bag X6, Florida Park 1710, South Africa

E-mail: bothaae@unisa.ac.za

Abstract. Solvation of bio-molecules in water is severely affected by the presence of co-solvent within the hydration shell of the solute structure. Furthermore, since solute molecules can range from small molecules, such as methane, to very large protein structures, it is imperative to understand the detailed structure-function relationship on the microscopic level. For example, it is useful know the conformational transitions that occur in protein structures. Although such an understanding can be obtained through large-scale molecular dynamic simulations, it is often the case that such simulations would require excessively large simulation times. In this context, Kirkwood-Buff theory, which connects the microscopic pair-wise molecular distributions to global thermodynamic properties, together with the recently developed technique, called finite size scaling, may provide a better method to reduce system sizes, and hence also the computational times. In this paper, we present molecular dynamics trial simulations of biologically relevant low-concentration solvents, solvated by aqueous co-solvent solutions. In particular we compare two different methods of calculating the relevant Kirkwood-Buff integrals. The first (traditional) method computes running integrals over the radial distribution functions, which must be obtained from large system-size NVT or NpT simulations. The second, newer method, employs finite size scaling to obtain the Kirkwood-Buff integrals directly by counting the particle number fluctuations in small, open sub-volumes embedded within a larger reservoir that can be well approximated by a much smaller simulation cell. In agreement with previous studies, which made a similar comparison for aqueous co-solvent solutions, without the additional solvent, we conclude that the finite size scaling method is also applicable to the present case, since it can produce computationally more efficient results which are equivalent to the more costly radial distribution function method.

1. Introduction

The solvation thermodynamics of solutes in water are extremely sensitive to the presence and concentration of co-solvents such as alcohols, osmolytes and salts [1, 2, 3, 4]. Their presence in the hydration shell of the solute in an aqueous co-solvent solution usually determines how well the solute is solvated by the solution [1]. In general, the free energy of solvation depends on the relative proportions of co-solvent and water in the mixture, and the more negative the free energy of solvation, the greater the extent to which the solute is solvated by the solution.

Understanding the effects of co-solvents on the solvation of solutes in aqueous co-solvent solutions, via calculations of the solvation free energy, for instance, may shed light on why proteins tend to unfold or denature in the presence of alcohols and urea [3]. It may also help to explain the salting in and out effects on hydrophobic solutes, due to favourable interactions between the solute and large monovalent ions of low charge density on the one hand, and unfavourable interactions between the solute and small ions of high charge density, on the other [2, 4].

In a recent study, Ganguly and Van der Vegt [5] have compared the efficiency of the so-called finite-size scaling method, first introduced by Schnell et al. [6], to the more traditional method of calculating the Kirkwood-Buff integrals (KBIs) via the radial distribution functions. Ganguly and Van der Vegt [5] consider the full range of concentrations of methanol-water and urea-water systems and conclude that the finite size scaling method (also called the fluctuation or sub-box method, see Refs. [7, 8, 5]) can provide accurate KBIs for system sizes as small as 2000 molecules, which is about a factor of ten smaller than the largest system size used in the present work, which is the minimal size required for the accurate computation of the radial distribution functions in the limit of large distances [9].

In this paper we further explore the validity of the two, above-mentioned methods by performing molecular dynamics (MD) simulations of another very important test case; namely, that of the solvation of methane (a hydrophobic solute) in a methanol (co-solvent) and water (solvent) mixture. In agreement with the previous results [5], which were obtained for for aqueous co-solvent solutions only, our results demonstrate that the finite size scaling method may also be used to improve the efficiency in aqueous co-solvent systems with the addition of a solute, such as methane.

2. Theory

2.1. Kirkwood-Buff theory of solutions

Macroscopic thermodynamic properties of solute-water-co-solvent systems can be directly obtained from microscopic molecular distributions. More specifically, the Kirkwood-Buff theory of solutions [10, 11] connects fluctuations in the grand canonical ensemble to macroscopic thermodynamic properties through the so-called Kirkwood-Buff integrals (KBIs) between components i and j of the solution

$$G_{ij} = 4\pi \int_0^{\infty} \left(g_{ij}^{\mu VT}(r) - 1 \right) r^2 \mathrm{d}r \qquad (1)$$

$$= V \left(\frac{\langle N_i N_j \rangle - \langle N_i \rangle \langle N_j \rangle}{\langle N_i \rangle \langle N_j \rangle} - \frac{\delta_{ij}}{\langle N_j \rangle} \right) \qquad (2)$$

where $\langle\, \rangle$ denotes the time and ensemble average, V is the volume of the simulation domain, N_i is the number of particles of component i, and $g_{ij}^{\mu VT}(r)$ is the radial distribution function in the grand canonical (μVT) ensemble. G_{ij}, the KBI between component i and j, is a local quantity that measures the deviation of the intermolecular distribution from that of a random one, i.e., an ideal gas. Hence, it is also a measure of the affinity of these two components for each other in the solution environment.

The macroscopic quantity that is of interest in this work is the free energy of solvation of the solute in the methanol-water mixture. This quantity can be expressed in terms of the G_{ij} and the solution component number densities ρ_i as

$$\left(\frac{\partial \Delta \mathrm{G_c}}{\partial \rho_c} \right)_{p,T} = \frac{-RT\,(G_{cc} - G_{cw})}{1 + \rho_c\,(G_{cc} - G_{cw})} \qquad (3)$$

for a binary system of water (w) and co-solvent (c), and as

$$\left(\frac{\partial \Delta \mathrm{G_s}}{\partial x_c} \right)_{p,T} = \lim_{\rho_s \to 0} \frac{RT\,(\rho_w + \rho_c)^2}{\eta} (G_{sw} - G_{sc}) \qquad (4)$$

for a ternary system in the limit of infinite dilution of the solute [10]. Here ΔG_s is the free energy of solvation of the solute, x_c is the co-solvent mole fraction, R is the universal gas constant, T the temperature, and $\eta = \rho_w + \rho_c + \rho_w \rho_c (G_{ww} + G_{cc} - 2G_{cw})$.

This paper outlines two independent methods for computing the right-hand side of Eq. (1) and Eq. (2) and compares the results obtained from each method. The first method involves generating the $g_{ij}^{\mu VT}(r)$ at every concentration by MD simulation of a system which is large enough to permit the accurate evaluation of the integral on the right-hand side of Eq. (1). The second method, that of finite size scaling, permits the evaluation of the right hand side of Eq. (2), as described in the next section.

2.2. *Finite Size Scaling*

The KBI between component i and j for a given methanol concentration can be calculated efficiently by an innovative particle counting and extrapolation scheme referred to as finite size scaling [6, 7, 8]. This scheme involves computing the KBI by extrapolation to a system of theoretically infinite size which produces more accurate values for G_{ij}, even at relatively small system sizes. The first step involves repeating a count of the number of particles of type i and j that fall within a certain sub-volume, also called a volumetric pixel (voxel). A large number of voxels of a fixed volume are randomly positioned within the simulation domain, without touching the boundary. By counting the numbers of particles in each voxel at each time step, the averages required in Eq. (2) are obtained. This process is repeated for a range of different voxel volumes. In theory, the ideal voxel volume should be larger than the correlation length of the system, and much smaller than the overall volume of the simulation domain, so that the simulation domain can still act as a particle reservoir to ensure that the ensemble of voxels do approximate a grand canonical ensemble. This means that there is an optimal range of voxel volumes for which the KBI will scale linearly with $1/L$, where L is a linear dimension of the voxel (in the case of a sphere, L would be the diameter). In practice, a good approximation to a grand canonical ensemble can be obtained by using a simulation domain with approximately 20000 molecules or more. (See, for example, Fig. 2 of Ref. [9].)

For a given concentration, a plot of G_{ij} as a function of $1/L$ will show the expected linear relationship over the optimal range of voxel volumes. The KBI value corresponding to a (theoretically) infinite system size may then be obtained by linearly extrapolating the value of G_{ij} back to the ordinate ($1/L = 0$) of the graph, where L is infinite. The KBIs thus determined can then be compared to those obtained directly from Eq. (1).

3. Methodology

Five mixtures of methanol and water were prepared with the concentrations of methanol given by $x_c = 0.00, 0.24, 0.50, 0.75$ and 1.00 mole fraction. The solvent co-solvent system always contained a total of 20000 molecules. An additional 250 methane molecules were added as solute.

In order to simulate the mixtures, the GROMACS molecular dynamics simulation package was used [12]. The methanol and water molecules were modeled using the GROMOS43a1 and SPC water force fields, respectively. All atom simulations were carried out in the NpT ensemble with pressure maintained by the Berendsen Barostat at a pressure of 1 atm and a coupling time of 0.5 ps [13]. The integration time step used was 1 fs for all concentrations. All the simulations were allowed to equilibrate for 15 ns, after which the trajectory was extended by a further 15 ns. Only the last 15 ns of the trajectories were used to calculate the KBIs. The electrostatic interactions in the all atom simulations were handled by means of the particle mesh Ewald.

Calculations of the KBIs, based on the particle fluctuation method (Eq. (2)), were implemented in the Python programming language. Several Python scripts were developed to extract the coordinates of each atom from the system trajectories and then to count the number of each atomic species within the voxels.

4. Simulation results

Initially, to test the water and methanol force fields, a detailed comparison of the computed KBIs was made with the experimentally determined KBIs from Refs. [14, 15]. Figure 1 shows the KBIs, as calculated from the RDFs (solid lines) for a 50% methanol-water mixture. The experimental values are given by the horizontal dashed and dotted lines for EXP 1 (Ref. [14]) and EXP 2 (Ref. [15]), respectively. In this simulation there were 10000 methanol and 10000 water molecules. Figure 1 shows that the KBIs reach a reasonably well-defined plateau beyond

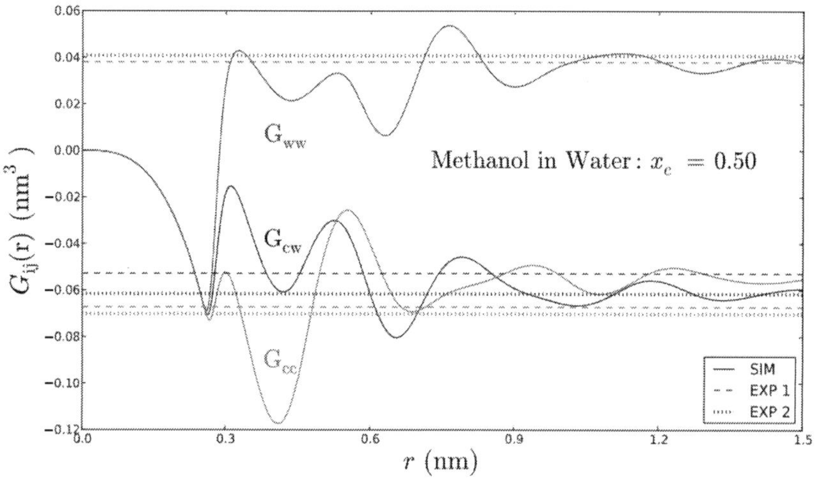

Figure 1. (Color online) The Kirkwood-Buff Integrals (KBIs) calculated via the radial distribution functions (solid curves) compared to the experimental values in references [14] (EXP 1) and [15] (EXP 2).

$r = 1.5$ nm, which is a good indication that the chosen system size for this work (20000 molecules in total) is sufficiently large. We arrived at this size by starting with a smaller system size and then doubling the system size until convergence in the KBIs was observed. Thus our system is sufficiently large to approximate the thermodynamic limit. In spite of this, however, we could not obtain agreement (within the experimental uncertainties) for concentrations of methanol below about 15% and we attribute this discrepancy to the well-known inaccuracies in the force fields employed here [16, 17, 18]. For the present work, in which our main aim is to compare the KBIs obtained from the same force field via two different methods, such inaccuracies in the force fields should not matter, and in view of the difficulties associated with re-parameterization of the force fields [18], we have chosen to ignore this problem in the present work.

In Fig. 2 we illustrate how the fluctuation method was used to obtain the KBIs via Eq. (2). In Fig. 2 the diameter L of the sample sub-volume (in this case a sphere) was varied from about half the total simulation box size, down to one twentieth of the simulation box size. As expected from theory [8], the KBI value (in this case for water-water) scales linearly over a certain range of sample system voxels. The KBI for a system of infinite size is therefore well approximated for each concentration by extrapolating to best line fit to infinite L, i.e. to the intercept corresponding to $1/L = 0$. Figures 5 (a) to (e) show comparisons of the KBIs obtained via the fluctuation method (Eq. (2)) and the radial distribution functions (Eq. (1)) for five different concentrations of the methane, methanol and water system. In Fig 5 (a), which is for pure water, the agreement between the two methods is satisfactory. Even though the system

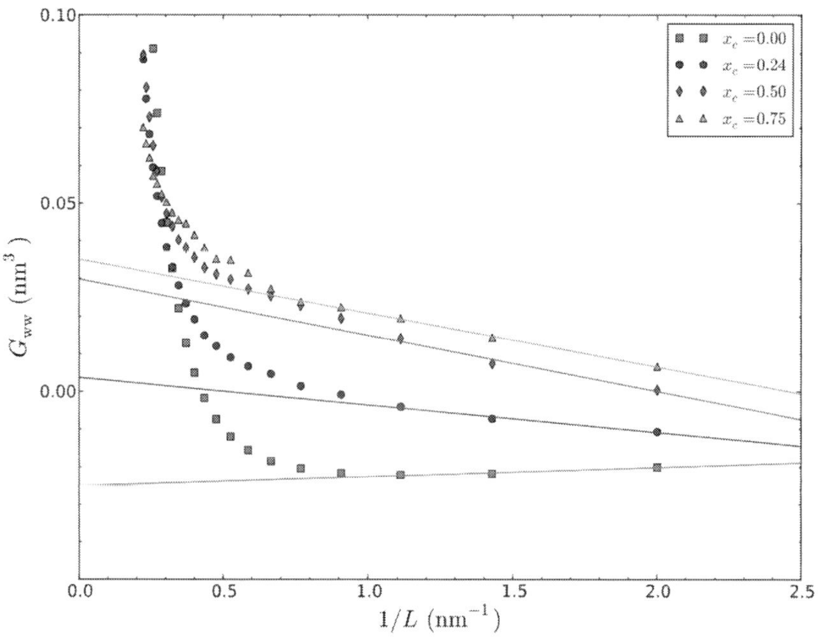

Figure 2. (Color online) Illustration of the method of finite size scaling for four different concentrations of co-solvent for the methane (250 molecules), methanol (co-solvent) and water system.

contains only 250 methane molecules, the RDF method still produces a KBI which shows a well-defined plateau, which indicated that in the calculation of the KBI, via the RDF method, the 15 ns trajectory was sufficiently long for the purposes of these calculations. In Fig. 5 (b) the two methods do not agree very well and the reason for this is not clear at present. One possibility could be that the inaccuracies in the methanol force field at lower concentrations produce unphysical clustering. The resolution of such clustering effects via the fluctuation method may require the use of much longer system trajectories (longer than the 15 ns used here).

5. Discussion and conclusion

We have developed a new, biologically relevant test case in order to compare two important methods for the calculation of Kirkwood-Buff integrals from molecular dynamic simulations. Specifically, we have compared the two methods for simulations of the hydration of methane (a hydrophobic solute) in an aqueous co-solvent solution. Unlike the traditional method, which requires the computation of radial distribution functions in the limit of large distances, the main advantage of the finite-size scaling method is that it can make use of relatively small systems to obtain results that are comparable to those of larger systems that would require substantially longer simulation times. This advantage is due to the analytical finite-size scaling relation [6], which allows the results from successively smaller sub-boxes to be extrapolated to systems which approach the thermodynamic limit.

We note that the finite size scaling method which we have employed here made use of cubic

Journal of Physics: Conference Series **574** (2015) 012092 doi:10.1088/1742-6596/574/1/012092

sub-volumes to determine the particle fluctuations. This geometry becomes problematic when the linear dimensions of the sub-volume voxel are on the order of the sizes of the molecules themselves [8]. These so-called nook and corner effects can be avoided by using spherical sub-volumes instead [6, 7, 8]. Although we have not done so in the present work, we thus expect that the agreement between the two methods discussed here could be improved by using spherical sub-volumes instead of cubes. Finally we note that problems with the employed force fields

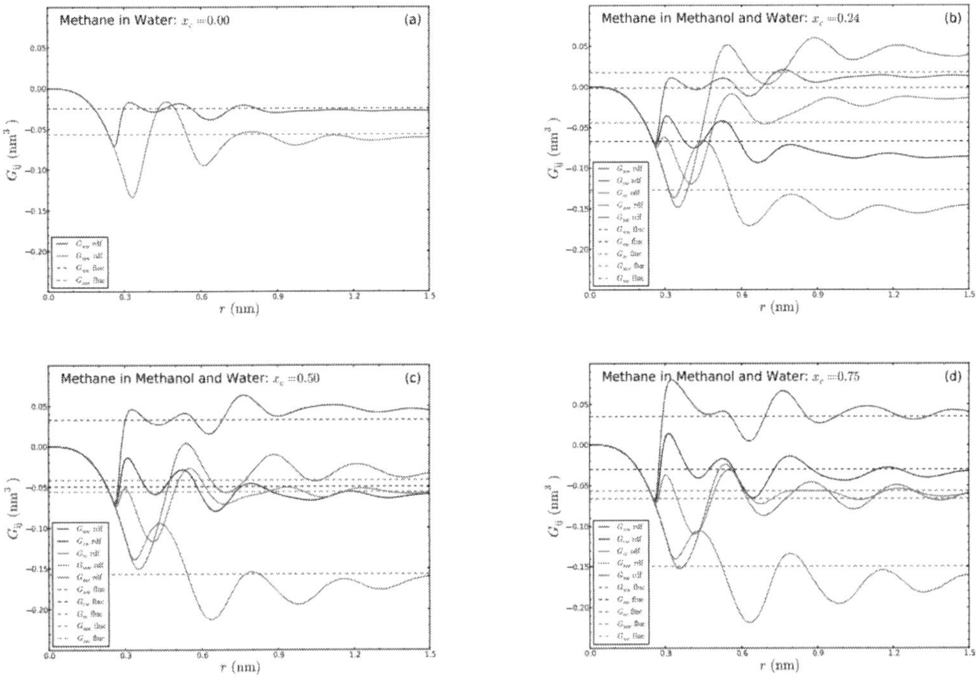

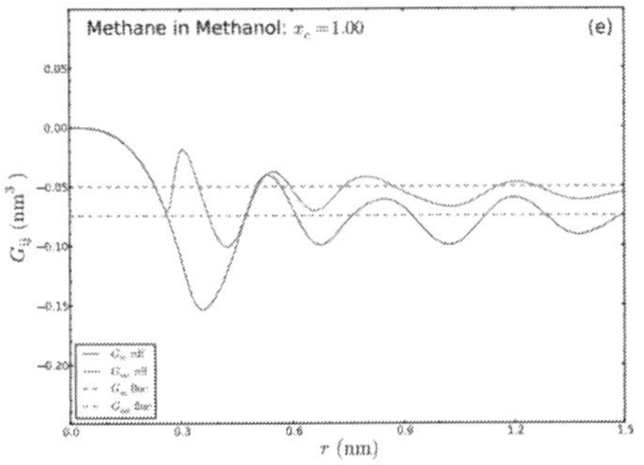

Figure 3. (Color online) Comparison of the KBIs obtained via the fluctuation method (Eq. (2)) and the radial distribution functions (Eq. (1)) for five different concentrations of the methane, methanol and water system. The correspondence between the labels in the legend (as read down the list in the legend) and the curves plotted in each figure is from top to bottom.

have prevented us from making a meaningful comparison of the solvation free energies with the experimental results. The difficulties associated with obtaining force field parameters that are applicable over a wide range of concentrations and to a variety of different systems are non-trivial [16, 17, 18], and since chemical potentials (and hence solvation free energies) are very sensitive to force field parameters, further refinement of the force fields would be required to facilitate a quantitative comparison of these quantities obtained from MD simulations and experimental results. However, the present work clearly demonstrates that finite size scaling can also be used in the new context of an aqueous co-solvent solution with the addition of a solute, to reduce simulation times.

Acknowledgments
AEB would like to thank D. Mukherji and N. F. A. van der Vegt for their helpful suggestions to improve this work.

References
[1] Roccatano D, Colombo G, Fioroni M and Mark A E 2002 *Proceedings of the National Acadamy of Sciences* **99** 12179
[2] Sadeghi R and Jahani F 2012 *J. Phys. Chem. B* **116** 5234
[3] Bennion B J and Daggett V 2003 *Proceedings of the National Acadamy of Sciences* **100** 5142
[4] van der Vegt N F A and van Gunsteren W F 2004 *J. Phys. Chem. B* **108** 1056
[5] Ganguly P and van der Vegt N F A 2013 *J. Chem. Theory Comput.* **9** 1347
[6] Schnell S K, Liu X, Simon J M, Bardow A, Bedeaux D, Vlugt T J H and Kjelstrup S 2011 *J. Phys. Chem. B* **115** 10911
[7] Liu X, Schnell S K, Simon J M, Bedeaux D, Kjelstrup S, Bardow A and Vlugt T J H 2011 *J. Phys. Chem. B* **115** 12921
[8] Schnell S K, Vlugt T J H, Simon J M, Bedeaux D and Kjelstrup S 2012 *Molecular Physics* **110** 1069
[9] Mukherji D, van der Vegt N F A, Kremer K and Site L D 2012 *J. Chem. Theory Comput.* **8** 375
[10] Ben-Naim A 2006 *Molecular Theory of Solutions* (New York: Oxford University Press)
[11] Kirkwood J G and Buff F P J 1951 *Chem. Phys.* **19** 774
[12] Lindahl E, Hess B and van der Spoel D 2001 *J. Mol. Mod.* **7** 306
[13] Berendsen H J C, Postma J P M, van Gunsteren W F, Dinola A and Haak J R 1984 *J. Phys. Chem.* **81** 3684
[14] Perera A, Sokolić F, Almásy L and Koga Y 2006 *J. Phys. Chem.* **124** 124515
[15] Matteoli E and Lepori L 1984 *J. Chem. Phys.* **80** 2856
[16] Weerasinghe S and Smith P E 2003 *J. Chem. Phys.* **118** 5901
[17] Weerasinghe S and Smith P E 2005 *J. Phys. Chem. B* **109** 15080
[18] Hawlicka E and Rybicki M 2012 *Molecular Dynamics - Theoretical Developments and Applications in Nanotechnology and Energy* ed Wang L (Rijeka: InTech) p 399

Structural investigations of fat fractals using small-angle scattering

Eugen M Anitas[1,2]

[1]Joint Institute for Nuclear Research, Dubna 141980, Moscow region, Russia
[2]Horia Hulubei National Institute of Physics and Nuclear Engineering, RO-077125 Bucharest, Romania

E-mail: anitas<at>theor.jinr.ru

Abstract. Experimental small-angle scattering (SAS) data characterized, on a double logarithmic scale, by a succession of power-law decays with decreasing values of scattering exponents, can be described in terms of fractal structures with positive Lebesgue measure (fat fractals). Here we present a theoretical model for fat fractals and show how one can extract structural information about the underlying fractal using SAS method, for the well known fractals existing in the literature: Vicsek and Menger sponge. We calculate analytically the fractal structure factor and study its properties in momentum space. The models allow us to obtain the fractal dimension at each structural level inside the fractal, the number of particles inside the fractal and about the most common distances between the center of mass of the particles.

1. Introduction

An important technique for investigating the structure of various types of materials is small-angle scattering (SAS) [1,2] which yields the differential elastic cross section per unit solid angle as a function of the momentum transfer. It is known that many disordered systems show the property of self-similarity across nano- and microscales [3], such as various types of elastomeric membranes [4, 5], cements [6], semiconductors [7], magnetic [8, 9] or biological structures [10–12], and therefore the concept of fractal geometry [3, 13] coupled with elements of SAS technique can give new insights regarding the structural characteristics of such fractal systems [14–20]. One of the main parameters which can be obtained is the fractal dimension D [3, 13]. For a mass fractal it is given by the scattering exponent of the power-law SAS intensity $I(q) \propto q^{-D}$ where $0 < \mathrm{D} < 3$.

However, a number of experimental SAS data shows a succession of power-laws whose scattering exponents take arbitrarily decreasing values [21–23]. Such a behavior was not clearly understood until recently [24], when it was shown that such type of successions corresponds to a fat fractal structure. In the later work, the main structural characteristics have been obtained by calculating the mono- and polydisperse structure and form factor for a system consisting of fat Cantor fractals (known also as ϵ-Cantor sets [25]). Although the fattened version of regular Cantor set (known also as thin Cantor sets) provides a simple and intuitive picture of the investigated properties, real physical systems have a more complex structure whose properties can be investigated by considering more complex structural models.

The aim of this paper is to extend the fat Cantor fractal model to fractal structures which can resemble more closely real physical systems and to show how we can extract information about the fractal dimensions at each level, total number of particles and about the most common distances in the fractal. To this end, we calculate the structure factor of fattened versions of Vicsek [26] and Menger sponge [3], as they are often considered as structural models for various natural or artificial systems.

2. Fat fractals

An intuitive way to understand fat fractals is the following: we start with an initial cube (here is used a cube, but any other shape could be used as well; $m = 0$, m being the fractal iteration number) which is divided into 27 smaller cubes with side length $1/3$ from the initial cube. Generally, in the first iteration ($m = 1$) some cubes (out of the 27 smaller ones) are kept and the others are removed. In particular, for the Vicsek fractal we keep the eight cubes in the corners plus the middle one and remove all the others, while for the Menger sponge we remove the smaller cubes in the middle of each face together with the cube in the center of the larger cube, and keep all the others. We repeat the same operation on each of the remaining cubes (9 for Vicsek, and respectively 20 for Menger), thus leaving 81 cubes, and respectively 400 cubes, of side length $1/3^2$ ($m = 2$) and so on. Therefore, at the m-th iteration the side length of the remaining cubes is $1/3^m$. The number of remaining cubes then is given by

$$N_m = \begin{cases} 9^m, & for\ Vicsek\ fractals, \\ 20^m, & for\ Menger\ fractals, \end{cases} \tag{1}$$

The thin version of these fractals is obtained in the limit $m \to \infty$, and have the same fractal dimensions $\log 9/\log 3 = 2$ (for Vicsek) and respectively $\log 20/\log 3 = 2.73$ (for Menger), independent on the iteration number. In this paper, the "fattened" version of these thin fractals is obtained by keeping the cubes instead of side length $1/3$ (for $m = 1, 2, 3$), then $1/3^2$ (for $m = 4, 5, 6$), then $1/3^3$ (for $m = 7, 8, 9$), etc. The resulting fractal is topologically equivalent to the thin version, but the holes decrease in size sufficiently fast so that, when $m \to \infty$, the fractal has nonzero and finite volume, and fractal dimension 3. Here, in the construction process we consider the same rules as already described in [24]: the side length of the initial cube is l_0 (called zero-order iteration or initiator) and specify it in Cartesian coordinates as a set of points satisfying the conditions $-l_0/2 \leq x \leq l_0/2$, $-l_0/2 \leq y \leq l_0/2$, $-l_0/2 \leq z \leq l_0/2$. The origin lies in the cube center, and the axes are parallel to the cube edges.

The iteration rule (*generator*) is to replace the initial cube by cubes of edge $\beta_s^{(1)} l_0$ ($m = 1$; whenever the quantity (...) appears in the exponent, it is to be interpreted as an index and not a power). For the Vicsek fractal the position of one cube is at the origin and the center of the eight cubes are shifted from the origin by the vectors $\boldsymbol{a}_j = \{\pm\beta_t^{(1)} l_0, \pm\beta_t^{(1)} l_0, \pm\beta_t^{(1)} l_0\}$ with all the combinations of the signs, where $\beta_t^{(1)} \equiv (1 - \beta_s^{(1)})/2$ and $\beta_s^{(1)}$ is a dimensionless positive parameter for the first iteration, obeying the condition $0 < \beta_s^{(1)} < 1/3$, in order to avoid overlapping of the cubes. For the Menger sponge the position of the eight cubes are shifted from the origin by the vectors $\boldsymbol{a}_j = \{\pm\beta_t^{(1)} l_0, \pm\beta_t^{(1)} l_0, \pm\beta_t^{(1)} l_0\}$ and the positions of the 12 cubes are shifted by the vectors $\{\pm\beta_t^{(1)} l_0, \pm\beta_t^{(1)} l_0, 0\}$, $\{\pm\beta_t^{(1)} l_0, 0, \pm\beta_t^{(1)} l_0\}$ and $\{0, \pm\beta_t^{(1)} l_0, \pm\beta_t^{(1)} l_0\}$ with all the combinations of the signs. The second and third iterations ($m = 2, 3$) are obtained by performing an analogous operation to each cube of the first iteration and with the same scaling factor $\beta_s^{(1)}$. For each subsequent iterations we repeat the same operation but for $m = 4, 5$ and $m = 6$ we take the scaling factor $\beta_s^{(2)}$, for $m = 7, 8$ and $m = 9$ we take the scaling factor $\beta_s^{(3)}$ and so on. If one consider that the edge of the removed parallelepiped at iteration m is $\gamma_m = \alpha^{p_m}$ where $0 < \alpha < 1$ and the exponent p_m is defined as

$$p_m \equiv \begin{cases} 1, & m = 1, 2, 3 \\ \cdots & \\ k, & m = 3k - 2, 3k, \end{cases} \tag{2}$$

where $m = 1, 2, \cdots$, then $\beta_s^{(m)} = (1 - \gamma_m)/2$, the side length of each cube by $l_m = l_0/2^m \prod_{i=1}^{m}(1 - \gamma_i)$ and the components of the $\boldsymbol{a}_j$ vectors are given by $\beta_t^{(m)} = l_{m-1}(1 + \gamma_m)/4$. Here, we consider $\alpha = 0.7$ in order to avoid superposition of cubes at least up to $m = 6$.

From their construction one can see that the fat fractals are built from a succession of exact self-similar fractals having different scaling factors at different scales. Since the scaling factor depends on

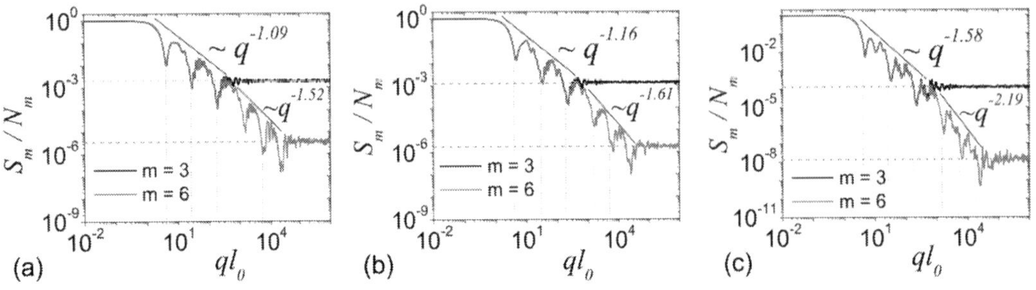

Figure 1. Monodisperse fractal structure factor (Eq. (4)) for the third and sixth iterations for a) Cantor (included for comparison, see details in [24]), b) Vicsek and c) Menger sponge, in units of N_m (Eq. (1)), at $\alpha = 0.7$. The position of minima indicated by vertical dotted-lines are estimated from Eq. (5)). The horizontal lines represent the asymptotes of the structure factor.

the iteration, each scale will have a different fractal dimension given by [17] $D_{\rm m} = -3\ln 2/\ln \beta_s^{(m)}$, and each of this scale gives a power-law decay of structure factor $S(q) \propto q^{-D_{\rm m}}$ [17, 19].

3. Results and discussions

For structures corresponding to a generalized Vicsek fractal [19] and Menger sponge, the generative function, which gives the positions of the centers of scattering cubes at each iteration, reads as

$$G_i(\boldsymbol{q}) = \begin{cases} \dfrac{1}{9}\left(1 + C_i(\boldsymbol{q})\right), & for\ Vicsek\ fractals, \\ \dfrac{1}{20}\left(C_i(\boldsymbol{q}) + 4(L_i^{xz}(\boldsymbol{q}) + L_i^{xy}(\boldsymbol{q}) + L_i^{yz}(\boldsymbol{q})\right), & for\ Menger\ fractals, \end{cases} \tag{3}$$

where $G_0(\boldsymbol{q}) \equiv 1$, $C_i(\boldsymbol{q}) = 8\cos(q_x u_i)\cos(q_y u_i)\cos(q_z u_i))$, $L_i^{xy}(\boldsymbol{q}) = \cos(q_x u_i)\cos(q_y u_i)$, $L_i^{xz}(\boldsymbol{q}) = \cos(q_x u_i)\cos(q_z u_i)$, $L_i^{yz}(\boldsymbol{q}) = \cos(q_y u_i)\cos(q_z u_i)$ and $u_i = l_0 \beta_{\rm t}^{(i)} \prod_{j=1}^{i-1} \beta_{\rm s}^{(j)}$. Then, the structure factor is obtained (Figs. (1) and (2)) from [19]

$$S(q) = N_m \left\langle \prod_{i=1}^{m} |G_i(\boldsymbol{q}u_i)|^2 \right\rangle . \tag{4}$$

The position of minima are obtained when the cubes inside the fractal interfere out of phase, and therefore using $2u_m = \pi/q$, for $k = 1, \cdots, m$ we obtain (vertical lines in Fig. (1)

$$q_k l_0 \simeq \frac{\pi}{2\beta_{\rm t}^{(k)} \prod_{i=1}^{k} \beta_{\rm s}^{(i)}} . \tag{5}$$

A more accurate description of a real physical system is obtained by considering an ensemble of fractals with different sizes taken at random and distributed according to a log-normal distribution function with relative variance $\sigma_{\rm r}$ [17]. The results are shown in Fig. (2), where $\sigma_{\rm r} = 0.4$.

One can be clearly observe that beyond the last power-law decay (or generalized power-law decay in the monodisperse case) we have $q \gtrsim 1/u_m$ and therefore, in this region the fractal structure factor given by Eq. (4) is $S(q) \simeq 1$ and the asymptotic values are $1/N_m$ (Figs. (1) and 2)), as for the case of thin fractals [17, 19].

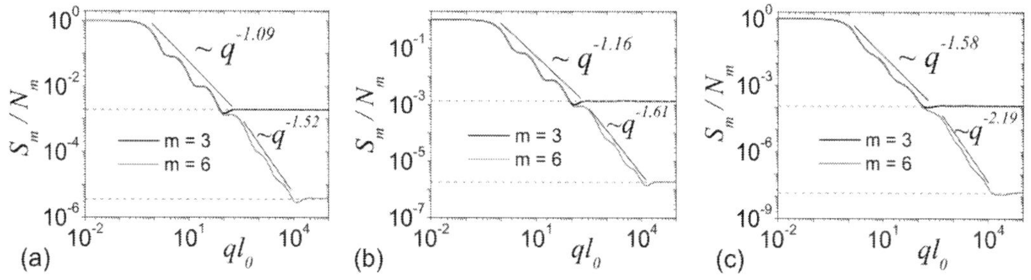

Figure 2. Polydisperse fractal structure factor (Eq. (4) together with the log-normal distribution [19]) for the third and sixth iterations for a) Cantor (included for comparison, see details in [24]), b) Vicsek and c) Menger sponge, in units of N_m (Eq. (1)), at $\alpha = 0.7$ and relative variance $\sigma_r = 0.4$. The horizontal lines represent the asymptotes of the structure factor.

4. Conclusions

We have extended the fat Cantor fractal model to other well known types of fractal structures described in literature (Vicsek and Menger sponge) which can model more closely real physical systems, and have presented an analytically expression for the fractal structure factor in mono- and polydisperse case. We have obtained the fractal dimensions at each level (from the slope of each power-law decay), total number of particles (from the asymptote of the structure factor) and the minima positions in the monodisperse factor (using Eq. 5). The model can be used to describe the structure of nano- and micro clusters and it can be applied to experimental SAS data showing a succession of power-law decay with decreasing values of scattering exponents ("concave"-like scattering curves).

References

[1] Glatter O and Kratky O 1982 *Small-angle X-ray Scattering* (London: Academic Press)
[2] Feigin L A and Svergun D I 1987 *Structure Analysis by Small-Angle X-Ray and Neutron Scattering* (NY: Plenum press)
[3] Gouyet J F 1996 *Physics and Fractal Structures* (Berlin: Springer)
[4] Balasoiu M et al 2008 *Optoelectronics and Advanced Materials - Rapid Communications* **2** 730–734
[5] Anitas E M, Balasoiu M, Bica I, Osipov V A and Kuklin A I 2009 *Optoelectronics and Advanced Materials - Rapid Communications* **3** 621–624
[6] Das A, Mazumder S, Sen D, Yalmali V, Shah J G, Ghosh A, Sahu A K and Wattal P K 2014 *J. Appl. Cryst.* **47** 421–429
[7] Cho K, Biswas P and Fraundorf P 2014 *J. Ind. Eng. Chem.* **20** 558
[8] Craus M L, Islamov A K, Anitas E M, Cornei N and Luca D 2014 *Journal of Alloys and Compounds* **592** 121–126
[9] Naito T et al 2014 *Eur. Phys. J. B* **86** 410
[10] Yadav I, Kumar S, Aswal V K and Kohlbrecher J 2014 *Phys. Rev. E* **89** 032304–1
[11] Gebhardt R 2014 *J. Appl. Cryst.* **47** 29–34
[12] Maji J, Bhattacharjee S M, Seno F and Trovato A 2014 *Phys. Rev. E* **89** 012121
[13] Mandelbrot B 1983 *The Fractal Geometry of Nature* (USA: W.H. Freeman)
[14] Schmidt P W and Dacai X 1986 *Phys. Rev. A* **33** 560–566
[15] Martin J E and Hurd A J 1987 *J. of Appl. Cryst.* **20** 61–78
[16] Schmidt P W 1991 *J. of Appl. Cryst.* **24** 414–435
[17] Cherny A Y, Anitas E M, Kuklin A I, Balasoiu M and Osipov V A 2010 *J. Appl. Cryst.* **43** 790–797
[18] Cherny A Y, Anitas E M, Kuklin A I, Balasoiu M and Osipov V A 2010 *J. Surf. Invest.* **4** 903–907
[19] Cherny A Y, Anitas E M, Kuklin A I and Osipov V A 2011 *Phys. Rev. E* **84** 036203–1–036203–11
[20] Cherny A Y, Anitas E M, Osipov V A and Kuklin A I 2014 *J. Appl. Cryst.* **47** 198–206
[21] Zhao J, Shi D and Lian J 2009 *Carbon* **47** 2329–2336
[22] Headen T F, Boek E S, Stellbrink J and M S U 2009 *Langmuir* **25** 422–428
[23] Golosova A A et al, 2012 *J. Phys. Chem. C* **116** 15765–15774
[24] Anitas E M 2014 *Eur. Phys. J. B* **87** 139
[25] Aliprantis C D and Burkinshaw O 1998 *Principles of Real Analysis* (Academic press)
[26] Vicsek T 1989 *Fractal Growth Phenomena* (Singapore: World Scientific)

Helical Disruptions in Small Loops of DNA

Marco Zoli

School of Science and Technology - CNISM
University of Camerino, I-62032 Camerino, Italy

E-mail: marco.zoli@unicam.it

Abstract. The thermodynamical stability of DNA minicircles is investigated by means of path integral techniques. Hydrogen bonds between base pairs on complementary strands can be broken by thermal fluctuations and temporary fluctuational openings along the double helix are essential to biological functions such as transcription and replication of the genetic information. Helix unwinding and bubble formation patterns are computed in circular sequences with variable radius in order to analyze the interplay between molecule size and appearance of helical disruptions. The latter are found in minicircles with $<$ 100 base pairs and appear as a strategy to soften the stress due to the bending and torsion of the helix.

It is known that the helicoidal conformation of DNA is essentially determined by the hydrophobicity of purine and pyrimidine bases and by the bond angles in the flexible sugar-phosphate backbone while sequence of the bases and environmental conditions due to the solvent also contribute to the molecule shape [1]. As each strand bears a negative charge (e) for each phosphate group and the rise distance between adjacent nucleotides is ~ 3.4Å, the bare double helix has a high linear charge density of $\sim 0.6\, e\, \text{Å}^{-1}$. Although the effective charge density is reduced by the counterions in the solvent, the electrostatic strands repulsion is key to the stability of the helix and also affects the inter-helical chiral interactions in those condensed phases of DNA assemblies (such as liquid crystals) which underlie the impressive growth of DNA-based structures recently witnessed in materials science.

Base pairing and base stacking are the fundamental interactions which control the synthesis of DNA and determine the thermodynamic stability both of single helices and of helix aggregates. However even stable duplexes at room temperature show local openings, temporary bubbles, which are intrinsic to the biological functioning as they permit the transcription and replication of the genetic code. Such bubbles are due to the strong fluctuational effects on the hydrogen bonds between complementary strands and cause the local unwinding of the double helix which ultimately leads to a state of negative supercoiled DNA for almost all living beings. While these processes are qualitatively understood, quantitative predictions of (energetically) optimal helical configurations for specific systems are scarce. We contribute to fill this gap by introducing a new path integral computational method which readily applies to loops of DNA as those found in bacterial plasmids, viral genomes and also mammalian cells.

1. Mesoscopic Model

Let's arrange N base pairs (bps) on a circle with radius R such that, $2\pi R/N \sim 3.4$ Å, as depicted in Fig. 1. When all bps centers of mass lie on the circumference, which represents the molecule backbone, the system is in the ground state.

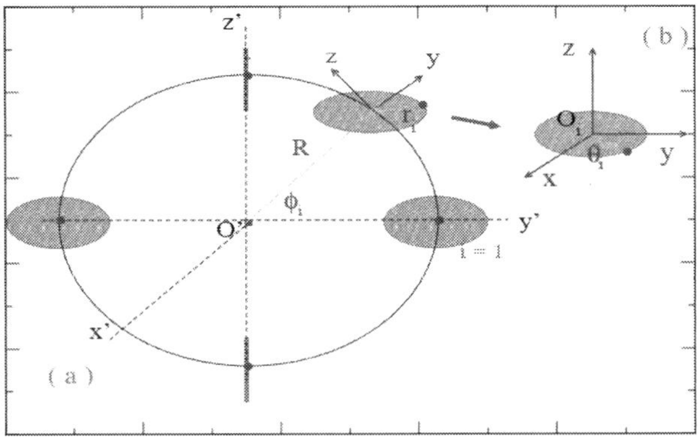

Figure 1. (Color online) (a) Helicoidal model for circular DNA with bending planes. The blue filled circles are the centers of mass of the base pairs stacked along the molecule backbone with a rise distance 3.4 Å. In the ground state all *bps* lie on the circumference with radius R. The red-shaded areas are spanned by the fluctuational vectors whose amplitude is measured by $|r_i|$ for the $i-$ *bp*. The azimuthal angle ϕ_i measures the bending of the $i-$ *bp* plane with respect to the (x', y') plane, **x'** being normal to the sheet plane. (b) Local reference system for the $i-$ *bp*. θ_i is the twist around the molecule backbone. The z-axis is tangent to the ground state circle.

Say $\mathbf{r}_i$, the inter-strand fluctuation for the i-base pair ($i = 1, ..N$) with respect to the ground state. Hence, we define the vector $\mathbf{t}_i$:

$$(t_i)_x = |\mathbf{r}_i| \cos\phi_i \cos\theta_i; \quad (t_i)_y = (R + |\mathbf{r}_i| \sin\theta_i)\cos\phi_i; \quad (t_i)_z = (R + |\mathbf{r}_i|)\sin\phi_i \,. \tag{1}$$

The ground state is recovered once all *bps*-fluctuations vanish hence, $t_i = R$, $\forall i$. The polar angle, $\theta_i = (i-1)2\pi/h + \theta_S$, measures the $i-$ *bp* twisting around the molecule backbone, with $h = N/Tw$ being the number of *bps* per helix turn and Tw is the twist number accounting for the coiling of the individual strands around the helical axis [1]. The azimuthal angle, $\phi_i = (i-1)2\pi/N + \phi_S$, defines the bending between adjacent *bps* along the stack. As the polynucleotide chain has a direction due to the chemistry of the intra-strand bonds, a distribution of values for the twist (θ_S) and the bending (ϕ_S) of the first *bp* in the sequence is weighed in the computation. The fluctuational orbits defined by $i = 1$ and $i = N + 1$ overlap consistently with the closure condition holding for the DNA ring.

2. Space-Time Mapping

My previous path integral analysis of DNA [2, 3, 4, 5, 6, 7, 8, 9, 10] were based on the ansatz that the *bps* displacements could be treated as one dimensional paths $x(\tau_i)$, $|\mathbf{r}_i| \to x(\tau_i)$, with the imaginary time $\tau_i \in [0, \beta]$ and β being the inverse the temperature.

Here I introduce a more general (albeit more CPU time consuming) space-time mapping technique, which does not pin a base pair to a specific τ_i thus avoiding the somewhat arbitrary partition of the β length in N intervals:

$$|\mathbf{r}_i| \to x_i(\tau); \quad |\mathbf{t}_i| \to \eta_i(\tau); \quad \tau \in [0\,, \beta]\,. \tag{2}$$

The paths $x_i(\tau)$ are expanded in Fourier series, the constraint $x_{i=1}(\tau) = x_{i=N+1}(\tau)$ for a DNA loop is implemented and, for the model in Fig. 1, the *bps* fluctuations in the path integral formalism are given by:

$$\eta_i(\tau) = \left[R^2 + x_i(\tau)^2 + 2R|x_i(\tau)|f(\theta_i, \phi_i)\right]^{1/2}$$
$$f(\theta_i, \phi_i) = \sin\theta_i \cos^2\phi_i + \sin^2\phi_i\,. \qquad (3)$$

This new mapping technique lets τ as an integration variable and permits to set a realistic rise distance between adjacent nucleotides along the stack. Written in terms of the $\eta_i(\tau)$, the Hamiltonian contains: *i)* a Morse potential, $V_M[\eta_i(\tau)]$, describing the effective hydrogen bond interaction between *bps* mates, *ii)* a solvent term, $V_{sol}[\eta_i(\tau)]$, accounting for hydrogen bond recombination with the counterions dissolved in water, *iii)* a (two particles) stacking potential, $V_S[\eta_i(\tau), \eta_{i-1}(\tau)]$, between adjacent bases along the strand. The potential parameters are widely discussed in Refs. [4, 5, 6]. Accordingly, the classical action $A[\eta_i]$ is a τ-integral of the base pair *kinetic plus potential* energies [9] and the partition function Z_N for a sequence with N *bps* reads:

$$Z_N = \prod_{i=1}^{N} \oint Dx_i \sum_{\theta_S, \phi_S} \exp\left\{-\beta A[\eta_i]\right\}$$
$$A[\eta_i] = \sum_{i=1}^{N} \int_0^{\beta} d\tau \left[\frac{\mu}{2}\dot{\eta}_i^2(\tau) + V_M[\eta_i(\tau)] + V_{sol}[\eta_i(\tau)] + V_S[\eta_i(\tau), \eta_{i-1}(\tau)]\right], \qquad (4)$$

where $\mu = 300\, amu$ is the *bp* reduced mass. The measure Dx_i is a multiple integral over the path Fourier coefficients while $\oint$ indicates that the paths $x_i(\tau)$ are closed trajectories. The integration over the two particles potential greatly enhances the computational times with respect to the previous method [10] but it offers a more realistic model for the base stacking in sequences of any length.

3. Free Energy of Minicircles

There has been a widespread interest in the properties of small sequences of DNA following measurements of high cyclization probabilities which have pointed to an intrinsic flexibility for fragments of ~ 100 *bps* or less [11]. It has been suggested that the formation of local disruptions in the helix of DNA loops may be the mechanism which permits to release the torsional stress and energetically favors the stability of bent molecules. Certainly, the breaking of some *bps* and the opening of fluctuational bubbles change the average pitch of the helix, that is the number of *bps* per helix turn.

In the theoretical framework synthesized by Eq. (4), I compute the free energy ($F = \beta^{-1} \ln Z_N$) of three heterogeneous loops with different length, $N = 18, 66, 86$, but similar content of AT-*bps* and GC-*bps*, $\sim 50\%$ each. The rise distance is pinned to the experimental value. The $66-$ and $86-$ *bps* loops have been prepared as described in Ref.[12] whereas the $N = 18$ *bps* is a toy sequence here introduced for comparison. For each loop, we simulate a broad set of values for the twist Tw and, for any Tw (i.e., h), Z_N is calculated by summing over an ensemble of fluctuational paths representing a large number of molecule configurations, about 10^7 paths for each base pair in Eq. (4).

Hence, the free energy is obtained as a function of the helical repeat h. The computational time for a simulation, e.g. for the $N = 66$ sequence, is about eight days on a workstation (Intel Xeon E5-1620 v2, 3.7GHz processor). The room temperature results for F/N are plotted in

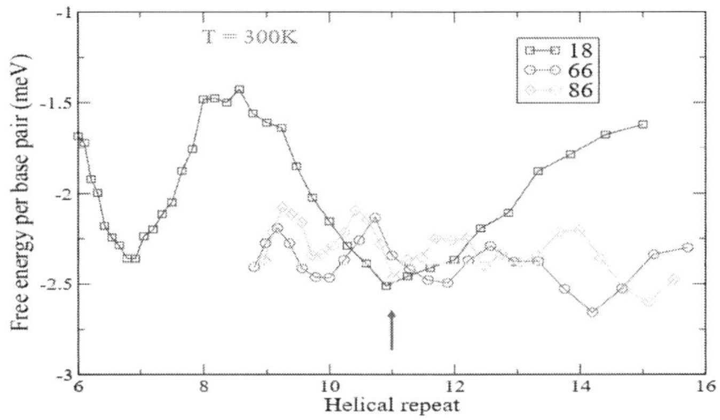

Figure 2. (Color online) Free energy per base pair for three loops with $N = 18, 66, 86$. The free energies are computed, at room temperature, as a function of the number of *bps* per helix turn.

Fig. 2. While the shortest loop shows a free energy minimum also at $h \sim 7$, the minima for all loops are remarkably found for h in the range $\sim (10 - 12)$ in line with the well known values of helical pitch in DNA sequences [1]. Even more interestingly, the 66− and 86− minicircles also show free energy minima at larger h due to a spontaneous unwinding of the complementary strands. This Tw reduction is consistent with the observations [12] of *bps* disruptions occuring in small loops as a consequence of the strong rotational deformations of the helix. The bending stress decreases by increasing the radius of the loop.

Then, for an ensemble of molecules with N *bps*, the free energy minimization evaluates the thermodynamically stablest values of helical repeat, that is an average property of the molecule ensemble. Furthermore, the path integral method can also determine the probabilities for the formation of fluctuational bubbles and select those base pairs along the DNA sequence for which hydrogen bond breaking is more likely to occur. By tuning the system temperature and the input parameters which control the counterions concentration in the solvent, we thus obtain a general and reliable computational scheme for the modeling of heterogeneous DNA loops in various ambient conditions.

References

[1] Calladine C R, Drew H R (1992) *Understanding DNA*, Academic Press, San Diego, USA
[2] Zoli M (2009) *Phys. Rev. E* **79** 041927
[3] Zoli M (2010) *Phys. Rev. E* **81** 051910
[4] Zoli M (2011) *J. Chem. Phys.* **135** 115101
[5] Zoli M (2011) *Eur. Phys. J. E* **34** 68
[6] Zoli M (2012) *J. Phys.: Condens. Matter* **24** 195103
[7] Zoli M (2013) *J. Phys.: Conf. Ser.* **410** 012038
[8] Zoli M (2013) *J. Chem. Phys.* **138** 205103
[9] Zoli M (2014) *J. Theor. Biol.* **354** 95-104
[10] Zoli M (2014) *Soft Matter* **10** 4304-4311
[11] Cloutier T E, Widom J (2004) *Mol. Cell* **14** 355-362
[12] Du Q, Kotlyar A, Vologodskii A (2008) *Nucl. Acids Res.* **36** 1120-1128

Modeling and simulation of defects detection in conductive multi-layered pieces by the eddy current technique

S Bennoud [1]* and M Zergoug[2]

[1]. Laboratory of Aircrafts, University of Blida 1, Blida, Algeria.
[2]. Center of Welding and Control CSC, Cheraga, Algeria.

bensalimen2006@yahoo.fr

Abstract. It has been shown that the eddy current method is one of the most effective techniques for the detection and characterization of surface and near-surface defects in conductive mediums especially in aluminum alloy. It is one of the most applied methods in industries which require a maximum of reliability and security (aerospace, aeronautics, nuclear, Etc).
In this study, a code to solve electromagnetic problems by employing the finite element method is developed.
The suggested model can simulate the probe response to the presence of a defect hidden in a multi-layered structure or a riveted structure on aluminum alloy. The developed code is based on the discretization in three dimensions of the Maxwell's equations in harmonic mode by the finite element method based on the combined potential formulations. That will enable us to interpret the results, to present them in graphical form and to carry out simulations for various applications

Keyword: Non destructive testing, eddy current, finite element method, aluminum parts, numerical simulations, inversing problem.

1. Introduction

The insurance of operation of the aircraft in the event of suspect or presence of damage requires the determination of critical elements in its structure. These elements should be identified during the design of the aircraft, and also during in-service exploitation.

Riveted joints have been frequently used in aeronautics for a long time. They present an important subject for much recent researches. This is due to the fact that very often rivet holes are places where fatigue cracks, corrosion and other defects occur; the last are crucial for the fatigue life and safe operation requirements [1, 2]. These defects are critical problems of in-service aircraft structures which lead to a degradation of structure integrity and fatigue resistance and directly affect the airworthiness of an aircraft and even result in loss of fuselage skin sections [3].

Eddy current testing can be used for a variety of applications such as detection of cracks (discontinuities), measurement of metal thickness, detection of metal thinning due to corrosion and erosion, determination of coating thickness and the measurement of electrical conductivity and magnetic permeability. Eddy current technique is an excellent method for detecting surface and near surface defects when the probable defect location and orientation is well known. In aeronautics for example, eddy current testing is used for over 50% of all applications for the detection of hidden defects in fuselage skins and multi-layers. [4]- [6].

In this work, we measured the magnetic signal due to the defects inside a thick aluminum plate and compared our results with the theoretical calculation. From calculated parameters, we deduce the real and imaginary components of the impedance which makes possible to determine the characteristic parameters of a crack in various metallic parts. With this method, we try to detect crack defects in the bottom layer of lap-joined aluminum plates.

2. Modeling

Cracks and corrosion frequently develop in the fuselage, often in hidden layers close to rivets (figure1).

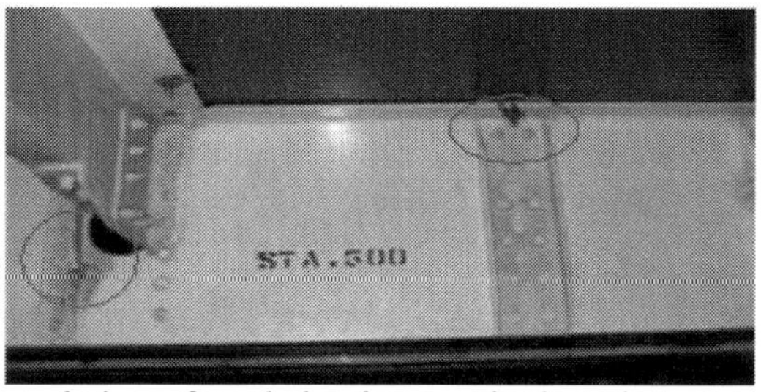

Figure 1. Internal view of crack development in riveted structure (longeron).

Modeling and simulations of eddy current technique using the numerical models of the finite element method (FEM) in order to found codes able to solve Maxwell's equations have been developed in different papers in the lasted years [7]-[9].

In this paper, the related direct problem consisting of modeling the impedance coil response over a multilayered structure used in eddy current testing is treated. Following the approach developed in [10, 11] for planar multilayer medium with constant electrical and magnetic properties, the numerical formulation of FEM is well established and the two main steps for establishing the model are determining the vector potential ***A*** and then calculating the coil impedance **Z**.

For a homogeneous extended conductor placed in a time-varying external magnetic field, with a frequency that is low enough that displacement currents and finite propagation velocity effects can be neglected. Using the formulation in potentials based on the introducing of two potentials, the magnetic vector potential ***A*** and the electric scalar potential *V*, the eddy current problem in terms of the magnetic vector potential can be described mathematically by the following equation (more information and details have been shown in [4, 5, 7, 9]):

$$\nabla^2 A + K^2 A = -\mu J \tag{1}$$

where A represents the magnetic vector potential, μ is the magnetic permeability of the tested domain, J is the excitation current density, $K^2=-j\omega\mu(\sigma+j\omega\varepsilon)$, ω is the angular frequency of the excitation current, σ is the electrical conductivity and ε is the dielectric constant.

Based on the Galerkin's method where irregular meshes of tetrahedral elements were used, the finite element formulation of (1) can be developed.

After some usual mathematical manipulations, the approximation of nodal values results in equations and elemental contributions of the solution can be calculated and summed into a global system of equations:

$$[K]\{A\} = \{Q\} \tag{2}$$

where [K] is the N×N banded symmetric complex global matrix (N is the total number of nodes) , and {Q} and {A} are respectively the Nx1 complex source matrix and the N×1 complex vector of unknowns.

Khalestki algorithm is applied to resolve this system of equations, taking advantage of the symmetry and bandwidth, to solve for A at the nodes of the finite element mesh.

From the ***A*** values, other quantities can be calculated such as flux densities. The dissipated energy ***P*** in the conductor and stored energy ***W*** in the whole solution domain can be calculated [4, 5, 7, 9].

The change in resistance **R** is associated with dissipated energy **P** and the change in inductance **L** is associated with the stored energy **W.**

The measurement model for coil impedance can be expressed in terms of changes in resistance and inductance in its complex form such as:

$$Z=R+j\,\omega\,L \tag{3}$$

3. Application and results

In the paper, we want to show how the developed program could be used in simulations of the non-destructive testing of aeronautical pieces.

For the simplicity we consider only an axisymmetric problem and the algorithm will be applied to some plated aluminum samples.

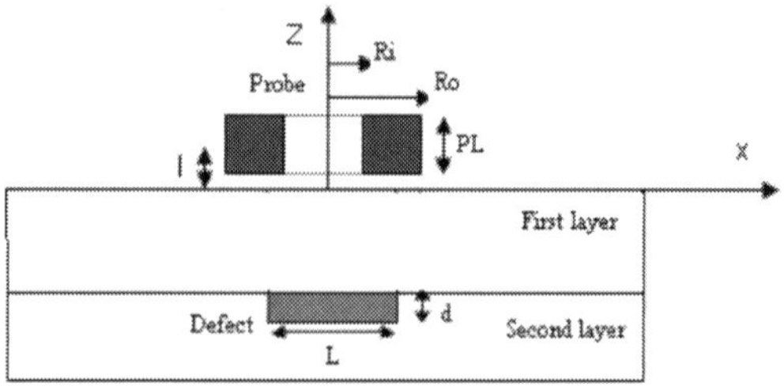

Figure 2. Test sample of second layer cracks.

The test sample shown in figure 2 is made of two aluminum layers. Each layer is 200 mm x 200 mm and 2.5 mm thick, the electrical conductivity is 17 MS/m and the relative permeability is 1.

A rectangular shaped defect is located on the surface of the second layer with length (*L*) of 10 mm, width (w) of 0.2 mm and depth (d) of 1 mm.

The coil properties and test parameters are inner radius (Ri) of 6 mm, outer radius (Ro) of 12 mm, probe length (PL) of 12 mm, number of turns (N) of 900, liff-off (l) of 0.1 mm, skin depth of 6 mm and frequency of 500 Hz.

In order to detect cracks with a known orientation in the sample (scan was carried out along the defect length that correspond to the axis Ox), it is convenient if the current can be injected (or induced) in two orthogonal directions. For a sample with rectangular geometry (see figure 1), it is easier to inject a uniform current longitudinally than transversely. In this simulation, the results obtained when a transverse current is induced in the test sample by a probe, carrying a 50 mA 500 Hz current that was located below the test sample are discussed.

Figure 3 shows the relationship of the impedance change and the positions of the probe. It can be seen that the impedance curve has a peak when defect is present, and the distance between the peak-start and peak-end points is just the approximate length of the defect.

The maximum amplitude appears as the probe is positioned over the centre of the defect.

The developed code can carry out to study the influence of certain parameters. In this part, simulation work is carried out for defects, the test parameters are fixed (the same parameters quoted in previous paragraph) and the frequency is modified, using the developed program.

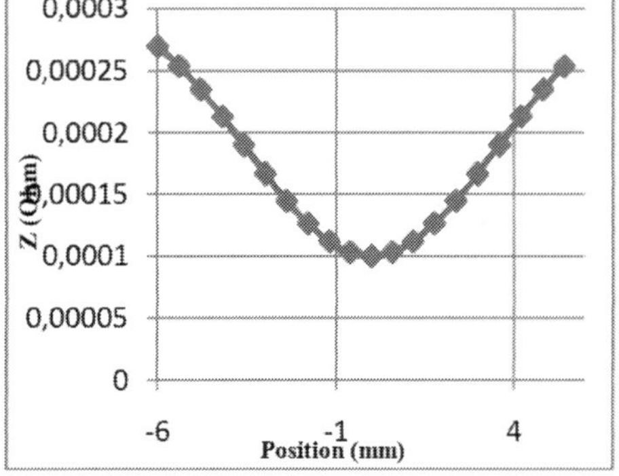

Figure 3. Impedance versus probe position.

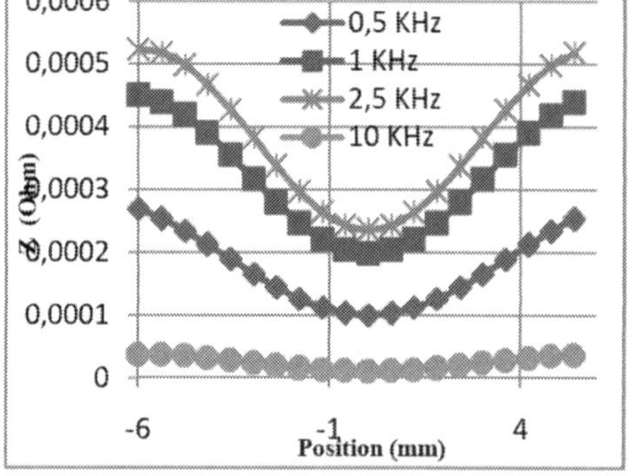

Figure 4. Defects detection of various frequencies.

For cracks located deeper in the conductor, the optimum excitation frequency moves towards lower values (figure 4).

Factors such as the type of material, surface finish and condition of the material, the design of the probe, and many other factors can affect the accuracy of the inspection.

4. Conclusion

Eddy current techniques are widely used for detection of surface cracks in metallic structures. These techniques have limited success in the detection of deep and subsurface defects that require low frequency eddy currents.

In this work, we calculated and measured the normal component of the magnetic fields due to a uniform current injected into a two aluminum plates assembled together that contains defects. From this parameter, we deduce the real (resistance) and imaginary (inductance) components of the impedance which makes possible to determine the characteristic parameters of a crack in various metallic parts.

Obtained results show that this method is adapted to selected problem. And from simulation, we see that the parameters of defects can be characterized according to the impedance amplitude.

Furthermore, the results also show that feasible to use the developed code to provide the forward solutions for the eddy current inversion problem which is sizing of inspected defects by scanning multi-layered structures, and to carry out simulations for other geometries of materials with different proprieties in order to study the limits of the chosen technique.

5. References

[1] Müller R P G, 1995, *An experimental and analytical investigation on the fatigue behavior of fuselage riveted lap joints. The significance of the rivet squeeze force, and a comparison of 2024-T3 and Glare 3.* (Ph.D. Thesis, Delft University of Technology).

[2] Rans C D, 2007, *The role of rivet installation on the fatigue performance of riveted lap joint.* (Ph.D. Thesis, Carleton University Ottawa, Ontario).

[3] Brooks C L, Honeycutt K and Prost-Domsky S, 2000, Case studies for life assessments with age degradation. *The Fourth Joint DoD/FAA/ NASA Conference on Aging Aircraft*, (May 2000, St. Louis,MO, USA).

[4] Bennoud S and Zergoug M, 2014, Modeling and simulation for 3D eddy current testing in conducting materials, *International Journal of Mechanical, Aerospace, Industrial and Mechatronics Engineering* **8(4)**, 757-760.

[5] Bennoud S, Zergoug M and Allali A, 2014, Numerical simulation for cracks detection using the finite elements method, *International Journal of Multiphysics (IJM)*. **8(1)**, 1-10.

[6] Helifa B, Oulhadj A, Benbelghit A, Lefkaier I K, Boubenider F and Boutassouna D, 2006. Detection and measurement of surface cracks in ferromagnetic materials using eddy current testing. *NDT & E international*, **39(1)**, 384-390.

[7] Monk P, 2003, *Finite element methods for Maxwell's equations,* (Oxford University Press, New York).

[8] Tikhonova D, Zentnerb J and Tkacheva A, 2004, Algorithm for simulation of electromagnetic processes using combined finite and boundary elements method, *Journal of Magnetism and Magnetic Material*, 662-664.

[9] Bennoud S and Zergoug M, 2014, Finite element modeling and analysis of electromagnetic field distribution for different shapes of metallic materials, *WULFENIA Journal*, **21(7)**, 318-326.

[10] Antimirov M Ya, Kolyshkin A A, and Vaillancourt R, 1997, *Mathematical models for eddy current testing*. Montreal: CRM.

[11] Theodoulidis T, and Kriezis E E, 2006, *Eddy current canonical problems (with applications to nondestructive evaluation)*. New York: Tech Science.

Establishing the number of distinct stabilizer bases for a quantum qudit error-correcting code

C M Wilmott

Department of Mathematics and Physics, Nottingham Trent University, Nottingham NG11 8NS, UK

E-mail: colin.wilmott@ntu.ac.uk

Abstract. The class of quantum codes called stabilizer codes is increasingly well-understood. The premise of the stabilizer formalism is that a quantum code can be efficiently described by a subgroup of its error group, and, interestingly, the stabilizer formalism permits correspondences with classical linear codes. In this paper, we examine one such correspondence, and we shall use this to establish the number of distinct stabilizer codes that exist for a fixed parametrisation.

1. Introduction

Coding theory is a branch of mathematics which seeks optimal solutions to problems concerning the safe and accurate transfer of information. Shannon (1948) established the topic of coding theory with his seminal paper on the mathematics of communication. Hamming (1950) then introduced the concept of an error-correcting codes with his work on the correction of errors on magnetic storage media. In the intervening decades, our society has become phenomenally technology-orientated and coding theory has played a prominent role in this change.

In recent years, coding theory has evolved beyond its original classical setting and is considered within a quantum theoretical perspective. Quantum stabilizer codes were introduced independently by Gottesman (1997) and Calderbank *et al* (1998), and rank among the most widely studied of all quantum error-correcting codes. The premise of the stabilizer formalism is that a quantum code can be described by a subgroup of its error group, and it is this subgroup which we refer to as the stabilizer of a quantum code. In this paper, we shall provide a brief overview of the stabilizer formalism before establishing the number of distinct stabilizer codes for a fixed parametrisation.

2. Preliminaries

Consider a d-dimensional Hilbert space $\mathbb{C}^d$ and fix each orthonormal basis state of the space to correspond to an element of ring $\mathbb{Z}_d$ of integers modulo d. As such, we have the basis $\{|0\rangle, |1\rangle, \ldots, |d-1\rangle\} \subset \mathbb{C}^d$ which we call the computational basis whose elements correspond to the column vectors of the identity matrix $\mathbb{I}_d$. A qudit is a d-dimensional quantum state $|\psi\rangle \in \mathbb{C}^d$ which can be written as $|\psi\rangle = \sum_{i=0}^{d-1} \alpha_i |i\rangle$, where $\alpha_i \in \mathbb{C}$ and $\sum_{i=0}^{d-1} |\alpha_i|^2 = 1$. The state space of an n-qudit is formed by taking the n-fold tensor product of the principal space $\mathbb{C}^d$, $(\mathbb{C}^d)^{\otimes n}$, which, correspondingly, possesses a set of orthonormal basis states that can given by $|i_1\rangle \otimes |i_2\rangle \otimes \cdots \otimes |i_n\rangle = |i_1 i_2 \ldots i_n\rangle$ for $i_j \in \mathbb{Z}_d$. The general state of a qudit in the n-fold space $\mathbb{C}^{d^n}$ is

$$|\psi\rangle = \sum_{(i_1 i_2 \ldots i_n) \,\in\, \mathbb{Z}_d^n} \alpha_{(i_1 i_2 \ldots i_n)} |i_1 i_2 \ldots i_n\rangle , \quad (1)$$

where $\alpha_{(i_1 i_2 \ldots i_n)} \in \mathbb{C}$ and $\sum |\alpha_{(i_1 i_2 \ldots i_n)}|^2 = 1$.

A basis for the set of bounded operators acting on $\mathbb{C}^d$ is given by the set of well-known generalised Pauli matrices $\mathcal{E} = \{X_i Z_j \mid (i,j) \in \mathbb{Z}_d \times \mathbb{Z}_d\}$. Any error E acting on a qudit state can be written as a linear combination of the generalised Pauli matrices. The generalised Pauli group, or error group, $\mathcal{G}$ is a group of order d^4 generated by $\mathcal{E}$ and τI with center $\zeta(\mathcal{G}) = \langle \tau I \rangle$. For an n-qudit system, any element $E_{i,j}$ of $\mathcal{G}^n$ may be written as $E_{i,j} = \tau^\alpha (X_{i_1} Z_{j_1}) \otimes (X_{i_2} Z_{j_2}) \otimes \cdots \otimes (X_{i_n} Z_{j_n})$ where $\alpha \in \mathbb{Z}_d$ and $((i_1,j_1),(i_2,j_2),\ldots,(i_n,j_n)) \in \mathbb{Z}_d^n \times \mathbb{Z}_d^n$. Since the group $\mathcal{G}^n/\langle \tau I \rangle$ is isomorphic to vector space $\mathbb{F}_d^{2n}$ with order d^{2n}, this allows us to write a correspondence between $\mathcal{G}^n/\langle \tau I \rangle$ and $\mathbb{F}_d^{2n}$ as

$$\begin{aligned} E_{ij} &= \otimes_{z=1}^n X_{i_z} \cdot \otimes_{z=1}^n Z_{j_z} \\ &\equiv (i_1, i_2, \ldots, i_n | j_1, j_2, \cdots, j_n) = (i|j). \end{aligned} \quad (2)$$

A quantum code consists of an encoding function e from the Hilbert space $(\mathbb{C}^d)^{\otimes k}$ to the Hilbert space $(\mathbb{C}^d)^{\otimes n} \equiv \mathbb{C}^{d^n}$, e: $\mathbb{C}^{d^k} \to \mathbb{C}^{d^n}$ where k and n are integers and $k < n$. In a manner similar to the classical case, we define the codewords of a quantum code to be those states contained in the image of e, Im(e). The length of the code is given by n while k denotes the number of encoded message qudits of the code. The extra $n - k$ qudits introduce additional information that allows the encoded qudits to be stored in a redundant manner which can then be later used to detect transmission errors. A code $\mathcal{Q}$ is a quantum $[\![n,k]\!]_d$ code over $\mathbb{C}^d$ if it is a subspace of dimension d^k in $\mathbb{C}^{d^n}$. For our purposes, it is not necessary to introduce minimum distance.

3. Stabilizer Codes

Let $\mathcal{Q}$ be a quantum error correcting code. The stabilizer $\mathcal{S}$ of $\mathcal{Q}$ is a formalism that describes a quantum code in terms of error operators acting on $\mathcal{G}^n$. More precisely, the stabilizer of $\mathcal{Q}$ is defined to be a set of operators $\mathcal{M} \in \mathcal{S}$ of $\mathcal{G}^n$ for which the condition

$$\mathcal{M} |\psi\rangle = |\psi\rangle \quad (3)$$

is satisfied for all codewords $|\psi\rangle$ (Gottesman 1997). Thus, the stabilizer maintains a common +1-eigenspace of codespace $\mathcal{Q}$. Stabilizer codes themselves were originally considered over dimension two, and in this situation, errors are said to either commute or anti-commute with a codeword. It is exactly this property that helps stabilizer codes elicit a relatively straight-forward error detection procedure. In particular, the set of all operators that commute with the stabilizer is called the centralizer. Error detection is then explained by the fact that any error which lies outside of the centralizer necessarily anti-commutes with some elements of the stabilizer. Note that since stabilizer codes were initially based on qubit systems, the stabilizer is necessarily an Abelian subgroup while the centralizer is referred to as the normalizer of the group. We shall however be considering more general systems based on qudits.

Definition 1 *A stabilizer code $\mathcal{Q}$ is a subspace of $\mathbb{C}^{d^n}$ that satisfies the relation*

$$\mathcal{Q} = \bigcap_{\mathcal{M} \in \mathcal{S}} \{|\psi\rangle \in \mathbb{C}^{d^n} \mid \mathcal{M} |\psi\rangle = |\psi\rangle\} \quad (4)$$

for some subgroup $\mathcal{S}$ of $\mathcal{G}^n$.

The above definition is a generalisation of the usual definition of stabilizer codes based on qubits. Since $\mathcal{S}$ describes the set of operators that leave each state in the quantum code invariant, $\mathcal{S}$ is therefore said to stabilize the code $\mathcal{Q}$.

4. Stabilizer Equivalence

4.1. A classical coding problem

In classical coding theory, an interesting problem is to determine the number of different generator bases that can be constructed for a linear code for fixed parameters. We will concern ourselves with this problem before asking if a comparable statement can be made for stabilizer codes — these being quantum analogue of linear classical codes.

Let F be a field. A classical code $\mathcal{C}$ of length n is a linear code if and only if $\mathcal{C}$ is a subspace of the vector space F^n. If $\mathcal{C}$ has dimension k over F, then we say that $\mathcal{C}$ is an $[n, k]$ linear code over F. Therefore, $\mathcal{C}$ can be specified by a basis consisting of a minimal set of vectors $v_1, v_2, \ldots, v_k$ such that $\mathcal{C} = \{\sum_{i=1}^{k} \alpha_i v_i \mid \alpha_i \in F\}$. Setting $F = \mathbb{F}_2$, we have the following.

Theorem 1 *(Hoffman et al 1991) A binary linear code of dimension k has precisely*

$$\frac{1}{k!}\prod_{i=0}^{k-1}(2^k - 2^i)$$

different generator bases.

4.2. A quantum coding problem

Motivated by theorem 1, let us now consider the problem of determining the number distinct $[\![n, k]\!]_d$ stabilizer codes for fixed n, k and d. We have the following.

Theorem 2 *The number of stabilizer sets in the Hilbert space $\mathbb{C}^{d^n}$ that maintain a +1-eigenspace of dimension d^k is given by*

$$d^{\frac{n-k-1(n-k)}{2}}\prod_{i=0}^{n-k-1}\left(d^{2(n-i)} - 1\right).$$

Proof. The error group associated with an n-qudit system is the n-fold product $\mathcal{G}^n = \{E_{ij} \mid (i, j) \in \mathbb{Z}_d^n \times \mathbb{Z}_d^n\}$. To count the number of elements within $\mathcal{G}^n$ that maintain a +1-eigenspace, note that the first such element $\mathcal{M}'_1$ can be chosen in $d^{2n} - 1$ ways. Let $\mathcal{M}'_2$ be another element of $\mathcal{G}^n$ that commutes with $\mathcal{M}'_1$ but is independent from $\mathcal{M}'_1$. Then it can be shown that there exists an $N'_2 \in \mathcal{G}^n$ that commutes with $\mathcal{M}'_1$ but fails to commute with $\mathcal{M}'_2$ (Gottesman 1997). Hence, M'_2 maintains the +1-eigenspace of $\mathcal{M}'_1$. Thus, the number of choices for $\mathcal{M}'_2$ that maintain a +1-eigenspace with $\mathcal{M}'_1$ is $\frac{d^{2n}}{d} - d$. In a similar fashion, we note that for $\mathcal{M}'_l \neq \{\mathcal{M}'_1, \mathcal{M}'_2, \ldots, \mathcal{M}'_{l-1}\}$, a suitable N'_l can be found such that N'_l commutes with $\mathcal{M}'_1, \mathcal{M}'_2, \ldots, \mathcal{M}'_{l-1}$ but does not commute with $\mathcal{M}'_l$. Hence, the number of choices for $\mathcal{M}'_l$ that preserve the +1-eigenspace of the set $\{\mathcal{M}'_i\}$, $i = 1, \ldots, l-1$, is $\frac{d^{2n}}{d^{l-1}} - d^{l-1}$. Therefore, the total number of ways of choosing $n-k$ stabilizers $\mathcal{M}'_i$ for a code $\mathcal{Q}$ is given as

$$\begin{aligned}\prod_{i=0}^{n-k-1}\left(\frac{d^{2n}}{d^i} - d^i\right) &= d^{\sum_{i=0}^{n-k-1} i}\prod_{i=0}^{n-k-1}\left(d^{2(n-i)} - 1\right) \\ &= d^{\frac{n-k-1(n-k)}{2}}\prod_{i=0}^{n-k-1}\left(d^{2(n-i)} - 1\right). \end{aligned} \quad (5)$$

Theorem 3 *An $[\![n, k]\!]_d$ stabilizer code $\mathcal{Q}$ has precisely*

$$d^{\frac{(n-k-1)(n-k)}{2}}\prod_{i=0}^{n-k-1}\left(d^{(n-i)} - 1\right) \quad (6)$$

possible choices for stabilizer sets which maintain its +1-eigenspace.

Proof. The codewords associated with a stabilizer code $\mathcal{Q}$ are given by

$$|\overline{c_1 \dots c_k}\rangle = \overline{X}_1^{c_1} \dots \overline{X}_k^{c_k} \frac{1}{|\mathcal{S}|} \sum_{\mathcal{M} \in \mathcal{S}} \mathcal{M} \, |00 \dots 0\rangle \tag{7}$$

(Gottesman 1997). As the stabilizer code of dimension d^k is defined in terms of a +1-eigenspace $\mathcal{S}$, there are d^{n-k} stabilizer elements that satisfy this definition in a non-trivial manner. There are a further d^k elements associated with the logical encoding process $\bar{X}$ and, thus, there are a total of d^n elements to choose from in order to maintain the code's +1-eigenspace. Counting the numbers of ways that the stabilizer elements $\mathcal{M}_i, i = 1, \dots, n-k$, can be chosen for a particular stabilizer code, we note there are $d^n - 1$ non-trivial choices that can be made for the first stabilizer $\mathcal{M}_1$. $\mathcal{M}_2$ can then chosen independently in $d^n - d$ ways, followed by $d^n - d^2$ choices for $\mathcal{M}_3$. In particular, $\mathcal{M}_l$ can be chosen to preserve the +1-eigenspace of $\mathcal{Q}$ in $d^n - d^{l-1}$ ways. Since we require $n-k$ stabilizers in total, there are

$$\begin{aligned} \prod_{i=0}^{n-k-1} \left(d^n - d^i\right) &= d^{\sum_{i=0}^{n-k-1} i} \prod_{i=0}^{n-k-1} \left(d^{(n-i)} - 1\right) \\ &= d^{\frac{(n-k-1)(n-k)}{2}} \prod_{i=0}^{n-k-1} \left(d^{(n-i)} - 1\right) \end{aligned} \tag{8}$$

choices for elements $\mathcal{M}_i \in \mathcal{S}$ that maintain the +1-eigenspace of $\mathcal{Q}$.

For fixed parameters n and k and d, we can now establish the number of distinct $[\![n,k]\!]_d$ stabilizer codes within $\mathbb{C}^{d^n}$.

Theorem 4 *The number of distinct* $[\![n,k]\!]_d$ *stabilizer codes over* $\mathbb{C}^{d^n}$ *is*

$$\prod_{i=0}^{n-k-1} (d^{(n-i)} + 1).$$

Proof. The quotient obtained from equations (5) and (8) is given by

$$\prod_{i=0}^{n-k-1} \left(\frac{d^{\frac{n-k-1(n-k)}{2}} d^{2(n-i)} - 1}{d^{\frac{n-k-1(n-k)}{2}} d^{(n-i)} - 1} \right) = \prod_{i=0}^{n-k-1} \left(d^{(n-i)} + 1\right) \tag{9}$$

and the result now follows.

References

Calderbank A R, Rains E M, Shor P W and Sloane N J A (1998) Quantum Error Correction via Codes over GF(4) *IEEE Trans. Inform. Theory* **44** pp 1369-87

Gottesman D (1997) Stabilizer Codes and Quantum Error Correction *Ph.D. Thesis Calif. Inst. Tech. Pasadena CA*

Hamming R W (1950) Error Detecting and Error Correcting Codes *Bell System Technical Journal* **29** pp 147-60

Hoffman D G, Leonard D A, Lidner C C, Phelps A T, Rodger C A and Wall J R 1991 *Coding Theory: The Essentials* (Marcel Dekker) p X

Shannon C E (1948) A Mathematical Theory of Communication *Bell System Technical Journal* **27** pp 379-423

From Fibonacci to the mathematics of cows and quantum circuitry

C M Wilmott

Department of Mathematics and Physics, Nottingham Trent University, Nottingham NG11 8NS, UK

E-mail: colin.wilmott@ntu.ac.uk

Abstract. The Fibonacci sequence is a famously well-known integer sequence from the thirteenth century which has transcended its original motivation. It possesses many interested and varied applications within architecture, engineering and science. Less well known is the Narayana sequence which itself has interesting and wide-ranging Fibonacci-type connections. In this paper, we shall recall Narayana's original motivation that gives rise to the sequence bearing his name. We also provide an interesting application of this sequence to the construction to quantum gate circuitry used in quantum computation.

1. Introduction

A celebrated feature of mathematics is the way in which it perpetually demonstrates good bridge-building capabilities between seemingly unrelated topics of study. An example of this extraordinary feature relates to the class of sequences called integer sequences which derives its name from the property that its sequence terms are integers. The terms of an integer sequence may be specified explicitly with respect to a formula, or implicitly through a recurrence relation (Sloane 1973). Integer sequences can be analyzed by a variety of techniques, some of which include the application of a data compression algorithm (Bell *et al* 1990), computation of the discrete Fourier transform (Loxton 1989), and evaluation of its generating function (Wilf 2005). Additionally, there are also a large number of transformation methods which can be applied to integer sequences, including the Euler transform, exponential transform and Möbius transform.

The Fibonacci sequence is an integer sequence that itself displays good bridge-building capabilities. Each term in this sequence is produced by adding together the two previous terms and the sequence takes its name from the famous thirteenth century mathematician Fibonacci, whose use he explained in his 1202 Liber abaci (Sigler 2002). The Fibonacci sequence is a sequence that is well-understood. The famous astronomer Johannes Kepler observed that the ratio of consecutive Fibonacci terms approached the famous golden ratio. It is through this connection that the Fibonacci sequence is widely recognized in art and architecture. Interestingly, and equally well-known is that the Fibonacci sequence also helps explain patterns that arise within biological contexts. The applications and important consequences that arise from the Fibonacci sequence demonstrate the this sequence has transcended it original motivation (Liber abaci). We now ask whether there exist other Fibonacci-like sequences possessing important applications which transcend the original motivations of the sequence in question.

Figure 1. Circuit descriptions for the CNOT gate types. (a) The CNOT1 gate; the control system $|m\rangle \in \mathcal{H}_{\mathcal{A}}$ remains unchanged after application whereas the state of the target system $|n\rangle \in \mathcal{H}_{\mathcal{B}}$ is transformed under modular arithmetic to the state $|n \oplus m\rangle$ with $m, n \in \mathbb{Z}_d$. (b) The CNOT2 gate in which the roles of systems $\mathcal{H}_{\mathcal{A}}$ and $\mathcal{H}_{\mathcal{B}}$ are reversed.

2. Narayana's Cows

Narayana was a fourteenth century Indian mathematician who gives name is given to a Fibonacci-inspired problem based on an idealized population of cows. For the sake of completeness, we shall state this problem (Waldschmidt 2009): A cow produces one calf every year. Then, beginning in its fourth year, each calf produces one calf at the beginning of each year. How many cows are there altogether after seventeen years? The Narayana sequence overlaps closely with that of Fibonacci; indeed, its nth-term is defined as

$$a(n) = a(n-1) + a(n-3) \tag{1}$$

with initial conditions $a(0) = a(1) = a(2) = 1$. For what follows, it will be useful to enumerate initial values of the Narayana sequence:

$$1, 1, 1, 2, 3, 4, 6, 9, 13, 19, 28, \ldots \tag{2}$$

The online encyclopedia of integer sequences records some interesting combinatorial connections offered by the Narayana sequence (Sloane). However, we shall present an application of the Narayana sequence which relates to the construction of optimal gate circuits that can be used for quantum computation. We now present preliminary material which will serve as a basis for our study.

3. Quantum Computation

Crucial for successful quantum computation is the requirement that engineers can build robust multiple-qubit quantum gates. The most elementary of all multiple-qubit quantum gates are given by two-qubit controlled unitary operators, and a classic example of these is the controlled-NOT (CNOT) gate. In terms of a classical perspective, the CNOT gate is the quantum analogue of the classical XOR gate. Barenco *et al* 1995 have shown that any multiple-qubit quantum operation may be restricted to compositions of single-qubit gates and the CNOT gate. It is for this reason that the CNOT gate has acquired the special status as the hallmark of multi-qubit control. Now, researchers in quantum computation have done considerable work in optimizing quantum circuitry networks. An important milestone in this direction related to general two-qubit operations requiring at most three CNOT gates (Vatan and Williams 2004). A crucial aspect of this result is the demand that the qubit SWAP gate requires at least three CNOT gates. Consequently, the SWAP gate has taken a prominent position in many quantum circuitry designs.

3.1. Elementary quantum gates

Let $\mathcal{H}$ represent the d-dimensional complex Hilbert space $\mathbb{C}^d$, and let us fix each orthonormal basis state of the d-dimensional Hilbert space to map to an element of the ring of integers modulo d, $\mathbb{Z}_d$. This yields the basis $\{|0\rangle, |1\rangle, \ldots, |d-1\rangle\} \subset \mathbb{C}^d$ whose elements correspond to the d column vectors of the identity matrix $\mathbb{I}_d$ is called the computational basis. We say a *qudit* is a d-dimensional quantum state $|\psi\rangle \in \mathcal{H}$ which can be expressed as $|\psi\rangle = \sum_{i=0}^{d-1} \alpha_i |i\rangle$ where

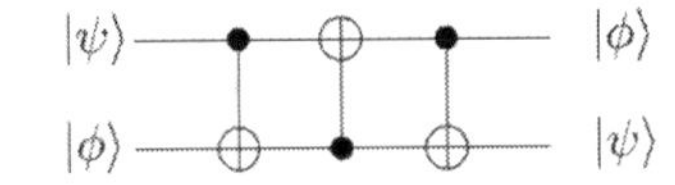

Figure 2. The SWAP gate illustrating the cyclical permutation of two qubits. System $\mathcal{A}$ begins in the state $|\psi\rangle$ and ends in the state $|\phi\rangle$ while system $\mathcal{B}$ begins in the state $|\phi\rangle$ and ends in the state $|\psi\rangle$.

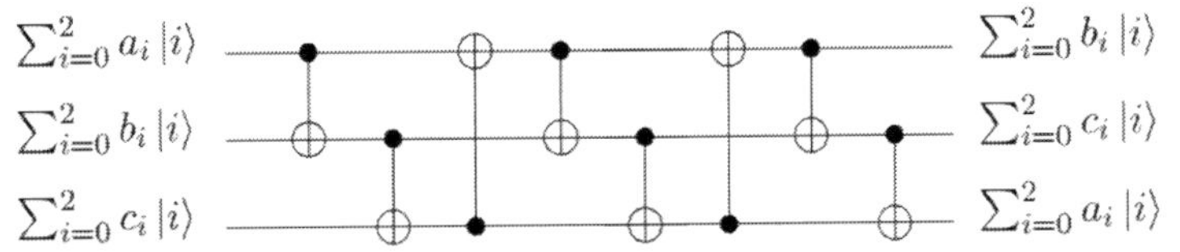

Figure 3. A qutrit SWAP gate that cyclically permutes the states of three qutrits. This gate is composed of eight two-qutrit CNOT gates

$\alpha_i \in \mathbb{C}$ and $\sum_{i=0}^{d-1} |\alpha_i|^2 = 1$. Given d-dimensional Hilbert spaces $\mathcal{H}_\mathcal{A}$ and $\mathcal{H}_\mathcal{B}$, consider the set of $d^2 \times d^2$ unitary transformations $U \in U(d^2)$ that act on the two-qudit quantum system $\mathcal{H}_\mathcal{A} \otimes \mathcal{H}_\mathcal{B}$. Let $U_{\text{CNOT1}} \in U(d^2)$ represent the generalized CNOT gate that has control qudit $|\psi\rangle \in \mathcal{H}_\mathcal{A}$ and target qudit $|\phi\rangle \in \mathcal{H}_\mathcal{B}$. The action of U_{CNOT1} on the set of basis states $|m\rangle \otimes |n\rangle$ of $\mathcal{H}_\mathcal{A} \otimes \mathcal{H}_\mathcal{B}$ is given by

$$U_{\text{CNOT1}} |m\rangle \otimes |n\rangle = |m\rangle \otimes |n \oplus m\rangle\,, \qquad m, n \in \mathbb{Z}_d, \tag{3}$$

with $\oplus$ denoting addition modulo d. Similarly, let $U_{\text{CNOT2}} \in U(d^2)$ denote the generalized CNOT gate having control qudit $|\phi\rangle \in \mathcal{H}_\mathcal{B}$ and target qudit $|\psi\rangle \in \mathcal{H}_\mathcal{A}$. The action of U_{CNOT2} on the set of basis states $|m\rangle \otimes |n\rangle$ of $\mathcal{H}_\mathcal{A} \otimes \mathcal{H}_\mathcal{B}$ is written

$$U_{\text{CNOT2}} |m\rangle \otimes |n\rangle = |m \oplus n\rangle \otimes |n\rangle\,, \qquad m, n \in \mathbb{Z}_d. \tag{4}$$

Fig. 1 provides the quantum gate circuitry representation for the respective CNOT types while Fig. 2 illustrates the well-known SWAP gate that permutes the states of two qubits, for $d = 2$.

4. The Narayana sequence and a qutrit SWAP gate

Let $d = 3$ and consider the following problem. Given three qutrit quantum systems: system $\mathcal{A}$ in the state $|a\rangle$; system $\mathcal{B}$ in the state $|b\rangle$; and system $\mathcal{C}$ in the state $|c\rangle$. Using only instances of the two-qutrit CNOT gate, determine the gate that implements a SWAP of the input states so that system $\mathcal{A}$ ends in the state $|b\rangle$, system $\mathcal{B}$ ends in the state $|c\rangle$, and system $\mathcal{C}$ ends in the state $|a\rangle$. Our construction method is presented in Fig. 3.

Construction Method: Let k and l be positive integers. For non-negative integers j, consider the function

$$f(j) = \binom{j}{k} \bmod l. \tag{5}$$

Integer functions of this type are periodic and we will make special use of this fact in the construction. In particular, we shall focus on the set of modular binomial coefficients given by

$$a_j = \sum_{i=0}^{j/3} \binom{j-2i}{i} \bmod 3 \tag{6}$$

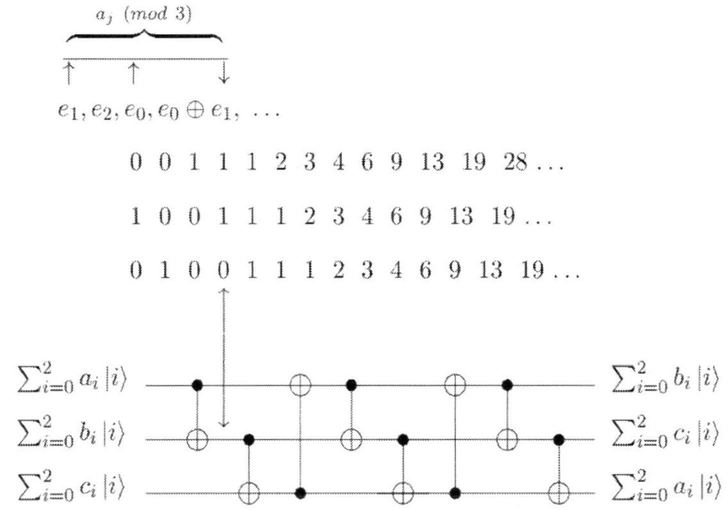

Figure 4. A quantum network composed entirely in terms of CNOT gates illustrating a cyclic SWAP of three 3-dimensional states. The columns of the array above describe the state of the system following the application of respective CNOT gates. In particular, system $\mathcal{A}_1$ is in the state $|e_0 + e_1\rangle_1$ following the application of the first CNOT gate. As the columns of array are periodic (modulo 3), a repeated application of generalized CNOTs on successive pairs of quantum systems induces a network design that eventually permutes the states of 3 qutrits

where we show how these coefficients, the terms of the Narayana sequence, can be used to construction of a regular generalized SWAP gate for qudits.

Fig. 4 describes our circuit design for a generalized SWAP gate for three 3-dimensional quantum states. The design is outlined in terms of a regular sequence of CNOT gates (modulo 3) that act on successive pairs of quantum states. The columns of the array in Fig. 4 describe the states of the target quantum systems after the corresponding CNOT gates have been applied. The up-down arrow between the array and network indicates the correspondence between system $\mathcal{A}_1$ being in the state $|e_0 + e_1\rangle_1$, where $e_0 + e_1$ is calculated modulo 3, and the application of the first CNOT gate. The sum $e_0 + e_1$ is represented as the dot product of the row vector (e_0, e_1, e_2) and the corresponding column vector $(1, 1, 0)^{\mathrm{T}}$. A similar case holds for subsequent CNOT applications. As a result of taking the dot product between regularly repeating column vectors representing three translations of the Narayana sequence modulo 3 and the row vector (e_0, e_1, e_2), we can trace the Narayana sequence to an ordering of eight qutrit CNOT gates which induce the required cyclic SWAP of original qutrit input states.

References

Barenco A, Bennett C H, Cleve R, DiVincenzo D P, Margolus N, Shor P, Sleator T, Smolin J A and Weinfurter H 1995 *Phys. Rev. A* **52** 5 3457

Bell T C, Cleary J G and Witten I H 1990 *Text Compression* (Englewood Cliffs, NJ)

Liber abaci `http://www-groups.dcs.st-andrews.ac.uk/~history/Quotations/Fibonacci.html`

Loxton J H 1989 *Spectral Studies of Automata. Irregularities of Partitions* (ed G Halasz and V T Sos) (New York: Springer-Verlag) pp 115-128

Sloane N J A 1973 *The Handbook of integer sequences* (Boston, MA: Academic Press)

Sloane N J A `https://oeis.org/A000930`

Sigler L E (2002) *Fibonacci's Liber Abaci* (Springer-Verlag)

Vatan F and Williams C 2004 *Phys. Rev. A* **69** 3 032315

Waldschmidt M `https://www.imj-prg.fr/~michel.waldschmidt/AgendaArchives.html`

Wilf H S 2005 *generatingfunctionology* (A K Peters/CRC Press; 3rd ed)

Wilmott C and Wild P 2014 *Quantum Inf. Process.* **13** 6

The Effect of a Non-Isotropic Flux of Very High Energy Cosmic Rays on the values of Mean Shower Maxima

Pantea Davoudifar; Keihanak Rowshan Tabari

Research Institute for Astronomy and Astrophysics of Maragha (RIAAM), P.O.Box:55134-441, IRAN

E-mail: dfpantea@riaam.ac.ir

Abstract. In our previous works we described a statistical method to interpret the results of extensive air shower simulations. For an isotropically distributed flux of cosmic rays, we used this method to deduce diagrams of mean values of shower maxima versus energy decades. To have a more realistic result, we considered the effect of a non-isotropic flux of cosmic rays at different energy ranges. This effect was considered as a weight factor deduced from a set of observed data. We discussed about the effect of this weight factor on our final resulted diagrams of mean shower maxima and for different interaction models compared the resulted distributions of very high energy cosmic ray's mass composition.

1. Introduction

Entering the *geomagnetic field*, charged particles and ultra high energy γ-rays ($E_\gamma \geq 10^{19}eV$) are affected by the *Earth magnetic field*. For instance, the deflection of low energy charged particles in the *Earth magnetic field*, cause the *azimuthal asymmetry* or interaction of ultrahigh energy γ-rays with *geomagnetic field* cause *pre-showering* effect. In an array of cosmic ray's detectors with a horizontal geometry, the *event rate* is also distorted due to *azimuthal event rate modulation* in the *geomagnetic field*.

Obviously, considering a set of simulated Extensive Air Showers (EASs), there is an statistical distribution of *depth of shower maxima*, X_{max}, over *zenith angle* intervals [1]. Using CORSIKA (version 7.37) [2] simulation, we studied this distribution for different hadronic interaction models and re-produced the diagrams of mean *depth of shower maxima* versus energy decades. The resulted diagrams then were used to study the cosmic ray mass composition, in which an *isotropic* flux of cosmic rays were supposed [1].

Anyhow the *zenith angle* dependence of the *arrival rate* of showers of *size*$>N$, $I(>N, X, \theta)$, for most of the experiments were considered to have a form of: $I(>N, X, \theta) = I(>N, X, 0)cos^n\theta[m^{-2}s^{-1}deg^{-1}]$ in which X is the vertical depth of experiment [3]. Though, such a phenomena has not a direct effect on the calculated [1] mean *depth of shower maxima*, $\langle X_{max}\rangle$; but it may cause a *tolerance* on the average values of X_{max} when the statistical fluctuations over *zenith angle* intervals are considered. In this paper, the order of magnitude of this possible *tolerances* are investigated.

2. The Method

For Yakutsk array, different fits to experimental data are available for the energy ranges of $E < 10^{18}eV$, $E > 10^{18}eV$, $E > 10^{19}eV$ and $E > 4 \times 10^{19}eV$ [4, 5]. For this array, it was also shown that the *azimuthal effect* on *event rate* at $E > 10^{17}eV$ is approximately the same in the whole energy range [5].

At the top of atmosphere and in the energy range of $\sim 10^{18} - 5 \times 10^{19}eV$, one may consider the *event rate* per degrees as:

$$n(N, 0, \theta) = n(N, 0, 0)cos^n(\theta)[s^{-1}deg^{-1}] \tag{1}$$

in which $n(N, 0, \theta) \equiv I(> N, 0, \theta) \times (1\ m^2)$ is the *event rate* per degree at the top of atmosphere with a *vertical depth* equal to 0. As a result, using data from Yakutsk array (for energies higher than $10^{19}eV$) [6] the expected *event rates* for different *zenith angle* intervals are given in table 1.

Table 1. Expected Event Rates in Different Zenith Angle Intervals. The values are multiplied by a factor of ~ 26.

$0-2°$	$8-10°$	$18-20°$	$24-26°$	$30-32°$	$34-36°$	$44-46°$	$58-60°$
100	1645	3120	3988	4593	4950	5195	4912

Simulating the total number of $\mathcal{N}$ showers for each *zenith angle* interval and dividing the *atmospheric depth* to intervals of 10 gr/cm^2, the *occurrence frequencies* of *depth of shower maxima* were counted. We used the second set of formulas were presented previously [1]:

$$\langle X_{max} \rangle = \sum_{i=1,\ldots,n} \omega_i X_{max,i} \tag{2}$$

in which n is the number of intervals (i.e. $\sum_{i=1,\ldots,n} frequency_i = \mathcal{N}$) and for energies in the range of $\sim 10^{19} - 5 \times 10^{19}$, $\mathcal{N}$ were chosen from table 1. Here ω_i is a weight factor equivalent to $\omega_i \equiv \frac{frequency_i}{\mathcal{N}}$. No functional fits are considered as the number of simulated showers are now different. As before, *zenith angle* dependency may be removed using a function of the form of:

$$X_{max}(\theta) = a. \cos\theta + b \tag{3}$$

and $\langle X_{max} \rangle$ has the value of $\sim a + b$.

Omitting the *zenith angles* of lower than $\sim 10°$, it is possible to use less simulations. But as the most uncertainties occur in this range, an important part of data may be lost. So we did this study for 5 fixed energies ($1 \times 10^{18}eV$, $5 \times 10^{18}eV$, $1 \times 10^{19}eV$, $5 \times 10^{19}eV$ and $1 \times 10^{20}eV$) for p and Fe primaries. Interaction models were QGSJETII (QGSJET-II-04) [7] and FLUKA [8]. Thinning parameters was set to 10^{-4}. No SLANT option was considered and using of Gaisser-Hillas curves [9] were avoided.

3. Discussion and Results

The effect of this *non-isotropic event rate* is a decrease in the values of statistical errors by increasing *zenith angle*. In fact, a fit of formula [3] to the distribution of mean values of X_{max} versus *zenith angle* now is possible by establishing a *standard deviation*, λ_θ, for each point (figure 1) which is now related to variable values of $\mathcal{N}(\theta)$. For the studied energies these values were ranged from 10 *down to* 2 for a primary proton of $5 \times 10^{19}eV$, and 9 *down to* 2 for a primary proton of $1 \times 10^{18}eV$. In comparison, the *standard deviation* had a constant value for an isotropic distribution of the *event rates*.

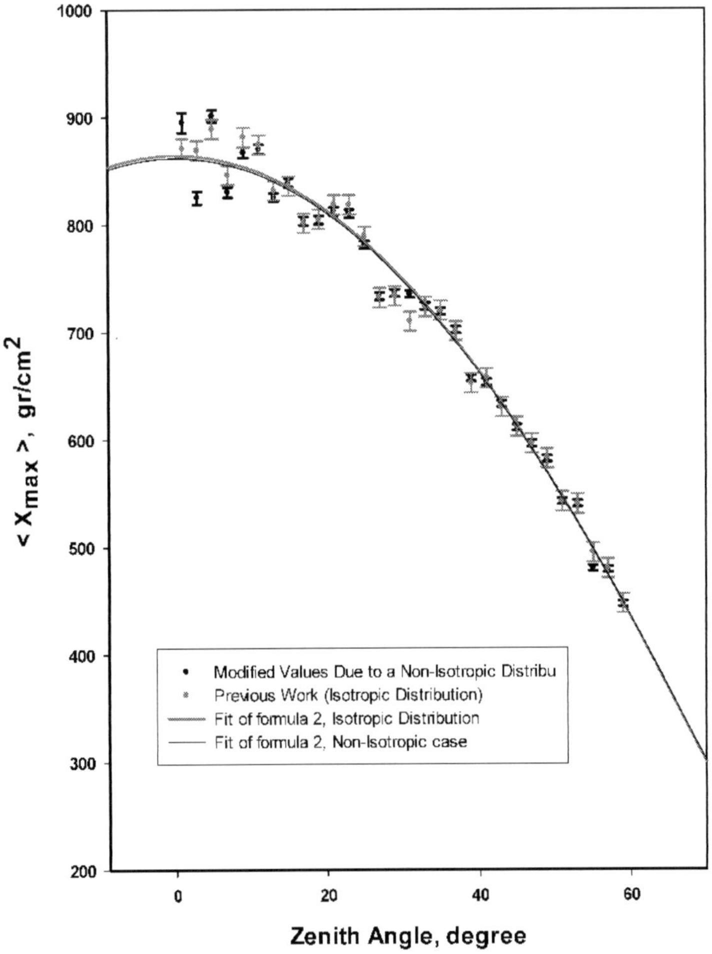

Figure 1. Distribution of *depth of shower maxima* in different *zenith angle* intervals for a 100 EeV proton (QUGSJET-II-04 in Combination with FLUKA were used).

Anyhow, in the case of a *Non-Isotropic* distribution of *event rates*; when the distribution of *depth of shower maxima* experiences higher fluctuations for lower *zenith angle* intervals, a fit of the form of formula 3 eliminates this effect and the resulted mean value of *depth of shower maxima* differs not much than 2% of its previous value [1]. For instance, for a 100 EeV proton using a *non-isotropic* distribution caused an upper value of $\langle X_{max} \rangle$ by 5 gr/cm^2 which is less than 1% of its previous calculated value for an *isotropic* distribution.

The percentage of these differences with respect to previous calculated values [1] were presented in table 2.

This research shows that in the method we used [1] for calculating diagrams of $\langle X_{max} \rangle$ versus energy decades, the use of a *non-isotropic event rate* of primaries may affect the results on *mass composition* of primaries only slightly. In fact in the energy range of $10^{19} eV$ and above, these values (i.e. the values of $\langle X_{max} \rangle$) may be shifted to larger values of $\langle X_{max} \rangle$ by less than 8 gr/cm^2

Table 2. The calculated percentage of differences, $\mathcal{P}\%$ with respect to previously calculated values for different primaries.

primary type	$1 \times 10^{18} eV$	$5 \times 10^{18} eV$	$1 \times 10^{19} eV$	$5 \times 10^{19} eV$	$1 \times 10^{20} eV$
p	2%	2%	1%	1%	1%
Fe	2%	2%	2%	1%	1%

for a proton primary and 7 gr/cm^2 for an Iron nuclei (see table I of [1]), and as these values are much less than the magnitude of experimental uncertainties[10, 11, 12, 13, 14, 15, 16], this has not a considerable effect on *mass composition* studies at the highest energies. Same result were attained for SYBILL 1.6 and SYBILL 2.1 [1, 17]

References

[1] P. Davoudifar, and K. Rowshan Tabari, 2013, *J. Phys.: Conf. Series* **410** 012087
[2] D. Heck et al., (1998), Report FZKA 6019, Forschungszentrum Karlsruhe, http://www-ik.fzk.de/corsika/physics description/corsika phys.html
[3] P.K.F. Grieder, 2010, *Extensive Air Showers, High Energy Phenomena and Astrophysical Aspects* (Springer Heidelberg Dordrecht London New York) Volumw I, p 255
[4] N.N. Efimov, A.A. Mikhailov, M.I. Pravdin, 1983, *Proc. of the 18th Int. Cosmic Ray Conf. (Bangalore)*, vol 2, p.149
[5] A.A. Ivanov, et al., 1999, *Proc. of the 26th Int. Cosmic Ray Conf. (Salt Lake City)*, vol 1, p.403
[6] N.N. Efimov, et al, 1988, *Catalogue of Highest Energy Cosmic Rays, Giant Extensive Air Showers*, World Data Center C2 for Cosmic Rays Institute of Physical and Chemical Research, Wako, Saitama, Japan; Online catalogue, http://eas.ysn.ru/catalog/yakutsk-array-data
[7] N.N. Kalmykov and S.S. Ostapchenko, 1993, Physics of Atomic Nuclei, 56, 346; N.N. Kalmykov, S.S. Ostapchenko, and A.I. Pavlov, 1994, Bulletin of Russian Academi of Science (Physics), 58, 1966; 1997, Nuclear Physics B (Proc. Suppl.), 52B, 17; S.S. Ostapchenko, 2011, Physical Review D, 83, 014018
[8] A. Ferrari, P.R. Sala, A. Fasso', and J. Ranft, 2005, CERN-2005-010, INFN TC-05/11, SLAC-R-773; G. Battistoni, S. Muraro, P.R. Sala, F. Cerutti, A. Ferrari, S. Roesler, A. Fasso', and J. Ranft, 2007, *Proc. of the Hadronic Shower Simulation Workshop 2006*, Fermilab 6–8 September 2006; M. Albrow, R. Raja eds., *AIP Conf. Proc.* 896, 31-49
[9] T.K. Gaisser, A.M. Hillas, 1977, *Proc. of the 15th Int. Cosmic Ray Conf., Plovdiv, Bulgaria*, 8, 353
[10] V.M. Grigoryev, et al., 1990, *Proce. of the 21th Int. Cosmic Ray Conf.*, 9, 206
[11] I.V. Doronina, et al., 1990, *Proc. of the 21th Int. Cosmic Ray Conf.*, 3, 150
[12] M.N. Dyakonov,et al., 1991, *Proc. of the 22nd Int. Cosmic Ray Conf.*, 4, 351
[13] M.N. Dyakonov et al., 1993, *Proc. of 23rd Int. Cosmic Ray Conf., Calagary, Canada*, vol 4, 303
[14] S. Knurenkov et al., 200), *Proc. of the 27th Int. Cosmic Ray Conf., Hamburg, Germany*, HE 1.3, 177
[15] A.A. Ivanov, et al.,(200), *Proc. of the 29th Int. Cosmic Ray Conf., Pune, India*, 6, 241
[16] S.P. Knurenko and A. Sabourov,(201), Astrophysics and Space Science Transactions, 7, 251
[17] J. Engel, T.K. Gaisser, P. Lipari, and T. Stanev, (1992), Physical Review D, 46, 5013; R.S. Fletcher, T.K. Gaisser, P. Lipari, and T. Stanev, (1994), Physical Review D, 50, 5710; R. Engel, (1999), *Proc. of 26th Int. Cosmic Ray Conf.*, 1, 415

Group theory analysis of the rhombohedral magnetic structures in rare earth iron garnets revisited

M Lahoubi
Department of Physics, Faculty of Sciences, *Laboratory L.P.S.*,
Badji-Mokhtar Annaba University, PO Box-12, 23000, Annaba, Algeria

E-mail: mlahoubi@hotmail.fr

Abstract. The representation analysis of Bertaut is used to enumerate all possible magnetic structures in rare earth iron garnets ($RE_3Fe_5O_{12}$) for the RE and Fe ions on the *6e*, *6e'* and *12f*, *2b*, *6d* Wyckoff positions respectively in the highest rhombohedral subgroup $R\bar{3}c$ of the cubic space group $Ia\bar{3}d$. The basis vectors of the one-dimensional irreducible representation A_{2g} lead to noncollinear magnetic structures of the RE moments which produce a better description than those based on the models of Wolf *et al.* Some results of the "double umbrella" structure found in our recent neutron powder diffraction studies of terbium iron garnet are reported.

1. Introduction

The rare earth iron garnets with general formula $\{RE^{3+}{}_3\}[Fe^{3+}{}_2](Fe^{3+}{}_3)O_{12}$ or REIG (where RE is a trivalent rare earth ion or the yttrium Y) are ferrimagnetic materials where the magnetic ions are distributed over the three crystallographic sites of the paramagnetic cubic space group $G = Ia\bar{3}d$ - (O_h^{10}) No. 230: RE ions occupy the dodecahedral 24*c*(222) Wyckoff position while the Fe ions are on the octahedral 16*a*($\bar{3}$) and tetrahedral 24*d*($\bar{4}$) Wyckoff positions. Below T_N (554 ± 6 K) [1] until room temperature, the Néel model [2] is generally accepted and the RE with the Fe sublattices form an antiparallel alignment along [111], the easy axis (EA) of magnetization. At low temperature, noncollinear magnetic structures of the RE magnetic moments appear due to two types of anisotropic interactions, RE-Fe superexchange and crystalline field represented by an exchange $\tilde{\boldsymbol{G}}$ and a magnetic $\tilde{\boldsymbol{g}}$ tensors respectively in the hypothesis of the effective spin Hamiltonian model of Wolf *et al.* [3]. As a results of this, the sublattice of the 24*c* RE ions split into six magnetically nonequivalent sublattices, C_j and C'_j with (j = 1-3) due to the symmetry (222). The RE moments lie in the principal planes of the magnetic $\tilde{\boldsymbol{g}}$-tensor but tilt away from [111] according to three possible arrangements based on the relative magnitudes of the components, g_x, g_y, g_z. This microscopic description which has been applied mainly in the previous neutron powder diffraction (NPD) studies seems now insufficient if we compare with our results obtained recently for DyIG and TbIG by using symmetry analysis [4-7].

2. Results and discussions

The "representation analysis" developed by Bertaut [8] is used. This group theory is based on the systematic reduction of a magnetic representation Γ of a space group G_k into irreducible representations (*irreps*). Γ is equal to the direct product of the permutation matrix representation Γ^{perm}

and an axial vector matrix representation $\widetilde{V}$: $\Gamma^{perm} \otimes \widetilde{V}$. Since the wave vector $\boldsymbol{k}$ = [000] is observed below T_N in the NPD analyses of all REIG, the little group G_k is obviously isomorphic to the full cubic symmetry $G = Ia\bar{3}d$ and then, the ten *irreps* related to the point group $G_0 = m\bar{3}m\ (O_h)$ are concerned, Γ_{ig} and Γ_{iu} (i = 1-5) where the subscripts g and u refer to representations which are even and odd respectively under inversion **I**. The cubic models based on the linear combinations of spins, ***F***, ***G***, ***C***, ***A*** [8] within the single three-dimensional *irrep* $\{\Gamma_{4g} = T_{1g}\}$ which look reasonable are indeed a first approximation if one takes into account the rhombohedral distortions observed at low temperature [5 and refs. herein]. Using the method of the "symmetry lowering device" of Bertaut, the highest subgroup to be considered is $G' = R\bar{3}c - (D_{3d}^6)$ No. 167. The following generators of this space group will be chosen: the identity $\mathbf{E} = (1|\tau_0)$, the inversion $\mathbf{I} = (\bar{1}|\tau_0)$, a ternary axis $\mathbf{3} = (3|\tau_0)$ and a diagonal binary axis $\mathbf{2_d} = (2_{X\bar{Y}}|\tau_1)$ perpendicular to the glide plane $c = \mathbf{I2_d}$. We apply the identification method of Olbrychski [9] which has been generalized by Bertaut [10]. Using the well known relation $(\alpha|\tau_\alpha).(\alpha'|\tau_{\alpha'}) = (\alpha\alpha'|\alpha\tau_{\alpha'}+\tau_\alpha)$, between symmetry operators usually described by the notation of Koster-Seitz, the space group $R\bar{3}c$ is then defined with the following relations between generators

$$\mathbf{3}^3 = \mathbf{2_d}^2 = \mathbf{I}^2 = (1|0,0,0) = (\mathbf{2_d 3})^2 = (\mathbf{3}^{-1}\mathbf{2_d})^2 ;\ \mathbf{3I} = \mathbf{I3} ;\ \mathbf{2_d I} = \mathbf{I2_d} \qquad (1)$$

Table 1. Transformations properties of the rhombohedral coordinates X, Y, Z and the S_X, S_Y, S_Z spin components in the six proper rotations and character table of the space group $R\bar{3}c$ with $\boldsymbol{k}$ = [000] and τ_0 = 0,0,0; τ_1 = 1/2,1/2,1/2; $\varepsilon = e^{2\pi i/3}$.

	E	**3**	**3²**	**2d**	**2d3**	**2d3²**
	$(1\|\tau_0)$	$(3\|\tau_0)$	$(3^2\|\tau_0)$	$(2_{X\bar{Y}}\|\tau_1)$	$(2_{Y\bar{Z}}\|\tau_1)$	$(2_{Z\bar{X}}\|\tau_1)$
X	X	Z	Y	$\bar{Y}+1/2$	$\bar{X}+1/2$	$\bar{Z}+1/2$
Y	Y	X	Z	$\bar{X}+1/2$	$\bar{Z}+1/2$	$\bar{Y}+1/2$
Z	Z	Y	X	$\bar{Z}+1/2$	$\bar{Y}+1/2$	$\bar{X}+1/2$
S_X	S_X	S_Y	S_Z	$-S_Y$	$-S_X$	$-S_Z$
S_Y	S_Y	S_Z	S_X	$-S_X$	$-S_Z$	$-S_Y$
S_Z	S_Z	S_X	S_Y	$-S_Z$	$-S_Y$	$-S_X$
$\Gamma_{1g}=A_{1g}$	+1	+1	+1	+1	+1	+1
$\Gamma_{2g}=A_{2g}$	+1	+1	+1	−1	−1	−1
$\Gamma_{3g}=E_g$	$\begin{pmatrix}1 & .\\ . & 1\end{pmatrix}$	$\begin{pmatrix}\varepsilon^2 & .\\ . & \varepsilon\end{pmatrix}$	$\begin{pmatrix}\varepsilon & .\\ . & \varepsilon^2\end{pmatrix}$	$\begin{pmatrix}. & 1\\ 1 & .\end{pmatrix}$	$\begin{pmatrix}. & \varepsilon\\ \varepsilon^2 & .\end{pmatrix}$	$\begin{pmatrix}. & \varepsilon^2\\ \varepsilon & .\end{pmatrix}$
Symbol	κ	β	γ	η	δ	ζ
$\Gamma_{1u}=A_{1u}$	+1	+1	+1	+1	+1	+1
$\Gamma_{2u}=A_{2u}$	+1	+1	+1	−1	−1	−1
$\Gamma_{3u}=E_u$	$+\kappa$	$+\beta$	$+\gamma$	$+\eta$	$+\delta$	$+\zeta$

Table 1 shows transformations properties of the rhombohedral coordinates X, Y, Z under only the six proper rotations in which we have added the characters of the six *irreps* of $R\bar{3}c$ with $\boldsymbol{k}$ = [000]. These *irreps* are those of the point group D_{3d}: Γ_{jg} and Γ_{ju} (j = 1-3). The following characters of the two-dimensional *irrep* $\Gamma_{3g} = E_g$ are deduced and then, their matrix representatives constructed

$$\chi(\kappa) = 2;\ \chi(\beta) = \chi(\gamma) = -1;\ \chi(\eta) = \chi(\delta) = \chi(\zeta) = 0 \qquad (2)$$

At first, we study the axial vector matrix representation $\widetilde{V}$ of dimension 3 which will be constructed. The magnetic moment or a spin being axial vector, transforms as a polar vector under rotation but remains invariant under the inversion so that we have: $\widetilde{V}(\mathbf{I}\alpha) = \widetilde{V}(\alpha)$. The transformation properties of the S_X, S_Y, S_Z spin components under the symmetry operators of the space group $R\bar{3}c$ are also shown on table 1 and then, the characters of $\tilde{V}$ are easily evaluated

$$\chi^{\tilde{V}}(\mathbf{E})=\chi^{\tilde{V}}(\mathbf{I})=3;\chi^{\tilde{V}}(\mathbf{3})=\chi^{\tilde{V}}(\mathbf{3}^2)=\chi^{\tilde{V}}(\mathbf{I3})=\chi^{\tilde{V}}(\mathbf{I3}^2)=0 \quad (3)$$

$$\chi^{\tilde{V}}(\mathbf{2_d})=\chi^{\tilde{V}}(\mathbf{2_d3})=\chi^{\tilde{V}}(\mathbf{2_d3}^2)=\chi^{\tilde{V}}(\mathbf{I2_d})=\chi^{\tilde{V}}(\mathbf{I2_d3})=\chi^{\tilde{V}}(\mathbf{I2_d3}^2)=-1 \quad (4)$$

The reduction of the axial vector matrix representation $\tilde{V}$ can be expressed as: $A_{2g} \oplus E_g$. The rhombohedral unit cell contains only the half ions of the cubic unit cell: while the Fe ions originally in the 24*d* and 16*a* sites, occupy respectively one and two nonequivalent sites, namely, 12*f*(1) and 2*b*($\bar{3}$), 6*d*($\bar{1}$), the twelve RE ions are located on two crystallographic sites 6*e* and 6*e*′. The transformations properties of theses twelve RE ions which are numbered in the same manner that we have used previously for these ions on the 24*c* Wyckoff position in $Ia\bar{3}d$ [5, 6] are shown in table 2.

Table 2. Transformations of permutation of the positions for the twelve RE ions in sites 6*e* and 6*e*′ under the symmetry operators of the space group $R\bar{3}c$.

	(1)	(2)	(3)	(4)	(5)	(6)	(7)	(8)	(9)	(10)	(11)	(12)
E	(1)	(2)	(3)	(4)	(5)	(6)	(7)	(8)	(9)	(10)	(11)	(12)
3	(1,2,3)				(4,5,6)			(7,8,9)			(10,11,12)	
3^2	(1,3,2)				(4,6,5)			(7,9,8)			(10,12,11)	
2$_d$	(1, 2)		(3)	(4,5)		(6)	(7,8)		(9)	(10,11)		(12)
2$_d$**3**	(1)	(2,3)		(4)	(5,6)		(7)	(8,9)		(10)	(11,12)	
2$_d$**3**^{2}	(1,3)	(2)		(4,6)	(5)		(7,9)	(8)		(10,12)	(11)	
I	(1,7)	(2,8)	(3,9)	(4,10)	(5,11)	(6,12)						
I3		(1,8,3,7,2,9)			(4,11,6,10,5,12)							
I3^2		(1,9,2,7,3,8)			(4,12,5,10,6,11)							
I2$_d$	(1, 8)	(2,7)	(3,9)	(4,11)	(5,10)	(6,12)						
I2$_d$**3**	(1, 7)	(2,9)	(3,8)	(4,10)	(5,12)	(6,11)						
I2$_d$**3**^{2}	(1, 9)	(2,8)	(3,7)	(4,12)	(5,11)	(6,10)						

It can be seen that the two symmetry operators, **I3** and **I3**^{2} allow for a decomposition of these ions into two distinct sublattices respectively, each sublattice having six equivalent ions; the lowered local symmetry (2) from (222) facilitates obviously such decomposition. Thus, the six ions of each sublattice permute between them under the six transformations of permutation related with the three binary axes (**2**$_d$, **2**$_d$**3**, **2**$_d$**3**^{2}) and three glide planes (*c*, *c*3, *c*3^2). Therefore, we find that the non equivalence of the crystallographic properties of the sublattices (C_j) and (C'_j) is due essentially to the remarkable role of the symmetry in the permutation operations of $R\bar{3}c$. The permutation matrix representation Γ^{perm} with the dimension 12 may be constructed and non-zero characters which correspond to points conserved in the permutation, i.e., to site symmetry elements, can be found

$$\chi^{\Gamma^{perm}}(\mathbf{E})=12;\chi^{\Gamma^{perm}}(\mathbf{2_d})=\chi^{\Gamma^{perm}}(\mathbf{2_d3})=\chi^{\Gamma^{perm}}(\mathbf{2_d3}^2)=4 \quad (5)$$

We finally get the complete reduction of Γ^{perm} over the six *irreps* Γ_{jg} and Γ_{ju} with j = 1-3. Thus, the transformation-induced matrix representation $\Gamma(6e, 6e')$ is defined as the direct product: $\Gamma^{perm} \otimes (A_{2g} \oplus E_g)$. Using the simple rules of decomposition, $\Gamma(6e, 6e')$ is simply decomposed into

$$\Gamma(6e, 6e') = 2(A_{1g} \oplus 2A_{2g} \oplus 3E_g \oplus A_{1u} \oplus 2A_{2u} \oplus 3E_u) \quad (6)$$

The transformation-induced matrix representations $\Gamma(6e)$ and $\Gamma(6e')$ with the dimension 6 related to each site $6e$ and $6e'$ respectively are equivalents and have a similar expression

$$\Gamma(6e) = \Gamma(6e') = A_{1g} \oplus 2A_{2g} \oplus 3E_g \oplus A_{1u} \oplus 2A_{2u} \oplus 3E_u \quad (7)$$

The basis vectors (*bv*) of all the *irreps* are obtained by the projection operator technique via the corresponding matrix representatives. We know that $\Gamma_{4g} = T_{1g}$ associated to the point group O_h of $Ia\bar{3}d$ is reduced in $R\bar{3}c$ to $\Gamma_{2g} = A_{2g} \oplus E_g$ and only the one-dimensional *irrep* A_{2g} will be considered after [4-7]. We shall show how to construct the *bv* of the *irrep* A_{2g} for the RE $6e$ site. This appears two times in the decomposition of the $\Gamma(6e)$ transformation-induced matrix representation so that, two different sets of linearly independent *bv* should have to be build. Let the function Ψ be a component or linear combination of components of a magnetic moment or spin. In most cases, it sufficient to take $\Psi = S_{1X}$, i.e., by using S_{1X} as the starting function. With the transformations of the S_X, S_Y, S_Z spin components (table 1) and that of the transformation positions (table 2), we get: $\Psi_{11}^{2g}(I) = \sum_\alpha [D^{A_{2g}}(\alpha)]*\alpha S_{1X} = (S_1 + S_7)_X + (S_2 + S_8)_Y + (S_3 + S_9)_Z$ with $[D^{A_{2g}}(\alpha)]*$, the matrix representative of the symmetry element α in the *irrep* A_{2g} and the asterisk indicating complex conjugation; $\Psi_{11}^{2g}(II) = [(S_1 + S_7)_Y + (S_1 + S_7)_Z] + [(S_2 + S_8)_Z + (S_2 + S_8)_X] + [(S_3 + S_9)_X + (S_3 + S_9)_Y]$.

Table 3. Magnetic structures of the RE ions in the *irrep* A_{2g} of $R\bar{3}c$.

RE ions	General spin configuration	Designation
{24*c*} 6*e* (C_j)	$S_1 = S_7 = (F \mp A, F \pm A, F \pm A)$ $S_2 = S_8 = (F \pm A, F \mp A, F \pm A)$ $S_3 = S_9 = (F \pm A, F \pm A, F \mp A)$	*Noncollinear ferromagnetic structure* called "umbrella" *e* around [111]
6*e'* (C'_j)	$S_4 = S_{10} = (F' \mp A', F' \pm A', F' \pm A')$ $S_5 = S_{11} = (F' \pm A', F' \mp A', F' \pm A')$ $S_6 = S_{12} = (F' \pm A', F' \pm A', F' \mp A')$	*Noncollinear ferromagnetic structure* called "umbrella" *e'* around [111]

We get as well for the RE $6e'$ site by using $\Psi = S_{4X}$ as the starting function: $\Psi_{11}^{2g}(IV) = (S_4 + S_{10})_X + (S_5 + S_{11})_Y + (S_6 + S_{12})_Z$ and $\Psi_{11}^{2g}(V) = [(S_4 + S_{10})_Y + (S_4 + S_{10})_Z] + [(S_5 + S_{11})_Z + (S_5 + S_{11})_X] + [(S_6 + S_{12})_X + (S_6 + S_{12})_Y]$. The linear combinations of these *bv* lead to the following resultant basis functions

$$\text{Site}(6e)\quad \boldsymbol{V}_{cf}(A_{2g}) = \Psi_{11}^{2g}(I) + \Psi_{11}^{2g}(II);\ \boldsymbol{V}_{ncf}(A_{2g}) = \mp[\Psi_{11}^{2g}(I) - \Psi_{11}^{2g}(II)] \quad (8)$$

$$\text{Site}(6e')\quad \boldsymbol{V'}_{cf}(A_{2g}) = \Psi_{11}^{2g}(IV) + \Psi_{11}^{2g}(V);\ \boldsymbol{V'}_{ncf}(A_{2g}) = \mp[\Psi_{11}^{2g}(IV) - \Psi_{11}^{2g}(V)] \quad (9)$$

where $\boldsymbol{V}_{cf}(A_{2g})$, $\boldsymbol{V'}_{cf}(A_{2g})$ and $\boldsymbol{V}_{ncf}(A_{2g})$, $\boldsymbol{V'}_{ncf}(A_{2g})$ are identified respectively as a *collinear ferromagnetic mode* along the [111] axis and a *noncollinear ferromagnetic mode* along the three $[\bar{1}11]$, $[1\bar{1}1]$, $[11\bar{1}]$ axes. From the general spin configuration described by the two *noncollinear ferromagnetic structures* around the [111] axis called "umbrella" *e* and *e'* and gathered on table 3, four rhombohedral models of "double umbrella" type were deduced [5-7]. The RE moments of the sublattices (C_j) and (C'_j) with j = 1-3 remain in the three principal glide planes: each plane contains the

[111] axis and one of the rhombohedral directions $\{[\bar{1}11], [1\bar{1}1], [11\bar{1}]\}$ respectively with their related constraints while the principal axes x and y of the local axes (g_x, g_y, g_z) of the magnetic $\tilde{g}$-tensor may deviate from the cubic [100] and [011] and equivalent directions of these planes and the projections of the spins $\boldsymbol{S}_j$ ($\boldsymbol{S}_{j+6}$) and $\boldsymbol{S}_{j'}$ ($\boldsymbol{S}_{j'+6}$) are not necessarily more equal and the predictions of Wolf *et al.* [3] will not be sufficient. The *bv* of the Fe ions in sites 12*f*(1), 2*b*($\bar{3}$) and 6*d*($\bar{1}$) are then deduced using the same steps and the Néel ferrimagnetic configuration of the Fe spins can be easily described. The example we are dealing with is TbIG where a detailed description of our refinements of the NPD studies are already reported [5-7]. We use the spherical coordinates (m_j, θ_j, ϕ_j) and (m'_j, θ'_j, ϕ'_j) respectively: (ϕ_j; ϕ'_j) are the angles associated to a variation in the (111) plane and are fixed with one specific condition ($\phi_1 = 180°$ and $\phi'_1 = 0°$) among the four possibilities (ϕ_j and $\phi'_j = 0°$ or 180°); the angles (θ_j; θ'_j) with respect to the [111] axis and the modulus (m_j; m'_j) will be the only variable parameters. From the thermal variations of the fitted parameters illustrated in figure 1, it can be show at first that the Néel model is not valid below 160 K and the "double umbrella" starts to develop at this "opening temperature" identified as T_J [11]. Secondly, the anomalous behavior observed near 54 K is close the low-temperature point T_B predicted previously by Belov at 58 K [12] for TbIG.

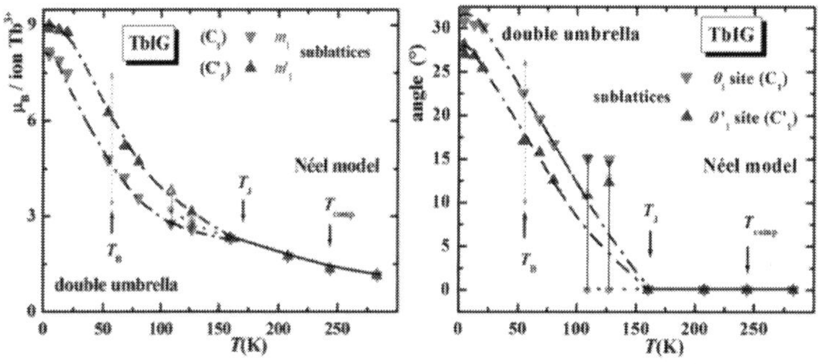

Figure 1. The parameters (m_1; m'_1) and (θ_1; θ'_1) versus T obtained from the NPD patterns of TbIG. The dash lines are guides for the eyes. The vertical lines are the error bars with the refined results found with the Néel model.

3. Conclusion

The representation analysis of Bertaut applied to the space group $R\bar{3}c$ has allowed to describe the rhombohedral magnetic structures of the RE moments in REIG. It was found with the case of TbIG that this method leads to more accurate results than those found using the predictions of Wolf *et al.*

4. References

[1] Pauthenet R. 1958 *Ann. Phys.* **3** 424
[2] Néel L 1948 *Ann. Phys.* **3** 137
[3] Wolf W P *et al.* 1962 *J. Phys. Soc. Japan* **17 Suppl. B-I** 443
[4] Lahoubi M *et al.* 2010 *J. Phys.: Conf. Ser.* **200** 082018
[5] Lahoubi M 2012 Temperature evolution of the double umbrella magnetic structure in terbium iron garnet *Neutron Diffraction* ed I Khidirov Intech Chapter 10 pp 203−230
[6] Lahoubi M 2012 *J. Phys.: Conf. Ser.* **340** 012068
[7] Lahoubi M and Ouladdiaf B 2014 *J. Magn. Magn. Mater.* (in press)
[8] Bertaut E F 1968 *Acta Cryst.* A **24** 217
[9] Olbrychski K 1963 *Phys. Status Solidi* **3** 1868
[10] Bertaut E F 1969 *C. R. Acad. Sci. Paris* B **268** 281
[11] Nelson T J and Mayer D C 1971 *I.E.E.E. Trans. on Magn.* **7** 616
[12] Belov K P 1996 *Sov. Phys. Usp.* **39** 623

Approximate analytical channel potential model of poly-Si thin film transistors operated in the strong inversion region under the high gate and low drain biases

Zhen Zhu[1] **and Junhao Chu**[2]

[1]School of Electronics and Information Engineering, Suzhou Vocational University, Suzhou 215104, People's Republic of China

[2]School of Information Science and Technology, East China Normal University, Shanghai 200241, People's Republic of China

E-mail: manuscript.z.zhu@gmail.com and jhchu@mail.sitp.ac.cn

Abstract. An approximate analytical channel potential model of polycrystalline silicon thin film transistors operated in the strong inversion region under the high gate and low drain biases is proposed. Thus, the linear relationship between the channel potential and the drain voltage is derived in the strong inversion region under the above bias condition when the polysilicon layer is ultrathin. This model agrees with the two-dimensional-device simulation results under different gate voltages, different drain voltages and different channel lengths. By comparing the relative errors between the model and the simulation results, it presents that this model is more suitable under the higher gate voltage V_g or the lower drain voltage V_d, regardless of the channel length. And this approximate analytical model is helpful in solving the two-dimensional-device problem by one-dimensional Poisson's equation since the drain bias is taken into account in the channel potential.

1. Introduction

In recent years, polycrystalline silicon thin film transistors (poly-Si TFTs) have been the essential electronic devices in flat panel displays (FPDs) [1]. For their applications, an indispensable circuit simulation is a key issue for their large scale circuitry. As a result, an accurate and efficient compact model of the poly-Si TFTs is required in the simulation. The surface-potential-based model structure is a choice for the new generation compact MOSFET models and the approach is recognized as both the most physical and the most difficult to implement. In the surface-potential-based models, the key issue is the calculation of the surface potential.

In order to calculate the surface potential, if we start from the basic Poisson's equation, the carrier density is needed. In order to well characterize it, the distribution of the channel potential, which is defined as bulk-referenced quasi-Fermi potential, is required. However, there are few discussions about the model of the channel potential. Or just a given value is assigned to the channel potential without discussing its dependence on the applied biases and the device structure parameters [2-4]. And if the channel potential is arbitrarily given, the relevant model is essentially a one-dimensional analysis.

In this paper, an approximate analytical channel potential model of poly-Si TFTs with the ultrathin polysilicon layer in the strong inversion region under the high gate and low drain biases is presented and is verified by the two-dimensional-device simulator MEDICI under various gate voltages, drain voltages and channel lengths. It is helpful in solving the two-dimensional-device problem by one-dimensional Poisson's equation since the drain bias is taken into account in the channel potential.

2. Model formulation

In the vicinity of the surface, for n-type poly-Si TFTs, the channel potential $V_c(y)$ is defined as $V_c(y) \equiv \varphi_n(y) - \phi_{f0}$, where y is a arbitrary point along the channel, $\varphi_n(y)$ is the electron quasi-Fermi potential at the arbitrary position along the channel in the vicinity of the surface and ϕ_{f0} is the Fermi potential at the bulk [5].

According to the previous work [6], for n-type poly-Si TFTs with the ultrathin polysilicon layer, the drain current can be expressed as:

$$I_{ds}(y) = \frac{W\mu_s}{y}\int_0^{V_c(y)}[qN_{Aa}\cdot t_{channel} + \frac{qN_T / L_g \cdot t_{channel}}{1 + g_A \cdot \exp(\frac{E_t / q + \phi_{f0} - \Psi_s}{\phi_t})} - C_{ox}(V_g - V_{fb} - \Psi_s)]dV_c \tag{1}$$

where I_{ds} is the drain current, W is the channel width, μ_s is the surface mobility, q is the electron charge, N_{Aa} is the active acceptor concentration, $t_{channel}$ is assumed as $\frac{1}{e}\cdot\sqrt{\frac{2\Psi_s\varepsilon_s}{qN_{Aa}}}$, Ψ_s is the surface potential, ε_s is the silicon permittivity, N_T is the acceptor type grain-boundary surface state (areal) density, L_g is the grain size, E_t is the energy level difference between the trap level and the intrinsic level, ϕ_t is the thermal voltage, C_{ox} is the unit gate oxide capacitance, V_g is the gate voltage, V_{fb} is the flat band voltage.

Meanwhile, in the strong inversion region, the first-order approximation of the surface potential is $\Psi_s = 2\phi_{f0} + V_c$, where V_c is equal to 0 V and the drain bias V_d at the source and drain electrode, respectively [7].

So when in the strong inversion region under the high gate and low drain biases, in the integrand of (1), all terms related to Ψ_s become neglectable compared with that with V_g. So the integrand of (1) can be determined by the term with V_g. And it is almost constant depending on V_g.

Therefore, (1) can be approximately rewritten as

$$I_{ds}(y) = -\frac{W\mu_s}{y}[C_{ox}(V_g - V_{fb})]V_c(y) \tag{2}$$

At $y=L$,

$$I_{ds}(L) = -\frac{W\mu_s}{L}[C_{ox}(V_g - V_{fb})]V_d \tag{3}$$

With (2) and (3), the approximate analytical channel potential model

$$V_c(y) = \frac{y}{L}V_d \tag{4}$$

is derived.

3. Result and discussion

The approximate analytical channel potential model of poly-Si TFTs in the strong inversion region under the high gate and low drain biases has been compared with the two-dimensional-device simulator MEDICI via different gate voltage V_g, drain voltage V_d and channel length L. And in the simulator, the model [8, p 3-111, 3-113] SRH (Shockley-Read-Hall recombination with fixed lifetime), ARORA (a concentration and temperature mobility model based on the work of Arora et al.) and SRFMOB (an effective mobility is calculated at semiconductor-insulator surfaces using an effective electric field) are involved. According to the MEDICI User Guide, if there is an Ohmic contact, the minority and majority carrier quasi-Fermi potentials are equal and are set to the applied bias of that electrode [8, p 2-73]. Since the source side is biased at 0 V, the quasi-Fermi potential at the source electrode, which equals to the bulk Fermi potential, is equal to 0 V. Therefore, the electron quasi-Fermi potential in the MEDICI is the same as the channel potential in this paper.

In figure 1, the correlation of the channel potential with the relative position (y/L) in the channel is shown under the various gate voltage V_g = 15 V, 20 V and 30 V while the drain voltage V_d = 1 V and the channel length L= 10 μm. And the model presents the well agreement with the simulation results. Moreover, since the term with the gate voltage plays a more important role in (1) under the higher gate bias, the model agrees with the simulation better.

In figure 2, the correlation of the channel potential with the relative position in the channel is shown under the different V_d = 0.1 V, 0.5 V and 1 V while V_g = 15 V and L= 10 μm. The model presents the well agreement with the simulation results. Furthermore, the smaller drain bias is, the more excellent agreement it achieves since the term related to Ψ_s in (1) turns to be less.

In figure 3, the correlation of the channel potential with the relative position in the channel is shown under the various L= 5 μm, 10 μm and 25 μm while V_g = 30 V, V_d = 1 V. The model presents the well agreement with the simulation results and is less affected by the different channel lengths.

In order to have an explicit insight on the differences between the model and the simulation results, the relative errors are compared under the different V_g, V_d and L in figure 4. In figure 4 (a), the relative error decreases to 2% under the higher gate voltage V_g = 30 V. The relative error is within about 2% under the lower drain voltage V_d = 0.1 V shown in figure 4 (b). And no obvious differences can be found in various channel lengths in figure 4 (c). It can be concluded that under the higher V_g and lower V_d, this approximate analytical model (4) can better depict the channel potential along the channel, regardless of the channel length.

4. Conclusion

The approximate analytical channel potential model of poly-Si TFTs with the ultrathin polysilicon layer in the strong inversion region under the high gate and low drain biases is obtained. The channel potential shows a linear dependence on the drain bias with the ratio factor y/L. Compared with the two-dimensional-device simulation results, this model is proved under different gate voltages, drain voltages and channel lengths. And the relative error decreases to about 2% under the higher gate voltage and the lower drain voltage respectively and is less affected by the different channel lengths. It presents that this model is more suitable under the higher V_g and lower V_d, regardless of the channel length. And this approximate analytical model is helpful in solving the two-dimensional-device problem by one-dimensional Poisson's equation since the drain bias is taken into account in the channel potential.

References

[1] Jung S.-H., Nam W.-J., Lee J.-H. and Han M.-K. 2006 *IEEE Electron Device Lett.* **27** 40-2
[2] Qureshi S and Siddiqui M J 2002 *Semicond Sci Technol* **17** 526-33
[3] Chen R, Zheng X, Deng W and Wu Z 2007 *Solid-State Electron.* **51** 975-81
[4] Deng W and Huang J 2011 *IEEE Electron Device Lett.* **32** 647-9
[5] Arora N 2007 *Mosfet Modeling for VLSI Simulation: Theory And Practice* (Singapore: World Scientific) p 169
[6] Zhu Z and Chu J 2012 *Int J Mod Phys B* **26** 1250069
[7] Van Langevelde R and Klaassen F M 2000 *Solid-State Electron.* **44** 409-18
[8] Taurus Medici 2010 *Medici User Guide, Version D-2010.03* (Synopsys Inc.)

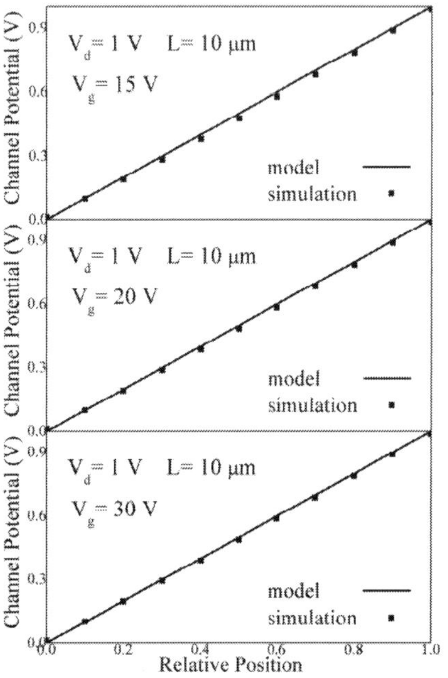

Figure 1. The correlation of the channel potential with the relative position (y/L) in the channel under the different gate voltages.

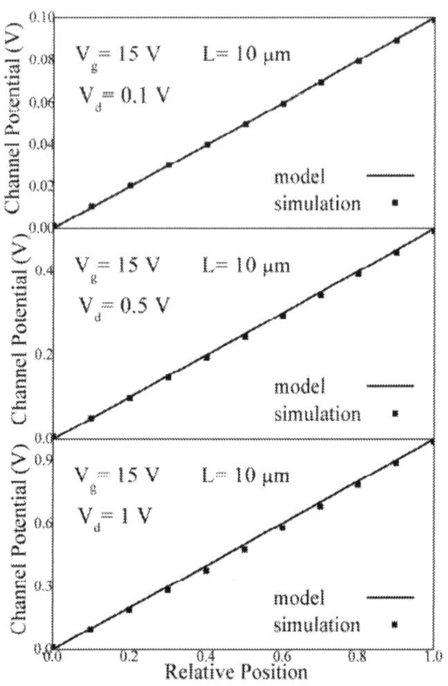

Figure 2. The correlation of the channel potential with the relative position (y/L) in the channel under the different drain biases.

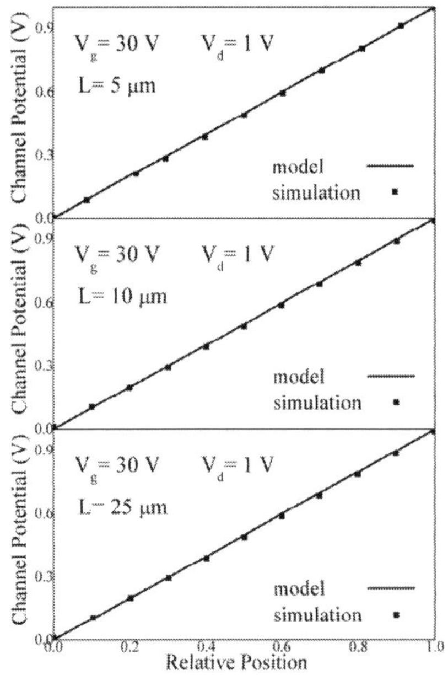

Figure 3. The correlation of the channel potential with the relative position (y/L) in the channel under the various channel lengths.

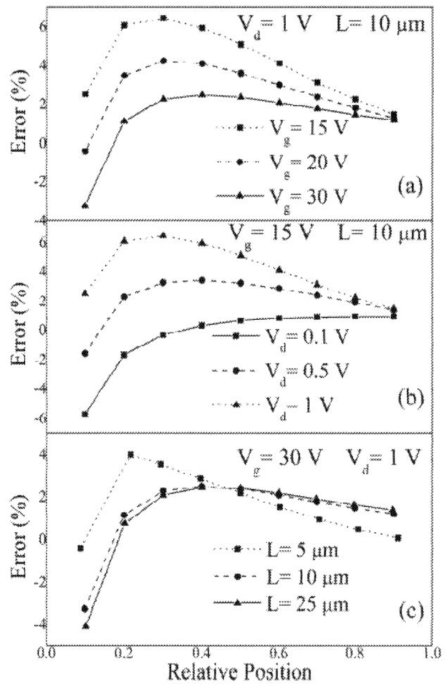

Figure 4. Relative errors between the model and the simulation are compared under the various gate, drain voltages and channel lengths.

Surface Tension, Pressure Difference and Laplace Formula for Membranes

Hiroshi Koibuchi
Department of Mechanical and Systems Engineering, National Institute of Technology, Ibaraki College, Nakane 866, Hitachinaka, Ibaraki 312-8508, Japan
E-mail: koibuchi@mech.ibaraki-ct.ac.jp

Andrey Shobukhov
Faculty of Computational Mathematics and Cybernetics, Lomonosov Moscow State University, 119991, Moscow, Leninskiye Gory, MSU, 2-nd Educational Building, Russia

The surface tension γ and the pressure difference Δp for spherical membranes are calculated using Monte Carlo simulation technique. We study the so-called tethered and fluid surface discrete models that are defined on the fixed-connectivity (tethered) and dynamically triangulated (fluid) lattices respectively. Hamiltonians of the models include a self-avoiding potential, which makes the enclosed volume well defined. We find that there is reasonable accuracy in the technique for the calculation of γ using the real area A if the bending rigidity κ or A/N is sufficiently large. We also find that γ becomes constant in the limit of $A/N \to \infty$ both in the tethered and fluid surfaces. The property $\lim_{A/N\to\infty} \gamma = \text{const}$ corresponds to certain experimental results in cell biology.
Keywords: Membranes, Triangulated surfaces, Surface tension, Laplace pressure, Monte Carlo

1. Introduction

The surface tension γ [1, 2, 3] and the pressure difference $\Delta p = p_{\rm in} - p_{\rm out}$, where $p_{\rm in}(p_{\rm out})$ is the pressure inside (outside) the surface, potentially reflect the microscopic interactions of the constituent molecules. Therefore, it is interesting to study γ and Δp by means of stochastic simulation of mechanical processes on the triangulated surface models such as Helfrich-Polyakov (HP) [4, 5]. The HP model is a coarse grained mathematical model for the microscopic interactions of lipid molecules [6]. The frame tension τ is defined via the macroscopic surface energy, which equals τA_p, where A_p is the projected area. This A_p may be regarded as the area contained within the boundary $\Gamma(\subset \mathbf{R}^2)$ fixed in $\mathbf{R}^3$ [7, 8]. For this reason, it is widely accepted that on the surfaces spanning Γ the frame tension τ is correctly calculated by using the projected area A_p of Γ rather than the real area A of the surface.

However, A_p is not always well-defined on the surfaces without boundary. In fact, we can use only the real area A in both simulations and data analyses. Therefore we have the surface tension γ of the HP model defined on spherical surface which has no boundary and the frame tension τ evaluated by Δp and the Laplace formula. Thus, we should prove that γ may be correctly defined through the real area A by comparing it with τ. γ and τ are expected to be identical in the limit of $\kappa(\sim 1/k_BT) \to \infty$ [8].

In this paper, we calculate the surface tension γ for spherical membranes and check whether γ is consistent with the frame tension τ, which is obtained from Δp and the Laplace formula.

2. Model

The partition function Z is defined by

$$Z = \sum_{\mathcal{T}} \int' \prod_{i=1}^{N} d\mathbf{r}_i \exp\left[-S(\mathbf{r}, \mathcal{T})\right], \tag{1}$$

where $\sum_{\mathcal{T}}$ denotes the sum over all possible triangulations $\mathcal{T}$. This $\sum_{\mathcal{T}}$ is included in Z only for the fluid surface model; the fixed-connectivity model is defined by Z without $\sum_{\mathcal{T}}$. The prime in $\int' \prod_{i=1}^{N} d\mathbf{r}_i$ denotes that the center of mass of the surface is fixed to the origin of $\mathbf{R}^3$, and the surface has no boundary Γ. The discrete Hamiltonian $S(\mathbf{r}, \mathcal{T})$ is defined on a triangulated sphere with $\mathbf{r}_i (i = 1, \cdots, N)$, which are the positions of vertices, and $\mathcal{T}$ such that

$$\begin{aligned} &S(\mathbf{r}, \mathcal{T}; \kappa, A_0, \Delta p) = S_1 + \kappa S_2 + U_{\rm Fix}(A_0) - \Delta p V + U_{\rm S}, \\ &S_1 = \sum_{ij} (\mathbf{r}_i - \mathbf{r}_j)^2, \quad S_2 = \sum_{ij} (1 - \mathbf{n}_i \cdot \mathbf{n}_j), \end{aligned} \tag{2}$$

where S_1 is the Gaussian bond potential and S_2 the bending energy with the bending rigidity $\kappa[1/k_BT]$. The symbol $\mathbf{n}_i$ denotes the unit normal vector of the triangle i, and $\sum_{ij}$ of S_1 and S_2 denote the sum over all nearest neighbor vertices and triangles respectively. The potential $U_{\rm Fix}(A_0)$, which fixes the surface area A to a constant A_0 without using the boundary Γ, is defined by

$$U_{\rm Fix}(A_0) = \begin{cases} \infty & (|1 - A/A_0| > \epsilon_A = 1/N_T), \\ 0 & (\text{otherwise}), \end{cases} \tag{3}$$

where N_T is the total number of triangles $N_T = 2N - 4$. Note that the surface is allowed to have only in-plane deformation if ϵ_A is exactly zero, therefore ϵ_A should be fixed to a small positive number. The energy $-\Delta p V$ with the enclosed volume V is well-defined only for the self-avoiding (SA) surface models [9]. Thus, the energy $-\Delta p V$ must be included in the Hamiltonian together with the SA potential $U_{\rm S}$ defined by $U_{\rm S} = \sum_{\Delta\Delta'} U_{\rm S}(\Delta, \Delta')$ where $\sum_{\Delta\Delta'}$ denotes the sum over all non-nearest neighbor triangles and

$$U_{\rm S}(\Delta, \Delta') = \begin{cases} \infty & (\text{triangles } \Delta\Delta' \text{ intersect}), \\ 0 & (\text{otherwise}). \end{cases} \tag{4}$$

This potential is considered as an extension of the one of Doi-Edwards model for polymers [10].

The partition function Z is independent of the change in the integration variables $\mathbf{r} \to \mathbf{r}'$. This property is called the scale invariance, which can be expressed by $\partial_\alpha \log Z(\alpha\mathbf{r})|_{\alpha=1} = 0$. The surface tension γ is obtained from this property as follows: It is easy to see that S_2 and $U_{\rm S}$ are scale independent and that $S_1(\alpha\mathbf{r}) = \alpha^2 S_1(\mathbf{r})$, $\Delta p V(\alpha\mathbf{r}) = \alpha^3 \Delta p V(\mathbf{r})$. Since $U_{\rm Fix}(A_0; \alpha\mathbf{r}) = U_{\rm Fix}(\alpha^{-2} A_0; \mathbf{r})$, we see that the corresponding derivative $\partial_\alpha \log Z(A_0; \alpha\mathbf{r})|_{\alpha=1}$ can be written as $-2A_0 \partial_{A_0} \log Z(A_0; \mathbf{r})$. Note that the partition function for the macroscopic membranes is given by

$$Z_{\rm mac} = \exp\left[-(\gamma A - \Delta p V)\right], \tag{5}$$

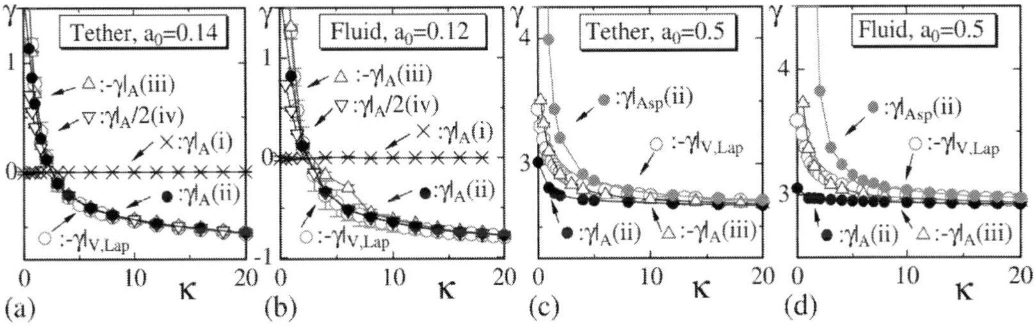

Figure 1. γ vs. κ at (a) $a_0(=A_0/N_T)=0.14$ (tethered), (b) $a_0=0.12$ (fluid), (c) $a_0=0.5$ (tethered) and (d) $a_0=0.5$ (fluid). The symbols ○ (×, •, △, ▽) denote the results obtained by the constant volume (area) simulations. The lattice size is $N=1442$.

where V is the enclosed volume. We have

$$\gamma|_A = (2\langle S_1\rangle_{A,\Delta p} - 3\Delta p\langle V\rangle_{A,\Delta p} - 3N)/(2A_0), \quad (6)$$

where $\langle S_1\rangle_{A,\Delta p}$ and $\langle V\rangle_{A,\Delta p}$ are obtained by the constant area simulations.

The pressure difference Δp in $Z_{\rm mac}$ of Eq. (5) can also be calculated using almost the same model and procedure as those described above for $\gamma|_A$. Indeed, the constraint $U_{\rm Fix}(A_0)$ in Eq. (2) can be replaced by $U_{\rm Fix}(V_0)$ to fix the enclosed volume V to V_0; and the energy $-\Delta pV$ should be removed from the Hamiltonian. In this model, Δp is not an input parameter but an output. Thus, the Hamiltonian is given by replacing A_0 with V_0 in Eq. (3) such that $S(\mathbf{r},\mathcal{T};\kappa,V_0)=S_1+\kappa S_2+U_{\rm Fix}(V_0)+U_{\rm S}$.

From the scale invariance of Z in Eq. (1) and $Z_{\rm mac}$ in Eq. (5), we obtain

$$\Delta p|_V = (3N - 2\langle S_1\rangle_V)/(3V_0). \quad (7)$$

The symbol $\Delta p|_V$ is used for the calculated pressure to distinguish it from the input parameter Δp in Eq. (2). It must be emphasized that $\Delta p|_V$ does not suffer from the problem encountered in the calculation of $\gamma|_{A_p}$.

3. Simulation results

The variables $\mathbf{r}$ and $\mathcal{T}$ are updated by the canonical Metropolis MC technique, where $\mathcal{T}$ is updated by using the standard bond flip technique. We use the surface of size $N=1442$. The total number of MCS is about 5×10^6 including 5×10^5 thermalization MCS.

3.1. Dependence of $\gamma|_A$ on κ

We firstly perform MC simulations by varying κ with constant V_0 to check whether the MC results of $\gamma|_A$ are consistent with those of $\Delta p|_V$. The constant V_0 is fixed by the triangle areas so that the mean value $a_0=A_0/N_T$ becomes $a_0=0.14$ (tethered model) and $a_0=0.12$ (fluid model). From the MC results $\Delta p|_V$ of the tethered and fluid models we obtain $\gamma|_{V,{\rm Lap}}$ using the Laplace formula

$$\gamma|_{V,{\rm Lap}} = (2/R)\Delta p|_V, \quad R=(3V_0/4\pi)^{1/3}. \quad (8)$$

These results are plotted (○) in Figs. 1(a)–(d). This $\gamma|_{V,{\rm Lap}}$ is completely different from $\gamma|_A$ because $\gamma|_{V,{\rm Lap}}$ ($\gamma|_A$) is obtained by the constant volume (area) simulations. Indeed, the sign of $\gamma(=\gamma|_A)$ of the surface expanded by its area in general is different from $\gamma(=\gamma|_{V,\mathrm{Lap}})$ of the same surface expanded by its volume. This comes from the definition of Z_{mac} in Eq. (5). Note also that the surface shape remains spherical in the constant V_0 simulations even in the limit of $\kappa\to 0$ for both tethered and fluid models.

Next, we perform the simulations with constant area A_0 and Δp such that $A_0=\langle A\rangle_V$ and

$$(\mathrm{i})\Delta p = -\Delta p|_V, \quad (\mathrm{ii})\Delta p = 0, \quad (\mathrm{iii})\Delta p = -2\Delta p|_V, \quad (\mathrm{iv})\Delta p = \Delta p|_V, \tag{9}$$

where $\langle A\rangle_V$ and $\Delta p|_V$ are the output of the first simulations. Under these four different conditions, we obtain the surface tensions $\gamma|_A$ from Eq. (6). We predict that the simulation results $\gamma|_A(\kappa)$ satisfy

$$(\mathrm{i})\gamma|_A = 0(\times), \quad (\mathrm{ii})\gamma|_A = -\gamma|_{V,\mathrm{Lap}}(\bullet), \quad (\mathrm{iii})\gamma|_A = \gamma|_{V,\mathrm{Lap}}(\triangle), \quad (\mathrm{iv})\gamma|_A = -2\gamma|_{V,\mathrm{Lap}}(\nabla). \tag{10}$$

These equalities correspond to the conditions in Eq. (9) (see Fig. 1).

The predictions in Eq. (10) come from the expectations

$$V_0 = \langle V\rangle_{A,\Delta p}, \quad \langle S_1\rangle_V = \langle S_1\rangle_{A,\Delta p}, \tag{11}$$

which are independent of Δp. It is easy to prove the predictions in Eq. (10) using Eq. (11). These equations are physically almost clear because the enclosed volume V_0 is uniquely determined by its surface area A_0 if the surface is a sphere. Figure 1 (b) shows that (iii) $\gamma|_A=\gamma|_{V,\mathrm{Lap}}$ is slightly broken in the region $\kappa\simeq 5$. The reason of this deviation is that the surface shape becomes prolate under the negative pressure $\Delta p=-2\Delta p|_V$ at $\kappa\simeq 5$ [11, 12].

The prediction (ii) of Eq. (10) in the small κ region is broken as we see in Figs. 1(c),(d). This deviation appears only because of the difference between A_0 and A_p expected in that region. Indeed, $\gamma|_{\mathrm{A}}$ ($\gamma|_{\mathrm{A_{sp}}}$) is slightly smaller (larger) than $-\gamma|_{V,\mathrm{Lap}}$, where $\gamma|_{\mathrm{A_{sp}}}$ is calculated by replacing A_0 with $A_{\mathrm{sp}}=4\pi\left(3\langle V\rangle_A/4\pi\right)^{2/3}$ in Eq. (6). This discrepancy implies that A_p is in the range $A_{\mathrm{sp}}<A_p<A_0$, because $\gamma|_{V,\mathrm{Lap}}$ in Eq. (8) is expected to be correct due to the fact that the surface remains spherical in the volume constant simulation in the whole region of κ as mentioned above. Note that the prediction (i) of Eq. (10), which is not depicted, is satisfied also in the whole region of κ including $\kappa=0$ for $a_0=0.5$.

3.2. Dependence of γ on A/N

Now we proceed to an additional check for the MC results of $\gamma|_A$ and $\gamma|_{V,\mathrm{Lap}}$ obtained by the area and volume constant simulations, respectively, by varying $A_0/N_T(=a_0)$ with fixed κ. In contrast to the check described above, these two different simulations can be performed independently. Thus we obtain $\Delta p|_V$, from which $\gamma|_{V,\mathrm{Lap}}$ is calculated by using the Laplace formula in Eq. (8). Therefore we expect that the simulation results satisfy (see Figs. 2(a) (h))

$$-\gamma|_{V,\mathrm{Lap}}(\bullet) = \bar{\gamma}|_{A,\Delta p}(\bigcirc), \quad \text{with} \quad \bar{\gamma}|_{A,\Delta p} := \left(2\langle S_1\rangle_{A,\Delta p} - 3N\right)/(2A_0). \tag{12}$$

The expectation in Eq. (12) is obtained by using the relations (as functions of a_0)

$$A_0 = \langle A\rangle_V, \quad \langle S_1\rangle_{A,\Delta p} = \langle S_1\rangle_V. \tag{13}$$

These relations also explain why we expect the surface areas to be the same in both simulations at least for sufficiently large a_0. The relations in Eq. (13) are weaker than those of Eq. (11) in the sense that the surface shape of the area constant simulation is not necessarily identical to the one of the volume constant simulation: stomatocyte, cup-like, and dumbbell, even branched-polymer, are allowed. We note that the surface shape of the latter simulation must be spherical

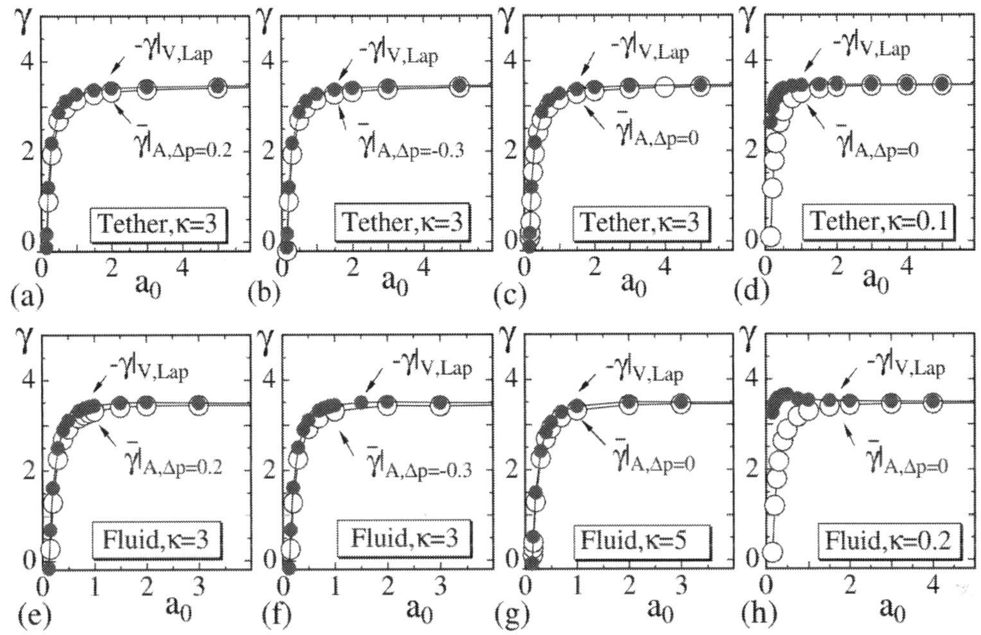

Figure 2. $-\gamma|_{V,\mathrm{Lap}}$ (•) and $\bar{\gamma}|_{A,\Delta p}$ (○) vs. $a_0(=A_0/N_T)$ for the tethered and fluid models. Δp=0.2, −0.3, 0, 0 are assumed for the constant area simulations. The lattice size is $N=1442$.

because the Laplace formula for a sphere is assumed to yield $\gamma|_{V,\mathrm{Lap}}$. This requirement is always fulfilled as mentioned above.

The result that $\gamma|_{A,\Delta p=0}$ (or $-\gamma|_{V,\mathrm{Lap}}$) becomes constant in the limit of $a_0 \to \infty$ in both tethered and fluid surfaces corresponds to the experimental fact observed in biological cells [13, 14].

This work is supported in part by the Grant-in-Aid for Scientific Research (C) Number 26390138. We acknowledge the support of the Promotion of Joint Research 2014, Toyohashi University of Technology. We are grateful to K. Osari and S. Usui for the computer analyses.

References

[1] Jan Ambjörn, Bergfinnur Durhuus, and Thordur Jonsson, Phys. Rev. Lett. **58**, 2619 (1987).
[2] J. Ambjörn, A. Irbäck, J. Jurkiewicz, B. Petersson, Nucl. Phys .B**393**, Issue 3, 571 (1993).
[3] J.F. Wheater, J. Phys. A Math. Gen. **27**, 3323 (1994).
[4] W. Helfrich, Z. Naturforsch **28c**, 693 (1973).
[5] A.M. Polyakov, Nucl. Phys. B **268**, 406 (1986).
[6] D. Nelson, in Statistical Mechanics of Membranes and Surfaces, Second Edition, edited by D. Nelson, T. Piran, and S. Weinberg, (World Scientific, 2004) p.1.
[7] F. David and S. Leibler, J. Phys. II Frans **1**, 959 (1991).
[8] Jean-Baptiste Fournier and Camilla Barbetta, Phys. Rev. Lett. **100**, 078103 (2008).
[9] M. Bowick, A. Cacciuto, G. Thorleifsson, and A. Travesset, Euro. Phys. J. E **5**, 149 (2001).
[10] M. Doi and F. Edwards, *The Theory of Polymer Dynamics*, (Oxford University Press, 1986).
[11] G. Gompper and D.M. Kroll, Phys. Rev. E **51**, 514 (1995).
[12] H.Koibuchi and A.Shobukhov, Physica A **410**, 54 (2014).
[13] Ramsey A. Foty, Gabor Forgacs, Cathie M. Pfleger, and Malcolm S. Steinberg, Phys. Rev. Lett. **72**, 2298 (1994).
[14] Ramsey A. Foty, Cathie M. Pfleger, Gabor Forgacs and Malcolm S. Steinberg, Development **122**, 1611 (1996).

Nematodynamics modelling under extreme mechanical and electric stresses

Antonino Amoddeo

Department of Civil, Energy, Environmental and Materials Engineering Università "Mediterranea" di Reggio Calabria Via Graziella 1, Feo di Vito, I-89122 Reggio Calabria, Italy

E-mail: antonino.amoddeo@unirc.it

Abstract. Nematic liquid crystals confined in asymmetric π-cells and subjected to intense electrical and mechanical stresses undergo strong distortions which can be relaxed by means of the order reconstruction, a fast switching mechanism connecting topologically different textures, assuming bulk and/or surface characteristics depending on both amplitude of the applied electric fields and anchoring angles of the nematic molecules on the confining surfaces. In the frame of the Landau-de Gennes order tensor theory, we provide a numerical model implemented with a moving mesh finite element method appropriate to describe the nematic order dynamics, allowing to map the switching properties of the nematic texture.

1. Introduction

Thermotropic nematic liquid crystals (NLC) are elongated molecules which exhibit orientational order, schematically represented as rods having cylindrical symmetry. In most electro-optical phenomena, the average orientation of the NLC long axis can be described by a dimensionless unit vector, the director $\boldsymbol{n}$, while the degree of order along such direction is accounted for by the scalar order parameter S: as NLCs possess a unique optical axis and cylindrical symmetry, they are said uniaxial [1]. Applying to a nematic texture distortions over length scales comparable to the biaxial coherence length ξ_b [2], the uniaxial symmetry is broken and the system presents two distinct optical axes. As a consequence, local biaxial domains grow, lowering the degree of order, and bistable behaviours become possible. Such phenomena attract great attention due to their direct influence on the optical an switching properties of NLC, but in this case the description requires the coupling of the $\boldsymbol{n}$ and S parameters within the Landau-de Gennes theory based on the order tensor $\mathbf{Q}$ [1].

In a π-cell a thin film of nematic material is confined between two flat glass plates, which internal faces are treated so as to align the nematics on both surfaces with strong anchoring energy and a pretilt angle θ: the two plates are arranged in an anti-parallel configuration generating a splayed texture [2], a set-up admitting the existence of a second topologically distinct texture where the nematic is π-twisted (or π-bent), and the energy barrier between them prevents any spontaneous relaxation from each other. Applying an electric field perpendicularly to the boundary plates the initial splay texture can be switched into the topological different bend one by means of the biaxial order reconstruction, by which the local order of the nematic phase is changed without any macroscopic director rotation.

Recently, using the Finite Element Method (FEM), we implemented the Moving Mesh Partial Differential Equation (MMPDE) numerical technique [3-6], in order to solve the Partial Differential

Equations (PDEs) system arising from the modelling of the order dynamics inside a π-cell containing a 5CB NLC. In such technique, the grid points in which the simulated domain is discretized are moved according to a monitor parameter towards regions where the gradient of the order, $\nabla\mathbf{Q}$, is growing: keeping constant the nodal connectivity and the number of mesh points inside the domain, the latter are clustered into regions of high spatial variability where more detail is required, thus improving the resolution; further, being the domain regions where $\nabla\mathbf{Q}$ is growing *apriori* unknown, a constant number of nodes allows computational saving, since a smaller number of mesh points is needed with respect to simulations performed on uniform grids. We present a numerical investigation of the biaxial order dynamics of a 5CB NLC confined in a asymmetric π-cell performed using the MMPDE technique in order to solve the time-dependent **Q** tensor governing equations, monitoring the evolution of both biaxiality and **Q**-eigenvalues, for applied electric fields with amplitudes extending far above the threshold value required to induce order reconstruction transitions in the bulk [2], imposing variable degrees of asymmetry.

2. Model and method

In the frame of the Landau-de Gennes theory, orientation and order of biaxial nematics are globally described by the symmetric, second-rank, traceless tensor in the orthonormal basis of its eigenvectors $\{\boldsymbol{e}_1, \boldsymbol{e}_2, \boldsymbol{e}_3\}$ [1,7]

$$\mathbf{Q} = \sum_{i=1}^{3} s_i \boldsymbol{e}_i \otimes \boldsymbol{e}_i \tag{1}$$

where the $\boldsymbol{e}_i$ lie along the directions of the preferred molecular orientation, and are assumed as pointing along the x, y and z direction of the laboratory frame of reference, respectively; the associated eigenvalues, s_i, represent the degree of order along such directions, and the isotropic, uniaxial or biaxial phases of nematics can be distinguished considering the s_i. When all the three eigenvalues vanish the calamitic molecules are in the isotropic phase, and the optical behaviour of the material is as an ordinary isotropic fluid; the uniaxial nematic phase presents a unique optical axis, described by the eigenvector $\boldsymbol{e}_i=\boldsymbol{n}$ associated to the maximum eigenvalue s_{max} giving the scalar order parameter $S=(3/2)s_{max}$, while in the biaxial nematic phase, all the eigenvalues are different.

The **Q** dynamical evolution model requires the minimization of the free energy density functional $\boldsymbol{F}$ inside the cell containing the NLC [1], which contributions are assumed to originate from bulk only, since surface terms are neglected because of the fixed boundary conditions imposed, thus leading to infinite anchoring strength. Neglecting ion effects [2] and assuming small **Q** distortions, thermotropic, elastic and electric contributions, respectively F_t, F_d and F_e, can be expressed in powers of **Q**

$$\boldsymbol{F} = \int_V \left(F_t + F_d + F_e\right) dV \tag{2}$$

The balance between the dissipation function D and the free energy variation [8,9] gives the generalized Euler-Lagrange equation (3), where the subscript "$_{,j}$" denotes differentiation with respect to the spatial coordinates x_1, x_2 and x_3, summation over repeated indices is assumed, and the order tensor is expressed in terms of the independent parameters q_i, $i\in[1,5]$,

$$\frac{\partial D}{\partial \dot{q}_i} + \frac{\partial F_t}{\partial q_i} + \frac{\partial F_e}{\partial q_i} + \frac{\partial F_d}{\partial q_i} - \frac{\partial}{\partial x_j}\left(\frac{\partial F_d}{\partial q_{i,j}}\right) - \frac{\partial}{\partial x_j}\left(\frac{\partial F_e}{\partial q_{i,j}}\right) = 0, \qquad i=1\ldots5 \tag{3}$$

The PDEs system (3), together with the governing equation for the electric potential U

$$\nabla\cdot\boldsymbol{D} = \nabla\cdot\left(-\varepsilon_0\boldsymbol{\varepsilon}\nabla U + \boldsymbol{P}_S\right) = 0 \tag{4}$$

where $\boldsymbol{D}$, ε_0, $\boldsymbol{\varepsilon}$ and $\boldsymbol{P}_S$ are, respectively, the displacement field, the vacuum dielectric permeability, the dielectric tensor and the spontaneous polarization vector, represent a system of six coupled PDEs solved imposing Dirichlet boundary conditions on the boundary surfaces of the π-cell, giving the dynamical evolution of the system. Full details about model and development can be found in [3-6]. The degree of biaxiality can be obtained from its invariant measure [2]

$$\beta^2 = 1 - 6\frac{tr\left(\mathbf{Q}^3\right)^2}{tr\left(\mathbf{Q}^2\right)^3} \in [0,1] \tag{5}$$

considering that a uniaxial nematic texture has $\beta^2=0$, while the maximum of biaxiality, $\beta^2=1$, corresponds to a biaxial nematic phase.

Starting from the discretization of the one dimensional integration domain, equidistributing in each subinterval an appropriate monitor function, a functional which can depend on the gradient of the unknown solution or on the solution error, a quality control on the mesh map is attained, resulting in a quite efficient method capable of accurate representations of sharp solution features. Denoting $u(z,t)$ a solution of a system PDEs over a physical domain $\Omega_p = [0,1]$ with z physical coordinate, if ξ is a computational coordinate of a fixed computational domain $\Omega_c = [0,1]$, we define a one-to-one mapping from physical space $\Omega_p \times (0,T]$ to computational space $\Omega_c \times (0,T]$ as $\xi = \xi(z,t)$, $z \in \Omega_p$, $t \in (0,T]$, and its reverse from computational space $\Omega_c \times (0,T]$ to physical space $\Omega_p \times (0,T]$ as $z = z(\xi,t)$, $\xi \in \Omega_c$, $t \in (0,T]$. The equidistribution of the monitor function $M(u(z,t))$ over the one-dimensional domain gives the mesh equation in differential form [3,4],

$$M(z(x,t))\frac{\partial}{\partial \xi}z(x,t) = \int_0^1 M(u(s,t))\,ds = C(t) \tag{6}$$

and we have demonstrated [4] that the choice of the monitor function tested by Beckett et al. [10], ensures a good quality control of the meshes and final convergence of the FEM solution. Hence we put

$$M(u(z,t)) = \int_0^1 \left(\left|\frac{\partial u(z,t)}{\partial z}\right|\right)^{1/2} dz + \left(\left|\frac{\partial u(z,t)}{\partial z}\right|\right)^{1/2}, \tag{7}$$

having the advantages of being naturally smoothed, with no need for external interventions.

The FEM discretization of the multidimensional time-dependent problem is done via the method of lines, while the adaptive solution process involves (i) a mesh generation using the equidistribution principle based on the numerical solution at the current time step and (ii) the PDEs solution on the new generated grid and then updated in time.

The PDEs solution was carried out replacing in (7) the unknown $u(z,t)$ with $tr(\mathbf{Q}^2)$, a quantity which is rapidly varying when the degree of order is not constant. At t=0 s a rectangular electric pulse is applied perpendicularly to the plates for a duration Δt = 0.25 ms. The imposed pretilt angles, both measured with respect to the plates are, on the lower boundary surface θ_L = 19°, while on the upper boundary plate θ_U was in the [-19°,0°] range with steps variable from 0.1° to 1° as needed. The investigated amplitudes of the applied electric pulse E ranged from 15 V/μm up to 90 V/μm, with steps variable from 1 to 5 V/μm. The dynamical evolution of the $\mathbf{Q}$ tensor inside the cell was sampled with a time step size δt= 0.1 μs, using the physical parameters typical for 5CB NLC at ΔT=-1°C [2], discretizing the physical domain with a mesh of 285 grid points.

3. Results and discussion

It has been demonstrated [3], that the switching properties of NLCs confined in a π-cell depend on both E and θ_U, once fixed θ_L. To build a phase diagram of the nematic switching regime, monitoring the biaxiality arising in the cell over several points of the (E,θ_U) plane is required. The results of the biaxiality computations extended to a wide range of (E, θ_U) pairs are summarized in the phase diagram of figure 1, where the boundaries among bulk switching (BS), surface switching (SS), bulk plus surface switching (BSS) or no switching (NS) regimes,

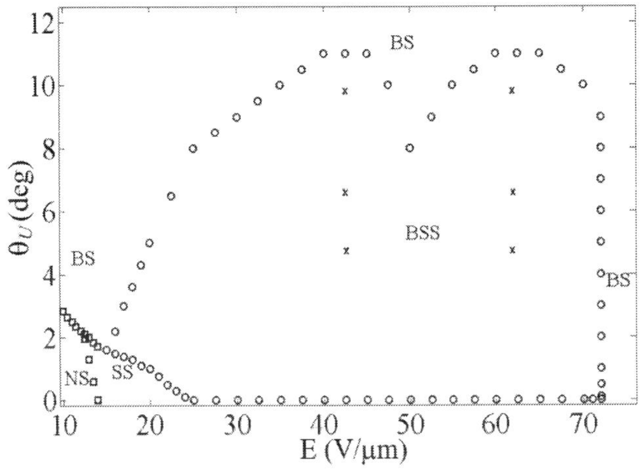

Figure 1. Phase diagram of the biaxial order reconstruction for $\theta_L = 19°$: on the horizontal axis are reported the applied electric pulse amplitudes E, and on the vertical axis are represented the opposite of the anchoring angles θ_U.

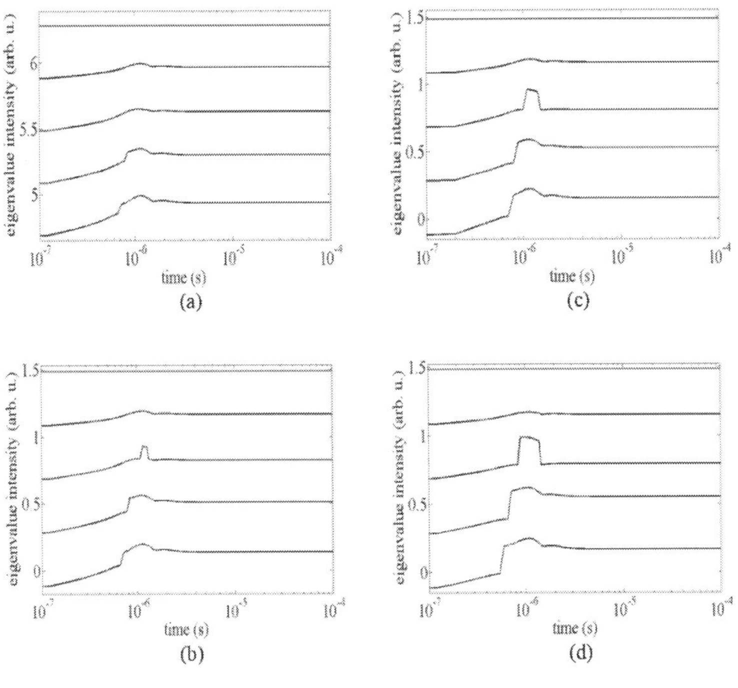

Figure 2. Temporal evolution of the s_3 eigenvalue computed on the first five mesh point starting from the upper boundary surface (top curve), obtained applying a pulse with amplitude E=42.5 V/μm, for θ_U values of -4° (a), -6° (b), -9.5° (c) and -14° (d), where on the horizontal axis time is reported in logarithmic scale.

obtained imposing $\theta_L = 19°$, are indicated by circles. Each point in the (E,θ_U) plane corresponds to a computed biaxiality map (see [3]), for which we read off and referred the corresponding texture as obeying to one of the above switching mechanisms. The points obtained for 10 V/μm $\leq E \leq$ 14 V/μm and 0° $\leq \theta_U \leq$ 3°, represented as squares, well reproduce that presented in [3]. It results a partition of the (E,θ_U) plane in four regions, in each of that are mapped textures homogeneous as regards their switching regime. A large portion of the phase diagram concerns the coexistence of both surface and bulk order reconstruction transitions, delimited by a bi-lobed boundary, and above E=70 V/μm, whatever θ_U is, the surface transition disappears: hence we expect that, for (E,θ_U) pairs falling in each of the lobed region of the phase diagram, the relaxation mechanism of the nematic texture switches over differently, along the cell thickness. In figure 1, with "**x**" are marked (E, θ_U) pairs for which we monitored the temporal evolution of the s_3 eigenvalue, as it is very sensitive to order variations along the electric field direction [6], i.e. the z-direction, adding also (E=42.5 V/μm, θ_U =-14°) and (E=62 V/μm, θ_U =-14°), not shown for scale reasons. The results obtained applying a pulse with amplitude E=42.5 V/μm are shown in figure 2. As the nematic molecules adsorbed on the confining plate are fixed by the infinite anchoring energy, the degree of order along $\boldsymbol{e}_3$ does not change and s_3 remain constant on the upper plate, for each of the anchoring angles. For θ_U =-4°, figure 2(a), during the time evolution s_3 increase smoothly on the first two grid points below the upper boundary plate, second and third curve from the top, a behaviour typical of second order transitions characterizing biaxial phenomena, in which the strong distortion relaxes through a wealth of biaxial states without any director rotation [6]; on the fourth grid point, instead, the discontinuity marks a first order transition, as the director $\boldsymbol{n}$ reorients along the z-direction under the effect of the applied electric pulse, then a predominant uniaxial order establishes. Imposing θ_U =-6°, figure 2(b), the temporal evolution of s_3 on the second grid point below the surface shows a sharp structure superimposed over the smooth evolution: the nematic director starts to reorient but it is suddenly pulled back, denoting a dynamical evolution of the biaxiality mediated by a metastable uniaxial state; on the remaining two points the s_3 evolution shows a first order transition. Increasing θ_U up to -14°, the duration of the metastable state increases and all the features become sharper and faster, see figures 2(c)-2(d). For E=62 V/μm, on going from θ_U =-4°to θ_U =-14°, in the s_3 temporal evolution the metastable uniaxial state is not present, and a stable director reorientation occurs on going from the upper boundary surface towards the bulk; the relative evolution is not shown for space reasons.

4. Conclusions

Our technique allowed us to map a phase diagram of the nematic order reconstruction phenomenon, following the biaxiality evolution inside the cell as a function of the (E, θ_U) pairs, a prohibitive task if performed on uniformly discretized domains. Our results show that intense electrical and mechanical stresses induce strong distortions of the texture which can be relaxed by growing biaxiality, ending with the coexistence of bulk and surface order reconstruction transitions in a bi-lobed region of the phase diagram. The evolution of the s_3 **Q**-tensor eigenvalue puts in evidence that textures can relax the electrical and mechanical stresses differently just underneath the upper boundary surface: the low energy side lobe includes textures which in such region have a layer where the biaxial evolution is mediated by a metastable uniaxial state; on the contrary, textures mapped within the high energy side lobe present a stable evolution from second to first order transition.

References

[1] de Gennes P G and Prost J 1993 *The Physics of Liquid Crystals* (Oxford: Clarendon Press)
[2] Barberi R, Ciuchi F, Durand G, Iovane M, Sikharulidze D, Sonnet A and Virga E G 2004 *Eur Phys J* E **13** 61–71
[3] Amoddeo A, Barberi R and Lombardo G 2012 *Phys Rev* E **85** 061705-10
[4] Amoddeo A, Barberi R and Lombardo G 2010 *Comput Math Appl* **60** 2239–52
[5] Amoddeo A, Barberi R and Lombardo G 2011 *Liq Cryst* **38** 93-103
[6] Amoddeo A, Barberi R and Lombardo G 2013 *Liq Cryst* **40** 799-809
[7] Virga E G 1994 *Variational Theories for Liquid Crystals* (London: Chapman and Hall)
[8] Sonnet A M, Maffettone P and Virga E G 2004 *J Non-Newton Fluid* **119** 51-9
[9] Sonnet A M and Virga E G 2001 *Phys Rev* E **64** 031705-10
[10] Beckett G, Mackenzie J A, Ramage A and Sloan D M 2001 *J Comput Phys* **167** 372-92

Mathematical Tools for Discovery of Nanoporous Materials for Energy Applications

M Haranczyk and R L Martin
Lawrence Berkeley National Laboratory, Berkeley, CA 94720, USA

E-mail: mharanczyk@lbl.gov

Abstract. Porous materials such as zeolites and metal organic frameworks have been of growing importance as materials for energy related applications such as CO_2 capture, hydrogen and methane storage, and catalysis. The current state-of-the-art molecular simulations allow for accurate *in silico* prediction of materials' properties but the computational cost of such calculations prohibits their application in the characterisation of very large sets of structures, which would be required to perform brute-force screening. Our work focuses on the development of novel methodologies to efficiently characterize and explore this complex materials space. In particular, we have been developing algorithms and tools for enumeration and characterisation of porous material databases as well as efficient screening approaches. Our methodology represents a ensemble of mathematical methods. We have used Voronoi tessellation-based techniques to enable high-throughput structure characterisation, statistical techniques to perform comparison and screening, and continuous optimisation to design materials. This article outlines our developments in material design.

1. Introduction

Crystalline porous materials have been exploited in industrial applications for many years, for instance as catalysts for oil refinement, water softeners, and membranes for separations. They are also promising candidates for economically-competitive application to emerging energy challenges such as vehicular natural gas storage and carbon dioxide capture. Porous materials exhibit complex networks of internal empty space, which permit the diffusion and adsorption of guest chemical species. The size, shape and connectivity of a material's pore structure determine the interactions which occur between the guest and the host material, and in turn, its suitability for a particular application.

However, identifying optimal materials for a particular application remains a great challenge. Zeolites, perhaps the most well-known class of porous material, consist of tetrahedral arrangements of silicon, bridged by oxygen atoms; despite their homogeneous chemistry, approximately 200 unique zeolite topologies are known to exist, while over two million hypothetically achievable topologies have been computationally enumerated [1]. Furthermore, new classes of so-called advanced porous materials have emerged in recent years; materials such as metal-organic frameworks (MOFs)[2] or porous polymer networks (PPNs)[3] permit nearly unlimited structural variation through control of both crystal topology and chemistry. On-going efforts to computationally enumerate advanced porous materials have so far resulted in over 100 thousand material designs, and soon next generation efforts will lead to exceeding the number of predicted zeolite structures. With such a vast search space of possible materials, it is clear that exhaustive synthesis is not a feasible strategy for materials discovery, and so there is a pressing need for computational methods for high-throughput screening and discovery of porous materials.

We have been developing techniques for the characterisation, enumeration, screening and design of porous materials. The core of our approaches involves a computational geometry technique - the Voronoi tessellation [4] - and focus on characterisation of a material's void space and its properties relevant to chemical applications, e.g. internal surface area. In a parallel effort, we have been developing the capability to construct three-dimensional structure models of advanced porous materials from a set of given molecular building blocks and a corresponding resulting material topology (also referred to as a *net*; determining how building blocks are connected in three-dimensional space). Our developed capabilities are being combined into higher-level hybrid approaches to perform material discovery and design, e.g. to perform a material's property optimisation with respect to its building blocks. In the following, we will briefly discuss the developed methodologies and capabilities as well as illustrate their application in the design of a high internal surface area material; in particular, two analogues of PPN materials based on the diamond topology.

2. Methods

2.1 Characterisation of porous materials

The core of our approaches involves the Voronoi tessellation, which is employed to translate structural information (e.g. position of atoms, size and shape of the periodic unit cell) into a periodic graph representation of the spaces between atoms and their connectivity. This graph representation (Voronoi network or Voronoi diagram) can be efficiently analysed to identify probe-accessible regions of void space (with respect to particular probe sizes) and their related properties.

In the Voronoi tessellation (also referred to as the Voronoi decomposition), the space surrounding n atoms is divided into n irregular polyhedral cells such that the cell for a given atom comprises of the space that is closer to that atom than any other. Mathematically, if the atoms have positions $\mathbf{x}_i$ for $i= 1,\ldots,n$ and the Euclidean distance between two positions $\mathbf{x}$ and $\mathbf{y}$ is written as $d(\mathbf{x},\mathbf{y})$, then the Voronoi cell for atom i is the space $\mathbf{x}$ that satisfies

$$d(\mathbf{x},\mathbf{x}_i) < d(\mathbf{x},\mathbf{x}_j) \quad (1)$$

for all j not equal to i. The Voronoi cells tessellate the space, and each Voronoi cell face is the perpendicular bisector between two neighbouring atoms. Edges of the Voronoi cells are lines that are equidistant from three neighbouring atoms, whereas the vertices of the cells, the Voronoi nodes, are positions equidistant from four neighbouring atoms. Taken together, the Voronoi edges and nodes form a three-dimensional graph that represents the pores and channels. The nodes of the graph are the local maxima of the function $D(\mathrm{x})=\min\{d(\mathbf{x},\mathbf{x}_i)\colon i= 1,\ldots,n\}$ that gives the minimum distance from a position $\mathbf{x}$ to an atom, and the edges of the graph are the least restrictive paths from one local maximum to the next. Due to this correspondence, it can be shown that determining the (guest-probe) accessibility of the structure is mathematically equivalent to examining the accessibility of the Voronoi network.

In the context of crystalline porous materials, the volume in question is the periodic structural unit, which comprises multiple atoms. As a consequence of the Voronoi tessellation, the Voronoi network is obtained. Our open source porous materials analysis suite, Zeo++ [5], exploits this graph abstraction of a material's porosity to efficiently calculate restricting pore diameters [5], internal surface areas [5] and other geometric characteristics (e.g. as pore size distributions[6]) of materials in a high-throughput manner. The corresponding algorithms rely on various mathematical methods. For example, Dijkstra-like graph propagation [7] is employed to identify diffusion paths in the periodic structure, and their corresponding restricting pore diameters. Monte Carlo sampling is used to calculate surface area and accessible volume etc.

2.2 Periodic structure assembly

When exploring the space of advanced porous materials, it is important to have the capability to predict the material structure(s) resulting from a set of given building blocks. We have developed an approach for modelling crystalline porous structures that emphasizes realism in terms of symmetry and stability [8]. Our new computational tools are based upon guaranteeing a pre-specified topology, by utilizing topology-specific symmetry operations to position building units in accordance with a particular underlying net (e.g., those provided by the Reticular Chemistry Structure Resource (RCSR)[9], which has assigned to each a three-letter name, e.g. Figure 1)). Our structure assembly method has been also implemented in our Zeo++ code.

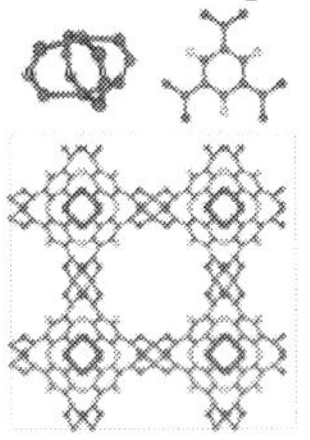

Figure 1. Illustration of crystal structure assembly. Combination of a square-coordinated building unit (Cu_2 paddlewheel) with a trigonal organic linker results in a three-dimensional periodic structure exhibiting the **tbo** net.

2.3 Materials Design by Optimisation

The outlined structure assembly and characterisation tools can be combined into a technique for the automated design of advanced porous materials using mathematical optimisation. At each iteration of optimisation, our structure assembly algorithm builds three-dimensional framework models, for which material properties, e.g. surface area and pore volume, can be calculated using algorithms implemented in Zeo++. Based on such properties, the next iteration is executed, and the procedure is continued until convergence.

In our current approach, the organic molecular building units of advanced porous materials are abstracted as geometric shapes (a union of rods and discs described by a set of radii and length parameters, the variables in optimisation, see Figure 2). This abstraction serves to enable optimisation in the continuous space of geometric shapes, rather than the discrete space of chemistry, and can therefore make use of rapid continuous optimisation algorithms. We utilize gradient-based optimisation to automatically tune properties of the material, for example, maximize its internal surface area, an important property in storage and separation applications. Here, for a ligand defined by a set of properties p_i (the vector of parameters describing the shape and size of organic ligand), the ligand $p_{i+1} = p_i + s\nabla g(p_i)$ is built in the optimisation step, where g() defines the surface area of the assembled material structure, and s is a scalar. Our approach utilizes a number (typically ten) random start points to mitigate the problem of converging to a local minimum, and is based on a radius/length step size of 0.1 Å. Moreover, the search space is typically constrained to the domain corresponding to parameters observed in experimental structures, and can be further extended to explore the frontiers of the material space. Having identified idealized molecular shapes with respect to a particular property, the real chemistry which most closely fits the shape can be identified, and properties of the resulting system determined.

Our work to date has demonstrated the potential of this strategy with a prototype method implemented for MOF materials[10]. Our prototype work focused on gravimetric surface area as the optimized property [10]. Calculating gravimetric surface area (i.e. surface area per mass of the material) requires defining the "molecular" mass of the abstract structure model, which is predicted via a statistical model requiring the surface area of the building block involved as input. This statistical model is trained using a set of hydrocarbon-backbone dicarboxylic acid molecules, the most common form of organic building block in MOF assembly [10]. In our most recent work, we have also explored optimisation of volumetric surface area[11] (i.e. surface area per volume of material) as well as multiobjective optimisation combining both surface area functions[12]. This initial work was focused on MOF materials. The following Results section will present an example of automated design of high surface area porous polymer network (PPN) materials based on the diamond (**dia**) net.

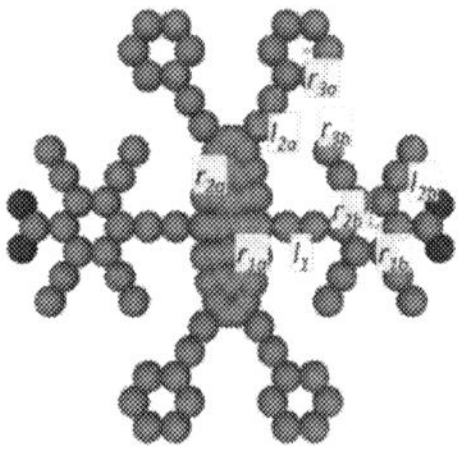

Figure 2. Abstract space-filling molecule constructed from the nine geometric parameters

3. Results

Here, we provide an example of our optimisation-based design technique applied to porous polymer network (PPN) materials, based on linear linkers and tetrahedral cores (adamantane or silicon, i.e. analogues of PPN-3 and PPN-4, see Figure 3)[13]. PPN materials are typically non-crystalline, however recent work has suggested that idealized **dia** net crystal models can be used to reproduce experimental adsorption isotherms[14].

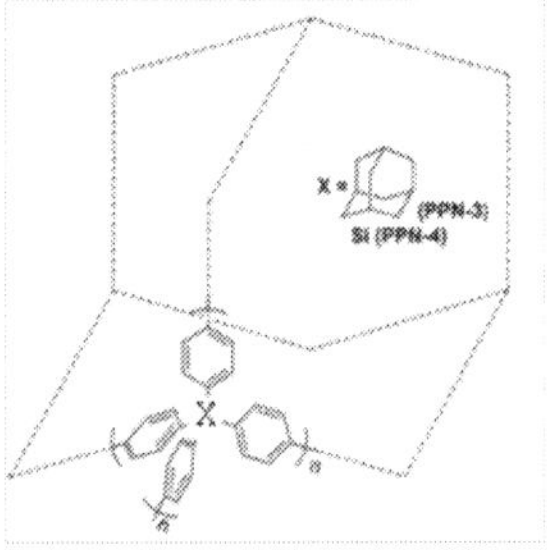

Figure 3. Schematic connectivity for **dia** net in PPN materials. The corner building blocks used to synthesize PPN-3 and PPN-4 are also presented.

Accordingly, we have generated optimized linker shapes for these PPN materials within the **dia** net, with constrained optimisation (limiting the search algorithm to the simplest linker complexity bounds as described in our previous work [11]). The objective function is the product of gravimetric surface area (GSA) and volumetric surface area (VSA), a metric which we refer to as GVSA; optimizing for this objective function leads to framework designs which exhibit the optimal compromise between these two typically competing objectives. Our previous results for the **dia** net, albeit for MOF materials rather than PPNs, highlighted that the optimal GVSA for this net arises from maximizing GSA at the expense of VSA[12]; this behaviour is also observed here for PPNs, and further exaggerated due to the lighter (non-metal) cores in PPN materials. In order to design real frameworks based on optimal abstract templates, it is necessary to project the final linkers onto real molecules. Projecting these abstract shapes onto the closest real chemistry corresponds to a slight extension of the organic linkers from the shape provided by the optimal template; this projection causes an increase in GSA and a corresponding decrease in VSA (see Table 1). We note that our designed analogues of PPN-3 and PPN-4 exhibit much higher gravimetric surface area than the current record holder material obtained in experiment (NU-110 with GSA of ca. 6700 m^2/g) and one obtained by enumeration of possible materials based on a linker library (hmof-5036495 with GSA of ca. 6700 m^2/g). Our GVSA-optimized GSA is a lower bound with respect to the maximum possible GSA.

Table 1. Calculated surface areas for optimal abstract frameworks and corresponding real frameworks generated by projecting onto the closest real chemistry. All descriptors were calculated using the

Zeo++ software package.

	PPN-3 analogue		PPN-4 analogue	
	Abstract	Real	Abstract	Real
Volumetric surface area (m^2/cm^3)	1515.92	1294.86	1759.24	1705.2
Gravimetric surface area (m^2/g)	8473.64	9933.75	8904.82	9993.47

4. Conclusions

We have developed algorithms and a software tool (Zeo++) that enable advanced porous material structure assembly and characterisation as well as efficient material design. We used Voronoi tessellation-based techniques to enable high-throughput structure characterisation. They provide the Voronoi network representation of the void space inside a material, which is used to derive material descriptors such as limiting pore diameters, accessible surface area and pore size distributions. We demonstrated how these structure characterisation tools can be combined with structure assembly capabilities to perform material structure optimisation with respect to its properties. This optimisation is performed within the framework of an abstract material model, in which molecular building blocks are represented by geometric objects (i.e. union of rods and disks). A material structure is assembled from these geometric building blocks, and its property is calculated. Then, in an iterative procedure the structure is modified following the gradient of property with respect to the structure parameters describing the building blocks. Multiple starting points are considered to mitigate the local minimum problem. We demonstrated such optimisation-based material design using two analogues of porous polymer network (PPN) material materials.

Acknowledgements

This work was partially supported by CAMERA: The Center for Applied Mathematics for Energy Research Applications at Lawrence Berkeley National Laboratory supported by the U.S. Department of Energy under Contract No. DE-AC02- 05CH11231.

References

[1] R. Pophale, P.A. Cheeseman and M.W. Deem: Phys. Chem. Chem. Phys. Vol. 13 (2011), p.12407.
[2] H.-C. Zhou, J.R. Long and O.M. Yaghi: Chem. Rev. Vol. 112 (2012), p. 673.
[3] T. Ben, H. Ren, S.Q. Ma, D.P. Cao, J.H. Lan, X.F. Jing, W.C. Wang, J. Xu, F. Deng and J.M. Simmons: Angew. Chem. Int. Ed. Vol. 48 (2009), p. 9457.
[4] G.Z.Voronoi: J. Reine Angew. Math. Vol. 134 (1908), p. 199.
[5] T.F. Willems, C.H. Rycroft, M. Kazi, J.C. Meza and M. Haranczyk: Microporous Mesoporous Mater. Vol. 149 (2012), p. 134.
[6] M. Pinheiro, R.L. Martin, C.H. Rycroft, A. Jones, E. Iglesia and M. Haranczyk: J. Mol. Graph. Model. Vol. 44 (2013), p. 208.
[7] E.W. Dijkstra: Numerische Math. Vol. 1 (1959), p. 269.
[8] R.L. Martin and M. Haranczyk: Cryst. Growth. Des. (2014).
[9] M. O'Keeffe, M.A. Peskov, S.J. Ramsden and O.M. Yaghi: Acc. Chem. Res. Vol. 41 (2008), p. 1782.
[10] R.L. Martin and M. Haranczyk: Chem. Sci. Vol. 4 (2013), p. 1781.
[11] R.L. Martin and M. Haranczyk: J. Chem. Theory Comput. Vol. 9 (2013), p. 2816.
[12] R.L. Martin and M. Haranczyk: Cryst. Growth Des. vol. 13 (2013), p. 4208.
[13] D. Yuan, W. Lu, D. Zhao and H.-C. Zhou: Adv. Mater. Vol. 23 (2011), p. 3723.
[14] R.L. Martin, M.N. Shahrak, J.A. Swisher, J. P. Sculley, H.-C. Zhou, B. Smit and M. Haranczyk: J. Phys. Chem. C vol. 117 (2013), p. 20037.

Improvements in the sensibility of MSA-GA tool using COFFEE objective function

A R Amorim[1], G F D Zafalon[1], L A Neves[1], A R Pinto[2], C R Valêncio[1], J M Machado[1]

[1] Department of Computer Science and Statistics (DCCE), São Paulo State University (UNESP), São José do Rio Preto, Brazil.
[2] Department of Control Engineering and Automation, Federal University of Santa Catarina, Blumenau, Brazil.

E-mail: anderson.rici@sjrp.unesp.br

Abstract. The sequence alignment is one of the most important tasks in Bioinformatics, playing an important role in the sequences analysis. There are many strategies to perform sequence alignment, since those use deterministic algorithms, as dynamic programming, until those ones, which use heuristic algorithms, as Progressive, Ant Colony (ACO), Genetic Algorithms (GA), Simulated Annealing (SA), among others. In this work, we have implemented the objective function COFFEE in the MSA-GA tool, in substitution of Weighted Sum-of-Pairs (WSP), to improve the final results. In the tests, we were able to verify the approach using COFFEE function achieved better results in 81% of the lower similarity alignments when compared with WSP approach. Moreover, even in the tests with more similar sets, the approach using COFFEE was better in 43% of the times.

1. Introduction

Some studies performed in biology have brought many improvements, specially in human genetics in the last years. However, these studies produce a huge amount of data, which must be refined for a more accurate analysis and after to propose some inferences.

Nowadays, the use of computational tools is necessary to provide a good genomic analysis, which can be classified as Bioinformatics. Thus, Bioinformatics is a computer science branch to solve biological problems, mainly related to multiple sequence alignments (MSA) and pattern recognition [1].

In this work, we proposed the implementation of COFFEE objective function in the multiple sequence alignment tool MSA-GA. Thus, it is possible to reach alignments with more biological significance for some sequence sets, specially those ones with lower similarity.

This work is organized as follows: in the section 2 a brief review about sequence alignment and genetic algorithm is provided. In the section 3 are described the materials and methods with the special attention to the implementation of the objective function in the MSA-GA tool. Some analysis and results are presented in the section 4. Finally, in the section 5, the conclusions and future perspectives are presented.

2. Sequence Alignment and Genetic Algorithms

The exact score of a sequence alignment can be obtained using dynamic programming algorithms, as Needleman-Wunsch [2]. However, this type of algorithm has high computational costs, becoming the alignments with more than two sequences unfeasible. Thus, to reduce the computational complexity, the multiple sequence alignment algorithms were developed, which can be based in many heuristics as Progressive Alignment, Simulated Annealing, Genetic Algorithms, and others [3].

Genetic Algorithms might be used to solve MSA problems through the Evolutionary Theory, where the participants are submitted to processes of mutation, recombination and gene selection to evolve the candidate alignments, which are measured by an objective function. Generally, in this heuristics, the alignment module is independent of score function, therefore the objective function can be implemented without changes in the alignment routines.

3. Materials and Methods

The MSA-GA [4] is a tool for MSA, which uses genetic algorithm as its base. The choice of this tool to the development of this work was done due to its good results when compared with other MSA tools. Moreover, the MSA-GA has a good modularity which allows the implementation of new objective functions in a feasible way. The MSA-GA uses the Weighted Sum-of-Pairs (WSP) as the objective function. However, the WSP has some limitations where regions with low similarity can cause some distortions in the final alignments.

In this context, a new score method proposed by Notredame [5], named COFFEE (Consistency based Objective Function For alignmEnt Evaluation), was implemented in the MSA-GA tool. The choice of this objective function was performed due to its approach is based in consistency, which smooths the noise caused by regions with low similarity, resulting in alignments with more biological significance even in sequences with low similarity.

Basically, the COFFEE function needs two components: the reference pairwise alignment set, which is called library, and an objective function which analyzes the consistence between a MSA and the pairwise alignments. In this work, the Needleman-Wunsch algorithm was used to build the library. To reduce the noise caused by sequences with low similarity, each pairwise alignment has an weight, which is related to the number of matches in the aligned sequences belonged to the library. The consequence is that the final alignment priviledges closer sequences instead of more distant ones.

Thus, the function score routine is based in the comparison between each pair of aligned residues with those belonged to the library, as presented in the Figure 1. Basically, for each column of multiple alignment, a residue matrix is declared. If the pair of residues is found in the pairwise alignment, the matrix cell is filled with the weight of the alignment.

The score of each column is reached by sum of matrix elements, divided by sum of the alignment weights in the library. Finally, the general score of consistency is equal to the sum of scores of columns, divided by the numbers of pairs of multiple alignment.

The COFFEE function in defined by Equation (1), where N is the number of sequences, LEN(A(i,j) is the length of the alignment, SCORE(A(i,j)) is the number of pairs of aligned residues shared between A(i,j) and the library, and W(i,j) is the weight of pairwise alignment.

$$COFFEE\ score = \frac{\left[\sum_{i=1}^{N-1}\sum_{j=i+1}^{N} W_{i,j} \times SCORE(A_{i,j})\right]}{\left[\sum_{i=1}^{N-1}\sum_{j=i+1}^{N} W_{i,j} \times LEN(A_{i,j})\right]} \quad (1)$$

4. Tests and Results

To analyze the improvements of this work, we used the test case sets from BAliBase [6]. This benchmark offers many sequence sets divided into different reference categories, where there is the possibility of an accurate comparison of obtained MSA with reference alignments. This comparison is performed using the BAliScore tool, which gives a score of biological significance, where 0 is the worst and 1 is the best alignment.

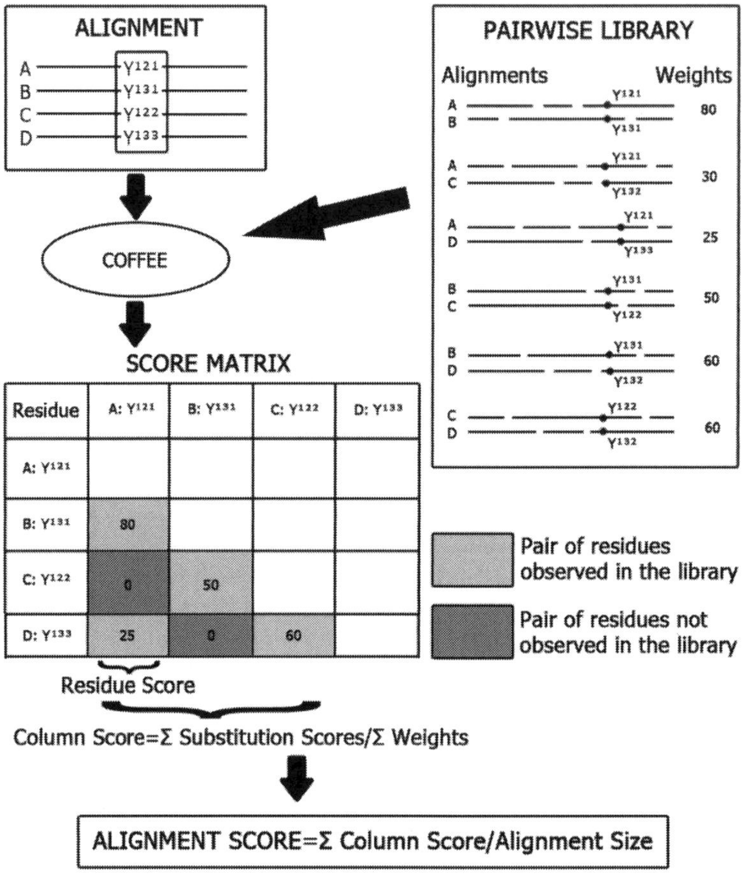

Figure 1. COFFEE function scheme.

In the tests, the cases of Reference 1 were selected, all with equidistant sequence sets with different conservation. The parameters used by genetic algorithm, in both approaches, were patterns of MSA-GA. Thus, in the Table 1 are shown the obtained scores of COFFEE comparing with WSP.

Table 1. COFFEE and WSP scores in the BAliScore.

Sequence Set	MSA-GA (COFFEE)	MSA-GA (WSP)	Sequence Set	MSA-GA (COFFEE)	MSA-GA (WSP)	Sequence Set	MSA-GA (COFFEE)	MSA-GA (WSP)
<25% identity			**20% ~ 40% identity**			**>35% identity**		
1idy	**0,4788**	0,4538	1ycc	**0,6520**	0,5488	1amk	**0,9176**	0,8874
1tvxA	0,2532	**0,2748**	1ad2	0,7408	**0,7498**	1aho	0,7386	**0,7848**
Kinase	**0,3532**	0,2940	1aym3	**0,8650**	0,8526	1csp	**0,8754**	0,8028
1r69	**0,1290**	0,1176	1fieA	**0,8058**	0,7902	1ar5A	0,8142	**0,8376**
1ubi	**0,5756**	0,5752	1ldg	**0,8634**	0,8546	1ad3	0,5872	**0,6394**
1wit	**0,5792**	0,5548	1sesA	**0,5476**	0,5416	1gpb	**0,8058**	0,7902
1ped	0,3906	**0,3968**	3cyr	**0,6302**	0,6088	1krn	0,8964	**0,9052**
2myr	**0,1926**	0,1548						

Thus, it can be noticed the COFFEE function was able to improve the sensibility of the final alignment of MSA-GA, where in 81% of the cases, the new strategy proposed here has reached better results in sequence sets of low similarity (<25% identity, 20% ~ 40% identity) , when compared with WSP as showed in Figure 2(a). Moreover, the COFFEE function has reached good results in sequence sets with high similarity (>35% identity), with better results in 43% of the cases, when compared with WSP, as can be seen in Figure 2(b).

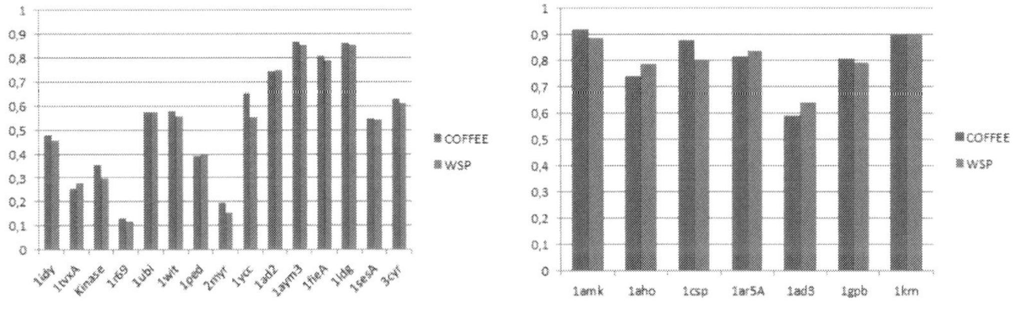

(a) (b)

Figure 2. BAliScore score: (a) sets with low similarity, (b) sets with high similarity.

5. Conclusion

The novelty proposed in this work is very important, because, through the implementation of COFFEE function into the MSA-GA, this tool was able to reach alignments with more biological significance, even for sequence sets with low similarity and also good results for those with high similarity. Moreover, some characteristics of genetic algorithm allow a parallel implementation of it, which will be conducted using multithreading strategy to reduce the execution time of the tool.

6. Acknowledgments

This work was financially supported by São Paulo Research Foundation (FAPESP), under grant 2013/08289-0.

References

[1] Edgar R C and Batzouglou S 2006 Multiple sequence alignment *Current Opinion in Structural Biology* **16** 368–373

[2] Needleman S B and Wunsch C D 1970 A general method applicable to the search for similarities in the amino acid sequence of two proteins *Journal of molecular biology* **48** 443-453

[3] Zafalon G F et al. 2013 Improvements in the score matrix calculation method using parallel score estimating algorithm *Journal of Biophysical Chemistry* **4** 47-51

[4] Gondro C and Kinghorn B P 2007 A simple genetic algorithm for multiple sequence alignment *Genetics and Molecular Research* **6** 964-982

[5] Notredame C, Holm L and Higgins D G 1998 COFFEE: an objective function for multiple sequence alignments *Bioinformatics* **14** 407-422

[6] Thompson J D, Koehl P, Ripp R and Poch O 2005 BAliBASE 3.0: latest developments of the multiple sequence alignment benchmark *Proteins: Structure, Function, and Bioinformatics* **61** 127-136

Analytic approximations to the energy eingenvalues of the quadratic Zeeman effect in two dimensions for hydrogenlike atoms

Citti Flavio[1], Castro Enrique[2], Martin Pablo[2,3], Cortés-Vega Luis[4]

1 Materials Engineering Department, Simón Bolivar University, Apartado 89000, Caracas 1080A, Venezuela
2 Physics Department, Simón Bolivar University, Apartado 89000, Caracas 1080A, Venezuela
3 Physics Department, Antofagasta University, Casilla 170, Antofagasta, Chile
4 Mathematics Department, Antofagasta University, Casilla 170, Antofagasta, Chile

E-mail: flaviocitti@gmail.com, ecastro@usb.ve, pablo.martin@uantof.cl, luis.cortes@uantof.cl

Abstract. Analytic energy eigenvalues approximations have been found for the 2-D quadratic Zeeman effect in hydrogenlike atoms. Here a new technique has been used, the so called multipoint quasirational technique (MPQA), which is an extension of the two-point quasirational approximation (TPQA) technique, previously used for the same problem. In our previous calculations, two expansions were used, one for small values of the magnetic field and other for large values, that is, expansions around zero and the infinity were determined. Later a bridge between both expansions was built using simple auxiliary functions as well as rational ones. In the present paper expansions around intermediate points were also determined, and used. In this way, the corresponding analytic bridge function is determined, taken care also of these new expansions, so the technique is a multiple point quasirational approximation (MPQA). The calculations are carried out for the ground state $1s$ and the excited states $2p^-$ and $3d^-$, respectively. These results are now much better than those obtained in previous works.

1. Introduction

Approximate analytic solutions for the quadratic Zeeman effect in 2-D have been found for hydrogenlike atoms in a magnetic field in 2-D for systems where the Coulomb energy scale is about the magnetic interaction are considered [1, 2]. This is the case, for instance, of conduction electrons in GaAS. Since neither the Coulomb energy, neither the magnetic energy can be considered small parameters, the linear approximation, denoted as Zeeman effect, is not a good approximation and the quadratic Zeeman effect should be taken in account. The characteristic parameter γ is of order one, where γ is the ratio between magnetic and Coulomb energies, that is, $\gamma = (\frac{\mu_B B}{R_0^*}) = (\frac{\epsilon^2 \hbar^3 B}{ce^3 m^{*2}})$, here B is the induction magnetic field, R_0^* is the effective Rydberg constant, m^* is the effective mass of the conduction electrons, $\hbar$ is the Planck constant, $(-e)$ is the electron charge, c is the speed light and μ_B is the Bhor magneton. The Hamiltonian with energy in units of R_0^* and position in units of effective Bhor radius $a_0^* = \frac{m^* e^4}{2\epsilon^2 \hbar^2}$ is

$$H = -\nabla^2 + \gamma L \; - \frac{2}{\;} + \gamma^2 \rho^2, \tag{1}$$

where ∇^2 is two-dimensional Laplacian (i.e., $a_0^* \approx 100\mathring{A}$ for GaAs) given by

$$\nabla^2 = \frac{\partial^2}{\partial \rho^2} + \frac{1}{\rho}\frac{\partial}{\partial \rho} - \frac{m^2}{\rho^2}, \quad m = 0, \pm 1, \pm 2, \ldots .$$

where m is the azimuthal number , the magnetic field is considered to be in z-direction, and the symmetric gauge $A = B(-y, x, 0)/2$ is used. Considering the solution of (1) as $\psi(\rho) = \frac{1}{\sqrt{2\pi}} R(\rho) e^{im\phi}$, then the Schrdinger equation becomes the second order differential equation [1]:

$$-\frac{d^2}{d\rho^2} - \frac{1}{\rho}\frac{d}{d\rho} + \frac{m^2}{\rho^2} - \frac{2}{\rho} + m\gamma + \frac{\gamma^2\rho^2}{4}\Big) R(\rho) = E\, R(\rho). \quad (2)$$

No analytic solutions are known for equation (2). Several methods have been used to obtain approximate solutions, such as Padé Method [1], the two-point quasi-rational approximants TPQA [3], $1/N$ shifted technique [7] and numerical computations [4]. Here a more modern technique will be used, the so called MPQA method of approximation where several points will be used simultaneusly, including the value at the points and several derivative [5, 9].

2. Multipoint Quasi-Rational Approximants

In this method there is not need to use the asymptotic expansion, thought this can be also used, and instead of that several intermediate point between zero and infinite will be considered.The numerical value of the function and its derivatives of first and second order will be used, and they will be determined by numerical computations. Once the form of the approximant has been decided, their parameters will be determined by equalizing its values and derivatives to those obtained by numerical computations as will be explained bellow. In our case, the functions to approximate are the eigenvalues of equation (2), and the form of the approximants will be

$$\widetilde{E}(\gamma) = \frac{P_L(\gamma)}{Q_M(\gamma)} = \frac{\sum_{k=0}^{L} p_k \gamma^k}{1 + \sum_{i=0}^{M} q_i \gamma^i}. \quad (3)$$

As in Padé method, $P_L(\gamma)$ and $Q_M(\gamma)$ are polinomials in γ of degree L and M, respectively. The coefficients p_k and q_i of the approximate are determined by imposing that $\widetilde{E}(\gamma)$ and $E(\gamma)$, and its first and second derivatives should coincide in some chosen points γ_r $(r = 1, 2, \ldots, n)$ as follows

$$Q_M(\gamma_r)\widetilde{E}(\gamma_r) - P_L(\gamma_r) = 0 \quad (4)$$

$$\frac{dQ_M(\gamma_r)}{d\gamma}\widetilde{E}(\gamma_r) + Q_M(\gamma_r)\frac{d\widetilde{E}(\gamma_r)}{d\gamma} - \frac{dP_{M+1}(\gamma_r)}{d\gamma} = 0 \quad (5)$$

$$\frac{d^2Q_M(\gamma_r)}{d^2\gamma}\widetilde{E}(\gamma_r) + 2\frac{dQ_M(\gamma_r)}{d\gamma}\frac{d\widetilde{E}(\gamma_r)}{d\gamma} + Q_M(\gamma_r)\frac{d^2\widetilde{E}(\gamma_r)}{d\gamma^2} - \frac{d^2P_L(\gamma_r)}{d\gamma^2} = 0. \quad (6)$$

The number of equations must be equal to the number of parameters to be determined. The relation between L and M must be determined first in order to produce the largest accuracy with the same number of parameters. Several cases were tested and the best results were obtained when $L = M + 1$. Now, increasing the degree of the polynomials the accuracy increases. Sometimes as in our case, it is better to leave a free parameter, by using a number of equations one unit lesser than the number of unknowns, and determining the free parameter by minimizing the error between the approximant and the numerical solution. In the present work the coefficient q_M was taken as the free parameter. The Table 1 shows the coefficients for $M = 4$.

Coeficcients	$\tilde{E}_{1s}$	$\tilde{E}_{2p^-}$	$\tilde{E}_{3d^-}$
p_0	0.160000	−0.444450	−4.000002
p_1	59.9455	−6.58034	−3.069952
p_2	1882.11	−32.3552	−1.254528
p_3	6363.85	−33.2901	−0.1089466
p_4	−3991.15	16.0552	0.02184713
p_5	−4641.90	6.63643	0.0008673304
q_0	1	1	1
q_1	−390.215	−12.5657	0.7674965
q_2	−6438.80	50.4000	0.3370095
q_3	−16263.9	50.1129	0.04541118
q_4	−5648.44	7.94572	0.000987733

Table 1: Values of the coefficients of the parameters $p's$ amd $q's$ for the MPQA ($L = 5$, $M = 4$) for the $1s$, $2p^-$ and $3d^-$ states, respectively.

3. Results and Conclusions

The accuracy of the MPQA depends first on the accuracy with which the eigenenergies and its derivatives were determined by numerical computation. The numerical calculations of the eigenvalues was performed with the shooting method using the equation (2). The derivatives were determined by central differences. The calculated values for the coefficients of the MPQA, for the $1s$, $2p^-$ and $3d^-$ states, are shown in Table 1. In figure 1, skeths of the eingenvalues as a fuction of $\gamma' = \frac{1+\gamma}{\gamma}$ as shown.. The number of parameters to determine were equal to those used in [3], using the TPQA method. This was necessary in order to obtain an adequate comparison.

States	MPQA	MPQA	MPQA	TPQA	TPQA	TPQA
	ϵ_{max}	$\epsilon^{\%}$	γ'_{ϵ}	ϵ_{max}	$\epsilon^{\%}$	γ'_{ϵ}
$1s$	4.7×10^{-6}	1.2×10^{-4}	0.5	1.7×10^{-4}	2.7×10^{-3}	0.95
$2p^-$	5.6×10^{-6}	1.3×10^{-3}	0	4.2×10^{-5}	1.0×10^{-2}	0.5
$3d^-$	3.8×10^{-6}	1.9×10^{-3}	0.35	1.2×10^{-5}	8.6×10^{-3}	0.05

Table 2: Maximum errors for the MPQA and TPQA of higher order for the three electron states considered.

In Table 2, the MPQA and TPQA method are compared. Since the TPQA approximants are the best analytic approximation published until now, there is not need to compare with other ones, as for instance, MacDonald et, al. [1]. The best way to do it is by looking at the maximum errors for the parameters to be determined. The accuracy of the MPQA is high and better than those previously published by Martin P. et, al., showing better results and arround one or two orders of magnitude lower error than the TPQA method. The MPQA technique looks like a worthwhile technique to obtain approximate eigenvalues for some potentials in Quantum Mechanics. The range of applications seems ampler than those of other approximation techniques, and the accuracy obtained is better for the same number of parameters to determine. An additional advantage is that the form of the approximants can be simpler than that of the TPQA method.

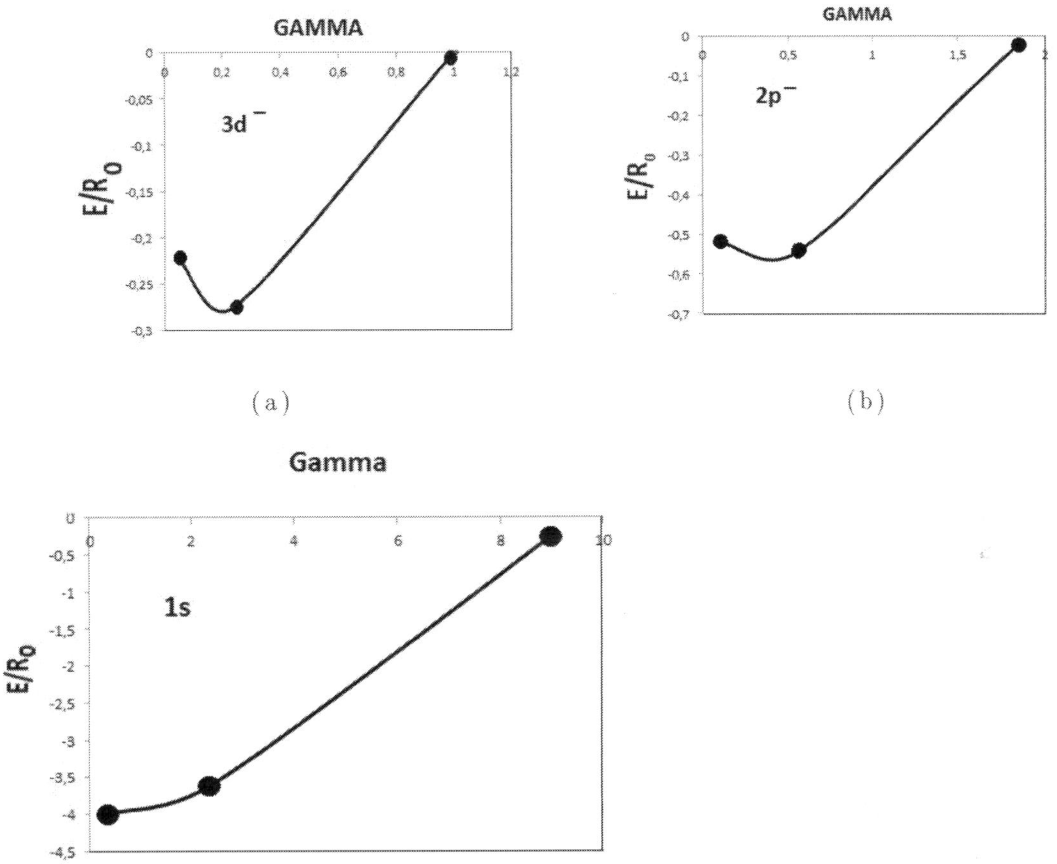

(a) (b)

(c)

Figure 1. 2D-hydrogenic energy curves in a magnetic field for the states $3d^-$ excite state, see plot (a); $2p^-$ excite state, see plot (b) and $1s$ ground state, see plot (c). The three circles in each curve show the chosen points used to get the numeric information required to determine the MPQA coefficients

4. References

[1] McDonald A. H. and Richie D. S., "Hydrogenic energy levels in two dimensions at arbitrary magnetic fields". Phys. Rev. B **33**, 8336 (1986)

[2] Martin P., Castro E., Paz J.L and De Freitas A., "Multipoint quasi-rational approximants in Quantum Chemistry", chapter 3 in New developments in Quantum Chemistry, Paz, J.L. and Hernandez A., Editors. (Trans work Research Network Trivandrum-695023, Kevala India, 2009), pp. 55-77

[3] Martin P., Rodrguez N. J. J. and Marquez J. L., Two-dimensional hydrogenlike atoms in the presence of a magnetic field: Quasifractional approximations. Phys. Rev. B **45**, 8359 (1992)

[4] Villalba V. M. and Pino R.,"Analytic Computation of the enrgy Levels of a Two-dimensional Hydrogenic Donor in a Constant Magnetic Field". Physica Scripta **58**, 605 (1998)

[5] Castro E. , Paz J. L. and Martn P. , Analytical approximations to the eigenvalues of the Morse potential with centrifugal terms. Journ. Mol. Structure: THEOCHEM **769**, 15 (2006)

[6] Castro E., Martn P. and Paz J. L., Precise spectra for the H2 molecule by a new approximate technique. Phys. Letters A **364**, 135 (2007)

[7] Mustafa O., The shifted-1/N-expansion method for two-dimensional hydrogenic donor states in an arbitrary

[8] P. Martin, J.L. Paz, G. Donoso, L. C. Ladera and J. Puerta,"Multi-Point Quasi-Rational Approximants: Applications to the Special Functions, Physics and Chemistry", Boletin Academia Ciencias Fsicas, Matemáticas y Naturales, Vol. LXXIL., pp. 53-74 (2012)

[9] P. Martin, E. Castro and J.L. Paz "Multi-point quasi-rational approximants for the energy eingenvalues two-power potentials, Rev. Mexicana Fsica, **58**, 301-307 (2012)

High Performance Computing of Meshless Time Domain Method on Multi-GPU Cluster

Soichiro Ikuno[1], **Susumu Nakata**[2], **Yuta Hirokawa**[3], **and Taku Itoh**[1]
[1]School of Computer Science, Tokyo University of Technology, Hachioji, Tokyo 192-0982, Japan
[2]Institute of Science and Engineering, Ritsumeikan University, Kusatsu, Shiga 525-8577, Japan
[3]Graduate School of Systems and Information Engineering, University of Tsukuba, Tsukuba, Ibaraki 305-8573, Japan

E-mail: ikuno@nal.ikulab.org

Abstract. High performance computing of Meshless Time Domain Method (MTDM) on multi-GPU using the supercomputer HA-PACS (Highly Accelerated Parallel Advanced system for Computational Sciences) at University of Tsukuba is investigated. Generally, the finite difference time domain (FDTD) method is adopted for the numerical simulation of the electromagnetic wave propagation phenomena. However, the numerical domain must be divided into rectangle meshes, and it is difficult to adopt the problem in a complexed domain to the method. On the other hand, MTDM can be easily adept to the problem because MTDM does not requires meshes. In the present study, we implement MTDM on multi-GPU cluster to speedup the method, and numerically investigate the performance of the method on multi-GPU cluster. To reduce the computation time, the communication time between the decomposed domain is hided below the perfect matched layer (PML) calculation procedure. The results of computation show that speedup of MTDM on 128 GPUs is 173 times faster than that of single CPU calculation.

1. Introduction

The finite difference time domain (FDTD) method is often applied for the electromagnetic wave propagation simulation. FDTD has an advantage in terms of parallelization and treatment of problems because the method is based on complete explicit finite difference method (FDM). However, the numerical domain must be divided into rectangle meshes. Therefore, it is difficult to treat the problem in a complex shaped numerical domain.

On the other hand, the meshless approach such as the element-free Galerkin method, the meshless local Petrov-Galerkin method and the radial point interpolation method (RPIM) [1] do not require finite elements or meshes of a geometrical structure. In particular, meshless approaches based on RPIM are applied to time dependent problems, and the method is called meshless time domain method (MTDM) [2].

In recent years, various researches of the general purpose computing on GPU (GPGPU) have been proposed [3, 4]. In previous study, we investigated the speedup of MTDM using a single GPU as well, and the parallelized method on a single GPU achieved about eight times faster than that of CPU with OpenMP [4].

The purpose of the present study is to implement MTDM method for analyzing electromagnetic wave propagation phenomena in a long waveguide on multi-GPU, and to

Table 1. Evaluation environment.

CPU	Intel E5 2.6 GHz x2 (Sandy-Bridge EP)
CPU memory	128 GB
GPU	NVIDIA M2090 ×4
GPU memory	24 GB
Infiniband	Mellanox Connect-X3 dual port QDR
compiler version and options	Intel 14.0.0 (-O3 -xhost -openmp)
CUDA version and options	5.5 (-O3 -use_fast_math -arch=sm_20)
MVAPICH2	2.0 (GPU Optimized, not GDR)
Intel MPI	4.1.3.049

investigate the details of the performance of the method. Note that the supercomputer HA-PACS (Highly Accelerated Parallel Advanced system for Computational Sciences) at University of Tsukuba is adopted for evaluations.

2. Meshless Time-Domain Method, MTDM

The basic concept of the MTDM is same as standard FDTD method, and the governing equations are completely same as FDTD as well. The equations are discretized in respect to time using Leap-Flog scheme. On the other hand, physical quantities on scattered node $\boldsymbol{x}_i$ $(i = 1, 2, \ldots n)$ in analytic domain $u(\boldsymbol{x}_i)$ (such as a electric field $E(\boldsymbol{x}_i)$ and a magnetic field $H(\boldsymbol{x}_i)$) is discretized in respect to space using the shape function $\phi_i(\boldsymbol{x}_j)$ which is derived from RPIM as follows.

$$u(\boldsymbol{x}_i) = \sum_{j=1}^{n} \phi_i(\boldsymbol{x}_j) u_i = u_i. \quad (1)$$

Note that the shape functions obtained by RPIM satisfy the Kronecker's delta function property, i.e.,

$$\phi_i(\boldsymbol{x}_j) = \begin{cases} 1; & i = j, \\ 0; & i \neq j. \end{cases} \quad (2)$$

In the present study, a two dimensional long straight waveguide with width 5 cm is adopted for evaluation. The perfectly matched layer (PML) and the perfect electric conductor (PEC) are used for absorbing boundary condition on edges and sides of the waveguide, respectively. In addition, the number of nodes for width direction is fixed as 51.

3. Parallelization Strategies

As we mentioned above, the supercomputer HA-PACS at University of Tsukuba is adopted for evaluations. HA-PACS is constructed by 332 nodes, and the peak performance of the system is about 1.166 PFLOPS. Dual CPU (Intel E5 2.6 GHz, Sandy-Bridge EP) and four GPUs (NVIDIA M2090) are implemented on a unit node. Additionally, nodes are connected by the Infiniband. The detail of the evaluation environment is shown in Table 1.

The major modification to the present study from the previous study is shown in below [4].

- Parallelized by using multi-GPU, and 32 nodes is used for evaluation (i.e. 128 GPUs).
- Optimize the memory access in the code by using CUDA Coalesced Access.
- Read-Only Cache (ROC) for Kepler generation is adopted for inner product calculation.

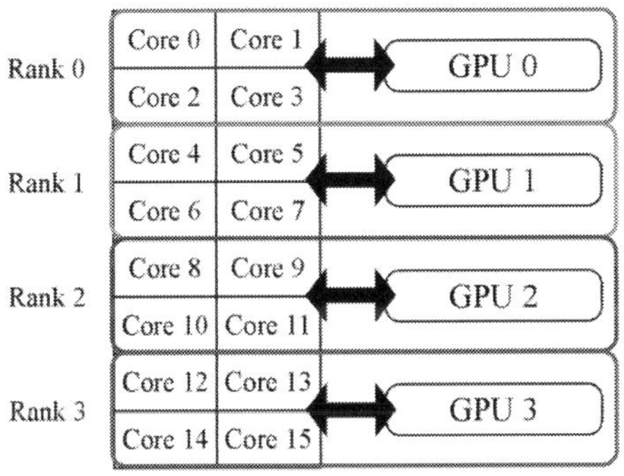

Figure 1. The schematic view of the parallelization strategy using MPI on unit node of HA-PACS.

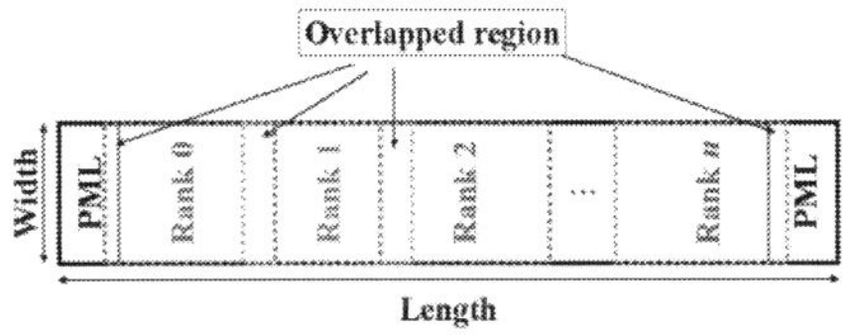

Figure 2. The schematic view of the analytic model of the straight waveguide.

- Texture memory on GPU is adopted for PML calculation.

In Figure 1, we show the schematic view of the parallelization strategy using Message Passing Interface (MPI) on a unit node of HA-PACS. As we mentioned above, two CPUs are implemented on a unit node, and a CPU has four cores. Thus, we divided into four Ranks in unit node as shown in Figure 1. Each Rank is communicated by using MPI even if GPUs are implemented on same node, and CUDA and OpenMP are used for parallelization.

The analytic domain is decomposed into n cells that including overlapped region for task distribution as shown in Figure 2. Each Rank is associated a cell, and data are communicated between neighboring cells using Peer to Peer communication (GPU Direct) to reduce the communication time [5]. Note that PML calculation is achieved at same time of data communication for overlapped region to hide the transmission time of the data.

4. Results and Discussions

Let us investigate the performance of the method using multi-GPU. To this end, the weak scaling and the strong scaling are calculated. In the weak scaling calculation, the length for each Rank is fixed as 10 m, and the node number of the length direction for MTDM is fixed as 10001 on a unit Rank. On the other hand, in the strong scaling calculation, the length of the waveguide is fixed as 100 m, and the node number of the length direction for MTDM is fixed as 100001.

In Figure 3, we show the weak scaling (the performance efficiency) of MTDM using the multi-GPU. We can see from this figure that the performance efficiency of MTDM using multi-GPU is about 97 %. This result indicates that the communication time between node of HA-PACS becomes suppressed under 3 %.

Finally, we calculate the strong scaling of MTDM using multi-GPU, and the values of speedup are shown in Figure 4 and the value of execution time is shown in Table 2. Note that the origin of the speedup (i.e. speedup = 1) is performed on a unit CPU, and the code is parallelized by using OpenMP. We can see from this figure that the value of the speedup increases lineally from a unit node to 16 node. However, the rate of increase slows down at 32 node. This is because associated node number of MTDM for unit Rank becomes about 40000 points, and the calculation cost becomes relatively smaller than the communication time.

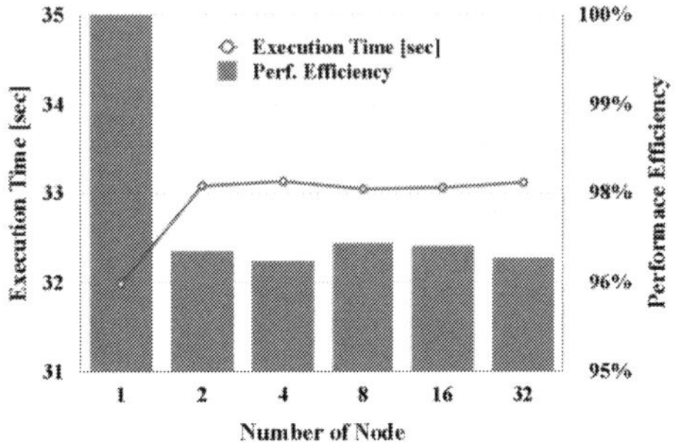

Figure 3. The performance efficiency and the execution time of MTDM using multi-GPU.

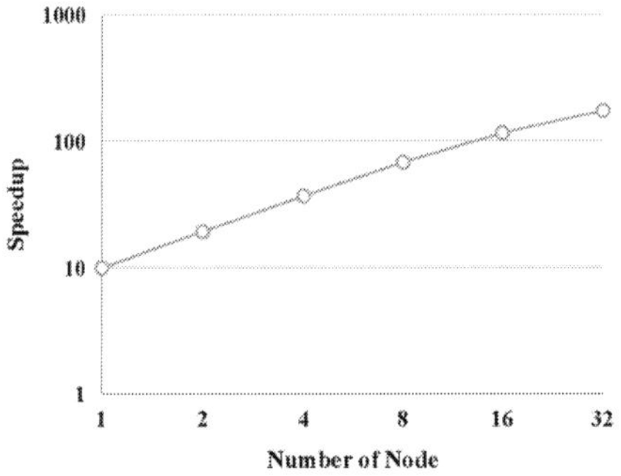

Figure 4. The speedup of MTDM using multi-GPU.

Table 2. Execution time of MTDM using multi-GPU.

Number of node	Execution time [sec]
CPU (origin of speedup)	7733.5391
1	780.70538
2	402.84217
4	210.99577
8	114.43655
16	67.028203
32	44.689656

Acknowledgment

This work was supported in part by Japan Society for the Promotion of Science under a Grant-in-Aid for Scientific Research (C) No. 26390135. And the numerical calculations for the present work was carried out under the "Interdisciplinary Computational Science Program" in Center for Computational Sciences, University of Tsukuba.

References

[1] J. G. Wang, G. R. Liu, *Int. J. Numer. Meth. Engng*, Vol. 54, No. 11, pp. 1623-1648, 2002.

[2] T. Kaufmann, C. Engstrom, C. Fumeaux, R. Vahldieck, *IEEE Trans. Microwave. Theo. Tech.*, Vol. 58, No. 12, pp. 3399-3408, 2010.

[3] N. Godel, S. Schomann, T. Warburton, and M. Clemens, "GPU Accelerated AdamsBashforth Multirate Discontinuous Galerkin FEM Simulation of High-Frequency Electromagnetic Fields", *IEEE Trans. Magn.*, **46**, No. 8, (2010) 2735-2738.

[4] S. Ikuno, Y. Fujita, Y. Hirokawa, T. Itoh, S. Nakata, and A. Kamitani "Large-Scale Simulation of Electromagnetic Wave Propagation Using Meshless Time Domain Method With Parallel Processing", *IEEE Trans. Magn.*, **49**, No. 5, (2013) 1613-1616.

[5] S. Ikuno, Y. Kawaguchi, N. Fujita, T. Itoh, S. Nakata, and K. Watanabe "Iterative Solver for Linear System Obtained by Edge Element: Bariable Preconditioned Method With Mixed Precision on GPU" *IEEE Trans. Magn.*, **48**, No. 2, (2012) 467-470.

Journal of Physics: Conference Series **574** (2015) 012107 doi:10.1088/1742-6596/574/1/012107

Precise Approximate Solution for the Bohm Sheath Potential

P. Martin[1,2], L. Cortés-Vega[3], F. Maass-Artigas[1]

1 Physics Department, Antofagasta University, Casilla 170, Antofagasta, Chile
2 Physics Department, Simón Bolivar University, Apartado 89000, Caracas 1080A, Venezuela
3 Mathematics Department, Antofagasta University, Casilla 170, Antofagasta, Chile

E-mail: pmartin@usb.ve or pablo.martin@uantof.cl, luis.cortes@uantof.cl, fernando.maass@uantof.cl

Abstract. The Poisson equation for the plasma sheath potential near a wall, leads to a non-linear differential equation, whose analytic solution is not known. The usual approximation taking only the first non-null term does not give good accuracy. Other approximations taken two additional terms are better, but they fail to give good accuracy in the intermediate region. Here a new analytic approximated solution is presented with much higher accuracy, and more precise results, not only near and far away of the wall, but also in the transition region. Two figures showing these new analytic solutions as a function of the relevant parameters are presented. The advantages of the present solution compared with those of pervious works are shown.

1. Introduction

The plasma potential near a wall has been treated in several ways and the most classic is the Bohm's treatment [1-3]. The Bohm's model is given by a non-linear second order differential equation of the form

$$\frac{\partial^2\varphi}{\partial x^2} = \frac{ne}{\epsilon_0}\left(exp\left(\frac{\epsilon\varphi}{k_BT}\right) - \sqrt{\frac{\frac{1}{2}mv^2}{\frac{1}{2}mv^2 - \epsilon\varphi}}\right), \tag{1}$$

where v is a characteristic ion velocity, φ is the potential, n is the ion and electron plasma density far away from the wall, k_B is the Boltzmann constant, m is the mass, T is the electron and ion temperature, and x is the distance to the wall. No analytic solution is known of Eq. (1), furthermore a solution of this equation will slow the position as a function of the potential, but the most usual need is the potential as a function of the distance to the wall [2,3]. In this contex, the procedure usually used to obtain φ as a function of x, is to expand all the functions as a power of φ and to keep only the term with the lower power in φ. However the accuracy of this approximation is low, and furthermore the results for $K \leq \frac{1}{2}$, are not suitable. A better aproximation will be shown here, which is also simpler than previously presented one [4].

2. Theoretical Model

It is convenient to use dimensionless quantities and variables, so using dimensionless variables:

$$\tilde{y} = \frac{x}{\lambda_D}, \quad K = \frac{\frac{1}{2}mv^2}{k_BT}, \quad \frac{1}{\lambda_D^2} = \frac{2ne^2}{\epsilon_0 k_BT},$$

where λ_D is the plasma Debye length, K is a dimensionless constant, and $\tilde{y}$, ϕ are dimensionless variables. So, Eq. (1) becomes

$$-2\frac{d^2\phi}{d\tilde{y}^2} = e^{-\phi} - \frac{1}{\sqrt{1+\frac{\phi}{K}}}. \tag{2}$$

As it is well known, by using the new variable

$$\psi = \frac{d\phi}{d\tilde{y}},$$

can be reduces (2) to the integral equation:

$$\psi^2(\phi) - \psi^2(0) = \int_0^\phi \left(\frac{1}{\sqrt{1+\frac{\phi}{K}}} - e^{-\phi}\right) d\phi. \tag{3}$$

It is convenient to consider the plasma potential as the reference, that is, ϕ is zero for $y = \infty$, and then $\psi(0)$ is also zero. In a more precise way, ϕ will be zero at the end of the sheath and beginning of the presheath. In this way, the sheath potential can be calculated by the non-linear first order equation

$$\frac{d\phi}{d\tilde{y}} = \sqrt{F(\phi)},$$

where

$$F(\phi) = 2K\sqrt{1+\frac{\phi}{K}} - 2K + e^{-\phi} - 1. \tag{4}$$

Now, by using a Taylor expansion for $F(\phi)$, it is obtained

$$F(\phi) = \phi^2\left[\left(\frac{1}{2} - \frac{1}{4K}\right) + \phi\left(\frac{1}{8K^2} - \frac{1}{6}\right) + \phi^2\left(\frac{1}{24} - \frac{5}{64K^3}\right) + \phi^3\left(\frac{7}{128K^4} - \frac{1}{120}\right) + \ldots\right].$$

So, $F(\phi)$ can be written as

$$F(\phi) = \left(\frac{2K-1}{4K}\right)\phi^2\left[1 + \phi\left(\frac{3-4K^2}{12K^2-6K}\right) + \phi^2\left(\frac{18K^3-15}{96K^3-48K^2}\right) + \ldots\right]. \tag{5}$$

The usual solution is obtained by keeping the first term of the expansion, giving

$$\phi_2(y) = \phi_w\, exp\left(-\frac{1}{2}\sqrt{\frac{2K-1}{K}}y\right) = \phi_w exp\left(-\lambda y\right),$$

where ϕ_w is the potential at the wall and ϕ_2 is our notation for this approximation. However the accuracy of this solution is very low, and it is not suitable for $K \le \frac{1}{2}$. In order to find a simple and better approximantion the following procedure will be followed. First, in order to get a simple integration for ϕ, the function $F(\phi)$ is approximated by a fourth degree polynomial, that is,

$$\frac{d\phi}{d\tilde{y}} = \sqrt{F(\phi)} = -\sqrt{\lambda^2\phi^2(1+\frac{\alpha}{2}\phi)^2} = -\lambda\phi(1+\frac{\alpha}{2}\phi),$$

where the procedure to find α will be shown bellow.

Integrating now the above equation, it is found

$$-\lambda y = \int_{\phi_w}^{\tilde{\phi}} \frac{d\phi}{\phi\left(1 + \frac{\alpha}{2}\phi\right)} = ln\left(\frac{\tilde{\phi}}{1 + \frac{\alpha}{2}\tilde{\phi}}\right) - ln\left(\frac{\phi_w}{1 + \frac{\alpha}{2}\phi_w}\right),$$

from which we obtain

$$\tilde{\phi}(y) = \frac{\phi_w}{e^{\lambda y}\left(1 + \frac{\alpha}{2}\phi_w\right) - \frac{\alpha}{2}\phi_w}.$$

The simplest procedure to find α is to compaire with the expansion in serie (5), giving

$$\alpha_1 = \frac{\left(\frac{1}{24K^2} - \frac{1}{6}\right)}{\lambda} = \frac{(3 - 4K^2)}{(12K^2 - 6K)}.$$

This value of α, denoted here by α_1, does not give the best approximation. A second procedure is to choose α in such a way that the slope in $y = 0$ coincides with the slope of the exact function defined in Eq. (5). This procedure leads to

$$\alpha = \frac{2}{\phi_w^2}\left[\sqrt{\frac{4K\left(2K\sqrt{1 + \frac{\phi_w}{K}} + e^{-\phi_w} - (2K + 1)\right)}{(2K - 1)}} - \phi_w\right] \tag{6}$$

3. Results

In Figure 1 and Figure 2, respectively, are shown the potentials for $K = 1$ and $K = 0.6$, obtained by:

(1) Numerical calculation, $\phi(y)$;

(2) The approximation, $\tilde{\phi}(y)$, calculating α by Eq. (6);

(3) The usual approximation, $\phi_2(y)$.

The accuracy of the approximant here found is always higher than previous ones.

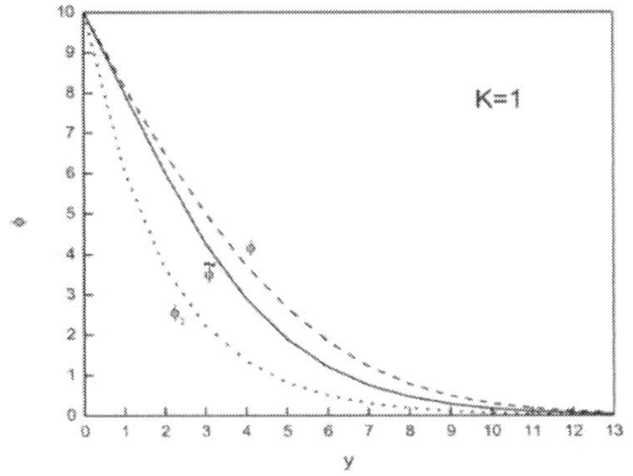

Figure 1. *Here, the potential as a function of the distance is shown for $K = 1$.*

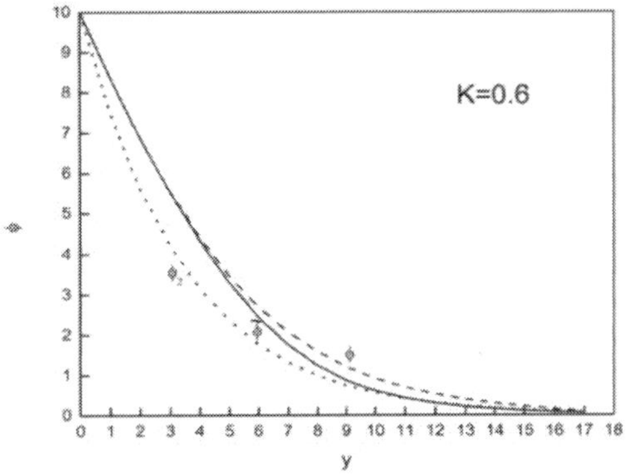

Figure 2. *Here, the potential as a function of the distance is shown for* $K = 0.6$.

4. Conclusion

An analytic approximate expression for the plasma potential near the wall has been found. This allows the direct calculation of the sheath potential as a function of the distance to the wall. This potential is given as a combination of exponential and rational functions. The accuracy of this approximant is higher than previous one [4] and enough for most experimental works.

References

[1] Bhom D. "The characteristics of electric discharges in magnetic fields", A.Guthrie, R Wakerling Ed. (McGraw Hill, New York, 1949) Chap. 3.
[2] Hazeltine R.D and Waelbroeck F. "The framework of plasma physics", Perseus Books, Reading Massachuset., 1998, pp. 81-84.
[3] Stangeby P.C. and McCracken G.M., Nuclear Fusion, **30**, 1228 (1990).
[4] Martin P. and Cereceda C., 29th EPS Conference on Plasma Phys. and Contr. Fusion, Montreux, 17-21 June 2002 ECA Vol. 26B, P-4.009 (2002)

Acknowledgments

Work supported by : 1) Decanato de la Facultad de Ciencias Bsicas, Universidad de Antofagasta, Antofagasta, Chile, 2) Grant G-**22** (P. Martin), Decanato de investigaciones, Universidad Simón Bolivar, Caracas, Venezuela, and 3) FONDECYT-CONICYT Regular Project # **1121103**, Chile (L. Cortés-Vega).

In memoriam, we thank also to Carlos Cereceda, since part of the present work is a continuation of his ideas published toghether with P. Martin (see here in References).

Effective mass Schrödinger equation with Thomas-Fermi potential

G Ovando, J J Peña and J. Morales
Universidad Autónoma Metropolitana - Azcapotzalco,
CBI - Area de Física Atómica Molecular Aplicada,
Av. San Pablo 180, 02200 México, D. F., MEXICO
E-mail: gaoz@azc.uam.mx

Abstract. The exactly-solvable position-dependent mass Schrödinger equation (PDMSE) for the Thomas-Fermi potential is presented. The PDMSE is transformed into a standard Schrödinger equation (SSE) with constant mass by means of a point canonical transformation scheme. By proposing an exponential type potential of the SSE it is possible to determine a PDMSE with the Thomas-Fermi potential. The resulting PDMSE is carried to the Sturm-Liouville form and the corresponding theory is developed for the particular resulting problem in order to obtain closed solutions in terms of Bessel functions of the first kind. Beyond the case considered in this work, the approach is general and can be useful in the study of electronic properties of non-uniform materials in which the carrier effective mass depends on the position as well as in the search of new solvable potentials suitable for quantum systems.

Keywords and Phrases: Position-dependent mass, Point canonical transformation, Thomas-Fermi potential, Schrödinger equation.

1. Introduction.

The quantum problem of position-dependent mass comes from the effective mass approximation [1] used for studying the motion of carrier electrons in nanostructured systems. The one-dimensional position-dependent mass Schrödinger equation (PDMSE) occurs in the quantum treatment of the electronic properties of semiconductors, liquid crystals, quantum dots and non uniform materials in which the carrier effective mass depends on the position [2]-[5]. The exactly solvable PDMSE has attracted considerable theoretical attention as demonstrated by already published methods on the subject such as the kinetic energy operator [6], Lie algebras [7],[8], supersymmetry [9] and path integration [10] approaches. Besides, the point canonical transformation method (PCTM) is an approach providing exact eigenvalues and eigenfunctions [11],[12] and can also be used to solve the PDMSE [13],[14]. In section II we describe the fundamentals of PCTM used to transform the PDMSE in variable x into a standard Schrödinger equation (SSE), in a new variable u, where the mass no longer depends on the position. The potential associated to the PDMSE and the potential involved in the SSE are connected through a Riccati-type relationship which includes a superpotential intrinsically determined by the mass distribution. To attain the purpose of this work, in section III, we assume that the superpotential is constant and consequently it is recognized an exactly-solvable PDMSE with the Thomas-Fermi potential for a specific mass function. This important potential has been studied previously under position-dependent mass conditions [15],[16]. In this work, orthogonal eigenfunctions and normalization constants are calculated in closed form by developing the particular details involved in the Sturm-Liouville theory for the corresponding SSE. It is proved that the eigenstates are those of definite parity with eigenvalues expressed through the order of the Bessel functions of the first kind. Straightforwardly, the proposal can be used with another former potential in the search of those PDMSE that could be useful in quantum applications where, for example, the chemical composition is a point to consider.

2. Position-dependent mass Schrödinger equation.

The PDMSE for known $\psi_n(x)$ in the Levy-Leblond representation is given by [17]

$$-\frac{\hbar^2}{2m(x)}\psi_n''(x) - \left(\frac{\hbar^2}{2m(x)}\right)'\psi_n'(x) + (E_n - V(x))\psi_n(x) = 0, \tag{1}$$

where E_n is the energy spectra, $m(x)$ is the position-dependent mass function and $V(x)$ is the potential function. With the aim to transform the equation above into a Schrödinger equation for constant mass, it is proposed the point canonical transformation

$$x = F(u) = g^{-1}(u), \tag{2}$$

such that

$$u = g(x) = \int^x \sqrt{M(t)}dt\ , \tag{3}$$

($M(x) = m(x)/m_0$ being the dimensionless form of the mass function and m_0 the mass of the carrier particle) and the similarity transformation

$$\psi_n(x) = \varphi(g(x))\sqrt{g'(x)}. \tag{4}$$

With this elements, the PDMSE given in Eq.(1) is transformed into the SSE given by

$$-\frac{\hbar^2}{2m_0}\varphi_n''(u) + U(u)\varphi_n(u) = E_n\varphi_n(u), \tag{5}$$

where the potentials are related by the Riccati-type equation

$$U(u) = V(F(u)) + \frac{\hbar^2}{2m_0}\left(W^2(u) + \frac{dW\ (u)}{du}\right), \tag{6}$$

with

$$W(u) = -\frac{1}{4}\frac{d}{du}\ln\left(M(F(u)\right), \tag{7}$$

which determines the position-dependent mass distribution in the form

$$M(x) = \exp\left(-4\int^{g(x)} W(t)\ dt\right). \tag{8}$$

Previous Eq. (6) shows that $W(x)$ is an essential function because it plays the role of a link between the PDMSE and the SSE. We can use $W(x)$ as in the case of the Witten superpotential [18], that is, to generate solvable cases for the PDMSE. In this case the scheme depends on the proposed solvable potential for the SSE [14]. The proposal of constant values for the function $W(x)$ has allowed to select an appropriate mass distribution Eq. (8) as well as an exactly-solvable exponential-type potential for the SSE in order to solve analytically the PDMSE for the Thomas-Fermi potential, as it will be seen in the next section. Eventually, the same procedure can be extended for other potential models.

3. PDMSE for the Thomas-Fermi Potential.

Previous considerations have carried us to consider the mass distribution

$$M(x) = \frac{1}{(\alpha \mid x \mid +1)^2}, \tag{9}$$

that gives respectively the transformation $g(x)$ and its corresponding inverse function $F(u)$, see Eqs.(3,4), in the form

$$g(x) = \frac{1}{\alpha} \ln(\alpha \mid x \mid +1), \qquad x \in (-\infty, \infty), \tag{10}$$

$$F(u) = \pm \frac{e^{\alpha u} - 1}{\alpha}, \qquad u \in (0, \infty). \tag{11}$$

In consequence, $W(u) = \alpha/2$, with α constant. Also, from the Eq.(6), the potentials are relationed by

$$V(F(u)) = U(u) - U_0, \tag{12}$$

with $U_0 = \frac{\hbar^2 \alpha^2}{8 m_0}$. Let us propose as solvable potential for the SSE Eq.(5) the exponential-type potential given by

$$U(u) = -V_0 \, e^{-\beta u} + U_0, \tag{13}$$

for which the potential of the PDMSE is

$$V(x) = -V_0(\alpha \mid x \mid +1)^{-\beta/\alpha}, \tag{14}$$

and which corresponds, in the special case of $\beta = 4\alpha$, to the Thomas-Fermi potential [15]

$$V(x) = -\frac{V_0}{(\alpha \mid x \mid +1)^4}. \tag{15}$$

To find the solution of this potential we solve the SSE Eq.(5) with the exponential potential given in Eq.(13), which is

$$-\frac{\hbar^2}{2m_0} \varphi_\rho''(u) - V_0 \, e^{-4\alpha u} \varphi_\rho(u) = \epsilon_\rho \varphi_\rho(u), \tag{16}$$

with $\epsilon_\rho = E_\rho - U_0$. Solutions of Eq.(16) are given in terms of the Bessel functions of the first kind as

$$\varphi_\rho(u) = N_\rho \, J_\rho(a e^{-2\alpha u}), \tag{17}$$

with $a = \sqrt{\frac{m_0 V_0}{2\hbar^2 \alpha^2}}$, $\epsilon_\rho = -16 U_0 \rho^2$, N_ρ the normalization constant and ρ the order of the Bessel functions that will be determined under condition to have an orthogonal system of eigensolutions. Let us consider the new variable

$$t = e^{-2\alpha u} = \frac{1}{(\alpha |x| + 1)^2}, \tag{18}$$

defined in the interval $t \in (0, 1)$, and the eigenfunctions expressed through this variable

$$y(t) = \varphi_\rho(u(t)), \tag{19}$$

in such a way that Eq. (16) is reduced to the Sturm-Liouville form

$$\left(t\ y'(t)\right)' + \left(a^2 t - \frac{\rho^2}{t}\right)\ y(t) = 0. \tag{20}$$

So, by considering ρ_j and ρ_k two distinct eigenvalues with respective eigensolutions $y_j(t)$ and $y_k(t)$, such that

$$\left(t\ y_j'(t)\right)' + \left(a^2 t - \frac{\rho_j^2}{t}\right)\ y_j(t) = 0, \tag{21}$$

$$\left(t\ y_k'(t)\right)' + \left(a^2 t - \frac{\rho_k^2}{t}\right)\ y_k(t) = 0, \tag{22}$$

and performing the usual combination of Sturm-Liouville theory, that is, the product of first equation by $y_k(t)$ minus the product of the second equation by $y_j(t)$, then arranging derivatives, we obtain

$$\left[t\left(y_k'(t)\, y_j(t) - y_k(t)\, y_j'(t)\right)\right]' + y_k(t)\, y_j(t)\, \frac{\rho_j^2 - \rho_k^2}{t} = 0, \tag{23}$$

from which we can evaluate the integral over the range of variable t of the product $y_k(t)\, y_j(t)$, that is

$$\left(\rho_k^2 - \rho_j^2\right) \int_0^1 y_k(t)\, y_j(t)\, \frac{dt}{t} = y_k'(1)\, y_j(1) - y_k(1)\, y_j'(1). \tag{24}$$

This equation allows to recognize conditions for the orthogonality of the eigenfunctions $\psi_{\rho_j}(x)$, $\psi_{\rho_j}(x)$ because of the identity

$$\int_{-\infty}^{\infty} \psi_{\rho_j}(x)\ \psi_{\rho_j}(x)\ dx = \int_0^{\infty} \varphi_{\rho_j}(u)\ \varphi_{\rho_j}(u)\ du = \frac{1}{2\alpha}\int_0^1 y_j(t)\, y_k(t)\, \frac{dt}{t}. \tag{25}$$

As previously mentioned $y_k(t) = N_\rho\ J_\rho(a\ t)$, then, according to Eq. (24) the orthogonal eigenfunctions fulfill the condition $J_\rho(a\ ) = 0$ for $\rho_j \neq \rho_k$, or $J_\rho'(a\ ) = 0$ for $\rho_j \neq \rho_k$. The first one means $\psi_\rho(0) = 0$, which is an odd parity condition for odd parity eigenfunctions; the second one imply $\psi_\rho'(0) = 0$, which is an even parity condition for even parity eigenfunctions. This situation ensures the orthogonality of all the eigenfunctions, that is, between those of the same parity with distinct eigenvalues $\rho_j \neq \rho_k$ and the orthogonality of eigenfunctions of different parity because of its integral in a symmetric interval $(-\infty, \infty)$, see Eq. (25). Consequently, due that Bessel functions of first kind alternate its points where the function is zero and where its derivative is zero, this means that the odd parity eigenfunctions and the even parity eigenfunctions alternate each other. Moreover, the ground state is gotten from a function of the second condition $J_\rho'(a\ ) = 0$ that is nonzero in the interval $t \in (0, 1)$ except for the point $t = 0$ (corresponding to $|x| \to \infty$) assuring that the ground state has no nodes. So that, the even parity eigenstates are those corresponding to eigenvalues ρ_0, ρ_2, ρ_4,..and so on and the odd parity eigenstates are those of eigenvalues ρ_1, ρ_3, ρ_5,..etc. In addition,we have to remark that given the fixed distance a, only a finite number of Bessel functions will satisfy the limiting conditions $J_\rho(a\ ) = 0$ or $J_\rho'(a\ ) = 0$.

Consequently, the wavefunctions for the PDMSE with the Thomas-Fermi potential, from Eq.(4) are given by

$$\psi_{\rho_n}(x) = \frac{N_{\rho_n}}{\sqrt{\alpha\,|x| + 1}} J_{\rho_n}\left(\frac{a}{(\alpha\,|x| + 1)^2}\right), \tag{26}$$

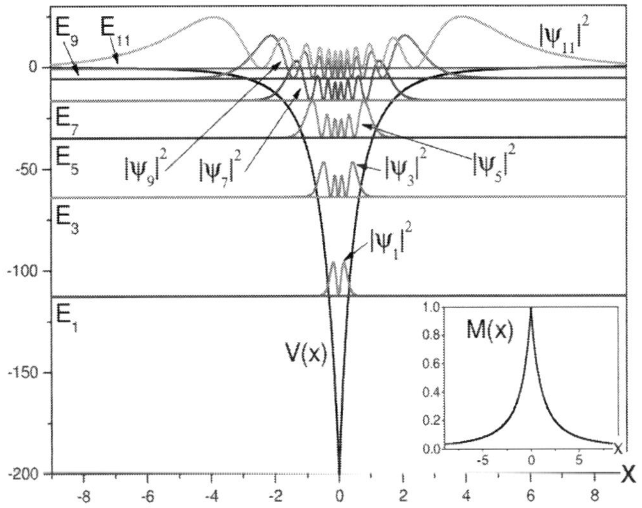

Figure 1. PDM Thomas-Fermi potential $V(x)$, Energy E_j, probability densities $|\psi_j|^2$, for $j = 1, 3, 5, ...11$, and PDM distribution $M(x)$.

with energy spectra

$$E_{\rho_n} = U_0 \left(1 - (4\rho_n)^2\right). \tag{27}$$

These features: the potential, energy levels, probability densities as well as the position-dependent mass distribution are shown in Figure 1 for the odd parity solutions.

Finally, the normalization of the eigenfunctions can be performed by evaluating the integrals coming from Eq.(24) after taking $\rho_k = \nu$, $\rho_j = \mu$, $y_k(t) = J_\nu(at)$ and $y_j(t) = J_\mu(at)$, that is

$$\int_0^1 J_\nu(at)\; J_\mu(at)\,\frac{dt}{t} = \frac{J_\nu{}'(a)\,J_\mu(a) - J_\nu(a)\,J_\mu{}'(a)}{\nu^2 - \mu^2}; \tag{28}$$

in the limit when ν tends towards μ, it yields

$$\int_0^1 J_\mu^2(at)\,\frac{dt}{t} = \underset{\nu\to\mu}{Lim}\frac{J_\nu{}'(a)\,J_\mu(a) - J_\nu(a)\,J_\mu{}'(a)}{\nu^2 - \mu^2}, \tag{29}$$

for which evaluation we use l'Hôpital's rule and Bessel properties [19] to obtain

$$\int_0^1 J_\mu^2(at)\,\frac{dt}{t} = \begin{cases} \frac{J_{\mu+1}(a)}{2\mu}\frac{\partial}{\partial\mu}J_\mu(a) & \text{if} \quad J_\mu(a) = 0 \\ \frac{J_\mu(a)}{2\mu}\frac{\partial}{\partial\mu}J_\mu{}'(a) & \text{if} \quad J_\mu{}'(a) = 0 \end{cases} \tag{30}$$

which determine the normalization constants of Eq. (26)

$$N_{\rho_n} = \begin{cases} \sqrt{\frac{4\alpha\rho_n}{J_{\rho_n}(a)\frac{\partial}{\partial\rho_n}J'_{\rho_n}(a)}} & n = 0, 2, 4... \\ \sqrt{\frac{4\alpha\rho_n}{J_{\rho_n+1}(a)\frac{\partial}{\partial\rho_n}J_{\rho_n}(a)}} & n = 1, 3, 5... \end{cases} \tag{31}$$

Concluding remarks

In this work, we have used again the general algorithm to find exactly solvable potentials for the effective mass Schrödinger equation of a previous work [14]. According to the proposal, the choice of the equivalent of Witten superpotential allowed obtain the position-dependent mass distribution to indentify an exactly-solvable effective mass Schrödinger equation for the potential $V(x) = (\alpha \mid x \mid +1)^{-\beta/\alpha}$, which has as a particular case the Thomas-Fermi model when $\beta = 4\alpha$. Although we performed calculations for this particular case, our work could be adapted for any value of β. This feature makes a difference between our work and the one already published by Axel *et al* [16]. Moreover, in the present work, we considered the normalization and orthogonality of the wavefunctions following the standard procedures of the Sturm-Liouville theory. The elements exposed in this work show the powerful of the general PCTM approach in the search of solvable models that can be useful in different theoretical studies of quantum chemistry and solid state physics.

Acknowledgements

This work was partially supported by the projects UAM-A-CBI-2232004 and 009. We are grateful with the SNI-Conacyt-Mexico for the stipend received.

References

[1] Slater J C 1949 *Phys. Rev.* **76** 1592
[2] Geller M R, Kohn W 1993 *Phys. Rev. Lett.* **70** 3103
[3] Serra L, Lipparini E 1997, *Europhys. Lett.* **40** 667
[4] Bastard G 1989 *Wave Mechanics Applied to Heterostructures*, Les Ulis, Les Éditions de Physique
[5] Barranco M, Pi M, Gatica S M, Hernández E S, Navarro J 1997 *Phys. Rev.* B **56** 8997
[6] Bagchi B, Gorain P, Quesne C, Roychoudhury R 2004 *Mod. Phys, Lett.* A **19** 2765
[7] Roy B 2005 *Europhys. Lett.* **72** 1
[8] Dong S H, Peña J J, Pacheco-García C, García-Ravelo J 2007 *Mod. Phys. Lett.* A **22** 1039
[9] Gonul B, Gonul B, Tutcu D, Ozer O 2002 *Modern Phys. Lett.* A **17** 2057
[10] Mandal B P 2000 *Int. J. Mod. Phys.* A **15** 1225
[11] Dutt R, Khare A, Varshni Y P 1995 *J. Phys.* A: Math. Gen **28** L107
[12] Cooper F, Khare A, Sukhatme U 2001 *Supersymmetry and Quantum Mechanics* (Singapore, World Scientific)
[13] Peña J J, Ovando G, Morales J, García-Ravelo J, García J 2008 *Int. J. Quant. Chem.* **108** 1750
[14] Peña J J, Ovando G, Morales J, García-Ravelo J, Pacheco-García C 2007 *Int. J. Quant. Chem.* **107** 3039
[15] Ioriatti L 1990 *Phys. Rev.* B **41** 8340
[16] Schulze-Halberg A, García-Ravelo J, Pacheco-Garcia C, Peña J J 2013 *Annals of Phys.* **333** 323
[17] Levy-Leblond J M 1995 *Phys. Rev.* A **52** 1845
[18] Witten E 1981 *Nucl. Phys.* B. **188** 513
[19] Bowman F 1958 *Introduction to Bessel functions* (New York, Dover Pub. Inc.)

Journal of Physics: Conference Series **574** (2015) 012109 doi:10.1088/1742-6596/574/1/012109

Detection of finger gesture using singular spectrum transformation

Y. Tamura[1], T. Umetani[1] and H. Nakamura[2]

[1] Department of Intelligence and Informatics, Konan University, Okamoto, Higashinada, Kobe, 658-8501, Japan

[2] Department of Helical Plasma Research, National Institute for Fusion Science, Oroshi, Toki, 509-5292, Japan

E-mail: tamura@konan-u.ac.jp

Abstract. The purpose of this study is to detect finger movement using a singular spectrum transformation method. Human gesture recognition is essential for realizing natural user interfaces. However, constructing a robust, easily installable interface is extremely di cult. Our proposed method uses singular spectrum transformation to classify finger movements. This method robustly classifies gestures and behavior.

1. Introduction

Gesture recognition of finger movement is among the most important issues for realizing natural user interfaces. Touch devices, such as those installed on smart phones, have become widely used. These devices are controlled by fingers. However, a natural interface requires a special device and is applied to a two-dimensional (2D) display. On the other hand, devices such as Microsoft Kinect and Leap motion sensors are operable in 3D space and have wider usability than touch devices. Unfortunately, the user's indication is less easily detected by 3D interfaces since its accuracy is lower and the degree of freedom is higher than that of the touch devices. Moreover, the raw data generated by 3D sensors is largely obscured by noise.

This study proposes a finger movement detection and classification method based on singular spectrum transformation (SST) [1]. The SST method has been applied in diverse fields such as the earth sciences, physics, and financial research. In human interface research, SST has been applied to the segmentation of human motions [2]. The present study applies SST to finger gesture recognition in noisy environments.

2. System configuration

Our system consists of a data glove and a PC. We used the data glove because it collects data at higher frequency than a Kinect or Leap motion sensor. The 5DT data glove is shown in Fig. 1. This data glove is installed with a single flexure sensor in each finger. The flexure of each finger is measured at 60 Hz, higher than the measuring frequency of the Kinect sensor.

3. Method

Figure 2 presents a flowchart of our proposed system.

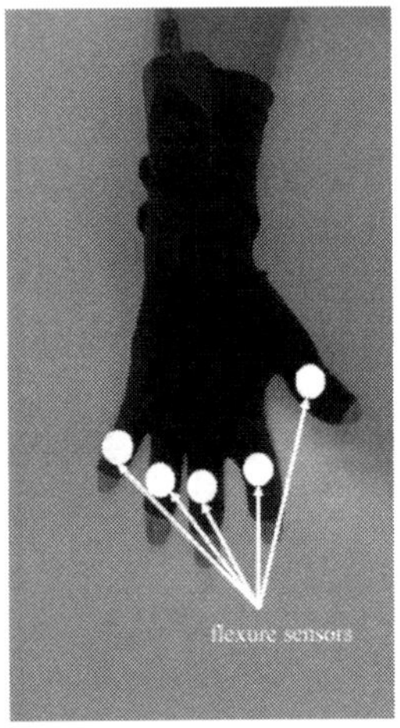

Figure 1. Data glove fitted with 5 flexure sensors

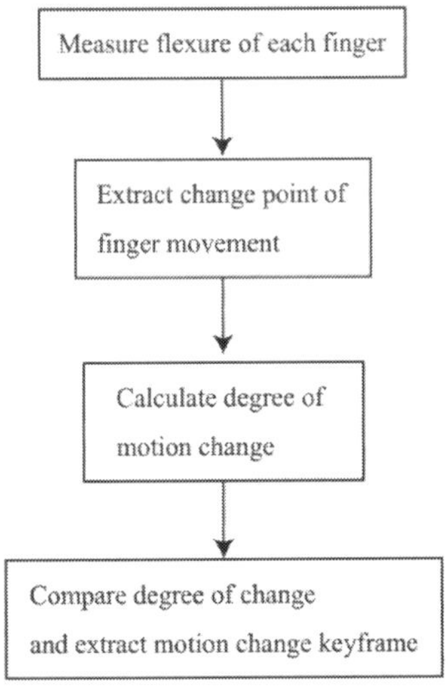

Figure 2. Flowchart of our finger movement detection system

In this study, we detect finger movement and the change-point of finger motions by an SST-based method. Since SST is a nonparametric decomposition method, it is suitable for analyzing unsteady datasets. The SST operates as follows: Consider a one-dimensional time series dataset $x(1), x(2), x(3), \cdots, x(t)$. A w-component vector $\mathbf{j}$ is defined as

$$\mathbf{j}(t-1) = [x(t-w)\ \ x(t-w+1)\cdots x(t-1)]^T. \tag{1}$$

In addition, the n-column Hankel matrix (i.e., the trajectory matrix) $\mathbf{H}(t)$ is given by

$$\mathbf{H}(t) = [\mathbf{j}(t-n)\ \ \mathbf{j}(t-n+1)\cdots\mathbf{j}(t-1)]. \tag{2}$$

The Hankel matrix can be decomposed by singular value decomposition as

$$\mathbf{H}(t) = \mathbf{U}(t)\mathbf{W}(t)\mathbf{V}(t)^T. \tag{3}$$

By applying this method to the time series of finger flexure data in the past and future datasets, the change-points of the motion are extracted from the dataset. The past dataset comprises data from time $(t-n-w-1)$ to $(t-1)$, whereas the future dataset contains data from $(t+s)$ to $(t+s+w+n)$, where s is the time shift from the past dataset, and w and n in Eqs. 1 and 2, respectively, are window range parameters.

Motion change is clearly detected by a "degree of change" factor, defined as follows. We first construct a hyperplane $\mathbf{A_p}(t)$ using the n largest left-singular vectors $(\mathbf{u_{p1}}, \mathbf{u_{p2}}, \cdots \mathbf{u_{pn}})$ in the past matrix $\mathbf{U_p}(t)$ of Eq. 3. We also define the largest left-singular vector $\mathbf{u_f}(t)$ in the future matrix $\mathbf{U_f}(t)$. If the past and the future datasets are similar, the inner product between vector $\mathbf{u_f}(t)$ and the vector formed by projecting $\mathbf{u_f}(t)$ onto the hyperplane $\mathbf{A_p}(t)$ is approximately 0. In the opposite scenario, this inner product approximates to 1. Thus, we define the factor $S(t)$ as follows:

$$S(t) = 1 - \mathbf{u_f}(t)\mathbf{P}(t) \tag{4}$$

$$\mathbf{P}(t) = \mathbf{A_p}(t)\mathbf{A_p}(t)^T, \tag{5}$$

where all components of matrix $\mathbf{A_p}$ are orthonormal bases.

4. Results

In this section, we show the result of the extraction using the SST method. The user wears a data glove on the right hand and sits on a chair next to a desk. The user taps the desk, and the changing finger movements are detected by the flexure sensor.

The original SST is a non-parametric method; however, it is necessary to define the parameters w, n, and s in Eqs. 1 and 2. Here we define $w = n$ and $s = -(w + n)/2$. In this definition of s, the central datum of the future data becomes t and the window range becomes $2w$. Moreover, in this experiment, the hyperplane $\mathbf{A_p}$ was obtained only from $\mathbf{u_{p1}}, \mathbf{u_{p2}}$.

4.1. Parameter dependency

The parameter w must be set sufficiently long so that all gesture signals are included. On the other hand, if w is too short, large amounts of noise are processed in the output data. Therefore, we tested the dependency of output of w. To this end, we set $w = 30, 60, 90$. Since the refresh rate of the data glove is 60Hz, the longest time range is 1.5 s.

The user was then requested to tap the desk 10 times. The results are shown in Fig. 3. Note that the 10th tapping signal cannot be analyzed in $w = 60$ and $w = 90$ since the window range is wider than the time range of this signal. Therefore, if nine peaks are detected, we infer that the method has detected all the gestures.

The analysis reveals that $w = 60$ is a suitable window size under this condition, $w = 30$ is slightly oversensitive whereas $w = 90$ misses some change-point signals. Therefore, we set $w = 60$ in the following experiment.

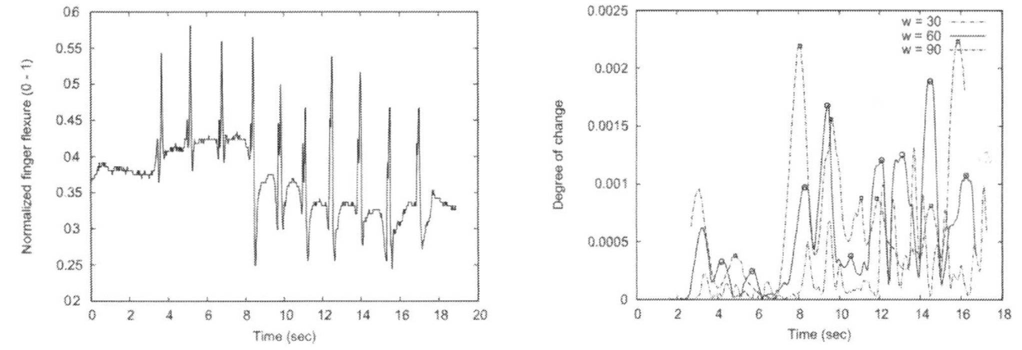

Figure 3. Dependency of output on the parameter w, which determines the window size of signal processing. Left image: Raw data of finger flexure. The rate of finger flexure is normalized from 0.0 to 1.0. Right image: The result of SST calculation. Circles and squares indicate the change-points for $w = 60$ and $w = 90$, respectively.

4.2. Influences of noise

In the previous section, the signal was relatively clear and most of the change-points in the data were distinguishable by human eyes. However, visual detection is much more difficult in a noisy environment. Therefore, we added white noise to the data and reapplied the SST method. Figure 4 shows the result of this experiment. The original signal is normalized from 0.0 to 1.0 and superimposed with white noise. The noise ratio is 0.1, 0.3 or 0.5. This result shows that noise in the input data obscures the change-points to some extent; however, the essential features of the original signal are preserved, especially at noise level 0.5.

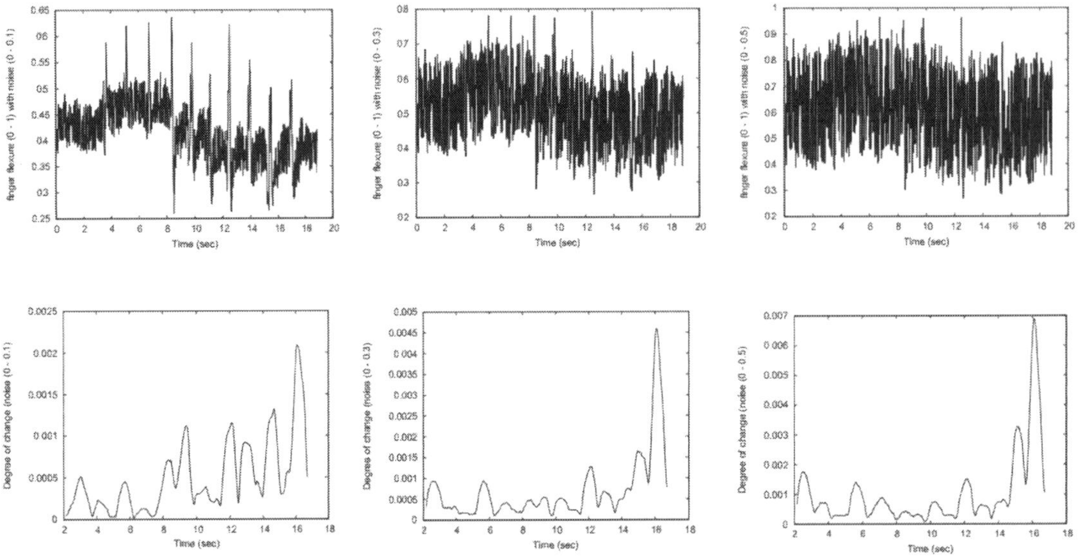

Figure 4. Experiment of signal detection in a noisy environment. The upper panels show the signal superimposed with white noise at a ratio of 0.1 (left), 0.3 (center), and 0.5 (right). The lower panels show the corresponding SST results.

5. Conclusion

We detected the change-points of finger motions using a data glove and an SST data processing method. We showed that this method detects finger movements in a noisy environment. Generally, useful data are difficult to extract from raw data and some parameters must be defined by a machine learning method. However, our results demonstrate the ability of the SST method to remove continuous noisy data and extract objective data with only a few specified parameters. Although all the change-points were not stably detected in a noisy environment, the salient features of the original signal were largely preserved. This implies that the SST method is applicable to finger detection in a low-accuracy system. However, a suitable algorithm is required that automatically extracts the change-points from the SST results. Moreover, the signals in this experiment were very peaky since the tapping was performed over short time periods. Therefore, the method requires further testing on long-term gestures.

Acknowledgement

This study is supported by KAKENHI (26330237).

References

[1] J. B. Elsner and A. A. Tsonis 1996 *Plenum Press*
[2] H. Nakanishi, S. Kanata, H. Hattori, T. Sawaragi and Y. Horiguchi 2011 *Journal of Advanced Computational Intelligence and Intelligent Informatics* **15** 8

Mathematical modeling of a water-in-oil emulsion droplet behavior under the microwave impact

Y I Fatkhullina[1], **A A Musin**[1], **L A Kovaleva**[1] **and I S Akhatov**[1,2]

[1] Bashkir State University, Center for Micro and Nanoscale Dynamics of Dispersed Systems, 32 Zaki Validi Street, Ufa 450074, Russia
[2] North Dakota State University, Department of Mechanical Engineering, Fargo, ND, USA

E-mail: liana-kovaleva@yandex.ru

Abstract. The problem of microwave (MW) electromagnetic radiation impact on a single water-in-oil droplet is considered. The system of heat equations within the droplet and in the surrounding liquid, incompressible Navier-Stokes equations within the droplet and in the surrounding liquid, and equation of state are considered. The formulated problem is solved numerically using TDMA (Tri-diagonal-matrix algorithm), SIMPLE algorithm and VOF method (volume of fluid method for the dynamics of free boundaries) in Euler coordinates. The results in the form of the dependence of the temperature within the droplet and in the surrounding liquid on the time of microwave impact and streamlines thermal convection are represented; dependence of the velocity of droplet's moving on the power of the of the microwave impact is shown. The obtained results can help to establish criteria for the efficient applicable of the microwave method for the water-in-oil emulsions destruction.

Introduction

The formation of high stability of water-in-oil emulsions is one of the negative factors in extracting and processing of oil, their preparation and transportation, as well as liquidation/recycling of oil-sludge barns. The use of conventional techniques to destroy the emulsions yields no positive results. The use of microwave radiation is one of the perspective method for the destruction of water-in-oil emulsions [1-3].

The emulsions are regarded as a heterogeneous system comprising two immiscible liquids such as oil and water, with one dispersed in the other in the presence of surfactants [4]. Besides, tiny solid rock particles (clay, quartz, salts, etc.) suspended in that system can also act as emulsion stabilizers [5]. High stability of water-in-oil emulsions is conditioned by the oil high-molecular polar components which cover water droplets forming the so called armor envelope, and that prevents the coalescence of the droplets [6,7].

To study the MW actions on the water droplet surrounded by a dielectric liquid the system of heat and Navier-Stokes equations within the droplet and in the surrounding liquid is considered. These equations are based on fundamental laws of motion of multi-phase medium. Formulation of the problem.

It is assumed that the water droplet with radius r_0 surrounded by a hydrocarbon liquid in a gravitational and MW fields. Spherical droplet is in the center of a cylindrical vessel with the height h and the radius r_1 (fig.1.). The surface tension effect is modeling. In addition, it is considered, that the infinitely thin of armor envelope is formed on the surface of the droplet. It is assumed, that droplet doesn't deform and keeps spherical shape in all time of influence. The inertial and ponderomotive forces are neglected. It's taken into account that the dissipation of the microwave energy is within a water droplet only. The motion of each component of the system is described by a system of thermal convection equations in linear Boussinesq approximation [8]. The law of conservation of energy and the conditions of dynamic equilibrium are executed on the interface.

TDMA (Tri-diagonal-matrix algorithm), SIMPLE algorithm and VOF method (volume of fluid method for the dynamics of free boundaries) were used for the numerical solution of this problem [9]. The surface tension converts to the bulk force using continuum surface force (CSF) model. The model interprets surface tension as a continuous, three-dimensional effect across an interface, rather than as a boundary value condition on the interface [10].

The direction and magnitude of bulk force are determined by the density gradient and curvature of the surface. In view of the assumptions made about the safety of a spherical droplet bulk forces are always directed to the center of the droplet and the curvature of the surface of the droplet is constant and equal to $2/r_0$.

Electromagnetic field energy is converted on heating of the water, and is modeled as a distributed heat source [2], magnitude of which is determined from the expression:

$$q_0 = \frac{\omega\varepsilon_0\varepsilon' \, tg\delta}{2}\left|E\right|^2 \tag{1}$$

where ω is the frequency of electromagnetic radiation; ε_0 is the dielectric constant; ε' is the dielectric permittivity; $\tan\delta$ is the dielectric loss tangent; E is the electrical intensity.

Then the dimensionless system of equations of thermal convection is considered in the following dimensionless form:

$$\begin{aligned}\frac{\partial\rho u}{\partial t}+\frac{1}{r}\frac{\partial}{\partial r}(r\rho uu)+\frac{\partial}{\partial z}(\rho vu) &= -\mathrm{Eu}\frac{\partial P}{\partial r}+\frac{1}{\mathrm{Re}}\frac{1}{r}\frac{\partial}{\partial r}(r\eta\frac{\partial u}{\partial r})\\ &+\frac{1}{\mathrm{Re}}\frac{\partial}{\partial z}(\eta\frac{\partial u}{\partial z})-\frac{1}{\mathrm{Re}}\eta\frac{u}{r^2}+\frac{2}{\mathrm{We}}\frac{\partial C}{\partial r}\end{aligned} \tag{2}$$

$$\begin{aligned}\frac{\partial\rho v}{\partial t}+\frac{1}{r}\frac{\partial}{\partial r}(r\rho uv)+\frac{\partial}{\partial z}(\rho vv) &= -\mathrm{Eu}\frac{\partial P}{\partial z}+\frac{1}{\mathrm{Re}}\frac{1}{r}\frac{\partial}{\partial r}(r\eta\frac{\partial v}{\partial r})\\ &+\frac{1}{\mathrm{Re}}\frac{\partial}{\partial z}(\eta\frac{\partial v}{\partial z})+\frac{2}{\mathrm{We}}\frac{\partial C}{\partial z}+\frac{\mathrm{Gr}}{\mathrm{Re}^2}\rho T\end{aligned} \tag{3}$$

$$\begin{aligned}\frac{\partial\rho cT}{\partial t}+\frac{1}{r}\frac{\partial}{\partial r}(r\rho cuT)+\frac{\partial}{\partial z}(\rho cvT) &= \frac{1}{\mathrm{Pr}}\frac{1}{\mathrm{Re}}\frac{1}{r}\frac{\partial}{\partial r}\left(rk\frac{\partial T}{\partial r}\right)\\ &+\frac{1}{\mathrm{Pr}}\frac{1}{\mathrm{Re}}\frac{\partial}{\partial z}\left(k\frac{\partial T}{\partial z}\right)+\mathrm{Os}\frac{1}{\mathrm{Re}}\frac{1}{\mathrm{Pr}}q\end{aligned} \tag{4}$$

$$\frac{\partial\rho}{\partial t}+\frac{1}{r}\frac{\partial}{\partial r}(r\rho u)+\frac{\partial\rho v}{\partial z}=0, \tag{5}$$

where ρ, c, μ, k are density, specific heat, dynamic viscosity and thermal conductivity of the medium; u and v are the velocity vector; p is the pressure; T is the temperature; β is the coefficient of volume expansion; C is the concentration (the volume fraction); q is the dimensionless density of distributed heat sources. Dimensionless numbers are defined as follows:

$$\mathrm{Pr} = \frac{c_0\eta_0}{k_0} \text{ is the Prandtl number}$$

$$\mathrm{Gr} = \frac{\beta g r_0^3 T_0 \rho_0}{\eta_0^2} \text{ is the Grashof number}$$

$$\mathrm{Os} = \frac{r_0^2 q_0}{T_0 k_0} \text{ is the Ostrogradskii number}$$

$$\mathrm{We} = \frac{\rho_0 v_0^2 r_0}{\sigma} \text{ is the Weber number}$$

$$\mathrm{Eu} = \frac{P_0}{\rho_0 v_0^2} \text{ is the Euler number}$$

$$\mathrm{Re} = \frac{v_0 r_0 \rho_0}{\eta_0} \text{ is the Reynolds number}$$

Due to the symmetry the problem was solved in a cylindrical coordinate system (r,h,φ) with the origin at the center of the drop (Fig. 1). Coordinate axes are oriented so that the direction of the z-axis coincides with the direction of the vector of electrical intensity $\mathbf{E} = (0, E_z, 0)$ and the vector of gravitational acceleration $\mathbf{g} = (0, g_z, 0)$. Thus, there is no dependence on the coordinate φ.

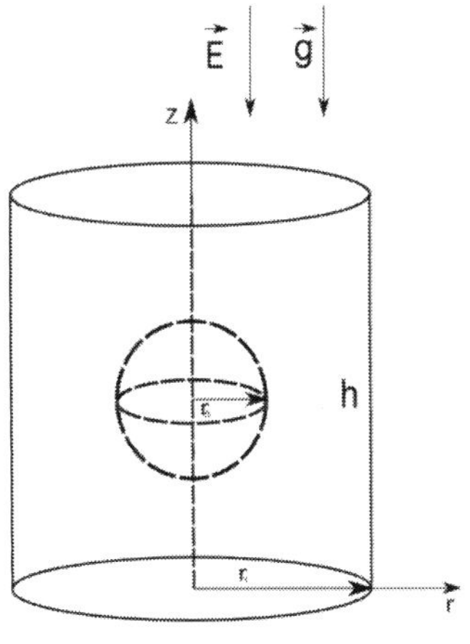

Figure 1. The computational domain scheme.

Accordingly the boundary conditions can be written as follows:

$$T(r,z,t=0) = T_0 \qquad u,v(r,z,t=0) = 0 \tag{6}$$

$$\frac{\partial T(0,z,t)}{\partial r} = 0 \quad \frac{\partial T(R_{\max},z,t)}{\partial r} = 0 \quad \frac{\partial T(r,Z_{\max},t)}{\partial z} = 0 \quad \frac{\partial T(r,0,t)}{\partial z} = 0 \tag{7}$$

$$u\big|_{R=0} = 0,\ u,v\big|_{R=R\max} = 0,\ u,v\big|_{Z=0} = 0,$$
$$u,v\big|_{Z=Z\max} = 0,\ \frac{\partial v}{\partial r}\Big|_{R=0} = 0 \tag{8}$$

Calculation results

Profiles of temperature along the equator and the poles in the droplet and surrounding liquid in case $q=5*10^7$ W/m^3 and temperature coefficient $\gamma=1.5$ 1/K are shown in fig. 2-3.

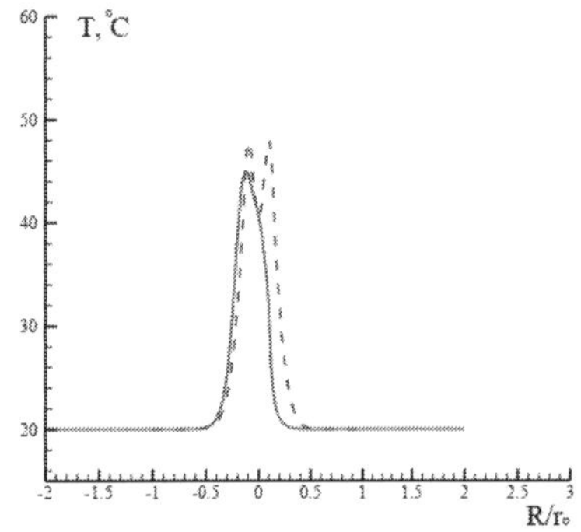

Figure 2. Profiles of temperature along the equator (dashed line) and the poles (solid line) in the droplet and surrounding liquid in case t=10 sec.

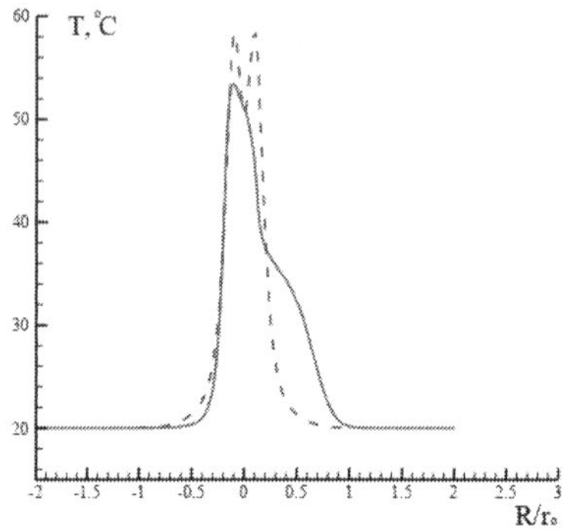

Figure 3. Profiles of temperature along the equator (dashed line) and the poles (solid line) in the droplet and surrounding liquid in case t=40 sec.

In the case of the MW influence the field energy dissipates directly within the droplet. Along the equator the droplet heats irregularly, one can observe two characteristic peaks. This is due to the fact that the convective flow (fig. 4) carries heating, thereby the center of droplet is heated slower than the edge (fig. 5). Also, both poles of the droplet are heated differently too. Surrounding liquid is heated primarily from the top this is due to the fact that the droplet falls down (fig. 5). Graph of droplet velocity depending on the time is shown in fig. 6.

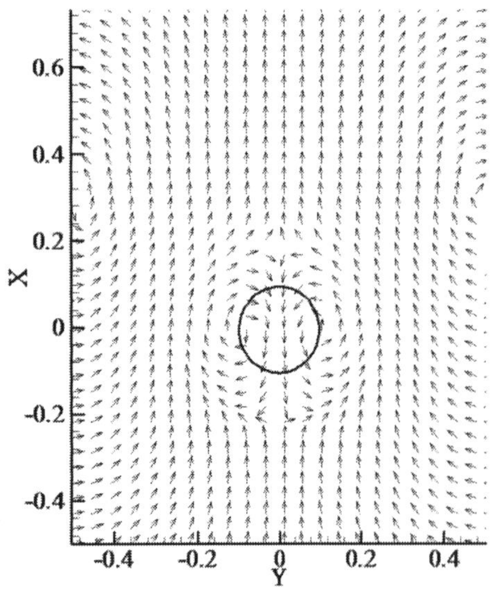

Figure 4. Streamlines in the droplet and the surrounding liquid under and MW heating.

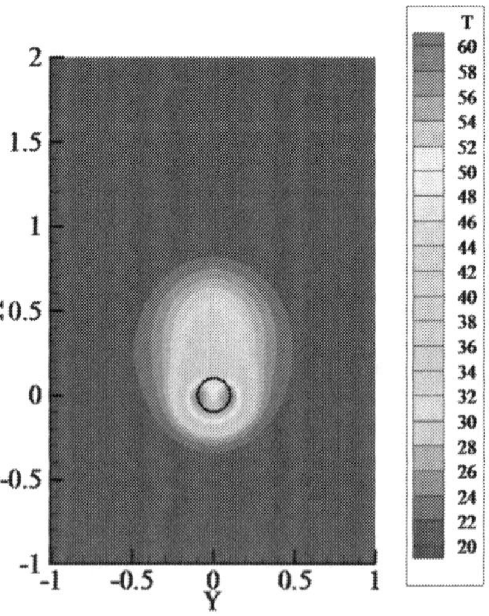

Figure 5. Distribution of temperature inside and outside the water droplet in case t=40 sec.

Journal of Physics: Conference Series **574** (2015) 012110 doi:10.1088/1742-6596/574/1/012110

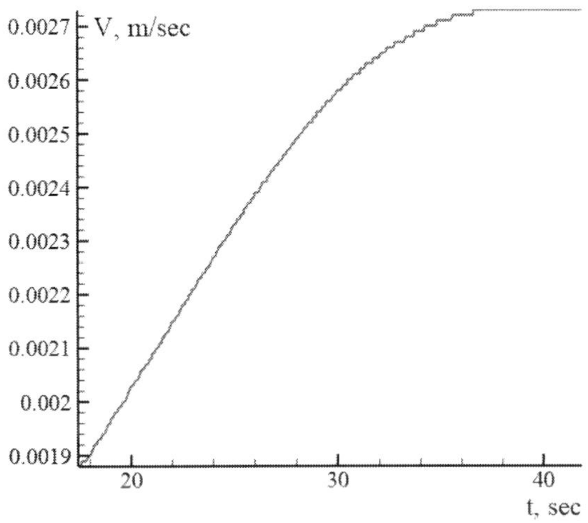

Figure 6. Droplet velocity depending on the time

Droplet velocity is established by the time t = 36 sec.
Profiles of temperature along the equator and the poles in the droplet and surrounding liquid in case $q=5*10^7$ W/m^3 and temperature coefficient γ=0.8 1/K are shown in fig. 7-8.

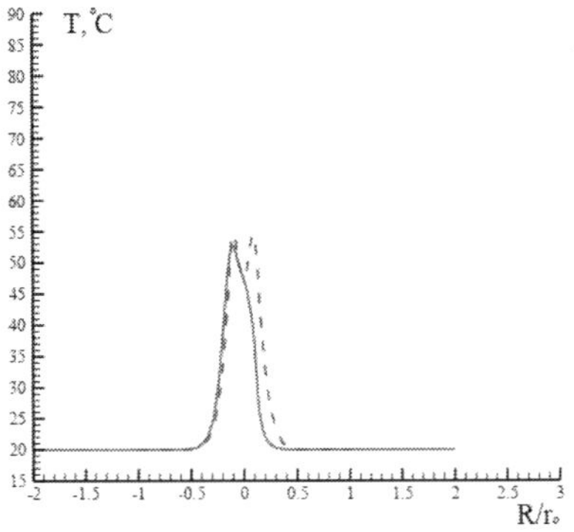

Figure 7. Profiles of temperature along the equator (dashed line) and the poles (solid line) in the droplet and surrounding liquid in case t=10 sec.

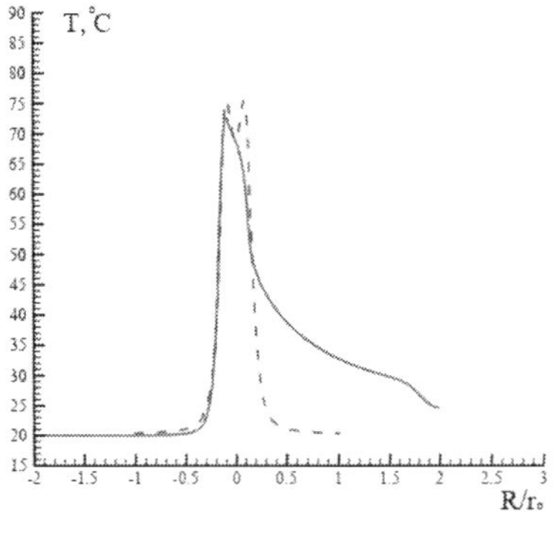

Figure 8. Profiles of temperature along the equator (dashed line) and the poles (solid line) in the droplet and surrounding liquid in case t=100 sec.

The droplet of water is heated by the previous case, but over time the heated region on top of the drop is increased significantly (fig. 7-8). This is due to the fact that in this case, the droplet doesn't fall, so the convective flow (fig. 9) carries the heated liquid from bottom to top. This is clearly seen in fig. 10.

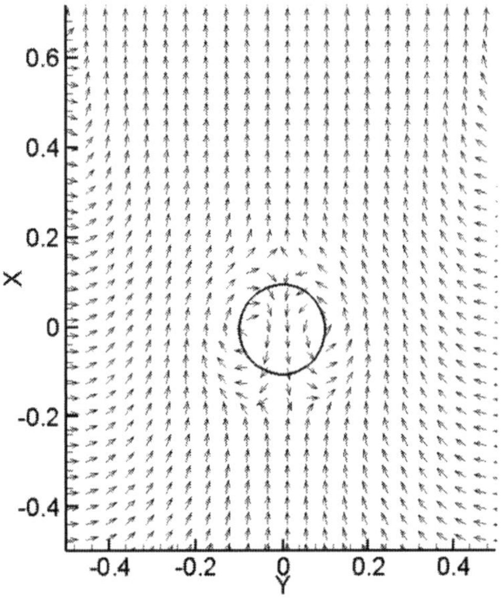

Figure 9. Streamlines in the droplet and the surrounding liquid under and MW heating.

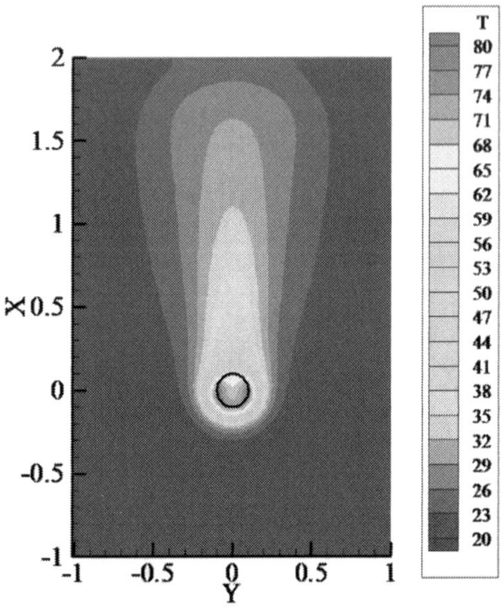

Figure 10. Temperature field in the drop and the surrounding liquid in case t=100 sec.

It should also be noted that the maximum temperature doesn't exceed 75 degrees and doesn't change with time due to temperature stabilization.

Conclusions

The numerical results show, that the singularity heating of the droplet and droplet movement in the liquid are different at various temperature coefficients. In case of decrease of the temperature coefficient one can see a more intense heating of the surrounding liquid with motionless droplet. When γ=1.5 1/K the droplet up down with established velocity V=0.00271 m/sec. The influence of the temperature coefficient is caused by the fact that the increasing of the temperature coefficient leads to more intense heating the medium, reducing its viscosity, as a result of which droplets fall downwards under the action of gravity. Streamlines in both cases are the same, but fluid moves clockwise to the right and counter-clockwise to the left.

Acknowledgments

This research is supported by the Ministry of Education and Science of the Russian Federation (government contract No.3.1251.2014/K and grant No. 11.G34.31.0040) and the RFBR grant № 14-01-97005.

References

[1] Kovaleva L A , Zinnatullin R , Minnigalimov R Z 2009 *Russia. J. Oil-field business.* **5** 54
[2] Zinnatullin R R , Kovaleva L A , Minnigalimov R Z 2009 *Russia. J. Oil facilities.* **7** 75
[3] Noïk C, Chen J, and Dalmazzone C 2006 SPE 103808 **12**
[4] Leal-Calderon F , Bibette J , Schmitt V 2007 *Springer* 228
[5] Gamal M , Mohamed A M O and Zekri A Y 2005 *J. Petroleum Science and Engin.* **16** 209
[6] Kovaleva L A , Minnigalimov R Z and Zinnatullin R R 2011 *J. Energy & Fuels* **8** 3731

[7] Kovaleva L ,Musin A , Zinnatullin R , Akhatov I S 2011 *Proceedings of ASME 2011 International Mechanical Engineering Congress & Exposition, IMECE2011-62935* (Denver, Colorado, USA)
[8] Boussinesq J 1903 *Theorie analytique de la chaleur (Gauthier-Villars, Paris* vol 3*)*
[9] Patankar C V 1980 *Numerical Heat Transfer and Fluid Flow* . (*McGraw-Hill, New York*)
[10] Brackbill U , Kothe D B, and Zemach C 1992 *J. Comp. Physics* **21** 335

Asymptotic Analysis of the Wnt/β Signaling Pathway

D T Maris and D A Goussis

Department of Mechanics, School of Applied Mathematics and Physical Sciences, National Technical University of Athens, 157 73 Athens, Greece

E-mail: dmaris@mail.ntua.gr, dagoussi@mail.ntua.gr

Abstract. The *Wnt/β-catenin* pathway is a signal transduction pathway made of proteins, which plays an important role in oncogenesis. Ethan Lee and and co-workers introduced in 2003 a detailed mathematical model of this pathway, incorporating the kinetics of protein-protein interactions, protein synthesis/degradation and phosphorylation/dephosphorylation. The fast/slow dynamics of Lee's system are examined here, by employing the *Computational Singular Perturbation (CSP)* algorithm. *CSP* reproduces the results of the classical singular perturbation analysis in an algorithmic fashion, producing an approximation of (i) the low dimensional *Slow Invariant Manifold (SIM)*, where the solution evolves and (ii) the reduced model that governs the flow there. The temporal variation of the dimensions of the *SIM* will be presented and the components of the pathway that are responsible (i) for the generation of the *SIM* and (ii) for driving the system on it will be identified.

1. Introduction

Models representing multiscale biochemical networks are usually non-linear and complex, making the interpretation of their dynamics a difficult task [1]. Among others, due to the generated equilibria between the fast components of such models, their long term evolution is determined by the slower components of the system [2]. The identification of these components is a very important step, since they generate and regulate the long term dynamics and response of the model.

Here, the most significant components of the *Wnt* signaling pathway model, that was introduced by *Lee et al.* in Ref. [3], will be identified. The *Wnt* signalling pathway is well known in the biology community due to its involvement to differentiation and to tissue development. It plays an important role to some kinds of cancers, such as colorectal cancer [4, 5] and in stem-cell maintenance [5, 6]. The Wnt pathway is hard to model due two lack of quantitive experimental data, and most of the mathematical modeling research is based upon the detailed model [3]. It has been analyzed extensively, among others by studying its robustness [7, 8] and by constructing reduced models on the basis of four *Partial Equilibrium* approximations [9].

The detailed model of *Lee et al.* will be considered here and it will be shown here that is amenable to further reduction, beyond the four *Partial Equilibrium* approximations employed in [3]. The number of the new equilibria that develop will be identified, as the system evolves in time, along with the components of the system that are responsible the most for their development. These identifications will be based on the *Computational Singular Perturbation* (CSP) algorithm, which reproduces the results of the classical singular perturbation analysis [10, 11]. A brief presentation of CSP will be given first and the discussion of the results will follow.

2. The Computational Singular Perturbation method

Consider a system of N nonlinear ordinary differential equations (ODE's):

$$\frac{d\mathbf{y}}{dt} = \mathbf{g}(\mathbf{y}) = \mathbf{S}_1 R^1(\mathbf{y}) + ... + \mathbf{S}_K R^K(\mathbf{y}) \tag{1}$$

where $\mathbf{y}$ is the N-dimensional *state vector* and $\mathbf{g}(\mathbf{y})$ is the N-dimensional *vector field*. It is assumed that each of the K additive terms $\mathbf{S}_k R^k$ correspond to a distinct component of the physical process; e.g., in a kinetic pathway $\mathbf{S}_k$ is the N-dimensional stoichiometric vector of the k-th reaction and R^k the corresponding rate. Suppose that Eq. (1) exhibits M time scales, which are dissipative (tend to drive the system to equilibrium) and much faster than the rest. *Geometrical Singular Perturbation Theory* proves that under the action of these M-fast time scales the solution is attracted on a $(N-M)$-dimensional SIM, where it evolves according to the slow time scales [12, 13]. Equation (1) can then be cast in the CSP form:

$$\frac{d\mathbf{y}}{dt} = \mathbf{a}_r \mathbf{f}^r + \mathbf{a}_s \mathbf{f}^s \tag{2}$$

where $\mathbf{a}_r = (\mathbf{a}_1,\ \mathbf{a}_2,\ ...\ \mathbf{a}_M)$ and $\mathbf{a}_s = (\mathbf{a}_{M+1},\ \mathbf{a}_{M+2},\ ...\ \mathbf{a}_N)$ are the CSP vectors which span the M-fast and the $(N-M)$-slow subspace of the tangent space, respectively. The amplitudes $\mathbf{f}^r = \mathbf{b}^r\mathbf{g}$ and $\mathbf{f}^s = \mathbf{b}^s\mathbf{g}$ are computed using the dual CSP vectors $\mathbf{b}^r = (\mathbf{b}^1,\ \mathbf{b}^2,\ ...\ \mathbf{b}^M)^T$ and $\mathbf{b}^s = (\mathbf{b}^{M+1},\ \mathbf{b}^{M+2},\ ...\ \mathbf{b}^N)^T$, where $\mathbf{b}^i \cdot \mathbf{a}_j = \delta^i_j$. When the M-fast time scales become exhausted the vector field $\mathbf{g}(\mathbf{y})$ lies entirely on the slow subspace, having no component in the fast one:

$$\mathbf{f}^r(\mathbf{y}) \approx \mathbf{0}^r \qquad\qquad \frac{d\mathbf{y}}{dt} \approx \mathbf{a}_s \mathbf{f}^s \tag{3}$$

The CSP basis vectors $\mathbf{a}_r$ and $\mathbf{a}_s$ along with their duals $\mathbf{b}^r$ and $\mathbf{b}^s$ are approximated by two iterative procedures; (i) the $\mathbf{b}^r$-refinement, which improves the accuracy of $\mathbf{a}_s$ and $\mathbf{b}^r$ vectors and (ii) the $\mathbf{a}_r$-refinement, which improves the accuracy of the $\mathbf{a}_r$ and $\mathbf{b}^s$ vectors. One $\mathbf{a}_r$-refinement assures stability of the reduced system in Eq. (3) and more than one $\mathbf{b}^r$-refinements improve the accuracy of the approximation of the SIM and of the solution from the reduced system [11, 14].

The construction of the reduced system allows for the acquisition of significant physical understanding of complex biological networks by employing the various CSP tools [2, 15]. For example, on the SIM where the M-fast time scales are exhausted, M equilibria develop among the various components of the system. Also, a number ($\geq M$) of variables relate to these M fast time scales. These components and variables are identified by the *CSP Participation Index* P_k^m and *CSP Pointer* $\mathbf{D}_m$ defined as:

$$P_k^m = \frac{f_k^m(\mathbf{y})}{\sum_{j=1}^{K} |f_j^m(\mathbf{y})|} \qquad\qquad \mathbf{D}_m = diag[\mathbf{a}_m \mathbf{b}^m] \tag{4}$$

where $m = 1, ..., M$ denotes the corresponding fast CSP mode. By definition $|P_1^m| + ... + |P_K^m| = 1$ and $D_m^1 + ... + D_m^N = 1$ [16]. A relatively large value of P_k^m denotes significant contribution of the k-th component to the m-th equilibrium and a relatively large D_m^n denotes strong relation of the n-th variable to the m-th fast time scale.

3. The Wnt model

As it is shown in Fig.(1), the Wnt pathway includes 17 reactions (steps) and 15 proteins and protein complexes (species). The most important features of this model are the core complexes participating in *β-catenin* phosphorylation and destruction. Except from *β-catenin*, these complexes include *GSK3β*, *APC* and *Axin*. The steps that assemble or decompose these complexes are the forward reactions 4-10. Some steps of the pathway are presented: (a) the synthesis of *Axin* and of *β-catenin* (reactions 14 and 12), (b) the degradation of *Axin* (reaction 15), (c) the *Axin*-dependent degradation of *β-catenin* (reaction 13), and (d) the steps participating in the Destruction Core Cycle in order for phosphorylated *β-catenin* to be degraded (reactions 8, 9, 10, 11) [3].

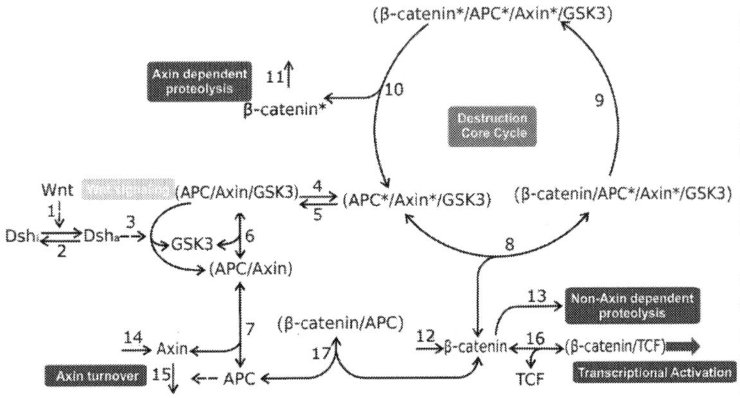

Figure 1. The 17 reaction scheme of the Wnt pathway by *Lee et al.*; single arrows characterize one way reactions, double arrows denote binding equilibria, broken arrows represent: *Dsh* activation by Wnt-stimulation, Dsh-mediated initiation of the release of *GSK3β* from the destruction complex and degradation of *Axin* through *APC* mediation (arrows 1,3,15); asterisks denote phosphorylation

The reversible binding steps involving *APC* (step 7), *Axin* (step 8), *TCF* (step 16) and *β-catenin* (step 17) along with their forward steps are considered very fast, so that the *Partial Equilibrium* approximation is valid [3]. Application of these four approximation yields a 15 dimensional ODE system, where rate equations in partial equilibrium are expressed through the remaining rates $V_4 - V_6$, V_9 and $V_{12} - V_{15}$. A complete list of the ODE system and the rate equations can be viewed in *Lee et al.* [3].

Table 1. Rates in Partial Equilibrium (left) and the existing conservation equations (right). X_i, $i = 1, ..., 15$ are the concentrations of proteins and protein complexes of the model and V_j, $j = 1, ..., 17$ are the rate equations. Concentrations marked with ^{o} are total (constants). Notation as in *Lee et al [3]*.

$V_{7f} - V_{7b} \approx 0$	$X_1 + X_2 = Dsh^o$
$V_{8f} - V_{8b} \approx 0$	$X_3 + X_4 + X_5 + X_8 + X_9 = GSK3^o$
$V_{16f} - V_{16b} \approx 0$	$X_3 + X_4 + X_6 + X_7 + X_8 + X_9 + X_{15} = APC^o$
$V_{17f} - V_{17b} \approx 0$	$X_{13} + X_{14} = TCF^o$

4. Reactions contributing to the formation of the *SIM*

Here, the case without Wnt stimulation ($Wnt = 0$) will be examined, so that the rates V_1, V_2 and V_3 are inactive and the X_1 and X_2 species do not participate. The results with Wnt stimulation are extracted with the same manner. The 15 dimensional Wnt model with 4 conservation equations and 4 Partial Equilibrium approximations, see Table (1), was analyzed with the *CSP* algorithm. Table (2) lists in column 3 the proteins and protein complexes that are identified by the *CSP Pointer* at a point in time where $M = 4$, see Fig.(2). The influence of the 7 reactions in the formation of the 8-dimensional *SIM* is assessed by the reaction in column 4 that were identified by the *Participation Index (PI)*.

(a) $M = 1$: the pointer identifies mainly *(β-catenin*/APC*/Axin*/GSK3)* and the PI suggests that reactions V_9 and V_{10} equilibrate, so that $k_{10}X_9 = V_{10} \approx V_{11} = k_{11}X_{10}$. We can assume that it is amenable to *Quasi Steady State* approximation $V_9 - V_{10} \approx 0$.

(b) $M = 2$: the pointer identifies mainly *(APC/Axin)* and the PI indicates that V_{6f} and V_{6b} equilibrate, so that $k_6X_5X_6 = V_{6f} \approx 10V_{6b} = k_6^- X_4$. We can assume that the PEA is valid.

(c) $M = 3$: the pointer identifies *β-catenin** and the PI indicates that there is an equilibria between reactions V_{10} and V_{11} so that $k_{10}X_9 = V_{10} \approx V_{11} = k_{11}X_{10}$. We can assume that *β-catenin** is amenable to *QSSA*, $V_{10} - V_{11} \approx 0$.

(d) $M = 4$: the pointer identifies mainly *(APC/Axin/GSK3)* and the PI produces the equilibration $k_4X_4 = V_4 \approx V_5 = k_5X_3$, given the equilibration $V_{6f} \approx V_{6b}$ and that $V_3 = 0$ we can assume that *(APC/Axin/GSK3)* is amenable to QSSA, $-V_3 - V_4 + V_5 + V_6 \approx 0$.

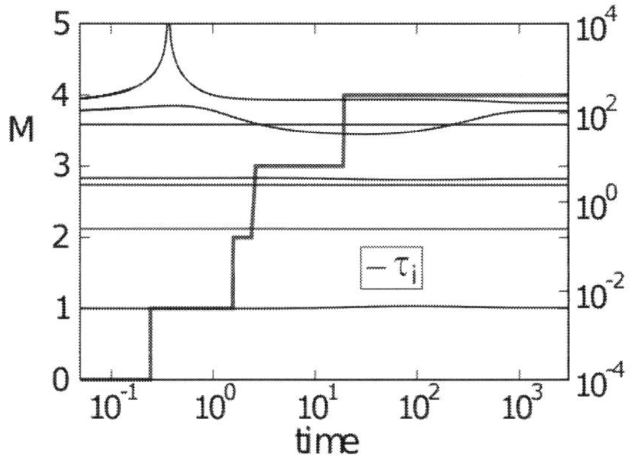

Figure 2. The evolution of the seven bounded time scales of the modified model are examined and it is shown that the number of fast time scales M increases with time; the time scales τ_i, $i = 1, ..., 7$ are denoted in black; M (blue) shows the increase of the exhausted time scales with time for one one $\mathbf{a}_r$-refinement and one $\mathbf{b}^r$-refinement.

Table 2. Proteins and protein complexes indicated by *CSP Pointer* are shown in column 3 (largest pointed marked bold); reactions exhibiting *Participation Indices* greater than 5% are listed in column 4 (those providing PI's greater than 15% are marked by bold); reactions with + sign equilibrate with those with the - sign.

M	Amplitude of exhausted modes	Proteins and protein complexes identified by CSP Pointer	Reactions
1	f^1	**(β-catenin*/APC*/Axin*/GSK3)**, (β-catenin/APC*/Axin*/GSK3)	**9**, **-10**
2	f^2	**(APC/Axin)**, (APC*/Axin*/GSK3), Axin	**-6f**, **6b**
3	f^3	**β-catenin***	**10**, **-11**
4	f^4	**(APC/Axin/GSK3)**, (APC*/Axin*/GSK3)	-4, 5,**-6f**, **6b**

5. Conclusions

The system of ODE's by *Lee et al.* exhibits multiscale behavior and it has amenable to singular perturbation analysis. It was demonstrated here that the algorithmic methodology of CSP can readily be employed, without the need (i) to cast the system in the proper non-dimensional form mechanism and (ii) to find the proper non-dimensional small parameters. The underlying physical insight for the control of the system was thus obtained without any effort. It was shown that as time evolves there exist 4 fast modes, see Fig 2. The fastest mode relates to the equilibration of the phosphorylation of *β-catenin* and its release through the destruction complex. The second mode relates to *(APC/Axin)* and the equilibria that forms between binding and releasing *GSK3*. The third mode involving *β-catenin** indicates that the rate *β-catenin** polyubiquitinated is the same as it is released from the destruction complex. The last mode involves the protein complex *(APC/Axin/GSK3)*. Given the equilibria that forms when *(APC/Axin)* binds and releases *GSK3*, an equilibrium develops between the phosphorylation and dephosphorylation of *APC* and *Axin* in the *(APC/Axin/GSK3)* complex.

6. Acknowledgments

The work of D T Maris and D A Goussis has been supported by the European Union (European Social Fund - ESF) and Greek national funds through the Operational Program "Education and Lifelong Learning" of the National Strategic Reference Framework (NSRF) - Research Funding Program: *ARISTEIA*, 2012-2015.

7. References

[1] Schnell S, Grima R and Maini P 2007 *American Scientist* **95** 134–142
[2] Kourdis P D, Steuer R and Goussis D A 2010 *Physica D: Nonlinear Phenomena* **239** 1798 – 1817
[3] Lee E, Salic A, Krüger R, Heinrich R and Kirschner M W 2003 *PLoS Biol* **1** 116–132
[4] Giles R H, van Es J H and Clevers H 2003 *Biochemica et Biophysica Acta - Reviews on Cancer* **1653** 1 – 24
[5] Reya T and Clevers H 2005 *Nature* **434** 1 - 24
[6] Daniel P and Hans C 2005 *Experimental Cell Research* **306** 357 - 363
[7] Krüger R and Heinrich R 2004 *Genome Informatics* **15** 138–148

[8] Heinrich R 2005 *Systems Biology* (*Topics in Current Genetics* vol 13) ed Alberghina L and Westerhoff H (Springer Berlin Heidelberg) pp 259–275
[9] Mirams G, Byrne H and King J 2010 *Journal of Mathematical Biology* **60** 131–160
[10] Lam S H and Goussis D A 1988 *Proc. Comb. Inst.* **22** 931–941
[11] Valorani M, Goussis D A, Creta F and Najm H N 2005 *Journal of Computational Physics* **209** 754 – 786
[12] Fenichel N 1979 *Journal of Differential Equations* **31** 53 – 98
[13] Kaper T 1999 *Proceedings of Symposia in Applied Mathematics* **56** 84 – 131
[14] Lam S H and Coussis D A 1991 *Reduced Kinetic Mechanisms and Asymptotic Approximations for Methane-Air Flames* (*Lecture Notes in Physics* vol 384) ed Smooke M (Springer Berlin Heidelberg) pp 227–242
[15] Kourdis P D and Goussis D A 2013 *Mathematical Biosciences* **243** 190–214
[16] Lam S H and Goussis D A 1994 *Intl. J. of Chem. Kinetics* **26** 461–486

The initiation of methane/air autoignition the important chemical components for various initial conditions

D M Manias and D A Goussis

Department of Mechanics, School of Applied Mathematics and Physical Sciences, National Technical University of Athens, 157 73 Athens, Greece

E-mail: dimanias@mail.ntua.gr, dagoussi@mail.ntua.gr

Abstract. A reactive system will be considered, the slow dynamics of which is characterized by time scales that are of explosive character; i.e., the components of the system that generate them tend to lead the system away from equilibrium. In particular, the initiation of a methane/air mixture autoignition will be considered for various initial temperatures and pressures. The species that relate to the explosive time scales and reactions that are responsible for the generation of these time scales will be identified. The analysis will be based on the *Computational Singular Perturbation (CSP)* algorithm, which is employed for the construction of the the reduced system that governs the long range (slow) evolution of the system. It will be demonstrated that an excellent agreement with the existing literature is obtained.

1. Introduction

Reactive systems introduce a wide range of time scales, the fastest of which (i) are of *dissipative* character and (ii) tend to constrain the evolution of the system on a low-dimensional surface in phase space; i.e., on the *slow invariant manifold* (SIM). On this surface, the system is characterized by slower time scales, leading the system to equilibrium or away from it. The case where some of these slow time scales are of *explosive* character will be considered here. These are time scales that are generated by components of the system that tend to lead it away from equilibrium. Such a situation arises in autoignition processes or even in flames [1 4]. The influence of these time scales can be assessed only when it is possible to construct the reduced (non-stiff) system that governs the flow on the SIM.

In the past, this reduced system was constructed with the traditional singular perturbation techniques, so only simple reactive systems were analyzed and the characteristic *explosive* time scale was then identified [5, 6]. However, the traditional techniques cannot handle the large mechanisms that are of interest today. Nowadays, the reduced system is constructed by algorithmic methodologies [7 10] so that the complexity of the reactive system is not an obstacle anymore and reactions responsible for the development of *explosive* time scales can be identified with a simple numerical tool [11, 12].

In this study, the autoignition process of homogeneous CH_4/air mixtures will be considered [13, 14]. The species and reactions that relate the most to the initiation of this process will be identified, by taking into account a wide range of initial temperatures and pressures and by employing the CSP algorithm. Recently, the reactions that favor or oppose the complete autoignition process and the related species were identified [8] for a single set of initial conditions. The analysis, here, is restricted to the start of the process, because the emphasis is placed on the influence of the initial values of temperature and pressure.

2. The Computational Singular Perturbation method

Consider a system consisting of N ordinary differential equations (ODE's):

$$\frac{d\mathbf{y}}{dt} = \mathbf{g}(\mathbf{y}) = \mathbf{S}_1 R^1(\mathbf{y}) + ... + \mathbf{S}_K R^K(\mathbf{y}) \tag{1}$$

where $\mathbf{y}$ is the N-dim. *state vector* and $\mathbf{g}(\mathbf{y})$ is the N-dim. *vector field*. The K additive terms in the RHS of Eq.(1) represent the various components (e.g. reactions) in the model. In chemical kinetics, $\mathbf{y}$ is the N-dim. vector of the species' mass fractions, $\mathbf{S}_k$ is the N-dim. stoichiometric vector of the k-th reaction and $R^k(\mathbf{y})$ is the corresponding reaction rate. Eq.(1) can be cast in CSP form as:

$$\frac{d\mathbf{y}}{dt} = \sum_{n=1}^{N} \mathbf{a}_n f^n, \qquad f^n = \mathbf{b}^n \cdot \mathbf{g}(\mathbf{z}) = \sum_{k=1}^{K} (\mathbf{b}^n \cdot \mathbf{S}_k) R^k \tag{2}$$

where $\mathbf{a}_n$ is the N-dim. CSP basis vector of the n-th mode, $\mathbf{b}^n$ is the N-dim. dual basis vector ($\mathbf{b}^i \cdot \mathbf{a}_j = \delta^i_j$) and f^n is the related amplitude [15, 16]. Assuming that the system in Eq. (2) exhibits $M < N$ time scales that are of dissipative nature and much faster than the rest ($\tau_1 < \cdots < \tau_M << \tau_{M+1} < \cdots < \tau_N$), yields:

$$f^m \approx 0 \quad (m = 1, M) \qquad\qquad \frac{d\mathbf{z}}{dt} \approx \sum_{n=M+1}^{N} \mathbf{a}_n f^n \tag{3}$$

The first of these two relations is an algebraic one and defines the SIM, while the second governs the slow evolution of the process on the SIM. The evolution Eq. (??) is free of the fast time scales (τ_1 to τ_M), so that its dynamics are characterized by the fastest of the slow time scales. The identification and analysis of the slow characteristic time scales, among them the explosive ones, is allowed by the availability of this equation [17 19].

The decomposition in fast/slow processes of the vector field $\mathbf{g}(\mathbf{y})$ allows the acquisition of significant physical understanding of complex chemical systems and it is possible by the use of CSP-related algorithmic tools. In particular, the contribution of each of the K reactions to the time scale τ_n and to the amplitude of each CSP mode f^n can be assessed by employing the *Time Scale Participation Index* (TPI) J^n_k and the *Amplitude Participation Index* (API) P^n_k, respectively, while the relation of the n-th mode to the i-th component of $\mathbf{y}$ can be assessed by employing the *CSP Pointer* (Po) D^n_i:

$$J^n_k = \frac{c^n_k}{|c^n_1| + ... + |c^n_K|} \qquad P^n_k = \frac{\mathbf{b}^n \cdot \mathbf{S}_k R^k}{|\mathbf{b}^n \cdot \mathbf{S}_1 R^1| + ... + |\mathbf{b}^n \cdot \mathbf{S}_K R^K|} \qquad D^n_i = a^i_n b^n_i \tag{4}$$

where, by definition $\sum_{k=1}^{K} |J^n_k| = \sum_{k=1}^{K} |P^n_k| = \sum_{i=1}^{N} a^i_n b^n_i = 1$ [10 12]. J^n_k measures the relative contribution of the k-th reaction to the n-th eigenvalue λ_n (therefore, to the time scale τ_n), while P^n_k measures the relative contribution of the rate of the k-th reaction to the amplitude of the n-th CSP mode f^n. Finally, D^n_k measures the relation of the n-th mode to the i-th component of $\mathbf{y}$.

CSP algorithmic tools will be employed next, in order to investigate the explosive modes that develop at the initiation of the autoignition process of various CH_4/air mixtures.

3. The Physical Problem

The homogeneous adiabatic autoignition at constant volume of a homogeneous stoichiometric CH_4/air mixture is considered. The reduced chemical kinetics mechanism employed here was derived from the GRI 3.0 mechanism[20] after ignoring the nitrogen chemistry reactions.

The t=0 case will be considered only. The purpose of this investigation is to examine the initiation of the process, by considering various initial pressures and temperatures. Here, four different values of the initial pressure p_o will be examined, while three different values for the initial temperature T_o will be accounted for each p_o. The slow dynamics of the system considered here exhibits two explosive time scales. Only the fastest one, say τ_e, will be considered, since this is the one that characterizes the autoignition process.

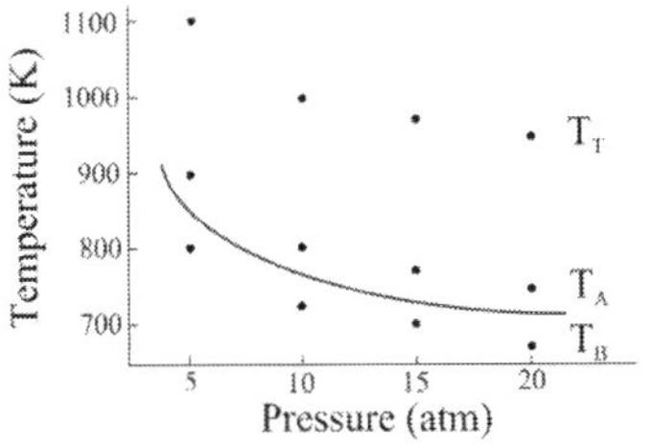

Figure 1: Explosion limit of methane/air [13, 14]. The twelve bullets denote the initial conditions for the temperature T_o and pressure p_o considered here, located below and above the explosion limit; subscripts A, B and T denote temperature right below (B), right above (A) and far above (T) the explosion limit.

4. Analysis of the fast explosive mode

In all cases considered, the reaction contributing the most towards the temperature evolution is reaction $HO_2 + CH_3 \leftarrow O_2 + CH_4$ $(118b)$, followed in a distance by reaction $H + CH_3(+M) \leftarrow CH_4(+M)$ $52b$. API data show that at $t = 0$ the amplitude f^e of the fast exhausted mode is entirely due to reaction $(118b)$, in all cases considered.

	5 atm	10 atm	15 atm	20 atm
T_T	CH_3 : +0.74	CH_3 : +0.65	CH_3 : +0.61	CH_3 : +0.56
	HO_2 : +0.14	HO_2 : +0.17	HO_2 : +0.17	HO_2 : +0.17
T_A	CH_3 : +0.35	CH_2O : +0.90	CH_2O : +0.92	CH_2O : +0.93
	CH_2O : +0.31	H_2O_2 : +0.08	H_2O_2 : +0.07	H_2O_2 : +0.06
T_B	CH_2O : +0.92	CH_2O : +0.96	T : +0.95	T : +0.98
	H_2O_2 : +0.07	H_2O_2 : +0.03		

Table 1: Species, the mass fractions of which are identified by the $CSP\ Pointer$. When $T_o \geq$ 900K, CH_3 relates the most to the fast time scale. When 900K> $T_o >$ 700K, the major pointed variable changes to CH_2O radical. In the cases $T_o \leq$ 700K, the pointed variable is T. The presence of the O-H chemistry has a negligible influence here.

According to Table 1, when $T_o >$ 900K (top row) the variable that relates the most to the fast explosive time scale is the CH_3 radical, followed by the HO_2 radical. This finding is full agreement with the TPI data which identify reaction $CH_3 + O_2 \rightarrow O + CH_3O$ $(155f)$ as the one contributing the most to τ_e, followed by reactions $2CH_3(+M) \rightarrow C_2H_6(+M)$ $(158f)$ and $CH_3 + H_2O_2 \leftarrow HO_2 + CH4$ $(157b)$. Clearly, the identified variables are reactants of the reactions that generate τ_e. At $T =$ 900K (second row, left entry) TPI data show that reaction $CH_2O + O_2 \rightarrow HO_2 + HCO$ $(32f)$ contributes significantly to τ_e along with reactions (155f, 158f, 157b). The contribution of (32f) is greater to that of (157b), so that CH_2O joints CH_3 as the variables identified by the $CSP\ Pointer$.

When 900K> $T_o >$ 700K (three entries on the right in second row and two entries on the left in third row), Table 1 shows that the $CSP\ Pointer$ identifies CH_2O as the one related the most to τ_e. TPI data show that reactions $(158f)$, $(32f)$ and $CH_3 + O_2 \rightarrow OH + CH_2O$ $(156f)$ are the ones contributing the most to τ_e. Of these reactions, $(32f)$ involves the species CH_2O, the mass fraction of which was identified by the $CSP\ Pointer$ as the one related the most to τ_e.

In the two cases where $T_o <$ 700K (two entries on the right in the third row), Table 1 indicates that the variable associated the most with τ_e is the temperature T. TPI data show that the reactions contributing the most to τ_e are $(158f)$, $(156f)$ and $(118b)$. As said preciously, reaction $(118b)$ is the one that drives the evolution of the temperature.

From the results displayed in Table 1 we can conclude that the fast exhausted time scale τ_e at $t = 0$ is mainly determined by the rate of oxidation of CH_3 when T_o is high and of (ii) CH_2O when T_o is low. At very low temperatures the variable that relates to the explosive mode is the temperature. The effect of pressure does not seem to be that much important in determining the nature of the initiation of the autoignition process.

5. Conclusions

Algorithmic asymptotic analysis was employed in order to identify the species and the reactions in which they participate as reactants that relate the most to the explosive time scale that characterizes an autoignition process. A homogeneous CH_4/air mixture was considered at various initial temperatures and pressures. Only the starting ($t = 0$) point was considered, since the aim of the present analysis was the assessment of the initial operating conditions (temperature and pressure) on the initiation of the process.

Although in all cases considered reaction $HO_2 + CH_3 \leftarrow O_2 + CH_4$ (118b) was shown responsible for initiating the process, its reactants (CH_4 and O_2) were not shown to relate to τ_e at $t = 0$. Instead, the reactants (e.g., CH_3, CH_2O, HO_2) of reactions that become active later on in the process were shown to relate at this starting point.

The predominant role of mainly CH_3 and CH_2O and secondary of HO_2 at high and low, respectively, initial temperatures is well documented in the existing literature regarding the action of the various reactions after the initiation of the process [13, 14, 21]. Here it was shown that the consumption of these intermediate species relate the most to τ_e at $t = 0$, although the rates of the consuming reactions there are practically zero, since the mass fraction of at least one of the reactants are zero at $t = 0$. In addition, it was shown that at very low temperatures, τ_e is mostly related to T.

The fact that the CH_3, CH_2O and HO_2 consuming reactions (155f), (156f), (158f), (32f) and (157b) determine τ_e at $t = 0$ does not contradict the fact that their rates is zero there because the mass fraction of these species is zero [8]. Keeping in mind that τ_e measures the time frame in which the autoignition takes place, these findings suggest that, τ_e is mainly dependent to the mass fraction in the initial mixture of the other reactants (O_2 for (155f), (156f) and (32f), M for (158f) and CH_4 for (157b)). In other words, the magnitude of τ_e is determined by the initial concentrations of O_2, M and CH_4 to the extend the reactions (155f), (156f), (158f), (32f) and (157b) consume CH_3, CH_2O and HO_2.

6. Acknowledgments

The work of D M Manias and D A Goussis has been supported by the European Union (European Social Fund - ESF) and Greek national funds through the Operational Program "Education and Lifelong Learning" of the National Strategic Reference Framework (NSRF) - Research Funding Program: "ARISTEIA" 2012-2015.

References

[1] Lee J C, Najm H N, Lefantzi S, Ray J, Frenklach M, Valorani M and Goussis D 2005 *Computational Fluid and Solid Mechanics* ed Bathe K J (Elsevier, Amsterdam) pp 717 720

[2] Kazakov A, Chaos M, hao and Dryer F L 2006 *The Journal of Physical Chemistry A* **110** 7003 7009

[3] Lu T, Yoo C, Chen J and Law C 2010 *Journal of Fluid Mechanics* **652** 45 64

[4] Najm H N, Valorani M, Goussis D A and Prager J 2010 *Combustion Theory and Modelling* **14** 257 294

[5] Williams F 1986 *Physica D: Nonlinear Phenomena* **20** 21 34

[6] Buckmaster J, Clavin P, Lin A, Matalon M, Peters N, Sivashinsky G and Williams F 2005 *Proceedings of the Combustion Institute* **30** 1 19

[7] Goussis D and Maas U 2011 *Fluid Mechanics and its Applications* **95** 193 220

[8] Diamantis D J, Kyritsis D C and Goussis D A 2014 *Proceedings Combustion Institute* in press

[9] Kourdis P D, Palasantza A G and Goussis D A 2013 *Computers and Mathematics with Applications* **65** 1516 1534

[10] Shan R and Lu T 2014 *Combustion and Flame* **161** 1716 1723

[11] Goussis D and Skevis G 2005 *Computational Fluid and Solid Mechanics* ed Bathe K J (Elsevier, Amsterdam) pp 650 653

[12] Goussis D A and Najm H N 2006 *SIAM Multiscale Modeling and Simulation* **5** 1297 1332

[13] Law C K 2006 *Combustion Physics* (New York: Cambridge University Press)

[14] Glassman I and Yetter R 2008 *Combustion* (Elsevier Science)

[15] Lam S H and Goussis D A 1988 *Proceedings Combustion Institute.* **22** 931 941

[16] Lam S H and Goussis D A 1994 *International Journal of Chemical Kinetics* **26** 461 486

[17] Trevino C 1989 *Prog. Aeronaut. Astronaut.* **131** 19 43
[18] Sanchez A L, Linan A and Williams F A 1997 *J. Eng. Math.* **31** 19 130
[19] Peters N, Paczko G, Seiser R and Seshadri K 2002 *Combust. Flame* **128** 38 59
[20] 2000 *GRI-Mech; release 3.0* http://www.me.berkeley.edu/gri_mech
[21] Westbrook C and Dryer F 1984 *Progress in Energy and Combustion Science* **10** 1 57

Numerical study on natural convection and entropy generation in squares and corrugated cavities

Souad Morsli[1], Amina Sabeur-Bendehina[2]

[1,2] Laboratoire d'Energie et Propulsion Navale, Faculté de Génie Mécanique, USTO MB BP 1505 el M'Naouer Oran Algeria

Email: sabeuramina@hotmail.com

Abstract. The present work deals with the study of the entropy generation in the natural convection process in both square cavities and square cavities with hot wavy walls through numerical simulations for different undulations and Rayleigh numbers, while keeping the Prandtl number constant. The results show that the heat transfer rate is notably affected by the shape of the hot wall geometry compared with the square one. It has been found in this investigation that the mean Nusselt number in the case of heat transfer in a cavity with wavy walls is lower, as compared to heat transfer in a cavity without undulations. Based on the obtained dimensionless velocity and temperature values, the distributions of the local entropy generation due to heat transfer and fluid friction, the local Bejan number, and the local entropy generation are determined and plotted for different undulations and Rayleigh number. The study is performed for Rayleigh number $10^5 < Ra < 10^7$ irreversibility coefficients $10^{-4} < \varphi < 10^{-2}$. The total entropy generation is found to increase with increasing undulation number

1. Introduction

In the past few decades, many researchers have investigated convective heat transfer in different configurations for practical engineering and industrial applications such as, heat exchangers, environmental comfort, and electronic cooling. By investigating the physical, geometric and the total entropy generation of processes, researchers can design systems with minimum loss of available energy, and enhance the thermal performance. Computational fluid dynamics and numerical codes make the calculation and simulation of local and total entropy generation easy in the complicated thermal models. Heat transfer in the system results in thermodynamic irreversibility, which generates the entropy. The entropy generation minimization is the method of optimization of the thermodynamic imperfections and fluid flow irreversibilities. Consequently, investigation of entropy generation and minimization in the heating systems provides an insight view of the thermodynamic irreversibilities involved. Bejan [1,2] has spent much effort to determine the gap between thermodynamics, heat transfer, and fluid mechanics. He has applied the second law of thermodynamics to determine the entropy generations due to heat and flow transport and consequently minimize the entropy generation. Narusawa [3] studied convective pattern change of natural convection in a rectangular cavity utilizing the second law of thermodynamics. He found that entropy generation from the perturbed temperature and velocity fields depends on aspect ratio of the cavity, the critical Rayleigh number and the ratio of entropy generation by viscous friction to that of thermal transport. Magherbi et al. [4] analyzed the entropy generation for laminar natural convection in transient state. They showed the effect of Rayleigh number and the irreversibility factor on the maximum entropy generation and the entropy generation in steady state condition. Famouri and Hooman [5] investigated numerically the entropy generation of natural convection in a partitioned cavity, with adiabatic horizontal and isothermally cooled vertical walls. They reported that, while entropy production due to fluid friction has small contribution to total entropy generation, the heat transfer irreversibility increases monotonically with both the Nusselt number and the dimensionless temperature difference.. Kaluri and Basak [6] studied entropy generation inside square cavities with distributed heated sources at wide ranges of Pr and Ra, using finite element method. They obtained that the heat transfer irreversibility dominates during conduction regime while irreversibility due to the fluid friction dominates in the convection regime.Ilis et al. [7] studied numerically the entropy generation in rectangular cavities with the same area but different aspect ratios. They found that for high value of Rayleigh number, with increasing aspect ratio the entropy generation due to fluid friction and the total entropy generation firstly increased, reaching to a maximum and then decreased. Bouabid et al. [8] investigated natural convection in an inclined rectangular cavity numerically. Their results showed that entropy generation tends towards asymptotic values for lower Grashof numbers, whereas it takes oscillation behaviour for higher values. Sheikhzadeh and Nikfar [9] investigated numerically laminar natural convection focused on entropy generation in a square cavity with a rectangular obstacle filled with either pure water or Cu-water nanofluid to simulate the effects of changing aspect ratio of the obstacle on fluid flow, heat transfer and entropy generation. Sabeur et al. [10] studied the influence of non-uniform boundary conditions on natural convection in inclined rectangular cavities differentially heated.The aim of the present work is to study the entropy generation in natural convection processes in a square cavity .The contribution of this work is the analysis of the variation of the entropy generation as a function of the Rayleigh number, undulation number, and irreversibility coefficient.

2. Mathematical Model

A two-dimensional heat transfer problem is considered in both cavities; the first one is square with plates walls while the second is square with wavy hot wall maintained at the constant temperature T_h. The cold wall is opposite to the latter and has a constant temperature T_c , while the other sides are insulated. The Rayleigh number is varied up to Ra = 10^7, while the Prandtl number is fixed to be Pr = 0.71. Figure 1 shows the geometry of the cavity under consideration and the coordinate system. The thermo-physical properties of the fluid in the flow model are assumed to be constant, except for density variations responsible for the body force term in the momentum equation. The Boussinesq approximation is assumed for the fluid properties related to the density and the temperature. The governing equations for the steady natural convection flow include the equations of conservation of mass, momentum, and energy:

$$\frac{\partial U}{\partial X}+\frac{\partial V}{\partial Y}=0$$
$$U\frac{\partial U}{\partial X}+V\frac{\partial U}{\partial Y}=-\frac{\partial P}{\partial X}+\Pr\left(\frac{\partial^2 U}{\partial X^2}+\frac{\partial^2 U}{\partial Y^2}\right) \quad (1)$$
$$U\frac{\partial V}{\partial X}+V\frac{\partial V}{\partial Y}=-\frac{\partial P}{\partial Y}+\Pr\left(\frac{\partial^2 V}{\partial X^2}+\frac{\partial^2 V}{\partial Y^2}\right)+Ra\Pr\theta$$
$$U\frac{\partial \theta}{\partial X}+V\frac{\partial \theta}{\partial Y}=\left(\frac{\partial^2 \theta}{\partial X^2}+\frac{\partial^2 \theta}{\partial Y^2}\right)$$

Figure 1. Geometry of the square cavity without and with undulation: (1) cold wall; (2) hot wall; (3) insulated walls.

. The shape of the wavy vertical wall is taken to be sinusoidal: $f(y)=\left[1-Amp(1-\cos 2\pi ny)\right]$ (2)

n and Amp are the number of undulations and the amplitude, respectively. The amplitude is fixed in this study at 0.05. The heat transfer rate due to convection in an enclosure is obtained from the Nusselt number calculation. On the wavy wall, the mean Nusselt number is expressed as follows:

$$\overline{Nu}=\frac{1}{s}\int_0^s\frac{\partial\theta}{\partial n}ds \quad (3)$$

The dimensionless local entropy generation due to heat transfer and the fluid friction (S_{LHT} and S_{LEF} respectively) for a two-dimensional heat and fluid flow in the Cartesian coordinates can be written as

$$S_{LHT} = \left(\frac{\partial \theta}{\partial X}\right)^2 + \left(\frac{\partial \theta}{\partial Y}\right)^2 \quad (4) \qquad S_{LEF} = \varphi\left\{2\left[\left(\frac{\partial U}{\partial X}\right)^2 + \left(\frac{\partial V}{\partial Y}\right)^2\right] + \left(\frac{\partial U}{\partial Y} + \frac{\partial V}{\partial X}\right)^2\right\} \quad (5)$$

The local entropy generation S_l is the sum of S_{LHT} and S_{LHT} The local Bejan number indicates the strength of the entropy generation due to heat transfer irreversibility

3. Grid validation

3.1 Square Cavity

Validation was carried out with respect to the benchmark case in square cavity as well as to those benchmark quality results in the literature. The grid independence test is performed successively sized grids 40 ×40, 50 ×50, 67 ×67 and 83 ×83. Table 1 indicates the average Nusselt numbers throughout the cavity for Rayleigh 10^5. It is clearly seen that the agreement is excellent with the deviation of less than 1%. So the grid 67 ×67 (Figure 2) is chosen for further computations.

Table1.Validation of the Grid

Grid	$\overline{Nu}$
40 ×40	**4.48**
50 ×50	**4.52**
67 × 67	**4.518**
83 × 83	**4.537**
Davis [11]	**4.519**

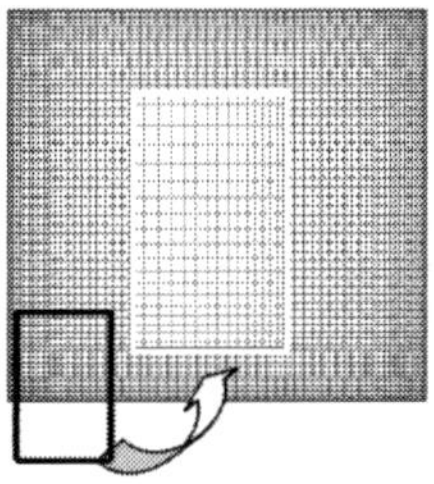

Figure 2. Mesh distribution in the square cavity

3.2 Cavity with the wavy wall

In order to study the precision of calculations; various grids were tested. Table 2 shows the various grids used for the Rayleigh number equal to 10^5 .The values of average Nusselt number are also presented. The choice of the grid 67 ×80 (Figure 3) which is used in the rest of the study is justified by the fact that the difference between the found values is lower than 4 %.

Table2. Validation of the Grid

Grid	$\overline{Nu}$
29 × 34	**3.907**
40 × 48	**3.820**
67 × 80	**3.760**
83 × 100	**3.720**

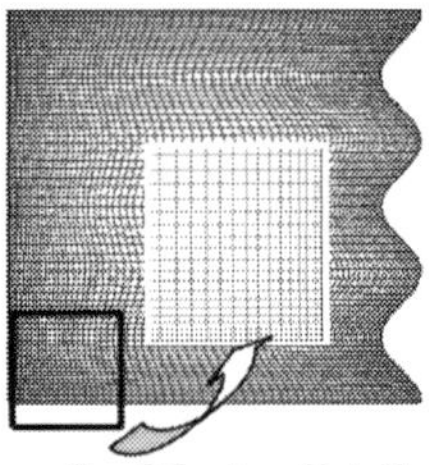

Figure3. Mesh distribution in the undulated cavity

4. Numerical Procedure

After transformation of coordinates, the Equations governing the flow and the energy are solved using a finite volume method with pressure-correction method as introduced by Patankar [12].The detailed information about the numerical procedure and convergence criteria can be found in Ref [10]. From the known temperature and velocity fields given by solving equations (1), the local entropy generation for the entire cavity is easily obtained after knowing the temperature gradient at the walls of the cavity.

5. Results and Discussions

Due to the several cases studied, the work was carried out in this paragraph only for the square cavity and the square cavity with four undulations, Rayleigh number equal to 10^7, and for $\phi = 0.0001$

Figure 4 illustrates the streamlines and isotherms in the square cavity. It was clearly seen that the thermal effects create a convection cell in the central region of the cavity turning it counterclockwise. It should be noted that the fluid remains a single cell by increasing the Rayleigh number; however stratification is obtained for high Rayleigh numbers.For low numbers of Rayleigh, the transfer mode is the dominant heat conduction. The increase of the Rayleigh intensifies air circulation loop in the

Journal of Physics: Conference Series **574** (2015) 012113 doi:10.1088/1742-6596/574/1/012113

boundary layer close to vertical walls. The flow then passes from laminar (10^3 <Ra <10^6) to turbulent (Ra> 10^6). It is noticed that, when the Rayleigh number values are above 10^6, an important flow along the hot and cold walls which extends along the horizontal walls. The total section of the cavity is occupied by a recirculation zone of low intensity. As and when the Rayleigh number increases (predominantly convection) the main flow is concentrated in the active walls of the cavity, and at the same time promoting the shearing effect, which lead to the formation of more and more cells. Figure 5 shows the isotherms, streamlines, contours of the local entropy generation due to the heat transfer and fluid friction, contours of local entropy generation and contours of Bejan number for Rayleigh number equal to 10^7. It is clearly seen that the undulated wall has an influence on the geometrical form taken by the cell as it is noticed on the different streamline patterns. In particular, near the hot wall, the streamlines converge toward the wall after each crest and diverge after each trough. This be behavior has a repercussion on the heat transfer by convection near the wall. Indeed, when the normal velocity component approaches the streamline to the wall, the heat transfer increases. It was showed the thermo-convective flow accelerating more and is represented by the form of one diagonally stretched cell. Near the crest, the streamlines move near the wall just after the crest. The isotherm distribution shows the same feature near the latter region as presented in the figure (4). This observation proves the diminution of the thermal boundary layer thickness just after the crest. The flow stays stratified in the core region of the cavity. Near and along the hot wall, the boundary layer thickness increases then decreases under the effect of the undulation. It is clear that the undulated wall has an influence on the geometrical form taken by the cell as it is noticed on the different streamline patterns.

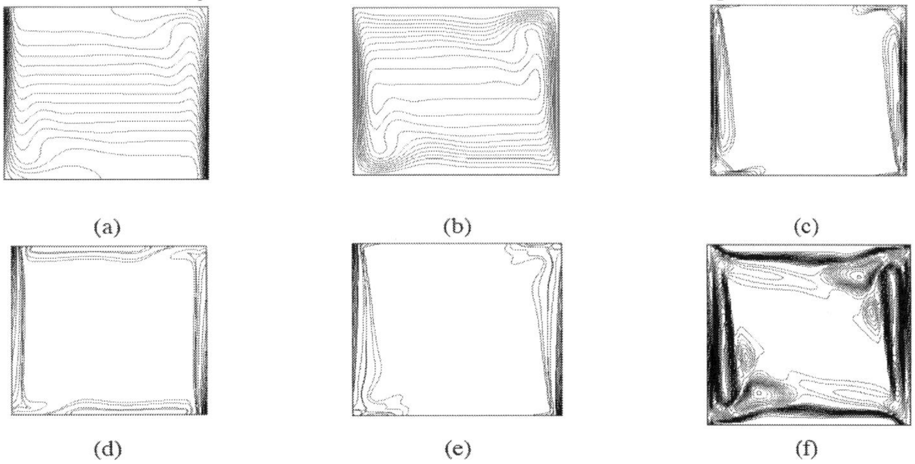

Figure 4. (a) Isotherms, (b)streamlines, (c) local entropy generation due to the heat transfer , (d)local entropy generation due to the fluid friction, (e) local entropy generation (f) Bejan Number in the square cavity (Ra=10^7) and $\phi = 0.0001$

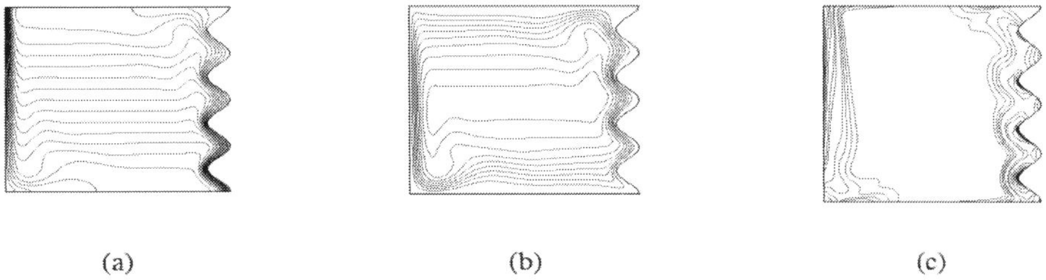

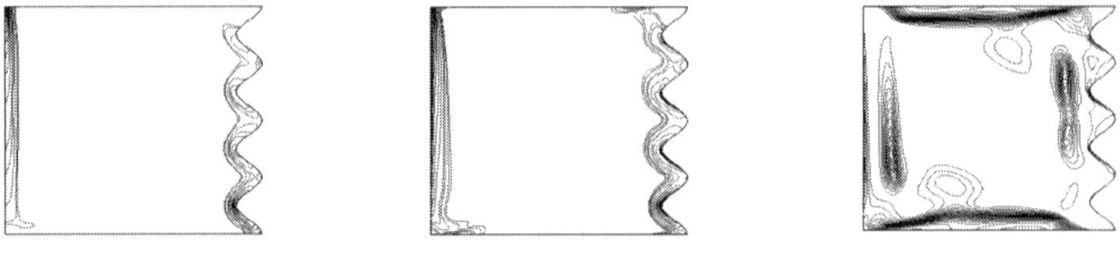

(d) (e) (f)

Figure 5.(a) Isotherms, (b)streamlines, (c) local entropy generation due to the heat transfer , (d) local entropy generation due to the fluid friction, (e) local entropy generation (f) Bejan Number in the undulated Cavity (Ra=10^7) and $\phi = 0.0001$

Figure 6 shows the variation of the total entropy generation with Rayleigh number for all cavities carried out and for ϕ=0.0001. Distribution pattern of entropy generation are very much similar for all undulations and for the square cavity. It is clearly seen that the increasing of Rayleigh number induces a linear increasing of entropy generation value. It could be noticed that for increasing undulation number, the total entropy generation value also increases.

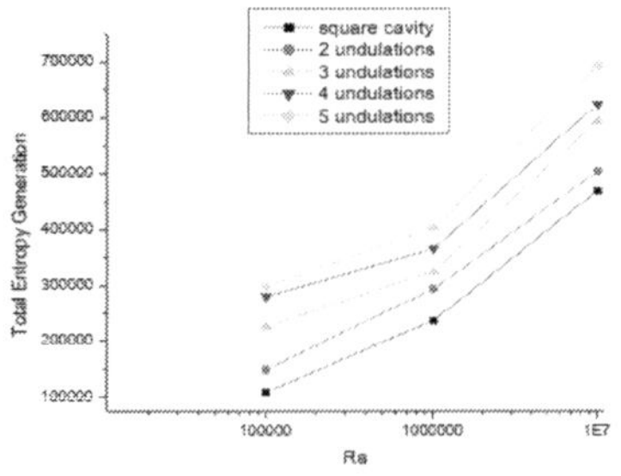

Figure 6.Variation of total entropy generation with Rayleigh number for all cavities and for $\phi = 0.0001$

6.Conclusion

This study has performed a numerical investigation into the natural convection heat transfer performance and entropy generation in two kinds of cavities , the first one is a square and the second one is a wavy wall square cavity filled with air. In modeling the cavity, it has been assumed that the left wall has a wavy surface and is heated along the wavy region only, while the right wall is flat and has a constant low temperature. In addition, it has been assumed that the upper and lower cavity walls are both flat and insulated. The simulations have examined the effects of Rayleigh number and wavy surface geometry parameters on the mean Nusselt number, total entropy generation and Bejan number, respectively. The numerical results support the following major conclusions

- The mean Nusselt number and total entropy generation both increase as the Rayleigh number increases.
- As the amplitude and wavelength of the partially-heated wavy surface increase, the mean Nusselt number reduces and the total entropy generation increases.
- The Bejan number increases as the amplitude and wavelength of the-heated wavy surface increase.
- The mean Nusselt number increases and the entropy generation reduces as the peak in the wavy surface approaches the horizontal center plane of the cavity

7. References

[1] A. Bejan 1980, *Second law analysis in heat transfer*, Energy **5 721–732.**
[2] A. Bejan 1996., *Entropy generation minimization*, CRC Press, Boca Raton, Fl,
[3] Narusawa, U 1999. *The second-law analysis of convective pattern change in a rectangular cavity*, J. Fluid Mech., **392**, pp. **361–377**
[4] Magherbi, M., Abbassi, H. and Ben Brahim, A 2003. *Entropy generation at the onset of natural convection*, Int. J. Heat Mass Transfer, **46**, pp. **3441–3450**
[5] Famouri, M. and Hooman, K. 2008 *Entropy generation for natural convection by heated*

partitions in a cavity, Int. Commun. Heat Mass Transfer, **35,** pp. **492–502**
[6] Kaluri, R.S. and Basak, T. 2011 *Analysis of entropy generation for distributed heating in processing of materials by thermal convection*, Int. J. Heat Mass Transfer, **54,** pp. **2578–2594**.
[7] Ilis, G.G., Mobedi, M. and Sunden, B. 2008 ''*Effect of aspect ratio on entropy generation in a rectangular cavity with differentially heated vertical walls*'', Int. Commun. Heat Mass Transfer, **35**, pp. **696–703**
[8] Bouabid, M., Magherbi, M., Hidouri, N. and Ben Brahim, A. 2011 Entropy *generation at natural convection in an inclined rectangular cavity*, Entropy, **13**, pp. **1020–1033**
[9] Sheikhzadeh G. A. and M. Nikfar 2013 *Aspect ratio effects of an adiabatic rectangular obstacle on natural convection and entropy generation of a nanofluid in an enclosure*, Journal of Mechanical Science and Technology **27** (11) **3495~3504**
[10] Sabeur-Bendehina A., Imine O. and Adjlout L. 2011 "*Laminar Free Convection in Undulated Cavity with Non Uniform Boundary Conditions*," Comptes Rendus Mecanique **339** (1), **42–57**
[11] De Vahl Davis, G. 1983 Natural Convection of air in a square cavity: a benchmark numerical Solution, International Journal of Numerical Methods in Fluids **3, 249-264**
[12] Patankar S.V 1980, Numerical heat transfer and fluid flow, Series in computational methods in mechanics and thermal sciences **Mc Graw-Hill book company**

Multi objective optimization of line pack management of gas pipeline system

A Chebouba
Laboratory of reliability of the oil equipments and the materials, Faculty of Hydrocarbon and Chemistry, University of Boumerdès, Algeria

E-mail: chebouba@yahoo.fr

Abstract. This paper addresses the Line Pack Management of the "GZ1 Hassi R'mell-Arzew" gas pipeline. For a gas pipeline system, the decision-making on the gas line pack management scenarios usually involves a delicate balance between minimization of the fuel consumption in the compression stations and maximizing gas line pack. In order to select an acceptable Line Pack Management of Gas Pipeline scenario from these two angles for "GZ1 Hassi R'mell-Arzew" gas pipeline, the idea of multi-objective decision-making has been introduced. The first step in developing this approach is the derivation of a numerical method to analyze the flow through the pipeline under transient isothermal conditions. In this paper, the solver NSGA-II of the modeFRONTIER, coupled with a matlab program was used for solving the multi-objective problem.

1. Introduction

This work is concerned with the minimization of energy cost for gas pipeline under transient conditions. The purpose of the work (Dupont and Rachford) [1] was to develop a mathematical programming approach to reducing fuel cost in long gas transmission lines with time varying demands. They use descent method together with Lagrange Multipliers to solve given non linear problem. An algorithm described in [2] based on sequential quadratic programming and takes account of the structure of the pipe flow equations by means of a reduced gradient technique which eliminates most of the variables from the quadratic sub-problem. Goldberg [3] solved the Wong's problem in the framework of the genetic algorithms. The GA method is a biologically inspired optimization and search technique developed by Holland [4]. Chebouba et al. [5] applied an ant optimisation algorithm to a linear system. In this paper, a genetic algorithm (NSGA II) [6] is used to solve a bi-objective problem.

2. Transient Flow in Pipeline Systems

The pipeleg is the most important component of the system because it defines the major dynamic characteristics. Isothermal unidirectional flow is usually assumed when modeling the flow of gas through a pipeleg, they result in the following equations, and the equation of continuity is

$$\frac{\partial m}{\partial x}+\frac{\partial(\rho A)}{\partial t}=0 \qquad (1)$$

The equation of motion is the transient form, accounting for the change of parameters in time:

$$\frac{\partial p}{\partial x} + \rho g \sin\alpha + \frac{\lambda v^2 \rho}{2d} + \rho\frac{\partial v}{\partial t} = 0 \tag{2}$$

The equation of state for a gas flow regarded as isothermal is,

$$\frac{p}{\rho} = z\frac{R}{M}T$$

the number of fundamental equations reduces to three and equation may be written in the simpler form

$$\frac{p}{\rho} = B^2 \tag{3}$$

Where B is the isothermal speed of sound, Equations (1) and (3) imply

$$F_1 = \frac{B^2\partial m}{A\partial x} + \frac{\partial p}{\partial t} = 0 \tag{4}$$

where mass flow is $m = \rho A v = \frac{p}{B^2}Av$

By equations (2) and (3), and the above definition of m,

$$F_2 = \frac{1\partial p^2}{2\partial x} + \frac{p\partial m}{A\partial t} + \frac{p^2 g}{B^2}\sin\alpha + \frac{\lambda B^2 m^2}{2dA^2} = 0 \tag{5}$$

We transform the system of partial differential equations into one of algebraic equations by the method of finite differences as performed in the implicit method [7].

4. Compressor Unit and Stations

4.1. Single centrifugal compressor unit

The primal quantities related to a centrifugal compressor unit are inlet volumetric flow rate Q, speed S, adiabatic head H, and adiabatic efficiency η. The relationship among these quantities can be well described by the following equations:

$$S = -B_H Q + \left((B_H Q)^2 - 4A_H\left(C_H Q^2 - (H/9.81)\right)^{0.5}\right)/(2A_H) \tag{6}$$

$$\eta = (C_E(Q/S)^2 + B_E(Q/S) + A_E)/100 \tag{7}$$

where A_H, B_H, C_H, A_E, B_E, and C_E are constants which depend on the compressor unit and are typically estimated by applying the least squares method to a set of collected data of the quantities Q, S, H, and η. Since, the preferred variables from the pipeline modeling perspective are mass flow rate m, suction pressure p_s and discharge pressure p_d, the relationships between (H,Q) and (m,p_s,p_d) are the following:

$$H = \frac{Z_s R T_s}{\omega}\left[\left(\frac{p_d}{p_s}\right)^{\omega} - 1\right] \tag{8}$$

$$Q = Z_s R T_s \frac{m}{p_s}\frac{1}{n} \tag{9}$$

where $\omega = \frac{\gamma - 1}{\gamma}$, the specific heat ratio γ, the gas compressibility factor at suction conditions Z_s, and the gas constant R, are positive parameters. n is an integer number which represents the number of compressor to put in service.

5. Objective functions

5.1 Total power Minimization

To find out how to run the compressor station so as to achieve a given value $(m/n,p_s,p_d)$, we proceed to map that point back to the original operating space by first computing H and Q from equations (8) and (9), respectively, and then solving for S and η in equations (6) and (7), respectively.
The power w consumed by a compressor station is given by

$$W = \frac{mH}{\eta\eta_m}$$

where η_m is a mechanical efficiency.

The functional to be minimized is the total power consumed by the system when it operates during the time period $[t=0\ t=tp]$ and is expressed over such time period as follows:

$$W_{Total} = \sum_{t=0}^{t=tp}\left(\frac{m_t H_t}{\eta_t \eta_m}\right)$$

5.2. Line pack maximization

The key consideration in pipeline operations is line pack, which is defined as the volume of natural gas between the compressor discharge pressure and the customer end-point delivery pressure. Gas pipelines not only serve as transportation links between producer and consumer, but they also represent potential storage units for safety stocks.

Moreover, adequate line pack allows pipeline operators to handle short-term fluctuations in demand and supply properly. On the other hand, when the line pack level is high, the energy consumption might be unnecessarily high. Total line pack maximization is expressed as follows [8]:

$$Maximise\ (LP_{TOT}) = \sum_{t=0}^{t=tp} 7.855\ 10^{-7}\left(\frac{T_{SC}}{P_{SC}}\right)\left(\frac{P_{avg(t)}}{Z_{avg(t)}T}\right)D^2\ L$$

7. SOFTWARE

All these equations were integrated in a computer program, built in matlab. This program was used for hydraulic modeling of the pipeline systems

On the other hand, for the optimization part, we use, in this work, the genetic algorithm optimization software NSGA II [6] available in modeFRONTIER by ESTECO [9].

modeFRONTIER allows optimization analysis to be achieved to changing the values assigned to the input variables of Matlab software. The output from matlab software can then be described as the objectives and constraints of the problem.

8. Application

In this case, we consider the last pipeleg of the pipeline. The initial conditions can be characterized as the steady state studied in the last section. The principal parameters used for the last pipeleg of the gas pipeline Hassi R'Mel- Arzew (Algeria) are listed in table 1.

Table 1

Length of the pipeleg (km)	Diameter (m)	gas constant J/(kg-°k)	contract pressure (bar)	Time horizon (h)	Step time (h)	Mass flow equation at load (kg/s)
110	0.990	435	42	24	1.5	300*(1-0.7*sin((2*3.14*t/3600)))

As with any meta heuristics, many parameters need to be set to have a good performance of NSGA II algorithm. The model performance was tested against variations of cross probability and mutation probability. A value of 0.3 for cross probability and one of 0.1 for mutation probability seem to be the best choice for our problem.

The calculations of simulation were realized by means of programs made in Matlab. We have determined, for a period of 24 hours, the discharge pressures, compressor speed and adiabatic head in the last compressor station which is directly connected with the point of delivery. This solution represents the optimal programming run of this station on a forecast of consumption over a period of 24 hours. We have chosen sixteen 1.5-h times steps. The results, for the minimum of power used by turbo compressors, are listed in tables 2 and 3. We can notice from table 3, that the pressure at delivery point never drops below the contract pressure (42 bars).

Table 2

Time step	Discharge pressure in bars	**Time step**	Discharge pressure in bars
Time step 1	57.35	Time step 9	55.9
Time step 2	65.39	Time step 10	65.01
Time step 3	55.9	Time step 11	56.1
Time step 4	66.57	Time step 12	64.9
Time step 5	56.24	Time step 13	55.86
Time step 6	64.52	Time step 14	65.98
Time step 7	56.96	Time step 15	61.55
Time step 8	66.04	Time step 16	55.86

Table 3

Time step	Pressure at delivery point in bars	**Time step**	Pressure at delivery point in bars
Time step 1	45.51	Time step 9	42.02
Time step 2	42.03	Time step 10	42.26
Time step 3	42.00	Time step 11	43.06
Time step 4	42.51	Time step 12	42.00
Time step 5	42.05	Time step 13	44.36
Time step 6	42.06	Time step 14	42.09
Time step 7	42.88	Time step 15	42.27
Time step 8	43.74	Time step 16	42.03

10. Conclusion

The objectives, in this paper, were to maximize the system line pack as the first, and to minimize total power as the second objective function. Also, since there was approximately a direct relation between the power consumed and CO2 emission, the Pareto points with minimum power resulted in minimum carbon dioxide emission and vice versa. At the end, we can provide a decision aid tool to operators of gas pipeline networks for making appropriate decision to determine discharge pressures and number of turbo compressors to put in service. In the future, to improve these results, we plan to use a hybrid method combining the genetic algorithm optimization software NSGA II and a local search.

References

[1] Dupont T F and Rachford H. H. Jr 1987 Optimization of Power Usage In Transient Gas Transmission Lines *PSIG Annual Meeting* Tulsa Oklahoma October

[2] Furey B P 1993 A Sequential Quadratic Programming Based Algorithm for optimization of Gas Networks *Automatica* **6 29**

[3] Golberg D E 1989 *Genetic Algorithm in search, Optimization, and Machine Learning* (Addison-Wesly)

[4] Holland J H 1975 *Adaptation in Natural and Artificial System* (Ann Arbor the University of

[5] Chebouba A, Yalaoui F, Smati A, Amodeo L, Younsi K and Tairi A 2009 Optimization of natural gas pipeline transportation using ant colony optimization *Computers & Operations Research* **36** 1916 – 1923

[6] Deb K, Partap A, Agarwal S and Meyarivan T 2002 A Fast and Elitist Non Dominated Sorting Genetic Algorithm NSGA-II *IEEE Transactions on Evolutionary Computation* **6** 182-197

[7] Streeter V L and Wylie E B 1970 Natural gas pipeline transients *Soc. Pet. Eng. J.* **10** 357-364

[8] Aliani Kashani A H and Molaei R 2014 Technico-economical and environmental optimization of natural gas network operation *Chem. Eng. Res. Des Research http//dx.doi.org/10.1016/j.cherd.2014.02.006*

[9] ModeFRONTIER software package, developed by ESTECO (Italy), distributed by SIRENHA, 1, Rue de la Noe - BP 42105 - 44321 Nantes Cedex – FRANCE, 2003

Toward the estimation of background fluctuations under newly-observed signals in particle physics

Federico Colecchia

Brunel University London, Kingston Lane, Uxbridge, Middlesex UB8 3PH, UNITED KINGDOM

E-mail: federico.colecchia@brunel.ac.uk

Abstract. When the number of events associated with a signal process is estimated in particle physics, it is common practice to extrapolate background distributions from control regions to a predefined signal window. This allows accurate estimation of the expected, or average, number of background events under the signal. However, in general, the actual number of background events can deviate from the average due to fluctuations in the data. Such a difference can be sizable when compared to the number of signal events in the early stages of data analysis following the observation of a new particle, as well as in the analysis of rare decay channels. We report on the development of a data-driven technique that aims to estimate the actual, as opposed to the expected, number of background events in a predefined signal window. We discuss results on toy Monte Carlo data and provide a preliminary estimate of systematic uncertainty.

1. Introduction

The task of data analysis in particle physics often deals with data sets comprising collision events that contain the signature of a scattering of interest as well as background events that correspond to uninteresting processes mimicking the signal. When estimating the number of signal events, background probability density functions (PDFs) are often extrapolated from control regions to a predefined signal window and subsequently used in template fits. However, this can only provide an estimate of the expected, or average, number of background events under the signal, and cannot take into account the effect of fluctuations. In practice, when the number of signal events, S, is much higher than the size of the typical fluctuations on the number of background events, $\sigma_B = \sqrt{\langle B \rangle}$, $\langle B \rangle$ being the expected number of background events in the signal window, the discrepancy between B and $\langle B \rangle$ can be neglected.

Nonetheless, when the number of signal events is low enough, the difference can be sizable. This can occur in the early stages of data analysis following the discovery of a new particle, or in the analysis of low-cross section processes. In such cases, the expected number of background events in the signal window can be a biased estimate of the actual number.

We report on the development of a data-driven technique that aims to estimate the actual number of background events under an observed signal, as opposed to the expected number. Our algorithm makes it possible to decompose an input mixture of signal and background events, e.g. a collection of events that pass all selection criteria corresponding to the end-point of a given analysis. This allows the shape of the background PDF to be estimated from the data, thereby taking into account the effect of statistical fluctuations. The development of this technique was influenced by a number of statistical methods, most notably the Gibbs Sampler [1] for mixture model decomposition, Expectation Maximisation [2], and Data Augmentation [3].

2. The algorithm

The algorithm that we use to decompose the input mixture of signal and background events is related to a method that we have proposed with reference to a different application to data analysis at high-luminosity hadron colliders [4, 5].

The PDF of the underlying statistical model has the form $F = \alpha_0 f_0(x) + \alpha_1 f_1(x)$, where α_0 and α_1 are the fractions of background and signal events in the input data set, respectively, with $\alpha_0 + \alpha_1 = 1$, and where f_0 (f_1) is the background (signal) PDF. In the context of this study, the variable x is interpreted as the invariant mass of a set of final state particles.

A notable feature of our approach, when compared to classical mixture models where predefined subpopulation PDF shapes are typically enforced a priori, is the nonparametric definition of the subpopulation PDFs, f_j. At every iteration of the algorithm, individual events are mapped to signal or background on a probabilistic basis, and the estimate φ_j of the subpopulation PDF f_j at that iteration is obtained by means of spline interpolation[1] of the histograms of x corresponding to those events that are mapped to signal or background at that iteration. This allows the algorithm to estimate generic deviations of the PDF shapes from the corresponding control sample templates due to fluctuations in the data. The shapes of the signal and background distributions in the data set analysed are ultimately estimated as splined histograms averaged over a predefined number of iterations.

The pseudocode of the algorithm is given below, subscripts "sig" and "bkg" relating to signal and background, respectively. The value of quantity v at iteration t is denoted by $v^{(t)}$ throughout.

(i) **Initialization:** Set $\alpha_{bkg} = \alpha_{bkg}^{(0)} = \alpha_{sig} = \alpha_{sig}^{(0)} = 0.5$, where $\alpha_{bkg} = \alpha_0$ and $\alpha_{sig} = \alpha_1 = 1 - \alpha_{bkg}$. Initial estimates $\varphi_j^{(0)}$ of the subpopulation PDFs f_j, $j = 0, 1$, are given by splined one-dimensional histograms of x obtained from high-statistics control samples.

(ii) **Iteration t:**

(a) Generate $z_{ij}^{(t)}$ for all events i and distributions j according to $P(z_{ij}^{(t)} = 1|\alpha_j^{(t-1)}, \varphi_j^{(0)}, x_i) = \frac{\alpha_j^{(t-1)}\varphi_j^{(0)}(x_i)}{\alpha_0^{(t-1)}\varphi_0^{(0)}(x_i)+\alpha_1^{(t-1)}\varphi_1^{(0)}(x_i)}$. Both the nonparametric treatment of the PDFs and the use of $\varphi_j^{(0)}$ instead of $\varphi_j^{(t-1)}$ to map individual events to signal or background distinguish this implementation from the classical Gibbs sampler for mixture models.

(b) Set $\alpha_j^{(t)} = \sum_{i=1}^{N} z_{ij}^{(t-1)}/N$, $j = 0, 1$.

We used a total number of 6,000 iterations, and averaged the PDF estimates, φ_j, over the last 4,000. These settings allowed the algorithm to reach convergence in all runs performed in this study, and no significant difference in the results was observed by changing them.

A more detailed description of this implementation of the algorithm can be found in [6]. The execution time was ~ 50 s per run on the data sets analysed using a 2 GHz Intel Processor with 1 GB RAM, which we consider reasonable for offline use.

3. Results

We illustrate this technique on a toy Monte Carlo data set obtained superimposing a gaussian signal with a first-order polyomial background. In the following, we will interpret the signal distribution as an invariant mass distribution corresponding to the decay of a particle with mass

[1] We have used the alglib C++ library [7] with this implementation of the algorithm.

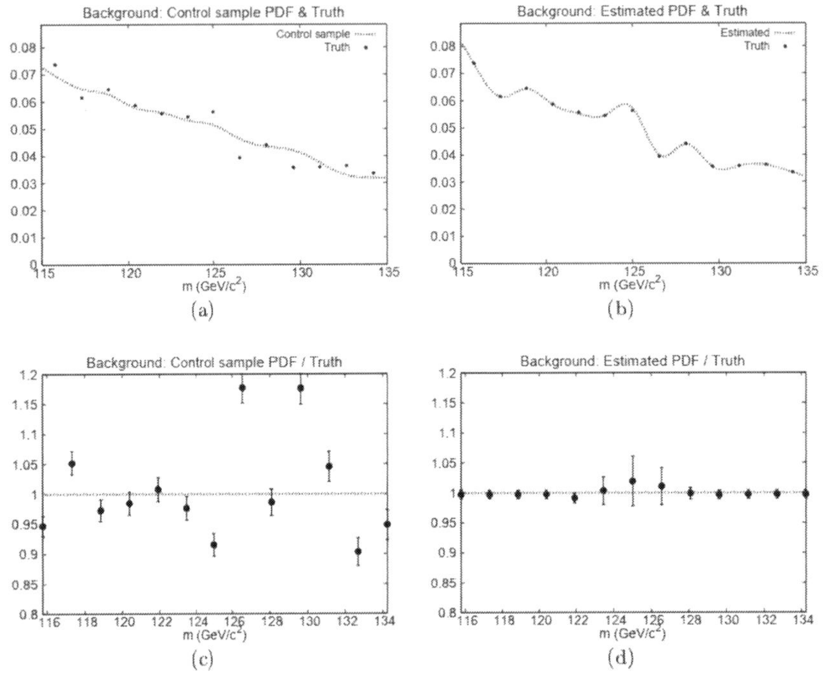

Figure 1: (a) True background distribution (points) superimposed with the PDF obtained from the high-statistics control sample (curve). (b) The same true background distribution (points) superimposed with the background PDF estimated from the data using the algorithm (curve). (c) Ratio between the background PDF obtained from the control sample and the true distribution. (d) Ratio between the background PDF estimated using the algorithm and the true distribution.

$m = 125$ GeV/c^2 and width 1 GeV/c^2. We superimposed $S = 200$ signal events to a total of 4,200 background events in the region 115 GeV/c^2 $< m <$ 135 GeV/ c^2, corresponding to an average of $\langle B \rangle = 1,600$ background events in the signal region, which is defined by 120 GeV/c^2 $< m <$ 130 GeV/c^2.

Due to statistical fluctuations in the data, different samples correspond to different numbers of background events in the signal window. In this study, the standard deviation on the number of background events with 120 GeV/c^2 $< m <$ 130 GeV/c^2 is $\sigma_B = \sqrt{\langle B \rangle} = 40$ events, which is sizable when compared to the number of signal events generated, $S = 200$. This illustrative scenario is not dissimilar from the early stages of data analysis following the observation of a Higgs boson in the $\gamma\gamma$ final state at the Large Hadron Collider (LHC) at CERN [8, 9].

High-statistics control samples were generated corresponding to 30,000 signal and 30,000 background events, and were used to obtain initial conditions on the signal and background PDF shapes. The function of the algorithm is essentially to iteratively refine those initial conditions based on the data, thereby taking into account the effect of statistical fluctuations. As a consistency check, the estimated fraction of background events in the input data set, $\hat{\alpha}_0$, was found to be in agreement with the true value within 2% in all runs used in this study.

The performance of the algorithm in terms of estimating the shape of the background PDF in the data set analysed is illustrated in figure 1. Figure 1 (a) displays the true background distribution (points) superimposed with the PDF obtained from the high-statistics control sample (curve). The discrepancies due to statistical fluctuations in the data are apparent. The points in figure 1 (b) show the same true background distribution as in figure 1 (a), but in this case the superimposed curve is the PDF estimated from the data using the algorithm, averaged over the last 4,000 iterations from a total of 6,000.

The ratio between the background control sample PDF and the true PDF is displayed in figure 1 (c), which again highlights the effect of fluctuations. The corresponding ratio between estimated and true PDF is shown in figure 1 (d), and shows a significantly-improved agreement.

It is worth recalling that, for the purpose of this study, what we are interested in is the shape of the background PDF. In fact, our objective is to estimate the actual number of background events under the signal as opposed to the expected number. The signal-related plots corresponding to figure 1 showed good agreement between the estimated and the true distribution, and were used together with the estimated fraction of background events in the data in order to check the consistency of the results obtained using the algorithm.

The plots in figure 1 refer to a run of the algorithm on a data set with $B = 1,571$ background events in the signal region 120 GeV/c^2 $< m <$ 130 GeV/c^2. The corresponding number of events estimated with that run of the algorithm was $\hat{B} = 1586.5$.

The algorithm was also run on multiple toy Monte Carlo data sets, corresponding to different numbers of background events in the signal window. Our preliminary estimate of the uncertainty on $\hat{B}$, i.e. on the estimated number of background events under the signal, is ~ 50 events. Work is underway to reduce this uncertainty below the size of typical background fluctuations in the data, $\sigma_B = \sqrt{\langle B \rangle} = 40$ events. Our studies suggest that the uncertainty on $\hat{B}$ is dominated by the uncertainty on the estimated fraction of background events in the data set, $\hat{\alpha}_0$. In fact, when the algorithm is run with α_0 kept fixed at the corresponding true value, the uncertainty on $\hat{B}$ drops from 50 to 12 events.

The results obtained running the algorithm on the different input data sets are summarised in table 1, where B_{gen} denotes the true number of background events in the signal window at generation, $\hat{B}$ is the corresponding number estimated using the algorithm, and $\Delta B = \hat{B} - B_{gen}$. The quantities $\hat{B}^*$ and ΔB^* in the table have a similar meaning as $\hat{B}$ and ΔB, but the values were obtained running the algorithm with α_0 kept fixed at its true value. The average and standard deviation of $\hat{B}$ across the runs are referred to as $\langle B \rangle$ ($\langle B \rangle^*$) and σ_B (σ_B^*), respectively.

4. Conclusions and outlook

We have reported on the development of a data-driven technique that aims to estimate the actual, as opposed to the expected, number of background events under an observed signal in particle physics. Established methods that rely on the extrapolation of background distributions from control regions to a predefined signal window allow a precise estimation of the expected, or average, number of background events under the signal. However, the actual number of background events in the signal window can deviate from the average due to statistical fluctuations in the data. Although the discrepancy is often negligible when compared to the number of signal events, it is not necessarily so in the early stages of data analysis following the discovery of a new particle, or more generally in the analysis of low-cross section processes.

We have described an algorithm that uses the data to estimate the shape of the background distribution in a predefined signal window, e.g. using the end-point of a given analysis i.e. a collection of events that pass all selection criteria. Control samples are used only to provide initial conditions for the background PDF, but the PDF shape is otherwise estimated directly from the same data set that contains the observed excess of signal events. We have discussed results on toy Monte Carlo data, with reference to an illustrative scenario that is not dissimilar from the early stages of data analysis following the discovery of a Higgs boson in the $\gamma\gamma$ channel.

We have provided a preliminary estimate of the uncertainty associated with the estimated number of background events in the signal window at the level of 50 events, out of a total average number $\langle B \rangle = 1,600$. Although we consider these results encouraging, the uncertainty is still larger than the size of the typical background fluctuations in the data, which is given

Run	B_{gen}	$\hat{B}$	ΔB	$\hat{B}^*$	ΔB^*
1	1536	1618.2	82.2	1549.3	13.3
2	1569	1645.0	76.0	1592.3	23.3
3	1579	1615.2	36.2	1584.7	5.7
4	1625	1637.7	12.7	1630.2	5.2
5	1558	1579.7	21.7	1548.0	-10.0
6	1576	1602.5	26.5	1588.2	12.2
7	1571	1586.5	15.5	1579.1	8.1
8	1584	1628.6	44.6	1584.8	0.8
9	1597	1664.1	67.1	1604.4	7.4
10	1644	1621.9	-22.1	1640.4	-3.7
11	1631	1688.9	57.9	1636.6	5.6
12	1573	1661.4	88.4	1586.9	13.9
13	1626	1655.6	29.6	1616.3	-9.7
14	1583	1641.2	58.2	1592.8	9.8
15	1613	1663.2	50.2	1635.8	22.8
16	1593	1663.7	70.7	1606.1	13.1
17	1583	1604.9	21.9	1585.2	2.2
18	1603	1646.8	43.8	1586.8	-16.2
19	1624	1667.8	43.8	1630.4	6.4
20	1580	1604.5	24.5	1575.4	-4.6
	$\langle\Delta B\rangle = 42.5$	$\sigma_B = 27.3$	$\langle\Delta B^*\rangle = 5.3$	$\sigma_B^* = 10.4$	

Table 1: Results obtained running the algorithm on different toy Monte Carlo data sets. The quantities B_{gen} and $\hat{B}$ refer to the true and to the estimated number of background events in the signal region, respectively, and $\Delta B = \hat{B} - B_{gen}$. The quantities $\hat{B}^*$ and ΔB^* have a similar meaning as $\hat{B}$ and ΔB, but the values were obtained keeping α_0 fixed at its true value. The average and standard deviation of $\hat{B}$ across the runs are represented by $\langle B\rangle$ ($\langle B\rangle^*$) and σ_B (σ_B^*), respectively.

by $\sigma_B = \sqrt{\langle B\rangle} = 40$ events. Work is underway to improve the performance of the algorithm in this respect. It should also be emphasised that, since the above uncertainty is expected to depend significantly on B, the assessment of the performance of this method will have to take into account the specifics of the analysis in question.

Acknowledgments
The author wishes to thank the High Energy Physics Group at Brunel University for a stimulating environment, and particularly Prof. Akram Khan, Prof. Peter Hobson and Dr. Paul Kyberd for fruitful conversations, as well as Dr. Ivan Reid for help on technical issues. Particular gratitude also goes to the High Energy Physics Group at University College London, especially to Prof. Jonathan Butterworth for his precious comments. The author also wishes to thank Prof. Trevor Sweeting and Dr. Alexandros Beskos at the UCL Department of Statistical Science for fruitful discussions. Finally, particular gratitude goes to Prof. Carsten Peterson and to Prof. Leif Lönnblad at the Department of Theoretical Physics, Lund University.

References

[1] Geman S and Geman D 1984 *IEEE T. Pattern Anal.* **6**(6) 721-41
[2] Dempster A P, Laird N M and Rubin D B 1977 *J. Roy. Statist. Soc. Ser.* B **39**(1):1-38
[3] Tanner M A and Wong W H 1987 *J. Amer. Statistical Assoc.* **82** (398):528-540
[4] Colecchia F 2013 *J. Phys.: Conf. Ser.* **410** 012028
[5] Colecchia F 2012 *J. Phys.: Conf. Ser.* **368** 012031
[6] Colecchia F 2013 arXiv:1311.2300 [physics.data-an]
[7] ALGLIB (www.alglib.net), Sergey Bochkanov
[8] The ATLAS Collaboration 2012 *Phys. Lett.* B **716**(1):1-29
[9] The CMS Collaboration 2012 *Phys. Lett.* B **716**:30-61

Fuzzy Clustering of Color Textures using Skew Divergence and Compact Histograms: Segmenting Thin Rock Sections

Hélio Siebra* **Bruno M. Carvalho*** [1] **Edgar Garduño**†

* DIMAp - Universidade Federal do Rio Grande do Norte, Brazil
† DCC - IIMAS - Universidad Nacional Autónoma de México, Mexico

E-mail: heliosiebra@hotmail.com, bruno@dimap.ufrn.br, edgargar@gmail.com

Abstract. Digital image segmentation is the process of assigning distinct labels to different objects in a digital image, and clustering techniques can be used to achieve such segmentations. However, many traditional segmentation algorithm fail to segment objects that are characterized by textures whose patterns cannot be successfully described by simple statistics computed over a very restricted area. In this paper we present a fuzzy clustering algorithm that achieves the segmentation of images with color textures by employing a distance function based on the Skew Divergence, that is based on the well-known Kullback-Leibler Divergence. In order for such a distance to produce good results when applied to color images, we reduced the dimensionality of the image's histogram, thus eliminating the sparsity of the color histogram and speeding up the execution of the algorithm. We performed experiments on thin rock sections and compared our results to the segmentations obtained by the Fuzzy C-Means and by another fuzzy segmentation technique, showing the superiority of our approach.

1. Introduction

Digital image segmentation is the process of assigning distinct labels to different objects in a digital image, i.e., the task of finding which pixels belong together. However, the level of detail indicated by the labeling depends on the application at hand. To perform object identification on digital or continuous images humans make use of high-level reasoning and knowledge, as well as of several visual hints. This task becomes particularly hard for a computer program when the images is affected by noise and/or inhomogeneous illumination or when the objects in the images are distinguished not by intensity itself, but by textural properties.

As mentioned by Forsyth and Ponce [1], textures are everywhere, easy for a human to recognize, but hard to define them precisely. Usually, textures are described as the feel or shape of a surface or substance, and in the areas of imaging they are characterized by statistical distributions of the observed intensities. Textures can be classified as structural or non-structural [2], where structural textures have a repeated pattern associated with a texture periodicity [3], and non-structural or statistical textures do not possess this characteristic and are globally or locally described by statistical distributions. Since visual textures can perceived as such depending on the observation scale, these distributions can change according to it.

[1] We would like to thank the support of the grant FAPERN PRONEM No 610006/2010.

The ubiquity of textures in nature makes the capacity of their segmentation in images a very important one. Visual problems in which the segmentation, description and identification of objects with textural properties in digital images or videos permeate several application areas such as Medicine, Security, Agriculture and Biology. In this work, we choose a problem in the field of Geology, the segmentation of thin rock samples, to show the applicability and efficiency of the proposed method.

The rest of this work is organized as follows. Section 2 briefly describes related techniques that can be used to segment objects with textural properties in monochromatic and color images, while Section 3 talks about the proposed method by describing the distance function used in the clustering process, as well as the procedure used for reducing the histogram dimensionality. Section 4 presents and discusses the experimental results, while Section 5 highlights the contributions and outlines possible future works.

2. Segmentation and Clustering of Textures

In this work, we are concerned with the segmentation of images that contain objects which are characterized by statistical textures, a much harder problem than when dealing with structural textures. Several different techniques have been employed to perform texture segmentation. One example of these techniques is the one proposed by Unser and Eden [4], where they extract local texture properties using local linear transforms that were optimized for maximal texture discrimination. Then, they reduce the texture features to a single component that is then thresholded to produce the segmentation. A very different approach is proposed by Hofmann et al. [5] that also performs unsupervised texture segmentation, but does it by relying on statistical tests as a measure of homogeneity. The texture segmentation task is then formulated as a clustering problem and solved by computing dissimilarities using multi-scale Gabor filters.

There are several model-based approaches [6, 7] that attack the texture segmentation problem by modeling the intensity field of textures as a Gauss-Markov or as a Markov random field to represent the local spatial dependencies between pixel intensities. However, these techniques are usually very computationally intensive because they require a large number of iterations to converge. These techniques would become even more computationally demanding if they were applied to the segmentation of color images.

There are also approaches based on fuzzy segmentation algorithms, such as the Fast-MOFS, proposed in [8]. The main advantage of this class of algorithm is its flexibility, since it can be easily adapted to work with different grids [9], or to incorporate color information in the computation of grades of membership [10]. This method has been successfully used in the segmentation of images acquired using varied modalities, such as MRI [8], CT [8] and EM [11].

The Fast-MOFS algorithm was later refined into the FAST-MOFS2 [12], providing for a simpler and more efficient implementation. However, the fuzzy affinity functions employed lacked the ability of gathering enough information for proper texture characterization. In order to cope with that, we introduced the concept of adaptive affinity functions [13], that allowed the algorithm to decide, at running time, on the appropriate size of the neighborhoods used. Even with the increased capacity for segmenting textures, this method still has problems when applied to images which contain objects that are characterized by more complex textures or that are formed by several isolated groups of pixels.

Here we analyze the behavior of three fuzzy clustering methods for segmenting color images. The first one is the conventional FCM (Fuzzy C-Means). As mentioned by [14], since the FCM does not consider any spatial information in the image context, it usually is very sensitive to artifacts such as noise or non-uniform illumination.

The second method is the one proposed by Samet et al. [15], that works by selecting a region which contains the mineral that is to be segmented in the rock thin section. Then, the RGB channels of these samples are median filtered and the three resulting histograms are used as membership functions. These membership functions are used to calculate the membership values of all pixels to the mineral from which the sample was taken. Pixels whose associated RGB values satisfy an user-defined fuzzy rule are considered to have strong membership to the mineral object. These pixels plus the pixels deemed to have weak membership to the mineral object, but that are connected to a pixel with strong membership, become part of the mineral object. Another samples can then be used to improve the quality of the segmentation.

The third one is a clustering algorithm that employs fuzzy affinity functions used with the FAST-MOFS2 [12] family of algorithms. These fuzzy affinity functions use the Skew divergence to compute the distance between the distributions of gray levels surrounding two pixels.

3. Skew Divergence of Color Histograms

Since we are interested in the characterization and further segmentation of objects which possess textural properties, specifically non-structural or statistical textures, it is that. In the fields of Statistics and Information Theory, the very well-known Kullback-Leibler (KL) divergence [16], also known as relative entropy is a non-symmetric measure of the distance or difference between two probability distributions q and r. Specifically, in Information Theory, the KL divergence of r from q, denoted by $KLD(q||r)$, is a measure of the information lost when r is used to approximate q, and is given by

$$KLD(q||r) = \sum_{y} q(y)(\log q(y) - \log r(y)). \tag{1}$$

The KL divergence is widely used to measure how far away are two distributions. However, the KL divergence is not defined when $r(y) = 0$, and that happens quite often when we are dealing with distributions that are characterized by histograms with few counts. For example, if we are dealing with monochromatic images with 256 gray levels and neighborhoods of size 9×9, we are characterizing a distribution by using a histogram with 256 bins and whose total count is 81.

In order to deal with this problem, Lee [17] proposed an asymmetric divergence, the Skew divergence, that can be seen as an approximation of the KL divergence and is given by

$$SD_{\alpha}(q, r) = D(r||\alpha q + (1 - \alpha)r), \tag{2}$$

for $0 \leq \alpha \leq 1$, where α can be seen as a degree of confidence in the distribution q and $(1 - \alpha)$ as the amount of smoothing of the distribution q over r. This allows us to work nicely with sparse data problems, such as the low count histograms of small neighborhoods.

This problem is amplified when we work with color images. If we are working with a color image where the intensity of each channel is represented by a byte, then we would be dealing with histograms with 256^3 bins or possible colors. This implies that we would be dealing with very sparse data, aside from the fact that the computation time of the KL or Skew divergences would be much larger than in the monochromatic case.

A solution to this problem is to perform a requantization, using a smaller number of bits to represent each channel. Usually, if this reduction is not very large, a very similar color image is produced. Similarly to what was performed in [18], we perform a further reduction of the dimension of the histogram by discarding colors whose bins have smaller counts than a specified value c. The counts associated with the discarded colors are associated to the closest color that was not discarded. Since we have to group colors that are perceptually similar, i.e., look closer for a human observer, we used the CIELAB color space, that is more perceptually uniform than the RGB color space.

Our method performs the clustering of color images which contain textural objects by computing the Skew divergence between the distributions of neighborhoods surrounding seed pixels and the other pixels. For a particular pixel p, we compute the Skew divergence between it and all seed pixels of all objects. After this, the smallest divergence value indicates the winner seed pixel, which will indicate the object to which the pixel p will be associated. These divergence values are normalized to the range $[0, 1]$, with 1 indicating that the pixel p has an identical distribution to a particular seed pixel. This divergence value can be seen as the fuzzy grade of membership of p to the object to which it belongs.

4. Experiments and Discussion

In order to show the applicability of the method proposed in this work, we choose to use it to segment images of thin sections of rock. Thin rock sections are produced from small slabs of rock that are glued with epoxy to a glass slide and grounded down until they reach the appropriate thickness (approximately $30\mu m$ thick). Usually, this grinding method involves using an interference color chart and a mineral color as a gauge to infer the thickness of the section. Typically quartz is used since it is one of the most abundant minerals.

These thin sections are then viewed using an optical microscope and are mainly used to investigate the optical properties of the minerals present in the sample. This is a common task in the field of petrology that is used to gather information about the origin and evolution of the parent rock or about the geometrical arrangement of the different minerals present in the sample. Usually, the segmentation of rock thin section images is not a trivial task due to the intricate patterns that the minerals can be arranged in the samples.

Here, we show the segmentations of two thin rock sections, one with 607×354 (Figure 1a) and the other with 607×486 (Figure 1f). The images were segmented manually with the aid of an image processing interface to produce the ground truth segmentations shown in Figures 1b and 1g. The experiments were run on a computer with an Intel Core 2 Duo E7400 running at 2.80Ghz and 4Gb of RAM memory.

Based on experiments performed with different images, we requantized each color channel selected, so that each color channel is represented by 3 bits, i.e., 8 different intensities per channel. Then, every color whose associated histogram bin had less counts than 0.1% of the number of pixels of the image was associated to the closest color according to the distance between colors computed in the CIELAB color space. This color space reduction is performed prior to the seed pixel selection, so, once it is done, the Skew divergence computation is performed as it was done in a monochromatic image.

The neighborhood size used was 9, i.e., each histogram used for computing the Skew divergence contains a total of 81 counts. Once the histograms of two neighborhoods are available, they are used to compute the Skew divergence, by using $\alpha = 0.67$. In both cases, the total number of bins of the histograms was smaller than 100, and so, the Skew divergence computation was faster than if we were dealing with complete histograms on monochromatic images. The running times for the clustering of images 1a and 1f were approximately $3.3s$ and $4.3s$, respectively.

The specific accuracy values of the segmentations for Figures 1a and 1b were 66.31% (Fuzzy C-Means), 78.65% (Samet et al. [15]) and 94.22% (proposed method), and 70.69% (Fuzzy C-Means), 87.72% (Samet et al. [15]) and 95.59% (proposed method). We can see from these results that the proposed method is clearly superior to the other two tested methods. We can also see in Figures 1d and 1i that the segmentations produced by the method of Samet et al. [15] does not label all pixels. This happens because the fuzzy membership values computed for all objects are smaller than the threshold used by the method. We can also see that the Fuzzy C-Means results are noisier than the other two and misclassify several regions of the two images.

On the other hand, while the proposed method produces the most accurate segmentation in both examples, it tends to overlook small areas originally classified as an object. That happens when there are objects with parts that are thinner than the support of the filter used as a neighborhood for computing the Skew Divergence.

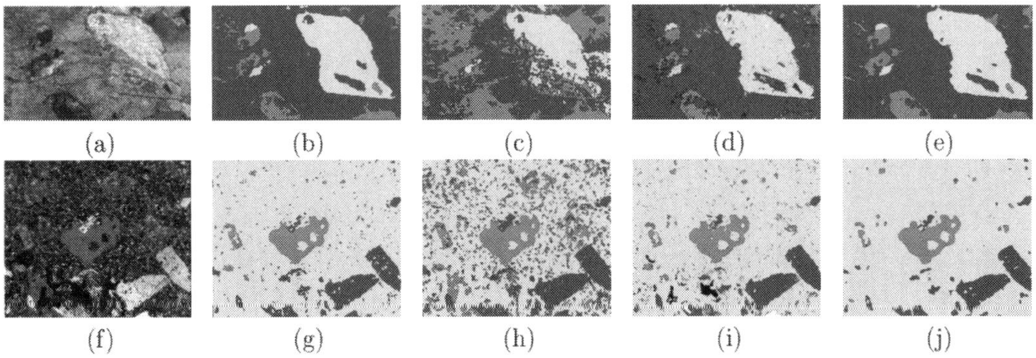

Figure 1. Original images of the rock samples (a and f) and its segmentation ground truth (b and g). Segmentations produced by using FCM (c and h), Samet et al. method (d and i) and the proposed method (e and j).

5. Conclusion

In this work we proposed a new fuzzy clustering method for segmenting color images that have objects which are characterized by textural properties. The method works by reducing the dimensionality of color histograms through requantization of the RGB color channels followed by the elimination of colors that do not appear often in the image. This reduction allows us to use the Skew divergence to calculate the distance between the distributions of two pixels' neighborhoods.

The efficacy of the proposed method was tested in the application of segmenting thin rock samples. The results show that our method is more accurate than the traditional Fuzzy C-Means and another fuzzy clustering method [15].

Future work will include the usage of the method proposed here to other applications, such as the segmentation of dermoscopic lesions or natural images. We also plan to analyze the robustness of this method when applied to noisy data and the automation of the segmentation for specific applications.

References

[1] Forsyth D A and Ponce J 2011 *Computer Vision: A Modern Approach* 2nd ed (Prentice Hall)
[2] Haralick R M 1979 *Proc. of The IEEE* **67** 786–804
[3] Jan S R and Hsueh Y C 1998 *Pat. Recogn. Lett.* **19**(5-6) 439–446
[4] Unser M and Eden M 1989 *IEEE Trans. Pat. Anal. Mach. Intel.* **11** 717–728
[5] T Hofmann J P and Buhmann J M 1998 *IEEE Trans. Pat. Anal. Mach. Intell.* **20** 803–818
[6] Manjunath B S and Chellappa R 1991 *IEEE Trans. Pat. Anal. Mach. Intell.* **13**(5) 478–482
[7] Robertson R, Villebrun E and Hoeher P 1995 *IEEE Int. Conf. on Comm.* vol 2 pp 1009–1013
[8] Carvalho B M, Herman G T and Kong T Y 2005 *Discr. Appl. Math.* **151** 55–77
[9] Carvalho B M, Garduño E and Herman G T 2001 *ICAPR '01* pp 339–348
[10] Carvalho B M, Oliveira L M and Andrade G S 2006 *DGCI* (*LNCS* vol 4245) (Springer) pp 494–505
[11] Carvalho B M, Oliveira L M and Garduño E 2006 *ISBI* (IEEE) pp 1024–1027
[12] Carvalho B M, Garduño E, Oliveira L M and Gomes R B 2014 *Pat. Anal. and Appl.* **17** 249–264
[13] Carvalho B M, Souza T S and Garduño E 2012 *ACM SAC* pp 51–53
[14] Vargas D M, Funes F J G and Silva A J R 2013 *Pat. Recogn. Lett.* **34** 400–413
[15] Samet R, Amrahov S and Ziroglu A 2012 *3rd Int. Conf. on IPTA* pp 402–406
[16] Kullback S and Leibler R A 1951 *The Annals of Mathematical Statistics* **22** 79–86
[17] Lee L 1999 *Proc. of Ann. Meet. of the ACL on Comput. Linguist.* ACL '99 (Stroudsburg, PA, USA) pp 25–32
[18] Cheng M M, Zhang G X, Mitra N J, Huang X and Hu S M 2011 *Proc. of IEEE CVPR* pp 409–416

Algorithms for Calculating Alternating Infinite Series

Hector Luna Garcia[1] and Luz Maria Garcia
Área de Física Atómica Molecular Aplicada, CBI, UAM-A, Av. San Pablo No. 180, Col. Reynosa, Azcapotzalco, México, D. F. 02200, México

E-mail: 5553972854.hl@gmail.com

Abstract. This paper are presented novel algorithms for exact limits of a broad class of infinite alternating series. Many of these series are found in physics and other branches of science and their exact values found for us are in complete agreement with the values obtained by other authors. Finally, these simple methods are very powerful in calculating the limits of many series as shown by the examples.

1. Introduction

In a previous paper [1] some novel algorithms which made use of polygamma functions for exact limits of a large branch of infinite series were introduced; moreover, the Laplace transform is used to find the sum of some of the infinite series. However, because of their importance in physics, remained pending deal the exact limits of alternating series which converge. The infinite series, including the alternate, are an important part of a course of mathematical analysis. In particular, alternating series $\sum_{n=0}^{\infty}\frac{(-1)^n}{(2n+1)}$ is convergent and is the Leibnitz definition for the irrational π number. To consider the infinite series in the literature is natural to introduce some convergence criteria due to Cauchy, Kummer, D'Alembert, and Gauss authors [2], which ensure us the convergence of infinite series. In addition, special forms of the term of the alternating series there are some criteria or tests that help quickly conclude its convergence: absolute convergence, ratio test and the root criteria [3]. However, the application of these criteria does not allow us to know their limits. In this paper, we present some algorithms to compute the exact limit of a broad class of alternating series which converge. This work is divided in four parts: This introduction, Section 2 illustrates the method; in Section 3 the method is generalized. Finally, Section 5 presents conclusions and future projects.

2. Special Cases of Alternantes Series

2.1 The infinite Series $\sum_{n=1}^{\infty}\frac{(-1)^{n-1}}{a^n}$.

This series is denoted as:

[1] Hector Luna Garcia, lghm@correo.azc.uam.mx.

$$S(b)=\sum_{n=1}^{\infty}\frac{(-1)^{n-1}}{b^n}, \quad (1)$$

where $b>1$ is an integer. Using the Feynman identity,

$$1/A=\int_0^{\infty}e^{-Ax}dx \ , \quad (2)$$

where $A>0$, the infinite series in equation (1) becomes

$$S(b)=\sum_{n=1}^{\infty}\int_0^{\infty}(-1)e^{-b^n x}dx. \quad (3)$$

Introducing the sum into the integral in equation (4), and changing $y=e^{-b^n x}$ variables, one has

$$S(b)=\int_1^0\sum_{n=1}^{\infty}\left(-\frac{1}{b}\right)^n dy=\int_1^0\lim_{N\to\infty}\left[\left(-\frac{1}{b}\right)-\left(-\frac{1}{b}\right)^N\right]\left[1-\left(-\frac{1}{b}\right)\right]^{-1}dy. \quad (4)$$

Considering now $N\to\infty$, the result is

$$S(b)=\int_0^1\frac{1}{(1+b)}dy=\frac{1}{(1+b)}. \quad (5)$$

Equation (5) is the most general expression result of this section. As an application, we present the following examples:

$$\sum_{n=1}^{\infty}\frac{(-1)^{n-1}}{2^n}=\frac{1}{3},\ \sum_{n=0}^{\infty}\frac{(-1)^n}{2^n}=\frac{2}{3},\text{ and, }\sum_{n=0}^{\infty}\frac{(-1)^{n-1}}{5^n}=-\frac{5}{6}.$$

2.2 The Infinite Series $\sum_{n=1}^{\infty}\frac{(-1)^n}{a^{2n}}$

Following the same algorithm as in the previous subsection

$$S(2n,b)=\sum_{n=1}^{\infty}\frac{(-1)^n}{b^{2n}}=-\frac{1}{(1+b^2)}. \quad (6)$$

As an application, the authors present the following examples

$$\sum_{n=1}^{\infty}\frac{(-1)^n}{3^{2n}}=-\frac{1}{10},\text{ and, }\sum_{n=0}^{\infty}\frac{(-1)^n}{2^{2n}}=\frac{4}{5}.$$

2.3 The Infinite Series $\sum_{n=1}^{\infty}\frac{(-1)^n}{n+a}$.

For this series denoted, $S(n,a)$ one has

$$S(a)=\sum_{n=1}^{\infty}(-1)^n\int_0^{\infty}e^{-(n+a)x}dx, \quad (7)$$

where we have used the Feynman's formula (2).
Then, interchanging the order of addition and integral, the equation (7) becomes

$$S(a)=\int_0^{\infty}e^{-ax}\sum_{n=1}^{\infty}\left(-e^{-x}\right)^n dx=\int_0^{\infty}e^{-ax}\lim_{x\to\infty}\left[\left(-e^{-x}\right)-\left(-e^{-x}\right)^N\right]\left[1-\left(-e^{-x}\right)\right]^{-1}dx. \quad (8)$$

In the limit as $N\to\infty$, the result is

$$S(a)=\int_0^\infty \left[e^{-ax}\left(-e^{-x}\right)\right]\left(1+e^{-x}\right)^{-1}dx, \tag{9}$$

Changing $z=e^{-x}$ variables, the equation (9) finally is expressed as:

$$S(a)=\int_1^0 u^a\left(1+u\right)^{-1}du. \tag{10}$$

At this point, it is important to mention that integral given by equation (9), cannot be written in terms of Eulerian functions [4] (polygamma functions), due to the plus sign in the binomial denominator. In contrast to those discussed in the Ref. [1] previously mentioned. However, the equation (9) is the most general expression result of this section. As one application, we present the following examples

$$\sum_{n=0}^{\infty}\frac{(-1)^n}{2n+1}=\frac{\pi}{4},\ \sum_{n=1}^{\infty}\frac{(-1)^{n+1}}{n}=\ln 2,\ \sum_{n=1}^{\infty}\frac{(-1)^{n+1}}{3n-2}=\frac{1}{3}\left(\ln 2+\frac{\pi}{\sqrt{3}}\right),\text{ and,}$$

$$\sum_{n=1}^{\infty}\frac{(-1)^{n+1}}{4n-3}=\frac{1}{4\sqrt{2}}\left(\pi+2\ln\left(\sqrt{2}+1\right)\right).$$

This series is found in Ref [5].

3. Series of type $\sum_{n=1}^{\infty}\frac{(-1)^{n+1}}{(n+b)^2}$.

This series is denoted as

$$S(2,b)=\sum_{n=1}^{\infty}\frac{(-1)^{n+1}}{(n+b)^2}. \tag{11}$$

The Feynman identity applied to this case is

$$1/A^2=\int_0^\infty xe^{-Ax}dx. \tag{12}$$

Following the same procedure as was used in section 2, leads to the following result:

$$S(2,b)=\int_0^\infty\left[xe^{-(a+1)x}\right]\left(1+e^{-x}\right)^{-1}dx=\int_1^0(\ln z)z^a(1+z)^{-1}dz, \tag{13}$$

where $z=e^{-x}$, the variable change is used.

Applying equation (13) the already reported result is obtained

$$\sum_{n=1}^{\infty}\frac{(-1)^{n+1}}{n^2}=\frac{\pi^2}{12}.$$

3.1 Series of type $S(2k,n)=\sum_{n=1}^{\infty}\frac{(-1)^{n-1}}{n^{2k}}$, with $k=1,2,\ldots$.

The Feynman identity applied to this case is

$$1/A^m=\frac{1}{(m-1)!}\int_0^\infty x^{m-1}e^{-Ax}dx. \tag{14}$$

Following the same procedure as was used in section 2, leads to the following result:

$$S(2k,n)=\frac{1}{(2k-1)!}\int_0^\infty x^{2k-1}e^{-x}\left(1+e^{-x}\right)^{-1}dx=\frac{1}{(2k-1)!}\int_0^1(-\ln z)^{2k-1}(1+z)^{-1}dz. \tag{15}$$

Applying equation (15), we have

$$\sum_{n=1}^{\infty}\frac{(-1)^{n-1}}{n^2}=\frac{\pi^2}{12},$$

whose value was found in section 2. As another examples, we find

$$\sum_{n=1}^{\infty}\frac{(-1)^{n-1}}{n^4}=\tfrac{1}{6}\int_1^0(\ln z)^3(1+z)^{-1}\,dz=\tfrac{1}{6}\left(\tfrac{7\pi^4}{120}\right)=\tfrac{7\pi^4}{720},$$

$$\sum_{n=1}^{\infty}\frac{(-1)^{n-1}}{n^6}=\tfrac{1}{120}\int_1^0(\ln z)^5(1+z)^{-1}\,dz=\tfrac{1}{120}\left(\tfrac{31\pi^4}{252}\right)=\tfrac{31\pi^6}{30240}.$$

This series is found in Ref [5].

3.2 Series of type $S(2k+1,n)=\sum_{n=1}^{\infty}\frac{(-1)^{n+1}}{(2n-1)^{2k+1}}$, with $k=1,2,\ldots$.

Following the same procedure, leads to the following result:

$$S(2k+1,n)=\tfrac{1}{2^{2k+1}(2k)!}\int_0^{\infty}x^{2k}e^{-\frac{x}{2}}\left(1+e^{-x}\right)^{-1}dx=\tfrac{1}{2^{2k+1}(2k)!}\int_0^{\infty}(\ln z)^{2k}z^{-\frac{1}{2}}(1+z)^{-1}\,dz. \quad (16)$$

Changing $z=w^2$ variables, the equation (16) finally is expressed as:

$$S(2k+1,n)=\tfrac{1}{2^{2k}(2k)!}\int_0^1\left(\ln w^2\right)^{2k}\left(1+w^2\right)^{-1}dw \quad (17)$$

Finally one has the following examples

$$\sum_{n=1}^{\infty}\frac{(-1)^{n+1}}{(2n-1)^3}=\tfrac{1}{8}\int_0^1\left(\ln w^2\right)^2\left(1+w^2\right)^{-1}dw=\tfrac{1}{8}\left(\tfrac{\pi^3}{4}\right)=\tfrac{\pi^3}{32},$$

$$\sum_{n=1}^{\infty}\frac{(-1)^{n+1}}{(2n-1)^5}=\tfrac{1}{384}\int_0^1\left(\ln w^2\right)^4\left(1+w^2\right)^{-1}dw=\tfrac{1}{384}\left(\tfrac{5\pi^5}{4}\right)=\tfrac{\pi^5}{1536}, \text{ and,}$$

$$\sum_{n=1}^{\infty}\frac{(-1)^{n+1}}{(2n-1)^7}=\tfrac{1}{46080}\int_0^1\left(\ln w^2\right)^6\left(1+w^2\right)^{-1}dw=\tfrac{1}{46080}\left(\tfrac{61\pi^7}{4}\right)=\tfrac{61\pi^5}{184320}.$$

This series is found in Ref [5].

5. Conclusions

The limits of a broad class of alternating series which converge were obtained using some novel algorithms. These algorithms are general and can be used to evaluate many de these series that are important in different branches of the physics, as well as in other relevant science applications. Moreover, we can apply these algorithms to other types of alternating series, not considered in this article.

References

[1] H. Luna Garcia, and Luz Maria Garcia, *Journal of Mathematics and Systems Science* **3:**110-113(2013).

[2] G.B. Arfken, H.J. Weber, and F.E. Harris. *Mathematical Methods for Physicists: A Comprehensive Guide*, 7th ed., Elsevier Academic Press, USA, 2012, pp. 321-344 (Chapter 5).

[3] M. Spivak. *Calculus*, 2nd ed., Editorial Reverté, España, 1992, pp. 641-678 (Chapter 22)

[4] M. Abramowitz, I.A. Stegun, Handbook of Mathematical Functions: with Formulas, Graphs, and Mathematical Tables, Dover Publications, USA, 1964,556-566(Chapter 15)

[5] I.S. Gradshteyn, I.M. Ryzhik, A. Jeffrey, D. Zwillinger, Table of Integrals, Series and Products, 6th ed., Academic Press, USA, 2000, pp. 321-347(0 Introduction).

Ab initio electronic band structure study of the valence bands of II–VI $C(2 \times 2)$ reconstructed surfaces

A. Rubio–Ponce[1] and D. Olguín[2]

[1] Departamento de Ciencias Básicas, Universidad Autónoma Metropolitana–Azcapotzalco, Av. San Pablo 180, 02200 México, D. F.

[2] Departamento de Física, Centro de Investigación y de Estudios Avanzados del Instituto Politécnico Nacional, A.P. 14740, C.P. 07300, México, Distrito Federal, Mexico

Abstract. The structural and electronic properties of CdTe(001), CdSe(001), and ZnSe(001) C(2 × 2) reconstructed surfaces have been investigated through the use of first-principles calculations. To simulate the surface, we employed the slab model. Using the experimentally determined lattice parameters as inputs, we relaxed the internal atomic positions of the outer atomic layers. We demonstrate that our model appropriately reproduces both the surface structural parameters and the known electronic properties found for these semiconductor compounds in bulk. Finally, we discuss our results of the projected bulk bands and the surface and resonance states found for these surfaces.

1. Introduction

Type II–VI semiconductors have continued to receive increasing attention because of their applications in optoelectronic devices for detecting and stimulating emission in the IR (mainly CdHgTe) and blue spectral regions (ZnSe). In this context, CdTe is used as a buffer for the growth of $Cd_xHg_{1-x}Te$ heterostructures. The (001) surfaces of II–VI compounds exhibit different reconstructions as functions of temperature and the growth conditions [1, 2, 3, 4, 5]. Knowledge of the atomic structures of these surfaces is crucial for understanding the energy band structure and, consequently, for the tailoring of heterostructures with desired electronic properties. The CdTe(001) surface was experimentally studied in early works [1, 6, 7]. From the theoretical perspective, this system has been studied using an empirical approach [8, 9, 10] and using first-principles calculations [11, 12]. Recently, a DFT study for CdSe(001) was conducted [13]. Furthermore, ZnSe(001) has also been the focus of extensive studies in the literature [3, 14, 15].

To contribute to the understanding of the electronic properties of the II–VI surface compounds, in this paper, we report a study on the CdTe(001) – , CdSe(001) –, and ZnSe(001) – $C(2 \times 2)$ reconstructed surfaces. For this purpose, a slab that included seventeen atomic layers and a vacuum of four atomic layers was constructed. The top view and side view of the (001) $C(2 \times 2)$ reconstructed surface is schematically illustrated in figure 4 of Ref. [14].

In a previous study, we showed the consistency of our results in terms of the size slab [12]. In the present work, we report the projected bulk bands in the valence band region, and we discuss the surface and resonance states found at these energies.

This paper is organized as follows: Section 2 briefly describes the method used in our calculations. In Section 3, we discuss our results. Finally, our conclusions are presented in Section 4.

2. Methods

The method used in our calculations was the full-potential linearized augmented plane-wave method (FP–LAPW), as implemented in the WIEN2k code [16]. The exchange–correlation energy was calculated using the GGA correction of Perdew *et al.*, [17] . In our calculations, the employed atomic electronic configuration was Zn: [Ar] 4s 3d; Se: [Ar] 4s 3d 4p; Cd: [Kr] 5s 4d; and Te: [Kr] 5s 4d 5p, in which the Zn–3*p* and Cd–4*p* states were treated as valence–band states using the local orbital extension of the LAPW method [16]. During the calculations, a step analysis was carefully performed to ensure the convergence of the total energy in terms of the variational parameters; here, we selected the following values: 9 for $R_{MT}K_{max}$ (where R_{MT} is the muffin-tin radius and K_{max} is the plane-wave cutoff) and 15 Ry for the cutoff energy parameter (G_{max}) for CdTe(001) were used, whereas values of 8.5 for $R_{MT}K_{max}$ and 12.0 Ry for G_{max} were used for CdSe(001); for ZnSe(001), the values used were 8 for $R_{MT}K_{max}$ and 15 Ry for G_{max}. Additionally, an appropriate set of k–points in the irreducible sector of the Brillouin zone was used; here, a value of 28 k–points for all cases were taken into account. Finally, we relaxed all of the atoms in the slab using Hellmann–Feynman forces until the force component was less than 0.5 mRy/au.

From the slab construction, we can distinguish different atomic sites for the cations (Cd, Zn) and for the anions (Te, Se). Thus we can identify the cation–1, which is located on the surface, and cation–3 and cation–5, which are located on the third and fifth atomic layers, respectively. Then, for the nearest-neighbor layer to the surface, we identify the anion–2 atoms, and the anion–4 and anion–6 atoms are located on the fourth and sixth atomic layers, respectively. This identification of the different atoms will be useful in the discussion of our results (see Fig. 4 in Ref. [14]).

3. Results

3.1. Geometry optimization

From the structural optimization procedure, we observed that all of the inner atomic layers are fixed, except for the two outermost atomic layers. That is, we found that the surface cation1 layer moves downward while the second layer, the anion2–atoms, moves upward in such a way that both atomic planes are located quite near each other. In fact, the relaxed anion2–cation1–anion2 angle is significantly different from the bulk value (see Table 1). Thus, the relaxed nearest-neighbor distances are quite different from their corresponding bulk values. The new nearest-neighbor bond lengths are shorter than their corresponding bulk values. As the anion2–layer moves upward, the bond length between the second atomic layer and the third atomic layer increases. Previous studies have shown that during the relaxation of the II–VI (001) $C(2 \times 2)$ surface, the cation layer relaxes downward while the anion layer relaxes upward; [1, 11, 13, 14, 15] our obtained results are in agreement with these previous studies.

Table 1 presents our calculated bond length values for the different systems studied in this work, and for comparison, we also include the values measured or calculated in other works. As shown, our calculated values for the different bond lengths are in good agreement with the measured and calculated values reported in the literature. However, note that the calculated LDA lattice parameters are generally smaller that the GGA ones, as is shown for the case of CdSe(001) [13].

3.2. Partial density of states

We have verified that our slab model appropriately reproduces the known bulk electronic density of states of the investigated systems. To prove this ability of our model, we compared the calculated partial density of states (PDOS) of the inner atoms of the slab with the bulk PDOS.

Figure 1 shows the calculated PDOS for the different systems investigated in this work. As shown in this figure, the calculated PDOS for the upper valence band (VB) for the atoms located

Table 1. Calculated and reported values for the nearest-neighbor (nn) bond length, the bond length between the second and third atomic layers (nnn), the anion–anion (A2–A2) interlayer distance, and the anion2–cation1–anion2 angle for the different systems studied. The figures in brackets are the bulk values, and all of the bond lengths are reported in Angstroms (Å).

System	Present work				Other works			
	nn	nnn	A2–A2	angle	nn	nnn	A2–A2	angle
CdTe(001)	2.66 (2.81)	3.03	3.89 (4.58)	167.06^o	2.62^a	2.81^a	3.92^a	
					2.63^b		3.93^b	173$^{o\,b}$
CdSe(001)	2.55 (2.62)	2.80	3.58 (4.28)	154.33^o	2.44^c			
ZnSe(001)	2.33 (2.45)	2.57	3.52 (4.01)	149.40^o	2.23^d	2.49^d	3.53^d	

^{a}Measured [1]; ^{b}Calculated [11]; ^{c}Calculated [13]; ^{d}Measured [14]

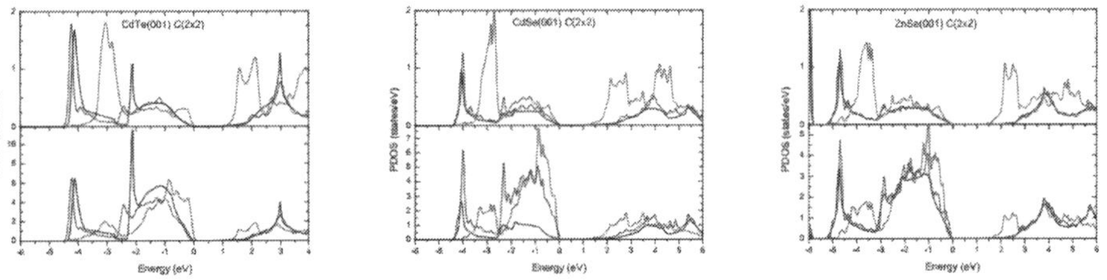

Figure 1. (Color online) Calculated PDOS for the different systems investigated in this work. The calculated bulk PDOS is shown as a solid line, the PDOS for the central layer atoms of the slab is shown as a blue line, and the PDOS for the atoms located on the surface is shown as a red line. The upper panel shows the cation PDOS, and the lower panel shows the anion PDOS.

in the central layer of the slab is almost similar to that obtained from the bulk calculations. The same agreement is observed for the PDOS of the lower conduction band (CB). In the case of CdTe, the energy difference for the main peaks in the VB is less than 0.11 eV in both the Cd– and Te–atomic layers.

However, we observed that differences are observed mainly for the atoms located in the outermost surface layers. These differences are associated with the existence of surface and resonance states, as discussed below.

In all of the cases shown in the figure, we observe that the surface cation PDOS exhibits important features at approximately 3.5 eV below the Fermi level and in the lower part of the CB. Furthermore, important features due to the anion second atomic layer are also obtained. In the VB energy region shown, the anion PDOS spreads more than the cation case.

3.3. Surface- and Resonance-states

The electronic states at the surface of a single crystal are strictly two–dimensional and show a dispersion that differs significantly from the bulk states. It is convention to distinguish between surface states (empty surface states), which lie in the projected band–gap, and surface resonances (occupied surface states) [21], which lie within the bulk band structure projected on the surface. These states originate once we break the translational symmetry to create the surface.

We have shown in Fig. 1 that our results for the CdTe(001) and ZnSe(001) surfaces are quite similar to those for the CdSe(001) case; thus, and to save space, we will only discuss the CdSe(001) $C(2\times2)$ case in detail; further details for the other surfaces will be provided elsewhere [23].

Figure 2 shows the calculated electronic bands in the first Brillouin zone of the

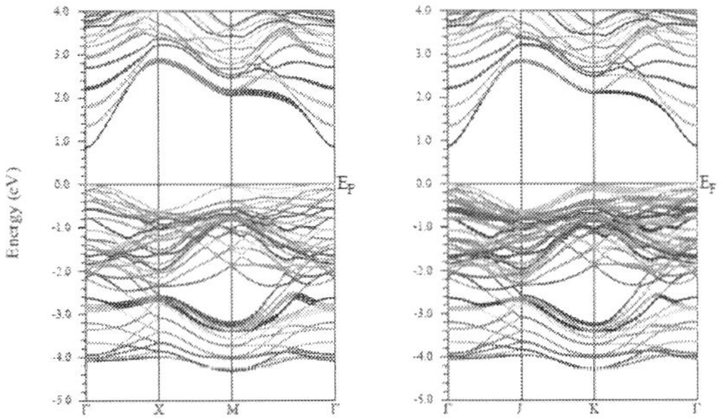

Figure 2. (Color online) Projected surface bulk bands for CdSe(001) $C(2 \times 2)$. The bands are shown in the irreducible Brillouin zone of the unreconstructed (1×1) surface.

unreconstructed (1×1) surface, in which the contributions of the surface–layer atoms to the surface states are indicated with circles. A larger circle sizes corresponds to more charge within the atomic sphere of the respective layer [16], and thus indicates an important contribution to the electronic bands of the atoms on the surface layers compared with the atoms located in the inner layers.

As indicated in Fig. 1, we obtained a number of surface and resonance states associated with the atoms located in the two outer layers of the slab.

In Fig. 2, we identify a series of resonance states associated with both the external Cd1 atoms (left panel) and the nearest-neighbor Se2 atoms (right panel).

For Cd1, we obtained a resonance state located at approximately 3.0 eV below the VB maximum. The state shows a wavy dispersion shape. Then, in the range from approximately 0.8 to 2.0 eV above the VB maximum, we obtain a surface state. The surface state follows the dispersion of the minimum of the CB. We observed that the obtained surface and resonance states have the symmetry of the Cd $s-$ and $p-$orbitals hybridized with the Se $p-$orbitals [23].

The right panel of the figure shows that in the upper VB energies, there are two surface states associated with the Se2 atom. At the upper VB, we obtained a surface state that follows the dispersion of the VB maximum; this state mainly has Se p_z-character. Then, we obtained a resonance state at approximately -0.1 eV. The state has anion p_x-character. Both states span over the shown Brillouin zone.

We also found that the Se2 atom contributes to the mentioned cation surface states, as noted in Fig. 1.

At the upper CB energies, there are some traces of resonance states; however, we will omit further discussion of these states.

Although we have identified that the cation lower CB and the anion upper VB states are surface states [24], previous works have claimed that for the II–VI $C(2 \times 2)$ surfaces, there are only resonance states in the VB [11, 15].

4. Conclusions

We investigated CdTe(001) –, CdSe(001) –, and ZnSe(001) – $C(2 \times 2)$ reconstructed surfaces using first-principles calculations. We found that our slab model reproduces the experimentally measured structural bond lengths, and the bulk partial density of states was also well reproduced in our model. The surface and resonance states reported from experimental results and in previous calculations were also found in our calculations.

Although recent *ab initio* calculations do not support the existence of surface states on the studied surfaces, we found a band gap surface state that shows an important dispersion as a function of **k**, which was also observed in previous empirical calculations. We will report a detailed investigation on the different features found in this work in a future study.

References

[1] Veron M B, *et al.*, 1995 *App. Phys. Lett.* **67**, 3957.
[2] Park C H and Chadi D J, 1995 *Phys. Rev. B* **49**, 16467.
[3] Ohtake A, Handa T, Yasuda T, Arai K, and Yao T, 1999 *Phys. Rev. B* **60**, 8326.
[4] Volkmann T, Ahr M, and Biehl M, 2004 *Phys. Rev. B* **69**, 165303.
[5] García A and Northrup J E, 1994 *Appl. Phys. Lett.* **65**, 708.
[6] Niles D W and Höscht H, 1992 *Phys. Rev. B* **43**, 1492.
[7] Gawlik K -U, *et al.* 1992 *Acta Physica Polonica A* **82**, 355.
[8] Olguín D and Baquero R, 1994 *Phys. Rev. B* **50**, 1980.
[9] Olguín D and Baquero R, 1995 *Phys. Rev. B* **51**, 16 891.
[10] Rodríguez F, Camacho A, Quiroga L, and Baquero R, 1990 *physica status solidi b* **160**, 127.
[11] Gundel S, Fleszar A, Faschinger W, and Hanke H, 1999 *Phys. Rev. B* **59** , 15261.
[12] Olguín D, 2007 *Rev. Mex. de Física* **53–S7**, 194.
[13] Zhu L, Yao K L, Liu Z L, and Li Y B, 2009 *J. Phys.: Condens. Matter* **21**, 095001.
[14] Weigand W, Müller A, Kilian L, Schallenberg T, Bach P, Schmidt G, Molenkamp L W, Bunk O, Johnson R L, Kumpf C, and Umbach E, 2003 *Phys. Rev. B* **68**, 241314.
[15] Plucinski L, Johnson R L, Fleszar A, Hanke W, Weigand W, Kumpf C, Heske C, Umbach E, Schallenberg T, and Molenkamp L W, 2004 *Phys. Rev. B* **70**, 125308.
[16] Blaha P, Schwarz K, and Luitz J, WIEN97, Vienna University of Technology 1997. [Improved and updated Unix version of the original copyrighted WIEN code, which was published by P. Blaha, K. Schwarz, P. Sorantin, and S. B. Trickey, *Comput. Phys. Commun.* **59**, 339 (1990).]: for a list of work published with the code see http://wein2k.at
[17] Perdew J P and Wang Y, 1992 *Phys. Rev. B* **45**, 13244; Perdew J P, Burke S, and Ernwerhof M, 1996 *Phys. Rev. Lett.* **77**, 3865.
[18] Yim W M and Stofko E J, 1972 *J. Electrochem. Soc.* **119**, 381.
[19] Zhang S B, Wei S -H, and Zunger A, 1995 *Phys. Rev. B* **52**, 13975.
[20] Chen Z, Eich D, Reuscher G, Waag A, Fink R, and Umbach E, 1999 *Phys. Rev. B* **60**, 8915.
[21] Hybertsen M S and Louie S G, 1998 *Phys. Rev. B* **38**, 4033.
[22] Plucinski L, Weigand W, Kumpf C, Heske C, Kosuch R, Schallenberg T, Molenkamp L W, Umbach E, and Johnson R L, 2005 *Surface Science* **585**, 95.
[23] In a future work we will present a full discussion of the different features found for other II–VI semiconductors.
[24] Using the empirical tight-binding approach and the surface Green function matching method, the lower CB surface state was found in previous work for the CdTe(001), ZnTe(001), and HgTe(001) ideal surfaces. [10] There it was found that the surface state does not show dispersion and was located almost in the mid band gap of the studied compounds. Also, it was arguded that for a relaxed and reconstructed surface, the surface state should present some dispersion.

Basic idea of Corbino-type single-electron transistor

Akira Suzuki and Hisao Taira

Department of Physics, Faculty of Science, Tokyo University of Science
1-3 Kagurazaka, Shinjyuku-ku, Tokyo 162-8601, Japan

E-mail: asuzuki@rs.kagu.tus.ac.jp

Abstract. We have formulated the transmission probability of an electron in a Corbino quantum disk by taking into account charging effect. The geometrical potential of the Corbino disk has a singularity at the centre of the disk. In order to avoid this singularity problem, we have to reformulate the Schrödinger equation in the Riemannian manifold. The Schrödinger equation describing the motion of the electron in the Corbino disk must be expressed by introducing a momentum operator reformed by the metric tensor. In order to obtain a Hermitian momentum operator, we must deform the Hilbert space by introducing a new wave function. This deformation leads to the extra potential term in the Schrödinger equation, which depends on the metric, *i.e.*, the geometry of the disk. It should be noted that the charging energy of confining electrons in the Corbino disk should depend on the geometry of the disk. We discuss the quantum tunneling of an electron confined in the Corbino disk in order to investigate the effect of both geometrical potential and charging energy of confining electrons in the Corbino disk by using the Wentzel-Kramers-Brillouin (WKB) method. It is expected that the charging energy, which depends on the effective confining potential, plays an important role in the transmission probability. This suggests that the formulated transmission probability is applicable to the analysis of the single-electron transistor.

1. Introduction

Recent technological advances enable us to fabricate a Corbino-type quantum disk in a two-dimensional (2D) electron system [1 5]. From the quantum mechanical viewpoint, it is interesting that a quantum centrifugal potential must be introduced in this system due to its geometry [6 8]. In general, it is necessary to reformulate the Schrödinger equation in terms of the Riemannian geometry to describe the motion of an electron constrained on the curved surface including the Corbino disk. The Schrödinger equation is generally given by [9 11]

$$-\frac{\hbar^2}{2m}\sum_{i,j=1}^{2}\frac{1}{\sqrt{g}}\frac{\partial}{\partial q_i}\left(\sqrt{g}g^{ij}\frac{\partial\sigma}{\partial q_j}\right)-\frac{\hbar^2}{8m}(\kappa_1-\kappa_2)^2\sigma=E\sigma\,, \tag{1}$$

where $g = \det[g^{ij}]$, $g^{ij} = g_{ij}^{-1}$'s denote the metric tensors, and q_i, q_j the generalized coordinates. The most intriguing point is that a geometrical potential term appears through $-\hbar^2(\kappa_1-\kappa_2)^2/(8m)$, where κ_1 and κ_2 are the principal curvatures of the surface at (q_1, q_2). This geometrical potential is uniquely determined if the configuration of the surface is given, in other words, the geometrical potential depends on the curvature of the surface. After we apply this equation to the Corbino disk, the *geometrical potential diverges at the center of the disk* since its functional form is proportional to $1/r^2$, where r is a radial component in the polar

Journal of Physics: Conference Series **574** (2015) 012119 doi:10.1088/1742-6596/574/1/012119

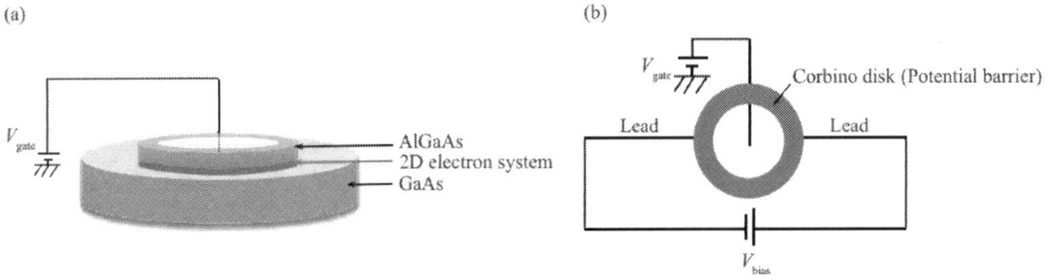

Figure 1. (Color online) (a) Three-dimensional illustration of the 2D electron system. Red lines indicate 2D electron system which is formed at the interface between GaAs and AlGaAs layers. A gate voltage V_{gate} can change the amount of charge in the Corbino disk. (b) Corbino disk with two leads. Shaded area indicates the potential barrier. An electron can transmit through the Corbino disk due to the nonzero bias voltage V_{bias}.

coordinates system. Previous studies showed that this potential singularity can be safely avoided by introducing a new wave function so as to obtain a Hermitian momentum operator $-i\hbar d/dr$ in the deformed Hilbert space [6 8]. These results have led us to the general formulation to describe the electronic states in the Corbino disk.

Corbino disk is a kind of the quantum dot whose transmission probability of an electron is tunable by changing the gate voltage as seen in a quantum point contact [12,13]. The amplitude of the charging energy is comparable with that of the centrifugal potential when the size of the system is to be of the order of micrometers. By applying the gate voltage, the transmission probability of an electron is also changed since the total charge in the Corbino disk associated with the Coulomb potential is tunable. This means that the charging energy should be taken into account to describe the quantum tunneling of the electron.

In this paper, we reformulate the Schrödinger equation by using the Riemannian geometrical method. The charging energy whose amplitude can be changed by the gate voltage is introduced in the potential term of the Schrödinger equation. In our model, the 2D Schrödinger equation can be reduced to the one dimensional one. Thus, the formula of the transmission probability of the electron confined in the Corbino disk is obtained by the Wentzel-Kramers-Brillouin (WKB) method.

2. The Schrödinger equation for an electron in a Corbino disk

Let us extend the general formula Eq. (1) to the case for an electron confined in the Corbino disk. Figure 1(a) shows a 2D electron system formed in the interface between e.g., GaAs and AlGaAs crystals in the Corbino disk [14]. Electron moves parallel to the interface of GaAs and AlGaAs layers. The Corbino disk is the concentric 2D circular disk (see Fig. 1(b)). Since the electron confined in the Corbino disk has a charge e (magnitude), a gate voltage V_{gate} can change the amount of charge $Q = \sum_{i=1}^{N} e_i$, where N is the total number of electrons in the Corbino disk. This means that the *static charging potential* $Q^2/(2C)$ induced by the presence of the electrons in the Corbino disk can be changed by applying V_{gate}, where C is the static capacitance of the Corbino disk. This phenomenon could be useful for the application to the single-electron transistor by using the Corbino disk. When we apply the Corbino disk to a functional device such as the transistor, we need the expression for the transmission probability of an electron in order to investigate its performance.

In order to obtain the expression of the transmission probability, we have to solve the Schrödinger equation Eq. (1) in the Corbino geometry. In this geometry, it is convenient to

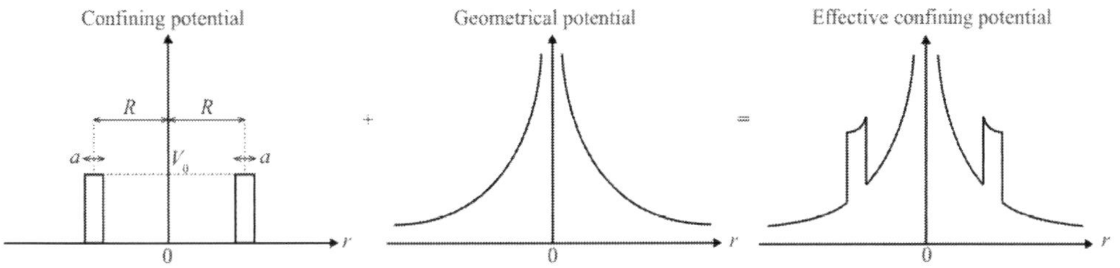

Figure 2. Schematic illustration of the confining potential barrier of the Corbino disk (left), geometrical potential induced by the Corbino disk (center), and effective potential $V_{\rm eff}(r)$ (right). No electron can transmit when the bias voltage is absent.

work in the polar coordinates, that is, (q_i, q_j) is replaced by (r, θ). In this transformation, the solution of Eq. (1) specific to the Corbino geometry (see Fig. 1) is found by introducing $\sigma(r,\theta) = R_{n,\ell}(r)\exp(i\ell\theta)$ in the separation of variables. The radial Schrödinger equation is then obtained in the following form:

$$\left(-\frac{d^2}{dr^2} + V_{\rm eff}(r)\right) R_{n,\ell}(r) = 0\,, \tag{2}$$

where $V_{\rm eff}(r)$ is the effective confining potential given by

$$V_{\rm eff}(r) = \frac{1}{r^2}\left(\ell^2 - \frac{1}{4}\right) - (k_{n,\ell})^2 + \frac{2m}{\hbar^2}\frac{Q^2}{2C} + \frac{2m}{\hbar^2}V_{\rm conf}(r)\,. \tag{3}$$

Here, $k_{n,\ell} = \sqrt{2mE_{n,\ell}}/\hbar$ with the eigenenergy $E_{n,\ell}$, where $n\,(=1,2,\cdots)$ and $\ell\,(=0,\pm1,\pm2,\cdots)$ are the principal and angular quantum numbers, respectively. $V_{\rm conf}(r)$ is the potential barrier of the Corbino disk. The term $-1/(4r^2)$ is arising from the geometrical potential given in Eq. (1) [9 11]. It should be noted that the effective confining potential $V_{\rm eff}(r)$ is formed by the confining potential barrier by the Corbino disk and the geometrical potential induced by the Corbino disk (See Fig. 2). Spatial profile of the effective potential is symmetric about the vertical axis at $r = 0$ when the bias voltage $V_{\rm bias}$ is absent. It should be noted that the charging potential energy $Q^2/2C$ is included in the effective confining potential in order to take account of the charging effect in the Corbino geometry.

3. The solution of the radial Schrödinger equation and transmission probability

Let us derive the formula for the transmission probability of the electron confined in the Corbino disk by applying the WKB method. Potential barrier $V_{\rm conf}(r)$ of the Corbino disk is set to be constant, i.e., $V_{\rm conf}(r) = V_0$ in the region of $R - a/2 \le r \le R + a/2$, where $R - a/2$ and $R + a/2$ denote the radii of inner and outer disk, respectively, and a is the width of the disk. An electron cannot transmit through the Corbino disk when $V_{\rm bias} = 0$ while they can transmit when $V_{\rm bias} \neq 0$ since the effective potential becomes asymmetric by applying the bias voltage $V_{\rm bias}$ (see Fig. 3).

According to the WKB method, the eigenenergy of the electron $E_{n,\ell}$ should be larger than V_0 to obtain the transmission probability. The eigenfunction obtained by solving Eq. (2) is given in terms of the Bessel function $J_n(k_{n,\ell}r)$ and the Neumann function $N_n(k_{n,\ell}r)$ [15]:

$$R_{n,\ell}(r) = \frac{a_{n,\ell}}{\sqrt{r}}J_n(k_{n,\ell}r) + \frac{b_{n,\ell}}{\sqrt{r}}N_n(k_{n,\ell}r)\,, \tag{4}$$

where the constants $a_{n,\ell}$ and $b_{n,\ell}$ are determined by the boundary conditions of the Corbino disk. The associated eigenenergy is

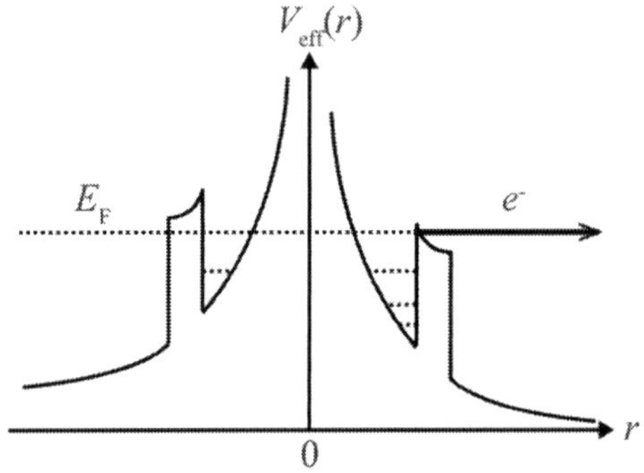

Figure 3. Schematic illustration of the effective confining potential $V_{\text{eff}}(r)$ under the nonzero bias voltage. An electron can transmit if the electron's energy equal to the Fermi energy E_F of the left lead.

$$E_{n,\ell} = \frac{\hbar^2}{2m} (k_{n,\ell})^2 , \tag{5}$$

where $k_{n,\ell}$ is numerically obtained. We finally obtain the formula of the transmission probability T for an electron in the confining potential of the Corbino disk [16] as in the following form:

$$T = \exp\left[-\frac{2}{\hbar} \int_{R-2/d}^{r_0(E_{n,\ell})} \sqrt{\frac{1}{r^2}\left(n^2 - \frac{1}{4}\right) + 2m\frac{V_0 + Q^2/(2C) - E_{n,\ell}}{\hbar^2}} dr \right] , \tag{6}$$

where

$$r_0(E_{n,\ell}) = \sqrt{\frac{\hbar^2(4n^2-1)}{8m(E_{n,\ell} - V_0 - Q^2/(2C))}} . \tag{7}$$

It should be noted that the formula given in Eq. (6) contains the charging potential $Q^2/(2C)$. This potential is comparable to the energy of $E_{n,\ell} - V_0$ by tuning the gate voltage when the size of the system is less than micrometer order. Experimental observation of our theoretical result given in Eq. (6) is accomplished by measuring the conductance G of the system. According to the Landauer formula, the conductance reads

$$G = \frac{2e^2}{h} T, \tag{8}$$

where T is given by Eq. (6). We hope our theoretical prediction shall raise both the theoretical and experimental interests in the Corbino disk-based single-electron transistor.

4. Conclusion

We have formulated the expression of the transmission probability for an electron confined in the Corbino disk by applying the WKB method. The charging energy is introduced to the potential term in the expression of the WKB formula. This formula elucidates the electronic tunneling phenomenon of the Corbino disk and serves as a basic equation for the single-electron transistor by using the Corbino disk. When the size of the system becomes the micrometer order, the charging potential is comparable with the geometrical potential. This means that it is important to incorporate the charging potential in the formulation of the transmission probability when one studies single-electron transistors.

References

[1] Gridin V, Sergeenkov S and Ausloos M 1996 *Solid State Communications* **98** 623–627
[2] Rycroft S, Doyle R, Fuchs D, Zeldov E, Drost R, Kes P, Tamegai T, Ooi S and Foord D 1999 *Physical Review B* **60** R757–R760
[3] Kawaji S 2008 *Proceedings of the Japan Academy. Series B, Physical and Biological Sciences* **84** 199–231
[4] Okuma S, Yamazaki Y and Kokubo N 2009 *Physical Review B* **80** 220501:1–220501:4
[5] Okuma S, Yamazaki Y and Kokubo N 2010 *Physica C: Superconductivity* **470** S776–S778
[6] Suzuki A and Matsutani S 1996 *Il Nuovo Cimento B Series 11* **111** 593–606
[7] Souma S and Suzuki A 1998 *Physical Review B* **58** 4649–4655
[8] Souma S and Suzuki A 1999 *Physical Review B* **60** 15928–15936
[9] Da Costa R C T 1981 *Physical Review A* **23** 1982–1987
[10] Taira H and Shima H 2007 *Surface Science* **601** 5270–5275
[11] Taira H and Shima H 2010 *Journal of Physics A: Mathematical and Theoretical* **43** 354013:1–354013:11
[12] Zazunov A, Yeyati A L and Egger R 2011 *Physical Review B* **84** 165440:1–165440:8
[13] Amasha S, Rau I, Grobis M, Potok R, Shtrikman H and Goldhaber-Gordon D 2011 *Physical Review Letters* **107** 216804:1–216804:5
[14] Kabir N, Yoon Y, Knab J, Chen J Y, Markelz A, Reno J, Sadofyev Y, Johnson S, Zhang Y H and Bird J 2006 *Applied physics letters* **89** 132109
[15] Arfken G B and Weber H J 2012 *Mathematical Methods for Physicists, Seventh Edition: A Comprehensive Guide* 7th ed 627–692 (Academic Press) ISBN 9780123846549
[16] Tomita I and Suzuki A 2006 *Journal of applied physics* **99** 126104:1–126104:3

Cluster analysis of word frequency dynamics

Yu.S. Maslennikova, V.V. Bochkarev, I.A. Belashova

Kazan Federal University, Kremlevskaya str.18, Kazan 420018, Russia

E-mail: JSMaslennikova@kpfu.ru, vbochkarev@mail.ru

Abstract. This paper describes the analysis and modelling of word usage frequency time series. During one of previous studies, an assumption was put forward that all word usage frequencies have uniform dynamics approaching the shape of a Gaussian function. This assumption can be checked using the frequency dictionaries of the Google Books Ngram database. This database includes 5.2 million books published between 1500 and 2008. The corpus contains over 500 billion words in American English, British English, French, German, Spanish, Russian, Hebrew, and Chinese. We clustered time series of word usage frequencies using a Kohonen neural network. The similarity between input vectors was estimated using several algorithms. As a result of the neural network training procedure, more than ten different forms of time series were found. They describe the dynamics of word usage frequencies from birth to death of individual words. Different groups of word forms were found to have different dynamics of word usage frequency variations.

1. Introduction

Literature is one of the main forms of artistic expression. Its temporal dynamics, both in content and style, allows us to track the historical aspects of the evolution of culture. To date, a lot of research has been conducted on the basis of the quantitative analysis of literary texts and frequency dictionaries. Through such research, the behavior word set of a language is approached as a kind of complex system. Continuing this line of investigation we here explore the stages of birth, evolution and death of individual word-forms. Significant progress in the understanding of the statistics of use of word-forms was achieved by Baayen in numerous works, e.g. [1, 2]. It should be noted, that processing techniques, the metric selected and the databases used play essential roles in the quantitative analysis of the evolution of language, including that of separate word-forms. By way of introduction we will review some of the most important studies in this field.

The authors of [3] conducted their study using a 108-word database from an online discussion group and a 1011 item word collection of digitized books, provided by Google Corporation. The model, presented in the article shows the existence of a strong relation between changes in word dissemination and changes in frequency. Google Books digital library and The Google Books Ngram corpus provide new opportunities for the quantitative analysis of a language evolution [4]. For example, using this database the authors of [5] showed that Heaps' law, which describes the number of distinct words in a document (or set of documents) as a function of the document length, does not always adequately describe the empirical facts: the exponent in the Heaps' law is not constant, it may vary with time, and, additionally, may exhibit oscillations, sometimes of a quasi-periodic character. This may mean that the influx of new words constantly varies. There are also papers devoted to research on narrowly defined lexical categories such as names. In [6], the authors studied the dynamics of the frequency of names that parents choose for their children, showing that names that quickly come into fashion quickly go out of fashion. This interesting observation contributes to shedding light on the genesis and extinction of different cultural tastes in general.

Ref. [7] presents an analysis of 10^7 words from the English, Spanish, and Hebrew parts of the Google Ngram database. The authors study trends in the rate of birth and death of words. For new words a peak in usage fluctuations peak is usually observed around 40 years after a word has gained wide circulation. Pronounced changes in the dynamics in times of war shows that the dynamics depends on co-evolutionary social, technological and political factors. The authors also showed that trends in birth and death of individual words generally have a similar shape. Ref. [8], however, shows the existence of difference between individual words dynamics. The authors of that paper analyzed the frequency of usage of function words using artificial neural networks. Here we propose an analysis of a broader database of words using an improved approach to clustering based on Kohonen neural networks.

2. Database

Our data are drawn from the 2012 version of the Google Books Ngram corpus of English words [4]. We did some filtering, selecting only 1-grams exclusively containing letters of the Latin alphabet and not more than one apostrophe. For each selected word-form we calculated a series of relative frequencies. While the interactive Google Books Ngram Viewer normalizes frequencies by the total number of 1-gram in the corpus, we normalized frequencies by the number of 1-grams retained after filtering.

Through this selection frequency charts for 6.78 million word forms was obtained. We prepared a reduced sample of the word forms that are among the 100,000 most frequent over the entire period which we focus on. This is the period from 1800to. For the period before 1800 the corpus is relatively small and the small sample size influences variation too much.. Moreover, the British database suffers from the additional problem of changes in fonts in years preceding 1800, something which leads to a large number of errors of optical character recognition.

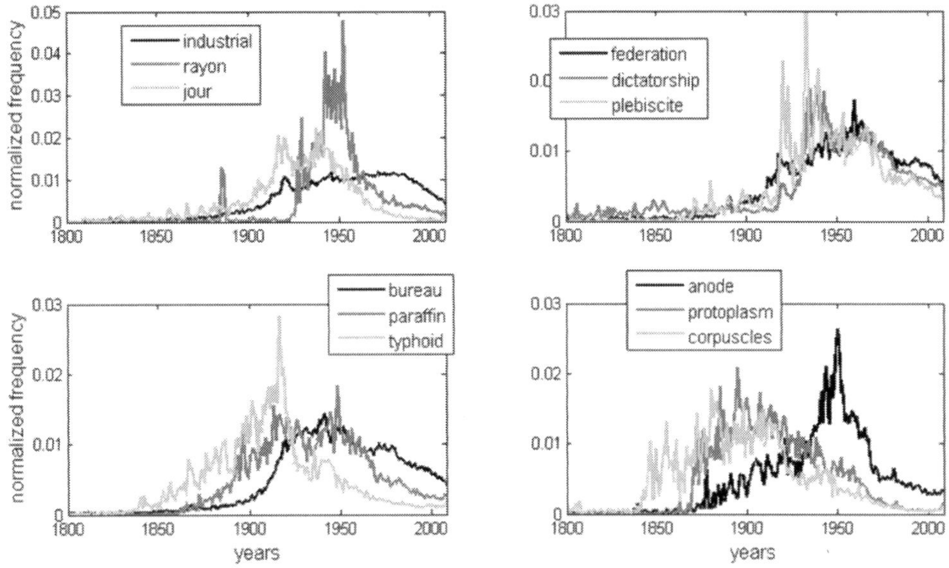

Figure 1. Dynamics of normalized frequencies for the 12 selected words.

There is a great variety in the types of frequency dynamics which can be observed. Even if we could overcome the technical difficulties involved in clustering the entire volume of the vocabulary, would still be difficult to visualize and interpret all types encountered. Therefore, in this paper we restrict the analysis to a narrow class of words which display clear 'bursts' in their frequency graphs, i.e., words for which we can distinguish a phase of increase, a clear maximum and a phase of decrease in frequency for the period under consideration). Examples of the frequency graphs for such words are shown in Fig.1. This shows the dynamics of the normalized frequencies for the 12 selected words: *'industrial'*, *'rayon'*, *'jour'*, *'federation'*, *'dictatorship'*, *'plebiscite'*, *'bureau'*, *'paraffin'*, *'typhoid'*, *'anode'*, *'protoplasm'*, *'corpuscles'* (normalization was performed according to the formula: $f_{norm}(t) = f(t) / \sum_t f(t)$). The selection of words with of the 'bursting' type was as accomplished in the follows way. For each word we calculated the center and the half-width of the frequency dependency according to the following formulas:

$$t_c = \frac{\sum_t t \cdot f(t)}{\sum_t f(t)}, \qquad \sigma_t^2 = \frac{\sum_t (t - t_c)^2 f(t)}{\sum_t f(t)}.$$

We selected word-forms that satisfy two requirements: first, we took only the words with a half-width σ_t of no more than 60 years. Secondly, to be able to observe all phases of the evolution we selected word forms for which $1850 \le t_c \le 1958$. Among the 100,000 most frequent words 4,445 satisfied these conditions whereas 48,969 word forms in the total corpus satisfy the conditions. First a comparison of different measures and clustering methods was performed on the short list (4445 word forms). Then, after determining the optimal approach, we extended clustering to the complete list of 48,969 word forms.

3. Clustering method

We used hierarchical clustering methods as well as clustering using the neural network of the Kohonen layer [9]. Hierarchical methods have the advantage of simple interpretation of the clustering results, and do not require a prior setting of the number of clusters. The advantage of the Kohonen layer is that it requires less computation.

To solve the clustering problem we need to choose the measure of the difference so that it is dependent only on the shape of the frequency dependencies compared and will not change with variation in absolute values and time shift. Achieving this requirement is possible with following method. Choosing some measure of the difference $D(x,y)$, we evaluate its value with different time shifts for a series compared against another one, just as is done in the calculation of the cross-correlation function. Thereafter, the difference between series x and y is calculated:

$$\min_{\tau} D(x(t+\tau), y(t))$$

We tested the following distance metrics:

- Correlation metric (via Pearson's coefficient);
- Correlation metric (via Spearman's coefficient);
- Distance metric based on L_1 norm. In this case, the distance measure is defined as the L_1-norm of the difference between vectors, normalized to unit length (according to $f_{norm}(t) = f(t) / \sum_t f(t)$) of frequencies;
- Kullback–Leibler metric. This measure arises naturally when deciding on a frequency vector's membership to the certain cluster based on the maximum likelihood method (it is necessary to take into account that the sampling frequencies have a Poisson distribution).

We used the following algorithms for computing distances between clusters: unweighted average distance (UPGMA), furthest distance, shortest distance, weighted average distance (WPGMA). Selection of the clustering method was performed by visual analysis. We checked how similar frequency dependencies are combined into one cluster. We also gave preference to methods which provided a more uniform grouping of words.

Let's consider the example of calculating the distance between two different words. Figure 2a shows dynamics of normalized frequency for words 'bolshevist' and 'isolationists'. Let's use the correlation metric based on Pearson's correlation coefficient. Figure 2b shows how the correlation distance depends on a shift of the first curve relative to the second one. We can see that the zero shift corresponds to the correlation coefficient 0.137 and a sufficiently large distance that is equal to 0.863. The minimum value of the distance that is equal to 0.097 (Pearson's coefficient equals 0.903), achieved with a shift for 22 years. It can be seen that curves are similar using the given shift. Therefore, for further analysis we use the found distance (0.097) as a distance between two considered words.

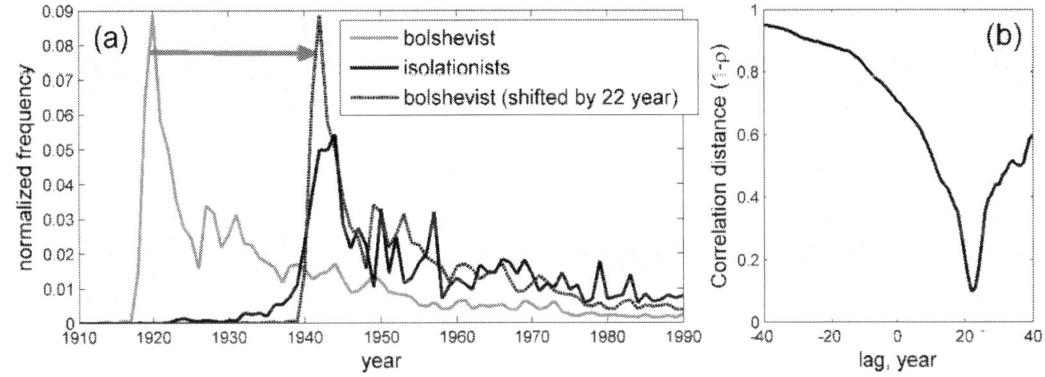

Figure 2. (a) Normalized frequencies for words 'bolshevist' and 'isolationists', frequency plot for the word 'bolshevist' shifted by 22 years is shown by dotted line; (b) Correlation distance at different shift values.

4. Results

Concerning the results of clustering, first of all, we note a wide variety of identified types of dynamics of frequency of word usage. When using any of the measures discussed above, the differences cannot be reduced to a classification into less than 70-100 clusters. Otherwise, with an increase in the threshold of clustering, the algorithm starts combining very different frequency trends. Figure 3 shows the normalized average frequency for 6 clusters containing the largest number of words. The dependencies presented were obtained by clustering the short list of words with a hierarchical method using the L_1-norm. In total, 6 clusters shown include 89.5% of the words. The remaining 10.5% of the words are distributed over 94 clusters of small size, which confirms that there is a great variability of types of frequency dynamics.

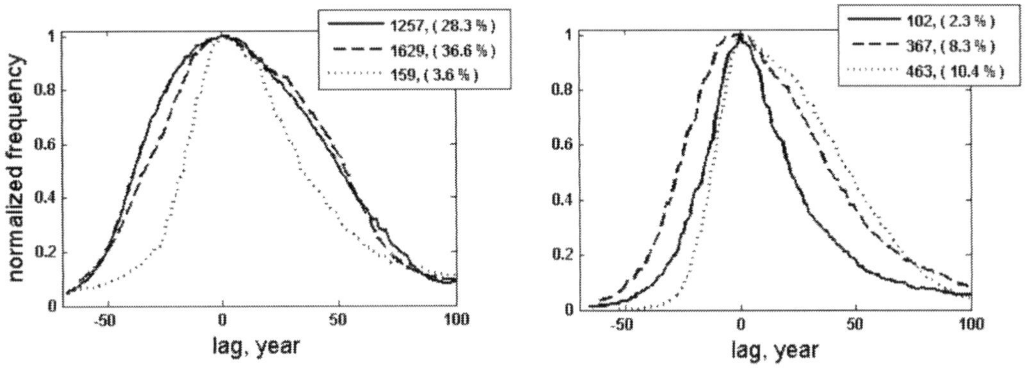

Figure 3. The normalized average frequency for 6 clusters containing the largest number of words.

The asymmetry of most dependencies is noteworthy: the phase of decrease in the frequency is longer than the phase of increase in the frequency. This impression can be confirmed by direct calculations. For each word from the complete list (48,969 word-forms) we evaluated the maximum and the point of intersection of the graph with the half of the maximum. Thus, for each word we estimated the duration of the growth phase of frequency (T_1) and the phase of decline of frequency (T_2). The median value of the ratio T_2 / T_1 was equal to 1.723, with T_2 greater than T_1 in 71% of the cases. Similar asymmetry was shown in frequency dynamics of child names in the paper [6] that we early considered. The duration of frequency increase phase is probably determined by rate of information exchange in society and social interaction. The duration of frequency decrease phase is determined by rate of falling a word into disuse and replacing it by frequently used words. As these two processes are substantially different, there are no a priori reasons to expect that the duration of these two phases will be equal. In fact, it can be seen that the second phase requires much more time.

This work was supported by the Russian Foundation for Basic Research (grant № 12-06-00404a).

References

[1] Baayen R H and Lieber R 1996/1997 Word frequency distributions and lexical semantics. *Computers and the Humanities* **30(4)** 281-291
[2] Baayen R Word Frequencies. Word Frequency Distributions 2001 *Text, Speech and Language Technology* **18** 1-38
[3] Eduardo G Altmann and Zakary L Whichard Adilson E. Motter Identifying Trends in Word Frequency Dynamics 2013 *J Stat Phys* **151** 277–288.
[4] The Google Books Ngram Corpuses, available at http://books.google.com/ngrams/datasets
[5] Bochkarev Vladimir V, Lerner Eduard Yu, Shevlyakova Anna V Deviations in the Zipf and Heaps laws in natural languages 2014 *Journal of Physics: Conference Series* **490**
[6] Berger J et al. How adoption speed affects the abandonment of cultural tastes 2009 *PNAS* **106(20)** 8146-50
[7] Petersen A M, Tenenbaum J, Havlin S, Stanley H E Statistical Laws Governing Fluctuations in Word Use from Word Birth to Word Death 2012 *Scientific Reports* **2** 313
[8] Maslennikova Yu S, Bochkarev V V, Voloskov D S Modelling of word usage frequency dynamics using artificial neural network 2014 *J. Phys.: Conf. Ser.* **490**
[9] Kohonen T Self-Organizing Maps. Third, extended edition 2001 *Springer*

Evaluation of outliers in acquired brain MR images

S Moldovanu[1,2], M (Vişan) Pungă[1,3], L Moraru[1],
[1]Faculty of Sciences and Environment, Department of Chemistry, Physics and Environment, Dunărea de Jos University of Galaţi, 47 Domnească St., 800008, Romania
[2]Dumitru Moţoc High School, 15 Milcov St., 800509, Galaţi, Romania
[3]Aurel Vlaicu High School, 1 Decembrie 1918 St., 800511, Galaţi, Romania

E-mail: moraru.luminita@ugal.ro

Abstract. Pre-processing is an important stage in the analysis of magnetic resonance images (MRI), because the effect of specific image artefacts, such as intensity inhomogeneity, noise and low contrast can adversely affect the quantitative image analysis. The image histogram is a useful tool in the analysis of MR images given that it allows a close relationship with important image features such as contrast and noise. The noise and variable contrast are elements that locally modify the quality of images. The key issue of this study derives from the fact that the spatial histogram can contain outliers indicating corrupted image information through the disorder of the bins. These aberrant errors should be excluded from the studied data sets. Here, the outliers are evaluated by using rigorous methods based on the probability theory and Chauvenet (CC), Grubbs (GC) and Peirce's (PC) criteria. In order to check the quality of the MR images, the Minkowsky (MD), Euclidean (ED) and cosine (CD) distance functions were used. They act as similarity scores between the histogram of the acquired MRI and the processed image. This analysis is necessary because, sometimes, the distance function exceeds the co-domain because of the outliers. In this paper, 32 MRIs are tested and the outliers are removed so that the distance functions generate uncorrupted and real values.

1. Introduction

Magnetic Resonance Image (MRI) that is widely used in research as well as in clinical practice [1], texture analysis or segmentation images [2-3].

The outliers detection techniques used in the image processing domain include: satellite imagery, hyperspectral images, digit recognition, spectroscopy, mammographic image, and video surveillance [4]. The sources of these outliers are motion, changes in an image during the acquisition process, or regions which appear abnormal on the static image. The main goal of the current research is that of restoring and enhancing the structural neuro-MR images by outliers removing operation in order to make accessible subtle details of tissues that are usually highlighted with tiresome intensity windowing methods. In a broad sense, outliers are patterns that deviate from expected normal behaviour. The high accuracy analysis of neuro-MR images is often hampered and various correction methods were developed. Commonly, they are categorized into low pass filtering, statistical modelling or surface fitting, or methods focused on mathematical modelling [5]. The statistical analysis of the outliers is a way of improving the quality of a corrupted image.

Generally, the MR images are analysed in four or five stages, as follows: a) the first stage deals with the image acquisition; b) in the pre-processing stage, various filters are used in order to eliminate the noise or the artefacts acquired in the first stage; c) then the segmentation of the images (supervised, unsupervised) could be accomplished [6]; d) structural approaches when the texture is represented by well-defined primitives (microtexture); or e) hierarchy spatial arrangements (macrotexture) could follow [2]. If the first two stages are compulsory for every image analysis, the choice of the last three stages depends on the research goal.

Practically, what this study aims at is including a new processing stage between the acquisition and the pre-processing stages. This is called detection of outliers (DO). Outliers can occur in any histogram image distribution. In a histogram image, the outliers occur because a bin is distant from other bins or the bins follow a sample kurtosis. In order to determine the outliers' percentage from an image the following tests were used: Chauvenet's criterion (CC) [7], Grubbs' criterion (GC) [8] and Peirce's criterion (PC) [9]. Chauvenet's criterion is based on the idea of finding a probability band, centred on the means of the normal distribution [7]. Peirce's method is a more general approach and it is applied in the case of more suspicious data values. The PC estimates the maximum allowable deviation of a measured value from the data means to the standard deviation. Grubbs' test is known as the maximum normed residual test and it is based on normal distribution.

In order to check for the quality of the restored image after outliers correction, the following distance-based methods were chosen: the Minkowsky (MD) [10], the Euclidean (ED) [10] and cosine (CD) distances [11]. All these functions measure the difference between the original image histogram and the histogram image corrupted by noise or by a modified contrast.

2. Materials and methods

The current study focuses on outliers detection tests and distance-based methods as tools for the qualitative restoration of the neuro-MR images [12]. In order to identify the outliers from the image histograms the following criteria were chosen.

1. Chauvenet's criterion is widely used in astronomy, nuclear technology, epidemiology, molecular biology and in many fields of Physical sciences. It makes an arbitrary assumption regarding the data rejection, namely all the data which lie within a "maximum deviation range" are assumed to be probable [7].
2. Peirce's criterion is a better alternative to Chauvenet's criterion, because it does not make any arbitrary assumptions [9].
3. Grubbs' criterion is a robust test for the detection of outliers for a normal distribution. Before applying GC it is necessary to check if the data could be approximated by a normal distribution and the data under examination are in ascending order [8].
4. Distance-based methods Texture analysis is a major step in texture classification and a sound analysis uses descriptors such as smoothness, coarseness and regularity. The histogram is a tool used to graphically interpret a digital image in tonal distribution. Three distance measures Euclidian [10] Minkowski [10] and the cosine distance [11] were applied between the histograms of the original neuro-MR image and the modified image by contrast and noise were used.

3. Experimental data

The neuro-MR images used in this study were retrieved from the Harvard Medical School database (http://www.med.harvard.edu/AANLIB/home.html). The images belong to a healthy aging patient.

During the proposed DO stage, the percentage of the outliers is established in order to verify the correctness of the subsequent measurements. This stage can be implemented between the acquisition and pre-processing stages. The logical steps to follow in the current study are:

(i) The images are tested with CC, PC and GC and the percentage of the outlier in the histogram image is computed.

(ii) For each image, the contrast has been varied from 0.1 to 0.9, with step of 0.1. The level 1 means the image itself. For all modified images, the CC, PC and GC were tested and the percentage of outliers was computed. The histograms of the original image and the images with 0.3 and 0.6 modified contrast level are shown in figure 1.

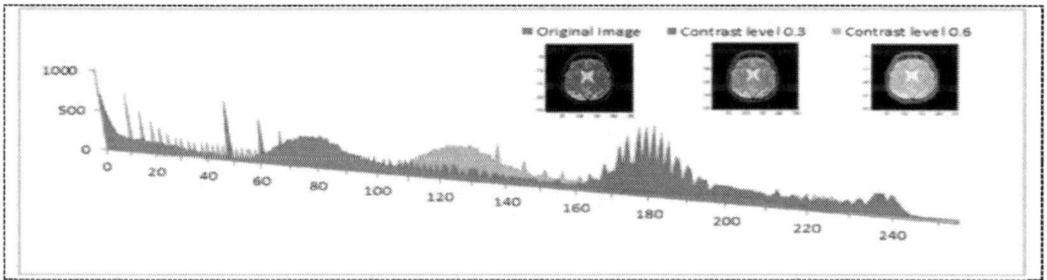

Figure 1. The histograms of the original and contrast modified images.

(iii) The Gaussian noise with variance ranges from 0.01 to 0.09 (with a step of 0.01) was added. Then, the CC, PC and GC were tested once again.
(iv) The pixels identified as outliers by CC, PC and GC are removed from the dataset.
(v) The similarity between images is tested by computing the distance functions between the original and modified images. Figure 2.a shows the representative distance functions ED, MD and CD, for different levels of the contrast. Label 1 denotes the distance functions computed for the original image (with outliers) and label 2 is for images without outliers. Figure 2.b presents only the ED, MD and CD metrics for images with outliers and added noise, because a perfect overlapping with the metrics for images without outlier and added noise has been found.

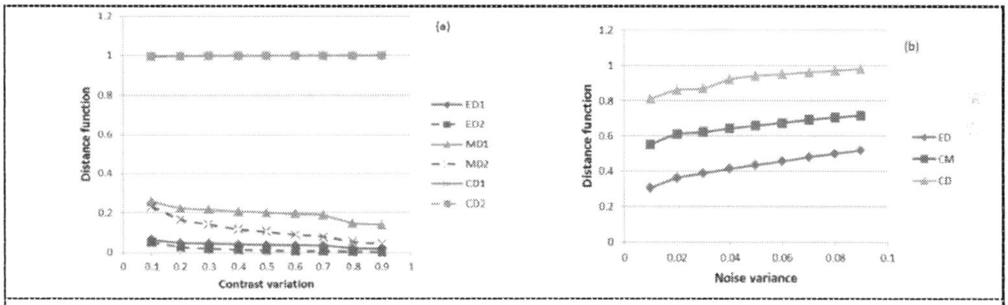

Figure 2. The distance functions for: (a) Contrast variation; (b) Noise variance

In the case of contrast level variation, the experimental data show that the CC, GP and PC gave different values of outlier percentage. The outlier percentage takes higher values for 0.2 and 0.3 contrast levels and lower values for levels of 0.8 and 0.9, in the case of CC. The percentage is constant for PC. In case of noise level variance, the outliers are not detectable by CC and PC (the last criterion detects outliers only for the noise levels of 0.08 and 0.09). The GC detects a relatively lower percentage of outliers.

4. Discussions

Nowadays, the challenge is to detect the outliers in neuro-MR images. An experimental study was conducted in order to compare the ability of Chauvenet's, Peirce's and Grubbs' criteria in outlier detection. Specifically, the contrast is increased in the range from 0.1 to 0.9 and Gaussian noise is added with variance ranging from 0.01 to 0.09. For each experimental image the percentage of outliers is computed. The quality of restoration process is verified by using the distance metrics between images with and without outliers.

When referring back to outliers in image, noise raises a sensitive issue. Typically, noise in the data tends to be similar to the actual outliers and it becomes difficult to remove the outliers from images. The current study highlights the idea that noise is a feature which cannot influence the outlier percentage in neuro-MR images.

The data from Figure 2a show a linear relationship of the similarity distance functions for both analysed cases (with/without outliers). The behaviour of the outliers related to the contrast variation is arguable. Thus, if the contrast level increases the distance between histogram images decreases, and the outliers percentage decreases, too. The ED1 distance is higher that ED2 indicating a higher similarity between images when the outliers were removed. The expected results are also confirmed by MD1 and MD2. On the contrary, CD1 and CD2 are highly overlapping and this indicates that the cosine distance is insensitive to the outliers.

The perfect overlapping of the EDs, MDs, and CDs in the case of noisy images (see Figure 2.b) leads to the conclusion that the analysis of the noise into image as an outlier can hardly be said to be useful. In this case it is necessary that the nature of noise in a MR image is taken into account. Noise is not an error. Of course, noise is not a true signal but its values are closer to the true signal values.

The experimental results of the current study show that the DO stage is necessary in the neuro-MR images processing, because, in the first instance, the user must know if the images contain outliers, the percentage of these outliers and how much they can influence the experiment results.

Conclusions

This study has brought together various outlier detection techniques. Three criteria for the outliers percentage detection were studied and three distance-based methods were employed for the detection of the similarity between images when the outliers were removed. The MR acquisition process is prone to various artefacts that affect the image features as contrast or noise. In case of contrast variation, the distance functions allow the conclusion that the images are affected by outliers. On the other hand, noise is not an error and the outliers in noisy images cannot be identified. Therefore, the necessity of this new proposed stage "Detection of outlier (DO)" should be in-depth studied for further applicability in the analysis of the complex systems as neuro-MR images.

Acknowledgements

The author Simona Moldovanu would like to thank the Project PERFORM, ID POSDRU/159/1.5/S/138963 of "Dunărea de Jos" University of Galati, Romania for support.

The author Mirela (Visan) Punga's work has been funded by the S.O.P.H.R.D. 2007-2013 of the Ministry of European Funds through the Financial Agreement POSDRU/159/1.5/S/132397.

References

[1] Symms M, Jäger H, Schmierer K and Yousry T, 2004 *J. Neurol. Neurosurg. Psychiatry* **75** 1235
[2] Materka A, 2004 *Dialogues Clin. Neurosci.* **6** 243
[3] Coltheart M 2008 *Scholarpedia* **33** 644
[4] Singh K and Upadhyaya S 2012 *IJCSI* **9** 694
[5] Hou Z 2006 *Int. J. Biomed. Imaging* **2006** 1
[6] Balafar M, Ramli A, Saripan M and Mashohor S 2010 *Artif. Intell. Rev.* **33** 261
[7] Zerbet A and M Nikulin 2003 *Commun. Stat-Theor. M.* **32** 573
[8] Ben-Gal I 2005 A Complete Guide for Practitioners and Researchers Mining and
[9] Knowledge Discovery Handbook *Outlier detection* ed Maimon O and Rockach L (Dordrecht :Kluwer Academic Publishers) chapter 1 p 1-15
[10] Wilson D and Martinez T 1997 *J. Artif. Intell. Res.* **6** 1
[11] Spruill M and Marcus C 2007 *Electron. Comm. Prob.* **12** 234
[12] Smith M, Martinez T and Giraud-Carrier C 2014 *Mach. Learn.* **95** 225

Applying Enhancement Filters in the Pre-processing of Images of Lymphoma

Sérgio Henrique Silva[1], **Marcelo Zanchetta do Nascimento**[2], **Leandro Alves Neves**[3], **Valério Ramos Batista**[4]

[1] *Faculty of Mechanical Engineering, Federal University of Uberlândia (UFU), Uberlândia, MG, Brazil*

[2] *Faculty of Computer Science, Federal University of Uberlândia (UFU), Uberlândia, MG, Brazil*

[3] *Institute of Biosciences, Literature and Exact Sciences, Department of Computer Science and Statistics, São Paulo State University (UNESP), São José do Rio Preto, SP, Brazil*

[4] *Mathematics, Computer Science and Cognition Centre, Federal University of ABC (UFABC), Santo André, SP, Brazil*

E-mail: marcelo.zanchetta@gmail.com

Abstract. Lymphoma is a type of cancer that affects the immune system, and is classified as Hodgkin or non-Hodgkin. It is one of the ten types of cancer that are the most common on earth. Among all malignant neoplasms diagnosed in the world, lymphoma ranges from three to four percent of them. Our work presents a study of some filters devoted to enhancing images of lymphoma at the pre-processing step. Here the enhancement is useful for removing noise from the digital images. We have analysed the noise caused by different sources like room vibration, scraps and defocusing, and in the following classes of lymphoma: follicular, mantle cell and B-cell chronic lymphocytic leukemia. The filters Gaussian, Median and Mean-Shift were applied to different colour models (RGB, Lab and HSV). Afterwards, we performed a quantitative analysis of the images by means of the Structural Similarity Index. This was done in order to evaluate the similarity between the images. In all cases we have obtained a certainty of at least 75%, which rises to 99% if one considers only HSV. Namely, we have concluded that HSV is an important choice of colour model at pre-processing histological images of lymphoma, because in this case the resulting image will get the best enhancement.

1. Introduction

The non-Hodgkin lymphoma is a diversified group of neoplasm that appears in the lymphocyte. This kind of lymphoma is estimated to happen in 10 out of 100 thousand people, but this number has been increasing about 4% yearly. According to [1], some of the reasons for this increase are the improvement in diagnostic methods, preciser morphological classifications and better health information systems.

This neoplasm does not follow a characteristic histological pattern, and it also grows with an altering cyclical behaviour. Diagnosing the non-Hodgkin lymphoma is therefore a difficult task [2]. Normally, diagnoses and prognoses are based upon histological analyses of the lymph nodes. The analyses are carried out by specialists that evaluate morphological and non-morphological characteristics of samples observed through the microscope. The main pieces of information they collect are: arrangement and quantity of centrocytes and centroblasts, internal characteristics of the cells, and tissue patterns [3]. The main neoplasms analysed by specialists are: Follicular Lymphoma (FL), Mantle Cell Lymphoma (MCL) and B-cell Chronic Lymphocytic Leukemia, also known as Chronic Lymphoid Leukemia (CLL).

Light microscopes enable the specialists to get fairly detailed information from the lymph nodes. These microscopes are equipped with digital cameras that render pictures on which image processing techniques can be applied in order to build a Computer-Aided Diagnosis (CAD). Such computational systems help specialists achieve a preciser diagnosis [4].

Together with that cutting-edge equipment, in the literature we cite [5] and [6] as works devoted to building CAD schemes applied to images of lymph nodes. Nonetheless, specialists still face a relevant problem that compromises a reliable diagnosis: the presence of *noise* in many digital images.

Indeed, the first problem is room vibration occasionally caused by external factors. Even if not perceptible they compromise the quality of the digital images rendered by camera and light microscope, which are highly sensitive. The second are scraps in the samples, and the third is defocusing, because it is not always easy to find the best calibration of the microscope to photographing. These and other problems were discussed in [7] but there the authors do not propose filters to fix them.

Of course, specialists can detect such problems in digital images. The ones that present best quality have been selected in databanks like [8] for scientific research. They are used as reference for many studies, and one of them is the analysis of filter performance. Filters of general purpose were discussed in [9]. Many of them have been implemented in Matlab toolboxes. Our work is devoted to evaluating three of these filters applied to images of lymphoma at the pre-processing step. The filters are Gaussian, Median and Mean-Shift.

In order to study their performance we have taken 48 images from [8] as reference. They were perturbed by an image manipulation program in order to simulate the aforementioned noises. The three filters were then applied to different colour models of the perturbed images (RGB, Lab and HSV). Afterwards, we carried out a quantitative analysis between the original and filtered images by means of the Structural Similarity (SSIM) Index. In all cases we have obtained a certainty of at least 75%, which rises to 99% if one considers only HSV. The 48 images consist of 16 of FL, 16 of MCL and 16 of CLL.

2. Methodology

2.0.1. Image Databank We have used the image databank [8], which was assembled after many studies carried out by researchers at the *National Cancer Institute* and the *National Institute on Aging*, both in the United States. A total of 30 histological slides were photographed, all of lymph nodes stained with Haematoxylin and Eosin. The aforementioned 48 images were obtained from this setting.

2.0.2. Colour Models With three colour models we could evaluate the influence of noise and of enhancement filters in the images. RGB is based on a cartesian cube whose points (R, G, B) have each of their coordinates varying in the interval $[0, 1]$. HSV is defined by hue, saturation and value (of brightness), whereas Lab consists of lightness and the two components (a, b) of chrominance. See [9] for details. According to [10–12] among all colour models these three have been the most successful ones when applied to histological images.

2.0.3. Noise We have followed [9] in order to simulate noise. Defocusing was obtained via *cyclic convolution*. The *sliding neighbourhood operation* on the image produced fuzziness, which in practice is caused by room vibration. Finally, salt and pepper simulated scraps.

For defocusing we applied a radius of 10 pixels, and for fuzziness a kernel of 21×21 pixels. Salt and pepper were sprinkled to a density of 0.02.

2.0.4. Enhancement Filters The Median Filter replaces each pixel of an image with a median of its surrounding pixels. In our work we have applied a punched square A of 3×3 pixels in order to determine its centre x by

$$x = \text{Med}\{A\}, \tag{1}$$

where Med is the median of the elements in A.

In the Gaussian filter the coefficients are obtained from a two-dimensional Gaussian function. The discrete Gaussian function with zero mean and standard deviation σ is given by

$$G(x, y) = \frac{1}{2\pi\sigma^2} \exp\left(-\frac{x^2 + y^2}{2\sigma^2}\right), \tag{2}$$

where (x, y) is the input mask position and $G(x, y)$ is the resulting new value for the (x, y)-pixel. Finally, the Mean-Shift filter locates the maxima of a density function through a non-parametric feature-space technique. Details are omitted here but the reader can find them in [9].

3. Results

Figures 1, 2 and 3 illustrate the SSIM of a filtered image for each colour model.

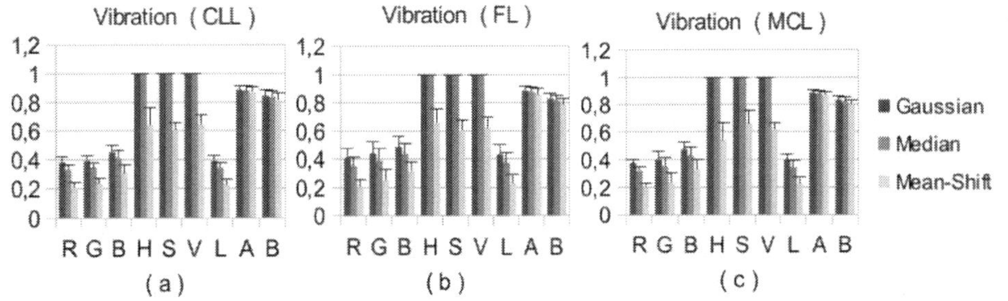

Figure 1: Evaluation of the filters of room vibration for (a) CCL, (b) FL and (c) MCL.

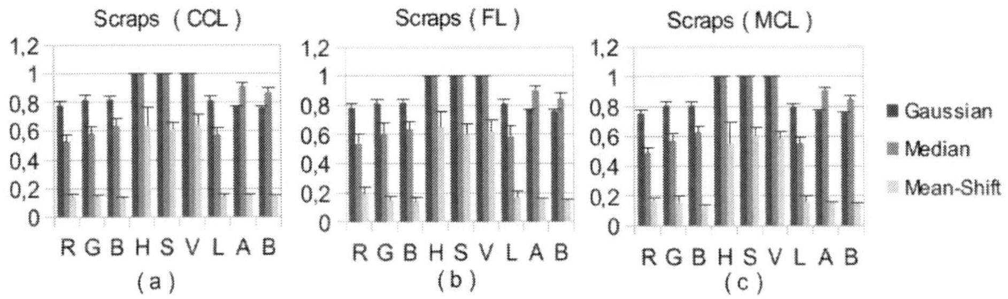

Figure 2: Values of SSIM obtained from filtering of scraps for (a) CCL, (b) FL and (c) MCL.

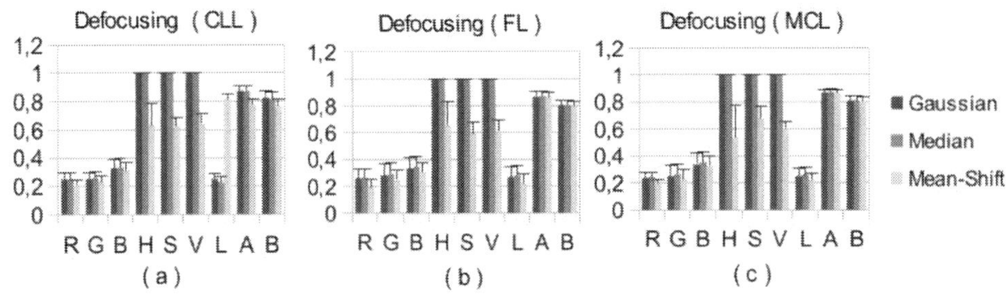

Figure 3: Values of SSIM obtained from filtering of defocusing of (a) CCL, (b) FL and (c) MCL.

An SSIM of almost 100% was achieved by the Gaussian and Median filters for both H and S channels in the three classes of lymphoma. This contrasts with channel V, of which the performance was no greater than 70%. By the way, in all but one of our experiments the Mean-Shift filter showed poorer performance than either the Gaussian or the Median filter, for all colour models and channels. The only exception was for the noise due to defocusing of CLL in the Lab-model, where the SSIM achieved 80% (much higher than the approximate 20% of both Gaussian and Median filters).

Filtering with RGB for room vibration gave an SSIM below 60% in all three classes of lymphoma. Nevertheless, for some channels of the other colour models the SSIM was quite above that.

In the case of noise due to scraps some channels of RGB and Lab achieved an SSIM around 80%, but HSV still had the best performance. The lowest values of SSIM came from RGB when applied to filtering noise due to defocusing. These remained below 40% in any of the three channels.

4. Conclusions

We have presented a study of comparisons between the performance of enhancement filters for different noises in histological images. In our study we considered the classes of cancer in these images separately. The chosen colour models (RGB, HSV and Lab) are the ones that have been mostly used in the literature for histological images. This is because of the successful results that they have been showing for such kind of images. In our study we concluded that the HSV model has the best performance at filtering noise of different kinds for all the images of lymphoma.

References

[1] Boyle P, Levin B *et al.* 2008 *World cancer report 2008.* (IARC Press, International Agency for Research on Cancer)
[2] Wright D H, Addis B J and Leong A 2011 *Diagnostic Lymph Node Pathology-9780340990681-182, 45*
[3] Swerdllow S, Campo E and Harris N L 2008 *WHO classification of tumours of haematopoietic and lymphoid tissues* (France: IARC Press, 2008)
[4] Gurcan M, Boucheron L, Can A, Madabhushi A, Rajpoot N and Yener B 2009 *Biomedical Engineering, IEEE Reviews in* **2** 147–171
[5] Oger M, Belhomme P and Gurcan M N 2012 *Computerized Medical Imaging and Graphics*
[6] Belkacem-Boussaid K, Prescott J, Lozanski G and Gurcan M N 2010 *SPIE Medical Imaging* (International Society for Optics and Photonics) pp 762436–762436
[7] Khan A, Rajpoot N, Treanor D and Magee D 2014 *IEEE Transactions on Bio-medical Engineering* **61** 1729–1738
[8] Orlov N V, Chen W W, Eckley D M, Macura T J, Shamir L, Jaffe E S and Goldberg I G 2010 *Trans. Info. Tech. Biomed.* **14** 1003–1013 ISSN 1089-7771 URL `http://dx.doi.org/10.1109/TITB.2010.2050695`
[9] Gonzalez R C and Woods R E 2006 *Digital Image Processing (3rd Edition)* (Upper Saddle River, NJ, USA: Prentice-Hall, Inc.) ISBN 013168728X
[10] Sertel O, Kong J, Lozanski G, Shana'ah A, Catalyurek U, Saltz J and Gurcan M 2008 *Acoustics, Speech and Signal Processing, 2008. ICASSP 2008. IEEE International Conference on* (IEEE) pp 597–600
[11] Kong H, Gurcan M and Belkacem-Boussaid K 2011 *Medical Imaging, IEEE Transactions on* **30** 1661–1677
[12] Irshad H, Veillard A, Roux L and Racoceanu D 2014 *Biomedical Engineering, IEEE Reviews in* **7** 97–114 ISSN 1937-3333

CFD-Exergy analysis of the flow in a supersonic steam ejector

M Boulenouar[1] and A Ouadha[2,3]

[1] Département de Génie Maritime, Faculté de Génie Mécanique,
Université des Sciences et de la Technologie Mohamed BOUDIAF d'Oran,
B.P. 1505 Oran El-M'naouar, 31000 Oran, Algérie
[2] Laboratoire d'Energie et Propulsion Navale, Faculté de Génie Mécanique,
Université des Sciences et de la Technologie Mohamed BOUDIAF d'Oran,
B.P. 1505 Oran El-M'naouar, 31000 Oran, Algérie

E-mail:ah_ouadha@yahoo.fr

Abstract. The current study aims to carry out a CFD-exergy based analysis to assess the main areas of loss in a supersonic steam ejector encountered in ejector refrigeration systems. The governing equations for a compressible flow are solved using finite volume approach based on *SST k-ω* model to handle turbulence effects. Flow rates and the computed mean temperatures and pressures have been used to calculate the exergy losses within the different regions of the ejector as well as its overall exergy efficiency. The primary mass flow rate, the secondary mass flow rate and the entrainment ratio predicted by the model have been compared with the experimental data from the literature.

1. Introduction

Understanding the entrainment process and the mechanisms responsible for the losses in an ejector is essential in order to construct more efficient ejector refrigeration systems. Methods used to improve its efficiency present complicated issues for both experimental and numerical investigations. Experimental techniques usually do not cover the whole operating parameters range of an ejector. These techniques can be successfully complemented using appropriate numerical simulations.

It seems that the literature devoted to ejectors and ejector systems is very extensive. Several experimental, analytical and numerical studies have been carried out in this area. A review paper published recently by Chen et al. [1] delineates progress and development of using ejectors in refrigeration and air conditioning applications using a number of refrigerants. The reviewed studies are reported and categorized in several topics including, refrigerant selections, mathematical modeling and numerical simulation of ejector system, geometric optimizations, operating conditions optimizations and combinations with other refrigeration systems.

Although that HFC and natural refrigerants have dominated most of published studies [2-6], steam as the working fluid stills attracting an increasing interest. In recent years a considerable volume of work has been published on the use of CFD for the prediction of the flow in steam ejectors [7-10].

From the above literature revue, it seems that less interest has been devoted to exergy analysis of ejectors. The majority of research studies have been focused on the determination of the performance

[3] To whom any correspondence should be addressed.

of ejectors in terms of the entrainment ratio using experimental and numerical approaches. Only few studies on exergy analysis have been found in the open literature to the best of the authors' knowledge, although there are many studies related to ejectors [11-12].
The current study aims to carry out a CFD-exergy based analysis to assess the main areas of loss in a supersonic steam ejector encountered in ejector refrigeration systems. The governing equations for a compressible flow are solved using finite volume approach based on *SST k-ω* model to handle turbulence effects. Flow rates and the computed mean temperatures and pressures have been used to calculate the rate of exergy at the ejector inlets and outlet as well as the exergy losses within the ejector. Furthermore, the exergy efficiency of the ejector has been calculated.

2. Mathematical model

2.1. Description of the problem

Figures 1*a* and 1*b* illustrate a tridimensional view of the ejector and a schematic representation showing its main dimensions, respectively. This configuration has been studied experimentally and numerically by Ruangtrakoon et al. [9,13]. The steam ejector consists of a primary nozzle, a secondary nozzle, a mixing chamber, a throat and a diffuser as shown in figure 1.*a*. The primary fluid at high pressure and temperature from the generator flows through a converging-diverging nozzle to reach supersonic velocity. It enters into a throat which is connected to a secondary inlet and an outlet. The primary flow leaves the nozzle at low pressure and drives a secondary flow of vapor refrigerant from the evaporator. The resulted mixing stream is recompressed in a mixing chamber involving a decrease in the flow velocity due to complex interactions between the mixing layer and the resulted shock waves. The combination of the shock waves and the subsonic flow through the diffuser increases the pressure to attain the condenser pressure. In this work, the value of NXP, the nozzle exit position, is set to 23 mm according to the recommendations of the authors of the reference cited above. Details of the main dimensions of the ejector are shown in figure 1.*b*.

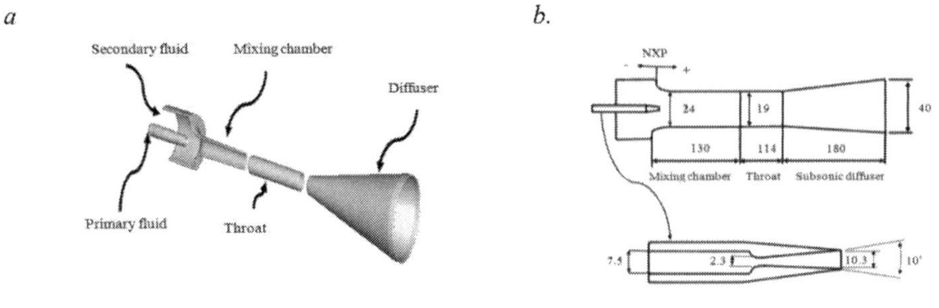

Figure 1. Physical problem : *a.* Tridimensional view; *b.* Schematic representation and dimensions in mm.

2.2 Governing equations

The working fluid (steam), assumed compressible and Newtonian, is governed by the compressible steady state form of mass, momentum and energy conservation equations. In compressible flows, the conservation equations are associated to a law relating pressure, density and energy. Typically, a perfect gas is assumed.
A literature survey reveals that several RANS turbulence models have been adopted in predicting the flow in ejectors. In the present study, the Menter's shear stress transport *SST k-ω* turbulence model has been used. It is a two-equation model based on the Boussinesq hypothesis. The detailed description of the model is beyond the scope of the present study and can be found in literature.

2.3 Boundary conditions

To accurately reproduce the physics of the problem observed during experimental tests carried out by Ruangtrakoon et al. [13], the primary fluid inlet and the secondary fluid inlet have been set as pressure-inlet type. At the exit of the ejector, a pressure-outlet type has been imposed. All wall surfaces of the ejector are considered adiabatic. In absence of information about turbulence levels in the experimental tests, a turbulence intensity of 5% has been specified at the inputs and 10% at the output of the ejector.

3. Numerical approach

3.1. Computational domain and mesh

Due to the symmetry of the problem, numerical simulations have been performed on a two-dimensional axisymmetric domain. The mesh has been created using multi-block technique by means of *Gambit*. Equidistant hexahedral cells have been used in order to reduce all the errors associated with cells extrusion and distortion.

It is necessary to check the quality of the numerical mesh in order to guarantee a grid independent solution. In the present study, the primary mass flow rate, the secondary mass flow rate and the entrainment ratio have been checked for grid independency by performing simulations on four non-uniform meshes with 8 000, 16 000, 32 000 and 64 000 cells. The numerical uncertainties for the three parameters are presented in table 3. The maximum numerical uncertainty levels between meshes 3 and 4 are less than 1.52%. Based on these uncertainties, the mesh consisting of 32 000 cells has been considered in the present study.

Table 1. Mesh sensitivity study.

Mesh	Number of cells	$\dot{m}_p$ (kg/h)	δ (%)	$\dot{m}_s$ (kg/h)	Δ (%)	Rm	δ (%)
1	8 000	4.636	- 0.086	1.226	1.712	0.264	-1.894
2	16 000	4.640	0.905	1.247	1.443	0.269	0.743
3	32 000	4.598	1.522	1.229	1.302	0.267	-0.375
4	64 000	4.528	-	1.213	-	0.268	-

3.2. Numerical solver

CFD codes, such as Fluent, provide wide computational flexibility and several options for turbulent compressible flow simulations. Fluent solves the three-dimensional Favre averaged equations of motion discretized using a control volume approach for flow, pressure, temperature and turbulence. Reynolds stress terms are modeled using one of several turbulence models available in the code. The density-based implicit solver, more suitable for supersonic flows, has been adopted. A second order upwind scheme has been used to discretize the convective terms. Exergy calculations are carried out in post processing after the flow field solution convergence.

4. Results and discussions

4.1. Numerical model validation

Before numerical results are presented, comparisons of computed results with experimental measurements are first carried out to assess the reliability of the numerical model used. The obtained primary mass flow rates, secondary mass flow rates and the entrainment ratios in the numerical computations at a fixed generator temperature of 120°C are in good agreement with the experimental data [13] as shown in table 2. The observed discrepancies may be attributed to the numerical modeling.

Table 2. Comparison of the numerical and experimental results.

	Exp. [13]	Present work	Error (%)
$\dot{m}_p$ (kg/h)	4.530	4.600	1.5
$\dot{m}_s$ (kg/h)	1.240	1.230	0.6
Rm	0.273	0.267	2.2

4.2. Operational characteristics

For a fixed geometry of the ejector, the entrainment ratio of an ejector is influenced by operating conditions. The entrainment ratio decreases with increasing generator temperature or decreasing evaporator temperature as shown in figure 2. However, for fixed generator and evaporator temperatures (pressures), the entrainment ratio has somewhat different behavior with the condenser temperature (pressure). According to the ejector discharge pressure (condenser pressure), three different operation modes can be encountered: the critical operation mode, sub-critical operation mode and back flow operation mode. Under critical operation mode, the entrainment ratio remains almost constant until a critical pressure value. During this operation mode, the entrainment of the secondary flow by the primary flow induces chocked phenomena. For pressures larger than the critical value, the entrainment ratio falls rapidly due to the fast decrease of the secondary mass flow rate. It is the sub-critical mode. For further large ejector discharge pressure, the entrainment ratio tends towards negative values induced by the inversion of the secondary flow into the secondary nozzle. The ejector operates in the backflow mode. It is also observed that the increase in the generator temperature results in an increase in the critical condenser temperature as shown in figure 5.*a* (dashed line). However, the critical condenser temperature is independent of the evaporator temperature.

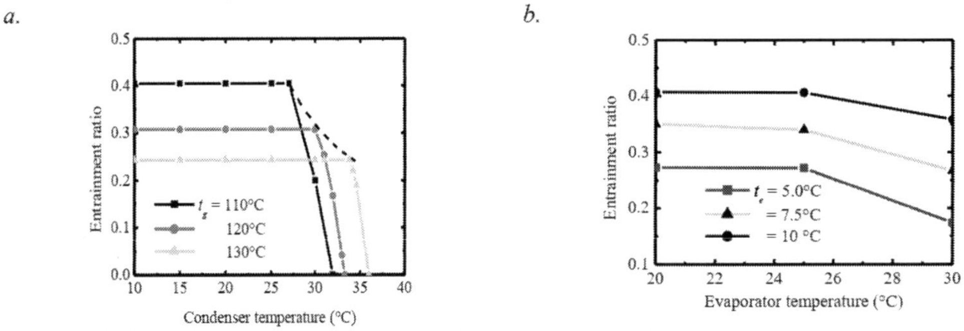

Figure 2. Effect of operating conditions on the entrainment ratio : *a.* Variable generator temperature; Variable evaporator temperature.

4.3. Flow fields

Figure 3 shows the Mach number displayed as contours in the symmetry plane and as profile along the ejector centreline. At first glance, the Mach number varies in both axial and radial directions as illustrated in the contours. At the nozzle, the centreline Mach number remains constant until the nozzle throat, and then it increases rapidly to reach a value of 4 at the exit of the nozzle. In the mixing chamber, it increases further to attain its maximum value of 4.54. Then, the Mach number decreases with noticeable fluctuations indicating the presence of shock waves. The latter extenuate moving axially to the entry of the diffuser. In the diffuser the flow becomes subsonic.

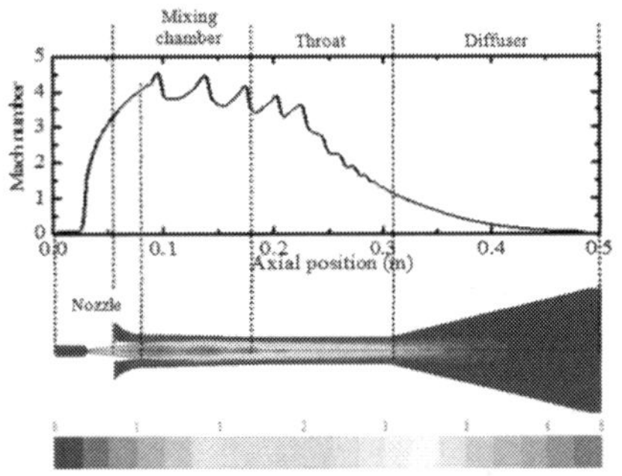

Figure 3. Mach number evolution in the symmetry plane and along the axis.

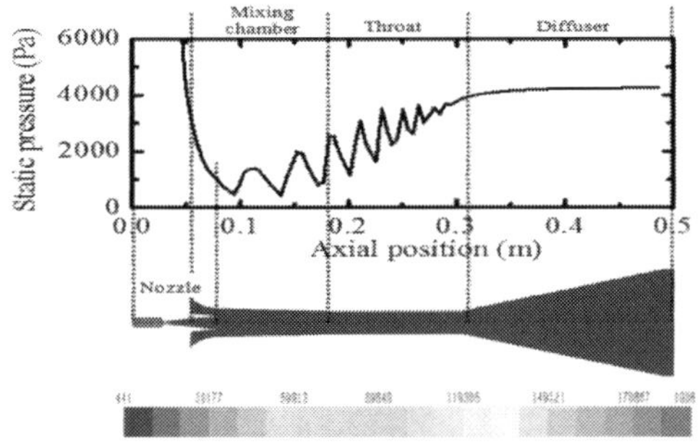

Figure 4. Static pressure evolution in the symmetry plane and along the axis.

The fluctuations in the centerline pressure reveal also the presence of shock waves commonly encountered in nozzle supersonic flows. The lowest values of the static pressure are displayed in the vicinity of the entry of the mixing chamber where the working fluid reaches its highest velocity. The mixing of the primary and the secondary flows results in a momentum exchange between the two flows. This process decreases the velocity of the working fluid in the mixing section. Thus, a noticeable increase in pressure in this section is observed as shown in figure 4.
The shock phenomena can also be observed in the velocity field displayed as contours in a symmetry plane and as a profile along the ejector centerline (figure 5). The distortion in the velocity profiles indicates the presence of shock waves. The velocity increases in the primary nozzle to reach its maximum value of 1094 m/s at the exit of the primary nozzle. Then series of shock waves, shown as discontinuities in the velocity profile, appear in the mixing chamber and the throat. The mixing process makes the velocity to decrease gradually. As the fluid flows toward the exit of the throat, the velocity decreases rapidly to attain its minimal value at the exit of the ejector.

4.4. Exergy analysis

The exergy calculations have been performed using the average return steam temperature and pressure values from the CFD analysis. Results, in terms of percentage of exergy losses in the different regions of the flow domain in the ejector, are given in figure 6. It appears that the main dominant exergy losses occur in the nozzle. They represent 43.29% of the total exergy losses. The contributions of exergy losses in the throat, the diffuser and the mixing chamber account for 33.75, 16.25 and 6.71% of the total exergy losses respectively. These losses are mainly due to the large velocity and pressure gradients. An appropriate re-design of the regions with large exergy losses may contribute to potential enhancements. Unlike the entrainment ratio (the traditional criteria measuring performance of ejectors), the exergy efficiency is firmly based on both the first and second laws of thermodynamics. For the actual design and under the operating conditions considered, the exergy efficiency is 77. 69%.

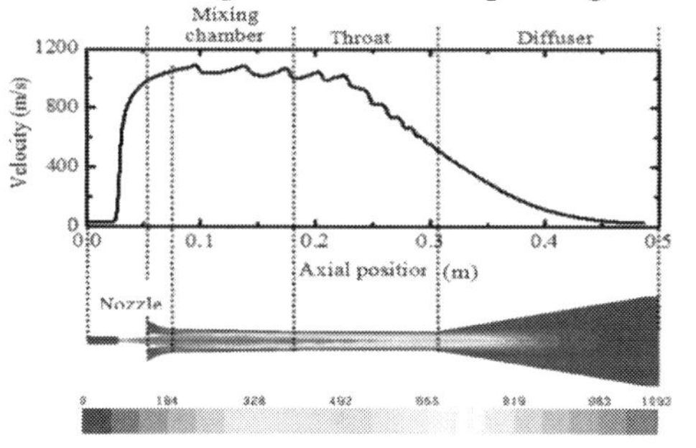

Figure 5. Velocity evolution in the symmetry plai and along the axis.

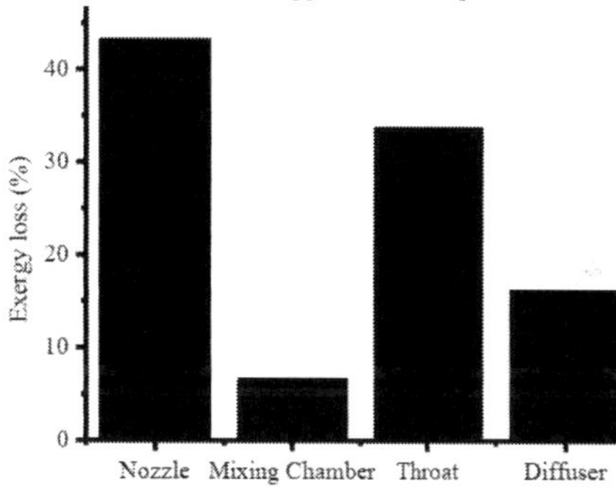

Figure 6. Percentages of exergy losses generated the different regions of the ejector.

5. Conclusions

A computational study on the performance and exergy loss mapping of a supersonic steam ejector is carried out. Reasonable predictions of the primary mass flow rate, the secondary mass flow rate and the entrainment ratio have been found in comparison with the literature experimental data. The present results show that the performances of the ejector are largely influenced by the operating conditions of the evaporator, the condenser and the generator. Furthermore, the exergy analysis indicates that the main source of irreversibilities is the nozzle due to high levels of pressure and velocity gradients. The calculated exergy efficiency provides an indication of the potential for improvement that is more fundamental than that shown by the entrainment ratio.

References

[1] Chen X, Omer S, Worall M and Riffat S 2013 Recent developments in ejector refrigeration technologies, *Renewable and Sustainable Energy Reviews* 19 629
[2] Banasiak K, Hafner A and Andresen T 2012 Experimental and numerical investigation of the influence of the two-phase ejector geometry on the performance of the R744 heat pump *Int. J. Refrig.* 35 1617
[3] Li C, Li Y and Wang L 2012 Configuration dependence and optimization of the entrainment performance for gas-gas and gas-liquid ejectors *Appl. Thermal Eng.* 48 237.
[4] Colarossi M, Trask N, Schmidt D P and Bergander M J 2012, Multidimensional modeling of condensing two-phase ejector flow *Int. J. Refrig.* 35 290
[5] Lucas C, Koehler J, Schroeder A and Tischendorf C 2013, Experimentally validated CO_2 ejector operation characteristic used in a numerical investigation of ejector cycle *Int. J. Refrig.* 36 881
[6] Banasiak K, Palacz M, Hafner A, Bulinski Z, Smołka J, Nowak A J and Fic A 2014 A CFD-based investigation of the energy performance of two-phase R744 ejectors to recover the expansion work in refrigeration systems: An irreversibility analysis *Int. J. Refrig.* 40 328
[7] Varga S, Oliveira A C, Ma X, Omer S A, Zhang W and Riffat S B 2011 Experimental and numerical analysis of a variable area ratio steam ejector *Int. J. Refrig.* 34 1668
[8] Yang X, Long X and Yao X 2012 Numerical investigation on the mixing process in a steam ejector with different nozzle structures *Int. J. Thermal Sciences* 56 95
[9] Ruangtrakoon N, Thongtip T, Aphornratana S and Sriveerakul T 2013 CFD simulation on the effect of primary nozzle geometries for a steam ejector in refrigeration cycle Int. *J. Thermal Sciences* 63 133
[10] Allouche Y, Bouden C and Varga S 2014 A CFD analysis of the flow structure inside a steam ejector to identify the suitable experimental operating conditions for a solar-driven refrigeration system *Int. J. Refrig.* 39 186
[11] Trela M, Kwidzinski R, Butrymowicz D and Karwacki J 2010 Exergy analysis of two-phase steam–water injector *Appl. Thermal Eng.* 30 340
[12] Cai Q, Tong M and Bai X 2012 Exergy analysis of two-stage steam-water jet injector *Korean J. Chemical Eng.* 29 513
[13] Ruangtrakoon N, Aphornratana S and Sriveerakul T 2011 Experimental studies of a steam jet refrigeration cycle: Effect of the primary nozzle geometries to system performance *Exp. Thermal Fluid Science* 35 676

Analytical Models for the Interactions between a Gas Mixture and Quartz Micro Balance Sensor Array Based on Polymers

Abdelaziz Abbas
Department of characterisation and Instrumentation, Welding and NDT research Center , route de Dely-Brahim , BP 64 ,311704 Chéraga , Alger, Algérie

E-mail:abbasabdelaziz@yahoo.de

Abstract. The aim of this work is to develop analytical models for the thermodynamic equilibrium at the interfaces (gas mixture / Quarz Micro Balance sensor arrays based on conducting polymers). Differential equations, which describe the change in the partial sensitivities of the sensor array elements depending on the gas mixture components concentrations, and the sensor array parameters, have been developed. Moreover, the responses of the sensor array as a function of the concentrations of the gas mixture components have been modeled.

1.Introduction

Acoustic sensors behave as a mass transducer, namely the response of the sensor to a vapor is proportional to the mass of the analyte. Because of the lack of perfect selectivity, sensor arrays are often implemented for analyse of gas mixture. Many scientific papers concerning the evaluation of acoustic wave sensor array data by using of pattern recognition techniques have been written [1]. However, to our best knowledge, no study has explored the physicochemical behavior at the interfaces gas mixture/ sensor array. Except the empirical formula of Clifford [2] , there is no model that describes the physicochemical behavior of sensor array / gas mixture multi- systems. Further improvement of the sensor array properties requires a better understanding of the processes that occur at the surfaces of the sensor array elements. Therefore, it is important to develop models for the sensor array responses. In previous work, we presented models of the interactions between a mixture of vapors and a sensor arrays consisted of chemical sensors [3] and Metal oxide sensors [4]. In the present report, we extend the application of the developed models to the interaction between **Q**uarz **M**icro **B**alance (QMB) sensor array and a mixture of vapors.

2. Theoretical Aspects

2.1 Interaction between a single QMB- sensor and a gas

If we consider the interaction between a QMB-sensor and a gas , a perturbation of the mass, Δm, on the surface of the sensor causes shifts in the oscillator frequency Δf, according to the equation of Sauerbrey [5-6]

$$\Delta f = f - f^{\text{air}} = -\frac{2(f^{\text{air}})^2}{\sqrt{\mu_Q \rho_Q}} \frac{\Delta m}{A} = K m^{\text{ads}} \tag{1}$$

where Δf in Hertz is the resonant frequency shift due to the added mass; f stands for the measured frequency in the presence of a gas ; f^{air} symbolizes the fundamental resonant frequency of the unloaded device is typically 5 or 10MHz; Δm (or m^{ads}) in g is the added mass (or the mass of adsorbed gas); A in cm^2 is the piezo-electrically active area (area of electrode surface); μ_Q in g cm^{-2} is the shear modulus of quartz ; ρ_Q in g cm^{-3} is the density of quartz. For a given sensor, the parameters fair, A, μ_Q , ρ_Q are constant , therefore they are summarized together with f^{air} in one constant K.

2.2 Interaction between a single QMB- sensor and a mixture of vapors

If we consider the interactions between a single QMB- sensor and a mixture of q-vapors; the frequency of the sensor can be expressed as follows

$$f^{*} = f^{\text{air}} + (K\sum_{j=1}^{q} m_{j}^{\text{ads}}) = f^{\text{air}} + (K\sum_{j=1}^{q}(M_{j}\ n_{j}^{\text{ads}})) \tag{2}$$

where f^* is the frequency of the single sensor , when it interacts with the gas mixture ; m_j^{ads} stands for the adsorbed mass of the jth vapor; n_j^{ads} in mole is the number of adsorption sites that have been occupied by the jth component ; M_j in g mole^{-1} is the molar mass of the jth component of the gas mixture. Most QMB- sensors use polymers (or conducting polymers) as sensitive layer. One of defining features of polymer surfaces and coatings is the chemical regularity of the surface. While many materials can be irregular mixtures of different components, the polymer surfaces tend to be chemically uniform; because of this, adsorption of molecules onto polymer surfaces can be easily modeled by the Langmuir isotherm. If we make the assumptions that there is only one type of active and immobile site, for which all molecules actively compete (competitive adsorption) and that the concentrations of the gas mixture components are small, the number of occupied adsorption sites of the gas mixture component j may be described by the well-known multi-components Langmuir- isotherm [7-8]; by using this model for all gas mixture components , we get

$$f^{*} = f^{\text{air}} + (K\sum_{j=1}^{q}(M_{j}\ n_{j}^{\max}\frac{b_{j}\gamma_{j}n_{j}}{1+(\sum_{j=1}^{q}b_{j}\gamma_{j}n_{j})})) \tag{3}$$

where $n_j^{\max}$ in mole is the maximal number of adsorption sites that can be occupied by the jth gas mixture component; n_j is the initial number of mole of the gas j , γ_j stands for the ratio n_j^{eq}/n_j ; n_j^{eq} stands for the mole number of the free particles of the gas j at thermodynamic equilibrium; b_j is a dimensionless parameter; it describes the thermodynamics at the sensor / gas j interface.

2.3Interactions between a sensor array and a mixture of vapors

We consider a system with m sensors and q different chemical species. The frequency f of the multi-sensor can be represented by a composite state function; namely, it depends on the frequency values of the individual sensors, which in turn depend on the gas mixture component concentrations and temperature. A small variation df of the frequency of the sensor array may be determined with the chain rule. Furthermore, the response of each chemical sensor must be a state function; this is an essential precondition for the operational and practically useful sensors .Under this condition df_i^* is a total differential; it can be expressed as follows

$$\mathrm{d}f = \sum_{i=1}^{m}\mathrm{d}f_{i} = \sum_{i=1}^{m}\frac{\partial f}{\partial f_{i}^{*}}\mathrm{d}f_{i}^{*} = \sum_{i=1}^{m}\sum_{j=1}^{q}\lambda_{i}\,S^{f_{ij}^{*},n_{j}}\,\mathrm{d}n_{j} \tag{4}$$

where f_i stands for the frequency of ith sensor array element ; f_i^* is the frequency of the single sensor i, when it interacts with the gas mixture; $\lambda_i = df/df_i^*$ is a dimensionless coefficient that describes the influence of the sensitivities of the sensor array elements and the influence of the gas mixture composition on the response of the sensor array; f_{ij}^* is the contribution of gas mixture component j in the frequency of the single sensor i; $S^{f_{ij}^*,n_j}$ symbolizes the partial sensitivity of the ith sensor array element, when it interacts with the component j of the gas mixture. We assume and show later that the following relationship is valid

$$S^{f_{ij}^*,n_j} = \left(\frac{\partial f_i^*}{\partial n_j}\right)_{n_2,\ldots,n_j,\ldots,n_q} = \frac{df_{ij}^*}{dn_j} = \frac{f_{ij}^*}{n_j} = \frac{f_{ij}^* + df_{ij}^*}{n_j + dn_j} = \text{constante} \quad (5)$$

The indefinite integral of the relationship (4) by making use of the assumption (5) gives

$$f = \int \sum_{i=1}^{m}\sum_{j=1}^{q} \lambda_i S^{f_{ij}^*,n_j} dn_j = \sum_{i=1}^{m}\sum_{j=1}^{q} \lambda_i S^{f_{ij}^*,n_j} \int dn_j = \sum_{i=1}^{m}\sum_{j=1}^{q} \lambda_i S^{f_{ij}^*,n_j} n_j \quad (6)$$

The differentiation of eq.6 using the rules of algebra and the comparison with eq.4, leads to the following differential equation

$$\sum_{i=1}^{m}\sum_{j=1}^{q} \lambda_i n_j dS^{f_{ij}^*,n_j} = 0 \quad (7)$$

The relationship (7) is a Gibbs Duhem like equation ; it connects the change of the partial sensitivities of the sensors with the concentrations of the gas mixture components and the multi-system coefficients λ_i . To find an analytical model for the sensor array response, we consider the total differential in eq.4 and rewrite it by introducing a new dimensionless coefficient β_i , we get

$$df = \sum_{i=1}^{m} df_i = \sum_{i=1}^{m} \lambda_i df_i^* = \sum_{i=1}^{m}\sum_{j=1}^{q} \lambda_i d(f_i^{air} + f_{ij}^*) = \sum_{i=1}^{m} d(f_i^*)^{\beta_i} \quad (8)$$

where β_i is a dimensionless coefficient. It is given by

$$\beta_i = \frac{f_i^*}{f_i}\lambda_i = \frac{d\ln f_i}{d\ln f_i^*} \quad (9)$$

The indefinite integral of (8) leads to the relationship

$$f = \sum_{i=1}^{m} (f_i^*)^{\beta_i} = \sum_{i=1}^{m}\sum_{j=1}^{q} (f_i^{air} + (S^{f_{ij}^*,n_j} n_j))^{\beta_i} = \left(f^{air} + (K\sum_{j=1}^{q} M_j n_j^{max} \frac{b_j \gamma_j n_j}{1 + (\sum_{j=1}^{q} b_j \gamma_j n_j)}) \right)^{\beta_i} \quad (10)$$

which represents the frequency of the sensor array as a function of the initial concentrations of the gas mixture components.

3.Results and discussion

The differential equations (7) and the function (10) were developed using the properties of functions of state. Therefore, they apply to all cross-reactive chemical sensors .In this paper, we defined a novel dimensionless coefficient λ_i. This variable has a similar meaning as the fugacity coefficient when we consider the interactions between the particles of real gas. At macroscopic scale, the assumption (5) is not valid, since the ratio df_{ij}^*/dn_j (slope of the tangent of the graph of the function $n_j \xrightarrow{g} f_{ij}^* = g(n_j)$ at the point (n_j, f_{ij}^*) is unequal f_{ij}^*/n_j (slope of the straight line f_{ij}^* = constant. n_j). The assumption (5) is fulfilled, if we go to small scales over and divide the system (sensor) in small discrete subsystems (sub-sensors). In the integration of f_i^* in eq. 7, one can imagine the sensor divided in an infinite number of discrete sub sensors, and the gas mixture M decomposed into infinitely many equivalent sub mixtures dM. Each sub-sensor interacts independently of the other sub sensors, with a sub mixture, and shows now a constant sensitivity with respect to the components of the sub mixtures. The response of the sensor is then obtained by integrating the responses of the discrete sub-sensors. Although the decomposition of the system in multi-system and the use of the relationships (5) are virtual and cannot be realized experimentally, these models are necessary and helpful for the development of our differential equations.

4.Conclusion

In this paper, novel differential equations and their solutions are developed. The eq.7 can be useful for checking experimental data for consistency. The function (10) shows that the Clifford model is a special case of a rather general result.

5.References

[1] Hierlemann A , Weimar U, Kraus G, Schweizer-Berberich M and Goepel W 1995 *Sensors & actuators B.* **26-27** 126
[2] Althainz P, Goschnick J, Ehrmann S and Ache H J 1996 *Sensors & Actuators. B* **33** 72
[3] Abbas A and Bouabdallah A 2007 *Journal of Mathematical Chemistry.* **42 (4)** 1023
[4] Abbas A and Bouabdellah A 2010 *Sensors &. Actuator . B* **145** 620
[5] Araya-Kleinsteuber B and Lowe C R 2007 *Handbook of Biosensors and Biochips* (Weinheim :J. Wiley & Sons , pp 1-21.
[6] Khot L R, Panigrahi S and Lin D 2011 *Sensors & Actuators. B* **153** 1
[7] Zhou X, Yi H, Tang X, Deng H and Liu H 2012 *Chemical Engineering Journal.* **200–202** 399
[8] Denayer J F, Baron G V, Jacobs P A, Martens J A 2000 *Phys. Chem. Chem. Phys.* **2** 1007

MR image intensity inhomogeneity correction

Mirela (Vişan) Pungă[1,2], Simona Moldovanu[1,3] and Luminita Moraru[1]
[1]Faculty of Sciences and Environment, Department of Chemistry, Physics and Environment, Dunărea de Jos University of Galaţi, 47 Domnească St., 800008, Romania
[2]Aurel Vlaicu High School, 1 Decembrie 1918 St., 800511, Galaţi, Romania
[3]Dumitru Motoc High School, 15 Milcov St., 800509, Galaţi, Romania

E-mail: moraru.luminita@ugal.ro

Abstract. MR technology is one of the best and most reliable ways of studying the brain. Its main drawback is the so-called intensity inhomogeneity or bias field which impairs the visual inspection and the medical proceedings for diagnosis and strongly affects the quantitative image analysis. Noise is yet another artifact in medical images. In order to accurately and effectively restore the original signal, reference is hereof made to filtering, bias correction and quantitative analysis of correction. In this report, two denoising algorithms are used; (i) Basis rotation fields of experts (BRFoE) and (ii) Anisotropic Diffusion (when Gaussian noise, the Perona-Malik and Tukey's biweight functions and the standard deviation of the noise of the input image are considered).

Keywords: MR images; Basis rotation fields of experts; Anisotropic Diffusion; bias field correction; image quality assessment

1. Introduction

Generally, pertaining to neurological research, MR technology is one of the best and most reliable ways of studying the brain. Various methods and procedures for the quantitative study of different brain pathologies were developed since the MRI technique has been elaborated [1-16]. In spite of the huge progress recorded by the means of this technique, different artifacts such as noise, poor homogeneity in the radiofrequency, and intensity inhomogeneity or bias field may be present in the images. These artifacts are corrected by using various mathematical algorithms which mainly focus on noise removal, inhomogeneity correction, contrast enhancement or edge emphasizing [1-7].

The aim of this paper is that of providing an accurate insight into the effective restoration methods of the original signal. Statistical method, diffusion method and intensity inhomogeneity or bias field correction method are used. In addition, the quality of the processed images is also an important factor that is being taken into consideration. This work particularly focuses on brain MR images, although such artifacts appear in all MRI techniques.

[1] Luminiţa Moraru

2. Methology

2.1. Experimental dataset

The data sets used in this study are freely available for download from 'The whole brain atlas' database of the Harvard Medical School's Whole Brain Atlas. The MR images measure (256 × 256). The total number of the images selected in our database was 42. The following diseases were investigated: cerebrovascular diseases/fatal stroke; carcinoma; Creutzfeld-Jacob disease and Alzheimer disease (figure 1). (http://www.med.harvard.edu/AANLIB/home.html)

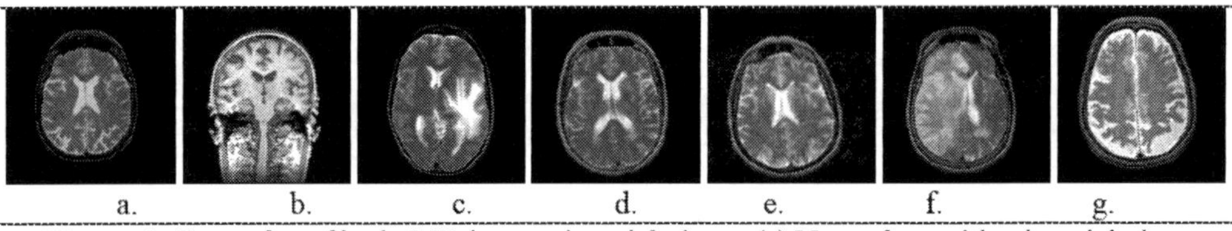

Figure 1. Examples of brain MR images in axial views. (a) Normal cognitive in axial view (NCAx); (b) normal cognitive in sagittal view (NCSag); (c) carcinoma (Car); (d) acute stroke (ASk); (e) Creutzfeld-Jacob disease (CJ); (f) fatal stroke (FSk); (g) Alzheimer's disease (AzD).

The experimental methodology was developed by using the Matlab ver. R2010a software (The Mathworks Inc., Natick, MA).

2.2 Anisotropic diffusion

The concept of anisotropic diffusion (AD) has been introduced by Perona and Malik [8] and it refers to the restoration of the images against the destruction of the features of the images due to noise artifacts. The AD model uses the image intensity as a 'concentration' and noise as a 'little concentration inhomogeneity'. The conductance function, the gradient threshold parameter and the stopping time of the iterative process are the parameters that govern the anisotropic diffusion process and they were presented in [8]. The following conductance functions are used: the Gaussian filter and the results of the filtration operation are referred to as AD1; $g_1 = exp\left[-\left(\frac{x}{K\sqrt{2}}\right)^2\right]$ denotes the Perona and Malik conductance function [9]. K is the gradient magnitude threshold parameter that controls the rate of the diffusion. The results of this filtration algorithm are referred to as AD2. The second conductance function used in our study is Tukey's biweight function [10] $g_2(x) = \begin{cases} 0.67\left[1-\left(\frac{x}{S}\right)^2\right]^2, & x \le S \\ 0, & otherwise \end{cases}$ where the gradient threshold parameter is chosen as $S = K\sqrt{5}$. S defines the threshold between the image gradients corresponding to the noise and those attributed to the true edges. The experimental results of this filter are referred to as AD3. In [9], an estimation of the standard deviation σ of the Gaussian noise of the original image as a stopping time parameter was used. Two cases were analyzed: the standard deviation of the noise of the input image is known a-priori (referred to here as AD4) and the standard deviation of the noise is estimated automatically (referred to here as AD5).

2.3 Basis rotation fields of experts

The Basis Rotation algorithm is part of Field of Experts models (BRFoE) was set up in [11] in order to improve the maximum likelihood filters performance in denoising task. They claimed that BRFoF filter shows a very good denoising performance. The optimization of the filters is done by using the principle of maximum likelihood. The main improvement brought by the BRFoE is broader in frequency.

2.4. Bias field correction

The intensity inhomogeneity or bias field in MR images represents a smooth variation of signal intensity across the image. In [12] the intensity inhomogeneities are derived and estimated by using the local intensity clustering property. It allows an energy functional to be defined, which is integrated in a level set formulation. In a two-phase level set formulation, intended for the reading of brain MR images, the intensities within each brain tissue become quite homogeneous in the bias corrected images. For a precise evaluation and comparison of the algorithms the following objective quality metrics are used: the mean squared error (MSE), the peak signal to noise ratio (PSNR), the average difference (Av Diff) and normalized absolute error (NAE). As correlation-based measurements the normalized correlation (NC) and the structural content (SC) are considered [13-15].

3. Results and discussions

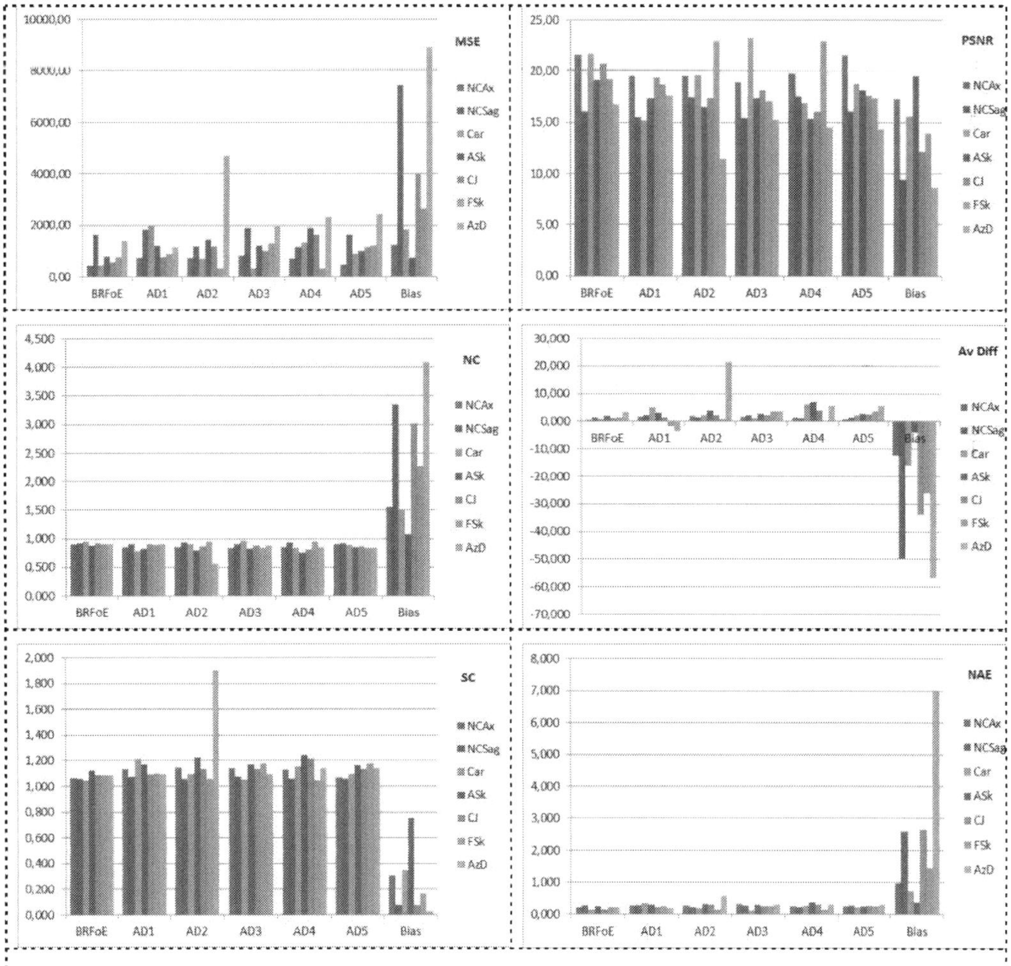

Figure 3. Objective Quality Measures for different algorithms and various brain MR images

Figures 1 to 6 comparatively show the computed quality metrics. It can be observed that, the BRFoE algorithm operates better in the case of noise removal for all MR analyzed images. In the case of the anisotropic diffusion algorithm, Tukey's biweight function (or AD3) and the standard deviation of the noise automatically estimated as $\sigma = 0.002$ (or AD5) are the best de-noising methods. The metrics that were most sensitive to the bias field and to image set variation were NC and SC.

Regarding the image type, the Alzheimer disease is the most responsive to the current analysis as it strongly correlates these metrics but in a negative way. These images are strongly affected by noise and their structural content is affected by de-noising and bias field correction operations.

The case of bias correction is a special one. The bias field is quite smooth and most parts cover the anatomical structure. The bias removal was conducted under the regularization of the level set formulation. This algorithm performs well in the case of for normal cognitive in axial view, carcinoma and acute stroke MR images. During the bias field correction a part of the pixels were removed from the bias corrected images. The pixel distribution value was checked and it was found that the range of the intensity of the gray level became narrower because some pixels, especially in the range from 190 to 255 were removed. This clearly affects the objective quality measures. Thus, NAE is ranging from 0.3606 for the acute stroke to 6.977 for Alzheimer disease. Similar data has been also found for NC (as a global measure, which compares the total weight of the corrected image and original image) and SC (that provides the closeness between the original and decoded image) measurements.

4. Conclusions

The experimental results indicate that the objective quality measurements such as MSE, PSNR and NC are close to the optimum values. Special attention was paid to the bias correction technique that shows a very high variation due to the number of pixels variation performed by the algorithm.

Acknowledgements

The author Mirela (Visan) Punga's work has been funded by the Sectoral Operational Programme Human Resources Development 2007-2013 of the Ministry of European Funds through the Financial Agreement POSDRU/159/1.5/S/132397.
The author Simona Moldovanu would like to thank the Project PERFORM, ID POSDRU/159/1.5/S/138963 of "Dunărea de Jos" University of Galati, Romania for support.

References

[1] Birkholz H 2011 *J. Comput. Appl. Math.* **235** 2502
[2] Luo J Y, Zhu C P and Magnin I E 2005 *IEEE Trans. Med. Imag.* **24** 1067
[3] Studholme C, Cardenas V, Song E, Ezekiel F, Maudsley A and Weiner M 2004 *IEEE Trans. Med. Imag.* **23** 99
[4] Styner M, Brechbühler C, Székely G and Gerig G 2000 *IEEE Trans. Med. Imag.* **19** 153
[5] Vovk U, Pernus F and Likar B 2007 *IEEE Trans. Med. Imag.* **26** 405
[6] Belaroussi B, Milles J, Carme S, Zhu YM and Benoit-Cattin HB 2006 *Med. Image Anal.* **10** 234
[7] Li C, Huang R, Ding Z, Gatenby C, Metaxas D, Gore J and D. Metaxas 2008 *Med Image Comput. Comput. Assist. Interv.* 2008 **5242 11** 1083
[8] Perona P and Malik J 1990 *IEEE Trans. Pattern Anal. Mach. Intell.* **12** 629
[9] Tsiotsios C and Petrou M 2013 *Pattern Recognit.* **46** 1369
[10] Black M J, Sapiro G, Marimont D, Heeger D 1998 *IEEE Trans. Image Process.* **7** 421
[11] Weiss Y and Freeman WT 2007 *Int. Conf.IEEE Computer Society (Minneapolis:MN/IEEE* p 1
[12] Chunming L, Huang R, Ding Z, Gatenby C J and Metaxas D N, Gore J C 2011 *IEEE Trans. Image Process.* **20** 2007
[13] Wang Z, Bovik A C, Sheikh H R and Simoncelli E P 2004 *IEEE Trans. Image Process.* **13** 600
[14] Sakuldee R and Udomhunsakul S 2007 *PWASET* **26** 434
[15] Poobal S, Ravindran G 2011 *IJEST* **3** 525

On the Representation Theory of the Ultrahyperbolic BMS group $\mathcal{UHB}(2,2)$. I. General Results

Evangelos Melas
Technological Educational Institution of Patras, Department of Management, Greece
E-mail: evangelosmelas@yahoo.co.uk

Abstract. The Bondi–Metzner–Sachs (BMS) group B is the common asymptotic group of all asymptotically flat (lorentzian) space–times, and is the best candidate for the universal symmetry group of General Relativity (G.R.). B admits generalizations to real space–times of any signature, to complex space–times, and supersymmetric generalizations for any space–time dimension. With this motivation McCarthy constructed the strongly continuous unitary irreducible representations (IRs) of B some time ago, and he identified $B(2,2)$ as the generalization of B appropriate to the to the 'ultrahyperbolic signature' $(+,+,-,-)$ and asymptotic flatness in null directions. We continue this programme by introducing a new group $\mathcal{UHB}(2,2)$ in the group theoretical study of ultrahyperbolic G.R. which happens to be a proper subgroup of $B(2,2)$. In this short paper we report on the first general results on the representation theory of $\mathcal{UHB}(2,2)$. In particular the main general results are that the *all* little groups of $\mathcal{UHB}(2,2)$ are compact and that the Wigner–Mackey's inducing construction is exhaustive despite the fact that $\mathcal{UHB}(2,2)$ is not locally compact in the employed Hilbert topology. At the end of the paper we comment on the significance of these results.

1. Introduction

The best candidate for the universal symmetry group of General Relativity (G.R), in any signature, is the so called Bondi–Metzner–Sachs (BMS) group B. These groups have been described [1] for all possible signatures and all possible complex versions of GR as well. The original BMS group B was discovered by Bondi, Metzner and Van der Burg [2] for asymptotically flat space–times which were axisymmetric, and by Sachs [3] for general asymptotically flat space–times, in the usual Lorentzian signature.

In earlier papers [1, 4, 5, 6, 7] it has been argued that the IRs of the BMS group and of its generalizations in complex space–times as well as in space–times with Euclidean or Ultrahyperbolic signature are what really lie behind the full description of (unconstrained) moduli spaces of gravitational instantons. Kronheimer [8, 9] has given a description of these instanton moduli spaces for *Euclidean* instantons. However, his description only partially describes the moduli spaces, since it still involves *constraints*. Kronheimer does not solve the constraint equations, but it has been argued [1, 7] that IRs of BMS group (in the relevant signature) give an *unconstrained* description of these same moduli spaces.

A terminological remark is in order. In an arbitrary dimension, an ultrahyperbolic metric is a metric with signature (s,t) or (t,s) with $s \geq t$ and for which $t \neq 0, 1$. The cases of $t = 0, 1$ are called Riemannian and Lorentzian (or hyperbolic) metrics respectively. In dimension four, there remains only the case $(2, 2)$. McCarthy in [1] introduced $B(2,2)$ as the BMS group appropriate to the 'ultrahyperbolic signature' and asymptotic flatness in null directions. The representation theory of $B(2,2)$ was initiated in [4] and [10]. In the present paper we introduce a new group $\mathcal{UHB}(2,2)$, unnoticed in [1], in the group theoretical study of Ultrahyperbolic G.R., and report some general results on its representation theory derived in [11]. We also mention in passing some more results on the representation theory of $\mathcal{UHB}(2,2)$ derived in [12] and [13].

The fact that $\mathcal{UHB}(2,2)$ is a proper subgroup of $B(2,2)$ does not render the study of the representations of $\mathcal{UHB}(2,2)$ superfluous. In general, the following holds: Let G be a group, H be a subgroup and $T(g)$ be an irreducible representation of the group G. Let $T_H(g)$ be the restriction of $T(g)$ on the subgroup H. In general, the representation $T_H(g)$ is *not* irreducible. Moreover, there are irreducibles of H which *cannot* be extended to the whole group G. An example of this phenomenon is actually provided by the original BMS group. The Poincare group P is a subgroup of the BMS group B. Not all irreducibles of P are obtained by restricting the irreducibles of B to P (here, it is understood that both B and P are endowed with the Hilbert topology). There are irreducibles of P, namely the continuous spin irreducible representations, which *cannot* be extended to the whole group B, which in the Hilbert topology has no continuous spin irreducibles, all its irreducibles carry discrete spin.

In Section 2 we define the group $\mathcal{UHB}(2,2)$ and sketch the inducing construction which we use in order to construct the IRs of $\mathcal{UHB}(2,2)$. In Section 3 we state the main general results on the representation theory of $\mathcal{UHB}(2,2)$ and comment on their significance.

2. The group $\mathcal{UHB}(2,2)$ and the inducing construction

Recall that the ultrahyperbolic version of Minkowski space is the vector space R^4 of row vectors with 4 real components, with scalar product defined as follows. Let $x, y \in R^4$ have components x^μ and y^μ respectively, where $\mu = 0, 1, 2, 3$. Define the scalar product $x.y$ between x and y by

$$x.y = x^0y^0 + x^2y^2 - x^1y^1 - x^3y^3. \tag{1}$$

Then the ultrahyperbolic version of Minkowski space, sometimes written $R^{2,2}$, is just R^4 with this scalar product.

In [11] it is shown that

Theorem 1 *The group $\mathcal{UHB}(2,2)$ can be realised as*

$$\mathcal{UHB}(2,2) = L^2(\mathcal{P}, \lambda, R) \circledS_T G^2 \tag{2}$$

with semi–direct product specified by

$$(T(g,h)\alpha)(x,y) = k_g(x)s_g(x)k_h(w)s_h(w)\alpha(xg, yh), \tag{3}$$

where $\alpha \in L^2(\mathcal{P}, \lambda, R)$ and $(x, y) \in \mathcal{P}$. For ease of notation, we write $\mathcal{P}$ for the torus $\mathrm{T} \simeq P_1(R) \times P_1(R)$, $P_1(R)$ is the one–dimensional real projective space, and $\mathcal{G}$ for $G \times G$, $G = SL(2, R)$. In analogy to B, it is natural to choose a measure λ on $\mathcal{P}$ which is invariant under the maximal compact subgroup $\mathrm{SO}(2) \times \mathrm{SO}(2)$ of $\mathcal{G}$. $L^2(\mathcal{P}, \lambda, R)$ is the separable Hilbert space of real–valued functions defined on $\mathcal{P}$.

Moreover, if $g \in G$ is

$$\begin{bmatrix} a & b \\ c & d \end{bmatrix}, \tag{4}$$

then the components x_1, x_2 of $\mathbf{x} \in R^2$ transform linearly, so that the ratio $x = x_1/x_2$ transforms fraction linearly. Writing xg for the transformed ratio,

$$xg = \frac{(\mathbf{x}g)_1}{(\mathbf{x}g)_2} = \frac{x_1a + x_2c}{x_1b + x_2d} = \frac{xa + c}{xb + d}. \tag{5}$$

The factors $k_g(x)$ and $s_g(x)$ on the right hand side of (3) are defined by

$$k_g(x) = \left\{ \frac{(xb+d)^2 + (xa+c)^2}{1+x^2} \right\}^{\frac{1}{2}}, \tag{6}$$

$$s_g(x) = \frac{xb+d}{|xb+d|}, \tag{7}$$

with similar formulae for yh, $k_h(y)$ and $s_h(y)$.

It is well known that the topological dual of a Hilbert space can be identified with the Hilbert space itself, so that we have $L^{2'}(\mathcal{P}, \lambda, R) \simeq L^2(\mathcal{P}, \lambda, R)$. In fact, given a continuous linear functional $\phi \in L^{2'}(\mathcal{P}, \lambda, R)$, we can write, for $\alpha \in L^2(\mathcal{P}, \lambda, R)$

$$(\phi, \alpha) = <\phi, \alpha> \tag{8}$$

where the function $\phi \in L^2(\mathcal{P}, \lambda, R)$ on the right is uniquely determined by (and denoted by the same symbol as) the linear functional $\phi \in L^{2'}(\mathcal{P}, \lambda, R)$ on the left. The representation theory of $\mathcal{UHB}(2,2)$ is governed by the dual action T' of $\mathcal{G}$ on the topological dual $L^{2'}(\mathcal{P}, \lambda, R)$ of $L^2(\mathcal{P}, \lambda, R)$. The dual action T' is defined by:

$$< T'(g,h)\phi, \alpha > = < \phi, T(g^{-1}, h^{-1})\alpha > \cdot \tag{9}$$

A short calculation gives

$$(T'(g,h)\phi)(x,y) = k_g^{-3}(x) s_g(x) k_h^{-3}(y) s_h(y) \phi(xg, yh). \tag{10}$$

Now, this action T' of $\mathcal{G}$ on $L^{2'}(\mathcal{P}, \lambda, R)$, given explicitly above, is like the action T of $\mathcal{G}$ on $L^2(\mathcal{P}, \lambda, R)$, continuous. The 'little group' L_ϕ of any $\phi \in L^{2'}(\mathcal{P}, \lambda, R)$ is the stabilizer

$$L_\phi = \{(g,h) \in \mathcal{G} \mid T'(g,h)\phi = \phi\}. \tag{11}$$

By continuity, $L_\phi \subset \mathcal{G}$ is a closed subgroup.

In the inducing construction, described in detail in [11], [12] and [13], attention is confined to measures on $L^{2'}(\mathcal{P}, \lambda, R)$ which are concentrated on single orbits of the $\mathcal{G}-$action T'. These measures give rise to IRs of $\mathcal{UHB}(2,2)$ which are induced in a sense generalising [14] Mackey's [15, 16, 17, 18, 19, 20]. Remarkably [11] this inducing construction gives rise to *all* the IRs of $\mathcal{UHB}(2,2)$ when $\mathcal{UHB}(2,2)$ is equipped with the Hilbert topology. In a nutshell the inducing construction is realized as follows: Let $\mathcal{O} \subset L^{2'}(\mathcal{P}, \lambda, R)$ be any orbit of the dual action T' of $\mathcal{G}$ on $L^{2'}(\mathcal{P}, \lambda, R)$. There is a natural homomorphism $\mathcal{O} = \mathcal{G}\phi_o \simeq \mathcal{G}/L_{\phi_o}$ where L_{ϕ_o} is the 'little group' of the point $\phi_o \subset \mathcal{O}$. Let U be a continuous irreducible unitary representation of L_{ϕ_o} on a Hilbert space D. Every coset space $\mathcal{G}/L_{\phi_o}$ can be equipped with a unique class of measures which are quasi−invariant under the action T of G^2. Let μ be any one of these. Let $D_\mu = L^2(\mathcal{G}/L_{\phi_o}, \mu, D)$ be the Hilbert space of functions $f : \mathcal{G}/L_{\phi_o} \to D$ which are square integrable with respect to μ. From a given ϕ_o and any continuous irreducible unitary representation U of L_{ϕ_o} on a Hilbert space D a continuous irreducible unitary representation of $\mathcal{UHB}(2,2)$ on D_μ can be constructed. The representation is said to be induced from U and ϕ_o. Different points of an orbit $\mathcal{G}\phi$ have conjugate little groups and give rise to equivalent representations of $\mathcal{UHB}(2,2)$.

3. Results and their significance

The main general results on the representation theory of $\mathcal{UHB}(2,2)$ proved in [11] are the following

(i) All the little groups L_{ϕ_o} of $\mathcal{UHB}(2,2)$ are compact

(ii) The Wigner–Mackey's inducing construction is exhaustive

Regarding the first result we note that the all the little groups L_{ϕ_o} of $\mathcal{UHB}(2,2)$, being compact, are up to conjugation subgroups of the maximal compact subgroup $\mathrm{SO}(2) \times \mathrm{SO}(2)$ of $\mathcal{G}$. They include groups which are finite as well as groups which are infinite, both connected and not–connected [11]. Therefore the construction of the IRs of $\mathcal{UHB}(2,2)$ involves at the first instance the classification of all the compact subgroups of $\mathrm{SO}(2) \times \mathrm{SO}(2)$. This task, far from being trivial, was undertaken in [12] and [13].

Regarding the second result we note that the Wigner–Mackey's inducing construction is exhaustive despite the fact that $\mathcal{UHB}(2,2)$ is not locally compact in the employed Hilbert topology [11]. This result is rather important because other group theoretical approaches to quantum gravity which invoke Wigner–Mackey's inducing construction (see e.g. [21]) are typically plagued by the non–exhaustiveness of the inducing construction which results precisely from the fact that the group in question is not locally compact in the prescribed topology. Exhaustiveness is not just a mathematical nicety: If the inducing construction is not exhaustive one cannot simply know if the most interesting information or part of it is coded in the irreducibles which cannot be found by the Wigner–Mackey's inducing procedure. These results, compactness of the little groups and exhaustiveness of the inducing construction, not only are they significant for the group theoretical approach to quantum gravity advocated here, but also they have repercussions for the other approaches to quantum gravity [11].

References

[1] P. J. McCarthy: *Phil. Trans. R. Soc. Lond. A* Vol. 338 (1992), p.271
[2] H. Bondi, M.G.J Van Der Berg and A.W.K Metzner: *Proc. R. Soc. Lond. A* Vol. 269 (1962), p.21
[3] R.K.Sachs: *Proc. R. Soc. Lond. A* Vol. 270 (1962), p.103
[4] P. J. McCarthy and E. Melas: *Nucl. Phys. B* Vol. 653 (2003), p.369
[5] E. Melas: *J : Phys : A : Math : Gen.* Vol. (2006), p.3341
[6] E. Melas: *Nucl. Phys. B. (Proc. Suppl.)* Vol. 104 (2002), p.212
[7] E. Melas: Vol. 45/3 (2004), p.996
[8] P. B. Kronheimer: *J. Diffl. Geom.* Vol. 29 (1989), p.665
[9] P. B. Kronheimer: *J. Diffl. Geom.* Vol. 29 (1989), p.685
[10] E. Melas: *J. Phys. A : Math. Gen.* Vol. 39 (2006), p.3341
[11] E. Melas, *Representations of the Ultrahyperbolic BMS group $\mathcal{HB}$. I. General Results* (unpublished).
[12] E. Melas, *Representations of the Ultrahyperbolic BMS group $\mathcal{HB}$. II. Determination of the representations induced from infinite little groups* arXiv:1312.0532
[13] E. Melas, *Representations of the Ultrahyperbolic BMS group $\mathcal{HB}$. III. Determination of the representations induced from finite little groups* arXiv:1402.1428
[14] P. J. McCarthy: *Proc. R. Soc. Lond. A* Vol. 343 (1975) p.489
[15] E. Wigner: *Annals of Mathematics* Vol. 40 (1939) p.149
[16] G. W. Mackey: *Induced representations of groups and quantum mechanics* Benjamin (1968)
[17] G. W. Mackey: *The theory of group representations* Chicago notes (1955)
[18] D. J. Simms: *Lie groups and quantum mechanics* Bonn notes Berlin: Springer (1968)
[19] C. J. Isham: *Topological and quantum aspects of quantum theory* in Dewitt B.S. & Stora R. (Editors), Relativity groups and topology II Elsevier (1984) p.1059
[20] G. W. Mackey: *Unitary group representations in physics, probability and number theory* London Benjamin (1978)
[21] C. J. Isham and A. C. Kakas: *Class. Quantum Grav.* Vol. 1 (1984) p.633

High performance computing and quantum trajectory method in CPU and GPU systems

Joanna Wiśniewska[1], **Marek Sawerwain**[2], **Wiesław Leoński**[3]

[1] Institute of Information Systems, Faculty of Cybernetics, Military University of Technology, Kaliskiego 2, 00-908 Warsaw, Poland
[2] Institute of Control and Computation Engineering, University of Zielona Góra, Licealna 9, 65-417 Zielona Góra, Poland
[3] Quantum Optics and Engineering Division, Institute of Physics, University of Zielona Góra, Licealna 9, Zielona Góra 65-417, Poland

E-mail: `jwisniewska@wat.edu.pl`

Abstract. Nowadays, a dynamic progress in computational techniques allows for development of various methods, which offer significant speed-up of computations, especially those related to the problems of quantum optics and quantum computing. In this work, we propose computational solutions which re-implement the quantum trajectory method (QTM) algorithm in modern parallel computation environments in which multi-core CPUs and modern many-core GPUs can be used. In consequence, new computational routines are developed in more effective way than those applied in other commonly used packages, such as Quantum Optics Toolbox (QOT) for Matlab or QuTIP for Python.

1. Introduction

The Quantum Trajectory Method (QTM) is still very popular method applied for investigation in the field of open quantum systems. However, its application requires a special attention in its reimplementation in modern hardware systems, especially those built with use of GPUs and modern multi-core CPUs.

The proposed implementation of QTM is also a hybrid method, in which power of traditional CPU and modern GPU is combined to obtain more effective routines applicable in simulations of open quantum systems' dynamics. It should be pointed out that discussed here solution which is based on GPU, is several times faster than other implementations of QTM prepared for traditional serial CPU systems. The proposed implementation allows for considerable reduction of calculation time and this speed-up depends on numerical method chosen for solving ordinary differential equations (ODE) appearing in considered problem. We will shortly discuss the fourth order Runge-Kutta method (RK4) and the fourth order backward differentiation formula (BDF4) which can be simultaneously applied in proposed method.

2. Quantum Trajectory Method

To describe the system's evolution with use of QTM we need an effective Hamiltonian H_{eff} defined with use of collapse operators C_n and system's Hamiltonian $H_{\text{eff}} = H_{\text{sys}} - \frac{i\hbar}{2}\sum_n C_n^+ C_n$.

The operator C_n allows to determine the probability of appearance of so-called *quantum jump*. Such jump is modelled by the collapse operator acting on a current quantum state of the system, whereas the probability can be defined as: $\delta p = \delta t \sum_n \left\langle \psi(t)|C_n^\dagger C_n|\psi(t)\right\rangle$.

If such quantum jump occurs, the system's state for the moment of time $(t + \delta t)$ just after the collapse operation can be written as:

$$|\psi(t+\delta t)\rangle = C_n\,|\psi(t)\rangle / \sqrt{\left\langle \psi(t)|C_n^\dagger C_n|\psi(t)\right\rangle}. \tag{1}$$

Moreover, if more than one collapse operator are applied in considered model, the probability of using i-th operator can be written as:

$$P_i(t) = \left\langle \psi(t)|C_i^\dagger C_i|\psi(t)\right\rangle / \delta p. \tag{2}$$

To ensure the probabilistic choice of adequate application of collapse operator in the process of simulation, a proper random numbers generator should be applied. Moreover all considered calculations correspond to operations performed on matrices and vectors. Applied matrices are usually band matrices, so they may be treated as sparse matrices to economize memory usage and to speed-up the calculations. In consequence, the compressed sparse row format will be used because of many matrix-vector multiplication – it also gives an additional speed-up.

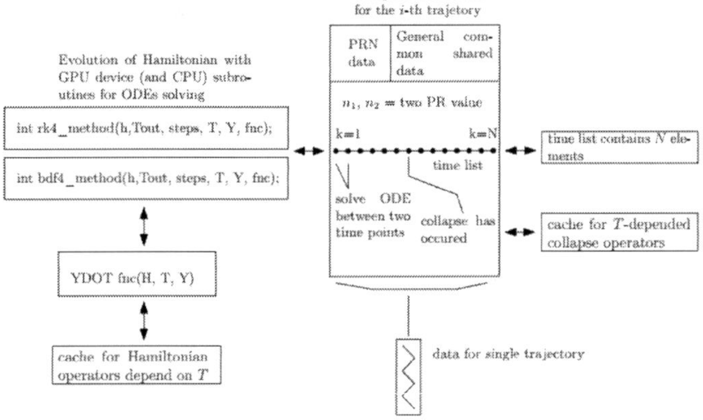

Figure 1. The general scheme of generation of a single quantum trajectory with GPU or CPU computation system (PRN – pseudo-random numbers; H – Hamiltonian data; RK4, BDF4 – methods for ODEs solving; T – time variable; Y – actual state of the system)

The above essential remarks constitute the idea of a following algorithm for a single quantum trajectory simulation which can be presented in a form of four computational steps:

(A) a random value $r \in (0,1)$ is computed, where r denotes the probability of quantum jump occurrence,

(B) the Schrödinger equation is integrated with application of the Hamiltonian H_{eff} to get state's vector at for the time t in such way to make the state's vector norm equal or greater to r: $\langle\psi(t)|\psi(t)\rangle \geq r$,

(C) a quantum jump occurrence causes the system's state projection at the moment t, to one of the states given by the equation (1). The operator C_n is selected to meet the following relation for adequate n: $\sum_{i=1}^{n} P_i(t) \geq r$ and $P_i(t)$ is specified by the equation (2),

(D) the projected state of obtained wave-function is a new initial value corresponding to the moment of time t; next, the new value of r is randomly selected and then, the procedure repeats the process of quantum trajectory simulation starting from the step (B) – more precisely: the simulation is performed again but starts from previously given value of t.

The presented approach refers to solutions described in the other publications [1, 2].

3. Some implementation notes and performance of numerical experiments

To implement QTM, the procedures defining rseudo-random numbers generator (PRNG) and some methods for solving ODEs are required. The presented solution uses the PRNG procedure defined in cuRand package [5] which is a part of CUDA Toolkit. Solving of ODEs will be realized by numerical methods for solving systems of differential equations – two methods will be used: RK4 and BDF4. Both methods are implemented in two versions – for CPUs and GPUs systems.

After performing a single QTM algorithm only one trajectory is obtained which is not sufficient to describe the state of quantum system. The QTM is a Monte Carlo method, so it necessary to generate more trajectories. Such trajectories have to be averaged to obtain the final trajectory which describes the dynamics of analysed system. Since the trajectories can be calculated without any correlations to one another with use of CPUs and GPUs, as well, this tactics leads to increasing the accuracy of calculations and reduction of computation time.

To ensure the high efficiency of numerical procedures, C++ and CUDA C++ programming languages were applied. Moreover, to simplify the use of proposed toolbox, special Python package was prepared. As an example, below we present a fragment of a script which allows for simulation of quantum system's dynamics with use of QTM (e.g. for 100 trajectories):

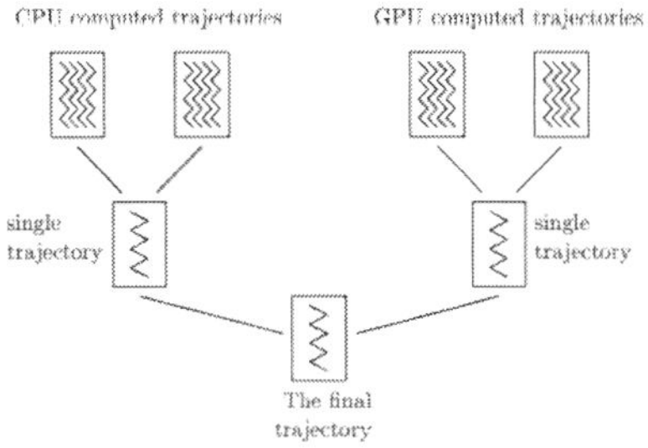

Figure 2. Generating and averaging trajectories with use of CPUs and GPUs. Moreover, using GPUs for averaging trajectories reduces the time of calculations.

```
import qcs
import matplotlib

N0=N1=N2=6
A0=qcs.TensorProduct(qcs.DestroyOperator(N0),qcs.Eye(N1),qcs.Eye(N2))
...
T = qcs.linspace(0, 10, 100) ; Psi0 = # initial state
H = 1.0j*(A0*a1.Adj()*A2.Adj()-A0.Adj()*A1*A2)
E0 = # expectation operator
C01 = # collapse operator 1 ; C02 = # collapse operator 2
expectdata = qcs.QuantumTrj(Psi0, T, "const", H, 100, E0, [C01, C02])
matplotlib.plot(T, expectdata)
```

Although the Hamiltonian H was assumed to be time-independent (the parameter `const-hamiltonian` was declared), we can define and apply time-dependent Hamiltonians, as well (`timedependent-hamiltonian`).

To compare the efficiency of proposed procedures with those already implemented in QuTIP package, we have performed simulations of time-evolution of optical parametric amplifier, described by the following trilinear Hamiltonian [4]:

$$\hat{H} = K(\hat{a}\hat{b}^{\dagger}\hat{c}^{\dagger} - \hat{a}^{\dagger}\hat{b}\hat{c}). \tag{3}$$

The operators $\hat{a}$, $\hat{b}$ and $\hat{c}$ are usual boson annihilation operators and correspond to the pump, signal and idler fields, respectively, whereas K denotes coupling constant. To perform simulations we assume that $K = 1$ and the initial state of considered system is $|\psi_0\rangle = |\alpha\rangle_a|0\rangle_b|0\rangle_c$ (coherent state for the pump mode (a) and vacuum states for signal (b) and idler (c) modes). Thus, Tab.(1) shows the computation times for simulations performed with use of QuTIP package, and our procedures using RK4 and BDF4 methods, for $(6 \times 6 \times 6)$-dimensional initial state. We see that if we use RK4 method, our procedures are $\sim$ 10 times faster then those implemented in QuTIP. For BDF4 method the speed-up is naturally smaller, but our package is still $\sim$ 3.5 times faster than QuTIP. Tab.(1) also shows the calculations' efficiency for time-dependent Hamiltonian:

$$H = H_0 - f(t) \cdot H_1 \quad \text{where} \quad f(t) = \alpha \cdot \exp\left(-(t/\sigma)^2\right), \tag{4}$$

Journal of Physics: Conference Series **574** (2015) 012127 doi:10.1088/1742-6596/574/1/012127

Table 1. The duration of simulations (in seconds) for exemplary Hamiltonians. The data refer to simulations with use of RK4 and BDF4 methods implemented in our toolbox (parallel CUDA technology NVIDIA graphic adapter Geforce 680 was used). Results are compared to those obtained with use of QuTIP package (computing system with Intel Core i7 950 3.00 Ghz with only one core was used – using two cores for given number of trajectories will shorten the simulation's time twice).

No. of trj.	QuTIP	GPU 680 RK4	GPU BDF4 680
Timings for Hamiltonian given by eq. (3)			
100	5.57	1.10	5.24
200	10.99	2.32	7.13
500	27.56	3.20	9.48
Timings for Hamiltonian given by eq. (4)			
100	12.33	1.42	7.91
200	37.02	3.02	11.10
500	60.08	3.85	13.22

where H_0 represents a constant part of the Hamiltonian and H_1 is time-dependent. The Hamiltonian H is defined with use of function $f(t)$ where α and σ are scalar values. In the presented solution the time-dependent Hamiltonian is calculated before the trajectories are computed for the points given by time variable – such approach speeds-up the calculations of trajectories (the missing Hamiltonians are computed during the simulation and built-in mechanism of caching of the Hamiltonians avoids recalculating previously obtained H). The duration of simulations, showed in Tab.(1), convict that application of GPUs and multi-core systems allows calculating more trajectories in shorter time. Furthermore, increasing the number of trajectories do not results in significant increase of computations' time as it was for the case of CPU systems. It should also be noticed that when time-dependent Hamiltonian was applied, the time of simulation with use of QuTIP package last about 50% longer.

4. Conclusions

We presented solution that allows to reduce the time of calculations as compared with other known toolkits. One should remember that discussed exemplary problem was represented by the Hamiltonian described by matrices with leading dimension $6 \times 6 \times 6$. For simpler unitary Hamiltonians which matrix representations are sized: 5×5, 15×15, GPU based computations can be performed faster than in CPU systems: twenty, fifty times, respectively. In cases of bigger quantum systems such speed-up will be three or even five times greater. Moreover, additional integration of presented solution with Python environment allows for more efficient application of our routines to simulate QTM, e.g. we can combine these routines with QuTIP package or use them in Matlab environment together with QOT package.

References

[1] J R Johansson P D Nation F Nori 2013 Comp. Phys. Comm. **184**(4) 234
[2] B M Garraway P L Knight 1994 Phys. Rev. A **50**, 25482563
[3] Sze M Tan 1999 J. Opt. B: Quantum Semiclass. Opt. **1** 424
[4] F Dell'Anno S De Siena F. Illuminati 2006 Phys. Rep. **428** 53
[5] NVIDIA, 2014, CURAND Toolkit Documentation. `http://docs.nvidia.com/cuda/curand/index.html`

Application of nonlinear methods to the study of ionospheric plasma

A.A. Chernyshov[1], **M.M. Mogilevsky**[1] **and B.V. Kozelov**[2]

[1] Space Research Institute of Russian Academy of Sciences, 84/32 Profsoyuznaya, 117997, Moscow, Russia

[2] Polar Geophysical Institute of Russian Academy of Sciences, Akademgorodok 26a, 184209, Apatity, Murmansk region, Russia

E-mail: achernyshov@iki.rssi.ru

Abstract. Most of the processes taking place in the auroral region of Earth's ionosphere are reflected in a variety of dynamic forms of the aurora borealis. In order to study these processes it is necessary to consider temporary and spatial variations of the characteristics of ionospheric plasma. Most traditional methods of classical physics are applicable mainly for stationary or quasi-stationary phenomena, but dynamic regimes, transients, fluctuations, self-similar scaling could be considered using the methods of nonlinear dynamics. Special interest is the development of the methods for describing the spatial structure and the temporal dynamics of auroral ionosphere based on the ideas of percolation theory and fractal geometry. The fractal characteristics (the Hausdorff fractal dimension and the index of connectivity) of Hall and Pedersen conductivities are used to the description of fractal patterns in the ionosphere. To obtain the self-consistent estimates of the parameters the Hausdorff fractal dimension and the index of connectivity in the auroral zone, an additional relation describing universal behavior of the fractal geometry of percolation at the critical threshold is applied. Also, it is shown that Tsallis statistics can be used to study auroral ionosphere

1. Introduction

Dynamic modes, transients, fluctuations, self-similarity in open dissipative systems which include the magnetosphere-ionosphere system may be considered by the methods of nonlinear dynamics. Therefore, the development and application of investigation methods of nonlinear processes characteristics in such systems is now an actual task, particularly for the study of auroral structures of the Earth's ionosphere. Special interest is the development of methods for describing the spatial structure and temporal dynamics of the auroral zone based on the concepts of percolation theory and fractal geometry [1]. Over the past two decades, the fractal description is convenient and productive tool for solving various physical problems including problems of space physics, for instance, the study of processes on the Sun, solar wind, near-Earth stretched tail, the auroral structures. The main advantage of the fractal method is the versatility of the results and the independence of the nature of fractal structures in the region under consideration. In physical problems, the estimations of fractal characteristics of real objects are understood as the characteristics of the scaling properties of these objects which are valid for a certain range of scales. Also, Tsallis non-extensive statistical theory (q-statistics) [2], introduced as a extension of Boltzmann - Gibbs statistics, was recently used for the description of complex process in the space plasma dynamics. For these reasons, in the present short review to describe of properties of auroral ionosphere, we use a geometric approach based on fractal theory and percolation theory as well as q-statistics. It is demonstrated that nonlinear methods are perspective and useful approach for studying near-Earth plasma.

2. Fractal model of ionospheric conductivity

We consider the E-region of nightside Earth's ionosphere at altitudes of 80 – 150 km and at latitudes where the major part of energetic particles precipitation is observed and these particles result in auroras. In this region, particle precipitation is the main cause of ionization in the nightside and, consequently, of increased conductivity. In E-layer of ionosphere, the electrons are magnetized and the ions are nonmagnetized. In this case, the expressions for the Hall and Pedersen conductivities are simplified and take the following form: $\sigma_P = q^2 n/m_i \nu_{in}$ and $\sigma_H = qn/B$ respectively.

The spatial fluctuations of the conductivity are described under the assumption that, according to experimental data, they obey a power-law distribution. We consider the ionosphere to be a "spongy" fractal medium the structure of which in the absence of other ionization sources is determined by the structure of precipitation of auroral particles. The rate of ionization caused by particle precipitation varies smoothly along the magnetic field lines; therefore, the nontrivial fractal structure can form only in the spatial distribution transverse to the magnetic field [3, 4]. Locally, in a given field line, the conductivity is determined by the parameters of the precipitating electron flux. Here we consider a quasi-stationary case assuming that changes of precipitation structure occur more slowly than a recombination processes in the considered layer of the ionosphere.

Diffusion processes on percolating fractal structures are significantly non-Gaussian. Due to holes, constrictions on the fractal set, the motion of a particle on it can change. Moreover, since the holes reveal themselves on a range of scales, a change takes place on all these scales. In order to evaluate this effect, we estimate both the Pedersen conductivity and Hall conductivity by means of a characteristic transverse length scale a. The minimum possible value of a is determined by the mean free path of the particles in the auroral zone ionosphere. It is necessary to know the fractal dimension d^f and the connectivity index of θ to describe the auroral ionosphere. We use the expression for the coefficient of turbulent transport that describes the kinetics of random processes on fractal geometry [1]:

$$D \sim \frac{< r^2 >}{\tau} (\frac{\tau}{t})^\mu \tag{1}$$

In the expression (1), $\tau = a/v_T$ is the collision time, $< r^2 >$ is the mean-square distance of the particle, μ is the fractal exponent of diffusion, which is expressed as $\mu = d_s/d^f = 2/(2+\theta)$ where d_s is a spectral fractal dimension [5]. Fractal dimension d_s of a fractal set is the ratio of the Hausdorff dimension d^f to the minimal Hausdorff dimension of paths $d_s = d^f/d_\theta = 2d^f/(2+\theta)$. The index μ is in fact the inverse Hausdorff dimension of geodesic lines d_θ [6], namely $\mu = 1/d_\theta$. To estimate the conductivity, we need an expression for the carrier concentration (density): $n \sim a^{d^f - d}$, where d is the Euclidean dimension. Thus, the Pedersen conductivity dependence of the characteristic spatial scale a_p (the mean free path of particle) becomes:

$$\sigma_p \sim \frac{q^2 n}{m_i \nu_{in}} \propto n\tau \sim n\frac{a_p}{D} \propto a_p^{d_P^f - d} * a_p * a_p^{-\mu+1} \sim a_p^{2-\mu+d_P^f-d} \tag{2}$$

Here, d_P^f is the fractal dimension that specifies the Pedersen current in (2), that is, the motion of ions; a_p is the characteristic length scale of the Pedersen current (mean free path of ions). In obtaining estimate (2), it is assumed that time for percolation case is estimated as $\tau \propto a_p/D$ in the expression (2) rather than $\tau \sim a_p^2/D$ which is valid for classical regular case.

Now we obtain an estimate for the Hall conductivity through the spatial scale a_H:

$$\sigma_H \sim \frac{qn}{B} \propto n \propto a_H^{d_H^f - d} \tag{3}$$

where d_H^f is the fractal dimension that characterizes the Hall current which is determined by the motion of electrons.

An important special case is the region of intense field-aligned currents. Since the field-aligned current $j_{||}$ generates transversal currents in the ionosphere and the corresponding transverse electric fields $E_\perp$, we have $j_{||} \simeq \nabla_\perp(\Sigma \cdot E_\perp)$. If a is the transverse characteristic scale, the change of potential drop in the ionosphere at this scale is $\Delta\varphi_\perp \sim j_{||}a^2/\Sigma$, and the change of field-aligned parallel potential drop is $\Delta\varphi_{||} \sim j_{||}$. Also, we use semi-empirical estimations of the conductivity depending on the flux and the kinetic energy of precipitating electrons [3]. Thus, we have second estimation (see [3] for detail information). Equating the exponents of the two estimates for the Pedersen conductivity, we obtain: $\frac{1}{3}d_P^f - \frac{1}{3}\theta = 2 - \mu + d_P^f - d$. If it is considered that the Euclidean dimension is $d = 2$, the consistency condition is one of two equations needed to unambiguously determine the parameters d_P^f and θ_P.. To close the system, criticality condition of percolation threshold (the Alexander-Orbach conjecture) ensuing from the universal value theorem can be taken as the second equation [5, 6]. Criticality condition relates uniquely the fractal parameters of the system: the Hausdorff dimension d^f and the connectivity index θ near the nonequilibrium quasi-stationary states. The parameter Λ characterizes the geometry of the percolation transition and determines the minimal fractional number of degrees of freedom that a particle must have to pass through a region under consideration in the process of random walks. The set of points visited by the particle then forms a percolating fractal network with the spectral dimension $d_s = \Lambda \approx 1,327$. Criticality condition leads to the independence of the fractal geometric characteristics from microscopic properties of the medium. In this case, this condition allows us to estimate the critical values of fractal parameters that are necessary for the closure of the magnetosphere-ionosphere current system. This follows from the inequalities: $(2d^f)/(2+\theta) \geq \Lambda$. Taking into account that $\mu = d_s/d^f = 2/(2+\theta)$, we deduce for the fractal dimension d_P^f and the connective index θ_P: $d_P^f \geq 1.38$, $\theta_P \leq 0.09$. The obtained values fully describe the fractal geometry of Pedersen current near the percolation threshold. For percolation of Pedersen current, enhanced conductivity regions should fill in only a fractal subset with the fractal dimension $1.38 \leq d_P^f \leq 2$.

In the same way we find the estimates of fractal characteristics of the Hall conductivity in the ionosphere: $d_H^f \leq 1.85$, $\theta_H \leq 0.77$. The fractal topology of the Hall current of auroral ionosphere region on the percolation threshold is given by obtained fractal parameters. It is interesting to note that the limit threshold fractal dimension d_H^f is very close to the Hausdorff dimension 1.89 of the set, titled "Cantor cheese" (two-dimensional "Cantor cheese "is also sometimes called "Sierpinski carpet"). Sierpinski carpet has remarkable property because it specifies the maximum Hausdorff dimension of the fractal set, all of whose points can be visited without self-crossings [6]. It should be noted that since the directions and heights of the Pedersen and Hall currents are different, the same precipitation can create structures with different dimensions for the Pedersen and Hall conductivity [4].

The obtained threshold values of the fractal Hausdorff dimension and the connectivity index are characteristic of a path-connected set for Pedersen conductivity. A set is called path-connected if along with any pair of points it also contains a path connecting these points [6]. The Hausdorff dimension of a path-connected fractal set has $d^f \geq d_\theta \geq 1$ and $\theta > 0$ that take place for Pedersen conductivity. Note that for fractal sets are not path-connected if the condition $0 \leq d^f < d_\theta < 1$ holds as well as $\theta < 0$. For the Hall conductivity using different empirical relationships, there is such a state where a negative connectivity index $\theta < 0$ but fractal dimension $d^f \geq 1$. Such interesting class of fractal objects are called as asymptotically path-connected set [4, 6].

It is well known that evaluation of indices that characterize the self-similarity in the experiment is usually performed by one-dimensional series, namely, time series of measurements (for example, along the trajectory of the satellite) or section of a two-dimensional distribution (for example, the auroral glow). In these cases, we are dealing with a statistically self-affine series. Fractal properties of the self-affine series are commonly described by Hurst index H, which is associated with the fractal dimension of the set d^f as $H = 2 - d^f$. The index H defined by the power dependence of the increment variance on the scale of $\sigma^2(\varepsilon) = |\varepsilon|^{2H}$ or by the power-law form of the power spectrum $f^{-\alpha}$, where f is the (spatial) frequency and $\alpha = 2H + 1$. Thus, the fractal structure characteristics of the Pedersen conductivity $d_P^f \geq 1.38$ have following value:$H_p = 2 - d_P^f \leq 0.62$ hence $\alpha = 2H_p + 1 \leq 2.24$. Values of the Hurst index H and the index α obtained from the various experimental data in the auroral region (satellite and ground-based data) agree well with our theoretical fractal parameters (detailed study of these comparisons are given in [4]). Correspondence of analytical estimates and different experimental results indicates availability and validity of the fractal and percolation methods. It can be concluded that fractional calculus is an effective approach modeling of phenomena in auroral zone plasma induced by nonlinear processes.

3. q-statistics

We demonstrate for the first time that q-statistics [2] can be used to study the Earth's auroral region. In order to show that q-statistics may apply for study of auroral ionosphere, we use pulsating aurora event on 2011-12-03 from 22:00UT which were observed by Apatity all-sky camera [7]. The obtained data (where background was subtracted and bright stars were deleted) were divided into ten time intervals and for each interval we calculated the value of parameter q. The parameter q is estimated using maximum likelihood. Specifically the maximum likelihood estimates (MLEs) have been determined from a solution of the stationary points of the log likelihood function. Parameter q indicates a departure from Gaussian distribution. In the limit $q \to 1$ we have usual Gaussian statistics.

Our study shows that when bright auroras are observed, there is a strong deviation from unity of the parameter q. Otherwise, the value of q tends to unity, that is, to Gaussian statistics. This is due to the fact that during auroral glow ionospheric processes are substantially non-Gaussian and strong intermittency occurs. Also, it is established a good correlation between the change in values of q and flatness. While aurora glows are observed, flatness and parameter q grow and vice versa. In fact, it is seen that the Tsallis statistics can be used to study of non-Gaussian process, intermittency, along with flatness and/or probability density function (PDF).

Acknowledgments

This work was supported by Program 22 of the Presidium of RAS and Grants of President of Russian Federation (MK-267.2014.5) and RFBR (13-05-90436, 14-02-31848).

References

[1] Mandelbrot B 1982 *The Fractal Geometry of Nature* (Times Books) ISBN 9780716711865
[2] Tsallis C 1988 *Journal of Statistical Physics* **52** 479–487
[3] Chernyshov A A, Mogilevsky M M and Kozelov B V 2013 *Plasma Physics Reports* **39** 562–571
[4] Chernyshov A A, Mogilevsky M M and Kozelov B V 2013 *Journal of Geophysical Research (Space Physics)* **118** 4108–4118
[5] Alexander S and Orbach R 1982 *Journal de Physique Lettres* **43** 625–631
[6] Zelenyi L M and Milovanov A V 2004 *Physics-Uspekhi* **47** 749–788
[7] Kozelov B V, Pilgaev S V, Borovkov L P and Yurov V E 2012 *Geosci. Instrum. Method. Data Syst.* **1** 1–6

The peculiarities of magnetization processes in layered high-temperature superconductors with ferromagnetic defects under applying of transport current and external magnetic field

V A Kashurnikov, A N Maksimova and I A Rudnev
National Research Nuclear University (Moscow Engineering Physics Institute), Moscow Russia

Abstract. The Monte-Carlo method was used for study of magnetization processes in 2D layered high temperature superconductors (HTSC) with internal ferromagnetic defects. The magnetization was treated under application of transport current and external dc magnetic field. The voltage-current characteristics (I-V curve) were calculated in presence of external dc magnetic field. A novel S-type I-V curve of the superconductor/ferromagnet system in external magnetic field was demonstrated. It was shown that the S-type nonlinearity is due to the local reversal magnetization of magnetic particles by the field of vortices. The H-T phase diagram which demonstrates the region of existence I-V curve nonlinearity was obtained. The conditions for electromagnetic generation at the region of nonlinearity were found and the frequency of such a generation was estimated.

1. Introduction

The value of the critical current is the most important characteristic of the type II superconductors for various electrical engineering applications. The critical current strongly depends on the properties of intrinsic defects. When the current exceeds the critical value, vortices of opposite polarities start to move under the action of Lorentz force and annihilate close to the geometrical centre of the slab. These processes give rise to the power losses, appearance of the voltage on the sample and the transition of the superconductor to the normal state. Therefore it is crucial to investigate the influence of defect structure on transport properties of type II superconductors. Transport properties, phase transitions in vortex lattice, magnetization processes and voltage-current characteristics have been widely investigated in [1-5] for HTSC with nonmagnetic defects by using Monte-Carlo method.

The effect of ferromagnetism on the superconducting state attracted considerable attention in recent years. The ferromagnetic impurities in high-temperature and low-temperature superconductors were investigated to improve current-carrying ability, and the interaction of ferromagnetic phases with superconductivity has been used for solve the problem of the coexistence of superconductivity and ferromagnetism in a mixed vortex state. An enhancement of vortex pinning by a magnetic nanoparticle is expected due to rise of the energy of magnetic particles in the field of vortex in addition to conventional nonmagnetic part of interaction. HTSC samples on ferromagnetic substrates and HTSC with intrinsic ferromagnetic defects were investigated in the papers [6,7] in which the magnetization curves were obtained. The nonlinear interaction between ferromagnetic particles and superconductor was demonstrated. The force of interaction of a single vortex line with a spherical magnetic nanoparticle of arbitrary radius was calculated [8]. Analysis of the behavior of ferromagnetic Gd in Nb [9] showed that the pinning force is due to the hysteretic loss for magnetization reversal.

The aims of this paper are investigation of transport properties of HTSC sample with ferromagnetic defects as internal pinning centers and a correct calculation of voltage-current characteristics in the presence of external dc magnetic field. Here, we represent the results of calculation of VACH by using the Monte-Carlo method. Orientation of ferromagnetic nanoparticles' easy axes and values of magnetic anisotropy parameter were taken into account, the case of periodic lattice of ferromagnetic defects was analyzed for specificity.

2. Model

To calculate an equilibrium distribution of vortices, we numerically minimized the Gibbs thermodynamic potential of the 2D vortex system with a varying number of pancake vortices; this potential has the form (see [4,7], where the model and the algorithm are described in details):

$$G = sN\varepsilon + \sum_{i<j} U_{in-plane}(r_{ij}) + s\sum_{i,j} U_p(r_{ij}) + \sum_{i,j} U_{surf}(r_{ij}^{(im)}), \tag{1}$$

where ε is a self-energy of vortex, s is a thickness of a superconducting layer and N is the number of vortices in the system, the second term describes pair interaction of vortices, the third term takes into account the interaction of vortices with pinning centers and the fourth term describes the interaction of vortices with the surface and the external field. Pancakes in a plane interact with the long-range potential in the following form:

$$U_{in-plane}(r_{ij}) = s\frac{\Phi_0^2}{8\pi^2\lambda^2}K_0\left(\frac{r_{ij}}{\lambda}\right), \tag{2}$$

where r_{ij} is a distance between vortices and $\lambda = \lambda(T) = \lambda(T=0)/\sqrt{1-(T/T_c)^{3.3}}$ is the magnetic field penetration depth, $K_0(x)$ is a Bessel function and T_c is a superconducting critical temperature.

Then we take into account an ensemble of ferromagnetic nanoparticles in the bulk of superconductor. The magnetic part of the pinning force was calculated as an energy of interaction of magnetic dipole of ferromagnetic defect with the field of Abrikosov vortex. Gibbs potential of the ferromagnetic particle has a form:

$$U_p = U + U_{pm} + U_{pn}, \quad U_{pm} = -\mu H_v, \tag{3}$$

μ is the projection of the magnetic moment onto the direction of the external field, H_v is the field produced by the vortex at the point of location of the particle. In our calculations, we choose $\mu \approx 10^4\mu_0$. U_{pn} is the nonmagnetic part of the interaction, U is a part of particle's energy, not associated with vortices. μ_0 is the Bohr magneton. U value includes the particle's energy in the field of Meissner current and the energy of magnetic anisotropy (in [7,10] the magnetization processes in assembly of ferromagnetic nanoparticles in the field of Abrikosov vortices are described in details). The magnetization of ferromagnetic nanoparticles depends on the mutual orientation of particle's easy axes and the direction of external magnetic field. In this paper easy axes are parallel to an external field.

The interaction of vortex with Meissner current is an inverse work of Lorentz force on a vortex to move it from the surface deep into the sample:

$$U_m = \frac{\Phi_0}{4\pi}\left[H_0\left(1-\frac{ch(x/\lambda)}{ch(d/2\lambda)}\right) + H_I\left(\frac{sh(x/\lambda)}{sh(d/2\lambda)} \mp 1\right)\right] \tag{4}$$

H_0 - is an external magnetic field, $H_I = 2\pi I/c$ - magnetic field at the sample's surface produced by transport current, I is a total transport current through the sample cross section, d – sample's width. One should take "-" before unity when vortex appears at $x>0$ (right side) and "+" otherwise [4].

Calculations were done at 1-10 K, for typical characteristics of $Bi_2Sr_2CaCu_2O_{8-\delta}$: $\lambda(0) = 180$ nm, $\xi(0) = 2$ nm, $T_c = 84$ K. The anisotropy parameter was chosen $K = 0.1\cdot 10^7$ erg/cm^3. Ferromagnetic nanoparticles have the size $\sim \xi$ which corresponds to magnetic moment $\mu \sim 10^4\mu_B$.

3. Results and discussion

The current-voltage characteristics were obtained by using method developed in [5]. When a viscous flow of vortex lines is established, the dynamic component of the losses per unit volume

$W = jE$ may be obtained by counting the number of annihilated vortex-antivortex pairs. In our calculations, we defined the critical current as the current at which the voltage per unit length in the sample is E=0.1 $\mu V/cm$. Figure 1 represents the voltage-current characteristics for the sample with ferromagnetic defects at different values of external dc magnetic field.

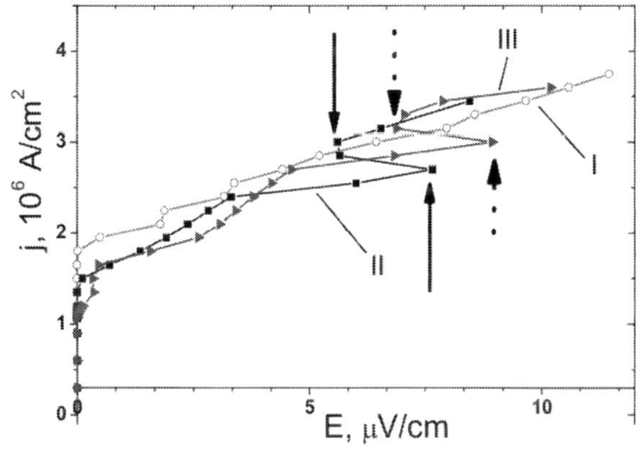

Figure 1. The voltage-current characteristics at fixed concentration of magnetic defects $c = 1{,}06 \cdot 10^9 cm^{-2}$ and different values of external field. I – H=200 Gs, II – H=400 Gs, III – H=500 Gs. $T = 1K$, much smaller than $T_c = 84K$ and Curie temperature. The regions of S-type nonlinearity are marked with arrows.

As it is replotted in Figure1, the S-type nonlinearity appears while external magnetic field increases. In contrast to the case with nonmagnetic defects, the orientation of particles magnetic moments is not fixed and depends on local magnetic fields and the field of transport current. As we can see from figure1, the nonlinearity that appears at lower current when magnetic field is smaller and the nonlinearity is absent below some critical value. The analysis of vortex configurations and magnetization of defects shows that the instabilities on the ac-characteristics correlate with drastic changes in magnetization of defects. Before the nonlinearity the vortex concentration is not enough for reorientation of magnetic moments and the effective concentration of pinning centers is small.

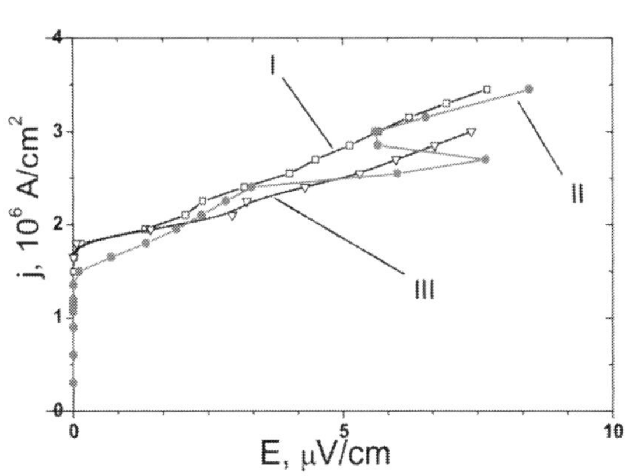

Figure 2. I-V curves at different concentrations of nonmagnetic defects: I – nonmagnetic, $c = 7.5 \cdot 10^8 cm^{-2}$, II - magnetic, $c = 1.06 \cdot 10^9 cm^{-2}$, III - nonmagnetic, $c = 4 \cdot 10^8 cm^{-2}$. I-V curve for nonmagnetic defects is shown for comparison (2), $H = 400Gs$, the depth of potential well of nonmagnetic defects was chosen equal to effective depth of magnetic defects, so that $\mu \frac{\Phi_0}{2\pi\lambda^2}\left(\ln\frac{\lambda}{\xi} - 0{,}28\right) = 0{,}02$ eV.

After nonlinearity the magnetic moments are parallel to the magnetization of vortices on the left side and the effective concentration of pinning centers increases. Figure 2 shows two I-V curves for different concentrations of nonmagnetic defects (with pinning force nearly equal to the pinning force of magnetic defects). The $j(E)$ curve is always higher for larger defect concentration and at the region of instability the transition from I-V curve for small concentration to I-V curve for higher concentration occurs. This effect is close to the one of the change of defect concentration in systems with high magnetic anisotropy examined in [7]. The analysis of influence of external parameters shows that the nonlinearity shifts to left and concurrently becomes straight at increasing temperature and above certain temperature is absent. Figure 3 shows the H-T phase diagram of the region of existence I-V curve nonlinearity. The temperature threshold at $T \approx 6K$ is due to magnetization reversal processes in assembly of magnetic particles. At higher temperature the magnetization of the particles becomes more reversible and the reorientation of magnetic moments occurs at lower magnetic fields.

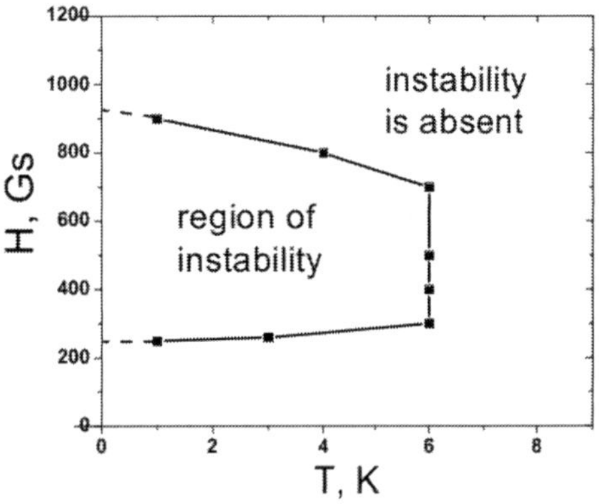

Figure 3. The H-T phase diagram of the region of existence I-V curve nonlinearity. The threshold temperature corresponds to the case when T is of order of the energy of magnetic anisotropy i.e. thermal fluctuations of magnetic moments become essential. $KV \sim T$, from which at the chosen value of magnetic anisotropy parameter $T \approx 6K$.

4. Conclusion

In summary, the magnetization of layered HTSC with internal ferromagnetic defects was calculated by Monte-Carlo method taking into account the field of transport current. The voltage-current characteristics in presence of external dc magnetic field were obtained. A novel, S-type of I-V curve was demonstrated. It was shown that the nonlinearity is caused by the local reversal magnetization of magnetic particles by the field of vortices which leads to an abrupt change in effective concentration of pinning centers. The influence of temperature and value of external dc magnetic field on the shape of $j(E)$ curve was investigated, the H-T phase diagram for the region of existing the nonlinearity was obtained.

Acknowledgements

The work was supported by RFBR, Grant **14-22-00098**.

[1] Kashurnikov V A, Rudnev I A, Gracheva M E, Nikitenko O A 2000 *JETP* **90** (1) 173
[2] Zyubin M V, Rudnev I A, Kashurnikov V A 2003 *JETP* **96** (6) 1065
[3] Kashurnikov V A, Rudnev I A, Zyubin M V 2001 *Supercond. Sci. and Techn.* **14**(9) 695
[4] Odintsov D S, Rudnev I A, Kashurnikov V A 2006 *JETP* **103** (1) 66-76
[5] Rudnev I A, Odintsov D S, Kashurnikov V A, 2008 *Phys. Lett. A* **372** 3934-3936
[6] Ilyina E A et al 2010 *Physica C* **470** 877-879
[7] Kashurnikov V A, Maksimova A N, Rudnev I A 2014 *Physics of the Solid State* **56** (5) 894-911
[8] Snezhko A, Prozorov T, Prozorov R 2005 *Phys. Rev. B: Condens. Matter* **71** 024527
[9] Palau A, Parvaneh H, Stelmashenko N A 2007 et al *Phys. Rev. Lett.* **98** 117003
[10] Prozorov R, Yeshurun Y, Prozorov T, Gedanken A 1999 *Phys. Rev. B* **59** 6956

Sharpening technique to decrease comb passband deviation

Gordana Jovanovic Dolecek

Department of Electronics, INAOE, E. Erro 1, 72740 Puebla , Mexico

E-mail: gordana@ieee.org

The decimation filter is necessary to prevent aliasing during the process of decreasing sampling rate, which is called decimation. Due its simplicity, the most popular decimation filter is the comb filter. However, this filter exhibits a high passband droop in the passband magnitude characteristic, which must be avoided to prevent the distortion of the decimated signal. In this paper we present the sharpening technique to improve comb passband characteristic. In order to decrease complexity introduced by sharpening, we also consider two-stage structure where the sharpening is applied only at the second stage. Different examples are included to illustrate the benefits of the proposed method.

1. Introduction

The current trend in the wireless radio receivers is to move the analog-to-digital converters (A/D) as close as possible to the antenna to do most of the signal processing in the digital domain. This imposes demands on design of innovative low-power analog-to-digital converters (A/D), known as the oversampling $\Sigma\Delta$-A/D converters. The oversampling $\Sigma\Delta$-A/D converter samples the analog information with a frequency much larger than the Nyquist frequency, which is the minimum sampling frequency required for preserving the quality of the information. The rate of the oversampled signal must be decreased to the Nyquist frequency to be efficiently processed by the DSP (Digital Signal processing). This process is performed in a digital format and is called down-sampling or decimation. Down-sampling may introduce the aliasing effects that deteriorate the signal and have to be eliminated by the decimation filter, as shown in Figure1, where M is the decimation factor [1].

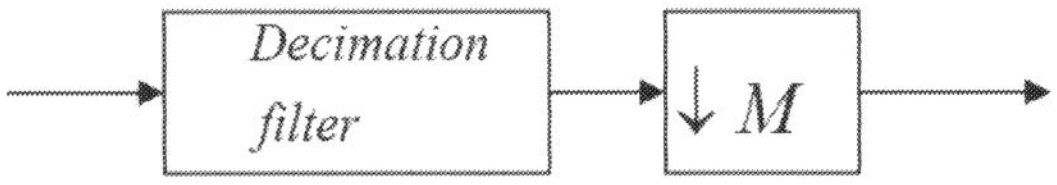

Figure 1. Decimation.

The decimation of a highly oversampled signal is usually accomplished using a cascade of two or more stages. Because the filter in the first stage must be very simple, it is usually used comb filter, which does not require the multipliers. The passband magnitude characteristic of the decimation filter must be flat to avoid the distortion of the decimated signal. However, the comb passband characteristic exhibits a high passband droop which must be compensated. The main objective of this work is to propose a novel procedure to obtain a flat passband comb characteristic considering a wide passband region, defined by the passband frequency,

$$\omega_p = \pi/(2M). \tag{1}$$

The rest of the paper is organized in the following way. Next section briefly describes the magnitude characteristic of comb filters. Section 3 presents the compensation filter [2] and the sharpening technique. The proposed procedure is described in Section 4 and illustrated with examples. Two-stage structure is elaborated in Section 5 and compared with the comb, method [2] and the one-stage sharpened structure.

2. Comb filters

The system function of a comb filter can be expressed, either in the recursive form,

$$H(z) = \left[\frac{1}{M}\frac{1-z^{-M}}{1-z^{-1}}\right]^K, \tag{2}$$

or in recursive form,

$$H(z) = \left[\frac{1}{M}\sum_{k=0}^{M-1} z^{-k}\right]^K, \tag{3}$$

where M is a decimation factor and K is order of the filter and presents the number of the cascaded comb filters.

The magnitude response of the filter is in the sin$x$/x form :

$$\left|H(e^{j\omega})\right| = \left|\frac{\sin(\omega M/2)}{M\sin(\omega/2)}\right|^K. \tag{4}$$

Figure 2 shows the magnitude characteristics of the comb filter for M=18 and K=1, 3, and 5. The passband zooms, shown in Figure 2 demonstrate that the comb filter exhibits a high passband droop which is increases with the increasing of the parameter K.

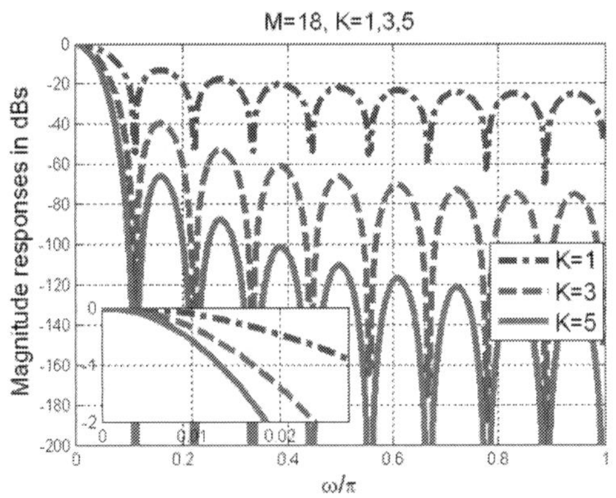

Figure 2. Magnitude responses of the comb filter for M=18 and the different values of K.

Different methods have been proposed to decrease the passband droop of the comb filters, using simple filters called compensators [2]-[8]. The method proposed here is the improvement of the method proposed in [2], which is briefly presented in next session, along with the brief review of the sharpening method.

3. Wideband comb compensator [2] and sharpening technique

3.1. Wideband compensator [2]

The compensation filter is a cascade of two compensators, both working at low rate:

$$G_1(z^M) = -2^{-(b+2)}\left[1-(2^{b+2}+2)z^{-M}+z^{-2M}\right], \tag{5}$$

where b is the integer which depends on the comb parameter K, and

$$G_2(z^M) = [-2^{-4}[z^{-M}-(2^4+2)z^{-2M}+z^{-3M}]]^{K_1}, \tag{6}$$

where K_1 depends on the comb parameter K in the following way:

$$K_1 = \begin{cases} K & for \quad 1 < K \leq 3 \\ K-1 & for \quad K > 3 \end{cases}. \tag{7}$$

Both filters can be moved to a low rate, which is M times less than the high input rate, thus becoming:

$$G_1(z) = -2^{-(b+2)}\left[1-(2^{b+2}+2)z^{-1}+z^{-2}\right]. \tag{8}$$

$$G_2(z) = [-2^{-4}[z^{-1}-(2^4+2)z^{-2}+z^{-3}]]^{K_1}. \tag{9}$$

The compensator [2] is the cascade of compensators $G_1(z)$ and $G_2(z)$:

$$G(z)= G_1(z)\, G_2(z), \tag{10}$$

where $G_1(z)$ and $G_2(z)$ are given in (8) and (9), respectively, and the values of parameters b are given in [2].

The compensated comb is the cascade of comb and compensator, given by:

$$H_c(z) = H(z)G(z^M), \tag{11}$$

where $H(z)$ is given in (2) and the compensator $G(z)$ is given in (10).

Example 1:

In this example we illustrate the method [2] taking K=3, and M=21. The parameters of compensator [2] are: K_1=2 and b=3. Figure 3 shows the corresponding overall magnitude responses of comb and compensated comb filters along with the passband zooms.

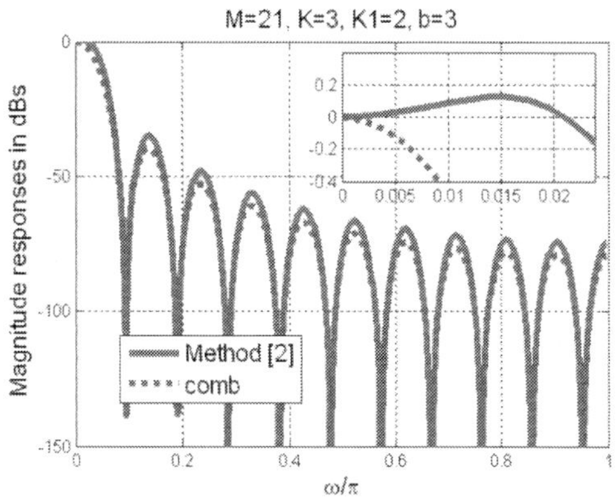

Figure 3. Magnitude responses of comb and compensated comb filters.

Note that the comb passband droop is significantly reduced. However, our goal here is to decrease the passband droop even more and practically obtain the flat passband characteristic. To this end we propose to use sharpening technique which is briefly described in the continuation.

3.2. Sharpening technique

The sharpening technique [9] uses the amplitude change function (ACF) which is a polynomial relationship of the form $H_{sh} = f(H)$ between the amplitudes of the sharpened and the prototype filters, H_{sh} and H, respectively, both the pass-band and stop-band of a linear-phase FIR filter. The improvement in the gain response near the passband edge H=1, or near the stopband edge H=0, depends on the order of tangencies m and n of the ACF at H=1, or at H=0, respectively.

The expressions for the mth and n^{th} order tangencies of the ACF at H=1 and H=0, respectively, are given as, [9]

$$H_{sh} = H^{n+1}\sum_{s=0}^{m}\frac{(n+s)!}{n!s!}(1-H)^s = H^{n+1}\sum_{s=0}^{m}C(n+s,s)(1-H)^s, \tag{12}$$

where $C(n+s,s)$ is the binomial coefficient.

The most simple polynomial for the passband improvement, (m=1, n=0) is in the form H_{sh}=2H-H^2.

4. Sharpened compensated filter

We propose to use simple sharpening polynomial H_{sh}=2H-H^2 to improve the compensated comb passband characteristic.

The compensated comb is defined by,

$$H_{sh}(z) = 2H_c(z) - H_c^2(z), \tag{13}$$

where $H_c(z)$ is given in (11).

Example 2:

In this example we compare magnitude responses of the compensated comb from Example 1, the sharpened compensated comb, and the comb filters. Figure 4 shows the overall magnitude responses and the passband zooms.

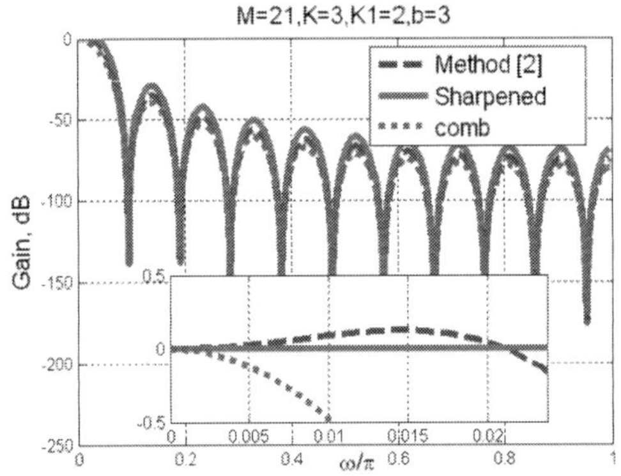

Figure 4. Magnitude responses of comb, compensated comb, and sharpened compensated comb filters.

Note that the passband characteristic of the sharpened compensated comb is practically flat. However, the sharpening is performed at high input rate which results in a higher complexity and the higher power consumption. In order to solve this problem we consider a two-stage structure in next session.

5. Two-stage sharpened compensated filter

Consider that the decimation factor can be factorized in the form M=M_1M_2. In this case we can split decimation into two stages, where in the first stage is the comb filter,

$$H_1(z) = \left[\frac{1}{M_1}\frac{1-z^{-M_1}}{1-z^{-1}}\right]^K, \tag{14}$$

decimated by M_1, while in the second stage is the comb filter,

$$H_2(z) = \left[\frac{1}{M_2}\frac{1-z^{-M_2}}{1-z^{-1}}\right]^K, \tag{15}$$

decimated by M_2.
The compensator (10) and the sharpened section $H_{sh2}(z)$:

$$H_{sh2}(z) = 2H_{c2}(z) - H_{c2}^2(z), \tag{16}$$

where

$$H_{c2}(z) = H_2(z)G(z^{M_2}), \tag{17}$$

are at the second stage.
The corresponding structure is given in Figure 5.

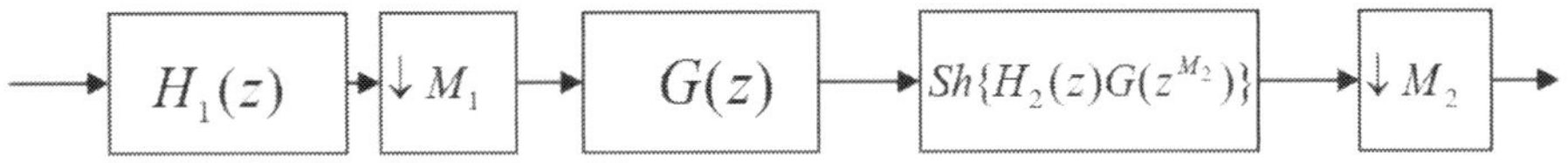

Figure 5. Two-stage sharpening structure.

Knowing that the sharpening is at the second stage, the choice of M_1 and M_2 has to satisfy

$$M_1 < M_2. \tag{18}$$

The choice of the parameters b and K_1 is obtained by MATLAB simulation:

$$K_1 = K-1, \text{ for } K=2,\ldots,6. \tag{19}$$

$$b = \begin{cases} 3 & for \quad K = 2,3,4 \\ 4 & for \quad K = 6 \end{cases}. \tag{20}$$

The method is illustrated in the following example.
Example 3:
Consider M=32 and K=4. According to (18) we chose M_1=4 and M_2=8. From (19) and (20) we get the parameters K_1=3, and b=3, respectively.
Figure 6 compares magnitude responses of method [5], sharpened one stage structure, and the sharpened two-stage structure.

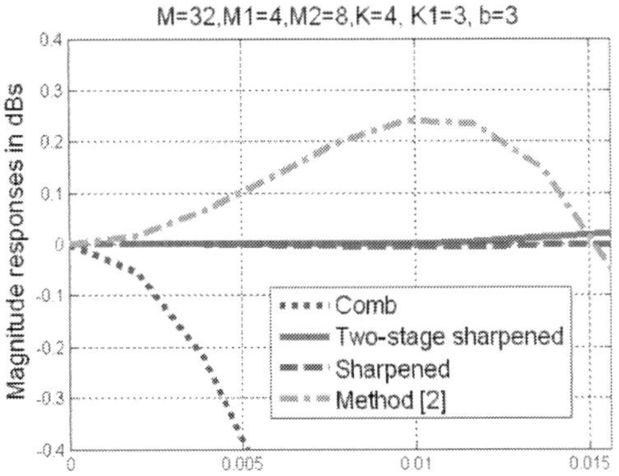

Figure 6. Passband responses for comb, Method [2], and one and two-stage sharpened.

Note that there is a small deviation compared with the one-stage sharpening. The deviation decreases if $M_1 \ll M_2$, as shown in Fig 7, for M_1=2 and M_2=16. However, in this case the sharpening section works at the rate which is only two times less than the high input rate. Consequently, the choice of M_1 and M_2 is a trade-off between the complexity and less deviation in the compensated comb passband.

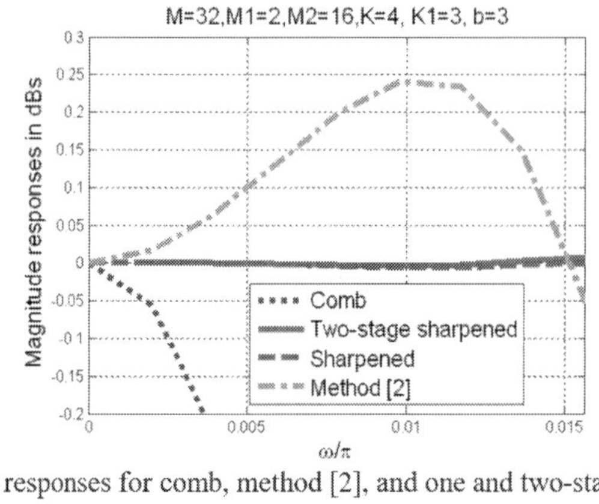

Figure 7. Passband responses for comb, method [2], and one and two-stage sharpened structures.

6. Concluding remarks

This paper presents the use of sharpening technique to obtain a flat passband comb magnitude characteristic. We consider the cascade of comb filter with narrowband and wideband compensators [2], which results in a passband deviation of approximately 0.25 dB. Applying the simple sharpening polynomial to the compensated comb, the flat passband comb magnitude characteristic is obtained. In order to decrease the complexity introduced by sharpening, we also proposed a two-stage structure, in which the sharpening is realized in the second stage. In this case the decimation factor must be presented as a product of two integers. As a result, the magnitude characteristic in the passband has only a non-significant deviation, in comparison with the passband response of the sharpened one-stage structure. The choice of the decimation factors is a trade-off between the decreased complexity and the increased passband deviation compared with the one-stage sharpened structure.

Acknowledgement
This work is supported by Conacyt Mexico under grant No.179587.

References

[1] Jovanovic Dolecek G 2002 *Multirate Signal Processing: Design and Application* (Hershey: IGP).
[2] Jovanovic Dolecek G 2013 Proc *International Conference on Advances in Computing, Communications and Informatics (ICACCI)* Mysore India On Wideband Comb Compensator **978-1-4673-6217-7/13/©2013 IEEE** 1987-1990.
[3] Kim S Lee W C Alm S Choi 2006 *Digital Signal Processing* Design of CIC roll-off compensation filter in a W-CDMA digital receiver **16**, 846-854.
[4] Jovanovic Dolecek G Mitra S K *Electronics Letters* Simple method for compensation of CIC decimation filter **44** 1162-1163.
[5] Fernandez-Vazquez A Jovanovic Dolecek G 2009 *IEEE Transactions on Circuits and Systems II: Express Brief* A general method to design GCF compensation filter **56** 409-413.
[6] Jovanovic Dolecek G 2009 *Electronics Letters* Simple wideband CIC compensator **45**, 1270-1272.
[7] Molnar G Vucic M 2011 *IEEE Transactions on Circuits and Systems II: Express Brief* Closed-form design of CIC compensators based on maximally flat error criterion **58** 926-30.
[8] Jovanovic Dolecek G Fernandez-Vazquez 2014 A *International Journal for Electronics and Communications (AEUE)* Trigonometrical Approach to Design a Simple Wideband Comb Compensator **68** 437–441.
[9] Kaiser J F Hamming R W 1977 *IEEE Transactions on Acoustics, Speech, and Signal Processing,* Sharpening the response of a symmetric non-recursive filter **ASSP-25** 415-422.

Two reliable approaches involving Haar wavelet method and Optimal Homotopy Asymptotic method for the solution of fractional Fisher type equation

Santanu Saha Ray

National Institute of Technology Rourkela, Department of Mathematics, Rourkela-769008, Orissa, India

E-mail: santanusaharay@yahoo.com, saharays@nitrkl.ac.in

Abstract. In this article, two reliable techniques, Haar wavelet method and optimal homotopy asymptotic method (OHAM) are presented. Haar wavelet method is an efficient numerical method for the numerical solution of fractional order partial differential equation like Fisher type. The approximate solutions of the fractional Fisher type equation are compared with the optimal homotopy asymptotic method as well as with the exact solutions. Comparisons between the obtained solutions with the exact solutions exhibit that both the featured methods are effective and efficient in solving nonlinear problems. However, the results indicate that OHAM provides more accurate value than Haar wavelet method.

1. Introduction

Fractional calculus is a field of applied mathematics which deals with derivatives and integrals of arbitrary orders [1,2]. In the last few decades, fractional calculus has been extensively investigated due to their broad applications in mathematics, physics and engineering such as viscoelasticity, diffusion of biological population, signal processing, electromagnetism, fluid mechanics, electrochemistry and so on. For this we need a reliable and efficient technique for the solution of fractional differential equations.

Wavelet theory is a relatively new and an emerging area in mathematical research. Approximation by orthogonal wavelets bases are becoming more popular for numerical solutions of fractional differential equations due to their excellent properties such as ability to detect singularities, orthogonality, flexibility to represent a function at different level of resolution and compact support. One of the pretty feature of wavelets is its ability to convert the given differential and integral equations to a system of linear or nonlinear differential equations, which can be solved by numerical methods.
The generalized time-fractional Fisher's biological population diffusion equation is given by

$$\frac{\partial^\alpha u}{\partial t^\alpha} = \frac{\partial^2 u}{\partial x^2} + F(u), \quad u(x,0) = \phi(x) \tag{1.1}$$

where $u(x,t)$ denotes the population density and $t > 0$, $x \in R$, $F(u)$ is a continuous nonlinear function satisfying the following conditions $F(0) = F(1) = 0$, $F'(0) > 0 > F'(1)$. The derivative in eq. (1.1) is the Caputo derivative of order α.
Definition: The fractional derivative introduced by Caputo [1], in the late sixties, is called Caputo

Fractional derivative. The Caputo fractional derivative ${}_0\mathrm{D}_t^{\alpha}$ of a function $f(t)$ is defined as [1]

$$ {}_0\mathrm{D}_t^{\alpha} f(t) = \frac{1}{\Gamma(n-\alpha)} \int_0^t \frac{f^n(\tau)}{(t-\tau)^{\alpha-n+1}} d\tau, \quad (n-1 < \alpha \leq n, \quad n \in N) \tag{1.2} $$

The notion of L_2 error in Haar wavelet method as well as in case of OHAM, the fine tuning of convergence region by using optimal values of convergence control parameters justify that both the methods are reliable. Our aim in the present work is to implement Haar wavelet method and optimal homotopy asymptotic method (OHAM) in order to demonstrate the capability of these methods in handling nonlinear equations of arbitrary order, so that one can apply it to various types of nonlinearity.

2. Function Approximation

Any function $y(t) \in L^2([0,1))$ can be expanded into Haar wavelets by [3–5]

$$ y(t) = c_0 h_0(t) + c_1 h_1(t) + c_2 h_2(t) + \ldots \tag{2.1} $$

where $c_j = \int_0^1 y(t) h_j(t) dt$.
If $y(t)$ is approximated as piecewise constant in each subinterval, the sum in eq. (2.1) may be terminated after m terms and consequently we can write discrete version in the matrix form as

$$ \mathbf{Y} \approx \sum_{i=0}^{m-1} c_i h_i(t_l) = C_m^T . H_m, \tag{2.2} $$

where $\mathbf{Y}$ and C_m^T are the m-dimensional row vectors.
Here H is the Haar wavelet matrix of order m defined by $H = [h_0, h_1, \ldots, h_{m-1}]^T$, $h_0, h_1, \ldots, h_{m-1}$ are the discrete form of the Haar wavelet bases.
The collocation points are given by $t_l = A + (l - 0.5)\Delta t, \quad l = 1, 2, \ldots, m$.

3. Operational matrix of the general order integration

The integration of $H_m(t) = [h_0(t), h_1(t), \ldots, h_{m-1}(t)]^T$ can be approximated by [5]

$$ \int_0^t H_m(\tau) d\tau \cong Q H_m(t), \tag{3.1} $$

where Q is called the Haar wavelet operational matrix of integration which is a square matrix of $m-$dimension.
The Haar wavelet operational matrix Q^{α} for integration of the general order α is given by

$$ Q^{\alpha} H_m(t) = J^{\alpha} H_m(t) = [J^{\alpha} h_0(t), J^{\alpha} h_1(t), \ldots, J^{\alpha} h_{m-1}(t)]^T = [Q h_0(t), Q h_1(t), \ldots, Q h_{m-1}(t)]^T \tag{3.2} $$

where

$$ Q h_0(t) = \begin{cases} \frac{t^{\alpha}}{\Gamma(1+\alpha)}, & t \in [A,B] \\ 0, & \text{elsewhere}, \end{cases} \quad \text{and} \quad Q h_i(t) = \begin{cases} 0, & A \leq t < \zeta_1(i) \\ \phi_1, & \zeta_1(i) \leq t < \zeta_2(i) \\ \phi_2, & \zeta_2(i) \leq t < \zeta_3(i) \\ \phi_3, & \zeta_3(i) \leq t < B, \end{cases} \tag{3.3} $$

where

$$ \phi_1 = \frac{(t-\zeta_1(i))^{\alpha}}{\Gamma(\alpha+1)}, \quad \phi_2 = \frac{(t-\zeta_1(i))^{\alpha}}{\Gamma(\alpha+1)} - 2\frac{(t-\zeta_2(i))^{\alpha}}{\Gamma(\alpha+1)}, \quad \phi_3 = \frac{(t-\zeta_1(i))^{\alpha}}{\Gamma(\alpha+1)} - 2\frac{(t-\zeta_2(i))^{\alpha}}{\Gamma(\alpha+1)} + \frac{(t-\zeta_3(i))^{\alpha}}{\Gamma(\alpha+1)} $$

for $i = 1, 2, \ldots, m$, $m = 2^J$ and J is a positive integer, called the maximum level of resolution. Here j and k represent the integer decomposition of the index i. i.e. $i = k + 2^j - 1$ and $1 \leq k < 2^j + 1$.

4. Application of Haar wavelet to Fractional Fisher type Equation

Consider the nonlinear diffusion equation of the fractional Fisher type [6, 7]

$$\frac{\partial^\alpha u}{\partial t^\alpha} = \frac{\partial^2 u}{\partial x^2} + u(1-u)(u-a), \quad 0 < \alpha \leq 1, \quad 0 \leq x \leq 1, \quad 0 < a < 1 \tag{4.1}$$

with the initial condition

$$u(x,0) = \frac{1}{1+\exp\left(\frac{-x}{\sqrt{2}}\right)} \tag{4.2}$$

When $\alpha = 1$, the exact solution of eq. (4.1) is given by [8, 9]

$$u(x,t) = \frac{1}{1+\exp\left[-\left(\frac{x+ct}{\sqrt{2}}\right)\right]}; \quad c = \sqrt{2}\left(\frac{1}{2}-a\right) \tag{4.3}$$

Let us divide both space and time interval $[0,1]$ into m equal subintervals; each of width $\Delta = \frac{1}{m}$.
Haar wavelet solution of $u(x,t)$ is sought by assuming that $\frac{\partial^2 u(x,t)}{\partial x^2}$ can be expanded in terms of Haar wavelets as

$$\frac{\partial^2 u(x,t)}{\partial x^2} = \sum_{i=1}^{m}\sum_{j=1}^{m} c_{ij} h_i(x) h_j(t) \tag{4.4}$$

Integrating eq. (4.4) twice w.r.t. x from 0 to x we get

$$u(x,t) = \sum_{i=1}^{m}\sum_{j=1}^{m} c_{ij} Q^2 h_i(x) h_j(t) + q(t) + x p(t) \tag{4.5}$$

Again $q(t) + xp(t)$ can be approximated using Haar wavelet function as

$$q(t) + xp(t) = \sum_{i=1}^{m}\sum_{j=1}^{m} r_{ij} h_i(x) h_j(t) \tag{4.6}$$

Substituting eq. (4.6) in eq. (4.5) we get

$$u(x,t) = \sum_{i=1}^{m}\sum_{j=1}^{m} c_{ij} Q^2 h_i(x) h_j(t) + \sum_{i=1}^{m}\sum_{j=1}^{m} r_{ij} h_i(x) h_j(t) \tag{4.7}$$

The nonlinear term presented in eq. (4.1) can be approximated using Haar wavelet function as

$$u(1-u)(u-a) = \sum_{i=1}^{m}\sum_{j=1}^{m} d_{ij} h_i(x) h_j(t) \tag{4.8}$$

Therefore,

$$\begin{aligned}&\left(\sum_{i=1}^{m}\sum_{j=1}^{m} c_{ij} Q^2 h_i(x) h_j(t) + \sum_{i=1}^{m}\sum_{j=1}^{m} r_{ij} h_i(x) h_j(t)\right)\left(1 - \sum_{i=1}^{m}\sum_{j=1}^{m} c_{ij} Q^2 h_i(x) h_j(t) + \sum_{i=1}^{m}\sum_{j=1}^{m} r_{ij} h_i(x) h_j(t)\right)\\ &\left(\sum_{i=1}^{m}\sum_{j=1}^{m} c_{ij} Q^2 h_i(x) h_j(t) + \sum_{i=1}^{m}\sum_{j=1}^{m} r_{ij} h_i(x) h_j(t) - a\right) = \sum_{i=1}^{m}\sum_{j=1}^{m} d_{ij} h_i(x) h_j(t)\end{aligned} \tag{4.9}$$

Substituting eqs. (4.4) and (4.8) in eq. (4.1) and applying J^α to both sides, we will have

$$u(x,t) - u(x,0) = J_t^\alpha\left[\sum_{i=1}^{m}\sum_{j=1}^{m} c_{ij} h_i(x) h_j(t)\right] + J_t^\alpha\left[\sum_{i=1}^{m}\sum_{j=1}^{m} d_{ij} h_i(x) h_j(t)\right] \tag{4.10}$$

Substituting eqs. (4.2) and (4.7) in eq. (4.9) we get

$$\sum_{i=1}^{m}\sum_{j=1}^{m} c_{ij}Q^2h_i(x)h_j(t) + \sum_{i=1}^{m}\sum_{j=1}^{m} r_{ij}h_i(x)h_j(t) - \frac{1}{1+\exp\left(\frac{-x}{\sqrt{2}}\right)} = \sum_{i=1}^{m}\sum_{j=1}^{m} c_{ij}h_i(x)Q_t^{\alpha}h_j(t) + \sum_{i=1}^{m}\sum_{j=1}^{m} d_{ij}h_i(x)Q_t^{\alpha}h_j(t) \quad (4.11)$$

Now substituting the collocation points $x_l = \frac{l-0.5}{m}$ and $t_k = \frac{k-0.5}{m}$ for $l,\ k = 1,2,...,m$ in eqs. (4.6), (4.9) and (4.11), we have $3m^2$ equations in $3m^2$ unknowns in c_{ij}, r_{ij} and d_{ij}. By solving these system of equations using mathematical software, the Haar wavelet coefficients c_{ij}, r_{ij} and d_{ij} can be obtained.

5. Application of OHAM to Fractional Fisher type Equation

The OHAM was introduced and developed by Merinca et al. [10]. In OHAM, the control and adjustment of the convergence region are provided in a convenient way.
We construct a homotopy $\varphi(x,t;p) : \Omega \times [0,1] \longrightarrow R$ for eq. (4.1) which is given by

$$(1-p)\frac{\partial^{\alpha}\varphi(x,t;p)}{\partial t^{\alpha}} = H(p)\left[\frac{\partial^{\alpha}\varphi(x,t;p)}{\partial t^{\alpha}} - \frac{\partial^2\varphi(x,t;p)}{\partial x^2} - \varphi(x,t;p)\left[1-\varphi(x,t;p)\right]\left[\varphi(x,t;p)-a\right]\right] \quad (5.1)$$

where $p \in [0,1]$ is an embedding parameter, $H(p)$ is a nonzero auxiliary function for $p \neq 0$ and $H(0) = 0$. When $p = 0$ and $p = 1$, we have $\varphi(x,t;0) = u_0(x,t)$ and $\varphi(x,t;1) = u(x,t)$, respectively.
Thus as p varies from 0 to 1, the solution $\varphi(x,t;p)$ approaches from $u_0(x,t)$ to $u(x,t)$.
Here

$$\varphi(x,t;p) = u_0(x,t) + \sum_{i=1}^{\infty} u_i(x,t)p^i \quad (5.2)$$

$$H(p) = pC_1 + p^2C_2 + p^3C_3 + ... \quad (5.3)$$

$$N(\varphi(x,t;p)) = N_0(u_0(x,t)) + \sum_{k=1}^{\infty} N_k(u_0,u_1,...,u_k)p^k \quad (5.4)$$

Substituting eqs. (5.2), (5.3) and (5.4) in eq. (5.1) and equating the coefficients of like powers of p and again solving those equations, we obtain the second order approximate solution using the following formula

$$u(x,t) = u_0(x,t) + u_1(x,t) + u_2(x,t) \quad (5.5)$$

The optimal values of the convergence control constants C_1 and C_2 can be obtained using collocation method. Substituting in eq. (4.1), we get the following expression for the residual $R_n(x,t;C_1,C_2,C_3,...)$. If $R_n(x,t;C_1,C_2,C_3,...) = 0$, then $\tilde{u}(x,t;C_1,C_2,C_3,...)$ is the exact solution. Generally such case does not arise for nonlinear problems. The convergence of the $n-$th approximate solution depends upon unknown constants $C_1,C_2,C_3,...$.
When the convergence control constants $C_1,C_2,C_3,...$ are known by the above mentioned methods then the approximate solution of (4.1) is well determined.

6. Numerical results and discussion

The following Table 1 shows the comparison of the approximate solutions of fractional Fisher type equation (4.1) obtained by using Haar wavelet method and OHAM at different values of x and t taking $\alpha = 0.5$. The following Table 2 exhibits the L_2 and L_∞ error norm for fractional Fisher type equation at different values of t and $\alpha = 1$. It can be easily observed from Table 2 that the solutions obtained by OHAM are more accurate than that of Haar wavelet method.

Table 1. The approximate solutions of fractional Fisher type equation (4.1) using Haar wavelet method and three terms for second order OHAM at various points of x and t taking $\alpha = 0.5$

x	$t=0.2$		$t=0.4$		$t=0.6$		$t=0.8$	
	u_{Haar}	u_{OHAM}	u_{Haar}	u_{OHAM}	u_{Haar}	u_{OHAM}	u_{Haar}	u_{OHAM}
0.1	0.531396	0.55521	0.550389	0.570645	0.570167	0.582442	0.586081	0.592358
0.3	0.557098	0.589616	0.582811	0.604511	0.607055	0.615806	0.625465	0.625237
0.5	0.586931	0.623164	0.617982	0.637394	0.645433	0.648094	0.665485	0.656966
0.7	0.613502	0.655571	0.644668	0.669033	0.671439	0.679069	0.690761	0.68733
0.9	0.643461	0.686593	0.669891	0.699213	0.69255	0.708537	0.709084	0.68733

Table 2. L_2 and L_∞ error norm for Fisher type equation t.

Time(sec)	HWM		OHAM	
	L_2	L_∞	L_2	L_∞
0.2	0.0377811	0.0489923	1.50470E-5	2.47294E-5
0.4	0.00380168	0.00797543	5.05627E-5	8.90920E-5
0.6	0.020685	0.0293499	9.97048E-5	1.79900E-4
0.8	0.0281228	0.0394803	1.69576E-4	2.85670E-4

7. Conclusion

In this paper, the classical as well as fractional Fisher type equation has been solved by using Haar wavelet method. The obtained results are then compared with exact solutions as well as with relatively new optimal homotopy asymptotic method (OHAM). These results have been cited in the tables demonstrated in order to justify the accuracy and efficiency of the proposed schemes. The Haar wavelet technique provides quite satisfactory results for the fractional Fisher type equation (4.1). The main advantages of this Haar wavelet method is they transfer the whole scheme into a system of algebraic equations for which the computation is easy and simple. OHAM allows fine tuning of convergence region and rate of convergence by suitably identifying convergence control parameters $C_1, C_2, C_3, \ldots$. The results obtained by OHAM are more accurate as its convergence region can be easily adjusted and controlled. The main advantages of these schemes are their simplicity, applicability and less computational errors. Although the obtained results indicate that optimal homotopy asymptotic method provides more accurate value than Haar wavelet method, however the accuracy of the wavelet method may be improved with the increase of level of resolution. Comparison shows that these two methods are reliable, appropriate and applicable for solving nonlinear fractional Fisher's equation.

References

[1] Podlubny I 1999 *Fractional Differential Equations* (New York: Academic Press)
[2] Hilfer R 2000 *Applications of fractional calculus in physics* (Singapore: World Scientific)
[3] Saha Ray S 2012 *Appl. Math. Comput.* **218** 5239–48
[4] Saha Ray S and Patra A 2013 *Appl. Math. Comput.* **220** 659–67
[5] Chen C F and Hsiao C H 1997 *IEE Proc-Control Theory Appication* **144**(1) 87-94
[6] Wazwaz A M 2009 *Partial Differential equations and Solitary Waves Theory* (Berlin: Springer)
[7] Fisher R A 1937 *Ann. Eugen.* **7**(4) 355–69
[8] Liu Y 2011 *Journal of Fractional Calculus and Applications* **1**(2) 1-8
[9] Wazwaz A M and Gorguis A 2004 *Appl. Math. Comput.* **154** 609–20
[10] Marinca V, Herisanu N, Bota C and Marinca B 2009 *Appl. Math. Lett.* **22**(2) 245–51

Trans-dimensional Monte Carlo sampling applied to the magnetotelluric inverse problem

Eric Mandolesi[1] **and Nicola Piana Agostinetti**[1]

[1]Dublin Institute for Advanced Studies, 5 Merrion Square, Dublin 2, Republic Of Ireland

E-mail: eric@cp.dias.ie

Abstract. The data required to build geological models of the subsurface are often unavailable from direct measurements or well logs. In order to image the subsurface geological structures several geophysical methods have been developed. The magnetotelluric (MT) method uses natural, time-varying electromagnetic (EM) fields as its source to measure the EM impedance of the subsurface. The interpretation of these data is routinely undertaken by solving inverse problems to produce 1D, 2D or 3D electrical conductivity models of the subsurface. In classical MT inverse problems the investigated models are parametrized using a fixed number of unknowns (i.e. fixed number of layers in a 1D model, or a fixed number of cells in a 2D model), and the non-uniqueness of the solution is handled by a regularization term added to the objective function. This study presents a different approach to the 1D MT inverse problem, by using a trans-dimensional Monte Carlo sampling algorithm, where trans-dimensionality implies that the number of unknown parameters is a parameter itself. This construction has been shown to have a built-in Occam razor, so that the regularization term is not required to produce a simple model. The influences of subjective choices in the interpretation process can therefore be sensibly reduced. The inverse problem is solved within a Bayesian framework, where posterior probability distribution of the investigated parameters are sought, rather than a single best-fit model, and uncertainties on the model parameters, and their correlation, can be easily measured.

1. Introduction

MT method is an electromagnetic sounding tomographic technique used to infer the resistivity distribution of the subsurface by measurements of orthogonal components of horizontal EM fields on the Earth surface. The MT inverse problem is known to be both unstable and highly non-linear. Despite these irregular traits, linearized inversion of MT data are routinely undertaken and lead to affordable and robust results [1]. The theoretical problem with this approach is in the inverse problem set up. The linearized version of the problem, even if solved in a 1D layered domain, requires in fact several key choices to be made, for example the number of parameters used to describe the subsurface, the presence of a regularization term in the objective function, the trade-off balance between a regular model and a model that fit the data better and so on. In the past, several authors used Bayesian inversion to interpreter MT data (cf. e.g. [2]). In this paper we present a trans-dimensional procedure in which the number of parameters to be estimated is an unknown itself. From a Bayesian perspective the inverse problem is not solved by a single model that fits the dataset, rather the solution of the inverse problem is given in a statistical form, returning the posterior probability distribution (PPD) of models that fit the

Table 1. Synthetic model used in the algorithm test. Layers are numbered top to bottom.

Layer number	Layer thickness (km)	Resistivity ($\Omega \cdot m$)
1	1.0	1000
2	1.0	100
3	∞	1000

dataset. The PPD is given by Bayes's rule as

$$p(\vec{m}|\vec{d},\mathcal{I}) = \frac{p(\vec{m}|\mathcal{I})p(\vec{d}|\vec{m},\mathcal{I})}{p(\vec{d}|\mathcal{I})} \tag{1}$$

in which p is the probability distribution, $\star|\star$ denotes the conditional dependence, $\vec{d}$ is the array of measured data and $\vec{m}$ is the array that stores the model parameters. Details about the quantities used in the Equation 1 can be found in Piana Agostinetti [3] and references therein. The reverse jump Markov chain Monte Carlo (rjMCMC) algorithm used here is formally identical to the one described by Piana Agostinetti [3].

As the MT method in 1D is mainly sensitive to conductance rather than conductivity (cf. e.g. [1]), the implicit parsimony in the model complexity that characterizes rjMCMC solutions helps the interpretation of data that cannot be adequately described by a simple model and that requires regularization terms or complex model to be fitted.

2. Numerical Experiment

The algorithm has been validated by the simulation of an inversion using data from a synthetic model, summarized in Table 1. The synthetic model is built as a stack of three layers: a resistive basement, a conductive layer and a resistive topmost layer. This model presents two the classic challenges for a MT inverse problem: retrieving the bottom of a conductor and fit the data with a model that is not overparametrized.

First we tested the prior probability distribution sampler – setting manually the likelihood to 1– to check that the candidate models are effectively drawn from the prior probability distribution. Let us pretend we miss a good *a priori* information about the parameters distribution, in this case we purpose as prior distribution for the resistivity ρ an uniform distribution in the interval $-0.5 < \log_{10}(\rho) < 3.5$. Frequency distribution of the sampled values for electrical resistivity, shown in Figure 1, demonstrate that the sampler draws this parameter from a uniform distribution, in the correct limits. Even the number of interfaces, and their location in depth, plotted respectively in Figure 3 and 5 result uniformly distributed.

We thus simulated data from the synthetic model, we added 3% random noise and test the algorithm, using 191 Markov chains with 500.000 models per chain.

3. Results and discussion

Overall the algorithm retrieved the (supposed unknown) synthetic model, with the resistivity distribution shown in Figure 2. The top interface of the conductor is well resolved, as expected, while the depth of the deeper resistor results blurred. This is due to the lack of sensitivity of the method on resistive region beneath conductors. Figure 4 proves the implicit parsimony of the algorithm showing that frequently the data can be fitted using three interfaces (i.e., two layers), one less than the true value. Even the distribution of the interfaces in depth (Figure 6 results overall correct, with the distribution peaks centered at 1 km and 2 km depth.

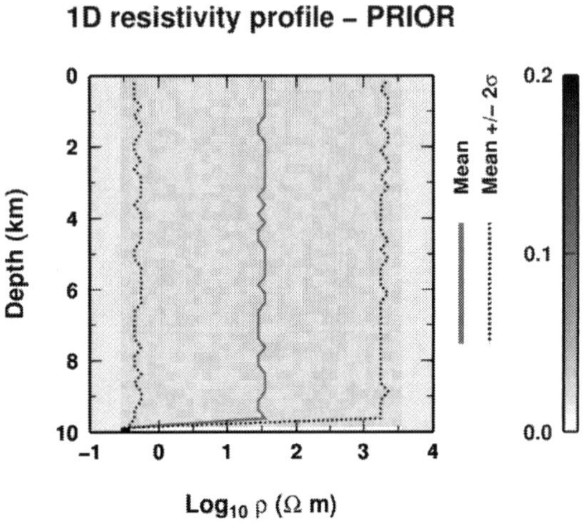

Figure 1. Prior sampling of the resistivity.

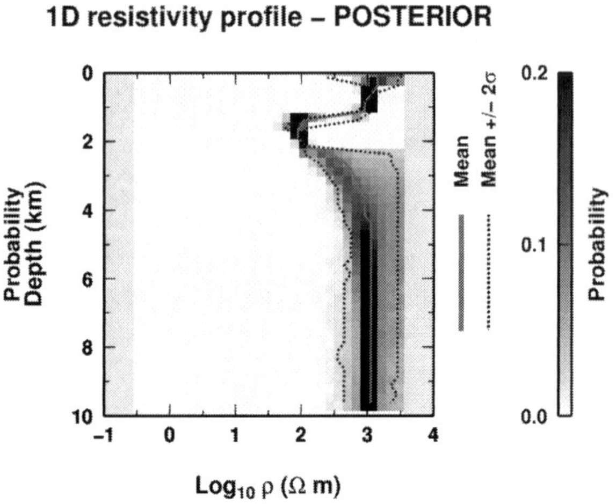

Figure 2. Posterior sampling of the resistivity.

Finally, Figure 7 and 8 show the data fit obtained using respectively the prior and posterior resistivity distributions. Without any affordable *a priori* information, it s natural that the prior

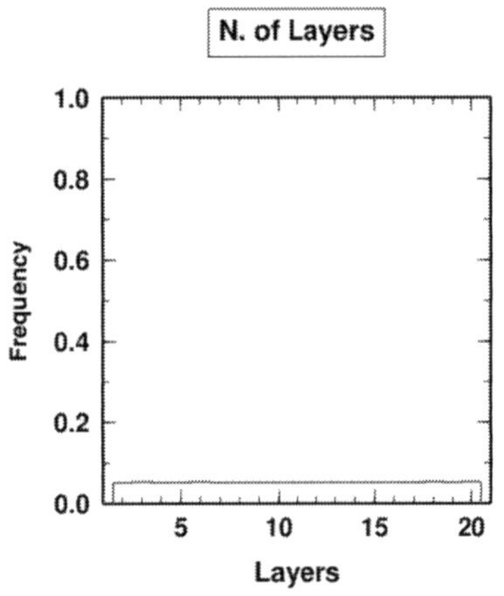

Figure 3. Prior sampling of the number of interfaces.

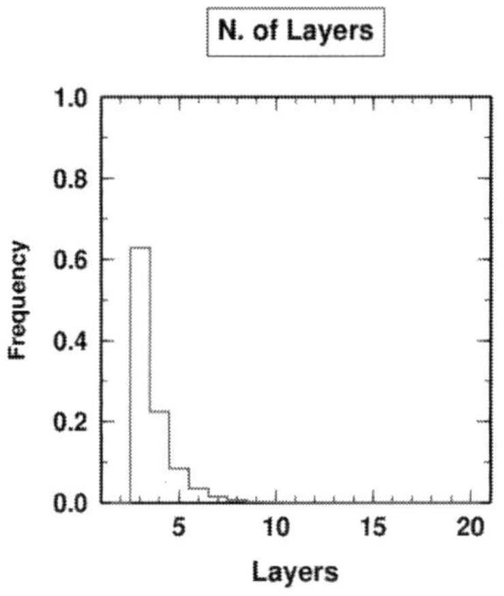

Figure 4. Posterior sampling of the number of interfaces.

distribution does not fit the data (red crosses). On the contrary the posterior distribution fits the dataset with accuracy, with the exception due to the MT method. The high frequency data are not fitted with accuracy as the topmost part of the model can be constrained only by higher frequency signals, in fact in Figure 2 is clearly shown the ambiguity in the corresponding region of the distribution.

In general the algorithm behaves as expected, retrieving the correct distribution and fitting the data using the simplest possible model. The trans-dimensional nature of the method allows to avoid the artificial use of en explicit regularization term and the superimposed selection of the appropriate number of parameters that is driven only by the data. Nevertheless, in order to validate the purposed method, other tests with both more complex synthetic models and real data are needed.

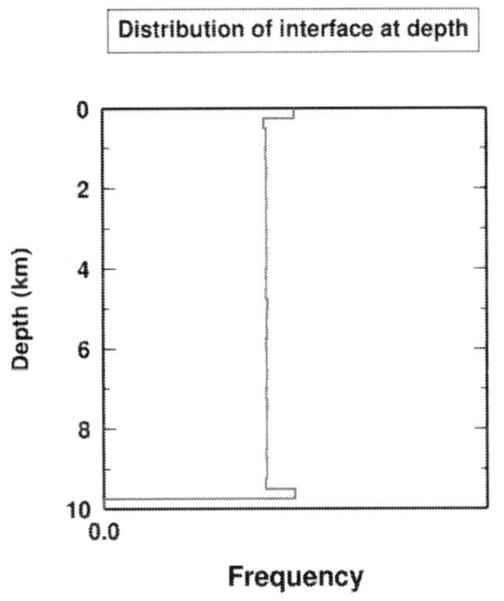

Figure 5. Prior sampling of the depth of the interfaces.

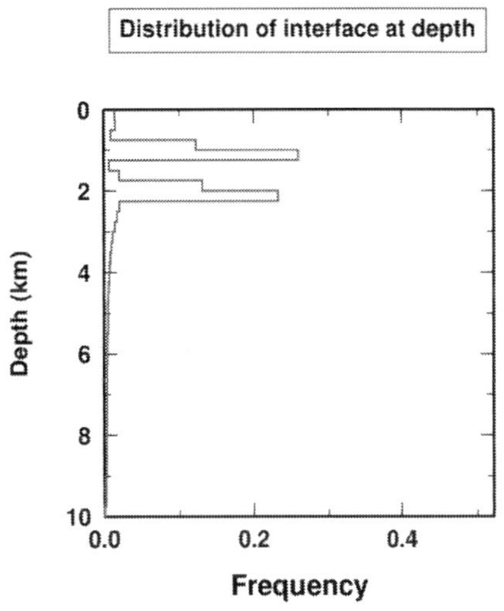

Figure 6. Posterior sampling of the depth of the interfaces.

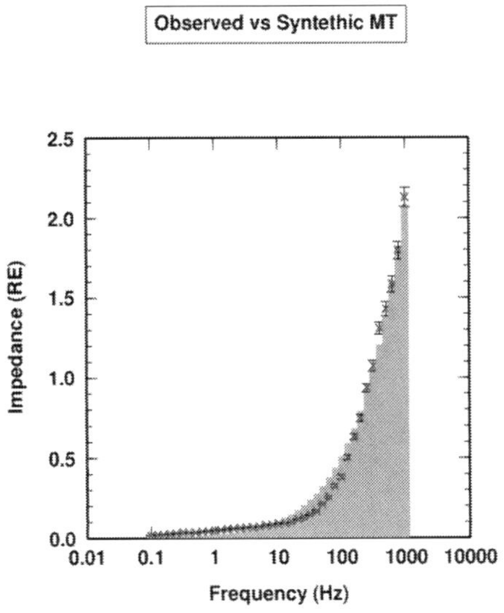

Figure 7. Prior sampling of the resistivity.

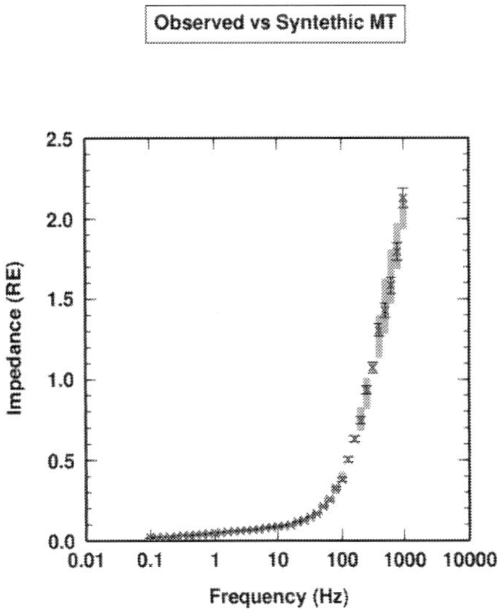

Figure 8. Posterior sampling of the resistivity.

[1] Chave A D and Jones A G 2012 *The magnetotelluric method: Theory and practice* (Cambridge University Press)
[2] Grandis H, Menvielle M and Roussignol M 1999 *Geophysical Journal International* **138** 757–768
[3] Agostinetti, N Piana and Malinverno A 2010 *Geophysical Journal International* **181** 858–872

Classification of Histological Images Based on the Stationary Wavelet Transform

M Z Nascimento[1], L Neves[2], S C Duarte[3], Y A S Duarte[3], V Ramos Batista[3]

[1] *Faculty of Computer Science (FACOM), Federal University of Uberlândia (UFU), Uberlândia, MG, Brazil*
[2] *Institute of Biosciences, Literature and Exact Sciences, Department of Computer Science and Statistics, São Paulo State University (UNESP), São José do Rio Preto, SP, Brazil*
[3] *Mathematics, Computer Science and Cognition Centre, Federal University of ABC (UFABC), Santo André, SP, Brazil*

E-mail: `nascimento@facom.ufu.br`

Abstract. Non-Hodgkin lymphomas are of many distinct types, and different classification systems make it difficult to diagnose them correctly. Many of these systems classify lymphomas only based on what they look like under a microscope. In 2008 the World Health Organisation (WHO) introduced the most recent system, which also considers the chromosome features of the lymphoma cells and the presence of certain proteins on their surface. The WHO system is the one that we apply in this work. Herewith we present an automatic method to classify histological images of three types of non-Hodgkin lymphoma. Our method is based on the Stationary Wavelet Transform (SWT), and it consists of three steps: 1) extracting sub-bands from the histological image through SWT, 2) applying Analysis of Variance (ANOVA) to clean noise and select the most relevant information, 3) classifying it by the Support Vector Machine (SVM) algorithm. The kernel types Linear, RBF and Polynomial were evaluated with our method applied to 210 images of lymphoma from the National Institute on Aging. We concluded that the following combination led to the most relevant results: detail sub-band, ANOVA and SVM with Linear and RBF kernels.

1. Introduction

Lymphoma is the term used for cancer when it happens in the lymph system, and it is classified as Hodgkin and non-Hodgkin (NHL). They are further subdivided into over 60 different subtypes [1]. This neoplasm can happen at any age. According to [1] in Canada lymphoma is the commonest cancer in people under 30, and it happens to one out of every 10 children diagnosed with cancer.

The most recent numbers of 2014 in the US are 731,277 people diagnosed with lymphoma, of which 558,340 are NHL [2]. In fact, the number of NHL-cases has been increasing. As for any kind of cancer, an early diagnosis can lead to a successful treatment and also a better quality of living for the patient.

Diagnoses and prognoses of lymphoma are obtained through the analysis of microscopic histological images carried out by a specialist. Despite the good knowledge of experienced pathologists, their diagnoses can be biased by several factors like subjectiveness, limitations of the human vision, and fatigue after several working hours. In order to reduce bias, a good alternative is the help of Computer Aided Diagnosis (CAD). These systems have been showing expressive success at classifying histological images correctly [3–5]. However, to the best of our knowledge in the specific case of lymphoma there is a lack of techniques devoted to extracting characteristics that help classify its subtypes.

In our work we present a method to extract characteristics of NHL and to classify three of its subtypes, namely Follicular Lymphoma (FL), Mantle Cell Lymphoma (MCL) and Chronic Lymphoid Leukemia (CLL). Our method is based on the Stationary Wavelet Transform (SWT) [6], on the Analysis of Variance (ANOVA) and on a Support Vector Machine (SVM) algorithm.

2. Materials and Methods

Three steps constitute the method: 1) extraction of sub-bands of the image with the SWT; 2) application of ANOVA to clean noise and select relevant information; 3) classification of lymphoma subtypes through SVM. Three different kernel types (Linear, RBF and Polynomial) were applied with our method. Its diagram is depicted in Figure 1.

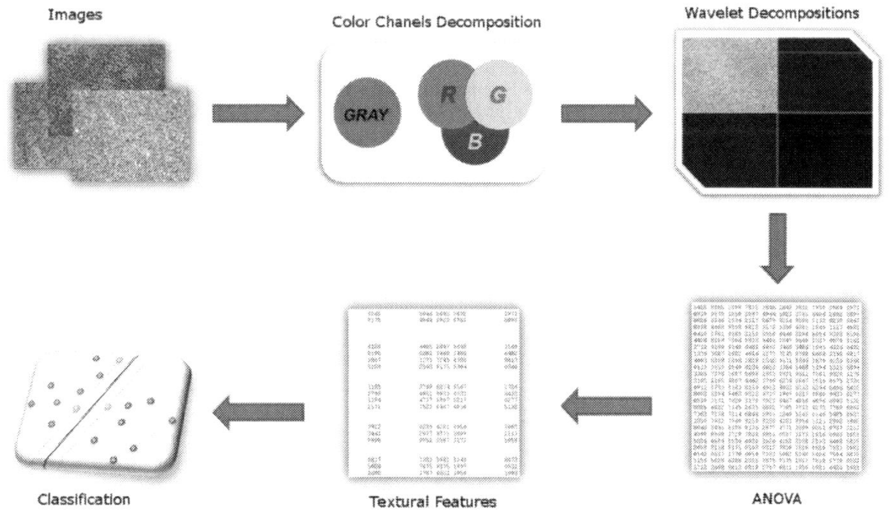

Figure 1: Diagram of the method.

3. Image Database

The database [7] was constructed by researchers of the *National Cancer Institute* and of the *National Institute on Aging*, both in the US. It contains images taken from 30 histological slides with lymph nodes stained with Haematoxylin and Eosin. They show 10 cases of each kind of lymphoma studied in our work (FL, MCL and CLL). The database contains 375 images, of which we have taken 210 at random: 70 of FL, 70 of MCL and 70 of CLL.

4. Extraction of Characteristics

We have chosen the SWT to extract sub-bands of the image because they are not downsampled by that transform, which is the case of the DWT. The SWT represents an image in the time-frequency domain (see [6] for details). In this way we kept information on low and high frequencies and obtained better results than with the DWT.

Our SWT was implemented with Quadrature Mirror Filters, which consist of recursive high- and low-pass filtering. For the multiresolution analysis we applied the wavelet function Daubechies 8 to a single decomposition level. This resulted in four sub-bands of coefficients: a sub-band of approximation (the original image in low resolution), and three of details (horizontal, vertical and diagonal).

4.1. Reduction of Attributes and Classification

ANOVA is a statistical test that compares the averages of multiple groups. As a second step one applies ANOVA to the three sub-bands of detail separately. In this way we can clean noise and select relevant information [8, 9] in order to compare the results between the sub-bands.

The SVM classifier is based on machine learning techniques and statistical learning theory. Briefly saying, SVM tries to construct hyperplanes that maximize the separation between sets of data in order to group them into classes. It will fail if the sets are not linearly separable, but in this case SVM maps data through a *kernel function* within a space of characteristics of higher dimension that eventually makes data linearly separable [10].

As a third and last step the sub-bands that resulted from ANOVA were submitted to the SVM. Here we used the *n-fold cross validation method* with $n = 10$. It consists of taking 90% of the data as training set and the remaining 10% as test set. This is done 10 times by exchanging data between the 90% and the 10%, in such a way that in the end all data will have taken part in the test. We have used three different kernel functions in order to compare the final results: Linear, RBF and Polynomial.

5. Results

Table 1 summarizes the second step of our method. The third step ~~of our method~~ is summarized in Table 2, where we see that the Linear Kernel had the best performance. Figures 2(a-d) show the graphical confusion matrices for this kernel.

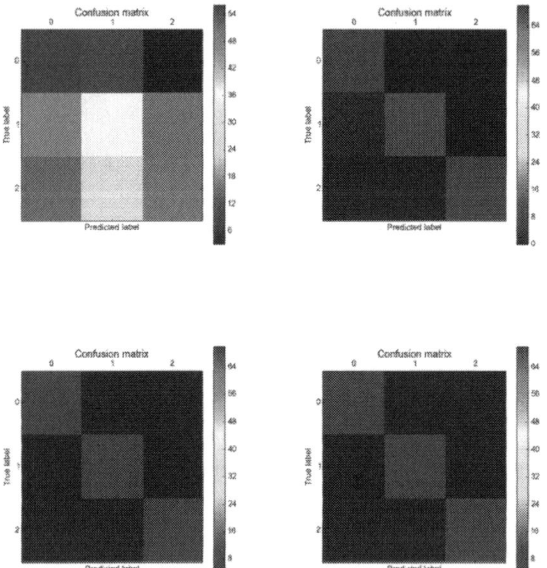

Figure 2: Confusion Matrices for (a) Approximation, (b) Horiz. (c) Vert. and (d) Diag. Details.

Table 1: Amount of attributes after ANOVA and their percentage reduction.

Sub-band	Attributes	ANOVA	Reduction (%)
Approximation	1,443,520	1,429,769	0.95
Horizontal Details	1,443,520	74,745	94.80
Vertical Details	1,443,520	75,321	94.80
Diagonal Details	1,443,520	34,236	97.60

Table 2: Results after applying SWT.

		Without Reduction		Reduction by ANOVA	
Sub-band	SVM-*Kernel*	Accuracy	Deviation	Accuracy	Deviation
Approximation	Linear	0.519	0.079	0.519	0.079
	RBF	0.514	0.044	0.514	0.044
	Polynomial	0.533	0.077	0.533	0.077
Horiz. Details	Linear	0.329	0.015	0.990	0.020
	RBF	0.433	0.065	0.948	0.035
	Polynomial	0.333	0.000	0.333	0.000
Vert. Details	Linear	0.333	0.000	1.000	0.000
	RBF	0.490	0.075	0.933	0.051
	Polynomial	0.333	0.000	0.333	0.000
Diag. Details	Linear	0.490	0.032	1.000	0.000
	RBF	0.319	0.032	0.886	0.060
	Polynomial	0.333	0.000	0.338	0.015

6. Conclusion

We presented a method to extract attributes through the analysis of textures based on SWT and ANOVA. The best results were obtained when we applied ANOVA to the vertical and diagonal coefficients, namely 100% of accuracy.

References

[1] Cancer Research UK 2014 Different types of non Hodgkin lymphoma we beat Cancer
[2] Leukemia & Lymphoma Society Team 2014 lls.org the Leukemia & Lymphoma Society (LLS) is the world's largest voluntary health agency dedicated to blood cancer URL `http://www.lls.org`
[3] Sertel O, Syst J K *et al.* 2009 *Journal of Signal Processing Systems* **55**
[4] Al-Kadi O S 2010 *Pattern Recognition* **43** 2043–2053 ISSN 0031-3203
[5] Krishnan M M R, Pal M, Bomminayuni S K, Chakraborty C, Paul R R, Chatterjee J and Ray A K 2009 *Computers in Biology and Medicine* **39** 1096–1104 ISSN 0010-4825 URL `http://www.sciencedirect.com/science/article/pii/S0010482509001668`
[6] Mallat S 1999 *A Wavelet Tour of Signal Processing* (Academic Press, San Diego, CA)
[7] on Aging N I 2008 Lymphoma - NIA Intramural Research Program Laboratory of Genetics URL `http://ome.grc.nia.nih.gov/iicbu2008/lymphoma/index.html`
[8] Vieira S 2006 *Análise de variância: ANOVA* (Atlas)
[9] Surendiran B, Sundaraiah Y and Vadivel A 2009 *Advances in Recent Technologies in Communication and Computing, 2009. ARTCom '09. International Conference on* pp 175–177
[10] Osta H, Qahwaji R and Ipson S 2008 *Systems, Signals and Devices, 2008. IEEE SSD 2008. 5th International Multi-Conference on* pp 1–6

Approximate Best Proximity Points of Cyclic Self-maps in Metric Spaces and Cyclic Asymptotic Regularity

M. De la Sen[1]
Institute of Research and Development of Processes,
University of the Basque Country,
Campus of Leioa, Bizkaia , Barrio de Sarriena 48940,Spain

E-mail: manuel.delasen@ehu.es

Abstract. This manuscript investigates some relations and properties between 2-cyclic self-mappings. In particular, a formal development is given to link some concepts and their basic related properties like those of approximate best proximity points, approximate best proximity point property and cyclic asymptotic regularity.

1. Introduction

Fixed point theory is receiving important attention in the last decades because of its applicability to many physical and engineering problems. See [1-13] and references therein. In [4-5] and other cited papers by the same authors and also some references therein, the problem of existence of approximate fixed points of functions in metric spaces is described and formalized as well as its links with asymptotic regularity of such functions including cases where contractive conditions are fulfilled . It is not required either that the metric space be complete by the same reasons. This paper extends such a formalism to 2 -cyclic self-mappings. In this way, a simple formal development is given which brings together the concepts of approximate best proximity points of 2 -cyclic self-mappings, approximate best proximity (respectively, partial best proximity) point property and cyclic asymptotic regularity of 2 -cyclic self-mappings. Let us remember that 2-cyclic self-mappings are defined on the union of two nonempty subsets of the set X where (X, d) is a metric space. Some related properties are proved while certain particular results are derived for contractive cyclic self-mappings f on $A \cup B$ with A and B being nonempty, in general disjoint, subsets of X. There are other interesting directly induced results available for composite self-mapping f^2 in $A \cup B$. It can be pointed out that the obtained results are applicable very easily to the stability and approximate stability of dynamic systems since equilibrium points of such systems are also fixed points of the mapping defining the trajectory solution from given initial conditions.

2. Problem statement

Let (X, d) be a metric space and let $f : A \cup B \to A \cup B$ be a 2 -cyclic self-mapping on the union of two nonempty subsets A and B of X. Since there are only two subsets involved, the self-mapping will be referred to simply as a cyclic self-mapping. The following definition will be then used:

[1] To whom any correspondence should be addressed.

Definitions 2.1. Let (X, d) be a metric space and let $f: A \cup B \to A \cup B$ be a 2-cyclic self-mapping on the union of two nonempty subsets A and B of X. Then:

(1) $x \in A \cup B$ is a ε-best proximity point of f (in A or in B) for a given $\varepsilon \in \mathbf{R}_{0+}$ if $d(x, fx) \le D + \varepsilon$,

$$\text{where } D = d(A, B) = \inf_{x \in A,\, y \in B} d(x, y).$$

(2) $x \in A$ is a ε-best proximity point of f in A for a given $\varepsilon \in \mathbf{R}_{0+}$ if $d(x, fx) \le D + \varepsilon$.

It turns out that $x \in A \cup B$ is an ε- best proximity point of f if and only if

$$x \in BP_{\varepsilon}(f) = \{ x \in A \cup B : d(x, fx) \le D + \varepsilon \}$$

Also $x \in A$ is an ε- best proximity point of f in A if and only if

$$x \in BP_{A\varepsilon}(f) = \{ x \in A : d(x, fx) \le D + \varepsilon \}$$

The following results are direct to prove:

Proposition 2.2. Let (X, d) and $f: A \cup B \to A \cup B$ be a metric space and a 2-cyclic self-mapping (referred to in the sequel as cyclic self-mappings) where A and B are nonempty subsets of X. Then, if $x \in A \cup B$ is an ε-best proximity point of f then it is an ε_1-best proximity point of f for any real $\varepsilon_1 \ge \varepsilon$.

Proposition 2.3. Let (X, d) be a metric space and let $f: A \cup B \to A \cup B$ be a 2-cyclic self-mapping on the union of two bounded nonempty subsets A and B of X. Then $x \in A \cup B$ is an ε-best proximity point of f then fx is an ε'- best proximity point for some $\varepsilon' \in \mathbf{R}_{+}$, then, $fx \in BP_{\varepsilon'}(f)$.

Definitions 2.4. Let (X, d) be a metric space and let $f: A \cup B \to A \cup B$ be a 2-cyclic self-mapping on the union of two nonempty subsets A and B of X. Then,

(1) $f: A \cup B \to A \cup B$ has the approximate best proximity point property if $BP_{\varepsilon}(f) \ne \varnothing$ for all $\varepsilon \in \mathbf{R}_{0+}$.

(2) $f: A \cup B \to A \cup B$ has the approximate best proximity point property in A if $BP_{A\varepsilon}(f) \ne \varnothing$ for all $\varepsilon \in \mathbf{R}_{0+}$.

(3) $f: A \cup B \to A \cup B$ has the ε_0-partial approximate best proximity point property if $BP_{\varepsilon}(f) \ne \varnothing$ for all real $\varepsilon \ge \varepsilon_0$ and a given $\varepsilon_0 \in \mathbf{R}_{0+}$.

(4) $f: A \cup B \to A \cup B$ has the ε_0-partial approximate best proximity point property in A if $BP_{A\varepsilon}(f) \ne \varnothing$ for all real $\varepsilon \ge \varepsilon_0$ and a given $\varepsilon_0 \in \mathbf{R}_{0+}$.

(5) $f^2: A \cup B \to A \cup B$ has the ε_0-partial approximate fixed point property if $FP_{\varepsilon}(f) = \{ x \in A \cup B : d(x, f^2 x) \le \varepsilon \} \ne \varnothing$; $\forall \varepsilon (\in \mathbf{R}_{+})$ for some given $\varepsilon_0 \in \mathbf{R}_{0+}$.

(6) $f^2: A \cup B \to A \cup B$ has the ε_0-partial approximate fixed point property in A if $FP_{A\varepsilon}(f) = \{ x \in A : d(x, f^2 x) \le \varepsilon \} \ne \varnothing$; $\forall \varepsilon (\in \mathbf{R}_{+})$ for some given $\varepsilon_0 \in \mathbf{R}_{0+}$.

(7) $f^2: A \cup B \to A \cup B$ has the approximate fixed point property if it has 0-partial approximate fixed point property.

Definitions 2.5. Let (X, d) be a metric space and let A and B be nonempty subsets of X with $d(A, B) = D$. Then,

(1) $f: A \cup B \to A \cup B$ is cyclic asymptotically regular if it is cyclic and

$$d\left(f^n x, f^{n+1} x\right) \to D \text{ as } n \to \infty; \; \forall x \in A \cup B$$

(2) $f: A \cup B \to A \cup B$ is cyclic asymptotically ε_0-regular,respectively, ε_0-regular in A, if it is cyclic and

$$d\left(f^n x, f^{n+1} x\right) \to D + \varepsilon_0 \text{ as } n \to \infty; \; \forall x \in A \cup B \text{ respectively } \forall x \in A$$

Proposition 2.6. Let (X, d) be a metric space and let A and B be nonempty subsets of X with $d(A, B) = D$. Then any strictly contractive cyclic self-mapping $f: A \cup B \to A \cup B$ is cyclic asymptotically regular, and equivalently, it has the approximate best proximity point property.

Proof: It is direct since, if $f: A \cup B \to A \cup B$ then

$$d\left(f^2 x, fx\right) \le K d(fx, x) + (1-K) D \tag{2.1}$$

for some $K \in [0, 1)$; $\forall x \in A \cup B$. Thus, it is cyclic asymptotically regular since

$$D \le d\left(f^{n+1} x, f^n x\right) \le K^n d(fx, x) + \left(1 - K^n\right) D \tag{2.2}$$

and $d\left(f^{n+1} x, f^n x\right) \to D$ as $n \to \infty$; $\forall x \in A \cup B$. Also, since $f: A \cup B \to A \cup B$ is cyclic asymptotically regular then there is $n_0 = n_0(\varepsilon) > 0$ for any given $\varepsilon \in \mathbf{R}_+$ such that $D \le d\left(f^{n+1} x, f^n x\right) \le D + \varepsilon$ so that $BP_\varepsilon(f) \ne \varnothing$; $\forall \varepsilon \in \mathbf{R}_+$, $\forall x \in A \cup B$. As a result, $f: A \cup B \to A \cup B$ has the approximate best proximity point property. Equivalently, if $BP_\varepsilon(f) \ne \varnothing \Leftrightarrow \left(D \le d\left(f^{n+1} x, f^n x\right) \le D + \varepsilon; \forall \varepsilon \in \mathbf{R}_+, \forall x \in A \cup B\right)$ then $\lim_{n \to \infty} d\left(f^{n+1} x, f^n x\right) = D$; $\forall x \in A \cup B$ so that $f: A \cup B \to A \cup B$ is cyclic asymptotically regular.

Lemma 2.7. Let (X, d) be a metric space and let A and B be nonempty subsets of X with $d(A, B) = D$. If $f: A \cup B \to A \cup B$ is cyclic asymptotically regular then it has the approximate best proximity point property.

Lemma 2.8. Let (X, d) be a metric space and let A and B be nonempty bounded subsets of X with $d(A, B) = D$. If $f^2: A \cup B \to A \cup B$ is asymptotically regular then $f: A \cup B \to A \cup B$ has the ε_0-partial best proximity point property for some threshold $\varepsilon_0 \in \mathbf{R}_{0+}$ with $\varepsilon_0 \le min\,(diam\, A, diam\, B)$.

Proof: One has $\left|d\left(f^{2n} x, f^{2n+1} x\right) - d\left(f^{2n+2} x, f^{2n+1} x\right)\right| \to 0$ as $n \to \infty$ so that $d\left(f^{2n} x, f^{2n+1} x\right) \to D_0 (\ge D)$ as $n \to \infty$ (2.3)
; $\forall x \in A \cup B$. Thus, $D_0 \le D + min\,(diam\, A, diam\, B) < \infty$ since A and B are bounded and each $f^{2n} x$, $f^{2n+1} x$ is one of them in A and the other one in B. Then, fix $x_0 = f^{2n} x$ there is $n_0 \in \mathbf{Z}_{0+}$ such that $d(x_0, f x_0) \le D + \varepsilon$ so that $BP_\varepsilon(f) \ne \varnothing$ for any real $min\,(diam\, A, diam\, B) \ge \varepsilon \ge \varepsilon_0$ and some $\varepsilon_0 \in \mathbf{R}_{0+}$.

Theorem 2.9. Let (X, d) be a complete metric space and let A and B be nonempty closed subsets of X with $d(A, B) = D$. Assume that A is approximatively compact with respect to B. Then $f: A \cup B \to A \cup B$ is cyclic asymptotically regular iff $f^2: A \cup B \to A \cup B$ is asymptotically regular.

Proof: Note that, since $BP_\varepsilon(f)\neq\varnothing\,;\forall\varepsilon\in\boldsymbol{R}_{0+}$, then $f:A\cup B\to A\cup B$ has 0-best proximity points in A and in B for any $\varepsilon\in\boldsymbol{R}_{0+}$ so that, in particular, $BP_0(f)\neq\varnothing$ and $f:A\cup B\to A\cup B$ has 0-best proximity points. Since A is approximatively compact with respect to B, the set $\{y\in B:d(x,A)=D\}$ is nonempty and, also, if $d(y,x_n)\to d(y,A)=D$ for some $y\in B$ and some sequence $\{x_n\}\subset A$, then there is a convergent subsequence $\{x_{n_k}\}\subset A$ of $\{x_n\}$.

a) First, it is proved that if $f:A\cup B\to A\cup B$ is cyclic asymptotically regular then $f^2:A\cup B\to A\cup B$ is asymptotically regular. Assume that $f:A\cup B\to A\cup B$ is cyclic asymptotically regular so that it has the approximate best proximity point property so that $BP_\varepsilon(f)\neq\varnothing$; $\forall\varepsilon\in\boldsymbol{R}_{0+}$ and $d(f^{n+1}x,f^nx)\to D$ as $n\to\infty$; $\forall x\in A\cup B$. Now, take $x\in A$. Since A is approximatively compact with respect to B, then $\{y\in B:d(x,A)=D\}\neq\varnothing$, and there is a convergent subsequence in A, $\{x_{2n_k}\}$ of $\{x_{2n}\}$ with the properties $d(x_n,x_{n+1})\to D$, $d(x_{2n},y)\to D$, $d(x_{2n_k},y)\to D$ as $n\to\infty$ for $x\in A$, and $\{x_{2n_k}=f^{2n_k}x\}\to z(\in A)$ with $x_{n+1}=fx_n=f^{n+1}x(\in A\cup B)$, $x_{2n}=f^{2n}x(\in A)$ since $x\in A$. Proceed by contradiction by assuming that $f^2:A\cup B\to A\cup B$ is not asymptotically regular. Then, there is $\varepsilon\in\boldsymbol{R}_{0+}$ and a sequence of positive integers $\{n_k\}$ such that $d(f^{2n_k+2}x,f^{2n_k}x)>\varepsilon$; $\forall x\in A$ with $z\in A$ and $fz\in B$ being best proximity points of A and B which are then 0-best proximity points (note that if $z\in A$ is a 0-best proximity point then $d(z,fz)=D$ so that $fz\in B$ is also a 0-best proximity point). Then, the following contradiction follows:

$$0<\varepsilon\leq \liminf_{k\to\infty} d\left(f^{2n_k+2}x,f^{2n_k}x\right)\leq \liminf_{k\to\infty} d\left(\lim_{k\to\infty}f^{2n_k+2}x,\lim_{k\to\infty}f^{2n_k}x\right)=0 \tag{2.4}$$

For $x'\in B$, we can repeat all the above reasoning for $x=fx'\in A$. In conclusion, $f^2:A\cup B\to A\cup B$ is asymptotically regular and $BP_\varepsilon(f)\neq\varnothing\Rightarrow\left[F_\varepsilon(f^2)=\{x\in A\cup B:d(x,f^2x)\leq\varepsilon\}\neq\varnothing\right]$ for any given $\varepsilon\in\boldsymbol{R}_{0+}$ if $f:A\cup B\to A\cup B$ is cyclic asymptotically regular.

b) Now, the converse implication is proved, that is, if $f^2:A\cup B\to A\cup B$ is asymptotically regular then $f:A\cup B\to A\cup B$ is cyclic asymptotically regular, equivalently,

$$\left[F_\varepsilon(f^2)=\{x\in A\cup B:d(x,f^2x)\leq\varepsilon\}\neq\varnothing\right]\Rightarrow BP_\varepsilon(f)\neq\varnothing \text{ for any given } \varepsilon\in\boldsymbol{R}_{0+}$$

or, equivalently, we prove its equivalent contrapositive logic proposition, that is, $BP_\varepsilon(f)=\varnothing\Rightarrow F_\varepsilon(f^2)=\varnothing$ for any given $\varepsilon\in\boldsymbol{R}_{0+}$. Assume on the contrary that $BP_\varepsilon(f)=\varnothing\Rightarrow F_\varepsilon(f^2)\neq\varnothing$. Then, $d(x,fx)\geq D+\varepsilon$; $\forall x\in A\cup B$ and $d(x,f^2x)<\varepsilon_1$ for some $\varepsilon\in\boldsymbol{R}_+$, some $x\in A\cup B$ and any $\varepsilon_1\in\boldsymbol{R}_+$; $\forall x\in A\cup B$. Note that, although $BP_\varepsilon(f)=\varnothing$ for $\varepsilon\in\boldsymbol{R}_+$ is being assumed, $BP_0(f)\neq\varnothing$ so that there are $z\in A$ and $fz\in B$ such that $d(z,fz)=D$ since $f:A\cup B\to A\cup B$ has 0-best proximity points in A and in B. As a result, one has for some $x\in A\cup B$:

$$d(f^2x,fx)+\varepsilon_1>d(x,f^2x)+d(f^2x,fx)\geq d(x,fx)\geq D+\varepsilon \tag{2.5}$$

By applying the above chain of inequalities to the 0-best proximity points z and fz satisfying $d(z,fz)=D$. It is proved that $d(f^2z,fz)=D$. Assume not. Then, the sequence of points (z,fz,f^2z) is generated through $f : A \cup B \to A \cup B$. If $f^2 z = z$ then $d(f^2 z, fz) = d(z, fz) = D$ holds. Assume that $f^2 z \neq z$ with $d(z, f^2 z) < \varepsilon_1$; $\forall \varepsilon_1 \in \boldsymbol{R}_{0+}$. Then, it follows from (2.5) that

$$D < d(fz, f^2 z) \le d(z, fz) + d(z, f^2 z) < \varepsilon_1 + D \tag{2.6}$$

fails for $\varepsilon_1 = 0$ so that $d(f^2 z, fz) = d(z, fz) = D$. Thus, $d(f^2 z, fz) = d(z, fz) = D$ then $D + \varepsilon_1 > D + \varepsilon$ from (2.2) and $\varepsilon_1 > \varepsilon$. This constraint fails for $\varepsilon \in \boldsymbol{R}_+$ and $0 \le \varepsilon_1 \le \varepsilon$ which contradicts that ε_1 is arbitrary.

Theorem 2.10. Let (X, d) be a metric space and let A and B be nonempty bounded closed subsets of X with $d(A, B) = D$. Assume that the cyclic self-mapping $f : A \cup B \to A \cup B$ satisfies:

$$d(f^2 x, fx) \le K d(fx, x) + (1 - K)(D + \delta(x)) \tag{2.7}$$

for $x \in A \cup B$ and some $K \in [0, 1)$, where $\delta(x) = \varepsilon_{0A}$ if $x \in A$ and $\delta(x) = \varepsilon_{0B}$ if $x \in B$. Then, $f : A \cup B \to A \cup B$ is cyclic asymptotically ε_{0A}-regular in A and cyclic asymptotically ε_{0B}-regular in B and it has both the ε_{0A}-partial best proximity point property in A and the $\varepsilon_{0B} (\neq \varepsilon_{0A})$-partial best proximity point property in B. Also, $f^2 : A \cup B \to A \cup B$ has not the approximate fixed point property and, equivalently, it is not cyclic asymptotically regular.

3. Concluding remarks

In this paper, a formal study is given which compares the close concepts of approximate best proximity points of 2 -cyclic self-mappings and approximate best proximity point property and cyclic asymptotic regularity. Some related properties are proved while particular results are given for contractive cyclic self-mappings. There are other derived results available in the paper concerned with composite self-mappings of the original mapping.

References

[1] Cho YJ, Kadelburg Z, Saadati R and W. Shatanawi W 2012 *Discrete Dynamics in Nature and Society* *2* **012** Article ID184534 doi: 10.1155/2012/184534

[2] Nashine HK and Khan MS 2010 *Applied Mathematics Letters* **23** pp 121-12, 2010, doi: 10.1016/j.aml.2009.06.025

[3] Khan MS , Rao KPR and Rao KRK 2011 *Journal of Advances Studies in Topology* **2** pp 37-45

[4] Dey D A, Laha AKR and Saha M 2013 **2013** Article ID 962058 4 pages doi: dx.doi.org/10.1155/2013/962058

[5] Dey D and Saha M 2013 *Acta Mathematica Universitatis Comenianae* **LXXXII** pp 119-123

[6] De la Sen M 2009 *Fixed Point Theory and Applications* **2009** Article ID 826438 doi: 10.1155/2009/826438, 2009

[7] De la Sen M and Agarwal RP 2011 *Fixed Point Theory and Applications* **2011**:102 doi:10.1186/1687-1812-2011-102

[8] De la Sen M, Agarwal RP and Nistal R 2013 *Journal of Nonlinear and Convex Analysis*, **14** pp 661-686

[9] De la Sen M 2013 *Abstract and Applied Analysis* **2013** Article ID 183174

[10] De la Sen M and Ibeas A 2013 *Applied Mathematics and Computation*. **219** pp 10655-10667

[11] De la Sen M 2010 Fixed Point Theory and Applications **2010** Article ID 572057

[12] De la Sen M 2009 Fixed Point Theory and Applications **2009** Article ID 410187

[13] De la Sen M 2011 Fixed Point Theory and Applications **2011** Article ID 867932

Automatic Method to Classify Images Based on Multiscale Fractal Descriptors and Paraconsistent Logic

E Pavarino[1], L A Neves[1], M Z Nascimento[1], M F Godoy[3,4], P F Arruda[4], D S Neto[5]

[1] São Paulo State University (UNESP), Department of Computer Science and Statistics (DCCE), Brazil
[2] Federal University of Uberlândia, FACOM, Brazil
[3] FAMERP - São José do Rio Preto, Brazil
[4] Transdisciplinary Center for Study of Chaos and Complexity (NUTECC), Brazil
[5] Hospital de Base de São José do Rio Preto, Department of Pathology, Brazil

E-mail: pavarinodu@gmail.com

Abstract. In this study is presented an automatic method to classify images from fractal descriptors as decision rules, such as multiscale fractal dimension and lacunarity. The proposed methodology was divided in three steps: quantification of the regions of interest with fractal dimension and lacunarity, techniques under a multiscale approach; definition of reference patterns, which are the limits of each studied group; and, classification of each group, considering the combination of the reference patterns with signals maximization (an approach commonly considered in paraconsistent logic). The proposed method was used to classify histological prostatic images, aiming the diagnostic of prostate cancer. The accuracy levels were important, overcoming those obtained with Support Vector Machine (SVM) and Best-first Decicion Tree (BFTree) classifiers. The proposed approach allows recognize and classify patterns, offering the advantage of giving comprehensive results to the specialists.

1. Introduction

The biological vision system is one of the most important alternative for human beings to explore the world, smoothly performing complex functions, such as analysis, recognition and classification of patterns. Many researches try to produce computational systems with the same efficiency of the biological system [1]. This task is complex, especially to transpose one of the most evident problem, the quantification and classification of information represented in different domains, such as intensity of gray levels, borders, contours and texture. A computer-aided diagnosis (CAD) can be an alternative. The desired specificity and sensibility for a CAD can be obtained if produced with techniques inspired in natural phenomena, capable to treat spatial and temporal data efficiently [2]. Thus, it is possible to highlight fractal approaches to quantify attributes [3,4] and paraconsistente logic to obtain classifications compatible with human reasoning [5,6].

In this context, an automatic method is presented to classify images from fractal descriptors, such as multiscale fractal dimension and lacunarity, and paraconsistent logic. As a first application, the proposed method was used to classify histological prostatic images, aiming the diagnostic of prostate cancer. The proposed approach allows recognize and classify patterns in the studied context, offering the advantage of giving comprehensive models to the specialists.

2. Methodology

The proposed method was organized as follows: formalization of multi-scale fractal dimension and lacunarity, which were used to quantify images given as input; definition of the classifier core, using the values of the descriptors as references; definition of the quantitative evaluation applied to validate the proposed method; and, description of the features from prostatic images representing normal, hyperplastic and tumor/cancer groups (the first application).

From each image given as input, the first step consisted in segmentation and quantification of the regions of interest. The method chosen to segment images was presented by [4]: the algorithm is an automatic technique and does not require specific knowledge about the explored context. Then, each segmented region was quantified from multiscale fractal dimension and lacunarity techniques. The fractal dimension was applied as a measure of the complexity of the organization of pixels in the region of interest or image [2-4]. The multiscale approach was obtained by overlapping a mesh of squares in the analyzed image. The objective was to obtain the number of frames needed to cover it. Therewith, considering an image I given as input, the number of squares that contains part of the form I, $N_l(I)$, depended of the size of the l box. This relation allowed estimating the fractal dimension D, equation (1).

$$D = -\lim_{l \to 0} \frac{\ln(N_l(I))}{\ln(l)}. \tag{1}$$

The multiscale approach requires a limit, which is the count of frames, $N_l(I)$, performed for different values of l (l_0=max(height, width) and $l_i+1=l_i / 2$). The approximation of a straight line was obtained by the regression (log-log graphic) of $N_l(I)$ (number of occupied boxes) by l (size of the side of this box). This allowed to define D=–θ as the fractal dimension of I in the multiscale approach.

The lacunarity was applied to quantify the pixel's organization in a determined region of the image (quantifying how the space is filled). The method used was the *Gliding-box* [2-4]. The lacunarity complemented the evaluations made by fractal dimension, because shapes with the same value of fractal dimension can present different values of lacunarity. The process was started with a box of size l positioned in the upper left corner of the image and counting the number of image points. This process was repeated to all rows and columns of the image, producing a distribution of frequency of the image mass. The number of boxes with size l containing a value of mass M of the interest region was designated by $n(M,l)$. The total counted boxes in the image was designated by the probability $P(M,l)$, equation (2). The first (A^1) e second (A^2) moments of this distribution were determined in the equations (3) e (4). The lacunarity L for a box of size l is defined in the equation (5).

$$P(M,l) = \frac{n(M,l)}{N(l)}. \tag{2}$$

$$A^{(1)} = \sum MP(M,l). \tag{3}$$

$$A^{(2)} = \sum M^2 P(M,l). \tag{4}$$

$$L(l) = \frac{A^{(2)}}{\left(A^{(1)}\right)^2}. \tag{5}$$

The classifier allows to treat the descriptors as feature vector (L-dimensional vector) of the type: A of order [n, d], where n is the number of analyzed images and d is the quantity of considered descriptors. In this work two descriptors (fractal dimension and lacunarity) were used: d=2. So, each pair of columns from the vector/matrix A contains the limiting values of the intervals of each descriptor considered. These values were verified to determine the interval that best represents each descriptor, inclusive treating possible overlaps. Therefore, the core classifier was composed by treating two types of overlaps between the intervals A e B: (1) A e B overlapped, but A $\not\subset$ B and B $\not\subset$ A; and (2) A e B overlapped, but A $\subset$ B and/or B $\subset$ A. For both types of overlapping, the classifier uses a numeric value α called "cut-off point (threshold value)", which is responsible for indicating the most appropriate value to classify the classes. The difference between the cases (1) and (2) it is the manner that the value of α is defined. Being S the overlapping interval between A and B, in overlaps of type (1), α was obtained as follows: (a) minA < minB and minB < maxA < maxB → S = [minB, maxA] → α = minB. Thus, A e B will be redefined as A = [minA, α[and B = [α, maxB]; and, (b) minB < maxA e min A < maxB < maxA → S = [minA, maxB] → α = maxB. Thus, B and A will be redefined as B = [minB, α[e A = [α, maxA]. For overlaps of type (2), in which the approach (1) does not apply, the value of α was defined as a cut-off point calculated from the Receiver Operating Characteristics (ROC) [7].

From the intervals calculated previously, the classifier core compares each row of the matrix, that represents a case to be studied, and assigns binary values called degree of certainty (*Gc*) and degree of uncertainty (*Gi*): this degrees define if a given row *i* (case) belongs or not to the interval of certain descriptor *d*. If it belongs, $Gc(i, d) = 1$ and $Gi(i, d) = 0$; otherwise, $Gc(i, d) = 0$ e $Gi(i, d) = 1$. Then, for each case *i*, the values of *Gc* where maximized (applying logic gate OR) and the values of *Gi* where minimized (applying logic gate AND): this procedure is commonly explored in paraconsistent logic [6,7]. The classification of each row *i* was performed comparing the degrees of certainty and uncertainty, allowing to identify the class of a specific case, made as follows: being $Gc^X(i,d)$, $Gi^X(i,d)$, $Gc^Y(i,d)$ e $Gi^Y(i,d)$ the degrees of certainty and uncertainty of an image *i* belong to the groups X e Y, according a descriptor *d*=2, then *i* will be classified as belonging to a class X if, and only if, $Gc^X(i,d) > Gc^Y(i,d)$ e $Gi^X(i,d) < Gi^Y(i,d)$.

As a first application context, the proposed method was tested using images with recognized scientific importance: detection of prostate cancer [4]. The bank of prostatic images used in this work was composed of 99 histological slides, obtained from 35 patients' prostates, carriers of prostate cancer, aged between 50 and 75 years (64.4 ± 5.9 years). The extraction of each prostate allowed producing histological slides from fragments removed from normal, hyperplastic and tumor areas of the same prostate. The microscopic images were acquired digitally, using a trinocular microscopic Olympus BX41, with planacromatic objective of 10x coupled to a digital camera Color Samsung SCC-31, with an adapter Olympus U-TV1X-2. Each slide was photographed digitally with 40x of magnification and saved in the jpeg file format. The images constituted the groups: normal, hyperplastic and tumor/cancer, with 33 cases each.

The discriminative power from the proposed method was evaluated applying known artificial intelligence algorithms, such as support vector machine (SVM) and Best-First Decision Tree (BFTree) [8]. The features matrixes, with values of fractal dimension and lacunarity, were compared two by two (normal *versus* hyperplastic, normal *versus* cancer and hyperplastic *versus* cancer), considering all the classifiers and the segmented regions stroma and lumen. This procedure resulted in 396 comparisons. The performance between groups and techniques were measured from accuracy, which is defined as the proportion of true results (both true positives (TP) and true negatives (TN)) in relation to all positive (P) and all negatives (N), equation (6) [7,8].

$$Accuracy = \frac{TP+TN}{P+N}. \tag{6}$$

3. Results

The performance of the method was evaluated from the previously described strategies and using prostatic images segmented in regions of stroma and lumen (application context). The values of accuracy are presented in figure 1, involving the regions of stroma and lumen of the groups normal (N), hyperplastic (H) and tumor (T). The values were obtained from the features matrixes composed by values of multiscale fractal dimension and lacunarity, which were used as input to the classifiers SVM, BFTree and that proposed in this work.

The results indicate that the proposal is relevant, providing accuracy values higher than those obtained from other recognized classifiers. The developed method overcame the SVM classifier in all the realized testes, considering the regions of interest stroma and lumen. This has not occurred regarding the classifier BFTree, which overcame the described approach in the comparisons between the classes: normal and hyperplastic. Nevertheless, the developed method showed better performance for the other testes (normal *versus* tumor and hyperplastic *versus* tumor), considering the regions of stroma and lumen, too. This is an important advantage, demonstrating that the presented technique allowed better distinction between the classes that requires therapeutic procedures totally different in the clinical practice. Another important observation for the clinical practice is that, in a 40x magnification, the developed method indicated lumen as the most relevant region to distinguish the classes normal, hyperplastic and tumor. Regarding the descriptors used, the fractal dimension was decisive in 66.67% of the tests to classify the groups from lumen, mainly between normal *versus* hyperplastic and hyperplastic *versus* tumor. This percentage was obtained with lacunarity too, but for comparisons with the region of stroma and highlighting to classify the groups normal *versus* tumor and hyperplastic *versus* tumor.

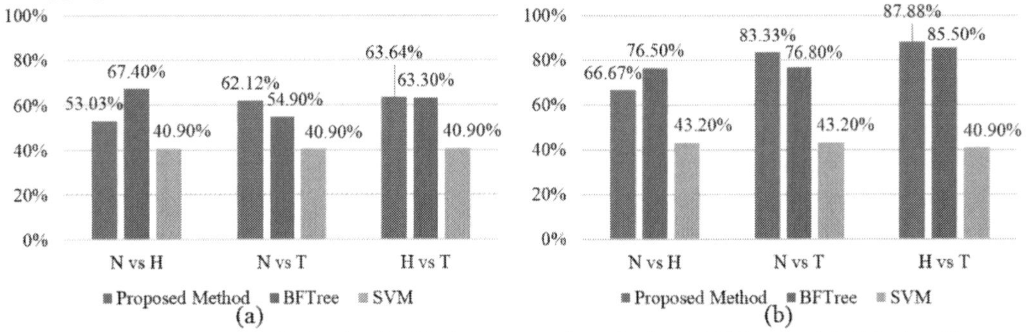

Figure 1. Accuracy levels obtained with the proposed method and BFTree and SVM classifiers, considering lacunarity and fractal dimension as descriptors. The results obtained with stroma are represented in (a) and with lumen in (b).

4. Conclusion

This work presented a new technique to classify images using fractal descriptors (fractal dimension and lacunarity) and paraconsistent logic. In a first context, the proposed model was subjected to tests with prostatic images with 40x of magnification to prove your efficiency. Through the values adopted as reference for the studied classes (normal, hyperplastic and tumor/cancer), the classifications reached higher levels of accuracy than other recognized classifiers in the majority of tests performed. The highest accuracy was approximately 88% and the method indicated lumen as the most relevant region to distinguish groups. By the results achieved, the proposed technique can be considered as a viable alternative to classify signals, with the advantage of making the results understandable to specialists.

Acknowledgements

This work was financially supported by PROPe/UNESP (Pró-Reitoria de Pesquisa/UNESP).

References

[1] Jain A K and Duin R 2000 Statistical pattern recognition: A review *IEEE Trans. on Patt. An. and Mac. Intel.* **22** 1 4–37

[2] Neves L A et al 2011 Maximum entropy, fractal dimension and lacunarity in quantification of cellular rejection in myocardial biopsy of patients submitted to heart transplantation *J. of Phys.: Conf. Series* **285** 1

[3] Wan-rong S, Bian-zhang Y, Xiao-jing Z and Zheng-hui, Z 2005 Fractal feature extraction of marrow cell images *Proceeedings of the 2005 IEEE, Engineering in Medicine and Biology 27th Annual Conference, Shanghai, China*

[4] Neves L A et al 2014 Multi-scale lacunarity as an alternative to quantify and diagnose the behavior of prostate cancer *Expert Systems with Applications* **41.11** 5017-29

[5] Da Costa N C, Krause D and Bueno O 2007 Paraconsistent Logics and Paraconsistency *Handbook of the Philosophy of Science, Philosophy of Logic, Elsevier*

[6] Lopes H F S, Abe J M and Anghinali R 2009 Application of Paraconsistent Artificial Neural Networks as a Method of Aid in the Diagnosis of Alzheimer Disease *J. Med. Syst.*

[7] Fawcett T 2006 An introduction to ROC analysis *Pattern recognition letters* **27.8** 861-74

[8] Duda R O, Hart P E and Stork D G 2012 Pattern classification *John Wiley & Sons*

Computer Simulation of Fire Dynamics in Industrial Hall

J Glasa[1,3] and V Steckova[2]

[1] Institute of Informatics, Slovak Academy of Sciences, Dubravska cesta 9, 845 07 Bratislava, Slovakia

[2] Faculty of Civil Engineering, Slovak University of Technology, Radlinskeho 11, 813 68 Bratislava, Slovakia

E-mail: jan.glasa@savba.sk

Abstract. In this paper, computer simulation of smoke spread dynamics in industrial hall is investigated. A set of simulations of fire in three industrial halls with the same geometry varying in the height of ceiling is realized using the FDS fire simulator, version 6. The obtained simulation results are described focusing on the impact of the ceiling height and fire barriers on the fire course and smoke spread dynamics.

1. Introduction

Advances in Computational Fluid Dynamics (CFD) stimulate the development of fire models capable to describe fire in environments with complex geometry incorporating a wide variety of physical phenomena related to fire. Several fire simulators based on CFD fire models are now available for instance the CFX, SMARTFIRE and FDS systems. The aim of this paper is to demonstrate the use of FDS (Fire Dynamics Simulator) [5] for analysis of smoke dynamics in industrial halls in the case of fire. FDS models fire-driven fluid flows. It solves numerically a form of Navier-Stokes equations appropriate for low-speed (Ma < 0.3), thermally-driven flow with an emphasis on smoke and heat transport from fire. The core algorithm is an explicit predictor-corrector scheme, second order accurate in space and time. Turbulence is treated by means of Large Eddy Simulation (for larger spaces) or Direct Numerical Simulation (for very fine mesh resolution). FDS comprises models for main relevant physical and chemical processes related to fire such as combustion, pyrolysis, radiation transport, suppression, etc. In December 2013, the 6th version of FDS was released. For multi-mesh calculations and multi-processor computer systems, parallel versions of FDS are also available. FDS has been used for simulation of fires in various structures such as theatre, supermarket, office building, airport boarding passage and nuclear power plant. However, there are no papers in the literature studying the impact of the height of ceiling of industrial halls on the smoke spread dynamics in case of fire. In this paper, we utilize our actual experience with FDS [1, 2, 3, 6] and use the system for such analysis.

2. Computer simulation of fire in industrial hall using FDS

We consider three industrial halls with the same geometry that differ from each other by the ceiling height (figure 1). *The first hall* is 3 m high. It is compartmented into two rooms by separating wall of the 2.2 m height and 10 cm thickness in which a closed 2 m high alloy door is located. The vertical smoke barrier mounted on the ceiling above the separating wall is 30 cm high and 10 cm thick. The

[3] To whom any correspondence should be addressed.

distance between the barrier and separating wall is 0.5 m. All walls, ceiling and barrier are composed from concrete. The floor is made from inert material. There are not any vents providing interaction between fire and external environment. Two 2.2 m high alloy doors in the hall are closed. Symmetrically arranged manufacturing desks composed from inert material are 0.9 m high. *The second hall* is 6 m high. The distance between the fire barrier and separating wall is 3.5 m. *The third hall* is 9 m high. The distance between the barrier and separating wall is 6.5 m. Other parameters of the halls are the same. The floor, walls and ceiling are represented by computational domain borders with the corresponding boundary conditions. The manufacturing desks, separating wall, doors and fire barrier are represented by obstructions (table 1). A form of technological accident leading to ignition of flammable technical material which fell down from manufacturing desk is considered. We assume that smoke will enter the next room and that the vertical barrier and separating wall will slow down the fire. The fire source is represented by a 0.5 x 1 x 0.2 m burning block with 4 burning faces of total fire area of 1 m^2 and 650 kW/m^2 HRRPUA boundary condition. In all scenarios, the same fire source located at the same place is considered.

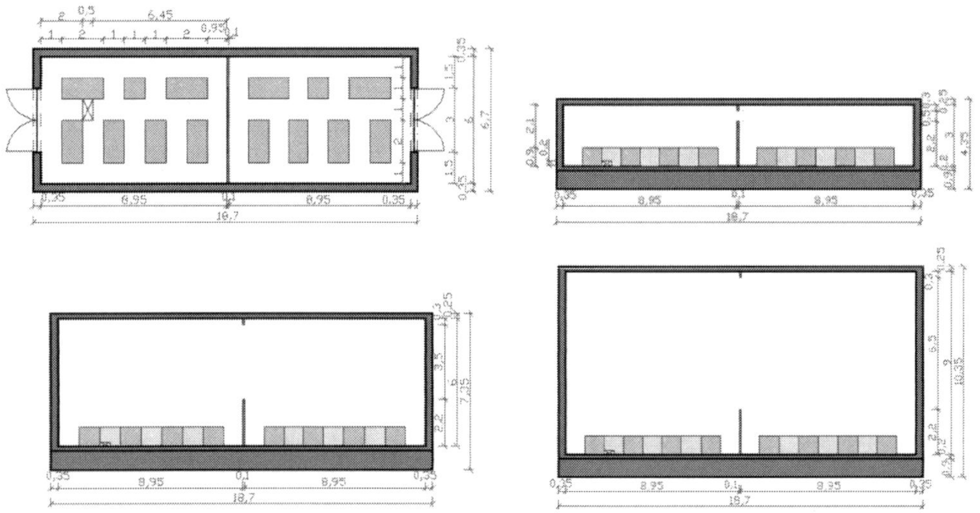

Figure 1. Scheme of 3 industrial halls considered: ground plan and side view.

Table 1. Simulations parameters: N_O, N_V and N is the total number of OBSTs (obstructions), VENTs and computational mesh cells, respectively; N_C is the number of cells per CPU core; N_x, N_y and N_z is the number of cells in x-, y- and z-direction, respectively; T is the total computational time.

Scenario	N_O	N_V	N	N_C	N_x	N_y	N_z	T
Hall 1	22	6	2592000	432000	60	120	60	12 hrs 15 min 15 s
Hall 2	22	6	5184000	864000	60	120	120	24 hrs 9 min 15 s
Hall 3	22	6	7776000	1296000	60	120	180	37 hrs 18 min 49 s

Table 2. Smoke spread: T_C, T_{LW}, T_{BW}, T_{FW}, T_{SW} and T_{FB} is the time of reaching the ceiling, left-, back-, front-, and separating walls, and the fire barrier, respectively.

Scenario	T_C	T_{LW}	T_{BW}	T_{FW}	T_{SW}	T_{FB}
Hall 1	1.9 s	2.9 s	3.1 s	4.0 s	6.7 s	6.7 s
Hall 2	3.3 s	4.1 s	4.0 s	5.1 s	35.5 s	7.4 s
Hall 3	5.1 s	4.0 s	4.5 s	6.6 s	40.5 s	8.8 s

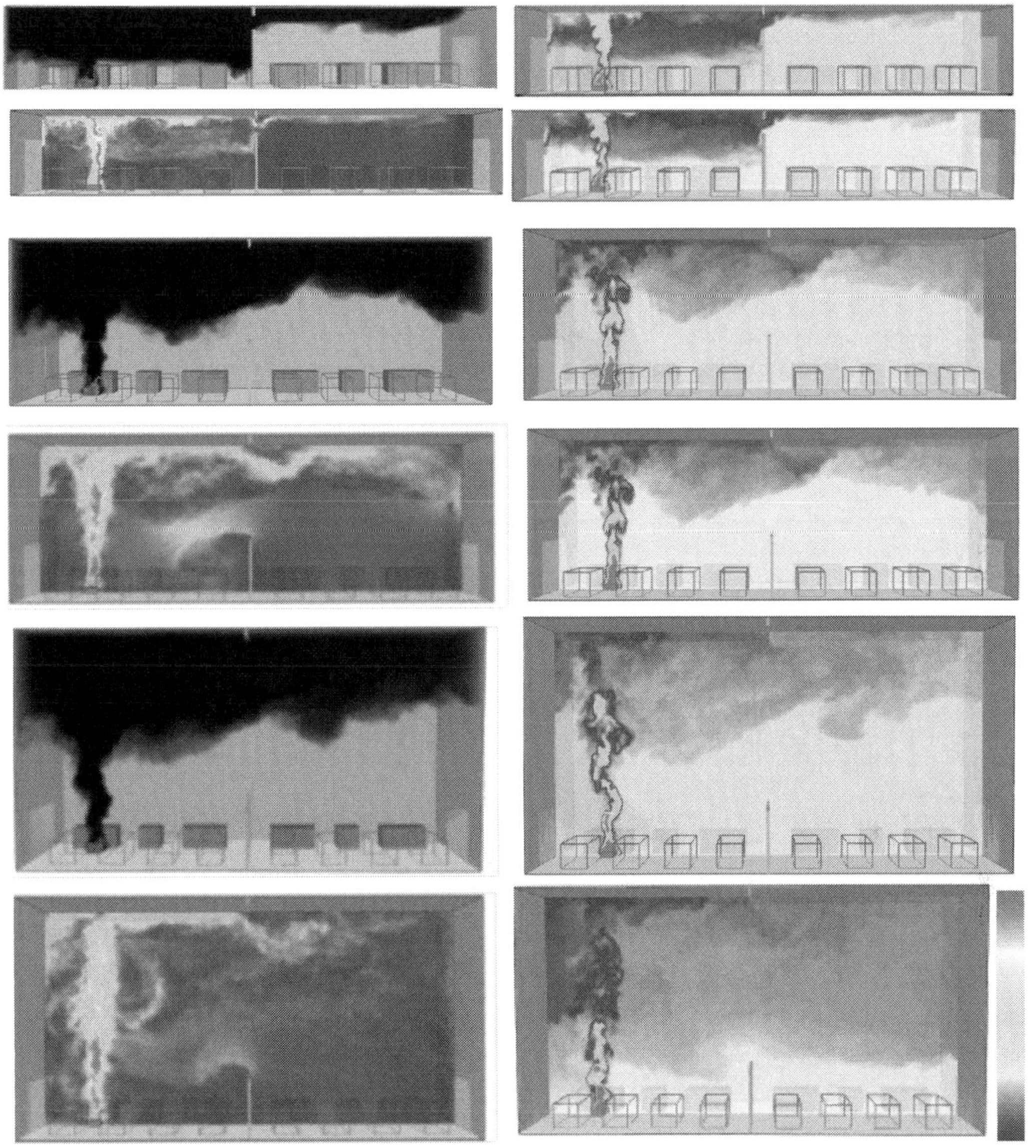

Figure 2. Fire simulation at the 23rd s: smoke spread dynamics and slices of gas temperature, velocity and soot concentration for 3 halls considered; the same colour scheme is used for temperature, velocity and concentration varying between 20 and 820°C, 0 and 5 m/s, and 0 and 0.015 mol/mol, respectively.

The calculations were realized in parallel on a 6-core personal computer (Intel i7-990x, 3.46 GHz CPU, 24 GB RAM) using all 6 CPU cores. In all 3 scenarios, the 18 m long and 6 m wide computational domain was divided into 6 computational meshes of 5 cm resolution (see table 1); all meshes were fine in the sense of mesh sensitivity study [4, 5]. The course of fire is illustrated in figure 2 and table 2. At first, smoke propagated upwards and then stratified under the ceiling. After reaching vertical barriers (the back, left and front walls, and fire barrier), turbulent clouds of smoke started to form spreading along the barriers downwards. The clouds size depended on velocity of gases movement. Later, a smoke layer height equalizing was observed which formed a new layer propagating backwards under the original layer. Main tendencies of spread of different smoke layers are illustrated in figure 2. In the *1st hall*, the fire barrier slowed entering smoke into the 2nd room. Moreover, only smaller part of smoke entered through the vent between the fire barrier and separating wall and began to stratify under the ceiling in the 2nd room. The best part of smoke moved downwards along the separating wall in the 1st room and formed a turbulent cloud of smoke (at the 12th s it reached the 1.6 m depth). Later, an equalizing of smoke layer height was observed. At the end of the 1st min, the 1st room was completely filled by smoke and a smoke stratification comprised of 3 relatively laminar layers under the ceiling was observed in the 2nd room. In this case, the fire barrier markedly decelerated the smoke spread from the 1st to the 2nd room. Since the distance between the fire barrier and separating wall was small, overall circulation between the 1st and 2nd rooms was limited. We observed a weak flow coming from the 2nd room moving downwards along the separating wall in the 1st room which started to rise and stabilized in a roughly horizontal direction. The main tendencies of fire course in the *2nd and 3rd halls* were similar; however, the vent between the fire barrier and separating wall was substantially bigger and significantly affected the smoke dynamics. Higher volume of cold air in halls with higher ceiling decreases the temperature of hot gases released from fire by phased hot/cold air mixing. Higher volume of oxygen supported combustion and caused that the temperature in proximity to fire source was markedly higher. The simulation results indicate that the ceiling height has a marked impact on the fire spread and fire safety in industrial halls. In both cases, the distance between the fire barrier and separating wall was big enough to enable circulation between the 1st and 2nd rooms. Because of small height of the fire barrier, smoke released form the fire source passed beneath the fire barrier with very limited delay only. The separating wall was too small to markedly decelerate the circulation between the 1st and 2nd room and inlet of fresh air to the fire source, and affected the smoke spread only a little. To provide sufficient decrease of smoke spread from the 1st to the 2nd room in the 2nd and 3rd halls, it would be necessary to reduce the size of vent between the fire barrier and separating wall, increasing the fire barrier depth or separating wall height.

3. Conclusions

In this paper, the impact of the ceiling height and fire barriers on the smoke dynamics in industrial halls in case of fire is illustrated using the FDS fire simulator. A great potential of the CFD-based fire model for analysis of fire-induced thermal flows is demonstrated. The simulation results indicate that FDS is able to capture main tendencies of smoke dynamics in the case of industrial hall fire and describe specific physical phenomena related to the type of fire; therefore, FDS has potential for the use for analysis of safety risks and suggestion of fire safety measures.

4. Acknowledgment

This work was partially supported by National Science Agency VEGA (contract No. 2/0184/14).

5. References

[1] Glasa J, Valasek L, Halada L and Weisenpacher P 2014 *Journal of Physics: Conference Series* **490** 012067 doi:10.1088/1742-6596/490/1/012067

[2] Glasa J, Valasek L, Weisenpacher P and Halada L 2012 *International Journal on Recent Trends in Engineering and Technology* **7** (2) 51-56

[3] Glasa J, Valasek L, Weisenpacher P and Halada L 2013 *Journal of Physics: Conference Series* **410** 01203 doi:10.1088/1742-6596/410/1/012013

[4] Hill K, Dreisbach J, Joglar F, Najafi B, McGrattan K, Peacock R and Hamins A 2007 Verification and Validation of Selected Fire Models for Nuclear Power Plant Applications *NUREG* **1824** U. S. Nuclear Regulatory Commission Washington USA

[5] McGrattan K, Hostikka S, McDermott R, Floyd J, Weinschenk C and Overholt K 2013 Fire Dynamics Simulator (Version 6), User's Guide *NIST Special Publication* **1019** Gaithersburg

[6] Weisenpacher P, Glasa J, Halada L, Valasek L and Sipkova V 2014 Parallel computer simulation of fire in road tunnel and people evacuation *Computing and Informatics* in press

Feedback stabilization of an oscillating vertical cylinder by POD Reduced-Order Model

Gilles Tissot, Laurent Cordier, Bernd R. Noack
PPRIME Institute, CEAT 43 route de l'aérodrome, 86036 Poitiers, France
E-mail: Gilles.Tissot@univ-poitiers.fr, Laurent.Cordier@univ-poitiers.fr, Bernd.Noack@univ-poitiers.fr

Abstract. The objective is to demonstrate the use of reduced-order models (ROM) based on proper orthogonal decomposition (POD) to stabilize the flow over a vertically oscillating circular cylinder in the laminar regime (Reynolds number equal to 60). The 2D Navier-Stokes equations are first solved with a finite element method, in which the moving cylinder is introduced via an ALE method. Since in fluid-structure interaction, the POD algorithm cannot be applied directly, we implemented the fictitious domain method of Glowinski *et al.* [1] where the solid domain is treated as a fluid undergoing an additional constraint. The POD-ROM is classically obtained by projecting the Navier-Stokes equations onto the first POD modes. At this level, the cylinder displacement is enforced in the POD-ROM through the introduction of Lagrange multipliers. For determining the optimal vertical velocity of the cylinder, a linear quadratic regulator framework is employed. After linearization of the POD-ROM around the steady flow state, the optimal linear feedback gain is obtained as solution of a generalized algebraic Riccati equation. Finally, when the optimal feedback control is applied, it is shown that the flow converges rapidly to the steady state. In addition, a vanishing control is obtained proving the efficiency of the control approach.

1. Flow configuration

The cylinder wake is a common generic configuration to test control methods which can be further implemented in more complex engineering applications. The stabilization of this flow consists in targeting and maintaining the unstable steady solution by control. Here, we consider the cylinder wake in fluid-structure interaction and suppose that the control is introduced by vertical oscillations of the cylinder, as sketched in figure 1. Let $\boldsymbol{u}$ be the velocity field and $\boldsymbol{u}^{\text{steady}}$ be the targeted steady state, the objective is to determine the best cylinder vertical velocity $V_c(t)$, taken as control parameter, such that we minimize

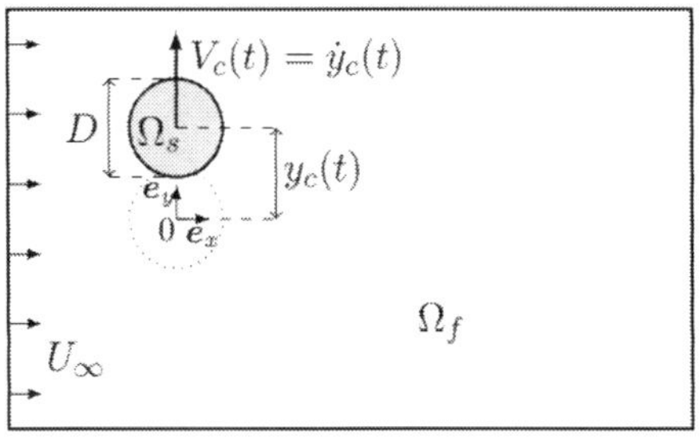

Figure 1. Cylinder wake in vertical oscillation.

$$\mathcal{J}(V_c(t)) = \int_0^T \|\boldsymbol{u}-\boldsymbol{u}^{\text{steady}}\|^2 \,\mathrm{d}t + \alpha \int_0^T y_c^2(t)\,\mathrm{d}t + \beta \int_0^T V_c^2(t)\,\mathrm{d}t$$

where y_c is the cylinder vertical position. The cost functional $\mathcal{J}$ is regularized to avoid very large, then unphysical, controls. Indeed, the role of the parameter α is to penalize too large cylinder displacements and then to enforce the cylinder to stay near the central position, while β penalizes too strong control actions.

2. Linear control framework

The determination of the control law is done with a linear control theory approach [2]. This framework relies on the existence of a linear model $E\frac{\mathrm{d}\boldsymbol{z}}{\mathrm{d}t} = A\boldsymbol{z} + B\boldsymbol{c}$, where $\boldsymbol{z}$ is the state of the system and $\boldsymbol{c}$ the control law to be determined. Considering that the state of the system is fully and perfectly known, the linear quadratic regulator (LQR) method is considered. The objective of the LQR problem is to determine the best control law $\boldsymbol{c}$, such that the cost functional

$$\mathcal{J}^{LQR}(\boldsymbol{z}, \boldsymbol{c}) = \frac{1}{2}\int_0^T \|F\boldsymbol{z}(t)\|^2 \,\mathrm{d}t + \frac{1}{2}\int_0^T \|\boldsymbol{c}(t)\|_R^2 \,\mathrm{d}t \tag{1}$$

is minimized. In (1), F is a matrix defining the objective to minimize, and R penalizes the control $\boldsymbol{c}$ through the norm $\|\cdot\|_R^2 = (R\cdot,\cdot)$ where $(\cdot,\cdot)$ denotes the inner product. The feedback control law that minimizes the value of $\mathcal{J}^{LQR}$ is $\boldsymbol{c} = K\boldsymbol{z}$ where the feedback gain K is found after solving a Riccati equation related to (1). For an infinite time horizon, the minimization of (1) leads to the *generalized algebraic Riccati equation* (GARE) given by

$$A^*XE + E^*XA - E^*XBR^{-1}B^*XE + FF^* = 0, \tag{2}$$

where A^* is the adjoint matrix of A. The optimal feedback gain is then $K = -R^{-1}B^*XE$ where X is solution of (2). For large-scale problems, as the one arising from spatial discretization of the Navier-Stokes equations with finite element methods, the resolution of the GARE given by (2) is still unfeasible [3]. One way is to obtain low-rank solutions of GAREs based on simulations of linear systems coupled with Proper Orthogonal Decomposition (POD) [4]. An alternative is to derive a reduced-order model (ROM) for the dynamics [5].

3. Flow modelling for optimal control in fluid-structure interaction

3.1. Fictitious domain approach

For the fluid-structure interaction problem, we consider a Lagrange-multiplier-based fictitious-domains method as in [1]. For that, let the whole computational domain $\Omega = \Omega_f(t) \cup \Omega_s(t)$ be constituted by the fluid domain $\Omega_f(t)$ and by an immersed body $\Omega_s(t)$, with $\Omega_f(t) \cap \Omega_s(t) = \emptyset$. The principle is to consider the whole domain as a fluid, governed by the Navier-Stokes equations, and to treat the solid in $\Omega_s(t)$ as a fluid subject to an additional constraint. The full system is then defined as

$$\begin{cases} \dfrac{\partial \boldsymbol{u}}{\partial t} + (\boldsymbol{u}.\nabla)\boldsymbol{u} = -\nabla p + \dfrac{1}{Re}\Delta \boldsymbol{u} \quad \forall \boldsymbol{x} \in \Omega \\ \nabla.\boldsymbol{u} = 0 \quad \forall \boldsymbol{x} \in \Omega \\ \boldsymbol{u}(\boldsymbol{x}, t) = \begin{pmatrix} 0 \\ V_c(t) \end{pmatrix} \quad \forall \boldsymbol{x} \in \Omega_s(t). \end{cases} \tag{3}$$

The cylinder velocity constraint in $\Omega_s(t)$ has still to be enforced. The system (3) can be numerically solved by enforcing the constraint with Lagrange multipliers [1]. The linearization of (3) around the steady state $\boldsymbol{u}^{\text{steady}}$ leads to a high-fidelity model that is too large for the direct resolution of (2). For this reason, a step of model reduction is required before searching for an optimal control law.

3.2. Model reduction in fluid-structure interaction

For enforcing the boundary conditions, the strategy is to first derive a reduced-order model by Galerkin projection of (3) onto the POD modes [2], and then to apply to this model the constraint of the domain motion expressed in the POD subspace [6]. The weak form of (3) obtained by Galerkin projection onto the POD modes provides a drastic reduction of the number of degrees of freedom. The Navier-Stokes equations being satisfied on the full domain Ω, (3) is projected onto the N_{gal} first POD modes which are properly defined on Ω [7]. For the cylinder motion constraint, $N_c = 3$ arbitrary test functions vanishing on $\Omega_f(t)$ are used. These functions are chosen to enforce the streamwise and transversal mean velocities and the mean rotation in $\Omega_s(t)$. Finally, it leads to a N_c-dimensional time-dependent linear system of constraints which is enforced by the introduction of a Lagrange multiplier $\boldsymbol{\lambda}(t) = (\lambda_1, \ldots, \lambda_{N_c})^T$. The reduced system can be written as a low-order constant-linear-quadratic descriptor system given by:

$$\begin{cases} \dfrac{\mathrm{d}a_i}{\mathrm{d}t}(t) = C_i + \displaystyle\sum_{j=1}^{N_{\text{gal}}} L_{ij}a_j(t) + \sum_{j=1}^{N_{\text{gal}}}\sum_{k=1}^{N_{\text{gal}}} Q_{ijk}a_j(t)a_k(t) + \sum_{j=1}^{N_c} \left(\mathcal{G}^*(y_c(t))\right)_{ij} \lambda_j(t) \quad i = [1, \ldots, N_{\text{gal}}] \\ \displaystyle\sum_{j=1}^{N_{\text{gal}}} \left(\mathcal{G}(y_c(t))\right)_{ij} a_j(t) = h_i(V_c(t), y_c(t)) \quad i = [1, \ldots, N_c], \end{cases} \tag{4}$$

where a_i $(i = 1, \cdots, N_{\text{gal}})$ are the time coefficients of the POD modes.

4. Results

4.1. POD Reduced-Order Model

Snapshots of the cylinder undergoing vertical oscillations have been generated using a moving mesh implemented in COMSOL Multiphysics 4.3 through an ALE formulation. The forced numerical simulation is initialized with the steady state solution $\boldsymbol{u}^{\text{steady}}$. For the cylinder, the velocity V_c and the position y_c are chosen such that the system first undergoes from $t = 0$ to $t = 100$ a transient regime from the steady state to an established unactuated state. Finally, the flow is actuated from $t = 100$ to $t = 200$ at the natural frequency. After interpolation on a regular mesh and subtraction of the steady state, the POD analysis is performed on the data [2]. A Galerkin projection of the Navier-Stokes equations onto the first 47 POD modes is performed. According to section 3.2, we then add the constraints of mean vertical, horizontal and tangential velocities inside the solid domain. Figure 2 represents the time coefficients obtained as solutions of (4) by comparison with the temporal POD eigenfunctions. We observe that the ROM's dynamics in the transient, established and actuated regimes are very close to the original dynamics. In conclusion, the constrained ROM (4) is considered to represent with a sufficient accuracy the oscillating vertical cylinder.

4.2. Feedback control

After linearization of (4) around the steady state, the optimal feedback gain expressed in the POD subspace is determined as solution of the low-order GARE (2). By construction, this feedback gain is designed to stabilize the ROM (4), and not the high-fidelity model (3). However, this feedback gain should be a good approximation of the optimal linear feedback gain solution of the high-order GARE (2) related to (3). To check this hypothesis, we expressed this gain in the high-order space and used the result to control the high-fidelity model (3). This control design leads to a stabilization of the cylinder wake at $Re = 60$ as shown in figure 3.

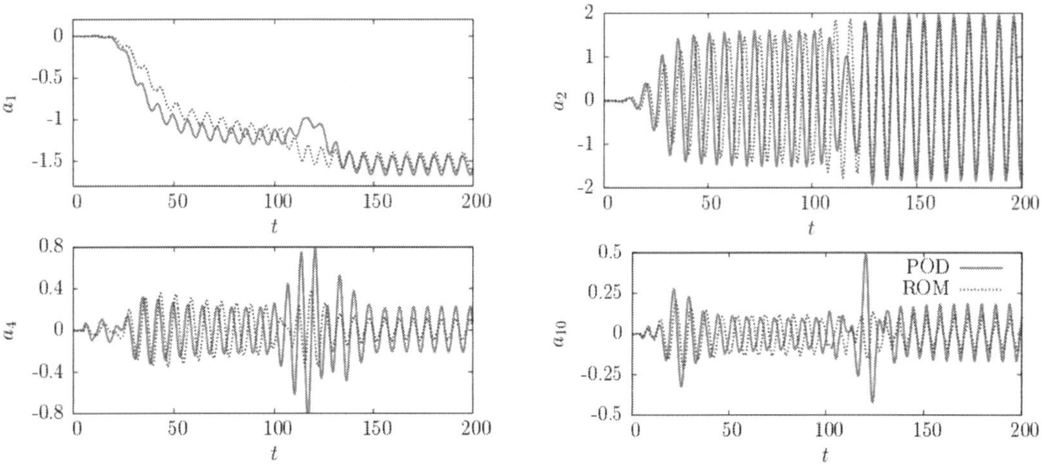

Figure 2. POD ROM of the vertical oscillating cylinder at Re = 60. The time coefficients obtained by projection on the POD modes are in red, whereas the solutions of the POD ROM (4) are represented in blue.

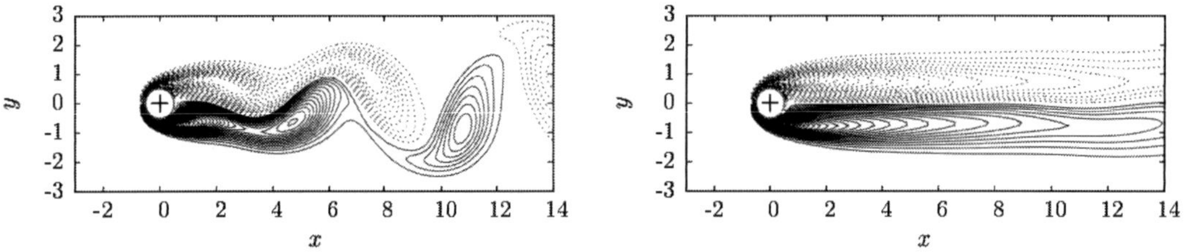

(a) Natural flow.

(b) Flow controlled by cylinder vertical oscillation.

Figure 3. Linear feedback stabilization of the wake behind a circular cylinder at Re = 60. Contour lines of vorticity. The dashed lines correspond to negative values.

5. Conclusion

An efficient feedback control has been developed in a moving domain configuration. The domain motion information has been incorporated in the POD ROM enforcing an additional constraint by Lagrange multipliers. This strategy allows the use of physical model-based control for large-scale problems in fluid-structure interaction. We have successfully tested this control design on the stabilization of a circular cylinder wake by vertical oscillations at $Re = 60$.

References

[1] Glowinski R, Pan T W, Hesla T I and Joseph D D 1999 *International Journal of Multiphase Flow* **25** 755–794
[2] Tissot G 2014 *Réduction de modèle et contrôle d'écoulements* Ph.D. thesis Université de Poitiers
[3] Benner P and Saak J 2010 *Preprint SPP1253-090, DFGSPP1253*
[4] Kramer B and Singler J R 2014 *Submitted for publication*
[5] Ravindran S S 2007 *Computer Methods in Applied Mechanics and Engineering* **196** 2555–2569
[6] Kalashnikova I and Barone M F 2012 *International Journal for Numerical Methods in Engineering* **90** 1337–1362
[7] Liberge E and Hamdouni A 2010 *Journal of Fluids and Structures* **26** 292–311

Inference of Transcriptional Network for Pluripotency in Mouse Embryonic Stem Cells

S Aburatani[1]
Team Leader, Computational Biology Research Center (CBRC), National Institute of Advanced Industrial Science and Technology,
AIST Tokyo Waterfront Bio-IT Research Building, 2-4-7 Aomi, Koto-ku, Tokyo, 135-0064, Japan

E-mail: s.aburatani@aist.go.jp

Abstract. In embryonic stem cells, various transcription factors (TFs) maintain pluripotency. To gain insights into the regulatory system controlling pluripotency, I inferred the regulatory relationships between the TFs expressed in ES cells. In this study, I applied a method based on structural equation modeling (SEM), combined with factor analysis, to 649 expression profiles of 19 TF genes measured in mouse Embryonic Stem Cells (ESCs). The factor analysis identified 19 TF genes that were regulated by several unmeasured factors. Since the known cell reprogramming TF genes (Pou5f1, Sox2 and Nanog) are regulated by different factors, each estimated factor is considered to be an input for signal transduction to control pluripotency in mouse ESCs. In the inferred network model, TF proteins were also arranged as unmeasured factors that control other TFs. The interpretation of the inferred network model revealed the regulatory mechanism for controlling pluripotency in ES cells.

1. Introduction

Clarification of the mechanism of pluripotency in mouse embryonic stem cells (ESCs) is an important theme in biology and regenerative medicine. Four transcription factors with pluripotency ability have been found, and these factors are known as the "Yamanaka factors" [1,2]. Since gene regulatory networks determine cell differentiation and specification [3], in order to decipher the mechanism of the pluripotency process in ESCs, the structures of the regulatory networks between genes should be elucidated. Although the regulatory networks for the genes encoding the important factors for pluripotency have been investigated [4,5,6], the global mechanism of pluripotency in ESCs has remained unclear. Many genes function in ESCs to control cell differentiation, and their expression is controlled by transcription factors (TFs). Therefore, an analysis of the regulatory network between TFs, which are related to pluripotency in mouse ESCs, will reveal the framework of the regulatory mechanism during the pluripotency process.

In order to clarify the regulatory network between TFs, I applied Structural Equation Modeling to huge expression profiles, which were measured in mouse ESCs. In previous investigations, the regulatory relationships between some TFs and their target genes were identified by ChIP-seq technology, gene expression data, and other experimental approaches. According to the former investigations, some aspects of the regulatory mechanism between TFs were clarified [7,8]. In a previous investigation, I utilized an improved SEM approach to infer the regulatory networks during developmental stages in model organisms [9,10]. In my SEM approach, both the previous results and un-measured factors can be included in the network model.

In this study, I improved the method for assuming the initial model for SEM calculations, for inferring the regulatory network between TFs in mouse ESCs. I applied my approach to 19 TF genes with regulated expression in mouse ESCs. Using my method, four signal-transduction pathways were estimated as regulators of initiators for pluripotency, and some TFs were identified as being regulated by these signal transduction pathways.

2. Materials & Methods

2.1. Expression data and selected genes

We compiled the expression profiles measured in mouse ESCs from the GEO Database (http://www.ncbi.nlm.nih.gov/geo/). Among the 1211 Series data (GSE) in GEO, four Series data (GSE16375, GSE26520, GSE42135 and GSE43597) were selected as the expression profiles measuring the expression of genes under various conditions: 342 conditions in GSE16375, 246 conditions in GSE26520, 42 conditions in GSE42135 and 26 conditions in GSE43507. For the SEM calculation, the downloaded data were normalized by a logarithmic transformation and the Z score for each condition, and the normalized data for 656 conditions were combined as a matrix. From the empirically detected transcription factors in mouse/human ESCs, 19 transcription factors are known to be important in ESCs. In this study, I utilized these 19 transcription factor genes to infer the regulatory network between them.

2.2. Factor Analysis

In the SEM calculation, un-measured regulatory factors can be included in the network mode as latent variables. In order to assume the network model, the optimal number of latent variables in the model was estimated by performing factor analysis. By using factor analysis, the variability among the expression profiles of the 19 TF genes can be described by a linear combination of a smaller number of latent variables and error variables, as follows:

$$x_i - u_i = \sum_{l=1}^{n} \alpha_l f_l + \varepsilon_i$$

where x_i is the vector of the expression levels of gene i, and u_i is the mean of x_i. Note that n is the number of latent variables f, α_l is the regression weight of f_l, and ε_i is an independent error term with zero mean and finite variance. Linear models expressed by equation (1) were constructed for each gene, and arranged in a matrix form. In this matrix form, the variance-covariance matrix S between genes can be expressed by the structurized matrix M, and each element of matrix M is expressed by parameters. A comparison between S and M allows the values of the parameters, defined by equation (1), to be estimated. In order to clarify the possible number of latent variables, the exploratory factor analysis with the Kaiser criterion states was performed. In the Kaiser criterion, the number of latent variables is equal to the number of eigenvalues of the covariance matrix that are greater than one. Furthermore, the causal relationships from the estimated latent variables to 19 TF genes were estimated by confirmatory factor analysis. The confirmatory factor analysis was performed with an assumed model with all possible causalities between estimated variables, and the relevant causalities were selected from the absolute values of the factor loadings.

2.3. Exploratory modeling for the initial model

According to the empirical knowledge of the regulatory relationships between TFs, three TFs are considered to be the initiators of sequential transcriptional regulation in ESCs. To construct an initial model that includes this knowledge within the network model, the three latent variables were arranged as translated transcription factor proteins: POU5F1(Oct4), NANOG and SOX2 [1,2,3]. These latent variables are the targets that control their corresponding genes. In order to estimate the effects of these three latent variables on the other TF genes, I improved my developed four-step procedure for exploratory model assumption [9].

In our four-step procedure, all combinations of the regulatory relationships from one latent variable to the other TF genes were tested by SEM. As one of the constraints of the SEM calculation, at least one regulatory arrow from the latent variable to the other variables should be fixed. In this case, the fixed regulation was not tested. In this study, I constructed models with different fixed regulatory arrows, and applied the four-step procedure to determine the best fitting model structure under the restriction. The structures of the fitted models were compared to estimate the remaining regulatory relationships from the latent variable to the other TF genes. This process was applied to all three latent variables. Finally, the remaining regulatory relationships from the latent variables to the other TF genes were combined, to construct an initial model for the SEM calculation.

2.4. Structural Equation Modelling (SEM)

After the estimation of the relationships between the variables for an initial model, I applied the SEM calculation to determine the structure of the network model. According to the initial model, all variables were defined by linear models, as follows:

$$l = \mathrm{A}l + \mathrm{B}g + h$$

$$g = \Pi l + \Gamma g + e$$

in which l is a vector of p latent variables; g is a vector of q measured gene expressions, A is a $p \times p$ matrix representing the regulatory weight between latent variables, Б is a $p \times q$ matrix of effectiveness from the gene to the latent variables, Г is a $q \times p$ matrix representing the regulation from the latent variables to the genes, and П is a $q \times q$ matrix representing the relationships between the genes. The error terms are denoted by the vectors h and e, respectively. From these linear models, the covariance matrix of model S_m, which is the matrix-valued function of the parameters, is given by

$$s_m = G \begin{bmatrix} I - A & -B \\ -C & I - D \end{bmatrix}^{-1} Cov_e \left[\begin{bmatrix} I - A & -B \\ -C & I - D \end{bmatrix}^{-1} \right]^t G^t$$

where G denotes a vector combined with p zero elements and q one elements, and Cov_e denotes the covariance matrix of the error terms. SEM is based on a covariance analysis to estimate the parameters for fitting S_m to the covariance matrix S, which is calculated from the observed data. The SEM software package SPSS AMOS 21.0 (IBM, USA) and the R package (sem) were used for the SEM calculation. To optimize the network model, I applied my developed iteration algorithm to the initial model [9,10].

3. Results & Discussion

According to the factor analysis, 4 factors were estimated as regulatory factors for 19 TF genes in mouse ESCs. The 19 TF genes were divided into 4 clusters by their regulators. Furthermore, the well known pluripotency-related genes are regulated by different factors: Pou5f1 is regulated by 2 factors (F1 and F4), Sox2 is regulated by another factor (F2), and Nanog is regulated by the remaining factor (F3). In general, these genes are considered to be controlled by LIF and Wnt signal transduction. Thus, the estimated regulatory factors were regarded as the effects of signal transduction.

The four factors were arranged as latent variables in the initial model, and the regulatory relationships between factors and genes were also displayed as directed arrows, depending on their factor loading values. In order to include some knowledge within the initial model, I arranged three more latent variables in the initial model. These three latent variables indicated the TF proteins of POU5F1, NANOG and SOX2, which are known to be initiators of pluripotency by regulating other TF genes. By my developed exploratory modeling for the initial model, the effects of the three initiators were estimated: the POU5F1 protein regulated 10 other TF genes, the NANOG protein regulated 8

Journal of Physics: Conference Series **574** (2015) 012138 doi:10.1088/1742-6596/574/1/012138

other TF genes, and the SOX2 protein regulated only 3 TF genes. These estimated effects from the three TF proteins were also included within the initial model.

The inferred network model is displayed in Figure 1a. Interestingly, the latent variable of the SOX2 protein was deleted from the network model during the model optimization process. Furthermore, the estimated regulatory factors (F1, F2, F3 and F4) were not related to each other. Figure 1b shows the schema of the inferred network. The strong associations between variables (weight >= 0.5) were extracted from the inferred network. Since F1 was a regulator of Pou5f1 expression, F1 was considered to be the effect of LIF signal transduction. In the lower class in F1, the Pou5f1 gene affected the Nanog gene and the POU5F1 protein. Actually, Pou5f1 is known to regulate Nanog gene expression. Furthermore, the effect from POU5F1 to Sox2 in the inferred model agreed with the known function of POU5F1.

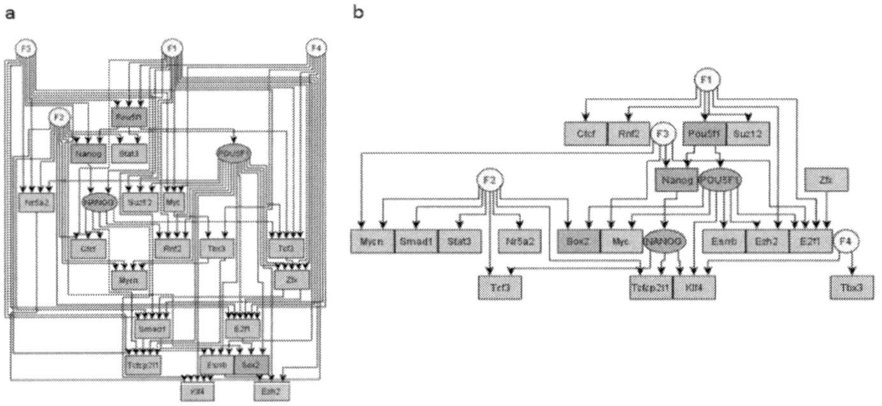

Figure 1. Inferred networks. (a) Whole network model. (b) Network with strong associations. Blue: estimated factors, pink: known factors for pluripotency, yellow: important TFs in ESCs

4. Conclusion

Here, I inferred the transcriptional regulation between 19 TFs, which are considered to be important for pluripotency in mouse ESCs. By including unobserved variables, the signal transduction effects were clarified. Our network model revealed the hierarchical regulation by well known initiators for pluripotency.

5. References

[1] Takahashi K and Yamanaka S 2006 *Cell* 126 663

[2] Okita K, Ichisaka T, Yamanaka S 2007 *Nature* 448 313

[3] Davidson E H 2006 *The Regulatory Genome: Gene Regulatory Networks in Development and Evolution* (Burlington, MA: Academic Press)

[4] Jaenisch R and Young R 2008 *Cell* 132 567

[5] Zeng F, Baldwin D A and Schultz R M 2004 *Dev. Biol.* 272 483

[6] Xie D, Chen C C, Ptaszek L M, Xiao S, Cao X, Fang F, Ng H H, Lewin H A, Cowan C and Zhong S 2010 *Genome Research* 20 804

[7] Boyer L A, Lee T I, Cole M F, Johnstone S E, Levine S S, Zucker J P, Guenther M G, Kumar R M, Murray H L, Jenner R G, Gifford D K, Melton D A, Jaenisch R and Young R A 2005 *Cell* 122(6) 947

[8] Kim J, Chu J, Shen X, Wang J and Orkin S H 2008 *Cell* 132 1049

[9] Aburatani S 2012 *Bioinformation* 8(14) 652

[10] Aburatani S, Fujibuchi W, Yamane J, Nagano R, Sone H, Imanish S and Ohsako S 2012 *Int. J. on Adv. in Life Sciences* 5 103

Analysis on the use of Multi-Sequence MRI Series for Segmentation of Abdominal Organs

M A SELVER[1], E SELVİ[2], E KAVUR[2], O DICLE[3]

[1]Dokuz Eylül University, Department of Electrical and Electronics Engineering, Tınaztepe Campus, 35160, Buca, İzmir, TURKEY

[2]Dokuz Eylül University, Institute of Natural and Applied Sciences, Tınaztepe Campus, 35160, Buca, İzmir, TURKEY

[3]Dokuz Eylül University, Faculty of Medicine, Department of Radiology, Tınaztepe Campus, Balçova, İzmir, TURKEY

E-mail: alper.selver@deu.edu.tr, esref.selvi@gmail.com, oguz.dicle@deu.edu.tr

Abstract. Segmentation of abdominal organs from MRI data sets is a challenging task due to various limitations and artefacts. During the routine clinical practice, radiologists use multiple MR sequences in order to analyze different anatomical properties. These sequences have different characteristics in terms of acquisition parameters (such as contrast mechanisms and pulse sequence designs) and image properties (such as pixel spacing, slice thicknesses and dynamic range). For a complete understanding of the data, computational techniques should combine the information coming from these various MRI sequences. These sequences are not acquired in parallel but in a sequential manner (one after another). Therefore, patient movements and respiratory motions change the position and shape of the abdominal organs. In this study, the amount of these effects is measured using three different symmetric surface distance metrics performed to three dimensional data acquired from various MRI sequences. The results are compared to intra and inter observer differences and discussions on using multiple MRI sequences for segmentation and the necessities for registration are presented.

1. Introduction

Magnetic Resonance Imaging (MRI) can provide very detailed and contemporary mappings of a patient through the use of different sequences. Therefore, MRI has become a unique tool in radiology for diagnostic imaging. The abdominal organs can be analyzed more precisely using these various sequences. Once an abdominal organ is segmented in each sequence, the results can be fused to extract the information of interest instead of using a single data. Moreover, segmentation of abdominal organs is a very challenging field of application due to the overlapping intensity ranges of the organs, variations in human anatomy and pathology. Therefore, these different MRI sequences can also be used to improve the segmentation accuracy. Since it is necessary to analyze and visualize abdominal organs (i.e. liver, right/left kidneys, spleen, pancreas, gall bladder) for several medical procedures, the analysis of these sequences are necessary in order to test their usability for data fusion.

Multiple MR sequences, which are acquired during the routine clinical practice, are being used by radiologists to help diagnosis. These sequences are sequentially acquired and therefore affected by two kinds of differences: i) differences in acquisition parameters such as contrast mechanisms and pulse sequence designs, ii) differences in image characteristics such as pixel spacing, slice thicknesses and dynamic range. To be able to use the information coming from these various MRI sequences in computational techniques such as segmentation and fusion, patient movements and respiratory motions should be at acceptable levels. For instance, these levels should preferably be lower than inter- and intra-observer variations on manual organ border delineation. In this study, the amount of these effects is measured using three different symmetric surface distance metrics that are performed to evaluate three dimensional data acquired from various MRI sequences. The results are compared to intra- and inter- observer differences reported in the literature and discussions on using multiple MRI sequences for segmentation and the necessities for registration are presented.

2. Datasets

The analyses are performed on MRI DICOM series, which are collected from the PACS of Dokuz Eylül University Radiology Department. The data sets were acquired using a 1.5 Tesla Philips MRI system, which produces 12 bit images having 256 x 256 pixels. Four different MRI sequences (i.e. T1-DUAL (in-phase and out-phase), T2-SPIR, THRIVE, SSH) are used to test the proposed method.

The sequences are used to scan the same part of the body but each sequence is obtained from a different combination of radiofrequency pulses and gradients. Since, the name of the sequences depends on the vendor, the above mentioned sequence acronyms are explained below [1, 2].

SPIR is the abbreviation for "Spectral Pre-saturation Inversion Recovery" and it is being used as a hybrid imaging sequence. The SPIR sequence uses T2-weighted contrast mechanism for imaging. It relies on selective suppression of fat protons [3]. The pre-saturation pulse is applied separately to each slice selection gradient. This sequence requires sensitive adjustment of calibration and a very homogenous magnetic field. The above mentioned features of SPIR makes it a preferred sequence to study liver, because the liver parenchyma can be analyzed very well with suppression of the fat content inside the parenchyma. Especially, mass lesions belong to parenchyma are more visible in this sequence. Because of being T2-weighted, it is possible to navigate the vessels within liver since they appear hyper-intense. The liver border appearance gets visually clearer, because of the suppression of the fat tissue around the liver. The adjacent abdominal organs and tissues such as gall bladder, duodenum contents, pancreas and right kidney become more separable from liver with their high-valued signal. One more important contribution of the SPIR sequence is its low sensitivity to motion. This feature provides minimization of the artifacts that adversely affects image quality in liver studies.

T1-DUAL (in-phase and out-phase) is a fat suppression sequence, which uses the difference in T1 times of fat and water protons. The signal is acquired twice: first when water and fat protons are in phase and second, when they are out of phase (while exciting protons are returning to their first position). For 1.5 Tesla devices the in-phase time, which water and fat protons are in same direction is 4,6 millisecond and the out-phase time, which fat and water protons are opposite directions, is 2,3 milliseconds. By determining TE (Time of Echo) value with this information, fat suppression is accomplished by subtracting corresponding frequencies of fat and water signals. This sequence is very useful to understand the fat content in lesions. Since T1-DUAL is a T1-weighted sequence, it is very effective to identify blood and tissues that are rich in protein. This sequence also helps determining the level of liver lubrication. In out-phase images, the border of the organs appear to be black, due to the sudden change in the amount of fat and water at the organ boundaries that cancels the acquired signal. This property of T1-DUAL is sometimes used for border delineation algorithms [4].

SSH refers to single shot imaging and it is a very useful sequence for the abdominal analysis, which are affected by the motion caused due to vessels, intestinal gas and diaphragm. By using this sequence, steady liquids can be studied very well, which allows analysis of biliary tracts.

THRIVE is a fast imaging sequence that is obtained by scanning the liver consecutive in a short time with thin slices to get respond of a known lesion to a contrast agent. THRIVE provides observation of the signal increase originating from the contrast agent instead of image quality.

(a)

(b)

(c)

Figure 1. Images acquired from different MRI sequences used in this study. (a) liver, (b) right kidney, (c) spleen (From left to right: SSH, T1-DUAL, T2-SPIR, WATS, THRIVE)

In this study, MR images obtained from 21 patients are used. The details about the characteristics of these data sets are given in Table 1. Four different sequences are acquired for each patient except one, who has 3 different sequences. While T2-SPIR, SSH and T1-DUAL sequences are common for each 21 patient, 20 patient has either THRIVE (7) or T1-WATS (13) sequences. For T2-SPIR, which is common for all 21 patients, slice thickness value changes between 7.7 mm and 9 mm and has an average value equal to 8.6 mm. Moreover, the x-y spacing in this sequence changes between 1.63 mm and 1.89 mm with an average of 1.53 mm. The number of slices for T2-SPIR sequence is 26 as minimum, 36 as maximum and 30 as average. On the other hand T1-DUAL sequences include two different series mentioned above. Each series has same x-y spacing, slice thickness and number of slices. For T1-DUAL sequences, slice thickness has a value between 5.5 mm and 9 mm with an average of 7.84 mm. The x-y spacing value in this sequence is between 1.44 mm and 1.89 mm and the average value is 1.61 mm. While the average number of slices is 32.8, minimum number is 26 and maximum number is 50. SHH sequence has slice thickness between 5.5 mm and 9 mm with an average of 7.84 mm. For SSH, sequences x-y spacing changes between 1,34mm and 1,82mm and has an average as 56mm. While average number of slices is 32.2, it is 50 at most and 25 at least. T1-WATS sequence, which is acquired for 13 patients, has slice thickness value between 5.5 mm and 9 mm like T1-DUAL and SSH sequences with an average of 7.9 mm. While x-y spacing in T1-WATS sequences changes between 1.36 mm and 1.67 mm, the average value of x-y spacing is 1.43mm. Finally, THRIVE sequences, which only have 7 data sets, has differences from the other sequences. The slice thickness is smaller and does not change between patients for THRIVE. All THRIVE series have 2.5 mm slice thickness. Since the slice thickness is smaller, THRIVE sequence has more number of slices in one series. Number of slices is 110 at most and 80 at least. Average number of slices is 92.9. For THRIVE, x-y spacing changes between 1.56 mm and 1.97 mm with an average 1.67 mm.

Slice thickness is one of the priority factors for the volume segmentation. Although it seems like that small slice thickness is better for volume segmentation, it is shown that small slice thickness is not always better [5].

Table 1. The list of the data sets and the MR sequences used in this study.

MRI Sequence	# of images	Slice Thickness			x- / y- spacing			Number of Slices		
		Min	**Max**	**Average**	**Min**	**Max**	**Average**	**Min**	**Max**	**Average**
T2-SPIR	21	7,7	9	8,6	1,36	1,89	1,53	26	36	30
T1-DUAL	21	5,5	9	7,84	1,44	1,89	1,61	26	50	32,8
THRIVE	7	2,5	2,5	2,5	1,56	1,97	1,67	80	110	92,9
T1-WATS	13	5,5	9	7,9	1,36	1,67	1,43	26	40	30,6
SSH	21	5,5	9	8,1	1.34	1,82	1,56	25	50	32,2

3. Method

As mentioned in the earlier sections, the aim is to evaluate the similarity (i.e. position and shape) of the same organ in different sequences of the same patient so that the effects of patient movement and respiratory motions can be analyzed. The comparison of the data sets acquired for the same patient using different sequences is performed using Symmetric Surface Distance (SSD) metrics, which are designed to compare two objects in three dimensions [6]. SSD is based on the distance between surface voxels of two volumes, which correspond to the segmented (V_S) and the reference (V_R) data. Surface voxels can be defined as the voxels having at least one non-object voxel within their 18-neighbourhood. SSD is obtained by calculating the Euclidean distance for each surface voxel of V_R to the closest surface voxel of V_S.

$$d\left(v_r, S(V_S)\right) = \min_{\substack{v_r \in S(V_R) \\ \forall q \in S(V_S)}} \left(\left\| v_r - q \right\|\right) \quad (4.1)$$

Here, $s(V_S)$ is the set of surface voxels of V_S and $s(V_R)$ is the set of surface voxels of V_R.
There are three different SSD measures, each of which aim to evaluate a different characteristic of the difference between the segmented and the reference objects.
The first metric is the Average Symmetric Surface Distance (ASSD), which is calculated by averaging the SSD values of each voxel and determined as follows:

$$ASSD(V_R, V_S) = \frac{1}{\left|S(V_S)\right| + \left|S(V_R)\right|}\left(\sum_{v_r \in S(V_R)} d\left(v_r, S(V_S)\right) + \sum_{v_s \in S(V_S)} d\left(v_s, S(V_R)\right)\right)(mm) \quad (4.2)$$

For perfect segmentation, ASSD is equal to 0. ASSD metric is not affected much by local errors even if they are big. Therefore, it provides information on the similarity of two objects in general.
The second metric is the Root Mean Square SSD (RMSSD), which is calculated similar to ASSD but with an important difference. The square of the distances between surface voxels are calculated. This difference results with exaggeration of the large distances, while smaller errors do not change much. RMSSD can be calculated using:

$$RMSSD(V_R, V_S) = \sqrt{\frac{1}{\left|S(V_S)\right| + \left|S(V_R)\right|}\left(\sum_{v_r \in S(V_R)} d^2\left(v_r, S(V_S)\right) + \sum_{v_s \in S(V_S)} d^2\left(v_s, S(V_R)\right)\right)}(mm) \quad (4.3)$$

Similar to ASSD, RMSSD should be 0 for the perfect segmentation. The RMSSD also provide a general information about the error rate, but it is also gives a better opinion about the distribution of the distance error when it is compared with ASSD.
The third metric extracted using SSD is the Maximum Symmetric Surface Distance (MSSD), which is calculated as:

$$MSSD(V_R,V_S)=\max\left\{\max_{v_r\in S(V_R)}\{d(v_r,S(V_S))\},\ \max_{v_s\in S(V_S)}\{d(v_s,S(V_R))\}\right\}(mm) \quad (4.4)$$

MSSD is equal to zero for the perfect segmentation, but the information provided by MSSD is significantly different than ASSD and MSSD. By calculating the maximum values of the error, it provides the important information about error margin. This information is critically important for the operations, in which the worst case scenario is of particular interest. A good example is surgical planning using the segmented objects.
To determine correlation of different sequences, the voxel dimensions should be converted in millimeters because of the differences x-y spacing and slice thickness.

On the other hand, assessing all there measures helps us to understand specific conditions. For example, for a good segmentation but a local error can be detected by looking these metrics. In this condition MSSD should be high while ASSD is low. If MSSD is close to ASSD, then it can be sign that is because of a shift on one volume in this condition.

4. Simulations and Results

Table 2 and Table 3 present the calculated SSD values for each organ among the selected MRI sequences presented in Table 1.

Table 2. SSD results for liver and spleen (mm)

liver	ASSD			
	SSH	T1DUAL	T2SPIR	THRIVE
SSH	0	6,4±2,8	6,3±2,9	7,4±5,5
T1DUAL	6,4±2,8	0	5,17±1,81	5,24±6,23
T2SPIR	6,3±2,9	5,17±1,81	0	7,37±5,62
THRIVE	7,4±5,5	5,24±6,23	7,37±5,62	0

liver	MSSD			
	SSH	T1DUAL	T2SPIR	THRIVE
SSH	0	30,9±9,6	30,9±9	32±10,4
T1DUAL	30,9±9,6	0	24,6±8,1	22,4±9,1
T2SPIR	30,9±9	24,6±8,1	0	28,6±10,9
THRIVE	32±10,4	22,4±9,1	28,6±10,9	0

liver	RMSSD			
	SSH	T1DUAL	T2SPIR	THRIVE
SSH	0	8,4±3,4	8,1±3,4	9,34±6,2
T1DUAL	8,4±3,4	0	6,64	6,55±6,76
T2SPIR	8,1±3,4	8,1±3,46	0	9,1±6,2
THRIVE	9,4±6,2	6,55±6,76	9,1±6,2	0

spleen	RMSSD			
	SSH	T1DUAL	T2SPIR	THRIVE
SSH	0	8,9±10,1	7,1±11	10±11,6
T1DUAL	8,9±10,1	0	9,2±12,1	11,3±12,8
T2SPIR	7,1±11	9,2±12,1	0	5,9±6,1
THRIVE	10±11,6	11,3±12,8	5,9±6,1	0

spleen	ASSD			
	SSH	T1DUAL	T2SPIR	THRIVE
SSH	0	6,1±6,7	5,4±8,4	7,6±9
T1DUAL	6,1±6,7	0	7,5±9,3	8,9±10
T2SPIR	5,4±8,4	7,5±9,3	0	4,8±5,1
THRIVE	7,6±9	8,9±10	4,8±5,1	0

spleen	MSSD			
	SSH	T1DUAL	T2SPIR	THRIVE
SSH	0	27,4±2,8	24,9±31	29,6±31,3
T1DUAL	27,4±2,8	0	32,7±34	34±34,4
T2SPIR	24,9±31	32,7±34	0	17,5±11,5
THRIVE	29,6±31,3	34±34,4	17,5±11,5	0

Table 3. SSD results for right and left kidneys (mm)

Right kidney	ASSD			
	SSH	T1DUAL	T2SPIR	THRIVE
SSH	0	2±1,8	1,9±0,4	5,6±4,5
T1DUAL	2±1,8	0	2,7±0,7	5,2±3,9
T2SPIR	1,9±0,4	2,7±0,7	0	6,7±5,6
THRIVE	5,6±4,5	5,2±3,9	6,7±5,6	0

Right kidney	MSSD			
	SSH	T1DUAL	T2SPIR	THRIVE
SSH	0	7,5±0,5	9,5±1,7	17,4±9,28
T1DUAL	7,5±0,5	0	11,67±2,9	15,4±7,4
T2SPIR	9,5±1,7	11,67±2,9	0	19,2±13,4
THRIVE	17,4±9,28	15,4±7,4	19,2±13,4	0

Right kidney	RMSSD			
	SSH	T1DUAL	T2SPIR	THRIVE
SSH	0	2,6±0,2	2,6±0,6	7±5,3
T1DUAL	2,6±0,2	0	3,6±1,2	6,2±4,5
T2SPIR	2,6±0,6	3,6±1,2	0	8,4±6,7
THRIVE	7±5,3	6,2±4,5	8,4±6,7	0

Left kidney	ASSD			
	SSH	T1DUAL	T2SPIR	THRIVE
SSH	0	2,4±1	2,1±1	3,4±3,4
T1DUAL	2,4±1	0	2,5±0,8	3,1±3,6
T2SPIR	2,1±1	2,5±0,8	0	3,8±3,5
THRIVE	3,4±3,4	3,1±3,6	3,8±3,5	0

Left kidney	MSSD			
	SSH	T1DUAL	T2SPIR	THRIVE
SSH	0	12,8±4,2	10,7±2,2	14,15±6,5
T1DUAL	12,8±4,2	0	12,3±3,1	13,4±8,2
T2SPIR	10,7±2,2	12,3±3,1	0	13,2±6,6
THRIVE	14,15±6,5	13,4±8,2	13,2±6,6	0

Left kidney	RMSSD			
	SSH	T1DUAL	T2SPIR	THRIVE
SSH	0	3,12±1,3	2,7±1	4,4±3,9
T1DUAL	3,12±1,3	0	3,32±0,9	4±4,2
T2SPIR	2,7±1	3,32±0,9	0	4,6±4,1
THRIVE	4,4±3,9	4±4,2	4,6±4,1	0

As seen on table 1 and table 2, SSD values between different sequences are different from each other. Liver is the biggest organ in the abdominal region and therefore, correlation between livers from different sequences can give us a strong opinion about the correlation of the abdominal region. ASSD values between different sequences are close to each other and they vary between 5.2 mm and 7.4 mm. MSSD values between different sequences are also close to each other like ASSD values. MSSD has values between 22.4 mm and 32 mm. As mentioned before, RMSSD values are in the same order as ASSD with a slight difference that exaggerates the bigger values. The RMSS values are between 8.4 mm and 9.2 mm.

Spleen is the second biggest organ in the abdomen. Although it is smaller than the liver, SSD values of the spleen are close to the SSD values of the liver. The SSD values of the spleen also have a wide range varying between from 4.8 mm to 8.9 mm. The MSSD values for the spleen have a wide range and they change from 17.4 mm to 34.4 mm. The RMSSD values for the spleen in different MRI sequences change between 7.1 mm and 11.3 mm. Moreover, the standard deviations of the all values are much greater than the values of the liver.

When the results for the right kidney are analyzed, it can be seen that the ASSD values between SSH, T1DUAL and T2SPIR are very small, which lie between 1.9 mm and 2.7 mm. On the other hand, THRIVE has greater ASSD values with the other sequences as 5.2 mm to 6.7 mm. The results of the MSSD is not very different from ASSD results. The MSSD values for sequences except THRIVE have small values that are from 7.5 mm to 11.6 mm. However, the MSSD values between the THRIVE and the other sequences are greater than the other ones as they change from 15.4 mm to 19.2 mm.

ASSD, MSSD and RMSSD values of the left kidney are close to the values of the right kidney. The ASSD values are between 2.1 mm and 3.8 mm, the MSSD values are between 10.7 mm and 14.1 mm. Finally, the RMSSD values lie between 2.7 mm and 4.7 mm.

Conclusion

The values of the liver are found close enough for all of the sequences compared to intra- and inter-observer manual delineation differences. Therefore, it can be concluded that liver can be segmented in different sequences and then these results can be combined in order to obtain more accurate results. On the other hand, spleen has more ASSD, MSSD and RMSSD values with greater standard deviation compared to its size. This decreases the reliability of the design of such a combination system for the spleen. Thus, it would be hard to combine the segmentation results of the spleen collected from different MRI sequences. The kidneys have ASSD, MSSD and RMSSD that are close enough compared to their size and they can be used for generating such combinations.

In conclusion, the SSD values of the liver and kidneys are close enough to each other in order to use for combining multi-Sequence MRI series for segmentation.

Acknowledgement

This study is supported by The Scientific and Technological Research Council of Turkey (TUBITAK) under grant number 112E032.

References

[1] Brown M.A, Semelka R-C. 1999 *MR imaging abbreviations, definitions, and descriptions: A review Radiology*; **213**; 647-662
[2] WestbrookC 2002 *MRI at a Glance*. (Oxford, England: Blackwell Science)
[3] Westbrook C and Kaut C 1998 *MRI in practice*. (Oxford, England: Blackwell Science)
[4] Kellberg J Karlsson A-K Stokland E and Svenssson P-A 2010 *Journal of Magnetic Resonance Imaging* **32** 204-210
[5] Hermoye L et al *Radiology* 2004 **234** 171-178
[6] Heimann T et al 2009 *IEEE Transactions on Medical Imaging* **28** 1251-1265

Spherical Target Recognition using Time-Domain Multiscale Approximation of Scattered Signals

M M Taygur, M A Selver[1], E Y Zoral[1]

[1]Department of Electrical and Electronics Engineering, Dokuz Eylul University, İzmir, TURKEY

Email: mtaygur@gmail.com, alper.selver@deu.edu.tr, yesim.zoral@deu.edu.tr

Abstract. Classification of similar shaped objects from scattered electromagnetic waves is a difficult problem to solve, as it heavily depends on the aspect angle. Eliminating the effects of the aspect angle is possible by extracting distinguishable features from the scattered signals. These features should be robust to noise effects especially at SNR levels, where noise effects become dominant on the scattered signal. In this paper, we propose a target classification method, which uses a structural feature set extracted from scattered signal. Prior to feature extraction, a multi-scale approximation is performed using hierarchical radial basis function network topology to suppress the effects of noise on scattered signal. After principle component analysis, k-fold cross validation based experiments is performed. Results show that spherical targets are recognized successfully up to -10dB SNR.

1. Introduction

The classification of similar targets from scattered electromagnetic signals is a challenging problem due to the strong dependency to aspect angle [1]. In order to classify a target effectively, distinguishable features should be extracted and adequately processed to gain independency from the adverse effects of aspect angle [2]. Moreover, these features should be robust to noise, which can alter scattered signal characteristics significantly.

Recently, structural properties of time domain scattered signals are used effectively in target recognition [3]. As these properties are based on hill-valley detection, corresponding features are very sensitive to noise. In this study, noisy scattered signals are approximated via a set of hierarchical radial basis functions. Opposed to well-known wavelet multi-resolution analysis using the discrete wavelet transform [4], proposed approach reconstructs the signal from low frequency components to high and this allows better extraction of features from noisy scattered signals.

At each layer of the hierarchy, a number of Gaussian units are used to approximate to the valleys/hills in the scattered data. The structural parameters of these units (i.e. centers, widths and amplitudes) are predefined for each layer and the number of these units is determined using a performance criterion. Finally, a combination of the selected units is used to create an approximation. For evaluation of the proposed technique in target classification, cross validation learning strategy [5] is used. The results show that, up to -10 dB SNR, proposed method can effectively be used for target classification.

Journal of Physics: Conference Series **574** (2015) 012140 doi:10.1088/1742-6596/574/1/012140

2. Method

2.1. Generation of Data

The scattered electric fields can be expressed in terms of Hertz and Debye potentials as in [2]. In this study, the scattered signals are calculated by the analytical solutions using these potentials. These analytical expressions are extracted for a plane wave excitation which is linearly polarized in x-direction and propagates in z-direction (Figure 1).

The far field scattered responses are computed using MATLAB 7.1 in frequency domain over a bandwidth from zero to 12 GHz at 873 frequency sample points with frequency resolution of 13.75 MHz which can be regarded as resonance region. These responses are also obtained at $\varphi = \pi/2$ plane, with a radial distance of 72 cm from the sphere center, for twelve different Bistatic Aspect Angles (BAA), 180-$\theta = \theta b = 10°, 20°, \ldots 180°$ degrees as shown in Figure 1.

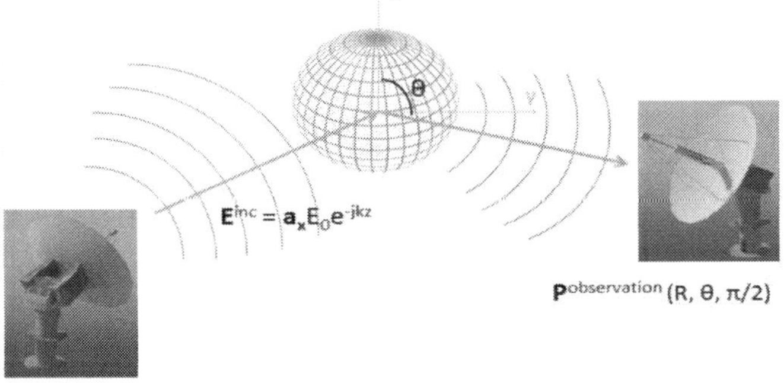

Figure 1. The geometry of the problem used to generate electromagnetic signals scattered from the spherical targets

After getting the scattered signals in frequency domain, inverse fast Fourier transformation was applied to get time-domain scattered fields. This allows scattered signals up to 5 ps resolution, which can be used to observe the frequency range below 12 GHz.

The resulting time signals, which are used throughout the simulation, have 1024 sample points with a total time span of 5.115 ns (Figures 2 and 3). The noisy scattered time domain signals at all the aspect angles stated above are obtained at the signal-to noise ratio (SNR) levels of 10, 0 and -10 dB to be used for classifier design and for performance testing.

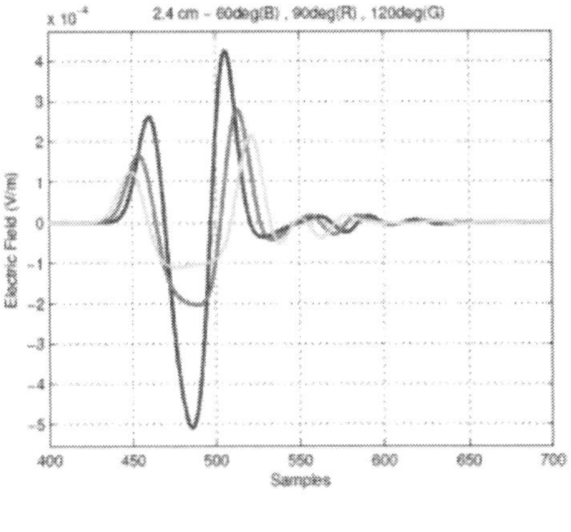

Figure 2. The scattered time domain signals for the dielectric sphere of radius 2.4 cm at BAA of 60° (blue), 90° (red) and 120° (green)

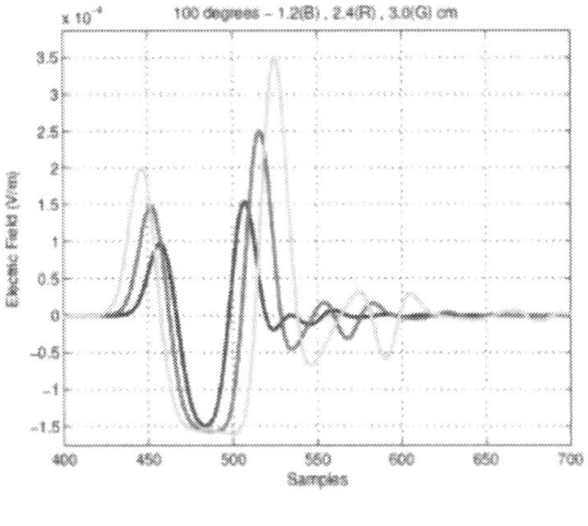

Figure 3. The scattered time signals for the dielectric sphere of radius 1.8cm (blue), 2.4cm (red) and 3.0cm (green) at BAA of 100°

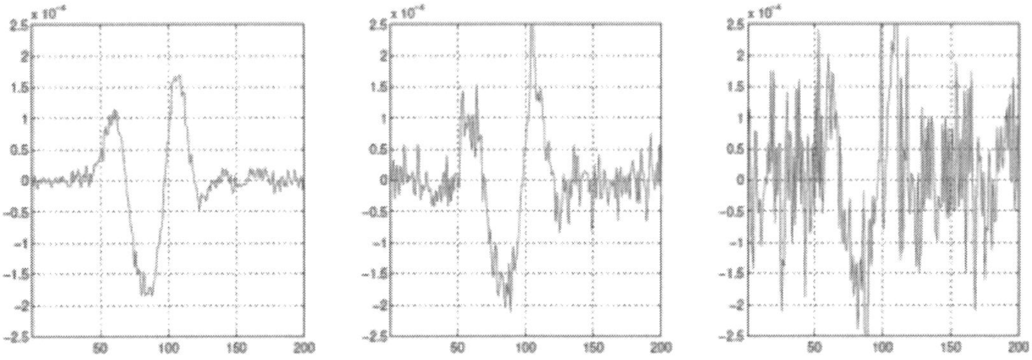

(a) (b) (c)

Figure 4. Scattered signals at 90° and 1.8cm distance with 10, 0, and -10dB SNR.

2.2. Structural Feature Extraction Strategy

Typical scattered signals in Figure 2 and Figure 3 show that these signals can be divided into four sub-waves. Analysis of these sub-waves shows that the amplitude and duration of these four sub-waves differ depending on the radius of the target. Thus, if aspect angle effects can be eliminated, these sub-wave properties (i.e. structural features) can be used to classify similar spheres with different radii.

Considering the fluctuating waveform, which is the common aspect for all signals we dealt with, extraction is built upon analyzing the 'peaks' (both negative and positive) in the signal. For each peak in the signal, 5 features are extracted. In addition, distances between the maxima/minima points of adjacent peaks are calculated. Hence, 6n-1 feature is obtained for a single scattered wave, where n is the number of detected peaks in the signal. The features are illustrated in Figure 5.

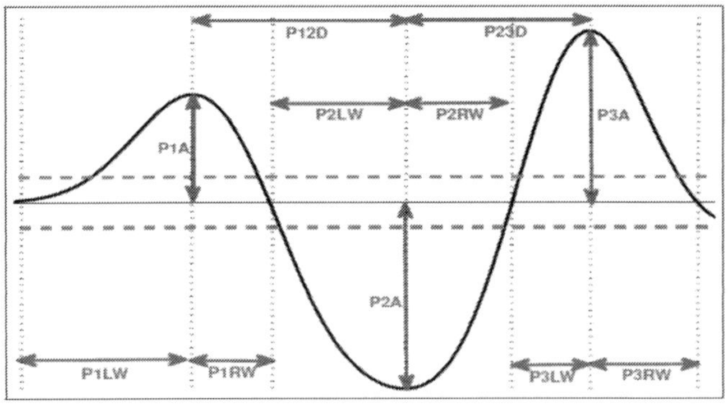

Figure 5. Feature extraction scheme: Amplitude of each peak (e.g. P1A), Left width of the peak (e.g. P1LW), iii. Right width of the peak (e.g. P1RW), iv.Tangent of peak from left (e.g. P1A/P1RW), v. Tangent of peak from right (e.g. P1A/P1RW), vi.Distances between peaks (e.g. P12D).

2.3. Hierarchical Use of Gaussian Units For Multi-scale Approximation of Scattered Signal

Extracting a feature set from noisy data is based on estimating local extremum points and zero-crossings by means of certain rules. Thus, the representation capability of the features relies on finding these important points on the signal. Since, it has been experimentally shown that the features extracted from a low-SNR signal provide better recognition performance [3], the noise effects over the signal should be suppressed prior to feature extraction.

The approximation function, f(x), to the scattered signal using bell shaped Gaussian functions can be defined as;

$$f(\boldsymbol{x}) = \sum_{j=1}^{N} w_j . g(\boldsymbol{x} - \boldsymbol{c}_j; \boldsymbol{\sigma}_j)$$

where $\boldsymbol{c}_j \in \mathbb{R}$ corresponds to the center of the jth Gaussian unit, $\boldsymbol{\sigma}_j \in \mathbb{R}$ to the width and $w_j \in \mathbb{R}$ corresponds to the amplitude of the jth Gaussian unit. In its most general form, the jth Gaussian unit in the ith layer is defined by $g_{ij}(.) = exp\left(-(\boldsymbol{x} - \boldsymbol{c}_{ij})^T.(1/\boldsymbol{\sigma}_{ij})^2.(\boldsymbol{x} - \boldsymbol{c}_{ij})\right)$.

The proposed approximation technique performs an approximation of f(x) using the sum of K approximations $\{l_i(.)\}_{i=1,2,...,K}$;

$$f(\boldsymbol{x}) = \sum_{i=1}^{K} l_i(\boldsymbol{x})$$

Herein, approximation layers, $l_i(.)$'s, are the sets of Gaussian functions, which have predefined centers, widths and amplitudes for that layer. In order to accomplish a multi-scale task, the Gaussian units become denser in position, narrower in width and smaller in amplitude as the levels proceed. Thus, each layer, $l_i(.)$, consists of a linear combination of Gaussian units, which are selected through a performance criteria to be used for that layer.

Consider that $l_i(.)$ Is composed of M_i Gaussian untis. Then, an approximation layer can be defined as,

$$l_i(\boldsymbol{x}) = \sum_{j=1}^{M_i} w_{ij}.g(\boldsymbol{x} - \boldsymbol{c}_{ij}; \boldsymbol{\sigma}_{ij})$$

As a result, the approximation uses $M = \sum_{j=1}^{M_i} M_i$ Gaussian units to approximate the scattered signal.

After approximating the function by using the Gaussian units found for a layer (Figure 4b), a residual is calculated point-wise, i.e. for each point $\boldsymbol{x}^n$ of the original scattered signal;

$$r_1(\boldsymbol{x}^n) = f(\boldsymbol{x}^n) - l_1(\boldsymbol{x}^n)$$

The next approximation layer considers the residual signal found in the previous layer as the new function to be approximated. The new Gaussian units having predefined center, width and amplitude for that layer is inserted. The performance criterion is used again to select which ones to use the corresponding residual signal. The general expression of a residual scattered signal to be approximated as i[th] layer is given as;

$$r_{i-1}(\boldsymbol{x}^n) = f(\boldsymbol{x}^n) + \sum_{j=1}^{i-1} w_{ij}.g(\boldsymbol{x} - \boldsymbol{c}_{ij}; \boldsymbol{\sigma}_{ij}) = r_{i-2}(\boldsymbol{x}^n) - l_{i-2}(\boldsymbol{x}^n)$$

To approximate the scattered signal, this procedure continues for several layers until the approximation error approaches a predefined value. The algorithm to accomplish this task and its pseudo code is given below;

1. Signal is divided into frames, which have equal lengths.
2. For each frame, Gaussian curves are added into a new array, which has the same length with the frame defined in 1. Gaussian curves are placed in the middle of the frame symmetrically and designed to have a standard deviation, which is equal to 0.125 of the frame length. Amplitude is chosen to be equal to 1/100 of maximum value of the signal.
3. Inside the frame, Gaussian curves are added until the error between approximation and original signal is smaller than a predefined error criteria.

Journal of Physics: Conference Series **574** (2015) 012140 doi:10.1088/1742-6596/574/1/012140

```
f[n]: Input signal
approx[n] ← 0
for f ← 2^8 to 2^6 by 2^0.1
  for no_frame ← 1 to #(f[n])/f
    gauss_prototype ← gauss(f/2,no_frame/4)
    do
      approx[no_frame.(1:f)] ← approx[no_frame.(1:f)] + gauss_prototype
    until |f[no_frame.(1:f)] – approx[no_frame.(1:f)]|< |f[no_frame.(1:f)] -
          approx[no_frame.(1:f)]-gauss_prototype|
return approx.
```

3. Application & Results

Four spherical targets, which have radius of 1.8cm, 2.4cm, 3.0cm, and 3.6cm, respectively, are used to test the method developed in this study. Feature vectors extracted for each sphere for 18 BAA, which generates a data set of 72 samples. The same procedure is also applied for scattered signal, which have added noise at SNR= 10, 0 and -10 dB levels.

Since this is a small data set, cross-validation techniques [5] are used to measure the performance of the technique. Among several possibilities, K-fold partitioning is chosen [5]. K-fold technique partitions the data set into K parts and performs K experiments, in which K-1 folds are used for training and the remaining fold is used for testing. The advantage of K-fold cross validation is that it prevents over-fitting by systematically using all the examples in the data set for both training and testing. K=9 is chosen as it is experienced to provide higher performance.

For classification, a Multi-Layer Perceptron (MLP) network [6], which is trained using back-propagation with adaptive learning rate, is employed. The MLP has one hidden layer with 6 neurons with tangent sigmoid activation functions and an output layer with 4 neurons with linear activation functions. The network goal is chosen to be 0.001 and the maximum number of iterations is determined as 15000 epochs. The adaptive learning rate is initialized to 0.01.

The results are shown in Tables 1-6. Tables 1 and 2 compares correct classification percentage (accuracy) for classification of the noisy data and the approximated data, respectively. The results show that, although the performances of both strategies are close to each other for high SNR levels, the correct classification percentage using the approximated data outperforms the use of original scattered signal as SNR level decreases. For instance, at -10dB SNR level, the classification performed using the features extracted from approximated scattered signal performs 4.40% and 6.43% higher accuracy for spheres 1 and 2, while the accuracy for 3 and 4 remains the same. Similar results are also achieved for Selectivity (Tables 3-4) and Specificity (Tables 5-6) metrics.

Table 1. Target recognition results (Accuracy) using features extracted from raw scattered data

Accuracy %	Sphere-1	Sphere-2	Sphere-3	Sphere-4
SNR=∞	95.83	92.55	87.53	88.57
SNR=10 dB	94.03	86.94	82.13	89.31
SNR=0 dB	88.19	74.81	68.66	76.39
SNR=-10 dB	75.60	61.53	69.95	73.75

Table 2. Target recognition results (Accuracy) using features extracted from approximated data

Accuracy %	Sphere-1	Sphere-2	Sphere-3	Sphere-4
SNR=∞	96.25	94.82	92.13	93.10
SNR=10 dB	95.79	94.82	91.11	92.64
SNR=0 dB	93.61	90.28	84.26	88.24
SNR=-10 dB	80.00	67.96	69.76	73.70

Table 3. Target recognition results (Sensitivity) using features extracted from raw scattered data

Accuracy %	Sphere-1	Sphere-2	Sphere-3	Sphere-4
SNR=∞	94.25	82.22	77.40	73.88
SNR=10 dB	90.74	72.03	69.63	72.40
SNR=0 dB	77.03	47.40	42.22	49.44
SNR=-10 dB	50.74	30.92	41.29	38.70

Table 4. Target recognition results (Sensitivity) using features extracted from approximated data

Accuracy %	Sphere-1	Sphere-2	Sphere-3	Sphere-4
SNR=∞	94.63	89.07	87.03	81.85
SNR=10 dB	94.07	93.14	80.74	80.74
SNR=0 dB	87.96	84.81	69.63	70.36
SNR=-10 dB	65.18	42.59	30.74	38.33

Table 5. Target recognition results (Specificity) using features extracted from raw scattered data

Accuracy %	Sphere-1	Sphere-2	Sphere-3	Sphere-4
SNR=∞	96.35	95.98	90.86	93.08
SNR=10 dB	95.12	91.91	86.29	94.39
SNR=0 dB	91.91	83.95	77.46	85.37
SNR=-10 dB	83.88	71.72	79.50	85.43

Table 6. Target recognition results (Specificity) using features extracted from approximated data

Accuracy %	Sphere-1	Sphere-2	Sphere-3	Sphere-4
SNR=∞	96.79	96.72	93.82	96.85
SNR=10 dB	96.35	95.37	94.56	96.60
SNR=0 dB	95.46	92.06	89.13	94.19
SNR=-10 dB	84.93	76.41	78.76	85.49

References

[1] Secmen M and Turhan-Sayan G 2009 *IET Radar Sonar and Navigation* **3** 583-595
[2] Turhan-Sayan G 2005 *IEEE Trans. On Antennas and Propagation* **53** 766-776
[3] Selver M A, Zoral E Y and Dorak S 2013 Aspect independent spherical target recognition using structural features of scattered signals *Conf. on the Computation of Electromagnetic Fields (Budapest, Hungary 30 June – 4 July 2013)*
[4] Mallat S G 1999 *A Wavelet Tour of Signal Processing* San Diego CA Academic Press 2nd ed.
[5] Raudis S J and Jain A K 1991 *IEEE Trans. On PAMI* **13** 252-264
[6] Haykin S 1998 *Neural Networks: A Comprehensive Foundation* Upper Saddle River NJ: Prentice Hall 2nd ed.

Ferromagnetic/superconducting interface in a hybrid nanoscopic disc

J D González[1],[2], J L Aguilar[1], J Barba-Ortega[2]

[1] Departamento de Física, Universidad del Magdalena, Santa Marta, Colombia

[2] Departamento de Física, Universidad Nacional de Colombia, Bogotá, Colombia

E-mail: jedgonzalezac@unal.edu.co

Abstract. The oscillations exhibited by the magnetization of superconducting nano and mesoscopics structures as a function of the applied magnetic field has attracted attention in the last few years. Nano/Mesoscopic hybrid systems in contact with ferromagnets exhibit interesting transport properties related to the influence of the exchange field on the density of states of clean ferromagnetic structures in contact with superconductor. In this work we study the vortex configurations in a superconducting disk with one central defect inside. The sample is surrounded by a ferromagnetic medium. We calculate the spatial distribution of the Cooper pairs and the phase of the superconducting order parameter and curves of magnetization and vorticity as a function of the magnetic applied field. We found that the first vortex penetration field decrease when a ferromagnetic/superconducting interface is used.

1. Introduction

We studied physical phenomena which appear when two mutually exclusive states of matter, superconductivity and ferromagnetism, are combined in a unified ferromagnet-superconductor hybrid (FSH) system. In the hybrid systems fabricated from materials with different and even mutually exclusive properties, a strong mutual interaction between subsystems can dramatically change properties of the constituent materials, in consequence, the spatial distribution of the Cooper pairs, the phase of the superconducting order parameter, vorticity and magnetization. The interplay of superconductivity and ferromagnetism has been thoroughly studied experimentally and theoretically [1, 2] for homogeneous system. In this paper, we solved the time dependent Ginzburg Landau (TDGL) equations for a superconducting thin disc with central pentagonal barrier (red), surrounded with a ferromagnetic material (blue) Fig.1.

2. Theoretical Formalism

The general form of the time dependent Ginzburg-Landau equations in the zero electric potential gauge are given by [3]–[5]:

$$\frac{\partial \psi}{\partial t} = -\left(i\nabla + \mathbf{A}\right)^2 \psi + \psi\left(1 - |\psi|^2\right) \tag{1}$$

$$\frac{\partial \mathbf{A}}{\partial t} = Re\left[\bar{\psi}\left(-i\nabla - \mathbf{A}\right)\psi\right] - \kappa^2 \nabla \times \nabla \times \mathbf{A} \tag{2}$$

In Eqs. (1) and (2) dimensionless units were introduced as follows: $|\psi|$ is the order parameter

Figure 1. Layout of the studied sample: Superconducting thin disc with central pentagonal barrier (red), surrounded with a ferromagnetic material (blue).

in units of $\psi_\infty(0) = \sqrt{-\alpha(0)/\beta}$, where $\alpha(0)$ and β are two phenomenological constants; lengths in units of the coherence length $\xi(0)$; time in units of $t_0 \equiv \pi\hbar/8K_BT_c$; A in units of $H_{c2}(0)\xi(0)$, where $H_{c2}(0)$ is the second critical field, temperature in units of the critical temperature T_c [4]. For a very thin disk of variable thickness, Eq. (1) can be rewritten as [6]−[9]:

$$\frac{\partial\psi}{\partial t} + \frac{1}{F}(i\nabla + \mathbf{A}_0)\cdot F(i\nabla + \mathbf{A}_0)\psi - \psi + \psi^3 = 0 \qquad (3)$$

where $F(r,\theta)$ is just a function which describes the thickness of the sample [10, 11]. In this case, the magnetic field can be taken nearly uniform inside the superconductor $\mathbf{H}_0 = \nabla \times \mathbf{A_0}$, so that Eq. (2) do not need to be solved.

3. Results and Discussion

We consider $F(r,\theta) = 1$ everywhere, except at defects position in the disk which are simulated by using $F = 1.2$ for the pentagonal barrier, see Fig.1. The parameters used in our numerical simulations were: $\kappa = 0.885438$, $T = 0.0$, $b = 0$ simulating the ferromagnetic boundary. The applied magnetic field was ramped in steps of $\Delta H = 10^{-4}$. Fig.2 (left) shown the magnetization as a function of increasing applied magnetic field, this curve exhibit the first clear jump at the field $H_0 = 0.8041H_{c2}(0)$ corresponding to the first entrance of vortices into the sample. The magnetization drops to zero for a value of the applied magnetic field $H_0 = 1.2464H_{c2}(0)$, therefore, our sample reaches the normal state. The Fig. Fig.2(right) shows the number of vortices entering whereas the applied magnetic field is increased. Following the jumps from the left to the right, in this order, we can see how many vortices there is in the sample for different values of $H_0/H_{c2}(0)$. When higher values of magnetization is reached, the contour plot of the phase of the superconducting order parameter is useful to determine the number of vortices into the sample. It is possible, by counting the phase variation in a closed path around this region. Values of the phase close to zero are given by blue regions and close to 2Φ by red regions. In Fig. 2 (right)(inset) shows the contour plot of phase of order parameter, additionally of the number of vortices in the superconductor, also is possible see their positions in the sample. For a value of $H_0 \approx 0.8149H_{c2}(0)$ we can see that the vortices are in outermost part of the disk, due to the repulsion caused by the defect, for this reason the vortices are not found inside. But by increasing the applied magnetic field, the pressure generated by the number of vortices in the superconductor exceeds the repulsion due to pentagonal barrier and the vortices can penetrate into the defect as shown in the graph of phase for a value of $H_0 = 1.0546H_{c2}(0)$. The panels of Fig. 3 (line up) show several spatial pattern of the order parameter for a disk with a pentagonal

Journal of Physics: Conference Series **574** (2015) 012141 doi:10.1088/1742-6596/574/1/012141

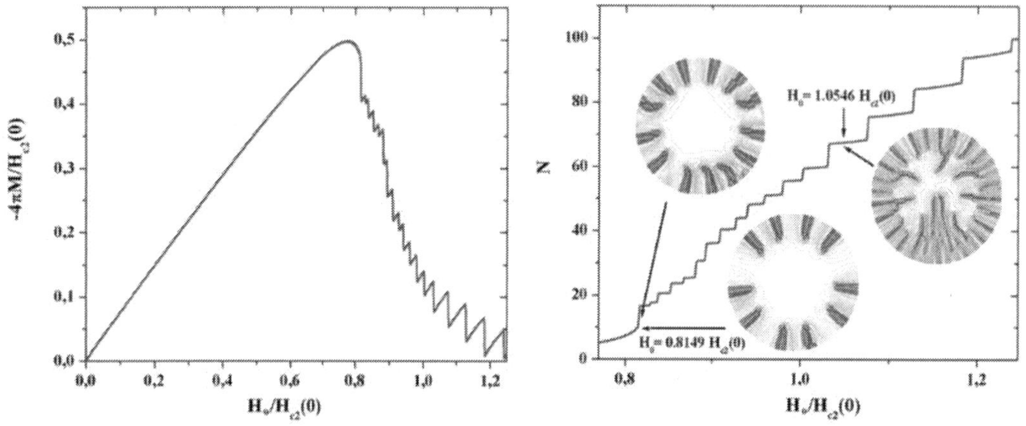

Figure 2. a). Magnetization curve (left) and vorticity N (right) as a function of the external applied magnetic field (the phase of the superconducting order parameter inset)

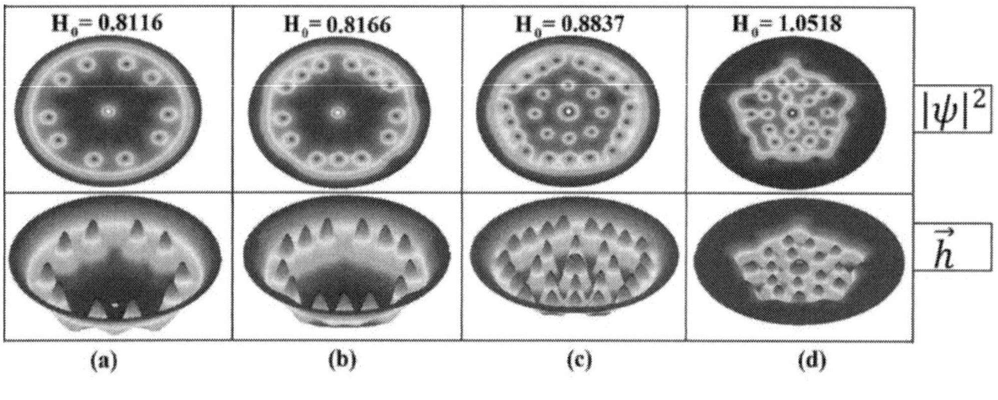

Figure 3. (Color online) The vortex structure in the indicated magnetic field for proposed sample. (line up) Square modulus of the order parameter $|\psi|^2$, (line down) Magnetic induction $\vec{h}$.

barrier by using $b = 0$. Following the panels from the left to the right, from (a, d) we can see that initially at $H_0 = 0.8116H_{c2}(0)$ pairs of vortices are located in every side of pentagono, whereas the blue region at the perimeter is the part of the disk which is reaching to normal state. When $H_0 = 0.8166H_{c2}(0)$ is applied the number of vortices grows a 33 percent aprox., despite the small increment en H_0, however, at $H_0 = 0.8837H_{c2}(0)$ its number reach almost 100 percent outside of defect and the vorticity $N_d = 8$ is obtained inside it. Fig. 3 (line down) shown the snapshot of the magnetic induction in which the red and/or blue regions indicate high and/or low values. We can see in the panel (a) and (b) that the blue color is centred on the disk and this confirm the Meissner effect on the pentagonal defect, so its behaviour is like type I superconductor. However, when the applied field reaches great values, the vortices penetrate the defects and the peaks of magnetic induction can even be counted (panel (c) and (d)).

4. Conclusions

We studied the effect of the geometry of a pentagonal barrier on the thermodynamical properties of a mesoscopic superconducting disk solving the time dependent Ginzburg-Landau equations. We take the value for the deGennes parameter $b = 0$, simulating a sample with its lateral surface sorrounded by a ferromagneto. Our results have shown that the vortex configuratin depend strongly of the chosen boundary condition and the nature of the defects, it is seen in the order parameter figure, for small changes in the applied magnetic field, the ferromagnet boundary encourages the entry of vortices.

5. Acknowledgments

This work was partially financed by the Universidad del Magdalena (Fonciencias) and the Colombian Agency COLCIENCIAS through doctoral scholarships 567.

6. References

[1] Fischer O *Magnetic Superconductors in Ferromagnetic Materials*, Vol. **5**, edited by Buschow K H J and Wohlfarth E P (Amsterdam, North-Holland, 1990), pp465549.
[2] Bulaevskii L N, Buzdin A I, Kulic M L and Panyukov S V 1985 *Adv. Phys.* **34** 175.
[3] Gropp W D, Kaper H G, Leaf G K, Levine D M, Palumbo M, Vinokur V M 1996 *J. Comput. Phys.* **123** 254.
[4] Tinkham M *Introduction to Superconductivity* 1996 McGraw-Hill, New York.
[5] Milosevic M V, Geurts R 2010 *Physica C* **470** 791.
[6] Sardella E, Filho P N L, Malvezzi A L 2008 *Phys. Rev. B* **77** 104508.
[7] Berdiyorov G R, Milosevic M V, Baelus B J, Peeters F M 2004 *Phys. Rev. B* **70** 024508.
[8] Du Q, Gunzburger M D 1993 *Physica D* **69** 215.
[9] Berdiyorov G R, Misko V R, Milosevic M V, Escoffier W, Grigorieva I V, Peeters F M 2008 *Phys. Rev. B* **77** 024526.
[10] Barba-Ortega J, Sardella E, Albino Aguiar J, Peeters F M 2013 *Physica C* **487** 47.
[11] Barba-Ortega J, Gonzalez J D and Sardella E 2014 *J. Low. Temp. Phys.* **174** 96.

Organization model for Mobile Wireless Sensor Networks inspired in Artificial Bee Colony

Guilherme Freire Roberto[1], Luis Fernando Castilho Maschi[2], Daniel Fernando Pigatto[3], Kalinka Regina Lucas Jaquie Castelo Branco[3], Leandro Alves Neves[1], Carlos Montez[4], Alex Sandro Roschildt Pinto[4]
[1]Sao Paulo State Univerity "Julio de Mesquita Filho", Sao Jose do Rio Preto, SP, Brasil
[2]Federal Institute of Sao Paulo Campus Catanduva, SP, Brasil
[3]University of Sao Paulo, Sao Carlos, SP, Brasil
[4]Federal University of Santa Catarina, Blumenau e Florianopolis, SC, Brasil

E-mail: guilhermefroberto@gmail.com

Abstract. The purpose of this study is to find a self-organizing model for MWSN based on bee colonies in order to reduce the number of messages transmitted among nodes, and thus reduce the overall consumption energy while maintaining the efficiency of message delivery. The results obtained in this article are originated from simulations carried out with SINALGO software, which demonstrates the effectiveness of the proposed approach. The BeeAODV (Bee Ad-Hoc On Demand Distance Vector) proposed in this paper allows to considerably reduce message exchanges whether compared to AODV (Ad-Hoc On Demand Distance Vector).

1. Introduction

The current industrial automation systems are typically characterized by a high degree of spatial distribution. Due to the complexity of production operations, solutions that help improving the productivity and efficiency of these systems are current challenges in the area. The use of WSN to monitor and control industrial applications has some advantages, such as facility of deployment and configuration, distance control and cost savings in infrastructure maintenance. However, the physical layout of large equipment and standards of operation make a very hostile environment for the characteristics of a WSN, and may have a large number of nodes failures, which makes its measurements unreliable [1]. Another issue to be addressed is the fact that in industrial environments wireless networks operating in the 2.4 GHz band are highly used, directly interfering in the transmission process of WSN. According to [2] in large and dense WSN networks, we can use fusion techniques of competitive data. The technique described in [3] defines that not all nodes have to operate at the same time, thus being possible to disconnect the radio module when the nodes are properly configured and in sleep state, therefore increasing the WSN life time and reducing the use of the communication channel. Real-time applications require constant monitoring of events in the environment for rapid decision-making, but the large number of messages exchanged between all nodes generates bumps and miscommunications, directly interfering the performance of the WSN, as described in [4]. The approach of this paper aims to reduce energy consumption by introducing the idea of mobile nodes, using techniques of bio-inspired algorithms and simulating the operation of a bee hive, where mobile nodes can efficiently communicate reducing power consumption.

2. Problem statement

Mobile Wireless Sensor Networks (MWSN) are an interesting alternative to circumvent the limited communication range of the nodes in a WSN. In [5] the authors present the BeeAdHoc, a routing algorithm for mobile ad hoc networks. It is based on the principles of a bee colony and uses mainly two types of agents, scouts and foragers, for doing routing in mobile ad hoc networks. This is a reactive routing algorithm that consumes less energy if compared to existing routing algorithms in the literature, because it operates with less control packets. The results presented in [5] show that BeeAdHoc consumes less energy as compared to DSR [6, 7], AODV [8] and DSDV [9]. When compared, the gains of BeeAdHoc and AODV, one realizes that there is a performance improvement. A comparison of the most used protocols in WSN was presented in [10]. However, this performance gain can be increased with the proposed self-organization for MWSN inspired by bee colony that will be presented in this article. It maintains network performance by reducing the number of messages exchanged between nodes, which makes it more effective to reduce approximately 30% of exchanged messages. The solution also reduces the overall energy consumption of MWSN.

3. Related works

The MWSN is based on routing and a high exchange of messages to maintain connectivity between nodes. Thus, if a node experiences problems (such as physical damage, interference with the media or lack of battery) and can no longer perform its task in the network, other nodes can rearrange their positions supplying the absence of the failed node. Several projects are being conducted in this area. It may be noted that the field of MWSN applications is broader than that of static networks, and with it, new challenges have emerged for the improvement of this technology, among them, the need to add an algorithm for mobility, minimize energy consumption (since mobile nodes consume more battery than fixed nodes) and synchronize the data collected [11]. One of the biggest challenges to overcome is the efficiency in the exchange of messages between nodes and base stations. For this new data transmission protocols are being developed and researched by proposing solutions to ensure such efficiency and communication security. Algorithms such as AODV are being prepared for the pursuit of energy efficiency for the realization of such communication as shown in [8, 12]. Bio-inspired algorithms, specifically regarding communication among colonies of insects such as bees and ants [13, 14], are seen as possible solutions to the communication in MWSN [15]. We can quote the ABC [16] and the ACA [12] as examples of bio-inspired algorithms that improve communication among mobile nodes. BeeAdHoc algorithm consumes less energy than the traditional AODV and other algorithms found in the literature, such as DSDV and DSR [5].

4. Algorithm description

The MWSN organization model presented in this project is inspired by the hierarchical operation of a bee colony. In nature, bees are organized in different roles in the hive. In this proposal, it is assumed three different functions: the queen, attendants and scouts. In Figure 1 is presented an overview of the organizational model proposed for the network. In a similar approach, the fixed base station acts as the queen, the attendants area is equivalent to the hive space, and the nodes located in this area correspond to the attendant bees. The search area corresponds to the region where the scout bees look for food for the hive. In the proposed organization model, when a node needs to transmit information ("food"), it sends a message to the base station. This message is then transmitted among explorer nodes until it is received by an attendant node. The attendant node that receives the message has to send it to the base station. In Fig. 2 is shown the overall implementation of the network, from initial configuration to the BeeAODV algorithm itself.

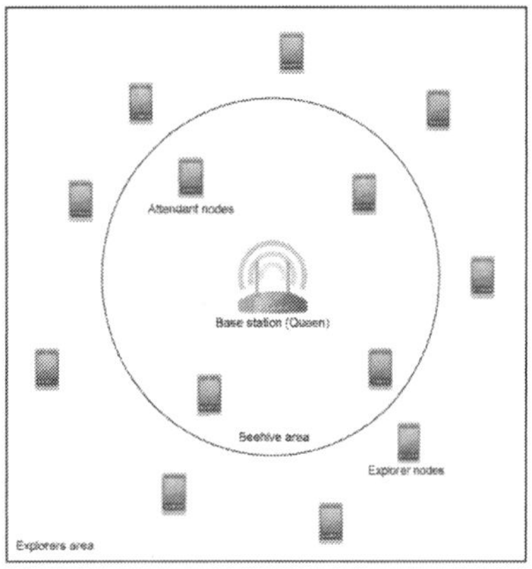

Figure 1. Overview of the organizational model proposed for the network.

```
Read k
Nodes are arranged in the initial distribution
Adjusting transmission power
node[0] = Queen
for n = 1 TO n < number_of_nodes, n + + do

    if (distance_between_node[n] AND Queen) < range_transmission then
        node[n] = NodeAttendant;
    else
        node[n] = Explorer
    end if
end for
while Network running do

    for n = 1 TO n< number_of_nodes, n + + do

        if n = 0 and existence_time_of_routing_table > 1second then
            Send Hello messages
        else

            if node[n] detects changes in the environment then
                node[n] sends messages destined to the Queen
            end if
        end if
    end for
end while
```

Figure 2. Operation of BeeAODV algorithm.

5. Methodology

For evaluating the efficiency of the algorithm, it was performed simulations using SINALGO software. The tests were conducted in a simulated rectangular area of 300 x 600 m, with a base station located in the center. The communication module adopted in the simulations was XBee, which uses the IEEE 802.15.4 communication protocol. When set at maximum power, it reaches a transmission range of approximately 90 m. The nodes were initially arranged in a distribution grid and then we performed a comparison between the traditional AODV algorithm and the organizational model proposed in this paper. Besides the base station that was present in all test cases with 50, 75, 100, 125, 150 and 200 nodes were simulated. Each simulation took 5 minutes to be executed. The data collected was the total number of transmitted messages and the time at which the node remained disconnected to the base. The "Hello Messages" were sent with 1 second interval.

6. Results

In Fig. 3 is shown the performance of the proposed organizational model BeeAODV compared to AODV, where we can observe that performance with a smaller amount of nodes is almost identical, but with the increase of nodes BeeAODV protocol becomes more efficient by sending a minor amount of messages. The time that the nodes are disconnected from the base station can be analyzed in Fig. 4, where, due to reduced transmission power, the algorithm proposed BeeAODV maintains some stability regardless of the amount of nodes, while the reliability of AODV tends to be 100% as the number of nodes is increased, once it is operated with maximum power transmission. Finally, in Fig. 5 it is shown that, statistically, the BeeAODV protocol is more stable in the exchange of "Hello Messages", when compared to AODV. The standard deviation of the number of messages in the simulations is significantly lower in BeeAODV model.

7. Conclusion

Simulation results point out that there was a significant reduction in the number of messages generated by the network. Consequently, this positively affects network performance and service life. Also due to the reduction in nodes transmission power, the overall energy consumption is even smaller, without compromising its reliability. The proposed algorithm BeeAODV proved to be a stable algorithm.

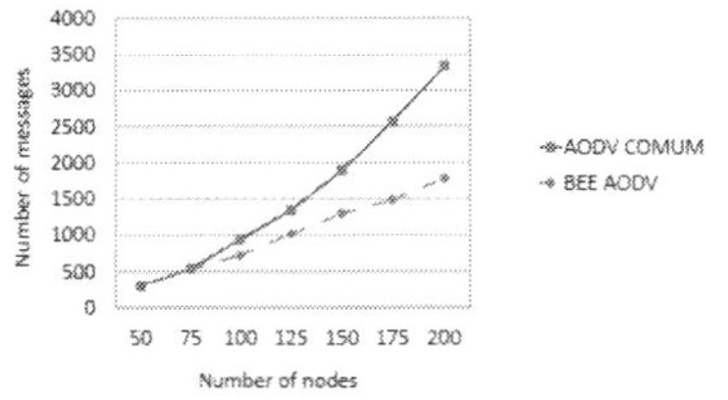

Figure 3. Comparison of the number of "Hello Messages" exchanged (AODV and BeeAODV).

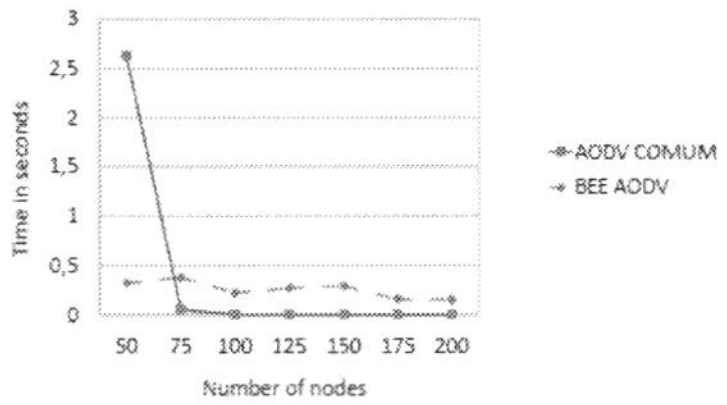

Figure 4. Comparison of time that the node is disconnected from the base station between BeeAODV and AODV protocols.

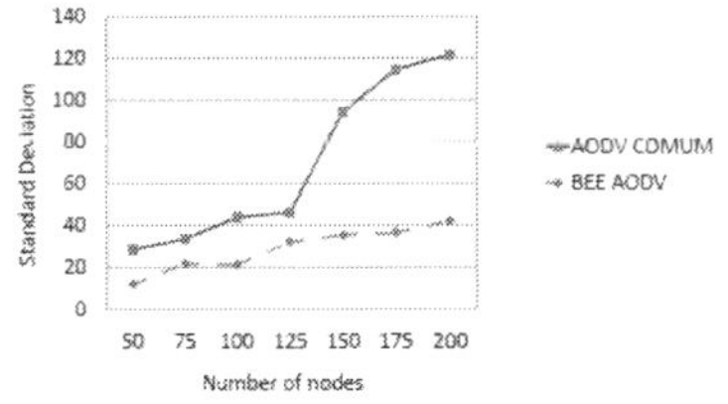

Figure 5. Default deviation for "Hello Messages" sent in simulations.

7.1. Acknowledgments

The authors would like to thank the institutions where they conduct their research for supporting this project and also the financial support provided by FAPESP (Process No. 2012/16171 -6).

References

[1] Gungor V and Hancke G 2009 *IEEE Transactions on Industrial Electronics* **56** 4258–4265 ISSN 0278-0046
[2] Elmenreich W 2002 157
[3] Tian D and Georganas N D 2002 *Proceedings of the 1st ACM international workshop on Wireless sensor networks and applications - WSNA '02* 32
[4] AbHalim A H and Zen K 2008 *2008 International Conference on Computer and Communication Engineering* (IEEE) pp 1290–1293 ISBN 978-1-4244-1691-2
[5] Wedde H F, Farooq M, Pannenbaecker T, Vogel B, Mueller C, Meth J and Jeruschkat R 2005 *Proceedings of the 2005 conference on Genetic and evolutionary computation - GECCO '05* 153
[6] Johnson D B, Maltz D A and Broch J 2001 *In Ad Hoc Networking, edited by Charles E. Perkins, Chapter 5* 139–172
[7] Cavalli A, Grepet C, Maag S and Tortajada V 2004 *24th International Conference on Distributed Computing Systems Workshops, 2004. Proceedings.* (IEEE) pp 768–773 ISBN 0-7695-2087-1
[8] Sundararajan T V P, Rajesh Kumar K and Karthikeyan R K 2009 *Control, Automation, Communication and Energy Conservation, 2009. INCACEC 2009. 2009 International Conference on* 1–5
[9] Khan K U R, Zaman R U, Reddy A V, Reddy K A and Harsha T S 2008 *2008 Second UKSIM European Symposium on Computer Modeling and Simulation* (IEEE) pp 506–511
[10] Royer E M 1999 *IEEE Personal Communications* **6** 46–55 ISSN 10709916
[11] Yick J, Mukherjee B and Ghosal D 2008 *Computer Networks* **52** 2292–2330 ISSN 13891286
[12] Hong J, Lu S, Chen D and Cao J 2008 *22nd International Conference on Advanced Information Networking and Applications (aina 2008)* 1054–1061
[13] Farooq M and Caro G A D 2008 *Swarm Intelligence: Introduction and Applications, Springer*
[14] Aksa K and Benmohammed M 2012 *International Journal of Networks and Communications* **2** 27–32 ISSN 2168-4936
[15] Dressler F and Akan O B 2010 *Computer Networks* **54** 881–900 ISSN 13891286
[16] Bansal J C, Sharma H and Jadon S S 2013 *International Journal of Advanced Intelligence Paradigms* **5** 123 ISSN 1755-0386

GPU Computing in Bayesian Inference of Realized Stochastic Volatility Model

Tetsuya Takaishi
Hiroshima University of Economics, Hiroshima 731-0192, JAPAN
E-mail: tt-taka@hue.ac.jp

Abstract. The realized stochastic volatility (RSV) model that utilizes the realized volatility as additional information has been proposed to infer volatility of financial time series. We consider the Bayesian inference of the RSV model by the Hybrid Monte Carlo (HMC) algorithm. The HMC algorithm can be parallelized and thus performed on the GPU for speedup. The GPU code is developed with CUDA Fortran. We compare the computational time in performing the HMC algorithm on GPU (GTX 760) and CPU (Intel i7-4770 3.4GHz) and find that the GPU can be up to 17 times faster than the CPU. We also code the program with OpenACC and find that appropriate coding can achieve the similar speedup with CUDA Fortran.

1. Introduction

Since volatility of asset returns plays an important role to manage financial risk, measuring values of volatility is an important task in empirical finance. Since one can not directly observe volatility in the financial markets, one needs to use a certain estimation technique such as volatility modeling. The first promising volatility modeling was introduced in 1982 by Engle[1]. The model introduced by him is called the AutoRegressive Conditional Heteroscedasticity (ARCH) model. Soon the ARCH model was extended by Bollerslev to the generalized ARCH (GARCH) model[2]. An alternative to the GARCH model is the stochastic volatility (SV) model[3] which allows the volatility to be a stochastic process. Recently an extended SV model which utilizes the realized volatility (RV)[4, 5] constructed by a sum of finely sampled intraday returns was proposed. This SV-type model is called the realized SV (RSV) model[6]. An advantage of the RSV model over the conventional SV model is that it uses the RV as additional information and thus could estimate the daily volatility more accurately. Since it is difficult to evaluate the likelihood function of the SV models usually the maximum likelihood method is not convenient for the parameter estimation of the SV model. The standard estimation algorithm for the SV model is the Markov Chain Monte Carlo (MCMC) method based on the Bayesian approach. Various MCMC algorithms for the SV model have been proposed and tested[7]-[12].

In this study we perform the MCMC method on graphics processing unit (GPU). GPU computing is able to perform computations in parallel that results in speeding up the computational performance and applied for various scientific computations, e.g. Ising model simulations[13]-[16]. Since in general MCMC algorithms are performed sequentially, not all MCMC algorithms can be easily parallelized. We use the hybrid Monte Carlo (HMC) algorithm[17] that can be implemented in parallel. We perform the HMC algorithm on GPU and CPU for the parameter estimations of the RSV model and compare thier computational performance. The GPU we used is NVIDIA's GeForce GTX 760 and coding for GPU was done using the PGI CUDA Fortran[18]. We also code the program using OpenACC[19] which enables us to do directive based programing and investigate performance results.

2. Realized Stochastic Volatility Model

The realized stochastic volatility (RSV) model introduced by Takahashi *et al.*[6] is written as

$$y_t = \exp(h_t/2)\epsilon_t, \quad \epsilon_t \sim N(0,1), \tag{1}$$
$$\ln RV_t = \xi + h_t + u_t, \quad u_t \sim N(0,\sigma_u^2), \tag{2}$$
$$h_{t+1} = \mu + \phi(h_t - \mu) + \eta_t, \quad \eta_t \sim N(0,\sigma_\eta^2), \tag{3}$$

where y_t and RV_t for $t = 1, ..., T$ are a daily return and a daily realized volatility RV at time t respectively, and h_t is a latent volatility defined by $\ln \sigma_t^2$. The model parameters that we have to estimate are $\theta = \phi, \mu, \xi, \sigma_\eta^2, \sigma_u^2$. We apply the Bayesian inference for parameter estimations and perform the Bayesian inference by the MCMC method. The most time consuming part of the MCMC approach for the SV-type models is volatility update[7]. In order to improve the volatility update several MCMC approaches have been developed, e.g. multi-move sampler[8, 9] and HMC algorithm[10, 11, 12]. In this study we use the HMC algorithm for the volatility update of the RSV model.

3. Hybrid Monte Carlo Algorithm

The HMC algorithm[17] appeared for the large scale MCMC simulations of the lattice Quantum Chromo Dynamics (QCD) calculations[20] for the first time and has been the standard MCMC algorithm of the lattice QCD calculations. The HMC algorithm combines the molecular dynamics (MD) simulation and the Metropolis accept/reject test. Since the HMC algorithm is a global algorithm that variables we consider are updated simultaneously. This means that the variables can be updated in parallel. For the RSV model we update volatility variables by the HMC algorithm. The basic HMC algorithm is as follows. First, candidates for next volatility variables in Markov chain are obtained by solving the Hamilton's equations of motion in fictitious time τ,

$$\frac{dh_i}{d\tau} = \frac{\partial H}{\partial p_i}, \tag{4}$$
$$\frac{dp_i}{d\tau} = -\frac{\partial H}{\partial h_i}, \tag{5}$$

where p_i for $i = 1, ..., T$ is a conjugate momentum to h_i. The Hamiltonian H is defined by $H(p,h) = \frac{1}{2}\sum_i^n p_i^2 - \ln f(h,\theta)$, where $f(h,\theta)$ is the conditional posterior density of the RSV model[6]. Since in general eqs.(4)-(5) can not be solved analytically, they are integrated numerically through the MD simulation. The conventional integrator for the MD simulation in the HMC algorithm is the 2nd order leapfrog integrator[17] given by

$$h_i(\tau + \delta\tau/2) = h_i(\tau) + \tfrac{\delta\tau}{2} p_i(\tau), \tag{6}$$
$$p_i(\tau + \delta\tau) = p_i(\tau) - \delta\tau \tfrac{\partial H}{\partial h_i}, \tag{7}$$
$$h_i(\tau + \delta\tau) = h_i(\tau + \delta\tau/2) + \tfrac{\delta\tau}{2} p_i(\tau + \delta\tau), \tag{8}$$

where $i = 1, ..., T$ and $\delta\tau$ stands for the step size. The higher order or other integrators[21]-[27] can be also used in the HMC algorithm if necessary. Eqs.(6)-(8) are repeatedly performed k times and then the total integration length l becomes $l = k \times \delta\tau$. After the MD simulations we obtain new volatility and conjugate momentum variables, denoted as $h_i' = h(\tau + l)$ and $p_i' = p(\tau + l)$. The new volatility variables h_i' are accepted at the Metropolis step with a probability $\sim \min\{1, \exp(-\Delta H)\}$ where $\Delta H = H(p', h') - H(p, h)$.

Table 1. GTX 760 Specifications[28]

GPU Engine Specs	
CUDA Cores	1152
Base Clock (MHz)	980
Boost Clock (MHz)	1033
Memory Specs	
Memory Speed	6.0 Gpbs
Memory Config	2048MB
Memory Interface	GDDR5
Memory Interface Width	265-bit
Memory Bandwidth (GB/sec)	192.2

Table 2. Fitting parameters

	A	C
GPU (CUDA Fortran)	1.13×10^{-5}	2.25×10^{-7}
CPU (Intel i7-4770 3.4GHz)	-1.42×10^{-6}	3.87×10^{-6}

4. GPU Coding Environment

We used the NVIDIA GeForce GTX760 for GPU computing. Table 1 shows the specifications of the GTX760[28]. The original HMC code of the RSV model for a single CPU was developed in [29]. Our codes for GPU were developed using the PGI CUDA Fortran[18] with CUDA 6.0 drivers. The PGI CUDA Fortran also supports OpenACC[19]. All programs are developed with single precision. We also execute a code on CPU (Intel i7-4770 3.4GHz) to compare performance between GPU and CPU.

5. HMC Algorithm by CUDA Fortran

To perform the HMC algorithm on GPU we make a code by CUDA Fortran. Each equation of eqs.(6)-(8) is assigned to a kernel executed on the GPU as follows.

$$h_i(\tau+\delta\tau/2) = h_i(\tau) + \tfrac{\delta\tau}{2} p_i(\tau), \qquad \Leftarrow \text{Kernel 1} \tag{9}$$

$$p_i(\tau+\delta\tau) = p_i(\tau) - \delta\tau \frac{\partial H}{\partial h_i} \qquad \Leftarrow \text{Kernel 2} \tag{10}$$

$$h_i(\tau+\delta\tau) = h_i(\tau+\delta\tau/2) + \tfrac{\delta\tau}{2} p_i(\tau+\delta\tau) \qquad \Leftarrow \text{Kernel 3} \tag{11}$$

For instance eq.(9) which integrates h_i for $i = 1, ..., T$, denoted by "Kernel 1" is coded by CUDA Fortran and executed on the GPU in parallel for $i = 1, ..., T$. After the execution of the Kernel 1, Kernels 2 and 3 are also executed. The Kernels 1-3 form an elementary step of the MD simulation. In this study we executed the elementary step 10000 times and measure an average execution time of the elementary step. In order to make a comparison between GPU and CPU we also measure an average execution time of the elementary step on CPU.

Fig.1 shows the average execution time of the elementary step as a function of the number of volatility variables or the size of time series where B has a relationship as $T = 512 \times B$. In the GPU computing we set the thread size to 512. From Fig.1 we recognize that the average execution time increases almost linearly with B for both GPU and CPU. We fit the results with a linear function of $f(B) = A + C \times B$ for both GPU and CPU. The fitting results are summarized in Table 2. Here we define the speedup of GPU over the CPU by $Gain = f_{CPU}(B)/f_{GPU}(B)$. Fig.2 shows the Gain as a function of B. The Gain increases with B and goes to $C_{CPU}/C_{GPU} = 17.2$ in the limit of $B \to \infty$.

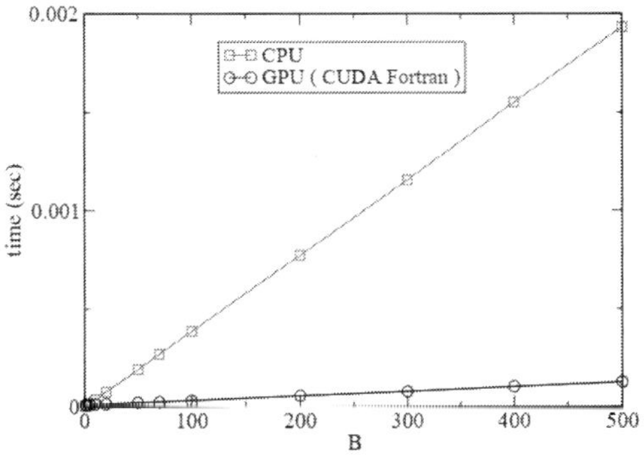

Figure 1. Average execution time of the elementary step on GPU and CPU as a function of B.

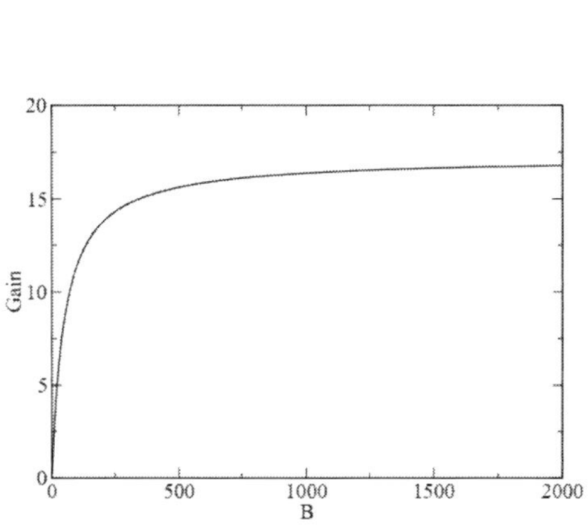

Figure 2. Gain defined by $f_{CPU}(B)/f_{GPU}(B)$ as a function of B.

6. HMC algorithm by OpenACC

OpenACC enables us to do directive based programming for GPU that can greatly reduce coding effort. We insert OpenACC directives into the existing Fortran program and actually in this study we used the program developed for the RSV model in [29]. The following is the schematic diagram for the OpenACC coding.

!$acc data copy(h,p)

!$acc kernels

$$h_i(\tau+\delta\tau/2) = h_i(\tau) + \frac{\delta\tau}{2}p_i(\tau) \tag{12}$$

$$p_i(\tau+\delta\tau) = p_i(\tau) - \delta\tau\frac{\partial H}{\partial h_i} \tag{13}$$

$$h_i(\tau+\delta\tau) = h_i(\tau+\delta\tau/2) + \frac{\delta\tau}{2}p_i(\tau+\delta\tau) \tag{14}$$

!$acc end kernels

!$acc end data

Eqs.(12)-(14) between "!$acc kernels" and "!$acc end kernels" are automatically translated to a GPU code and performed on GPU. The data directive "!$acc data copy(h,p)" specifies the variables (here h and p) that are used in the GPU code between "!$acc data copy(h,p)" and "!$acc end data". This avoids unnecessary data transfer between CPU and GPU. Actually in order to measure an average execution time as in the previous section we repeat 10000 times the code from "!$acc kernels" to "!$acc end kernels" and if no data directive is inserted to the program we can not achieve the speedup as the code by CUDA Fortran. Fig.3 shows the average execution time as a function of B. The squares (circles) indicate the average execution time of the OpenACC code with (without) the data directive. We find that the average execution time of the OpenACC code with the data directive is similar with that of the CUDA Fortran. On the other hand, without the data directive the average execution time takes more time than that with the data directive.

7. Conclusion

We have performed the HMC algorithm in the Bayesian estimation of the RSV model on GPU (GTX 760) using CUDA Fortran. It is found that the GPU can be up to 17 times faster than the CPU (Intel i7-4770 3.4GHz) when the size of time series is big. We have also coded a HMC program for GPU with the OpenACC that enable us to do directive based programming for

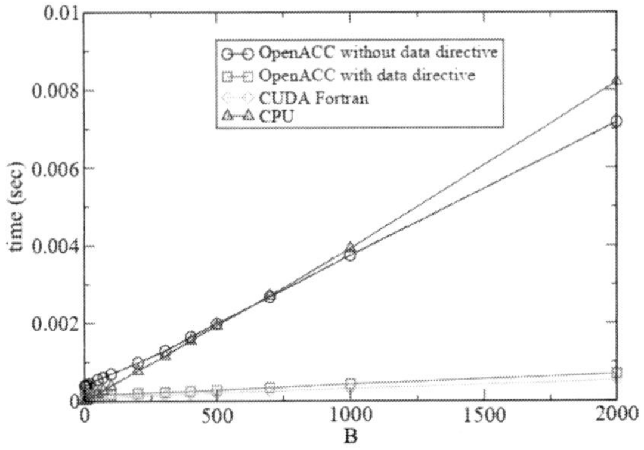

Figure 3. Average execution time of the elementary step of GPU computing by OpenACC with (without) data directive.

GPU computing and found that the OpenACC program with appropriate coding can achieve the similar speedup with CUDA Fortran.

Acknowledgement

Numerical calculations in this work were carried out at the Yukawa Institute Computer Facility and the facilities of the Institute of Statistical Mathematics. This work was supported by JSPS KAKENHI Grant Number 25330047.

References

[1] Engle R F 1982 *Econometrica* **50** 987-1007
[2] Bollerslev T 1986 *Journal of Econometrics* **31** 307-327
[3] Taylor S J 1986 *Modelling Financial Time Series* John Wiley & Sons, New Jersy
[4] Andersen T G and Bollerslev T 1998 *International Economic Review* **39** 885-905
[5] Andersen T G, Bollerslev T, Diebold F X and Labys P 2001 *J. Am. Statist. Assoc.*A **96** 42-55
[6] Takahashi M, Omori Y and Watanabe T 2009 *Compt. Stat. & Data anal.* **53** 2404-2425
[7] Kim S, Shephard N and Chib S 1998 *Review of Economic Studies* **65** 361-393
[8] Shephard N and Pitt M K 1997 *Biometrika* **84** 653-667
[9] Watanabe T and Omori Y 2004 *Biometrika* **91** 246-248
[10] Takaishi T 2008 *Lecture Notes in Computer Science* **5226** 929-936
[11] Takaishi T 2009 *Journal of Circuits, Systems, and Computers* **18** 1381-1396
[12] Takaishi T 2013 *J. Phys.: Conf. Ser.* **423** 012021
[13] Preis T, Virnau W and Schneider J J 2009 *J. Comput. Phys.* **228** 4468-4477
[14] Block B, Virnau P and Preis T 2010 *Comput. Phys. Commun.* **181** 1549-1556
[15] Weigel M and Yavors'kii T 2011 *Physics Procedia* **15** 92-96
[16] Komura Y and Okabe Y 2012 *J. Comput. Phys.* **231** 1209-1215
[17] Duane S, Kennedy A D, Pendleton B J and Roweth D 1987 *Phys. Lett. B* **195** 216-222
[18] https://www.pgroup.com/resources/cudafortran.htm
[19] http://www.openacc.org/
[20] Montvay I and Münster G 1994 *Quantum Fields on a Lattice* Cambridge University Press
[21] Forest E and Ruth R D 1990 *Physica D* **43** 105-117
[22] Sexton J C and Weingarten D H 1992 *Nucl. Phys. B* **380** 665-677
[23] Yoshida H 1990 *Phys. Lett. A* **150** 262-268
[24] Takaishi T 2000 *Comput. Phys. Commun.* **133** 6-17
[25] Takaishi T 2002 *Phys. Lett. B* **540** 159-165
[26] Omelyan I P, Mryglod I M and Folk R 2003 *Comput. Phys. Commun.* **151** 272-314
[27] Takaishi T and de Forcrand Ph 2008 *Phys. Rev. E* **73** 036706
[28] http://www.geforce.com/hardware/desktop-gpus/geforce-gtx-760
[29] Takaishi T 2014 *J. Phys.: Conf. Ser.* **490** 012092

Journal of Physics: Conference Series **574** (2015) 012144 doi:10.1088/1742-6596/574/1/012144

A mathematical model and simulation results of plasma enhanced chemical vapor deposition of silicon nitride films

S A Konakov [1], **V V Krzhizhanovskaya** [1,2,3]

[1] St. Petersburg State Polytechnical University, 195251, St. Petersburg, Russia
[2] University of Amsterdam, 1098 XH Amsterdam, The Netherlands
[3] ITMO University, 197101, St. Petersburg, Russia

E-mail: konakov.st@gmail.com

Abstract. We developed a mathematical model of Plasma Enhanced Chemical Vapor Deposition (PECVD) of silicon nitride thin films from SiH_4-NH_3-N_2-Ar mixture, an important application in modern materials science. Our multiphysics model describes gas dynamics, chemical physics, plasma physics and electrodynamics. The PECVD technology is inherently multiscale, from macroscale processes in the chemical reactor to atomic-scale surface chemistry. Our macroscale model is based on Navier-Stokes equations for a transient laminar flow of a compressible chemically reacting gas mixture, together with the mass transfer and energy balance equations, Poisson equation for electric potential, electrons and ions balance equations. The chemical kinetics model includes 24 species and 58 reactions: 37 in the gas phase and 21 on the surface. A deposition model consists of three stages: adsorption to the surface, diffusion along the surface and embedding of products into the substrate. A new model has been validated on experimental results obtained with the «Plasmalab System 100» reactor. We present the mathematical model and simulation results investigating the influence of flow rate and source gas proportion on silicon nitride film growth rate and chemical composition.

1. Introduction

Silicon nitride thin films are used in solar cells, nano- and microelectronics, and other micro-technologies [1]. Plasma enhanced chemical vapor deposition (PECVD) at low-pressure RF glowing discharge is the most applicable technology for silicon film production because a high-quality hydrogenated amorphous silicon nitride (SiN_xH_y) can be obtained on large-area surface at low temperature [2]. Chemical composition and physical properties of the deposited films depend on process parameters in a highly non-linear way. It is extremely difficult to predict film properties before deposition, and experiments are expensive and time-consuming. Multiphysics computer simulation is the most efficient way for technology optimization.

The goal of our research is the development of a reliable model and investigation of correlation between process parameters and chemical composition of SiN_xH_y. We modeled a real PECVD reactor, and used experimental results and our previous studies [3, 4, 5] to build the chemical kinetics model. For process optimization at the macroscale level we have chosen a continuum modeling approach. Other researchers also used this approach to simulate silicon nitride PECVD from silane (SiH_4), ammonia (NH_3), nitrogen (N_2) and argon (Ar) [6, 7, 8], but they did not consider all four source gases together, as presented in our work. Recent progress in silicon nitride atomic structure investigation by Fourier transform infrared (FTIR) spectroscopy [9, 10] allowed us to develop a new model of surface chemistry mechanism.

2. Mathematical model

A continuum modeling approach is justified at Knudsen numbers $Kn \sim 5 \cdot 10^{-3}$ calculated for our conditions. The PECVD processes require modeling gas dynamics, heat and mass transfer, electromagnetism and chemical kinetics. Mass conservation law (1) and Navier-Stokes equations for a transient laminar flow of a compressible chemically reacting gas mixture (2) describe the gas dynamics in the process chamber:

$$\frac{\partial \rho}{\partial t} + \Delta \cdot (\rho \boldsymbol{V}) = 0, \tag{1}$$

$$\frac{\partial (\rho u_i)}{\partial t} + \nabla \cdot (\rho \mathbf{V} u_i) = -\frac{\partial p}{\partial x_i} + \nabla \cdot (\mu \nabla u_i) \qquad i = x, y, z, \tag{2}$$

where ρ is density, t is time, V is velocity vector, u_i is projection of velocity vector on i axis, μ is the kinematic viscosity. Heat transfer equation can be written as

$$\frac{\partial (\rho h_0)}{\partial t} + \nabla \cdot (\rho \boldsymbol{V} h_0) = \nabla \cdot (k_{eff} \nabla T) + \frac{\partial p}{\partial t} + S_h, \tag{3}$$

where h_o is enthalpy, k_{eff} is thermal conductivity, T is temperature, p is pressure, S_h is the heat source due to chemical reactions. A model of capacitively coupled plasma (CCP) is based on Maxwellian electron energy distribution function. Electron balance is given by

$$\frac{\partial n_e}{\partial t} + \nabla \cdot \boldsymbol{\Gamma_e} = S, \tag{4}$$

where n_e is electron density, Γ_e is electron flux and S is a source of electrons due to chemical reactions. Energy balance in plasma can be written as

$$\frac{3}{2}\frac{\partial}{\partial t}(n_e T_e) + \nabla\left(\frac{5}{2} T_e \cdot \boldsymbol{\Gamma_e} - \chi \nabla T_\varepsilon\right) = P - P_c \tag{5}$$

where T_e electron temperature, χ is a constant for electron termodiffusion, P is power adsorbed in plasma and P_c is an energy loss due to electron impact reactions.The spatial distribution of electrostatic potential φ is described by the Poisson equation:

$$-\nabla \cdot \varepsilon_r \nabla \varphi = e\left(\textstyle\sum_i q_i n_i - n_e\right), \tag{6}$$

where ε_r is relative permittivity, q_i is charge of ions, n_i is ion density, e is elementary charge. Mass transfer equation for the ith species can be written as

$$\frac{\partial}{\partial t}(\rho Y_i) + \frac{\partial}{\partial x_j}(\rho u_j Y_i) = \frac{\partial}{\partial x_j} J_{ij} + M_i \omega_i, \tag{7}$$

where ρ is density, Y_i is a mass fraction of ith species, J_{ij} is projection of diffusion flux vector on j^{th} axis, M_i is molar mass of ith species, ω_i is chemical reaction rate.

3. Model of chemical kinetics in plasma

The model of chemical processes in plasma describes reactions in the bulk and on the surface. For the gas phase processes we considered 24 species: source gases Ar, N_2, NH_3, SiH_4, ions N_2^+, NH_3^+, NH_4^+, Ar^+; molecules and radicals H, NH_2, NH, SiH_3, SiH_2, H_2, Si_2H_6; aminosilanes $H_2Si(NH_2)_2$, $HSi(NH_2)$, $HSi(NH_2)_2$, $HSi(NH_2)_3$, $Si(NH_2)_2$, $Si(NH_2)_3$, $Si(NH_2)_4$; excited state of argon Ar* and electron e^-. The model of 37 bulk chemical reactions is based on [7, 8]. The cross-sections of electron impact reactions are taken from [11].

A deposition model consists of three stages: (1) adsorption to the surface, (2) diffusion along the surface to the next reactant and (3) embedding of products into the substrate. We considered the most stable species that can be adsorbed on the surface: $H_2Si(NH_2)$, $HSi(NH_2)$, $HSi(NH_2)_2$, NH_2, $Si(NH_2)_2$, $Si(NH2)_3$, SiH_2, SiH_3. Sticking coefficients defining the probability of adsorption were taken from [7, 8]. The surface reaction model was developed based on the experimental data of Fourier transform infrared spectroscopy (FTIR) [9, 10]. An example of the spectrum decomposition into the Gaussian components is shown in Figure 1. It was found that silicon nitride films are composed of four functional groups with chemical bonds Si_3–Si–N, H_2–Si–N_2, Si_3N_4 and H–Si–N_3. We suggested 7 reactions between the adsorbed components producing these molecular groups and found reaction rates by matching the simulation results with experimental data. Another 14 reactions describe adsorption to the surface and ions neutralization.

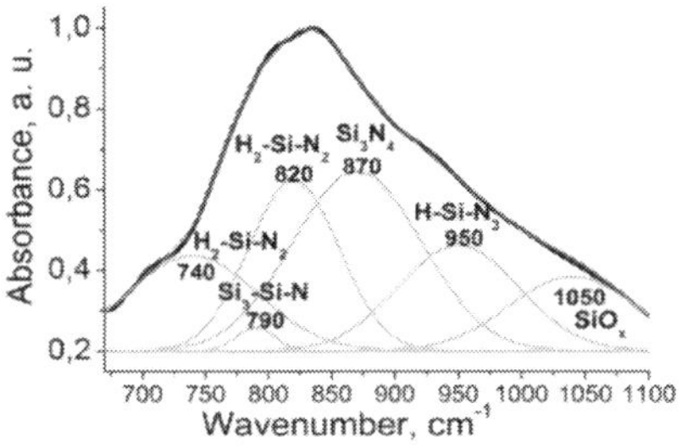

Figure 1. Infrared spectrum and Gaussian components for the Si–N absorption bands. Experiement.

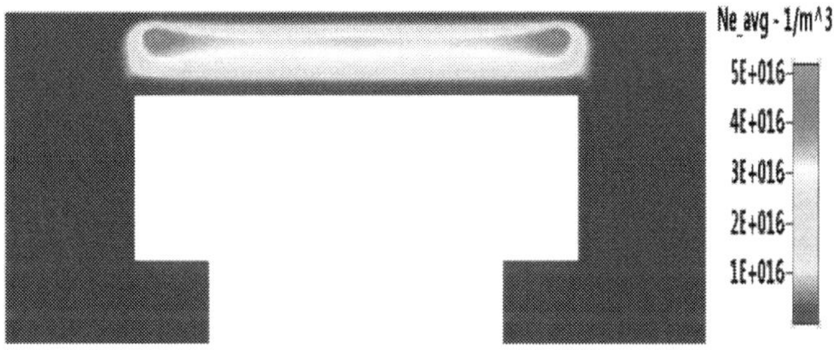

Figure 2. Simulation result: distribution of electron density in a PECVD reactor. Flow rate of SiH_4-Ar mixture is 400 sccm, N_2 is 600 sccm and NH_3 is 20 sccm.

4. Implementation and results

A finite volume method on a regular rectangular computational mesh was used for solving equations (1)-(7) by the CGS-Pre iterative solver. 20 time steps were performed in each RF discharge cycle with a maximum of 500 numerical iterations per time step. Typical simulation time was 18 hours per experiment. We simulated a two-dimensional geometry of «Plasmalab System 100» PECVD reactor by Oxford Instruments Plasma Technology. The frequency of CCP discharge was 13.56 MHz, power 20 W, temperature 350 °C, and pressure 0.65 Torr.

We carried out two series of experiments: In the first one we varied the total gas flow rate, keeping the ratio of source gases constant. In the second series we varied the flow rate of silane-argon mixture under the constant flow rate of other gases: 600 sccm of N_2 and 20 sccm of NH_3. An example of 2D distribution of electron density in a simulated PECVD reactor is shown in Figure 2. A comparison of simulation results with experimental data is presented in Figures 3 and 4.

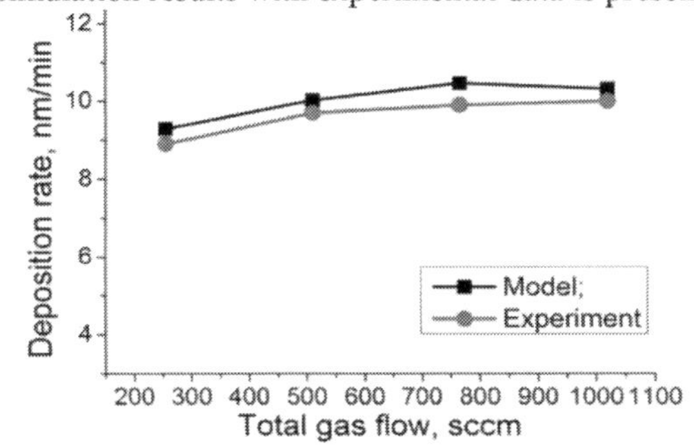

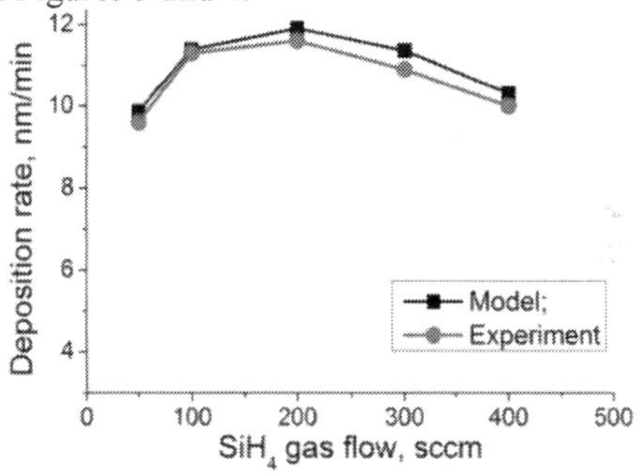

Figure 3. Influence of the total gas flow on deposition rate. Simulations and experiments.

Figure 4. Influence of the silane flow on deposition rate. Simulations and experiments.

Maximum discrepancy between the simulated deposition rate and experimental data is less than 5%, but further model fine-tuning is still possible. Two important observations can be made: (1) The deposition rate grows a little (by 10%) with a four-fold increase in the total gas flow rate at fixed source gas proportion. (2) The influence of silane flow rate is non-linear, with a maximum of deposition rate at 200 sccm. At lower silane flow rates, the deposition rate is lower because there are fewer silicon atoms for producing aminosilanes. At higher flow rates, low deposition rate is explained by higher film density and possibly by higher silane formation and removal by the flow.

Chemical composition of the films in the second series is shown in Figures 5 and 6. Simulation results are in good agreement with the experimental data. The Si_3–Si–N fraction increases with growing SiH_4 flow rate (Figure 5), thus increasing the Si/N ratio (Figure 6). The Si_3N_4 and H_2–Si–N_2 fractions depend non-linearly on the silane flow rate, thus allowing to find process parameters that provide certain film properties.

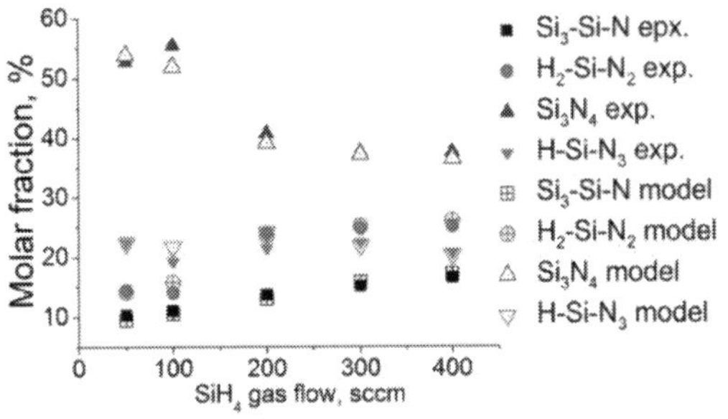

Figure 5. Influence of silane gas flow on film composition: fractions of the four groups of chemical bonds. Simulations and experiments.

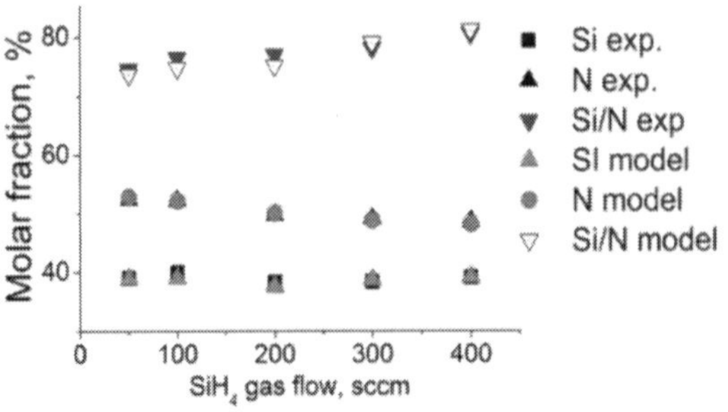

Figure 6. Influence of silane gas flow on film composition: fractions of atoms in the film. Simulations and experiments.

5. Conclusions

We developed and validated a mathematical model of plasma enhanced chemical vapor deposition (PECVD) of hydrogenated silicon nitride SiN_xH_y thin films. The model is based on Navier-Stokes equations for a transient laminar flow of a compressible chemically reacting gas mixture, together with the mass transfer and energy balance equations, Poisson equation for electric potential, electrons and ions balance equations. A new model of surface chemistry has been suggested, based on the latest experiments and spectral analysis of films deposited under different process conditions. The model correctly predicts the deposition rate and film chemical composition, thus it can be used for optimization of PECVD technology.

Acknowledgements. This work was supported by the Government of Russian Federation (contract 11.G34.31.0019 under Leading Scientist Programme and Grant 074-U01 under "5-100-2020" Programme) and by the Georgius-Agricola-Scholarship program by TU Dresden. We specially thank Dr. Lars Rebohle from the Institute of Ion-Beam Physics and Materials Research of Helmholtz-Zentrum Dresden-Rossendorf for providing experimental facilities and for his great support.

References

[1] Ghodssi R, Lin P 2011 *MEMS Materials and Processes Handbook* (New-York, Springer)
[2] Iliescu C, Tay F, Weil J *2006 J. Micromech. Microeng.* **16** 869–874
[3] Klimovich S, Gorbachev Y E, Krzhizhanovskaya V V 2010 *ASME Proceedings*. Paper DETC2010-28455 http://dx.doi.org/10.1115/DETC2010-28455
[4] Krzhizhanovskaya V V, Ivanov D A, Gorbachev Y E, Smirnov A S 2009. *ASME Proceedings*. Paper DETC2009-87039 http://dx.doi.org/10.1115/DETC2009-87039
[5] Krzhizhanovskaya V V, Sloot P M A, Gorbachev Y E 2005, *SIMULATION* **81,** no. 1, p. 77-85 http://dx.doi.org/10.1177/0037549705051972
[6] Kushner M J 1992 *J. Appl. Phys.* **71** 4173-89
[7] Dollet A, Coudere J P, Despax B 1995 *Plasma Sources Sci. Technol* **4** 94-106
[8] Bavafa M, Ilati H, Rashidian B 2008 *Semicond. Sci. Technol* **23** 095023
[9] Xu X, Zhou D, He Q, Jiang J, Fan T, Huang L, Ao T, He S 2013 *Appl. Phys. A.* **111** 867-876
[10] Scardera G, Puzzer T, Conibeer G, Green M A 2008 *J. Appl. Phys.* **104** 104310
[11] NIST Electron-Impact Cross Sections for Ionization and Excitation database http://www.nist.gov/pml/data/ionization/
[12] Xu X, Zhang L, Huang L, He Q, Fan T, Yang Z, Jiang J, Li Z *2012 J. Non-Crystalline Solids* **358** 99–106

Modeling chemical vapor deposition of silicon dioxide in microreactors at atmospheric pressure

S A Konakov[1], **V V Krzhizhanovskaya** [1,2,3]

[1] St. Petersburg State Polytechnical University, 195251, St. Petersburg, Russia
[2] University of Amsterdam ,1098 XH Amsterdam, The Netherlands
[3] ITMO University, 197101, St. Petersburg, Russia

E-mail: konakov.st@gmail.com; Valeria.Krzhizhanovskaya@gmail.com

Abstract. We developed a multiphysics mathematical model for simulation of silicon dioxide Chemical Vapor Deposition (CVD) from tetraethyl orthosilicate (TEOS) and oxygen mixture in a microreactor at atmospheric pressure. Microfluidics is a promising technology with numerous applications in chemical synthesis due to its high heat and mass transfer efficiency and well-controlled flow parameters. Experimental studies of CVD microreactor technology are slow and expensive. Analytical solution of the governing equations is impossible due to the complexity of intertwined non-linear physical and chemical processes. Computer simulation is the most effective tool for design and optimization of microreactors. Our computational fluid dynamics model employs mass, momentum and energy balance equations for a laminar transient flow of a chemically reacting gas mixture at low Reynolds number. Simulation results show the influence of microreactor configuration and process parameters on SiO_2 deposition rate and uniformity. We simulated three microreactors with the central channel diameter of 5, 10, 20 micrometers, varying gas flow rate in the range of 5-100 microliters per hour and temperature in the range of 300-800 °C. For each microchannel diameter we found an optimal set of process parameters providing the best quality of deposited material. The model will be used for optimization of the microreactor configuration and technological parameters to facilitate the experimental stage of this research.

1. Introduction and microreactor design

Microreactor technology as part of microfluidics offers many advantages compared to the macroreactors: high efficiency of heat and mass transfer, precise control of process parameters and easy migration from lab experiments to commercial production. Synthesis of nanomaterials with unique properties is the most interesting application of microreactors, for instance monodisperse nanoparticle production from liquid phase [1, 2] or nanowires and nanorods deposition in a channel of microfluidic chip [3]. A relatively new direction in microreactor technology is Chemical Vapor Deposition (CVD), a popular method for nanomaterial manufacturing. In this paper we investigate the problem of microreactor CVD technology for silicon dioxide (SiO_2) deposition. We present a mathematical model and computer simulation results.

Silicon dioxide can be deposited from different chemical mixtures: silane with oxygen, dichlorosilane with oxygen, or TetraEthyl OrthoSilicate (TEOS). These mixtures differ in technological process conditions and quality of deposited materials. TEOS offers several advantages, most notably. (1) thermal decomposition can be operated at atmospheric pressure and (2) silicon atom in TEOS is in the same chemical state as in the deposited SiO_2, therefore it is not necessary to oxidize it, which is beneficial from the chemical point of view. Oxygen is however added to avoid contamination of silicon dioxide with carbon, since oxygen reacts with the products of TEOS pyrolysis. Ozone is often used for reducing the temperature of TEOS decomposition [4, 5], but it is difficult to supply ozone in microreactors because of its high reactivity. Without ozone, production of silicon dioxide from TEOS is efficient at 700-900°C.

Our design of a microreactor is based on a macroreactor scheme [6]. The microreactor has two channels for gas supply: central channel for TEOS+Argon mixture and lateral ring channel for oxygen (see Figure 1). Argon was used as a carrier gas diluting TEOS. Varying the ratio between TEOS and O_2 flow rates, we can regulate the distribution of gas flow, deposition rate and localization. The goal of technology optimization is maximizing the deposition rate and TEOS utilization and minimizing the deposition area.

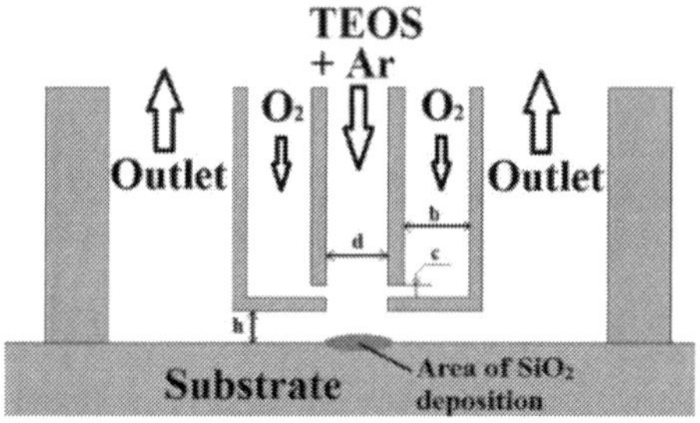

Figure 1. Microreactor scheme. h=10 μm; b=20 μm; c=5 μm; d=5, 10, 20 μm. Outlet ring width is 50 μm.

We simulated this CVD microreactor with three different diameters of the central channel: d = 5, 10 and 20 μm. For each microchannel diameter we found an optimal set of process parameters providing the best deposition localization.

2. Mathematical model and chemical reactions

Based on the calculated Knudsen numbers 10^{-3}–10^{-2}, we selected a continuum modeling approach. In this range, slip boundary conditions for velocity and thermal jump on reactor walls shall be applied [7]. In the next stage of our research we will investigate the effect of these boundary conditions. Our mathematical model is based on mass, momentum and energy balance equations for a laminar transient flow of a chemically reacting gas mixture at low Reynolds numbers [8]. A model of chemical reactions is based on [9, 10]. The rate of a chemical reaction between species A and B is defined by

$$R = k\cdot[A]\cdot[B], \tag{1}$$

where $[A]$ and $[B]$ are the species concentrations, k is the reaction rate constant. The constants depend on temperature, as described by Arrhenius' equation

$$k = A \cdot \exp\left(-\frac{E_a}{RT}\right), \tag{2}$$

where A is a pre-exponential factor, E_a is the activation energy, R is the universal gas constant, and T is the temperature. Chemical reactions in the gas phase and on the surface selected for our model are presented in Tables 1 and 2. Silicon substrate with SiO_2 and adsorbed –OH groups is modeled as SiG_3OH. This surface specie reacts with the gas phase species, leading to SiO_2 deposition.

It is important to note that to obtain a better localization of silicon dioxide it is necessary to use a diffusion-limited regime, where diffusion of reactants is the slowest stage in reaction mechanism [11]. In this case the uncertainty in reaction rate values is not critical for the accuracy of simulation.

Table 1. Gas phase reactions

N	Reaction	A, $cm^3 s^{-1}$	E_a, kcal/mol
1	SiE_4=> $Si(OH)E_3+C_2H_4$	$4.9\cdot10^{13}$	61.46
2	SiE_4+H_2O=> $Si(OH)E_3+C_2H_6O$	$1.0\cdot10^{11}$	25.00
3	$Si(OH)E_3$+ SiE_4=> $O(SiE_3)_2+C_2H_6O$	$1.0\cdot10^{11}$	30.00
4	$2Si(OH)E_3$=> $O(SiE_3)_2+H_2O$	$1.0\cdot10^{11}$	30.00
5	C_2H_5OH=>$C_2H_4+H_2O$	$4.9\cdot10^{14}$	65.50
6	$C_2H_4+3O_2$=>$2CO_2$+ $2H_2O$	$4.2\cdot10^{13}$	60.80

"E" is the ethoxy group (OC_2H_5);
"G" denotes a "glass" Si-O bond, contributing ½ of the O atom to the Si atom

Table 2. Surface reactions

N	Reaction	A, $cm^3 s^{-1}$	E_a, kcal/mol
1	$SiGE_3$=> $SiG(OH)E_2+C_2H_4$	$5.1\cdot10^{12}$	47.00
2	$SiE_4+SiG_3(OH)$=> SiO_2+SiGE_3 +C_2H_6O	$8.8\cdot10^{3}$*	43.80
3	$SiG(OH)_2E$=> $SiG_3(OH)+C_2H_6O$	$2.2\cdot10^{12}$	44.00
4	$Si(OH)E_3$+ $SiG_3(OH)$=> SiO_2+ $SiGE_3$+ H_2O	20.0*	12.00
5	$SiG(OH)E_2$=> $SiG(OH)_2E+C_2H_4$	$3.4\cdot10^{12}$	47.00
6	$SiG(OH)E_2$=> $SiG_3E+C_2H_6O$	$2.2\cdot10^{12}$	44.00
7	$SiG(OH)_2E$=> SiG_3E+H_2O	$2.2\cdot10^{12}$	44.00
8	SiG_3E=> $SiG_3(OH)$+ C_2H_4	$1.7\cdot10^{12}$	47.00

* - sticking coefficient was used to simulate SiO_2 adsorption

3. Results and discussion

We have studied the influence of source gases flow rates and microreactor central channel diameter on the deposition rate and localization. Due to the page limit, here we show only the results for 800°C temperature, atmospheric pressure, 10 µm central channel diameter, and 1 sccm of TEOS+Ar mixture (0.15 sccm of TEOS and 0.85 sccm of Argon).

To illustrate the influence of source gases flow rates, we present velocity distributions for three TEOS-to-Oxygen ratios (Fig. 2). With 1 sccm of TEOS+Ar fixed, case "a" had 0.01 sccm of oxygen, case "b" 1 sccm, and case "c" 5 sccm of O_2. Figure 3 shows $Si(OH)E_3$ distribution, the main product of TEOS decomposition, for the same three regimes (a,b,c).

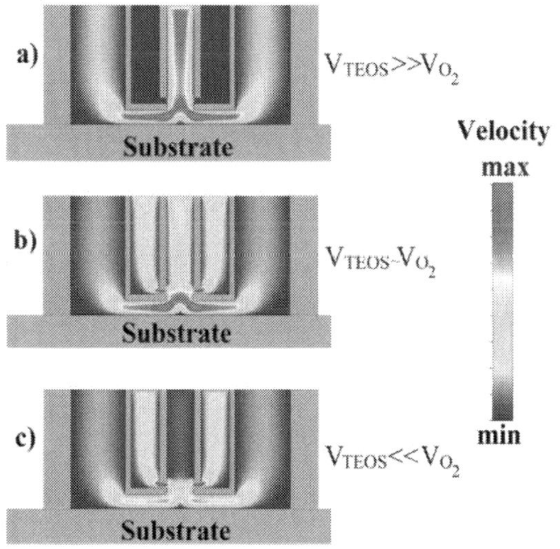

Figure 2. Gas velocity in microreactor. Three gas flow regimes: a) $V_{TEOS}>>V_{O2}$ b) $V_{TEOS}\sim V_{O2}$ c) $V_{TEOS}<<V_{O2}$.

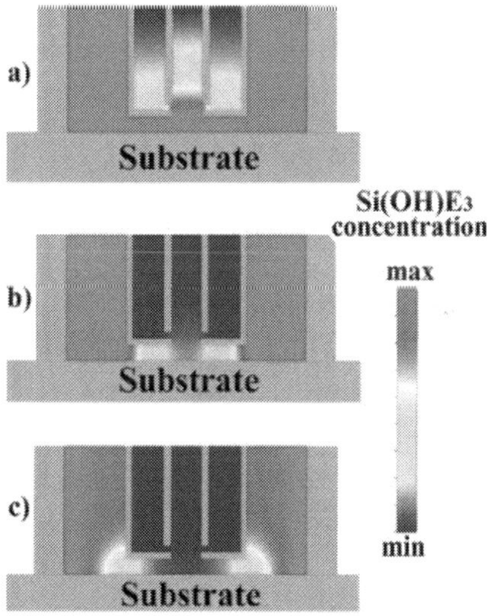

Figure 3. Distribution of $Si(OH)E_3$ concentration in microreactor

We observed that an increase of the oxygen flow (coming out of the ring near the substrate) confines the TEOS flow towards the central line of microreactor, thus the deposition area is reduced (see also Fig. 4). This is beneficial for the technology. But further increase in the oxygen flow reduces the rate of SiO_2 deposition, while we need to maximize it. Simulations will allow to find the balance between these two opposite trends, optimizing the process of micro-electro-mechanical systems (MEMS) production. Future work shall include studying also the influence of oxygen outflow direction and distance to the substrate.

For different diameters of the central TEOS channel or lateral O_2 ring channel, we can expect the same behavior, just at different values of the flow rates. We found the optimal TEOS:O_2 flow rate ratios for three diameters of the central channel, 1:55 for 5 µm channel; 1:100 for 10 µm channel; and 1:200 for 20 µm channel. It is important to note that the main factor in deposition localization is the ratio of flow velocities, rather than the total flow rates. It means that by varying the width of the oxygen channel we can control the flow streamlines, and thus the deposition area. Further study shall investigate the properties of the deposited SiO_2, minimizing its contamination with carbon.

Figure 4 shows the deposition rate distribution along the substrate for cases "a" and "b" at diameter d=10 µm. In case "c" deposition is extremely low, thus it is not plotted in this figure. The Full Width Half Maximum (FWHM) of deposition in case "b" (at higher oxygen flow rate) is 8 µm, smaller than in case "a", which is 12 µm. Case "b" represents a very good result, with the deposition width even smaller than the central channel diameter. Similar results were obtained in microreactors with central channel diameters of 5 and 20 µm.

Temperature dependence was investigated in the range from 300°C to 800°C (results are available in an extended report). For temperatures below 600°C the SiO_2 deposition stops, because TEOS thermal decomposition requires higher temperatures.

4. Conclusions and future plans

We developed a multiphysics mathematical model for simulation of silicon dioxide chemical vapor deposition from tetraethyl orthosilicate and oxygen mixture in a microreactor at atmospheric pressure. Several series of computational experiments have investigated the influence of TEOS to Oxygen flow rate ratios, process temperature and central channel diameter on deposition rate and localization. Simulation results demonstrated three regimes of gas flows in microreactors. For each microreactor we found the TEOS/O_2 ratios separating these three regimes, and an optimum combination of process parameters, minimizing deposition area and maximizing deposition rate.

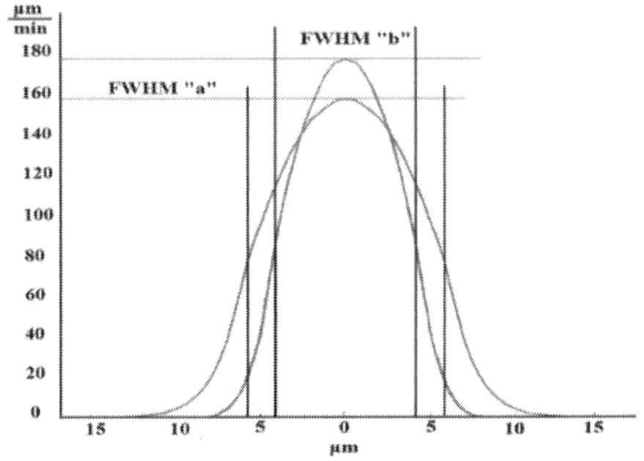

Figure 4. Distribution of deposition rate along the substrate for CVD microreactor with 10 μm central channel. Full Width Half Maximum (FWHM) for case "a" is 12 μm, for case "b" is 8 μm.

As the next step of this project, we will further adjust and validate the model. It includes studying the effect of velocity slip and thermal jump boundary conditions; investigating the influence of oxygen outflow direction and distance to the substrate; and optimizing the properties of the deposited material. For model validation, laboratory experiments will be carried out based on the preliminary results obtained in this work. The validated model will be used for designing microreactors for MEMS production technology.

Acknowledgements. This work was supported by the Government of Russian Federation: contract 11.G34.31.0019 under Leading Scientist Programme, Grant 074-U01 under "5-100-2020" Programme.

References

[1] Badilescu S, Packirisamy M 2012 *Polymers* **4** 1278-1310
[2] Wagner J, Köhler J M 2005 *Nano letters* **5** 685-691
[3] Kim J, Li Z, Park I 2011 *Lab on a Chip* **11** 1946-19
[4] Kim E J, Gill W N 1998 *Korean Journal of Chemical Engineering* **15** 56-63
[5] Zhou N, Krishnan A, Kudriavtsev V, Brichko Y 1997 *Fifth International Conference on Advanced Thermal Processing of Semiconductors*, RTP **97** 257-268
[6] Nieto J P, Jeannerot L, Caussat B 2005 *Chemical engineering science* **60**, 19, 5331-5340
[7] Nguyen N T, Wereley S T 2002 *Fundamentals and applications of microfluidics*. Artech House
[8] Krzhizhanovskaya V V, Sloot P M A, Gorbachev Y E 2005, *SIMULATION* **81,** no. 1, p. 77-85 http://dx.doi.org/10.1177/0037549705051972
[9] Coltrin M E, Ho P, Moffat H K, Buss R J 2000 *Thin Solid Films* **365** 251-263
[10] Wang, H. (1998). *A comprehensive kinetic model of ethylene and acetylene oxidation at high temperatures* (Doctoral dissertation, University of Delaware).
[11] Pierson H O 1999 *Handbook of chemical vapor deposition: principles, technology and applications*. William Andrew.

Journal of Physics: Conference Series **574** (2015) 012146 doi:10.1088/1742-6596/574/1/012146

Quantum Magnets and Matrix Lorenz Systems

J Tranchida[1,2], P Thibaudeau[1] and S Nicolis[2]

[1] CEA DAM/Le Ripault, BP 16, F-37260, Monts, France

[2] CNRS–Laboratore de Mathématiques et Physique Thdorique (UMR7350)
Fédération de Recherche "Denis Poisson" (FR2964)
Université "François Rabelais" de Tours
Parc Grandmont, 37200 Tours, France

E-mail:
julien.tranchida@cea.fr,pascal.thibaudeau@cea.fr,stam.nicolis@lmpt.univ-tours.fr

Abstract. The Landau–Lifshitz–Gilbert equations for the evolution of the magnetization, in presence of an external torque, can be cast in the form of the Lorenz equations and, thus, can describe chaotic fluctuations. To study quantum effects, we describe the magnetization by matrices, that take values in a Lie algebra. The finite dimensionality of the representation encodes the quantum fluctuations, while the non-linear nature of the equations can describe chaotic fluctuations. We identify a criterion, for the appearance of such non-linear terms. This depends on whether an invariant, symmetric tensor of the algebra can vanish or not. This proposal is studied in detail for the fundamental representation of $\mathfrak{u}(2) = \mathfrak{u}(1) \times \mathfrak{su}(2)$. We find a knotted structure for the attractor, a bimodal distribution for the largest Lyapunov exponent and that the dynamics takes place within the Cartan subalgebra, that does not contain only the identity matrix, thereby can describe the quantum fluctuations.

1. Introduction

Recent advances in magnetic materials and techniques allow manipulation of spin moments at nanoscale resolution. Thus, chaotic fluctuations become significant and their control have been the subject of both experimental and theoretical extensive studies [1]. One of the first direct observation of period doubling and chaos was spin-wave instabilities in yttrium iron garnet (YIG) and has been documented more than thirty years ago [2]. Using the ferromagnetic resonance technique, several routes to chaos have been found and explored including periodic-doubling cascades, quasi periodic and intermittent dynamics which exhibit complex magnetic behaviors [3, 4]. Recently, the phase diagram of a chaotic magnetic nanoparticle has been presented [5], which was obtained by monitoring the classical dynamics of its magnetization, modeled by the Landau-Lifshitz-Gilbert (LLG) equation.

What has received much less attention is the contribution of *quantum* fluctuations, that become significant at nanoscale resolution and are crucial for controlling qubit devices [6]. These might affect non-linear effects in new ways.

The challenge, therefore, is to describe the interplay between the two sources of fluctuations in a way that can lead to a deeper understanding of their effects and predict new features. To this end a model is proposed that displays both and allows us to distinguish them in a particularly clean fashion.

Let take a closer look to the LLG equation. The motion of the magnetization $\vec{M}$ in 3D space that spins around an effective pulsation $\vec{\omega}$ with a damping vector $\vec{D}$ and an external torque $\vec{T}$, is given by

$$\dot{\vec{M}} = \vec{\omega} \times \vec{M} - \vec{D} + \vec{T}. \tag{1}$$

This equation can be used to describe a particular magnetic system with 3 single-ion anisotropy axes $\vec{n}_i$ of intensity η_i, under a static magnetic field $\vec{B}$ and a diagonal Bloch-Blombergen damping $\vec{D} = (M_1/\tau_1, M_2/\tau_2, M_3/\tau_3)$ with the torque $\vec{T} = (0, 0, d)$ along the z-axis. In such a situation, $\vec{\omega} = \sum_{i=1}^{3} \eta_i(\vec{n}_i \cdot \vec{M})\vec{n}_i - \vec{\beta}$ with $\eta_i = 2K_i/\mu_0 M_s^2$, $\vec{\beta} = \vec{B}/\mu_0 M_s$, K_i is the anisotropy energy along the i-axis and M_s is the saturation magnetization. Equation (1) can be written in components as

$$\begin{aligned}
\dot{M}_1 &= (\eta_2 - \eta_3)M_2 M_3 - \beta_2 M_3 + \beta_3 M_2 - \frac{M_1}{\tau_1} \\
\dot{M}_2 &= (\eta_3 - \eta_1)M_3 M_1 - \beta_3 M_1 + \beta_1 M_3 - \frac{M_2}{\tau_2} \\
\dot{M}_3 &= (\eta_1 - \eta_2)M_1 M_2 - \beta_1 M_2 + \beta_2 M_1 - \frac{M_3}{\tau_3} + d
\end{aligned} \tag{2}$$

The linear transformation

$$\begin{aligned}
M_1 &= x \\
M_2 &= y \\
M_3 &= z - r - \sigma
\end{aligned} \tag{3}$$

when $\eta_1 = 2, \eta_2 = 1, \eta_3 = 1, \beta_1 = 0, \beta_2 = 0, \beta_3 = \sigma, \tau_1 = 1/\sigma, \tau_2 = 1, \tau_3 = 1/b, d = -b(r+\sigma)$, lets the Landau–Lifshitz–Gilbert equations (2) take the form of the well known prototypical system that displays the full repertoire of behaviors from regular to chaotic, along many routes, namely the Lorenz system [7]

$$\begin{aligned}
\dot{x} &= \sigma(y - x) \\
\dot{y} &= x(r - z) - y \\
\dot{z} &= xy - bz
\end{aligned} \tag{4}$$

whose typical solution (in the chaotic phase) is displayed in Fig. 1. Here $\dot{x} \equiv dx/dt$ and similarly

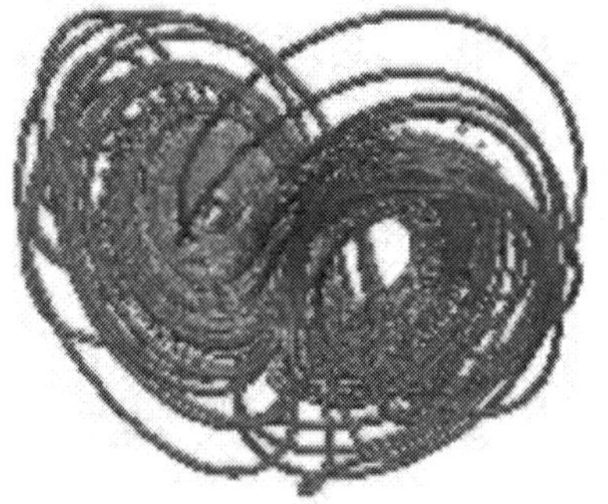

Figure 1. (Color online) Plot of the parametric solution of the time evolution of $\mathrm{Tr}(X)$, $\mathrm{Tr}(Y)$ and $\mathrm{Tr}(Z)$ with $\sigma = 10$, $r = 28$, $b = 8/3$. Classical Lorenz system in gray and its matrix version in red.

for the other variables. While these equations were deduced to describe classical fluid dynamics and the original parameters reflect this fact: σ is the Prandtl number, r the Rayleigh number and b is the aspect ratio of the "cell" (real or virtual), their scope is, in fact, much broader, as was realized from the work in the 70's [8]. These are, still, classical equations and the variables (x, y, z) are commuting quantities.

To describe the quantum fluctuations, a generalization of these equations to the case where the variables become operators, (X, Y, Z) has been considered in ref. [9], specifically, the model discussed in ref. [10], where the operators are described by square matrices, that are expanded in the generators of a given Lie algebra. As is usual in quantum mechanics, a prescription for the expressions that involve products of non-commuting variables is mandatory. The Weyl ordering [11] is adopted, so that the product XY of two operators, X and Y is replaced by its symmetric expression $XY \to \frac{1}{2}(XY + YX)$.[1] The X, Y, Z are now expanded in the generators, T^a, of a Lie algebra, $\mathcal{G}$, where $a = 1, 2, \ldots, \dim_{\mathrm{Ad}(\mathcal{G})}$, thus $X \equiv x^a T^a, Y \equiv y^a T^a$ and $Z \equiv z^a T^a$. A Lie algebra is defined by its structure constants, f^{abc}, that enter in the commutation relations, $[T^a, T^b] = \mathrm{i} f^{abc} T^c$ and the fact that this algebra is compact implies that $\mathrm{Tr}(T^a T^b) = \kappa\delta^{ab}$ with $\kappa \neq 0$ and Tr stands for the trace on the algebra.

Here we come upon the following subtlety: The Lie algebra we are considering here, *must* contain a $\mathfrak{u}(1)$ factor. The reason is that, otherwise, the equations are inconsistent, since the trace of the left hand side vanishes, but the trace of the right hand side does not: the trace of the non-linear combinations is non-zero.

Taking into account such requirements, the quantum counterparts to eqs. (4) take the following, intriguing, form

$$\begin{aligned} \dot{x}^a &= \sigma(y^a - x^a) \\ \dot{y}^a &= -y^a + r x^a - \mathsf{d}^{abc} x^b z^c \\ \dot{z}^a &= -b z^a + \mathsf{d}^{abc} x^b y^c \end{aligned} \tag{5}$$

These equations, herald the appearance of the invariant symmetric tensor

$$\mathsf{d}^{abc} \equiv \frac{1}{2\kappa}\mathrm{Tr}\left[\left\{T^a, T^b\right\} T^c\right] \tag{6}$$

of the Lie algebra. This tensor appears in gauge theories, since the gauge fields belong to the adjoint representation of the group and plays an important role in the classification of gauge anomalies [12]. Groups, for which this tensor vanishes identically are called "anomaly–safe" and in the present context, this means that eqs. (5) are *linear*, thus do not give rise to chaotic fluctuations. Only quantum fluctuations can appear. Groups, for which this tensor does not identically vanish, on the other hand, lead to non–linear equations, thus can describe both chaotic *and* quantum fluctuations.

For the case of abelian Lie groups, the structure constants vanish. In that case, eqs. (5) is equivalent to eqs. (4) up to a rescaling of all the variables proportional to the single non-vanishing element of d. This means that they share the same route to chaos, i.e. belong to the same universality class.

When the d tensor no longer vanishes, chaotic fluctuations and quantum effects mix in a non-trivial way. The simplest case is the $\mathfrak{u}(2) = (\mathfrak{u}(1)\times\mathfrak{su}(2))/\mathbb{Z}_2$ algebra, where all the components of d^{abc} are zero, except $\mathsf{d}^{0cc} = 1/2$ for $c \in \{0, 1, 2, 3\}$ with the corresponding circular permutations. This can be made particularly clear by writing the corresponding Lorenz system of equations as follows:

[1] It may be checked that another ordering prescription does not change, qualitatively, the results presented here. A more complete discussion will be presented elsewhere.

$$\begin{aligned}
&\mathfrak{u}(1)\left\{\begin{array}{l}
\dot{x}^0=\sigma(y^0-x^0)\\
\dot{y}^0=-y^0+rx^0-x^0z^0-2x^bz^b\\
\dot{z}^0=-bz^0+x^0y^0+2x^by^b
\end{array}\right.\\
&\mathfrak{su}(2)\left\{\begin{array}{l}
\dot{x}^a=\sigma(y^a-x^a)\\
\dot{y}^a=-y^a+rx^a-x^0z^a-x^az^0\\
\dot{z}^a=-bz^a+x^ay^0+x^0y^a
\end{array}\right.
\end{aligned}\tag{7}$$

The first set of the three equations highlights the fact that $\mathfrak{u}(1)$ is responsible for the chaotic fluctuations and the second set shows that the $\mathfrak{su}(2)$ components satisfy linear equations, with the $\mathfrak{u}(1)$ variables acting as sources.

The two sets of equations were integrated using an eighth-order Runge-Kutta scheme with a fixed time stepping scheme. The group invariants, $\mathrm{Tr}(X)$, $\mathrm{Tr}(Y)$ and $\mathrm{Tr}(Z)$, plotted in Fig. 1, define a subspace where a structure similar to the Lorenz "butterfly" appears, "decorated" by the $\mathfrak{su}(2)$ terms that make it "knotted". This last property is, in fact, expected [13], given that great circles on the $\mathfrak{su}(2)$ manifold, the 3–sphere, do have non-zero linking number, the well known Hopf invariant [14]. While knots are stable in the *non-chaotic* phase, since they describe periodic orbits, it will be interesting to investigate in more detail the import of the matrix structure of the equations on their properties.

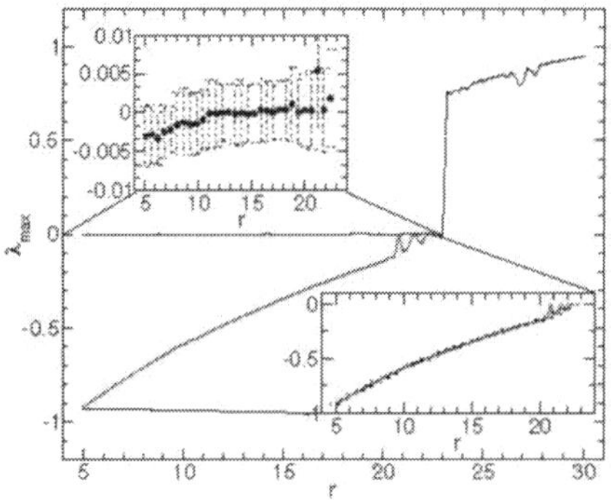

Figure 2. (Color online) Phase diagram of the largest Lyapunov spectra in the $\mathfrak{u}(2)$ matrix Lorenz system. Details of selected parts of the diagram are inserted to exhibit the 1σ error bars.

A striking feature of the largest Lyapunov exponent, as function of r, is its bimodal distribution, along the two group factors, $\mathfrak{u}(1)$ and $\mathfrak{su}(2)$, in the non-chaotic phase, $r < r_{\mathrm{crit}}$, whereas, in the chaotic phase, $r > r_{\mathrm{crit}}$, the distribution becomes unimodal, giving the same values for each group factor. While a mathematical proof for this result is not available, we stress that it provides a consistency check for the reliability of our numerical analysis, since the $\mathfrak{su}(2)$ factor cannot give rise to chaos by itself. The transition to chaos, at $r = r_{\mathrm{crit}}$, appears at the same value as for the classical Lorenz system, only much more abrupt in the matrix case.

Another way to characterize the quantum fluctuations is by studying the time evolution of the three commutators, $\mathrm{Tr}([X,Y])$, $\mathrm{Tr}([Y,Z])$ and $\mathrm{Tr}([X,Z])$. If all vanish, this means that all three matrices belong to the Cartan subalgebra. Preliminary results show that this does, in fact, occur [10] and our numerical results seem to confirm it. What is noteworthy in Fig. 3 is that the time evolution of the average of the three commutators taken over random initial conditions, collapses along the same curve. We shall now eliminate the possibility that the Cartan subalgebra

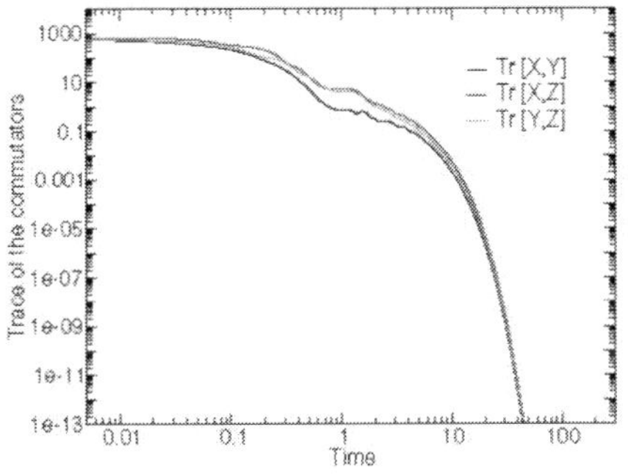

Figure 3. (Color online.) Plot of the average of the three commutators $\mathrm{Tr}([X,Y])$, $\mathrm{Tr}([Y,Z])$, $\mathrm{Tr}([X,Z])$ taken over random initial conditions for $r = 15$.

so obtained is proportional only to the identity. This is achieved by monitoring the invariants in the $\mathfrak{su}(2)$ subspace, namely $\mathrm{Tr}X^2 - (\mathrm{Tr}X)^2 = x_1{}^2 + x_2{}^2 + x_3{}^2$ and so on. Performing runs for different values of r, at fixed b and σ, we find that, in the non-chaotic phase, $r < r_{\mathrm{crit}}$, these quantities seem to converge to a single point, not the origin, whereas in the chaotic phase, $r > r_{\mathrm{crit}}$, they appear to describe a fuzzy region, of finite volume in phase space. That it does not collapse to the origin indicates the persistence of the quantum fluctuations in both phases.

A natural generalization of our results is towards a quantum Nambu description of spin systems, where it has been demonstrated that the Nambu mechanics leads to novel identities for extended Lorenz system with dissipation that are not obvious in an Hamiltonian approach [15]. Moreover this generalization may also provide insights into the origin of dissipation in such magnetic systems that has become a subject of topical research [16] and lead to new relations that are hard to guess from the Hamiltonian viewpoint.

Acknowledgments

JT acknowledges financial support from a joint Rŕegion Centre–CEA doctoral fellowship. SN would like to thank M Axenides and E G Floratos for stimulating conversations.

[1] Wigen P 1994 *Nonlinear Phenomena and Chaos in Magnetic Materials* (Singapore: World Scientific)
[2] Gibson G and Jeffries C 1984 *Phys. Rev. A* **29** 811–818
[3] Fernandez Alvarez L, Pla O and Chubykalo O 2000 *Phys. Rev. B* **61** 11613
[4] Mayergoyz I, Bertotti G and Serpico C 2009 *Nonlinear magnetization dynamics in nanosystems* (Elsevier)
[5] Bragard J, Pleiner H, Suarez O, Vargas P, Gallas J and Laroze D 2011 *Phys. Rev. E* **84** 037202
[6] Boissonneault M, Doherty A C, Ong F R, Bertet P, Vion D, Esteve D and Blais A 2014 *Phys. Rev. A* **89**(2) 022324
[7] Lorenz E N 1963 *J. Atmos. Sci.* **20** 130
[8] Feigenbaum M 1978 *J. Stat. Phys.* **19** 25
[9] Floratos E 2012 *Int. J. Bifurcation and Chaos* **22** 1250213
[10] Axenides M and Floratos E 2010 *JHEP* **1004** 36
[11] Balazs N and Jennings B 1984 *Phys. Rep.* **104** 347
[12] Georgi H and Glashow S L 1972 *Phys. Rev. D* **6** 429
[13] Ghrist R 1998 *Chaos, Solitons & Fractals* **9** 583–598
[14] Niemi A J and Sutcliffe P 2014 *Nonlinearity* **27** 2095
[15] Blender R and Lucarini V 2013 *Physica D* **243** 86–91
[16] Garanin D A 1997 *Phys. Rev. B* **55** 3050

Impact of Vehicles on Smoke Spread Dynamics in the Case of Fire in Road Tunnel

L Valasek[1], J Glasa, P Weisenpacher and L Halada

Institute of Informatics, Slovak Academy of Sciences, Dubravska cesta 9, 845 07 Bratislava, Slovakia

E-mail: lukas.valasek@savba.sk

Abstract. In this paper, the impact of vehicles in road tunnel on smoke spread in case of fire is illustrated using the FDS (Fire Dynamics Simulator) system. FDS is a CFD-based fire field model capable to simulate fire in various environments capturing a big variety of physical processes related to fire. A set of simulations of fire in a 300 m long two-lane single directional road tunnel with longitudinal ventilation is described focusing on the impact of vehicles on the smoke spread dynamics in the tunnel.

1. Introduction

Current fire models are capable to describe fire in various environments and incorporate a wide variety of physical phenomena related to fire. They utilize the knowledge of Computational Fluid Dynamics (CFD) and increasing computational power of modern computers. In this paper, smoke spread dynamics in a road tunnel in case of fire is investigated using the FDS (Fire Dynamics Simulator) system. FDS [9] solves numerically a form of Navier-Stokes equations appropriate for low-speed ($Ma < 0.3$), thermally-driven flow with an emphasis on smoke and heat transport from fire. The core algorithm is an explicit predictor-corrector scheme, second order accurate in space and time. Turbulence is treated by means of Large Eddy Simulation or Direct Numerical Simulation for larger spaces or very fine mesh resolutions, respectively. FDS covers other models for relevant physical and chemical processes related to fire such as pyrolysis, combustion, radiation, heat transport, suppression, etc. FDS allows to model people evacuation in structures and capture the impact of fire on the course of evacuation and people behaviour. Recently, the 6th version of FDS was released; parallel versions of FDS are also available for multi-mesh and multi-processor calculations. FDS has been used for simulation of fires in various structures, for instance in a cinema, supermarket, office building, industrial hall and nuclear power plant (see e.g. [4, 5, 6, 7, 8]). In this paper, we utilize our actual experience with FDS for simulation of fire in short road tunnel in order to study the impact of vehicles in a road tunnel on the smoke spread dynamics in case of fire. In our previous papers [10, 11] and [3, 1, 2], we studied the impact of parallelization of simulation calculation on efficiency and accuracy and various aspects of modelling of people evacuation, respectively.

2. Description of road tunnel and fire scenario

We consider a 10 MW fire in a single-directional 2-lane road tunnel with curved ceiling and longitudi-

[1] To whom any correspondence should be addressed

nal ventilation. The tunnel is 10 m wide, 300 m long and 7.2 m high with two couples of jet fans which are placed about 1 m under the tunnel ceiling at the distances of 32.6 m and 137.8 m from the left tunnel portal (see figure 1). The fans are placed 3 m far from each other. They have the 0.9 m effective diameter and 5.2 m length. A 2 m wide emergency exit placed 193 m far from the left tunnel portal is considered. The tunnel geometry is represented by 38 OBSTs and 5 VENTs.

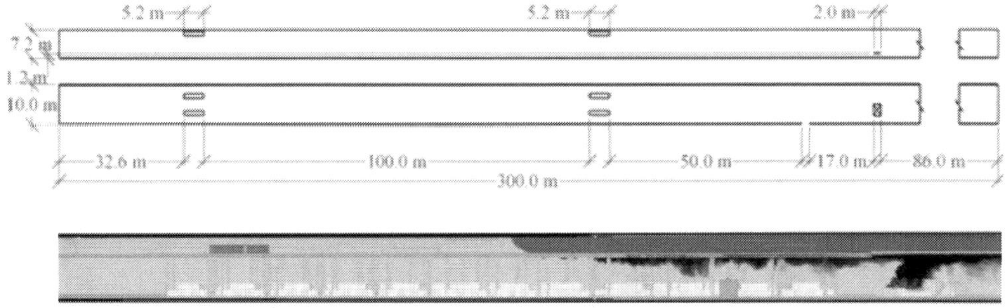

Figure 1. Scheme of road tunnel and fire source and their representation (simulation at the 27th s).

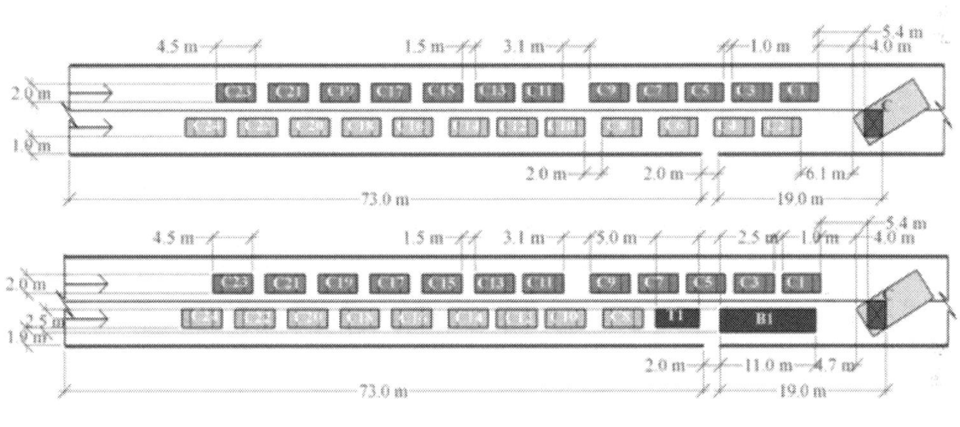

Figure 2. Traffic scheme for the first two scenarios: with 24 cars and with 21 cars, bus and transporter.

Table 1. Arrival times (AT) of particular vehicles in the first two scenarios: C1-C24 denote cars; B1 and T1 denotes bus and transporter, respectively.

Vehicles	C1	C2	C3	C4	C5	C6	C7	C8	C9	C10	C11	C12	C13	C14	C15	C16	C17	C18	C19	C20	C21	C22	C23	C24
AT (s)	3	4	5	6	7	8	9	10	11	12	13	14	15	16	17	18	19	20	21	22	23	24	25	26
Vehicles	C1	B1	C3		C5	T1	C7	C8	C9	C10	C11	C12	C13	C14	C15	C16	C17	C18	C19	C20	C21	C22	C23	C24
AT (s)	3	4	5		7	5	9	10	11	12	13	14	15	16	17	18	19	20	21	22	23	24	25	26

At the beginning, the 1 m/s steady air flow with the left direction is assumed in the tunnel, provided by setting the initial conditions for air flow in the tunnel. The fire source is represented by burning of 2 m x 3 m surface producing heat at 1666.667 kW/m^2 heat release rate per unit area since the beginning until the end of simulation (at the 240th s). The fire source is placed about 1.1 m above the road 212 m far from the left tunnel portal. We do not consider any flammable materials inside the tunnel. We assume the following ventilation action. At the beginning, all fans work with the 1 m/s velocity; at the 29th s, they start to work with linearly increasing velocity (from 1 m/s) reaching the 2.8 m/s velocity at the 30th s. Since that time, the fans velocity was not changed until the end of simulation. The 2.8 m/s flow velocity is close to the value of critical velocity. We assume 13°C ambient temperature in the tunnel at the beginning of simulation. We consider three different traffic situations: 24 cars (1st scenario); 21 cars, bus and transporter (2nd scenario); and the situation without any vehicles (3rd scenario), respectively. Vehicles are located in parallel with tunnel direction (figure 2) and arrive through the left tunnel portal at arrival times listed in table 1. At the times they are inserted in the computational domain and start to influence the air flow dynamics. Each car is represented by two OBSTs. The bus and transporter are represented by one OBST.

3. Smoke spread dynamics simulation

The described three fire simulations were realized on 6-core PC (Intel i7-3930K, 3.26 GHz, 64 GB RAM) using 5 CPU cores assigned to five 3D computational meshes of the 10 cm density (the total number of cells was 21600000; 4320000 cells per mesh). Total computational time of 4-minute simulation of the 1st, 2nd and 3rd scenario was 112.23, 113.00 and 110.06 hours, respectively.

Figure 3. Smoke dynamics at the 13th s, 31st s, 45th s and 100th s for scenario 1 (side view).

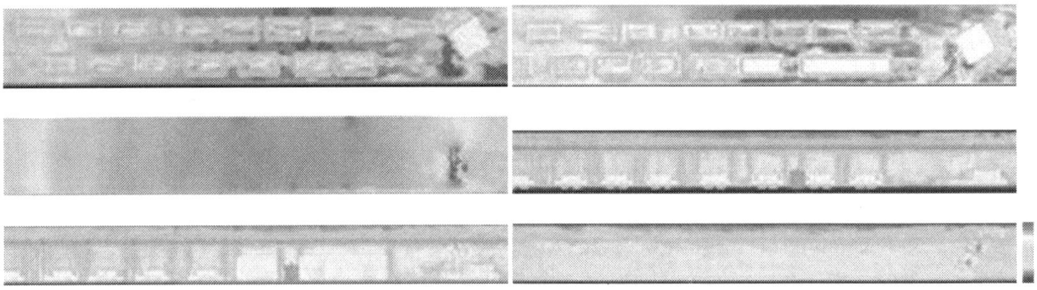

Figure 4. Horizontal and vertical slices of flow velocity at head level and at the center of tunnel tube at the 30th s for scenarios 1-3, respectively; color scheme is the same for both cases, the values vary from -3.9 to 4.6 m/s and from -0.8 to 2.1 m/s, respectively.

Table 2. Time T (in s) at which smoke reached the distances corresponding to the distances of odd vehicles in scenario 1 and fans from the fire source.

Distance to	C1	C3	C5	C7	C9	C11	C13	C15	C17	C19	C21	C23	Fan2	Fan1
T (s) in Scenario1	6.4	8.6	12.2	15.6	18.2	23.2	26.4	30.6	34.4	39.4	42.8	47.4	46.6	134.2
T (s) in Scenario2	6.8	9.8	13.4	16.4	19.6	24.0	27.0	31.8	35.8	39.8	43.6	48.0	47.0	132.4
T (s) in Scenario3	8.0	10.4	13.0	15.4	18.2	21.8	25.8	29.4	33.6	37.4	41.2	45.4	40.6	133.4

The course of fire was similar for all three scenarios (see figures 3 and 4, and table 2). At first, hot gases released from the fire source propagated upwards reaching the ceiling at the 3rd s. Then smoke spread under the ceiling in the direction to both portals being drifted slightly more towards the right portal because of the original air flow in the tunnel. The air flow supported later by ventilation blowing with the velocity close to the critical velocity caused that the smoke in upper part of the tunnel propagated to the left with decreased velocity. In lower part of the smoke layer a colder smoke, which was drifted to the right by air flow forced by ventilation, can be observed. Later, local turbulent clouds of smoke started to origin at lower part of the smoke layer. These clouds were induced by eddies caused by circumfluence of vehicles standing in the tunnel (vehicles act as obstructions against laminar air flow in the tunnel tube). The size of vehicles and their location cause small differences in smoke spread (see table 2 and figure 4); however, even these small differences can have significant impact on increase of safety risks for passengers in the tunnel and their intoxication as it was discussed in [1, 2].

3. Conclusion
In this paper, a set of simulations of fire in road tunnel is described. The simulations differ from each other by traffic situations considered. Detailed analysis of simulation results indicates that FDS is able to realistically capture the main features of road tunnel fire and study the influence of vehicles on smoke spread dynamics and increase of safety risks for passengers in tunnel in the case of fire.

4. Acknowledgment
This paper was partially supported by National Science Foundation VEGA (contract No. 2/0184/14).

5. References
[1] Glasa J and Valasek L 2014 Study of applicability of FDS+Evac for evacuation modeling in case of road tunnel fire *Research J. of Applied Sciences, Eng. and Techn.* **7**(17) 3603-15
[2] Glasa J, Valasek L, Halada L and Weisenpacher P 2014 Impact of Turned Cars in Tunnel on Modelling People Evacuation in Fire Conditions In Proceedings of 8th EUROSIM Congress on Modelling and Simulation, IEEE Computer Society (CPS) Cardiff 84-89
[3] Glasa J, Valasek L, Halada L and Weisenpacher P 2014 Modelling of impact of fire on safe people evacuation in tunnel *Journal of Physics: Conference Series* **490** 012067 doi:10.1088/1742-6596/490/1/012067
[4] Glasa J, Valasek L, Weisenpacher P and Halada L 2012 Use of PyroSim for simulation of cinema fire *International Journal on Recent Trends in Engineering and Technology* **7** (2) 51-56
[5] Glasa J, Valasek L, Weisenpacher P and Halada L 2013 Cinema fire modeling by FDS *Journal of Physics: Conference Series* **410** 01203 doi:10.1088/1742-6596/410/1/012013
[6] Halada L, Glasa J, Valasek L and Weisenpacher P 2012 The use of FDS for modeling fires in human structures *Journal on Law, Economy and Management* **2**(2) 27-35
[7] Hill K, Dreisbach J, Joglar F, Najafi B, McGrattan K, Peacock R and Hamins A 2007 Verification and Validation of Selected Fire Models for Nuclear Power Plant Applications *NUREG* **1824** U. S. Nuclear Regulatory Commission Washington USA
[8] McGrattan K, Hostikka S, McDermott R, Floyd J, Weinschenk C and Overholt K 2013 Fire Dynamics Simulator (Version 6) Technical Reference Guide (Validation) *NIST Special Publication* **1018** Gaithersburg
[9] McGrattan K, Hostikka S, McDermott R, Floyd J, Weinschenk C and Overholt K 2013 Fire Dynamics Simulator (Version 6) User's Guide *NIST Special Publication* **1019** Gaithersburg
[10] Weisenpacher P, Glasa J, Halada L, Valasek L and Sipkova V 2014 Parallel computer simulation of fire in road tunnel and people evacuation *Computing and Informatics* in press
[11] Weisenpacher P, Halada L and Glasa J 2011 Computer Simulation of Fire in a Tunnel Using Parallel Version of FDS In Proceedings of the Mediterranean Combustion Symposium, Associazione Sezione Italianadel Combustion Institute, Cagliari, 11 p.

Optimal models of extreme volume-prices are time-dependent

Paulo Rocha[1], **Frank Raischel**[2], **João Pedro Boto**[1] **and Pedro G. Lind**[3]

[1] Centro de Matemática e Aplicações Fundamentais Avenida Professor Gama Pinto, 2 1649-003 Lisboa, Portugal
[2]Instituto Dom Luiz, CGUL, University of Lisbon, 1749-016 Lisbon, Portugal
[3]ForWind and Institute of Physics, University of Oldenburg, DE-26111 Oldenburg, Germany

E-mail: paulorocha99@hotmail.com, raischel@cii.fc.ul.pt, boto@ptmat.fc.ul.pt, pedro.g.lind@forwind.de

Abstract. We present evidence that the best model for empirical volume-price distributions is not always the same and it strongly depends in (i) the region of the volume-price spectrum that one wants to model and (ii) the period in time that is being modelled. To show these two features we analyze stocks of the New York stock market with four different models: Γ, Γ-inverse, log-normal, and Weibull distributions. To evaluate the accuracy of each model we use standard relative deviations as well as the Kullback-Leibler distance and introduce an additional distance particularly suited to evaluate how accurate are the models for the distribution tails (large volume-price). Finally we put our findings in perspective and discuss how they can be extended to other situations in finance engineering.

1. Introduction

It has been acknowledged in financial distributions that the statistics of extreme events, leading to heavy-tails, as well as correlations between noise sources and other components need to be taken into account if one pretends to establish a model that describes accurately the underlying dynamical system. In this paper we are interested in the study of extreme events present in financial markets distributions. As study subject we shall consider the New York stock market (NYSM). We focus here on the evolution of volume-price, i.e. on capital exchanged, which should have more the character of a conserved quantity than the price per se. Having access to the overall distribution of volume-price provides information about the entire capital traded in the market. However, extreme events focus in what happens on the right tails of such distributions, the region comprising the highest volume-prices occurring in a given market.

In this paper we will address the problem of modelling such tails at high volume-prices in comparison with models for the full volume-price distribution.

Our data base is composed by approximately 2000 listed shares extracted from the NYSM and freely available at `http://finance.yahoo.com/`. The data has a sample frequency of 0.1 min^{-1}, starting in January 27th 2011 and ending in April 6th 2014, with a total of 976 days ($\sim 10^5$ data points).

Using this data we build the volume-price distribution of the listed companies every 10 minutes. We define the volume-price $s = pV$ as the product of the trading volume V and the

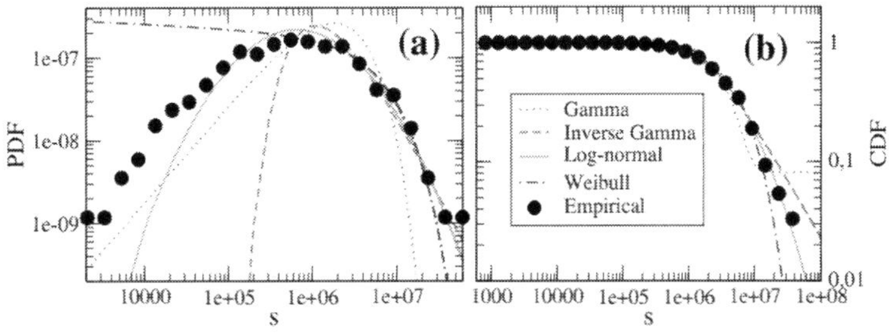

Figure 1. (a) Probability density function and (b) cumulative density function given by the four different distributions: Γ (Eq. 1), inverse Γ (Eq. 2), log-normal (Eq. 2), and Weibull (Eq. 4).

last trade price p. More details, concerning the data mining and collection can be found in Ref. [1].

We start in Sec. 2 by studying the shape of volume-price distribution and introducing four different typical models in finance to fit the empirical data[2, 1]. In Sec. 3, we present the error analysis of each models, using standard relative deviations and more sophisticated measures, namely the Kullback-Leibler distance. This measure is particularly suited to evaluate how close the (non-linear) models fit is from the empirical distributions. Section 4 concludes the paper.

2. Four models for volume-price distributions

We will consider four well-known[2] bi-parametric distributions to fit the empirical volume-price distribution, namely the Γ−distribution:

$$p_{\Gamma}(s) = \frac{s^{\phi-1}}{\theta^{\phi}\Gamma[\phi]}e^{-\frac{s}{\theta}} \tag{1}$$

the inverse Γ−distribution:

$$p_{1/\Gamma}(s) = \frac{\theta^{\phi}}{\Gamma[\phi]}s^{-\phi-1}e^{-\frac{\theta}{s}} \tag{2}$$

the Log-normal distribution:

$$p_{ln}(s) = \frac{1}{\sqrt{2\pi}\theta s}e^{-\frac{(\log s-\phi)^2}{2\theta^2}} \tag{3}$$

the Weibull distribution

$$p_w(s) = \frac{\phi}{\theta^{\phi}}s^{\phi-1}e^{-\left(\frac{s}{\theta}\right)^{\phi}} \tag{4}$$

Using the least square scheme, we fit the empirical cumulative density function (CDF) with each one of the four models above. Figures 1a and 1b show respectively the probability and corresponding cumulative density function of each model (lines) that fit the empirical distribution (bullets) at one particular 10-minute snapshot.

While the log-normal and the Γ-distribution are the best ones for the low range of volume-price values, all models, except Γ-distribution are closer to the right tail comprising the largest volume-prices observed. Therefore, one should not expect to obtain always the same optimal model, i.e. the smallest deviations or error measures are not observed always for the same model.

In Fig. 2 we show the typical evolution of one parameter during a short time-interval of approximately 3 days. Here one plots the parameter ϕ of the inverse Γ-distribution. Similar evolutions are observed with other parameters above.

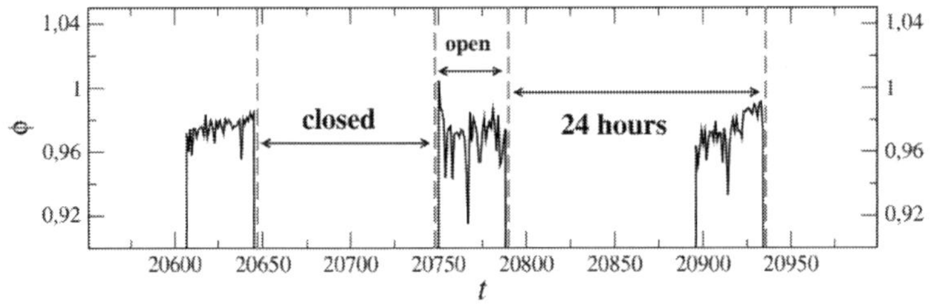

Figure 2. Illustrative time series of the parameter ϕ for inverse $\Gamma-$distribution (Eq. 2) during approximately three days. Similar plots are obtain for all parameters in the four models here considered.

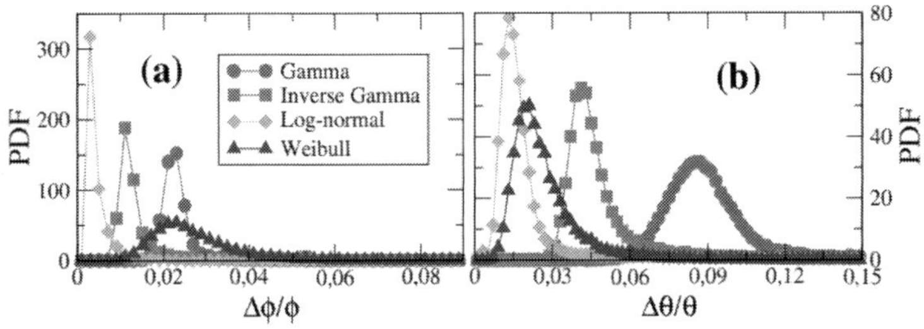

Figure 3. Probability density function of the resulting relative errors **(a)** $\Delta\phi$ and **(b)** and $\Delta\theta$, corresponding to the fitting parameters ϕ and θ respectively.

3. What is the best model for our data?

To evaluate how accurate a model is, we first consider the relative error, $\Delta\phi$ and $\Delta\theta$, of each parameter value, ϕ and θ respectively. In Figs. 3a and 3b one plots the distributions of the observed relative errors.

For both parameters the smallest relative deviation is observed for the log-normal distribution, with typical relative deviations of the order of 1%. The inverse Γ-distribution shows also acceptable deviations. The other two models are not as good as the log-normal and the inverse Γ-distribution. In a previous work, where the evolution of the mean volume-price $\langle s\rangle$ was considered separately and the models were used to fit the distribution of the normalized volume-price, $s/\langle s\rangle$, the optimal model according to relative deviations was only the inverse Γ-distribution.

The relative deviations do no take into account the importance of the volume-price spectrum in the distribution, weighted by the probability density function or another density function. To weight each value in the volume-price spectrum according to some density function we introduce here the *generalized* Kullback-Leibler (KL) distance:

$$D^{(F)}(P||Q) = \sum_i \ln\left(\left|\frac{P(i)}{Q(i)}\right|\right) F(i)\Delta x, \tag{5}$$

where $Q(i)$ is the empirical distribution, $P(i)$ is the modelled distribution and $F(i)$ is a weighting function. For $F(i) = P(i)$ one obtains the standard KL distance[3], where the logarithmic deviations are heavier weighted in the central region of the distribution than on the sides. Figure 4a shows the distribution of the values for this standard KL distance, $D^{(P)}$, obtained when considering each one of the four models. Once again one observes that the log-normal distribution is the one yielding smaller deviations.

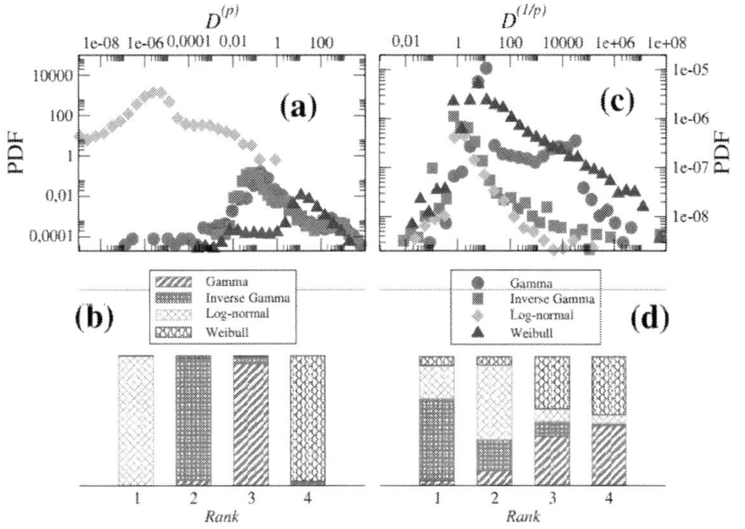

Figure 4. **(a)** PDF of the Kullback-Leibler distance $D^{(P)}$ test for the full spectrum of the volume-price. **(b)** Percentage of accuracy rankings for each model, using the Kullback-Leibler distance $D^{(P)}$. A model with rank 1 is more accurate then a model with rank 2. **(c)** PDF of the tail Kullback-Leibler distance $D^{(1/P)}$ (see text) using only the values of s larger the median of the distribution. **(d)** Percentage of accuracy rankings for each model, using the tail Kullback-Leibler distance $D^{(1/P)}$.

Since the $D^{(P)}$ distributions overlap, one may argue that the log-normal distribution might not be *always* the best model. To address this question we plot in 4b the ranking ordering all four models in their accuracy for each time step. Indeed, almost always the log-normal is the best model, followed by the inverse Γ-distribution.

For financial purposes, volume-price values are not equally important. For instance, a deviation from the observed distribution in the region of small volume-prices result in a smaller fluctuation of the amount of transactions than in the region of largest values, where the risk is the highest and therefore should be more accurately fitted.

To weight the largest volume-prices we first consider only the region of the distribution for s larger than the median and then consider in Eq. (5) $F(i) = 1/P(i)$. In this way, the largest values of the volume-price, i.e. those for which $P(i)$ is the smallest will be weighted heavier than the others. Figures 4c and 4d show, respectively, the distance distributions for this case and the corresponding rankings.

Interestingly, not only the best model is the inverse Γ-distribution but the dominance of one single model in each rank is not as strong as when considering the full distributions (Figs. 4a and 4b). This indicates that in NYSM the best model of the volume-price tail distribution is most probably the inverse Γ-distribution but the probability that another model is observed as the best one is significative.

4. Conclusions

In this paper we analyzed New York stock market volume-price distributions. By testing four different models, commonly applied in finance, to the empirical volume-price distributions, we found evidence that the best model is not always the same.

In particular, we show that it strongly depends on the region of the spectrum that one wants to model, being the log-normal the best model for low values of s and the inverse Γ for the tail distributions. Moreover, the best model seems to change depending on the period of time that the distribution is being modelled.

For further investigation, one can ask what is the reason behind this time dependency. A good study of this strange behaviour can eventually enable one to improve measures of risk and to provide additional insight in risk management.

All in all, our findings put in question common assumptions used when analyzing finance data. Namely, to assume that volume-price distributions are log-normal[4] seems to be valid only under particular restrictions and it seems not to be the optimal model when addressing extreme events. Furthermore, we provided evidence to claim that the non-stationary character of finance data is reflected not only in the fluctuations of the statistical moments defining the empirical distributions of observed values, but also in the functional shape of those distributions.

Acknowledgments
The authors thank João P. da Cruz for useful discussions and *Fundação para a Ciência e a Tecnologia*, *Deutscher Akademischer Auslandsdienst*, *Centro de Matemática e Aplicações Fundamentais* and German Environment Ministry for financial support.

References
[1] P. Rocha, F. Raischel, J. Cruz and P.G. Lind, "Stochastic Evolution of Volume-Price Distributions" submitted in 2014.
[2] S. Camargo, S.M.D. Queirós and Celia Anteneodo, Eur. Phys. J. B **86** 159 (2013).
[3] S. Kullback and R.A. Leibler, An. Math. Stat. **22** 7986 (1951).
[4] M.F.M. Osborne, Operations Research **7** 145-173 (1959).

Multiple Time Series Ising Model for Financial Market Simulations

Tetsuya Takaishi
Hiroshima University of Economics, Hiroshima 731-0192, JAPAN
E-mail: tt-taka@hue.ac.jp

Abstract. In this paper we propose an Ising model which simulates multiple financial time series. Our model introduces the interaction which couples to spins of other systems. Simulations from our model show that time series exhibit the volatility clustering that is often observed in the real financial markets. Furthermore we also find non-zero cross correlations between the volatilities from our model. Thus our model can simulate stock markets where volatilities of stocks are mutually correlated.

1. Introduction

The financial markets are considered to be complex systems where many agents are interacting at different levels and acting rationally or in some cases irrationally. Such financial markets produce a rich structure on time variation of various financial assets and the pronounced properties of asset returns has been classified as the stylized facts, e.g. see [1]. The most prominent property in the stylized facts is that asset returns show fat-tailed distributions that can not be explained by the standard random work model. A possible explanation for the fat-tailed distributions is that the return distributions are viewed as a finite-variance mixture of normal distributions, suggested by Clark[2]. In this view asset returns follow a Gaussian random process with a time-varying volatility. This view, using realized volatility[3, 4] constructed from high-frequency data, has been tested and it is found that the asset returns are consistent with this view[5, 6, 7, 8, 9, 10].

Bornholdt proposed an Ising model designed to simulate financial market as a minimalistic agent based model[11]. The Bornholdt model successfully exhibits several stylized facts such as fat-tailed return distributions and volatility clustering[11, 12, 13, 14]. Variants of the model have also been proposed and they exhibit exponential fat-tail distributions[15] or asymmetric volatility[16]. The view of finite-variance mixture of normal distributions has also been tested for the Bornholdt models and it is shown that returns simulated from the models are consistent with the view of the finite-variance mixture of normal distributions[17, 18].

So far the models studied only dealt with a financial market where a single asset is traded. In the real financial markets various financial assets are traded and they are correlated each other. Measuring correlations between assets is important to investigate stability of financial markets and many studies have been conducted to reveal properties of correlations in financial markets, e.g. [19, 20, 21, 22, 23]. In this study we propose an Ising model that extends the single Bornholdt model to a multiple time series model. Then we perform simulations of our model and show that the model can exhibit correlations between the return volatilities.

2. Multiple Time Series Ising Model

Let us consider a financial market where K stocks are traded and assume that each stock is traded by $N = L \times L$ agents on a square lattice. Each agent has a spin s_i and it takes two states $s_i = \pm 1$ corresponding to "buy" (+1) or "sell" (-1), where i stands for the ith agent. The decision of agents are made probabilistically according to a local field. In our model the local field $h_i^{(k)}(t)$ at time t is given by

$$h_i^{(k)}(t) = \sum_{\langle i,j \rangle} J s_j^{(k)}(t) - \alpha s_i^{(k)}(t)|M^{(k)}(t)| + \sum_{j=1}^{K} \gamma_{jk} M^{(j)}(t), \quad (1)$$

where $\langle i, j \rangle$ stands for a summation over the nearest neighbor pairs, k denotes the kth stock and $M^{(k)}(t)$ is the magnetization that shows an imbalance between "buy" and "sell" states, given by $M^{(k)}(t) = \sum_{l=1}^{N} s_l^{(k)}(t)$. J is the nearest neighbor coupling and in this study we set $J = 1$. The third term on the right hand side of eq.(1) describes the interaction with other stocks that is not present in the Bornholdt model. More precisely this interaction couples to the magnetization of other stocks and introduce an effect of imitating the states of other stocks. The magnitude of the interaction is given by the interaction parameters that form a matrix γ_{lm} having zero diagonal elements, i.e. $\gamma_{ll} = 0$. As in the Bornholdt model the states of spins are updated according to the following probability.

$$\begin{aligned} s_i^{(k)}(t+1) &= +1 \quad p = 1/(1+\exp(-2\beta h_i^{(k)}(t))), \\ s_i^{(k)}(t+1) &= -1 \quad 1-p. \end{aligned} \quad (2)$$

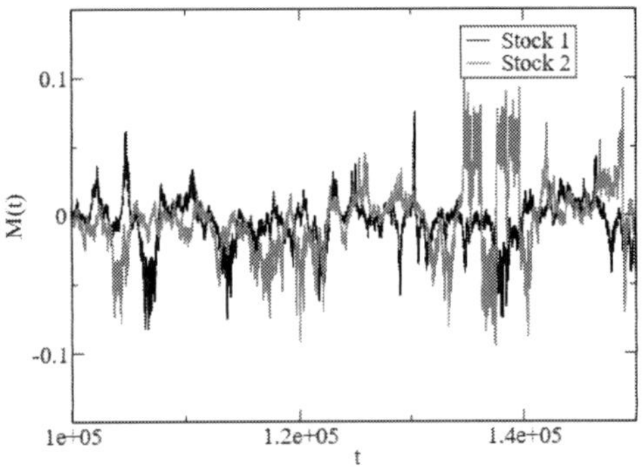

Figure 1. Time series of the magnetization for stocks 1 and 2 at $\gamma = 0.0$.

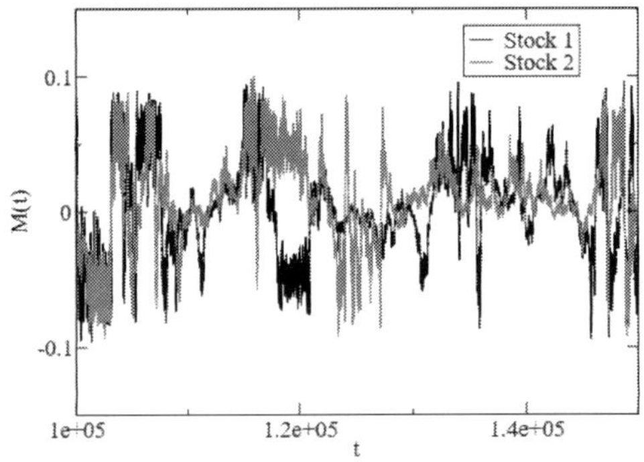

Figure 2. Time series of the magnetization for stocks 1 and 2 at $\gamma = 0.15$.

3. Simulations

In this study we perform simulations for $K = 2$, i.e. we simulate a stock market consisting of two stocks. The simulations are done on 120×120 square lattices with the periodic boundary condition. We set simulation parameters to $(\beta, \alpha, J) = (2.0, 30, 1)$. Here we assume symmetric γ, i.e. $\gamma_{12} = \gamma_{21}$, which means that the stocks we consider give the same interaction effect to other stocks each other. Here we make simulations for five values of γ, $\gamma = (0.0, 0.05, 0.07, 0.10, 0.15)$. The states of spins are updated randomly according to eq.(2). After discarding the first 10^4 updates as thermalization we collect data from 5×10^5 updates.

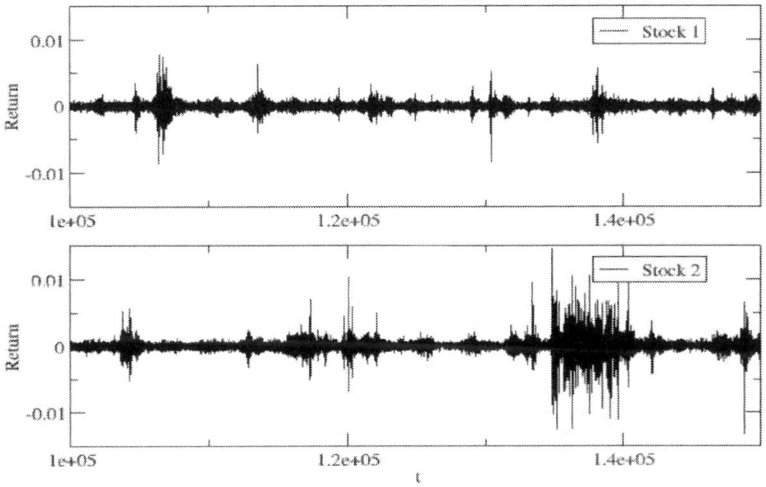

Figure 3. Time series of return for stocks 1 and 2 at $\gamma = 0.0$.

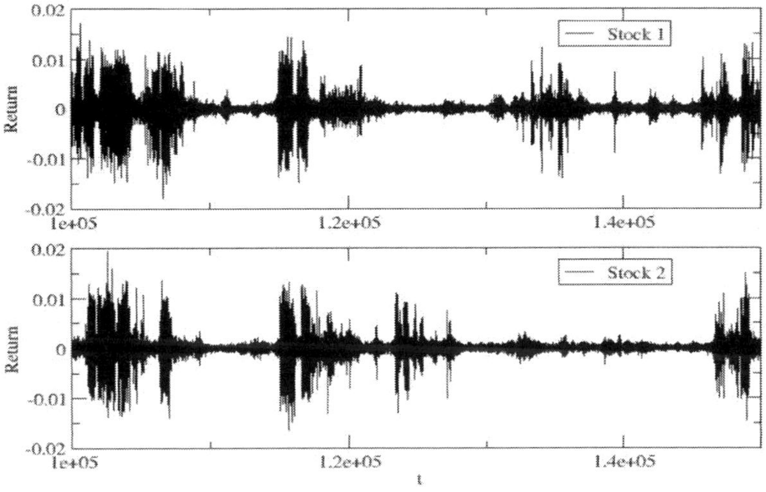

Figure 4. Time series of return for stocks 1 and 2 at $\gamma = 0.15$.

Fig.1 shows the time series of the magnetization for two stocks (1 and 2) at $\gamma = 0.0$. Simulating at $\gamma = 0.0$ means that two stocks are independent. On the other hand Fig.2 shows the time series of the magnetization at $\gamma = 0.15$ where the interaction between stocks are present and we find that synchronization occurs between the magnetizations. For other $\gamma > 0$ we also see similar synchronization between the magnetizations to a certain extent.

Such synchronization also occurs between the returns defined by $Return(t) = (M(t+1) - M(t))/2$ as in [13]. The time series of returns at $\gamma = 0.0$ and 0.15 are shown in Figs. 3 and 4 respectively. It is clearly seen that volatility clustering occurs in the return time series. At $\gamma = 0.0$, however, we see no synchronization between the volatility clusterings. On the other

Table 1. Cross correlation between the volatilities for various γ.

γ	0.0	0.05	0.07	0.10	0.15
cross correlation	8.2×10^{-3}	5.8×10^{-2}	8.4×10^{-2}	0.15	0.31

hand some volatility clusterings synchronize at $\gamma = 0.15$.

To quantify the strength of the synchronization we measure cross correlations between volatilities from the two stocks. The cross correlation between the lth and the mth volatilities is given by $\frac{\langle (v^{(l)}(t) - \langle v^{(l)}(t)\rangle)(v^{(m)}(t) - \langle v^{(m)}(t)\rangle)\rangle}{\sigma^{(l)}\sigma^{(m)}}$, where $v^{(l)}$ and $\sigma^{(l)}$ stands for the volatility and its standard deviation of the lth stock respectively. Here we define the volatility by the absolute value of return. Table 1 shows the values of cross correlations for various γ and we find that the cross correlation of volatility increases with γ.

4. Conclusion

We have proposed an Ising model which can simulate multiple financial time series and performed simulations of a stock market with two stocks. The interaction parameter γ in the model tunes the strength of correlations between stocks. We calculated cross correlation between volatilities of the two stocks and found that the cross correlation increases with γ. Therefore our model serves to simulate stock markets where volatilities of stocks are mutually correlated. Since we have demonstrated the model including only two stocks it may be desirable to further investigate the more realistic model that includes many stocks.

Acknowledgement

Numerical calculations in this work were carried out at the Yukawa Institute Computer Facility and the facilities of the Institute of Statistical Mathematics. This work was supported by JSPS KAKENHI Grant Number 25330047.

References

[1] Cont R 2001 *Quantitative Finance* **1** 223–236
[2] Clark P K 1973 *Econometrica* **41** 135-155
[3] Andersen T G and Bollerslev T 1998 *International Economic Review* **39** 885-905
[4] Andersen T G, Bollerslev T, Diebold F X and Labys P 2001 *J. Am. Statist. Assoc.*A **96** 42-55
[5] Andersen T G, Bollerslev T, Diebold F X and Labys P 2000 *Multinational Finance Journal* **4** 159–179
[6] Andersen T G, Bollerslev T, Diebold F X and Ebens H 2001 *Journal of Financial Economics* **61** 43–76
[7] Andersen T G, Bollerslev T and Dobrev, D 2007 *Journal of Econometrics* **138** 125-180
[8] Andersen T G, Bollerslev T, Frederiksen P and Nielsen M Ø 2010 *Journal of Applied Econometrics* **25** 233-261
[9] Takaishi T, Chen TT and Zheng Z 2012 *Prog. Theor. Phys. Supplement* **194** 43-54
[10] Takaishi T 2012 *Procedia - Social and Behavioral Sciences* **65** 968–973
[11] Bornholdt S 2001 *Int. J. Mod. Phys. C* **12** 667–674
[12] Yamano Y 2002 *Int. J. Mod. Phys. C* **13** 89-96
[13] Kaizoji T, Bornholdt S and Fujiwara Y 2002 *Physica A* **316** 441–452
[14] Krause S M and Bornholdt S 2011 *arXiv:1103.5345*
[15] Takaishi T 2005 *Int. J. Mod. Phys. C* **16** 1311–1317
[16] Yamamoto R 2010 *Physica A* **389** 1208-214
[17] Takaishi T 2013 *J. Phys.: Conf. Ser.* **454** 012041
[18] Takaishi T 2014 *JPS Conf. Proc.* 019007
[19] Plerou V, Gopikrishnan P, Rosenow B, Amaral L A N and Stanley H E 1999 *Phys. Rev. Lett.* **83** 1471-1474
[20] Plerou V, Gopikrishnan P, Rosenow B, Amaral L A N, Guhr T and Stanley H E 2002 *Phys. Rev. E* **65** 066126
[21] Utsugi A, Ino K and Oshikawa M 2004 *Phys. Rev. E* **70** 026110
[22] Kim D and Jeong H 2005 *Phys. Rev. E* **72** 046133
[23] Wang D, Podobnik B, Horvatić D and Stanley H E 2011 *Phys. Rev. E* **83** 046121

Modification of Gaussian mixture models for data classification in high energy physics

Michal Štěpánek, Jiří Franc, and Václav Kůs

Department of Mathematics, Faculty of Nuclear Sciences and Physical Engineering, Czech Technical University, Trojanova 13, 120 00 Prague 2, The Czech Republic

E-mail: `michal.stepanek@fjfi.cvut.cz`, `francji@fnal.gov`, `vaclav.kus@fjfi.cvut.cz`

Abstract. In high energy physics, we deal with demanding task of signal separation from background. The Model Based Clustering method involves the estimation of distribution mixture parameters via the Expectation-Maximization algorithm in the training phase and application of Bayes' rule in the testing phase. Modifications of the algorithm such as weighting, missing data processing, and overtraining avoidance will be discussed. Due to the strong dependence of the algorithm on initialization, genetic optimization techniques such as mutation, elitism, parasitism, and the rank selection of individuals will be mentioned. Data pre-processing plays a significant role for the subsequent combination of final discriminants in order to improve signal separation efficiency. Moreover, the results of the top quark separation from the Tevatron collider will be compared with those of standard multivariate techniques in high energy physics. Results from this study has been used in the measurement of the inclusive top pair production cross section employing D$\emptyset$ Tevatron full RunII data (9.7 fb^{-1}).

1. Introduction

One of the most important step in the data analysis in experimental high energy physics (HEP) is the discrimination phase and the separation of signal from the noise, called background. A lot of machine learning techniques have been adapted for the last decade and they have become an essential part of the analysis tools. The Model Based Clustering (MBC) method and the Expectation-Maximization algorithm (EM) are well known techniques in acoustic emission, image recognition and so on, but with no utilization in the HEP. The first attempt to use MBC in this area was done in the single top analysis at Tevatron D$\emptyset$ experiment [1] and the method was described for example in [2].

This paper extends the first attempts and describe the modification and the genetic optimization of the MBC method. The application from the analysis of the inclusive top pair production cross section in the $t\bar{t}$ lepton+jets channel at D$\emptyset$ experiment is presented and it results will be used to improve the analysis that has been done on previous data sample (5.3 fb^{-1}) [3].

2. Model Based Clustering method

MBC method is a statistical classification technique based on creating mixture model of training sample for subsequent separation of testing sample. Let $\boldsymbol{x} \in \mathbb{R}^{D\times N}$ represent a set of data of dimension D with N independent and identically distributed observations and Θ represent a parameter space. According to Bayes' theorem, the *a posteriori* probability that observation $\boldsymbol{x}_i$

belongs to the class of signal can be expressed as follows

$$P(\text{signal} \,|\, \boldsymbol{x}_i) = \frac{p(\boldsymbol{x}_i \,|\, \text{signal})P(\text{signal})}{p(\boldsymbol{x}_i \,|\, \text{signal})P(\text{signal}) + p(\boldsymbol{x}_i \,|\, \text{background})P(\text{background})},$$

where $P(\cdot \,|\, \cdot)$ denotes a conditional probability and $p(\cdot \,|\, \cdot)$ is a conditional probability density function. We can express the *a priori* probabilities to signal or background as the ratio of corresponding events in training MC samples. Distributions of signal and background are considered as a Gaussian distribution mixture model (GMM)

$$p(\boldsymbol{x} \,|\, \boldsymbol{\theta}) = \sum_{l=1}^{M} \alpha_l p_l(\boldsymbol{x} \,|\, \boldsymbol{\mu}_l, \boldsymbol{\Sigma}_l), \qquad \sum_{l=1}^{M} \alpha_l = 1,\ \alpha_l \geq 0,$$

where α_l denotes the weight of the l-th component, $\boldsymbol{\mu} \in \mathbb{R}^{D\times 1}$ is the vector of expected values, and $\boldsymbol{\Sigma} \in \mathbb{R}^{D\times D}$ is the covariance matrix. In order to avoid the overtraining, the quality of testing sample classification must be taken into account for the choice of the number of mixture components. Let $\boldsymbol{z}$ be the complete data containing both observed data $\boldsymbol{x}$ and their membership to mixture components. Parameters of the GMM can be estimated via the EM algorithm, whose k-th iteration ($k \in \mathbb{N}_0$) consists of two steps:

(i) *E-step*: Calculate the auxiliary function $Q(\boldsymbol{\theta}, \boldsymbol{\theta}^k) = \mathbf{E}[l_c(\boldsymbol{\theta} \,|\, \boldsymbol{z}) \,|\, \boldsymbol{x}, \boldsymbol{\theta}^k]$.

(ii) *M-step*: Find $\boldsymbol{\theta}^{k+1} \in \Theta$ maximizing $Q(\boldsymbol{\theta}, \boldsymbol{\theta}^k)$, i.e., $Q(\boldsymbol{\theta}^{k+1}, \boldsymbol{\theta}^k) \geq Q(\boldsymbol{\theta}, \boldsymbol{\theta}^k)$, $\forall \boldsymbol{\theta} \in \Theta$.

In the case of GMM for weighted data, i.e., each observation $\boldsymbol{x}_i$ is associated with a weight $\gamma(\boldsymbol{x}_i)$, new values of parameters α_l^{k+1}, $\boldsymbol{\mu}_l^{k+1}$, $\boldsymbol{\Sigma}_l^{k+1}$ are given by the solution of M-step [4].

Let us mention a few words about the shape of covariance matrix $\boldsymbol{\Sigma}$. Considering the product mixture model in the form

$$p_l(\boldsymbol{x}_i \,|\, \boldsymbol{\mu}_l, \boldsymbol{\Sigma}_l) = \prod_{d=1}^{D} p_l(x_{di} \,|\, [\boldsymbol{\mu}_l]_d, [\boldsymbol{\Sigma}_l]_{dd}),$$

i.e., employing solely diagonal covariance matrices, provides several advantages. First of all, choice of using diagonal covariance matrices improves the stability of the algorithm, in addition to simplifying the calculation of the matrix determinant and of the inverse matrix. Our studies suggest (see [4]) that the choice of the number of components generally has a stronger impact on the results than the form of the covariance matrix. This model is also directly applicable to incomplete data without estimating the missing values or their omission. Let us consider the product mixture model, $\mathcal{D}(\boldsymbol{x}_i)$ denote the subset of indices for the defined variables in a given vector $\boldsymbol{x}_i$, and $\mathcal{X}_d$ be the subset of vectors $\boldsymbol{x}_i$ with defined value of variable d,

$$\mathcal{D}(\boldsymbol{x}_i) = \{d \in \widehat{D} \,:\, x_{di} \text{ is defined}\}, \quad \mathcal{X}_d = \{\boldsymbol{x}_i,\ i \in \widehat{N} \,:\, d \in \mathcal{D}(\boldsymbol{x}_i)\}.$$

The formula for computation of the posterior probabilites of the missing data can be expressed in the form of

$$p(l \,|\, \boldsymbol{x}_i, \boldsymbol{\theta}) = \frac{\alpha_l \prod_{d\in\mathcal{D}(\boldsymbol{x}_i)} p_l(x_{di} \,|\, [\boldsymbol{\mu}_l]_d, [\boldsymbol{\Sigma}_l]_{dd})}{\sum_{l=1}^{M} \alpha_l \prod_{d\in\mathcal{D}(\boldsymbol{x}_i)} p_l(x_{di} \,|\, [\boldsymbol{\mu}_l]_d, [\boldsymbol{\Sigma}_l]_{dd})},$$

where $p_l(x_{di} \mid [\boldsymbol{\mu}_l]_d, [\boldsymbol{\Sigma}_l]_{dd})$ is the probability density function of a one-dimensional Gaussian distribution with mean $[\boldsymbol{\mu}_l]_d$ and standard deviation $[\boldsymbol{\Sigma}_l]_{dd}$. Analogously, the modification of the M-step is done by computation of the parameters of each dimension d only for vectors $\boldsymbol{x}_i \in \mathcal{X}_d$.

2.1. Genetic optimization of the Expectation-Maximization algorithm

Due to the fact that convergence of the EM algorithm is strongly dependent on the initial values, genetic optimization for finding maximum of the logarithmic likelihood function (i.e., the fitness function) is extremely useful. Firstly, the population $\mathcal{P}$ of P individuals $\mathcal{I}_n$, $n \in \widehat{P}$, is initialized by random starting values of the EM algorithm: α_l^0, $\boldsymbol{\mu}_l^0$, $\boldsymbol{\Sigma}_l^0$, $l \in \widehat{M}$. In the k-th iteration, new posterior probabilites $p(l|\boldsymbol{x}_i, \boldsymbol{\theta}^k)$, $i \in \widehat{N}$, are computed for each individual. Hence, the individual is associated with a $M \times N$ matrix containing current states, where the individual has evolved during the iteration process and by undergoing genetic-type operations. In this case, the gene of an individual is represented by index $i \in \widehat{N}$ associated with the probabilites of the membership to mixture components for one specific observation $\boldsymbol{x}_i$. Reproductive success is proportional to the fitness function employing rank selection, which assigns the values to individiuals according to the ranking of their fitness – the worst will have fitness 1, second worst fitness 2 etc. and the best will have fitness equal to the number of individuals in population $\mathcal{P}$.

Mutation is carried out using a biased mutation operation inspired by [5], which transforms the mutation operation into a guided search that can rapidly find the optimal combination of cluster assignments. It means that the probability of mutating to a cluster is related to the Mahalanobis distance from the data point to the center of the cluster, i.e., the closer a data point is to a cluster center, the higher the probability of mutating to this cluster. Genes of the individual are randomly selected with probability $p_{mut} \approx 0.1 - 0.5$.

In order to maintain the monotonic convergence property of the EM algorithm, the elitism must be employed. Elitism first copies a few best individuals to new population and the rest of evolution is accomplished by mutation. Moreover, a very common operation is parasitism, which can increase the variability of the population and thus improve performance of genetic algorithm. In every epoch, some of the worst individuals do not undergo the second selection and go directly into the new population.

2.2. Variable transformation

Variable transformation provides more suitable input for the GMM and also possibility of a combined separation via the MBC. Combination of classification methods means using their output as a new variable input, thereby creating a matrix whose columns are formed by the posterior probabilites of the membership to signal (or discriminants, in general) from different methods. Gaussianization using the Probability Integral Transformation (see [4]) can be expressed by mapping

$$x \mapsto \sqrt{2}\,\mathrm{erf}^{-1}\left(2 \cdot \int_{-\infty}^{x} f_X(t)\mathrm{d}t - 1\right),$$

where $\mathrm{erf}(\cdot)$ is the error function and $f_X(\cdot)$ is the probability density function of a random variable X. Suppose $x > 0$, the univariate Box-Cox transformation is defined by

$$x(\lambda) = \begin{cases} \dfrac{x^\lambda - 1}{\lambda} & \lambda \neq 0, \\ \ln(x) & \lambda = 0, \end{cases}$$

by which the variable is transformed to a normally distributed variable in terms of skewness and kurtosis. For both ways of gaussianization, worth noting that the transformation parameters can be estimated for signal or background set separately and we subsequently choose which one will be used for the gaussianization.

Journal of Physics: Conference Series **574** (2015) 012150 doi:10.1088/1742-6596/574/1/012150

The goal of the decorrelation is the transformation of a set of r.v. $\boldsymbol{X} = (X_1, \ldots, X_D)^T$ with a known covariance matrix $\boldsymbol{\Sigma}$ into a set of r.v. with the identity matrix as a covariance matrix. Suppose zero mean, decorrelation can be performed by

$$\boldsymbol{Y} = \mathbb{D}^{-\frac{1}{2}} \mathbb{U}^T \boldsymbol{X},$$

where $\mathbb{D}$ is a diagonal matrix with eigenvalues of $\boldsymbol{\Sigma}$ on the diagonal and $\mathbb{U}$ is composed of the corresponding eigenvectors.

3. Application of the MBC method in high energy physics

The presented MBC method with genetic modification was used in the analysis of the inclusive top pair production cross section employing DØ Tevatron full RunII data (9.7 fb^{-1}). Our task in this analysis was to separate signal from the background in the $t\bar{t}$-lepton+jets channel. This decay family has 6 subtask according to the lepton type (electron, or muon) and to the number of jets in the decay (2,3, or 4+). The DØ collaboration prepared the final selection in lepton+jets channel, which is described in [6]. The ratio of the signal differs from 1.5 % in 2 jetbins channels to nearly 45 % in 4 jetbins channels. The required Monte Carlo samples, corresponding to real blind data sets from the detector, were generated from Alpgen+Pythia generator and split into 3 groups - training, testing and validation (called yield). Each group matched the real data in accordance with the number of events (sum of weights of entries describing the simulation of one single decay).

The results of the separation for muon 4+ jets channel are presented in the figures 1 and 2. The former shows that the output discriminants from the blind data has the same distribution as the validation MC sample, the ROC curve for this channel, the optimal cut computed as the maximum of the ratio of true positive S and $\sqrt{S+B}$ (where B stands for false positive), and finally, the check of the overtraining. It is clear from the last subplot that the output discriminants of the signal and the background have the same distribution both in the training sample and in the testing sample.

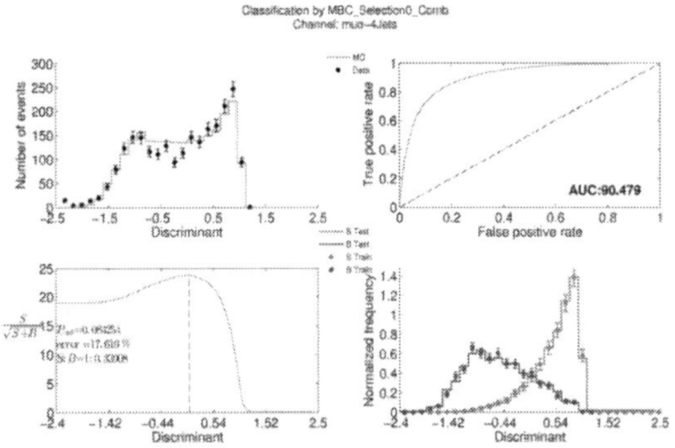

Figure 1. Control plots of results from muon 4+ Jets channel

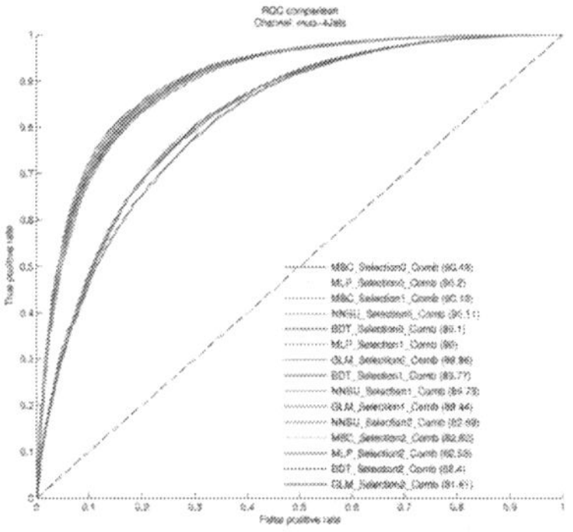

Figure 2. Results muon 4+ Jets channel: MBC, MBCcomb, ROC comparison.

The figure 2 shows the comparison of the quality of various methods ranked by the area under the ROC curve. Different variables selections were used during the analysis. It can be observed that the results of the MBC method is comparable with other well known standard multivariate techniques in high energy physics. The MBC method had even the best result in the presented channel (muon, 4+jets) displayed in the figure 2.

4. Discussion

The genetic modification of the MBC method and its application in high energy physics was presented. Although the field of high energy physics is one of the most difficult for machine learning and separation, the MBC method proved its quality and its results are at least comparable with other classification multivariate techniques. There are still some pending work and the MBC method can be improved via modification of the fitness function in terms of focus on testing sample separation efficiency, for instance.

Acknowledgments

This work has been supported by the MSMT (CZ) grant INGO II INFRA LG12020 and CTU (CZ) grant SGS12/197/OHK4/3T/14.

References

[1] DØ Collaboration 2013 Evidence for s-channel single top quark production in $p\bar{p}$ collisions at $\sqrt{s} = 1.96$ TeV *Preprint* arXiv:1308.6690
[2] Štěpánek M, Franc J, and Kůs V 2014, Model based clustering method as a new multivariate technique in high energy physics, Journal of Physics: Conference Series 490 - 012225.
[3] Abazov V M 2011 *et al*, D0 Collaboration, Measurement of the top quark pair production cross section in the lepton + jets channel in proton-antiproton collisions at $\sqrt{s} = 1.96$ TeV *Phys Rev. D.* Vol. 84, Issue 1.
[4] Štěpánek M 2014 Application of statistical methods for separation of signal in the production of top quark at the DØ experiment (CTU FNSPE)
[5] Wicker J E 2006 Applications of modern statistical methods to analysis of data in physical science (University of Tennessee)
[6] Abazov V M 2014 *et al*, D0 Collaboration, Measurement of differential $t\bar{t}$ production cross sections in $p\bar{p}$ collisions *ARXIV:1401.5785*

The effect of the number of states on the validity of credit ratings

P. Lencastre[1,2], **F. Raischel**[3], **P.G. Lind**[4]

[1] ISCTE-IUL, Av. Forças Armadas, 1649-026 Lisboa, Portugal
(e-mail: pedro.lencastre.silva@gmail.com)
[2] Mathematical Department, FCUL, University of Lisbon, 1749-016 Lisbon, Portugal
[3] Instituto Dom Luiz, University of Lisbon, 1749-016 Lisbon, Portugal
[4] ForWind and Institute of Physics, University of Oldenburg, Ammerländer Heerstrasse 136, DE-26111 Oldenburg, Germany

Abstract. We explicitly test if the reliability of credit ratings depends on the total number of admissible states. We analyse open access credit rating data and show that the effect of the number of states in the dynamical properties of ratings change with time, thus giving supportive evidence that the ideal number of admissible states changes with time. We use matrix estimation methods that explicitly assume the hypothesis needed for the process to be a valid rating process. By comparing with the likelihood maximization method of matrix estimation, we quantify the "likelihood-loss" of assuming that the process is a well grounded rating process.

1. Motivation and Scope

Credit ratings are a popular tool to evaluate credit risk and calculate the capital adequacy requirements for banks [8]. However, there is usually no explicit criteria to determine the number of states a rating scale should have. This is true for both the credit ratings published by the credit rating agencies and for the internal ratings used by banks [2]. It has been argued that the validity of the rating process as a measure of credit risk depends on it being Markov and time continuous [6]. If ratings follow criteria based on financial and economic variables, which are time continuous, then they should themselves be time continuous. If this fails, then it is possible that ratings are biased by concerns other than financial and economical ones.

In this paper, we discuss the possibility that the dynamical properties of the process change when we change the number of states. If the number of states is artificially large, there might be additional constrains to make the process stable or to ascribe the ratings. In turn, this can affect the validity the usual assumption that ratings are a time continuous process. On the other hand, if there are too few states, then it might be impossible to distinguish two issuers with the same rating based on their intrinsic credit risk. Furthermore, in this case, it is possible that the historical data of previous ratings might be a criterion to determine the different levels of credit risk within the same rating state, meaning that the process is not Markov. By using different techniques to estimate a transition matrix from a finite sample of data [5], one can evaluate the dependency of the dynamical properties of ratings with the number of states. Below, we compare transition matrices calculated under different assumptions and show that the quality of the time continuity and Markov assumptions changes considerably in time. We start, in Sec. 2, by describing the empirical data collected from Moody's. In Sec. 3 we present the theoretical

Journal of Physics: Conference Series **574** (2015) 012151 doi:10.1088/1742-6596/574/1/012151

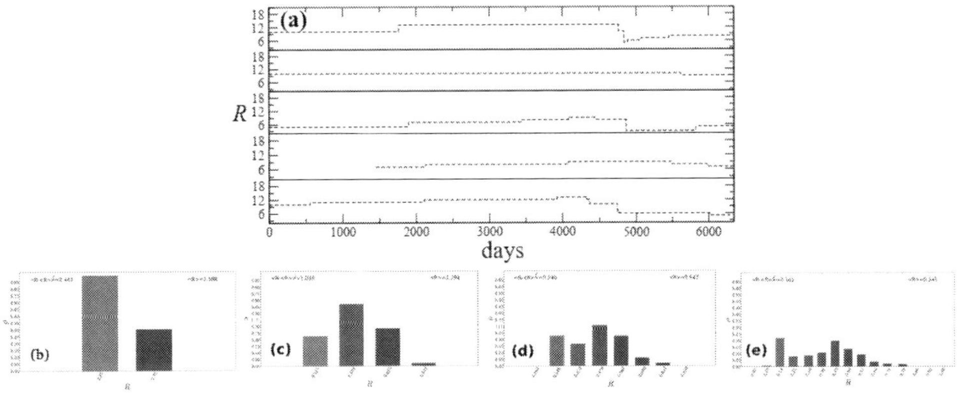

Figure 1. (a) Time series of the credit ratings of 10 individual entities from the time when they are first rated until the present. **(b-e)** Histogram of rating frequencies in the first of January of 2010 for different number of states $n_s = \{2, 4, 8, 15\}$.

background for estimating transition matrices and in Sec. 4 we describe how to test the validity of both the existence of a generator and Markov assumptions. Section 5 concludes the paper and presents some discussion of our results in the light of finance rating procedures.

2. Data description

The time series used in this paper was reconstructed by us[6] from data provided by Moodys, an influential credit rating agency, and publicly available in compliance with Rule 17g-2(d)(3) of US. SEC regulations [3]. Our data sample is the set of rating histories from the European banks, with a sample frequency of one day, starting in January 1st 2007 and ending in January 1st of 2013.

The rating class considered for our analysis is the so-called *Banking Financial Strength*[7], a measure of a bank's intrinsic credit risk including factors such as franchise value, and business and asset diversification [7] and excluding external factors, such as government support. In the Financial Strength rating class there are 15 credit rating grades, represented by letters from "A" to "E" with the two possible extra suffixes "+" and "-".

Figure 1a shows several examples of individual rating histories for each entity. It can be seen that rating changes occur infrequently, with an average of 0.43 transitions each year. We further label each state with a number ranging from $\frac{1}{2n}$ (default) to $1 - \frac{1}{2n}$ (highest rating), where n, $n > 1$ represents the number of accessible states.

In Figs.1b-e we plot the histogram of the ratings R on January 1st 2010, for $2, 4, 8$ and 15 states respectively. The histograms were constructed from the original 15-state histogram by merging pairs of adjacent states into one single state.

We define the rating increments as

$$T_i(t, \tau) = R_i(t) - R_i(t - \tau). \tag{1}$$

When $T_i(t) > 0$ (resp. < 0) it means that bank i saw its rating increased (resp. decreased) during the last τ period of time. Unless stated otherwise we will use always $\tau = 1$ year.

In Fig. 2 one sees that, though the first two statistical moments of $T(t, \tau)$ grow when we use less states, they have always a similar shape whatever number of admissible states is used.

3. Estimating rating transition matrices

There are three methods of extracting a transition matrix directly from a time series [4].

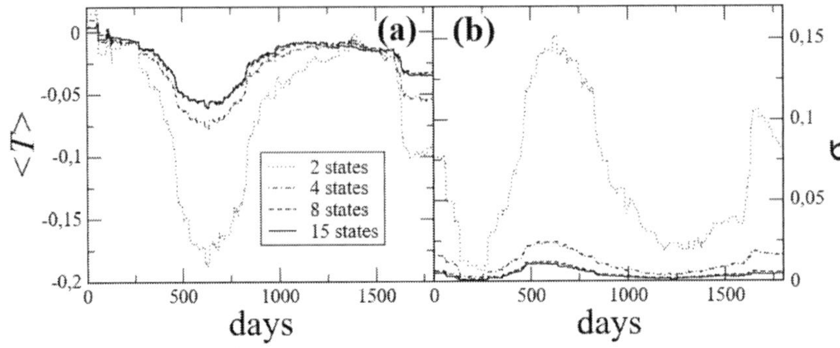

Figure 2. Comparing the sensitivity of the rating increment evolution with the number of admissible states. For 2, 4, 8 and 15 admissible states, one plots **(a)** the mean $\langle T(t,\tau)\rangle$ of the ratings increments, **(b)** the standard deviation $\sigma_T(t,\tau)$, both for $\tau = 1$ year.

One is simply the normalization of the transition number matrix $\mathbf{N}$, where each entry $N(t,\tau)_{i,j}$ gives the number of transition in the desired time interval $[t-\tau, t]$. The transition matrix is then obtained by normalizing the row-sums to one:

$$(\mathbf{T}(t,\tau))_{i,j} = \frac{(N(t,\tau))_{i,j}}{\sum_j (N(t,\tau)_{i,j}}. \tag{2}$$

This estimation method is called the Cohort Method[5] and is the default method.

Another method consists in estimating the logarithm of a generator matrix $\mathbf{Q}$ directly from empirical data[5, 1]. The off-diagonal elements are empirically estimated as

$$(Q(t,\tau))_{i,j} = \frac{(N(t,\tau))_{i,j}}{\int_{t-\tau}^{t} N_i(t\prime)dt\prime}, \tag{3}$$

where the number $N_i(t\prime)$ stands for the number of entities in state i at moment t. The diagonal elements are calculated by forcing the row-sums of $\mathbf{Q}$ to be zero. We then calculate the transition matrix as $\mathbf{T}'(t,\tau) = e^{\mathbf{Q}(t,\tau)}$. This estimation method is valid if and only if the underlying process is time-continuous, Markov and time-homogeneous[5].

Finally, the third method consists in using the Chapman-Kolmogorov Equation[9], which holds for any Markov process, given by

$$\bar{\mathbf{T}}(t,\tau) = \prod_{n=1}^{k} \mathbf{T}_{t-(k-i)\tau,\tau\prime}, \tag{4}$$

where $\tau\prime = \tau/k$. This corresponds to the multiplication of matrices with a smaller non-overlapping time-window of size $\tau\prime$.

4. Comparing different estimations

In this work we want to compare all three estimations, $\mathbf{T}$, $\mathbf{T}'$ and $\bar{\mathbf{T}}$ and quantify the difference between them. To compare them, we will use a so-called likelihood difference between the transition matrix of the default method $\mathbf{T}$ and the other two estimations:

$$d(\mathbf{T},\mathbf{T}_{other}) = \frac{\sum_{i,j}(N(t,\tau))_{i,j}\left(\frac{\log(T(t,\tau)_{i,j})}{\log((\mathbf{T}(t,\tau)_{other})_{i,j})}\right)}{\sum_{i,j}(N(t,\tau))_{i,j}} \tag{5}$$

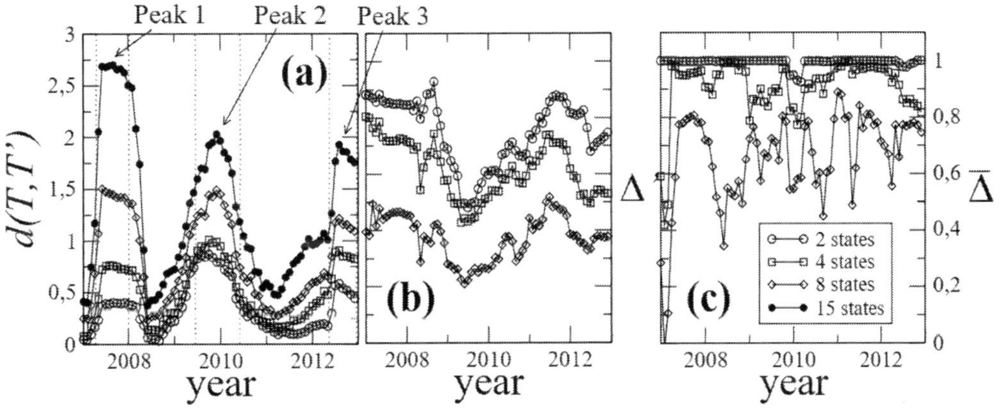

Figure 3. **(a)** Comparison between **T** and **T**′, calculated over a time-window of one year, using the log-likelihood distance of the Eq. 5. **(b)** Comparison between the likelihood distance with 15 states and a smaller number of states. **(c)** The same as (b) for the transition matrix $\overline{\mathbf{T}}$ computed from the Chapman-Kolmogorov equation.

where $\mathbf{T}_{other}$ represents either the matrix $\mathbf{T}'$ of the generator matrix method or the matrix $\bar{\mathbf{T}}$ in the Chapman-Kolmogorov method. This norm is particularly useful when one of the matrices maximizes of the likelihood function, which is the case for the cohort estimated matrix [5]. This way, the distance $d(\mathbf{T}_{other}, \mathbf{T})$ becomes an adequate measure of the "likelihood loss" by choosing a different estimation method.

We compare $\mathbf{T}$ with $\mathbf{T}\prime$, and $\overline{\mathbf{T}}$ with a fixed time-window of one year, and plot $d(\mathbf{T}, \mathbf{T}\prime)(t)$ and $d(\mathbf{T}, \overline{\mathbf{T}})(t)$ in Fig. 3. If the process is time-homogeneous, Markov and time-continuous, as one expects from a rating process [6], then the difference between $\mathbf{T}$, $\mathbf{T}\prime$ and $\overline{\mathbf{T}}$ should be white noise resulting from dealing with a finite sample.

However, when too few states are considered, e.g. only two states (good and bad), it might be possible to distinguish entities in the same state due to their history, making ratings not Markov. In general we expect the system to lose the Markov property when the "width" of each rating grade, a percentile of the maximum financial health, is bigger than the uncertainty associated to the uncertainty characteristic of the average entity's financial health. This might be counter-balanced by the fact that when we reduce the number of states we also lose information about the system's past. Nonetheless, when we lower the number of states, the "pure" white noise of dealing with a finite sample of data becomes lower, but the difference between matrices might become larger due to the fact that the process might not be Markov.

To see if the process is time-continuous, Markov, and time-homogeneous, we measure how correct it is to assume that a generator exists through the difference $d(\mathbf{T}, \mathbf{T}\prime)$. Results are shown in Fig. 3a. In Fig. 3b one plots the normalized difference between the distances for the standard 15 states and the lower cases:

$$\Delta(\mathbf{T}, \mathbf{T}_{other}) = \frac{d(\mathbf{T}, \mathbf{T}^{(15)}_{other}) - d(\mathbf{T}, \mathbf{T}^{(n)}_{other})}{d(\mathbf{T}, \mathbf{T}^{(15)}_{other})}, \qquad (6)$$

with $\mathbf{T}_{other} = \mathbf{T}\prime$ and with $n = 2, 4$ and 8. As a notation simplification, we use $\Delta\prime = \Delta(\mathbf{T}, \mathbf{T}\prime)$ and $\bar{\Delta} = \Delta(\mathbf{T}, \bar{\mathbf{T}})$.

When testing the validity of the assumption of the existence of a generator, the shape of $d(\mathbf{T}, \mathbf{T}\prime)$ preserves most features when we change the number of states, although the scale of the different graphics change. In fact, as can be seen from Fig. 3b, all three peaks in 2007, 2009 and 2012 remain visible and occur at approximately the same time and have the same duration. Their relative strength, however, is inverted. With 15 rating classes, the peak in 2007 is stronger than the other two, equally strong peaks in 2009 and 2012. The relative strength of the three peaks is somehow equal using the reduced sets of 8 and 4 states, whereas for two states the 2009 peak becomes strongest.

To test when the process is Markov we check when does the Chapman-Kolmogorov Equation (4) hold, and in Fig. 3c one plots a quantity $\bar{\Delta}$ using Eq. (6) but this time for using the matrices $\mathbf{T}$ and $\bar{\mathbf{T}}$. The peaks of $d(\mathbf{T}, \overline{\mathbf{T}})$ occur now at the same instants as the peaks of $d(\mathbf{T}, \mathbf{T}\prime$, namely at the beginning of 2007, 2009 and 2012.

Reducing the number of states preserves the shape of the graphic, and the years at which a greater deviation occurs, but greatly diminishes the overall scale and difference between $\mathbf{T}$ and $\overline{\mathbf{T}}$. In the limit of two states, almost all information is lost.

5. Conclusions

We have used publicly available Moody's time series of credit ratings and studied simple ways to compute the validity of the usual assumptions, regarding credit rating time series, as a function of the number of rating states.

In general, we have presented evidence that the accuracy of the time-continuous Markov time homogeneous assumptions vary not only in time but also with the number of states. Consequently, the choice of how many admissible states one should use to categorize a sample of rated entities should not be fixed in large time-spans.

Moreover, we also found that during periods when the process shows to be non-Markovian or during which no proper generator exists the choice of the number of admissible states is of importance, since the non-Markovianity and the unreliability of a generator is less pronounced when one chooses a number of admissible states smaller than the standard 15 states.

Using empirical data for which 15 states were prescribed it is not possible to extend this work to consider a number of admissible states larger than 15. One may conjecture that, the choice of 15 is better than a smaller number for detecting deviations from Markov time-continuous processes, but whether or not this is an optimal choice is an open question for future investigations.

References

[1] Theodore Charitos, Peter R de Waal, and Linda C van der Gaag. Computing short-interval transition matrices of a discrete-time markov chain from partially observed data. *Statistics in medicine*, 27(6):905–921, 2008.

[2] Jens Grunert, Lars Norden, and Martin Weber. The role of non-financial factors in internal credit ratings. *Journal of Banking & Finance*, 29(2):509–531, 2005.

[3] https://www.moodys.com/Pages/reg001004.aspx.

[4] Yusuf Jafry and Til Schuermann. Measurement, estimation and comparison of credit migration matrices. *Journal of Banking & Finance*, 28(11):2603–2639, 2004.

[5] David Lando and Torben M Skødeberg. Analyzing rating transitions and rating drift with continuous observations. *journal of banking & finance*, 26(2):423–444, 2002.

[6] Pedro Lencastre, Frank Raischel, Pedro G Lind, and Tim Rogers. Are credit ratings time-homogeneous and markov? *arXiv preprint arXiv:1403.8018*, 2014.

[7] S Moody. moodys rating symbols & definitions. *Moodys Investors Service, Report*, 79004(08):1–52, 2004.

[8] Samu Peura and Esa Jokivuolle. Simulation based stress tests of banks' regulatory capital adequacy. *Journal of Banking & Finance*, 28(8):1801–1824, 2004.

[9] Hannes Risken. *Fokker-Planck Equation*. Springer, 1984.

Use of principal component analysis for identification of temporal and spatial patterns in the dynamics of ionospheric equatorial anomaly

Yu.S. Maslennikova, V.V. Bochkarev, D.S. Voloskov

Kazan Federal University, Kremlevskaya str.18, Kazan 420018, Russia

E-mail: JSMaslennikova@kpfu.ru, vbochkarev@mail.ru

Abstract. In this paper we describe results of the principal components analysis of the dynamics of Total Electronic Content (TEC) data with the use of global maps presented by the Jet Propulsion Laboratory (NASA, USA) for the period of 2007–2011. We show that the result of the decomposition in principal components essentially depends on the method used for pre-processing the data, their representation (the used coordinate system), and the data centering technique (e.g., daily and seasonal components extracting). The use of momentarily co-moving frame of reference and other special techniques provide opportunity for the detailed analysis of the ionospheric equatorial anomaly. The covariance matrix of decomposition was calculated using Spearman's rank correlation coefficient that allows reducing statistical relationship between principal components.

1. Introduction

Principal component analysis (PCA), also called *empirical orthogonal functions* method (EOF) has been used for wide variety of geophysical data analysis tasks for many years. It should be noted, that for correct analysis of investigated data dynamics one should take into account data representation features and presented trends. For example in paper [1] the authors modelled TEC dynamics using EOF analysis, representing data in local time domain. As a result, the first mode with the highest energy contained equatorial anomaly information. The second mode reflected seasonal variations of ionosphere. In other words, the first high-energy modes, obtained using such approach, contain information only about well-known ionosphere dynamics features. An approach, proposed in paper [2] involves transition to local time domain and coordinate system, taking into account the angle of magnetic axis declination. As a result first mode contains daily TEC variations and the second mode contains seasonal variations. All these examples show, that while using PCA for geophysical data analysis, one should pay particular attention to factoring in the way, data is presented, prevailing periodicities and statistical properties of the data. This paper is dedicated to the development of the methods allowing to take these factors into account. Also for decomposition we use covariance matrix based on Spearman's rank correlation coefficient. It allows reducing statistical relationship between components of decomposition.

2. Total Electronic Content

Total electron content (or TEC) is an important descriptive quantity for the ionosphere of the Earth. TEC is the total number of electrons integrated between two points, along a tube of one squared meter cross section, i.e., the electron columnar number density. For ionosphere dynamics analysis global

TEC maps, provided by JPL (Jet Propulsion Laboratory, NASA, USA), were used for period from 2004 to 2011 years. TEC maps were presented in geographical coordinate system with spatial resolution of 2.5° for latitude and 5° for longitude and time resolution of 2 hours. TEC daily dynamics for 13.11.2006 is presented on the Fig.1. It shows that the maximum of TEC field is located in the subsolar point and migrates, following after motion of the Sun.

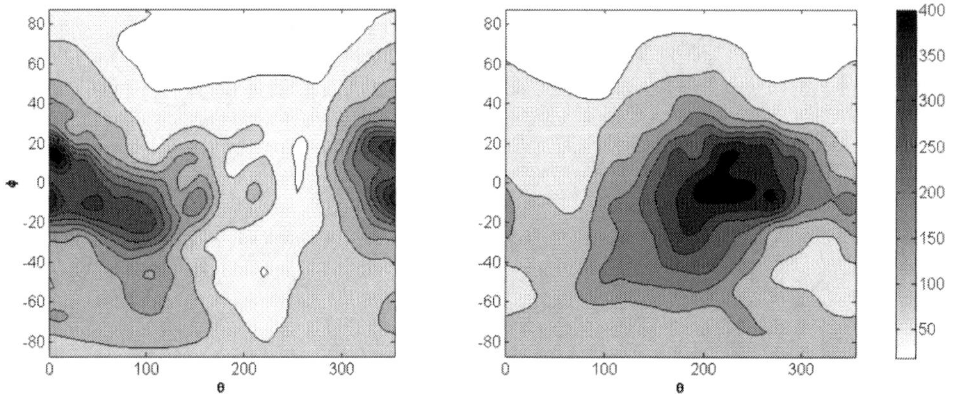

Figure 1. Global TEC maps at 00:00 and 12:00 of 13th November, 2006.

3. Principal component analysis

We can represent a matrix of TEC observations as a matrix $X_{i,j}$, where the index i indicates a point on the Earth's surface and index j corresponds to a time moment. Principal component analysis allows representing the matrix of TEC data in the following form: $X = \sum_k \sigma_k U_k V_k^H$. The orthonormalized vector U_k defines the spatial distribution of TEC variations that corresponds to the kth principal component. The orthonormalized vector V_k defines temporal evolution of the corresponding component. For centralized data, value σ_k^2 corresponds to the variance of the kth principal component.

For the matrix $X_{i,j}$ of TEC data we can estimate the covariance matrix $\mathbf{K_x}$ that defines a character of the time-spatial dependence of TEC global maps. The variance of the 1st component can be defined as follows: $\sigma_1 = \varphi_1^* \mathbf{K_x} \varphi_1$. The search of the best transform φ_1 is reduced to the maximization of the function σ_1 subject to the normalizing condition: $\left(\varphi_1^*, \varphi_1\right) = \sum_{j=1}^{p} \varphi_{1,j}^2 = 1$. This optimization problem can be solved with the help of the Lagrange function. Remaining components can be found by analogy. The calculation procedure is reduced to the maximization of the variance subject to conditions of normalizing conditions of components. The process should be performed until we find all eigenvalues and the corresponding eigenvectors of the covariance matrix $\mathbf{K_x}$, that represent variances and coefficients of linear combinations of principal components. More detailed information about principal component analysis is presented in [3].

It was shown in the paper [4] that the result of the principal component analysis essentially depends on the data pre-processing technique, used coordinate system and the data centering technique. Following the recommendations from [4], in current work we use momentarily co-moving frame of reference to reduce diurnal periodicity in TEC data and take into account transfer matrix of the spherical coordinate system.

4. Reducing statistical relationship between principal components

In principal component analysis U_k components can be found as eigenvectors of spatial covariance matrix. Using of covariance matrix is equivalent to estimating statistical connection between values in

Journal of Physics: Conference Series **574** (2015) 012152 doi:10.1088/1742-6596/574/1/012152

different spatial points using Pearson's correlation coefficient. However, Pearson's coefficient can be considered as correct estimation of statistical connection only when distribution of fluctuations values is close enough to normal distribution. If fluctuation distribution is different from normal one, components, obtained from principal component analysis could be not statistically independent

In order to assess distribution of $V_k(t)$ amplitude fluctuations, we found discrete wavelet-transform coefficients of these vector (Daubechies-5 wavelet was used). Distribution functions of wavelet coefficients were fitted using t-distribution function. v parameters of approximating t-distribution were following: 2.12 ± 0.13 for the 1st mode, and- $v=12.62\pm3.21$ for the 15th mode. Large value of t-distribution's v parameter for high modes shows that their distribution functions are close enough to normal ones whereas for the first modes deviation from the normal distribution is significant.

In such situation we can use Spearman's rank correlation coefficient to estimate statistical relationship. This coefficient is independent from the absolute values of the data and the distribution shape, taking into account the position of the element in ranked set. We used covariance matrix analogue, built the following way: $B_{ij} = \sigma_i \sigma_j \rho^S_{i,j}$, where σ_i –standard deviation of TEC fluctuations in i^{th} point and $\rho^S_{i,j}$ –Spearman's correlation coefficient between values in i^{th} and j^{th} points. Having built the B matrix we can find U_k components like it's eigenvectors.

The most restrictive quantitative characteristic of statistical interconnection is the mutual information measure I [5]. Having large enough datasets we found estimations of 2-dimensional distribution densities of $V_k(t)$ components amplitudes, after that we calculated the mutual information measure I. For more convenient interpretation one can apply the following functional transform to I: $R = \sqrt{1-2^{-I}}$.

Thus we obtain the value between 0 and 1 with properties which are similar to ones of correlation coefficient. If values have normal distribution, equivalent correlation coefficient R is equal to absolute value of Pearson's coefficient.

Fig. 2 represents mutual distribution functions of 1st and 2nd modes as well as 14th and 15th ones, obtained using the kernel density estimation method. From the Fig.2 one can see that 1st and 2nd components are statistically dependent, what means that decomposition is unsuccessful. Equivalent correlation coefficient between 1st and 2nd modes is $R_{12}=0.714$, between 1st and 3th modes is $R_{13}=0.539$, between 2nd and 3th ones – $R_{23}=0.416$, between 14th and 15th $R_{14,15}=0.121$.

For the same dataset decomposition $X = \sum_k \sigma_k U_k V_k^H$ was performed where covariance matrix was calculated using Spearman's correlation coefficient. In Fig.3 the comparison between results of classic PCA and proposed approach for the first 3 components is presented. Estimations of statistical connection between first modes of decomposition were also performed. The correlation between components decreased by more than 40% ($R'_{12}=0.39$, $R'_{13}=0.29$). Proposed above approach provides opportunity for analyzing of temporal patterns of TEC data. Fig. 4 shows dynamics of the first 2 components. After median filtering we can see 28-30 days periodic fluctuations.

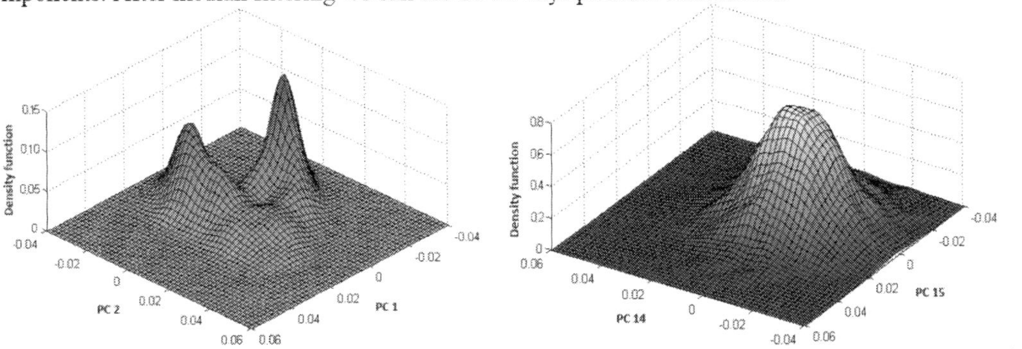

Figure 2. Mutual distribution functions for 1st and 2nd components (left side); for 14th and 15th components (right side).

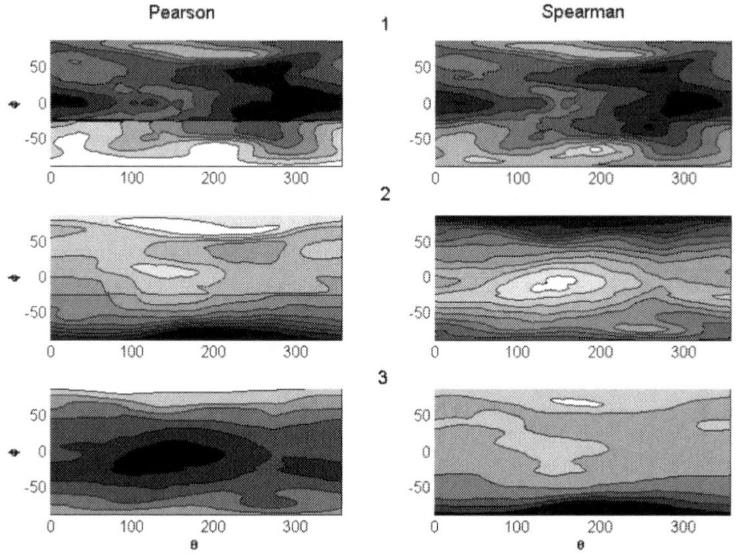

Figure 3. Spatial patterns of TEC data using classic PCA method (left side) and proposed approach using Spearman's correlation coefficient for the first 3 components (right side).

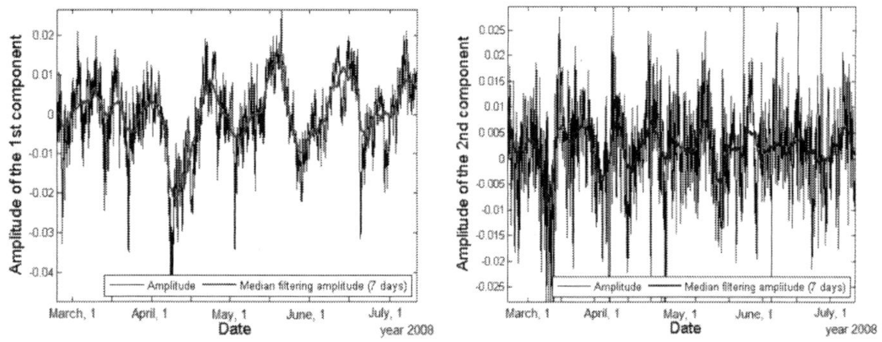

Figure 4. Temporal patterns of TEC data for the first 2 components.

The work is performed according to the Russian Government Program of Competitive Growth of Kazan Federal University.

References

[1] Wan, W., Ding, F., Ren, Z., Zhang, M., Liu, L., and Ning, B., Modeling the global ionospheric total electron content with empirical orthogonal function analysis, Science China Technological Sciences, 2012, vol. 55, pp. 1161– 1168.

[2] Zhang D, Ridley A J, Xiao Z, Hao Y A global model: Empirical orthogonal function analysis of total electron content 1999–2009 data 2012 *J. Geophys. Res.* p. 117

[3] Jolliffe I T Principal component analysis 2002 *Series: Springer Series in Statistics, 2nd ed., NY*

[4] Maslennikova Yu S, Bochkarev V V Principal Component Analysis of Global Maps of the Total Electronic Content 2014 *Geomagnetism and Aeronomy* **54(2)** 216–223

[5] Borda Monica Fundamentals in Information Theory and Coding 2011 *Springer*

Influence of Mobility Models in Precision Spray Aided by Wireless Sensor Networks

L B L Gonçalves[1], F G Costa[2], L A Neves[1], J Ueyama[2], G F D Zafalon[1], C Montez[3], A S R Pinto[3]

[1]Department of Computer Science and Statistic, São Paulo State University, São José do Rio Preto, SP, 15054-000, Brazil

[2]Institute of Mathematics and Computer Science, University of São Paulo, São Carlos, SP, 13566-590, Brazil

[3]Department of Control Engineering and Automation, Federal University of Santa Catarina, Florianópolis, SC, 88040-900, Brazil

E-mail: leandroblg@ieee.org

Abstract. Precision Spray is a technique to increase performance of Precision Agriculture. This spray technique may be aided by a Wireless Sensor Network, however, for such approach, the communication between the agricultural input applicator vehicle and network is critical due to its proper functioning. Thus, this work analyzes how the number of nodes in a wireless sensor network, its type of distribution and different areas of scenario affects the performance of communication. We performed simulations to observe system's behavior changing to find the most fitted non-controlled mobility model to the system.

1. Introduction

Precision Agriculture (PA) aim to increase the efficiency of resource management and productivity of crops, resulting in better and less expensive products [1]. A technique used for PA is Precision Spray (PS) of agricultural inputs, because the distribution of such materials affects the quality of grown products. The use of Unmanned Aerial Vehicles (UAVs) aided by Wireless Sensor Network (WSN) is an approach for PS in which the WSN report, ongoing, to UAV details of distribution, allowing to improve the efficiency of the process [2]. However, proper functioning of this PS technique depends the communication efficiency between UAV and WSN.

The aim of this work is to observe the behavior of network and find the mobility model, among Random Walk, Random Waypoint, Random Direction and Manhattan Grid, have better performance on this system.

This paper is organized as follow: In Section 2 is commented some theoretical base about concepts of PA, WSN and Mobile Sink (MS), in Section 3 are detailed the parameters of simulations, in section 4 we present the results and discuss about. Conclusions are shown in section 5, followed by references in section 6.

2. Concepts in precision agriculture aided by WSN and mobile sink

PA is one of most prosperous fields for use of WSN [3]. Since PA is based in a big volume of data from the crop, WSN may aim PA collecting data such as micrometeorological, soil conditions [4], presence of nutrients, vegetal health conditions [1] and agricultural inputs distribution [2].

A MS, also called Data Mule, is a concept initially developed for collecting data from nodes or groups of nodes unconnected of others. A MS travel the sensored area collecting data from nodes at its communication range. It consists of a node with improved energetic and data storage capabilities, attached to a mobility agent [5].

PA may be better executed by use of PS, because PS distributes more properly agricultural inputs at the crop. An approach for PS consists in to have a UAV as applicator vehicle and it be aided by a WSN, in which this WSN reports to UAV how the distribution has made, allowing ongoing corrections [2].

3. Materials and Methods

Movement of a MS affects how many and with which this MS is connected on a moment, may interfering directly on network efficiency. To know which mobility model has better performance on the system we made experiments by simulations. We used the network simulator OMNet++ [6] with the framework MiXiM. For generate the mobility models outputs we used the software BonnMotion [7].

Each simulation had duration of 600 seconds. We varied two parameters individually: The number of nodes and the scenario area, called experiments 1 and 2, respectively. The number of nodes was varied by 28, 45, 66, 91 and 120, for a area of 1.8x10^5 m^2. These values were chosen by complete a grid with equal distance between its intersections with same proportion of scenario, 1:2. The area was varied by 4.24x10^5, 2.64x10^5, 1.8x10^5, 1.3x10^5 and, 9.89x10^4 m^2, keeping the density values from experiment 1; and used 66 nodes, median of values of varied number of nodes on experiment 1. Each experiment was performed with random and grid node distributions, as shown on Figure1.

The main metric used is the Communication Efficiency (*Ef*). Ef is calculated by *Ef=Ru/Ts*, where *Ru* is the number of messages received by UAV and *Ts* is the sum of messages sent by nodes. The mobility models were parameterized so that its behaviors were as similar as possible, as commented in [8].

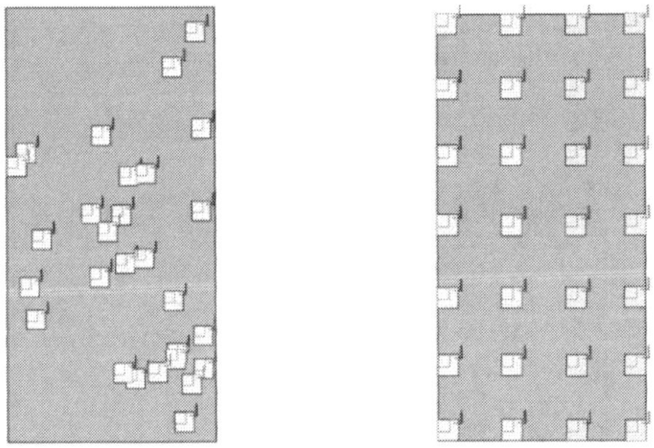

(a) **(b)**

Figure 1. Examples of node distributions with 28 nodes for experiments made: (a) Represents a possible random node distribution. (b) Represents a grid node distribution.

The follow assumptions were taken in this work: UAV had constant speed at 20 m/s, has not the capacity of static planning over a region; all nodes, including the UAV, are in the same plan; nodes work the same manner during all the simulation time; all energetic sources and buffers were considered unlimited and; do not existed elements that could affect the signal at specific parts of scenario (e.g. trees, buildings).

4. Results and discussion

Simulation results are presented on Figures 2, 3, 4 and 5. The results were calculated by me mean of 32 executions of each simulation with different random number seeds. Figures 2 and 4 correspond to random node distribution simulations, and Figures 3 and 5 correspond to grid node distribution simulations. Figures 2 and 3 correspond to number of nodes variation simulations, Figures 4 and 5 corresponds to area variation simulations.

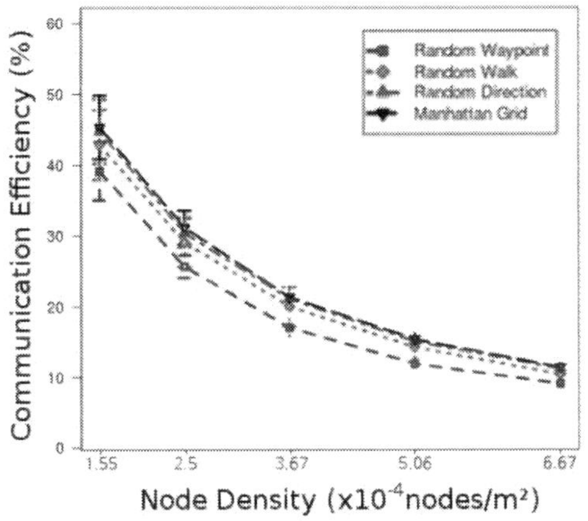

Figure 2. Communication efficiency for experiment 1 with random node distribution.

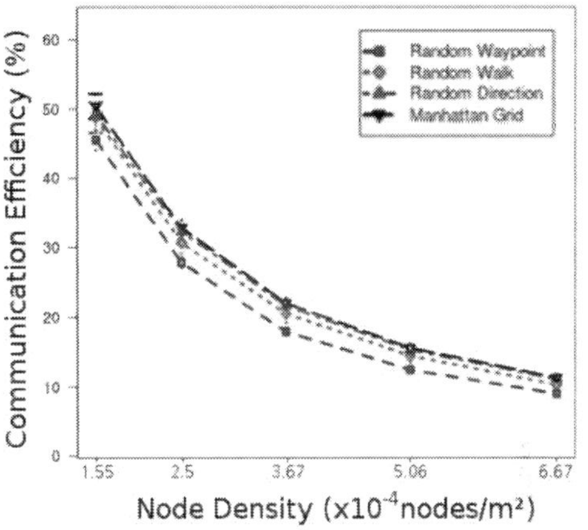

Figure 3. Communication efficiency for experiment 1 with grid node distribution.

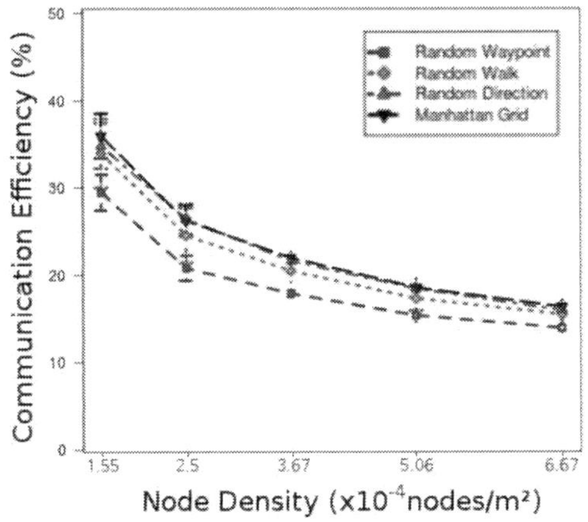

Figure 4. Communication efficiency for experiment 2 with grid node distribution.

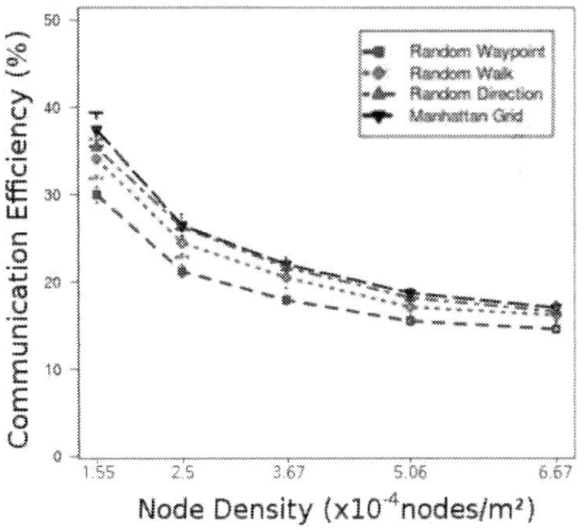

Figure 5. Communication efficiency for experiment 2 with grid node distribution.

Ef curves for random and grid node distribution had respectively average 22.75% and 23.74%. Thus, the average of Ef for random distributions (heterogeneous) present 95.8% of relative efficiency compared with average of grid distributions (homogeneous).

Network behavior for experiment 1 (Figures 2 and 3) had average Ef equals 24.14% and for experiment 2 (Figures 4 and 5) had average Ef equals 22.36%. It, initially, shows a better performance for node variation over area variation, however, the rates of efficiency decrease with increase of density in experiment 1 are from 15.74% until 3.85%, with average 8.82%, while for experiment 2 these rates are from 9.34% until 1.63%, with average 4.53%. It implies the average Ef decreases 1.94 times faster for experiment than experiment 2.

Since the curves for each mobility model presented similar behavior in all cases we may observe their average Efs. Average Efs for Random Waypoint, Random Walk, Random Direction and Manhattan Grid are, respectively, 20.56%, 23.24%, 24.35% and 24.83%. Thus, Manhattan Grid mobility model was the one with better average Ef, and then better fitted to the problem. Mobility models Random Waypoint, Random Walk and Random Direction had relative efficiencies compared to Manhattan Grid respectively 82.80%, 93.60% and 98.03%.

5. Conclusion

The node distribution slightly affected the network efficiency, with homogeneous network subtly more efficient than heterogeneous network. Node variation simulations, despite had been better average network performance than area variation simulations, had bigger efficiency degradation rate than area variation simulations. Manhattan Grid mobility model had better performance on the system. As this model executes grid mobility, Random Direction mobility model may be used for similar systems that require multi directional mobility, once it had almost the same efficiency.

6. References

[1] Lee W S, Alchanatis V, Yang C, Hirafuji M, Moshou D, Li C 2010 Sensing technologies for precision specialty crop production, *Computers and Electronics in Agriculture*, **74** 2-33

[2] Costa F G, Ueyama J, Braun T, Pessin G, Osorio F S, Vargas P A 2012 The use of unmanned aerial vehicles and wireless sensor network in agricultural applications, *Geoscience and Remote Sensing Symposium, 2012 IEEE International,* 5045-5048

[3] Díaz S E, Pérez J C, Mateos A C, Marinescu M C, Guerra B B 2011 A novel methodology for the monitoring of the agricultural production process based on wireless sensor networks, *Computers and Electronics in Agriculture*, **76** 252-265

[4] Matese A, Di Gennaro S F, Zaldei A, Genesio L, Vaccari F P 2009 A wireless sensor network for precision viticulture: The NAV system, *Computers and Electronics in Agriculture*, **69** 51-58

[5] Shah R C, Roy S, Jain S, Brunette W 2003 Data MULEs: modeling a three-tier architecture for sparse sensor networks, *Sensor Network Protocols and Applications, 2003. Proceedings of the First IEEE. 2003 IEEE International Workshop on*, 30-41

[6] Varga A 1999 Using the OMNeT++ discrete event simulation system in education, *IEEE Transactions on Education*, **42** 11

[7] Aschenbruck N, Ernst R, Gerhards-Padilla E, Schwamborn M 2010 BonnMotion: a mobility scenario generation and analysis tool, *Proceedings of the 3rd International ICST Conference on Simulation Tools and Techniques*, 1-10

[8] Camp T, Boleng J, Davies V 2002 A survey of mobility models for ad hoc network research, *Wireless Communications and Mobile Computing*, **2** 483-502

Universal quasi-equilibrium state of self-gravitating systems not depending on dimensions

Tohru Tashiro

Department of Physics, Ochanomizu University, 2-1-1 Ohtsuka, Bunkyo, Tokyo 112-8610, Japan

E-mail: tashiro@cosmos.phys.ocha.ac.jp

Abstract. We study a universality of self-gravitating systems in the quasi-equilibrium state. It is shown numerically that the two dimensional self-gravitating system in the quasi-equilibrium state has the same kind of density profile as the three dimensional one. We develop a model to describe this universal quasi-equilibrium state by using a special Langevin equation with a distinctive random noise to self-gravitating systems. We find that the density profile derived theoretically is consistent well with results of observations and simulations.

1. Introduction

There are many astronomical objects in which a number of elements are organized mainly by gravity. They are called *self-gravitating systems* (SGSs). It is known that, before the thermal equilibrium state, there is another stable state for SGS which we call *quasi-equilibrium state* (QES) [1]. Recent observations suggest that a density profile of SGS in QES has a universality not depending on dimensions. Globular clusters consisting of about hundreds of thousands of stars, a good instance of three dimensional SGS, are well-known to have a spherically symmetric number density. We shall point to a filamentary structure of molecular clouds as an example of two dimensional SGS. The *Herschel* Space Observatory have revealed the 27 filamentary structures in IC 5146 which is a reflection nebula in Cygnus and the number density of molecular clouds have a cylindrical symmetry around the axis of filament [2]. Therefore, the self-gravity generated by molecular clouds can be regarded as the two dimensional SGS: By using the mass density of molecular clouds $\rho_{\rm mc}(r)$ and the gravitational potential per unit mass $\phi_{\rm mc}(r)$ where r means a distance from the axis, the Poisson equation becomes

$$\frac{1}{r}\frac{\mathrm{d}}{\mathrm{d}r}\left\{r\frac{\mathrm{d}\phi_{\rm mc}(r)}{\mathrm{d}r}\right\} = 4\pi G\rho_{\rm mc}(r)\ , \tag{1}$$

which is formally equivalent to the Poisson equation of two dimensional SGS as discussed later. These SGSs of different dimensions have the following universal density profile in QES: The number density is constant around the center of SGS and decrease in a power law at large radius. Thus, the number density $\mathcal{N}(r)$ can be approximated well by

$$\mathcal{N}(r) \simeq \frac{\mathcal{N}(0)}{(1+r^2/a^2)^\kappa} \tag{2}$$

where r is the distance from the center of SGS and a is called a core radius. With regard to molecular clouds and globular clusters, $0.75 < \kappa < 1.25$ [2] and κ is around $3/2$ [1], respectively.

Obviously, the universal density cannot be derived by assuming the system is isothermal [3, 4]. Therefore, a new theory explaining the physical mechanism behind the universality is necessary. The author has developed a model to unveil the mechanism in three dimensional SGS [3, 4]. So, the aim of this paper is to describe QES universally by extending the dimension to μ which is 2 or 3.

2. Two dimensional gravity and N-body simulations of two dimensional SGS

2.1. Two dimensional gravity

The two dimensional gravitational potential per unit mass ϕ generated by mass source ρ ditributing circular-symmetrically satisfies the following Poisson equation in the two dimensional space.

$$\frac{1}{r}\frac{\mathrm{d}}{\mathrm{d}r}\left\{r\frac{\mathrm{d}\phi(r)}{\mathrm{d}r}\right\} = 4\pi G'\rho(r) \ . \tag{3}$$

This Poisson equation is formally equivalent to one governing molecular clouds which are distributed axisymmetrically.

When the mass source is a mass point with a mass m existing in the origin, $\rho(r) = m\delta(0)/\pi r$. Then, the Poisson equation is altered to

$$\frac{1}{r}\frac{\mathrm{d}}{\mathrm{d}r}\left\{r\frac{\mathrm{d}\phi(r)}{\mathrm{d}r}\right\} = \frac{4mG'\delta(0)}{r} \ . \tag{4}$$

Therefore, the gravity per unit mass from the mass point in the two dimensional space is unveiled to be

$$-\phi'(r) = -\frac{2G'm}{r} \ . \tag{5}$$

2.2. N-body simulations

In order to investigate QES of two dimensional SGS, we followed the time evolution of 10^4-body system in the two dimensional space interacting by Eq. (5) numerically.

A polytrope solution with a polytrope index n was adopted as the initial condition. Generally, the solution in the three dimensional space is well-known [1]. Here, we have extended it to the two dimensional space: The distribution function $f(r, v)$ in phase space can be shown as

$$f(r,v) \propto \begin{cases} \mathcal{E}^{n-1} & (\mathcal{E} < 0) \\ 0 & (\mathcal{E} \geq 0) \end{cases} \quad \left(\mathcal{E} \equiv \phi(R) - \left\{\frac{1}{2}v^2 + \phi(r)\right\}\right) \tag{6}$$

where R means a radius of the system. Then, we run the N-body simulation by varying an initial virial ratio V which is one of the kinetic energy to the gravity potential of the system [5].

Specifically, when $V = 0$, it is found that the number density in QES has a universality and the density around the center of the system can be fitted well by Eq. (2). Figure 1 denotes that the best-fit κ which is the index in Eq. (2) ranges from 0.8 to 1.1. The results have a good consistency with the observations of molecular clouds [2].

3. Fokker-Planck model

In the previous section, the number density of the two dimensional SGS in QES can be represented well by Eq. (2) with $0.8 < \kappa < 1.1$. On the other hand, the number density of most globular clusters, a good example of the three dimensional SGS, is well-known to be fitted by Eq. (2) with $\kappa \simeq 3/2$ (see Refs. in [3]). Furthermore, we have reported that the same density

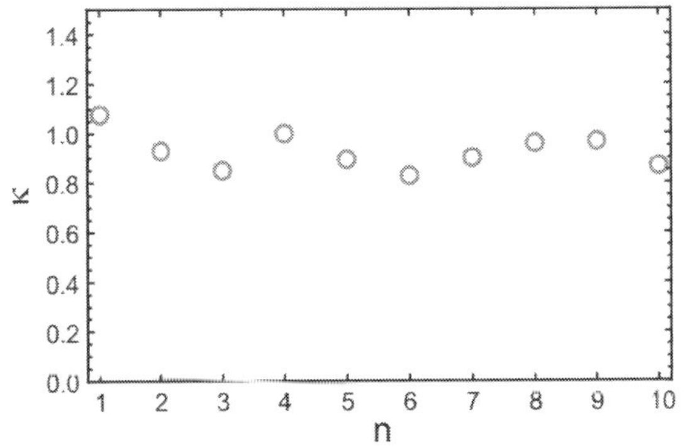

Figure 1. (color online) The optimum values of κ in Eq. (2) for fitting the number densities vs the initial polytope index n.

profile is obtained through N-body simulations of the three dimensional SGS [3, 4]. Therefore, we can conclude that the SGS has the universal density profile not depending on dimensions.

Here, let us describe this density profile universally by a special Fokker-Planck equation in the μ dimensional space. Before constructing the Fokker-Planck equation, we shall model forces influencing an element of the system. That is, we shall begin by constructing a Langevin equation.

We assume that the frictional force $-m\gamma\dot{\boldsymbol{r}}^{\mu}(t)$ and the random noise with constant intensity $\sqrt{2D}\boldsymbol{\xi}^{\mu}(t)$ which are essential for a many-body system to reach the thermal equilibrium state act on the element, where m is a mass of the element, $\boldsymbol{r}^{\mu}$ is a position vector, and $\boldsymbol{\xi}^{\mu}(t)$ means Gaussian-white noises. The index represents that the vector is in the μ dimensional space. The element is also influenced by a mean gravitational force $-F^{\mu}(r)$ along the radial direction of the system which is derived by differentiating $m\phi^{\mu}(r)$: $-F^{\mu}(r) = -m\partial_r\phi^{\mu}(r)$ where $\phi^{\mu}(r)$ is the mean gravitational potential in μ dimensional space. However, this is just a mean gravity. It is natural to consider that the element actually is influenced by a fluctuating gravity around the mean value. This means that another noise which prevents the system from reaching the thermal equilibrium state simply is added to the ordinary Langevin equation, and so this system goes to another stable state, that is, QES. Therefore, we can consider the noise distinctive to SGS.

If assuming the intensity of the noise to be constant, we can obtain the following Langevin equation in the over-damped limit:

$$m\gamma\dot{\boldsymbol{r}}^{\mu}(t) = -F^{\mu}(r)\left\{1+\sqrt{2\epsilon}\eta(t)\right\}\boldsymbol{e}_r^{\mu} - \frac{\partial}{\partial r}\frac{\epsilon}{2m\gamma}F^{\mu}(r)^2\boldsymbol{e}_r^{\mu} + \sqrt{2D}\boldsymbol{\xi}^{\mu}(t) \,, \tag{7}$$

where $\boldsymbol{e}_r^{\mu}$ is a unit vector along the radial direction in μ dimensional space. The second term on the right hand side of the above equation is a correction term in order to regard products as the Storatonovich product [6].

The corresponding Fokker-Planck equation is given by

$$\begin{aligned}\frac{\partial}{\partial t}P^{\mu}(r,t) = {} & \frac{D}{(m\gamma)^2}\left\{\frac{\partial^2}{\partial r^2}+\frac{\mu-1}{r}\frac{\partial}{\partial r}\right\}P^{\mu}(r,t) + \frac{1}{m\gamma}\frac{1}{r}\frac{\partial}{\partial r}rF^{\mu}(r)P^{\mu}(r,t) \\ & + \frac{\epsilon}{(m\gamma)^2}\left\{\frac{\partial^2}{\partial r^2}+\frac{\mu-1}{r}\frac{\partial}{\partial r}+\frac{\mu-1}{r^{\mu-1}}\frac{\partial}{\partial r}r^{\mu-2}\right\}F^{\mu}(r)^2P^{\mu}(r,t) \,. \end{aligned} \tag{8}$$

We shall treat the system as a circular ($\mu = 2$) or a spherical symmetric one ($\mu = 3$) including N elements. So, PDF P^{μ} is a function of the distance from the origin $r \equiv |\boldsymbol{r}^{\mu}|$ where the index μ are abbreviated for simplify. Note that a relation among PDF P^{μ}, the number density $\mathcal{N}^{\mu}$ and the mass density ρ^{μ} is as follows: $P^{\mu} = \mathcal{N}^{\mu}/N = \rho^{\mu}/(mN)$.

Let us use $\mathcal{P}^\mu(r,t) = J^\mu(r)P^\mu(r,t)$ instead of P^μ, where J^μ means Jacobian determinant[1].

$$\frac{\partial}{\partial t}\mathcal{P}^\mu(r,t) = \frac{D}{(m\gamma)^2}\left\{\frac{\partial^2}{\partial r^2} - \frac{\partial}{\partial r}\frac{\mu-1}{r}\right\}\mathcal{P}^\mu(r,t) + \frac{1}{m\gamma}\frac{\partial}{\partial r}F^\mu(r)\mathcal{P}^\mu(r,t) + \frac{\epsilon}{(m\gamma)^2}\frac{\partial^2}{\partial r^2}F^\mu(r)^2\mathcal{P}^\mu(r,t) \quad (9)$$

When the system reaches QES, $\partial_t \mathcal{P}^\mu_{\rm qe}(r) = 0$. Here, we integrate the Fokker-Planck equation by r. Owing to use $\mathcal{P}^\mu_{\rm qe}$, the integration becomes easier:

$$\left\{\frac{D}{(m\gamma)^2} + \frac{\epsilon F^\mu(r)^2}{(m\gamma)^2}\right\}\mathcal{P}^{\mu\,\prime}_{\rm qe}(r) - \left[\frac{D}{(m\gamma)^2}\frac{\mu-1}{r} - \frac{2\epsilon F^\mu(r)F^{\mu\prime}(r)}{(m\gamma)^2} - \frac{F^\mu(r)}{m\gamma}\right]\mathcal{P}^\mu_{\rm qe}(r) = \text{const.} \quad (10)$$

Since $P_{\rm qe}(0)$ and $P'_{\rm qe}(0)$ are bounded, $\mathcal{P}^\mu_{\rm qe}(0) = \mathcal{P}^{\mu\,\prime}_{\rm qe}(0) = 0$. Therefore, the constant of the right hand side of Eq. (10) becomes 0:

$$\mathcal{P}^{\mu\,\prime}_{\rm qe}(r) = -\frac{rF^\mu(r)\{m\gamma + 2\epsilon F^{\mu\prime}(r)\} - (\mu-1)D}{r\{D + \epsilon F^\mu(r)^2\}}\mathcal{P}^\mu_{\rm qe}(r)\ . \quad (11)$$

If using the number density in QES $\mathcal{N}^\mu_{\rm qe}(= N\mathcal{P}^\mu_{\rm qe}/J^\mu)$, we can obtain

$$\mathcal{N}^{\mu\,\prime}_{\rm qe}(r) = -\frac{rF^\mu(r)\{m\gamma + 2\epsilon F^{\mu\prime}(r)\} + (\mu-1)\epsilon F^\mu(r)^2}{r\{D + \epsilon F^\mu(r)^2\}}\mathcal{N}^\mu_{\rm qe}(r)\ . \quad (12)$$

By substituting the gravitational potential per unit mass $\phi^\mu(r)(= \frac{1}{m}\int {\rm d}rF^\mu(r))$ into the Poisson equation $\triangle\phi^\mu = 4\pi\mathcal{G}^\mu\rho^\mu_{\rm qe}$, an equation governing F^μ can be obtained as follows:

$$F^{\mu\prime}(r) + \frac{\mu-1}{r}F^\mu(r) = 4\pi\mathcal{G}^\mu m\rho^\mu_{\rm qe}(r) = 4\pi\mathcal{G}^\mu m^2\mathcal{N}^\mu_{\rm qe}(r)\ , \quad (13)$$

where $\rho^\mu_{\rm qe} = m\mathcal{N}^\mu_{\rm qe} = mNP^\mu_{\rm qe} = mN\mathcal{P}^\mu_{\rm qe}/J^\mu$ and $\mathcal{G}^\mu$ relates quantities in the previous sections as $\mathcal{G}^2 = G'$ and $\mathcal{G}^3 = G$.

Here, we nondimensionalize these equations by using the following units of length and force: $[\text{length}] = \sqrt{\frac{2(\mu+2)\mu^2 D}{4\pi\mathcal{G}^\mu m^2\mathcal{N}^\mu_{\rm qe}(0)\{\mu^2 m\gamma + 8\pi(\mu^2+4\mu+2)\epsilon\mathcal{G}^\mu m^2\mathcal{N}^\mu_{\rm qe}(0)\}}}$, $[\text{force}] = \sqrt{\frac{8\pi(\mu+2)D\mathcal{G}^\mu m^2\mathcal{N}^\mu_{\rm qe}(0)}{\mu^2 m\gamma + 8\pi(\mu^2+4\mu+2)\epsilon\mathcal{G}^\mu m^2\mathcal{N}^\mu_{\rm qe}(0)}}$. Then, equations (12) and (13) are altered to

$$\bar{\mathcal{N}}^{\mu\,\prime}_{\rm qe}(\bar{r}) = -\frac{2(\mu+2)\bar{r}\bar{F}^\mu(\bar{r})\{1 + 2q\bar{F}^{\mu\prime}(\bar{r})\} + 2(\mu-1)(\mu+1)q\bar{F}^\mu(\bar{r})^2}{\bar{r}\{\mu + 2(\mu^2+4\mu+2)q + 2(\mu+2)q\bar{F}^\mu(\bar{r})^2\}}\bar{\mathcal{N}}^\mu_{\rm qe}(\bar{r})\ , \quad (14)$$

and

$$\bar{F}^{\mu\prime}(\bar{r}) + \frac{\mu-1}{\bar{r}}\bar{F}^\mu(\bar{r}) = \mu\bar{\mathcal{N}}^\mu_{\rm qe}(\bar{r})\ , \quad (15)$$

where $q \equiv 4\pi\epsilon\mathcal{G}^\mu m^2\mathcal{N}^\mu_{\rm qe}(0)/(\mu m\gamma)$ and variables with overbars denote dimensionless. We should solve these equations with boundary condition $\bar{\mathcal{N}}^\mu_{\rm qe}(0) = 1$.

4. Results

The numerical solutions for $\mu = 2$ and 3 are shown in Fig. 2 by changing q. For comparison with observations, $1/(1+\bar{r}^2)$ and $1/(1+\bar{r}^2)^{3/2}$ are also plotted by a dashed curve in Fig. 2 (a) and (b), respectively. The dashed curves are typical best-fit ones for densities of molecular clouds or globular clusters. We can see that the best-fit curves are derived from our model by varying q appropriately.

[1] For $\mu = 2$ and 3, $J^2 = 2\pi r$ and $J^3 = 4\pi r^2$, respectively: We can depict it by $J^\mu = 2^{\mu-1}\pi r^{\mu-1}$.

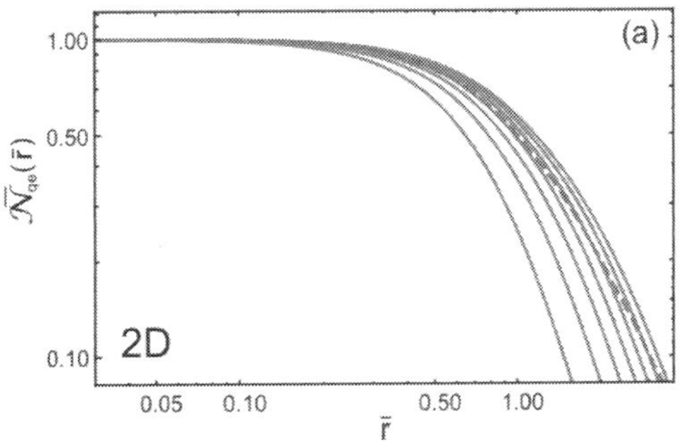

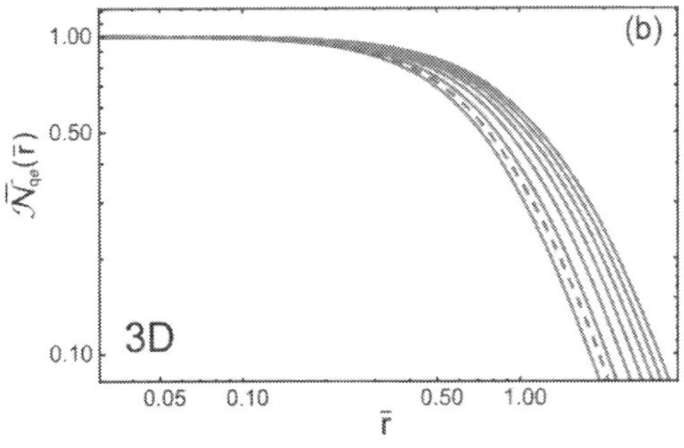

Figure 2. (color online) Numerical solutions of Eqs. (14) and (15) for $\mu = 2$ (a) and $\mu = 3$ (b). As the curve changes from the left to the right in (a), q gets larger from 0 to 0.24 in steps of 0.04. On the other hand, q gets larger from 0 to 0.12 in steps of 0.02 in (b). The (red) dashed curve in (a) and (b) means $1/(1+\bar{r}^2)$ and $1/(1+\bar{r}^2)^{3/2}$, respectively.

Finally, we examine the range of the index κ. Owing to adopt Eqs. (12) and (13) as the units of length and force, the number density around the center of the system in QES can be depicted by

$$\bar{\mathcal{N}}^{\mu}(\bar{r}) \simeq \frac{1}{(1+\bar{r}^2)^{\kappa(q)}} , \tag{16}$$

where κ is a function of q:

$$\kappa(q) = \frac{(\mu+2)\{1+(\mu+1)q\}}{\mu+2(\mu^2+4\mu+2)q} . \tag{17}$$

From this equation, because of $q \geq 0$, one can see the range of κ as follows:

$$\frac{(\mu+1)(\mu+2)}{2(\mu^2+4\mu+2)} < \kappa \leq \frac{\mu+2}{\mu} . \tag{18}$$

Therefore, with regard to $\mu = 2$ and $\mu = 3$, $\frac{3}{7} < \kappa \leq 2$ and $\frac{10}{23} < \kappa \leq \frac{5}{3}$, respectively. Both ranges include the observed indexes.

4.1. Acknowledgments

We would like to thank Prof. Takayuki Tatekawa for N-body simulations and members of astrophysics laboratory at Ochanomizu University for extensive discussions. This work was supported by JSPS Grant-in-Aid for Exploratory Research 12849926.

References

[1] Binney J and Tremaine S 2008 *Galactic Dynamics: (Second Edition)* (Princeton, NJ: Princeton University Press)
[2] Arzoumanian D *et al.* 2011 *A&A* **529** L6
[3] Tashiro T and Tatekawa T 2010 *J. Phys. Soc. Jpn* **79** 063001
[4] Tashiro T and Tatekawa T 2011 *Numerical Simulations of Physical and Engineering Processes* (Croatia: INTECH) p 301
[5] Teles T N, Levin Y, Pakter R and Rizzato F B 2010 *J. Stat. Mech.* **2010** P05007
[6] Sekimoto K 1999 *J. Phys. Soc. Jpn* **68** 1448

Ideal Fluids, the Quark Gluon Plasma, and Hadronic Gases

Nasser Demir

[1] Physics Department, Faculty of Science, Kuwait University, Khaldiya, Kuwait 13060

E-mail: nasser.demir@gmail.com

Abstract. A detailed comparison between two methods to calculate the shear viscosity coefficient of a hot hadronic gas is presented. We choose two systems in this comparison which are massless particles with current algebra cross section and a mixture comprised of pions with rho resonances. The two methods involved are the Green-Kubo method, applied using the Ultrarelativistic Quantum Molecular Dynamics (UrQMD) model to simulate the hadronic medium, and the Chapman-Enskog method. In addition, the effect of the resonance lifetime on the shear viscosity coefficient is investigated.

1. Introduction

Ultrarelativistic heavy ion collisions at the Relativistic Heavy Ion Collider (RHIC) and the Large Hadron Collider (LHC) are thought to have created a Quark Gluon Plasma (QGP) with characteristics of an ideal fluid [1, 8, 11]. As such, the characterization of the shear viscosity coefficient η and the shear viscosity to entropy density ratio η/s of hot, dense QCD matter is of significant interest.

One should note that the shear viscosity, especially in the context of a relativistic heavy ion collision, is a *time-dependent* quantity. While the partonic phase of such a collision is expected to have a very low value of η/s, after hadronization occurs η/s is expected to rapidly increase [6]. In order to quantify the viscosity of the deconfined phase of a relativistic heavy ion collision, it is necessary to separately determine the viscosity of the hadronic phase in order to quantitatively constrain the viscosity of the deconfined phase. Although several calculations for the hadronic shear viscosity exist, in order to validate a method and test the limits to which a calculation can be reliable, it is necessary to perform a systematic comparison between two different methods and see where the differences exist, such as in [13] , and that is the goal of this paper.

2. Theory

Two independent methods here are presented which are used to calculate the shear viscosity coefficient η. They are the Chapman Enskog Method and the Green Kubo method.

2.1. Calculating Shear Viscosity: The Chapman Enskog Method

The idea of the Chapman Enskog approximation is that the system, which is slightly disturbed from its equilibrium, will have a perturbed distribution function which can be expressed in terms of hydrodynamic variables [5].

For a single component gas, the first approximation to the shear viscosity takes the form of [13]

$$\eta = \frac{1}{10} k_B T \frac{\gamma_0^2}{c_{00}} \,, \qquad (1)$$

where $\gamma_0 = -10\,\hat{h}$, $\hat{h} = K_3(z)/K_2(z)$ and $z = mc^2/k_B T$ is called the relativity parameter. The term c_{00} contains the interaction in the system which is defined as

$$c_{00} = 16\left(w_2^{(2)} - \frac{1}{z} w_1^{(2)} + \frac{1}{3\,z^2} w_0^{(2)}\right) . \qquad (2)$$

The quantity $w_i^{(s)}$ is the relativistic omega integral given by

$$w_i^{(s)} = \frac{2\pi z^3 c}{K_2(z)^2} \int_0^\infty d\psi \, \sinh^7\psi \, \cosh^i\psi \, K_j(2z\cosh\psi) \times \int_0^\pi d\Theta \, \sin\Theta \, \sigma(\psi,\Theta) \, (1-\cos^s\Theta) \,, \quad (3)$$

where $j = 5/2 + 1/2(-1)^i$, $\sinh\psi = g/mc$ and $\cosh\psi = P/2mc$. The quantities g and P are the relative and center of mass momenta, respectively. The interaction in the system is represented by the differential cross section, $\sigma(\psi,\Theta)$.

2.2. Simulating the Hadronic Medium

To simulate the hadronic medium we utilize the Ultrarelativistic Quantum Molecular Dynamics (UrQMD) model. UrQMD is based upon the Boltzmann equation. Interactions in our simulation are based only upon scattering, and we neglect any interparticle potential. The criterion for a collision to occur is based upon the geometric interpretation of the cross section:

$$d_{min} \leq \sqrt{\frac{\sigma_{tot}}{\pi}}, \qquad (4)$$

where d_{min} is the transverse distance at closest approach between two particles. UrQMD is described in great detail in [2].

In order to use the Green-Kubo formalism to extract the shear viscosity coefficient, it is necessary to establish that our system reaches a state of thermal equilibrium. To force our system into equilibrium, we confine our hadronic medium to a cubic box with periodic boundary conditions in real space. This technique was also used in [3, 9, 10]. The input parameters of our system are: volume of the box, initial particle species, along with their chemical potentials. We initialize our systems to be in thermal and chemical equilibrium ($\mu_i = 0$). For the test case involving a gas of massless pions with energy dependent cross sections, the initial yields corresponding to a given temperature and chemical potential are calculated using the Boltzmann distribution function. For the test case involving a pion rho mixture, the initial particle yields are calculated using the Statistical Hadronization with Resonances (SHARE) model [12], with the resonance and decay tables modified to represent the degrees of freedom of the system.

2.3. Shear Viscosity Calculation Via Green Kubo Method

The second method to calculate the shear viscosity is the Green Kubo method, which amounts to finding the time integral of the correlations of the shear component of the energy-momentum tensor about the equilibrium state [7]. The energy-momentum tensor is related to the phase-space density of the system through the following equation

$$\pi^{\mu\nu} = \int d^3p \frac{p^\mu p^\nu}{p^0} f(x,p). \qquad (5)$$

The phase space density for a system of particles uniformly distributed in phase space is given by

$$f(x,p) = \frac{1}{V}\sum_{j=1}^{N}\delta(\vec{p}-\vec{p_j}). \quad (6)$$

This enables one to calculate π^{xy} directly

$$\pi^{xy} = \frac{1}{V}\sum_{j=1}^{N}\frac{p^x(j)p^y(j)}{p^0(j)}. \quad (7)$$

Note that since we assumed the particles were uniformly distributed in real space, the Green-Kubo formula for the shear viscosity coefficient becomes

$$\eta = \frac{V}{T}\int dt\langle\pi^{xy}(0)\pi^{xy}(t)\rangle. \quad (8)$$

In order to simplify evaluating this integral, we assume the correlation function $\langle\pi^{xy}(0)\,\pi^{xy}(t)\rangle$ has an exponential ansatz:

$$\langle\pi^{xy}(0)\,\pi^{xy}(t)\rangle \propto exp\left(-\frac{t}{\tau_\pi}\right) \quad (9)$$

This ansatz has also been used in [9]. If we assume the exponential dependence of the stress tensor correlation function on post-equilibriation time t, the Kubo formula reduces to

$$\eta = \frac{V}{T}\tau_\pi\langle\pi^{xy}(0)^2\rangle. \quad (10)$$

The volume V of the system is an input parameter, the temperature T is extracted from fitting momenta distributions to a Boltzmann distribution after checking that the momenta distributions in the system are isotropic, the correlator relaxation time τ_π is obtained from fitting the momentum correlation function to an exponential, and $\langle\pi^{xy}(0)^2\rangle$ is obtained from the intercept of the momentum correlation function. Hence, the shear viscosity coefficient can be calculated in this way, and in the next section we present the results of the shear viscosity coefficient for two systems.

3. Results

In this section we present the results for the calculation of the shear viscosity via both the Chapman-Enskog and Green-Kubo method for two systems. One system involves a pure chiral pion gas with an energy dependent cross section given by current algebra. The other system is a $\pi\rho$ mixture, where massive pions could scatter elastically or inelastically via an intermediate ρ resonance state. We also scale the resonance lifetime to see if there is any effect on the shear viscosity coefficient.

3.1. Massless Pions with Energy Dependent Cross Section (with Current Algebra)

The total (energy-dependent) cross section for the first system is given by

$$\sigma\left(\sqrt{s}\right) = \frac{5s}{72\pi f_\pi^4}, \quad (11)$$

where s is the Mandelstam variable representing the total energy in the center of mass frame and $f_\pi = 93$ MeV is the pion decay constant. The results for the shear viscosity coefficient are shown as a function of temperature in Figure 1, and we find that there is very good agreement between the calculations from the Green-Kubo method and the Chapman-Enskog method.

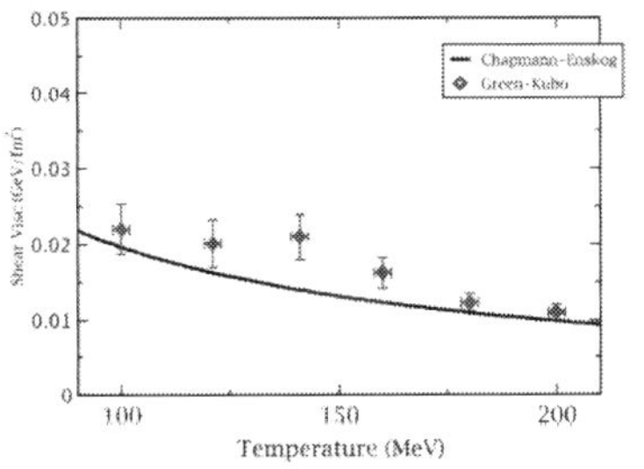

Figure 1. Shear viscosity coefficient as a function of temperature for an gas of chiral pions with energy dependent cross section. Chapmann Enskog calculation courtesy of A. Wiranata.

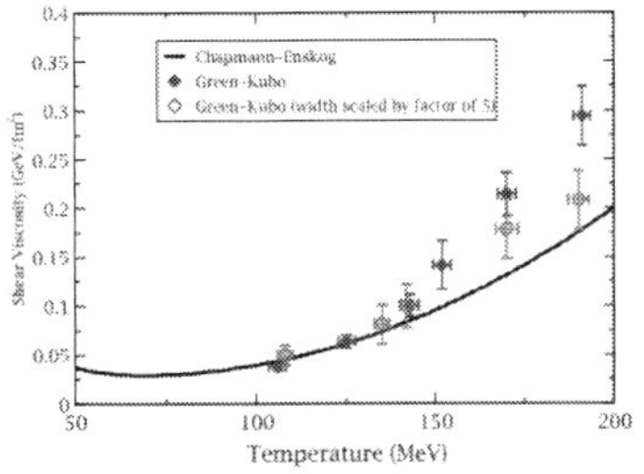

Figure 2. Shear viscosity coefficient versus temperature for a gas of pions scattering elastically and inelastically though an intermediate rho resonance state. Also shown is the effect of scaling the resonance lifetime by 5 in the UrQMD model. Chapmann Enskog calculation courtesy of A. Wiranata.

3.2. Pion Rho Mixture

The $\pi\rho$ mixture involves the two following processes:

- (a) $\sigma_{\pi\pi\to\pi\pi,elastic} = 17.778$ mb.
- (b) $\pi\pi \to \rho \to \pi\pi$ (inelastic)

The value of the cross section in (a) is based upon the additive quark model, which dictates that meson-meson and baryon-baryon cross sections are related via

$$\sigma_{MM} = \left(\frac{2}{3}\right)^2 \sigma_{BB}, \tag{12}$$

where M/B stands for meson/baryon, respectively. Since the elastic nucleon-nucleon cross section is known to be approximately 40 mb, this implies that $\sigma_{\pi\pi,elastic} = 17.778$ mb. The results for the shear viscosity coefficient are shown as a function of temperature in Figure 2. Also shown in Figure 2 are the results for the shear viscosity coefficient when the resonance lifetime is scaled by a factor of 5.

4. Discussion

As seen in Figure 1 there is good general agreement between the Chapman Enskog and Green Kubo methods for the case of chiral pions with energy dependent cross sections. In Figure 2, we find that, although there is good agreement between the Green-Kubo and Chapmann-Enskog methods in the temperature range $(100 < T < 150)$ MeV, the results from the Green-Kubo calculation are systematically higher than those for the Chapmann-Enskog method in the temperature range $(150 < T < 200)$ MeV. This is a noteworthy result, particularly when one realizes that many more inelastic scattering events through the intermediate resonance channel are expected in the temperature range $T > 150$ MeV than $T < 150$ MeV. In order to fully understand why the Green-Kubo method yields a higher result for the shear viscosity than the Chapmann-Eskog method does, it is necessary to understand the effect of resonance lifetimes in UrQMD on the shear viscosity coefficient. This is a subtle effect and the effect of timescales on transport coefficients in general has been investigated in the past [4]. Scaling the resonance lifetime by a factor of 5 yields no significant effect except in the higher temperature range. This interesting preliminary result should be continued by scaling the resonance lifetime by a factor of 0.2 to see if the trend is witnessed in the higher temperature range ($T > 175$) MeV in the opposite direction.

5. Conclusions

We have compared two methods to calculate the shear viscosity coefficient of hot hadronic matter, which is important to quantify when addressing the general question of transport coefficient behavior in the context of a relativistic heavy ion collision. The two methods used to calculate the shear viscosity were the Chapman-Enskog method and the Green-Kubo method using the UrQMD model to simulate the hadronic medium in equilibrium. When the results of the calculation for the two different methods were compared, excellent agreement was found between the Chapman Enskog and Green Kubo methods for the system of pure massless pions with an energy dependent cross section, whereas a systematic deviation occured between the Chapmann Enskog and Green Kubo methods in the temperature range $150 < T < 200$ MeV, where significant inelastic scattering is expected to occur. Preliminary results indicate changing the resonance lifetime does impact the shear viscosity coefficient at temperatures $T > 175$ MeV, and this result should be investigated further.

Acknowledgements

N. Demir wishes to thank Kuwait University for its support via decree number 1906 for the 2014-2015 fiscal year, which helped in completion of this work. He also thanks A. Wiranata for providing the Chapmann-Enskog calculations to compare against.

References

[1] B. B. Back et al. Centrality dependence of charged hadron transverse momentum spectra in d + au collisions at s(nn)**(1/2) = 200-gev. *Phys. Rev. Lett.*, 91:072302, 2003.

[2] S. A. Bass et al. Microscopic models for ultrarelativistic heavy ion collisions. *Prog. Part. Nucl. Phys.*, 41:225–370, 1998.

[3] M. Belkacem et al. Equation of state, spectra and composition of hot and dense infinite hadronic matter in a microscopic transport model. *Phys. Rev.*, C58:1727–1733, 1998.

[4] Pawel Danielewicz and Scott Pratt. Delays associated with processes in nuclear reaction simulations. *Phys. Rev.*, C53:249–266, 1996.

[5] Sybren R. De Groot. *Relativistic Kinetic Theory: Principles and Applications*. North-Holland Publishing Company, third edition, 1980.

[6] Nasser Demir and Steffen A. Bass. Shear-Viscosity to Entropy-Density Ratio of a Relativistic Hadron Gas. *Phys. Rev. Lett.*, 102:172302, 2009.

[7] R. Kubo. *Rept. Prog. Phys.*, 29:255, 1966.

[8] Berndt Muller and James L. Nagle. Results from the relativistic heavy ion collider. 2006.

[9] Azwinndini Muronga. Shear viscosity coefficient from microscopic models. *Phys. Rev.*, C69:044901, 2004.

[10] Shin Muroya and Nobuo Sasaki. A calculation of the viscosity over entropy ratio of a hadronic gas. *Prog. Theor. Phys.*, 113:457–462, 2005.

[11] Bjoern Schenke, Sangyong Jeon, and Charles Gale. Anisotropic flow in $\sqrt{s} = 2.76$ TeV Pb+Pb collisions at the LHC. *Phys.Lett.*, B702:59–63, 2011.

[12] Giorgio Torrieri, Jean Letessier, Johann Rafelski, and Steve Steinke. Statistical hadronization with resonances. *Acta Phys.Polon.*, B35:2911–2927, 2004.

[13] Anton Wiranata and Madappa Prakash. Shear Viscosities from the Chapman-Enskog and the Relaxation Time Approaches. *Phys.Rev.*, C85:054908, 2012.

Asymptotic solution of natural convection in a uniformly Joule-heating shallow cavity

Xiaohui Zhang[1] , Ahmed Kadhim Hussin[2]

1. College of Physics, Optoelectronics and Energy,Soochow University,Suzhou,China
E-mail: xhzhang@suda.edu.cn
2. College of engineering, Babylon University, Babylon,IRAQ

Abstract. The steady laminar two-dimensional Joule heating natural convection is investigated using asymptotical analysis, the fluid is in a rectangular cavity, the direct current contributes heat for heating the process medium by a pair of plate electrodes, the top wall is cooled with atmosphere and all the other walls are kept thermally insulated. The asymptotic solution is obtained in the core region in the limit as the aspect ratio, which is defined as the ratio of the vertical dimension of cavity to the horizontal dimension of cavity, goes to zero. The numerical experiments are also carried out to compare with the asymptotic solution of the steady two-dimensional Joule heating convection. The asymptotic results indicate that the expressions of velocity and temperature fields in the core region are valid in the limit of the small aspect ratio.

1. Introduction

Application of Joule heating in engineering can be found in various industrial processing,such as melting glasses in electric glass melters and for heating molten slag in electro-slag remelting process applied in the steel industry and nuclear waste treatment technology. Two types of body forces prevail in an electrically conducting liquid under Joule heating, they are the gravitational body force due to the non-uniform temperature field and the electromagnetic body force due to the interaction between self-induced magnetic field and moving charge carriers in the liquid.

With the development of the computer technology, many numerical simulations of Joule heating convection have been performed for rectangular geometries [1,2], However, to our knowledge, few asymptotic or analytical solutions of Joule heating convection have been reported, because the governing equations are nonlinear and coupled. The asymptotic or analytical solutions are very useful even for the newly rapidly developing computational fluid dynamics.

The purpose of this paper is to use the asymptotic analysis to examine convection and obtain an approximate analytical solution of the velocity and temperature field in uniformly Joule-heating shallow cavity.

2. Mathematical formulation of the problem

Consider the system described in Ref. [2] for simplicity, the fluid in the cube is heated by the electrode pair, which is assumed to be isopotential surfaces with an externally applied potential across them. All the boundaries of the pool are solid–fluid interfaces.

Mathematical model describing the state of an electrically conducting liquid pool in a cavity involves model equations for fluid flow, heat transfer and Maxwell's equations describing the electromagnetic field. Furthermore, if we examine the ratio $Ha^2/(Ra/Pr)$ we find that:

$$\frac{Ha^2}{Ra/\mathrm{Pr}} = \frac{\sigma\mu(\nu/L)}{\nu/\alpha}\cdot\frac{c_P\beta}{Lg} = \frac{\mathrm{Re}_m}{\mathrm{Pr}}\cdot\frac{c_P\beta}{Lg}$$

[1] To whom any correspondence should be addressed.

For the type of applications envisioned here $\mathrm{Re_m}$ (the magnetic Reynolds number, which can be defined as the ratio of convective to diffusive transport of electromagnetic field) is order of 10^{-4}, one then needs to estimate the magnitude of ratio (cp/β)/(g*L*), it gives a measure of the extent of buoyant force per unit temperature change for given thermal properties of the pool material, in order to appreciate the importance of the electromagnetic forces in the momentum balance, this parameter depends on the properties of the pool material and needs to be quite large in order for the effect of the induced magnetic field to be felt in the momentum balance. Otherwise it makes no sense to drop the effect of the induced magnetic field in Ohm's law while keeping it in the momentum balance.

Consequently if we define $U=\dfrac{\partial\psi}{\partial Y}$ $V=-\dfrac{\partial\psi}{\partial X}$,

Taking A as the aspect ratio $A=L/W$, Equations can be written as

$$\frac{\partial^2\psi}{\partial X^2}+\frac{\partial^2\psi}{\partial Y^2}=-\omega$$

$$\frac{\partial\psi}{\partial Y}\frac{\partial\omega}{\partial X}-\frac{\partial\psi}{\partial X}\frac{\partial\omega}{\partial Y}=\frac{\partial^2\omega}{\partial Y^2}-\frac{\partial^2\omega}{\partial X^2}+\frac{Ra}{Pr}\frac{\partial\theta}{\partial X} \tag{1}$$

$$\frac{\partial\psi}{\partial Y}\frac{\partial\theta}{\partial X}-\frac{\partial\psi}{\partial X}\frac{\partial\theta}{\partial Y}=\frac{1}{Pr}\left(\frac{\partial^2\theta}{\partial Y^2}+\frac{\partial^2\theta}{\partial X^2}\right)+\frac{1}{Pr}\frac{1}{A}$$

3. Core flow and end region flow

With introduction of the characteristic horizontal scale $X=O(A^{-1})$, and

$$\left.\begin{aligned}\theta&=\theta_0+A\theta_1+A^2\theta_2+\cdots,\\ \psi&=\psi_0+A\psi_1+A^2\psi_2+\cdots,\\ \omega&=\omega_0+A\omega_1+A^2\omega_2+\cdots,\end{aligned}\right\} \tag{2}$$

The systematic solution, valid for $A \ll 1$, i.e.

$$\psi=K_1\left(\frac{1}{24}Y^4-\frac{1}{12}Y^3+\frac{1}{24}Y^2\right) \tag{3}$$

$$\omega=-K_1\left(\frac{1}{2}Y^2-\frac{1}{2}Y+\frac{1}{12}\right) \tag{4}$$

$$\hat{\theta}=-K_1\frac{Pr}{Ra}\hat{X}-\frac{1}{2}AX^2+K_1^2A\frac{Pr^2}{Ra}\left(\frac{1}{120}Y^5-\frac{1}{48}Y^4+\frac{1}{72}Y^3\right)+A^2 \tag{5}$$

Where $\hat{X}=AX, \hat{\theta}=A\theta$.

As in the core region, the solution can be obtained as a regular perturbation expansion in the small parameter A of the form, when $X\to\infty$,the solution is $\theta_0=-Y^2+A$,and

$$\lim_{X\to\infty}\psi_1=-\frac{3}{2}\frac{Ra}{Pr}\left(\frac{1}{24}Y^4-\frac{1}{12}Y^3+\frac{1}{24}Y^2\right)$$

In order to clarify the physics meaning of the solutions above, we have plotted the isotherms and streamline contours in Fig. 1 and 2 from the above solutions. It should be noted that all of the results are plotted as square figures, hence, for cavities with small aspect ratios the horizontal length scale is substantially compressed. Fig. 1 reveals that the maximum liquid temperature occurs at the adiabatic bottom wall, this supports the present physics observation of internal heating condition, the upper part, the lower temperature. Centro-symmetry property (Y=0.5) is shown in Fig. 2 which is anticipated the convection roll in the cavity from physics point.

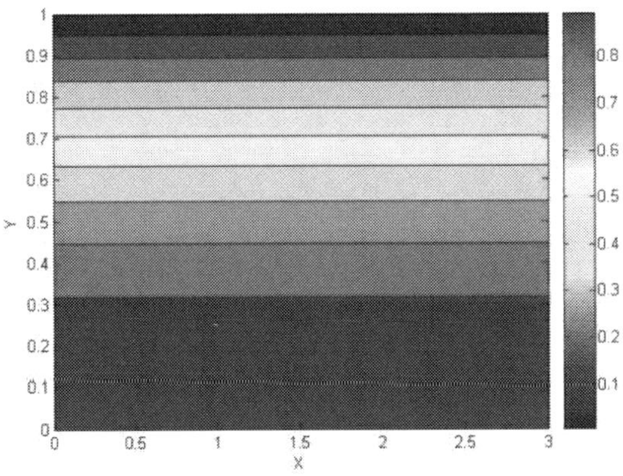

Fig. 1 At $O(1)$ isotherms $X = \infty$

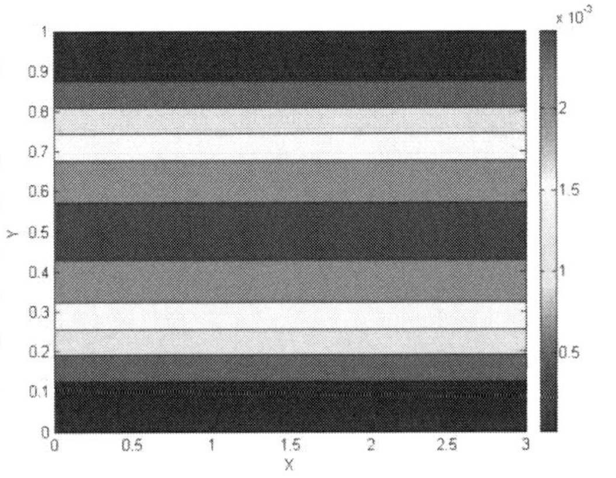

Fig. 2 Streamlines $X = \infty$

4. Numerical simulation and comparison

Numerical experiments are carried out to compare simulation results with the asymptotic solution of the steady two-dimensional Joule-heating convection. The validation of the code for the simulation was performed in our previous works [2], and there is no need to repeat here.

Fig. 3 is a plot of the theoretically and numerically determined velocity profiles u/K_1 on the centre-line $X = 1/2A$ as a function of Y. The numerical data are expressed with the dotted line and solid line is for the theoretical result in the figure.

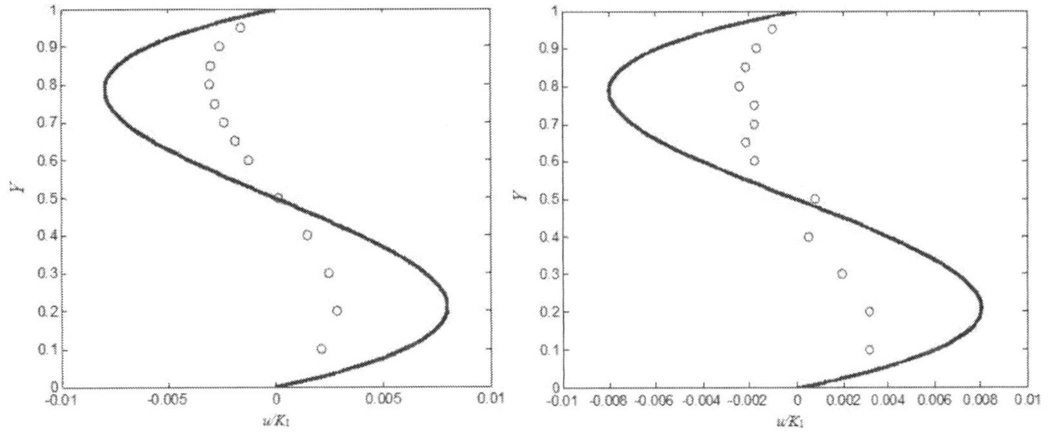

(a) A=0.5 (b) A=0.25

Fig. 3 Comparison of normalized horizontal velocity at X=1/2A with theoretical parallel-flow profile (solid line). Ra=15000 ,Pr=1

For both cases, the analytical solution and numerical results present the same symmetry characteristics, from the asymptotic profile, the flow obeys to asymmetry by the vertical mid-plane of the cavity, the horizontal velocity is zero on this plane. The flow field consists of anti-clockwise and clockwise convection rolls in the left and right cavity respectively. It is found that the asymptotical solutions are not suitable in the Y=0.2 and Y=0.8 plane, In the asymptotical solutions, the terms $O(A^2)$ are neglected, therefore, the error between the asymptotical solutions and the simulation results increases. Further consideration of core solution of equation (5) indicates that a plot $\theta + K_1 \frac{Pr}{Ra} X$ as a function of Y yield the fifth-order polynomial $f(Y)$.Fig. 4 is a plot of this function evaluated at the centre-line 1/2A.

Predicted isotherms show that the isotherms are more uniform close the liquid surface which means the conduction mechanism dominates. The numerical data in lower part of cavity is somewhat different from the theoretical profile, as the temperature in the lower part of cavity is not linear. In the asymptotical solutions, in order to get K_1 the term of fifth order of temperature is neglected, therefore, it is found that there is error between the asymptotical solutions and the simulation results.

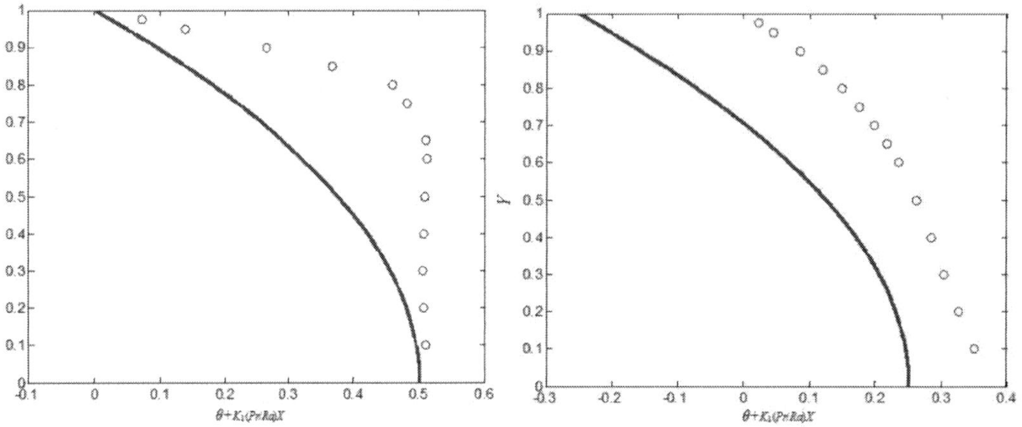

(a) A=0.5 (b) A=0.25

Fig. 4 Comparison of numerically determined temperature Θ at X= $1/2A$ with theoretical core temperature (solid line).

5. Conclusion

Approximate solutions of steady laminar two-dimensional Joule heating natural convection are obtained in the core region in the limit as the aspect ratio goes to zero.

Numerical experiments are carried out to compare simulation results with the asymptotic solution under the same conditions.

The asymptotical solutions show that the temperature and the velocity in the core region have the same trend as the numerical solutions respectively.

The asymptotical solutions show the flow patterns and isotherms are steady symmetric with respect to the mid-plane of cavity, which have agreement with the simulation results and physics meaning of convection.

The reason of discrepancies between the asymptotic results and the numerical simulations is that the higher order terms are neglected in the asymptotical solutions.

Acknowledgments

This work is supported by National Natural Science Foundation of China (Grant No. 51176132).

The corresponding author gratefully acknowledges the support of K.C.Wong Education Foundation and DAAD.

References:

[1] G. Sugilal, P.K. Wattal, K. Iyer, Convective behaviour of a uniformly Joule-heated liquid pool in a rectangular cavity, *International Journal of Thermal Sciences* 44, (2005), pp.915–925

[2] X.Zhang ,D.Angeli.Flow transitions in a Joule-heated cavity of a low-Prandtl number fluid,*International Journal of Thermal Sciences*, 50, (2011), pp.2063-2077

Gravitational radiation as radiation same level of electromagnetic and its generation in pulsed high-current discharge

S I Fisenko, I S Fisenko

Rusthermosynthesis JSC,
38, Ostozhenka Str., Moscow 119034, Russia

StanislavFisenko@yandex.ru

Abstract. The notion of gravitational radiation as a radiation of the same level as the electromagnetic radiation is based on theoretically proved and experimentally confirmed fact of existence of stationary states of an electron in its gravitational field characterized by the gravitational constant $K = 10^{42}G$ (G is the Newtonian gravitational constant) and unrecoverable space-time curvature Λ. This paper gives an *overview* of the authors' works [1, 2, 3, 4], which set out the relevant results. *Additionally*, data is provided on the broadening of the spectra characteristic radiation. The data show that this broadening can be explained only by the presence of excited states of electrons in their gravitational field. What is more, the interpretation of the new line of X-ray emission spectrum according to the results of observation of MOS-camera of XMM-Newton observatory is of interest. The given work contributes into further elaboration of the findings considering their application to dense high-temperature plasma of multiple-charge ions. This is due to quantitative character of electron gravitational radiation spectrum such that amplification of gravitational radiation may take place only in multiple-charge ion high-temperature plasma.

1. Introduction

Numerous attempts to detect experimentally gravitational radiation (as based on the notions of the general relativity theory and alternative theories) brought no result. Studying the development of the relativistic theory of gravitation (which is the relativistic theory of gravitation, and not any of its particular case, such as general relativity theory), the authors obtained a model of gravitational interaction at the quantum level, which has no analogues. [1,2,3,4].

2. Gravitational radiation of the relativistic theory of gravitation

From the 70's onwards, it became obvious [5] that in the quantum region the numerical value of G is not compatible with the principles of quantum mechanics. The essence of the problem of the generalization of relativistic equations on the quantum level was thus outlined: such generalization must match the numerical values of the gravity constants in the quantum and classical regions.

In the development of these results, as a micro-level approximation of Einstein's field equations, a model is proposed, based on the following assumption [1,2]:

"The gravitational field within the region of localization of an elementary particle having a mass m0 is characterized by the values of the gravity constant K and of the constant □ that lead to the

Journal of Physics: Conference Series **574** (2015) 012157 doi:10.1088/1742-6596/574/1/012157

stationary states of the particle in its proper gravitational field, and the particle stationary states as such are the sources of the gravitational field with the Newtonian gravity constant G".

Complexity of solving this problem compels us to employ a simpler approximation, namely: energy spectrum calculations in a relativistic fine-structure approximation. In this approximation the problem of the stationary states of an elementary source in the proper gravitational field will be reduced to solving the following equations:

$$f''+\left(\frac{\nu'-\lambda'}{2}+\frac{2}{r}\right)f'+e^{\lambda}\left(K_n^2e^{-\nu}-K_0^2-\frac{l(l+1)}{r^2}\right)f=0 \qquad (1)$$

$$-e^{-\lambda}\left(\frac{1}{r^2}-\frac{\lambda'}{r}\right)+\frac{1}{r^2}+\Lambda=\beta(2l+1)\left\{f^2\left[e^{-\lambda}K_n^2+K_0^2+\frac{l(l+1)}{r^2}\right]+f'^2e^{-\lambda}\right\} \qquad (2)$$

$$-e^{-\lambda}\left(\frac{1}{r^2}+\frac{\nu'}{r}\right)+\frac{1}{r^2}+\Lambda=\beta(2l+1)\left\{f^2\left[K_0^2-K_n^2e^{-\nu}+\frac{l(l+1)}{r^2}\right]-e^{\lambda}f'^2\right\} \qquad (3)$$

$$\left\{-\frac{1}{2}\left(\nu''+\nu'^2\right)-\left(\nu'+\lambda'\right)\left(\frac{\nu'}{4}+\frac{1}{r}\right)+\frac{1}{r^2}\left(1+e^{\lambda}\right)\right\}_{r=r_n}=0 \qquad (4)$$

$$f\left(\sqrt{\Lambda^{-1}}\right)=0 \qquad (5)$$

$$f(r_n)=0 \qquad (6)$$

$$\lambda(0)=\nu(0)=0 \qquad (7)$$

$$\int_0^{r_n} f^2r^2dr=1 \qquad (8)$$

Equations (1)—(3) follow from equations (9)—(10)

$$\left\{-g^{\mu\nu}\frac{\partial}{\partial x_\mu}\frac{\partial}{\partial x_\nu}+g^{\mu\nu}\Gamma^{\alpha}_{\mu\nu}\frac{\partial}{\partial x_\alpha}-K_0^2\right\}\Psi=0 \qquad (9)$$

$$R_{\mu\nu}-\frac{1}{2}g_{\mu\nu}R=-\kappa\left(T_{\mu\nu}-\mu g_{\mu\nu}\right), \qquad (10)$$

after the substitution of Ψ in the form of $\Psi=f_{El}(r)Y_{lm}(\theta,\varphi)\exp\left(\frac{-iEt}{\hbar}\right)$ into them and specific computations in the central-symmetry field metric with the interval defined by the expression [6]

$$dS^2=c^2e^{\nu}dt^2-r^2\left(d\theta^2+\sin^2\theta d\varphi^2\right)-e^{\lambda}dr^2 \qquad (11)$$

The following notation is used above: f_{El} is the radial wave function that describes the states with a definite energy E and the orbital moment l (hereafter the subscripts El are omitted), $Y_{lm}(\theta,\varphi)$ – are spherical functions, $K_n=E_n/\hbar c$, $K_0=cm_0/\hbar$, $\beta=(\kappa/4\pi)(\hbar/m_0)$.

Condition (4) defines rn,, whereas equations (10) through (7) are the boundary conditions and the normalization condition for the function f, respectively. Condition (4) in the general case has the form $R(K,r_n)=R(G,r_n)$. Neglecting the proper gravitational field with the constant G, we shall write down this condition as $R(K,r_n)=0$, to which equality (4) actually corresponds.

In its simplest approximation (from the point of view of the original mathematical estimates) the problem on steady states in proper gravitational field (with constants K and Λ) is solved by [1]. The solution of this problem provides the following conclusions.

a) With the numeric values $K \approx 5.1\times10^{31}$ Nm^2kg^{-2} and $\Lambda = 4.4\times10^{29}$ m^{-2} there is a spectrum of steady states of the electron in proper gravitational field (0.511 MeV ...0.681 MeV). The basic state is the observed electron rest energy 0.511 MeV.

b) These steady states are the sources of the gravitational field with the G constant.

c) The transitions to stationary states of the electron in proper gravitational field cause gravitational emission, which is characterized by constant K, i. e. gravitational emission is an emission of the same level as electromagnetic (electric charge e, gravitational charge $m\sqrt{K}$). In this respect there is no point in saying that gravitational effects in the quantum area are characterized by the G constant, as this constant belongs only to the macroscopic area and cannot be transferred to the quantum level *(which is also evident from the negative results of registration of gravitational waves with the G constant, they do not exist).*

Existence of such numerical value Λ denotes a phenomenon having a deep physical sense: introduction into density of the Lagrange function of a constant member independent on a state of the field. This means that the time-space has an inherent curving which is connected with neither the matter nor the gravitational waves. The distance at which the gravitational field with the constant K is localized is less than the Compton wavelength, and for the electron, for example, this value is of the order of its classical radius. At distances larger than this one, the gravitational field is characterized by the constant G, i.e., correct transition to Classical GR holds, (see Figure 1).

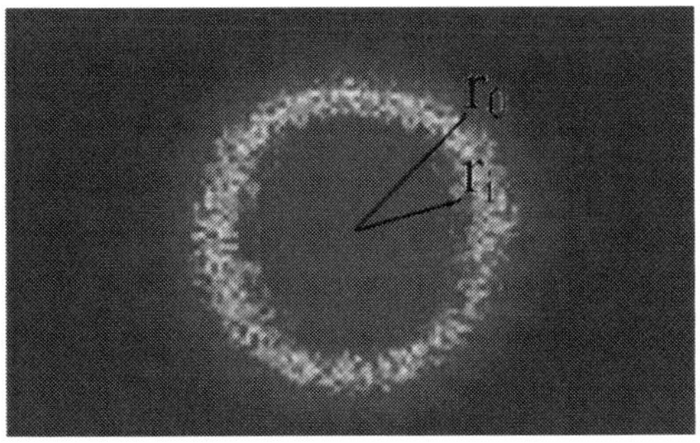

Figure 1. Electron in proper gravitational field in the ground stationary state.

r_i – radius of irremovable curvature,

r_0 – "classical" electron radius being a radius of the ground stationary state of electron in proper gravitational field

3. Spectral lines widening of the radiation of multiple-charge ions

Figs 2 show characteristic parts of micropinch soft X-ray radiation spectrumAs a mechanism of spectrum lines widening, a Doppler, radiation and impact widening were considered. Such adjustment according to said widening mechanisms does not lead to complete reproduction of the registered part of the micropinch radiation spectrum. This is the evidence (under the condition of independent conformation of the macroscopic parameters adjustment) of additional widening mechanism existence due to electron excited states and corresponding gravitational radiation spectrum part already not having clearly expressed lines because of energy transfer in the spectrum to the long-wave area.

That is to say that the additional mechanism of spectral lines widening of the characteristic electromagnetic radiation of multiple-charge ions (in the conditions of plasma compression by radiated gravitational field) is the only and unequivocal way of quenching electrons excited states at the radiating energy levels of ions and exciting these levels by gravitational radiation at resonance frequencies. Such increase in probability of ion transitions in other states results in additional spectral lines widening of the characteristic radiation. The reason for quick degradation of micropinches in various pulse high-currency discharges with multiple-charge ions is also clear. There is only partial thermolization of accelerated plasma with the power of gravitational radiation not sufficient for maintaining steady states [4].

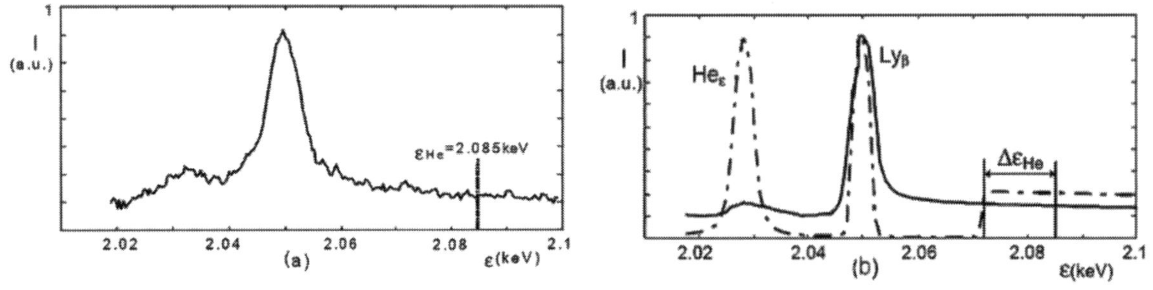

Figure 2. Experimental (a) and calculated (b) parts of a micropinch spectrum normalized for line Lyβ intensity in the area of the basic state ionization threshold of He-like ions. The firm line in variant (b) corresponds to density of 0.1 g/cm^3 , the dotted line – to 0.01 g/cm^3; it was assumed that T_e = 0.35 keV, (see [4]).

4. Thermonuclear plasma steady states generation

The problem of controlled fusion realization is directly connected to obtaining steady state of dense high-temperature plasma.

It can also be unambiguously stated that the present state of the art does not solve the problem of dense high-temperature plasma retention for a time required for the reaction of nuclear fusion but only solving the problem of heating plasma to the state when these reactions can exist. In the offered method of forming dense-high temperature plasma steady states for nuclear fusion a new fundamental concept is used, namely retaining plasma by radiated gravitational field as radiation of the same kind as electromagnetic:

Forming and accelerating binary plasma with multivalent ions by accelerating magnetic field in a pulse high-current discharge, injection of binary plasma from the space of the accelerating magnetic field:

Exciting stationary states of an electron in its own gravitational field in the range of energy up to 171 keV with following radiation under the condition of quenching lower excited energy levels of ion electron shell of a heavy component (including quenching excited state of electrons directly in nuclei of small sequential number as carbon) when retarding plasma bunch ejected from the space of the accelerating magnetic field. Cascade transitions from the upper levels are realized in the process of gravitational radiation energy transit to long-wave range.

The sequence of the operations is carried out in a two-sectional chamber of MAGO installation (developed in Experimental Physics Research Institute, Sarov, [7]; the structure of the installation is most suitable for the claimed method of forming steady states of the dense high-temperature plasma) with magnetodynamic outflow of plasma and further conversion of the plasma bunch energy (in the process of quenching) in the plasma heat energy for securing both further plasma heating and exciting gravitational radiation and its transit into a long-wave part of the spectrum with consequent plasma compression in the condition of radiation blocking and increasing [4].

5. Characteristic radiation spectra of stars

Two groups of researchers, independently of each other [8,9] reported that in the X-ray spectra of galaxy clusters there is a new line of radiation with an energy of 3.57 keV. This radiation should go from hot intergalactic gas that fills the cluster of galaxies, but, unlike other identified emission lines, this cannot be explained by any atomic transition.

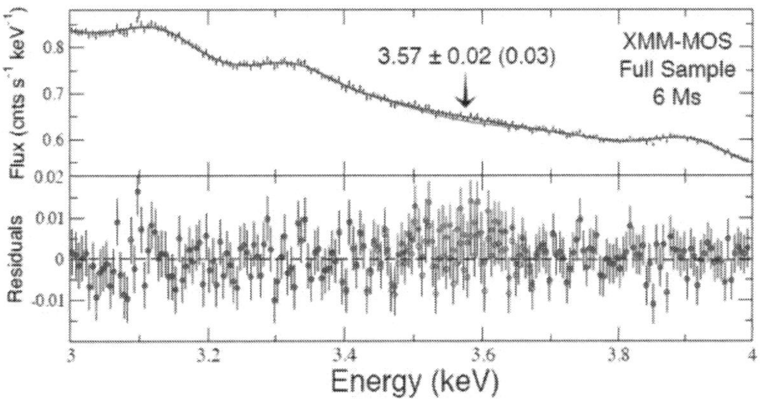

Figure 3. Above: MOS-camera spectrum in the range of 3 to 4 keV, XMM-Newton observatory .Separate dashes are the results of observation with errors; red curve is the best reproduction of the spectrum when only known emission lines of ions taken into account, blue curve is the result of addition of another previously unknown emission line. Bottom: deviation of observation data from the red and blue curves, [8].

The results of these measurements are shown in Figure 3. At the same time, the resonance spectra of stationary states of electrons in its gravitational field and the spectra of multiple-charge ions can produce not only the registered emission line, but also other lines of similar properties. We can expect the presence of such lines in detailed registration of the emission spectra of astrophysical objects in the energy range greater than 8 keV.

References

[1] Fisenko S I and Fisenko I S 2009 Gravitational interaction on quantum level and consequences thereof *The old and new concepts of physics 4* **6** 495–521
[2] Fisenko S I and Fisenko I S December 2012 The discrete energy spectrum of the gravitational radiation in the relativistic theory of gravitation *International Journal of Theoretical and Applied Physics (IJTAP) 2* **2** 32–39
[3] Fisenko S I and Fisenko I S April 2013 Method of forming stable states of dense high-temperature plasma *Journal of Modern Physics 4* **4** 481–485.
[4] Fisenko S I and Fisenko I S April 2014 *Journal of Applied Mathematics and Physics 5* **2**
[5] Siravam C and Sinha K 1979 The concept of "strong" gravity *Phys. Rep.* **51** 112–123
[6] Landau L D and Lifshitz E M 1976 *Field Theory* (Moscow: Nauka), pp 451–457
[7] Burenkov O M, Dolin Y N, Duday P V, Dudin V I, Ivanov V A, Ivanovsky A V, Karpov G V et al New Configuration of Experiments for MAGO Program *XIV Int. Conf. on Megagauss Magnetic Field Generation and Related Topics (Maui, Havaii, USA, 14–19 October 2012)* pp 95–99
[8] Bulbul E et al Detection of An Unidentified Emission Line in the Stacked X-ray spectrum of Galaxy Clusters *Preprint* astro-ph/1402.2301
[9] Boyarsky A, Ruchayskiy O, Iakubovskyi D, Franse J An unidentified line in X-ray spectra of the Andromeda galaxy and Perseus galaxy cluster *Preprint* astro-ph/1402.4119

A simple model of hysteresis behavior using spreadsheet analysis

A Ehrmann[1, 3] and T Blachowicz[2,3]
[1] Faculty of Textile and Clothing Technology, Niederrhein University of Applied Sciences, 41065 Mönchengladbach, Germany
[2] Institute of Physics – Center for Science and Education, Silesian University of Technology, 44-100 Gliwice, Poland
[3] VIARAM – Virtual Institute for Applied Research on Advanced Materials
E-mail: andrea.ehrmann@hsnr.de

Abstract. Hysteresis loops occur in many scientific and technical problems, especially as field dependent magnetization of ferromagnetic materials, but also as stress-strain-curves of materials measured by tensile tests including thermal effects, liquid-solid phase transitions, in cell biology or economics. While several mathematical models exist which aim to calculate hysteresis energies and other parameters, here we offer a simple model for a general hysteretic system, showing different hysteresis loops depending on the defined parameters. The calculation which is based on basic spreadsheet analysis plus an easy macro code can be used by students to understand how these systems work and how the parameters influence the reactions of the system on an external field. Importantly, in the step-by-step mode, each change of the system state, compared to the last step, becomes visible. The simple program can be developed further by several changes and additions, enabling the building of a tool which is capable of answering real physical questions in the broad field of magnetism as well as in other scientific areas, in which similar hysteresis loops occur.

1. Introduction

Each student of physics knows the typical ferromagnetic hysteresis loop in different shapes, depending on material, temperature, sample shape and other parameters. However, general models of hystereses often include sophisticated mathematical descriptions, such as the Bouc-Wen model [1,2], or are hard to bring into agreement with the physical situation, such as the Preisach hysteresis model [3].

What is offered here, instead, is a simple program, based on a spreadsheet analysis plus an easy macro code, to enable testing the influence of system parameters on the resulting hysteresis loops. The program is optimized for Microsoft Excel®, included in Microsoft Office 2007 or higher.

In this way, students can get a feel for the importance of the different parameters, for their effects on coercivities (loop width) and loop shapes. Our article is targeted on developing an intuitive approach of these influences, which is why – additional to automatically creating a complete hysteresis loop – a step-by-step mode is included, allowing for examination of each single step after changing the external field. While magnetism is surely one of the most important research areas for such a model to be applied [4], we have intentionally avoided all special mathematical magnetism-related descriptions in the model, to allow for using it for a broader spectrum of topics without unnecessary restrictions. The model also allows for examination of other systems showing hysteresis loops, such as of ferroelectric

spins [5], hysteretic phase transformations in Lennard-Jones fluids [6], adsorption / desorption from a fluid reservoir in a porous material [7], and further effects in physics, chemistry or material sciences.
The article is structured as follows: In section 2, the simple mathematical base of the program is described. Section 3 shows a small selection of possible results, calculated with our program, showing the influence of the different parameters. Section 4, finally, offers some additional ideas, allowing for the students to develop the program further, before the article is concluded in Section 5.

2. Mathematical Model

The following mathematical rules are the base for the program described here:
The system under investigation is subdivided into uniformly distributed cells. Each cell (depicting e.g. one magnetic moment or one ferroelectric spin) can have two discrete states, -1 or +1, as usual in Ising models [8]. Each cell is influenced by its four nearest neighbors, further interactions between diagonally neighboring cell or others are excluded in the basic version. Here, Ising regions of 16 x 16 cells have been chosen (Fig. 1), ten of which are used to calculate an average value of the respective system parameter (e.g. the overall magnetization) and an averaged hysteresis loop.
The first input parameter is the weight of the switching energy. Its value describes the relative importance of the switching energy, i.e. for the maximum value of 1, switching is independent of the neighbors, while for smaller values, the influence of the magnetization of the neighbors becomes larger. The rarely met value of 0 should not be used since it results in too many +1 values.
The second input parameter, the probability of the "wrong" magnetization, describes the influence of the temperature in a real system. For this probability = 0, a completely deterministic, stationary system is depicted. A probability of 0.5 describes – taking into account a magnetic system – a paramagnet, values between 0 and 0.5 result in ferromagnetic particles (softer ferromagnets for higher values). Values between 0.5 and 1 describe antiferromagnetic particles, which show a chessboard-like structure in the step-by-step processing for small external magnetic fields.

0	0	0	0	0	0	0	0	0	0	0	0	0	0	0	0	0	0
0	-1	-1	-1	1	1	-1	1	1	1	1	-1	-1	1	1	1	1	0
0	-1	-1	-1	-1	1	1	1	1	1	1	1	1	1	-1	1	-1	0
0	1	1	1	1	1	1	1	1	1	-1	-1	1	1	1	1	1	0
0	-1	-1	1	1	1	1	1	1	1	-1	-1	1	-1	1	1	1	0
0	1	1	1	1	-1	1	-1	1	-1	1	-1	1	-1	-1	1	1	0
0	-1	1	1	-1	-1	1	-1	-1	-1	-1	-1	-1	-1	-1	1	-1	0
0	-1	1	-1	-1	1	1	1	-1	-1	-1	-1	1	-1	-1	-1	-1	0
0	-1	1	-1	-1	1	1	1	-1	-1	-1	-1	1	-1	-1	1	1	0
0	-1	1	-1	-1	-1	1	1	-1	-1	-1	1	-1	-1	-1	1	1	0
0	1	1	-1	-1	1	1	1	-1	-1	-1	-1	-1	-1	-1	-1	1	0
0	1	1	1	1	1	-1	-1	-1	1	-1	1	1	1	1	1	1	0
0	1	1	1	1	1	-1	-1	1	1	-1	1	1	1	1	1	-1	0
0	1	-1	-1	-1	-1	-1	1	1	1	-1	-1	-1	1	1	1	-1	0
0	1	1	1	1	-1	1	1	1	-1	1	-1	-1	-1	-1	-1	-1	0
0	-1	1	1	1	-1	1	1	1	1	1	1	1	1	-1	-1	-1	0
0	1	-1	1	1	-1	1	-1	1	1	-1	-1	1	-1	-1	-1	-1	0
0	0	0	0	0	0	0	0	0	0	0	0	0	0	0	0	0	0

Figure 1. Example of the cells moments at zero external field in one 16 x 16 "particle".

The third input parameter, the value of the external switching field (i.e. the magnetic field in case of magnetic particles), can be chosen between ± 1. At these borders, the external switching field is stronger than all possible interactions within the particle. The hysteresis loops, however, only sweep up to switching fields of ± 0.1 (± 0.2), since in most cases, saturation is reached at these values. It must be mentioned that the field sweeping speed – i.e. the number of calculation steps per field unit controlled by the user – strongly influences the resulting hysteresis loops, thus the consequences of changes in different parameters should be examined with identical field sweeping speeds, i.e. with identical field step sizes. In the extreme case of a quasi-static loop (i.e. for very low sweeping speeds), this model is well-known to result in vanishing hysteresis loops [9,10]; a behavior which can be examined by the students themselves by means of more advanced spreadsheet models which are available on request from the authors.

In Fig. 2 the idea of simulation algorithm is described. In general, interaction with the four closest neighbors determines the resulting +1/-1 state, governed via the above mentioned parameters w_{cr}, P_{wm}, and H_{ext}, i.e. the weight of switching energy, the probability of the "wrong" state, and the external switching field, respectively, as well as a random number R between 0 and 1.

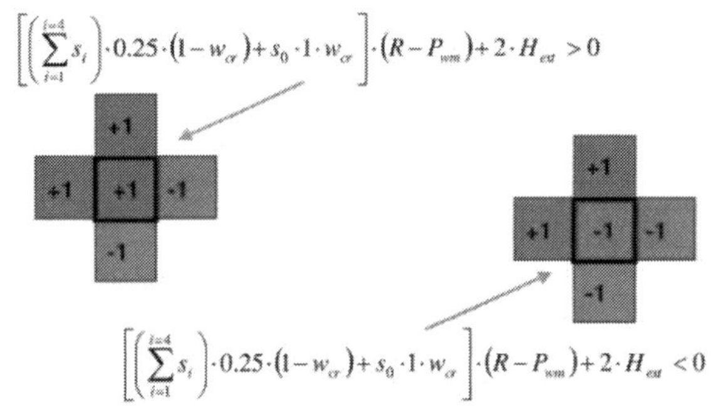

Figure 2. The two main conditions of switching for a cell s_0, surrounded by the four neighbors s_i.

The general condition of switching of a single cell s_0 is based on the following expression

$$\left[\left(\sum_{i=1}^{i=4} s_i\right)\cdot 0.25\cdot(1-w_{cr})+s_0\cdot 1\cdot w_{cr}\right]\cdot(R-P_{wm})+2\cdot H_{ext} \tag{1}$$

The easiest possibility to work with the program is to choose manual processing. After choosing the desired input values, a calculation step is manually started (by pressing F9 in Microsoft Excel®). Continuously starting new calculation steps – by keeping the respective key depressed – will result in a stable state of the simulated particles. Due to the included probability of the "wrong" state, it can be easily seen that each of the ten particles shown in the program looks different, even if they are all set to saturation (i.e. high external switching field to achieve completely blue or red particles) before. The overall system parameter under examination (e.g. magnetization) of the ten particles is steadily shown in the orange field "Magnetization".

For the hysteresis loops, a macro is needed. For this, the developer mode has to be switched on, by clicking (in Microsoft Excel®) on the Microsoft Office Button → Excel Options → Popular → Show Developer Tab in the Ribbon. Now the hysteresis loop can be carried out by pressing "Ctrl + m" for a "small" loop (external magnetic field in the range of ± 0.1) or "Ctrl + M" for a "large" loop (± 0.2). Depending on the computer speed, the recalculation will take a few seconds or a little longer time. Each hysteresis loop can be copied or exported as a graph, or the x-y-values can be copied into another document.

3. Results

In this section, a few results are shown to describe the influence of the different parameters.

Figure 3 depicts hysteresis loops calculated for a constant probability of the "wrong" state = 0.1 and a weight of the switching energy of 0.01, 0.1, and 0.5, respectively. While the coercive fields (the external fields for which the system parameter becomes 0, i.e. the intersection points with the x-axis) are broadest for a middle weight of the switching energy, larger weights lead to a change in the shape of the hysteresis loop.

In Fig. 4, the influence of the probability of a "wrong" state is shown for a constant weight of the switching energy. Apparently, the coercive fields are largest for small probabilities, while for larger probabilities, the system becomes softer, until a paramagnet-like state (completely reversible, without a visible hysteresis loop) is reached for a probability of 0.5.

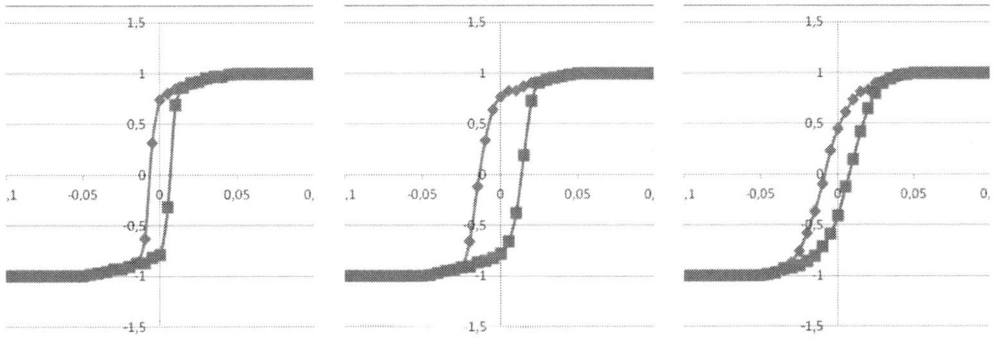

Figure 3. Hysteresis loops, calculated for a probability of "wrong" state $P_{wm} = 0.1$ and a weight of the switching energy $w_{cr} = 0.01$ (left panel), 0.1 (middle panel), and 0.5 (right panel).

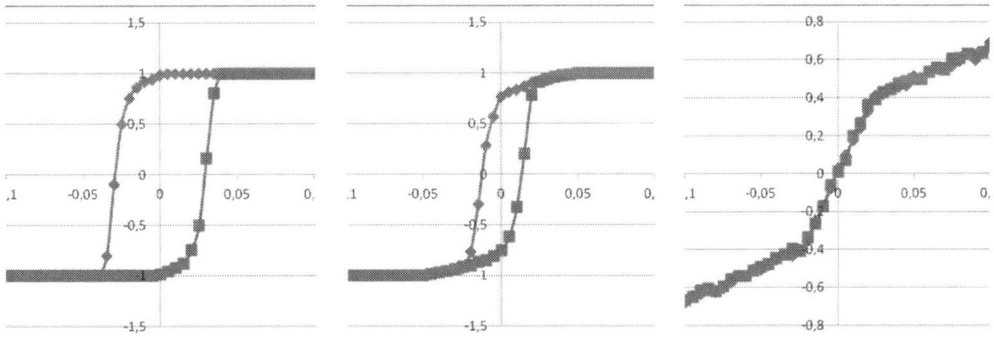

Figure 4. Hysteresis loops, calculated for a weight of the switching energy $w_{cr} = 0.1$ and a probability of "wrong" state P_{wm} of 0.01 (left panel), 0.1 (middle panel), and 0.5 (right panel, paramagnetic).

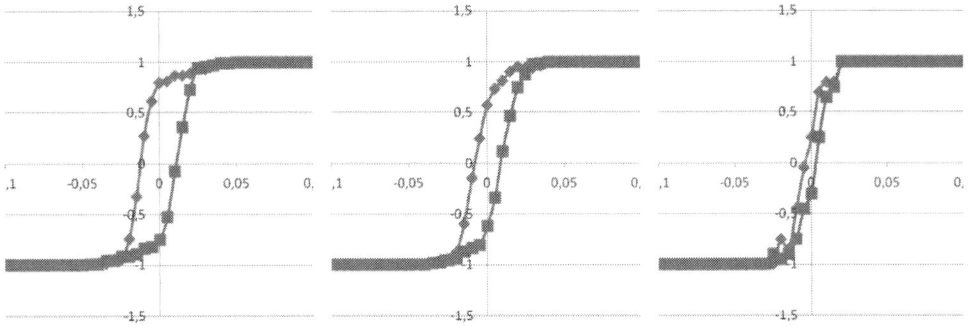

Figure 5. Hysteresis loops, calculated for a weight of the switching energy = 0.1 and a probability of "wrong" state of 0.1, for system dimensions of 4 x 16 cells (left panel), 2 x 16 cells (middle panel), and 1 x 4 cells (right panel).

Finally, Fig. 5 shows the influence of a more advanced change. Here, the system dimensions have been reduced, from 16 x 16 cells, as in the original program version, to significantly smaller dimensions. It should be mentioned here that, due to the circular reference which is necessary to introduce the switching energy, deleting columns or rows will always lead to errors in the neighboring cells of the particles. To overcome this problem, the desired number of cells should be copied from one of the unchanged particles into the changed one with the broken references. Smaller particles show, on the one hand, stronger deviations from "smooth" loops. Recalculation can always be used to identify if any unexpected loop shapes are real or just arbitrary, due to the small number of cells involved in the simulation. On the other hand, the hysteresis loops show smaller coercive fields for smaller particles – which is equivalent to real magnetic systems, in which the so-called super-paramagnetic limit exists for very small particles, which lose their ferromagnetic behavior below a certain number of magnetic moments per particle.

4. Further Developments

Besides the simple basic examples shown above, several other parameters can be tested by the students, since the program is completely open. For getting clearer hysteresis loops with more points on the sides of the loops, the macros can easily be changed from the recent version with equidistant field values to varied field values by simply exchanging the respective values (twice per block). In this way, it is possible to test the influence of the field sweeping speed on the resulting hysteresis loop.
Besides changing the dimensions of square or rectangular particles, it is also possible to test different forms, such as circles, donuts, triangles etc. or quite long, thin particles, e.g. to simulate hysteresis in yarn tension measurements.
While the recent program version offers the possibilities to simulate magnetic particles such as ferromagnets, paramagnets and antiferromagnets, a ferrimagnet could be created by, e.g., exchanging +1 by +2 and leaving the -1 unchanged. In the same way, non-magnetic hysteresis loops with different behavior can be defined.
A more sophisticated further development introduces additional layers, influencing their direct neighbors in out-of-plane direction and summing the magnetization over all layers. Another further development is the introduction of two additional possible states, by defining vectors (+1,+1), (+1,-1) etc., as often seen in magnetic systems with fourfold anisotropies or as to simulate multi-axial stress-strain measurements. Additionally, it is possible to allow for the magnetic moments to be oriented in each in-plane direction at random angles, as in the Stoner-Wohlfarth model [11].
As can be seen, the program offers a broad variety of simple and also more sophisticated possibilities to change parameters or calculation methods due to the students' own ideas.

5. Conclusion

An Ising particle with variable parameters has been built using Microsoft Excel® and a simple macro. While the step-by-step mode supports understanding of the parameters' influence and getting a feel for the changes in each step, the macro mode results in complete hysteresis loops.
As we have shown, the program can be developed further by easier or more sophisticated changes and additions, enabling the building of a tool which is capable of answering real physical questions in the broad field of magnetism as well as in other scientific areas, in which similar hysteresis loops occur.

6. References

[1] Bouc R 1967 *Proceedings of the Fourth Conference on Nonlinear Oscillation* (Prague, Czechoslovakia) p. 315
[2] Wen Y K 1976 *Journal of Engineering Mechanics* **102** 249-63.
[3] Mayergoyz I D 2003 *Mathematical Models of Hysteresis and their Applications: Second Edition (Electromagnetism)*. Academic Press. ISBN 978-0-12-480873-7.
[4] Beckmann B, Nowak U and Usadel K D 2003 *Phys. Rev. Lett.* **91** 187201
[5] Potter Jr. B G, Tikare V and Tuttle B A 2000 *J. Appl. Phys.* **87** 4415
[6] Gor G Yu, Rasmussen C J and Neimark A V 2012 *Langmuir* **28** 12100-7
[7] Sarkisov L and Monson P A *Langmuir* **16** 9857-60
[8] Ising E 1925 *Zeitschrift für Physik* **31** 253-8
[9] Goldsztein G H, Broner F and Strogatz S H 1997 *SIAM J. Appl. Math.* **57** 1163-85
[10] Acharyya M 1998 *Physica A: Statistical Mechanics and its Applications* **253** 199
[11] Stoner E C and Wohlfarth E P 1948 *Philosophical Transactions of the Royal Society A: Physical, Mathematical and Engineering Sciences* **240** 599–642

Shubnikov state in complex geometry Type-II Technetium device

J Barba-Ortega[1], J D Gonzalez[2], J L Aguilar [2]

[1] Departamento de Física, Universidad Nacional de Colombia, Bogotá, Colombia
[2] Grupo en Teoría de la Materia Condensada, Universidad del Magdalena, Santa Marta, Colombia

E-mail: jjbarbao@unal.edu.co

Abstract. We investigate numerically the flux quantum configurations and some thermodynamic properties of a superconducting Technetium film by using the link variables technique for one shape of circular geometry. The Technetium exhibits superconductivity properties indicated by to an extrapolated critical magnetic field value of $H_c(T=0) = 1410$ Oe, a Ginzburg-Landau parameter of $\kappa = 0.92$, and a critical temperature $T_c = 7.86K$, being the magnetic behavior of this material characteristic of a type-II superconductor with a weak-coupling superconductor of the BCS type. The studied sample is a circular sector with angular width $\theta = 7\pi/4$ surrounded by a dielectric material and submitted to external magnetic field applied perpendicular to its plane. We evaluate the magnetic moment density, Shubnikov state and free Gibbs energy as a function of the external magnetic field.

1. Introduction

The short-lived gamma ray-emitting nuclear isomer technetium is used in nuclear medicine for a wide variety of diagnostic tests. Technetium is used as a gamma ray-free source of beta particles. Long-lived technetium isotopes produced commercially are by products of fission of uranium-235 in nuclear reactors and are extracted from nuclear fuel rods. Experimental works reported the normal and superconducting-state parameters for technetium from heat-capacity measurements and showed that the thermodynamic properties of the superconducting state indicate that this element is an intermediate-coupling superconductor, also reported the superconducting transition temperatures of Molybdenum-Technetium, noting a similarity of Technetium and Rhenium with respect to alloy and intermetallic compound formation alloys [1, 2, 3, 4]. Several authors studied the magnetic and superconducting properties in mesoscopic bi dimensional and tri-dimensional samples, e.g. disk, cubes, circular sector and spheres; they found different features in the magnetization and the creation of a surface layer at higher magnetic fields that increase the number of vortices in the sample [5, 6]. Sardella *et. al* studied the superconducting state in a circular sector, founding that, when the superconducting area decrease, the nucleation field increases and the value of the first nucleation field is approximately the same independently of the form of the circular sector [7, 8]. Using the Ginzburg-Landau theory, the authors of this paper studied the superconducting matter in a semicircular film with general boundary conditions, they presented evidences that, with appropriate superconductor-superconductor interface, the superconductivity is strongly enhance or suppressed [9, 10]. In this work we studied the magnetic moment density, Shubnikov state and free Gibbs energy

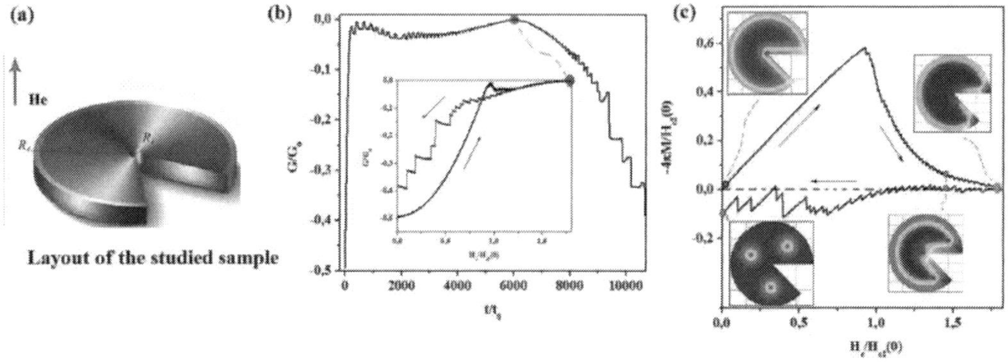

Figure 1. (a) Layout of the studied sample, (b) Free Gibbs energy loop as a function of the time t and external magnetic field H_e (insets) and (c) Magnetization loop as a function of H_e, contour plot of the order parameter (insets).

as a function of the external magnetic field for a Technetium circular sector with angular width $\theta = 7\pi/4$ surrounded by a dielectric material. We show that in our sample new and interesting vortex configurations are possible due to the competing interactions of the vortex-vortex repulsion and the geometry of the sample.

2. Theoretical Formalism

Let a thin perforated superconducting circular sector angular width $\theta = 7\pi/4$, external and internal radius $R_e = 8.2\xi(0)$, $R_i = 0.2\xi(0)$ respectively, $\kappa = 0.92$ at $T = 0$ (see Fig. 1(a)). Largest unit cells lengths were $a_r = (R_e - R_i)/N_r$ and $a_\theta = 2\pi/N_\theta$, taking $N_r = N_\theta = 128$ as the number of cells in the r and θ directions respectively. The superconducting state properties are described in the time dependent Ginzburg-Landau theory by the order parameter ψ, where $|\psi|^2$ represents the electron superconducting density, and the potential vector $\mathbf{A}$, related to magnetic induction by $\mathbf{B} = \nabla \times \mathbf{A}$ in the zero electrical potential gauge [11, 12, 13]:

$$\frac{\partial \psi}{\partial t} = -(i\boldsymbol{\nabla} + \mathbf{A})^2\psi + \psi - \psi|\psi|^2 \quad (1)$$

$$\frac{\partial \mathbf{A}}{\partial t} = J_s - \kappa^2 \boldsymbol{\nabla} \times \mathbf{B} \quad (2)$$

In Eqs. (1) and (2) dimensionless units were introduced as follows: $|\psi|$ is the order parameter in units of $\psi_\infty(0) = \sqrt{-\alpha(0)/\beta}$, where $\alpha(0)$ and β are two phenomenological constants; lengths in units of the coherence length $\xi(0)$; time in units of $t_0 = \pi\hbar/8K_BT_c; t_0 \approx 10^{-13}$; $\mathbf{A}$ in units of $H_{c2}(0)\xi(0)$, where $H_{c2}(0)$ is the second critical field, the supercurrent $J_s = \mathrm{Re}\left[\bar{\psi}(-i\boldsymbol{\nabla} - \mathbf{A})\psi\right]$ in $J_0 = \hbar c^2/8\pi e\xi$ units, the free Gibbs energy in units of $G_0 = H_{c2}^2 V/8\pi$, temperature in units of the critical temperature T_c [14]. The Eqs. (1) and (2) are supplemented by the superconducting/dielectric boundary condition $\mathbf{n} \cdot (-i\boldsymbol{\nabla} - \mathbf{A})\psi = 0$, where the current density normal to sample vanish at the interface.

3. Results and Discussion

The Figs. 1 (b-c) show the free Gibbs energy and magnetization curves as function of the external applied magnetic field H_e and time t. It is seen from these figures that the lower and upper critical fields $H_1 = 0.924$ and $H_2 = 1.800$ respectively, correspond at the times $t_{H_1} = 6000$ and $t_{H_2} = 12000$. The decrease in the Gibbs free energy in the downward branch of

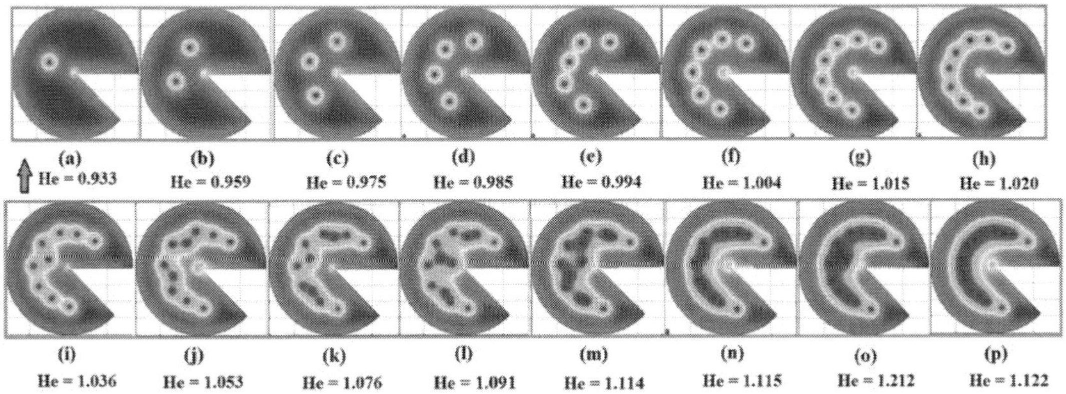

Figure 2. Spatial pattern of the order parameter $|\psi|^2$ in the upward branch of the magnetic field. Green arrows indicate if the magnetic field is increasing (up) or decreasing (down), (blue and red regions represent values of the modulus of the order parameter from 0 to 1).

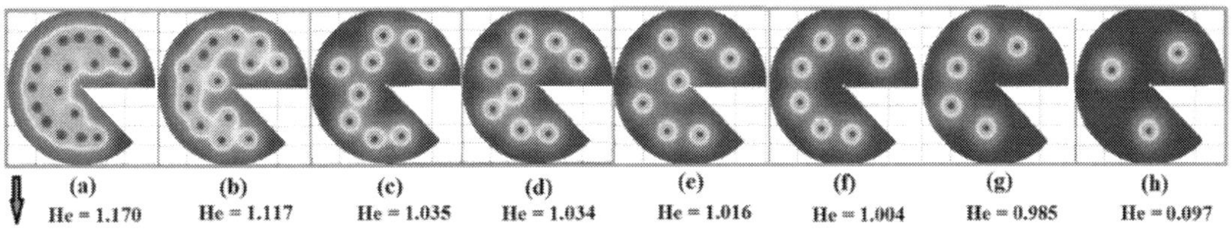

Figure 3. Spatial pattern of the order parameter $|\psi|^2$ in the downward branch of the magnetic field. Green arrow indicate if the magnetic field is increasing (up) or decreasing (down), (blue and red regions represent values of the modulus of the order parameter from 0 to 1).

the magnetic field is related to the fact that there are less vortices inside the sample. The jump in the magnetization and free energy could be correlated to structural transitions in the sample where the configuration of vortices is altered, is clear that these jumps signalling that one or more vortices have entered or exited the sample. The free energy increases with H_e until that the first critical field H_1 is reached, at this point the first vortex penetrates the sector at the center; so, is energetically more favorable for all vortices to nucleate at the center of the sample. In the insets of the Fig. 1 (c) is represented the square of the order parameter $|\psi|^2$ at the indicated magnetic fields for several vorticities N, (blue and red regions represent values of the modulus of the order parameter from 0 to 1). At $H_e = 0.5$ increasing the magnetic field the superconducting order parameter vanishes at the surface due to an imminent first vortex entrance; at $H_e = 1.45$, $N = 33$ vortices are present in the sample in both scenarios, increasing and decreasing the magnetic field, in this figure we can appreciate that the superconductivity remains in a region at the surface of the sample; finally at $H_e = 1.790$ closer to H_2, the superconducting region obey a geometry of the defect but the corner of the sample remains in the superconducting state. In the Fig. 2 we show the spatial pattern of the order parameter $|\psi|^2$ for a sector circular in the upward branch (downward branch, Fig. 3) of the magnetic field. Following the panels from the left to the right and from the top to the bottom, in this order, we can see that initially we have a single vortex state $N = 1$ at $H_e = 0.933$. The vortex transitions occurs at different values of H_e following a $N \rightarrow N+1$ dynamics in the upward branch of the magnetic field until $N = 43$ at H_2. Meanwhile, the order parameter phase $\Delta\varphi$ allows to determine the number of vortices

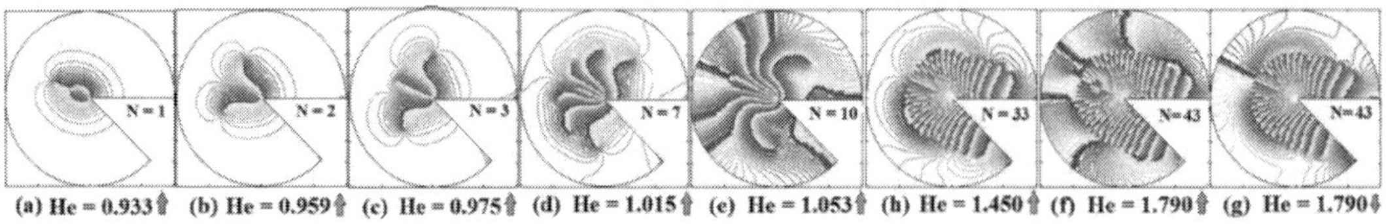

Figure 4. Plot of the order parameter phase $\Delta\varphi$ at the indicated magnetic field and vorticity N. (Green arrows indicate if the magnetic field is increasing (up) or decreasing (down), red and yellow regions represent values of the order parameter phase from 0 to 2π.

in a given region. If the vorticity in this region is N, then the phase changes by $\Delta\varphi = 2\pi N$ (red and yellow regions represent values of the order parameter phase from 0 to 2π) (See Fig. 4). When the magnetic field decrease occur a $N \to N-1$ vortex transition until $N = 15$ at $H_e = 0.556$. Now, under this magnetic field a new vortex transition take place, $N \to N-2$ until $N = 3$ at zero magnetic field. When the magnetic field is decreasing, we can see that the superconducting order parameter do not vanish at the surface, which is a direct consequence of the superconducting/dielectric interface.

4. Conclusions

In summary we solved the non lineal time dependent Ginzburg Landau equations for a superconducting Technetium film using the link variable method. We found interesting and novel vortex configurations in this sample. The nucleation field, superconducting/normal transition field, order parameter phase, magnetization and free Gibbs energy have been evaluated. We found a $N \to N+1$ ($N \to N-2$) vortex transition in the upward (downward) branch of the magnetic field, with $N = 3$ vortices trapped in the sample at $H_e = 0.0$. Finally near to H_2 the sample reaches the normal state except in the peaks or corners of the circular sector.

5. Acknowledgements

The authors would like to thank Edson Sardella for their very useful discussions. This work was partially financed by the Universidad del Magdalena (Fonciencias) and the Colombian Agency Colciencias through doctoral scholarships 567.

References

[1] R. J. Trainor and M. B. Brodsky, Phys. Rev. **B** 12, 4867 (1975).
[2] V. B. Compton, E. Corenzwit, J. P. Maita, B. T. Matthias, and F. J. Morin, Phys. Rev. **123**, 1567 (1967).
[3] X. Q. Hong and J. E. Hirsch, Phys. Rev. **B** 46, 14702 (1992).
[4] S. T. Sekula, R. H. Kernohan, and G. R. Love, Phys. Rev. **155**, 364 (1967).
[5] M. M. Doria, Antonio R. de C. Romaguera, and F. M. Peeters, Phys. Rev. **B** 75, 064505 (2007).
[6] B. Xu, M. V. Milosevic, and F. M. Peeters, Phys. Rev. **B** 77, 144509 (2008).
[7] E. Sardella, P. Lisboa-Filho, and A. Malvezzi, Phys. Rev. **B** 77, 104508 (2008).
[8] T. S. Alstrm, M. P. Srensen, N. F. Pedersen, S. Madsen, Acta. Appl. Math. **115** 63 (2011).
[9] J. Barba-Ortega, M. R. Joya and J. Albino Aguiar, J. Supercond. Nov. Magn. **26** 2253 (2013).
[10] J. Barba-Ortega, Edson Sardella, J. Albino Aguiar, Supercond. Sci. Technol. **24** 015001 (2011).
[11] W. D. Gropp, H. G. Kaper, G. K. Leaf, D. M. Levine, M. Palumbo, and V. M. Vinokur, J. Comput. Phys. **123**, 254 (1996).
[12] G. C. Buscaglia, C. Bolech and A. Lopez, *Connectivity and Superconductivity*, J. Berger, and J. Rubinstein Eds, Springer, (2000).
[13] M. V. Milosevic, R. Geurts, Physica C **470** 791 (2010).
[14] M. Tinkham, *Introduction to Superconductivity*, McGraw-Hill, New York, (1996).

Numerical Investigation of Grain Coarsening and Coalescence Model

Aliki D Muradova[1,2] and Dionisios T Hristopulos[2]

[1] Geostatistics Laboratory, School of Mineral Resources Engineering, Technical University of Crete, Chania, GR-73100, Greece

[2] Institute of Computational Mechanics and Optimization, School of Production Engineering and Management, Technical University of Crete, Chania, GR-73100, Greece

E-mail: aliki@mred.tuc.gr, dionisi@mred.tuc.gr

Abstract. A kinetic nonlinear model of mass transfer, grain coarsening and coalescence with potential applications in sintering processes is studied. The model involves nonlinear differential equations that determine the transport of mass between grains. The rate of mass transfer is controlled by the activation energy (an Arrhenius factor) leading to a nonlinear model of mass transfer and grain coarsening. The resulting dynamical system of coupled nonlinear differential equations with random initial conditions (i.e., initial grain mass configuration) is solved by means of the Runge-Kutta method. An analysis of the fixed points of the two-grain system is carried out, and the solution of the multi-grain system is studied. We incorporate coalescence of smaller grains with larger neighbors using a cellular automaton step in the evolution of the system.

1. Introduction

Many technological materials, including ceramics, are produced by means of nonequilibrium physical processes that generate phase changes. A typical example is sintering, a process during which a powder is transformed into a monolithic material through atomic diffusion mechanisms. The modeling of sintering is a long-standing problem which remains poorly understood to date. Early studies of the kinetics of phase change have focused on interactions between grains that produce a crystal aggregate [1, 2]. Recent approaches use the Discrete Element Method (DEM) to relax assumptions regarding the particle kinematics [4]. In the above, grain coarsening is accounted by considering the overlapping volume of neighboring spherical particles, increasing the radius of the larger particle while decreasing the radius of the smaller by the same amount.

Here we investigate numerically the kinetic, nonlinear model of mass transfer, grain coarsening, and coalescence introduced in [3]. The model comprises a system of nonlinear ordinary differential equations (ODEs) and incorporates local fluctuations in mass transfer rates due to variations in the degree of amorphization of the grains. The initial conditions for the system are the grain masses which are considered to be randomly distributed. The initial value problem is solved by means of the fourth order Runge-Kutta method. The ultimate goal is to use the model in order to predict changes in the grain size distribution as a function of key sintering parameters such as time and temperature, as well as the powder grain size distribution and the non-homogeneity of the transfer rates.

2. Nonlinear kinetic model

A collection of spherical grains such that the surface of each grain (particle) is in contact with at least one other particle is considered. We study a one-dimensional geometry in which N grains are assumed to be located along a line (the x-axis), and each grain has two neighbors, one on each side, except for the two boundary grains. The grains are indexed by i ($i = 1, 2, \dots, N$). Taking into account that the first grain ($i = 1$) does not have a left neighbor and the last grain ($i = N$) does not have a right neighbor, we rewrite the original model ([3]) as the following system of ODEs:

$$\dot{m}_1(t) = e^{-um_2(t)}m_2(t) - e^{-um_1(t)}m_1(t), \tag{1}$$

$$\dot{m}_i(t) = e^{-um_{i-1}(t)}m_{i-1}(t) + e^{-um_{i+1}(t)}m_{i+1}(t) - 2e^{-um_i(t)}m_i(t), \quad i = 2, 3, \dots, N-1, \tag{2}$$

$$\dot{m}_N(t) = e^{-um_{N-1}(t)}m_{N-1}(t) - e^{-um_N(t)}m_N(t). \tag{3}$$

In the above, $m_i(t)$ is the mass of i-th grain, t is time, N is the number of grains and u represents the normalized grain activation energy. For the mass of i-th grain we suppose $m_i(t) = l_d r_i^d(t)\rho$, where $r_i(t)$ is the radius of i-th grain and ρ is the mass density of the grains (assumed to be constant); for spherical grains $d = 3$ and $l_d = 4\pi/3$.

Equations (1) and (3) correspond to the mass transfer process for the boundary grains ($i = 1, N$). Equations (2) describe the mass transfer process for the grains in the bulk. We assume that the initial grain masses, i.e., the initial conditions for (1)-(3) are random numbers,

$$m_i(0) = m_{i0}, \quad i = 1, 2, \dots, N. \tag{4}$$

The initial conditions can be drawn from the normal, lognormal or some empirically determined probability distribution.

If we assume that the grains are arranged along a ring with cyclical boundary conditions so that $m_0(t) := m_N(t)$, $m_{N+1}(t) := m_1(t)$, then the original model (1)-(3) is written in the following simpler form, whereas the final result of the simulation is not significantly affected by the change,

$$\dot{m}_i(t) = e^{-um_{i-1}(t)}m_{i-1}(t) + e^{-um_{i+1}(t)}m_{i+1}(t) - 2e^{-um_i(t)}m_i(t), \quad i = 1, 2, \dots, N. \tag{5}$$

3. Analysis of the two-grain system

In this section we focus on the case $N = 2$ and present the results of numerical analysis, which will be helpful for investigating systems with a large number of grains. The initial masses are considered to take values at the nodes of a square lattice $[a, b] \times [a, b]$. According to the numerical experiments, a *diffusive regime* was observed for $a \leq 1$, $b \leq 1$: the two masses converged to equal values for all $u \leq 1$. These results were verified with lattice steps $h = 0.1, 0.02, 0.005$. In general, smaller u values lead to faster convergence. The *mass growth regime*, i.e., the accumulation of mass on the larger grains and reduction of the smaller, is obtained at large values of u. The critical value of u depends on the initial masses. For example, if $u = 0.1$ the mass growing regime sets is at $m_1(0) \approx 8.5$ and $m_2 \approx 12$ (and vice versa); $u = 1$ requires $m_1(0) \approx 0.85$ and $m_2 \approx 1.2$ and so on. In general, higher values of u imply that smaller initial mass values lead to the mass growth regime. However, for large values, e.g. approximately $u\,m_1(0)$, $u\,m_2(0) > 10$ the initial masses change extremely slowly.

For every initial distribution of grain sizes there exists a critical u (activation energy) that leads to mass growth. For some particular situations the masses change very slowly and a many time steps will be needed for convergence to a final state. By virtue of the numerical results and in light of the equations (1)-(3) (or for the system of equations (5)) it follows that very small or very large values of u lead to *almost stationary processes*; for small u the grains tend to equal mass distribution, whereas for large u the masses are practically "*trapped*" at their initial values.

4. Computational algorithm for grain mass transfer

The kinetic model (1)-(3) with the initial conditions (4) is discretized using the fourth order Runge-Kutta method (RK4). An analogous scheme is obtained for the system of equations (5) with the initial conditions (4). For the numerical implementation of the RK4 scheme we use the Matlab ordinary differential equation toolbox, function (*ode45 solver*). The statistical analysis of the resulting grain distributions is based on an ensemble of simulated states.

Below we describe an algorithm for solving the kinetic model including the possibility for grain coalescence.

Grain mass transfer simulation algorithm

(i) Set the simulation counter to one.

(ii) Choose an initial configuration of grain masses from a specified probability distribution (e.g., normal, lognormal or empirical distributions).

(iii) Solve (1)-(3), or equivalently (5), with the initial conditions (4) by means of a suitable numerical method (e.g. RK4).

(iv) Check the neighboring mass ratios and implement grain coalescence if needed. Namely, if $m_i/m_{i+1} < \delta$, where $0 < \delta < 1$ is the coalescence parameter, then the $(i+1)$-th grain becomes $m_{i+1} + m_i$ whereas the i-th grain is removed. If $m_i/m_{i+1} > \delta$ the i-th grain accumulates the mass of both grains whereas the $(i+1)$-th grain is removed.

(v) Reduce the number of the equations in the system (1)-(3), or equivalently (5), by the number of grains removed according to the rules in step (iv).

(vi) Check if the updated number of grains is greater than 1: If it is, go to the next step; otherwise, the simulation terminates.

(vii) Check the number of iterations: If it does not exceed the specified limit go to the next step; otherwise, the simulation terminates.

(viii) Solve the system of equations with the updated number of equations (grains).

(ix) Repeat steps (iv)-(viii) until the simulation stops. Increase the simulation counter by one.

(x) If the value of the simulation counter is less than the specified number of simulations go to step (ii); otherwise stop and calculate statistics.

5. Numerical simulation of multi-grain system

In this section we present numerical simulation results for the equation system (1)-(3), (4) for $N = 10\,000$ grains, different values of u and δ. We have verified that the numerical solution of the system conserves the total grain mass.

Example 1. Figure 1 shows the histogram of the initial masses which are drawn from the normal (Gaussian) distribution with mean $\mu = 10$ and standard deviation $\sigma = 2$. The maximum number of time steps used is 5 000, and the size of the time step is $\tau = 0.02$.

Figure 2 demonstrates the final mass distributions without coalescence. The plots show that small values of u lead to a diffusive regime which implies convergence of all the masses to the mean of the initial distribution (e.g., see plot 1 with ($u = 0.05$)). The next higher values of u lead to growth of some grains at the expense of other grains becoming smaller (Plots 2-6 with $u = [0.08, 0.1, 0.3, 0.6, 1]$). Finally, higher values of u lead to a quasi-static system in which the "final" state does not differ significantly from the initial histogram (Plots 7-9).

Figure 3 demonstrates the "final" mass distributions with coalescence ($\delta = 0.1$) after 5 000 steps. The number of remaining grains and the end of the procedure is $N_{\text{final}} = [10\,000, 21, 1, 3\,166, 6\,036, 9\,375, 9\,995, 9\,998, 9\,999]$. The plots of Figure 3 exhibit mass growth for $u = [0.08, 0.1, 0.3, 0.6, 1]$. For $u = 0.05$ the diffusive final state is observed, whereas for $u = 2, 3, 4$ the final masses are essentially unchanged with respect to the initial conditions.

Journal of Physics: Conference Series **574** (2015) 012160 doi:10.1088/1742-6596/574/1/012160

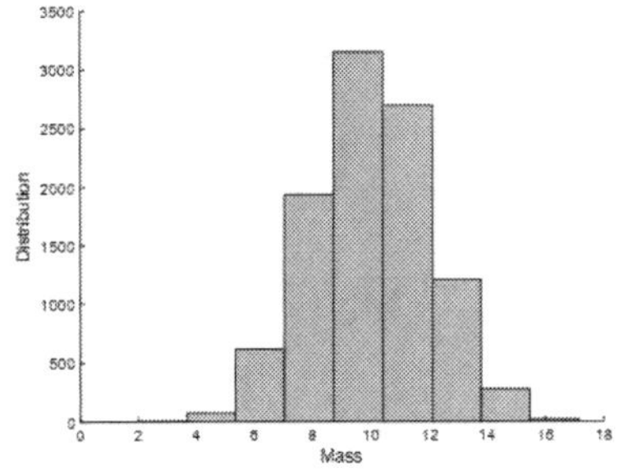

Figure 1. Histogram for initial configuration of grain masses for a system comprising $N = 10\,000$ grains drawn from the $N(0, 2)$ distribution.

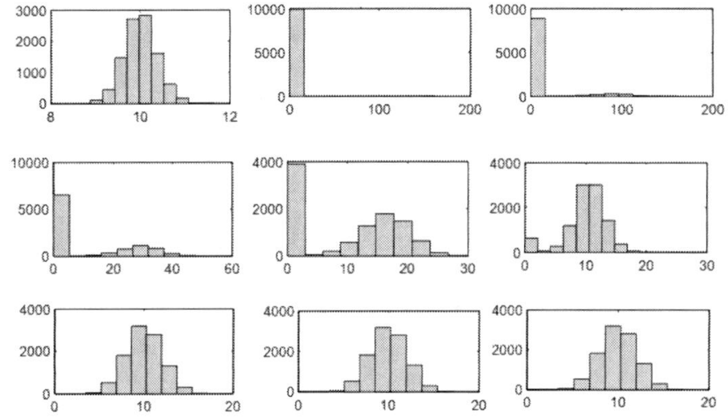

Figure 2. Mass distribution of system comprising $N = 10\,000$ grains without coalescence after 5 000 steps. The values of the activation energy are $u = [0.05, 0.08, 0.1, 0.3, 0.6, 1, 2, 3, 4]$.

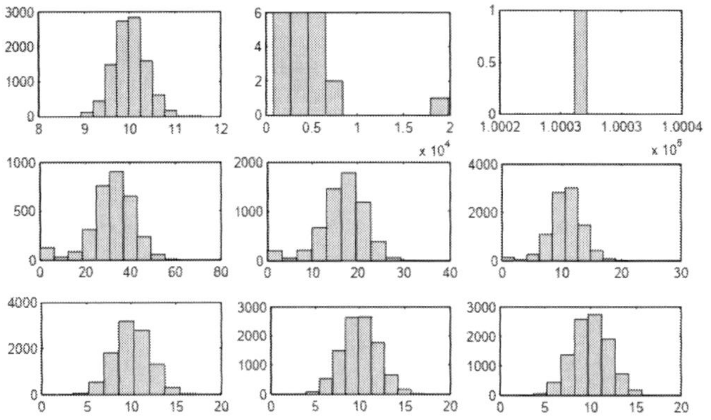

Figure 3. Mass distribution of system comprising $N = 10\,000$ grains with coalescence parameter $\delta = 0.01$ after 5 000 steps. The values of the activation energy are $u = [0.05, 0.08, 0.1, 0.3, 0.6, 1, 2, 3, 4]$.

Example 2. Here we investigate the system with different values of the ratio δ. Figure 4 shows the mean value of the grain radius at the final time (one check after 1000 time steps).

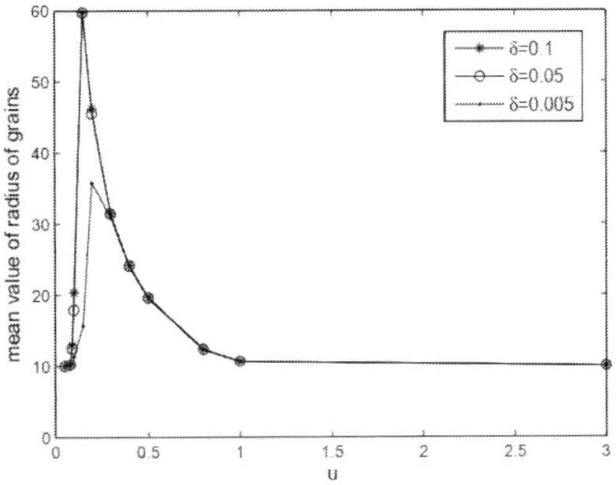

Figure 4. The dynamics of changing of the mean value of the radius of grains for $u = 0.05, 0.08, 0.09, 0.1, 0.15, 0.2, 0.3, 0.4, 0.5, 0.8, 1, 3$ with $\delta = \{0.1, 0.05, 0.005\}$ and the initial normal distribution ($\mu = 10,\ \sigma = 2$) after 5000 time steps.

6. Conclusions

Numerical simulations of a system of coupled, nonlinear ordinary differential equations that describe mass transfer between grains of inhomogeneous sizes have been presented. Three different types of behavior have been identified: a diffusive regime in which all the grains tend to the same mass in equilibrium, a growth-decay regime in which the bigger grains grow whereas the smaller ones lose mass, and a trapping regime in which the initial mass distribution practically does not change. The parameters that control these regimes involve the activation energy, the coalescence ratio and the initial mass distribution of the grains.

Acknowledgments

Acknowledgments This work has been funded by the project *NAMCO: Development of High Performance Alumina Matrix Nanostructured Composites*. NAMCO is implemented under the "THALIS" Action of the operational programme "Education and Lifelong Learning" and is co-funded by the European Social Fund and National Resources. The authors would like also to thank Prof. Athena Tsetsekou (National Technical University of Athens) for helpful discussions.

References

[1] Avrami M 1939 *J. Chem. Phys.* **7** 1103–1112
[2] Avrami M 1941 *J. Chem. Phys.* **9** 177–184
[3] Hristopulos D T, Leonidakis L and Tsetsekou A 2006 *Eur. Phys. J.* B **50** 83–87
[4] Martin C I, Schneider L C R, Olmos L and Bouvard D 2006 *Scripta Materialia* **55** 425-428

Mathematical modeling of a rigid circular membrane in a tri-material transversely isotropic full-space undergoing a prescribed displacement

L Mohtati, M Rahimian and M Eskandari-Ghadi
School of Civil Engineering, College of Engineering, University of Tehran, P. O. Box 11165-4563, Tehran, Iran.

E-mail: rahimian@ut.ac.ir

Abstract. A horizontal layer between two transversely isotropic half-spaces forms a tri-material full-space, involved in this paper. A mathematical formulation is presented to determine the response of a rigid circular membrane, which is laid down at an interface of the tri-material transversely isotropic full-space and is considered to be under a prescribed horizontal displacement. The governing equations are expressed in the cylindrical coordinate system. With the aid of a system of two scalar potential functions, the governing equations of motion can be uncoupled into two separated partial differential equations, which may be transformed to some ordinary differential equations by applying the Hankel integral transforms in the radial direction and Fourier series along the angular coordinate. After determining the unknown functions by imposing the relaxed boundary conditions, they are transformed to a set of four coupled integral equations, which are reduced to two coupled Fredholm-Volterra integral equations of the second kind, from which both displacement and the stress fields are computed. To confirm the accuracy of the numerical evaluation of the integrals involved, the numerical results are compared with the solutions exists for a transversely isotropic half-space. In order to investigate the degree of material anisotropy, some numerical evaluations are given for different combinations of transversely isotropic region.

1. Introduction

As da Vinci said, mechanics is the paradise of the mathematical sciences because by means of it one comes to the fruits of mathematics. Mechanics of an elastic medium undergoing force or displacement boundary conditions either applied on the surface or intrinsically, has been an interesting mathematical problem in the context of elasticity theory. There are some research works in this area for both static and dynamic cases [1-4]. Eskandari-Ghadi [5] introduced a complete set of scalar potential functions. With the use of these potential function, Moghaddasi et al.[2] have studied interaction of rigid disc in transversely isotropic half-space. In this study, a mathematical model describing a rigid membrane embedded in a tri-material transversely isotropic full space undergoing a prescribed displacement is presented. The aforementioned model is applicable to different practical examples including the interaction of foundations and the supporting soil.

2. Boundary-value problem and its solution

A tri-material full-space includes a horizontal layer between two different transversely isotropic half-spaces. Consider a cylindrical coordinate system attached to the domain of the boundary value problem, see figure 1. The axes of symmetries of all materials are assumed to be parallel to each other and normal to the interface of the middle layer and surrounding half-spaces. A rigid circular membrane located at the an interface of the tri-material transversely isotropic full-space is considered and the coordinate system oxyz is attached at the centre of the disc as shown in figure 1. The rigid membrane is undergoing a prescribed translation Δ in the horizontal plane in x-direction. The tri-material full-space can be defined as the upper half-space, the middle layer and the lower half-space denoted as Region I ($z \le -h$), Region II ($0 \le z \le -h$) and Region III ($z \ge 0$), respectively. A relaxed treatment of the mixed boundary-value problem and continuity conditions can be stated in terms of the components of the Cauchy stress tensor σ_{ij} and the displacement vector u_i, $(i, j = r, \theta, z)$ as follows:

$$u_r(r,\theta,0) = \Delta\cos\theta, \qquad u_\theta(r,\theta,0) = -\Delta\sin\theta, \tag{1}$$

$$\sigma_{zr}(r,\theta,0^-) - \sigma_{zr}(r,\theta,0^+) = P(r,\theta;0), \qquad \sigma_{z\theta}(r,\theta,0^-) - \sigma_{z\theta}(r,\theta,0^+) = Q(r,\theta;0), \tag{2}$$

For $r < a$ and $0 \le \theta \le 2\pi$, and

$$\sigma_{zr}(r,\theta,0^-) - \sigma_{zr}(r,\theta,0^+) = 0, \qquad \sigma_{z\theta}(r,\theta,0^-) - \sigma_{z\theta}(r,\theta,0^+) = 0, \tag{3}$$

For $r > a$ and $0 \le \theta \le 2\pi$, and

$$u_z(r,\theta,0^-) = u_z(r,\theta,0^+), \qquad u_r(r,\theta,0^-) = u_r(r,\theta,0^-), \qquad u_\theta(r,\theta,0^-) = u_\theta(r,\theta,0^-), \tag{4}$$

$$u_z(r,\theta,-h^-) = u_z(r,\theta,-h^+), \quad u_r(r,\theta,0-h^-) = u_r(r,\theta,-h^-), \quad u_\theta(r,\theta,-h^-) = u_\theta(r,\theta,-h^-), \tag{5}$$

$$\sigma_{zz}(r,\theta,0^-) - \sigma_{zz}(r,\theta,0^+) = 0, \qquad \sigma_{zr}(r,\theta,-h^-) - \sigma_{zr}(r,\theta,-h^+) = 0, \tag{6}$$

$$\sigma_{zz}(r,\theta,-h^-) - \sigma_{zz}(r,\theta,-h^+) = 0, \qquad \sigma_{z\theta}(r,\theta,-h^-) - \sigma_{z\theta}(r,\theta,-h^+) = 0, \tag{7}$$

For $r \ge 0$ and $0 \le \theta \le 2\pi$.

$P(r,\theta;0)$ and $Q(r,\theta;0)$ are the unknown tangential contact-load distribution acting on the disc in the radial and angular directions, respectively.

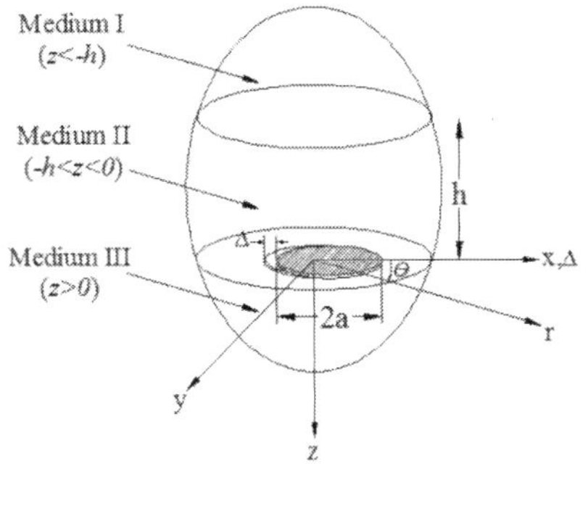

Figure 1. A rigid disc in a tri-material transversely isotropic full-space

Expressing the governing equations of motion in terms of displacement and in the absence of body forces and with the aid of the scalar potential functions F and χ by virtue of Fourier expansion with respect to the angular coordinate and utilizing the mth order Hankel with respect to the radial coordinates, the solutions for the potential functions in regions are [2,6]:

$$\tilde{F}_m^{Im}(\xi,z) = A_m^I(\xi)e^{\xi s_1^I z} + C_m^I(\xi)e^{\xi s_2^I z}, \qquad \tilde{\chi}_m^{Im}(\xi,z) = E_m^I(\xi)e^{\xi s_3^I z},$$

$$\tilde{F}_m^{Im}(\xi,z) = A_m^{II}(\xi)e^{\xi s_1^{II} z} + B_m^{II}(\xi)e^{-\xi s_1^{II} z} + C_m^{II}(\xi)e^{\xi s_2^{II} z} + D_m^{II}(\xi)e^{-\xi s_2^{II} z}, \quad \tilde{\chi}_m^{Im}(\xi,z) = E_m^{II}(\xi)e^{\xi s_3^{II} z} + F_m^{II}(\xi)e^{-\xi s_3^{II} z},$$

$$\tilde{F}_m^{Im}(\xi,z) = B_m^{III}(\xi)e^{-\xi s_1^{III} z} + D_m^{III}(\xi)e^{-\xi s_2^{III} z}, \qquad \tilde{\chi}_m^{Im}(\xi,z) = F_m^{III}(\xi)e^{-\xi s_3^{III} z}, \tag{8}$$

For half-space medium the regularity conditions have been considered and $s_0 = \sqrt{c_{66} / c_{44}}$, s_1 and s_2 are the roots of (9) which are not zero or pure imaginary numbers.

$$c_{33}c_{44}s^4+\left(c_{13}^2+2c_{13}c_{44}-c_{11}c_{33}\right)s^2+c_{11}c_{44}=0 \tag{9}$$

c_{ij} are the elasticity constants. Equations (2,4-7) form a system of twelve equations and twelve unknown coefficients $A_m^I,..,F_m^{III}$, which could be expressed in terms of the $X_m(\xi)=P_m^{m-1}(\xi)-iQ_m^{m+1}(\xi)$ and $Y_m(\xi)=P_m^{m-1}(\xi)+iQ_m^{m+1}(\xi)$. The boundary conditions imply that $X_1=Y_{-1}$, $X_{-1}=Y_1$, $X_m=Y_m=0$ for $m\neq 0$. With the aid of multiplying-factor method and Sonine integration introduced by Nobel [7], the remaining four conditions (1,3) result in four coupled integral equations as [2]

$$\int_0^\infty \xi^{-0.5}\left\{\left(1+H_1(\xi;h)\right)X(\xi)+\frac{l_2}{l_1}\left(1-H_2(\xi;h)\right)Y(\xi)\right\}J_{-\frac{1}{2}}(r\xi)d\xi=\frac{\Delta}{l_1}\sqrt{\frac{2}{\pi r}},\qquad r<a, \tag{10}$$

$$\int_0^\infty \xi^{-0.5}\left\{\left(1-H_2(\xi;h)\right)X(\xi)+\frac{l_1}{l_2}\left(1+H_1(\xi;h)\right)Y(\xi)\right\}J_{\frac{3}{2}}(r\xi)d\xi=0,\qquad r<a, \tag{11}$$

$$\int_0^\infty \xi^{-0.5}X(\xi)J_{-\frac{1}{2}}(r\xi)d\xi=0,\qquad r>a, \tag{12}$$

$$\int_0^\infty \xi^{-0.5}Y(\xi)J_{\frac{3}{2}}(r\xi)d\xi=0,\qquad r>a, \tag{13}$$

where $X(\xi)=X_1\xi$, $Y(\xi)=Y_1\xi$, $H_1(\xi;h)$, $H_2(\xi;h)$ are in terms of the coefficients $A_m^I,..,F_m^{III}$ and l_1, l_2 are the convergence factor of $H_1(\xi;h)$ and $H_2(\xi;h)$ respectively, which as ξ tends to infinity, $H_1(\xi;h)$ and $H_2(\xi;h)$ tends to zero.

Defining $\theta_A(r)=\sqrt{\pi r/2}\int_0^\infty \xi^{-0.5}X(\xi)J_{-\frac{1}{2}}(r\xi)d\xi$, $\theta_B(r)=\sqrt{\pi r/2}\int_0^\infty \xi^{-0.5}Y(\xi)J_{\frac{3}{2}}(r\xi)d\xi$ for $r<a$, these equations are reduced to two coupled Fredholm-Volterra integral equations of the second kind as (14-16) which can be numerically solved for $\theta_A(r)$ and $\theta_B(r)$.

$$\theta_A(r)+\frac{l_2}{l_1}\left[\int_r^a\frac{1}{\rho}\theta_B(\rho)d\rho-\theta_B(r)\right]+\int_0^a K_{11}(r,\rho;h)\theta_A(\rho)d\rho+\int_0^a K_{12}(r,\rho;h)\theta_B(\rho)d\rho=\frac{\Delta}{l_1}, \tag{14}$$

$$\theta_B(r)+\frac{l_2}{l_1}\left[\frac{1}{r}\int_0^r\theta_A(\rho)d\rho-\theta_A(r)\right]+\int_0^a K_{21}(r,\rho;h)\theta_A(\rho)d\rho+\int_0^a K_{22}(r,\rho;h)\theta_B(\rho)d\rho=0 \tag{15}$$

$$\begin{aligned}K_{ij}(r,\rho;h)&=\left((1-l_2/l_1)\delta_{ij}+l_2/l_1\right)\sqrt{r\rho}\times\\&\left[\int_0^\infty\xi\left(\left(H_1(\xi;h)+H_2(\xi;h)\right)\delta_{ij}-H_2(\xi;h)\right)J_{(-1)^i(i-\frac{1}{2})}(r\xi)J_{(-1)^j(j-\frac{1}{2})}(\rho\xi)d\xi\right],\qquad i,j=1,2,\end{aligned} \tag{16}$$

3. Numerical results

Because of complex kernel functions in the integrals involved in this paper, there is no existing analytical solution for the Fredholm-Volterra integral equations (14,15), so these integral equations can be solved numerically. For numerical evaluations, some materials are listed in table 1, and some different cases are defined in table 2. The case 1 is a homogenous half-space for embedded disc, which is considered as a benchmark. Depth-wise variations of normalized stress σ_{rz} and displacement U_{rr} at $r=0$ are illustrated in figure 2,3. All numerical results are dimensionless. The results are in an excellent agreement with the literature [2] and the effect of degree of anisotropy on the displacement U_{rr} and stress σ_{rz} is observed clearly in figure 2,3.

Table 1. Synthetic material engineering constants.

Mat	E(GPa)	E'(GPa)	G(GPa)	G'(GPa)	ν	ν'
1	50	50	20	10		
2	50	55	20	20	0.25	0.25
3	55	50	20	20	0.25	0.25
4	100	50	40	20	0.25	0.25
5	150	50	60	20	0.25	0.25

Table 2. Combination of materials for different cases.

Case	Region1	Region2	Region3
1	-	Mat4	Mat4
2	Mat1	Mat3	Mat5
3	Mat1	Mat3	Mat2

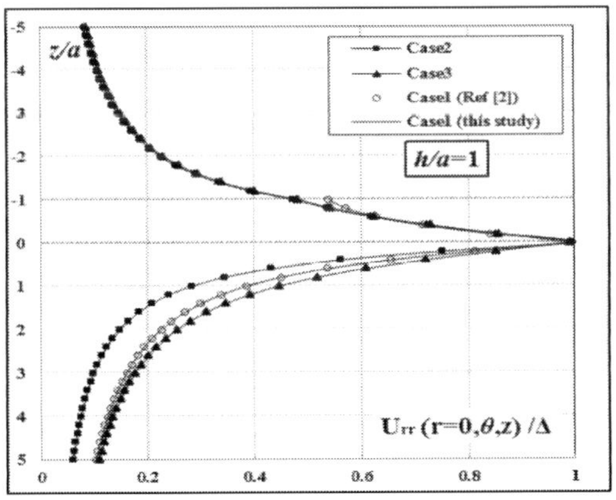

Figure 2. Depth-wise variation of normalized displacement U_{rr} at $r = 0$.

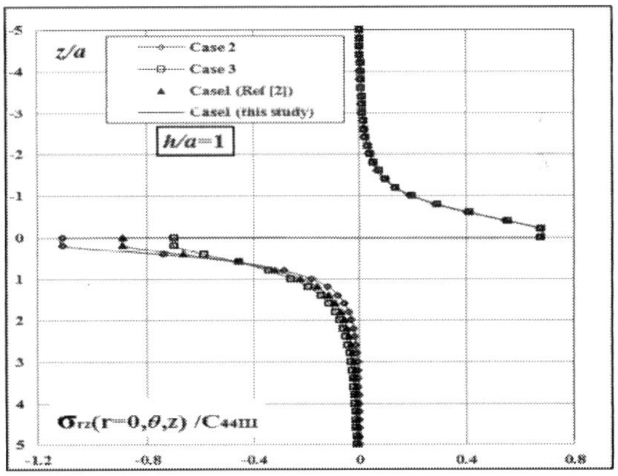

Figure 3. Depth-wise variation of normalized stress σ_{rz} at $r = 0$.

4. Conclusion

With the aid of Hankel integral transforms, Fourier series and a method of potentials the relaxed boundary-value problem is formulated as two coupled Fredholm-Volterra integral equations, which can be solved numerically. The numerical results match very well with the literature, which proves the accuracy of the numerical evaluation of the integrals involved. In order to investigate the degree of material anisotropy, some numerical evaluations are given for different combinations of transversely isotropic region.

Refrences

[1] Pak R Y S and Saphores J-D M 1992 *Q. J. Mech. Appl. Math.* **45** 435-449
[2] Moghaddasi H, Rahimian M, Khojasteh A, and Pak R Y S 2012 *Q. J. Mech. Appl. Math.* **65** 513-533
[3] Eskandari M, Shodja H M and Ahmadi S F. 2013 *Eur. J. of Mech A/Solids.* **39** 134-143
[4] Eskandari-Ghadi M, Nabizadeh M S and Ardeshir-Behrestaghi A. 2014 *Soil Dyn Earthq Eng.* **61** 135-139
[5] Eskandari-Ghadi M 2005 *J. Elast.* **81** 1-19
[6] Eskandari-Ghadi M, Sture S, Pak R Y S and Ardeshir-Behrestaghi A 2009 *Int. J. Solids Struct.* **46** 1121-33
[7] Noble B 1963 *Proc. Camb. Phil. Soc.* **59** 351-362

Journal of Physics: Conference Series **574** (2015) 012162 doi:10.1088/1742-6596/574/1/012162

Compton Operator in Quantum Electrodynamics

Hector Luna Garcia[1] **and Luz Maria Garcia**
Área de Física Atómica Molecular Aplicada, CBI, UAM-A, Av. San Pablo No. 180, Col. Reynosa, Azcapotzalco, México, D. F. 02200, México

E-mail: 5553972854.hl@gmail.com

Abstract. In the frame in the quantum electrodynamics exist four basic operators; the electron self-energy, vacuum polarization, vertex correction, and the Compton operator. The first three operators are very important by its relation with renormalized and Ward identity. However, the Compton operator has equal importance, but without divergence, and little attention has been given it. We have calculated the Compton operator and obtained the closed expression for it in the frame of dimensionally continuous integration and hypergeometric functions.

1. Introduction

This work considers the paper published in the first issue of the Revista Mexicana de Física [1], this paper considers the so-called divertex operator and which corresponds, in the space of time, to Feynman diagram shown in figure 1, where we have used the notation Lozano-Medina in the above mentioned work.

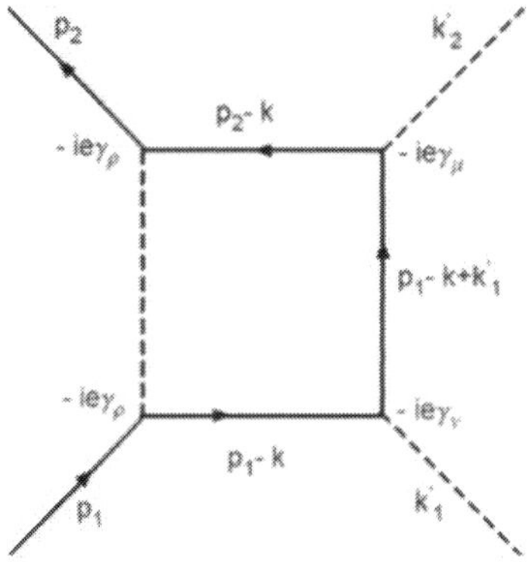

Figure 1. Feynman graph of the divertex operator.

The fundamental result that these authors reach, making the integration over k is

$$C_{\mu\nu} = \frac{-i\pi\alpha}{3(2\pi)^6}\int_0^1\int_0^1\int_0^1 u^2 v du dv dt \times \frac{1}{c^2}\left\{\Lambda^{(0)}_{\mu\nu} + \frac{1}{2c^2}\left[u\Lambda^{(1)}_{\mu\nu}\left(p_1^2+m^2\right)+\Lambda^{(1)}_{\mu\nu}\left(p_2^2+m^2\right)\right.\right.$$

[1] Hector Luna Garcia, lghm@correo.azc.uam.mx.

$$+(\not{p}_2+m)\Lambda^{(2)}_{\mu\nu}(\not{p}_1+m)+(\not{p}_2+m)\Lambda^{(3)}_{\mu\nu}+\Lambda^{(4)}_{\mu\nu}(\not{p}_1+m)+\Lambda^{(5)}_{\mu\nu}\Big]\Big]\Big\}, \tag{1}$$

where *u*, *v*, *t* are Feynman parameters and the meaning of c and Λ's are found in his article. As can be seen, remain to be determined parametric integrals, which is quite difficult because of its complexity.

In the present work, *ab-initio* avoids introducing the Feynman parameterization, even though it is implicit in the integrals over propagators no-massive, and is then employed: a) the method of integration dimensionally continuous of Giambiagi Bollini [2]; b) the recent application of Davydychev [3, 4] to obtain accurate results for Feynman integrals of *N* points, and to integrals of *N* points, but with different masses. The metric used is that of Bjorken and Drell [5].

2. Compton Operator

Applying Dyson rules in the metric mentioned and transferring the integration over d^4k to an integration on d^nk, leads to operator

$$C_{\mu\nu}=A_n\int d^nk\gamma_\rho\frac{\not{p}_2-\not{k}+m}{\left((p_2-k)^2-m^2\right)}\gamma_\mu\frac{\not{p}_1-\not{k}-\not{k}+m}{\left((p_1+k_1-k)^2-m^2\right)}\gamma_\nu\frac{\not{p}_1-\not{k}+m}{\left((p_1-k)^2-m^2\right)}\gamma_\rho\frac{1}{k^2-\lambda^2} \tag{2}$$

where λ^2 is introduced to treat the infrared divergence. However, as demonstrated in [6, 7], the dimensional regularization is powerful enough to normalize the divergence ultraviolet, infrared and mass singularity; therefore, in what follows the last term of (2) is $(k)^{-2}$; also constant A_n include both charges and other constants.

3. N-Dimensional Diracology

Following Lozano-Medina, is introduced:

$$\Gamma_{\alpha\beta\cdots\omega}=\gamma_\alpha\gamma_\beta\cdots\gamma_\omega, \tag{3}$$

and reduced matrix Γ when there is repeated indices and leads to

$$\gamma_\mu\Gamma_\alpha\gamma^\mu=(2-n)\Gamma_\alpha,\ \gamma_\mu\Gamma_{\alpha\beta\delta\varepsilon\lambda}\gamma^\mu=2\left(\Gamma_{\beta\delta\varepsilon\lambda\alpha}-\Gamma_{\alpha\delta\varepsilon\lambda\beta}-\Gamma_{\alpha\beta\lambda\varepsilon\delta}\right)-(n-4)\Gamma_{\alpha\beta\delta\varepsilon\lambda},$$
$$\gamma_\mu\Gamma_{\alpha\beta}\gamma^\mu=4g_{\alpha\beta}I+(n-4)\Gamma_{\alpha\beta},\ \gamma_\mu\Gamma_{\alpha\beta\delta\varepsilon}\gamma^\mu=2\Gamma_{\beta\delta\varepsilon}+2\Gamma_{\alpha\varepsilon\delta\beta}+(n-4)\Gamma_{\alpha\beta\delta\varepsilon},$$
$$\gamma_\mu\Gamma_{\alpha\beta\lambda}\gamma^\mu=-2\Gamma_{\lambda\beta\alpha}-(n-4)\Gamma_{\alpha\beta\delta\varepsilon}, \tag{4}$$

whereof, $C_{\mu\nu}$ can be written as

$$C_{\mu\nu}=A_n\int d^nk\frac{N_{\mu\nu}}{D}, \tag{5}$$

where

$$N_{\mu\nu}=\gamma_\rho(\not{p}_2-\not{k}+m)\gamma_\mu(\not{p}_1-\not{k}-\not{k}+m)\gamma_\nu(\not{p}_1-\not{k}+m)\gamma_\rho, \tag{6}$$

$$D=\left((p_2-k)^2-m^2\right)\left((p_1+k_1-k)^2-m^2\right)\left((p_1-k)^2-m^2\right). \tag{7}$$

From the diracology we have

$$\begin{aligned}N_{\mu\nu}=&(p_2-k)_\alpha(p_1+k_1-k)_\beta(p_1-k)_\lambda\Gamma_{\rho\alpha\mu\beta\nu\lambda\rho}+m(p_2-k)_\alpha(p_1+k_1-k)_\beta\Gamma_{\rho\alpha\mu\beta\nu\rho}\\&+m(p_2-k)_\alpha(p_1-k)_\lambda(p_1-k)_\lambda\Gamma_{\rho\alpha\mu\nu\lambda\rho}+m^2(p_2-k)_\alpha\Gamma_{\rho\alpha\mu\nu\rho}\\&+m(p_1+k_1-k)_\beta(p_1-k)_\lambda\Gamma_{\rho\mu\beta\nu\lambda\rho}+m^2(p_1+k_1-k)_\beta\Gamma_{\rho\mu\beta\nu\rho}\\&+m^2(p_1-k)_\lambda\Gamma_{\rho\mu\nu\lambda\rho}+m^3\Gamma_{\rho\mu\nu\rho}.\end{aligned} \tag{8}$$

4. Integration Symmetric

In the integral (5) and precisely the nature of (8), several terms cancel when calculating the integration on k, this is due to the symmetrical integration. The essential result is

$$\int d^n k f\left(k^2\right)\prod_{j=1}^{2r+1} k_{\mu_j} = 0, \tag{9}$$

$$\int d^n k f\left(k^2\right)\prod_{j=1}^{2r} k_{\mu_j} = \int d^n k f\left(k^2\right)\left(k^2\right)^r \times \left\{g_{\mu_{j_1},\mu_{j_2}} \cdots g_{\mu_{j_{2r-1}},\mu_{j_{2r}}}\right\} N_r \frac{\sigma^{j_1\cdots j_{2r}}}{r!2^r}, \tag{10}$$

where $\sigma^{j_1\cdots j_n} = \varepsilon^{j_1\cdots j_n}\left(-1\right)^{sign(\varepsilon)}$ and N_r is a constant to be determined.

The constant N_r is determined by contraction as already indicated with $g^{\sigma\rho}\int d^n k f\left(k^2\right)\prod_{j=1}^{2r+2} k_{\mu_j}$, and where $\rho = \mu_{2r+1}$ and $\sigma = \mu_{2r+2}$. In this way is obtained

$$N_r = \left(n+2r\right)N_{r+1} = 2^{-r}\Gamma\left(n/2\right)\left(\Gamma\left(n/2+r\right)\right)^{-1}.$$

Symmetric integration allows to express the Compton operator as follows

$$\begin{aligned} C_{\mu\nu} = A_n\Big[& p_{2\alpha}p_{1\lambda}\left(p_1+k_1\right)_\beta \Gamma_{\rho\alpha\mu\beta\nu\lambda\rho} + mp_{2\alpha}\left(p_1+k_1\right)_\beta \Gamma_{\rho\alpha\mu\beta\nu\rho} + mp_{2\alpha}p_{1\lambda}\Gamma_{\rho\alpha\mu\nu\lambda\rho} \\ & + m^2 p_{2\alpha}\Gamma_{\rho\alpha\mu\nu\rho} + mp_{1\lambda}\left(p_1+k_1\right)_\beta \Gamma_{\rho\mu\beta\nu\lambda\rho} + m^2\left(p_1+k_1\right)_\beta \Gamma_{\rho\mu\beta\nu\rho} \\ & + m^2 p_{1\lambda}\Gamma_{\rho\mu\nu\lambda\rho} + m^3 p_{1\lambda}\Gamma_{\rho\mu\nu\rho}\Big]\times I_1 + \frac{A_n}{n}\Big[\Big(g_{\alpha\beta}p_{1\lambda} + g_{\beta\lambda}p_{2\alpha} + g_{\alpha\lambda}\left(p_1+k_1\right)_\beta\Big)\Gamma_{\rho\alpha\mu\beta\nu\lambda\rho} \\ & + \frac{A_n}{n}\Big[\Big(g_{\alpha\beta}p_{1\lambda} + g_{\beta\lambda}p_{2\alpha} + g_{\alpha\lambda}\left(p_1+k_1\right)_\beta\Big)\Gamma_{\rho\alpha\mu\beta\nu\lambda\rho} + g_{\alpha\beta}\Gamma_{\rho\alpha\mu\beta\nu\rho} + g_{\beta\lambda}\Gamma_{\rho\mu\beta\nu\lambda\rho}\Big]\times I_2, \end{aligned} \tag{11}$$

where I_1 and I_2 are following Feynman integrals

$$I_1 = \int \frac{d^n k}{\left(\left(p_2-k\right)^2 - m^2\right)\left(\left(p_1+k_1-k\right)^2 - m^2\right)\left(\left(p_1-k\right)^2 - m^2\right)k^2}, \tag{12}$$

$$I_2 = \int \frac{d^n k}{\left(\left(p_2-k\right)^2 - m^2\right)\left(\left(p_1+k_1-k\right)^2 - m^2\right)\left(\left(p_1-k\right)^2 - m^2\right)}. \tag{13}$$

5. Reduction of Compton Operator

In this section we provide the expanded form of $C_{\mu\nu}$, using equations (3) and (4) where are the integrals I_1 and I_2; these integrals will be discussed in the next section. The result of this operator is

$$\begin{aligned} C_{\mu\nu} = A_n\Big\{ & 2\gamma_\mu\left(\not p_1+\not k_1\right)\gamma_\nu \not p_1 \not p_2 - 2\not p_2\left(\not p_1+\not k_1\right)\gamma_\nu \not p_1 \gamma_\mu - 2\not p_2\gamma_\mu \not p_1 \gamma_\nu\left(\not p_1+\not k_1\right) \\ & -\left(n-4\right)\not p_2\gamma_\mu\left(\not p_1+\not k_1\right)\gamma_\nu \not p_1 + m\Big[2\gamma_\mu\left(\not p_1+\not k_1\right)\gamma_\nu \not p_2 + 2\not p_2\gamma_\nu\left(\not p_1+\not k_1\right)\gamma_\mu \\ & +\left(n-4\right)\not p_2\gamma_\mu\left(\not p_1+\not k_1\right)\gamma_\nu + 2\Gamma_{\mu\nu}\not p_1\not p_2 + 2\not p_2\not p_1\Gamma_{\nu\mu} + \left(n-4\right)\not p_2\Gamma_{\mu\nu}\not p_1 \\ & +2\not p_1\gamma_\mu\left(\not p_1+\not k_1\right)\gamma_\nu + 2\gamma_\mu\not p_1\gamma_\nu\left(\not p_1+\not k_1\right) + \left(n-4\right)\gamma_\mu\left(\not p_1+\not k_1\right)\gamma_\nu \not p_1\Big] \\ & -m^2\Big[2\Gamma_{\mu\nu}\not p_2 + \left(n-4\right)\not p_2\Gamma_{\mu\nu} + 2\gamma_\nu\left(\not p_1+\not k_1\right)\gamma_\mu + \left(n-4\right)\gamma_\mu\left(\not p_1+\not k_1\right)\gamma_\nu \\ & +2\not p_1\Gamma_{\nu\mu} + \left(n-4\right)\Gamma_{\mu\nu}\not p_1\Big] + m^3\Big[4g_{\mu\nu}I + \left(n-4\right)\Gamma_{\mu\nu}\Big]\Big\}\times I_1 \end{aligned}$$

$$
\begin{aligned}
&+{}^{A_n}\!/_{n}\{8(\gamma_\mu p_{1\mu}-\gamma_\nu p_{1\mu})+2(n-4)(\Gamma_{\mu\nu}p_1-p_1\Gamma_{\mu\nu})+2(2-n)(\Gamma_{\mu\nu}p_2-p_2\Gamma_{\mu\nu})\\
&+4p_1\Gamma_{\mu\nu}+2(n-4)\Gamma_{\mu\nu}p_1-2np_2\Gamma_{\mu\nu}-8(p_1+k_1)g_{\mu\nu}-2(n-4)\Gamma_{\mu\nu}(p_1+k_1)\\
&-(n-4)(2-n)(\Gamma_{\mu\nu}p_1-p_2\Gamma_{\mu\nu})+2(n-4)\gamma_\nu(\Gamma_{\mu\nu}p_2-p_2\Gamma_{\mu\nu})(p_1+k_1)\gamma_\mu\\
&+(n-4)^2\gamma_\mu(p_1+k_1)\gamma_\nu+4(2-n)g_{\mu\nu}I+(2-n)(n-4)\Gamma_{\mu\nu}+4ng_{\mu\nu}I+(n-4)g_{\mu\nu}I\\
&+(n-4)^2\Gamma_{\mu\nu}+4(2-n)g_{\mu\nu}I+(2-n)(n-4)\Gamma_{\mu\nu}\}\times I_2. \qquad (14)
\end{aligned}
$$

Naturally, the physical case of Compton is obtained at the limit $n \to 4^+$.

6. Feynman-Davydychev Integrals

The Davydychev method for Feynman integrals is in Ref [3, 4]. In both cases, is based on the application of the following expressions:

a) From the Mellin-Barnes transform (for I_1 and I_2)

$$\frac{1}{(k^2-m^2+i\varepsilon)}=\frac{1}{(k^2+i\varepsilon)}\frac{1}{\Gamma(\nu)}\frac{1}{2\pi i}\int_{-i\infty}^{+i\infty}ds\left(\frac{-m^2}{k^2+i\varepsilon}\right)^s\Gamma(-s)\Gamma(\nu+s), \qquad (15)$$

and for the case of I_2

$$J^{(N)}\left(\{\nu_j\}\middle|\{p_j\};m\right)=\int\left(\prod_{j=1}^{N}\left((p_j+r)^2-m^2\right)^{\nu_j}\right)^{-1}d^nk. \qquad (16)$$

b) Lemma of Barnes

$$\frac{1}{2\pi i}\int_{-i\infty}^{+i\infty}ds\Gamma(A+s)\Gamma(B+s)\Gamma(C-s)\Gamma(D-s)=\frac{\Gamma(A+C)\Gamma(B+C)\Gamma(A+D)\Gamma(B+D)}{\Gamma(A+B+C+D)} \qquad (17)$$

The purpose of (17) is to remove the poles of the function $\Gamma(z)$ shown in the right side of the complex plane, being the residue $\mathrm{Res}[\Gamma(z)]=(-1)^i(i!)^{-1}$. The integral I_2 is founded applying the Eq. (4.6) of Ref [3]

$$
\begin{aligned}
I_2=J^{(3)}(1,1,1|q_1,q_2,q_3;m)&=\pi^{n/2}i^{1-n}(-m^2)^{n/2-3}\frac{\Gamma(3-n/2)}{\Gamma(3)}\sum_{a_{12}=0}^{\infty}\sum_{a_{13}=0}^{\infty}\sum_{a_{23}=0}^{\infty}\frac{1}{a_{12}!}\frac{1}{a_{13}!}\frac{1}{a_{23}!}\left(\frac{k_{12}^2}{4m^2}\right)^{a_{12}}\\
&\times\left(\frac{k_{13}^2}{4m^2}\right)^{a_{13}}\left(\frac{k_{23}^2}{4m^2}\right)^{a_{23}}\frac{(3-\frac{n}{2})_{a_{12}+a_{13}+a_{23}}(1)_{a_{12}+a_{13}}(1)_{a_{12}+a_{23}}(1)_{a_{13}+a_{23}}}{(\frac{3}{2})_{a_{12}+a_{13}+a_{23}}(2)_{a_{12}+a_{13}+a_{23}}}, \qquad (18)
\end{aligned}
$$

where $(C)_j$ is the Pochhammer symbol defined by $(C)_j=\Gamma(C+j)(\Gamma(C))^{-1}$. Also, $k_{23}^2=(k_1)^2$, $k_{12}^2=(p_1+k_1-p_2)^2$, and $k_{13}^2=(p_1-p_2)^2$.

Finally I_2 is written in terms of Lauricella's generalized hypergeometric function defined in Eq. (A.8) of Appendix A in Ref [3]:

$$I_2=\pi^{\frac{n}{2}}i^{1-n}(-m^2)^{\frac{n}{2}-3}\frac{\Gamma(3-\frac{n}{2})}{2}\times F_{2:0;0}^{1:3;0}\left(\begin{matrix}((3-\frac{n}{2}):1,1,1),(1:1,1),(1:1,1),(1:1,1)\\(\frac{3}{2}:1,1,1),(2:1,1)\end{matrix}\middle|\frac{k_{12}^2}{4m^2},\frac{k_{13}^2}{4m^2},\frac{k_{23}^2}{4m^2}\right) \qquad (19)$$

Applying Eq. (3.6) of Ref [3], the integral I_1 is obtained from $I_1=J^{(4)}(1,1,1,1|q_1,q_2,q_3,q_4;m_1,m_2,m_3,m_4)$, with $m_1=0$, $m_2=m_3=m_4=m$, and $q_1=0$,

$q_2 = -p_1$, $q_3 = -p_2$, and $q_4 = -p_1 - k_1$.
In this way is obtained

$$I_1 = \pi^{\frac{n}{2}} i^{1-n} \left(-m^2\right)^{\frac{n}{2}-4} \frac{\Gamma\left(4-\frac{n}{2}\right)}{\Gamma(4)} \sum_{a_{12}=0}^{\infty} \sum_{a_{13}=0}^{\infty} \sum_{a_{14}=0}^{\infty} \sum_{a_{23}=0}^{\infty} \sum_{a_{24}=0}^{\infty} \sum_{a_{34}=0}^{\infty} \frac{1}{a_{12}!} \frac{1}{a_{13}!} \frac{1}{a_{14}!} \frac{1}{a_{23}!} \frac{1}{a_{24}!} \frac{1}{a_{34}!}$$
$$\times \left(\frac{k_{12}^2}{m^2}\right)^{a_{12}} \left(\frac{k_{13}^2}{m^2}\right)^{a_{13}} \left(\frac{k_{14}^2}{m^2}\right)^{a_{14}} \left(\frac{k_{23}^2}{m^2}\right)^{a_{23}} \left(\frac{k_{24}^2}{m^2}\right)^{a_{24}} \left(\frac{k_{34}^2}{m^2}\right)^{a_{34}} \sum_{b_1=0}^{\infty} \sum_{b_2=0}^{\infty} \sum_{b_3=0}^{\infty} \frac{1}{b_1!} \frac{1}{b_2!} \frac{1}{b_3!} \left(1-\frac{m_1^2}{m^2}\right)^{b_1}$$
$$\times \left(1-\frac{m_2^2}{m^2}\right)^{b_2} \left(1-\frac{m_3^2}{m^2}\right)^{b_3} \frac{\left(4-\frac{n}{2}\right)_{a_{12}+a_{13}+a_{14}+a_{23}+a_{24}+a_{34}+b_1+b_2+b_3}}{(4)_{2a_{12}+2a_{13}+2a_{14}+2a_{23}+2a_{24}+2a_{34}+b_1+b_2+b_3}} (1)_{a_{12}+a_{13}+a_{14}+b_1} (1)_{a_{12}+a_{23}+a_{24}+b_2}$$
$$\times (1)_{a_{13}+a_{23}+a_{31}+b_3} (1)_{a_{14}+a_{24}+a_{34}}. \tag{20}$$

It follows from the values $m_1 = 0$, $m_2 = m_3 = m$ that in the summations over b_2 and b_3 remain only terms with $b_2 = b_3 = 0$ and the sum of b_1 leads hypergeometric function given by

$${}_2F_1\left(\begin{matrix} 4-\frac{n}{2}+a_{12}+a_{13}+a_{14}+a_{23}+a_{24}+a_{34}, 1+a_{12}+a_{13}+a_{14} \\ 4+2a_{12}+2a_{13}+2a_{14}+2a_{23}+2a_{24}+2a_{34} \end{matrix} \Big| 1\right). \tag{21}$$

Then, I_1 is given by

$$I_1 = \pi^{\frac{n}{2}} i^{1-n} \left(-m^2\right)^{\frac{n}{2}-4} \frac{\Gamma\left(4-\frac{n}{2}\right)}{\Gamma(4)} \sum_{a_{12}=0}^{\infty} \sum_{a_{13}=0}^{\infty} \sum_{a_{14}=0}^{\infty} \sum_{a_{23}=0}^{\infty} \sum_{a_{24}=0}^{\infty} \sum_{a_{34}=0}^{\infty} \frac{1}{a_{12}!} \frac{1}{a_{13}!} \frac{1}{a_{14}!} \frac{1}{a_{23}!} \frac{1}{a_{24}!} \frac{1}{a_{34}!}$$
$$\times \left(\frac{k_{12}^2}{m^2}\right)^{a_{12}} \left(\frac{k_{13}^2}{m^2}\right)^{a_{13}} \left(\frac{k_{14}^2}{m^2}\right)^{a_{14}} \left(\frac{k_{23}^2}{m^2}\right)^{a_{23}} \left(\frac{k_{24}^2}{m^2}\right)^{a_{24}} \left(\frac{k_{34}^2}{m^2}\right)^{a_{34}} (1)_{a_{12}+a_{13}+a_{14}} (1)_{a_{12}+a_{23}+a_{24}} (1)_{a_{13}+a_{23}+a_{24}}$$
$$\times (1)_{a_{14}+a_{24}+a_{34}} \, {}_2F_1\left(\begin{matrix} 4-\frac{n}{2}+a_{12}+a_{13}+a_{14}+a_{23}+a_{24}+a_{34}, 1+a_{12}+a_{13}+a_{14} \\ 4+2a_{12}+2a_{13}+2a_{14}+2a_{23}+2a_{24}+2a_{34} \end{matrix} \Big| 1\right). \tag{22}$$

Finally, the hypergeometric function can be reduced even more in terms of gamma functions using the Gauss's theorem.

$${}_2F_1\left(\begin{matrix} a,b \\ c \end{matrix} \Big| 1\right) = \frac{\Gamma(c)\Gamma(c-a-b)}{\Gamma(c-a)\Gamma(c-b)}. \tag{23}$$

7. Conclusions

By employing dimensional regularization or dimensionally continuous integration method and the use thereof for obtaining Feynman integrals, we obtain a complete analytical expression for the Compton operator studied by Lozano-Medina. In addition it is interesting to note that the prize Mexico Science and Technology 1992, was granted to Juan José Giambiagi precisely for the invention of the method called dimensional regularization [8].

References

1. J. Lozano y F. Medina, Rev. Mex. Fís., **I**(1), (1952) 44-51.
2. C. G. Bollini and J. J. Giambiagi, Nuovo Cimento, B **12**, (1972) 20-25.
3. A. I. Davydychev, J. Math. Phys., **32**, (1991) 1052-1060.
4. A. I. Davydychev, J. Math. Phys., **33**, (1992) 358-369.
5. J. D. Bjorken y S. Drell, LSC Relativistic Quantum Mechanics, Mc Graw-Hill, (1998).
6. W. J. Marciano, Phys. Rev., **12**, (1975) 3861-3871.
7. R. Mares Gallardo y H. Luna, Rev. Mex. Fis., **43**(5), (1997).
8. Excélsior, year 5, 6, september 1992. México.

Journal of Physics: Conference Series **574** (2015) 012163 doi:10.1088/1742-6596/574/1/012163

Generation of thrust and lift with airfoils in plunging and pitching motion

M.Moriche, O.Flores, M.García-Villalba

Departamento de Bioingeniería e Ingeniería Aeroespacial, Universidad Carlos III de Madrid

E-mail: mmoriche@ing.uc3m.es

Abstract.
We present fully resolved Direct Numerical Simulations of 2D flow over a moving airfoil, using an in-house code that solves the Navier-Stokes equations of the incompressible flow with an Immersed Boundary Method. A combination of sinusoidal plunging and pitching motions is imposed to the airfoil. Starting from a thrust producing case (Reynolds number, $Re = 1000$, reduced frequency, $k = 1.41$, plunging amplitude $h_0/c = 1$, pitching amplitude $\theta_0 = 30°$, phase shift $\phi = 90°$), we increase the mean pitching angle (in order to produce lift) and vary the phase shift between pitching and plunging (to optimize the direction and magnitude of the net force on the airfoil). These cases are discussed in terms of their lift coefficient, thrust coefficient and propulsive efficiency.

1. Introduction

Flight performance of small birds and insects is very attractive for aerodynamicists, specially in terms of maneuverability. These small animals produce thrust, lift and position control by flapping their wings and adjusting their body position. Similar strategies are being used in bio-inspired Micro Aerial Vehicles (MAVs), which are small unmanned aircrafts with sizes of the order of 15 cm. These flight devices have received a lot of attention in the last years, since equipped with a camera and some sensors they can be used for surveillance and reconnaissance missions in hazardous environments like bush fires or nuclear accidents. Due to their small size, the Reynolds number (Re) that characterizes MAVs flight is typically $10^2 - 10^4$, similar to that of small birds and insects [1]. Improving our understanding of flapping wings at low Re is a key factor to unlock the standardized design of bio-inspired MAVs, with the potential to achieve high maneuverability and efficiency in their flapping flight.

It is important to realise that this low Re combined with high amplitude flapping motions lead to an unsteady flight regime where massive flow separation is present. Under these conditions, unsteady aerodynamic mechanisms such as leading edge vortex, clap and fling and wake recapture are dominant [2]. The classical unsteady aerodynamic models for fixed wings are no longer valid, since they are designed for high Re, low amplitude motions.

Our capability to predict the aerodynamic forces for a flapping wing at low Re is very limited, and only Direct Numerical Simulations (DNS) are able to deliver reliable predictions. There are essentially no appropriate low order models, mainly because of the vast parameter space of flapping flight. Although there have been a large number of studies (see the reviews of [3] and [4]) the mapping of that parametric space is sparse.

In the present paper, we consider the pitching and plunging motion of 2D airfoils. In particular, we focus on the balance between the lift and thrust, and in how a thrust generation case can be converted into lift generation by changing the mean pitching angle.

2. Numerical method

The simulations presented here have been developed using an in-house code that solves the Navier Stokes equations for an incompressible flow, using a fractional step method in a staggered grid. The presence of solid boundaries (airfoil) is modelled using the direct-forcing Immersed Boundary Method proposed by Uhlmann [5]. The code is written in Fortran and uses MPI (for parallelization) and HYPRE libraries (for the solution of the linear systems). The code has been extensively validated, as reported by Moriche [6].

In the following, we present two-dimensional (2D) simulations of a NACA-0012 airfoil of chord c immersed in a free stream of velocity U_∞, flapping with a sinusoidal plunging and pitching motion. The computational domain has a size of $25c$ in the streamwise direction and $15c$ in the vertical direction. The number of grid points is 3200 in the streamwise direction and 1920 in the vertical direction. This number has been selected based on preliminary simulations to properly predict the aerodynamic forces. We employ a Dirichlet boundary condition at the inflow plane, free slip at the top and bottom boundaries and a convective boundary condition at the outflow plane.

The plunging motion is imposed by the vertical displacement of the quarter chord point of the airfoil

$$h(t) = h_0 \sin(2\pi f\, t)\,, \tag{1}$$

where h_0 is the plunging amplitude and f is the oscillation frequency. The pitching motion is imposed by the angle of the chord line of the airfoil with the streamwise direction

$$\theta(t) = \theta_m + \theta_0 \sin(2\pi f\, t + \phi)\,, \tag{2}$$

where θ_m is the mean pitching value, θ_0 the pitching amplitude and ϕ the pitching phase shift. Using c and U_∞ as characteristic dimensions, the airfoil kinematics is characterized by the following non-dimensional parameters: the Reynolds number $Re = U_\infty c/\nu$, the reduced frequency $k = 2\pi f\, c/U_\infty$, the plunging amplitude h_0/c, the pitching amplitude θ_0, the mean pitching value θ_m and the phase shift ϕ.

3. Results

We have performed a series of simulations in which Re, k, h_0/c and θ_0 are kept constant, while we vary θ_m and ϕ. The aim is to evaluate how these two parameters influence the lift production, tilting and modulating the net force produced by the airfoil. We start from a thrust producing case that is taken as reference, case C0, and which corresponds to one of the cases reported by Anderson [7], except for a somewhat lower Reynolds number: $Re = 1000$, $\theta_0 = 30°$, $h_0/c = 1$ and $k = 1.41$. We then vary θ_m in the range (0° - 20°) and ϕ in the range (30° - 110°), to produce a database with 15 different cases.

A summary of the results obtained is presented in figure 1 which shows the values of the averaged lift (c_l) and thrust (c_t) coefficients. The solid (dashed) lines join points that correspond to cases with the same θ_m (ϕ). Finally, the color indicates the propulsive efficiency defined as

$$\eta = \frac{U_\infty \int_0^T c_t(t)\, dt}{\int_0^T c_l(t)\, \dot{h}(t)\, dt}, \tag{3}$$

where $c_l(t)$ and $c_t(t)$ are the instantaneous lift and thrust coefficients, $\dot{h}(t)$ is the vertical velocity of the quarter chord point and $T = 1/f$.

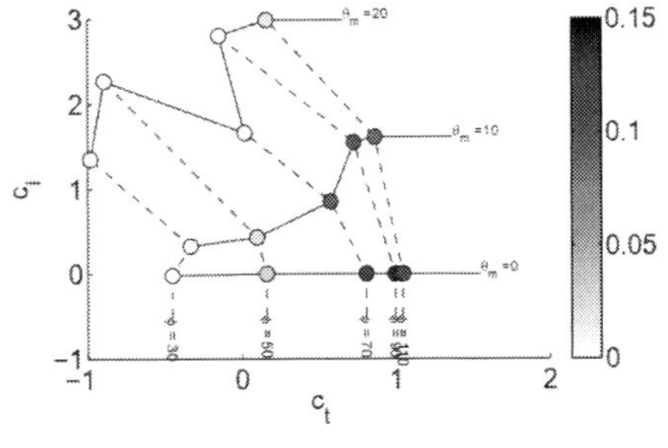

Figure 1. Averaged c_l versus averaged c_t. The color corresponds to the propulsive efficiency.

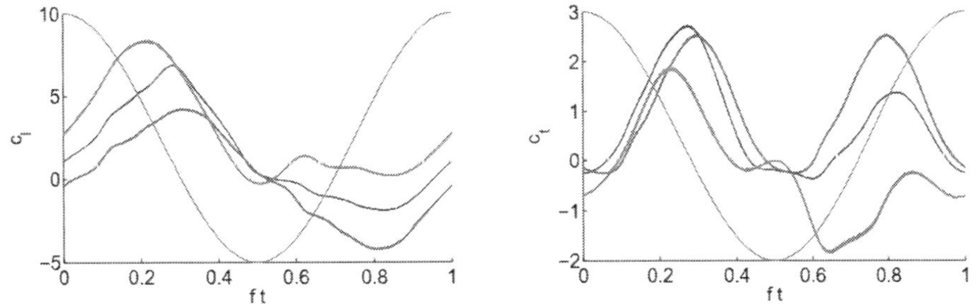

(a) (b)

Figure 2. Time history of force coefficients for selected cases with $\phi = 90°$. (a) c_l, (b) c_t. ——— $\theta_m = 10°$ (C1). ——— $\theta_m = 0°$ (C0). ——— $\theta_m = 20°$. The grey line corresponds to the vertical position of the quarter chord point in arbitrary units.

We can observe in figure 1 that for $\theta_m = 0°$, thrust is only produced for $\phi \gtrsim 50°$. Increasing ϕ up to the maximum value considered $\phi = 110°$, c_t increases but the propulsive efficiency reaches a maximum at $\phi = 90°$, note that this is the reference case (C0). Increasing θ_m leads, in general, to lift production at the expense of losing thrust. We select as optimal lift-production case (C1) the case with $\theta_m = 10°$, $\phi = 90°$, since the propulsive efficiency is comparable to C0, and c_l and c_t are of order unity.

In order to analyze the influence of θ_m and ϕ, figures 2 and 3 display the time history of the instantaneous force coefficients for selected cases. Figure 2 shows cases with $\phi = 90°$ and $\theta_m = 0°, 10°$ and $20°$. Recall that C0 ($\theta_m = 0°$) is the reference thrust producing case, and accordingly thrust is produced symmetrically both in the downstroke and the upstroke (see grey line for plunge position reference). In C1 ($\theta_m = 10°$), a similar amount of thrust is produced in the downstroke (compared to C0) and a somewhat lower amount of thrust is produced in the upstroke. Note that the lift is mainly produced in the downstroke, as expected. Further increasing θ_m to $20°$ we observe that much more lift is produced in the downstroke, while roughly no vertical force is produced in the upstroke. In terms of thrust, c_t is lower in the downstroke compared to C0 and C1. However, the main difference between $\theta_m = 20°$ and C0/C1 is that in the upstroke the former has a negative thrust (drag), resulting in a much lower averaged c_t.

Figure 3 shows cases with $\theta_m = 10°$ and $\phi = 70°, 90°$ and $110°$. It can be observed that overall, the effect of ϕ is less pronounced than the effect of θ_m. The cases with $\phi = 90°$ (C1) and $\phi = 110°$ are rather similar, since the latter produces more lift in the downstroke, which is lost

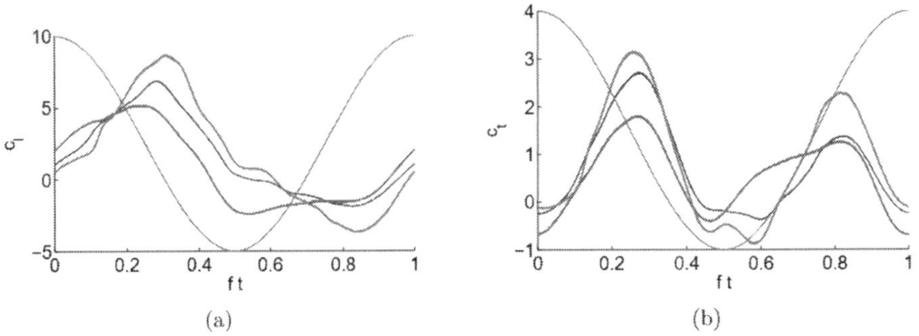

Figure 3. Time history of force coefficients for selected cases with $\theta_m = 10°$. (a) c_l, (b) c_t. ——— $\phi = 90°$ (C1). ——— $\phi = 70°$. ——— $\phi = 110°$. The grey line corresponds to the vertical position of the quarter chord point in arbitrary units.

in the upstroke. This results in a slightly lower propulsive efficiency that for C1. Concerning the thrust, the case with $\phi = 110°$ produces more thrust in both downstroke and upstroke, but this advantage is partially cancelled out by a worse performance in the downstroke to upstroke transition. Finally, the case with $\phi = 70°$ produces less lift and thrust than the other two cases, specially in the downstroke.

4. Conclusions

We have performed a series of DNS of low *Re* flow over a pitching and plunging airfoil, varying the mean pitching angle (θ_m) and the phase shift between the pitching and plunging motions (ϕ). We have identified an optimal lift-and-thrust producing case ($\theta_m = 10°$, $\phi = 90°$), with a propulsive efficiency comparable to the thrust producing case used as a reference. In addition we have studied the effect of variations in ϕ and θ_m with respect to this optimal lift-and-thrust producing case. The effect of θ_m is more pronounced, increasing the lift as θ_m increases from 0° to 20°. However, there is a clear degradation in the thrust performance for $\theta_m > 10°$, associated with drag production during the upstroke.

References

[1] W. Shyy, Y. Lian, J. Tang, D. Viieru, and H. Liu, *Low Reynolds number flyers*. Cambridge University Press, 2008.

[2] S. P. Sane, "The aerodynamics of insect flight," *Journal of Experimental Biology*, vol. 206, no. 23, pp. 4191–4208, 2003.

[3] M. F. Platzer, K. D. Jones, J. Young, and J. S. Lai, "Flapping wing aerodynamics: progress and challenges," *AIAA journal*, vol. 46, no. 9, pp. 2136–2149, 2008.

[4] W. Shyy, H. Aono, S. K. Chimakurthi, P. Trizila, C.-K. Kang, C. E. Cesnik, and H. Liu, "Recent progress in flapping wing aerodynamics and aeroelasticity," *Progress in Aerospace Sciences*, vol. 46, no. 7, pp. 284–327, 2010.

[5] M. Uhlmann, "An immersed boundary method with direct forcing for the simulation of particulate flows," *J. Comput. Phys.*, vol. 209, no. 2, pp. 448 – 476, 2005.

[6] M. Moriche, "Development and validation of a numerical solver for unsteady aerodynamics applications," Master's thesis, Univ. Carlos III de Madrid, 2013.

[7] J. M. Anderson, "Vorticity control for efficient propulsion.," tech. rep., DTIC Document, 1996.

Journal of Physics: Conference Series **574** (2015) 012164 doi:10.1088/1742-6596/574/1/012164

Bistability in self-activating genes regulated by non-coding RNAs

Jesus Miro-Bueno

Research Institute of the IT4Innovations Centre of Excellence, Faculty of Philosophy and Science, Silesian University in Opava, 74601 Opava, Czech Republic

E-mail: `jesus.mirobueno@fpf.slu.cz`

Abstract. Non-coding RNA molecules are able to regulate gene expression and play an essential role in cells. On the other hand, bistability is an important behaviour of genetic networks. Here, we propose and study an ODE model in order to show how non-coding RNA can produce bistability in a simple way. The model comprises a single gene with positive feedback that is repressed by non-coding RNA molecules. We show how the values of all the reaction rates involved in the model are able to control the transitions between the high and low states. This new model can be interesting to clarify the role of non-coding RNA molecules in genetic networks. As well, these results can be interesting in synthetic biology for developing new genetic memories and biomolecular devices based on non-coding RNAs.

1. Introduction

Non-coding RNA (ncRNA) molecules are strands of RNA that are not codified in proteins and are involved in important cell functions [1]. These ncRNA molecules are able to regulate gene expression by repressing the activity of messenger RNA (mRNA) molecules. The regulation of a genetic network by ncRNAs can produce bistability and oscillations [2, 3, 4, 5]. Bistability is important in cellular processes such as cell differentiation and signalling. A way to produce bistability in genetic networks is a self-activating gene, i.e., a single gene that activates its own transcription creating a positive feedback loop [6]. Cooperative reactions or multimers are necessary in the positive feedback in order to obtain bistability. An alternative way to produce bistability, without cooperative reactions or multimers, is a gene with positive feedback in which the protein is sequestered by a repressor molecule [7]. In this case, the bistable dynamics has been poorly studied if the sequestered molecule is the mRNA instead of the protein. Here, we present and study an ODE model that shows how to obtain bistability in a simple way in a self-activating gene repressed by ncRNA molecules. We also show how all the parameters involved in the model are able to control the transitions between the low and high states.

2. Results and Discussion

The model comprises a gene with positive feedback that is repressed by non-coding RNA molecules (Fig. 1A). The gene is transcribed into mRNA molecules. And the mRNAs are translated into proteins. These proteins are able to bind to the promoter of its own gene increasing the transcription rate. Therefore, a positive feedback loop is created in the system. On the other hand, ncRNA molecules are synthesized at constant rate in the system. These

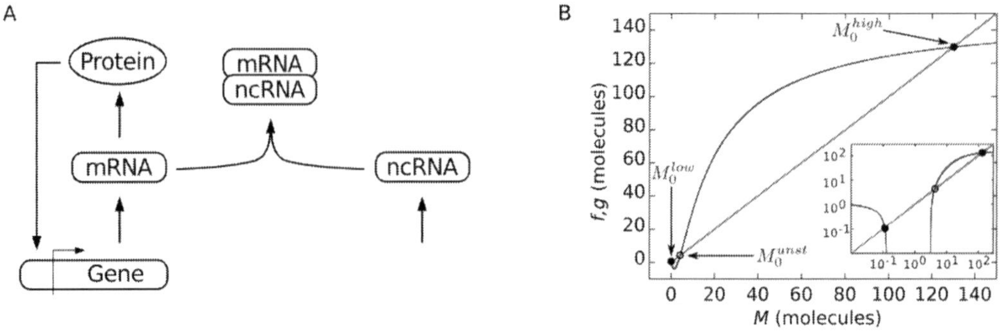

Figure 1. A) Diagram of the model. A protein binds to the promoter of its own gene increasing the transcription rate and creating a positive feedback loop. Non-coding RNAs bind to the mRNAs repressing the expression of the gene. B) Fixed points of the model. Inset: same plot in logarithmic scale. (Red and blue lines represent functions f and g, respectively). The rate values are: $k_1 = 1$ molecules^{-1} hour^{-1}, $k_{-1} = 100$ hour^{-1}, $k_2 = 1$ hour^{-1}, $k_3 = 500$ hour^{-1}, $k_4 = 1$ hour^{-1}, $k_5 = 10$ hour^{-1}, $k_6 = 1$ hour^{-1}, $k_7 = 1$ molecules^{-1} hour^{-1}, $k_{-7} = 50$ hour^{-1}, $k_8 = 10$ hour^{-1}, $k_9 = 350$ molecules hour^{-1} and $k_{10} = 1$ hour^{-1}.

ncRNA molecules recognize and bind to mRNA molecules forming a complex mRNA-ncRNA. In this complex, the mRNA can not be translated, and therefore, the gene expression is silenced. The positive feedback does not produce bistability by itself because of it does not include the formation of multimers or cooperative binding reactions. The repression of the gene expression by ncRNA molecules introduces in the system the necessary non-linearity to produce bistability. The biochemical reactions that described the gene with positive feedback and the repression carried out by the ncRNA molecules are as follows:

$$
\begin{array}{llll}
\text{Positive feedback} & & \text{Repression} & \\
G + P \xrightarrow{k_1} G_a & \text{(Activation)} & M + S \xrightarrow{k_7} C & \text{(Complex creation)} \\
G_a \xrightarrow{k_{-1}} G + P & \text{(Deactivation)} & C \xrightarrow{k_{-7}} M + S & \text{(Complex decay)} \\
G \xrightarrow{k_2} G + M & \text{(Slow transcription)} & C \xrightarrow{k_8} \phi & \text{(Complex degradation)} \\
G_a \xrightarrow{k_3} G_a + M & \text{(Fast transcription)} & \phi \xrightarrow{k_9} S & (S\ \text{synthesis}) \\
M \xrightarrow{k_4} \phi & \text{(mRNA degradation)} & S \xrightarrow{k_{10}} \phi & (S\ \text{degradation}), \\
M \xrightarrow{k_5} M + P & \text{(Translation)} & & \\
P \xrightarrow{k_6} \phi & (P\ \text{degradation}) & &
\end{array}
\quad (1)
$$

where G denotes the gene without P bound to its promoter, M denotes mRNA transcribed from G, P denotes the activator protein translated from M, G_a denotes the gene with P bound to its promoter, S denotes the ncRNA molecules and C denotes S bound to M. The description of the rates is as follows: k_1 is the binding rate of P to the promoter of G, k_{-1} is the unbinding rate of P from the promoter of G, k_2 is the basal transcription rate, k_3 is the activated transcription rate, k_4 is the degradation rate of M, k_5 is the translation rate, k_6 is the degradation rate of P, k_7 is the binding rate of S to M, k_{-7} is the decay rate of C into S and P, k_8 is the degradation rate of the complex C, k_9 is the synthesis rate of S and k_{10} is the degradation rate of S.

The dynamics of the biochemical reactions (1) follows the law of mass action and can be

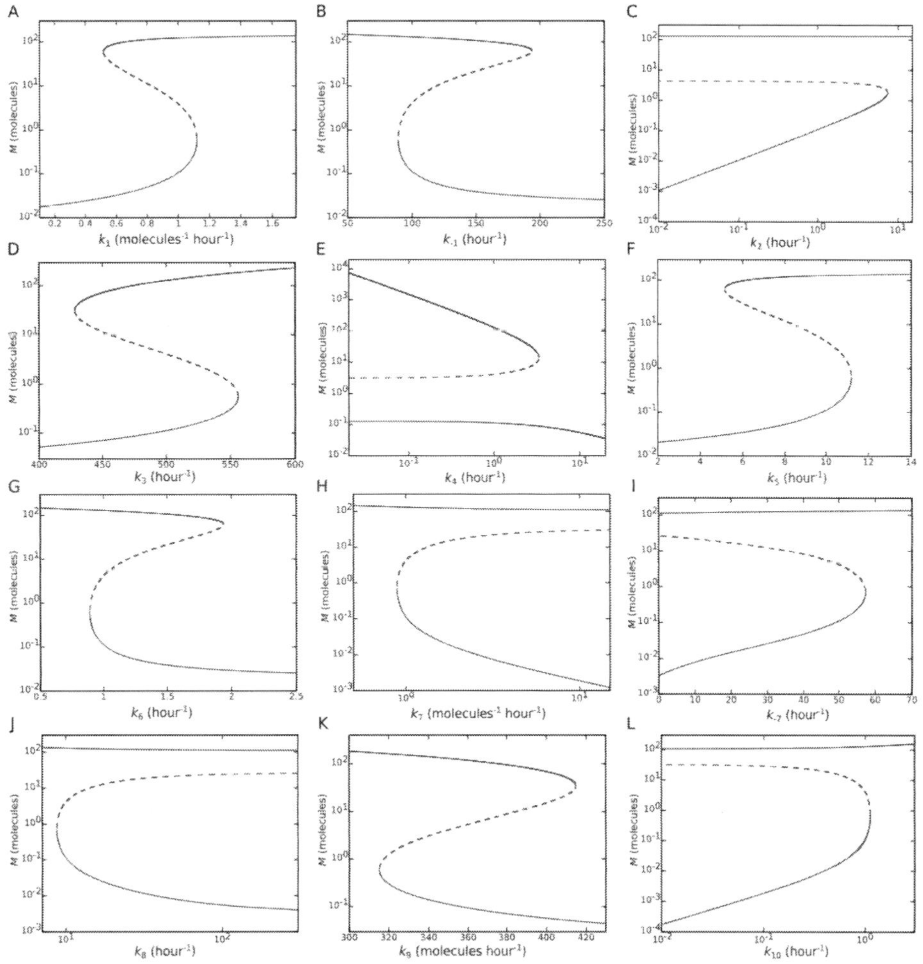

Figure 2. Bifurcation diagrams for all the reaction rates of the model. Solid and dashed lines represent stable and unstable points, respectively.

described by the following ordinary differential equations:

$$
\begin{aligned}
dG/dt &= -k_1GP + k_{-1}G_a & dP/dt &= -k_1GP + k_{-1}G_a + k_5M - k_6P \\
dG_a/dt &= k_1GP - k_{-1}G_a & dS/dt &= -k_7MS + k_{-7}C + k_9 - k_{10}S \\
dM/dt &= k_2G + k_3G_a - k_4M - k_7SM + k_{-7}C & dC/dt &= k_7MS - (k_{-7} + k_8)C,
\end{aligned} \tag{2}
$$

where all the biochemical species are measured in molecules. The equilibrium of the system is defined by the fixed points of the differential equations (2). These fixed points satisfy $dG/dt = dG_a/dt = dM/dt = dP/dt = dS/dt = dC/dt = 0$ [8]. Then, we have a system of equations and the solution for the number of mRNAs can be written as

$$M = \frac{1}{k_4}\left(\frac{k_2\alpha + k_3M}{\alpha + M} - k_9\frac{M}{\beta + M}\right), \tag{3}$$

where the parameters are as follows: $\alpha = k_{-1}k_6/k_1k_5$, $\beta = k_{10}(k_{-7} + k_8)/k_7k_8$. There is only one gene in the system, then we have assumed $G + G_a = 1$ molecule. A method to visualize the position of the fixed points consists on plotting the two members of the equation (3) as different functions in the same graph (Fig. 1B). The function for left and right members of the equation are denoted by $f = M$ and $g = \frac{1}{k_4}\left(\frac{k_2\alpha+k_3M}{\alpha+M} - k_9\frac{M}{\beta+M}\right)$, respectively. The points where the function f intersects the function g are the fixed points of the differential equations (2). The Fig. 1B shows three intersections, then there are three fixed points in the system. The middle point is an unstable fixed point and its value is M_0^{unst}=4.1 molecules. The other two points are stable fixed points and their values are M_0^{low}=0.1 and M_0^{high}=129.7 molecules. Therefore, the system exhibits bistability, i.e., two stable steady states: *high state* and *low state*. Depending on the initial conditions the system evolves toward the high or low state. The value of the parameters used for plotting are shown in the caption of the Fig. 1. We have used standard values for the reaction rates [9, 10].

The positions of the fixed points showed in the Fig. 1B depend on the values of the reaction rates. There is a dependency of the steady state on the direction of the parameter changes. The system responds differently to increase or decrease of the reactions rates due to there is a region of bistability. This phenomenon is known as hysteresis [8]. In the Fig. 2 are shown the bifurcation diagrams for all the reaction rates of the model. These reaction rates are able to control the transitions between the high and low states. When the value of the rates k_1, k_2, k_3, k_5, k_{-7} and k_{10} increase the bistable switches from the low to the high state, and when the rates decrease the bistable switches from the high to the low state (Figs. 2A,C,D,F,I,L). On the other hand, when the rates k_{-1}, k_4, k_6, k_7, k_8, and k_9 increase the bistable switches from the high to the low state, and and when the rates decrease the bistable switches from the low to the high state (Figs. 2B,E,G,H,J,K). The rate k_9, that is the synthesis rate of ncRNA molecules, is specially interesting for engineering new biomolecular devices in synthetic biology. This rate can be interpreted as the input of the bistable and can be easily adjust to switch the bistable between the two states.

3. Conclusions

In conclusion, we have shown that a self-activating gene repressed by ncRNA molecules can produce bistability in a simple way. We have studied how the reaction rates involved in the model can control the transitions between the high and low states. This simple model can be useful for clarifying the role of ncRNAs in cells and human diseases and for designing new genetic devices like, for example, memories in synthetic biology.

Acknowledgements

This work was supported by the European Regional Development Fund in the IT4Innovations Centre of Excellence project (CZ.1.05/1.1.00/02.0070) and EU project Development of Research Capacities of the Silesian University in Opava (CZ.1.07/2.3.00/30.0007).

References

[1] Mattick J S and Makunin I V 2006 *Hum. Mol. Gen.* **15** R17
[2] Miró-Bueno J M and Rodríguez-Patón A 2009 *Lecture Notes in Computer Science* **5518** 1170
[3] Liu D, Chang X, Liu Z, Chen L and Wang, R 2011 *PLoS ONE* **6** e17029.
[4] Zhou P, Cai S, Liu Z and Wang R 2012 *Phys. Rev. E* **85** 041916.
[5] Miró-Bueno J 2013 *Lecture Notes in Bioinformatics* **8130** 250
[6] Becskei A, Séraphin B and Serrano L 2001 *EMBO J.* **20** 2528
[7] François P and Hakim V 2004 *Proc. Natl. Acad. Sci. USA* **101** 580
[8] Strogatz S H 2001 *Nonlinear dynamics and chaos: with applications to physics, biology and chemistry* (Perseus publishing)
[9] Miró-Bueno J M and Rodríguez-Patón A 2011 *PLoS ONE* **6** e27414
[10] Legewie S, Dienst D, Wilde A, Herzel H and Axmann I M 2008 *Biophys. J.* **95** 3232

An efficient implicit direct forcing immersed boundary method for incompressible flows

S-G Cai[1], A Ouahsine[1], H Smaoui[1,2], J Favier[3] and Y Hoarau[4]

[1]Laboratoire Roberval, UMR 7337, Université de Technologie de Compiègne
Centre de Recherches de Royallieu, CS 60319, 60203 Compiègne Cedex, France
[2]Direction Technique Eau, Mer et Fleuves (CEREMA)
134, Rue de Beauvais, CS 60039, 60280 Margny-lès-Compiègne, France
[3]Laboratoire M2P2 UMR 7340, Aix Marseille Université, Marseille, France
[4]Laboratoire ICube UMR 7357, Université de Strasbourg, Strasbourg, France

E-mail: `shanggui.cai@utc.fr`, `ouahsine@utc.fr`

Abstract. A novel efficient implicit direct forcing immersed boundary method for incompressible flows with complex boundaries is presented. In the previous work [1], the calculation is performed on the Cartesian grid regardless of the immersed object, with a fictitious force evaluated on the Lagrangian points to mimic the presence of the physical boundaries. However the explicit direct forcing method [1] fails to accurately impose the non-slip boundary condition on the immersed interface. In the present work, the calculation is based on the implicit treatment of the artificial force while in an effective way of system iteration. The accuracy is also improved by solving the Navier-Stokes equation with the rotational incremental pressure-correction projection method of Guermond and Shen [2]. Numerical simulations performed with the proposed method are in good agreement with those in the literature.

1. Introduction

Immersed boundary method (IBM) is a type of non-body-conforming method, which significantly simplifies the flow simulations with arbitrary complex boundaries. Our work is based on the Uhlmann's explicit direct forcing immersed boundary method [1], which is known not to precisely satisfy the non-slip boundary condition on the interface. To improve this, we propose a low computer-consuming implicit scheme in conjunction with the rotational incremental pressure-correction projection method of Guermond and Shen [2] for solving the incompressible Navier-Stokes equations. Numerical simulations of a two-dimensional flow over a circular cylinder at two different Reynold numbers are performed, which are in good agreement with the previous experimental and numerical results in the literature.

2. Numerical method

2.1. Rotational incremental pressure-correction projection method

Consider the incompressible Navier-Stokes equations

$$\frac{\partial \mathbf{u}}{\partial t} + \nabla \cdot (\mathbf{u}\mathbf{u}) = -\nabla p + \frac{1}{Re}\nabla^2 \mathbf{u} \tag{1}$$

$$\nabla \cdot \mathbf{u} = 0 \tag{2}$$

where $\mathbf{u}$, p, Re are the non-dimensionalized velocity vector, pressure, and the Reynolds number, respectively. The equations are discretized with finite difference scheme on a staggered grid [3]. All the spatial derivatives are approximated by the second-order central difference. The second-order Adams-Bashforth scheme is used for the temporal discretization of convective term and the Crank-Nicolson scheme for the diffusive term. This results in the following set of equations:

$$\frac{\mathbf{u}^{n+1}-\mathbf{u}^n}{\Delta t}+\left[\frac{3}{2}\mathcal{H}(\mathbf{u}^n)-\frac{1}{2}\mathcal{H}(\mathbf{u}^{n-1})\right]=-\mathcal{G}p^{n+1}+\frac{1}{2Re}\mathcal{L}\left(\mathbf{u}^{n+1}+\mathbf{u}^n\right) \tag{3}$$

$$\mathcal{D}\mathbf{u}^{n+1}=0 \tag{4}$$

where $\mathcal{H}$, $\mathcal{G}$, $\mathcal{D}$, $\mathcal{L}$ are the discrete convection, gradient, divergence and diffusion operators, respectively.

The above equations are solved with the rotational incremental pressure-correction projection method [2], in which an auxiliary velocity $\mathbf{u}^*$ is estimated with the pressure value at previous time level in the viscous step, and the pseudo pressure ϕ is used to project the predicted velocity into the divergence-free field. The two sub-steps are performed as:

$$\frac{\mathbf{u}^{*}-\mathbf{u}^n}{\Delta t}+\left[\frac{3}{2}\mathcal{H}(\mathbf{u}^n)-\frac{1}{2}\mathcal{H}(\mathbf{u}^{n-1})\right]=-\mathcal{G}p^{n}+\frac{1}{2Re}\mathcal{L}(\mathbf{u}^{*}+\mathbf{u}^n) \tag{5}$$

$$\frac{\mathbf{u}^{n+1}-\mathbf{u}^*}{\Delta t}=-\mathcal{G}\phi^{n+1} \tag{6}$$

By taking the divergence of Eq. (6) and applying the incompressibility constraint (4), the pressure corrector is obtained

$$\mathcal{L}\phi^{n+1}=\frac{1}{\Delta t}\mathcal{D}\mathbf{u}^* \tag{7}$$

The pressure is then advanced by

$$p^{n+1}=p^n+\phi^{n+1}-\frac{1}{2Re}\mathcal{D}\mathbf{u}^* \tag{8}$$

Therefore, the preceding equations are solved in the order (5), (7), (6), (8) at each time step.

2.2. Direct forcing immersed boundary method

In the direct forcing immersed boundary method of Uhlmann [1], the force $\mathbf{F}^{n+1}$ is determined on the Lagrangian points such that the desired velocity $\mathbf{U}^d$ is satisfied on the embedded boundary

$$\mathbf{F}^{n+1}=\frac{\mathbf{U}^d-\mathbf{U}^*}{\Delta t} \tag{9}$$

where the upper-case letters indicate the quantities evaluated at the Lagrangian locations, and $\mathbf{U}^*$ is the estimated velocity by all the explicit terms in the momentum equation. Therefore, the non-slip boundary condition is not fully verified. As the Navier-Stokes equations are inherently implicit, we need to iterate the whole system such that $\mathbf{F}^{n+1}$ is calculated by the final divergence-free velocity $\mathbf{U}^{n+1}$, which, however, is too cumbersome to perform.

In the present study, we propose to evaluate the force after the viscous step. Instead of iterating the whole system, we iterate the force within a small system coupled with the pressure poisson equation (PPE), which is often the most time-consuming part. As implied by Ji [4] that a considerable amount of computational time could be saved if the PPE is solved only once at each iteration within a time step no matter whether it is converged, since the intermediate pressure is not the true value. We follow this idea and iterate the small system in combination with GAMG (geometric algebraic multi-grid) pre-conditioner and BICGSTAB solver for the PPE by the library PETSc [5]. The modified algorithm is as follows:

Prediction: $\frac{\mathbf{u}^* - \mathbf{u}^n}{\Delta t} + \left[\frac{3}{2}\mathcal{H}(\mathbf{u}^n) - \frac{1}{2}\mathcal{H}(\mathbf{u}^{n-1})\right] = -\mathcal{G}p^n + \frac{1}{2Re}\mathcal{L}(\mathbf{u}^n + \mathbf{u}^*)$;

Initialization: $\mathbf{u}^k = \mathbf{u}^*, k = 0$;

while *TRUE* **do**

$$\mathbf{U}^k(s) = \sum_{\mathbf{x}\in\Omega_f} \mathbf{u}^k(\mathbf{x})\delta_h(\mathbf{x} - \mathbf{X}(s))\Delta\mathbf{x};$$

$$\mathbf{F}^{k+1}(s) = \frac{\mathbf{U}^d(s) - \mathbf{U}^k(s)}{\Delta t};$$

$$\mathbf{f}^{k+1}(\mathbf{x}) = \sum_{\mathbf{X}\in\Gamma_b} \mathbf{F}^{k+1}(s)\delta_h(\mathbf{x} - \mathbf{X}(s))\Delta s;$$

$$\mathcal{L}\phi^{k+1} = \frac{1}{\Delta t}\mathcal{D}\mathbf{u}^* + \mathcal{D}\mathbf{f}^{k+1};$$

$$\mathbf{u}^{k+1} = \mathbf{u}^* - \Delta t\mathcal{G}\phi^{k+1} + \Delta t\mathbf{f}^{k+1};$$

if $\|\mathbf{u}^{k+1} - \mathbf{u}^k\| < \epsilon$ *and* $\|\phi^{k+1} - \phi^k\| < \epsilon$ **then**

$\mathbf{u}^{n+1} = \mathbf{u}^{k+1}$;

$\phi^{n+1} = \phi^{k+1}$;

break;

else

$k = k + 1$;

end

end

Correction: $p^{n+1} = p^n + \phi^{n+1} - \frac{1}{2Re}\mathcal{D}\mathbf{u}^*$.

The three-points discrete delta function of Roma et al. [6] is employed for the regularization and interpolation.

$$\delta_h(r) = \begin{cases} \frac{1}{3h}(1 - \sqrt{3(r/h)^2 + 1}), & |r| \le 0.5h, \\ \frac{1}{6h}(5 - 3|r/h| + \sqrt{-3(1 - |r/h|)^2 + 1}), & 0.5h \le |r| \le 1.5h, \\ 0, & \text{otherwise.} \end{cases} \quad (10)$$

3. Results

3.1. Flow over a stationary circular cylinder

In the present work, the two-dimensional simulations of flow around a circular cylinder at two different Reynolds numbers $Re = 30, 185$ are performed, with the initial uniform velocity $u_\infty = 1$ and the cylinder diameter $D = 1$. The flow is simulated in a domain of $[-10D, 40D] \times [-15D, 15D]$. Inlet boundary condition is set to uniform ($u = u_\infty, v = 0$), while $\partial\mathbf{u}/\partial x = 0$ is applied at the outlet plane. Free-slip conditions are assigned at lateral boundaries.

At $Re = 30$, the flow presents steady-state characteristics with a recirculating region in the wake of the cylinder. Whose dimensions are characterized by the length of the wake l, the distance a from the cylinder to the vortex center, the distance b between two symmetry vortex centers, and the angle θ of flow separation measured from x-axis. The wake properties in the present study is compared against previous experimental and numerical studies in Table 1 and proved to be in accord. The streamlines and vorticity contours for this flow are shown in Fig. 1. At higher Reynolds number $Re = 185$, periodic vortex shedding is reproduced, Fig. 1. The

Strouhal number $St = Df_s/u_\infty$ (where f_s is the shedding frequency) is 0.19, which is in close agreement with the result of Pinelli et al. [7].

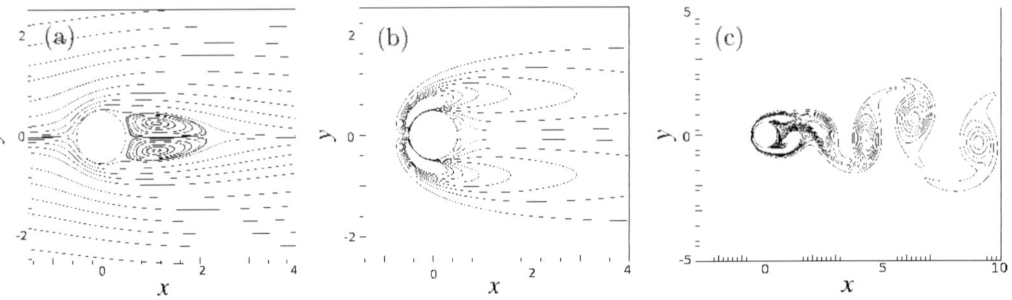

Figure 1: (a) streamlines at $Re = 30$, (b) vorticity contours at $Re = 30$, (c) vorticity contours at $Re = 185$, where all the contours levels are set from -3 to 3 in increments of 0.4.

Table 1: Comparison of experimental and numerical studies of steady-state wake dimensions for $Re = 30$, where experimental study is listed with (*).

	l/D	a/D	b/D	θ^o
Coutanceau and Bouard [8]*	1.55	0.54	0.54	50
Pinelli et al. [7]	1.70	0.56	0.52	48.05
Present	1.71	0.61	0.53	49.46

4. Conclusions

An implicit direct forcing immersed boundary method is developed with a small coupled system solved in an efficient way. The rotational incremental pressure-correction projection method is employed to enforce the incompressibility constraint. The accuracy of the proposed method has been proven by the numerical simulations and comparisons with the previous experimental and numerical result in the literature.

Acknowledgement

The first author acknowledge the China Scholarship Council for the financial support. The computations were performed on the computer cluster PILCAM2.

References

[1] Uhlmann M 2005 *J. Comput. Phys.* **209** 448 – 476
[2] Guermond J, Minev P and Shen J 2006 *Comput. Methods Appl. Mech. Engrg.* **195** 6011 – 6045
[3] Harlow F and Welch J 1965 *Phys. Fluids* **8** 2182–2189
[4] Ji C, Munjiza A and JJR W 2012 *J. Comput. Phys.* **231** 1797–1821
[5] Balay S, Gropp W D, McInnes L C and Smith B F 1997 *Modern Software Tools in Scientific Computing* ed Arge E, Bruaset A M and Langtangen H P (Birkhäuser Press) pp 163–202
[6] Roma A, Peskin C and Berger M 1999 *J. Comput. Phys.* **153** 509–534
[7] Pinelli A, Naqavi I, Piomelli U and Favier J 2010 *J. Comput. Phys.* **229** 9073 – 9091
[8] Coutanceau M and Bouard R 1977 *J. Fluid Mech.* **79** 231–256

Polarons on nonlinear lattice in the Su–Schrieffer– Heeger approximation Exact solution and multipeaked polarons.

T.Yu. Astakhova, V.A. Kashin, V.N. Likhachev, and G.A. Vinogradov*

Emanuel Institute of Biochemical Physics, Russian Academy of Sciences, Ul. Kosygina 4, Moscow 119991, Russian Federation

E-mail: *The corresponding author: gvin@deom.chph.ras.ru

Abstract.
We investigate the polaron dynamics on the nonlinear lattice with the cubic nonlinearity. The electron-phonon interaction is accounted in the Su-Schrieffer-Heeger approximation. An exact analytical solution is obtained in the continuum approximation at certain relation of parameters. The numerical simulation agrees with analytics very well. Moreover, colliding polarons recover their shapes and velocities after the elastic collision suggesting that the solution belongs to the exactly integrable system. When the continuum approximation is invalid (parameters of nonlinearity and electron-phonon interaction are not small), a new family of stable multipeaked polarons is found. These polarons are formed by the coupled solitons hold together by the electron-phonon interaction.

1. Introduction

The Su Schrieffer Heeger (SSH) approximation aimed at the accounting the electron-phonon interaction is know since 1979 [1]. Polarons as "self-trapped" charge carriers can explain many effects associated with the charge transport in nonmetallic materials. Special interest arose after the effective charge transport over long distances (tenth nanometers) was discovered in synthetic DNA and polypeptides [2-9] (see also reviews [2, 10-12]). E. Conwell with colleagues was the first who applied the SSH approximation in an attempts to describe the charge transfer in DNA [13, 14] using the polaron paradigm. This line of research was further extensively studied [15-20].

Obtaining the analytical solution for polarons is of primary interest as it allows to make qualitative and quantitative assessments of different properties. Polaron solutions on the harmonic lattice in the SSH approximation were found recently [21-23]. The solution has the hyperbolic secants form typical for the soliton solution.

In the present paper the analytical solution for polarons on the anharmonic α Fermi-Pasta-Ulam (FPU) lattice is derived at special relation between parameters of lattice nonlinearity α and electron-phonon interaction χ, and when they are both small. At larger parameters values a new family of multipeaked polarons is found in numerical modelling.

2. Exact solution in the continuum approximation

2.1. Theoretical model

We consider a lattice model of a molecular system (e.g. DNA duplex stack) consisting of N particles with free ends. The dimensionless hamiltonian consists of two contributions. One is classical lattice hamiltonian H_{lat} and the other accounts for the electron-phonon interaction H_{int}: $H = H_{\text{lat}} + H_{\text{int}}$, where the lattice hamiltonian reads

$$H_{\text{lat}} = \frac{1}{2}\sum_{j=1}^{N}\dot{x}_j^2 + \frac{1}{2}\sum_{j=1}^{N-1}(x_{j+1}-x_j)^2 - \frac{\alpha}{3}\sum_{j=1}^{N-1}(x_{j+1}-x_j)^3 \tag{1}$$

and α is the nonlinearity parameter. This α-FPU potential is chosen because it represents the series expansion up to the third order of such potentials as Morse, Lennard-Jones etc. The electron-phonon interaction is

$$H_{\text{int}} = \left\langle \mathbf{\Psi}^* \middle| \widehat{H}_{\text{ep}} \middle| \mathbf{\Psi} \right\rangle \tag{2}$$

where $\widehat{H}_{\text{ep}}$ in the matrix representation reads

$$\widehat{H}_{\text{ep}} = \begin{pmatrix} e_1 & t_1 & \dots & 0 & 0 \\ t_1 & e_2 & \dots & 0 & 0 \\ \dots & \dots & \dots & \dots & \dots \\ 0 & 0 & \dots & e_{N-1} & t_{N-1} \\ 0 & 0 & \dots & t_{N-1} & e_N \end{pmatrix} \qquad \text{where} \quad t_j = t_0 - \chi\,(x_{j+1} - x_j), \tag{3}$$

and the wave function $\mathbf{\Psi}$ is the N vector: $\mathbf{\Psi} = \psi_1, \psi_2, \dots, \psi_N$. If the lattice is comprised by equal particles then all on-site energies e_j are also equal and without loss of generality they can be put to zero, what means the electron energy point of reference.

It is more convenient to use variables $q_j \equiv (x_{j+1} - x_j)$. Then the coupled system of Newton's and Schrödinger equations arising from (1) (3) reads

$$\begin{aligned} \ddot{q}_j &= q_{j+1} - 2q_j + q_{j-1} - \alpha[(q_{j+1} - q_j)^2 - (q_j - q_{j-1})^2] - \\ &\quad \chi[(\psi_{j+1}^*\psi_{j+2} - 2\psi_j^*\psi_{j+1} + \psi_{j-1}^*\psi_j) + \text{c.c.}] \\ \dot{\psi}_j &= \frac{i}{\tilde{\hbar}}\left[(1 - \chi q_j)\psi_{j+1} + (1 - \chi q_{j-1})\psi_{j-1}\right]. \end{aligned} \tag{4}$$

where $\tilde{\hbar}$ is the dimensionless Planck's constant (in molecular systems $\tilde{\hbar} \sim 10^{-2}$).

2.2. The continuum approximation and the solution.

A general way of obtaining the solution of discrete equations like (4) is the usage of the continuum approximation. In the continuum approximation discrete variables are expanded into series, e.g. $q_{j\pm1} = q \pm a q_j' + \frac{a^2}{2}q'' \pm \frac{a^3}{3}q''' + \frac{a^4}{4}q'''' \pm \dots$ where superscripts mean spatial derivatives of the corresponding orders and a is a dimensionless parameter of expansion ($a = 1$). After the substitution of this expansions into (4) a system of partial differential equations (PDEs) is obtained:

$$\begin{aligned} q_{tt} &= \left(q_{xx} + \frac{1}{12}q_{xxxx}\right) - \alpha\left(q^2\right)_{xx} + 2\chi\left(\psi\psi^*\right)_{xx} \\ \psi_t &= \frac{i}{\tilde{\hbar}}[2(1 - \chi q)\,\psi + \psi_{xx}]. \end{aligned} \tag{5}$$

The PDEs system (5) does not belong to the class of exactly integrable equations and has no an exact solution. But if $\alpha = 2\chi$ then (5) coincides with the akharov-Shabat system [24] with the solution

$$q(x,t) = -\frac{A}{\cosh^2[d(x - v_{\text{p}}t)]}; \qquad \psi(x,t) = \frac{B\,\exp[i(kx + \omega t)]}{\cosh[d(x - v_{\text{p}}t)]}. \tag{6}$$

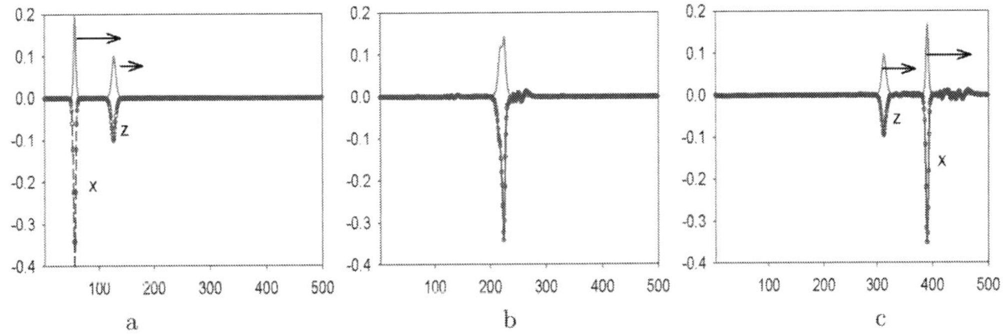

Figure 1. (Color online). Collision of two of polarons. a) At time moment $t = 0$ polarons are centered at sites $j_0^X = 56$ and $j_0^Z = 128$. The initial velocity of the left polaron is larger than the velocity of the right polaron . b) Polarons collision at $t = 200$. c) Polarons after the collision at $t = 400$: polaron overruns more slower polaron . Positive values modulus of the wave function, negative relative displacements. $N = 500$.

It can be shown that the soliton-like solution is one-parameteric and

$$d = \sqrt{\alpha A} = \sqrt{2\chi A}, \quad B = \sqrt{d/2}, \quad v_{\mathrm{p}} = \left(1 + \frac{2\alpha A}{3} - \sqrt{\frac{\chi^3}{A}}\right)^{1/2} \tag{7}$$

with A being the free parameter. In the phase of the wave wave function: $k \ll 1$; $\omega \gg 1$.

Note that if the electron-phonon interaction is absent, i.e. $\chi = 0$, then $q(x,t)$ is nothing else then soliton and its velocity coincides with the velocity of soliton on the α-FPU lattice [25]. But if the lattice is harmonic, i.e. $\alpha = 0$ then the polaron velocity coincides with the velocity on the harmonic lattice [21, 23]. Thus, the expression for the polaron velocity is correct in two limiting cases.

2.3. Numerical test of polaron stability

The solution (6) is checked in numerical simulation on the lattice with parameters $\alpha = 0.4$ and $\chi = 0.2$. The crucial experiment indicting the polaron stability is their collision. Figures 1 show the elastic collision of polarons. Initial conditions are chosen with parameters $A = 0.2$ and $A = 0.1$ for polarons ' ' and ' ', correspondingly. This numerical experiment has no physical meaning as it does not take into account the Coulomb interaction. Its primary goal is the demonstration of high polaron stability and the accuracy of the solution. Moreover, this result points to the fact that PDEs (5) belong to the exactly integrable system.

3. Polarons at arbitrary parameters values: multipeaked polarons.

The exact solution (6) is valid only if $\alpha = 2\chi$ and both parameters are small. In the general case these limitations are not fulfilled. Do polarons exist and are they stable at arbitrary parameters values? As the analytical solution is absent in this case, the numerical modelling is the only way to check this possibility. The parameter values $\alpha = 1.0$ and $\chi = 0.4$ are chosen for the more detailed analysis for definiteness. The results differ unessentially at other values of α and χ.

When the polaron velocity is comparable with or exceeds the sound velocity, new polaron shapes with the envelope consisting of few peaks emerge. Four examples for stable polarons with two, three, four and five peaks are shown in Fig. 2

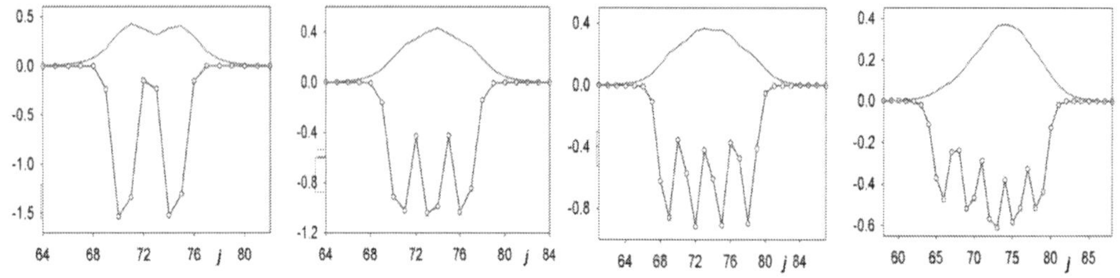

Figure 2. (Color online) Stable polarons with two, three, four and five peaks. Snapshots are shown at $t = 900$ when polarons travel along the lattice several hundreds lattice sites.

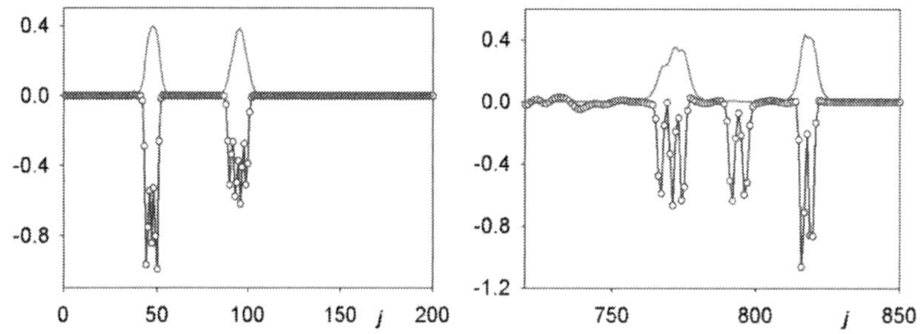

Figure 3. (Color online) Inelastic collision of two polarons. Left panel: initial positions of polarons. The left three peaked polaron has velocity larger then the right four-peaked polaron. Right panel: three-peaked and two-peaked polarons with two solitons in between after the collision at $t = 700$.

Multipeaked polarons were tested in collisions (see Fig. 3). The collision is inelastic in contrast to the elastic collision of "analytical" polarons. The results presented in Fig. 3 can help in elucidating the peaks nature. Indeed, there are seven peaks in total before collision (three-peaked polaron collides with the four-peaked polaron). Seven "peaks" are also observed after the collision. Five peaks belong to polarons (three peaked plus two peaked polarons). And two "peaks" are nothing else but solitons. Thus one can suspect that the multipeaked polaron might be comprised by solitons hold tightly by the electro-phonon interaction.

The following numerical experiment is performed to check this possibility. The two-soliton solution of the KdV equation is

$$q(x,t) = 1 + \exp\theta_1 + \exp\theta_2 + \left(\frac{a_1 - a_2}{a_1 + a_2}\right)\exp(\theta_1 + \theta_2), \tag{8}$$

where $\theta_i = a_i x - a_i^3 t$, $i = 1, 2$ and a_i are parameters. The initial relative displacements are (8) with $a_1 = 3.0$ and $a_2 = 2.95$. The initial wave function is the eigenfunction of the matrix (3) with the hopping integrals employing relative displacements (8). After quick self-organization, the two-peaked polaron is formed (Fig. 4). It looks very much like the earlier found two-peaked polaron (Fig. 2).

On contrary, if the polaron is "dressed down", i.e. the wave function is put to zero for a multipeaked polaron at any time instant, it decays into solitons. The number of solitons coincides with the number of peaks.

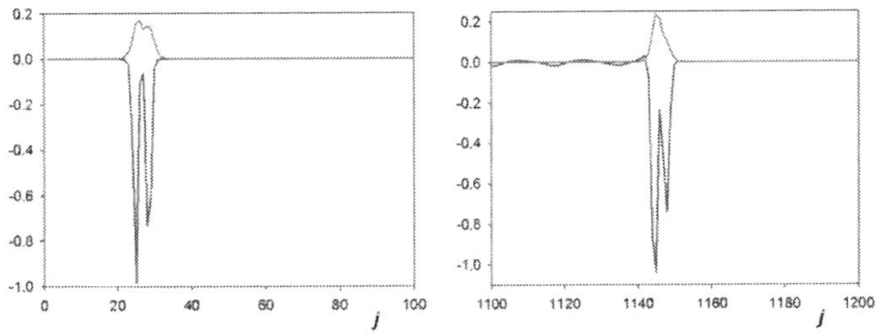

Figure 4. (Color online) Formation of two-peaked polaron. Left panel: initial condition two-closely located solitons are "dressed up" by the wave function. Right panel: snapshot of the initial condition evolution at $t = 1000$.

In conclusion: 1) we have found an exact solution for polarons on the α-FPU lattice, and 2) found a new family of multipeaked polarons formed by the coupled solitons hold together by the electron-phonon interaction.

[1] Su W P, Schrieffer J R and Heeger A J 1979 *Phys. Rev. Lett.* **42** 1698
[2] Genereux J C and Barton j K 2010 *Chem. Rev.* **110** 1642
[3] Genereux J C, Wuerth S M and Barton J K 2011 *J. Am. Chem. Soc.* **133** 3863
[4] Arikuma Y, Nakayama H, Morita T and Kimura S 2010 *Angew. Chem. Int. Ed.* **49** 1800
[5] Augustyn K E, Genereux J C and Barton J K 2007 *Angew. Chem. Int. Ed.* **46** 5731
[6] Barton J K, Olmon E D and Sontz P A 2011 *Coordin. Chem. Rev.* **255** 619
[7] Slinker J D, Muren N B, Renfrew S E and Barton J K 2011 *Nature Chemistry* **3** 228
[8] Henderson P T, Jones D, Hampikian G, Kan Y and Schuster G B 1999 *Proc. Natl. Acad. Sci. USA* **96** 8353
[9] Nunẽz M, Hall D B and Barton J K 1999 *Chem. Biol.* **6** 85
[10] Cordes M and Giese B 2009 *Chem. Soc. Rev.* **38** 892
[11] Astakhova T Yu, Likhachev V N and Vinogradov G A 2012 *Russian Chem. Rev.* **81** 994
[12] Lakhno V D 2010 *Intern. J. Quant. Chem.* **110** 127
[13] Conwell E M and Rakhmanova S V 2000 *Proc. Natl. Acad. Sci. USA* **97** 4556
[14] Rakhmanova S V and Conwell E M 2001 *J. Phys. Chem.* B **105** 2056
[15] Conwell E M and Basko D M 2001 *J. Am. Chem. Soc.* **123** 11441
[16] Conwell E M , Park J -H and Choi H -Y, 2005 *J. Phys. Chem.* B **109** 9760
[17] Conwell E M, McLaughlin P M and Bloch S M 2008 *J. Phys. Chem.* B **112** 2268
[18] Zhang G, Hu H, Cui S and Lv Z 2010 *Physica* B **405** 4382
[19] Zhang G, Cui P, Wu J and Liu Ch 2009 *Physica* B **404** 1485
[20] Wei J H, Liu X J, Berakdar J and Yan Y 2008 *J. Chem. Phys.* 2008 **128** 165101
[21] Carstea A S 2009 *Chaos Solitons and Fractals* **42** 923
[22] Li L, Li E and Wang M 2010 *Appl. Math. J. Chinese Univ.* **25** 454
[23] Astakhova T Yu , Likhachev V N and Vinigradov G A 2013 *Russ. J. Phys. Chem.* B **7** 521
[24] Dubrovin B A, Malanuyk T M, Krichever I M and Makhan'kov V G 1988 *Physics of Elementary Particles and Atomic Nuclei* **19** 579 (in Russian)
[25] Dauxious T and Peyrard M 2006 *Physics of Solitons* (Cambridge University Press, United Kingdom)

Journal of Physics: Conference Series **574** (2015) 012167 doi:10.1088/1742-6596/574/1/012167

Computational treatment of the neutrino signals produced in nuclear detectors

V. Tsakstara[1,2], T.S. Kosmas[1], J. Sinatkas[3]

[1]Theoretical Physics Section, University of Ioannina, GR 45110 Ioannina, Greece
[2] TEI of Western Macedonia, Dept. of Business Administration, GR-51100 Grevena, Greece
[3] TEI of Western Macedonia, Dept. of Information Engineering, GR-52100 Kastoria, Greece

E-mail: vtsaksta@cc.uoi.gr

Abstract.
We apply the convolution (folding) method in neutrino physics studies. We focus on its use to explore the response of some nuclear detectors to the energy spectra of laboratory neutrinos. After calculating the neutrino-nucleus cross sections (within the context of a nuclear model) for a neutrino detector, the obtained cross section values must be folded with a specific neutrino-energy distribution. In the present work we use the ν-distribution of pion-muon decay at rest neutrino beams created in muon factories like the Fermilab, J-PARC, and other laboratories. Due to the fact that neutrino-nucleus interactions are very weak, the evaluated cross sections are small ($\sim 10^{-42}\text{cm}^2$). Thus, one needs a very fine convolution tool to obtain accurate description of the ν-signals (laboratory, supernova neutrinos, etc.) recorded at some nuclear detectors.

1. Introduction

In recent years, there are two neutrino-beam facilities for measuring neutrino-nucleus cross sections: (i). The pion-muon decay at rest (DAR) neutrino facilities existing at Fermilab, USA, at J-PARC, Japan, etc., or proposed to operate near spallation neutron sources (ORLaND experiment, at ORNL, and European Spallation Source, Sweden) [1]. There, the neutrino beams may be intense neutrino pulses of energy spectra that could be unique for terrestrial studies of astrophysical reactions (solar neutrino reactions, etc.) and also suitable for investigation of other neutrino-mediated processes. (ii). The beta-beam neutrino sources which produce neutrino beams of low and intermediate energies through the use of boosted radioactive nuclei that decay by emitting $\beta^{\pm}$ particles and (anti)neutrinos. The neutrino beams of such facilities are intense, collimated and pure, appropriate for searching standard and non-standard ν-physics as well as for interpretating low-energy neutrino signals. For example, the analysis of supernova neutrino energy spectra, whenever observed, could be realized through proper measurements on low-energy beam spectra originating from boosted radioactive ions ^6He, ^{18}Ne, and other ions.

With the above neutrino-beam facilities the physics research that could be undertaken, is associated with the open issues of low- and intermediate-energy neutrinos within nuclear, particle physics, and astrophysics. Within this research there are some well motivated studies on prominent isotopes. They have the potential to act as neutrino detectors. Through these studies we may also investigate the structure properties of nuclei.

The energy spectra of the laboratory neutrinos (see Fig. 1) generated from muons decaying at rest (also the low-energy β-beam neutrinos), are similar to those of the neutrinos created in

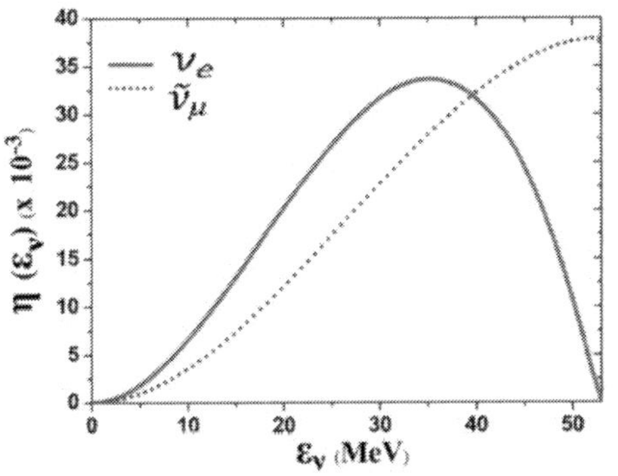

Figure 1. Energy-spectra of ν_e and $\tilde{\nu}_\mu$ neutrino beams, originating from the muon-decay reaction Eq. (1). The ν energy-distributions are described by the expressions of Eqs. (2).

the core collapse supernova, except in their high energy region where the Supernova neutrinos have a long tail in contrast to the sharp cutoff present in the spectrum of the laboratory neutrino beams. Despite this difference in the high energy region, the similarity in the energy range and spectrum of the supernova neutrinos with those coming from the muons decaying at rest, opens up the possibility of connecting the ground based neutrino-nucleus experiments with the study of supernova neutrino nuclear cross sections.

Such a type of ν-simulations, which could be useful in calibrating the terrestrial detectors proposed for the detection of supernova neutrinos, are going to be discussed in the present paper. This study is also useful in understanding the mechanism of r-process nucleosynthesis leading to the formation of heavy elements in the interstellar medium [3].

2. Energy-spectra of neutrino sources

2.1. Pion-muon decay at rest neutrino energy distributions

In the operating pion-muon decay at rest neutrino sources (Fermilab, at USA, J-PARC, at Japan) and the expected to operate neutrino facilities at the Neutron Spallation Sources (ORNL, USA, and Lund, Sweden), ν_e neutrinos, and $\tilde{\nu}_\mu$ anti-neutrinos are produced from the decay of muons according to the reaction

$$\mu^+ \to e^+ + \nu_e + \tilde{\nu}_\mu \, . \tag{1}$$

Because the decaying muons result from the decay of slow pions ($\pi^+ \to \mu^+ + \nu_\mu$), they have relatively low energies. In these facilities, the energy-spectra of ν_e and $\tilde{\nu}_\mu$ neutrinos (see Fig. 1) are approximately described by the normalized distributions [1, 2]

$$\eta_{\nu_e}(\varepsilon_\nu) = 96\varepsilon_\nu^2 M_\mu^{-4} \, (M_\mu - 2\varepsilon_\nu) \, , \qquad \eta_{\tilde{\nu}_\mu}(\varepsilon_\nu) = 16\varepsilon_\nu^2 M_\mu^{-4} \, (3M_\mu - 4\varepsilon_\nu) \, , \tag{2}$$

where $M_\mu = 105.6$ MeV, is the muon rest mass. The maximum energy of ν_e and $\tilde{\nu}_\mu$ in the later equations is $\varepsilon_\nu^{max} = 52.8 MeV = M_\mu/2$ [3, 4, 5]. The distribution of these ν_e neutrinos is known as Michel energy spectrum. It is worth noting that, the pion-muon decay at rest neutrino beams are not completely pure as, for example, the β-beam neutrinos.

From a simulation performance point of view, the analytic expressions of Eqs. (2) are convenient for the required numerical integration in the convolution procedure [4, 5]. On the other hand, their energy range and shape roughly resembles that of SN neutrino (alternative distributions are discussed in Refs. [3, 6]).

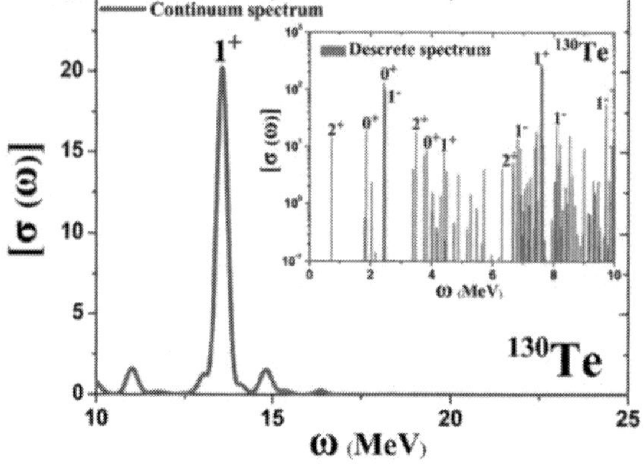

Figure 2. Total cross sections (in $10^{-42}cm^2$) for neutrino scattering on ^{130}Te isotope. In the inset contributions coming from the discrete spectrum are shown, while the solid line illustrates the contribution of the continuum nuclear spectrum.

The theoretical results of ν-nucleus cross sections (see Fig. 2 for the case of ^{130}Te-ν scattering) can be connected with the neutrino experiments and the neutrino sources discussed before. To this end, we have to carry out the folding (convolution) procedure on the calculated cross sections with the distribution $\eta(\varepsilon_\nu)$ of the neutrino source of interest. In this way we estimate the response of several nuclear isotopes to the corresponding neutrino spectrum.

3. The convolution procedure with laboratory neutrino distribution

The signal of a nuclear ν-detector (throughout its energy spectrum as a function of the incoming energy ε_ν created by a specific neutrino distribution $\eta(\varepsilon_\nu)$ is written as (see Ref. [7])

$$\sigma_{fold}(\varepsilon_\nu) = \sum_{\omega=\varepsilon_{thres}}^{\omega=\varepsilon_\nu} \sigma(\omega, \varepsilon_\nu)\eta(\varepsilon_\nu)\,. \tag{3}$$

For the total ν-nucleus cross section $\sigma_{tot}(\varepsilon_\nu)$, the signal on the detector of the ν spectral distribution $\eta(\varepsilon_\nu)$, is evaluated by

$$\sigma_{fold}^{sign}(\varepsilon_\nu) = \sigma_{tot}(\varepsilon_\nu)\eta^{sign}(\varepsilon_\nu)\,. \tag{4}$$

As a concrete example, in Fig. 3 we present results based on the latter expression for the cross sections of ν-scattering on ^{130}Te isotope. The folding is performed with the normalized energy distributions given by the expressions of Eq. (2). More specifically, with the distribution $\eta_{\nu_e}(\varepsilon_\nu)$ we obtain the signal of Fig. 3(left) and with the $\eta_{\tilde{\nu}_\mu}(\varepsilon_\nu)$ we obtain the signal of Fig. 3(right).

It is worth remarking that, difficulties arise due to the fact that, the nuclear excitation spectrum is partly discrete for low-lying transitions $\omega \leq 10MeV$ (see the inset of Fig.2) and partly continuum, i.e. above nucleon emission threshold, namely for $\varepsilon_\nu \succeq 10$ MeV (see Fig. 2). This imposes the use of discrete convolution for the discrete spectrum, and numerical integration for the continuum spectrum or the use of proper interpolation techniques [3, 6, 7]. We note that, in most of the previous similar convolutions, the discrete spectrum is not included in the integrals of Eqs. (3) and (4), but also in other convolution integrals. The latter observable is obtained by inserting the total neutrino-nucleus cross sections, $\sigma_{tot}(\varepsilon_\nu)$, in the convolution integral [2, 5, 7]

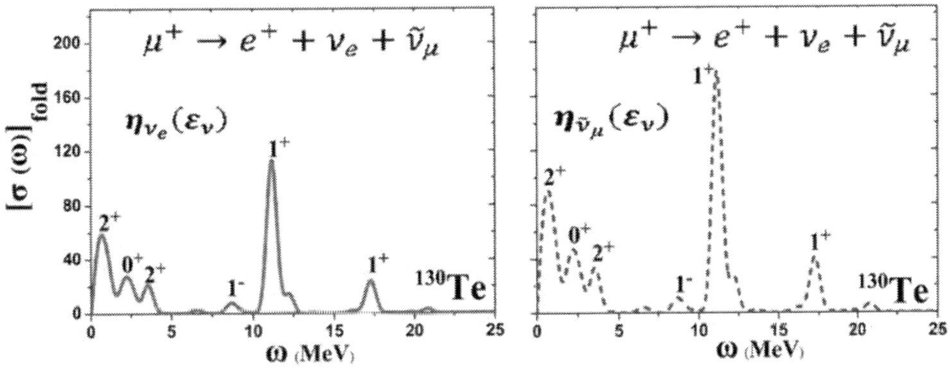

Figure 3. Folded total cross sections (in $10^{-42}cm^2$) for ν-^{130}Te scattering obtained by using the Michel spectrum $\eta_{\nu_e}(\varepsilon_\nu)$ (left panel), and the $\eta_{\tilde{\nu}_\mu}(\varepsilon_\nu)$ antineutrino spectrum (right panel).

which is a useful quantity like the known as flux averaged total cross section, $\langle\sigma_{tot}\rangle$, for various neutrino studies.

$$\langle\sigma_{tot}\rangle = \int_{\varepsilon_{thres}}^{\infty} \sigma_{tot}(\varepsilon_\nu)\eta(\varepsilon_\nu)d\varepsilon_\nu\,, \qquad (5)$$

(ε_{thres} is the threshold energy of the nuclear detector in question). For the sake of completeness, we mention that, we furthermore distinguish two special cases of the convolutions of Eqs. (3), (4) and (5): the radiochemical (σ_{rad}^{incoh}) which includes only the particle bound transitions (discrete spectrum), and the coherent flux-averaged cross sections (σ_{coh}^{fold}) which includes the ground-state to ground state transitions. These quantities are important to be computed for various neutrino sources.

4. Summary and Conclusions

In this work, we discussed the computational treatment of neutrino-nucleus cross section calculations performed with the use of the convolution method. In this way, the signal created at nuclear ν-detectors can be simulated by employing various ν-energy distributions. As a special case, the convolution involved in Eqs. (4) and (5) was carried out by employing the MERLIN package [8] which has been successfully used in similar studies in Refs. [3, 4, 5].

5. Acknowledgments

One of us (T.S.K) wishes to thank Dept. of Information Engineering, TEI of W. Macedonia for financial support.

References

[1] Louis W C 2009 *Prog. Part. Nucl. Phys.* C **63** 51.
[2] Kosmas T S and Oset E 1996 *Phys. Rev.* C **53** 1409.
[3] Tsakstara V and Kosmas T S 2011 *Phys. Rev.* C **84** 064620 and references therein.
[4] Tsakstara V and Kosmas T S 2011 *Phys. Rev.* C **83** 054612.
[5] Tsakstara V and Kosmas T S 2012 *Phys. Rev.* C **86** 044618.
[6] Jachowicz N, McLaughlin G C, Volpe C 2008 *Phys. Rev.* C **77**, 055501.
[7] Tsakstara V, Kosmas T S, Sinatkas J et. al. 2007 *Proc. Int. Conf. on Computational Methods in Science and Engineering (Corfu)* vol 2 (AIP) p 1383 and references therein.
[8] Lagaris I E, Likas A, Fotiadis D 1997 *Comput. Phys. Commun.* **104** 1; Kosmas T S, Lagaris I E 2002 *J. Phys.* G **28** 2907.

Using simulated annealing algorithms to solve the Schrödinger equation in muonic atoms

I.S. Kardaras[1] **and O.T. Kosmas**[2]

[1] Theoretical Physics Section, University of Ioannina, GR-45110 Ioannina, Greece
[2] Chair of Applied Dynamics,University of Erlangen-Nuremberg, Germany

E-mail: `iokard@uoi.gr`, `odysseas.kosmas@ltd.uni-erlangen.de`

Abstract. In many physical problems, the computation of exact wave functions for muons (particles about two hundred times heavier than electrons), bound in the extended Coulomb field created by the atomic nucleus, is required. Even though the problem is trivial under the assumption of point-like nuclear systems, the consideration of the nuclear finite-size necessitates the use of advantageous numerical techniques. In the case of non-relativistic bound muons, the solution of the Schrödinger equation is reliable, but for a relativistic description the solution of the Dirac equations for the bound muon is needed. In the present contribution, as a first step, we attempt to derive a method for solving the Schrödinger equation on the basis of simulated annealing algorithms. To this end, one may optimize appropriate parametric expressions for the wave function of a muon orbiting around complex nuclei by employing the simulated annealing method recently constructed to minimize multi parametric expressions in several physical applications.

1. Motivation

As is well known, when negative muons, μ^-, produced in a meson factory (e.g. at the Fermilab, USA, or at the J-PARC, Japan), slow down in matter, it is possible for them to be captured in atomic orbits (in this way muonic atoms are produced). Afterwards, fast electromagnetic cascades bring the muon of the muonic atom down to the innermost (1s or 2p) quantum orbits [1]. A bound in such an orbit muon may disappear either by decay known as muon decay in orbit or by capture by the nucleus. The main reaction-channel of the muon capture is the ordinary muon capture represented by the reaction [1]

$$\mu_b^- + (A, Z) \to (A, Z-1)^* + \nu_\mu \,, \tag{1}$$

where (A, Z) denotes the initial atomic nucleus with mass number A and proton number Z (A and Z are considered integer numbers). Even though processes like the reaction (1) have been the subject of extensive experimental and theoretical investigations started on the early 50's, recently the interest has been revived due to the important role they play in physical applications and astrophysics. In these studies mostly a mean value of the muon wave function, $\Phi_\mu(r)$, with r being the spherical coordinate, has been utilized. However, for a reliable description of the reaction (1) and of any reaction having the same initial state with it, i.e. a muon orbiting around an atomic nucleus (A, Z), the derivation of an exact muon wave function is necessary.

It is the purpose of our present work to construct an advantageous method that would provide an accurate muon wave functions by solving the Schrödinger equation (or the Dirac equations) which obeys a muon bound in the extended Coulomb field of the nucleus of such a muonic atom. Our method is exploiting the advantages of the simulated annealing algorithms [2].

2. Brief description of the Formalism

In several approaches the description of the muon capture process (1) is based on non-relativistic muons [1, 3]. Then, one needs to solve the Schrödinger equation which gives the radial part of the muon wave function $\Phi_\mu(r)$, or the known as the reduced radial wave function $U(r)$ defined as

$$U(r) = r\Phi_\mu(r) \,. \tag{2}$$

In the latter case, the Schrödinger equation is written as [4]

$$-\frac{\hbar^2}{2\mu}\frac{d^2}{dr^2}U(r) + V(r)U(r) = EU(r) \,. \tag{3}$$

where E is the total energy of the muon (energy eigenvalue). In the latter equation μ denotes the reduced mass of the muon-nucleus system given by

$$\frac{1}{\mu} = \frac{1}{m_\mu} + \frac{1}{Zm_p + (A-Z)m_n}, \tag{4}$$

where m_μ is the muon rest mass and m_p and m_n stand for the mass of the proton and neutron, respectively. In (3) $V(r)$ represents the extended Coulomb potential originating from the finite-size nuclear charge-density distribution $\rho(r)$ which for various spherically symmetric nuclei is available from electron scattering data [4]. For a point-like nucleus, Eq. (3) can be solved trivially. However, by taking into consideration the finite size of the nucleus, $V(r)$ is obtained from the expression

$$V(r) = -e^2 \int \frac{\rho(r')}{|r-r'|} d^3r' \,. \tag{5}$$

(e denotes the electric charge of the proton) Then, in order to solve the Schrödinger equation (3) one needs to apply special numerical integration techniques [4, 5, 6].

By multiplying (3) by $U(r)$ and integrating we obtain

$$-\frac{\hbar^2}{2m}\int U(r)\frac{d^2}{dr^2}U(r)dr + \int V(r)U(r)^2 dr = E\int U(r)^2 dr. \tag{6}$$

Integrating by parts the term with the second derivative (using boundary conditions the term $U(r)dU(r)/dr$ vanishes) and solving for the energy E the latter equation, we find

$$E = N_0 \int_0^\infty [\frac{\hbar^2}{2\mu}(\frac{dU(r)}{dr})^2 + V(r)U^2(r)]dr, \tag{7}$$

where N_0 is the normalization factor

$$N_0 = \left[\int_0^\infty U^2(r)dr\right]^{-1} . \tag{8}$$

In solving the second order differential equation (3), Eq. (7) acts as a constraint.

We mention that $U(r)$ obeys the boundary condition $U(r=0) = 0$ and has asymptotic behavior $U(r) \sim e^{-kr}$, with $k > 0$ (usually $0.1 \le k \le 0.6$). Equation (3) has been solved previously by assuming that $U(r)$ can be written as a superposition of appropriately parametrized sigmoid functions as explained below [4].

3. Solving the Schrödinger equation with artificial neural networks

At first we derive a grid from $r = 0$ up to a point $r = b$ (b is dependent on the nuclear system in question, $30 \leq b \leq 100fm$) where the wave function practically vanishes, and denote it by r_j, for $j = 1, 2, \ldots, n$. Then (3) must hold at every point r_j of the grid. This is equivalently expressed as

$$\sum_{j=1}^{n} \left[\left(-\frac{\hbar^2}{2\mu} \frac{d^2}{dr^2} + V(r_j) - E \right) U(r_j) \right]^2 = 0 \tag{9}$$

The artificial neural networks approach, chosen to solve the latter equation, consists in parameterizing $U(r)$ and then minimizing the left-hand side of (9) divided by the normalization $\int U(r)^2 dr$, so as to avoid the trivial solution $U(r) = 0$ everywhere [7]. To this aim we use the parametrization

$$U(r) = re^{-kr} N(r, \mathbf{u}, \mathbf{v}, \mathbf{w}), \qquad k > 0, \tag{10}$$

where $N(r, \mathbf{u}, \mathbf{v}, \mathbf{w})$ is a feed-forward artificial neural network with one hidden layer and one input unit (r). The biases are denoted by $\mathbf{u} = (u_1, u_2, \ldots, u_m)$, where m is the number of hidden units. The weights to the hidden layers are denoted by $\mathbf{w} = (w_1, w_2, \ldots, w_m)$ and the weights to the output by $\mathbf{v} = (v_1, v_2, \ldots, v_m)$. The hidden layer units have sigmoid activations of the form $f(x) = (1 + e^{-x})^{-1}$. Thus, the artificial neural network is defined as

$$N(r, \mathbf{u}, \mathbf{v}, \mathbf{w}) = \sum_{i=1}^{m} u_i f(w_i r + v_i). \tag{11}$$

To obtain the precise expression for the reduced radial wave function $U(r)$, we insert this form in equation (10). By combining the latter equation with (9) we get

$$\mathcal{C} = \sum_{j=1}^{n} \left[\left(-\frac{\hbar^2}{2\mu} \frac{d^2}{dr^2} + V(r_j) - E \right) \left(r_j e^{-kr_j} \sum_{i=1}^{m} u_i f(w_{ji} r_j + v_i) \right) \right]^2 = 0. \tag{12}$$

In the present paper, inspired by the simulated annealing method of Ref. [2], we will derive an algorithm to optimize the latter expression which would play the role of a total cost function $\mathcal{C}$.

4. Optimization using simulated annealing techniques

We consider that the initial radial part of the μ^- wave function $U(r)$ is divided into several line segments, the number of which depends on the nuclear system in question. In this way, any radial wave function may be represented by a set of way-points $U_i = U(r_i)$. The total cost of the initial radial wave function $U(r)$ is readily calculated [see Eq. (12)]. Then, at any iteration step, every way-point of the wave function $U(r)$ is moved by an elementary length perpendicular to the line which connects the initial and the end points of the segment. This elementary length is specified by the Coulomb potential and the nuclear system (region in which the nuclear charge density $\rho(r)$ is, practically, non-zero). Every displacement has a positive or negative contribution to the total cost and it is accepted even if it has a positive contribution to the total cost with a probability, however, which depends on a parameter T (the temperature-parameter of the algorithm) that plays the same role as the temperature of a physical system. In simulated annealing methods the probability distribution is described by [2]

$$p(E) \equiv Prob(E) \approx exp(-E/T) \tag{13}$$

(the so-called Boltzmann probability distribution) where the temperature T and energy E are measured in MeV). The physical meaning of the latter equation is that a system in thermal equilibrium at temperature T has its energy states probabilistically distributed according to their energy E. Even at low temperature, there is a chance for the system to get out of a local energy minimum in favor of finding a better, more global energy minimum. In other words, the system sometimes goes uphill as well as downhill, but the lower the temperature, the less likelihood for any significant uphill transition. Initially, the temperature-parameter T of the algorithm is high, but as the algorithm proceeds, the temperature is approaching to zero. At this point, only movements with negative contributions are accepted. This method, known as simulated annealing, is used to prevent the algorithm from being trapped in local minima through the above mechanism it proceeds further to find the global minimum (or minima) [2].

4.1. Description of the code for the minimization process

As mentioned before, simulated annealing mimics the annealing process to solve an optimization problem. The temperature parameter T [see Eq. (13)] controls the search which typically starts off high and is slowly "cooled" or lowered in every iteration. At each iteration a new point is generated and its distance from the current point is proportional to the temperature. If the new point has a better function value it replaces the current point and iteration counter is incremented. It is possible, however, to accept and move forward with a worse point, but the probability of doing so is directly dependent on the temperature T (this step, sometimes, helps identify a new search region in hope of finding a better minimum and protects the algorithm from being trapped in local minima).

Our objective function $\mathcal{C}$ is parameterized by the arguments $\mathbf{u}, \mathbf{v}, \mathbf{w}$ that act as constants during the optimization. These parameters can change to create a family of objective functions (they are not varied as part of the minimization). We should note that, there is no systematic way to decide if the calculated radial wave function $U(r)$ is the optimal one. However, in our cases the energy E is very critical and, thus, the optimal wave function $U(r)$ is close to that corresponding to the minimum energy. We are currently working on the derivation and checks of a MATLAB code for the minimization process and we expect to obtain results for concrete examples soon.

5. Summary and Conclusions

Exact bound-muon wave functions can be computed within the context of simulated annealing algorithms by solving the Schrödinger equation and can be compared with those obtained by previous numerical methods. Afterwards, improved wave functions for relativistic muons, coming out of similar solutions of the two-component Dirac equation could be obtained in a straightforward way by extending our present method.

References

[1] Measday D F 2001 Phys. Rep. **354** 243.
[2] Kosmas O T and Vlachos D 2012 Computers & Operations Research **39** 576.
[3] Suzuki T, Measday D F, and Roalsvig J P 1987 Phys. Rev. **C 35** 2212.
[4] Kosmas T S and Lagaris I E 2002 J. Phys. **G 28** 2907.
[5] Zinner N T, Laganke K, Vogel P Phys. Rev. **C 74** 024326, and References therein.
[6] Kardaras I S 2010 MSc. Thesis, Univ. of Ioannina.
[7] Tsoulos I G 2008 Appl. Math. and Comput. **203** 598.

Journal of Physics: Conference Series **574** (2015) 012169 doi:10.1088/1742-6596/574/1/012169

First-principles Study of Structural Properties of $Mg_xZn_{1-x}O$ ternary alloys

J. Gao, G. J. Zhao, X. X. Liang and T. L. Song
Department of Physics, School of Physical Science and Technology, Inner Mongolia University, Hohhot 010021, P. R . China

Email: stzhaogj@imu.edu.cn

Abstract. First-principles calculations have been carried out to investigate the structural and electronic properties of $Mg_xZn_{1-x}O$ ternary alloys. The calculations are performed using the full potential linearized augmented plane wave (FP-LAPW) method within the density functional theory(DFT). We conclude that the structural properties of these materials, in particular the composition dependence on the lattice constant and the band gap is found to be linear. The *a*-axis length in the lattice gradually increases, while the *c*-axis length decreases with the increase in Mg doping concentration, and be corresponded with the Vegard's law linear rule. The lattice parameters of the $Mg_xZn_{1-x}O$ ternary alloys are consistent with experimental data and other theoretical results. We found in the conduction band portion, the Mg 2p 2s states are moved to high energy region as the Mg content increases, so the band gap increases.

1. Introduction

During the past few years, the II-VI compound semiconductors have received considerable interest from both experimental and theoretical points of view. This is due to their potential technological applications in light-emitting diodes (LEDs) and laser diodes (LDs)[1]. As a typical II-VI wide-gap semiconductor and oxide, ZnO have been of growing interest and have shown great potential in ultraviolet photoelectric devices application, especially in deep ultraviolet light emitters and detectors application, because of its wide band gap(3.37eV) and large exciton binding energy (60 meV) at room temperature[2-3]. It has wide range of technological applications such as transparent conducting[4] electrodes in solar cells and flat panel displays, surface acoustic wave devices and gas sensors[5-7]. All the time, there are two important requirements in fabricating ZnO laser diodes are p-type doping and band gap engineering in alloy semiconductors to create barrier layers and quantum wells which facilitate radiative recombination by carrier confinement. The addition of impurities among the wide band gap semiconductors often induces dramatic changes in their structural and optical properties. Ternary $Mg_xZn_{1-x}O$ compounds are ideal materials for the development of deep ultraviolet photoelectric devices because they have many particular advantages, such as availability of lattice-matched single-crystal substrates, wide tunable band gap(3.3 to 7.8eV), low growth temperature (100-750 °C), high radiation hardness[8-9]. In addition, because the ionic radius of Mg^{2+}(0.57Å) is similar to that of Zn^{2+}(0.60Å), Zn can be substituted by Mg without much lattice distortion[10], Furthermore, the bonding strength of Mg-O is stronger than that of Zn-O, therefore MgZnO is expected to have higher lattice stability than ZnO. Moreover, the accomplishment of UV LEDs based on MgZnO has been demonstrated very recently[11].

In order to understand the nature of these materials better we present the results of first-principles calculations of ternary alloys $Mg_xZn_{1-x}O$ in wurtzite structure. In this work, we study the electronic and structural properties of $Mg_xZn_{1-x}O$ ternary alloys in wurtzite phase over a wide range of compositions $0 \leq x \leq 1$ using the full potential linearized augmented plane wave (FP-LAPW) method within the density functional theory.

2. Calculation Method

The calculations presented in this work were performed using first-principles, plane-wave pseudopotential approach within the framework of density-functional theory(DFT) implemented in the Cambridge Serial Total Energy Package(CASTEP) codes[12]. For the exchange-correlation potential, we used the generalized gradient approximation(GGA) of Perdew and Wang, known as PW91[13]. The plane-wave cutoff energy is set to be 400eV in the present calculations. The special points sampling integration over the Brillouin zone is employed by using the Monkhorst-Pack method [14] with a 4×4×5 special k-point mesh. The Mg $2p^6 3s^2$, Zn $3d^{10} 4s^2$ and O $2s^2 2p^4$ electrons are treated as valence states. The Brodyden-Fletcher-Goldfar-Shanno(BFGS) minimization scheme[15] was used in geometry optimization. The tolerances for geometry optimization were set as the difference in total energy being within 2×10-5eV/atom, the maximum ionic Hellmann-Feynman force within 0.05eV/Å, the maximum ionic displacement within 0.002Å and the maximum stress within 0.1GPa. Figure 1 depicts the schematics of $Mg_xZn_{1-x}O$ alloys.

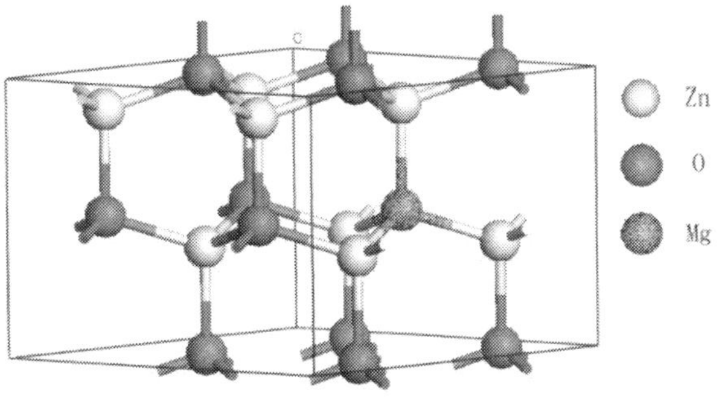

Figure1. The crystal structure of wurtzite $Mg_xZn_{1-x}O$ alloys.

3. Results and discussions

3.1 Structural properties

In the present work, we model the alloys at some selected compositions with the ordered structures described in terms of periodically repeated supercells with sixteen atoms per unit cell. For the considered structures, we perform the structural optimization by minimizing the total energy with respect to the cell parameters and also the atomic positions. Our calculated values for the equilibrium lattice constants for $Mg_xZn_{1-x}O$ alloys are given in Table 1. Usually, in the treatment of alloy problems, it is assumed that the atoms are located at ideal lattice sites and the lattice constants of alloys should vary linearly with compositions x according to the so-called Vegard's law[16]. So the lattice constant of ternary $Mg_xZn_{1-x}O$ alloys can be expressed by Vegard's law.

$$a\left(Mg_xZn_{1-x}O\right) = xa_{MgO} + \left(1-x\right)a_{ZnO} \quad (1)$$

Figure 2 shows the variation of the calculated equilibrium lattice constants versus concentration for $Mg_xZn_{1-x}O$ alloys. From the figure 2 we can conclude that the a-axis length in the lattice gradually increases, while the c-axis length decreases with increasing Mg content, consistent with experimental results and other theoretical results[17]. The linear equation of lattice parameter a and c are obtained using the linear fitting method. We reach the conclusion that the results are very close to that of other papers' conclusion [18].

$$\begin{aligned} a &= 3.28162 + 0.06653x \\ c &= 5.31564 - 0.15373x \end{aligned} \tag{2}$$

Table 1. Calculated lattice parameter a and c for ZnO(WZ) and MgO(RS) and their alloys at equilibrium volume. Compared with other theoretical calculations and experimental results.

	Lattice constant a (Å)			Lattice constant c (Å)		
x	This work	Exp.	Other calculations	This work	Exp.	Other calculations
0	3.280	3.249[c]	3.280[a]	5.280	5.206[c]	5.296[a]
			3.275[b]			5.270[b]
0.125	3.291			5.291		
0.25	3.295		3.297[b]	5.288		5.248[b]
0.375	3.303			5.269		
0.5	3.308		3.313[b]	5.275		5.22[b]
0.625	3.324			5.218		
0.75	3.327		3.325[b]	5.215		5.203[b]
0.875	3.337			5.191		
1	3.359		3.333[b]	5.119		5.175[b]

[a] Ref [19], [b] Ref [17], [c] Ref [20]

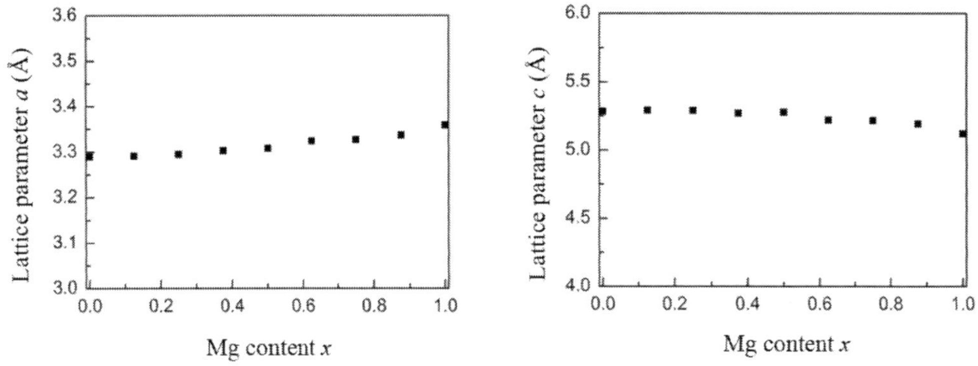

Figure 2. Calculated equilibrium lattice constants of $Mg_xZn_{1-x}O$ alloys for different Mg concentrations.

3.2. Electronic properties

3.2.1. Band structures

We have calculated the energy band for the $Mg_xZn_{1-x}O$ alloys along the high directions in the first Brillouin zone at the calculated equilibrium lattice constants. Figure 3 shows the variation of band gap with Mg content in hexagonal $Mg_xZn_{1-x}O$ crystals with $0 \le x \le 1$. In fact, it is well known that the GGA usually underestimates the experimental energy band gap and this is an intrinsic feature of the density functional theory (DFT), DFT being a ground-state theory is not suitable for describing excited-state properties, such as the energy gap. However, it is widely accepted that GGA (LDA) electronic band structures are qualitatively in good agreement with the experiments as regards the ordering of the energy levels and the shape of the bands.

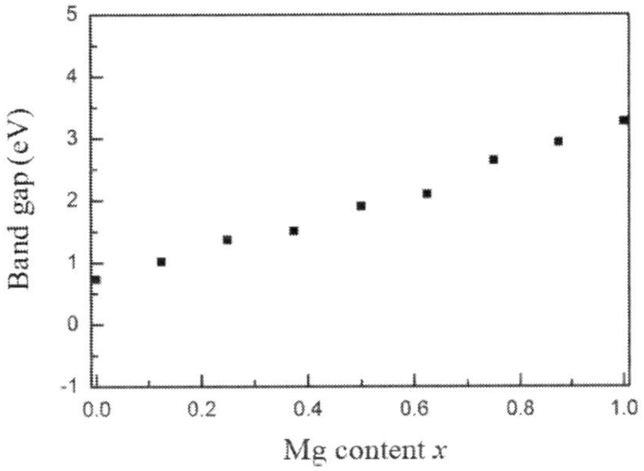

Figure 3. Composition dependence of the calculated band gap for $Mg_xZn_{1-x}O$ alloys.

From the Figure 3 we found that with the increase of components the energy gap of the alloys is also increased, it is clearly seen that the calculated band gap exhibits strong composition dependence for $Mg_xZn_{1-x}O$ alloys. This is different from conventional III-V alloys which show a weakly compositional dependent energy gap.

3.2.2. Density of states

We also calculated the partial and total density of states(DOS) of $Mg_xZn_{1-x}O$ in the WZ structures. Due to the close similarity between the results obtained for these alloys, the DOS is given only for the$Mg_{0.75}Zn_{0.25}O$ compound as shown in Figure 4. From the partial DOS, we find that the anion (O) s states are strongly localized in the energy range from -17eV to -15eV and the upper valence band is derived mainly from the hybridization of Zn d and O p states. In the conduction band, the Mg, Zn and O states become partially occupied, while the conduction band is mainly dominated by cation states (Mg). In order to interpret the effect of Mg on the electronic structure, we have offered the changing curve of the density of states with different concentration.

It clearly indicates that in the conduction band portion the Mg 2p 3s states are moved to high energy region as the Mg content increases, and the Mg 2p state is more and more dominant in Figure 5. Therefore it leads to an increase in the band gap.

4. Conclusions

In this study, we have presented a complete theoretical analysis of the structural and electronic properties of $Mg_xZn_{1-x}O$ alloys by the first-principles density functional calculations. The structural parameters of $Mg_xZn_{1-x}O$ alloys are fully relaxed and have been optimized. The calculated results show the calculated lattice constants scale linearly with composition, showing the validity of Vegard's linear rule in the definition of lattice constants of $Mg_xZn_{1-x}O$ alloys. We have investigated the composition dependence of the lattice constant and band gap. It can be expected that some of our calculated results will be useful for the device applications of $Mg_xZn_{1-x}O$ alloys and can be verified in the future experiments.

5. Acknowledgments

One of the authors (J. Gao) is supported by the Foundation of enhancing comprehensive strength project of Inner Mongolia University(No.14020202), and the research is supported in part by the PhD Progress Foundation of Higher Education Institutions of China(20111501110003), Natural Science Foundation of Inner Mongolia (No.2011MS0105) and Talent Development Foundation of Inner Mongolia.

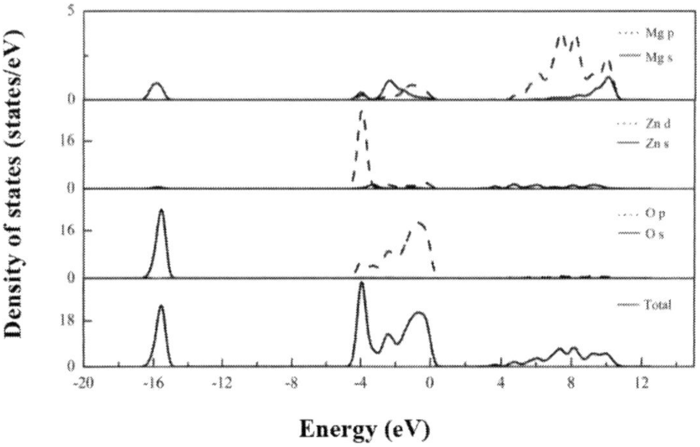

Figure 4. The partial density of states of $Mg_xZn_{1-x}O$ with $x = 75\%$. The position of Fermi level is located at 0 eV.

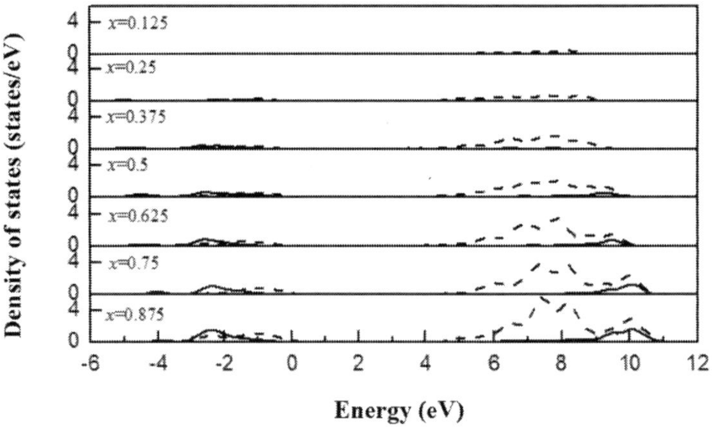

Figure 5. The partial density of states of Mg 2p(dashed line), 3s(solid line) bands in hexagonal $Mg_xZn_{1-x}O$ crystals($x = 0.125, 0.375, 0.5, 0.625, 0.75, 0.875$ respectively).

References

[1] Li X P, Zhang B L, Zhu H C, Dong X, Xia X C, Cui Y G, Ma Y and Du G T 2008 *J. Phys.* D: Appl. Phys. **41** 035101
[2] Ko H J, Chen Y F, Zhu Z, Yao T, Kaobayyashi I and Uchiki H 2000 *Appl. Phys. Lett.* **76** 1905
[3] Park W I, Jun Y H and Jung S W 2003 *Appl. Phys. Lett.* 82 964
[4] Zhao Q, Zhang H Z, Zhu Y W, Feng S Q, Sun X C, Xu J 2005 *Appl. Phys. Lett.* **86** 203115
[5] Wong L M, Chiam S Y, Huang J Q, Wang S J 2010 *J. Appl. Phys.* **108** 033702
[6] Liao L, Lu H B, Li J C, Liu C 2007 *Appl. Phys. Lett.* **91** 173110

[7] Ahn M W, Park K S,Heo J H, Park J G, Kim D W and Choi K J 2008 *Appl. Phys.Lett.* **93** 263103
[8] Hierro A , Tabares G, Ulloa J M , Muñoz E, Nakamura A, Hayashi T, Temmyo J 2009 *Appl. Phys. Lett.* **94** 232101
[9] Ghosh R, Basak D J 2007 *Appl. Phys.* **101** 113111.
[10] Ku C J, Duan Z Q, Reyes P I, Lu Y C, Xu Y 2011 *Appl. Phys. Lett.* **98** 123511
[11] Nakahara K, Akasaka S, Yuji H and Tamura K 2010 *Appl. Phys. Lett.* **97** 013501
[12] Payne M C, Teter M, Allan D C and Joannopoulos J D 1992 *Rev. Mod. Phys.* **64** 1045
[13] Perdew J P, Chevary J A, Vosko S H, Jackson K A, Pederson M R, Singh D J and Fiolhais C 1992 *Phys. Rev.* B **46** 6671
[14] Monkhorst H J and Pack J D 1976 *Phys. Rev.* B **13** 5188
[15] Fischer T H and Almlof J 1992 *J. Phys. Chem.* **96** 9768
[16] Denton A R and Ashcrof N W 1991 *Phys. Rev.*A **43** 3161
[17] Chang Y S, Chien C T, Chen C W, Chu T Y, Chiang H H, Ku C H, Wu J J, Lin C S, Chen L C, Chen K H 2007 *J. Appl. Phys.* **101** 033502
[18] Ohtomo A, Kawasaki M, Ohkubo I, Koinuma H, Yasuda T, and Segawa Y 1999 *Appl. Phys. Lett.* **75** 980
[19] Wang Z J, Li S C, Wang L Y, Liu Z 2009 *Chin. Phys.* B **18** 2992
[20] Kim W J, Leem T H, Han M S, Park I M, Ryu Y R and Lee T S 2006 *J. Appl. Phys.* **99** 096104

Specific heat amplitude ratios for generic competing systems

Marcone I. de Sena Jr. and Marcelo M. Leite

Laboratório de Física Teórica e Computacional, Departamento de Física, Universidade Federal de Pernambuco, 50670-901, Recife, PE, Brazil

E-mail: marcone.sena@df.ufpe.br, mleite@df.ufpe.br

Abstract. *The specific heat amplitude ratios for generic anisotropic as well as isotropic Lifshitz critical behaviors for the N-vector model are computed at one-loop level using the ϵ_L-expansion. In the anisotropic case, we show that the universal result obtained reduces very easily to that from the simpler m-axial universality class result. In the isotropic case, if n is the number of neighbors coupled via competing interactions, we demonstrate that the ratio vanishes close to $n = 4$ and becomes negative for $n > 4$ when it is calculated exactly, which is rather odd. The evaluation using the orthogonal approximation is shown to yield positive results for arbitrary n. Explicit computations for the case $n = 2, d = 3, N = 1$ yield an exact amplitude ratio equal to 0.06, with the approximate amplitude ratio being 1.17. We discuss two physical mechanisms to pick out one of the amplitude ratio values. We propose an experiment in homopolymer-diblock copolymer blends in order to determine the amplitude ratio.*

1. Introduction

Competition represents the balance between attractive and repulsive microscopic short range forces in macroscopical systems. Their description is different from ordinary physical criticality where only either attractive or repulsive interactions take place. The main distinction is the need to including additional parameters, the number and kinds of "competing axes" which indicate how far the attractive-repulsive couplings extend themselves along certain space directions [1, 2]. (In the language of magnetic systems n is the number of neighbors interacting via competing forces.)

Specific heat amplitude ratio is an example of a universal critical property which has not been investigated yet for this sort of system. (A recent study of the susceptibility amplitude ratio of this universality class can be found in [3].) In this work we compute this amplitude ratio for anisotropic as well as isotropic systems. In the anisotropic cases we show how general competition universality classes reduce to those from the m-axial Lifshitz criticalities, whereas a similar property is demonstrated for generic isotropic systems too.

Although (second character) m-axial isotropic Lisfhitz points [4] have been reported in the literature [7, 8, 9, 10] from theoretical and experimental results in unconventional polymers, recent results taking into account the effect of fluctuations on these systems [11, 12] make the subject far from being completely understood. Besides, the appearance of the isotropic case $n = 6$ at mean-field level has been discussed in mixtures of $(AB - BC)$ diblock copolymers [13].

Hence, we found convenient to propose an experiment in polymers in the attempt to compare our results with real isotropic physical systems with competition.

2. Anisotropic Amplitude ratio

We begin with the renormalized free energy density at one-loop level at the fixed point for the N-vector model describing the generic anisotropic Lifshitz critical behavior. This object is written in terms of the two factor scale, namely, the renormalized temperature (t) and order pameter (M) and is defined up to a polynomial of second order in t. Furthermore, we choose the free energy such that it coincides with the case without competition, *i.e.*, $m_n = 0$ [5], for which $\epsilon_L(= 4 + \sum_{n=2}^{L} \frac{(n-1)}{n} m_n - d)) \equiv \epsilon = 4 - d$. Defining the quantity $y = u_n^* M^2$ (u_n^* is the dimensionless coupling constant at the fixed point useful to our purposes in what follows), the free energy density at one-loop order can be written as

$$F(t,y) = \frac{1}{2u_n^*}\Big(ty + \frac{y^2}{12}\Big) + \frac{1}{2}\int d^{d-\sum_{n=2}^{L} m_n}q \prod_{n=2}^{L} d^{m_n}k_{(n)} \Bigg[(N-1)ln\left(1 + \frac{t+\frac{y}{6}}{q^2 + \sum\limits_{n=2}^{L} k_{(n)}^{2n}}\right)$$
$$+ln\left(1 + \frac{t+\frac{y}{2}}{q^2 + \sum\limits_{n=2}^{L} k_{(n)}^{2n}}\right) - \frac{1}{q^2 + \sum\limits_{n=2}^{L} k_{(n)}^{2n}}\Big[N\Big(t+\frac{y}{6}\Big) + \frac{y}{3}\Big]\Bigg] + \frac{1}{4} I_{SP_n}\Big[N\Big(t+\frac{y}{6}\Big)^2 + \frac{2y}{3}\Big(t+\frac{y}{3}\Big)\Big]$$
$$+O(\epsilon_n^2). \quad (1)$$

The integral I_{SP_n} is the one-loop contribution to the one-particle ($1PI$) irreducible four-point vertex part computed at fixed external momenta and is defined by

$$I_{SP_{n'}}(K_n^2 = \delta_{nn'}\kappa_n^2) = \int \frac{d^{d-\sum_{n=2}^{L} m_n}q \prod\limits_{n=2}^{L} d^{m_n}k_{(n)}}{\Big(q^2 + \sum\limits_{n=2}^{L} k_{(n)}^{2n}\Big)\Big[(q+K_{(1)})^2 + \sum\limits_{n=2}^{L}(k_{(n)} + K_{(n)})^{2n}\Big]}. \quad (2)$$

Note that at this symmetry point, only the external momenta associated to the n' subspace is fixed at nonvanishing values, whereas all other external momentum scales are set to zero. The specific heat above an below the critical temperature is given by

$$C_{\pm}(t) = -\frac{\partial^2 F(t,M)}{\partial t^2} - \frac{N\nu_n}{2\alpha_L}\Big[\kappa_n \frac{\partial I_{SP_n}(\kappa_n)}{\partial \kappa_n}\Big]_{\kappa_n^{2n}=1}, \quad (3)$$

and above T_L we should take the derivatives and set $M = 0(y = 0)$. Below T_L, we should employ the value of M (y) at the coexistence curve, namely $y = u_n^* M^2 = -6t$. The integral can be computed in a closed analytical form using the orthogonal approximation [1, 2] and is given by $I_{SP_n}(K_{(n)}) = \frac{K_{(n)}^{-n\epsilon_L}}{\epsilon_L}[1 + h_{m_L}\epsilon_L]$ ($h_{m_L} = 1 + \frac{[\psi(1)-\psi(2-\sum_{n=2}^{L}(\frac{m_n}{2n}))]}{2}$). Using this result in conjunction with the fixed point and critical exponents up to two-loop level computed using this approximation given, respectively, by the following expressions [2]

$$u_n^* = \frac{6\epsilon_L}{(N+8)}\Big[1 + \epsilon_L\Big[-h_{m_L} + \frac{(9N+42)}{(N+8)^2}\Big]\Big], \quad (4a)$$

$$\nu_n = \frac{1}{n}\Big(\frac{1}{2} + \frac{(N+2)}{4(N+8)}\epsilon_L + \frac{1}{8}\frac{(N+2)(N^2+23N+60)}{(N+8)^3}\epsilon_L^2\Big). \quad (4b)$$

$$\alpha_L = \frac{(4-N)}{2(N+8)}\epsilon_L - \frac{(N+2)(N^2+30N+56)}{4(N+8)^3}\epsilon_L^2, \quad (4c)$$

is sufficient to compute the ratio up to one-loop order. After performing the integrals resulting from the derivatives of the free energy with respect to t we identify the specific heat up to one-loop order with the expressions $C_{\pm}(t) = A_{\pm}|t|^{-\alpha_L}$, which yields

$$\frac{A_+}{A_-} = \frac{N}{4} 2^{\alpha_L}(1 + \epsilon_L). \tag{5}$$

Notice that in the above expression the exponent is expressed up to first order in ϵ_L. This formula reduces to the case $m_2 = m$, $m_3 = ... = m_L = 0$ corresponding to the ratio of the m-axial Lifshitz universality class [6].

3. Exact isotropic amplitude ratio

We write the free energy density for the N-vector model describing the generic isotropic Lifshitz critical behavior using the same ingredients as before, with minor modifications. The expansion parameter is $\epsilon_n = 4n - d$ ($d = m_n$). The free energy is such that it coincides with the case without competition, *i.e.*, $n = 1$ [5], for which $\epsilon_1 \equiv \epsilon = 4 - d$. The free energy density at one-loop order reads

$$F(t,y) = \frac{1}{2u_n^*}\Big(ty + \frac{y^2}{12}\Big) + \frac{1}{2}\int d^{m_n}k\Big[(N-1)ln\Big(1 + \frac{t+\frac{y}{6}}{(k^2)^n}\Big) + ln\Big(1 + \frac{t+\frac{y}{2}}{(k^2)^n}\Big)$$
$$-\frac{1}{k^{2n}}\Big[N\Big(t + \frac{y}{6}\Big) + \frac{y}{3}\Big]\Big] + \frac{1}{4}I_{SP_n}\Big[N\Big(t+\frac{y}{6}\Big)^2 + \frac{2y}{3}\Big(t + \frac{y}{3}\Big)\Big] + O(\epsilon_n^2). \tag{6}$$

Similarly as dicussed in the anisotropic case, the integral I_{SP_n} and the condition on the specific heat are both defined by the expressions

$$I_{SP_n}(K) = \Big[\int \frac{d^{m_n}k}{k^{2n}(k+K)^{2n}}\Big]_{K^2=\kappa_n^2}, \tag{7a}$$

$$C_{\pm}(t) = -\frac{\partial^2 F(t,M)}{\partial t^2} - \frac{N\nu_n}{2\alpha_n}\Big[\kappa_n\frac{\partial I_{SP_n}(\kappa_n)}{\partial \kappa_n}\Big]_{\kappa_n^{2n}=1}. \tag{7b}$$

Above T_L the derivatives are taken and set $M = 0 (y = 0)$ afterwards, whereas below T_L we set $y = u_n^* M^2 = -6t$. Now, the exact result $I_{SP_n}(K) = \frac{K^{-\epsilon_n}}{\epsilon_n}[1 + D(n)\epsilon_n]$ ($D(n) = \frac{1}{2}[\psi(2n) + \psi(1)] - \psi(n)$) together with the fixed point and critical indices up to two-loop order given, respectively, by [2]

$$u_n^* = \frac{6\epsilon_n}{(N+8)}\Big[1 + \epsilon_n\Big[\frac{1}{(N+8)^2}\frac{(-1)^n\Gamma(2n)^2(2N+4)}{\Gamma(n+1)\Gamma(3n)} + (20N+88)(1-D(n)) - D(n)$$
$$+\frac{(20N+88)}{(N+8)^2}\Big(\sum_{p=1}^{2n-2}\frac{1}{2n-p} - 2\sum_{p=1}^{n-1}\frac{1}{2n-p}\Big)\Big]\Big], \tag{8a}$$

$$\nu_n = \frac{1}{2n} + \frac{(N+2)}{4n^2(N+8)}\epsilon_n + \frac{(N+2)}{4n^2(N+8)^3}\epsilon_n^2\Big[(-1)^n(N-4)\frac{\Gamma(2n)^2}{\Gamma(n+1)\Gamma(3n)} + \frac{(N+2)(N+8)}{2n}$$
$$+(14N+40)D(n)\Big], \tag{8b}$$

$$\alpha_n = \frac{(4-N)}{2n(N+8)}\epsilon_n - \frac{(N+2)}{4n^2(N+8)^3}\epsilon_n^2\Big[(-1)^n\frac{4n(N-4)}{\Gamma(n+1)}\frac{\Gamma(2n)^2}{\Gamma(3n)} + (N-4)(N+8) + 4n(14N$$
$$+40)D(n)\Big], \tag{8c}$$

are required to evaluate the exact ratio up to one-loop order. From the identification $C_{\pm}(t) = A_{\pm}|t|^{-\alpha_n}$, one learns that

$$\frac{A_+}{A_-} = \frac{N}{4} 2^{\alpha_n}(1 + 2D(n)\epsilon_n). \tag{9}$$

This formula reduces to the case $n = 1$ from ref.[19], since $D(1) = 1/2$. Nevertheless, as n increases $D(n)$ becomes negative for $n \geq 2$. The choice $\epsilon_n > 0$ leads naturally to *vanishing and even negative values for the amplitude ratio.* Accordingly, perturbation theory in the framework of the ϵ_n-expansion is not a systematic correction to the mean field solution in the case of this amplitude ratio, since neither zero nor negative values are allowed in the later situation. Could this indicate a failure of the ϵ_n-expansion for isotropic systems with n higher enough? Let us compute the amplitude ratio using the generalized orthogonal approximation [1, 2] and try to figure out possible explanations to this odd behavior.

4. Isotropic amplitude ratio in the orthogonal approximation

Now we just have to compute the one-loop integral at the symmetry point, the fixed point and the critical exponents using the orthogonal approximation. The integral at the symmetry point is given by $I_{SP_n}(K) = \frac{K^{-\epsilon_n}}{\epsilon_n}[1+\frac{\epsilon_n}{2n}]$. The fixed point and critical exponents in this approximation can be obtained from ref.[2] but we do not bother to write them down explicitly. Substitution of these values above and below T_L as explicated before lead to the orthogonal approximation version of the specific heat amplitude ratio, namely

$$\frac{A_+}{A_-} = \frac{N}{4} 2^{\alpha_n}(1 + \frac{\epsilon_n}{n}). \tag{10}$$

This result only agrees with the exact computation for $n = 1$. But the result from last equation has the virtue that it does represent an expansion around the mean field value ($\epsilon_n = 0$) since the positiveness is guaranteed at this loop order.

5. Discussion

In fig. 1 we plot simultaneously the exact amplitude ratios for fixed ϵ_n. Inpired by the traditional use of the ϵ-expansion for three-dimensional systems in order to compare with experiments, we set $\epsilon_n = 1$ for the various values of the space dimension (i.e., below the critical dimension) in the amplitude with the variation of (N, n). The asymptotes of the orthogonal approximation amplitude curves are defined by the $n \to \infty$ limit. The inset indicates how different values of N produce a slight deviation in the cancellation of the amplitude ratio as a function of n. For all values of $N \neq 0$, $n \geq 4$ correspond to negative values of the specific heat amplitude ratio in the exact approach.

The amplitude ratio await for experimental appplications in generic anisotropic systems yet to be discovered. From the experimental viewpoint, at least the traditional second character case $m_2 \equiv m$ can be compared with theoretical predictions. Suggested theoretically in the context of copolymer-homopolymer ternary blends [7, 8], the first mean-field isotropic experimental realization was reported on symmetric diblock copolymer-homopolymer blends (of polyethylene (PE) and polyethylenepropylene (PEP)) using small angle neutron scattering ($SANS$) for the case $(N, d, n) = (1, 3, 2)$ [9]. However, a subsequent study showed that the phase behavior of symmetric PE-PEP/PE/PEP mixtures points out to the destruction of the mean-field Lifshitz point [10]. Another polymer system consisting of deuterated polybutadiene (dPB)and polystyrene (PS) homopolymers has also been analyzed with $SANS$, though its phase diagram was constructed from the temperature and diblock copolymer concentration [11, 12]. The PE-PEP/PE/PEP and dPB-PS/dPB/PS mixtures agree between each other with respect to the disappearance of the Lifshitz point caused by fluctuations, except that in the latter the

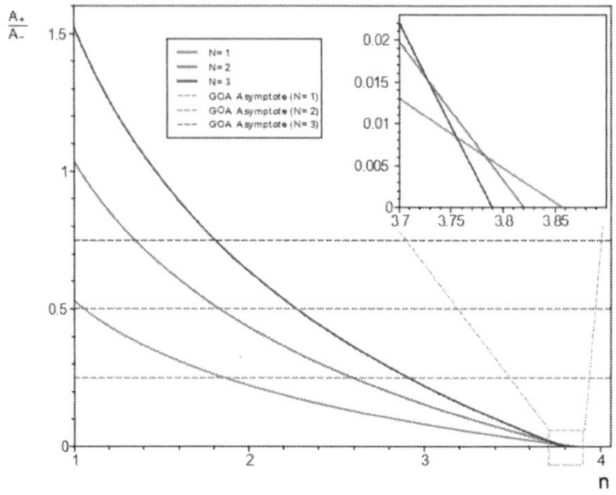

Figure 1. The amplitude ratio plotted as a function of n and N (different collors). The horizontal lines correspond to the value $n \to \infty$ using the generalized orthogonal approximation (GOA). The inset shows how the amplitude vanish and start to take negative values for different values of N.

fluctuations were explicitly taking into account in the isotropic Lifshitz critical region and produced nontrivial effects on the susceptibility amplitudes and associated critical exponent [12]. The universality class ($N = 1, d = 3, n = 2$) in ref.[12] encountered the susceptibility exponent $\gamma_2 = 1.55 \pm 0.15$ from the isotherm at 69.5^0C with concentration of diblock copolymer $\Phi_{DB} = 0.071$. Surprisingly, the exact two-loop value determined from ref.[2] using $\epsilon_2 = 5$ yields $\gamma_2 = 1.50$ which is quite close to the experimental value. Unfortunately, the setup of this experiment was not able to encounter universal values for the susceptibility amplitude ratio.

In the case of the specific heat amplitude ratio we propose the following experiment: using a scanning adiabatic calorimeter [14], determine the mean field Lifshitz temperature T_L in the case of the homopolymer-diblock copolymer blend, for instance in the dPB-PS/dPB/PS system. Next, in the lamellar modulated phase one should measure the amplitude ratio in a certain temperature $T = T_L - \Delta T$ as well as performing the measurement of the amplitude in the disordered phase for $T = T_L + \Delta T$ and dividing the latter result by the former. The result could be checked in principle by measuring the amplitude at $T = T_L - \Delta T$ within the uniformly ordered phase with different diblock copolymer concentration followed by a similar determination at $T = T_L + \Delta T$. Inside the Lifshitz critical region, the two results should be the same. After that, the output should be compared with our theoretical predictions in the present work by setting $d = 3(\epsilon_2 = 5), N = 1, n = 2$: $\left[\frac{A_+}{A_-}\right]_{exact} = 0.06$, whereas $\left[\frac{A_+}{A_-}\right]_{GOA} = 1.17$, which is quite big a difference to clearly distinguish the correct experimental result in comparison with both theoretical predictions and rule out one of the hypotheses raised herein.

If the would be experimental result confirms the exact ratio, then we can conclude for its smallness that this rules out isotropic Lifshitz systems for $n > 3$. In that case, $n = 4$ represents effectively a long range system in the isotropic competitive world, since it is well known that systems with long range order originate negative specific heat [17, 18], even though no pathology occurs to the exponents, leading to a new criterion for stability of isotropic competing systems. On the other hand, if the other possibility prevails, then the propagation of fluctuations must be restricted and the orthogonal approximation is a way out to the issue of thermodynamical instability in higher character isotropic Lifshitz critical behavior.

Acknowledgments
MISJ would like to thank financial support by CNPq from Brazil.

References
[1] Leite M M, 2004 *Phys. Lett. A* **326** 281
[2] Leite M M, 2005 *Phys. Rev. B* **72** 224432
[3] Farias C F and Leite M M, 2012 *J. Stat. Phys.* **148** 972
[4] Hornreich R M, Luban M and Shtrikman S, 1975 *Phys. Rev. Lett.* **35** 1678
[5] Brézin E, Le Guillou J C and Zinn-Justin J, 1976 *Phase Transitions and Critical Phenomena* vol 6, ed C Domb and M S Green (London: Academic) p125
[6] Leite M M, 2003 *Phys. Rev. B* **68** 052408
[7] Broseta D and Fredrickson G H, 1990 *J. Chem. Phys.* **93** 2927
[8] Holyst R and Schick M, 1992 *J. Chem. Phys.* **96** 7728
[9] Bates F S, Maurer W, Lodge T P, Schulz M F, Matsen M W, Almdal K and Mortensen K, 1995 *Phys. Rev. Lett.* **75** 4429
[10] Bates F S, Maurer W W, Lipic P M, Hillmyer M A, Almdal K, Mortensen K, Fredrickson G H and Lodge T P, 1997 *Phys. Rev. Lett.* **79** 849
[11] Pipich V, Schwahn D and Willner L, 2005 *Phys. Rev. Lett.* **94** 117801
[12] Pipich V, Schwahn D and Willner L, 2005 *J. Chem. Phys.* **123** 124904
[13] Olmsted P D and Hamley I W, 1999 *Europhys. Lett.* **45** 83
[14] Flewelling A C, DeFonseka R J, Khaleeli N, Partee J and Jacobs D T, 1996 *J. Chem. Phys.* **104** 8048
[15] Thirring W, 1970 *Z. Phys.* **235** 339
[16] Ramírez-Hernández A, Larralde H and Leyvraz F, 2008 *Phys. Rev. Lett.* **100** 120601
[17] Posch H A and Thirring W, 2005 *Phys. Rev. Lett.* **95** 251101
[18] Lederhendler A and Mukamel D, 2010 *Phys. Rev. Lett.* **105** 150602
[19] Brézin E, Le Guillou J C and Zinn-Justin J, 1974 *Phys. Lett. A* **47** 285

Low voltage operation of electro-absorption modulator promising for high-definition 3D imaging application using a three step asymmetric coupled quantum well structure

Byung Hoon Na,[1] Gun Wu Ju,[1] Hee Ju Choi,[2] Soo Kyung Lee,[3] Jin Myeong Jeon,[2] Yong Chul Cho,[4] Yong Hwa Park,[4] Chang Young Park,[4] and Yong Tak Lee[1, 2, 3,*]

[1]School of Information and Informations, Gwangju Institute of Science and Technology (GIST), 123, Cheomdangwagi-ro, Buk-gu, Gwangju 500-712, Republic of Korea
[2]Department of Physics and Photon Science, GIST, Republic of Korea
[3]Department of Nanobio Materials and Electronics, GIST, Republic of Korea
[4]Advanced Device Lab, Samsung Advanced Institute of Technology, Giheung, Yongin, 446-712, Republic of Korea

E-mail: ytlee@gist.ac.kr

Abstract. In this paper, we propose a transmission type electro-absorption modulator (EAM) operating at 850 nm having low operating voltage and high absorption change with low insertion loss using a novel three step asymmetric coupled quantum well (3 ACQW) structure which can be used as an optical image shutter for high-definition (HD) three dimensional (3D) imaging. Theoretical calculations show that the exciton red shift of 3 ACQW structure is more than two times larger than that of rectangular quantum well (RQW) structure while maintaining high absorption change. The EAM having coupled cavities with 3 ACQW structure shows a wide spectral bandwidth and high amplitude modulation at a bias voltage of only -8V, which is 41% lower in operating voltage than that of RQW, making the proposed EAM highly attractive as an optical image shutter for HD 3D imaging applications.

1. Introduction

Optical image shutter is an important component for realizing high-definition (HD) three dimensional (3D) imaging systems. Low voltage operation, high speed operation and wide bandwidth are the key requirements of such an optical image shutter. Over the past years, several optical image shutters such as Kerr cells [1], liquid crystal or mechanical shutter [2] and micro channel plate (MCP) [3] were proposed. They however have several drawbacks. Of them, Kerr cells suffer from extremely high operating voltage while the modulation speed of liquid crystal or mechanical shutter is limited to few kHz. MCP is not only expensive and bulky but also requires high voltage for its operation. These shutters are therefore not suitable for reliable high-definition (HD) 3D imaging system. Recently, surface normal electro-absorption modulator (EAM) which can function as optical image shutter operating at wavelength near 850 nm based on quantum-confined Stark effect (QCSE) was developed [4]. The reliable operation of EAM with high amplitude modulation over wide wavelength range was reported. However, the reported EAM still suffer from high operating voltage at high modulation speeds, which increases power consumption and the generated heat in the system which significantly degrades the modulation performance. In the past years, several band-gap engineering techniques has been extensively studied in GaAs/AlGaAs quantum well (QW) system such as stepped QW [5], symmetric [6], asymmetric [7] and partially doped coupled QW [8] to reduce the operating voltage. In case of symmetric and asymmetric coupled QW, these structures have a large residual absorption limiting high amplitude modulation. Doped coupled QW operate only at low temperatures and due to the high doping concentration in QW region, significant broadening of absorption spectra occurs leading to low contrast ratio which limits their usage, whereas stepped QW did not show any improvement in modulation performance.

In this paper, we present the design and simulation of transmission type EAM structure having high amplitude modulation with wide spectral bandwidth operating at low bias voltage. These improved features are achieved by using a novel three step GaAs/AlGaAs asymmetric coupled quantum well (3 ACQW) structure for the EAM. Theoretical calculations show that the exciton red shift of 3 ACQW structure is more than two times larger than that of rectangular quantum well (RQW) structure while maintaining high absorption variation. The EAM having coupled cavities with 3 ACQW structure show a wide spectral bandwidth and large amplitude modulation at a bias voltage of only -8V, which is 41% lower than that of RQW. These simulation results reveal that EAM with 3 ACQW structure are promising as optical image shutter for HD 3D imaging systems.

2. Design of 3 ACQW for low voltage operation and large absorption modulation

Low voltage operation can be achieved in coupled GaAs/$Al_xGa_{1-x}As$ QW by optimizing the coupling barrier thickness and height [9, 10, 11]. For an applied electric field, the electron and hole wave functions of coupled QW are shifted rapidly compared to that of RQW, resulting in lower operating voltage. However, a rapid shift of wave functions to opposite directions greatly reduces the overlap of electron and hole wave functions leading to a weak oscillator strength and therefore to a lower electro-absorption. Consequently, in spite of having low operating voltage, the coupled QW structure suffers from low absorption modulation. The key idea of our purposed structure lie in increasing the overlap of electron and hole wave functions while simultaneously reducing the operating voltage, which is attributed to asymmetric, thin and low height coupling barrier. All absorption spectrums were calculated using commercial APSYS program [12]. For the calculation, the Schrodinger equation was numerically solved with appropriate boundary conditions in order to obtain the eigen values (particle energies) and energy states (wave functions).

Figure 1 (a) and (b) respectively shows the calculated electron and hole wave functions of the RQW and 3 ACQW structures at different applied electric fields. The RQW structure consists of 8 nm thick GaAs well and a 4 nm thick $Al_{0.3}Ga_{0.7}As$ barrier, and the 3 ACQW structure consists of 3 nm thick, 6.6 nm thick and 2 nm thick GaAs wells separated by two 1 nm thick $Al_{0.2}Ga_{0.8}As$ barrier. The barrier is sufficiently thin so that the particle wave functions of adjacent wells can couple to each other. In case of RQW, the wave functions are not seriously distorted with applied fields. On the other hand, in the 3 ACQW with properly chosen position of the barrier, the wave functions are seriously distorted, leading to much larger dependence of electro absorption on applied field. As a result, the red shift of the 3 ACQW structure is almost two times larger than that of RQW structure. In addition, the absorption coefficient of 3 ACQW structure is larger than that of RQW due to large overlap of electron and hole wave functions.

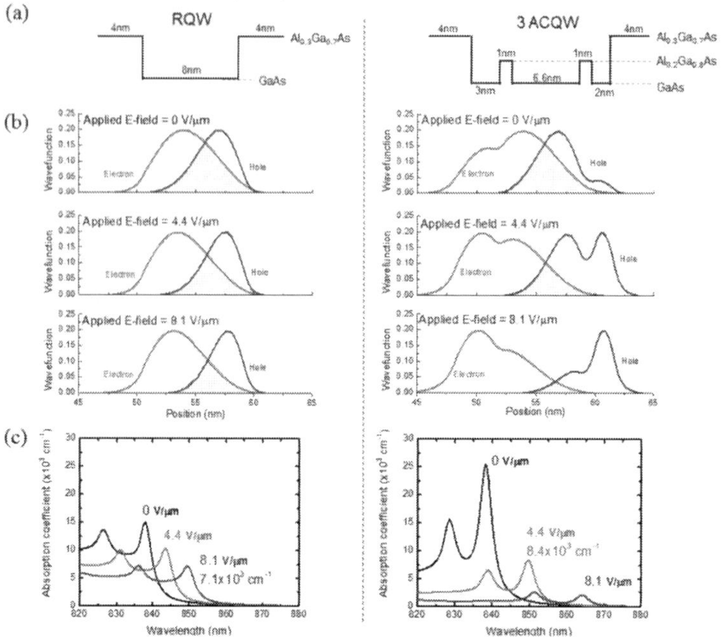

Figure 1 (a) Quantum well structure (b) electron and hole wave functions, and (c) absorption coefficient of RQW(left) and 3 ACQW(right) structures at an applied electric fields of 0, 4.4 and 8.1 V/μm. The yellow region indicates the wave function overlap..

3. Design of low voltage operation electro-absorption modulator

To confirm the feasibility of EAM with 3 ACQW for low voltage operation and high amplitude modulation, we have designed and fabricated the EAM with RQW and with 3 ACQW using a combination of coupled Fabry-Perot cavity with micro-cavity (CCMC) [13].

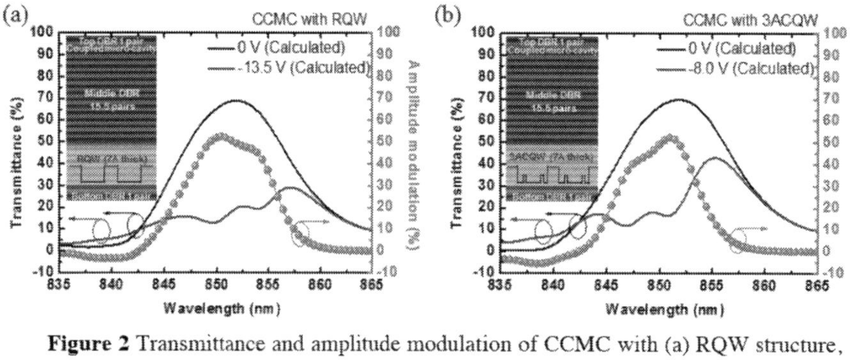

Figure 2 Transmittance and amplitude modulation of CCMC with (a) RQW structure, (b) 3 ACQW structure. The inset of (a) and (b) show the schematic of CCMC with RQW structure and 3 ACQW structure, respectively.

The structures shown in the insets of Figure 2 (a) and (b) have a 7-λ thick cavity sandwiched between one pair of bottom DBR and 15.5 pairs of middle DBR for both structures. On the top of middle DBR is the micro-cavity followed by one pair of top DBR. In the CCMC with RQW structure, the active layer consists of 60 and 78 pairs of 8 nm thick and 8.5 nm thick GaAs RQW, respectively. For CCMC with 3 ACQW structure, the active layer consists of 45 and 50 pairs of 6.6 nm thick and 6.0 nm thick GaAs QW, respectively which is placed in the middle of 3 ACQW structure. Figure 2 shows the calculated transmittance spectra and amplitude modulation for CCMC having RQW and 3

ACQW structures. It can be seen that for CCMC with 3 ACQW structure, the operating voltage is reduced by about 41% compared to CCMC with RQW structure, while maintaining a broad spectral bandwidth with high amplitude modulation over 50%.

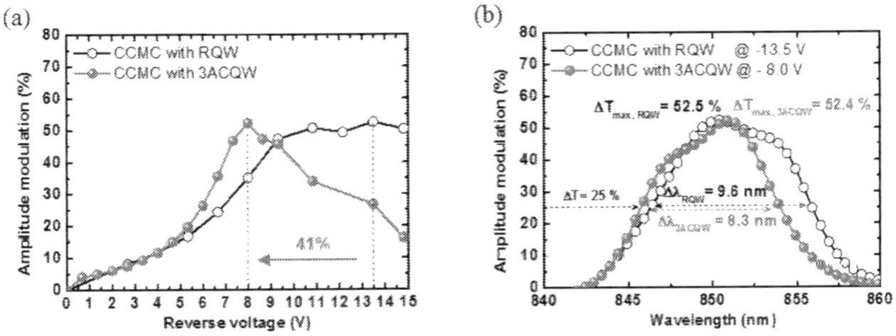

Figure 3 The amplitude modulation of CCMC with RQW and 3 ACQW (a) at operating wavelength for different applied bias voltages and (b) at operating voltage.

Figure 3 shows the amplitude modulation of CCMC with RQW and 3 ACQW as a function of applied bias and wavelength. From Fig 3(a), it can be seen that the amplitude modulation increases as the applied bias voltages increases and show a maximum when exciton peak wavelength coincides with the operating wavelength. The maximum amplitude modulation of CCMC with 3 ACQW structure occurs at a bias voltage of -8 V which is 41% lower compared to that of CCMC with RQW structure. For CCMC with 3 ACQW at operating voltage (-8 V), the amplitude modulation and spectral bandwidth (for $\triangle T = 25\%$) are 52.4% and 8.3 nm, respectively, as shown Figure 3 (b). Even though CCMC with RQW shows a higher spectral bandwidth (9.6 nm) compared to CCMC with 3 ACQW, the former requires a much larger bias of -13.5V. It is noteworthy that the operating voltage of CCMC with 3 ACQW was significantly reduced by 41% while simultaneously maintaining high amplitude modulation (>50%) and wide spectral bandwidth (>8 nm), demonstrating their capability to be used as optical shutter for high-definition imaging applications.

4. Conclusion

We have proposed a novel 3 ACQW structure for EAM to simultaneously achieve low voltage operation and high amplitude modulation. The proposed 3 ACQW structure showed the lowest operating voltage, while maintaining high absorption modulation. The CCMC with 3 ACQW structure also showed high amplitude modulation (>50%) and a wide spectral bandwidth of over 8 nm at the operating voltage of only -8 V, which is about 41% lower in operating voltage. This low voltage operation of EAM with 3 ACQW structure also resulted in a 65% lower power dissipation compared to that of EAM with RQW structure. These results indicate that EAM with the proposed 3 ACQW are highly promising for the realization of full HD 3D imaging system.

Acknowledgement

This research was supported by Samsung Advanced Institute of Technology (SAIT).

References

[1] R. Schwarte, H.-G. Heinol, Z. Xu, and K. Hartmann 1995 *Proc. SPIE* **2588** 126-134
[2] B. Saleh and M. Teich 1991 *John Wiley & Sons* ISBN 0-471-83965-5
[3] M. Kawakita, K. Iizuka, H. Nakamura, I. Mizuno, T. Kurita, T. Aida, Y. Yamanouchi, H. Mitsumine, T. Fukaya, H. Kikuchi, and F. Sato 2004 *Opt. Express* **12** 2781-2794
[4] Y. H. Park, Y. C. Cho, J. W. You, C. Y. Park, H. Yoon, S. H. Lee, J. O. Kwon, and S. W. Lee, B. H. Na, G. W. Ju, H. J. Choi, and Y. T. Lee 2013 *J. Micro-Nanolithogr. Mems. Momes.* **12** 023011

[5] T. K. Woodward, J. E. Cunningham, and W. Y. Jan 1995 *J. Appl. Phys.* **78** 1411-1414
[6] Y. C. Chan and K. Tada 1991 *IEEE J. Quantum Electron.* **27** 702-707
[7] H. Feng, J. P. Pang, M. Sugiyama, K. Tada, and Y. Nakano 1998 *IEEE J. Quantum Electron.* **34** 1197-1208
[8] M. F. Krol, R. P. Leavitt, J. T. Pham, B. P. McGinnis, and N. Peyghambarian 1995 *Appl. Phys. Lett.* **66** 3045-3047
[9] Y. J. Chen, E. S. Koteles, B. S. Elman, and C. A. Armiento 1987 *Phys. Review B* **36** 4562-4565
[10] D. Atkinson, G. Parry, and E. J. Austin 1990 *Semicond. Sci. Technol.* **5** 516-524
[11] N. Debbar, S. Hong, J. Singh, and P. Bhattacharya 1989 *J. Appl. Phys.* **65** 383-385
[12] Y. Chiu, S. Z. Zhang, V. Kaman, J. Piprek, and J. E. Bowers 2001 *Proc. SPIE* **4490** 1-10
[13] B. H. Na, G. W. Ju, H. J. Choi, Y. C. Cho, Y. H. Park, and Y. T. Lee, 2012 *Opt. Express* **20** 19511–19519

Modeling of the transitions between magnetic states of core/shell nanoparticles

I. G. Iliushin[1*], L. L. Afremov[1], S. V. Anisimov

[1]Department of Theoretical and Experimental Physics, School of Natural Sciences FEFU, 8 Suhanova st., Vladivostok, Russia

*corresponding author's e-mail: futted@gmail.com

Abstract. Core/shell nanoparticles can be used in industry, medicine and biophysics, due to their unique properties. Theoretical studies of core/shell nanoparticles are mainly based on the Stoner-Wohlfarth model, and the Monte-Carlo simulation is mostly used the Metropolis algorithm. This method is an extension of one-phase nanoparticle model and not entirely correct for solving the magnetic states of core/shell nanoparticles, which represents a magnetic core covered with a magnetic or nonmagnetic shell. We developed a model of core/shell nanoparticles based on the analysis of total energy consisting of anisotropy energy E_a, magnetostatic interaction energy E_m, exchange interaction energy E_{ex} and energy of magnetic moment of the grain in external magnetic field E_h: $E=E_a+E_m+E_{ex}+E_H$.

Keywords: Core/shell nanoparticles, magnetic characteristics, model hysteresis calculation.

Introduction

The thermal stability of magnetic media and magnetic storage elements becomes important in reducing the size of the magnetic structures [1]. The thermal stability requires estimates of the rate of transitions between equilibrium states of core/shell nanoparticles. To calculate the probability of a transition of the magnetic moments of nanoparticles between the two equilibrium states need to possess the most complete information about the magnitude of the barrier between these states. Energy landscape may contain several local minima and maxima, respectively, making it difficult calculation of the energy barrier which determines the optimal transition from one state to another. An illustration of what has been said may serve shown in fig. 1 "relief" energy dependence on the distribution of the magnetic moment in a ferromagnetic film [2].

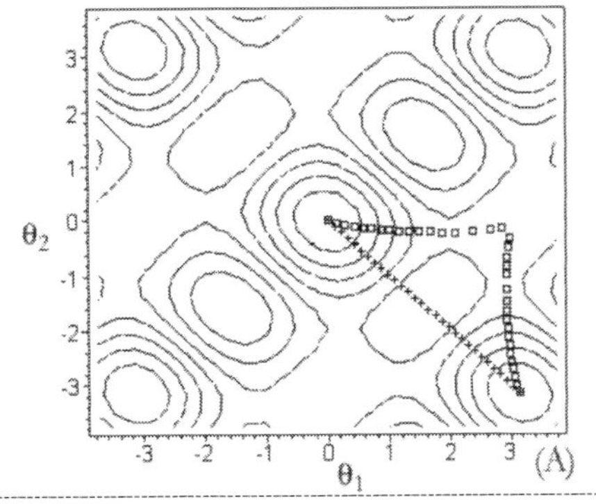

Figure 1. An illustration of the problem of finding a minimum barrier.

The problem is solved by means of various numerical methods are widely used in solid-state physics, theoretical chemistry and materials science. At the moment, there are many different search algorithms barriers. This method is optimal energy paths (MEPs), the method of elastic bands (NEB) and etc. Many of these methods are concentrated on finding the singularities of the function - the saddle points [2-5]. This method of calculation of the energy barrier separating equilibrium states we used in our work.

Model of Core/Shell nanoparticles

Basic assumptions of the model of the uniaxial core/shell nanoparticles, based on the model we developed two-phase particles [6], can be summarized as follows:

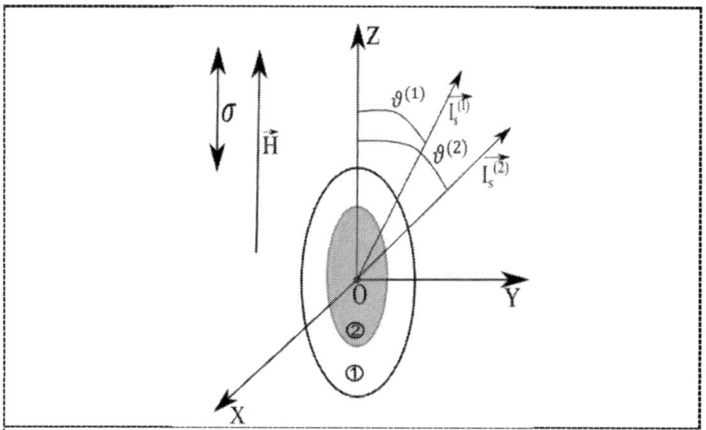

Figure 2. An illustration of the model of a two-phase particle

1. An uniformly magnetized nanoparticle (phase (1)) of volume V and the shape of an ellipsoid with elongation q_1 contains an uniformly magnetized ellipsoidal inclusion (phase (2)) with a volume $v = \varepsilon V$ and elongation q, the long axis of which coincides with the long axis of the particle oriented along the Oz axis (see Fig. 1).
2. Both uniaxial ferromagnets' axis of crystal anisotropy $\langle k_A \rangle^{(1)}$ и $\langle k_A \rangle^{(2)}$ are considered to be parallel to the Oz axis, and the vectors of spontaneous magnetization of the phases $I_s^{(1)}$ and $I_s^{(2)}$ are in the yOz plane containing long axis of magnetic phases and make angles $\vartheta^{(1)}$ and $\vartheta^{(2)}$ with the Oz axis, respectively.
3. Both external magnetic field H and uniaxial mechanical stresses σ are applied along the Oz axis.

In the accepted approximation, free energy of nanoparticle F located in an external field H can be written as the sum of

- crystallographic anisotropy energy:

$$E_A = -\frac{1}{4}\left\{\left(I_s^{(1)}\right)^2 k_A^{(1)}(1-\varepsilon)\cos 2\vartheta^{(1)} + \left(I_s^{(2)}\right)^2 k_A^{(2)}\varepsilon\cos 2\vartheta^{(2)}\right\}V\,, \tag{1}$$

- interaction energy of magnetic moment with its own magnetic field, which, in accordance with [67] can be represented as :
- $E_m =$

$$\left\{-\frac{(I_s^{(1)})^2}{4}(1-2\varepsilon)k_N^{(1)}\cos 2\vartheta^{(1)} - \frac{(I_s^{(2)})^2}{4}\varepsilon k_N^{(2)}\cos 2\vartheta^{(2)} + \frac{\varepsilon I_s^{(1)} I_s^{(2)}}{3}\left(k_N^{(2)} - k_N^{(1)}\right)\left(\sin\vartheta^{(1)}\sin\vartheta^{(2)} - 2\cos\vartheta^{(1)}\cos\vartheta^{(2)}\right)\right\}V, \tag{2}$$

- energy of magnetoelastic interaction with uniaxial mechanical stress σ:

$$E_\sigma = -\frac{1}{4}\left\{\left(I_s^{(1)}\right)^2 k_\sigma^{(1)}(1-\varepsilon)\cos 2\vartheta^{(1)} + \left(I_s^{(2)}\right)^2 k_\sigma^{(2)}\varepsilon\cos 2\vartheta^{(2)}\right\}V, \tag{3}$$

- exchange interaction energy across the border, which according to [7] can be defined as:

$$E_{ex} = -\frac{2A_{in}}{\delta}\cos(\vartheta^{(1)} - \vartheta^{(2)})s, \quad (4)$$

- interaction energy of magnetic moment with an external magnetic field

$$E_H = -H\left[I_s^{(1)}(1-\varepsilon)\cos\vartheta^{(1)} + I_s^{(2)}\,\varepsilon\cos\vartheta^{(2)}\right]V, \quad (5)$$

In the formula (1) – (5) $k_A^{(1,2)} = K_1/\left(I_s^{(1,2)}\right)^2$, $k_N^{(1,2)}$, dimensionless constants of anisotropy and phases' shape anisotropy are K_1 – the first anisotropy constant, and $k_\sigma^{(1,2)} = 2\left(\lambda_2^{(1,2)} + \lambda_3^{(1,2)}\right)\sigma/\left(I_s^{(1,2)}\right)^2$, $\lambda_2^{(1,2)}$, $\lambda_3^{(1,2)}$ – the second and third magnetostriction constants of a uniaxial crystal, V – the volume of a particle, s – the surface area that separates the phases, ε – the ratio of the inclusion volume to the volume of the whole particle, A_{in} - the constant of interfacial exchange interaction, δ - the width of the transition region with the order of a constant lattice. Note that the shape anisotropy constant $k_N = 2\pi(1-3N_z)$ is expressed through the demagnetizing factor along the long axis N_z depending only on the elongation of the ellipsoid q: $N_z = \left[q ln\left(q+\sqrt{q^2-1}\right) - \sqrt{q^2-1}\right]/(q^2-1)^{3/2}$.

Analysis of a two-phase particle's magnetic states is a procedure of minimization of the total free energy (F) of a nanoparticle:

$$F = \left\{-\frac{(I_s^{(1)})^2}{4}K^{(1)}\cos 2\vartheta^{(1)} - \frac{(I_s^{(2)})^2}{4}K^{(2)}\cos 2\vartheta^{(2)} + I_s^{(1)} I_s^{(2)}(-\mathcal{U}_1\sin\vartheta^{(1)}\sin\vartheta^{(2)} + \mathcal{U}_2\cos\vartheta^{(1)}\cos\vartheta^{(2)}) - H\left[I_s^{(1)}(1-\varepsilon)\cos\vartheta^{(1)} + I_s^{(2)}\,\varepsilon\cos\vartheta^{(2)}\right]\right\}V, \quad (6)$$

where the effective anisotropy constants $\mathcal{K}^{(1,2)}$ and the position of are determined by the following formulas:

$$\mathcal{K}^{(1)} = (1-\varepsilon)\left(k_A^{(1)} + k_\sigma^{(1)}\right) + (1-2\varepsilon)k_N^{(1)} + \varepsilon k_N^{(2)}, \quad (7)$$

$$\mathcal{K}^{(2)} = \varepsilon\left(k_A^{(2)} + k_\sigma^{(2)} + k_N^{(2)}\right)$$

The constants of an interfacial interaction $\mathcal{U}_1$ and $\mathcal{U}_2$ are expressed through the constants of magnetostatic and exchange interaction of the phases:

$$\mathcal{U}_1 = \varepsilon\left(\frac{\left(k_N^{(1)} - k_N^{(2)}\right)}{3} + \frac{2sA_{in}}{v\,\delta\, I_s^{(1)} I_s^{(2)}}\right), \mathcal{U}_2 = \varepsilon\left(\frac{2\left(k_N^{(1)} - k_N^{(2)}\right)}{3} - \frac{2sA_{in}}{v\,\delta\, I_s^{(1)} I_s^{(2)}}\right). \quad (9)$$

Minimization of free energy (6) by $\vartheta^{(1)}$ and $\vartheta^{(2)}$ leads to a system of equations determining static statesof phases' magnetic moments:

$$\frac{\mathcal{K}^{(1)}}{2}\sin 2\vartheta^{(1)} - j\left[\mathcal{U}_1\cos\vartheta^{(1)}\sin\vartheta^{(2)} + \mathcal{U}_2\sin\vartheta^{(1)}\cos\vartheta^{(2)}\right] + \frac{H}{I_s^{(1)}}(1-\varepsilon)\sin\vartheta^{(1)} = 0$$

$$\frac{j\mathcal{K}^{(2)}}{2}\sin 2\vartheta^{(2)} - \left[\mathcal{U}_1\sin\vartheta^{(1)}\cos\vartheta^{(2)} + \mathcal{U}_2\cos\vartheta^{(1)}\sin\vartheta^{(2)}\right] + \frac{H}{I_s^{(1)}}\varepsilon\sin\vartheta^{(2)} = 0 \quad (10)$$

The system of equations (10) to determine the ground and metastable states of the magnetic moment of the nanoparticle. As noted earlier [6] that neglecting thermal fluctuations in the absence of an external magnetic field two-phase particle can be in one of the four equilibrium states. Note that when$\mathcal{U}_2 > 0$ are stable *first* «(↑↑)-state» the magnetic moments of the two phase are parallel and oriented along the axis Oz; and *third* «(↓↓)-state» differs from the first state in antiparallel orientation of the phases magnetization regarding to the axis Oz, *second* «(↑↓)-state» and *fourth* «(↓↑)-state» metastable, which differs from the first state in antiparallel orientation of the phases magnetization regarding to the axis Oz. When $\mathcal{U}_2 < 0$ are stable *second* «(↑↓)-state» and *fourth* «(↓↑)-state», *first* «(↑↑)-state» and *third* «(↓↓)-state» – metastable. The probability of transition from one equilibrium state (i –th) to another (k –th) determined by the height of the potential barrier F_{ik}:

$$W_{ik} = f_0 exp\left(-\frac{F_{ik}}{k_B T}\right) \tag{11}$$

where $f_0 \sim 10^7 \div 10^{10} c^{-1}$ — characteristic frequency of the "attempts" to overcome the potential barrier, which is then assumed to be equal 10^{10} c^{-1}[7], k_B — Boltzmann's constant.

To calculate the energy barrier has been developed following algorithm:

1. With the system of equations (10) is determined by the entire range of solutions suspicious extremum.

2. Using standard conditions determine the extrema of the free energy $F = F(\vartheta^{(1)}, \vartheta^{(2)})$, that is positive and negative second derivatives $\partial^2 F(\vartheta^{(1)}, \vartheta^{(2)})/\partial^2\vartheta^{(1)}$ or $\partial^2 F(\vartheta^{(1)}, \vartheta^{(2)})/\partial^2\vartheta^{(2)}$ and the positivity of the quadratic form $[\partial^2 F(\vartheta^{(1)}, \vartheta^{(2)})/\partial^2\vartheta^{(1)}][\partial^2 F(\vartheta^{(1)}, \vartheta^{(2)})/\partial^2\vartheta^{(2)}] - -[\partial^2 F(\vartheta^{(1)}, \vartheta^{(2)})/\partial\vartheta^{(1)}\partial\vartheta^{(2)}]^2$ were decisions not related to the distributions of the magnetic moment with the minimum or maximum value $F(\vartheta^{(1)}, \vartheta^{(2)})$.

3. Solutions of the system (10) is not related to the decisions outlined in the result of the research described in the preceding paragraph, are checked for the possibility of implementing a saddle point.

4. Potential barrier height is determined by the difference $F_{ik} = (F_{ik\,max})_{min} - F_{i\,min}$, where $(F_{ik\,max})_{min}$ – minimum of several maximum values of the free energy separating i and k equilibrium states. Thus, using a relatively simple calculation, we can calculate the entire spectrum of energy barriers determining the probabilities of transitions from one equilibrium state to another.

Conclusion

A simple scheme for detecting the extrema, and the allocation of these saddle points allows you to quickly and with minimal errors to calculate the barriers between states, and the more and the populations of states.

Acknowledgement

Work supported by Ministry of Education and Science, project № 559.2014.

References

[1] R. Dittrich, T. Schrefl, D. Suess, W. Scholz, H. Forster, J. Fidler // A path method for finding energy barriers and minimum energy paths in complex micromagnetic systems. Institute for Solid State Physics, Vienna University of Technology, Wiedner Hauptstr. 8 - 10, A-1040 Vienna, Austria, pp. 18.

[2] T. Shimatsu, H. Uwazumi, Y. Saki, A. Otsuki, I. Watanabe, H. Muraoka, Y. Nakamura // *IEEE Trans. Mag.* **37** (2001) 1567.

[3] Graeme Henkelman, Gísli Jóhannesson and Hannes Jónsson Department of Chemistry // Methods for finding saddle point and minimum energy paths , University of Washington, Seattle, WA 98195-1700.

[4] Daniel Sheppard, Rye Terrell, and Graeme Henkelman Department of Chemistry and Biochemistry // Optimization methods for finding minimum energy paths , The University of Texas at Austin, Austin, Texas 78712-0165, USA.

[5] Graeme Henkelman, Blas P. Uberuaga and Hannes Jónsson // *Journal of Chemical Physics* Vol. 113, n. 22.

[6] Afremov L.L., Ilyushin I.G. // *Journal of Nanomaterials*, vol. 2013, Article ID 687613, 15 pages, 2013. doi: 10.1155/2013/687613.

[7] Brown W.F. // *Phys. Rev.*, 1963, **130**, p. 167.

AUTHOR INDEX

AUTHOR INDEX

AUTHOR INDEX

AUTHOR INDEX